ENERGY

1 B = 1.055 056 kJ

1 B = 778.169 ft·lbf

1 ft·lbf = 1.3558 J

1 J = 9.478×10^{-4} B

1 IT cal = 4.1868* J

1 cal = 4.1840* J

SPECIFIC ENERGY

1 B/lbm = 2.326 kJ/kg

1 kJ/kg = 0.4299 B/lbm

1 B/lbmol = 2.326 kJ/kmol

1 kJ/kmol = 0.4299 B/lbmol

SPECIFIC ENTROPY, SPECIFIC HEAT, GAS CONSTANT

1 B/lbm·R = 4.1868* kJ/kg·K

1 kJ/kg·K = 0.2388 B/lbm·R

1 B/lbmol·R = 4.1868* kJ/kmol·K

1 kJ/kmol·K = 0.2388 B/lbm·R

DENSITY

1 lbm/ft^3 = 16.018 kg/m^3

1 kg/m^3 = 0.062 428 lbm/ft^3

SPECIFIC VOLUME

1 ft^3/lbm = 0.062 428 m^3/kg

1 m^3/kg = 16.018 ft^3/lbm

POWER

1 B/s = 1.055 056 kJ/s

1 hp = 2545 B/h

1 hp = 550 ft·lbf/s

1 kW = 1.3410 hp

VELOCITY

1 mph = 1.467 ft/s

1 mph = 0.4470 m/s

1 ft/s = 0.3048* m/s

TEMPERATURE

$T[°C] = \frac{5}{9}(T[°F] - 32)$

$T[°F] = \frac{9}{5}T[°C] + 32$

$T[°C] = T[K] - 273.15$

$T[°F] = T[R] - 459.67$

1 K = 1.8 R or 1.8 T[K] = T[R]

*Exact value

ENGINEERING THERMODYNAMICS

SECOND EDITION

Engineering Thermodynamics

AN INTRODUCTORY TEXTBOOK

J. B. Jones, P.E.

Professor of Mechanical Engineering
Virginia Polytechnic Institute and State University

G. A. Hawkins, P.E.

Late Professor of Thermodynamics and Vice President for Academic Affairs
Purdue University

John Wiley & Sons, Inc.

New York Chichester Brisbane Toronto Singapore

Library of Congress Cataloging in Publication Data:

Jones, J. B. (James Beverly)
Engineering thermodynamics.

Includes indexes.
1. Thermodynamics. I. Hawkins, George A. (George Andrew), 1907– II. Title.
TJ265.J64 1986 621.402′1 85-17821
ISBN 0-471-81202-1

Printed in the United States of America

10 9 8 7 6 5 4 3 2 1

PREFACE

This textbook is intended for use in undergraduate engineering courses in thermodynamics. Its purpose is to help develop in the student (1) an understanding of the first law, the second law, and some physical property relationships and (2) competence in the application of these principles to engineering systems. The book is written directly to students; one indication of this is the frequent use of the personal pronoun "you." The intention is to free the teacher from the necessity of interpreting the text to students. The teacher's talents are more effectively used if classroom time is devoted to those teaching activities that require a face-to-face exchange of thoughts and therefore cannot be accomplished by a textbook.

This edition follows the general plan of the first edition where George Hawkins and I expressed our conviction that students need copious explanation and illustration during the early part of a thermodynamics course sequence. A person who has studied thermodynamics at length can state the basic principles concisely and from them deduce many far-reaching conclusions. A thorough familiarity with the subject reveals the simplicity as well as the importance of the fundamentals; but people who have achieved this familiarity may not recognize the need of the neophyte for repeated explanations and illustrations of the fundamentals. Experienced teachers realize this when in subsequent courses they find that too many students are weak precisely on the fundamentals. Consequently, the explanations in the early chapters of this book are often exhaustive, with special emphasis on various points that have been stumbling blocks for students. The early chapters include many fully solved example problems. As the students gain maturity in the subject, they need less explanation, and the instructor can rely more on their reasoning powers; thus, both the amount of explanation and the number of solved example problems decrease in later chapters.

Some derivations are somewhat abbreviated because they are better appreciated after the results have been used enough for the student to recognize their usefulness. Initial derivations that are exhaustive accomplish little with most students, but after one has gained facility with the results, a rigorous examination of derivations can be quite fruitful.

The order of material is based more on pedagogical considerations than on logical economy. (This point is called to the students' attention in Sec. 9·13.) Thus, after the introductory chapter and Chapter 2 on the first law, there are two chapters on physical properties. The problems following these chapters involve extensive applications of the first law and material from the introductory chapter, as well as physical property relationships. The student can thereby practice applying the first law to various systems before taking up the new material on the second law.

Separate chapters are devoted to the second-law statement, reversibility, the Carnot principle and cycle, and entropy in order to emphasize the sequence. Reversibility has meaning only in light of the second law; the Carnot principle and cycle can be established only after reversibility has been defined; and the definition of entropy involves a reversible process and also thermodynamic temperature, which is most easily explained after the Carnot principle is introduced. Chapter 9 then treats physical property relationships that follow from the first and second laws combined, and Chapter 10 introduces availability and irreversibility. A problem posed by the inclusion of availability and irreversibility calculations in an introductory course is that although the conclusions are simple as well as highly useful, the general derivation of these conclusions is somewhat involved. Some instructors may wish to use the conclusions without devoting appreciable time to their derivation.

Chapters 14 through 18 illustrate the application of the basic principles covered in earlier chapters to various systems. The emphasis on fundamentals is maintained by omitting special techniques that are applicable only to certain situations. An instructor can always introduce some of these techniques in the classroom when advisable. It is much easier, in fact, to introduce such special material as a supplement to the basic approach in the textbook than to convince students to rely on the basic approach if their textbook frequently uses more specialized methods.

Chapter 19 on binary mixtures introduces briefly a part of the subject not otherwise mentioned in the book except in connection with absorption refrigeration in Chapter 18. Some teachers may prefer to assign Chapter 19 before Chapter 18. Chapter 20 is a brief introduction to fuel cells and provides additional applications of principles introduced in Chapters 12 and 13.

The final chapter is a short one on the elements of heat transfer. Although heat transfer is not part of the subject of thermodynamics, all engineering students should have some knowledge of this field closely related to thermodynamics. In curricula that do not require a course in heat transfer and even in those that do require one, the chapter on heat transfer may be useful even though it is very brief.

Naturally, many decisions have been made on the selection of material presented. For the most part, in Chapters 1 through 15, no material is introduced unless it is used subsequently or ties in with other material in the book. Some valuable and highly useful concepts (for example, chemical potential, partial molal properties, and fugacity) are not introduced, because covering them to the extent that students could use them effectively would require many additional pages.

Over 1600 problems of varying difficulty are included, and answers to approximately one third are given in the appendix. Many problems are highly "cumulative"; that is, their solutions involve material from several previous chapters. Some simple problems are included so that for illustrating some specific point the teacher can use one of these rather than make up a problem on the spur of the moment and have students end up with solutions in their notes but not the statement of the problem. Some problems involve physical property data that must be obtained from other sources.

This edition includes many problems for which the use of a computer is either required or advisable. However, the problem statements themselves do not indicate this because an essential part of problem solving is the selection of appropriate calculation methods.

For many years to come, engineers in English-speaking countries must be familiar with both SI and English units. Consequently, most of the solved example problems use SI but some use English units. The analysis and early part of a solution are nearly always independent of the system of units used, so the book can be used for courses using either SI or English units alone as may be advisable in some curricula. Doing so is facilitated by the inclusion at the end of each of the first eight chapters of problems similar to each example problem but using the other system of units. The problems that involve units are approximately evenly divided between SI and English units.

The references are for *students*. They have been selected as ones that can be helpful to most students at their current stage in the study of thermodynamics. They may not be the most suitable references for an advanced student reviewing the subject or seeking the greatest rigor in its development. Some of the references use different symbols or conventions from those of this textbook, but students must recognize that they will often encounter this situation in practice.

Many users of the first edition have provided helpful comments. One who has been most helpful in this regard over the years is Robert C. Fellinger of Iowa State University, and I am grateful to him for many excellent suggestions.

Four colleagues have generously helped me by reviewing early versions of various chapter revisions: G. H. Beyer, E. F. Brown, and H. L. Wood of VPI&SU and Professor S. B. Thomason of Memphis State University. I am also indebted to the four reviewers of the nearly completed manuscript: R. C. Fellinger (Iowa State University), Robert J. Heinsohn (Pennsylvania State University), Richard K. Irey (University of Florida), and Michael J. Moran (Ohio State University). Each one made highly perceptive comments and several valuable suggestions. I have considered all of them in detail, and only after careful thought have I chosen not to follow a few of the suggestions.

In preparing this edition, I have been extremely fortunate in having the assistance of two people who have been unsurpassably competent and helpful co-workers. Even when the work was mountainous in extent and tedious or intricate, I appreciated how pleasant it was to work with them. Ada B. Simmons prepared the manuscript, exercising superb judgment not just in the typing, assembling, and organization but also in editorial improvements. Regina D. Rieves performed many of the calculations and read the entire draft manuscript critically. She also solved many of the examples and problems, gathered and organized data, and prepared some of the figures. I warmly thank Mrs. Simmons and Mrs. Rieves for the quality of their work and their enthusiastic dedication to the project.

All of the people at Wiley with whom I have worked on this edition—like their counterparts years ago on the first edition—have been most helpful, offering many sound suggestions and professional guidance.

Again on this edition my wife, Jane Hardcastle Jones, has supported me through long hours of intense work on what sometimes looked like a never-to-be-ended task, and I am grateful to her.

J. B. Jones

Blacksburg, Virginia
June 1985

George Andrew Hawkins

(*1907–1978*)

George Hawkins earned B.S.M.E., M.S.M.E., and Ph.D. degrees from Purdue University. He served on the Purdue faculty for forty-four years as Professor of Thermodynamics, Dean of Engineering (1953–67), Vice President for Academic Affairs (1967–71), and in other positions. His primary field of research was heat transfer. In 1940 he was awarded the Pi Tau Sigma Gold Medal as the outstanding young mechanical engineer within ten years of the baccalaureate degree. He made numerous contributions to engineering and engineering education through his wisdom and vigor in teaching, research, and administrative leadership. He served as president of the American Society for Engineering Education in 1970 to 1971. Among his many honors was the election to Honorary Membership in the American Society of Mechanical Engineers. His other books include *Elements of Heat Transfer*, Third Edition, written with Max Jakob (Wiley, 1957) and *Thermodynamics*, Second Edition (Wiley, 1951).

CONTENTS

9. Physical Properties

10. Availability and Irreversibility

11. Gas and Gas–Vapor Mixtures

17. Vapor Power Cycles

18. Refrigeration

19. Binary Mixtures

20. Fuel Cells: Direct Energy Conversion

21. Heat Transfer

SYMBOLS

A	area; Helmholtz function, $U - TS$
a	linear acceleration; specific Helmholtz function, $u - Ts$; velocity of a pressure wave
b	Darrieus function, $h - T_0 s$
C	a constant; number of components (in the phase rule)
C_p	molar specific heat at constant pressure
C_v	molar specific heat at constant volume
c	constant-temperature coefficient, $(\partial h/\partial P)_T$; velocity of sound
c_p	specific heat at constant pressure, $(\partial h/\partial T)_P$
c_v	specific heat at constant volume, $(\partial u/\partial T)_V$
E	stored energy
e	specific stored energy, E/m
F	force; maximum number of independent intensive properties (in the phase rule)
F_A	shape factor for radiant heat transfer
F_E	emissivity factor for radiant heat transfer
$\mathfrak{F}$	Faraday constant, 96.487×10^6 coulomb/kmol of electrons
f	number of degrees of freedom of a molecule
G	Gibbs function, $H - TS$
g	gravitational acceleration or the acceleration of a freely falling body; specific Gibbs function, $h - Ts$
g_c	dimensional constant
H	enthalpy, $U + PV$
ΔH_f	enthalpy (change) of formation
ΔH_R	enthalpy (change) of reaction
h	specific enthalpy, $u + Pv$; height of a fluid column; convective heat-transfer coefficient
I	irreversibility
i	specific irreversibility; I/m; electric current
k	ratio of specific heats, c_p/c_v; thermal conductivity
K_p	equilibrium constant
KE	kinetic energy
ke	kinetic energy per unit mass
L	length
M	molar mass; Mach number, V/c
$\dot{m}$	mass rate of flow
m	mass

m'	mass of a molecule (Chapter 4); mass of extracted steam per pound of steam entering turbine
N	number of moles
n	polytropic exponent
P	pressure; number of phases (in the phase rule)
P_R	reduced pressure
p_r	relative pressure
PE	potential energy
pe	potential energy per unit mass
Q	heat
$\dot{Q}$	time rate of heat transfer
g	heat transfer per unit mass
R	gas constant
R_u	universal gas constant
r	radius; compression ratio
r_c	cutoff ratio
S	entropy
s	specific entropy
T	absolute temperature, temperature
T_R	reduced temperature
t	temperature
U	internal energy; overall heat-transfer coefficient
ΔU_R	internal energy (change) of reaction
u	specific internal energy; velocity; velocity of a point on a rotor
V	volume, velocity, voltage (emf)
v	specific volume
v_r	relative specific volume
W	work
$\dot{W}$	power
X	mass fraction (Chapters 18 and 19)
x	quality; mole fraction; a property in general; distance in the direction of heat conduction
Y	stream availability
y	specific stream availability; a property in general
Z	compressibility factor, $Z = PV/mRT$
z	elevation

Greek letters

α	absorptivity
β	coefficient of performance; coefficient of volume expansion, $(\partial v/\partial T)_P/V$
γ	specific weight
ε	emissivity, "fuel cell efficiency"

η	efficiency; number of molecules
Θ	Debye's constant or characteristic temperature
θ	temperature on any nonthermodynamic scale
κ_s	isentropic compressibility, $-(\partial v/\partial P)_s/V$
κ_T	isothermal compressibility, $-(\partial v/\partial P)_T/V$
μ	Joule–Thomson coefficient, $(\partial T/\partial P)_h$
υ	stoichiometric coefficient (number of moles in chemical equation)
ρ	density; reflectivity
σ	Stefan–Boltzmann constant
τ	time; transmissivity
Φ	availability of a closed system
ϕ	specific availability of a closed system, $\frac{\Phi}{m}$; $\int \frac{c_p dT}{T}$; relative humidity
ω	humidity ratio

Subscripts

a	air
c	critical state; (see also g_c in list of symbols)
da	dry air
dg	dry gas
f	final state; saturated liquid; fuel
fg	difference between property of saturated liquid and property of saturated vapor at the same pressure and temperature
g	saturated vapor
H	high temperature (as in T_H and Q_H)
i	initial state; ice point; ideal or isentropic; intermediate (as intermediate pressure in multistage compression)
if	difference between property of saturated solid and saturated liquid at the same pressure and temperature
ig	difference between property of saturated solid and saturated vapor at the same pressure and temperature
L	low temperature (as in T_L and Q_L)
m	mixture
N	molar (as in v_N for molar specific volume)
O	base state; state of the atmosphere
R	reduced coordinate; energy reservoir
r	relative
s	steam point
t	total or stagnation
u	universal (in R_u)
v	vapor (in gas–vapor mixtures)

σ referring to an open-system boundary

1,2,3 referring to different states of a system or different locations in space

Superscripts

$^\circ$ standard state

$*$ state used in relating real gas and ideal gas properties; state at which $M(= V/c) = 1$

ENGINEERING THERMODYNAMICS

CHAPTER ONE

Fundamental Concepts and Definitions

The study of any science must begin with an understanding of certain definitions and conventions. A sound knowledge of these definitions and conventions at the outset, just like an understanding of the ground rules at the start of a baseball game, will prevent many later misunderstandings.

The beginning student of thermodynamics is sometimes dismayed by the emphasis placed on definitions; however, precise definitions, well understood, pave the way toward a sound understanding of the science.

This chapter introduces definitions and conventions that are used repeatedly in later chapters. Careful study of this chapter is therefore essential for ready understanding of topics which follow.

1·1 Thermodynamics

Thermodynamics is the science dealing with energy transformations, including heat and work, and the physical properties of substances that are involved in energy transformations.

The origin of thermodynamics may be traced to early studies of the performance of steam engines. Its range of application was then broadened to cover other types of heat engines, or devices which convert heat or fuel energy into work, and the early refrigerating machines. Since the middle of the nineteenth century the scope of thermodynamics has grown so greatly that today an understanding of thermodynamics is essential for much of the work of the engineer, physicist, and chemist.

1·2 Engineering Thermodynamics

Engineering thermodynamics is that part of the science that deals with all types of heat engines or power plants (stationary or vehicular), refrigeration, air conditioning, combustion, the compression and expansion of fluids, chemical processing plants, and the physical properties of the substances used in these and other applications.

The automobile engine is an example of a device that converts part of the chemical energy of a fuel into work. The remainder of the energy obtained from the fuel is discharged to the surroundings by way of the exhaust gases and the cooling water. By thermodynamic analysis it is possible to predict the amounts of fuel and air required by the engine to provide a certain work output, the amount of energy carried away by the cooling water and by the exhaust gases, and the effects of changing the engine compression ratio.

A household electric refrigerator is another example of a device that is designed in accordance with the principles of thermodynamics. Suppose the temperature to be maintained in the refrigerator compartment and the amount of heat to be removed from the compartment are known. The problem is to design a refrigerating unit which will perform this service. Typical questions which can be answered by the use of thermodynamics are: What will the normal power consumption be? Of the many different refrigerating fluids available, is there one which will result in a lower power requirement than the others? What pressure must the units be built to withstand? Can the pressure range be changed by using a different refrigerant? At what rate must air flow across the warm condenser coils during normal operation?

Steam engines were commercially successful long before the basic laws of thermodynamics were formulated, and the early builders of refrigerating machines and internal-combustion engines had little or no knowledge of thermodynamics as we now know it. However, while it was the practical problems concerned with these early machines that called for the development of thermodynamics, the resulting science has led the way toward improvements and new processes undreamed of by the pioneer builders and inventors. Thus engineering practice and thermodynamics have advanced together with now one and now the other in the lead. Today, the continuing work of engineers toward improvements and innovations in equipment and processes and the need for accurate predictions of performance make thermodynamics an indispensable engineering tool.

1·3 On Definitions in General

In order for people to communicate effectively with one another, they must agree on the meanings of the terms they use. That is, each term or word used must relate to the same (or at least to a similar) experience, object, or operation for each of the people involved. It is sometimes difficult to achieve such agreement on terms. We know that many words have multiple meanings. For example, consider the many different meanings of the nouns "run," "break," and "point" and of the verbs "run" and "draw." These obvious examples cause little misunderstanding, however, because people are aware that the words have multiple meanings and the intended meaning is usually made clear by the context.

Serious difficulties in communication and understanding result, however, when it is not realized that some of the words used have different meanings for the various people involved. This causes much trouble in the study of engineering and science. To many people a statement about "the power in a gallon of gasoline" may have some meaning, but as power is defined in physics and engineering, such a statement is nonsense. Also, people who use the words speed and velocity synonymously think it is nonsense to refer to the difference in velocity of an aircraft moving north at 500 mph and another one moving west at 500 mph, because they do not realize that, for sound reasons, the definition of velocity involves direction as well as magnitude. Thus the need for precise definitions is obvious.

A definition of a word must relate the word to some experience, where experience includes both events and objects. Now we shall investigate how definitions should be made in order to establish clearly the intended relation between a word and an experience.

The best way to explain the meaning of a word to someone who has no idea of its meaning is to point out as many objects or operations as possible to which the word relates. This would amount to a nonverbal definition.

For many reasons, definitions usually must be verbal; that is, a word is usually defined by means of other words. To be of value, a verbal definition must use only words which are understood or independently defined. Violations of this rule can be found in any poor dictionary. For example, suppose we were entirely ignorant of the meaning of the word *shrill.* We could refer to a dictionary and find

shrill: having a high-pitched tone or sound.

Now if we do not know the meaning of *high* as used with *pitch,* we look and find

high: (Music) acute in pitch.

We are still no better off than at first unless we learn the meaning of *acute,* so we turn again to the dictionary

acute: high or shrill.

We are obviously going in a circle, and as long as we use only verbal definitions we shall continue to go in circles. We have the same difficulty if we try to learn a foreign language by using a dictionary in that language alone. Each word we look up will be defined in terms of other words we do not know, and these other words in turn are defined by still others we do not know. We must start by learning the meaning of a few words in the foreign language in terms of words in a language we understand.

By the same reasoning, we can see that the meaning of words in our own language can be made clear by verbal definitions only if we start with some words which we directly relate to physical events or objects. These words are *undefined* verbally. Other words can then be defined in terms of these *undefined* words.

The idea that definitions of some words cannot be stated is surprising to some people, but the logical structure of any science must be based on the acceptance of certain verbally undefined terms as starting points. Precisely which terms are accepted as undefined is usually a matter of convenience. The undefined terms which will be used in this book are listed in Sec. 1·5.

Specifying a word as undefined does not mean that we are unable to explain its meaning to someone, but rather that we are unable to express its meaning only in terms of simpler words. If X is an undefined term (say mass, time, temperature, or length), we cannot complete a simple defining statement that begins, "X is . . . ". However, we can describe operations and experiments that demonstrate the effects of X and which allow us to assign numbers to the magnitude of X. In brief, we can tell *about* X but we cannot answer the question "What *is* X?"

An *operational* definition is one which specifies a method of measuring the defined quantity or specifies a test to be applied to an event or object to see whether it fits the definition. It gives directions for performing the operations that relate the defined word to nonverbal events or objects. It is therefore more valuable than a nonoperational definition.

An example of a nonoperational definition is

translational kinetic energy: energy of a body which is due to its translational motion.

An operational definition of the same term is

translational kinetic energy: the energy of a body which is evaluated by $mV^2/2$, where m is the mass of the body and V its translational velocity.

The terms energy, mass, and velocity which are used in these definitions must be independently defined or must be accepted as undefined terms. In the operational definition given above, the operations are partly paper-and-pencil operations; but they are nevertheless operations which result in a definite number for the translational kinetic energy.

In order for the definition to be of value, the specified measurement or test operations must be possible. We might define length in terms of operations with a yardstick or a foot rule. These operations would be possible for the measurement of the length of a tennis court, but they would not serve for measuring the dimensions of an atom. As far as operations are concerned, these two "lengths" are different kinds of quantities, since they cannot be measured by the same operations.

Summarizing this discussion of definitions:

1. Precise definitions are essential for effective communication and understanding.

2. Some words must be accepted as undefined on the verbal level.

3. All other words should be defined in terms of these undefined words in order to avoid "circularity" in definitions.

4. Definitions are most effective in relating words to events or objects if the definitions are operational. An operational definition includes directions for measuring the defined quantity or for testing an event or object to see if it fits the definition.

1·4 The Language of Mathematics

The language we use in everyday communication and in much of our thinking is not a precise language; that is, the meaning of a written statement may not be the same for all readers. Let us list some reasons for this.

1. As mentioned in the preceding section, many words have multiple meanings. (What connections are there among a heavy weight, a heavy line, a heavy opera, a heavy motor oil, and a heavy accent?)

2. Many words have "biased" meanings. For example, we think of steam as being "hot" and are therefore surprised when we first learn of a home being cooled by steam. Similarly, if we are told that someone put his finger into molten metal we think of the event as tragic until we consider that he might have put his finger into a cup of mercury at room temperature. Our conceptions of steam and molten metal are "biased" toward steam and molten metal which are at high temperatures.

3. We deal largely with two-valued properties. Our habit of classifying as good or bad, extrovert or introvert, hard or soft, and so forth tends to obscure the fact that these are all multivalued properties, that there are many degrees of hardness and of introversion and so forth.

We do not encounter these same difficulties when we use a mathematical language. One reason for the clarity of mathematical language is that it is an operational language. One way of interpreting the equation

$$x^2 - x + 4 = 0$$

is: "The number for which x stands is such that, if it is multiplied by itself and then this product is diminished by the number itself and increased by four, the result is zero." Thus the equation gives us directions for specific operations that will test whether any given number can be substituted for x.

A further advantage of a mathematical statement is that it is more concise than an equivalent word statement. Mathematical notation is sometimes referred to as a "shorthand." Consider how many words would be required to express the same information given by

$$\left(\frac{\partial v}{\partial T}\right)_P = 0 \qquad \text{or} \qquad T_1 = T_2 e^{\mu\alpha}$$

A third advantage of mathematics as a language is that it is precise. The result of a specified operation on x is the same regardless of who performs the operation or when or where or why it is performed. This point probably needs no supporting discussion except the observation that the variation in numerical answers obtained by different people performing the same operations does not result from any ambiguity of the mathematical statement. It results from variations in the degree of precision involved in carrying out the specified operations.

A final comment on mathematics is that the operations are performed on abstract numbers only. A useful equation expresses a relationship among physical quantities, and the solution of the equation yields a numerical value for some physical quantity. However, the mathematical operations involved in solving the equation are in no way influenced by these physical relationships.

1·5 The Undefined Terms

In this textbook, terms that will be accepted as undefined verbally are time, length, temperature, mass, and force. A relationship among some of these is given by Newton's second law

$$F = \frac{1}{g_c}\frac{d}{d\tau}(mV) \tag{1·1a}$$

where F is the net force acting on a body of mass m which has a velocity V, τ is time, and g_c is a dimensional constant whose value depends only on the units selected for the other quantities. The dimensional constant g_c is written because with force, mass, length, and time all undefined, there are four independent dimensions. (See Appendix A, Dimensions and Units.) It is also possible to choose *either* mass or force as an undefined term and then to define the other term by means of Newton's second law, although this will not be a *verbal* definition. Then the dimensions of the term defined by the equation are established by the equation and no dimensional constant is needed. That is, the equation is written

$$F = \frac{d}{d\tau}(mV) \tag{1·1b}$$

In such case, whether we select mass or force as undefined is a matter of convenience or personal preference. The selection may be entirely arbitrary. If mass, length, and time are selected as undefined and their dimensions are designated by M, L, and τ, respectively, then the dimensions of force are derived as ML/τ^2. If force (dimension F), length, and time are selected as undefined, then the dimensions of mass are derived as $F\tau^2/L$.

In the same manner, the selection of the other undefined terms is entirely arbitrary. As an illustration of this point, it would be possible to select as undefined quantities time, velocity, temperature, and mass or force. Length could then be defined in terms of time and velocity. Such a procedure may not appeal to us, but it could be used.

A few terms, such as matter and acceleration, which are used frequently can be defined in terms of the undefined words listed, but the definitions are not repeated here because they are usually covered at length in physics courses with which you should be familiar.

Temperature is sometimes defined in terms of molecular activity. Such a definition may have some value, but it does not lend itself to any test by physical operations nor does it describe a relationship between temperature and other observable physical events. Occasionally temperature is defined in terms of heat, and such definitions are acceptable if heat is defined independently or is accepted as an undefined term. In this book, however, we select the alternative approach and use temperature as an undefined term in the definition of heat.

1·6 Dimensions, Units, and Some Abbreviations

Several systems of dimensions are in use. They differ principally in the selection of primary dimensions from among the dimensions of force (F), mass (M), length (L), and time (τ). A system of dimensions that uses all four of these as primary dimensions is called an $FML\tau$ system. Others are referred to as $FL\tau$ and $ML\tau$ systems. For each of these, several systems of units are employed, and these are referred to as $FML\tau$ systems of units, $FL\tau$ systems, and so forth. For engineering work, the two most common systems of units are the SI (Système International d'Unités) and the English (also called English engineering, U.S. customary, and other names). These are described in Appendix A. At least these two systems are likely to be used in engineering practice for many years to come, so it is essential that engineers be able to use both of them. Therefore, both are employed in this textbook. Most of the text material and most of the example problem solutions use SI; some of the problems at the ends of chapters use SI, some use English units, and some use mixtures of the kind frequently encountered in practice. Sometimes essentially the same problem is presented in one system and then in another. Also, some use units such as bars and calories, which are part of neither of the two most common systems. This may be annoying, but it is typical of engineering practice and literature, including not only papers, journals, magazines, and books but also reports, specifications, and proposals.

An extensive table of conversion factors is given inside the covers of this book.

The different systems of units have different conventions. For example, Newtons per square meter in SI would be written as N/m^2 (or Nm^{-2}) or in terms of the defined unit, 1 pascal = 1 N/m^2; but in the English system, units are often abbreviated so that pounds force per square foot (lbf/ft^2) is written as psf, and gallons per minute is written as gpm. Periods are not used with abbreviations except in cases where the abbreviation without periods might be read as a word (such as the abbreviation ''in'' for inch). The abbreviations for singular and plural forms are the same, for example, 1 min and 4.2 min, not 4.2 mins.

It should be noted that SI is now well codified through agreements reached at a series of international conferences so that its use is characterized by a great consistency. On the other hand, no such codification or consistency in use is associated with the English system, so variations in symbols, units, conventions, and even the name of the system are frequently encountered.

1·7 Thermodynamic Systems

A thermodynamic system is defined as any quantity of matter or any region of space to which attention is directed for purposes of analysis. The quantity of matter or region of space must be within a prescribed boundary. This boundary may be deformable and may be imaginary.

If a system is defined as a *particular quantity of matter,* then the system always contains the same matter and there can be no transfer of mass across the boundary. However, if a system is defined as a *region of space within a prescribed boundary,* then matter may cross the system boundary. In order to distinguish between these two types of systems, the type that has no mass transfer across its boundary we call a *closed system* or *control mass.* An *open system* is a region of space within a boundary which matter may cross. This boundary may be moving. An open system is also called a *control volume* and its boundary is called a *control surface.*

Everything outside the system boundary is referred to as the *surroundings.* Usually the term surroundings is restricted to those things outside the system that in some way interact with the system or affect the behavior of the system.

A special case of a closed system is an *isolated system.* An *isolated system* is a system that in no way interacts with its surroundings. Notice that an isolated system must be a closed system, since the requirement that there be no interaction of the system with its surroundings prohibits any transfer of mass across the system boundary.

Examples of systems used in thermodynamic analysis range from tiny particles to complete, complex power plants and even to large regions in the earth's atmosphere. The most important step in the solution of a problem in thermodynamics is often the selection and careful specification of the system which is to be considered. The importance of this step will be emphasized frequently later in this book. A few examples of thermodynamic systems will now be considered.

In studying the flow of a gas through a pipe, the system might be defined as the gas within a certain length of pipe. For the study of gas compression, a gas trapped within a cylinder and being compressed by a piston can be considered as a system. In this case, part of the system boundary is movable. Notice that, although the boundary moves, as long as it always encloses the same material, the system under consideration is a closed system.

In studying the operation of an automobile engine, the entire engine can be considered as an open system. Such a system interacts with the surroundings in several ways. Air and gasoline are drawn in and hot exhaust gases are discharged. The turning of the output shaft against a resisting torque is an interaction between the system and its surroundings. The engine also causes a rise in temperature of the air blown across its radiator. In making a thermodynamic analysis of the entire engine, we must consider all these interactions with the surroundings.

In order to study particular aspects of the engine operation, other systems may be selected. For example, a convenient system to use in a study of the combustion process might be the contents of one cylinder during the time interval between the closure of the intake valve and the opening of the exhaust valve. During this time interval no mass enters or leaves the cylinder; hence the contents of the cylinder constitute a closed system. The interactions between this system and its surroundings include the electric impulse to the spark, the action of the gas against the moving piston, and the transfer of heat between the system and its surroundings.

For a study of the exhaust process, the system selected might be one cylinder while its exhaust valve is open. This system is bounded by the piston face, the cylinder walls, the cylinder head, the intake valve, and the exhaust-valve port opening. While the exhaust valve is open, mass flows across this boundary, so the system must be classified as an open system. The interactions between this system and its surroundings include the action of the gas against the moving piston, heat transfer, and the flow of matter from the system to the surroundings.

1·8 Ideal and Actual Systems

In solving physical problems, we often focus our attention not on the actual system at hand but instead on some idealized system which is similar to, but simpler than, the actual system. For example, in calculating the mechanical advantage of a system of ropes and pulleys (actual system), it is customary to start with the consideration of frictionless pulleys and weightless, nonstretching ropes (ideal system). An ideal system differs from the corresponding actual system in that it can be completely described in terms of a few characteristics. An actual system has many characteristics, some of which are highly pertinent to the behavior under study, many of which are immaterial, and some of which may have slight or unknown influence. The ideal system may be described in terms of only those characteristics that have a major influence on the actual system's behavior. Then this ideal system is analyzed under the assumption that no other system characteristics influence its behavior.

In order to obtain an accurate prediction of the behavior of an actual system, the result of the calculation on a corresponding ideal system must be adjusted to account for the differences between ideal and actual system behaviors. Nevertheless, an ideal system is helpful in solving the problem. Another example of the use of an ideal system is in the calculation of the stress in a simple beam. Part of this calculation is generally made by using an ideal beam which is assumed, among other things, to be homogeneous and undistorted by the application of a load. No actual beam meets these conditions, even though for many applications the behavior of the ideal beam simulates that of the actual beam so closely that no correction need be made when applying the result for one to the other.

Many other examples of the substitution of an ideal system for an actual one for the purposes of analysis can be given. It is important to recognize such a substitution when it is made in order to avoid the error of predicting the behavior of an actual system by an analysis of an ideal system which may behave quite differently. This is the error which frequently leads to statements such as "Theoretically yes, but actually no." No contradiction troubles us if we realize that we are probably talking about different systems when we speak of a "theoretical" result and an "actual" result.

1·9 Properties, States, and Processes

Properties

A *property* is any observable characteristic of a system. Examples of properties are pressure, temperature, modulus of elasticity, volume, and dynamic viscosity. We also consider as properties any combination of observable characteristics such as, for example, the product of pressure and temperature. Such properties can be thought of as *indirectly* observable characteristics of a system. Any number of such properties can be defined, but only a few are useful. Another type of property is the kind which cannot be directly observed and cannot be obtained by mathematical operations on other properties but can be defined only by means of the laws of thermodynamics. Two such properties, internal energy and entropy, will be introduced later.

States

The *state* or condition of a system is specified by the values of its properties. Since there are numerous relationships among the properties of particular systems, the values of a few properties will often identify a state completely because all other properties can be determined in terms of these few. Precisely how many properties are required to specify the state of a system depends on the complexity of the system.

If a system has the same values of its properties at two different times, the system is in identical states at these two times.

A system is said to be in an equilibrium state (or in equilibrium) if no changes can occur in the state of the system without the aid of an external stimulus. A test to see if a system is in equilibrium is to isolate the system and observe whether any changes in its state occur. Obviously the temperature must be the same throughout a system in equilibrium. Otherwise, when the system is isolated there would be a transfer of heat from one part of it to another to change the temperature distribution. Also, there can be no eddying motions of a fluid in a system in equilibrium, because, when the system is isolated, these motions will eventually cease and thus the state of the system will have changed without any external effects. In order to be in equilibrium, a system must be homogeneous or must consist of a finite number of homogeneous parts in contact. Such homogeneity is not sufficient to ensure equilibrium, however. For example, a system comprised of iron, water vapor, and air at room temperature consists of a finite number of homogeneous parts in contact, but it is not in equilibrium because it will certainly change state through the oxidation of the iron without any interaction with the surroundings. Notice that the properties of a system such as pressure, temperature, and velocity can each be represented by a single number only if the system is in a state of equilibrium. For this reason, equilibrium states are much easier to specify than nonequilibrium states and elementary thermodynamics is concerned chiefly with equilibrium states and the changes of systems from one equilibrium state to another.

Processes

A transformation of a system from one state to another is called a *process*. The *path* of the process is the series of states through which the system passes during the process. A *cycle* or *cyclic process* is a process (or a series of processes) which returns the system to the state it was in before the process began. The properties of the system vary during a cycle, but at the completion of a cycle all properties have been restored to their initial values. In other words, the net change in a property is zero for any cycle. This is concisely stated by

$$\oint dx = 0$$

where x is any property and the symbol $\oint$ indicates integration around a cycle.

Point and Path Functions

By definition, the properties of a system are characteristics that we can directly or indirectly observe while the system is in any given state. The system may have reached that state by means of any one of an infinite number of different processes, but the properties of the system in that state are independent of the history of the system. Thus, when a system changes from one state to another, the properties of the system undergo a change which depends only on the end states and not on the path followed between the two end states. Since the values of properties do depend only on the states of systems and not on the paths followed by systems in changing states, properties are called *state functions* or *point functions*.

A quantity whose value depends on the particular path followed in passing from one state to another is called a *path function*. The distinction between point functions and path functions is considered in detail in Sec. 1·19. The terms are introduced here so that one important characteristic of point functions can be introduced: Point functions (properties) can be used as coordinates on diagrams representing states of a system. Property diagrams are used widely in thermodynamics.

Consider a system for which two properties are sufficient to determine all other properties, and thus the state, of the system. Let these two properties be X and Y. (X and Y might stand for pressure and volume, for example.) Each state of the system can then be represented by a point on XY coordinates. In Fig. 1·1, point A represents the state for which $X = X_A$ and $Y = Y_A$. Any other property Z is a function of the state and hence of X and Y.

$$Z = f(X, Y) \qquad Z_A = f(X_A, Y_A)$$

At another state B, the properties have the values X_B, Y_B, and Z_B. The differences in values of the properties between the two states are

$$\Delta X = X_B - X_A \qquad \Delta Y = Y_B - Y_A \qquad \Delta Z = Z_B - Z_A$$

and no information regarding any path between A and B is needed in order to evaluate these differences. Conversely, measurements made on the system only at states A and B can give no information as to which of the many possible paths between A and B was actually followed.

Intensive, Extensive, and Specific Properties

If a homogeneous system is divided into two parts, the mass of the whole system is equal to the sum of the masses of the two parts. The volume of the whole is also equal to the sum of the volumes of the parts. On the other hand, the temperature of the whole is not equal to the sum of the temperatures of the parts. In fact, the temperature, pressure, and density of the whole are the same as of the parts. This brings us to the distinction between *extensive* and *intensive* properties.

If the value of a property of a system is equal to the sum of the values for the parts of the system, that property is an *extensive* property. Mass, volume, weight, and several other properties (energy, enthalpy, entropy) that will be introduced later are *extensive* properties. An *intensive* property is one which has the same value for any part of a homogeneous system as it does for the whole system. The measurement of an intensive property can be made without knowledge of the total mass or extent of the system. Pressure, temperature, and density are examples of intensive properties.

If the value of any extensive property is divided by the mass of the system, the

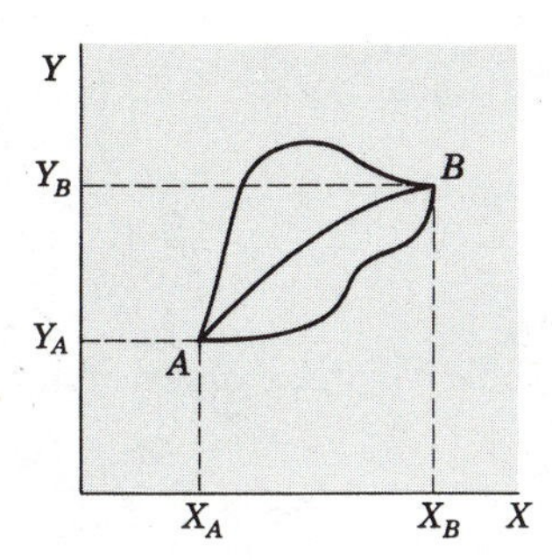

Figure 1·1 Property diagram showing three paths between states A and B.

resulting property is intensive and is called a *specific* property. For example, specific volume is obtained by dividing the volume (an extensive property) of a system by its mass. This ratio of volume to mass is the same for any part of a homogeneous system and for the system as a whole; therefore it is an intensive property. A capital letter is usually used as the symbol for an extensive property and the same lowercase letter stands for the corresponding specific property. Thus V is used for volume and v is used for specific volume.

$$v = \frac{V}{m}$$

where m is used for mass (an exception to the convention of using capital letters for extensive properties).

Macroscopic and Microscopic Viewpoints

If all possible physical measurements on a system indicate that it is in the same state it was in at some previous time, we refer to the two states as identical states. This does not mean that we believe each molecule to have the same location and velocity that it previously had. It means only that on a *macroscopic* level the states are identical; on the *microscopic* level we can make no direct measurements and consequently have no evidence for conclusions regarding individual molecules. The macroscopic viewpoint is used almost exclusively in thermodynamics. We occasionally use a molecular picture to enhance our understanding of some phenomenon, but it must be clearly understood that the laws of thermodynamics are based solely on macroscopic observations and in no way depend on molecular theory.

The explanation of physical phenomena on the basis of molecular behavior is the goal of *kinetic theory,* which is based on the application of the laws of mechanics to individual molecules. Another approach is that of *statistical thermodynamics,* in which the behavior of individual molecules is not treated but probability considerations are applied to the very large numbers of molecules that comprise any macroscopic quantity of matter. The approach to thermodynamics followed in this book is often called *classical thermodynamics* to distinguish it from statistical thermodynamics.

1·10 Density, Specific Volume, and Specific Weight

Density (ρ) is defined as the mass of a substance divided by its volume, or the mass per unit volume:

$$\rho \equiv \frac{\text{mass}}{\text{volume}} = \frac{M}{V}$$

Specific volume (v) is defined as the volume per unit mass or the reciprocal of density:

$$v \equiv \frac{V}{m} = \frac{1}{\rho}$$

Specific weight (γ) is defined as the weight of a substance divided by its volume, or the weight per unit volume:

$$\gamma \equiv \frac{\text{weight}}{\text{volume}} = \frac{w}{V}$$

The relation between specific weight and density can be developed by applying Newton's second law of motion to a body of fixed mass to get

$$F = ma \tag{1·2}$$

If the only external force acting on a body is the gravitational force that we call weight, the resulting acceleration is the gravitational acceleration g:

$$w = mg \tag{1·3}$$

This equation is the basic relation between weight and mass. If each side of this equation is divided by the volume of the body under consideration, the basic relation between specific weight and density is obtained:

$$\gamma = \rho g \tag{1·4}$$

(If an $FML\tau$ system of dimensions and units were being used, this equation would be $\gamma = \rho g / g_c$. This different approach is fully valid but is not used in this book. This point is discussed in Appendix A.)

Specific gravity is defined as the ratio of the density of a substance to some standard density. The standard density used with solids and liquids is often that of water at some specified temperature such as 0°C, 4°C (the temperature of maximum density for water at a pressure of one atmosphere), or 20°C. Specific gravity can also be considered as the ratio of the specific weight of a substance to some standard specific weight if both are measured in the same gravitational field. Since specific gravity is a dimensionless ratio, its numerical value is independent of any system of units.

1·11 Pressure

Pressure is defined as the normal force exerted by a system on a unit area of its boundary. The pressure may vary from place to place on the system boundary, even when the system is in equilibrium. For example, consider a system consisting of a fluid (either gas or liquid) in a closed tank. A simple force balance on the fluid shows that the pressure increases toward the bottom of the tank as a result of the weight of the fluid. In many applications the variation in pressure caused by gravity is negligible.

Most pressure-measuring instruments measure the difference between the pressure of a fluid and the pressure of the atmosphere. This pressure *difference* is called *gage pressure*. The absolute pressure of the fluid is then obtained by the relation

$$P_{\text{abs}} = P_{\text{atm}} + P_{\text{gage}}$$

If a fluid exists at a pressure lower than atmospheric pressure, its gage pressure is negative and the term *vacuum* is applied to the magnitude of the gage pressure. For

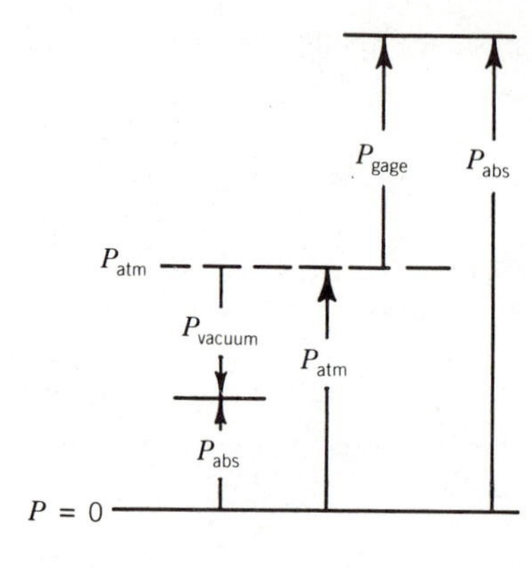

Figure 1·2 Relationships among absolute, gage, atmospheric, and vacuum pressures.

example, a gage pressure of -30 kPa is spoken of as a *vacuum* of 30 kPa. The relationships among absolute pressure, gage pressure, atmospheric (or barometric) pressure, and vacuum are shown graphically in Fig. 1·2.

A manometer is a simple instrument that indicates a pressure difference by balancing a measurable length of fluid column against the pressure difference. For a fluid in static equilibrium, the relationship between pressure and elevation within the fluid is given by the basic equation of fluid statics,

$$dP = -\gamma \, dz \tag{1·5}$$

where γ is the specific weight of the fluid and z is the elevation. The minus sign results from the convention of measuring z positively upward. Thus as z increases in a fluid, P decreases. A derivation of this equation can be found in textbooks on fluid mechanics. In general, the specific weight is a function of pressure and temperature. However, for liquids, which are only slightly compressible, the specific weight can be assumed constant with respect to pressure so that the basic equation can be integrated to give

$$\Delta P = -\gamma \, \Delta z$$

or, if magnitudes only are being considered and h stands for the height of a fluid column,

$$\Delta P = \gamma h$$

As a result of the use of manometers and the direct proportionality between pressures and manometric fluid heights, pressures are often expressed in units such as inches of mercury, inches of water, or millimeters of mercury. Vacuum readings, no matter how they are obtained, are generally expressed in inches or millimeters of mercury. A pressure that is less than atmospheric pressure by 200 mm of mercury is spoken of as a "vacuum of 200 mm mercury" and is written "200 mm Hg vac."

Example 1·1. Determine the barometric pressure in kilopascals (kPa) if the barometer reading is 76.0 cm of mercury. The specific gravity of mercury is 13.6.

Solution. The basic equation of fluid statics will be used. A barometer measures the difference between atmospheric pressure and essentially zero pressure, so

$$P = P - 0 = \Delta P = \gamma h$$

Using a water density of 1000 kg/m³ and a gravitational acceleration of 9.81 m/s²,

$$\begin{aligned} P &= \gamma h = \rho g h = (\text{sp gr})_{\text{Hg}}\rho_{\text{water}}gh \\ &= 13.6(1000)9.81\left(\frac{76}{100}\right) = 101\ 000\ \frac{\text{kg·m}}{\text{s}^2} \\ &= 101\ 000\ \text{N/m}^2 = 101\ 000\ \text{Pa} = 101\ \text{kPa} \end{aligned}$$

(Note that the data used do not justify more than three significant digits in the result.)

Example 1·2. Determine the pressure difference represented by (*a*) 1 in. of mercury, (*b*) 1 ft of water.

Solution. (*a*) For 1 in. of mercury, using a water specific weight of 62.4 lbf/ft³,

$$\Delta P = \gamma h = 13.6(62.4)\frac{1}{12}\left(\frac{1}{144}\right) = 0.491\ \text{psi}$$

(*b*) For 1 ft of water,

$$\Delta P = \gamma h = 62.4(1)\frac{1}{144} = 0.433\ \text{psi}$$

If the equation involving density were used,

$$\Delta P = \gamma h = \rho g h$$

and if a numerical value for density of 62.4 lbm/ft³ were substituted directly,

$$\Delta P = \rho g h = 62.4\left[\frac{\text{lbm}}{\text{ft}^3}\right]32.174\left[\frac{\text{ft}}{\text{s}^2}\right]1[\text{ft}] = 2008\ \text{lbm/ft·s}^2$$

At first glance these units do not appear to be those of pressure, but they are. To obtain the units desired, use the *conversion factor* (in an $FL\tau$ or $ML\tau$ English system of dimensions)

$$32.174\ \text{lbm·ft/lbf·s}^2 = 1$$

so that

$$\begin{aligned} \Delta P &= 2008\left[\frac{\text{lbm}}{\text{ft·s}^2}\right] \div 32.174\left[\frac{\text{lbm·ft}}{\text{lbf·s}^2}\right] = 62.4\ \text{lbf/ft}^2 \\ &= 0.433\ \text{psi} \end{aligned}$$

(If one were using an $FML\tau$ system of dimensions and units, the equation would have been written initially as $\Delta P = \rho g h/g_c$, where g_c is a dimensional constant, not a conversion factor. This is a fully valid approach, but it is not used in this book.)

Example 1·3. A pressure gage reads 20.0 psi when the barometer stands at 28.2 in. of mercury. Compute the absolute pressure of the fluid in the gage.

Solution. In the solution of Example 1·2 above it was shown that a pressure of 1 in. of mercury is equivalent to 0.491 psi.

$$\begin{aligned} P_{\text{barometric}} &= 28.2[\text{in. Hg}]0.491[\text{psi/in. Hg}] = 13.9\ \text{psia} \\ P &= P_{\text{barometric}} + P_{\text{gage}} = 13.9 + 20.0 = 33.9\ \text{psia} \end{aligned}$$

Example 1·4. A vacuum gage reads 10 in. of mercury when the atmospheric pressure is 29.4 in. of mercury. Compute the absolute pressure.

Solution

$$P = P_{atm} + P_{gage} = 29.4 + (-10) = 19.4 \text{ in. Hg abs} \quad \text{or} \quad 9.53 \text{ psia}$$

1·12 Temperature

The familiar sense perceptions of hot and cold are qualitative indications of the *temperature* of a body. No definition of temperature is given here because temperature is one of the terms (such as mass, length, and time) that we will consider as undefined verbally. We cannot make a simple, direct statement that defines temperature in terms of words that are either independently defined or accepted as undefined. We will, however, specify some operations by which numerical values can be assigned to various temperatures.

It is customary to speak of a hot body as having a higher temperature than a cold body. Our sense of touch readily indicates which of two bodies of the same material is "hotter" or at the higher temperature, but numerical values cannot be assigned to various temperatures on the basis of physiological sensations alone. Fortunately, when the temperature of a body changes, several other properties also change. Any one of these temperature-dependent properties might be used as an indirect measurement of temperature. For example, both the volume and the electrical resistance of a bar of steel increase as the steel gets hotter. Many other temperature-dependent properties of materials can be brought to mind, and several different ones are actually used in the measurement of temperature. Before the measurement of temperature can be discussed further, the concept of equality of temperature must be introduced.

If a hot body and a cold body are brought into contact with each other while isolated from all other bodies, the hot body becomes colder or the cold body becomes hotter, or both of these changes occur.* Finally, all changes in the properties of the bodies cease. The bodies are then at the same temperature and are said to be in *thermal equilibrium* with each other. It should be noted that such *equality of temperature* is possible even though the bodies have equal values of no other properties. Two bodies may be at the same temperature although the mass of one is many times that of the other.

It is a matter of experience that *two bodies each in temperature (or thermal) equilibrium with a third body are in temperature equilibrium with each other.* (This statement is sometimes called the zeroth law of thermodynamics.) In view of this fact, it is possible to determine if two bodies are at the same temperature without bringing them into contact with each other; it is necessary only to see if they are each in thermal equilibrium with a third body. The third body is usually what we call a thermometer. A thermometer is a body which has a readily measurable property which is a function of temperature. In a mercury-in-glass thermometer, the volume of the mercury depends on its temperature. In a resistance thermometer, the electrical resistance of the thermometer element is a temperature-dependent property. In order for a thermometer to indicate the temperature of another body, the thermometer and the other body must be in contact with each other

*If the hot body is 1 kilogram of steel at 30°C and the cold body is 10 kilograms of ice at 1 atmosphere, 0°C, then the temperature of only the hot body will change. You can readily think of other cases in which the temperature of only one of the bodies changes.

long enough and must be sufficiently isolated from other bodies so that they will attain temperature equilibrium with each other. The temperature of the thermometer is then the temperature of the other body.

1·13 Temperature Scales

We now consider the establishment of a numerical scale of temperature which will permit the temperature of a body to be specified quantitatively. One way to establish a temperature scale is first to assign numerical values to certain accurately reproducible temperatures. The reproducible temperatures so selected as reference points are often (1) the equilibrium temperature of ice and air-saturated water under a pressure of 101.325 kPa (14.696 psia) and (2) the equilibrium temperature of pure liquid water in contact with its vapor at 101.325 kPa. These two temperatures are referred to as the *ice point* and the *steam point*, respectively. They are used as reference temperatures because they are accurately reproducible in any laboratory. On the Celsius* scale, the ice point is assigned the value 0 and the steam point is assigned the value 100. On the Fahrenheit† scale the values 32 and 212 are assigned.

We establish a mercury-in-glass thermometer scale by first bringing such a thermometer to the ice point and marking the position of the mercury surface in the stem 0, and then bringing the thermometer to the temperature of the steam point and marking the position of the mercury surface 100. The stem is then divided into 100 equal parts or degrees. We can establish a resistance thermometer scale by first noting the electrical resistance of a wire at the ice point and at the steam point and then making a straight-line plot of temperature versus resistance. We can follow a similar procedure with a thermocouple to obtain a straight-line plot (based on two points) of temperature versus emf for a fixed cold-junction temperature. If these three thermometers are placed together in a fluid which is at the ice point, all will read 0. If they are placed together in a fluid at the steam point, all will read 100. Agreement at the ice and steam points results, of course, from the calibration of these thermometers at these two temperatures which are used to define the Celsius scale. If the three thermometers are placed in a fluid at a temperature *A* which causes the mercury-in-glass thermometer to read 50°C, the other two thermometers will generally indicate temperatures slightly different from 50°C. This means that, while the expansion of mercury between the ice point and temperature *A* is

*Anders Celsius (1701–1744), a Swedish astronomer, presented a paper before the Swedish Academy of Sciences in 1742 describing the centigrade thermometer scale with 100 degrees between the ice and steam points with the ice point designated as the zero of the scale. The present Celsius scale is very nearly the same as the original centigrade scale proposed by Celsius.

†Gabriel Daniel Fahrenheit (1686–1736), a German instrument maker, around 1714 was the first to use mercury-in-glass thermometers. Alcohol and linseed oil had earlier been used as thermometric fluids. Fahrenheit's scale was a modification of one proposed by Sir Isaac Newton with 0 for the ice point and 12 for normal human body temperature. Fahrenheit lowered the zero of the scale to the temperature of a salt–ice mixture and made the degree smaller so that body temperature was 96. His measurements showed the ice and steam points to be at 32 and 212, respectively, based on the reference points at 0 and 96. Subsequently, the 32 and 212 were adopted as reference points, and refinements in thermometers have revealed that the salt–ice-mixture minimum temperature and normal body temperature are not exactly 0 and 96 on the present Fahrenheit scale.

one half of its expansion between the ice and steam points, the changes in the thermometric properties of the other thermometers do not have the same one-to-two ratio for these two temperature differences. This sort of discrepancy is found even among thermometers based on the same thermometric property, say, the expansion of a liquid. (Strictly speaking, the volume is the thermometric property, but only the change of volume is measured.) If three thermometers, all calibrated at 0°C and 100°C read 50, 49, and 53 when at the same temperature, which one is "correct"? We cannot answer this question from the data given. We may be tempted to say that two of the thermometers are "nonlinear" with temperature, but notice that there is no reason to establish any one as a standard in preference to the others. Thus a shortcoming of any temperature scale defined in terms of the physical properties of a substance is apparent. In Chapter 7 we shall see how it is possible to establish a *thermodynamic scale of temperature* which is independent of the properties of any substance.

Still, for convenience and for the purpose of having some scale as a reference until we develop the thermodynamic scale, we must define a temperature scale in terms of a readily measurable property of some substance. For reasons which will be brought out in Secs. 4·3 and 7·3, one of the best temperature scales related to physical properties is based on the variation with temperature of the pressure or volume of certain gases at low pressures.

A constant-volume gas thermometer consists of a gas, usually hydrogen or helium, contained in a constant-volume vessel provided with a means for measuring the gas pressure. The gas pressure varies with the temperature of the gas. Constant-volume gas thermometers calibrated at the ice point and the steam point agree closely with each other at other temperatures, even though different gases may be used. This agreement becomes better as the pressure of the gases (hence the mass in a given volume) is reduced. Extrapolation shows that the agreement becomes exact as the pressure approaches zero. Temperatures are measured on the constant-volume gas thermometer scale by the use of a linear relationship between temperature and pressure as shown in Fig. 1·3. The equation of the line shown is

$$t = t_0 + \left(\frac{t_s - t_i}{P_s - P_i}\right) P = t_0 + \frac{t_s - t_i}{P_s/P_i - 1}\left(\frac{P}{P_i}\right)$$

where the subscripts i, s, and 0 denote respectively conditions at the ice point, at the steam point, and at the temperature at which the gas pressure is zero. The value of t_0 can be obtained by the following procedure: The ratio of the pressure at the steam point to

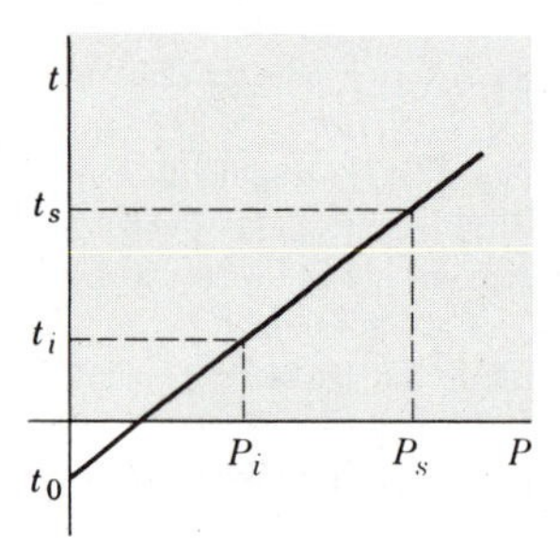

Figure 1·3 Constant-volume gas thermometer relationship.

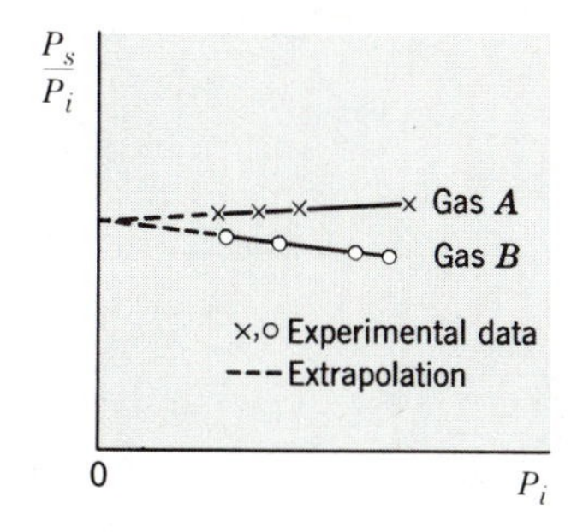

Figure 1·4 Determination of t_0 on a constant-volume gas thermometer.

the pressure at the ice point is measured for several different pressures at the ice point, the pressure at the ice point being varied by changing the amount of gas in the constant-volume thermometer. It is found that the measured pressure ratio P_s/P_i is a function of the ice-point pressure as shown in Fig. 1·4. In that figure, four measurements of the ratio are shown for some gas *A*. Measurements on some other gas *B* give slightly different values. It is an important fact that when the measurements are extrapolated to $P_i = 0$, the value of the ratio P_s/P_i is the same for all gases and is 1.3661. Using this value in the equation of the line in Fig. 1·3, substituting t_i and P_i as one possible pair of corresponding values of t and P, and using Celsius scale values for t_s and t_i

$$t_0 = t - \frac{t_s - t_i}{P_s/P_i - 1}\left(\frac{P}{P_i}\right) = 0 - \frac{100 - 0}{1.3661 - 1}(1) = -273.15°\text{C}$$

The *absolute Celsius gas thermometer* scale is defined as the scale on which (1) t_0 in the equation above is assigned the numerical value of zero, and (2) the interval between the ice and steam points is 100 degrees, that is, the degree is the same "size" as the Celsius degree. We will see later that the absolute Celsius gas thermometer scale is very close to the Kelvin scale, which is a thermodynamic temperature scale. Likewise, the absolute Fahrenheit gas thermometer scale is very close to the Rankine* thermodynamic temperature scale.

Notice that the definition of a gas thermometer absolute-temperature scale does not imply that a temperature equal to or lower than zero on that scale is unattainable, nor does it involve any assumption regarding the behavior of any substance at or near that temperature.

Despite its simplicity in principle, a constant-volume gas thermometer is difficult to use for precise measurements. Also, even though only two fixed points are required to establish a temperature scale, for practical purposes in calibrating thermometers, additional fixed points across a wider temperature range are needed. Therefore, use is made of the International Practical Temperature Scale, IPTS-68, adopted by the International Committee on Weights and Measures in 1968 as a revision of earlier temperature scales adopted

*William John MacQuorn Rankine (1820–1872), Scotch engineer and professor of civil engineering at the University of Glasgow, made several outstanding contributions to the development of thermodynamics and its engineering applications. He was the author of several books and more than 150 papers on thermodynamics, mechanics, canals, shipbuilding, steam engines, and water supply systems.

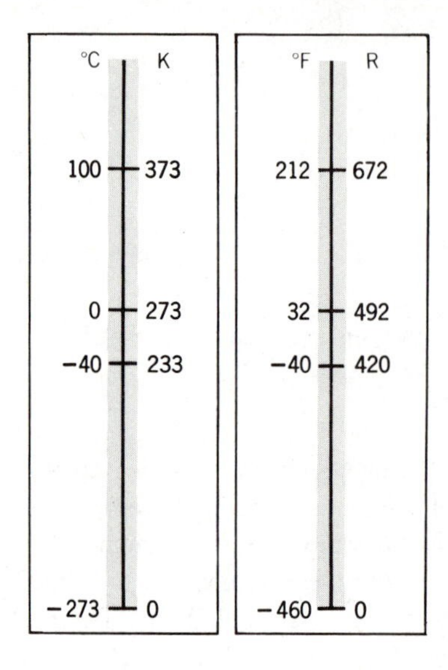

Figure 1·5 Comparison of temperature scales.

by that body. This scale specifies (1) numerical values for several easily reproducible temperatures (fixed points) over a wide temperature range and (2) temperature-measuring devices and interpolation methods in various temperature ranges. IPTS-68 agrees closely with the thermodynamic temperature scale discussed in Chapter 8.

Four temperature scales are encountered frequently in engineering practice: Fahrenheit, Rankine, Celsius, and Kelvin. The relationships among them are shown in Fig. 1·5 and are given by the following equations in which temperatures on the various scales are distinguished by subscripts and absolute temperatures are denoted by T instead of t.

$$T_R = t_F + 459.67 = \frac{9}{5} T_K$$

$$T_K = t_C + 273.15 = \frac{5}{9} T_R$$

The values 459.67 and 273.15 are replaced by the approximate values 460 and 273 in this textbook.

1·14 Work

Work is an interaction between a system and its surroundings. Work is done by a system on its surroundings if the sole external effect of the interaction could be the lifting of a body. The magnitude of work is the product of the weight of the body lifted and the distance it would be lifted if the lifting of the body were the sole external effect of the interaction. This definition points out that work involves both a system and something outside the system, whether it is called the surroundings or another system. In two systems A and B, interacting only with each other, the work done *by* system A is work done *on* system B and vice versa. The definition tells how to identify and measure the work done *by* a system. Work done *on* a system must be identified as work done *by* some other system. This roundabout method is necessary because the definition of work when reworded as "work is done on a system if the sole external effect could be the fall of a body" is *not true,* as is shown in Example 1·10.

Consider a system comprised of a compressed coil spring. As the spring expands against some part of its surroundings, the action of the spring on its surroundings could be reduced to the lifting of a weight. It does not matter whether the spring is actually being used to move an object against frictional resistance, to accelerate a body, or to push a plunger that in turn forces a fluid to flow through a small opening. The important fact is that the sole external effect *could be* the lifting of a weight while the spring undergoes the same process.

Consider a system that is a gas trapped in a cylinder behind a movable piston. If the gas expands, pushing the piston outward, the sole effect external to the gas *could be* the lifting of a body. If frictional effects are present in the surroundings, there may actually be effects other than the lifting of a body. For example, the temperature of some part of the surroundings may increase. However, if the frictional effects are reduced, the limiting case in which the sole external effect is the lifting of a weight is approached. The limiting condition of no friction in the surroundings is a useful concept in the identification and measurement of work. The use of this concept does not restrict us to the consideration of processes that involve no friction. It must be remembered that in deciding whether a certain interaction of a system with its surroundings is work, we ask, not if the sole external effect *is* the lifting of a weight, but rather: "*Could* the sole external effect be the lifting of a weight?" In seeking an answer to this question we consider as one possibility the limiting case of no friction in the surroundings.

A system comprised of a storage battery interacts with its surroundings by means of the electric current passing through its terminals. The electric current can be used to run a motor that in turn lifts a weight. For the limiting case of no frictional effects in the surroundings, the lifting of the weight would be the only effect outside the system, so the storage battery has done work on its surroundings. (The turning of the motor is not in itself an effect on the surroundings because, after the process is completed and the motor is brought to rest, it is in a state identical with its initial state. There has been no net change in the state of the motor. The weight, on the other hand, does show an effect of the process: The elevation of the weight after the process is different from the initial elevation.)

There is a widespread convention that work done *by* a system is expressed by *positive* numbers and work done *on* a system is expressed by *negative* numbers. Of course, since work is an interaction between systems, work done by one system must be done on some other system, so the same work is positive with respect to one system and negative with respect to another. Suppose that system A does 1000 J of work on system B. Following the convention described, we express this as

$$W_A = 1000 \text{ J} \qquad \text{or} \qquad W_B = -1000 \text{ J}$$

In this textbook, the work of a system is occasionally specified as work_{in} (meaning work done on the system) or work_{out} (meaning work done by the system) in order to avoid dependence on the sign convention. Where the subscript indicating direction is not used, however, the usual convention is followed: The term work, without a subscript indicating direction, means work done by the system to which the term is applied. As an illustration of both the sign convention and the use of directional subscripts, suppose again that system A does 1000 J of work on system B. We can then write

$$W_{in,A} = -1000 \text{ J} \qquad W_{in,B} = 1000 \text{ J}$$
$$W_{out,A} = 1000 \text{ J} \qquad W_{out,B} = -1000 \text{ J}$$
$$W_A = 1000 \text{ J} \qquad W_B = -1000 \text{ J}$$

Even though we speak of the work of a system, it is important to remember that work is an interaction between two systems.

The state of a system changes as work is done on or by the system. Work is not a characteristic which can be observed while a system is in a particular state. Thus work is not a property of a system. The change in value of a property is the same for all processes between two given end states; the amount of work done depends on the path of the process between the two end states. For this reason, work is called a *path function.* It was pointed out in Sec. 1·9 that properties are *point functions.* The distinction between point and path functions is the subject of Sec. 1·19.

Work done by an electric current is called electrical work. Work done by a magnetic field is called magnetic work. Work done by the action of a force at a moving boundary of a system is called mechanical work. The actions of the spring and the expanding gas discussed above are examples of mechanical work. *Mechanical work is work which is due to the action of a force on a moving boundary of a system. Its magnitude is equal to the product of the force and the displacement of its point of application in the direction of the force.* In many engineering applications of thermodynamics, mechanical work is the only form of work involved. However, thermodynamics deals with all forms of work; consequently, this section opened with a general definition which includes all forms of work that come within the broad scope of thermodynamics.

In this textbook of engineering thermodynamics, the term *work* without modifiers will generally mean mechanical work. Exceptions to this rule will be made clear by the context.

The definition of mechanical work does not require that the system boundary move in a direction normal to itself. Shear forces at the system boundary as well as normal forces are to be considered. An illustration of these two points is provided by a system comprised of the gas inside a rigid tank. A shaft extends through the wall of the tank, and on the end of the shaft inside the tank is a thin circular disk. When the shaft and disk turn at high speed, they do work on the gas even though (1) the volume of the gas remains constant, and (2) the boundary does not move in a direction normal to the boundary. The part of the gas near the shaft and disk is moving, however, and the shaft and disk exert a force on this moving part of the system. This illustration can also be made by taking as the system everything inside the tank, including the disk and that part of the shaft which is within the tank. An interaction between this system and its surroundings occurs where the shaft crosses the system boundary. Work is done on the system by means of the torque on the rotating shaft. If that part of the system boundary that is a plane cutting the shaft is considered to be moving (rotating) because the part of the system adjacent to it is moving, this interaction between the system and its surroundings comes within the given definition of mechanical work.

The definition of mechanical work specifies that its magnitude is equal to the product of a force and the displacement of the point of application of the force in the same direction. If the force F acts on a body which is displaced a distance s in the direction of the force, then the work done is given by

$$\text{work} = Fs$$

(If the line of action of the force and the total displacement are not in the same direction, then either F is taken as the force component in the direction of the displacement, or s is taken as the displacement component in the direction of the force.) The work done during a differential displacement ds is a differential amount of work, δ work, and is given by

$$\delta\ \text{work} = F\,ds \tag{1·6a}$$

The differential of a path function is written with the symbol δ instead of the symbol d. (This point is discussed further in Sec. 1·19.) For a finite displacement, the total amount of work done is given by

$$\text{work} = \int F\,ds \tag{1·6b}$$

Integration of this expression requires a knowledge of the functional relationship between F and s.

1·15 Work of a Frictionless Process of a Simple Compressible Closed System

Derivation of a Useful Expression from Mechanics

In some cases the work done in a process can be calculated by expressing $\int F\,ds$ in terms of a relationship between two properties of a system. This section and Sec. 1·17 discuss two of these cases.

We now develop an expression for the work done by a system during a process which satisfies the following conditions:

1. The system is closed.

2. There are no effects of electricity, magnetism, distortion of solids, motion, gravity, or capillarity. (The absence of solid distortion effects could also be specified as the absence of anisotropic stress or as the absence of any shear stress, because any one of these specifications implies the others.) A system that meets these conditions is often called a *simple compressible closed system.*

3. There are no frictional effects within the system. This precludes fluid shear forces on any part of the system.

4. The pressure is the same on all boundaries of the system; thus a single number represents the pressure of the system. (Conditions 3 and 4 can be established by requiring

that the system be at any instant in a state of equilibrium or infinitesimally close to an equilibrium state. A process carried out under such conditions is sometimes called a *quasistatic* process.)

The conditions listed can be summarized as a *quasistatic process of a simple compressible closed system.* (If we refer to the process as only *frictionless,* the additional restriction of condition 4 must be implied. In a later chapter another term, *reversible process,* will be introduced and will make the term *quasistatic* unnecessary.)

How can any expression developed for the condition of no gravity be of value? The answer to this question is that, although the relation to be developed is exactly correct only for this condition, it is sufficiently accurate for many situations encountered in practice where gravitational effects are negligible in comparison with other effects. If we are told that air at 300 kPa, 30°C, is stored in a 2-m diameter spherical tank, we generally do not ask where in the tank the air pressure is 300 kPa. The reason we do not ask is that the difference in pressure between air at the highest point and the lowest point in the tank is only about 0.06 kPa, and for most purposes this slight variation in pressure caused by gravity can be neglected. Similarly, the effects of motion such as the acceleration of part of the system by means of a moving piston are often negligible.

As an example of a system that satisfies the conditions listed above, consider a gas trapped in a cylinder and expanding against a piston as shown in Fig. 1·6. The gas expands from an initial state 1 to a final state 2. The piston moves slowly so that effects of motion on the system are negligible. The pressure is uniform throughout the system at any stage of the expansion, but the value of this uniform pressure changes as the expansion proceeds. At any stage of the expansion, the force on the piston is the product of the pressure of the gas and the area of the piston. Since this force acts in the direction of motion of the piston, the work done by the gas on the piston while the piston moves a distance ds is

$$\delta \text{ work} = F\, ds = PA\, ds$$

$A\, ds$ is the volume increase dV of the system as the piston travels the distance $ds,$ and so

$$\delta \text{ work} = P\, dV \tag{1·7a}$$

and the total amount of work done by the gas on the piston as the gas expands from state 1 to state 2 is

$$\text{work}_{1\text{-}2} = \int_1^2 P\, dV \tag{1·7b}$$

The *PV* Diagram

The area beneath a curve on pressure-volume coordinates is $\int P\, dV$. Therefore, the work of the process described above is represented by an area on a plot of the system pressure versus its volume as shown in the lower part of Fig. 1·6. The crosshatched area represents the work done by the gas as its volume increases by an amount dV. The total area beneath curve 1-2 is $\int_1^2 P\, dV$ and represents the work done by the system as it passes from state 1 to state 2.

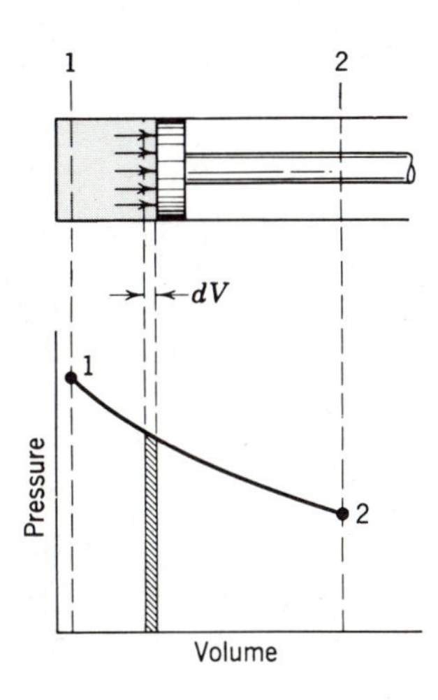

Figure 1·6 Pressure–volume diagram for a gas expanding in a cylinder.

A closed system can follow any one of many different paths as it changes from one particular state to another by means of frictionless processes. A plot of these paths on a *PV* diagram shows that, in general, the area beneath each path differs from that beneath other paths. This result was expected, because work is a path function and its value therefore depends on the path of the process between any two states and not on the states themselves.

The Sign Convention

Work done as the system expands is work done by the system, because the sole external effect could be the lifting of a body. When work is done by a system, $\int P\,dV$ is positive. When a system is compressed frictionlessly, work is done *on* the system and $P\,dV$ is negative. Thus the equation

$$\text{work} = \int P\,dV \tag{1·7c}$$

agrees with the sign convention of work out of (or by) the system as positive and work into (or on) the system as negative.

A Special Application

There are cases in which $\int P\,dV$ has significance even though shear forces at the system boundary are present. Consider a simple compressible system that is expanding or being compressed while simultaneously a shear force acts at some part of the boundary. Under these conditions, if the pressure is uniform over those parts of the boundary which are moving in such a manner as to change the volume of the system, then $\int P\,dV$ is the work done which is associated with the change in volume. It is not the total work done by the system, however. (See Example 1·7.) The total work done by a system is given by $\int P\,dV$ only for a quasistatic (and hence frictionless) process of a simple compressible closed system. (A gas stirred by a paddle wheel inside a rigid tank shows that work cannot be calculated by $\int P\,dV$ for all processes of simple compressible closed systems. For this system, $\int P\,dV$ is zero because the volume is constant; but the work is not zero as long as the paddle wheel is stirring the gas. Therefore, work $\neq \int P\,dV$.)

Absolute or Gage Pressure in $\int P\,dV$?

Absolute pressure must be used when the work of a system is evaluated by $\int P\,dV$. However, sometimes the result obtained by using gage pressure is significant. Referring to Fig. 1·7, the work done on the piston by the gas (the system) expanding frictionlessly is given by

$$\text{work} = \int P\,dV \tag{1·7c}$$

where P is the absolute pressure of the gas. Part of this work is used to push back the atmosphere, and the rest is used to move the piston against a resistance made up of the force F on the piston rod and the frictional force of the cylinder wall on the piston. (Notice that this frictional effect is outside the system; the system is expanding frictionlessly.) The atmosphere is assumed to be so large that its pressure P_0 is unaffected by changes in volume of the order of that caused by the expansion of the system. Therefore, the work done by the system on the atmosphere is given by

$$\text{work} = \int P_0\,dV = P_0 \int dV = P_0\,\Delta V$$

The work done on the piston rod and to overcome the cylinder wall friction is the work of the expanding gas minus the work done in pushing back the atmosphere.

$$\begin{aligned}(\text{total work of system}) - (\text{work done on atmosphere}) &= \int P\,dV - \int P_0\,dV \\ &= \int (P - P_0)\,dV \\ &= \int P_{\text{gage}}\,dV\end{aligned}$$

Generalizing, the use of gage pressure in $\int P\,dV$ for a closed-system frictionless process gives the work of the system excluding the work done on the surrounding atmosphere. We make little use of this fact because (1) we frequently deal with cyclic processes in which the net work on the atmosphere is zero, and (2) in our later work we find that it is the total work done by a system that we can most conveniently relate to various system properties and to other interactions of the system with its surroundings.

The Work of a Cyclic Frictionless Process of a Closed System

A closed system may pass by means of frictionless processes from a state 1 to a state 2 along a path such as 1-*a*-2 in Fig. 1·8. It may then complete a cycle by returning to state

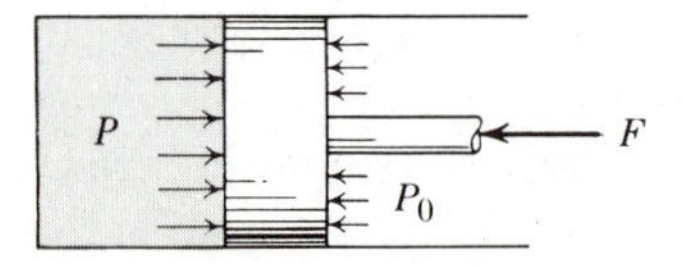

Figure 1·7 Effect of atmospheric pressure.

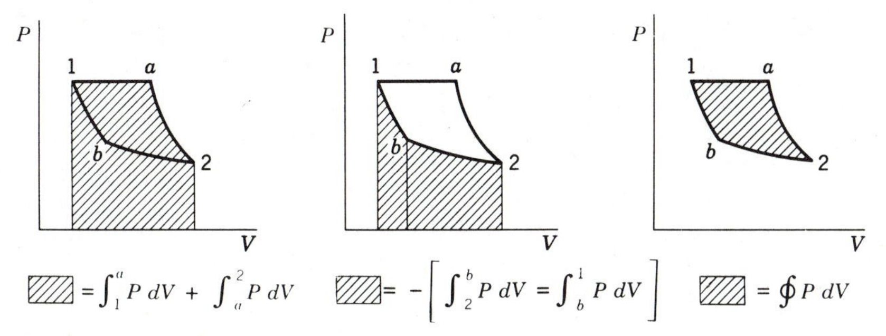

Figure 1·8 Cyclic, frictionless process of a closed system.

1. During the return process, the system may pass through different states from those it passed through during the initial processes 1-a-2. Let one of the states passed through during the return process be state b, as shown in Fig. 1·8.

During some of the processes of the cycle the system does work on its surroundings; during other processes work is done on the system. The net work of the system for the cycle is the sum of the work of all the processes making up the cycle. For the cycle made up of four processes,

$$\begin{pmatrix}\text{work of closed system}\\ \text{during frictionless cycle}\end{pmatrix} = \int_1^a P\,dV + \int_a^2 P\,dV + \int_2^b P\,dV + \int_b^1 P\,dV$$
$$= \oint P\,dV$$

where the symbol $\oint$ denotes integration along a closed or cyclic path.

The work done by the system during each process is represented by the area beneath the PV plot of the process, and Fig. 1·8 shows that the net work of the cycle is represented by the area enclosed within the diagram. Notice that the net work of a cycle is generally not zero, even though (by definition of a cycle) the system is returned to its initial state. This is another reminder that work is a path function and cannot be evaluated from only the end states of a process.

Indicator Diagrams

The pressure inside the cylinder of a reciprocating engine or compressor can be measured and plotted against the piston position or cylinder volume automatically. Because the device originally used for this purpose was called an indicator, the resulting pressure–volume record is called an indicator diagram. The area within the pressure–volume trace on an indicator diagram is proportional to the work done on the piston. However, the indicator diagram is not the same as the pressure–volume diagrams shown in this section because the PV diagrams shown here are for closed systems, whereas the cylinder of an engine or compressor is not a closed system during a complete cycle of operation.

Example 1·5. A fluid expands frictionlessly in a closed system from a volume of 0.100 m^3 to 0.160 m^3 in such a manner that the pressure is given by $P = CV^{-2}$, where C is a constant. The initial pressure is 300 kPa. Calculate the amount of work done.

Solution. Since a frictionless process of a closed system is involved and the relation between pressure and volume implies that the pressure is uniform throughout the system, the work done by the system is given by

$$\text{work} = \int_1^2 P\,dV$$

In order to integrate this expression, the relation between P and V must be used. Thus,

$$\text{work} = \int_1^2 P\,dV = \int_1^2 \frac{C\,dV}{V^2} = -C\left(\frac{1}{V_2} - \frac{1}{V_1}\right)$$

The numerical value of C could be computed and then substituted with the values of V_2 and V_1 into the equation above. However, less numerical computation is required if the value of C from the given PV relation is substituted: $C = PV^2 = P_1V_1^2 = P_2V_2^2$.

$$\text{work} = -C\left(\frac{1}{V_2} - \frac{1}{V_1}\right) = -P_1V_1^2\left(\frac{1}{V_2} - \frac{1}{V_1}\right) = -P_1V_1\left(\frac{V_1}{V_2} - 1\right)$$

$$= -300(0.100)\left(\frac{0.100}{0.160} - 1\right) = +11.3\text{ kJ}$$

The numerical value obtained is the work *out* or the work done *by* the system. If the numerical value obtained were negative, the conclusion would be that work had been done *on* the system.

Example 1·6. The cylinder shown in the figure is closed at its upper end and has a cross-sectional area of A m^2. Initially it contains V_1 m^3 of a gas at atmospheric pressure. The outer vessel, which contains a liquid with a free surface exposed to the atmosphere, is filled to the point of overflowing. How much work must be done on or by the gas if it expands until the liquid level in the cylinder is h m lower than that in the surrounding vessel?

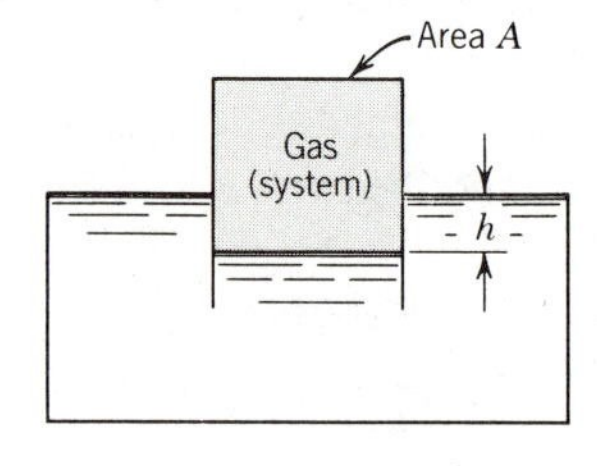

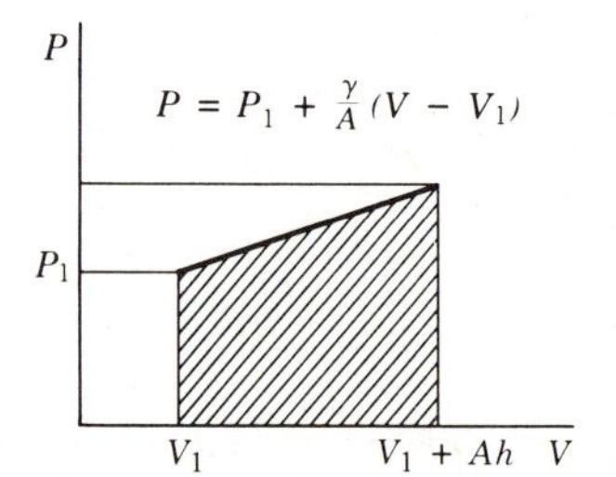

Example 1·6.

Solution. The gas trapped in the cylinder constitutes a closed system. Assuming that the expansion of the gas occurs slowly so that fricitonal effects within the gas are negligible and the pressure throughout the gas is uniform, the work done by the gas is given by

$$\text{work} = \int_{V_1}^{V} P\,dV$$

where the subscript 1 refers to the initial condition, and properties without subscripts are those at any stage of the expansion. In order to integrate this expression, we must find a relationship between P and V. Initially, since the gas in the cylinder is at atmospheric pressure, the liquid level is the same inside and outside the cylinder or $h = 0$. The volume of the gas at any stage of the expansion is given by

$$V = V_1 + Ah$$

and then, since V_1 and A are constants,

$$dV = A\,dh$$

The pressure can also be expressed in terms of h. Making use of the pressure–height relation for incompressible fluids,

$$\Delta P = \gamma h$$

where γ is the specific weight of the liquid, we have

$$P = P_1 + \gamma h$$

Since we now have both P and dV in terms of h, we can use h as the variable of integration:

$$\text{work} = \int_{V_1}^{V} P\,dV = \int_{h=0}^{h} (P_1 + \gamma h)A\,dh$$

Integration and substitution of the limits gives

$$\text{work} = A\left(P_1 h + \frac{\gamma h^2}{2}\right) = Ah\left(P_1 + \frac{\gamma h}{2}\right)$$

This work is represented by the crosshatched area on the PV diagram shown.

Example 1·7. Consider as a system the fluid contained in the cylinder as shown in the figure. The fluid expands from a volume of 1.40 to 1.60 cu ft while the pressure remains constant at 100 psia and while the paddle wheel does 3600 ft·lbf of work on the system. How much work is done by the system on the piston? What is the net amount of work done on or by the system?

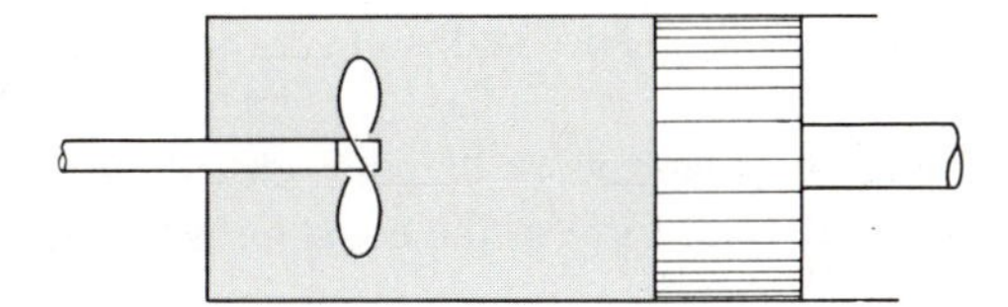

Example 1·7.

Solution. This process is obviously not frictionless because fluid shear forces are involved in the interaction between the paddle wheel and the gas (system). However, *if the action of the paddle wheel is such that the pressure at the piston face is uniform and of known value at each stage of the process,* the work done on the piston (or in other words the work associated with a change in volume of the system) can be calculated by

$$\text{work} = \int P\,dV$$

For the special case of a constant pressure, this relation becomes

$$\text{work} = P\int_1^2 dV = P(V_2 - V_1)$$

$$= 100(144)(1.60 - 1.40) = 2880 \text{ ft·lbf}$$

Since the paddle-wheel work is work done on the system, we write in accordance with the usual sign convention, work = −3600 ft·lbf. The net work of the system (meaning work done *by* the system in accordance with the convention) is

$$\text{net work} = \text{work}_{\text{piston}} + \text{work}_{\text{paddle}}$$
$$= 2880 + (-3600) = -720 \text{ ft·lbf}$$
$$\text{net work}_{\text{in}} = 720 \text{ ft·lbf}$$

Recall that these results are based on the assumption that the pressure on the piston face is 100 psia throughout the process.

Analogous Expressions for Work of Quasistatic Processes of Other Systems

Expressions for the work of quasistatic processes of several other kinds of systems are analogous to Eqs. 1·7 which apply to simple compressible closed systems.* For example, the work done by an elastic wire is given in differential form by

$$\delta W = -F\,dL \qquad \text{or} \qquad \delta W = -(\text{volume})\sigma\,d\varepsilon$$

where F is the tensile force, σ is the tensile stress, and ε is the unit strain (dL/L).

The work done by a surface film as it changes area in a quasistatic manner is given by

$$\delta W = -\mathcal{S}\,dA$$

where $\mathcal{S}$ is the surface tension and A is the area of the film.

The work involved in the quasistatic changing of the magnetization of a magnetic solid is given by

$$\delta W = -\mu_0\,\mathcal{H}\,d\mathcal{M}$$

where μ_0 is the permeability of free space, $\mathcal{H}$ is the magnetic field strength, and $\mathcal{M}$ is the magnetization.

Several other similar expressions for the quasistatic work of other closed systems exist, and it is significant that in each case the general form of the equation is

$$\delta \text{ work} = (\text{intensive property})\, d(\text{extensive property})$$

1·16 Steady Flow

The preceding section concerned a formulation for the mechanical work of a frictionless process of a simple compressible closed system. There is no analogous method for determining the work of a frictionless process of an open system except for the special case

*See, for example, Reference 1·1, pp. 108–113, or Reference 1·7, pp. 60–70. (Note, however, that the sign convention for work in Reference 1·7 is different from that in this textbook.)

of *steady flow* through the open system. This article will discuss steady flow, and the next one will present a means of calculating the work of a frictionless steady-flow process.

The flow through an open system is *steady flow* (and the system is often called a *steady-flow system*) if all properties at each point within the system remain constant with respect to time. This definition requires that the following particular conditions be met:

1. The properties of the fluids crossing the boundary remain constant at each point on the boundary.
2. The flow rate at each section where mass crosses the boundary is constant. (The flow rate cannot change as long as all properties, including velocity, at each point remain constant.)
3. The rate of mass flow into the system equals the rate of mass flow out of it, or the amount of mass enclosed by the boundary is constant. (If this were not true, the density in some parts of the system or the volume of the system would change in violation of the defining condition of steady flow.)
4. All interactions with the surroundings occur at a steady rate.

Notice that these four conditions are in terms of observations that can be made at the boundary of the system, so that no knowledge of the inner workings of a system is needed in order to determine whether steady-flow conditions prevail.

Some examples of steady-flow systems are a section of pipe line, a gas turbine, a welding torch, a carburetor, a boiler, and an air-conditioning unit. (If all the listed conditions are met except that the properties at various points within the system or at the system boundary vary cyclically with time, periodically returning to the same values, the flow may be treated as steady flow. For example, the flow through a reciprocating engine can sometimes be treated as steady flow.)

The mass rate of flow of a fluid passing through a cross-sectional area A is

$$\dot{m} = \frac{AV}{v} \tag{1·8a}$$

where V is the average velocity of the fluid in a direction normal to the plane of the area A, and v is the specific volume of the fluid.* For steady flow with fluid entering a system at a section 1 and leaving at a section 2,

$$\dot{m}_1 = \dot{m}_2 = \frac{A_1V_1}{v_1} = \frac{A_2V_2}{v_2} \tag{1·9}$$

This is the *continuity equation of steady flow*. This important relation is frequently used. It can readily be extended to any number of system inlets and outlets.

*A more general expression is

$$\dot{m} = \int \frac{u\,dA}{v} \tag{1·8b}$$

where u and v are values of velocity and specific volume which may vary across the flow cross section.

1·17 Work of a Frictionless Steady-flow Process

An expression for the work of a frictionless process of a *closed system* was derived in Sec. 1·15. A useful expression for the work of a frictionless *steady-flow process* will now be derived. The derivation procedure is: (*a*) make a free-body diagram of an element of fluid, (*b*) evaluate the external forces on the element, (*c*) relate the sum of the external forces to the mass and acceleration of the element, (*d*) solve the resulting relation for the force by which work is done on the fluid, and (*e*) apply the definition of work as $\int F\, ds$.

The absence of friction means that no shearing forces act on the fluid. An element of a fluid flowing under these conditions is shown in Fig. 1·9 where ΔL is to be considered small enough so that the variations of P, A, and V along ΔL are very nearly linear. This does not mean that the P, A, and V vary linearly with L, but only that ΔL is so small that a linear relation is a good approximation for the short distance ΔL. This approximation becomes better as ΔL becomes smaller, and later in the derivation an exact expression will be obtained by letting ΔL approach the differential length dL. The forces acting on this element in a direction parallel to the direction of flow are as follows:

1. The force of the adjacent fluid on the upstream face of the element, PA.
2. The force of the adjacent fluid on the downstream face of the element, $(P + \Delta P)(A + \Delta A)$.
3. The component of the weight of the element, weight $\cos\theta$ or $mg \cos\theta$.
4. The component of the normal wall forces in the direction of flow, $P_{avg}\, A_{surface} \sin \alpha$. Since the element is assumed to be small enough so that P varies almost linearly along ΔL, the average pressure is the arithmetic mean of P and $(P + \Delta P)$, and the normal wall force component in the direction of flow is $(P + \Delta P/2)A_{surface} \sin\alpha$ or $(P + \Delta P/2)\Delta A$.
5. The force on the element F_w, by which work is being done on the fluid. This force is applied to the fluid by means of some impeller which is not shown in the figure.

The sum or resultant of these forces is

$$\sum F = PA - (P + \Delta P)(A + \Delta A) - mg\cos\theta + \left(P + \frac{\Delta P}{2}\right)\Delta A + F_w$$

$$= -A\,\Delta P - \frac{\Delta P\,\Delta A}{2} - mg\cos\theta + F_w \tag{a}$$

Applying Newton's second law of motion, the sum of the external forces on the fluid element must equal ma. The mass of the element is $\rho(A + \Delta A/2)\Delta L$, and the acceleration is approximately $\Delta V/\Delta\tau$. Thus

$$\sum F = ma = \rho\left(A + \frac{\Delta A}{2}\right)\Delta L\,\frac{\Delta V}{\Delta\tau}$$

Noting that $\Delta L/\Delta\tau = V_{avg} = V + \Delta V/2$,

$$\sum F = \rho\left(A + \frac{\Delta A}{2}\right)\left(V + \frac{\Delta V}{2}\right)\Delta V \tag{b}$$

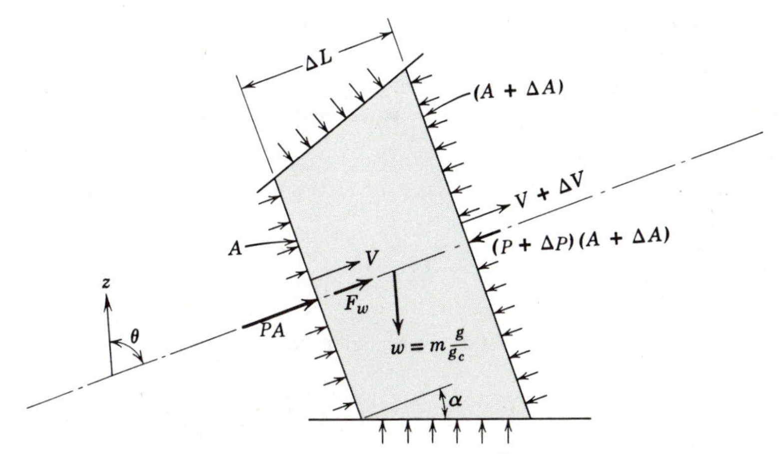

Figure 1·9 Element of fluid in frictionless steady flow.

Combining Eqs. a and b and neglecting differences of the order of $\Delta P\ \Delta A$ and $\Delta A\ \Delta V$,

$$-A\ \Delta P - mg\cos\theta + F_w = \rho AV\Delta V$$

or

$$F_w = A\ \Delta P + \rho AV\Delta V + mg\cos\theta$$

The work done on the element (or the work put into the element) is

$$\text{work}_{\text{in}} = F_w\ \Delta L = A\ \Delta L\ \Delta P + \rho A\Delta LV\Delta V + mg\ \Delta L\cos\theta$$

If differences of the order of $\Delta A\ \Delta L$ are neglected, then $A\ \Delta L$ is the volume of the element and $\rho A\ \Delta L$ is its mass. $\Delta L\cos\theta$ is Δz. Then

$$\text{work}_{\text{in}} = F_w\ \Delta L = (\text{volume})\ \Delta P + mV\Delta V + mg\Delta z$$

and per unit mass

$$\text{work}_{\text{in}} = v\ \Delta P + V\Delta V + g\ \Delta z$$

Now, if ΔL is made to approach dL, then the other differences also approach differentials, and the work (per unit mass) done on the fluid in the distance dL is

$$\text{work}_{\text{in}} = v\ dP + V\ dV + g\ dz \qquad (1·10a)$$

or for flow between sections a finite distance apart

$$\text{work}_{\text{in}} = \int v\ dP + \Delta\left(\frac{V^2}{2}\right) + g\Delta z \qquad (1·10b)$$

This is an important relation for the *mechanical work done on a unit mass of fluid in a frictionless steady-flow process*. Notice that it was derived from the principles of mechanics just as the analogous expression for the work of a frictionless process of a simple compressible closed system was derived.

Example 1·8. A fluid flowing at a steady rate of 3.0 lbm/s through an open system expands frictionlessly according to the relation Pv^2 = constant from an initial pressure of 45 psia to a final pressure of 15 psia. The density of the fluid entering the system is 0.25 lbm/cu ft. Changes in velocity and elevation are negligible. Calculate the power delivered by the fluid.

Solution. The power, or the rate at which work is done, can be calculated by multiplying the work done per pound by the mass rate of flow. Since the expansion is a frictionless steady-flow process, the work done by the fluid is given by

$$\text{work}_{\text{out}} = w = -\int_1^2 v\,dP - \frac{V_2^2 - V_1^2}{2} - g(z_2 - z_1)$$

Noting that the changes in velocity and elevation are negligible and that $Pv^2 = C = P_1v_1^2$,

$$w = -\int_1^2 v\,dP = -C^{1/2}\int_1^2 P^{-1/2}\,dP = -C^{1/2}(2)[P_2^{1/2} - P_1^{1/2}]$$

$$= -P_1^{1/2}v_1(2)(P_2^{1/2} - P_1^{1/2}) = -2P_1v_1\left[\left(\frac{P_2}{P_1}\right)^{1/2} - 1\right] = 2\frac{P_1}{\rho_1}\left[\left(\frac{P_2}{P_1}\right)^{1/2} - 1\right]$$

$$= -\frac{2(45)144}{0.25}\left[\left(\frac{15}{45}\right)^{1/2} - 1\right] = 21\,900 \text{ ft·lbf/lbm}$$

$$\text{power} = (\text{work})\,\dot{m} = 21\,900(3.0) = 65\,700 \text{ ft·lbf/s}$$

Using the definition 1 hp = 550 ft·lbf/s,

$$\text{power} = \frac{65\,700}{550} = 119 \text{ hp}$$

Example 1·9. Nitrogen flows steadily and frictionlessly through a nozzle at a rate of 0.82 kg/s. It enters the nozzle at 300 kPa, 350 K, with a specific volume of 0.346 m³/kg and a velocity of 160 m/s. The nitrogen expands within the nozzle according to the relation $Pv^{1.4}$ = constant and leaves at 150 kPa. The change in elevation between inlet and outlet is negligible. Calculate the cross-sectional area of the nozzle outlet.

Analysis. Since the mass rate of flow is known, the outlet area can be obtained from the continuity equation,

$$A_2 = \frac{\dot{m}v_2}{V_2}$$

provided the specific volume and the velocity at the outlet can be found.

The outlet specific volume v_2 can be found from $P_2v_1^{1.4} = P_2v_2^{1.4}$ since v_1 and both pressures are known.

In order to find the outlet velocity V_2, let us investigate the suitability of the expression derived from the principles of mechanics,

$$w = -\int_1^2 v\,dP - \frac{V_2^2 - V_1^2}{2} - g(z_2 - z_1)$$

Since there is no work done on or by a fluid passing through a nozzle, and the elevation change is negligible, the above expression reduces to

$$\frac{V_2^2 - V_1^2}{2} = -\int_1^2 v\,dP$$

V_1 is known and the right-hand side can be integrated because the functional relationship between P and v is known; so V_2 can be found from this expression. The values of V_2 and v_2 which can be found by the methods outlined above can then be used in the continuity equation to give the value of A_2 which is being sought. This analysis thus outlines the method of solution.

Solution. The relation $P_1 v_1^{1.4} = P_2 v_2^{1.4}$ can be solved for v_2:

$$v_2 = v_1 \left(\frac{P_1}{P_2}\right)^{1/1.4} = 0.346 \left(\frac{300}{150}\right)^{1/1.4} = 0.568 \text{ m}^3/\text{kg}$$

Since $Pv^{1.4} = C$, work $= 0$ for flow through a nozzle, and the elevation change is negligible, the expression

$$\frac{V_2^2 - V_1^2}{2} = -w - \int_1^2 v \, dP - g(z_2 - z_1)$$

reduces to

$$\frac{V_2^2 - V_1^2}{2} = 0 - C^{1/1.4} \int_1^2 P^{-1/1.4} \, dP - 0$$

$$= -P_1^{1/1.4} v_1 \left(\frac{1.4}{1.4 - 1}\right) (P_2^{1-1/1.4} - P_1^{1-1/1.4})$$

$$= -\frac{1.4}{0.4} P_1 v_1 \left[\left(\frac{P_2}{P_1}\right)^{1-1/1.4} - 1\right]$$

$$= -\frac{1.4}{0.4}(300)0.346 \left[\left(\frac{150}{300}\right)^{1-1/1.4} - 1\right] = 116 \text{ kJ/kg}$$

$$V_2 = \sqrt{2(116)\,1000 + V_1^2} = \sqrt{2(116)\,1000 + (160)^2}$$

$$= 508 \text{ m/s}$$

The factor of 1000 is necessary in the last line, as a check of units shows, because it is 1 newton (*not* 1 *kilo*newton) that is defined as the force required to accelerate 1 *kilo*gram of mass at a rate of 1 m/s^2.

The values found for v_2 and V_2 can now be used to find A_2:

$$A_2 = \frac{\dot{m} v_2}{V_2} = \frac{0.82(0.568)}{508} = 9.17 \times 10^{-4} \text{ m}^2$$

1·18 Heat

As discussed in Sec. 1·12, if two bodies at different temperatures are brought into contact with each other while isolated from all other bodies, they will interact with each other so that the temperature of one or both will change until both bodies are at the same temperature. This interaction between the bodies or systems is the result of only the temperature difference between them and is called *heat*. *Heat is an interaction between a system and its surroundings which is caused by a difference in temperature between the system and its surroundings.* Conventionally, we say that heat is added to a body which becomes hotter or is taken from a body which becomes colder.* The definition above provides a means of *recognizing* the interaction called heat (and particularly provides a means of distinguishing this interaction from the one called work). In addition, we must have some method of *measuring* heat.

Heat can be measured by the use of *standard systems* which can be made to change from one readily recognizable state to another by means of the transfer of heat under specified conditions. The magnitude of the heat transferred is then given by the number of such standard systems which can be made to undergo the specified state change as a result of the interaction. For example, suppose we wish to know how much heat is transferred from a block of steel which cools from the temperature of boiling water at one atmosphere (100°C) to the temperature of freezing water at one atmosphere (0°C). The standard systems we use might be 1-kg blocks of ice at 0°C. The steel block can be cooled by being isolated with one such block of ice after another so that each block of ice is melted but the temperature of the resulting liquid is not changed. The number of such standard systems (blocks of ice) so used is a measure of the amount of heat removed from the steel block. Of course, many different kinds of standard systems other than blocks of ice could be used.† (A word of caution is needed here. The conditions under which the standard systems interact with the steel block must be specified and precisely controlled. The blocks of ice, for example, can be made to change state even without heat transfer, so the state changes are related to heat transfer only if they are carried out under the same specified conditions in each instance.) The important conclusion is that it is possible to describe certain operations by means of which heat can be measured; hence, heat can be defined operationally.

The widely adopted sign convention for heat is that heat added *to* a system is expressed by *positive* numbers and heat taken *from* a system is expressed by *negative* numbers. Suppose that 100 J of heat is transferred from system A to system B. Following the convention described, we express this as

$$Q_A = -100 \text{ J} \qquad \text{or} \qquad Q_B = 100 \text{ J}$$

In this textbook, directional subscripts are used for heat just as for work. Where the subscript indicating direction is not used, the usual convention is followed: Q (without a subscript) stands for the net amount of heat *added to* the system to which the term is applied. As an illustration of both the sign convention and the use of directional subscripts, suppose again that heat is transferred from system A to system B in the amount of 100 J. We can express this in the following ways:

$$\begin{aligned} Q_{\text{in},A} &= -100 \text{ J} & Q_{\text{in},B} &= 100 \text{ J} \\ Q_{\text{out},A} &= 100 \text{ J} & Q_{\text{out},B} &= -100 \text{ J} \\ Q_A &= -100 \text{ J} & Q_B &= 100 \text{ J} \end{aligned}$$

*Unfortunately, conventional expressions such as "heat is added" and "heat is transferred," which stem from the now discredited caloric theory, tend to give the erroneous impression that heat is a substance. No such difficulty is encountered with the other interaction between systems, work.

†The original definitions of the units of heat, the calorie and the British thermal unit (Btu, abbreviated further as B), were based on standard systems comprised of specified masses of water at specified temperatures. The specified state change in each case was a temperature change of one degree on a particular scale of temperature. The calorie and the Btu are no longer so defined.

Heat, like work, is an interaction between systems. It is not a characteristic which can be observed while a system is in a particular state. It is not a property of a system. The amount of heat transferred to or from a system during a process cannot be determined from the end states alone. Instead, the amount of heat transferred depends on how the system was changed from one state to another. Heat, like work, is a *path function*.

Occasionally, beginning students in thermodynamics are confused because statements about "the heat in a gallon of gasoline" or "the heat in a radiator" are frequently heard in everyday conversation (conversation in which there is no concern with measurements or with consistency among definitions). The definition of heat used in thermodynamics makes such statements meaningless.

A full understanding of the differences between the two interactions, heat and work, depends on the second law of thermodynamics which is considered in Chapters 5, 6, and 7. Nevertheless, it is well to review at this point the essential difference between them which is established by their definitions: Heat is an interaction caused by a temperature difference between a system and its surroundings; work is done by a system if the sole effect of the system on its surroundings could be reduced to the lifting of a weight.

Consider a system consisting of air and an externally driven paddle wheel in a thermally insulated vessel.* The system and the surroundings are initially at the same temperature. If the paddle wheel is turned by an external motor, the temperature of the system increases. The interaction between the system and its surroundings is the result of the torque exerted on the rotating shaft and is not the result of a temperature difference between the system and its surroundings because (1) the system and surroundings were at the same temperature at the beginning of the process and (2) the vessel is thermally insulated to prevent any interaction resulting from a temperature difference. Thus the interaction is work, not heat. It is unfortunate and misleading that in everyday language it would often be said that the gas was "heated by the paddle wheel" or "heated by friction."

A process in which there is no heat transfer is called an *adiabatic* process. A system which exchanges no heat with its surroundings is called an adiabatic system.

Example 1·10. Referring to the figure, the air, oil, ice, and liquid water within the outer container are all initially at a temperature of 0°C. The outer container is thermally insulated so that no heat can be transferred through it. The weight is allowed to fall, turning the paddle wheel in the oil by means of the pulley. After some time has elapsed, it is observed that the air, oil, ice, and liquid water are again at 0°C but that some of the ice in the inner container has melted. Identify the work done on or by each system and the heat transferred to or from each of the systems which are identified as follows: System A is everything within the outer container,

*You may justifiably raise an eyebrow at some of the systems chosen for study and may wonder if thermodynamics is concerned only with peculiar systems which a practicing engineer is unlikely to encounter. The reply is that the systems used in proofs and explanations are occasionally unusual because, for the illustration of principles, it is desirable to use systems that involve only one interaction or effect at a time. Sometimes this calls for a highly simplified system (as the paddle-wheel system here), and sometimes it calls for an elaborately controlled system (as the standard systems mentioned earlier in this section, where pressure and temperature were both controlled and interactions with systems other than the steel block were prevented). These systems are chosen because each clearly illustrates some principle or definition; in turn, the application of the principle or definition to some more common or more complex system is then more readily understood.

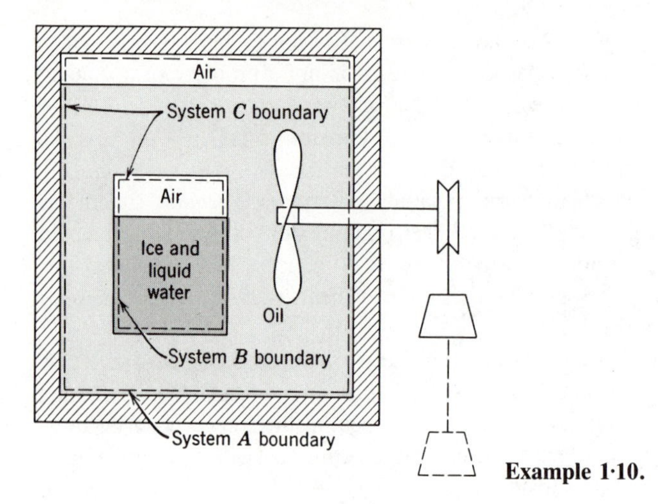

Example 1·10.

system B is everything within the inner container, and system C is everything within the outer container except system B. (System A = system B + system C.)

Solution. A study of the figure and the problem statement indicates that, as a result of the paddle-wheel action, as the weight falls the temperature of the oil (and the air above the oil) increases. The resulting temperature difference between the oil and the contents of the inner container (system B) causes a transfer of heat from the oil into the inner container. This transfer of heat melts some of the ice and causes the temperature of the oil and air to return to 0°C. The interactions involving the various systems are as follows:

System B. *Heat.* There is a transfer of heat into system B from system C inasmuch as there is an interaction which results from a temperature difference. *Work.* No mechanical work is done on or by system B inasmuch as its interaction with its surroundings involves no motion of the system boundary.

System A. *Heat.* No heat is transferred to or from system A because the system is thermally insulated from its surroundings. (Something occurs *within* system A as a result of a temperature difference *within* the system, but with regard to system A this is not heat transfer because heat is an interaction *between a system and its surroundings.*) *Work.* Work is done on system A by means of the shaft which turns the paddle wheel. Notice that, as far as the action of the weight and pulley system external to system A is concerned, the action within system A could be reduced to the lifting of a weight (by removing the oil and replacing the paddle wheel by a weight and pulley arrangement).

System C. *Heat.* Heat is transferred from system C to system B. No other heat is transferred to or from system C because that part of its boundary which is not common to system B is thermally insulated. *Work.* Work is done on system C by means of the shaft which turns the paddle wheel.

As an additional point of interest, consider system B and its surroundings. After the process described is completed, the temperature of everything outside system B is the same as at the beginning of the process, so the sole effect on the surroundings of system B is the fall of the weight. In Sec. 1·14 it was pointed out, however, that an interaction between a system and its surroundings in which the only effect on the surroundings is the *fall* of a body is not necessarily work. In fact, the entire interaction of system B with its surroundings is the result of a temperature difference between the system and its surroundings (even though the temperature difference is zero at both the beginning and the end of the process); so the interaction between system B and its surroundings is only a transfer of heat.

1·19 Point and Path Functions

Several references have already been made to point functions and path functions. The differences between them will now be discussed.

Consider a system composed of a gas trapped in a cylinder behind a piston, as shown in Fig. 1·10. Initially, the system is at state 1. At this state 1 the properties pressure, volume, and temperature can be measured. If the gas is then compressed to state 2, the properties can again be measured. The measurements at each state are entirely independent of those made at the other state. The change in a property such as pressure is $P_2 - P_1$, and the value of this change depends only on the states 1 and 2; it does not depend on the manner in which the system was changed from state 1 to state 2. For example, the pressure might first have been increased to its final value while the piston was stationary (a constant-volume process) and then held constant while the piston was moved to its final position (a constant-pressure process). This path is shown as 1-*a*-2 in Fig. 1·10. Three other possible paths connecting states 1 and 2 are shown. The change in any property between state 1 and state 2 is the same for any path, because the value of a property is a characteristic of a system in a given state and is independent of the path followed by the system in reaching that state.

On a diagram using properties as coordinates, each point represents a particular state of a system; that is, associated with each point on the diagram is a particular value of each property of the system. Conversely, each state of the system can be represented by a single point on the diagram. (If more than two properties must be specified to define the state of a system, as is often the case, then a two-dimensional diagram is insufficient.) Properties are therefore referred to as point functions.

Refer again to the system of Fig. 1·10 and suppose that all properties of the system were measured at state 1 and again at state 2. Is it possible from these measurements to

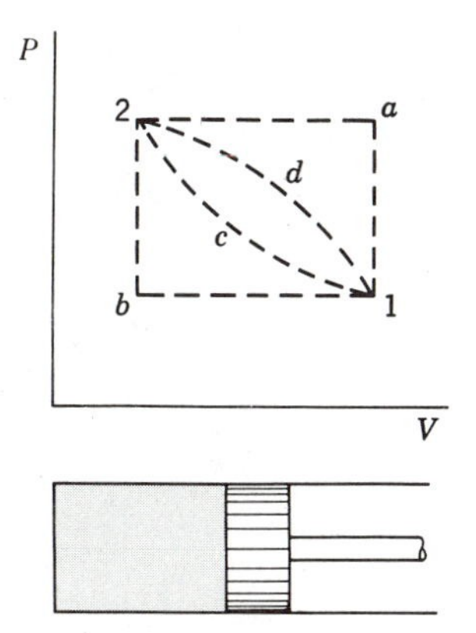

Figure 1·10 Four possible paths from state 1 to state 2.

determine the work done in compressing the gas from state 1 to state 2? The answer is no. Determination of the work done requires more than a knowledge of the end states of the process. To illustrate this, consider that the process is frictionless so that work can be evaluated by $\int P\,dV$. It is readily seen that the value of $\int P\,dV$ is different for each of the four paths shown in Fig. 1·10, and many more paths connecting states 1 and 2 are possible. Thus the amount of work done depends on which path is followed between states 1 and 2. Work is therefore called a path function.

Heat is also a path function, although the demonstration of this fact at this point is more involved that the demonstration that work is a path function.

Notice that there is no such thing as a value of work or of heat at state 1 or at any other state. Work and heat are interactions between systems, not characteristics of systems in particular states.

The mathematical notation for point and path functions and a test which distinguishes between differentials of point functions and those of path functions deserve attention. If x is a function of two independent variables, y and z, this fact is expressed by the notation

$$x = f(y, z)$$

and x is called a *point function,* because at each point on a plane of yz coordinates there is a discrete value of x. The differential dx of a point function x is called an *exact differential,* and

$$dx = \left(\frac{\partial x}{\partial y}\right)_z dy + \left(\frac{\partial x}{\partial z}\right)_y dz$$

This may be written as

$$dx = M\,dy + N\,dz$$

where

$$M = \left(\frac{\partial x}{\partial y}\right)_z \qquad N = \left(\frac{\partial x}{\partial z}\right)_y$$

Then

$$\frac{\partial M}{\partial z} = \frac{\partial}{\partial z}\left(\frac{\partial x}{\partial y}\right) = \frac{\partial^2 x}{\partial z\,\partial y} \qquad \frac{\partial N}{\partial y} = \frac{\partial}{\partial y}\left(\frac{\partial x}{\partial z}\right) = \frac{\partial^2 x}{\partial y\,\partial z}$$

and consequently, since the order of differentiation is immaterial,

$$\frac{\partial M}{\partial z} = \frac{\partial N}{\partial y}$$

In fact, this provides a test for exactness. *A differential in the form* $dx = M\,dy + N\,dz$ *is an exact differential (hence is the differential of a point function) if and only if*

$$\frac{\partial M}{\partial z} = \frac{\partial N}{\partial y}$$

A proof of this relation can be found in any elementary textbook on differential equations.

For an exact differential dx

$$\int_1^2 dx = x_2 - x_1$$

where $x_2 = f(y_2, z_2)$ and $x_1 = f(y_1, z_1)$. The value of $\int_1^2 dx$ is independent of the path followed on yz coordinates in going from (y_1, z_1) to (y_2, z_2). For a closed path or cycle the initial and end states (or points on a property diagram) are identical; so

$$\oint dx = 0$$

Let G be a path function, a quantity which depends on the path followed in going from state 1 (y_1, z_1) to state 2 (y_2, z_2). For such a quantity *no* relation of the form

$$G = F(y, z)$$

exists, because specifying a value of y and a value of z does not determine a value of G. The notation G_1 or G_2 should not be used, as this implies that there is a particular value of G at state 1 or state 2, and this is not true. The value of G corresponding to a particular path between states 1 and 2 is therefore not spoken of as a "change" in G, but simply as a value of G for that particular path.

This value of G is equal to the sum of the G values for any number of segments into which the path may be divided. If these segments are made smaller and smaller, the limiting value of G of one segment is δG. The symbol δ can be interpreted as "a very small amount of," whereas a corresponding loose interpretation of the differential operator d would be "the difference between two values very close together."

There may be a relationship of the form

$$\delta G = M' \, dy + N' \, dz$$

but, since G is a path function, δG is an *inexact differential* and

$$\frac{\partial M'}{\partial z} \neq \frac{\partial N'}{\partial y}$$

For any specific path between states 1 and 2, we may write

$$G_{1\text{-}2} = \int_1^2 \delta G$$

However, for a path function,

$$\int_1^2 \delta G \neq G_2 - G_1$$

because (1) a path function cannot be evaluated in terms of end states alone, and (2) there are no values such as G_1 and G_2 which can be assigned to states 1 and 2.

In order to illustrate a path function other than work or heat, consider the length of a line connecting two points, 1 and 2, on an xy coordinate plane (Fig. 1·11). Let the length of some path connecting 1 and 2 be L. There is no value of L for point 1 nor point 2, nor is there a single value for points 1 and 2 together. There is, in general, a different

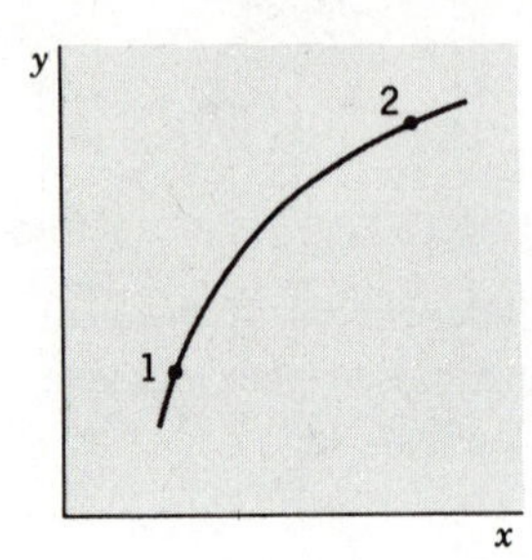

Figure 1·11 Path on a property diagram.

value of L for each of the many possible paths between 1 and 2. For any small segment of one of these paths, the limiting relationship is

$$(\delta L)^2 = (dx)^2 + (dy)^2$$

and for the entire path,

$$L_{1\text{-}2} = \int_1^2 \delta L = \int_1^2 \sqrt{1 + (dy/dx)^2}\, dx$$

This cannot be evaluated unless the relationship between y and x is known, that is, unless the path is specified. Knowledge of the end points alone is insufficient, because L is a path function and $L \neq f(x, y)$.

For a closed path on an xy coordinate plane, notice that

$$\oint dx = 0 \qquad \text{and} \qquad \oint dy = 0$$

but that

$$\oint \delta L \neq 0$$

Example 1·11. For some substances, $Pv = BT$, where P, v, T, and B are pressure, specific volume, temperature, and a constant, respectively. The quantities s and I are given by

$$\delta s \qquad \text{or} \qquad ds = \frac{C\, dT}{T} - \frac{v\, dP}{T}$$

and

$$\delta I \qquad \text{or} \qquad dI = \frac{D\, dT}{T} + \frac{P\, dv}{v}$$

where C and D are constants. Determine whether each of these quantities is a point function.

Solution. Since in each case the differential is equal to an expression of the form of $M\, dy + N\, dz$, we need only to use the test for exactness

$$\frac{\partial M}{\partial z} \stackrel{?}{=} \frac{\partial N}{\partial y}$$

In the case of δs or ds, $M = C/T$, $N = -v/T$, $z = P$, and $y = T$. Then for $Pv = BT$

$$\left(\frac{\partial(C/T)}{\partial P}\right)_T \stackrel{?}{=} \left(\frac{\partial(-v/T)}{\partial T}\right)_P$$

$$= \left(\frac{\partial(-B/P)}{\partial T}\right)_P$$

$$0 = 0$$

Hence, s is a point function. In the case of I, we have

$$\left(\frac{\partial(D/T)}{\partial v}\right)_T \stackrel{?}{=} \left(\frac{\partial(P/v)}{\partial T}\right)_v$$

$$0 \neq \frac{B}{v^2}$$

Hence I is not a point function.

Example 1·12. Evaluate $\int_A^B dR$ over each of the two paths (1) $y = x^2$, and (2) $y = 3x - 2$ connecting points A (1,1) and B (2,4) for each case:

$$(a)\ R = \int (x\,dy + y\,dx)$$

$$(b)\ R = \int (x\,dy - y\,dx)$$

Solution. (*a*) For the path $y = x^2$

$$\int_A^B dR = \int x\,dy + \int y\,dx = \int 2x^2\,dx + \int x^2\,dx$$

$$= 3\int x^2\,dx = x^3\Big]_{x=1}^{x=2} = 8 - 1 = 7$$

For the path $y = 3x - 2$

$$\int_A^B dR = \int x\,dy + \int y\,dx = \int 3x\,dx + \int (3x - 2)\,dx$$

$$= \int (6x - 2)\,dx = 3x^2 - 2x\Big]_{x=1}^{x=2} = 8 - 1 = 7$$

The same result for both paths should have been expected because application of the test for exactness shows that $x\,dy + y\,dx$ is exact; hence $R = f(x, y)$ and R has a particular value for each value of x and y. It is readily seen by inspection that $x\,dy + y\,dx = d(xy)$; therefore $R = xy$, and the simplest way to evaluate $\int_A^B dR$ is by

$$\int_A^B dR = R_B - R_A = x_B y_B - x_A y_A = 2(4) - 1(1) = 7$$

No knowledge of any particular path connecting A and B is necessary because R is a point function.

This particular function allows a simple graphical explanation of its independence of the path chosen between A and B. In each figure, the area between the curve and the x-axis

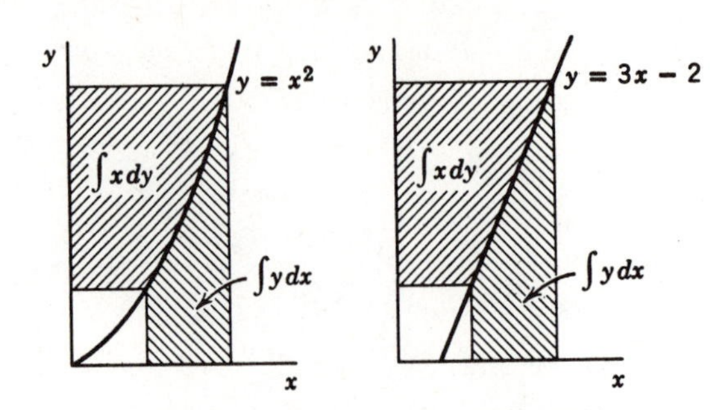

Example 1·12.

represents $\int y\,dx$. This area is not the same for the two paths. The same is true of the areas between the curves and the y axes which represent $\int x\,dy$. However, the *sum* of the two areas on one figure is the same as on the other figure, so that $\int x\,dy + \int y\,dx$ has the same value for either path.

(*b*) For the path $y = x^2$

$$\int_A^B dR = \int x\,dy - \int y\,dx = \int 2x^2\,dx - \int x^2\,dx$$

$$= \int x^2\,dx = \frac{x^3}{3}\bigg]_{x=1}^{x=2} = \frac{8}{3} - \frac{1}{3} = \frac{7}{3}$$

For the path $y = 3x - 2$

$$\int_A^B dR = \int x\,dy - \int y\,dx = \int 3x\,dx - \int (3x - 2)\,dx$$

$$= \int 2\,dx = 2x\bigg]_{x=1}^{x=2} = 4 - 2 = 2$$

Different results for the two paths should have been expected, since application of the test for exactness to $R = \int(x\,dy - y\,dx)$ shows that $R \neq f(x, y)$. The differential of R should then be δR instead of dR. On the figures above it is apparent that the *difference* in areas $\int x\,dy$ and $\int y\,dx$ does depend on the path followed between A and B.

1·20 Summary

Thermodynamics is the science that deals with energy transformations, including heat and work, and the physical properties of substances which are involved in energy transformations. Engineering thermodynamics is the part of the science which pertains to all types of heat engines, turbines, compressors, refrigeration, air conditioning, combustion, chemical processing plants, and the physical properties of substances used in these applications.

The study of any science should begin with the acceptance of certain terms as being undefined on the verbal level. Precise definitions of other terms must then be established in relation to the undefined terms. Definitions are most valuable if they are *operational*. An operational definition includes directions for measuring the defined quantity or for testing an event or object to see if it fits the definition.

In this textbook, terms which will be accepted as undefined verbally are time, length, temperature, mass, and force.

A thermodynamic system is defined as any quantity of matter or region of space within prescribed boundaries. A system which has no mass crossing its boundaries is called a *closed system*. An *open system* is a region of space within a boundary that mass may cross. Everything outside the system boundary is called the *surroundings*. A closed system that in no way interacts with its surroundings is called an *isolated system*. An *ideal system* is one which can be completely described in terms of a few characteristics.

A *property* is any characteristic of a system which is observable while the system is in any one condition or *state*. Some examples of properties are pressure, temperature, volume, and density.

A *process* is a change in a system from one state to another. The *path* of a process is the series of states through which the system passes during the process. A *cycle* or *cyclic process* is a process or series of processes that returns the system to the state it was in before the process began. The net change in any property of a system which executes a cycle is zero; that is, if x is a property

$$\oint dx = 0$$

An *intensive property* is one which has the same value for any part of a homogeneous system as it does for the whole system. An *extensive property* is one which has a value for a system equal to the sum of its values for the various parts of the system.

For a fluid in static equilibrium, the relationship between pressure P and elevation z within the fluid is given by the basic equation of fluid statics

$$dP = -\gamma\, dz \tag{1·5}$$

where γ is the specific weight of the fluid and z is measured positively upward.

Two bodies are at the *same temperature* or are in *thermal equilibrium* with each other if all properties remain unchanged when the two bodies are brought into conjunction with each other while isolated from all other bodies. Two bodies in thermal equilibrium with a third body are in thermal equilibrium with each other.

Work is an interaction between a system and its surroundings. Work is done by a system on its surroundings if the sole external effect of the interaction could be the lifting of a body. The magnitude of work is the product of the weight of the body lifted and the distance it could be lifted if the lifting of the body were the sole external effect of the interaction. This definition of work includes electrical work, magnetic work, and mechanical work.

Mechanical work is the interaction between a system and its surroundings that is due to the action of a force on a moving boundary of the system. Its magnitude is equal to the product of the force and the displacement, in the direction of the force, of the point of application. In this textbook, the term *work* without modifiers will mean mechanical work except in a few cases where the exceptions to the rule are specifically pointed out.

For a frictionless (or quasistatic) process of a closed simple compressible system

$$\text{work}_{\text{done by system}} = \int P\, dV \tag{1·7}$$

where P is the pressure of the system and V is its volume.

Steady flow is flow in which all properties at each point within an open system remain constant with respect to time. For steady flow with fluid entering a system at a section 1 and leaving at a section 2, the mass rate of flow $\dot{m}$ is given by

$$\dot{m}_1 = \dot{m}_2 = \frac{A_1 V_1}{v_1} = \frac{A_2 V_2}{v_2} \tag{1·9}$$

in which A is cross-sectional area, V is average velocity across the section, and v is the specific volume of the fluid. This is the continuity equation of steady flow.

For a frictionless steady-flow process, per unit of mass

$$\text{work}_{\text{done by system}} = -\int v\,dP - \Delta\left(\frac{V^2}{2}\right) - g\Delta z \tag{1·10}$$

Heat is an interaction between a system and its surroundings which is caused by a difference in temperature between the system and its surroundings. Heat can be measured by means of standard systems which are changed from one readily recognizable state to another by means of the transfer of heat under specified conditions. An *adiabatic* process is one in which no heat is transferred.

The interactions heat and work are *path functions.* They can be evaluated for a process only if the path of the process is known. Differentials of heat and work are *inexact differentials.* In general

$$\oint \delta W \neq 0 \qquad \oint \delta Q \neq 0$$

Suggested Reading

1·1 Dixon, J. R., *Thermodynamics I,* Prentice-Hall, Englewood Cliffs, N.J., 1975, Chapter 1 and Section 2-4.

1·2 Holman, J. P., *Thermodynamics,* 2nd ed., McGraw-Hill, New York, 1974, Chapter 1.

1·3 Kestin, Joseph, *A Course in Thermodynamics,* revised printing, McGraw-Hill, New York, 1979, Chapters 1 and 2.

1·4 Lewis, G. N., and M. Randall, *Thermodynamics,* 2nd ed., revised by K. S. Pitzer and Leo Brewer, McGraw-Hill, New York, 1961, Chapters 1 to 3.

1·5 Reynolds, William C., and Henry C. Perkins, *Engineering Thermodynamics,* 2nd ed., McGraw-Hill, New York, 1977, Chapter 1.

1·6 Van Wylen, Gordon J., and Richard E. Sonntag, *Fundamentals of Classical Thermodynamics,* 3rd ed., Wiley, New York, 1985, Chapters 1 and 2.

1·7 Zemansky, Mark W., Michael M. Abbott, and Hendrick C. Van Ness, *Basic Engineering Thermodynamics,* 2nd ed., McGraw-Hill, New York, 1975, Chapters 1 and 3.

On Definitions in General

1·8 Hayakawa, S. I., *Language in Thought and Action,* 4th ed., Harcourt Brace Jovanovich, New York, 1978, pp. 8–16 and Chapters 4 and 10.

1·9 Rapoport, Anatol, *Operational Philosophy,* Harper & Brothers, New York, 1954, Chapter 1.

Problems

1·1 Which of the following definitions are operational?

(a) *Potential energy*: energy of a body which is due to the position of the body.

(b) *Kinetic energy*: energy due to the motion of a body.

(c) *Kinetic energy*: energy possessed by a body which is evaluated by $mV^2/2$, where m is the mass of the body, and V is its velocity.

(d) *Work*: the act of overcoming a resistance through space.

(e) *Work*: work is done by a system when the system acts on its surroundings in such a way that the entire external effect can be reduced to the raising of a weight in the surroundings, and the magnitude of the work done is the product of the weight and the distance the weight is raised.

(f) *Heat*: disordered energy.

(g) *Mass*: the quantity of matter in a body.

(h) *Mass*: a property of matter which can be measured by means of a balance.

(i) *Mechanical work*: the interaction between a system and its surroundings which is due to the action of a force at a moving boundary of the system, and its magnitude is equal to the product of the force and the displacement of its point of application in the direction of the force.

1·2 Which of the following definitions are operational?

(a) *Specific volume*: the volume of a substance divided by its mass.

(b) *Specific volume*: the reciprocal of density.

(c) *Temperature*: the degree of hotness of a body.

(d) *Energy*: that which is capable of producing an effect.

(e) *Total enthalpy:* a property which is the sum of enthalpy and kinetic energy.

(f) *Property:* a statistical average of a characteristic of the molecular behavior of a substance.

1·3 If the specific volume of a gas is 1.7 cu ft/lbm, what volume will 3 lbm occupy?

1·4 A cylinder having a volume of 2 cu ft contains 20 lbm of gas. Determine the specific volume. If 14 lbm of gas escapes, compute the final specific volume and the final density.

1·5 A room with a volume of 170 m^3 contains 200 kg of air. One half of the air is drawn from the room and compressed into a tank with a volume of 7.0 m^3. Determine the density of the air in the tank and of the air remaining in the room.

1·6 Two pounds of gas occupy a volume of 14 cu ft. Compute the specific volume in cubic centimeters per gram (cc/g).

1·7 Determine the pressure in psia units on a rock submerged 80 ft below sea level. Assume that the specific gravity of sea water is 1.025. The barometric pressure is 14.7 psia.

1·8 Determine the pressure on the hull of a submarine submerged 50 m below sea level. Assume that the specific gravity of sea water is 1.025. The barometric pressure is 76.0 cm of mercury.

1·9 The absolute steam pressure at the entrance and exit of a turbine are 5500 kN/m^2 and 2.6 cm Hg, respectively. If the barometric pressure is taken as 75 cm of mercury, compute the gage pressure for the entering steam and the vacuum gage pressure for the steam at exit from the turbine.

1·10 The absolute steam pressure at entrance and exit of a turbine are 800 and 0.5 psia, respectively. If the barometric pressure is taken as 29.46 in. of mercury, compute the gage pressure of the entering steam and the vacuum gage pressure of the steam at exit from the turbine.

1·11 A vacuum gage reads 25 in. of mercury. Determine the absolute pressure in psia units if the barometer reads 29.7 in. of mercury.

1·12 A mercury manometer at sea level has the same deflection as one at an altitude of 15 000 ft. The temperatures are the same. Which one indicates the greater pressure difference? Explain.

1·13 Convert the following Celsius temperatures to Fahrenheit: (*a*) −30, (*b*) −10, (*c*) 0, (*d*) 200, (*e*) 1050.

1·14 Convert the following Fahrenheit temperatures to Celsius: (*a*) −60, (*b*) −40, (*c*) 5, (*d*) 540, (*e*) 2070.

1·15 On the Réaumur temperature scale the ice point is at zero and the steam point is at 80. What is the Réaumur temperature of absolute zero?

1·16 Determine the validity of the following procedures for converting between Celsius and Fahrenheit temperature scales. (*a*) Add 40 to the temperature on one of these scales, multiply by 5/9 or 9/5, and subtract 40 to obtain the temperature on the other scale. (*b*) Double the Celsius temperature scale value, subtract 10 percent of that result, and add 32 to obtain the Fahrenheit temperature. (*c*) Subtract 32 from the Fahrenheit temperature, halve this amount, and add one-ninth of the result to obtain the Celsius temperature.

1·17 A hoisting engine in a mine is raising a cage weighing 500 lbf at a rate of 200 fpm. What power is developed by the engine?

1·18 A 1000-ton train operates at a speed of 60 mph on a level track. If the frictional resistance is 15 lbf/ton, determine (*a*) the force required and (*b*) the power (in hp) developed by the engine.

1·19 A locomotive draws a train at a velocity of 4 mph up a grade which rises 1 ft vertically in each 100 ft measured horizontally. The total weight of the locomotive and cars is 4000 tons. If the frictional resistance is 4 lbf/ton, how much power is developed by the engine?

1·20 A locomotive draws a train at a velocity of 50 km/h up a grade which rises 30 cm vertically in each 30 m measured horizontally (a 1.0 percent grade). The total mass of the locomotive and cars is 3800 t. If the frictional resistance is 30 N/t, what power is developed by the engine?

1·21 A locomotive draws a train at a velocity of 80 km/h down a 0.20 percent grade. The total mass of the locomotive and cars is 2400 tonnes. Frictional resistance is 50 N/t. Determine the power delivered by the locomotive engine.

1·22 An aircraft with a total mass of 20 000 kg is in steady level flight at a velocity of 800 km/h. The thrust provided by the engine is 70 kN. Determine the power being used to propel the craft.

1·23 A tank 4 m long, 3 m wide, and 2 m deep is half full of water. How much work is required to raise all the water over the top edge of the tank?

1·24 A tank 10 ft long, 7 ft wide, and 6 ft deep is half full of water. How many foot-pounds of work would be required to raise all the water over the top edge of the tank?

1·25 A well hole having a diameter of 2 in. is to be cut into the earth to a depth of 200 ft. Determine the total work required to raise the earth material to the surface if the average weight of 1 cu ft is 115 lbf.

1·26 A $\frac{1}{2}$-hp electric motor has nothing attached to its shaft. The motor is operating at 3600 rpm and is connected to a 110-volt power line. What is the power output?

1·27 A steel coil spring stands on end on a table. Its linear spring constant is 0.98 N/mm. An object with a mass of 1.10 kg is placed on the spring so that it is compressed to a new equilibrium position. How much work is done on or by the spring as it is compressed? How much work is done on or by a similar spring with the same spring constant if it is hanging from a hook and an object of 1.10 kg mass is attached to the lower end of the spring to stretch it to a new equilibrium position?

1·28 Explain the difference: $dP/d\tau$ is the rate of change of pressure with time, but power, $\dot{W} = dW/d\tau$, cannot be defined as the rate of change of work with time.

1·29 In Sec. 1·14 it is stated that the magnitude of mechanical work is equal to the product of a force on a moving boundary of a system and the displacement of its point of application in the direction of the force. Show how the case of a rotating shaft passing through the system boundary is covered by this statement, with work being the product of a torque and an angular displacement.

1·30 Calculate the work done on a body having a mass of 140 kg to accelerate it from a velocity of 120 m/s to 150 m/s in (*a*) 10 s, (*b*) 1 min.

1·31 Calculate the work done on a body having a mass of 10 slugs to accelerate it from a velocity of 400 to 500 fps in (*a*) 10 s, (*b*) 1 min.

1·32 Calculate the work done in lifting a 7-slug body from an elevation of 640 ft above mean sea level to an elevation 60 ft higher in (*a*) 2 min, (*b*) 10 min.

1·33 Assuming that the gravitational acceleration g varies with altitude h above the earth's surface according to the relationship

$$g = \frac{a}{(b + h)^2}$$

where a and b are constants, sketch a curve of the weight of a given object versus h, and write an equation for the work done in lifting an object of mass m to an elevation h. Write the equation in terms of m, h, a, and b.

1·34 Compute the work performed when 4 cu ft of a gas expands to a volume of 9 cu ft under a constant pressure of 200 psia. Draw a pressure–volume diagram for the process.

1·35 Compute the work performed when 2 m^3 of a gas expands to a volume of 5 m^3 under a constant pressure of 1380 kN/m^2. Draw a pressure–volume diagram for the process.

1·36 A vertical U-tube has a circular cross section with a diameter of 1.0 cm. One leg of the U-tube is capped and the other is open to the atmosphere. The two ends are at the same elevation. Mercury in the U-tube traps nitrogen in the closed leg so that the mercury level is 30 cm below the capped end. In the other leg, the mercury is at the open end, so increasing the pressure of the nitrogen causes mercury to overflow. Atmospheric pressure is 95.0 kPa. The nitrogen is heated so that it expands until the level of the mercury in that leg is lowered to 40 cm below the capped end. Calculate the work done by the nitrogen as it expands. (You will undoubtedly neglect the effect of the change in mercury temperature near the nitrogen. If this effect were considered, would the calculated work be greater, smaller, or the same?)

1·37 A rigid steel tank contains 6 cu ft of oxygen at a pressure of 200 psia. If the oxygen is cooled so that its pressure is reduced to 180 psia, is any work performed? Assume that the volume of the tank remains constant. Draw a pressure–volume diagram for the process.

1·38 A toy balloon filled with a light hydrocarbon gas floats against the ceiling of a room. If the gas cools and the balloon contracts, is any work done on or by the gas? Draw a pressure–volume diagram for the cooling process.

1·39 A gas is trapped by a frictionless piston in a vertical cylinder having an inner diameter of 80 cm. The piston is connected to nothing else and so is supported by the gas pressure. The distance from the cylinder closed end to the piston face is 40 cm. The gas pressure is 250 kPa. The gas is then heated so that the piston is lifted 10 cm. Atmospheric pressure is 100 kPa. Calculate the work done by the gas. How much of this work is done in lifting the piston and how much in pushing back the atmosphere? Determine the mass of the piston.

1·40 In a closed system a gas is compressed frictionlessly from a volume of 0.100 m^3 and a pressure of 0.70 kPa to a volume of 0.025 m^3 in such a manner that $P(V + 0.030) =$ constant, where V is in m^3. Calculate the work.

1·41 A process occurs for which the pressure changes according to the equation $P = 288V + 900$. In this relation the pressure is expressed in psia units and the volume is expressed in cubic foot units. If the volume changes from 10 to 20 cu ft, compute the work performed.

1·42 Air expands in a cylinder according to the law $PV^{1.4} = C$ from an initial volume of 3 m^3 and pressure of 450 kN/m^2 to a final volume of 4 m^3. The symbol C represents a constant. Compute the work.

1·43 Carbon dioxide is compressed frictionlessly in a cylinder in such a manner that $PV^{1.2} =$ constant from $P_i = 100$ kPa, $V_i = 0.0040$ m^3, to $P_f = 500$ kPa. The subscripts i and f signify initial and final conditions respectively. Compute the work.

1·44 Air expands in a cylinder according to the law $PV^{1.4} = C$ from an initial volume of 5 cu ft and pressure of 70 psia to a final volume of 10 cu ft. Compute the work done. The symbol C represents a constant.

1·45 A gas expands according to the law $PV^a = C$, where a and C are constants. Derive a general equation for the work done when the gas expands from an initial absolute pressure and volume of P_1 and V_1 to a final absolute pressure and volume of P_2 and V_2 (*a*) in a closed system, (*b*) in a steady-flow system.

1·46 A spherical balloon has a diameter D_1 when the pressure of the gas inside it is P_1 and atmospheric pressure is P_0. The gas is heated, causing the balloon diameter to increase to D_2. The pressure of the gas is proportional to the balloon diameter. How much work is done by the gas? Express your answer in terms of P_1, D_1 and the ratio D_2/D_1. How much work is done on the atmosphere?

1·47 A vapor in a cylinder is compressed frictionlessly from 20 psia and 1 cu ft to 100 psia in such a manner that $PV =$ constant. Sketch the process on PV coordinates, and compute the work done.

1·48 A gas trapped in a cylinder behind a movable piston has an initial volume of 0.40 m^3, and its pressure is 159 kPa. It is made to undergo a process which follows the relationship $(V - 0.2)10^5 = (P - 300)^2$ until its pressure is 400 kPa. Sketch a PV diagram of this process and determine the work.

1·49 Liquid sulfur dioxide with a density of 1394 kg/m^3 enters an expansion valve at a rate of 32.6 kg/h with a velocity of 0.610 m/s. In passing through the valve, the sulfur dioxide partially vaporizes, so that it leaves the valve with a density of 11.9 kg/m^3. Inlet and outlet areas are equal, and the flow is steady. Calculate (*a*) the outlet area and (*b*) the outlet velocity.

1·50 Air with a density of 1.20 kg/m^3 enters a steady-flow system through a 30-cm diameter duct with a velocity of 3 m/s. It leaves with a density of 3.20 kg/m^3 through a 10-cm diameter duct. Determine the outlet velocity and the mass rate of flow.

1·51 Air with a density of 0.075 lbm/cu ft enters a steady-flow system through a 12-in. diameter duct with a velocity of 10 fps. It leaves with a density of 0.20 lbm/cu ft through a 4-in. diameter duct. Determine the outlet velocity and the mass rate of flow.

1·52 Consider a liquid in a circular pipe of radius r_0. At one instant, the velocity is a maximum u_{max} at the pipe axis and zero at the pipe wall. Determine the ratio V/u_{max}, where V is the average velocity at a given section, if the velocity u at any radius r is given by $u = u_{max}(1 - r/r_0)^2$.

1·53 Consider a liquid flowing steadily in a circular pipe of radius r_0. Determine the ratio V/u_{max}, where V is the average velocity across the pipe and u_{max} is the maximum velocity, if the velocity u at any radius is given by $u = u_{max}[1 - (r/r_0)^2]$.

1·54 A gas with a density of 0.079 kg/m³ flows at a velocity of 200 m/s through a cross-sectional area of 0.010 m². Determine the mass rate of flow.

1·55 Steam with a specific volume of 1.155 m³/kg flows at a rate of 3.30 kg/s through a cross-sectional area of 0.010 m². Determine the average velocity across the cross section.

1·56 Steam with a specific volume of 0.551 m³/kg has an average velocity of 200 m/s across a flow area of 0.0140 m². Determine the mass rate of flow.

1·57 Water flows steadily through a converging tube that is a truncated circular cone with an included angle of θ. At inlet, the diameter is D_1 and the water pressure and average velocity are P_1 and V_1. Determine the pressure P at any section a distance L downstream from the inlet in terms of P_1, V_1, θ, L, and the water density.

1·58 Give a specific example of (*a*) a closed-system process in which $\int P\,dv = 0$ and work $\neq 0$, and (*b*) a steady-flow process in which work $= 0$ and $\int v\,dP \neq 0$.

1·59 Consider a nutating disk or "wobble-plate" liquid meter. If the gears of the counter mechanism become hard to turn, will the pressure drop across the meter be affected for a given flow rate?

1·60 Air is compressed in a frictionless steady-flow process from 70 kN/m², −29°C, and a specific volume of 1 m³/kg to 100 kN/m² in such a manner that $P(v + 0.5) =$ constant, where v is in m³/kg. Inlet velocity is negligibly small, and discharge velocity is 100 m/s. The flow rate is 0.20 kg/s. Calculate the work required per kilogram of air.

1·61 Carbon dioxide is compressed frictionlessly in a steady-flow process in such a manner that $Pv^{1/2} =$ constant from 100 kPa to 500 kPa. The initial specific volume is 0.560 m³/kg. Compute the work per kilogram of gas.

1·62 Air is compressed in a frictionless steady-flow process from 10 psia, 60°F, and a specific volume of 19.25 cu ft/lbm to 15 psia in such a manner that $P(v + 5) =$ constant, where v is in cu ft/lbm. Inlet velocity is negligibly small and discharge velocity is 350 fps. The flow rate is 0.45 lbm/s. Calculate the work required per pound of air.

1·63 Refer to Fig. 1·9 and the derivation of Eq. 1·10 in Sec. 1·17. The pressure and cross-sectional area at one end of the fluid element considered are P and A; those at the other end are P and ΔP and A and ΔA. Carry out the derivation by using P and A as the pressure and area near the center of the element. Then, at the ends of the element the pressure will be $P + \Delta P/2$ and $P - \Delta P/2$, and so on.

1·64 Derive an expression for the work done in the elastic stretching of a steel wire in terms of modulus of elasticity and strain.

1·65 Determine the amount of work required to stretch a steel wire 1.00 mm in diameter and initially 500 mm long to a length of 510 mm at room temperature. (Use a modulus of elasticity of 200 × 10³ MPa.)

1·66 In everyday language the noun *heat* and the verb *to heat* are often used in sentences different from those which follow from the definition of heat used in engineering thermodynamics. One type of misuse leads to confusion between an interaction between systems and a property. The other type of misuse leads to a confusion between different types of interactions between systems. Give several examples of each type of misuse which may be heard in everyday conversation.

1·67 For each process and each system listed, indicate whether the work is more than, less than, or equal to zero, and do the same for heat, observing the usual sign convention for work and heat:

(a) A coil spring stands on end on a table. A book is placed on the spring, compressing it. Consider as the system (1) the book, (2) the spring, (3) the table.

(b) A paddle wheel turned by a motor stirs a liquid in an insulated vessel. Consider as the system (1) the liquid, (2) the paddle wheel.

(c) A gas in an insulated cylinder is compressed so that its pressure and temperature both increase. The system is the gas.

(d) A steel wire is bent back and forth until it becomes hot to the touch. The system is the wire.

(e) Carbon dioxide is compressed in a water-cooled compressor. Steady flow prevails. The system is (1) the carbon dioxide, (2) the cooling water, (3) a section of the compressor cylinder wall.

1·68 For each process and each system listed, indicate whether the work is more than, less then, or equal to zero, and do the same for heat, observing the usual sign conventions for work and heat:

(a) An automobile stands with bright sunlight on its left side. The temperature of the air in the tires is thereby increased. The system is (1) the air in the tires, (2) the automobile.

(b) A household refrigerator is driven by an electric motor. Steady-flow conditions prevail for several hours. The system is (1) the motor, (2) the entire contents of the refrigerator cabinet.

(c) An aircraft engine drives a propeller through a speed-reducing gearbox. Operation is steady. Power delivered to the propeller is 96 percent of the power delivered to the gearbox. The system is (1) the gearbox, (2) the propeller.

1·69 Someone has proposed the use of three new thermodynamic quantities: X, Y, and Z. Determine which ones are properties if they are defined as (1) $X = \int (P\,dv + v\,dP)$; (2) $Y = \int (P\,dv - v\,dP)$; and (3) $Z = \int (R\,dT + P\,dv)$, where $R = Pv/T =$ constant.

1·70 Evaluate F, G, and H as defined by

$$F = \int (x^2\,dy + y\,dx)$$

$$G = \int [(x - y)\,dy + (y^2 - 2x)\,dx]$$

$$H = \int [2y(x - 2y)\,dx + x(x - 8y)\,dy]$$

over each of the following paths from $x = 0$, $y = 0$, to $x = 1$, $y = 2$: (*a*) $y = 2x$, (*b*) $y = 2x^2$, and (*c*) $y = 2x^{1/2}$.

1·71 Evaluate I, J, and K as defined by

$$I = \int (xydx + xydy)$$

$$J = \int [(x + y - 20)\,dx - (x - 2y)\,dy]$$

$$K = \int 2ydx$$

over each of the following paths from $x = 20$, $y = 20$, to $x = 40$, $y = 60$: (*a*) $y = 2x - 20$, (*b*) $y = 0.025x^2 + 0.5x$, and (*c*) $y = 60 - 0.10(x - 40)^2$.

1·72 Make a list of the 10 numbered equations in this chapter. For each one, state the principle involved and the restrictions on its use. Which ones were you familiar with before studying this chapter and which ones are new to you?

CHAPTER TWO

The First Law of Thermodynamics

Thermodynamics is characterized by a great number and diversity of applications of its few basic principles. One of these basic principles, the first law of thermodynamics, is introduced in this chapter, and several examples of its use are presented. It is shown that the conservation of energy principle follows from the first law of thermodynamics. Further applications of the first law are made throughout the remaining chapters of the book.

2·1 The First Law of Thermodynamics

It was pointed out in Chapter 1 that, when a closed system passes through a cycle, usually

$$\oint \delta W \neq 0 \qquad \text{and} \qquad \oint \delta Q \neq 0$$

Many experiments show, however, that *if* the net work of a cycle is zero, then the net heat transfer of that cycle is also zero. Further experiments show that whenever there is a net work input to a closed system during a cycle, there must be a net heat transfer from the system. Conversely, a net work output from a closed system during a cycle is always accompanied by a net heat input. These experimental observations suggest that there is a relationship between work and heat. Let us consider some experiments which might be useful in a search for such a relationship.

Consider a closed system which is comprised of a gas in a rigid vessel fitted with a paddle wheel as shown in Fig. 2·1. If work is done on the gas by means of the paddle wheel (process A), the temperature of the gas will rise. Then heat must be removed from the gas (process B) in order to restore the gas to its initial state. During process A there is work done but no heat transfer; during process B there is heat transferred but no work done. During the complete cycle there is a net work input and a net heat transfer from the system. Even relatively crude experiments with this apparatus show that as the amount of work input is increased, the amount of heat that must be removed to restore the system to its initial state increases proportionately.

Another useful system is a gas that is trapped inside a cylinder fitted with a piston (Fig. 2·2). The piston is a thermal insulator (i.e., heat cannot pass through it) and the

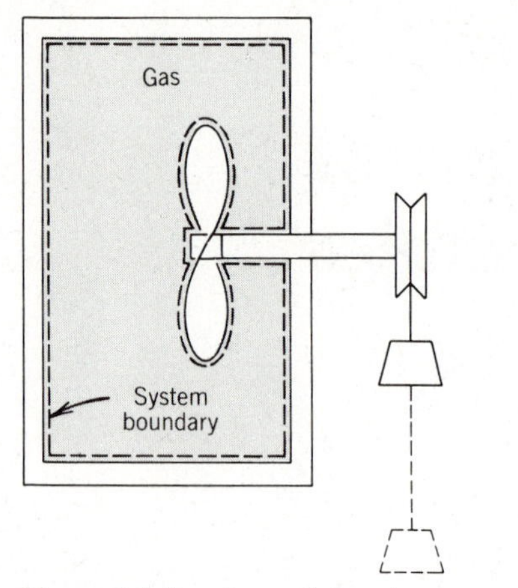

Figure 2·1 One form of closed system.

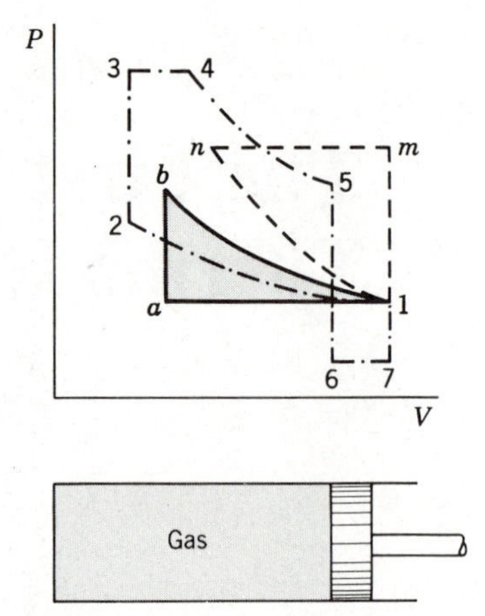

Figure 2·2 Three possible cycles of a closed system.

cylinder can be surrounded either by thermal insulating material or by liquid baths at various temperatures so that the transfer of heat to or from the gas can be readily controlled. The work done on or by the gas is determined from measurements of the force on the piston and the displacement of the piston. Many different cycles can be executed by this system. Three possible ones are shown on the *PV* diagram of Fig. 2·2. The work of the system can be positive, zero, or negative.

Many experiments on systems as shown in Figs. 2·1 and 2·2 and on many other types of systems indicate that, *whenever a closed system executes a cycle, the net work output of the system is proportional to the net heat input.* This conclusion from experiments is known as the first law of thermodynamics and is expressed by*

$$\oint \delta W \propto \oint \delta Q$$

or by

$$\oint \delta W = J \oint \delta Q \tag{2·1}$$

where J is a proportionality factor whose value depends only on the units selected for W and Q.

The dimensions and possible units of work are obvious from the definition of work given in Sec. 1·14. This is not the case with heat as defined in Sec. 1·18, and one might infer from that definition that suitable units might be devised in terms of numbers of standard systems undergoing prescribed state changes. In fact, the early units of heat such as the calorie and the British thermal unit (Btu, further abbreviated as B) were established in just such a manner. On the basis of the experimental result given by Eq. 2·1, many measurements were made of the proportionality factor which was called the "mechanical equivalent of heat," often represented by the symbol J. For example, in English units,

$$J = 778.169 \text{ ft·lbf/B}$$

Since there is a fixed proportionality between these units, one (say, ft·lbf) could be

*The notation $(\Sigma W)_{cycle}$ and $(\Sigma Q)_{cycle}$ is also used.

defined in terms of the other. Or heat and work can be expressed in the same units. Both can be measured in calories or Btus, and heat as well as work can be measured in newton-meters or foot-pounds force. This is now done. Consequently, J is simply a conversion factor.* Any equation expressing a physical relationship must hold regardless of the system of units used for the various quantities in the equation. The only requirement is that the units used be *consistent* and that the proper conversion factors be used in any numerical calculation in order to make them so. When this is recognized, it is unnecessary to write conversion factors as part of an equation. Therefore, the conversion factor J will not be written in equations in this book, it being understood that work and heat, for example, appearing in the same equation must be expressed in the same units. Thus Eq. 2·1 becomes

$$\oint \delta W = \oint \delta Q \qquad \text{or} \qquad \oint (\delta Q - \delta W) = 0 \tag{2·1}$$

We can then state the *first law of thermodynamics* as: *For any cyclic process of a closed system, the net work output of the system is equal to the net heat input.*

The first law of thermodynamics is a far-reaching principle of nature which is induced from the results of many experiments. It cannot be deduced or proved from any other principles of nature. The inductive reasoning process, by which we take the results of a finite number of experiments and extend them to cover all cases, always leaves some room for doubt as to the value of its conclusions. With regard to its truth and range of application, all we can say about the first law of thermodynamics is that many, many experimental measurements are in accord with it and all attempts to find an exception to it have so far failed.

Example 2·1. Even though a household electric refrigerator is covered by thermal insulating material, there is some transfer of heat from the surrounding room into the refrigerator compartment. In a certain refrigerator, heat is transferred inward through the walls at an average rate of 30 J/s. A watt-hour meter connected to the motor shows that 6.0 kW·h of electrical work was put into the motor during a 10-day period. What was the average rate of heat transfer away from the refrigerating unit in joules per second during the 10-day period if all conditions

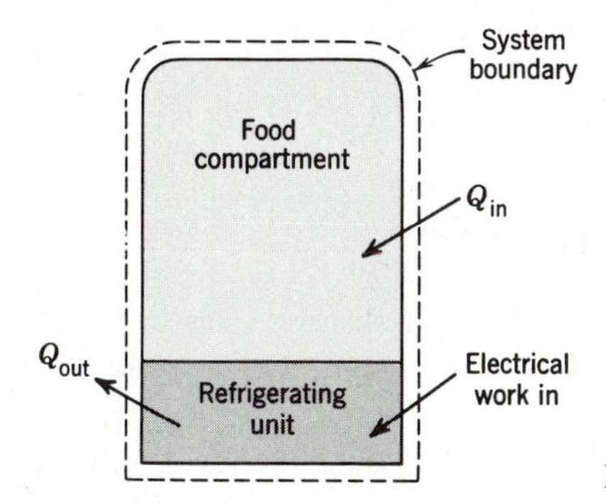

Example 2·1.

*Values of the conversion factor J in other units can readily be derived from the above value and the definitions of other units. Some approximate values frequently used in connection with the units of power, the horsepower (1 hp = 33 000 ft·lbf/min) and the kilowatt (0.746 kW = 1 hp), are

$$1/J = 1 = 3413\ \text{B/kW·h} = 2545\ \text{B/hp·h} = 42.4\ \text{B/hp·min} = 0.707\ \text{B/hp·s}$$

inside the refrigerator are the same at the end of the 10-day period as they were at the beginning of it?

Solution. If all conditions inside the refrigerator are the same at the end of the 10-day period as at the beginning of it, then the closed system comprised of the refrigerator (food compartment and refrigerating unit, including the motor) has undergone one or more complete cycles of operation, and the first law of thermodynamics as stated above can be applied to the system. The only work involved is the 6.0 kW·h of electrical work input to the motor. Heat is transferred into the system through the compartment walls and is transferred from the system by the heat exchanger which is part of the refrigerating unit. Heat is also rejected by the electric motor. Therefore, we integrate

$$\oint \delta Q = \oint \delta W$$

to

$$Q_{\text{in through walls}} - Q_{\text{out from heat exchanger and motor}} = W$$

Rearranging,

$$\begin{aligned} Q_{\text{out from heat exchanger and motor}} &= Q_{\text{in through walls}} + W_{\text{in}} \\ &= 30\left[\frac{\text{J}}{\text{s}}\right] 10[\text{d}]\ 24\left[\frac{\text{h}}{\text{d}}\right] 3600\left[\frac{\text{s}}{\text{h}}\right] \\ &\quad + 6.0[\text{kW}\cdot\text{h}]\ 3600\left[\frac{\text{s}}{\text{h}}\right] 1000\left[\frac{\text{J}}{\text{s}\cdot\text{kW}}\right] \\ &= 25\ 920\ 000 + 21\ 600\ 000 = 47\ 500\ 000\ \text{J} \end{aligned}$$

$$\text{Average rate of heat transfer} = \dot{Q} = \frac{Q}{\text{total time}} = \frac{47\ 500\ 000}{10(24)3600} = 55\ \text{J/s}$$

2·2 The First Law for Noncyclic Processes: Energy

The first law of thermodynamics as stated above applies only to cyclic processes of a closed system. We will now extend the application of this principle to noncyclic processes, that is, to processes which cause a net change in the state of a system. When a closed system undergoes a process which changes it from a state 1 to a state 2, it is usually true that

$$\int_1^2 \delta Q - \int_1^2 \delta W \neq 0$$

or

$$Q - W \neq 0$$

However, since $\oint (\delta Q - \delta W) = 0$, it can be readily shown that the value of $\int_1^2(\delta Q - \delta W)$ or $(Q - W)$ is the same for any path between states 1 and 2. The proof of this follows:

Consider a closed system which is changed from a state 1 to a state 2 by some process shown as path *A* on a property diagram such as the *PV* diagram of Fig. 2·3. It is then returned to state 1 via path *B*. The first law states that

$$\oint_{1\text{-}A\text{-}2\text{-}B\text{-}1} (\delta Q - \delta W) = 0$$

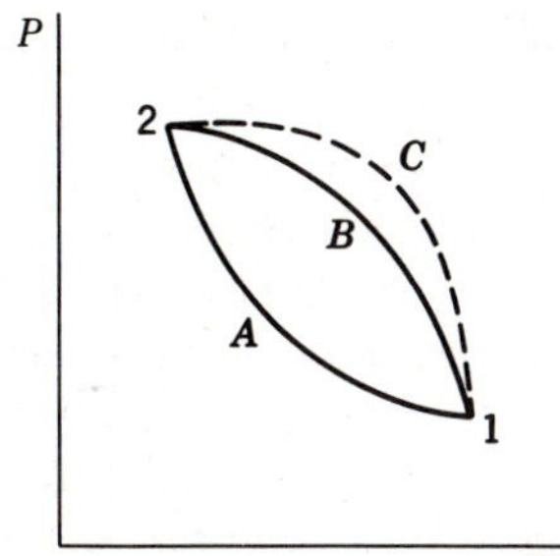

Figure 2·3 State change of a closed system.

or

$$\int_{1\text{-}A}^{2} (\delta Q - \delta W) + \int_{2\text{-}B}^{1} (\delta Q - \delta W) = 0 \tag{a}$$

If C is any other path by which the system could be restored from state 2 to state 1, then it is likewise true that

$$\int_{1\text{-}A}^{2} (\delta Q - \delta W) + \int_{2\text{-}C}^{1} (\delta Q - \delta W) = 0 \tag{b}$$

Comparing Eqs. a and b, we see that

$$\int_{2\text{-}B}^{1} (\delta Q - \delta W) = \int_{2\text{-}C}^{1} (\delta Q - \delta W)$$

Since B and C were *any* two paths between states 2 and 1, it follows that the value of $\int(\delta Q - \delta W)$ is the same for *all* paths between the two states. In other words, the value of $\int(\delta Q - \delta W)$ or $(Q - W)$ depends only on the end states of a process. *Thus* $\int(\delta Q - \delta W)$ or $(Q - W)$ *is a property*. (This could have been inferred directly from the statement of the first law in the form $\oint (\delta Q - \delta W) = 0$, since if the cyclic integral of any quantity is always zero, that quantity must be a property. The demonstration presented above, however, is generally more satisfying to the beginning student of thermodynamics.) This property is called *stored energy* and is denoted by the symbol E. Thus

$$\Delta E \equiv \int (\delta Q - \delta W) \tag{2·2a}$$

or

$$\Delta E \equiv Q - W \tag{2·2b}$$

This defining equation for ΔE is generally regarded as a useful statement of the first law.

Since E is a property, its differential is exact, and the differential form of the defining equation is

$$dE = \delta Q - \delta W \tag{2·2c}$$

Only the *change* in the value of E between two states can be evaluated by means of Eqs. 2·2. In fact, thermodynamics provides no information about absolute values of E for any system. It is only the change in E, however, which is important in engineering problems. Consequently, a value of $E = 0$ can be assigned to any particular state of a system. Then from measurements of heat and work during a process that transforms the system to another state, the value of E at the other state can be determined. Since the state for which $E = 0$ is selected arbitrarily, some of the other states will have negative values of E, and some will have positive values. Many different states of a closed system may have the same value of E, because usually more than one property must be specified in order to define the state of a system. For example, one state of a system may be defined by certain values of E and pressure, and another state may have the same value of E and a different value of pressure.

The definition of stored energy given by Eqs. 2·2 is an *operational* definition. Equation 2·2b, for example, states that the number that represents the change in the stored energy of a closed system during any process is obtained by subtracting the net amount of work done by the system from the net amount of heat added to the system. Work and heat have previously been defined operationally.

Because energy is a term in everyday use and most people have (or believe they have) some intuitive grasp of its meaning, nonoperational and less precise definitions are numerous and often satisfy people. For example, the term energy was used in the first section of this book without being defined, and probably very few readers were troubled by the term at that point. Nevertheless, in scientific work where precise meanings are essential, important terms such as energy should be either defined operationally or accepted at the start as verbally undefined.

Since heat and work are related to the stored energy of a system, they are often spoken of as forms of *energy in transition* or *transitional energy*.* Thus the term *energy* is a general term referring to stored energy (a property of a system) and to heat and work (interactions between systems). This is the reason why the property E has been referred to always as *stored* energy. E is sometimes called internal energy, but we will reserve this name for one particular form of stored energy in conformity with many reference works and published tables of thermodynamic properties.

With heat and work considered as forms of energy in transition, Eqs. 2·2 become statements of the *conservation of energy principle:* The increase in the energy stored in a system is equal to the net transfer of energy into the system. This is equivalent to the statement that energy can be neither created nor destroyed although it can be stored in various forms and can be transferred from one system to another. Thus it is often said that the first law of thermodynamics is the principle (or law) of conservation of energy. (In this book and many others, however, energy is defined by means of the first law, so the law of conservation of energy is a *consequence* of the first law.)

Notice should be taken of the form of the first law as applied to an isolated system. By definition of an isolated system, both work and heat must be zero, so that Eq. 2·2b reduces to

$$\Delta E_{\text{isolated system}} = 0 \tag{2·3}$$

*Heat and work are often *defined* as energy in transition. Possible definitions are: *Heat* is energy in transition between a system and its surroundings (or between two systems) as a result of a temperature difference between the system and its surroundings (or between the two systems). *Work* is energy in transition between a system and its surroundings by virtue of a force acting through a distance at the system boundary. (For generality, the force acting through a distance must include cases such as an electromotive force acting on a moving charge, etc.)

that is, the stored energy of an isolated system must remain constant. This does not mean that there can be no change of state of an isolated system; it does mean that an isolated system can exist only in those states which have the same stored energy as the initial state.

Although stored energy has been defined in terms of work and heat measurements involving a closed system and the first law is established from measurements made on a closed system, the principle of conservation of energy which is a consequence of the first law can be extended to apply to any type of system, open or closed. We shall see later that stored energy is a useful property of open systems as well as of closed systems.

Before looking further into the nature of E, let us summarize the preceding discussion of the first law of thermodynamics as follows:

1. The first law is a generalization based on the results of many experiments. It cannot be deduced from any other physical principles; it is entirely empirical.

2. As a result of the first law, energy can be defined in terms of heat and work, the operationally defined interactions between systems.

3. The first law is stated by any one of the following equations which refer to a closed system:

$$\oint \delta Q = \oint \delta W \tag{2·1}$$

$$Q - W = \Delta E \tag{2·2}$$

$$\Delta E_{\text{isolated system}} = 0 \tag{2·3}$$

More precisely, Eq. 2·1 is the empirical conclusion which we call the first law, Eq. 2·2 defines stored energy which the first law shows to be a property, and Eq. 2·3 is a consequence of the first two.

4. Equivalent word statements of the first law are:

(*a*) Whenever a closed system executes a cycle, the net work done by the system is equal to the net heat transfer to the system. (This is the statement closest to the experimental results that support the first law.)

(*b*) The net heat added to a closed system minus the net work done by the system is equal to the increase in the stored energy of the system. (This is the statement most convenient for application to many engineering problems.)

(*c*) The stored energy of an isolated system remains constant. (In a few engineering problems this is a convenient statement for direct application.)

5. One form of the first law is the conservation of energy principle (or the law of conservation of energy): Energy can be neither created nor destroyed although it can be stored in various forms and can be transferred from one system to another.

Example 2·2. A system is comprised of a certain mass of air contained in a cylinder fitted with a piston. The air expands from a state 1 for which $E_1 = 70$ kJ to a state 2 for which $E_2 = -20$ kJ. During the expansion, the air does 60 kJ of work on the surroundings. Determine the amount of heat transferred to or from the system during the process.

Solution. Applying the first law in the form

$$Q = E_2 - E_1 + W \tag{2·2}$$

to this system,

$$Q = -20 - 70 + 60 = -30 \text{ kJ}$$

Since Q without subscript stands for heat added to a system, the minus sign of the result indicates that 30 kJ of heat is transferred from the system. The result can be written also as $Q_{out} = 30$ kJ.

Notice that in this process heat and work are both taken from the system. The total decrease in the stored energy of the system is equal to the sum of the energy removed as heat and as work.

Example 2·3. Determine the final E value of a mass of water that is in an initial state for which $E = 20$ kJ and then undergoes a process during which 10 kJ of work is done on the water and 3 kJ of heat is removed from it.

Solution. Apply the first law to the system comprised of the water. The final amount of stored energy is equal to the initial amount of stored energy plus the energy added as work minus the energy removed as heat.

$$E_2 = E_1 + W_{in} - Q_{out} \tag{2·2}$$
$$= 20 + 10 - 3 = 27 \text{ kJ}$$

If the first law is applied stated as

$$E_2 = E_1 + Q - W \tag{2·2}$$

the substitution of numerical values (recalling that Q denotes Q_{in} and W denotes W_{out}) would be

$$E_2 = 20 + (-3) - (-10) = 27 \text{ kJ}$$

2·3 The Nature of E

Stored energy E is defined as

$$\Delta E \equiv Q - W \tag{2·2}$$

This is an operational definition which tells how the change in E of a closed system can be obtained from measurements of heat and work. Since E has been shown to be a property, the question arises as to whether E (or ΔE) values obtained by measurements of heat and work can then be correlated with other properties of a system. There would be several benefits from such a correlation. For one thing, ΔE for any process could then be determined from measurements of other properties at the two end states. This would increase the utility of the first law because, for any system for which E had already been correlated with other properties, it would give an independent method for evaluating ΔE in the relation

$$Q = \Delta E + W \tag{2·2}$$

Many engineering problems involve the use of Eq. 2·2 or one of the many special relations which can be derived from it. Often the value of either Q or W is specified for a process. The problem is to find out how much of the other form of transitional energy must be added to or removed from the system in order to bring about a particular change of state. In such a case, the advantage of having E previously correlated with other properties is apparent.

Fortunately, it is possible in certain cases to make the desired correlation between E and other properties. We shall now investigate some of these cases in which the value of ΔE can be determined by measurements of properties of a system at the beginning and the end of a process.

The procedure to be followed in two cases is to consider a system undergoing a process for which heat and work are measurable and during which a limited number of properties change. For each of these two cases the change in stored energy calculated from the heat and work measurements can be easily correlated with properties of the initial and final states of the system.

Potential Energy

Consider a system composed of a body that has its elevation changed while no other property of the body changes. Obviously a change in elevation can be brought about without a transfer of heat. The change in stored energy associated with the change in elevation is then equal to the work done on the body.

$$\Delta E_{\text{change in elevation}} = W_{\text{on body}} = \int_1^2 F\,dz$$

The force required in order to lift the body while producing no other effect on it is equal to the weight of the body, so the preceding equation becomes

$$\Delta E_{\text{change in elevation}} = \int_1^2 F\,dz = \int_1^2 (\text{weight})\,dz = \int_1^2 mg\,dz$$

where m is the mass of the body and g is the acceleration of gravity. Usually the variation in g with elevation is negligible, so that

$$\Delta E_{\text{change in elevation}} = mg \int_1^2 dz = gm(z_2 - z_1)$$

Experiments show that when a system undergoes any process involving a change in elevation, part of the change of stored energy is always attributable to the change in elevation alone, regardless of what other properties may change. This part of the stored energy which is a function of elevation alone is spoken of as a *form* of stored energy and is called *potential energy, PE*. Thus

$$\Delta PE = PE_2 - PE_1 = (\text{weight})(z_2 - z_1) = gm(z_2 - z_1) \qquad (2\text{·}4\text{a})$$

The elevation at which the potential energy of a body is zero may be selected arbitrarily. If z_0 is the elevation at which we assign $PE_0 = 0$, and PE is the potential energy at any other elevation, the equation

$$PE - PE_0 = gm(z - z_0)$$

reduces to

$$PE = gmz = (\text{weight})z \tag{2·4b}$$

where z is measured from the datum plane at an elevation of z_0. Thus potential energy is often defined as the energy stored in a system* as a result of its location in the earth's gravitational field, and its magnitude is the product of (1) the weight of the system and (2) the distance between the center of gravity of the system and some arbitrary horizontal datum plane.

Kinetic Energy

Consider a system composed of a body that experiences a change in velocity while its elevation, temperature, volume, and all other properties remain constant. Such a change in velocity can be brought about without heat transfer. Application of the first law then shows that the change in stored energy associated with the change in velocity is equal to the work done on the body in order to accelerate it:

$$\Delta E_{\text{change in velocity}} = W_{\text{on body}} = \int_1^2 F\,dL = \int_1^2 ma\,dL$$

Noting that $a = dV/d\tau$ and $V = dL/d\tau$, this becomes

$$\Delta E_{\text{change in velocity}} = m\int_1^2 \frac{dV}{d\tau}\,dL = m\int_1^2 V\,dV = \frac{m}{2}(V_2^2 - V_1^2)$$

Experiments show that when a system undergoes any process, part of the change of stored energy is always attributable to any change in velocity alone, regardless of what other properties may change. This part of the stored energy is called kinetic energy, KE. Thus

$$\Delta KE = KE_2 - KE_1 = m\left(\frac{V_2^2 - V_1^2}{2}\right) \tag{2·5a}$$

It is customary and convenient (but not mandatory) to assign a value of zero kinetic energy to a system with zero velocity relative to some arbitrary frame of reference. When this is done, the kinetic energy of a system in any state is

$$KE = \frac{mV^2}{2} \tag{2·5b}$$

*Strictly speaking, in the earth's gravitational field potential energy is stored in the *system* comprised of the earth and the relatively small system under study. A broader and more rigorous definition is: Potential energy is energy stored in a system as a result of gravitational forces between parts of the system. Potential energy changes are evaluated from measurements of work and heat in processes in which parts of the system move relative to each other and no other changes of state occur. When this definition is used, the energy stored in a system comprised of the earth and some other body is a special case that is sometimes referred to as geopotential energy. Throughout this book, "potential energy of a body" will mean "geopotential energy of the system composed of the earth and the body."

where V is the velocity of the system relative to some specified frame of reference. Thus kinetic energy is defined as the energy stored in a system by virtue of the motion of the system, and its magnitude is given by $mV^2/2$, where the symbols are as previously defined. If not all parts of the system have the same velocity, the kinetic energy is given by

$$KE = \frac{1}{2} \int V^2 \, dm \tag{2·5c}$$

where V is the velocity of any elemental mass dm, and the integration must be carried out over the entire mass of the system. For example, from Eq. 2·5c it can be shown that the kinetic energy of a solid body rotating about a fixed axis is $I\omega^2/2$, where ω is the angular velocity and I is the moment of inertia of the body about the axis of rotation.

Other Forms of Stored Energy

Several other forms of energy can be identified by procedures similar to those used to identify potential and kinetic energies. A form of energy related to magnetic fields can easily be identified, and others related to electrical effects and surface tension effects can be found. Even in the absence of all these effects as well as motion and gravity, a difference in the stored energy of a system can still be discerned by measurements of heat and work. The form of stored energy which is independent of electricity, magnetism, surface tension, motion, and gravity is called *internal energy* and is represented by the symbol U. Thus

$$U \equiv E - KE - PE - \text{(magnetic energy)} - \text{(electric energy)} - \text{(surface energy)}$$

The ease with which internal energy is correlated with other properties of a system depends on the nature of the system. This point is discussed in Chapters 4 and 9.

Although the principles of thermodynamics are independent of any assumptions regarding the structure of matter, it is often helpful to picture the internal energy of a substance, in the absence of chemical or nuclear reactions, as the summation of the kinetic and potential energies of the molecules of the substance. The kinetic energy of molecules is associated with translational, rotational, and vibratory motions of molecules. The kinetic energy of the molecules of a substance increases as the temperature of the substance increases. Molecular potential energy is associated with the attractive forces between molecules. These forces are large in a solid where molecules are close together, smaller in a liquid, and very small in a gas where molecules are separated from each other by distances which are large in terms of molecular dimensions. Molecular potential energy increases as the distance between molecules increases, so it is highest for gases and lowest for solids.

In a process which involves a chemical reaction, the change in the internal energy of a system is related to the changes in the internal structure of molecules, that is, to changes on the atomic level rather than on the molecular level. Nuclear reactions involve changes on the subatomic level, that is, changes within the atoms of a substance.*

*In connection with nuclear reactions and the conversion of matter to energy, the conservation of energy and conservation of mass principles must be generalized to treat matter as a form of energy. This subject is not covered in this book. Note, however, that the first law of thermodynamics as stated here and Eq. 2·1 do hold in any case.

Measurements of only heat and work provide no means of distinguishing among the molecular, atomic, and nuclear levels of energy storage, so all are included under the one category of internal energy.

Thermodynamics provides no information as to absolute values of internal energy; only changes in internal energy can be determined by measurements of heat and work. For convenience, the internal energy of a substance is often assigned the value of zero at some arbitrary reference state just as the potential energy of a body is often assigned the value of zero at some convenient and arbitrary horizontal datum plane.

The total change in the stored energy of a system is equal to the sum of the changes in all of the particular forms of stored energy:

$$\Delta E = \Delta PE + \Delta KE + \Delta U + \Delta(\text{magnetic energy}) + \Delta(\text{electric energy}) + \Delta(\text{surface energy}) \qquad (2{\cdot}6a)$$

In the absence of electricity, magnetism, and surface tension, the stored energy of a system is

$$E = U + KE + PE \qquad (2{\cdot}6b)$$

If all parts of the system move with the same velocity, and z is the vertical distance of the center of gravity of the system from an arbitrary horizontal datum plane,

$$E = U + \frac{mV^2}{2} + gmz \qquad (2{\cdot}6c)$$

Per unit mass of substance, this is written

$$e = u + \frac{V^2}{2} + gz \qquad (2{\cdot}6d)$$

2·4 Enthalpy

In many thermodynamic analyses the sum of internal energy U and the product of pressure and volume PV appears. Because this combination $(U + PV)$ occurs so frequently, it has been given a name, *enthalpy,* and is represented by the symbol H. Since U, P, and V are all properties, this combination of them is also a property. The defining relation is

$$H \equiv U + PV$$

or, per unit mass

$$h \equiv u + Pv$$

It should be noted that u represents a form of stored energy, but Pv does not; therefore, their sum is not a form of stored energy. It will be seen later that in certain applications enthalpy may be treated as energy, but this should not obscure the fact that *enthalpy is simply a useful property defined by an arbitrary combination of other properties and is not a form of energy.*

Since we cannot obtain absolute values of internal energy, we cannot obtain absolute values of enthalpy. Only changes in enthalpy are of importance to us, however.

In English units, internal energy values are often expressed in B/lbm, whereas typical engineering units for pressure and specific volume give the Pv term in ft·lbf/lbm. Consequently, careful attention is required when working with numerical values to ensure that consistent units are used.

2·5 The First Law Applied to (Stationary) Closed Systems

A closed system may move, but usually engineering systems analyzed as closed systems do not. Therefore, in this book when the term *closed system* is used and there is no indication to the contrary, it will be assumed that effects of gravity and motion are negligible.* This means that there will usually be no change in the kinetic energy or the potential energy of a closed system. If we also exclude effects of electricity, magnetism, and surface tension, then the total change in the stored energy of the system is the change in internal energy U. Thus the first law applied to a closed system under these conditions becomes

$$Q - W = \Delta U \tag{2·7}$$

or

$$Q - W = U_2 - U_1 \tag{2·7}$$

where the subscripts 1 and 2 refer respectively to the states at the beginning and the end of a process. Notice that Eq. 2·7 is a special case of Eq. 2·2. U is internal energy, the form of stored energy which is independent of gravity, motion, electricity, magnetism, and surface tension; E is stored energy in general.

Example 2·4. As a certain quantity of gas expands, it performs 26.0 kJ of work while its internal energy decreases by 20.0 kJ. Consider the gas as a closed system, and determine the heat transfer of the process.

Solution. There is no indication that forms of stored energy other than internal energy change during this process, so we may apply the first law in the form

$$Q = \Delta U + W$$

and substitute numerical values to obtain

$$Q = -20.0 + 26.0 = 6.0 \text{ kJ}$$

(If a negative value were obtained for Q, the conclusion would be that heat is removed from the system during the process.)

Example 2·5. A table of the properties of rubidium vapor shows that the internal energy and specific volume at 200 kPa, 800°C, are 714.3 kJ/kg and 0.471 m³/kg, respectively. One-half kilogram of rubidium initially at this condition in a closed rigid container is heated until its temperature is 1227°C and its internal energy is 808.2 kJ/kg. Determine the amount of heat added to the rubidium.

*Some engineers use the term "stationary system" or "nonflow system" to designate a closed system which is at rest or in a state of uniform horizontal motion.

Solution. The system is the $\frac{1}{2}$ kg of rubidium. It is a closed system. Applying the first law and noting that no work is done on or by the rubidium in a closed rigid container,

$$\begin{aligned} Q &= U_2 - U_1 + W \\ &= m(u_2 - u_1) + 0 \\ &= \tfrac{1}{2}(808.2 - 714.3) = 47.0 \text{ kJ/kg} \end{aligned}$$

Example 2·6. In a cylinder fitted with a piston is trapped 0.02 lbm of helium initially at 15 psia with a specific volume of 93.0 cu ft/lbm. The helium is compressed frictionlessly in such a manner that $Pv^{1.3}$ = constant until its pressure is 30 psia. The internal energy of helium is given in ft·lbf/lbm by $u = 1.51Pv$, where P is in psfa and v is in cu ft/lbm. Determine the heat transfer of the compression process.

Analysis. Q can be found from the first law,

$$Q = U_2 - U_1 - W_{\text{in}} = m(u_2 - u_1) - W_{\text{in}}$$

if the two terms on the right-hand side can be evaluated. u_1 can be found from $1.51P_1v_1$ and u_2 can be found from $1.51P_2v_2$ after v_2 is found from $P_1v_1^{1.3} = P_2v_2^{1.3}$.

For this frictionless process, $\text{work}_{\text{in}} = -\int P\,dV = -m\int P\,dv$, and $Pv^{1.3}$ = constant is the relation needed between P and v in order to perform the integration.

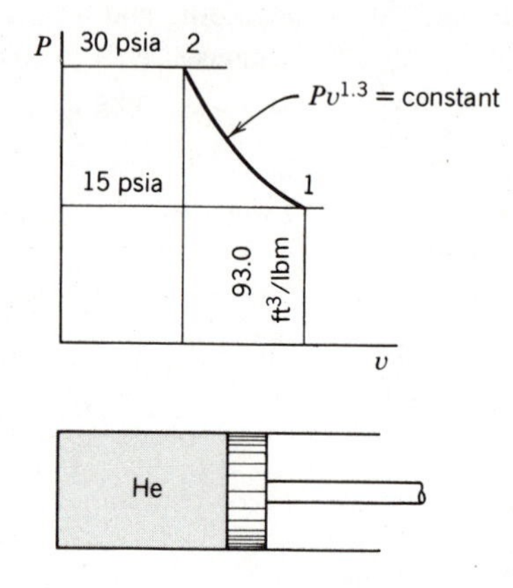

Example 2·6.

Solution. v_2 is found from $Pv^{1.3}$ = constant $= P_1v_1^{1.3} = P_2v_2^{1.3}$ as

$$v_2 = v_1\left(\frac{P_1}{P_2}\right)^{1/1.3} = 93.0\left(\frac{15}{30}\right)^{0.769} = 54.6 \text{ cu ft/lbm}$$

Then

$$\begin{aligned} u_2 - u_1 &= 1.51[P_2v_2 - P_1v_1] = 1.51(144)[30(54.6) - 15(93.0)]/778 \\ &= 67.9 \text{ B/lbm} \end{aligned}$$

The work input for this frictionless process of a closed system is

$$W_{in} = -\int_1^2 P\,dV = -m\int_1^2 P\,dv = -m\int_1^2 \frac{\text{constant}}{v^{1.3}}\,dv$$

$$= -m\int_1^2 \frac{P_1 v_1^{1.3}}{v^{1.3}}\,dv = -P_1 m_1 v_1^{1.3}\left(\frac{1}{-0.3}\right)\left(\frac{1}{v_2^{0.3}} - \frac{1}{v_1^{0.3}}\right)$$

$$= +mP_1 v_1\left(\frac{1}{0.3}\right)\left[\left(\frac{v_1}{v_2}\right)^{0.3} - 1\right] = 0.02(15)144(93.0)\frac{1}{0.3}\left[\left(\frac{93.0}{54.6}\right)^{0.3} - 1\right]$$

$$= 2320 \text{ ft·lbf} = 2.98 \text{ B}$$

Substituting the values of Δu and W into the first law formulation

$$Q = m(u_2 - u_1) - W_{in}$$

gives

$$Q = 0.02(67.9) - 2.98 = -1.62 \text{ B}$$

The minus sign indicates that heat was removed from the helium during the compression, because $Q_{out} = -Q = -(-1.62) = 1.62$ B.

Example 2·7. A mixture of ammonia liquid and vapor is heated slowly at constant pressure until all the liquid has evaporated. Properties of the ammonia at the two end states of the process are given in the accompanying table.

Property	Initial State (1)	Final State (2)
Pressure, psia	50	50
Temperature, °F	21.67	21.67
Specific volume, cu ft/lbm	2.86	5.71
Enthalpy, B/lbm	342.4	618.2

Calculate the amount of heat added per pound of ammonia, assuming that there are no frictional effects within the closed system.

Analysis. The amount of heat added can be found from

$$q = u_2 - u_1 + w \tag{2·7}$$

by finding the u values from $u_2 = h_2 - P_2 v_2$ and $u_1 = h_1 - P_1 v_1$ and by calculating the work of this frictionless process of a closed system from work $= \int_1^2 P\,dv$. Since P = constant, the expression for work is readily integrated to give $P(v_2 - v_1)$ or $(P_2 v_2 - P_1 v_1)$. When this is substituted into Eq. 2·7, the result is

$$q = u_2 - u_1 + P_2 v_2 - P_1 v_1 = h_2 - h_1$$

By analyzing the problem completely in this manner before beginning numerical calculations, we have saved ourselves the trouble of calculating the internal energy values. A little practice makes it possible to analyze a problem such as this entirely "in one's head," without use of pencil and paper, or perhaps with only sketchy notes on scratch paper. In any event, the problem *solution* which is to be presented to someone else or filed away for later use by the same engineer should be complete and should show each step clearly. The solution based on the above analysis follows.

Solution. The heat added is found from the first law as

$$q = u_2 - u_1 + w \tag{2·7}$$

For a frictionless process of a closed system, $w = \int_1^2 P\, dv$. Making this substitution in Eq. 2·7 and integrating for this case of P = constant $= P_1 = P_2$,

$$q = u_2 - u_1 + \int_1^2 P\, dv = u_2 - u_1 + P(v_2 - v_1)$$

$$= u_2 - u_1 + P_2 v_2 - P_1 v_1$$

Noting that $h \equiv u + Pv$, we obtain

$$q = h_2 - h_1 = 618.2 - 342.4 = 275.8 \text{ B/lbm}$$

2·6 The First Law Applied to Open Systems

From the first law or the law of conservation of energy, we can conclude* that for any system, open or closed, there is an "energy balance" as

$$\begin{pmatrix}\text{Net amount of energy} \\ \text{added to system}\end{pmatrix} = \begin{pmatrix}\text{net increase in stored} \\ \text{energy of system}\end{pmatrix} \tag{2·8a}$$

With both open and closed systems, energy can be added to the system or taken from it by means of heat and work. In the case of an open system, there is an additional mechanism for increasing or decreasing the stored energy of the system. When matter enters a system, the stored energy of the system is increased by an amount equal to the stored energy of the entering matter. The stored energy of a system is decreased whenever matter leaves the system because the matter leaving the system takes some stored energy with it. If we distinguish this transfer of stored energy of matter crossing the system boundary from heat and work, Eq. 2·8a becomes

$$\begin{pmatrix}\text{Net amount of} \\ \text{energy added to} \\ \text{system as heat} \\ \text{and all forms of} \\ \text{work}\end{pmatrix} + \begin{pmatrix}\text{stored energy} \\ \text{of matter} \\ \text{entering} \\ \text{system}\end{pmatrix} - \begin{pmatrix}\text{stored energy} \\ \text{of matter} \\ \text{leaving} \\ \text{system}\end{pmatrix} = \begin{pmatrix}\text{net increase} \\ \text{in stored} \\ \text{energy of} \\ \text{system}\end{pmatrix} \tag{2·8b}$$

If we investigate the flow of matter across the boundary of a system, we find that work is always done on or by a system where fluid flows across the system boundary. Therefore, the work term in an energy balance for an open system is usually separated into two parts: (*a*) the work required to push a fluid into or out of the system, and (*b*) all other forms of work.

In order to derive an expression for the work required to push a fluid into or out of a system, consider a system as shown in Fig. 2·4 with fluid entering at section 1 and leaving at section 2. In particular, consider the process by which a volume V_1 of fluid is

*For a rigorous demonstration of applying principles established for closed systems to open systems, see, for example, References 2·3 and 2·7.

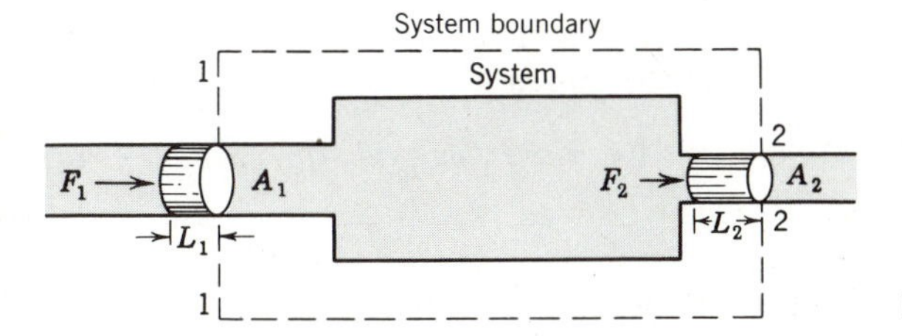

Figure 2·4 Open system.

pushed into the system at section 1. The fluid element of volume V_1 has a cross-sectional area A_1 and a length L_1 as it crosses the system boundary at section 1. The force acting on this element to push it across the system boundary is $F_1 = P_1A_1$. This force acts through a distance L_1 so that the work done in pushing the element across the system boundary is

$$\text{work} = F_1L_1 = P_1A_1L_1 = P_1V_1$$

This is work done on the system by the fluid outside the system, because the effect within the system could be reduced entirely to the lifting of a weight. In a similar manner, the work performed by the system to push an element of fluid out at section 2 is

$$\text{work} = P_2V_2$$

It is seen from this analysis that work must be done in causing fluid to flow into or out of a system. This work is called *flow work.* Other names such as flow energy and displacement energy are sometimes used, but they are not used in this book.*

Per unit mass crossing the boundary of a system, the flow work is Pv. If the pressure or the specific volume or both vary as a fluid flows across a system boundary, the flow work is calculated by integrating $\int Pv\,\delta m$, where δm is an infinitesimal mass crossing the boundary. The symbol δm is used instead of dm because the amount of mass crossing the boundary is not a property. The mass within the system is a property, so the infinitesimal change in mass within the system is properly represented by dm.

Since the work term in an energy balance for an open system is usually separated into two parts, (*a*) flow work and (*b*) all other forms of work, the term *work* without modifiers is conventionally understood to stand for all other forms of work except flow work, and the complete two-word name is always used when referring to flow work. In accordance with this convention, we can rewrite Eq. 2·8b to show flow work separately

$$Q - W + \sum_{\substack{\text{mass}\\\text{entering}}} (PV + E) - \sum_{\substack{\text{mass}\\\text{leaving}}} (PV + E) = E_f - E_i \qquad (2\cdot 8c)$$

*Disagreement on the name used for this quantity results from the fact that the PV term is generally derived as a work quantity; yet it is unlike other work quantities in that it is expressed in terms of a point function. Because it is so expressed, some engineers prefer to group it with stored energy quantities and sometimes speak of it as "transported energy" or "convected energy" instead of work; but it must be remembered that PV can be treated as energy only when a fluid is crossing a system boundary. For a closed system, PV does not represent any form of energy. Incidentally, arguments as to whether PV "really is" flow work or flow energy are fruitless. Both terms are successfully used.

where E_f and E_i are respectively the final and initial stored energies of the system. In differential form, where δm is an infinitesimal mass crossing the system boundary,

$$\delta Q - \delta W + [(e + Pv)\,\delta m]_{\text{in}} - [(e + Pv)\,\delta m]_{\text{out}} = dE \tag{2.8d}$$

2·7 The First Law Applied to Open Systems — Steady Flow

An open system through which fluid passes under steady-flow conditions is called a steady-flow system. We shall now apply the first law to steady-flow systems.

From the definition of steady flow (Sec. 1·16), it is seen that the stored energy of an open system must remain constant as long as steady-flow conditions prevail. Therefore, when the energy balance as written in Eq. 2·8b is applied to a steady-flow system, its right-hand side becomes zero, and we have

$$\begin{pmatrix}\text{Net amount of}\\ \text{energy added to}\\ \text{system as heat}\\ \text{and work}\end{pmatrix} + \begin{pmatrix}\text{stored energy of}\\ \text{matter entering}\\ \text{system}\end{pmatrix} - \begin{pmatrix}\text{stored energy of}\\ \text{matter leaving}\\ \text{system}\end{pmatrix} = 0 \tag{2·9a}$$

Let us consider a steady-flow system such as shown in Fig. 2·5 which has only two fluid streams crossing its boundary. Fluid enters at section 1 and leaves at section 2. (The system might be an air compressor, a gas turbine, a centrifugal water pump, a fan, a nozzle, or a section of pipe.) Rewriting Eq. 2·9a in terms of symbols already introduced, we have

$$Q - W + mP_1v_1 - mP_2v_2 + E_1 - E_2 = 0 \tag{2·9b}$$

where Q = net amount of heat added to the system while an amount of mass m passes through the system

W = net amount of work, excluding flow work, done by the system during the same time

mP_1v_1 = amount of flow work done on the system by the fluid entering

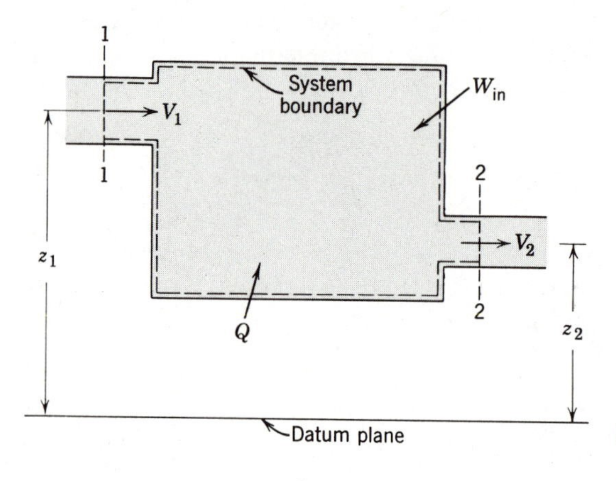

Figure 2·5 Steady-flow system.

mP_2v_2 = amount of flow work done by the system on the fluid leaving (v_1 and v_2 as well as P_1 and P_2 are generally unequal, but for steady-flow conditions the mass of fluid entering must equal the mass of fluid leaving.)

E_1 = stored energy of the fluid entering the system

E_2 = stored energy of the fluid leaving the system

Notice that Q and W may each be either positive or negative.

If each fluid stream crossing the system boundary has a uniform velocity and z is the vertical distance of the stream from the horizontal datum plane, then the stored energy of fluid crossing the system boundary is given by

$$E = me = m\left(u + \frac{V^2}{2} + gz\right) \tag{2·6c}$$

provided there are no effects of electricity, magnetism, and surface tension. m is the mass of fluid crossing the boundary and is the same for inlet and outlet for a steady-flow system with only one inlet and one outlet. Equation 2·9b may then be written, after rearranging slightly,

$$Q - W + mP_1v_1 - mP_2v_2 + m\left(u_1 + \frac{V_1^2}{2} + gz_1\right) = m\left(u_2 + \frac{V_2^2}{2} + gz_2\right) \tag{2·9c}$$

(Recall the sign convention introduced in Sec. 1·14: W stands for work done *by* the system, so that $W_{in} = -W$.) Per unit of mass passing through the system,

$$q - w + P_1v_1 - P_2v_2 + u_1 + \frac{V_1^2}{2} + gz_1 = u_2 + \frac{V_2^2}{2} + gz_2 \tag{2·9d}$$

Using the definition of enthalpy, $h \equiv u + Pv$, the energy balance equations above can be shortened. For example, equation 2·9d becomes after a slight rearrangement

$$q - w = h_2 - h_1 + \frac{V_2^2 - V_1^2}{2} + g\,(z_2 - z_1) \tag{2·9e}$$

This is a frequently used form of the energy balance applied to a steady-flow system. It can be written more concisely as

$$q - w = \Delta h + \Delta ke + \Delta pe$$

A form of Eq. 2·9e which is frequently seen is

$$q + h_1 + ke_1 + pe_1 = w + h_2 + ke_2 + pe_2$$

Equations 2·9b through 2·9e are all forms that apply only to steady-flow systems that have a single inlet and a single outlet. Application of Eq. 2·9a to a steady-flow system with fluid crossing the boundary at several places leads to

$$Q - W = \sum_{\substack{\text{all streams}\\ \text{leaving}}} m(e + Pv) - \sum_{\substack{\text{all streams}\\ \text{entering}}} m(e + Pv) \tag{2·9f}$$

In the absence of electricity, magnetism, and surface tension effects, we may replace e

by $u + V^2/2 + gz$. Then the combination $(u + Pv)$ can be replaced by h to give a convenient form that is frequently used:

$$Q - W = \sum_{\substack{\text{all streams}\\ \text{leaving}}} m\left(h + \frac{V^2}{2} + gz\right) - \sum_{\substack{\text{all streams}\\ \text{entering}}} m\left(h + \frac{V^2}{2} + gz\right) \tag{2·9g}$$

Equation 2·9e, which applies to a system with a single inlet and a single outlet, is a special case of Eq. 2·9g.

In all the above equations where the symbol m appears, it stands for the mass of the fluid. In many engineering applications it is more convenient to use the mass rate of flow $\dot{m}$. When this is done, all terms in the energy balance must be on a time basis. Work per unit time is called power $\dot{W}$, so Eq. 2·9g becomes

$$\dot{Q} - \dot{W} = \sum_{\substack{\text{all streams}\\ \text{leaving}}} \dot{m}\left(h + \frac{V^2}{2} + gz\right) - \sum_{\substack{\text{all streams}\\ \text{entering}}} \dot{m}\left(h + \frac{V^2}{2} + gz\right) \tag{2·9h}$$

The symbol $\dot{Q}$ is used for heat transfer per unit time or rate of heat transfer.

Equation 2·9 has been written above in several forms, and many other arrangements are useful in various applications. You must understand that these are all forms of the same basic relation, Eq. 2·9a, and it is neither necessary nor advisable to learn the several different forms independently. You should learn and thoroughly understand the relationship expressed by Eq. 2·9a, and from this relationship you should be able to formulate quickly any of the special forms which may be convenient for a particular application.

Notice that the subscripts used in several forms of Eq. 2·9 (pertaining to steady-flow systems) refer to different sections of the system boundary during the same time interval or at the same instant, whereas the subscripts 1 and 2 used in equations such as

$$Q - W = U_2 - U_1 \tag{2·7}$$

(pertaining to closed systems) refer to the same mass at different times. The viewpoint used in dealing with steady-flow systems is that of an observer outside the system making simultaneous measurements at all places where mass or energy crosses the system boundary. Another viewpoint for a steady-flow system is that of an observer who travels through the system from inlet to outlet with the flowing fluid and measures the properties of the same mass of material at inlet and outlet. The moving observer viewpoint is not used in this book. It is unsatisfactory in processes involving the mixing of two or more fluid streams as illustrated in Fig. 2·6 because each entering stream loses its identity as it is mixed with or perhaps reacts chemically with another stream. For example, if a stream of methane and a stream of oxygen enter a system and react with each other so that a stream composed of carbon dioxide, carbon monoxide, and water leaves the system, an observer who intended to travel through the system with the methane and measure its properties at inlet and outlet would be faced with an impossible task.

The application of the first law to steady-flow systems is illustrated by the following examples.

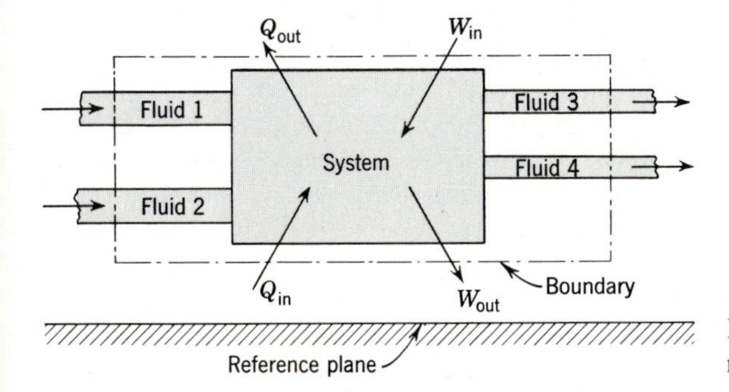

Figure 2·6 Steady-flow system with mixing of fluid streams.

Example 2·8. In a steady-flow process, the pressure of a fluid decreased between the entrance and exit of a machine from 200 to 40 psia, the specific volume increased from 3 to 10 cu ft/lbm, the internal energy decreased by 40 B/lbm, and 10 B/lbm of heat was added to the fluid between entrance and exit. Entrance and exit pipes were at the same elevation, and entrance and exit velocities were 10 and 15 fps, respectively. Determine the amount of work done on or by the fluid.

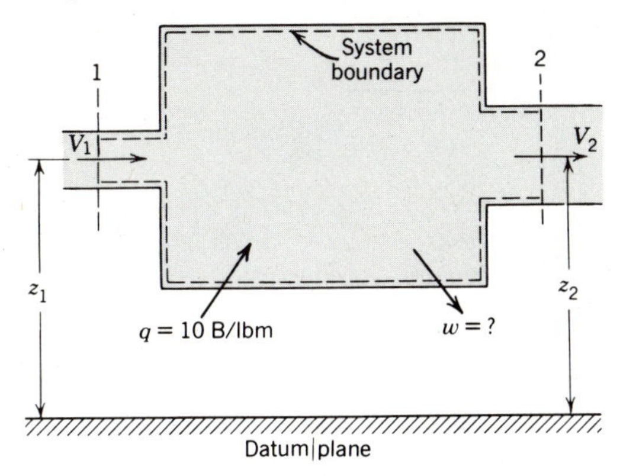

Example 2·8.

Solution. A diagram of the steady-flow system is made first. Next, the first law is applied to the system in the form of an energy balance solved for work per pound:

$$w = q + u_1 - u_2 + P_1 v_1 - P_2 v_2 + \frac{V_1^2 - V_2^2}{2} + g(z_1 - z_2) \qquad (2\cdot 9e)$$

(If a negative value is obtained for w, it will indicate that work is done *on* the system in this process.) Factoring $(V_1^2 - V_2^2)$ to $(V_1 - V_2)(V_1 + V_2)$ for convenience and substituting numerical values gives

$$w = 10\left[\frac{\text{B}}{\text{lbm}}\right] 778 \left[\frac{\text{ft}\cdot\text{lbf}}{\text{B}}\right] + 40\left[\frac{\text{B}}{\text{lbm}}\right] 778\left[\frac{\text{ft}\cdot\text{lbf}}{\text{B}}\right] + 200\left[\frac{\text{lbf}}{\text{in.}^2}\right] 144\left[\frac{\text{in.}^2}{\text{ft}^2}\right] 3\left[\frac{\text{ft}^3}{\text{lbm}}\right]$$

$$- 40\left[\frac{\text{lbf}}{\text{in.}^2}\right] 144\left[\frac{\text{in.}^2}{\text{ft}^2}\right] 10\left[\frac{\text{ft}^3}{\text{lbm}}\right] + \frac{(-5)(25)[\text{ft}^2/\text{s}^2]}{2(32.2)[\text{lbm}\cdot\text{ft}/\text{s}^2\cdot\text{lbf}]} + 0$$

$$= 7780 + 31\,120 + 86\,400 - 57\,600 - \text{negligible} + 0$$

$$= 67\,700 \text{ ft}\cdot\text{lbf/lbm}$$

The positive sign indicates that work was performed by the fluid (the system) on the surroundings.

Example 2·9. A compressor requires 5 hp to compress ammonia at a certain rate from 40 psia, 11.7°F, to 135 psia while heat is removed from the compressor by means of water jackets at a rate of 150 B/min. Neglecting changes in potential and kinetic energy, how much is the enthalpy of the ammonia changed per minute?

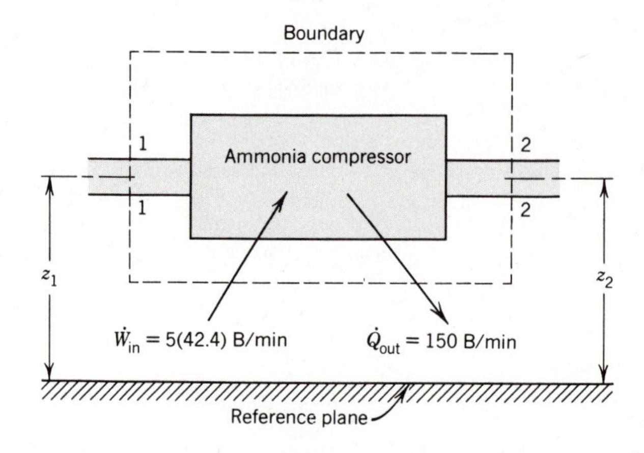

Example 2·9.

Solution. A sketch of the steady-flow system is made first. The enthalpy change per minute is the product of the mass rate of flow and the enthalpy change per pound, $\dot{m}\Delta h$; so an energy balance is written for the system in a form giving $\dot{m}\Delta h$ explicitly

$$\begin{aligned}\dot{m}\Delta h = \dot{m}(h_2 - h_1) &= \dot{m}(w_{in} + q - \Delta ke - \Delta pe)\\ &= \dot{W}_{in} + \dot{Q} - 0 - 0\\ &= 5(42.4) + (-150)\\ &= 212 - 150 = 62 \text{ B/min}\end{aligned}$$

The positive sign indicates that the enthalpy of the ammonia increased. The value 42.4 is a conversion factor: 42.4 B/min·hp = 1.

Example 2·10. Consider a gas turbine power plant with air as the working fluid. Air enters at 100 kPa, 20°C (ρ = 1.19 kg/m³), with a velocity of 130 m/s through an opening of 0.112 m² cross-sectional area. After being compressed, heated, and expanded through a turbine, the air leaves at 180 kPa, 150°C (ρ = 1.48 kg/m³), through an opening of 0.100 m² cross-sectional area. The power output of the plant is 375 kW. The internal energy and enthalpy of the air are given in kJ/kg by $u = 0.717T$ and $h = 1.004T$, where T is temperature on the Kelvin scale. Determine the net amount of heat added to the air in kJ/kg.

Analysis. A sketch of the steady-flow system is made first. It is sometimes convenient to write on the sketch the values of any physical quantities which are known and which appear to be pertinent. It looks as though we must use an energy balance in order to find q; so an energy balance is written for the system

$$q = w + h_2 - h_1 + \frac{V_2^2 - V_1^2}{2} + g(z_2 - z_1) \tag{a}$$

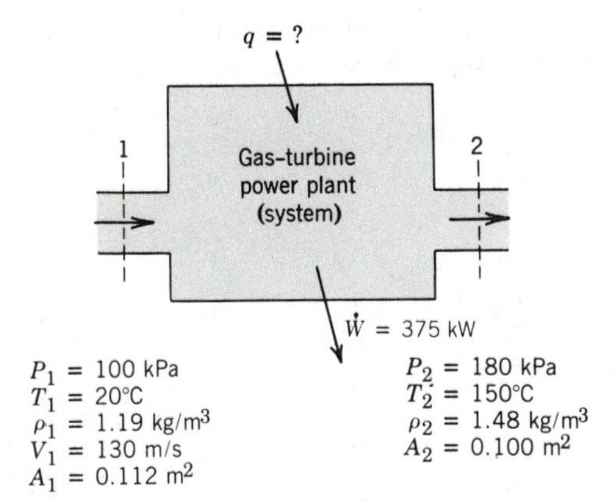

Example 2·10.

(We could use an energy balance with Δu and ΔPv as separate terms instead of using Δh, but the use of Δh will reduce the amount of numerical calculations needed.) Now all that must be done is to evaluate each of the terms on the right-hand side of the equation in order to obtain the value of q. The four terms will be considered in the order in which they appear in the energy balance above.

The work output (in kJ/kg) can be found from the power output and the mass rate of flow

$$w = \frac{\dot{W}}{\dot{m}} \tag{b}$$

$\dot{W} = 375$ kW, but the mass rate of flow must be calculated by

$$\dot{m} = \rho_1 A_1 V_1 \tag{c}$$

and ρ, A, and V at the inlet are all known. Thus Eqs. b and c together provide the value of w for use in Eq. a.

The change in enthalpy can be found from the given relationship $h = 1.004T$.

The change in kinetic energy requires a knowledge of both V_1 and V_2. $V_1 = 130$ m/s, but the value of V_2 must be obtained. Since A_2 and ρ_2 are known, we can use the continuity equation

$$\rho_2 A_2 V_2 = \rho_1 A_1 V_1 = \dot{m}$$

ρ_1, A_1, and V_1 are all known, and their product $\dot{m}$ was found earlier by means of Eq. c.

The change in potential energy can be found only if we have some information about the elevations of the inlet and outlet, and we do not have this information. However, if we notice the magnitudes of the other terms in the energy balance and then notice that an elevation change of 102 m is required to cause a change in potential energy of 1 kJ/kg (for $g = 9.81$ m/s^2), we should not hesitate to assume that the change in potential energy is negligibly small. (An elevation difference of 10 m between inlet and outlet of a gas-turbine plant would be unusually great, and it would correspond to a potential energy change of only 0.1 kJ/kg.)

Thus our analysis of the problem shows us (1) that we can solve the problem, and (2) the procedure to use in our solution.

An engineer tackling a problem usually analyzes it thoroughly in some manner such as presented above. For a simple problem like this, the analysis would not be written out as it is here, and it would be made so rapidly that one would hardly be aware that it had been made. The analysis might be outlined in writing in a form such as shown below.

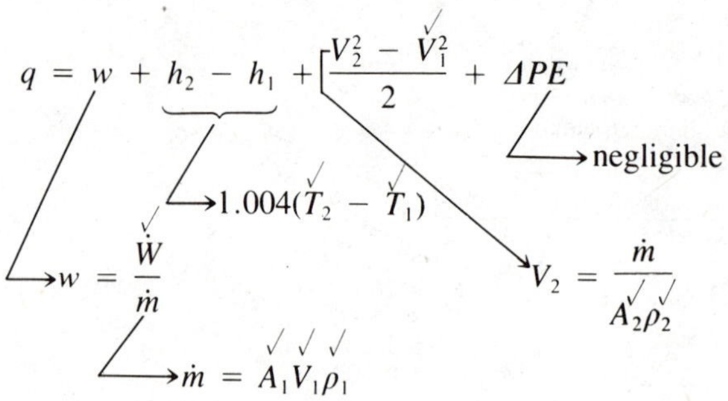

$$q = w + h_2 - h_1 + \frac{V_2^2 - \overset{\checkmark}{V_1^2}}{2} + \Delta PE$$

$\Delta PE \rightarrow$ negligible

$h_2 - h_1 \rightarrow 1.004(\overset{\checkmark}{T_2} - \overset{\checkmark}{T_1})$

$w \rightarrow w = \dfrac{\overset{\checkmark}{\dot{W}}}{\dot{m}}$

$\dot{m} \rightarrow \dot{m} = \overset{\checkmark}{A_1}\overset{\checkmark}{V_1}\overset{\checkmark}{\rho_1}$

$V_2 \rightarrow V_2 = \dfrac{\dot{m}}{\overset{\checkmark}{A_2}\overset{\checkmark}{\rho_2}}$

The check marks ($\surd$) over certain symbols indicate that the numerical values of those quantities are known.

This form of analysis provides a clear outline of the solution. It will be used frequently throughout this book.

Solution.

$$\dot{m} = A_1V_1\rho_1 = 0.112(130)1.19 = 17.3 \text{ kg/s}$$

$$w = \frac{\dot{W}}{\dot{m}} = \frac{375}{17.3} = 21.7 \text{ kJ/kg}$$

$$V_2 = \frac{\dot{m}}{A_2\rho_2} = \frac{17.3}{1.48(0.100)} = 117 \text{ m/s}$$

$$q = w + h_2 - h_1 + \frac{V_2^2 - V_1^2}{2} + \Delta pe$$

$$= w + 1.004(T_2 - T_1) + \frac{(V_2 - V_1)(V_2 + V_1)}{2} + 0$$

$$= 21.7 + 1.004(423 - 293) + \frac{(-13)247}{2(1000)}$$

$$= 151 \text{ kJ/kg}$$

Alternative solution. If you have at hand a computer with a program for solving simultaneous algebraic equations, you might approach this problem by entering into the program all the equations which apply. The program may be such that it lists the variables involved, or you may have to list them outside the program. In either case, the equations and variables might be listed as

	Variables	
Equations	Assigned	Unknown
$\dot{m} = \rho_1A_1V_1$	ρ_1, A_1, V_1	$\dot{m}$
$\dot{m} = \rho_2A_2V_2$	ρ_2, A_2	V_2
$\dot{W} = \dot{m}w$	$\dot{W}$	w
$u = 0.717T$	T	u
$h = 1.004T$	T	h
$q = w + h_2 - h_1 + \dfrac{V_2^2 - V_1^2}{2}$		q

Obviously, this set of six equations can be solved for the unknown values. (You might have entered the equation for u as $u_2 = 0.717T_2$ and $u_1 = 0.717T_1$, and this would have increased by one the number of equations and also the number of unknowns. The same could have been done with the equation for h. The equation $\rho_1 A_1 V_1 = \rho_2 A_2 V_2$ could not have been added to the above list, because it is not independent of the first two equations listed. Several similar points must be considered if you are using a general program for solving simultaneous equations.) For a problem as simple and straightforward as this one, however, a solution such as that shown in its entirety above is probably as convenient and as quick as any other at this stage of your study.

Example 2·11. Air is compressed in a frictionless steady-flow process from 90 kPa, 15°C ($v = 0.918$ m³/kg), to 130 kPa in such a manner that $P(v + 0.250) =$ constant, where v is in m³/kg. Inlet velocity is negligibly small, and discharge velocity is 110 m/s. Calculate the work required per kilogram of air.

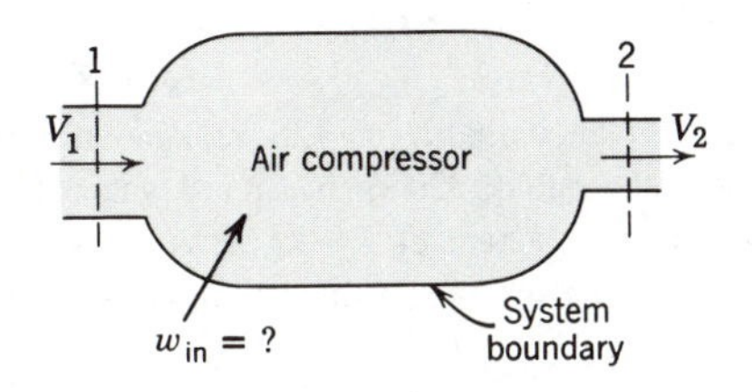

Example 2·11.

Analysis. Applying the first law to this steady-flow system gives an energy balance in the form

$$w_{in} = q_{out} + h_2 - h_1 + \frac{V_2^2 - V_1^2}{2} + g(z_2 - z_1)$$

When we begin to evaluate these terms, we see that the potential energy term is probably negligibly small and that we have values for both velocities which appear in the kinetic energy term. There appears to be no way to calculate q or Δh (unless one has already studied Chapter 4!); so we appear to be "stuck." The energy balance written above is valid, but we do not have all the information required in order to solve it for w_{in}. Therefore, we immediately seek some other means of solving the problem. Perhaps we can find w_{in} by means of the expression which is based on the principles of mechanics (Sec. 1·17),

$$w_{in} = \int_1^2 v\,dP + \frac{V_2^2 - V_1^2}{2} + g(z_2 - z_1)$$

We have noted before that we can evaluate the kinetic and potential energy terms. Since we have a functional relationship between P and v, we can integrate $\int v\,dP$, and so we will be able to determine w_{in}. (The unsuccessful attempt to use the energy balance is shown above as an illustration that no harm is done by making such a "false start" as long as we quickly recognize that we are not on the right track and immediately look for another method of solution).

Solution.

$$w_{in} = \int_1^2 v\,dP + \frac{V_2^2 - V_1^2}{2} + g(z_2 - z_1) = \int_1^2 v\,dP + \frac{V_2^2}{2} + 0$$

$$\int_1^2 v\,dP = \int_1^2 \left(\frac{C}{P} - 0.250\right) dP = C \ln \frac{P_2}{P_1} - 0.250\,(P_2 - P_1)$$

$$= P_1\,(v_1 + 0.250) \ln \frac{P_2}{P_1} - 0.250\,(P_2 - P_1)$$

$$= 90(0.918 + 0.250) \ln \frac{130}{90} - 0.250(130 - 90)$$

$$= 28.7 \text{ kJ/kg}$$

$$w_{in} = \int_1^2 v\,dP + (V_2^2/2) = 28.7 + \frac{(110)^2}{2(1000)} = 34.8 \text{ kJ/kg}$$

Example 2·12. A mixture of air and water vapor with an enthalpy of 126 kJ/kg enters the dehumidifying section of an air-conditioning system at a rate of 320 kg/h. Liquid water drains out of the dehumidifier with an enthalpy of 42 kJ/kg at a rate of 7.0 kg/h. An air–vapor mixture leaves with an enthalpy of 47 kJ/kg. Determine the rate of heat removal from the fluids passing through the dehumidifier.

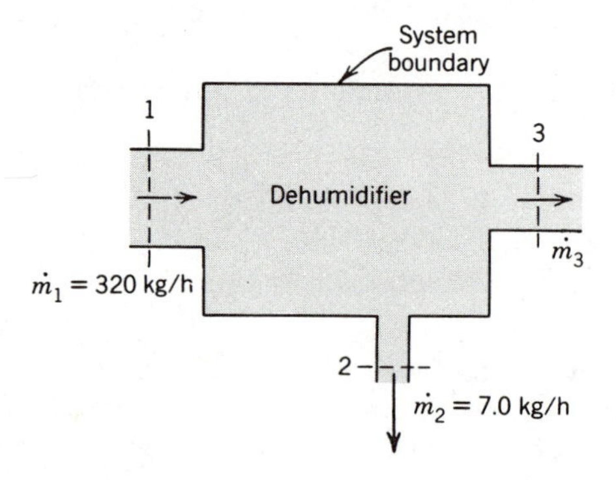

Example 2·12.

Solution. After a sketch of the system is made, the first law is applied in the form of an energy balance for steady flow

$$\dot{m}_1 h_1 = \dot{m}_2 h_2 + \dot{m}_3 h_3 + \dot{Q}_{out}$$

The assumption has been made that changes in kinetic and potential energies are negligibly small. There is no work done. The conservation of mass requires that $\dot{m}_3 = \dot{m}_1 - \dot{m}_2$. Rearranging and substituting numerical values gives

$$\dot{Q}_{out} = \dot{m}_1 - \dot{m}_2 - (\dot{m}_1 - \dot{m}_2)\,h_3$$

$$\dot{Q}_{out} = 320(126) - 7.0(42) - (320 - 7)47$$

$$= 25\,300 \text{ kJ/h}$$

Example 2·13. Brine with a specific gravity of 1.20 enters a pump at 14 psia, 15°F, through a 3-in.-diameter opening and is discharged at 45 psia through a 2-in.-diameter opening. The pump outlet is 3 ft above the inlet. The flow rate is 200 gpm. If the pumping were done frictionlessly, how much power would be required?

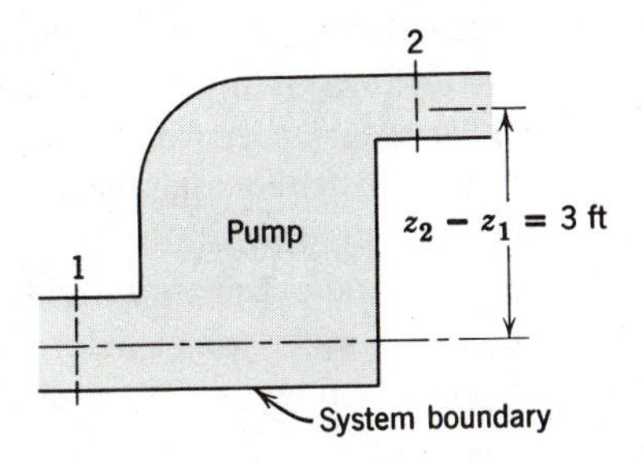

Example 2·13.

Analysis. This is an open system. Steady-flow conditions will be assumed. We can find the power input by multiplying the work per unit mass by the mass rate of flow. In order to find the work, we first consider the use of the first law. As in Example 2·11, however, we see that our ignorance of q and Δu stops us. (However, see problem 2·45.) Therefore, we turn to the relation from mechanics which applies to frictionless steady-flow processes:

$$w_{in} = \int_1^2 v\,dP + \frac{V_2^2 - V_1^2}{2} + g(z_2 - z_1)$$

Noting that brine is incompressible (v = constant) and that the velocities can be obtained by means of the continuity equation, we see that the work done on the brine can be calculated by this relation.

The flow rate is given in the usual unit of gallons per minute, a volume rate of flow. Since the specific gravity is known, the mass rate of flow can easily be obtained.

Solution.

$$\dot{m} = (\text{volume rate of flow})\,\rho$$

$$= \frac{200}{60}\left[\frac{\text{gal}}{\text{s}}\right]\frac{231[\text{in.}^3]}{[\text{gal}]}\frac{[\text{ft}^3]}{1728[\text{in.}^3]}(62.4)1.20\left[\frac{\text{lbm}}{\text{ft}^3}\right]$$

$$= 33.4 \text{ lbm/s}$$

$$V_1 = \frac{\dot{m}}{\rho A_1} = \frac{33.4(4)144}{1.20(62.4)\pi(3)^2} = 9.1 \text{ fps}$$

$$V_2 = \frac{\dot{m}}{\rho A_2} = \frac{\rho A_1 V_1}{\rho A_2} = \left(\frac{D_1}{D_2}\right)^2 V_1 = \left(\frac{3}{2}\right)^2 9.1 = 20.5 \text{ fps}$$

$$w_{in} = \int_1^2 v\,dP + \frac{V_2^2 - V_1^2}{2} + g(z_2 - z_1)$$

$$= v(P_2 - P_1) + \frac{(V_2 - V_1)(V_2 + V_1)}{2} + g(z_2 - z_1)$$

$$= \frac{1}{1.2(62.4)}(45 - 14)144 + \frac{11.4(29.6)}{32.2} + \frac{32.2}{32.2}(3)$$

$$= 59.7 + 5.2 + 3 = 67.9 \text{ ft·lbf/lbm}$$

$$\dot{W} = \dot{m}w_{in} = \frac{33.4(67.9)}{550} = 4.12 \text{ hp}$$

(Note that in the numerical evaluation of the potential energy term the value in the numerator is $g = 32.2$ ft/s^2 while the value in the denominator is the conversion factor $1 = 32.2$ lbm·ft/lbf·s^2).

2.8 The First Law Applied to Open Systems — General Formulations

We have pointed out (in Sec. 1·16) that steady flow is a *special case* of the flow through an open system. An open system that has mass crossing its boundaries under any conditions other than those of steady flow is sometimes called a transient-flow system or an unsteady-flow system. The properties of the mass crossing the boundaries may vary with time. The mass of matter in the system may also vary, since the mass of matter entering the system may not equal the mass of matter leaving it. Some examples of transient-flow processes are the filling or emptying of any vessel, the flow in the intake and exhaust manifolds of an automobile engine, the flow through an automobile-engine water jacket before the engine parts have reached steady operating temperatures, and the flow through a turbine while the shaft speed or power output is changing.

We have seen (Sec. 2·7) that the application of the first law to any open system results in an energy balance in a form such as

$$\begin{pmatrix}\text{Net amount of}\\ \text{energy added to}\\ \text{system as heat}\\ \text{and all forms of}\\ \text{work}\end{pmatrix} + \begin{pmatrix}\text{stored}\\ \text{energy}\\ \text{of matter}\\ \text{entering}\\ \text{system}\end{pmatrix} - \begin{pmatrix}\text{stored}\\ \text{energy}\\ \text{of matter}\\ \text{leaving}\\ \text{system}\end{pmatrix} = \begin{pmatrix}\text{net increase}\\ \text{in stored}\\ \text{energy of}\\ \text{system}\end{pmatrix} \qquad (2\cdot 8b)$$

It was also shown in Sec. 2·7 that the work term in an energy balance for an open system can be separated into two parts: (*a*) flow work, and (*b*) all other forms of work. In the general case of an open system, the properties of the matter crossing the system boundary vary with mass or with time. If the variations of P and v with mass crossing the system boundary at some section are known, the flow-work term for that section can be evaluated as $\int_0^{m_1} Pv\,\delta m$, where m_1 is the total amount of mass which crosses the boundary there. If the variations of P, v, and the mass rate of flow with time are known, it is more convenient to write the flow-work term for one section as $\int_{\tau_i}^{\tau_f} \dot{m}\,Pv\,d\tau$, where τ_i and τ_f are the times at which the process starts (*i*nitial time) and ends (*f*inal time). In a similar manner, the stored energy of fluid crossing the system boundary can be evaluated as $\int_0^m e\,\delta m$ or as $\int_{\tau_i}^{\tau_f} \dot{m}\,e\,d\tau$. Thus the terms in Eq. 2·8b can be evaluated as follows:

$$\begin{pmatrix}\text{Net amount of energy}\\ \text{added to system as heat}\\ \text{and work}\end{pmatrix} = \begin{cases} Q - W + \sum\limits_{\text{entering}} \int Pv\,\delta m - \sum\limits_{\text{leaving}} \int Pv\,\delta m \\ Q - W + \sum\limits_{\text{entering}} \int \dot{m}\,Pv\,d\tau - \sum\limits_{\text{leaving}} \int \dot{m}\,Pv\,d\tau \end{cases}$$

$$\begin{pmatrix}\text{Stored energy of matter}\\ \text{entering system}\end{pmatrix} = \sum_{\text{entering}} \int e\,\delta m = \sum_{\text{entering}} \int \dot{m}e\,d\tau$$

$$\begin{pmatrix}\text{Stored energy of matter}\\ \text{leaving system}\end{pmatrix} = \sum_{\text{leaving}} \int e\,\delta m = \sum_{\text{leaving}} \int \dot{m}e\,d\tau$$

$$\begin{pmatrix}\text{Net increase in stored}\\ \text{energy of system}\end{pmatrix} = E_f - E_i$$

where E_i and E_f are respectively the stored energy of the entire system in its initial state and in its final state. If the power input or output of the system is known as a function of time, W may be evaluated as $\int_{\tau_i}^{\tau_f} \dot{W}\, d\tau$.

For a system with fluid entering only at a section 1 and leaving only at a section 2, Eq. 2·8b can be written as

$$Q - W + \int (e_1 + P_1 v_1)\, \delta m_1 - \int (e_2 + P_2 v_2)\, \delta m_2 = E_f - E_i \qquad (2\cdot 8c)$$

In the absence of electrical, magnetic, and surface tension effects, $e = u + ke + pe$ and $e + Pv = h + ke + pe$, so that

$$Q - W + \int \left(h_1 + \frac{V_1^2}{2} + gz_1\right) \delta m_1 - \int \left(h_2 + \frac{V_2^2}{2} + gz_2\right) \delta m_2$$
$$= U_f - U_i + \frac{m_f V_f^2 - m_i V_i^2}{2} + g(m_f z_f - m_i z_i) \qquad (2\cdot 8e)$$

It must be remembered that the subscripts i and f denote that the values so marked are for the entire system before and after the process occurs.

The closed system and steady-flow energy balances presented in Secs. 2·2 and 2·7 can be considered as special cases of the general open-system energy balances presented in this article and Sec. 2·6.* To illustrate this, let us first rewrite Eq. 2·8 in differential form:

$$\delta Q - \delta W + (e_1 + P_1 v_1)\, \delta m_1 - (e_2 + P_2 v_2)\, \delta m_2 = dE \qquad (2\cdot 8d)$$

For the special case of a *closed system*, $\delta m_1 = \delta m_2 = 0$ and Eq. 2·8d becomes

$$\delta Q - \delta W = dE \qquad (2\cdot 2)$$

which was introduced in Sec. 2·2. For the special case of a *steady-flow* system, $dE = 0$, $\delta m_1 = \delta m_2$, and the properties (e, P, v) of the fluids crossing the system boundary do not vary with time. Thus Eq. 2·8 becomes

$$Q - W + (e_1 + P_1 v_1)m - (e_2 + P_2 v_2)m = 0 \qquad (2\cdot 9b)$$

which was introduced in Sec. 2·7 for the case of steady flow.

Frequently, moving the boundary of an open system simplifies an energy balance. This is usually true for the flow of a fluid into an open system from a region where the fluid is at rest. As a specific example, consider the flow from the atmosphere into the air compressor shown in Fig. 2·7. The fluid at section 1 has properties P_1, v_1, h_1, etc. and a velocity V_1. Consider an energy balance between section 1 and some section 0 in the atmosphere where the fluid on its way toward section 1 has a velocity that is very small, $V_0 \approx 0$. Certainly no work is done between 0 and 1. Further, let us neglect Δpe. Then

*Notice that the closed-system energy balance *can be considered* as a special case of the open-system energy balance, but that to *derive* the closed-system equation from the open-system equation is to reverse the logical structure we have been following. The first law is an inductive conclusion based on observations on *closed systems*. It can then be extended to open systems. Starting from the open-system result and deducing the closed-system energy balance simply reverses the development: so of course it works out, but it proves nothing.

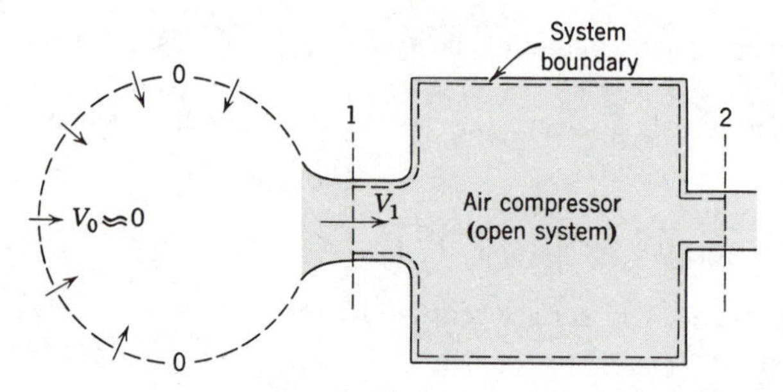

Figure 2·7 Flow into an open system from an atmosphere at rest.

if we assume that the flow between 0 and 1 is adiabatic and that there is no change in the total amount of energy or mass stored between 0 and 1, the first law gives us

$$h_0 = h_1 + \frac{V_1^2}{2}$$

Thus two terms, enthalpy and kinetic energy at section 1, which appear in an energy balance on the compressor can be replaced by the enthalpy of the fluid at rest in the surrounding atmosphere. This substitution is often made tacitly. For example, if we are applying the first law to a machine which draws air from a room, we usually make use of the pressure and temperature of the air at rest in the room in evaluating inlet conditions and therefore treat the inlet velocity as being zero. Actually, of course, right at the machine inlet the velocity is greater than zero; but, if we evaluate and use the kinetic energy at that point, then we must use it in conjunction with the enthalpy (which may depend on pressure and temperature) at that point, and this will be lower than the enthalpy in the room.

In case you are tempted at this point to memorize some of the many energy balance equations that have been displayed in the last three sections, we repeat: *Learn and understand the relationship*

$$\begin{pmatrix}\text{Net amount of}\\ \text{energy added to}\\ \text{system as heat}\\ \text{and all forms of}\\ \text{work}\end{pmatrix} + \begin{pmatrix}\text{stored}\\ \text{energy}\\ \text{of matter}\\ \text{entering}\\ \text{system}\end{pmatrix} - \begin{pmatrix}\text{stored}\\ \text{energy}\\ \text{of matter}\\ \text{leaving}\\ \text{system}\end{pmatrix} = \begin{pmatrix}\text{net increase}\\ \text{in stored}\\ \text{energy of}\\ \text{system}\end{pmatrix} \qquad (2{\cdot}8\text{b})$$

*From this, develop any other equations you may need for particular systems and processes.**

Example 2·14. Air at 700 kPa, 90°C, flows through a large pipeline. Connected to the line through a valve is an insulated tank with a volume of 0.200 m³ containing 0.230 kg of air at 100 kPa, 30°C, while the valve is closed. The valve is then opened and air flows into the tank until the pressure in the tank is 700 kPa. Properties of air are related by $h = 1.4u = 1.004T = 3.50Pv$,

*You may well ask, "If all these equations are not to be learned, why are they in the book?" The answer is that these equations are presented as *samples* of the many that can be obtained from the general energy balance. It is hoped that the study of these samples will help you in developing your own equations to fit the various special cases you encounter.

with h and u in kJ/kg, T in kelvins, P in kPa, and v in m^3/kg. Assuming that no heat transfer occurs between air in the tank and any part of the surroundings and that conditions are uniform throughout the tank at any instant, compute the final mass and the final temperature of air in the tank. Neglect changes in kinetic and potential energy.

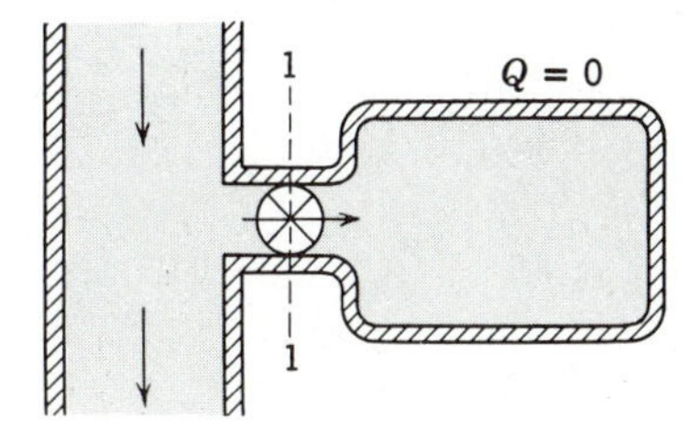

Example 2·14. Open-system analysis.

Solution. The region within the tank will be considered as the system. Since the quantity of air in the tank varies, this is a transient-flow system. A sketch is made. Since the properties of the air entering the system are constant, they can be tabulated along with the initial and final properties of air in the system as shown in the accompanying table.

	Air Initially in the Tank	Air Flowing into the Tank	Air Finally in the Tank
Pressure, kPa	$P_i = 100$	$P_1 = 700$	$P_f = 700$
Volume, m^3	$V_i = 0.200$	. . .	$V_f = 0.200$
Mass, kg	$m_i = 0.230$	$m_1 = m_f - m_i$	m_f
Temperature, °C	$T_i = 30$	$T_1 = 90$	T_f

Noting that no work (except flow work) is done during the process, we apply the first law to this adiabatic process of a transient-flow system to get the following energy balance

$$\begin{pmatrix}\text{Stored energy}\\ \text{of air initially}\\ \text{in tank}\end{pmatrix} + \begin{pmatrix}\text{stored energy of}\\ \text{air which flows}\\ \text{into tank}\end{pmatrix} + \begin{pmatrix}\text{flow work of}\\ \text{air which flows}\\ \text{into tank}\end{pmatrix} = \begin{pmatrix}\text{stored energy}\\ \text{of air finally}\\ \text{in tank}\end{pmatrix}$$

Neglecting the kinetic and potential energy terms and recalling that u_1, P_1, and v_1 are constant, we can express this energy balance as

$$m_iu_i + m_1(u_1 + P_1v_1) = m_fu_f$$

$$m_iu_i + (m_f - m_i)h_1 = m_fu_f$$

In this equation there are two unknowns, m_f and u_f. However, another relation between these two variables is given by $1.4u = 3.50Pv$.

$$1.4u_f = 3.50P_fv_f = 3.50P_f\frac{V_f}{m_f}$$

$$u_f = 2.50\frac{P_fV_f}{m_f}$$

Substituting this value of u_f into the preceding energy balance,

$$m_i u_i + (m_f - m_i)h_1 = 2.50 P_f V_f$$

$$m_f = \frac{2.50 P_f V_f - m_i u_i + m_i h_1}{h_1}$$

$$= \frac{2.50 P_f V_f - m_i(1.004/1.4)T_i + m_i(1.004)T_1}{1.004 T_1}$$

$$= \frac{2.50(700)0.200 - 0.230(1.004)\ (303/1.4 - 363)}{1.004(363)}$$

$$m_f = 1.05 \text{ kg}$$

$$T_f = \frac{3.50}{1.004} P_f v_f = \frac{3.50 P_f v_f}{1.004 m_f}$$

$$= \frac{3.50(700)0.200}{1.004(1.05)} = 465 \text{ K} = 192°\text{C}$$

These results and those of the following two solutions are based on the assumption that the tank is filled adiabatically. It is difficult to check these results experimentally because they are affected greatly by heat transfer to the tank wall, even if the outside of the tank is covered by an excellent insulating material. Also, heat transfer from the air to any thermometer placed in the tank will affect the results; yet such a transfer of heat is necessary in order for the thermometer to indicate the temperature of the air.

Solution with slightly different aproach. Some people prefer to start always with an energy balance in differential form such as (for the case of $Q = 0$ and $W = 0$)

$$\begin{pmatrix} \text{stored energy and} \\ \text{flow work added to} \\ \text{system by entering} \\ \text{mass } \delta m \end{pmatrix} = \begin{pmatrix} \text{increase in stored} \\ \text{energy of system} \\ \text{while mass } \delta m \text{ enters} \end{pmatrix}$$

$$(u_1 + P_1 v_1)\delta m_1 = dU$$

$$h_1 \delta m_1 = dU$$

If m is the mass in the tank, dm is the change in mass in the tank. Since mass enters at only section 1, $\delta m_1 = dm$.

$$h_1 dm = dU$$

Since $u = 2.50Pv$, then $U = mu = 2.50PV$, and $dU = 2.50(V\,dP + P\,dV) = 2.50V\,dP$.

$$h_1 dm = 2.50V\,dP$$

Integrating, noting that h_1 is constant,

$$h_1 dm = 2.50V\,dP$$

$$h_1(m_f - m_i) = 2.50V(P_f - P_i)$$

$$m_f = m_i + \frac{2.50V(P_f - P_i)}{1.004 T_1}$$

Substitution of numerical values gives $m_f = 1.05$ kg, and T_f can then be found as in the first solution.

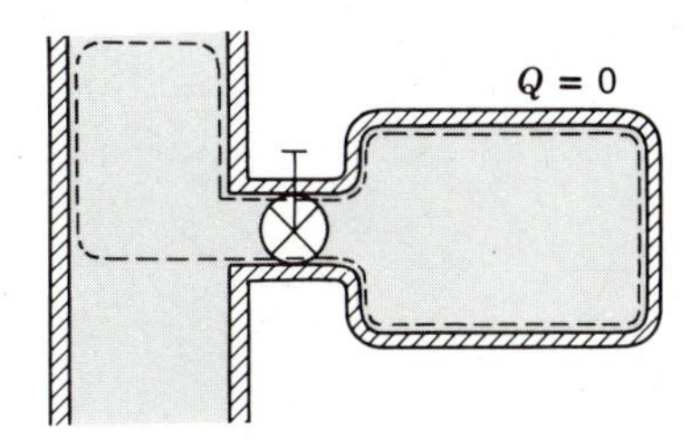

Example 2·14. Closed-system analysis.

Alternative solution. This problem can also be solved by the use of a closed-system analysis. Let the system be all the air which finally is in the tank. The system boundary is then an imaginary envelope that encloses all the air in the tank and all the air in the pipe that eventually flows into the tank. No mass crosses this boundary. As air flows from the pipe into the tank, the volume of that part of the system outside the tank is reduced to zero. Work is done on the system by the surrounding air in the pipe as the system volume is reduced. Properties of the system before and after the process occurs are listed in the accompanying table.

	Initial Condition	Final Condition
Mass	$m_f = m_i$ in tank $+ (m_f - m_i)$ in pipe	m_f (all in tank)
Internal energy	$m_i u_i + (m_f - m_i) u_1$	$m_f u_f$
Volume	$V_f + V_1 = m_f v_f + (m_f - m_i) v_1$	$V_f = m_f v_f$

The subscript 1 refers to conditions in the pipe.

Applying the first law to this system and neglecting changes in kinetic energy,

$$Q + W_{\text{in}} + U_i = U_f \tag{a}$$

The work done on the system is the work done by the surrounding gas in the pipe to compress that part of the system in the pipe from a volume of V_1 to a volume of zero. Assuming that *this decrease in volume of the system is brought about frictionlessly* and that *the pressure of that part of the system in the pipe remains constant*

$$W_{\text{in}} = -\int_{V_f+V_1}^{V_f} P\,dV = -P_1 \Delta V = +P_1 V_1 = P_1(m_f - m_i)v_1$$

Also, the process is adiabatic. Equation a becomes

$$0 + P_1(m_f - m_i)v_1 + m_i u_i + (m_f - m_i)u_1 = m_f u_f$$

$$m_i u_i + (m_f - m_i)h_1 = m_f u_f$$

This is the same energy balance obtained in the first solution that was based on an open-system analysis, and so the numerical solution from this point is identical for the open-system analysis and the closed-system analysis.

Example 2·15. A well-insulated tank with a volume of 0.600 m^3 contains 1.40 kg of air at 300 kPa. A small valve is then opened to let some of the air escape into the atmosphere which is at 101 kPa. Assume that conditions are uniform throughout the tank at any instant. Determine the amount of air left in the tank when its pressure reaches 150 kPa. (For air under these conditions, $h = 1.4u = 1.004T = 3.50Pv$, where u and h are in kJ/kg, T is in kelvins, P is in kPa, and v is in m^3/kg.)

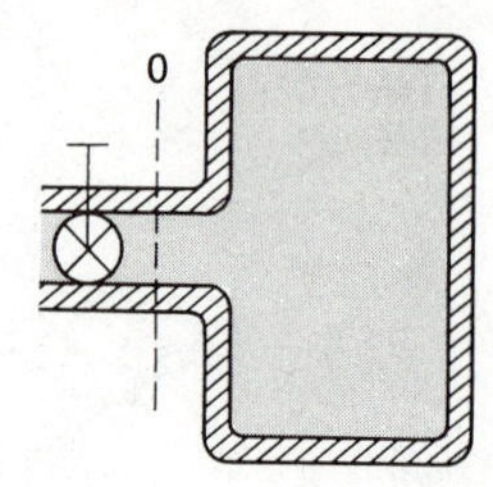

Example 2·15.

Solution. Define the open system as being the region within the tank. A sketch is made. The mass of air remaining in thc tank at 150 kPa will depend on its temperature, and its temperature will be influenced by the energy removed from the system by means of the outflowing air; so the first step is to set up an energy balance for the system. Since the properties of the air crossing the system boundary vary (unlike the case of Example 2·14), we must start with a differential form of the energy balance. As an infinitesimal mass of air, δm_o, flows out of the system, the mass of air within the system m changes by an amount dm. Since mass δm_0 leaves the system, $\delta m_0 = -dm$. Since $Q = 0$ and $W = 0$,

$$\begin{pmatrix}\text{decrease in stored energy}\\ \text{of air in tank}\end{pmatrix} = \begin{pmatrix}\text{stored energy of}\\ \delta m_0 \text{ leaving tank}\end{pmatrix} + \begin{pmatrix}\text{flow work of}\\ \delta m_0 \text{ leaving tank}\end{pmatrix}$$

$$-dU = e_0\,\delta m_0 + P_0 v_0\,\delta m_0 = (e_0 + P_0 v_0)\,\delta m_0$$

where a property without a subscript is a property of the air in the tank. Potential energy changes can be neglected. If we locate section 0 just upstream of the small valve opening so that ke_0 is negligibly small, $e_0 = u_0$ and the last equation becomes

$$-dU = (u_0 + P_0 v_0)\,\delta m_0$$

Since the properties of the air leaving are the same as those of the air in the tank, $u_0 = u$, $P_0 = P$, and $v_0 = v$. Recalling also that $\delta m_0 = -dm$

$$\begin{aligned} -dU &= -(u + Pv)\,dm \\ d(mu) &= (u + Pv)\,dm \\ m\,du + u\,dm &= (u + Pv)\,dm \\ m\,du &= pv\,dm \end{aligned}$$

Specific internal energy can be expressed in terms of Pv as $1.4u = 3.50\,Pv$ or $u = 2.50Pv$, so that $du = 2.5\,d(Pv)$, and the energy balance becomes

$$\begin{aligned} 2.5m\,d(Pv) &= Pv\,dm \\ \frac{dm}{m} &= 2.5\,\frac{d(Pv)}{Pv} \\ \ln\frac{m_f}{m_i} &= 2.5\ln\left(\frac{P_f v_f}{P_i v_i}\right) \\ \frac{m_f}{m_i} &= \left(\frac{P_f v_f}{P_i v_i}\right)^{2.5} = \left(\frac{P_f V_f m_i}{m_f P_i V_i}\right)^{2.5} \end{aligned}$$

But the tank volume is constant, $V_f = V_i$, so

$$\frac{m_f}{m_i} = \left(\frac{P_f}{P_i}\right)^{2.5}\left(\frac{m_i}{m_f}\right)^{2.5}$$

$$\frac{m_f}{m_i} = \left(\frac{P_f}{P_i}\right)^{2.5/3.5} = \left(\frac{150}{300}\right)^{0.714} = 0.609$$

$$m_f = 0.609m_i = 0.609(1.40) = 0.853 \text{ kg}$$

2·9 Heat Engines, Thermal Efficiency; Refrigerators, Heat Pumps, Coefficient of Performance

A heat engine is a system that receives heat and produces work while executing a cycle. A heat engine may be as simple as a gas confined within a cylinder fitted with a piston or as complex as an entire power plant. Two examples of heat engines are shown schematically in Fig. 2·8. In Fig. 2·8*a,* a gas trapped in the cylinder is heated at constant pressure (process 1-2), doing work on the piston. Then the gas is cooled while the piston is stationary (process 2-3), there being no work done on or by the gas during this process. Then the piston is moved inward, doing work on the gas in compressing it adiabatically to its final state.

In Fig. 2·8*b,* heat is added to water in the boiler in order to generate steam that then expands adiabatically through the turbine, doing work. The steam flows from the turbine into the condenser. Heat is removed from the steam in the condenser in order to condense the steam. The liquid leaving the condenser enters a pump that pumps it into the boiler to complete a cycle. Work is done on the liquid flowing through the pump. Even though each of the four pieces of equipment has fluid flowing into and out of it and hence must be treated as an open system, the four pieces of equipment and the connecting piping together always contain the same fluid. No matter enters or leaves this larger system, so it can be treated as a closed system.

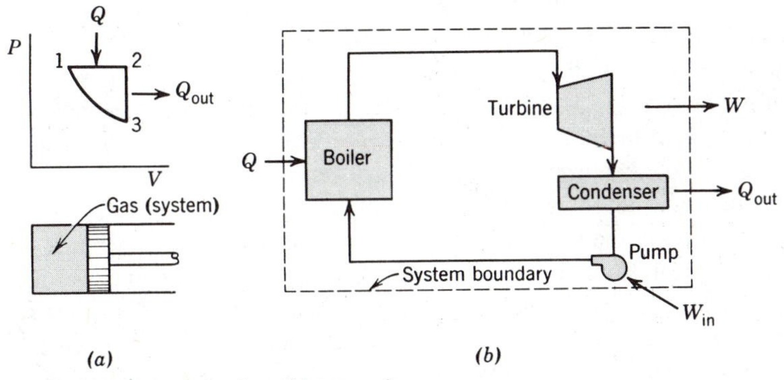

Figure 2·8 Two examples of heat engines.

Notice that in each of these heat-engine cycles some heat is rejected from the system. By the first law,

$$\oint \delta W = \oint \delta Q = Q_{\text{in}} - Q_{\text{out}}$$

where Q_{in} and Q_{out} stand respectively for the *gross* amount of heat added and the *gross* amount of heat rejected. Since Q_{out} in the cycles described is not zero, the net work done by the system is less than the gross amount of heat added. In other words, not all the heat added is converted into work. *Thermal efficiency* is defined as that fraction of the gross heat input to a system during a cycle that is converted into net work output, or

$$\eta \equiv \frac{\oint \delta W}{Q_{\text{in}}} = \frac{\text{net work output of cycle}}{\text{gross heat added}}$$

Applying the first law,

$$\eta \equiv \frac{\oint \delta W}{Q_{\text{in}}} = \frac{Q_{\text{in}} - Q_{\text{out}}}{Q_{\text{in}}} = 1 - \frac{Q_{\text{out}}}{Q_{\text{in}}}$$

Thus the thermal efficiency of the steam power plant cycle discussed above and shown in Fig. 2·8*b* is given by either

$$\eta = \frac{W_T - W_{\text{in},P}}{Q_{\text{boiler}}} \qquad \text{or} \qquad \eta = \frac{Q_{\text{boiler}} - Q_{\text{out, condenser}}}{Q_{\text{boiler}}}$$

Notice that thermal efficiency is used only with cycles; there is no such thing as the thermal efficiency of a process.

A system which absorbs heat from some low-temperature part of its surroundings and discharges heat to some other part of its surroundings at a higher temperature while executing a cycle is called either a *refrigerator* or a *heat pump*. The difference between a refrigerator and a heat pump is one of purpose. If the purpose is to remove heat from the low-temperature part of the surroundings to cool it or maintain it at a low temperature, the device is called a refrigerator. The discharging of heat to the higher-temperature region is incidental, although necessary, to the operation of the refrigerator. If the purpose is to discharge heat from the system to some part of the surroundings in order to achieve or maintain a higher temperature there, the device is called a heat pump. Some looseness in terminology should be noted, however. A device that during the winter draws heat from the cold atmosphere and discharges heat to a building to maintain it at a higher temperature can also be used in the summer to maintain a relatively low temperature in the building by drawing heat from the building and discharging heat to the warmer atmosphere. In accordance with the distinction made previously, the device operates as a heat pump during the winter and as a refrigerator during the summer, but the device is called simply a heat pump. The working fluid in either a refrigerator or a heat pump is called the *refrigerant*.

In one type of refrigerator or heat pump, called the vapor-compression type, shown in Fig. 2·9, a vapor is compressed and led into a condenser where it is condensed to a liquid by the removal of heat. The high-pressure liquid refrigerant then passes through an expansion valve into an evaporator where the pressure is low. The temperature of the refrigerant in the evaporator is low enough so that heat can be absorbed from some relatively low-temperature part of the surroundings in order to evaporate the refrigerant. The refrigerant then passes to the compressor to begin the cycle anew. In a household refrigerator, the low-temperature part of the surroundings is in the freezer compartment, and the high-temperature part of the surroundings is the air in the room where the refrigerator is operated. For a heat pump used to heat a house in winter, the low-temperature part of the surroundings is the outdoor atmosphere (or perhaps well water or earth), and the high-temperature part of the surroundings is the air in the house which is being heated.

An index of performance used with heat engines is *thermal efficiency*. For refrigerators and heat pumps, *coefficient of performance* is used as an index of performance. Neither index is applicable to a system or cycle other than the one for which it is defined. For example, it is meaningless to speak of "the thermal efficiency of a refrigerator."

The *coefficient of performance* of a *refrigerating* cycle is defined as the ratio of the heat absorbed from the low-temperature part of the surroundings to the net work input

$$\beta_R \equiv \frac{Q_{\text{in},L}}{-\oint \delta W}$$

The *coefficient of performance* of a *heat pump* cycle is defined as the ratio of the heat discharged to the high-temperature part of the surroundings to the net work input,

$$\beta_{\text{HP}} \equiv \frac{Q_{\text{out},H}}{-\oint \delta W}$$

Notice that in each case a coefficient of performance is defined as a ratio of the heat transfer that is the purpose of the device to the net work input. Also, notice that values of either coefficient of performance may be greater or less than one.

Application of the first law to any cycle shows that

$$W_{\text{in}} = Q_{\text{out}} - Q_{\text{in}}$$

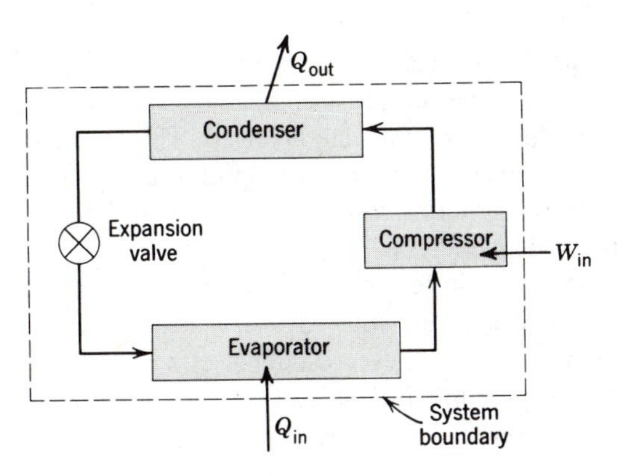

Figure 2·9 Refrigerator.

Notice, however, that the operation of an actual device may involve stray heat transfer with parts of the surroundings other than the higher-temperature and lower-temperature regions between which the device is intended to operate. In such a case, Q_{out} may not be the same as $Q_{out,H}$, and Q_{in} may not be the same as $Q_{in,L}$.

2·10 Limitation of the First Law

All the relations developed in the preceding sections are used in making an energy accounting for various types of systems. In addition to being able to write an energy balance, an engineer must be able to predict how much of the heat transferred to a system may be converted into useful work, or how much work is required to produce a certain refrigerating effect, or whether or not a system will undergo a specified change. These predictions cannot be made on the basis of the first law alone. For example, it is impossible to predict from the first law alone how much of the heat transferred to a heat engine may be converted into work. As far as the first law is concerned, all the heat transferred could conceivably be converted into work. In order to obtain information regarding these aspects of engineering analysis, the second law of thermodynamics must be employed. This law is discussed in detail in Chapters 5 through 8.

2·11 Historical Note on the First Law

The historical approach to the study of a science is usually not the most effective method of study for a person seeking a working knowledge of the science, whether that person seeks the working knowledge in order to apply it in the solution of existing problems or to use it as a basis for further investigation of the subject. The reason for this is that the development of a science rarely proceeds in an orderly logical manner. The history of a science is usually a story of a few giant strides and many smaller steps in the right direction interspersed with, and nearly obscured by, semiblind groping, mistakes, confusion, misunderstandings, and trips down attractive byroads that turn out to be dead ends.

A science begins to develop as a result of either (1) a practical need for the understanding of certain phenomena so that the behavior of systems can be predicted or controlled or (2) a curiosity on the part of someone simply to know why things behave as they do. The science grows as hypotheses are formulated, tested, modified, and accepted or discarded. At the same time, the vocabulary of the science is continually modified; for convenience and clarity, definitions of even the basic terms are changed. This is one of the reasons why the study of the "original sources" is often not helpful to a beginning student: the student is often confused by the use of terms which have undergone changes in meaning. This is especially true in thermodynamics.

The concept of energy was used in mechanics in the seventeenth century, restricted to what we now call kinetic energy, potential energy, and work. The seventeenth-century energy analyses in mechanics were of great value whenever frictional effects could be ignored, but they could not cope with any process in which there was a frictional effect because the relationship involved in a conversion of work to heat was unknown.

The development of the steam engine introduced a number of practical problems regarding the inverse conversion: heat to work. In 1698, Thomas Savery built a steam engine in London for pumping water. Thomas Newcomen built a much-improved engine in 1705, and for three-quarters of a century Newcomen engines were widely used. They were succeeded by the engines of James Watt,* who patented the separate condenser in 1769 as the first of many improvements. The practical need for means to predict and evaluate the performance of these early steam engines stimulated interest in the relationship between work and heat, two concepts that had developed independently. The idea of work had been developed through mechanics. Heat was the subject of an entirely different "science," and the caloric theory, which postulated that heat was an indestructible fluid, was widely accepted.

Origins of the caloric theory have been traced to the Greek philosophers, but the "modern" statement of it was formulated by William Cleghorn in 1779, and the term *caloric* was suggested by Lavoisier in 1787. Though many people today scoff at the caloric theory, it must be said in its defense that it provided adequate explanations for many physical phenomena, and for many years, as apparent exceptions to it were demonstrated, the theory was ingeniously modified to account for these phenomena. (For a discussion of the caloric theory, see Reference 2·12.)

Although several people had expressed the belief that the caloric theory was unsound and that heat might be produced without limit by means of work, the first experimental attack on the caloric theory was made by Count Rumford (Benjamin Thompson),† beginning in 1787. While directing the manufacture of cannon, Rumford observed that a large amount of heat was evolved from a cannon as it was bored, and that the amount of heat was not directly related to the amount of metal removed or to the size of the chips produced, as would be required by the caloric theory. Experiments showed him that more heat was evolved and more work was required when a dull boring tool was used instead of a sharp one. These and other observations convinced Rumford that the caloric theory

*James Watt (1736–1819), Scottish engineer, was trained in his youth as an instrument maker. While serving in this capacity at the University of Glasgow he became interested in the possibility of improving the performance of steam engines. His interest was intensified when he was called on to repair a model of a Newcomen engine. His first patent covered not only a separate condenser but also the condenser air pump, insulation of the engine cylinder, steam jacketing of the cylinder, and the use of steam on both sides of the piston. During the period of his early work on steam engines, Watt also engaged in engineering work pertaining to canals, harbors, and bridges. Watt and Matthew Boulton, a progressive manufacturer and businessman, formed a happy partnership that fully utilized the talents of both men. Engines built by Boulton & Watt, Birmingham, played an important part in the industrial growth of Great Britain during the nineteenth century.

†Benjamin Thompson (1753–1814) was born in Woburn, Massachusetts. At the age of 14 he is reported to have calculated a solar eclipse with an error of only a few seconds. At 19 he married a wealthy widow and immediately became prominent in Boston social circles. He remained loyal to the Crown when the American Revolution began and in 1776 prudently left Boston for London to become active in the British civil service. Five years later he was involved in a plot to sell British naval information to the French. For 11 years he was in Munich, holding various positions including minister of war and minister of police in the Bavarian government. He was made a count of the Holy Roman Empire and selected the title of Rumford from the name of an American township. His cannon-boring experiments were made while he was in Bavaria. To Rumford's credit are also prison reforms, stove and fireplace improvements, and numerous other items. He was a founder of the Royal Institution of London. To his discredit are several instances of fantastic duplicity and intrigue. His second wife was the wealthy widow of Antoine Lavoisier.

was unsound, but his views were not widely accepted. He reported his cannon-boring experiments in 1798 and his demonstration that bodies do not change weight when they are heated or cooled in 1799. Sir Humphry Davy reported in 1799 further experiments on the conversion of work to heat by means such as the rubbing together of two pieces of ice. Still the caloric theory was not generally abandoned, although more people began to question it. Marc Séguin, a French engineer, stated the theory of the equivalence of heat and work in 1839 and performed some experiments in order to support it, but his experiments were not convincing. Julius Robert Mayer, a German physician, stated the theory independently in 1842 and even calculated the mechanical equivalent of heat (the relationship between the foot-pound and the Btu) from data on the specific heat of gases, but he performed no experiments himself. The man who placed the equivalence of heat and work on a sound experimental basis was Joule, who between 1840 and 1849 carefully measured the mechanical equivalent of heat by several methods. His results were published in 1843 and 1849. The law of conservation of energy gained widespread acceptance after the publication in 1848 of a paper in which Hermann Helmholtz, then a surgeon in the Prussian army, clearly showed the applications of the law in several scientific fields.

Anyone reading the original papers referred to here or some histories of thermodynamics will perceive that much confusion was caused by the dual use of the term heat: (1) to designate an interaction between bodies and (2) to designate "something" stored in a body. (Expressions such as "the heat in a body" are frequently found.) There was also a confusion of purpose between trying to answer (*a*) the question of what relation heat has to other effects or how it is measured and (*b*) the nebulous question of what heat *is*. An operational approach tells us that the first question can be answered but that the second one cannot.

2·12 Summary

Heat and work were defined operationally and independently of each other in Chapter 1. An empirical relationship between heat and work has been found. As a result of this relationship, stored energy can be operationally defined as

$$\Delta E \equiv Q - W$$

where ΔE is the change in the stored energy of a closed system, Q is the net amount of heat added to the system, and W is the net amount of work done by the system. Also it can be shown that stored energy E is a property.

These facts are generally summarized in a statement of the first law of thermodynamics or the law of conservation of energy. Possible statements are the following.

1. Whenever a closed system executes a cycle, the net amount of heat added to the system during the cycle is equal to the net amount of work done by the system, or

$$\oint \delta Q = \oint \delta W \tag{2·1}$$

2. Energy can be neither created nor destroyed, although it can be stored in various forms and can be transferred from one system to another as heat or work or

$$\begin{pmatrix}\text{Net amount of}\\ \text{energy added to}\\ \text{system as heat}\\ \text{and all forms of}\\ \text{work}\end{pmatrix} + \begin{pmatrix}\text{stored}\\ \text{energy}\\ \text{of matter}\\ \text{entering}\\ \text{system}\end{pmatrix} - \begin{pmatrix}\text{stored}\\ \text{energy}\\ \text{of matter}\\ \text{leaving}\\ \text{system}\end{pmatrix} = \begin{pmatrix}\text{net increase}\\ \text{in stored}\\ \text{energy of}\\ \text{system}\end{pmatrix} \qquad (2{\cdot}8b)$$

Many other statements of the first law have been formulated, and several of them are in common use. If any one of the valid statements is accepted as a postulate, all of the others can be deduced from it.

Stored energy may be classified as

1. Potential energy ≡ the energy stored in a system as a result of its location in a gravitational field, its magnitude being the product of (*a*) the weight of the system and (*b*) the distance between the center of gravity of the system and some arbitrary horizontal datum plane.

2. Kinetic energy ≡ the energy stored in a system by virtue of the motion of the system, its magnitude being given by

$$\text{kinetic energy} \equiv \frac{mV^2}{2} \qquad (2{\cdot}5b)$$

where m is the mass of the system and V is its velocity relative to an arbitrary frame of reference. If not all parts of the system have the same velocity, a more general expression must be used (see Eq. 2·5c).

3. Internal energy U ≡ the form of energy stored in a system that is independent of gravity, motion, electricity, magnetism, and surface tension. The internal energy of a substance is related to the potential and kinetic energies of the molecules of the substance and to the internal structure of the molecules. In the absence of motion, gravity, electricity, magnetism, and surface tension, $U = E$. In the absence of electricity, magnetism, and surface tension, the stored energy of a system is

$$E = U + \frac{mV^2}{2} + gmz \qquad (2{\cdot}6c)$$

A useful property formed by the arbitrary combination of other properties is enthalpy, defined as

$$h \equiv u + Pv$$

Whenever a fluid flows across the boundary of a system, an amount of work equal to Pv is done on the system per unit of mass entering or is done by the system per unit of mass leaving. This work done in pushing fluid across the system boundary is called flow-work.

The beginning student of thermodynamics sometimes feels that there are too many equations to learn. Leafing through the pages of a thermodynamics textbook does give the impression that there are many equations printed. Closer inspection, however, shows that many of the equations are simply special forms of a few basic relations. These special forms are included not with the intent that you should learn each of them. They are presented only to show you a few of the many special forms that you can derive yourself from the basic relations in order to apply the basic principles to specific systems and processes. The broad scope of thermodynamics is indicated by the large number and great diversity of applications made of its basic principles. Because the different types of systems to which the basic principles of thermodynamics have been applied are so numerous and because there are so many more as yet untried applications to be made in the future, it would be futile to try to learn all the special forms of the basic relations which have been developed. Instead, an engineer needs a sound understanding of the basic principles, their ranges of application, and their limitations. The best way to achieve this understanding is through practice in the application of the principles to many different situations.

In order to show how few relations (other than definitions) are needed for the solution of the problems in this chapter and the preceding one, formulations of the basic principles for the three types of systems discussed (closed, steady flow, and general open) are presented in the following table. These are not in all cases the most general forms. For example, the conservation of mass and conservation of energy equations for steady flow

Formulations of Basic Principles

Conditions: Closed system: $\Delta KE = 0, \Delta PE = 0$

Conservation of mass: $m_1 = m_2$

Conservation of energy: $Q = U_2 - U_1 + W$ (2·7)

For *frictionless* processes, an expression for mechanical work of simple compressible systems:

$$W = \int P\,dV \tag{1·7}$$

Conditions: Open system: one inlet, one outlet

Conservation of mass: $m_f = m_i + m_1 - m_2$

Conservation of energy:

$$Q - W + \int (e_1 + P_1 v_1)\,\delta m_1 - \int (e_2 + P_2 v_2)\,\delta m_2 = E_f - E_i \tag{2·8}$$

Conditions: Open system, steady flow: one inlet, one outlet

Conservation of mass: $\dot{m}_1 = \dot{m}_2 = \rho_1 A_1 V_1 = \rho_2 A_2 V_2$ (1·9)

Conservation of energy:

$$q = w + h_2 - h_1 + \frac{V_2^2 - V_1^2}{2} + g(z_2 - z_1) \tag{2·9}$$

For *frictionless* processes an expression for mechanical work per unit mass of fluid:

$$w_{\text{in}} = \int v\,dP + \Delta ke + \Delta pe \tag{1·10}$$

and closed systems are actually special cases of the more general equations for an open system. However, the formulations listed here are convenient starting points in many engineering analyses because they can be readily generalized or restricted as the situation demands. In all cases the symbols and subscripts have been previously defined. In all cases effects of electricity, magnetism, and surface tension are assumed to be negligible. Remember that learning the restrictions on each equation and the precise meaning of the various terms is just as important as learning the equation itself.

Thermal efficiency, a performance parameter that applies to heat engine cycles, is defined as

$$\eta \equiv \frac{\oint \delta W}{Q_{\text{in}}} = \frac{\text{net work output of cycle}}{\text{gross heat added during cycle}}$$

Coefficient of performance, a performance parameter which applies to refrigeration and heat pump cycles, is defined for refrigerators as

$$\beta_R \equiv \frac{Q_{\text{in}}}{-\oint \delta W} = \frac{\text{heat absorbed from low-temperature region}}{\text{net work input of cycle}}$$

and for heat pumps as

$$\beta_{HP} \equiv \frac{Q_{\text{in}}}{-\oint \delta W} = \frac{\text{heat discharged to high-temperature region}}{\text{net work input of cycle}}$$

References and Suggested Reading

2·1 Cravalho, Ernest G., and Joseph L. Smith, Jr., *Engineering Thermodynamics,* Pitman, Boston, 1981, Chapters 2, 3, and 4.

2·2 Dixon, J. R., *Thermodynamics I,* Prentice-Hall, Englewood Cliffs, N.J., 1975, Chapter 8.

2·3 Fox, Robert W., and Alan T. McDonald, *Introduction to Fluid Mechanics,* 3rd ed., Wiley, New York 1985, Sections 4-2 and 4-8.

2·4 Holman, J. P., *Thermodynamics,* 2nd ed., McGraw-Hill, New York, 1974.

2·5 Keenan, J. H., *Thermodynamics,* Wiley, New York, 1941, Chapter IV.

2·6 Reynolds, William C., and Henry C. Perkins, *Engineering Thermodynamics,* 2nd ed., McGraw-Hill, New York, 1977, Chapter 2.

2·7 Shames, Irving H., *Mechanics of Fluids,* 2nd ed., McGraw-Hill, New York, 1982, Sections 4-8 and 5-11 through 5-13.

2·8 Wark, K., *Thermodynamics,* 4th ed., McGraw-Hill, New York, 1983, Chapter 2.

On the Historical Development of the First Law

2·9 Angrist, Stanley W., and Loren G. Hepler, *Order and Chaos,* Basic Books, New York, 1967, Chapter 3.

2·10 Baehr, Hans Dieter, *Thermodynamik,* 5th ed., Springer-Verlag, New York, 1984, Chapter 1.

2·11 Cardwell, D. S. L., *From Watt to Clausius: The Rise of Thermodynamics in the Early Industrial Age,* Heinemann Educational Books, London, 1971.

2·12 Roller, Duane, *The Early Development of the Concepts of Temperature and Heat—The Rise and Decline of the Caloric Theory,* Harvard University Press, Cambridge, Mass., 1950.

Problems

2·1 For each of the following five cases of processes of a closed system, fill in the blanks where possible:

	Q	W	E_1	E_2	ΔE
(a)	10 kJ	5 kJ	71 kJ		
(b)	10 kJ	−5 kJ	60 kJ		
(c)	25 kJ	−10 kJ		−10 kJ	
(d)		20 kJ	55 kJ		17 kJ
(e)	−10 kJ			65 kJ	−20 kJ

2·2 A closed system passes from state 1 to state 2 while 40 kJ of heat is added and 60 kJ of work is done. As the system is returned to state 1, 35 kJ of work is done on it. What is the heat transfer during process 2-1?

2·3 The internal energy of a closed system changes from 300 to 260 kJ/kg while the system performs 20 kJ/kg of work. Compute the heat transfer.

2·4 During the expansion of 55 kg of gas in a closed system, heat is added to the gas in the amount of 600 kJ. The internal energy decreases by 1500 kJ. Compute the work.

2·5 During a process of a closed system, 2500 kJ of heat is removed and there is a decrease of 500 kJ in the internal energy. Compute the work.

2·6 Compute the work of a closed-system process during which 30 kJ of heat is added to the system and the stored energy of the system increases by 100 kJ.

2·7 An expanding gas does 7000 kJ of work while it receives 5000 kJ of heat. Calculate the change in its stored energy.

2·8 During a process 1-2 of a closed system, 250 kJ of heat is added to the system and the stored energy of the system increases by 200 kJ. During the return process 2-1 which restores the system to its initial state, 100 kJ of work is done on the system. Determine the heat transfer of process 2-1.

2·9 A closed system does 60 kJ of work while 20 kJ of heat is removed. Then the system is restored to its initial state by means of a process in which 10 kJ of heat is added to it. Determine ΔE for this second process.

2·10 A system comprised of a vapor trapped in a cylinder behind a piston has an initial internal energy of 60.0 kJ. Heat is added in the amount of 12.0 kJ, and the system does 32.0 kJ of work. The mass of vapor in the system is 0.35 kg. Determine the final internal energy of the system.

2·11 One-half kilogram of gas is held in a rigid tank. By means of an impeller in the tank, an external motor does 50.0 kJ/kg of work on the gas while its internal energy increases from 120.0 kJ/kg to 160.0 kJ/kg. Determine the heat transfer in kJ/kg.

2·12 A closed system passes from state 1 to state 2 while 6290 cal of heat is added and 8810 cal of work is done. As the system is returned to state 1, work in the amount of 500 cal is done on it. What is the heat transfer during process 2-1?

2·13 During the expansion of 15 kg of gas in a closed system, heat is added to the gas in the amount of 443 kJ. The internal energy decreases by 1266 kJ. Compute the work.

2·14 During a process of a closed system, 252.0 kcal of heat is removed, and there is a decrease of 50.4 kcal in the internal energy. Compute the work.

2·15 An expanding gas does 8.20 kJ of work while it receives 10.60 kJ of heat. Calculate the change in its stored energy.

2·16 A closed system contains 30 g of a gas. How much heat (in joules) is added to the system in a process which produces 5000 N·m of work while the stored energy decreases by 200 J/g?

2·17 A gas in a cylinder fitted with a piston undergoes a cycle comprised of three processes. First, the gas expands at constant pressure with a heat addition of 40.0 kJ and a work output of 10.0 kJ. Then it is cooled at constant volume by a removal of 50 kJ of heat. Finally, an adiabatic process restores the gas to its initial state. Determine (*a*) the work of the adiabatic process and (*b*) the internal energy of the gas at each of the other two states if its internal energy in the initial state is assigned the value of zero.

2·18 A closed system passes from state 1 to state 2 while 25 B of heat is added and 35 B of work is done. As the system is returned to state 1, 20 B of work is done on it. What is the heat transfer during process 2-1?

2·19 The internal energy of a fluid in a closed system changes from 500 to 440 B/lbm while the fluid performs 30 000 ft·lbf/lbm of work. Compute the heat transfer.

2·20 During the expansion of 30 lbm of gas in a closed system, heat is added to the gas in the amount of 420 B. The internal energy decreases by 1200 B. Compute the work.

2·21 During a process of a closed system, 1000 B of heat is removed and there is a decrease of 200 B in the internal energy. Compute the work.

2·22 Compute the work of a closed-system process during which 50 B of heat is added to the system and the stored energy of the system increases by 300 B.

2·23 An expanding gas does 6000 ft·lbf of work while it receives 10 B of heat. Calculate the change in its stored energy.

2·24 During a process 1-2 of a closed system, 200 B of heat is added to the system and the stored energy of the system increases by 150 B. During the return process 2-1 which restores the system to its initial state, 80 B of work is done on the system. Determine the heat transfer of process 2-1.

2·25 During a certain process a closed system does 30 B of work while 10 B of heat is removed. Then the system is restored to its initial state by means of a process in which 4 B of heat is added to it. Determine ΔE for this second process.

2·26 A cylinder with a volume of 2 cu ft contains 0.5 lbm of a gas at 20 psia and 50°F. The gas is compressed without friction to 120 psia in such a way that P (in psia) = constant − $100V$, with V in cu ft. Its final temperature is 20°F. The internal energy of the gas is given by u (in B/lbm) = $0.20T$, with T in degrees Fahrenheit. Sketch the process on PV coordinates and compute (*a*) heat transferred and (*b*) enthalpy change per pound.

2·27 In a closed system 0.27 lbm of air expands frictionlessly from 30 psia, 140°F, until its volume is doubled. The initial volume is 2.0 cu ft, and the expansion follows the path PV = constant.

During the process heat is added to the air in the amount of 28.5 B/lbm. Determine the internal energy change in the air in B/lbm.

2·28 Air in a cylinder expands frictionlessly against a piston in such a manner that P^2V = constant. Initially the air is at 30 psia, 140°F, has a mass of 0.54 lbm and occupies a volume of 4 cu ft. The final volume is 16 cu ft. Heat is added to the air in the amount of 185 B/lbm. Sketch a *PV* diagram, and determine the change in internal energy per pound of air.

2·29 Ethane initially at 35 kPa with a volume of 0.120 m^3 is compressed frictionlessly in a cylinder until its volume is halved in such a manner that its pressure and volume are linearly related, $P = a + bV$. The final pressure is 80 kPa, and the change in internal energy of the ethane is 3.22 kJ. Determine the heat transfer.

2·30 Ethene in a closed system is compressed frictionlessly from 95 kPa to 190 kPa in such a manner that PV = constant. Initially, the density is 1.11 kg/m^3 and the volume is 0.045 m^3. During the compression, heat is removed from the ethene in the amount of 2.96 kJ. Determine the change in internal energy of the ethene.

2·31 Neon in a cylinder expands frictionlessly against a piston so that $P^{0.7}V$ = constant from initial conditions of 300 kPa, 90°C, and a volume of 0.024 m^3, to a final pressure of 120 kPa. The internal energy change is -2.59 kJ and the enthalpy change is -4.31 kJ. Determine the heat transfer.

2·32 A well-insulated tank filled with neon has an electric resistance heating element in it. Because current from an external storage battery flows through the heating element, the temperature of the gas rises. Determine the algebraic sign (>0, $=0$, <0) of the heat transfer and of the work of a system comprised of (*a*) just the neon, (*b*) the neon and the heating element. The heating element has a resistance of 5 ohms. If during the process the potential difference across the element is 12 V, and the process occurs in 15 min, determine, if possible, ΔE for the system comprised of (*c*) just the neon, (*d*) the neon and the heating element.

2·33 A single-phase electric transformer operates under steady conditions with 440 V across the primary and a current of 8.60 A with unity power factor. The secondary provides 30.0 A at 120 V, unity power factor. Determine the net heat transfer.

2·34 A robot arm is moved by means of an electric motor and a speed-reducing gear train mounted within the arm. During a complete cycle of arm operation, the work output from the gear train is 2.40 kJ. The arm completes seven cycles of operation per minute. The average efficiency of the motor and gear train combination is estimated to be 75 percent. In order to simulate the heat dissipation required, the motor and gear train are to be replaced by an electric resistance heating element for the purposes of a heat dissipation study. How much energy should be dissipated by the electric resistance heating element?

2·35 A gearbox driven by a gasoline engine has an efficiency of 92.0 percent when its output shaft delivers 120.0 hp at 240 rpm. The gearbox is cooled by means of a fan which blows room air over it. Under these steady operating conditions how much heat is given off by the gearbox? (Use your judgment as to the definition of gearbox efficiency.)

2·36 A closed thermally insulated room contains a household refrigerator. The electric motor of the refrigerator is connected to a 110-volt power line which passes through the wall. If the refrigerator is operated, will the average temperature of the air in the room increase, decrease, or remain constant if the refrigerator door is (*a*) open, (*b*) closed?

2·37 Consider an air bubble rising through an open tank of water. Does the size of the bubble change? What are the energy transfers across a boundary which encloses only the bubble? What are the energy transfers across a boundary which encloses all the water and the bubble?

2·38 Occasionally ones sees the relation $\delta Q = dU + P\,dV$ referred to as the first law of thermodynamics. Comment on this, stating the restrictions which apply to this equation.

2·39 Three blocks of steel in a rigid insulated container are initially at different temperatures. As they exchange heat to attain temperature equality, their stored energies change by -260, $+140$, and $+100$ kJ. Nothing else is in the container except air. Determine ΔE of the air.

2·40 Calculate the amount of work required to accelerate a 3000-lbm automobile from rest to a speed of 50 fps on a level highway if there are no frictional effects.

2·41 A certain body has a potential energy of 400 ft·lbf relative to datum plane A and a potential energy of -300 ft·lbf relative to datum plane B. The body has a mass of 35 lbm. The local acceleration of gravity is 31.6 ft/s^2. What is the relative location of plane A with respect to plane B?

2·42 Calculate the work required to accelerate a 140-kg motorcycle and 80-kg rider from rest to a speed of 30 m/s on a level highway if there are no frictional effects.

2·43 A body has a potential energy of 200 kJ relative to datum plane A and a potential energy of -150 kJ relative to datum plane B. The body has a mass of 10 kg. The local acceleration of gravity is 9.78 m/s^2. What is the relative location of plane A with respect to plane B?

2·44 For a closed system, $\oint \delta Q = \oint \delta W$. Show by an example that this relationship does not hold for an open system.

2·45 Prove that in a frictionless steady-flow process of an incompressible fluid, $q - \Delta u = 0$. Does this mean that such a process is adiabatic?

2·46 A body falls freely in a vacuum. Write an energy balance for this process, considering the body as the system and using a stationary frame of reference. Notice that there is a force on the body and the body is moving. Does this mean that work is done on or by the body? If so, with what other system is the work exchanged? If not, what is the meaning of the product of the body weight and the distance moved?

2·47 Comment on this proposal for destroying some energy: A steel coil spring is compressed, and in this manner more energy is stored in it than in the same spring when uncompressed. The compressed spring is held to a constant length by means of a glass thread tied around it and is then submerged in an acid which dissolves the spring but not the thread. Therefore, it is claimed, the energy used to compress the spring is lost and can never be recovered.

2·48 A craft moves in a north-to-south direction at constant speed and constant altitude above sea level. The acceleration of gravity at constant altitude changes with latitude; therefore the weight of the craft itself (excluding the fuel) changes, and consequently its potential energy changes. Explain what is wrong with this reasoning, or explain the energy transformation which is involved.

2·49 Derive an expression for the kinetic energy of a rotating solid body (*a*) from Eq. 2·5c, (*b*) from the work required to put the body in motion.

2·50 Derive an expression involving the modulus of elasticity for the change in stored energy of a wire which is stretched adiabatically within its elastic limit. (Be careful with the algebraic sign.)

2·51 Consider two masses in interstellar space. The attractive force between them varies inversely as the square of the distance separating them. How does the potential energy of the two masses vary with that distance?

2·52 Consider a storage battery which has a terminal potential of 12 V while it delivers a current of 10 A for 3 h. The stored energy of the battery decreases by 1205 B. Determine the heat transfer.

2·53 A dc electric motor operates steadily at 800 rpm when the torque applied to its shaft is 96 lbf·in. It draws a current of 50 A at 24 V. Determine the rate of heat transfer.

2·54 Consider two streams of a liquid flowing through identical circular pipes at the same mass flow rate. The velocity of one is uniform across the pipe cross section, and the velocity of the other varies across the pipe cross section in such a manner that the velocity u at any radius r is given by $u/u_{max} = 1 - (r/r_0)^2$, where u_{max} is the velocity at the pipe axis and r_0 is the pipe radius. Determine the ratio of the kinetic energy of the stream with the parabolic velocity distribution to that of the stream with the uniform distribution.

2·55 Seventy kilojoules of work is done by each kilogram of fluid passing through an apparatus under steady-flow conditions. In the inlet pipe which is located 30 m above the floor, the specific volume is 3 m^3/kg, the pressure is 300 Pa, and the velocity is 50 m/s. In the discharge pipe which is 15 m below the floor, the specific volume is 9 m^3/kg, the pressure is 60 Pa, and the velocity is 150 m/s. Heat loss from the fluid is 3 kJ/kg. Determine the change in internal energy of the fluid passing through the apparatus.

2·56 A steam nozzle is designed to pass 450 kg of steam per hour. The initial and final pressures are 1400 and 14 kPa. The initial and final velocities are 150 and 1200 m/s. Neglecting heat losses, compute the change in enthalpy.

2·57 A steam nozzle is designed to pass 1000 lbm of steam per hour. The initial and final pressures are 200 and 2 psia. The initial and final velocities are 500 and 4000 fps. Neglecting heat losses, compute the change in enthalpy.

2·58 A fluid passes through a turbine at a rate of 2.50 kg/s. The inlet and exit velocities are 30 and 120 m/s, respectively. The initial and final enthalpies are 2930 and 2675 kJ/kg, respectively, and the heat loss amounts to 45.0 kJ/s. Compute the power output.

2·59 A turbine is supplied with 20 000 lbm/h of a fluid. The inlet and exit fluid velocities are 6000 fpm and 24 000 fpm, respectively. If the initial and final enthalpy values are 1260 and 1000 B/lbm, respectively, and the heat loss amounts to 140 000 B/h, compute the power output.

2·60 Air exapnds through a nozzle from a pressure of 500 kPa to a final pressure of 100 kPa. The enthalpy decreases by 100 kJ/kg during the flow of air. If the entering velocity and heat-flow terms are neglected, compute the exit velocity.

2·61 Air expands through a nozzle from a pressure of 75 psia to a final pressure of 15 psia. The enthalpy decreases by 48 B/lbm during the flow of air. If the entering velocity and heat-flow terms are neglected, compute the exit velocity.

2·62 An air compressor takes in air at a pressure of 100 kPa and specific volume of 0.90 m^3/kg. The discharge pressure and specific volume are 690 kPa and 0.20 m^3/kg. The initial and final internal energy values for the air are 224 and 346 kJ/kg respectively. The cooling water around the cylinders removes 70 kJ/kg from the air. Neglecting the change in kinetic and potential energy terms, compute the work.

2·63 An air compressor takes in air at a pressure of 14.4 psia and specific volume of 13.7 cu ft/lbm. The discharge pressure and specific volume are 100 psia and 2.74 cu ft/lbm. The initial and final internal energy values for the air are 12 and 47 B/lbm, respectively. The cooling water around the cylinders removes 33 B/lbm of entering air. Neglecting the change in kinetic and potential energy terms, compute the work.

2·64 Steam expands through a nozzle from 1400 kPa to 14 kPa. The initial and final enthalpy values are 3300 and 2800 kJ/kg, respectively. Neglecting the initial velocity and heat losses, compute the final velocity.

2·65 Steam expands through a nozzle from a pressure of 200 psia to a final pressure of 2 psia. The initial and final enthalpy values are 1289 and 955 B/lbm respectively. Neglecting the initial velocity and heat losses, compute the final velocity.

2·66 The supply line to a steam radiator is located 2 ft above the discharge line. Data relative to the steam as it enters follow. The pressure equals 15 psia, the specific volume equals 26.3 cu ft/lbm, and the internal energy value is 1077 B/lbm. After condensing in the radiator, the steam leaves at a pressure of 15 psia. The specific volume and enthalpy at exit are 0.0167 cu ft/lbm and 181 B/lbm respectively. Neglecting the entering and leaving velocities, compute the heat released from the radiator per pound of steam entering.

2·67 An aircraft fuel with a specific gravity of 0.840 is to be pumped at a rate of 5.20 kg/s from a pressure of 40 kPa to a pressure of 800 kPa. The fuel temperature is approximately 5°C. Velocity change through the pump is very low. The required power input to the pump is estimated to be 140 percent of that required to do the pumping frictionlessly. State the assumptions you make and determine the required power input.

2·68 Benzene (specific gravity = 0.90) flows through a pump at a rate of 1.8 cfs. The inlet pressure is 10 in. of mercury vacuum. Outlet pressure is 24 psig, and the pump outlet is 3 ft above the inlet. Inlet cross-sectional area is 0.36 sq ft, and outlet area is 0.18 sq ft. The pump operates at 2400 rpm. Calculate the power input.

2·69 A hydraulic turbine is located slightly lower than the level of the water surface downstream of the dam in which the turbine is placed. The difference in water level between the upstream and downstream sides of the dam is 22.0 m. Water entering the turbine has a velocity of 3.2 m/s, and that leaving has a velocity of 1.6 m/s. The flow rate through the turbine is 1.16 m^3/s. For frictionless flow, determine the turbine power output.

2·70 The flow rate through a hydraulic turbine is 6 cfs. The pressure of the water is 30.6 psig at the inlet which has a cross-sectional area of 0.50 sq ft. At the outlet, 8 ft below the inlet, the velocity is 2.5 fps, and the pressure is 4 psig. Calculate the power imparted to the turbine by the fluid.

2·71 The water level in an open tank is maintained 3.20 m above the centerline of a nozzle through which water flows from the tank into the atmosphere. The exit diameter of the nozzle (and the diameter of the water jet leaving) is 13.0 mm. Determine the flow rate of the water jet. What assumptions do you make, and how do they affect the calculation?

2·72 Steam flows through a turbine at a rate of 10 000 kg/h. The inlet and outlet enthalpy values are 2800 and 2100 J/g respectively. Inlet and outlet velocities are 30 and 200 m/s respectively. Heat loss to the surroundings amounts to 150 000 kJ/h. Calculate the power output.

2·73 A gas flows steadily through a turbine at a rate of 1.40 kg/s, entering at 500 kN/m^2, 940 K (h_1 = 900 J/g, u_1 = 724 J/g), and leaving at 100 kN/m^2, 770 K (h_2 = 690 J/g, u_2 = 554 J/g, ρ_2 = 0.70 kg/m^3) through a cross-sectional area of 0.010 m^2. The inlet velocity is negligibly low. The power output is 240 kW. Calculate the heat transfer in joules per gram.

2·74 A gas flows steadily through a machine, entering at 35 psia, 115°F, with a density of 0.091 lbm/cu ft and negligible velocity and leaving at 16.7 psia, 40°F, with a density of 0.05 lbm/cu ft and a velocity of 500 fps through an opening of 0.4-sq-ft cross-sectional area. As the gas flows through the machine, its internal energy decreases by 30 B/lbm and its enthalpy decreases by 40 B/lbm. The power output is 400 hp. Determine the heat transfer per pound of gas.

2·75 Calculate the power required by a compressor if air flowing at a rate of 0.9 kg/s enters at 100 kPa, 5°C, with a velocity of 60 m/s, and leaves at 200 kPa, 70°C, with a velocity of 120 m/s.

The enthalpy of the air increases by 65.3 kJ/kg as it passes through the compressor; its internal energy increases by 46.6 kJ/kg. Heat transferred from the air to the cooling water circulating through the compressor casing amounts to 19 kJ/kg.

2·76 Calculate the power required by a compressor if air flowing at a rate of 2.0 lbm/s enters at 15.0 psia, 40°F, with a velocity of 200 fps, and leaves at 30.0 psia, 160°F, with a velocity of 400 fps. The enthalpy of the air increases by 28.8 B/lbm as it passes through the compressor; its internal energy increases by 20.5 B/lbm. Heat transferred from the air to the cooling water circulating through the compressor casing amounts to 8.0 B/lbm.

2·77 Air enters a machine at 100 kPa, 30°C (ρ = 1.150 kg/m^3), at low velocity and leaves at 100 kPa, 55°C (ρ = 1.062 kg/m^3), with a velocity of 90 m/s through an opening having a cross-sectional area of 0.00090 m^2. The enthalpy of the air increases 25.2 kJ/kg, and its internal energy increases 18.0 kJ/kg. A blower delivers 0.75 kW to the air as it passes through the machine. Calculate the heat added to or removed from the air in kJ/kg.

2·78 Methane flows frictionlessly through a turbine at a rate of 13.2 kg/h, expanding from 375 kPa to 110 kPa in such a manner that $Pv^{1.25}$ = 117, with P in kPa and v in m^3/kg. The inlet velocity is quite low, but the exit velocity is 60 m/s. The internal energy change of the methane passing through the turbine is −107.6 kJ/kg and the enthalpy change is −139.7 kJ/kg. Determine the heat transfer per kilogram of methane.

2·79 Air enters a wind tunnel at 96.0 kPa, 20°C, v = 0.876 m^3/kg, with low velocity and then flows steadily through a nozzle into the test section. If the expansion in the nozzle is such that $Pv^{1.40}$ = constant, to what pressure must the air be expanded to achieve a velocity of 220 m/s in the test section?

2·80 A gas flows steadily through a machine while expanding frictionlessly from 60 to 15 psia according to the relation $Pv^{1.25}$ = 32 000 with P in psf and v in cu ft/lbm. The enthalpy decreases 35 B/lbm, and the change in kinetic energy is negligible. Calculate the heat transferred in B/lbm.

2·81 Air flows through a gas turbine at a rate of 6.8 kg/s, entering at 400 kPa, 370°C, (u = 468.1 kJ/kg, h = 652.7 kJ/kg), and leaving at 100 kPa. The net increase in kinetic energy of the air passing through the turbine is 12 kJ/kg. There is no heat transfer. The power output is 820 kW. Determine the enthalpy of the air at the turbine outlet.

2·82 Air flows through a gas turbine at a rate of 15 lbm/s, entering at 60 psia, 700°F, (u = 201.6 B/lbm, h = 281.1 B/lbm), and leaving at 15 psia. The net increase in kinetic energy of the air passing through the turbine is 5.0 B/lbm. There is no heat transfer. The power output is 1100 hp. Determine the enthalpy of the air at the turbine outlet.

2·83 A gas enters a compressor at 100 kPa, 30°C (h = 303.4 kJ/kg, u = 216.5 kJ/kg, v = 0.870 m^3/kg) with negligible velocity and is discharged at 490 kPa, 260°C (h = 537.4 kJ/kg, u = 384.4 kJ/kg). The gas leaves the compressor with a velocity of 150 m/s. The power input is 2400 kW. The flow rate is 9.0 kg/s. Determine the heat transfer in kJ/kg.

2·84 A gas enters a compressor at 14 psia, 80°F (h = 129 B/lbm, u = 92 B/lbm, v = 14.3 cu ft/lbm) with negligible velocity and is discharged at 70 psia, 500°F (h = 231 B/lbm, u = 165 B/lbm). The gas leaves the compressor with a velocity of 500 fps. The power input is 3200 hp. The flow rate is 20 lbm/s. Determine the heat transfer in B/lbm.

2·85 The power output of a steam turbine is 3000 kW while the steam flow rate is 20 600 kg/h. Barometric pressure is 710 mm of mercury. Determine the amount of heat, in kJ/kg, added to or removed from the steam passing through the turbine if inlet and exhaust conditions are respectively: absolute pressures, 1400 and 7.0 kPa; temperatures, 300°C and 39°C; specific volumes, 0.182 and

19.71 m^3/kg; velocities, 60 and 150 m/s; internal energies, 2785.2 and 2338.2 kJ/kg; and enthalpies, 3040.4 and 2476.1 kJ/kg.

2·86 The power output of a steam turbine is 4000 hp while the steam flow rate is 56 000 lbm/h. Barometric pressure is 29.0 in. of mercury. Determine the amount of heat, in B/lbm, added to or removed from the steam passing through the trubine if inlet and exhaust conditions are as listed in the table.

	Inlet	Exhaust
Pressure, psia	200	1
Temperature, °F	560	102
Specific volume, cu ft/lbm	2.93	331
Velocity, fps	200	500
Internal energy, B/lbm	1193	1035
Enthalpy, B/lbm	1301	1096

2·87 A blower supplying air to a diesel engine takes in air at 90 kPa, 20°C (v = 0.934 m^3/kg, u = 209.3 kJ/kg) with low velocity at a rate of 12.3 m^3/min. It discharges air into the engine at 106 kPa, 30°C (v = 0.820 m^3/kg, u = 303.4 kJ/kg) with a velocity of 24 m/s. The flow is adiabatic. Calculate the power consumption of the blower.

2·88 A blower receives air at P = 14 psia, v = 14.3 cu ft/lbm, u = 13.6 B/lbm, V = 10 fps, and discharges it at P = 15 psia, v = 13.6 cu ft/lbm, u = 15.5 B/lbm, V = 50 fps. Flow is adiabatic and 1500 cfm enters the blower. Calculate the power consumption.

2·89 Carbon monoxide flows steadily through a machine at a rate of 0.40 kg/s, entering at 170 kPa. 17°C (ρ_1 = 1.97 kg/m^3, u_1 = 215.1 kJ/kg, h_1 = 301.2 kJ/kg), and leaving at 90 kPa, −7°C (ρ_2 = 1.14 kg/m^3, u_2 = 197.2 kJ/kg, h_2 = 276.2 kJ/kg). The inlet velocity is quite small. Discharge is through a cross-sectional area of 0.0070 m^2. Heat transfer into the gas as it passes through the machine is estimated as 1.2 kJ/s. Determine the power output.

2·90 Carbon dioxide flows steadily through a machine at a rate of 0.5 lbm/s, entering with negligible velocity at 20 psia, 80°F (u_1 = 67.2 B/lbm, h_1 = 91.6 B/lbm). Carbon dioxide leaves the machine at 60 psia, 200°F (ρ_2 = 0.0116 slug/cu ft, u_2 = 85.8 B/lbm, h_2 = 115.6 B/lbm), through an opening which has an area of 0.4 sq in. Power input to the machine is 25 hp. Determine the amount of heat transfer per pound of carbon dioxide.

2·91 A piece of electronic equipment fits into a case with dimensions of 9 by 16 by 22 cm. Power consumption under steady conditions is 600 W. The output signals carry a negligible amount of energy. Under steady operating conditions, the cooling air blown through the unit can experience an increase in internal energy of not more than 15 kJ/kg and an increase in enthalpy of not more than 21 kJ/kg. At what rate (kg/s) must cooling air be supplied?

2·92 During a test of an automotive engine, the amount of heat transferred to the cooling water is determined by measuring the amount of cooling water flowing through the engine and its temperature rise, since in the temperature range involved c_p for water is known. The inlet and outlet temperatures are measured at the points labeled T_i and T_o on the diagram. The flow rate is measured by the precision flow meter FM. All the piping is well insulated. After a test is completed, it is discovered that bypass valve B was partially open during the test. Explain why the data cannot be used or how they might still be used.

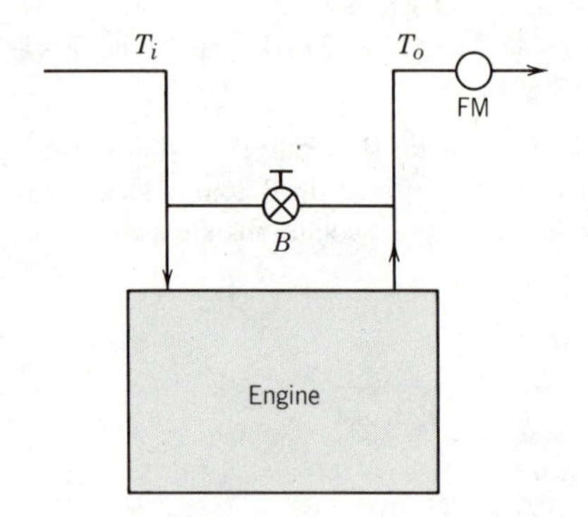

Problem 2·92.

2·93 A gaseous hydrocarbon fuel enters a burner at 1 atm, 25°C, with an enthalpy of −3556 kJ/kg at a rate of 18.0 kg/s. Air at 1 atm, 25°C, enters the burner with an enthalpy of 300 kJ/kg at a rate of 391 kg/s. The combustion products leaving the burner at 540°C have an enthalpy of −1693 kJ/kg. Determine the amount of heat transferred from the burner in kJ/kg of fuel.

2·94 A gas enters an insulated device at 200 kPa, 27°C, with an enthalpy of 214.4 kJ/kg and a velocity of 70 m/s, at a rate of 0.00230 kg/s. Within the device, the gas stream is separated into two streams, and measurements show that one stream leaves at 100 kPa with an enthalpy of 206.0 kJ/kg and a very low velocity at a rate of 0.0090 kg/s. If the other leaving stream also has a very low velocity, what is its enthalpy?

2·95 Two fluids flow through a heat exchanger. Fluid A flows at a rate of 3000 kg/h, entering with an enthalpy of −321.6 kJ/kg and leaving with an enthalpy of −344.0 kJ/kg. Fluid B flows at a rate of 165 kg/h and enters with an enthalpy of 42.0 kJ/kg. All fluid velocities are very low, and it is estimated that 4 percent of the heat removed from the higher temperature fluid B is lost as heat to the surroundings because the insulation of the heat exchanger is not perfect. Determine the enthalpy of fluid B as it leaves the heat exchanger.

2·96 An insulated chamber initially contains 0.35 kg of a gas having an internal energy of 220.0 kJ/kg. Additional gas having an internal energy of 260.0 kJ/kg and an enthalpy of 350.0 kJ/kg enters the chamber until the total mass of gas contained is 0.90 kg. Calculate the internal energy (kJ/kg) of the gas finally contained by the chamber.

2·97 A well-insulated tank of volume V contains a gas initially at a pressure P_i with a specific volume of v_i. The internal energy of the gas is given by $u = aPv$, where a is a constant. The gas leaks from the tank through a porous plug at a rate that varies with time. Determine the value of a required to have the mass of gas remaining in the tank at any time proportional to the gas pressure raised to the power of 0.80.

2·98 A well-insulated tank initially holds air at a low pressure P_i. Air from the surrounding atmosphere then leaks slowly into the tank through a porous plug. For air, $u = bPv$, where b is a constant. Show that the mass in the tank at any instant is given by $m = m_i\,(Pv + A)/(P_iv_i + A)$, where A is a constant that depends on the conditions of the surrounding atmosphere.

2·99 A gas in a closed tank of volume V is initially at $P = P_i$ and $T = T_i$. A small valve is opened, allowing gas to escape at a rate of $\dot{m} = KP/\sqrt{T}$, where K is a constant and P and T are the pressure and temperature of the gas at any instant. There is no heat transfer. The mass in the tank at any instant is $m = bP/T$, where b is a constant; and the internal energy of the gas is given by $u = cT$, where c is a constant. Determine the rate of temperature change with time just after the valve is opened in terms of b, c, K, V, and T_i.

2·100 Air which has properties such that $h = 1.4u = 1.004T = 3.50Pv$, where u and h are in kJ/kg, T is in kelvins, P is in kPa, and v is in m^3/kg, is initially trapped in a well-insulated rigid tank. A paddle wheel in the tank is turned by an external motor, and air is bled from the tank through a small valve to keep the temperature constant at 40°C. Assume that the pressure and temperature are uniform throughout the tank. Determine the mass rate of flow of air out of the tank at an instant when the pressure is 400 kPa and the power input is 0.037 kW.

2·101 An insulated tank contains air initially at 140 kPa, 15°C. An electrical heating coil in the tank steadily dissipates 100 W to the air in the tank while a pressure regulating valve bleeds air from the tank to the atmosphere at such a rate as to maintain the pressure constant in the tank. The tank volume is 0.110 m^3. For air, $h = 1.4u = 1.04T = 3.50Pv$, where h and u are in kJ/kg, T is in kelvins, P is in kPa, and v is in m^3/kg. Determine the temperature of the air in the tank 2 min after the initial conditions.

2·102 A liquid which is incompressible enters a device at a constant temperature T_1 and flows through the device at a constant mass flow rate $\dot{m}$. During the initial operating time period between $\tau = 0$ and $\tau = a$, the heat added to the liquid as it passes through the device is expressed as $\dot{Q} = \dot{Q}_0 + b\tau$, where $\dot{Q}_0$ and b are constants. The volume of liquid in the device remains constant. No work is done. The energy stored in the liquid within the device at any instant is $E = E_i + FT_2$, where E_i and F are constants and T_2 is the temperature of the liquid leaving at that instant. The enthalpy and internal energy per unit mass of liquid can be expressed as $h = CT$ and $u = DT$, where C and D are constants. Derive an expression for T_2 at any instant τ, with $0 < \tau < a$.

2·103 A small leak develops in a 20-cu ft insulated tank which initially contains air at 40 psia. What fraction of the air escapes by the time the pressure drops to 20 psia? (For air, $h = 1.4u = 0.24T = 0.0045Pv$, where h and u are in B/lbm, T is in Rankine, P is in psfa, and v is in ft^3/lbm.)

2·104 A 4-cu-ft tank is to be filled from a compressed air line in which air at 100 psia, 200°F, is flowing. Initially the tank contains air at room conditions, 14.7 psia, 80°F. Determine the final mass of air in the tank. Assume that the filling process is adiabatic and that for air, $h = 1.4u = 0.24T = 0.0045Pv$, where h and u are in B/lbm, T is in Rankine, P is in psfa, and v is in ft^3/lbm. List other assumptions you make as part of your solution.

2·105 Comment on the following definition of thermal efficiency: Thermal efficiency is the ratio of the energy outputs from a system to the energy inputs to it.

2·106 For actual refrigeration unit removing an amount of heat Q_L from a low-temperature region and discharging an amount of heat Q_H to a high-temperature region, the coefficient of performance is defined as $\beta_R \equiv Q_L/W_{in}$. Explain why it would not be satisfactory to define the coefficient of performance as $\beta_R \equiv Q_L/(Q_H - Q_L)$.

2·107 A power plant receives 300 GJ/h from burning fuel. Energy leaving with exhaust gases amounts to 50 GJ/h more than the energy introduced by air entering, cooling water absorbs 200 GJ/h, and heat loss to the surrounding atmosphere amounts to 10 GJ/h. Determine the plant thermal efficiency.

2·108 A closed system undergoes a cycle during which it receives 3050 kJ from a fluid at a very high temperature, rejects 550 kJ to part of the surroundings at 120°C, and also rejects heat to part of the surroundings at 17°C. The work output of the cycle is 680 kJ. Determine the thermal efficiency of the cycle.

2·109 A refrigerator removes heat steadily from a food compartment at 6°C at a rate of 800 kJ/min while its power input is 3.05 kW. Cooling water removes heat from the refrigerating unit at a rate of 870 kJ/min, and some heat is transferred from the refrigerating unit to the atmosphere. Determine (*a*) the heat loss to the atmosphere and (*b*) the coefficient of performance.

2·110 A household refrigerator operates in such a manner that when it is running, the driving electric motor requires a power input of 450 W. If it runs one third of the time and its coefficient of performance is 2.7, at what rate is heat discharged to the room that contains the refrigerator?

2·111 A building heated by electric resistance heaters requires an average of 21 kW during one winter month. If the building were to be heated by means of a heat pump that absorbs heat from a solar-heated pond at a temperature of 7°C and has a coefficient of performance of 3.8, (*a*) what would the power requirement be and (*b*) at what rate would heat be withdrawn from the pond?

2·112 For heat pumps and refrigerators having a ratio of work input to heat absorbed ranging from 0.2 to 2.0, what is the range of the ratio of their coefficients of performance, β_{HP}/β_R?

CHAPTER THREE

Physical Properties I

Chapter 1 covered several definitions and fundamental concepts. Chapter 2 introduced the first law of thermodynamics and its application to various systems. For the application of the first law to most engineering systems, a knowledge of the physical properties of various substances is required. This chapter and the following one are concerned with some of these physical properties, particularly the relationships among pressure, volume, temperature, and internal energy for various substances.

The study of physical properties could be deferred until after consideration of the second law of thermodynamics. Indeed, such a sequence of topics is nearly always followed in advanced textbooks and is generally a more satisfying sequence to those who have already completed some study of thermodynamics. Also, a more complete presentation of physical property relationships can be made when both the first and second laws are used. Nevertheless, physical properties are made the subject of the next two chapters in order to give you more opportunity to strengthen your grasp on the first law through additional applications using physical property relationships before taking up the second law in Chapter 5. Then after you have studied the second law in Chapters 5, 6, 7, and 8, Chapter 9 introduces a more general approach to physical properties using both the first and second laws.

3·1 Phases of Substances: Solids, Liquids, and Gases

It is a matter of everyday experience that substances can exist as solids, as liquids, and as gases. At normal room pressure and temperature, copper is a solid, mercury is a liquid, and oxygen is a gas; but each of these substances can appear in a different *phase* if the pressure or temperature is changed sufficiently.

A *phase* is any homogeneous part of a system that is physically distinct and is separated from other parts of the system by definite bounding surfaces. Ice, liquid water, and water vapor constitute three separate phases of the pure substance H_2O, because each is homogeneous and physically distinct from the others and each is clearly defined by the boundaries existing between them. At high pressures there are several different forms of ice, each of which constitutes a separate phase because each is clearly distinct from the others and separated from them by definite boundary surfaces. Three solid phases of sulfur can exist at room pressure and temperature. Graphite and diamond are two solid phases of carbon. In general, each solid in a system constitutes a separate phase, since each is homogeneous, physically distinct, and separated from the rest of the system by definite boundary surfaces. A solid solution consists of but a single phase no matter how many substances are involved. A liquid solution constitutes a single phase no matter how many substances are present as long as it is homogeneous. For example, a solution of salt and sugar in water consists of but one phase, even though this one phase consists of three separate constituents: salt, sugar, and water. If additional amounts of salt and sugar are added until the water is unable to dissolve any more, part will remain undissolved (in solid form). Under this condition, three phases will exist: solid salt, solid sugar, and liquid solution. If a liquid contains two distinct layers, two phases exist. For example, under certain concentrations ether and water form two separate layers; each layer is homogeneous and the two are separated by a bounding surface. If a vapor exists above a single- or double-layer liquid system, still another phase is present. A gas or a mixture of gases always constitutes a single phase because the mixture is homogeneous as a result of the intimate mixing of the molecules.

The study of physical properties of substances from the point of view of classical thermodynamics does not involve considerations of molecular structure and behavior. Nevertheless, some understanding of molecular phenomena is helpful in the study of physical properties, so occasionally in this book explanations based on molecular theory are presented. Let us now consider from the molecular point of view the characteristics of solids, liquids, and gases.

In crystalline solids, the units of the crystal structure are commonly atoms or ions rather than molecules, although molecules in certain types of crystals maintain their identity. In this discussion of solids, the word molecule is used to simplify comparison with liquids and gases. The molecules of a solid are very close together and are arranged in a three-dimensional pattern that is repeated with but minor irregularities throughout the solid. The arrangement of the molecules (or atoms or ions) in a solid is referred to as a *lattice*. In noncrystalline solids, such as glasses, the immediate neighbors of any molecule are generally arranged in a fixed pattern, but the regularity of the pattern does not extend throughout the solid as it does in crystalline solids. Crystalline solids are said to possess long-range order as well as short-range order, whereas noncrystalline solids possess only short-range order. In all solids the molecules are so close together that the forces of molecules on each other (which are repulsive forces at very close spacing and attractive forces when there is a greater separation between molecules*) are large. It is these forces which tend to keep the molecules in fixed positons within the lattice and that give solids their resistance to deformation. The molecules of a solid continually oscillate or vibrate about their equilibrium positions, but the amplitude of this motion is small, so that a

*The force between two molecules a distance r apart is often assumed to be given by the Lennard–Jones equation

$$F = \frac{a}{r^m} - \frac{b}{r^n}$$

with positive values corresponding to attractive forces. a and b are positive constants. Since the attractive force predominates as r increases, $n > m$.

molecule is never far from its equilibrium position. The velocity of the molecules during the oscillation depends on the temperature of the solid: The higher the temperature, the higher the velocity. As the temperature of a solid is increased, the molecular velocities and momentum increase until the attractive forces are partially overcome. Then groups of molecules shift position relative to other such groups and the solid becomes a liquid.

In a liquid, the molecular spacing is about the same as in a solid. The molecules are not held in relatively fixed positions, but in regions of molecular dimensions a semblance of crystalline order is retained. That is, each molecule retains an orientation with respect to some of its neighbors that it had in the solid crystalline phase. These small groups of molecules in which remnants of crystalline structure persist are not firmly held in position relative to neighboring groups. For this reason, a liquid in static equilibrium cannot withstand a shearing stress. The presence of a crystalline structure of short-range order in some liquids has been verified by measurements of the scattering of X rays. A number of the macroscopic properties of liquids can be accounted for on the basis of a liquid structure which involves an ordered arrangement of molecules in regions of molecular dimensions. The distances between molecules in a liquid are generally slightly greater than the distances between molecules in the solid phase of the same substance. One of the notable exceptions is water. Water, unlike most substances, expands on freezing, so that the molecular spacing is slightly greater in the solid phase than in the liquid phase.

A gas is composed of molecules which are relatively far apart. There is no regularity or permanence in their arrangement in space, for gas molecules are continually in motion, colliding with each other and rebounding to travel in new directions. For gases of low density, collisions are the only interactions of consequence between molecules, because they are so widely separated that intermolecular forces are small.

Statistical laws exist by which the behavior of great numbers of molecules in a gas may be predicted. It is not possible to study the motion of a single molecule by these laws, just as it is impossible to predict the length of life of a given person from the statistical averages established by life insurance companies. Bearing this point in mind, let us look at some numerical values obtained by means of kinetic theory.

In air at normal room conditions, there is a large range of molecular velocities, but an "average" value is about 500 m/s. On the average, a molecule of oxygen or nitrogen in such air travels about 0.00006 mm between collisions and experiences about 8 billion collisions per second. One cubic centimeter of normal room air contains about 2.5×10^{19} molecules; but, if this gives the impression that the molecules are jammed together, it should be noticed that the molecules themselves occupy only about 1/1000 of the volume which the gas fills.

In review, this discussion of the molecular picture of solids, liquids, and gases has pointed out that the molecules of a substance are restrained to some extent by the forces that act between individual molecules and groups of molecules. The closer the molecules are grouped together, the greater are the forces between them, and hence they have less freedom of motion. It has been pointed out (J. A. Eldridge, *The Physical Basis of Things,* McGraw-Hill, New York, 1934) that the molecules in a gas and in a liquid, respectively, may be compared with life in the country and in a congested city. In a gas the molecule exists in the wide open spaces where a molecule's a molecule and is more or less free to move about without much interference. In a liquid the molecules are more closely spaced, and their movement is greatly influenced by their neighbors. In a solid the congestion is even greater, and here the freedom of the individual molecule is even less.

3·2 Equilibrium of Phases of a Pure Substance

A *pure substance* is a substance which is chemically homogeneous and fixed in chemical composition. A system comprised of liquid and vapor phases of H_2O is a pure substance. Even if some of the liquid is vaporized during a process, the system will still be chemically homogeneous and unchanged in chemical composition. A mixture may be a pure substance. For example, air is a mixture of (principally) oxygen and nitrogen. In its gaseous phase it may be heated, cooled, compressed, or expanded without undergoing any change in its chemical composition. If it is cooled until part of it liquefies, it is no longer a pure substance, because the liquid will contain a higher fraction of nitrogen than the vapor does and the substance will no longer be chemically homogeneous. A mixture of oxygen and carbon monoxide is a pure substance as long as it remains fixed in composition. If some of the CO combines with some of the O_2 to form CO_2, the system is not a pure substance during the process because its chemical composition has changed. A system cannot be treated as a pure substance during any process that involves a chemical reaction.

The state of a pure substance at rest is usually specified completely by the values of two independent properties, provided that there are no electric, magnetic, solid distortion, or surface tension effects. Thus, any property of a system composed of a pure substance is usually a function of only two independent properties of the system, provided there are no effects of electricity, magnetism, etc. For example, if the pressure and temperature of air are specified, then the values of all other properties of the air such as density, internal energy, enthalpy, viscosity, and thermal conductivity are fixed. If the pressure and density of air are specified, then the temperature and all other properties have fixed values. Care must be exercised to see that the two properties selected to specify a state are *independent* properties. Two properties are independent if either one can be varied throughout a range of values while the other remains constant. It will be pointed out later that, when two phases of a pure substance exist together in equilibrium, the pressure and temperature of the mixture depend only on each other. Consequently, under these conditions pressure and temperature are not independent variables, and values of pressure and temperature do not completely specify the state of the system. Under these conditions it would be necessary to specify pressure and specific volume, pressure and enthalpy, temperature and specific volume, internal energy and temperature, or some other pair of independent variables in order to specify the state of the system completely.

The liquid phase of a substance can exist at many different pressures and temperatures. For example, liquid water a 1 atm (101.3 kPa or 14.7 psia) can exist in equilibrium at any temperature between 0°C and 100°C, inclusive; and liquid water at 15°C can exist in equilibrium under pressures ranging from less than 1 atm to several hundred atmospheres. Likewise, the solid phase of a substance can exist in equilibrium at many different temperatures and pressures, and the gas or vapor phase of a substance also exists under many combinations of pressure and temperature. In short, the pressure and temperature of *any one phase* of a substance can be varied independently over wide ranges. However, *when two phases of a pure substance coexist in equilibrium, there is a fixed relationship between their pressure and temperature*. At any given pressure there is but one temperature at which a certain two phases will exist together in equilibrium. Conversely, the two phases can coexist in equilibrium at a given temperature only if the pressure is a particular value that corresponds to that temperature. These corresponding values of pressure and temperature are spoken of as the saturation pressure for a given temperature and the saturation temperature for a given pressure.

Generalizing, *saturation conditions* are those conditions under which two or more phases of a pure substnace can exist together in equilibrium. Any phase of a substance existing under such conditions is called a *saturated* phase. For example, liquid water and water vapor in equilibrium with each other are spoken of as *saturated liquid* and *saturated vapor*. Ice in equilibrium with liquid or vapor or both is spoken of as a *saturated solid*. The presence of two or more phases is not required in order to have a saturated phase. A phase is saturated even if it exists alone as long as it is at a pressure and a temperature under which two or more phases *could* exist together in equilibrium.

As an illustration of these facts regarding equilibrium of two phases of a pure substance, consider the familiar substance *water*. Liquid water (a single phase) can exist in equilibrium at various temperatures while under a pressure of 1 atm. However, it cannot exist at a temperature higher than 100°C while under this pressure. On the other hand, water vapor (a single phase) can exist under a pressure of 1 atm only at temperatures of 100°C and higher. The only temperature at which *either* the liquid *or* the vapor phase can exist under a pressure of 1 atm is 100°C. This is also the only temperature at which the two phases can exist *together* in equilibrium at a pressure of 1 atm. Therefore, 100°C is the saturation temperature of water at 101.325 kPa (1 atm). Likewise, 101.325 kPa is said to be the saturation pressure of water at 100°C. If the temperature of liquid in a system comprised only of the pure substance water is slowly increased while the pressure is held constant at 1 atm, no vapor will form until the temperature reaches 100°C. Then, no matter how much vapor is formed, as long as both liquid and vapor are present in equilibrium and the pressure is held constant, the temperature will be 100°C. Not until all of the liquid has changed to vapor can the temperature rise above the saturation temperature of 100°C. If liquid water at some pressure *other* than 101.325 kPa has its temperature slowly increased, vapor will form only at some temperature *other* than 100°C. For liquid–vapor equilibrium, the higher the pressure, the higher the saturation temperature.

A fixed relationship between saturation pressures and temperatures is characteristic of all purc substances. The temperature at which a substance evaporates or condenses (or the temperature at which liquid and vapor phases can coexist in equilibrium) depends only on the pressure of the substance. The relationship between liquid–vapor saturation pressures and temperatures is shown for several substances in Fig. 3·1. These curves are called liquid–vapor saturation curves, vaporization curves, or condensation curves. The upper and lower ends of these curves, known respectively as the critical states and the triple states of the substances, will be discussed later. Figure 3·1 does not show accurately the lower ends of the liquid–vapor saturation curves, because for several of the substances the triple-state pressure is too low to be shown acccurately on a diagram of this scale.

The temperature at which any two phases of a pure substance can coexist in equilibrium

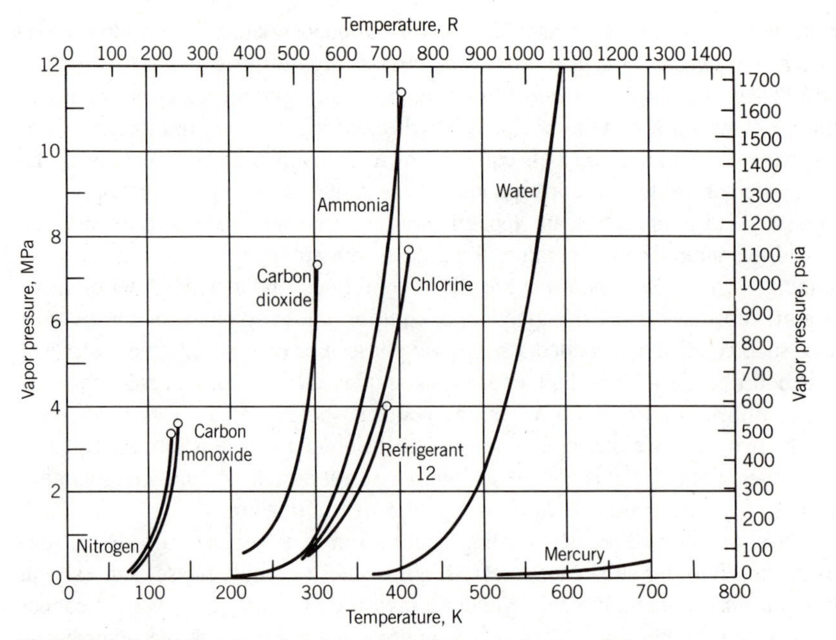

Figure 3·1. Liquid–vapor saturation curves for several pure substances.

depends on the pressure. This applies to solid–liquid and solid–vapor equilibrium as well as to the liquid–vapor equilibrium already discussed. Thus the melting or freezing temperature of a substance depends on the pressure, although the variation of this temperature with pressure is usually small. For most substances the freezing temperature increases as the pressure increases, so that the solid–liquid saturation curve or fusion line appears as in Fig. 3·2. For water and other substances which expand on freezing, increasing the pressure lowers the freezing point. This is shown by the familiar demonstration in which a wire weighted at both ends is passed over a cake of ice. The increased pressure caused by the wire lowers the freezing point slightly, causing the ice to melt directly below the wire. The liquid water formed passes above the wire and freezes because the pressure on it is again atmospheric. In this manner the wire will pass completely through the cake of ice which will freeze solid above the wire.

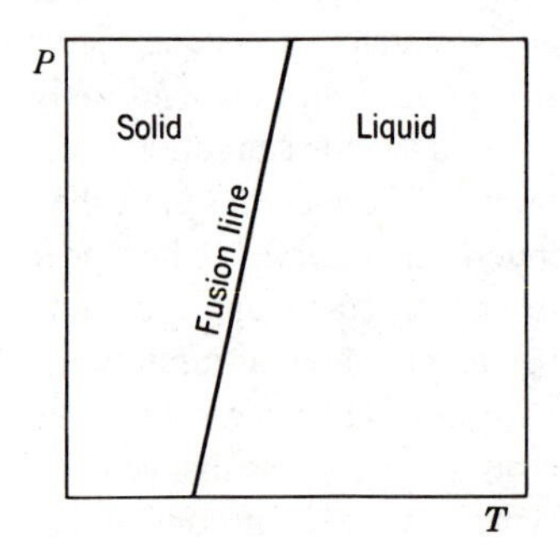

Figure 3·2. Solid–liquid saturation curve for a substance that contracts on freezing.

Figure 3·3 shows a pressure–freezing temperature chart for water. At any point on the line, the liquid and solid may exist together in equilibrium. The area to the left of the line represents the solid, and that to the right the liquid. For example, at a pressure of 40 MPa and a temperature of −5°C water can exist only in the solid form, ice. At 40 MPa and −2°C, only liquid water may exist.

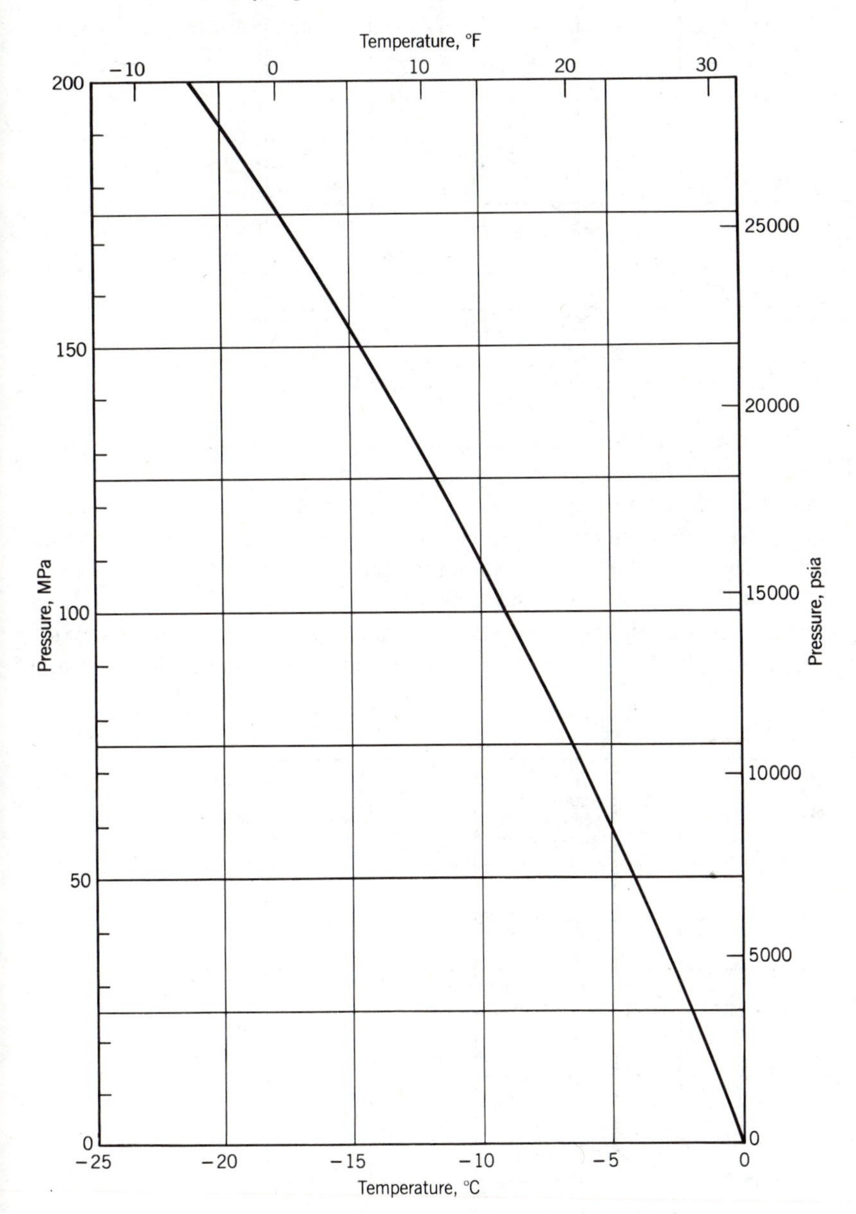

Figure 3·3. Solid–liquid saturation curve for water. Data from N. W. Dorsey, *Properties of Ordinary Water Substance*, Reinhold, New York, 1940.

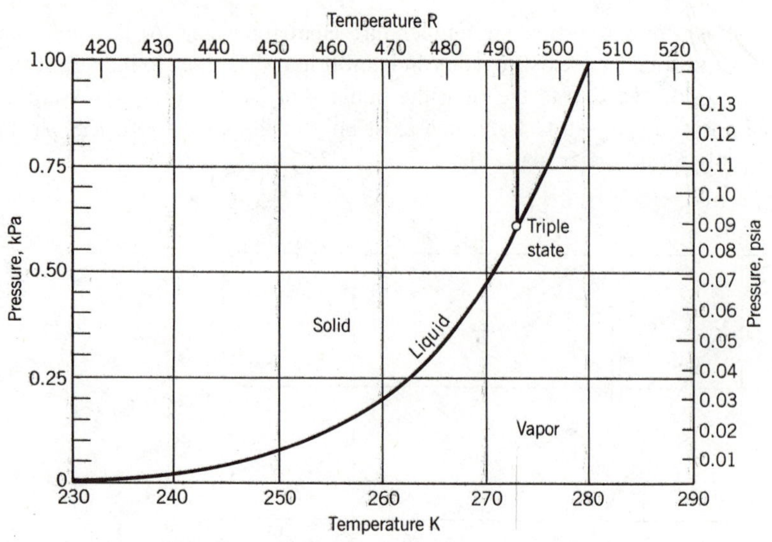

Figure 3·4. Saturation lines for water. Data from Reference 3·5.

For most substances (helium is an interesting exception), there is some pressure below which liquid cannot exist. Below this pressure the solid and vapor phases can coexist in equilibrium, and the relation between pressure and temperature for these solid–vapor saturation states is fixed for each substance. The transformation from a solid to a vapor is known as sublimation; so the solid–vapor saturation line on a *PT* diagram is often called a sublimation line. These curves are shown for water and for carbon dioxide in Figs. 3·4 and 3·5. At pressures below 517.8 kPa, solid and vapor phases of carbon dioxide can exist but the liquid phase cannot. Therefore, solid carbon dioxide (dry ice) at 1 atm

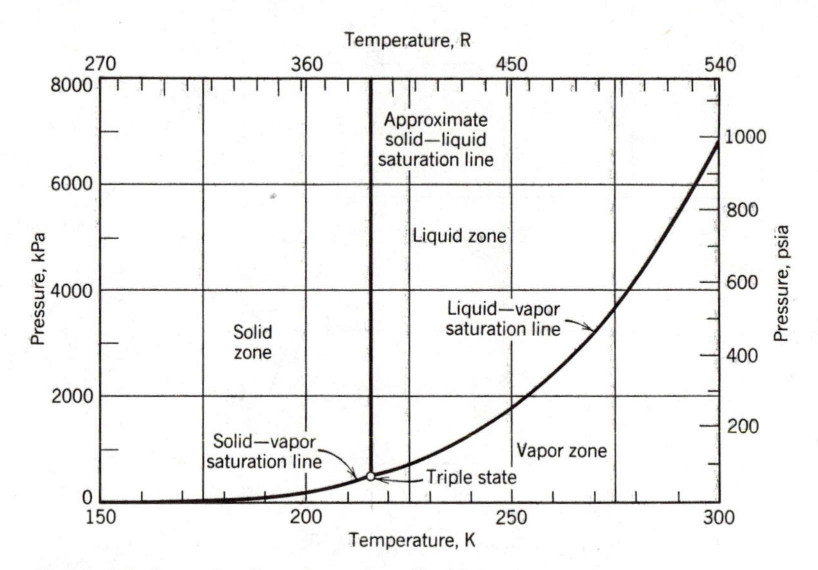

Figure 3·5. Saturation lines for carbon dioxide.

sublimes instead of melting. Another familiar example of sublimation is the transformation of moth balls and solid camphor from the solid to the gaseous phase. Ice disappears from a sidewalk during winter weather by the process of sublimation while the temperature remains well below 0°C. (Since the gaseous phase involved in this process is a mixture of air and water vapor, and the solid phase is ice only, this is not a case of equilibrium between phases of a pure substance. This point is discussed further in Chapters 11 and 19.)

3·3 Phase Diagrams

A pressure–temperature diagram showing more than one of the saturation lines (liquid–vapor, liquid–solid, solid–vapor, or other) of a pure substance is called a *phase diagram.* A phase diagram for a substance which contracts on freezing (as most substances do) is shown in Fig. 3·6, and Fig. 3·7 is a phase diagram for water. Figure 3·7 is a combination of appropriate parts of Figs. 3·1, 3·3, and 3·4. Any point on a saturation line of a phase diagram represents conditions of pressure and temperature under which two or more phases can coexist in equilibrium. All points not on a saturation line represent conditions under which only one phase of the pure substance can exist in equilibrium. Points between the solid–liquid and liquid–vapor lines represent liquid states, and this region of the phase diagram is referred to as the *liquid region.* Similarly, other areas are designated as the *solid region* and the *vapor region.* Liquid existing at a temperature lower than its saturation temperature (or, in other words, at a pressure higher than its saturation pressure) is called a *compressed liquid* or a *subcooled liquid* to distinguish it from a saturated liquid. Thus all points in the liquid region and not on the liquid–vapor saturation line represent states of compressed liquid. Water flowing from a drinking fountain and mercury in an open cup at room temperature are both compressed liquids. In a like manner, a solid at a temperature below its saturation temperature is called a *compressed solid,* and a vapor at a temperature above its saturation temperature is called a *superheated vapor.* These modifiers (compressed, subcooled, superheated) are generally used only in cases where their omission might reasonably mislead a reader into inferring the existence of saturation conditions.

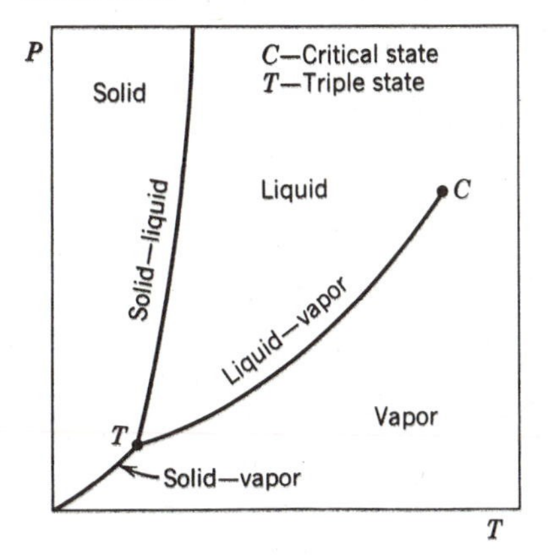

Figure 3·6. Phase diagram for a substance that contracts on freezing.

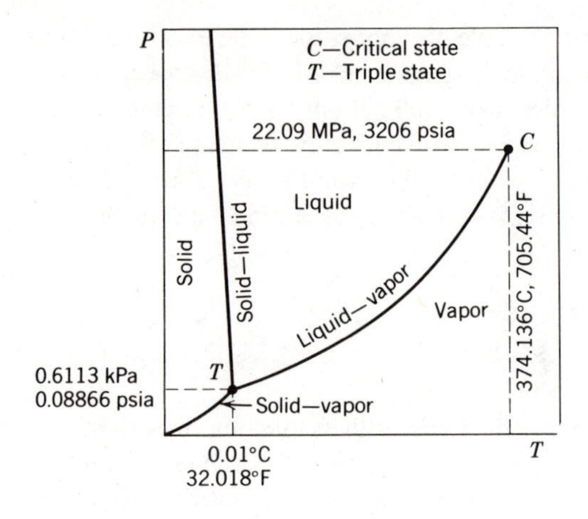

Figure 3·7. Phase diagram for water (scales distorted).

The intersection of the vaporization line, fusion line, and sublimation line on a phase (*PT*) diagram represents the conditions under which three phases can coexist in equilibrium and is called the *triple state, triple point,* or *triple-phase point.* These conditions are represented by a point only on a pressure–temperature diagram; on other property diagrams they are represented by a line or an area. If a substance can exist in more than three phases, there will be more than one triple state for that substance. For example, at high pressures several solid phases of water other than common ice have been observed, so there are several (at least seven, in fact) triple states of water. References to *the* triple state of water always pertain to the triple state at which liquid, solid, and vapor can exist. Triple-state data for several substances are given in Table 3·1.

In Figs. 3·6 and 3·7 the vaporization line extends from the triple point to a point *C* known as the *critical point,* representing the *critical state* of the substance. At pressures or temperatures higher than the critical-state value, no distinction can be made between liquid and vapor phases. The critical state is discussed in Sec. 3·7, and critical-state data for several substances are given in Table 3·2.

TABLE 3·1 Triple-State Data

	Pressure		Temperature	
Substance	kPa	psia	°C	°F
Ammonia	6.1	0.89	−78	−108
Carbon dioxide	517	75	−57	−70
Helium	5.1	0.75	−271	−456
Hydrogen	7.0	1.0	−259	−434
Nitrogen	12.5	1.81	−210	−346
Oxygen	0.15	0.022	−219	−361
Water	0.611	0.0886	0.01	32.02

TABLE 3·2 Critical-State Data

Substance	Pressure		Temperature	
	MPa	psia	°C	°F
Ammonia	11.3	1640	132	270
Carbon dioxide	7.39	1071	31	88
Helium	0.23	33	−268	−450
Hydrogen	1.3	188	−213	−400
Mercury	18.2	2650	899	1650
Nitrogen	3.4	493	−147	−233
Oxygen	5.0	731	−119	−182
Water	22.1	3206	374	705

3·4 Other Property Diagrams

Other property diagrams besides the pressure–temperature or phase diagrams are useful in studying the physical properties of pure substances and in analyzing various thermodynamic processes. Let us turn our attention first to *Pv* diagrams.

In Fig. 3·8, line *a-c,* called the saturated liquid line, is a plot of the specific volume of saturated liquid versus pressure. As the pressure increases (and consequently the saturation temperature increases), the specific volume of saturated liquid increases slightly. The volume scale of Fig. 3·8 has been distorted to magnify this increase. The region immediately to the left of the saturated liquid line represents compressed or subcooled liquid states.

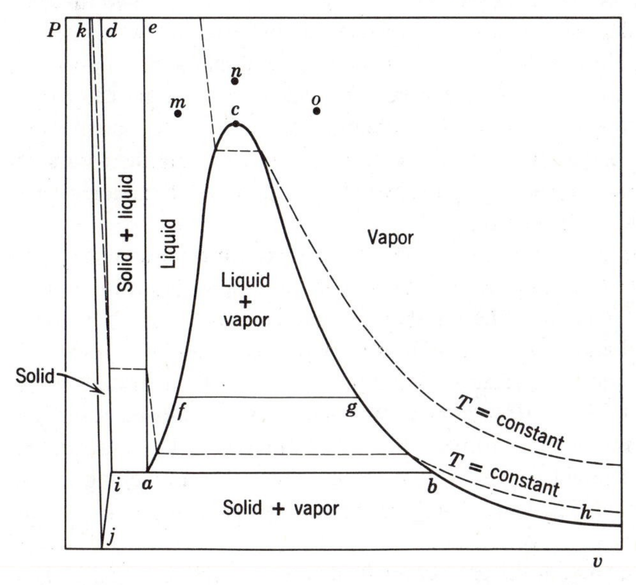

Figure 3·8. *Pv* diagram for a substance that contracts on freezing.

The line *c-h,* called the saturated vapor line, is a plot of the specific volume of saturated vapor versus pressure. As the pressure increases, the specific volume of saturated vapor decreases. The region immediately to the right of the saturated vapor line represents superheated vapor states.

Point *c,* which is common to the saturated liquid and saturated vapor lines, represents the state at which the specific volume is the same for saturated liquid and for saturated vapor. In fact, there is no distinction between liquid and vapor at this state which is called the *critical state* or *critical point.* On a pressure–temperature diagram, the liquid–vapor saturation line ends at the critical point (see Figs. 3·6 and 3·7) because at higher pressures and temperatures no distinction can be made between liquid and vapor states. There can be no continuation of the line which represents states in which two distinct phases can exist together in equilibrium. There is a question as to whether a substance in a state such as represented by point *n* in Fig. 3·8 should be called a liquid or a vapor. Arbitrary rules have been established, but it is best simply to recognize the fact that only one phase exists in the region where *n* is located, and this single phase has no properties which characterize it as a liquid instead of a vapor or vice versa. (Remember that we are considering now only pure substances.) Point *m* represents a state generally referred to as a liquid; point *o* represents a state generally referred to as a gas or vapor; but a transition from state *m* to state *o* or from *o* to *m* can occur with no discontinuities in properties occurring and without any phenomena of boiling or condensation.

On a *PT* diagram, saturated liquid and saturated vapor at the same pressure (and hence at the same temperature) as well as a mixture of the two phases are represented by the same point. These states are not represented by a single point on a *Pv* diagram because the specific volume of saturated vapor is greater than that of saturated liquid at the same pressure, and a mixture of the two phases has a specific volume between these two limiting values. Therefore, on a *Pv* diagram the two-phase (liquid and vapor) mixture states at any pressure lie on a horizontal line between the saturated liquid and saturated vapor lines. The lowest of such lines is the one at the lowest pressure under which liquid can exist. This is the *triple-phase* line. It is at the pressure of the triple state. The area bounded by this *triple-phase* line (*a-b* in Fig. 3·8), the saturated liquid line (*a-c*), and the saturated vapor line (*c-b*) is called the *wet region.* All states involving mixtures of liquid and vapor are represented by points within this region.

The location of a point in the wet region representing a mixture of liquid and vapor phases depends on the pressure (or temperature) and also on the proportion of liquid and vapor in the mixture. *Quality,* denoted by the symbol x, is defined as the fraction by mass of vapor in a mixture of liquid and vapor. The term quality has no meaning in regard to compressed liquid or superheated vapor states. The limiting values of quality are 0 for the saturated liquid alone and 1.0 or 100 percent for saturated vapor alone. Saturated vapor existing with no liquid present ($x = 100$ percent) is sometimes called "dry saturated vapor," and vapor mixed with liquid is called "wet vapor." This is a misleading usage because the properties of *the saturated vapor* itself are the same whether liquid is present or not.

Consider a liquid–vapor mixture at the pressure indicated by line *f-g* in Fig. 3·8. The liquid in the mixture is represented by point *f*; its specific volume is v_f. The vapor in the mixture is represented by point g; its specific volume is v_g. The mixture has a specific volume v_x that is greater than v_f but less than v_g; and v_x can be expressed in terms of v_f, v_g, and x. (The subscripts f and g are quite generally used in tables of properties to designate states of saturated liquid and saturated vapor, respectively.) The volume of the mixture is the sum of the volume of the liquid and of the vapor because the liquid, although it may be dispersed in very small drops, occupies a volume from which the vapor is excluded. (This is unlike the case of two vapors which, as a result of the intermingling of their individual molecules, occupy the same volume when mixed. That is, the volume of the vapor mixture equals the volume of each constituent.) Thus, letting m_L and m_V denote the mass of liquid and of vapor, respectively, the specific volume of the mixture is

$$v_x = \frac{V}{m} = \frac{V_L + V_V}{m_L + m_V} = \frac{m_L v_f + m_V v_g}{m_L + m_V}$$

$$= \frac{m_L}{m_L + m_V} v_f + \frac{m_V}{m_L + m_V} v_g$$

Using the definition of quality x,

$$x \equiv \frac{m_V}{m_L + m_V} \qquad \text{and} \qquad \frac{m_L}{m_L + m_V} = 1 - x$$

and consequently

$$v_x = (1 - x)v_f + xv_g \tag{3·1a}$$

or

$$v_x = v_f + x(v_g - v_f) \tag{3·1b}$$

The difference $(v_g - v_f)$ is denoted by the symbol v_{fg}, so that

$$v_x = v_f + xv_{fg} \tag{3·1b}$$

This shows that, when $x = 50$ percent, the point representing the state of the mixture is midway between points f and g; when $x = 20$ percent, the point is one fifth of the distance along f-g from point f, and so forth. A physical interpretation of Eq. 3·1b is that the volume of 1 kg of mixture is the volume of 1 kg of saturated liquid plus the increase in volume during the vaporization of x kg of substance. Equation 3·1b is generally more convenient for computations than Eq. 3·1a. A form that is convenient for precise computations when x is high is

$$v_x = v_g - (1 - x)v_{fg} \tag{3·1c}$$

A physical interpretation of Eq. 3·1c is that the volume of 1 kg of mixture is the volume of 1 kg of saturated vapor minus the volume decrease during the condensation of $(1 - x)$ kg of substance.

Saturated liquid at the triple-phase pressure is represented by point a in Fig. 3·8, and point b represents saturated vapor. The saturated solid at the same pressure (and temperature) is represented by point i. Points on the triple-phase line between i and a represent mixtures of solid and liquid, solid and vapor, or all three phases. Points between a and *b* represent solid–vapor, liquid–vapor, or solid–liquid–vapor mixtures. Notice that the triple state, which is represented by a point on a *PT* diagram, is represented by a line on a *Pv* diagram. On some other property diagrams, it is represented by an area.

The saturated solid at pressures higher than the triple-phase pressure is represented by points along the line *i-d,* called the saturated solid line. Liquid at the freezing temperature is represented by points along the line *a-e*. In accordance with the definition of saturation conditions, liquid at the freezing temperature could be called saturated liquid except that this would lead to confusion with liquid at the boiling temperature. Hence the line *a-e* is called the freezing liquid line. Mixtures of solid and liquid are represented by points in the region bounded by lines *a-e, i-d,* and *i-a*.

Line *i-j* is also a saturated solid line, but points along it represent states in which the solid can be in equilibrium with a vapor instead of with a liquid. Line *b-h* is likewise an extension of the saturated vapor line and solid–vapor mixtures are represented by points within the area bounded by *j-i-a-b-h*.

Line *j-k* is a plot of the specific volume of the solid versus pressure at some minimum temperature. Figure 3·9 (discussed below) makes this clear.

It must be kept in mind that the volume scale in Fig. 3·8 is very greatly distorted. For all substances the change in volume between points *a* and *b* is many times that between *i* and *a*; and the critical volume v_c is only slightly greater than v_a or v_i.

The broken lines in Fig. 3·8 are lines of constant temperature. Notice that in all two-phase regions constant-temperature lines coincide with constant-pressure lines. The pressure–temperature dependence of saturated phases was discussed in Sec. 3·2.

Two-dimensional property diagrams serve adequately in most thermodynamic analyses, but a better picture of the relationship among pressure, specific volume, and temperature of a substance is given by a three-dimensional *PvT* diagram such as that shown in Fig. 3·9. All equilibrium states of the substance are represented by points on the *PvT* surface. Figure 3·9 shows the *Pv* and *PT* diagrams, which have been discussed above,

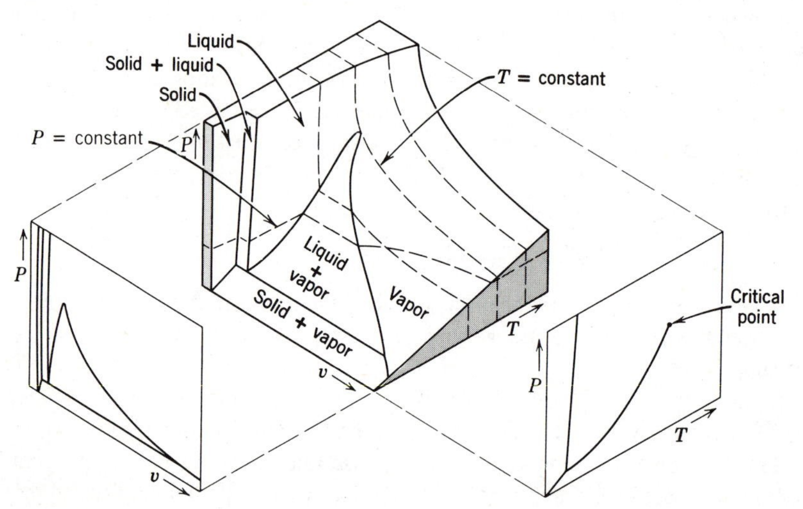

Figure 3·9. *PvT* surface for a substance that contracts on freezing.

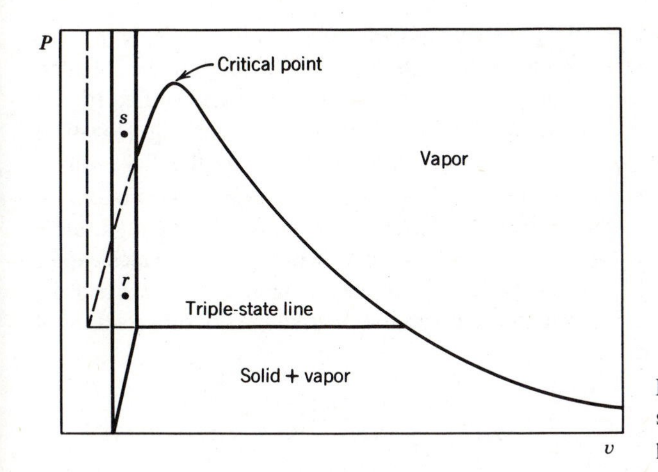

Figure 3·10. *Pv* diagram for a substance that, like water, expands on freezing.

as projections of the *PvT* surface. Points representing mixtures of two phases all lie in surfaces that have elements perpendicular to the *PT* plane. Since a straightedge held parallel to the *v* axis can contact any one of these surfaces all along one of its elements, these are called ruled surfaces. They project into lines on the *PT* diagram. Various surfaces other than the *PvT* surface are useful in some types of thermodynamic analyses.

The advantage of a *PvT* diagram over a two-dimensional diagram is shown by Figs. 3·10 and 3·11 which are for a substance which expands on freezing. Inspection of these

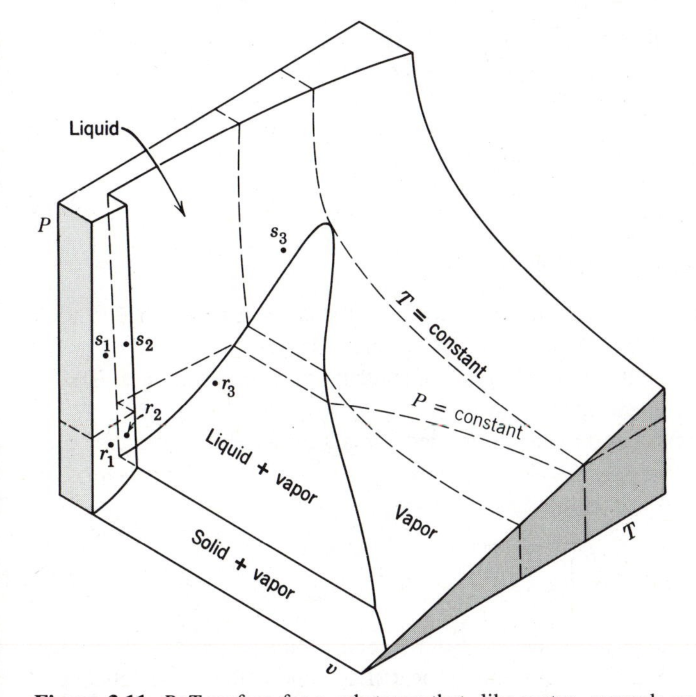

Figure 3·11. *PvT* surface for a substance that, like water, expands on freezing.

figures reveals that point r of the Pv diagram can represent solid (r_1), solid–liquid mixture (r_2), or a liquid–vapor mixture (r_3). Point s can represent a solid, a solid–liquid mixture, or a liquid. Thus the various regions on a Pv diagram that were easily identified in Fig. 3·8 overlap each other on a Pv diagram for a substance that expands on freezing. However, the regions are clearly shown on a PvT diagram, and each point on the PvT surface represents only one state of the substance.

A Tv diagram can also be projected from the PvT surface and is occasionally useful.

The value of any specific property of a two-phase mixture of a pure substance can be calculated from the values of that property for the individual saturated phases and the fraction of each phase present. For example, the specific internal energy of a mixture of liquid and vapor of quality x is

$$u_x = \frac{U}{m} = \frac{U_L + U_V}{m_L + m_V} = \frac{m_L u_f + m_V u_g}{m_L + m_V}$$

$$= (1 - x)u_f + xu_g = u_f + xu_{fg} \qquad (3\cdot2)$$

where the subscripts have the same meaning as in the example given above of the calculation of the specific volume of a mixture.

It can easily be shown that on a diagram of specific properties (u, v, h, etc.) any line that represents mixtures of two phases at a given pressure or temperature is a straight line. Using as an example a line on a uv diagram, the slope of such a line is

$$\frac{du_x}{dv_x} = \frac{\dfrac{du_x}{dx}}{\dfrac{dv_x}{dx}} = \frac{\dfrac{d}{dx}(u_f + xu_{fg})}{\dfrac{d}{dx}(v_f + xv_{fg})} = \frac{u_{fg}}{v_{fg}}$$

and, since u_{fg} and v_{fg} are constant at each pressure or temperature, a line representing mixture states at any pressure or temperature is straight.

3·5 Specific Heat and Latent Heat

The development of the science of calorimetry before the first law was established resulted in some unfortunate choices of names for certain physical properties. Some of these names which persist today occasionally confuse beginning students of thermodynamics. Two of these are *specific heat* and *latent heat.*

Several different specific heats can be defined for a substance, but the two most frequently used are called the *specific heat at constant pressure* c_p and the *specific heat at constant volume* c_v. These are defined as

$$c_p \equiv \left(\frac{\partial h}{\partial T}\right)_P \quad \text{and} \quad c_v \equiv \left(\frac{\partial u}{\partial T}\right)_v \qquad (3\cdot3)$$

where P and T are, of course, independent. Both c_p and c_v are properties. In modern thermodynamics they are not defined in terms of heat. Each one is the rate of change of a property with temperature while some other property is held constant. For example, c_p is the slope of a constant-pressure line on an hT diagram of a substance.

The properties c_p and c_v were originally given the name of specific heats because *under certain conditions* they relate the temperature change of a system to the amount of heat added to the system. Let us answer the question, "Just what are the conditions under which $\delta q = c_v\, dT$?" It has been pointed out that the state of a pure substance which is at rest and uninfluenced by electricity, magnetism, etc., is usually determined by any two independent properties. Thus any property of a system comprised of a pure substance is a function of any two independent properties. Using temperature and specific volume as independent properties, we can express the internal energy of the system as

$$u = f(T, v) \tag{a}$$

Then

$$du = \left(\frac{\partial u}{\partial T}\right)_v dT + \left(\frac{\partial u}{\partial v}\right)_T dv \tag{b}$$

and, using the definition of c_v,

$$du = c_v\, dT + \left(\frac{\partial u}{\partial v}\right)_T dv \tag{c}$$

and this relationship is valid for *any process* of a pure substance, provided only that $u = f(T, v)$. For the *special case* of a *constant-volume* process, c becomes

$$du = c_v\, dT \tag{d}$$

Thus, for a pure substance, c_v relates *internal energy* changes with *temperature* for a *constant-volume* process. The question now is, "Under what further conditions is *heat* related to *internal energy* changes only?" The answer is seen by noting that the first law as applied to a closed system,

$$du = \delta q - \delta w \tag{2·7}$$

reduces to

$$du = \delta q$$

for any process in which no work is done, or (since we intend to apply it to a constant-volume process) for any *frictionless constant-volume* process. Therefore, the answer to the original question is that, *for a frictionless constant-volume process,*

$$\delta q = c_v\, dT$$

This relationship was originally the defining equation for c_v which was defined explicitly as

$$c_v = \left(\frac{\delta q}{dT}\right)_{v=\text{const,frictionless}}$$

The definition as given by Eq. 3·3 is much more concise and less likely to be misinterpreted than this older one. In the same manner, the definition of c_p given by Eq. 3·3 is superior to

$$c_p = \left(\frac{\delta q}{dT}\right)_{P=\text{const,frictionless}}$$

which was formerly in wide use. It is unfortunate that the name *specific heat* has persisted for these properties, since they are properties which have much more important roles in thermodynamics than simply the roles of coefficients in the calculation of heat under special conditions.

The specific heats as defined by Eqs. 3·3 are sometimes called *instantaneous* specific heats to indicate that they can be evaluated at any one state of a system and to distinguish them from *mean* specific heats, which are defined as

$$\overline{c_p} \equiv \left(\frac{\Delta h}{\Delta T}\right)_P \quad \text{and} \quad \overline{c_v} \equiv \left(\frac{\Delta u}{\Delta T}\right)_v$$

Mean specific heats are related to instantaneous specific heats by

$$\overline{c_p} = \frac{\int_1^2 c_p \, dT}{T_2 - T_1} \quad \text{and} \quad \overline{c_v} = \frac{\int_1^2 c_v \, dT}{T_2 - T_1}$$

Notice that these relationships, as well as the defining equations for mean specific heats, involve the specification of a temperature interval, so that a mean specific heat is not a property in the thermodynamic sense because its value does not depend only on the state of the system.

Throughout this textbook, the term *specific heat* without modifiers always signifies *instantaneous* specific heat; *mean* specific-heat values are always so identified.

Latent heat is defined as the magnitude of the difference between the (specific) enthalpy of one phase of a pure substance at saturation conditions and the (specific) enthalpy of another phase of the pure substance at the same pressure and temperature. Letting the subscript i refer to saturated solid, f to saturated liquid, and g to saturated vapor, we distinguish among the various latent heats of a substance as

Latent heat of vaporization, $h_{fg} = h_g - h_f$ (at the same P and T)

Latent heat of fusion, $h_{if} = h_f - h_i$ (at the same P and T)

Latent heat of sublimation, $h_{ig} = h_g - h_i$ (at the same P and T)

(It is actually unnecessary to *specify* that the two enthalpy values in each of the three expressions above must be at the same pressure *and* temperature, because, if two saturated phases of a substance are at the same pressure, they must be at the same temperature and vice versa.)

Notice that latent heat as defined here is a property, and its definition does not refer to heat. It has been proposed that h_{fg} be called the *enthalpy of vaporization,* h_{if} the *enthalpy of fusion,* and so forth, but these terms have not been widely adopted. The term *latent heat,* like the term specific heat, evolved from the science of calorimetry that preceded the development of the first law of thermodynamics. This early science was beclouded by a lack of distinction between quantities which are properties and those which are not.

Application of the first law to phase changes of pure substances shows that the latent heat of a substance at a given pressure is the amount of heat which must be added to a unit mass of the substance to cause it to change phase in a *constant-pressure frictionless* process. This was formerly widely accepted as the definition of latent heat. The definition of latent heat in terms of properties is preferred.

The latent heats of a substance are functions of pressure. The latent heat of vaporization usually varies more with pressure than the latent heat of fusion or of sublimation does. A typical variation in the latent heat of vaporization is shown by the *Ph* diagram of Fig. 3·12. The distance along a constant-pressure line between the saturated liquid and saturated vapor lines is proportional to h_{fg} for that pressure.

Example 3·1. For a certain substance in the temperature range of 0 to 1000°C and at low pressures,

$$c_p = 0.400 + \frac{140}{T} + \frac{T}{1000}$$

where c_p is in kJ/kg·K and T is in kelvins. (Notice that because the definition of specific heat involves a temperature difference or differential, the units can be interchangeably kJ/kg·°C or kJ/kg·K. With $T[\text{K}] = T[°\text{C}] + 273$, we see that $dT_K = dT_C$. This is sometimes explained by saying that the size of a unit is the same on the Kelvin and Celsius scales, so a *change* or *difference* in temperature has the same value on both scales.) Calculate (*a*) $\overline{c_p}$ between 10 and 600°C, (*b*) $\overline{c_p}$ between 10 and 300°C, (*c*) c_p at 10, 300, and 600°C, and (*d*) h at 100 kPa, 400°C, if $h = 65.0$ kJ/kg at 100 kPa, 10°C.

Solution. (***a***) From the definition of mean specific heat, we have for the temperature range of 10°C (283 K) to 600°C (873 K)

$$\overline{c_p} \equiv \frac{\int_1^2 c_p\,dT}{T_2 - T_1} = \frac{\int_1^2 \left(0.400 + \frac{140}{T} + \frac{T}{1000}\right) dT}{T_2 - T_1}$$

$$= \frac{0.400(T_2 - T_1) + 140 \ln (T_2/T_1) + (T_2^2 - T_1^2)/2000}{T_2 - T_1}$$

$$= 0.400 + \frac{140}{T_2 - T_1} \ln \frac{T_2}{T_1} + \frac{T_2 + T_1}{2000} \qquad \text{(A)}$$

$$= 0.400 + \frac{140}{873 - 283} \ln \frac{873}{283} + \frac{873 + 283}{2000} = 1.25 \text{ kJ/kg·K}$$

(***b***) Substituting in equation A above the values of $T_1 = 283$ K and $T_2 = 573$ K we have

$$\overline{c_p} = 0.400 + \frac{140}{573 - 283} \ln \frac{573}{283} + \frac{573 + 283}{2000} = 1.17 \text{ kJ/kg·K}$$

(***c***) Substituting the proper temperatures into the given expression for c_p,

$$c_{p,10°\text{C}} = 0.400 + \frac{140}{283} + \frac{283}{1000} = 1.18 \text{ kJ/kg·K}$$

$$c_{p,300°\text{C}} = 0.400 + \frac{140}{573} + \frac{573}{1000} = 1.22 \text{ kJ/kg·K}$$

$$c_{p,600°\text{C}} = 0.400 + \frac{140}{873} + \frac{873}{1000} = 1.43 \text{ kJ/kg·K}$$

(*d*) Considering h as a function of P and T,

$$dh = \left(\frac{\partial h}{\partial T}\right)_P dT + \left(\frac{\partial h}{\partial P}\right)_T dP = c_p\, dT + \left(\frac{\partial h}{\partial P}\right)_T dP$$

Letting subscripts 1 and 2 denote the conditions at 100 kPa 10°C, and at 100 kPa, 400°C, respectively,

$$h_2 = h_1 + \int_1^2 dh = h_1 + \int_1^2 \left[c_p\, dT + \left(\frac{\partial h}{\partial P}\right)_T dP\right]$$

Since the pressure is the same at states 1 and 2, the second term in the integrand is zero, so that

$$h_2 = h_1 + \int_1^2 c_p\, dT = h_1 + \int_1^2 \left[0.400 + \frac{140}{T} + \frac{T}{1000}\right] dT$$

$$= h_1 + 0.400(T_2 - T_1) + 140 \ln \frac{T_2}{T_1} + \frac{T_2^2 - T_1^2}{2000}$$

$$= 65.0 + 0.400(673 - 283) + 140 \ln \frac{673}{283} + \frac{(673)^2 - (283)^2}{2000}$$

$$= 529 \text{ kJ/kg}$$

Example 3·2. One-half pound of a gas is contained at 30 psia, 140°F, in a closed, rigid vessel. By means of a paddle wheel, 560 ft·lbf of work is done on the gas while 3.0 B of heat is added. During this process, the temperature of the gas rises to 240°F. Calculate $\overline{c_v}$ of the gas.

Solution. The definition of $\overline{c_v}$ is

$$\overline{c_v} \equiv \frac{\int_1^2 c_v\, dT}{T_2 - T_1}$$

and *for a constant-volume process* (see Eq. d),

$$u_2 - u_1 = \int_1^2 c_v\, dT$$

so that *for a constant-volume process*

$$\overline{c_v} = \frac{u_2 - u_1}{T_2 - T_1}$$

For a closed system

$$u_2 - u_1 = \frac{U_2 - U_1}{m} = \frac{Q + W_{\text{in}}}{m}$$

Substituting this value of Δu into the expression above for $\overline{c_v}$

$$\overline{c_v} = \frac{Q + W_{\text{in}}}{m(T_2 - T_1)} = \frac{3.0 + 560/778}{0.5(240 - 140)} = 0.0744 \text{ B/lbm·°F}$$

3·6 Phase Changes

In order to clarify the nature of phase changes, several processes in which a pure substance changes phase will now be considered and will be represented on six different property diagrams (Fig. 3·12). You should make certain that you can sketch all of these property diagrams from the process descriptions, provided you have data at hand for plotting the saturation lines of the substance. Inspection of Fig. 3·12 shows that the property diagram that is best for showing a particular process depends on the process.

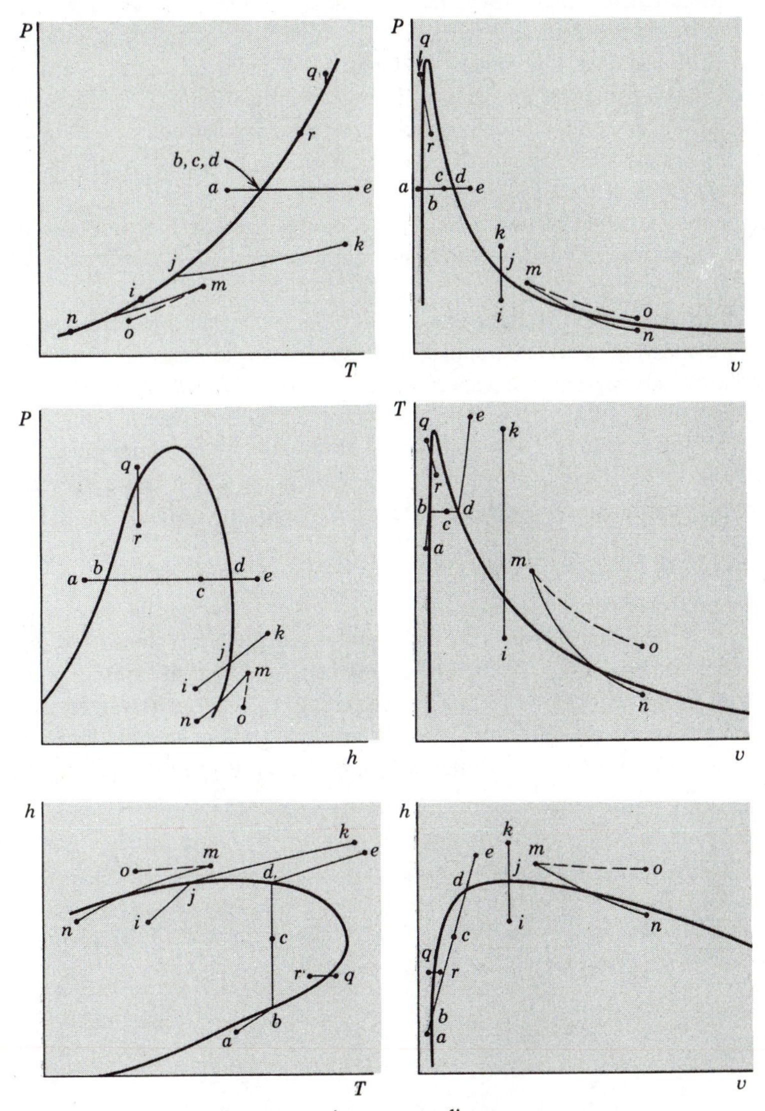

Figure 3·12. Phase changes on six property diagrams.

Let us consider first a process in which a pure substance is changed at constant pressure from a compressed liquid (state *a*) to a superheated vapor (state *e*). For all substances, the internal energy of superheated vapor is greater than that of compressed liquid at the same pressure; therefore, energy must be added to the substance during this process. The energy may be added as heat, as work, or as any combination of heat and work. Property diagrams of the process are the same whether the process occurs in an open or a closed system.

As energy is added to the compressed liquid at constant pressure, the temperature of the liquid rises. While liquid is present in equilibrium, the temperature cannot be higher than the saturation temperature, which is reached at point *b*. On the Tv diagram of Fig. 3·12, the reciprocal of the slope of line *a-b* is $(\partial v/\partial T)_P$, v times the coefficient of cubical expansion. This property has such low values for liquids that line *a-b* is nearly vertical on a Tv diagram. (For *water* under a pressure of 1 atm, specific volume is a minimum at about 4°C, and at lower temperatures the coefficient of cubical expansion of the liquid is negative; hence for water below about 4°C a constant-pressure line of 1 atm has a negative slope on a Tv diagram.)

On the hT diagram, the slope of line *a-b* is $(\partial h/\partial T)_P$, the specific heat at constant pressure. For liquids, c_p varies but slightly with temperature, so line *a-b* is nearly straight on the hT diagram.

After the liquid reaches state *b*, further energy addition at constant pressure causes some of the substance to evaporate. The vapor thus formed is at the same pressure and temperature as the liquid. The state of this vapor is represented by point *d*. If only enough energy to vaporize part of the substance is supplied, then the resulting state is a mixture of liquid at state *b* and vapor at state *d*. The properties of the mixture are represented by some point *c* which is located between *b* and *d* on any diagram which has a specific property as at least one coordinate. On such a diagram the ratio of the distance *b-c* to the distance *b-d* is equal to the quality of the mixture.

In order to evaporate one unit mass of saturated liquid to a saturated vapor at the same pressure, its internal energy must be increased by $u_g - u_f$ or u_{fg}, where the subscripts are as defined in Sec. 3·4. The fluid expands during the process. If it is in a closed system, it does some work on the surroundings so that the total energy added to the system is u_{fg} plus an amount equal to the work done by the system. If it is in an open system, any fluid leaving the system does more flow work on the surroundings than an equal amount of fluid entering does on the system so that the total energy added to the system is again greater than u_{fg}.

Example 3·3. One kilogram of saturated liquid propane at 100 kPa in a closed system is completely vaporized at constant pressure in a frictionless process. Calculate, in kilojoules per kilogram, (*a*) the amount of heat added and (*b*) the amount of work done. Properties of the saturated liquid and saturated vapor are as follows:

$$u_f = 68.19 \text{ kJ/kg} \qquad u_g = 451.13 \text{ kJ/kg}$$
$$v_f = 0.00175 \text{ m}^3\text{/kg} \qquad v_g = 0.4172 \text{ m}^3\text{/kg}$$
$$h_f = 68.36 \text{ kJ/kg} \qquad h_g = 492.85 \text{ kJ/kg}$$

Solution. Analysis of the problem shows that it is possible to solve part *b* first and that this result will be useful in the solution of part *a*.

(***b***) For this frictionless process of a closed system,

$$w = \int_1^2 P\,dv$$

and, since the pressure is constant

$$w = P\int_1^2 dv = P(v_2 - v_1) = P(v_g - v_f)$$

$$= 100(0.4172 - 0.00175) = 41.55 \text{ kJ/kg}$$

(***a***) Applying the first law to this closed system

$$q = u_2 - u_1 + \text{work} = u_g - u_f + \text{work}$$
$$= 451.13 - 68.19 + 41.55 = 424.5 \text{ kJ/kg}$$

This result may also be obtained by

$$q = u_2 - u_1 + w = u_2 - u_1 + \int_1^2 P\,dv$$

$$= u_2 - u_1 + P_2v_2 - P_1v_1 = h_2 - h_1 = h_g - h_f$$

$$= 492.85 - 68.36 = 424.5 \text{ kJ/kg.}$$

Notice that because work is done by the system during the vaporization process, the amount of heat added must be greater than the increase in the internal energy of the system.

Example 3·4. One kilogram of saturated liquid propane at 100 kPa in a closed system is completely vaporized at constant pressure by the addition of heat while 15.0 kJ of work is done on the propane by means of a stirrer. Physical properties are tabulated in Example 3·3. Calculate the amount of heat added in kJ/kg.

Solution. This process is not frictionless because the action of the stirrer involves fluid shear forces; therefore work $\neq \int P\,dv$. However, if the action of the stirrer is such that the pressure on any boundary of the system which moves so as to change the volume of the system remains uniform and constant, then the work done on this moving boundary (or the work associated with the change in volume of the system) is

$$\text{work}_{\Delta v} = \int P\,dv = P(v_2 - v_1)$$

and the net work is

$$\text{work} = \text{work}_{\Delta v} + \text{work}_{\text{stirrer}} = P(v_2 - v_1) + \text{work}_{\text{stirrer}}$$
$$= 100(0.4172 - 0.00175) + (-15.0) = 26.5 \text{ kJ/kg}$$

Then, applying the first law,

$$q = u_2 - u_1 + \text{work} = 451.13 - 68.19 + 26.5$$
$$= 409.4 \text{ kJ/kg}$$

(This example illustrates that latent heat of vaporization cannot be defined simply as the amount of heat required to change 1 lbm of a saturated liquid to saturated vapor in *any* constant-pressure process. Here, $q = 409.4$ kJ/kg $< h_{fg}$.)

Example 3·5. Saturated liquid propane at 100 kPa is completely vaporized at constant pressure in a steady-flow heat exchanger. Calculate the amount of heat added in kJ/kg. Physical properties are tabulated in Example 3·3.

Solution. Applying the first law to this steady-flow system, noting that no work is done in a heat exchanger, and assuming that changes in potential and kinetic energy are negligible, we have

$$q = h_2 - h_1 + \text{work} + \Delta pe + \Delta ke$$

$$= h_2 - h_1 + 0 + 0 + 0 = h_g - h_f = 492.85 - 68.36 = 424.5 \text{ kJ/kg}$$

(Notice that $q = h_{fg}$ *only* when work $+ \Delta pe + \Delta ke = 0$.)

The addition of energy to a saturated vapor at constant pressure causes an increase in its temperature, specific volume, internal energy, and enthalpy as shown by the lines *d-e* in Fig. 3·12. For all substances, $(\partial v/\partial T)_P$ is greater for the vapor than for the liquid at the same pressure and c_p, which is defined as $(\partial h/\partial T)_P$, is lower for the vapor than for the liquid. Thus the relative slopes of lines *a-b* and *d-e* on some of the diagrams can be explained.

Consider now the addition of energy to a liquid–vapor mixture of a pure substance that is held in a closed rigid container and initially has a specific volume greater than the specific volume at the critical point. Such an initial state is represented by point *i* on each diagram of Fig. 3·12. Addition of energy, either as heat or as some form of frictional work such as paddlewheel work, causes some of the liquid to evaporate as the pressure and temperature increase. At state *j* all of the liquid has been vaporized; only saturated vapor is present. Further addition of energy to the substance in the constant-volume container causes its state to change along line *j-k*.

The expansion of a vapor in a nozzle, a turbine, or an engine is often adiabatic or very nearly so. From any initial state, several adiabatic paths are possible. One such path is shown in Fig. 3·12 as *m-n*. Another is shown as *m-o*. For each of these paths, work is done by the fluid or there is a change in kinetic energy, or both these energy transformations occur.

A *throttling process* is an adiabatic steady-flow expansion in which no work is done and there is no change in kinetic energy. Application of the first law shows that the initial and final enthalpies are the same in a throttling process. Examples of a throttling process are the flow of a fluid through a porous plug and the flow through a pressure-reducing valve. A throttling process in which a phase change occurs is shown by line *q-r* in Fig. 3·12. Notice that this phase change is brought about without the use of heat or work.

In addition to those listed above, many other processes in which phase changes occur are possible and occur in engineering applications.

3·7 The Critical State

The critical state, the limiting state for the existence of saturated liquid and vapor, has already been introduced. The *critical pressure* is the highest pressure under which distinguishable liquid and vapor phases can exist in equilibrium. The *critical temperature* is the highest temperature at which distinguishable liquid and vapor phases can exist in equilibrium. As the critical state is approached from lower pressures and temperatures, the properties of saturated liquid and saturated vapor approach each other. Therefore, properties such as u_{fg}, h_{fg}, and v_{fg} reach the limiting value of zero at the critical point.

Since the properties, especially density, of liquid and vapor phases of a substance under normal ambient conditions are so different, it is sometimes difficult to visualize conditions under which the two phases are indistinguishable from each other. It is often helpful to consider the following experiment.

If a rigid transparent vessel is filled with a liquid-vapor mixture of a substance in a state represented by point *a* on Fig. 3·13, all the liquid will in time settle to the bottom of the vessel, and a meniscus can be seen separating the two phases. If the mixture is then heated, the pressure and temperature will increase. Inspection of Fig. 3·13 shows that the fraction of liquid, as well as the liquid specific volume, will increase, causing the meniscus to rise until it reaches the top of the vessel at point *b*. During further heating to point *m*, the vessel contains liquid only. If the vessel had originally been filled with a mixture represented by point *d*, heating would have caused the evaporation of the liquid, and the meniscus would have fallen until the last bit of liquid in the bottom of the vessel was vaporized as the substance reached the state represented by point *e*. Further heating to point *o* would superheat the vapor.

The most interesting case is the one in which the vessel is initially filled with a mixture at state *h* such that the mixture specific volume v_h is equal to the specific volume at the critical state. As the mixture is heated, it approaches the critical state with the liquid becoming less dense and the vapor becoming more dense. As the critical state is approached, the meniscus fades away because there are no longer two phases present. The single phase which remains may be considered as either a liquid or a vapor, and there is fully as much reason for one designation as for the other.

It appears that the amounts of liquid and vapor initially placed in the vessel must be very carefully controlled (or, in other words, that point *h* must be carefully selected) to

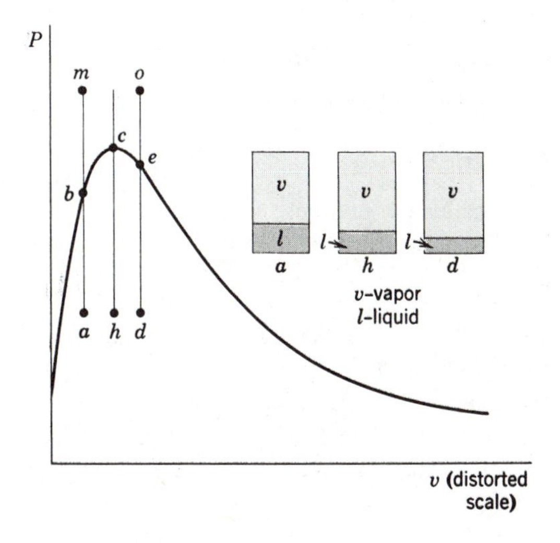

Figure 3·13. Critical-state experiments.

insure that constant-volume heating will cause the meniscus to disappear somewhere near the center of the tube instead of rising to the top as the entire sample is liquefied or falling to the bottom as the entire sample is vaporized. Actually experiments show that the meniscus between liquid and vapor phases disappears not only at the critical state but in a region around it where the two phase densities are nearly equal. This means that the disappearance of the meniscus can be demonstrated without making the substance pass exactly through its critical state. Conversely, the critical state cannot be precisely identified by the disappearance of the meniscus. These facts were not anticipated before the performance of such experiments. In fact, the investigation of properties near the critical state offers several eloquent examples of the importance of careful experimentation in conjunction with analytical studies.

3·8 Tables of Properties

An equation relating P, v, and T of a substance is called an *equation of state*. The relationships among P, v, and T for some substances are such that they can be represented by fairly simple equations of state, and one of these is discussed in the next chapter. Also, it is sometimes possible to express properties such as internal energy and enthalpy as simple functions of one or two of the state properties P, v, and T. Unfortunately, for many substances of importance in engineering the relationships among properties over a wide range cannot be expressed by simple equations. Therefore, the properties of these substances must be determined over the range of interest by measurements supplemented by calculations which ensure the self-consistency of interpolated or derived values. The results of these measurements and calculations are then presented in the form of tables or charts.

The properties which are usually tabulated are:

P and T, intensive properties, which are directly measurable and usually controllable.

v, which is useful in determining the size of equipment, flow rates, and velocities.

h and u, which are useful in applications of the first law (u is not used so much as h and is frequently not tabulated, but it can always be found by $u = h - Pv$).

s (entropy), which is introduced in Chapter 8.

Several other properties are occasionally tabulated, but those listed above are found in all tables of thermodynamic properties. The properties are usually tabulated for the following states:

1. *Saturated liquid*, for which either P or T is used as the argument or independent variable. Only one argument is needed, because P and T are dependent on each other for saturation states. (Notice that this does not mean that we can specify the state of a pure substance by only one variable. In addition to the value of pressure or temperature, we must also know the fact that the substance exists in a saturated liquid state.) Complete

tables often include one saturation-states table with integral (i.e., whole number) values of pressure and one with integral values of temperature as the argument in order to reduce and simplify the interpolation needed.

2. *Saturated vapor,* these properties usually being listed in the same table as the saturated liquid properties.

3. *Superheated vapor,* for which P and T are used as arguments because P and T are readily measurable independent variables, and two independent variables are required to specify the state of the substance.

Compressed liquid properties are seldom tabulated. They are functions of temperature and pressure but vary only slightly with pressure. Therefore, the properties of a compressed liquid are approximately equal to the properties of saturated liquid *at the same temperature*. When very high pressures are encountered or a high degree of precision is desired, this approximation may be unsatisfactory. Then the properties needed must be obtained from tables or must be calculated by means of data on specific heats, bulk moduli, and coefficients of expansion. The use of a compressed liquid table for water is illustrated later in this article.

Saturated solid (solid–vapor equilibrium) properties are occasionally tabulated in the same fashion as saturated liquid and vapor properties. Water and carbon dioxide are two substances that are occasionally encountered in solid–vapor equilibrium applications.

Liquid–vapor mixture and *solid–vapor mixture* properties are not tabulated because they can be readily calculated from the quality and the properties of their constituent phases by means of equations such as 3·1 and 3·2.

It was pointed out in Sec. 2·4 that thermodynamics provides no information as to absolute values of internal energy. We are concerned only with *changes* in internal energy. Therefore, the state at which internal energy or enthalpy has a value of zero may be selected arbitrarily. Steam tables generally assign $h = 0$ to saturated liquid at 0°C or 32°F;* tables for refrigerants often assign $h = 0$ to saturated liquid at -40°C or -40°F. Other datum temperatures are also used.

Several examples of the use of property tables will now be given, using some of the abbreviated or skeleton tables of properties in the appendix. For ease in solving many of the problems in this textbook and for most engineering work, the original tables or the extensive tables available in various reference works should be used to reduce the amount of interpolation needed. Tables from various sources are likely to be in various formats, and even the tables in the appendix of this book use various formats; but anyone proficient in the use of tables for one substance or tables in one format can quickly adapt to the use of property tables for other substances and in other formats. Water or steam is important to the engineer because of its widespread use as a working fluid and a heat transfer medium and its presence in atmospheric air and in the products of combustion of any fuel that contains hydrogen. Many of the examples and problems that follow involve water not

*Since the triple-point temperature of water is 0.01°C or 32.018°F, saturated liquid cannot exist at 0°C or 32°F. Nevertheless it is possible and quite convenient to extrapolate saturated liquid properties to 0°C or 32°F and to use this as the datum temperature. Obviously, for some states values of u and h are negative.

only because of the importance of water but also on account of the completeness of the available tables that cover a wide range of variables with small tabular differences.

Property tables for a few substances are given in the appendix in both SI and English units, and the units used in each table are indicated in the table heading by SI or E in parentheses following the table number. Tables A·1.1(E) and A·1.1(SI) are therefore tables of the same properties. When we refer in the text to these tables, we designate them simply as Table A·1.1, because the discussion applies to either table.

For water, the properties of saturated liquid and vapor are given in Tables A·1.1 and A·1.2 of the appendix. The primary difference between these two is that Table A·1.1 is based on integral values of temperature and Table A·1.2 on integral values of pressure. The first and second columns of Table A·1.1 give the saturation temperature and pressure, respectively. For example, the saturation pressure for 20°C is 2.339 kPa (and that for 50°F is 0.17811 psia). The other columns of the table are appropriately labeled. The units for each property are given with each table. Whenever values of internal energy are not given in such a table, they can be readily calculated from $u = h - Pv$.

Table A·1.2 is similar to Table A·1.1 except that the pressure and temperature columns are interchanged.

The data of Tables A·1.1 and A·1.2, in conjunction with equations such as Eqs. 3·1 and 3·2 (Sec. 3·4), can be used to determine the properties of liquid–vapor mixtures. These mixtures are commonly referred to as "steam," even though they contain liquid.

The properties of superheated steam are presented in Table A·1.3 of the appendix. The saturation temperature is shown in parentheses beside each pressure value.

Properties of *compressed* or *subcooled* liquid water are given in Table A·1.4. (As mentioned earlier, the two designations are entirely equivalent, but they stem from two different points of view. *Compressed* liquid is at a pressure higher than the saturation pressure corresponding to its temperature. Liquid in the same state is also called *subcooled* liquid because it is at a temperature lower than the saturation temperature corresponding to its pressure.) Although the compressed liquid table in this book is in the same format as the superheated vapor table for water, other formats are also used. For many substances, compressed liquid property tables are unavailable. In such cases, a useful approximation is to take the properties of compressed liquid as being the same as those of saturated liquid *at the same temperature*. A perusal of Table A·1.4 reveals the magnitude of the error involved in this approximation at various pressures and temperatures.

Properties of saturated solid and saturated vapor phases (that is, states along the solid–vapor saturation line or sublimation line on a *PT* diagram) have been tabulated for only a few substances. Table A·1.5 presents a few values for water.

In addition to tables of properties, charts are used for the same purpose. Data can usually be read more rapidly from charts than from tables at some sacrifice in accuracy. Coordinates commonly used for charts of physical properties are enthalpy–entropy, enthalpy–volume, temperature–entropy, and pressure–enthalpy. The use of some of these charts will be discussed following the introduction of the property entropy.

Thermodynamic property data are also available in computer data banks for ready retrieval. In fact, published tables and charts result from calculations based on property relationships (including some discussed in Chapter 9), experimental data, and various interpolation methods. Therefore, by means of a computer utilizing the same data and methods used to generate published tables, one can expect the same accuracy as provided by the tables. However, some computer data banks make use of approximate relationships and coarser interpolation methods, thereby sacrificing some accuracy in comparison with the more extensive programs used for generating most published tables.

Example 3·6. Ten kilograms of steam at 20 kPa has a volume of 52 m³. Determine the temperature and the enthalpy per kilogram.

Solution. From the data given it may not be possible to visualize at once the condition of the steam. Therefore we first calculate the specific volume

$$v = \frac{\text{volume}}{\text{mass}} = \frac{52}{10} = 5.2 \text{ m}^3/\text{kg}$$

and then compare this with the values given in Table A·1.2(SI) at 20 kPa for saturated liquid ($v_f = 0.001\ 017 \text{ m}^3/\text{kg}$) and for saturated vapor ($v_g = 7.649 \text{ m}^3/\text{kg}$). We see that $v_f < v < v_g$. Therefore, the steam is a wet vapor ($0 < x < 100$ percent) at a temperature of 60.06°C which corresponds to the saturation pressure of 20 kPa. In order to determine the enthalpy we first compute the quality. Using Eq. 3·1,

$$v_x = v_f + x(v_{fg})$$

and solving for x gives

$$x = \frac{(v_x - v_f)}{(v_g - v_f)} = \frac{5.2 - 0.001\ 074}{7.649 - 0.001\ 074} = 0.68 \quad \text{or} \quad 68 \text{ percent}$$

The enthalpy per kilogram (that is, the specific enthalpy) is

$$h = h_f + x(h_{fg}) = 251.40 + 0.68(2358.3) = 1855.0 \text{ kJ/kg}$$

Example 3·7. Steam at 400 psia has an enthalpy of 600 B/lbm. Determine the internal energy.

Solution. The first step is to determine whether the steam is superheated or is a liquid–vapor mixture. To do this, refer to Table A·1.2(E), and note that at 400 psia $h_g = 1205.5$ B/lbm. The steam under consideration has a lower enthalpy than saturated vapor at the same pressure; therefore it must be a liquid–vapor mixture rather than a superheated vapor. Knowing the pressure and the enthalpy, we can find the quality by rearranging

$$h = h_f + x(h_{fg})$$

to give

$$x = \frac{h - h_f}{h_{fg}} = \frac{600 - 424.2}{781.2} = 0.225$$

The internal energy is then given by

$$u = u_f + xu_{fg} = 422.8 + 0.225(696.7) = 579.6 \text{ B/lbm}$$

Notice that in these examples the terms *enthalpy* and *internal energy* are used in place of *specific enthalpy* and *specific internal energy*. This is common practice. However, the modifier is not omitted from *specific volume*.

Example 3·8. Determine the enthalpy and specific volume of steam at 5 MPa and 400°F.

Solution. From Table A·1.2(SI) or A·1.3(SI), the saturation temperature for a pressure of 5 MPa is 263.99°C. The temperature of the steam under consideration is greater than the saturation temperature; hence the steam is superheated. Table A·1.3 (SI) shows the following values for the enthalpy and specific volume

$$h = 3195.7 \text{ kJ/kg} \qquad \text{and} \qquad v = 0.057\,81 \text{ m}^3/\text{kg}$$

Example 3·9. Determine the enthalpy of steam at 1.0 MPa, 270°C.

Solution. Since the saturation temperature for a pressure of 1 MPa is 179.91°C, the steam under consideration is superheated. Sometimes it is necessary to interpolate between values if available tables are not extensive enough. In Table A·1.3 (SI), values at 1 MPa pressure are available only at temperatures of 250 and 300°C. A linear interpolation between these two tabulated temperatures gives

$$\begin{aligned} T &= 250°\text{C} \qquad & h &= 2942.6 \text{ kJ/kg} \\ T &= 270°\text{C} & & \\ T &= 300°\text{C} \qquad & h &= 3051.2 \text{ kJ/kg} \end{aligned}$$

for a value at 270°C of h = 2986.0 kJ/kg. (The Keenan, Keyes, Hill, and Moore tables, Reference 3·5, from which Table A·1.3 was abridged, give a value of 2986.5 kJ/kg. Therefore, some inaccuracy is seen in the use of linear interpolation.)

Example 3·10. Determine the specific volume and enthalpy of water at 20 MPa, 200°C.

Solution. Since 200°C is lower than the saturation temperature corresponding to 20 MPa, water at 20 MPa, 200°C, is a compressed liquid. From Table A·1.4 for a pressure of 20 MPa and a temperature of 200°C,

$$v = 0.001\,138\,8 \text{ m}^3/\text{kg} \qquad h = 860.5 \text{ kJ/kg}$$

For saturated liquid at 200°C, v_f = 0.001 157 m³/kg. Consequently the error in approximating the specific volume of this compressed liquid as the specific volume of saturated liquid at the same temperature would be less than 2 percent. In other words, water at 200°C in this pressure range is almost incompressible.

Example 3·11. Fill in the blanks in the following table for the properties of water. (The given property values are in boldface type.)

	P, psia	T, °F	x, %	v, cu ft/lbm	h, B/lbm
(a)	**40**	**400**	*M*	*12.623*	*1236.4*
(b)	**40**	*267.26*	**70**	*7.366*	*889.8*
(c)	*247.1*	**400**	*74.8*	**1.400**	*993.3*
(d)	**1000**	*700*	*M*	**0.608**	*1324.6*
(e)	**3000**	**200**	*M*	*0.016476*	*174.89*
(f)	**14.696**	**212**	*I*	*I*	*I*
(g)	*24.97*	**240**	*72.6*	*11.86*	**900**
(h)	*430*	*700*	*M*	**1.53**	**1361**
(i)	**0.0309**	**10**	*I*	*I*	*I*
(j)	*275*	**500**	*M*	*1.9047*	**1260.4**

Solution. Values found from the steam tables are inserted in the accompanying table in italics. In a few cases, simple computations similar to those in previous examples were made. *M* denotes *meaningless* and *I* denotes *indeterminate.*

Example 3·12. Fill in the blanks in the following table for the properties of ammonia. (The given property values are in boldface type.)

	P, psia	*T*, °F	*x*	*v*, cu ft/lbm	h, B/lbm
(*a*)	**200**	**150**	*M*	*1.740*	*671.8*
(*b*)	*5.55*	**−60**	**15**	*6.73*	**67.2**
(*c*)	**153**	**80**	*I*	*I*	*I*
(*d*)	*128.4*	**100**	*M*	*2.534*	**650**
(*e*)	**5**	*−63.11*	*20.3*	*10.03*	**100**
(*f*)	**260**	*130*	*M*	**1.22**	*647.8*

Solution. Values found from the ammonia tables in the appendix are inserted in the accompanying table in italics. *M* denotes *meaningless* and *I* denotes *indeterminate.*

Example 3·13. Saturated liquid ammonia at 200 psia enters an expansion valve in a refrigerating system and is throttled to 40 psia. What is the state of ammonia leaving the valve? Also, determine the internal energy of the ammonia leaving.

Solution. For the flow through a valve, work = 0. Assuming that this is a throttling process as defined in Sec. 3·6, $q = 0$ and the change in kinetic energy is negligibly small. The first law applied to the steady flow through the valve then reduces to

$$h_2 = h_1$$

At the inlet (section 1), the ammonia is saturated liquid, so, using Table A·2.2(E)

$$h_2 = h_1 = h_{f1} = 150.9 \text{ B/lbm}$$

This value of enthalpy is greater than h_f and less than h_g at 40 psia; therefore, the ammonia leaving the valve is a liquid–vapor mixture. Its quality is

$$x_2 = \frac{h_2 - h_{f2}}{h_{fg2}} = \frac{150.9 - 55.6}{559.8} = 17.0 \text{ percent}$$

This indicates that 17 percent of the ammonia evaporated during the throttling process. The final state is 40 psia, 17 percent quality.

Internal energy values are not included in Table A·2.2, so we calculate values of u_f and u_g from tabulated properties at 40 psia:

$$u_{f2} = h_{f2} - P_2 v_{f2} = 150.9 - \frac{40(144)0.02451}{778} = 150.7 \text{ B/lbm}$$

$$u_{g2} = h_{g2} - P_2 v_{g2} = 615.4 - \frac{40(144)7.047}{778} = 563.2 \text{ B/lbm}$$

Then

$$u_2 = u_{f2} + x_2(u_{g2} - u_{f2}) = 150.7 + 0.170(563.2 - 150.7) = 220.8 \text{ B/lbm}$$

(This calculation could also be made by calculating $v_2 = v_{f2} + x_2 v_{fg2}$ and then $u_2 = h_2 - P_2 v_2$.)

3·9 Summary

This chapter has dealt, largely in a qualitative manner, with some of the physical properties of pure substances. It has also introduced the use of property tables. In summary, some of the terms introduced in this chapter are listed or discussed here.

A *pure substance* is a substance that is chemically homogeneous and fixed in chemical composition.

A *phase* is any homogeneous part of a system that is physically distinct and is separated from other parts of the system by definite boundary surfaces.

Two phases of a pure substance can exist together in equilibrium only if their pressure and temperature bear a certain fixed relationship to each other. The corresponding values of pressure and temperature under which two phases can coexist in equilibrium are called the *saturation pressure* and the *saturation temperature*. Any phase existing under these saturation conditions is called a *saturated* phase.

Liquid existing at a temperature lower than the saturation temperature corresponding to its pressure (or, in other words, at a pressure higher than the saturation pressure corresponding to its temperature) is called *compressed liquid* or *subcooled liquid*. Vapor existing at a temperature higher than the saturation temperature corresponding to its pressure (or at a pressure lower than the saturation pressure corresponding to its temperature) is called *superheated vapor* or *gas*.

The condition of pressure and temperature under which three phases of a pure substance can coexist in equilibrium is called the *triple state, triple point,* or *triple-phase point* of the substance.

For every substance, at low pressures the density of the saturated vapor is much less than that of saturated liquid. As pressure increases, the difference in densities of the two phases decreases until at the *critical pressure* and the *critical temperature* the difference in density between the two phases becomes zero. All other properties also reach the same value for both saturated liquid and saturated vapor as the *critical state* is reached. The critical state is therefore the limiting condition of pressure and temperature under which separate liquid and vapor phases can be distinguished.

Various property diagrams are useful in showing the characteristics of a substance. The pressure–temperature diagram which shows the limiting pressures and temperatures for the existence of the various phases of a substance called a *phase diagram*. Three-dimensional property diagrams are occasionally used.

The *quality* of a liquid–vapor mixture is defined as the fraction by mass of vapor in a mixture of liquid and vapor. Properties of a liquid–vapor mixture may be found by relations such as

$$v = v_f + xv_{fg}$$
$$h = h_f + xh_{fg}$$

where x is the quality, the subscript f refers to saturated liquid, and the subscript fg refers to the difference between the value of a property for saturated vapor and for saturated liquid at the same pressure. Similar terminology is used for solid–vapor mixtures.

Specific heat at constant pressure c_p and specific heat at constant volume c_v are defined as

$$c_p \equiv \left(\frac{\partial h}{\partial T}\right)_P \quad \text{and } c_v \equiv \left(\frac{\partial u}{\partial T}\right)_v$$

Both c_p and c_v are properties. They are not defined in terms of heat. It can be shown by means of the first law and these definitions of specific heat that for both open and closed systems

$$\delta Q = mc_p\, dT \text{ for } \textit{frictionless} \text{ constant-pressure processes}$$
$$\delta Q = mc_v\, dT \text{ for } \textit{frictionless} \text{ constant-volume processes}$$

(It is from these special relations that c_p and c_v have been given the name specific heats, even though c_p and c_v are used in many other relations which do not involve heat.)

Latent heat of vaporization is defined as

$$h_{fg} \equiv h_g - h_f$$

where h_g and h_f are the specific enthalpies of saturated vapor and saturated liquid, respectively, at the same pressure. Similar terminology is used for other phases.

For many substances the relationships among properties over a wide range cannot be expressed by simple equations. Therefore, the properties of these substances must generally be obtained from tables based on experimental measurements and extensive thermodynamic calculations.

References

3·1 Van Wylen, Gordon J., and Richard E. Sonntag, *Fundamentals of Classical Thermodynamics,* 3rd ed., Wiley, New York, 1985, Sections 3.1 to 3.3 and 3.5,6.

3·2 Reynolds, William C., and Henry C. Perkins, *Engineering Thermodynamics,* 2nd ed., McGraw-Hill, New York, 1977.

3·3 Wark, K., *Thermodynamics,* 4th ed., McGraw-Hill, New York, 1983, Sections 4-1 to 4-5.

For Thermodynamic Property Data

3·4 Keenan, J. H., J. Chao, and J. Kaye, *Gas Tables,* 2nd ed., Wiley, New York, 1980 (English units), 1983 (SI units).

3·5 Keenan, J. H., F. G. Keyes, P. G. Hill, and J. G. Moore, *Steam Tables,* Wiley, New York, 1969 (English units), 1978 (SI units).

3·6 Potter, J. H., *Steam Charts,* ASME, New York, 1976.

3·7 Reynolds, W. C., *Thermodynamic Properties in SI,* Department of Mechanical Engineering, Stanford University, Palo Alto, Calif., 1979.

3·8 *Marks' Mechanical Engineers' Handbook,* 8th ed., McGraw-Hill, New York, 1978.

3·9 *Chemical Engineers' Handbook,* 5th ed., McGraw-Hill, New York, 1973.

3·10 *Handbook of Tables for Applied Engineering Science,* R. E. Bolz and G. L. Tuve, (eds.), CRC Press, Boca Raton, Fla., 1970.

3·11 *Tables of Thermal Properties of Gases,* Circular 564, National Bureau of Standards, Washington, D.C., 1955.

Problems

3·1 Which of the systems described below are composed of pure substances during the described process? (The containers and partitions are not parts of the systems.)

(a) A tank contains oxygen and nitrogen on opposite sides of a partition. The partition is removed (or punctured).

(b) A tank contains nitrogen at 130 kPa on one side of a partition and nitrogen at 100 kPa on the other side. The partition is removed.

(c) Air is contained in a tank. Also within the tank is a covered dish containing water. The cover is removed from the dish.

(d) A tank contains liquid water and steam. The tank is cooled so that some of the steam condenses.

3·2 Which of the systems described below are composed of pure substances during the described processes? (The containers and partitions are not parts of the systems.)

(a) A tank contains air, water vapor, and iron filings. Solid iron oxide is formed very slowly.

(b) A tank contains ice and water vapor. Some of the ice sublimes, but no liquid water is formed.

(c) A tank contains liquid water on one side of a partition and steam at a lower pressure on the other side. The partition is ruptured.

(d) Air flows over very cold tubes and some of it is condensed, with the resulting liquid being slightly richer in nitrogen than the gas is.

3·3 What are the saturation temperatures corresponding to the following pressures for nitrogen: 500, 1000, and 2000 kPa?

3·4 What are the saturation temperatures corresponding to the following pressures for nitrogen: 100, 200, and 400 psia?

3·5 What is the saturation temperature for chlorine for a pressure of (*a*) 4000 kPa, (*b*) 600 psia?

3·6 Is it possible to have ice and water at 70 MPa and -10°C?

3·7 Is it possible to have ice and water at a pressure of 20 000 psia and a temperature of 5°F?

3·8 At what temperature would ice at a pressure of 100 MPa melt?

3·9 At what temperature would ice at a pressure of 10 000 psia melt?

3·10 Would it be possible to maintain ice at 55 MPa and -2°C?

3·11 Would it be possible to maintain ice at 8000 psia and 28°C?

3·12 It has been suggested that the liquid–vapor saturation curve of a pure substance can be represented by an equation of the form $P = P_0 e^{-\alpha/T}$, where T is absolute temperature and P_0 and α are constants. How well does such an equation fit the data for water?

3·13 What are some of the substances besides water that expand on freezing? In what applications is this characteristic important?

3·14 Sketch a phase (*PT*) diagram for water, and then sketch on this diagram a few lines of (*a*) constant volume, (*b*) constant enthalpy.

3·15 On a *Pv* diagram, how many states of a system can be represented by a single point on the triple-phase line?

3·16 Sketch a *Tv* diagram, and identify the various lines and regions on it for (*a*) H_2O, (*b*) CO_2. Show constant-pressure lines for pressures below the triple-phase value, between the triple phase and critical values, and above the critical value.

3·17 A three-phase mixture of H_2O consists of 40 percent solid, 50 percent liquid, and 10 percent vapor by mass. Explain how the point representing this mixture is located on a *uv* diagram and on a *Ph* diagram.

3·18 Sketch a curve of *T* versus *u* for constant-pressure heating of a substance which passes through solid, liquid, and vapor phases.

3·19 A rigid vessel with a volume of 0.182 m³ is divided into two equal volumes by a partition. Initially, 7 g of hydrogen is held on one side of the partition, and the other side of the tank is evacuated. After 6 h it is discovered that the side of the tank initially evacuated holds 0.010 g of hydrogen because the partition is slightly pervious. Determine the average number of molecules per second passing through the partition.

3·20 A vessel having a volume of 0.32 m³ was sealed containing slightly compressed air 4200 years ago and placed in a tomb in Egypt. If, since that time, 0.30 g of the air has leaked from the vessel, determine the average rate (molecules per second) at which air has left the vessel.

3·21 A container of nitrogen leaks at a rate of 0.010 oz in 30 years. What is the average number of molecules escaping per second?

3·22 Show that, for a frictionless constant-pressure process and only for such a process, $q = \int c_p\, dT$ for either a closed system or for a steady-flow system.

3·23 Critical temperature is sometimes defined as the temperature above which the liquid phase of a substance does not exist. Comment on this.

3·24 In the solution of part *d* of Example 3·1 the statement is made that the second term in $\int_1^2 [c_p\, dT + (\partial h/\partial P)_T\, dP]$ is zero because $P_2 = P_1$. Is it generally true that $\int_1^2 f(x)dx = 0$ if $x_2 = x_1$? Explain.

3·25 For each of the five states of water, fill in the blanks in the following table with the property values or with *M* or *I* (for meaningless or indeterminate):

State	(a)	(b)	(c)	(d)	(e)
P, kPa	150	270.1	2500		5000
T, °C		130	300	150	100
x, %	70				
v, m³/kg				1.9364	
h, kJ/kg					
u, kJ/kg					

3·26 For each of the five states of water, fill in the blanks in the following with the property values or with *M* or *I* (for meaningless or indeterminate):

State	(a)	(b)	(c)	(d)	(e)
P, psia	14.696				4.0
T, °F				60	200
x, %				100	
v, ft³/lbm	5.909	48.50	3.679		
h, B/lbm					
u, B/lbm		588.63	1171.9		

3·27 For each of the five states of water, fill in the blanks in the following table with the property values or with M or I (for meaningless or indeterminate):

State	(a)	(b)	(c)	(d)	(e)
P, kPa			200		5000
T, °C	500	200		20	
x, %		30			
v, m³/kg			0.00106		0.03963
h, kJ/kg				2538.5	
u, kJ/kg	3086.6				

3·28 For each of the five states of water, fill in the blanks in the following table with the property values or with M or I (for meaningless or indeterminate):

State	(a)	(b)	(c)	(d)	(e)
P, psia	500			50	
T, °F	50	170		281.03	35
x, %		75			
v, ft³/lbm			27.84		1179.2
h, B/lbm			1168.7		
u, B/lbm					

3·29 Potassium at 1 atm boils at approximately 757°C. The specific volumes of the saturated liquid and vapor are 0.0015 and 1.991 m³/kg, respectively. Compute the specific volume of a mixture of 70 percent quality.

3·30 A closed tank contains 0.50 m³ of dry saturated steam at a gage pressure of 899 kPa. (*a*) What is its temperature? (*b*) How many kilograms of steam does the tank contain? Barometric pressure equals 101 kPa.

3·31 A tank having a volume of 0.050 m³ contains 80 percent saturated vapor by volume and 20 percent saturated liquid water at a temperature of 25°C. If the liquid and vapor are agitated until thoroughly mixed, what would be the quality of the mixture?

3·32 Steam at 1000 kPa and 95 percent quality is formed from saturated feedwater at 70°C. What is the difference in enthalpy between the two states?

3·33 One kilogram of water at 80°C is converted into steam at 340°C under 1500 kPa pressure. How much is the enthalpy changed?

3·34 Steam at 2600 kPa leaves a boiler carrying 4 percent moisture. After it passes the superheater, the pressure is unchanged but the temperature has risen to 400°C. What is the difference in enthalpy between the two states? What is the change in specific volume?

3·35 Cesium at 100 psia has a saturation temperature of 1740°F. Specific volumes of saturated liquid and saturated vapor are 0.00991 and 1.597 ft^3/lbm, respectively. Determine the specific volume of a mixture of 60 percent quality.

3·36 A closed tank contains 20 cu ft of dry saturated steam at a gage pressure of 140.3 psig. (*a*) What is its temperature? (*b*) How many pounds of steam does the tank contain? Barometric pressure equals 14.7 psia.

3·37 A tank having a volume of 1 cu ft contains 80 percent saturated vapor by volume and 20 percent saturated liquid water at a temperature of 80°F. If the liquid and vapor were agitated until thoroughly mixed, what would be the quality of the mixture?

3·38 One pound of water at 180°F is converted into steam at 660°F under 200 psia pressure. How much is the enthalpy changed?

3·39 Steam at 380 psia leaves a boiler carrying 4 percent moisture. After it passes the superheater, the pressure is unchanged but the temperature has risen to 800°F. What is the difference in enthalpy between the two states? What is the change in specific volume?

3·40 Water at 60°F stands 6 cm deep in an open container with an inside diameter of 2 cm. During a 96-h period, the water level drops 2 mm as a result of evaporation. Determine the net average number of molecules leaving per second during the 4-day period.

3·41 Pure water at room temperature is stored in a bottle, with the water surface standing in the bottle neck which has a diameter of 20 mm. during a period of 24 h, water evaporates so that the liquid level drops by 0.10 mm. Determine the average rate (molecules/s) at which molecules leave the liquid phase.

3·42 For each of the five states of ammonia, fill in the blanks in the following table with the property values or with *M* or *I* (for meaningless or indeterminate):

State	(a)	(b)	(c)	(d)	(e)
P, kPa	226.45	429.44			
T, °C		0		40	
x, %				2	
v, m^3/kg			0.8533		1.3654
h, kJ/kg	1398.0				1495.7
u, kJ/kg			1473.9		

3·43 For each of the five states of ammonia, fill in the blanks in the following table with the property values or with *M* or *I* (for meaningless or indeterminate):

State	(a)	(b)	(c)	(d)	(e)
P, psia	50				50
T, °F		100	200		50
x, %	35				
v, ft^3/lbm			8.185	0.02299	
h, B/lbm		670.7			
u, B/lbm				−10.63	

3·44 For each of the five states of refrigerant 12 (CCl_2F_2), fill in the blanks in the following table with the property values or with *M* or *I* (for meaningless or indeterminate):

State	(a)	(b)	(c)	(d)	(e)
P, kPa		2788.50			80.7
T, °C	0		50	−50	
x, %					
v, m³/kg		0.005258			0.1175
h, kJ/kg				34.63	
u, kJ/kg	80				

3·45 For each of the five states of refrigerant 12 (CCl_2F_2), fill in the blanks in the following table with the property values or with *M* or *I* (for meaningless or indeterminate):

State	(a)	(b)	(c)	(d)	(e)
P, psia		349.00	60		
T, °F			80		180
x, %	50				
v, ft³/lbm		0.10330		2.4491	
h, B/lbm	42.8957			114.080	
u, B/lbm					52.913

3·46 For each of the five states of carbon dioxide, fill in the blanks in the following table with the property values or with *M* or *I* (for meaningless or indeterminate):

State	(a)	(b)	(c)	(d)	(e)
P, kPa		20 000	5000	3659	10 000
T, °C	22		−53	2	
x, %					
v, m³/kg	0.001324				0.01929
h, kJ/kg		403.03			
u, kJ/kg					

3·47 For each of the five states of carbon dioxide, fill in the blanks in the following table with the property values or with *M* or *I* (for meaningless or indeterminate):

State	(a)	(b)	(c)	(d)	(e)
P, psia	608.9				
T, °F				5	5
x, %	0.25		100		97
v, ft³/lbm		0.0153	0.2045	0.0610	
h, B/lbm		−22.1			
u, B/lbm					

3·48 For each of the five states of butane (C_4H_{10}), fill in the blanks in the following table with the property values or with *M* or *I* (for meaningless or indeterminate):

State	(a)	(b)	(c)	(d)	(e)
P, kPa		2956			50
T, °C			307	7	147
x, %		30			
v, m³/kg	0.06101		0.1159		
h, kJ/kg	917.64			566.25	
u, kJ/kg					

3·49 For each of the five states of sulfur dioxide, fill in the blanks in the following table with the property values or with *M* or *I* (for meaningless or indeterminate):

State	(a)	(b)	(c)	(d)	(e)
P, psia					30
T, °F	100		100	100	
x, %					
v, ft³/lbm		0.01114	0.2223		3.189
h, B/lbm	198.6	22.64		197.9	
u, B/lbm					

3·50 Estimate the ratio of the amount of energy required for a man to shave with an electric shaver to that required for him to shave with a razor and hot water.

3·51 Estimate the ratio of the amount of energy used for water heating for a person to take a bath to that for a person to take a shower.

3·52 Using data from Table A·1.5, plot a curve of c_p versus temperature for ice in the temperature range of 0 to -40°C. Assume that the specific heat of ice in this temperature range is independent of pressure.

3·53 Using data from Table A·1.4, plot a curve of c_p versus temperature for water in the temperature range of 0 to 260°C at 5 MPa.

3·54 A certain house requires 48×10^6 kJ for heating during an entire winter. It is proposed that water heated by the sun during the summer be stored to provide the heating. If the water storage temperature at the beginning of the winter heating season is 70°C and the water can be used for heating until its temperature has been reduced to 35°C, what length of insulated cylindrical tank having an inside diameter of 2.0 m would be required to store the water?

3·55 The radiant heat incident upon a solar collector is 500 W/m², and 60 percent of this amount of energy is absorbed by the water flowing through the collector. The collector has a face area of 1.30 m². If the water enters at 30°C and leaves at 75°C, what is the water flow rate?

3·56 In a certain location, the solar radiation incident upon a collector amounts to 10 000 kJ/m²·day. It is estimated that 45 percent of this energy can be transferred to water flowing through the collector. If a dwelling requires 0.35 m³ of water at 60°C each day, and water supplied to the collector is at 8°C, determine the solar collector area required.

3·57 Sketch *Pv* and *PT* diagrams beside each other and with pressure scales aligned with each other for a substance which contracts upon freezing. Starting at a state 1 which is a compressed liquid, sketch a constant pressure line, 1-2-3-4-5, extending to a superheated vapor state and a constant temperature line, 1-6-7-8-9, also extending to a superheated vapor state. States 2 and 6 are saturated liquid, 3 and 7 are liquid–vapor mixtures, and 4 and 8 are saturated vapor. Label the designated states on the two diagrams.

3·58 Steam at 260°C contained in a cylinder fitted with a piston initially has a quality of 70 percent and a volume of 0.02 m^3. The steam is expanded at constant temperature until it is dry and saturated (that is, x_2 = 100 percent). Determine (*a*) the mass of steam in the cylinder and (*b*) the work done during the expansion process.

3·59 Dry saturated steam at 7200 kPa is contained in a rigid tank having a volume of 0.060 m^3. Determine the amount of heat which must be added (in kJ/kg) to increase the pressure to 8300 kPa.

3·60 A rigid container with a volume of 0.170 m^3 is initially filled with steam at 480 kPa, 340°C. It is cooled to 90°C. (*a*) At what temperature does a phase change start to occur? (*b*) What is the final pressure? (*c*) What mass fraction of the water is liquid in the final state? (*d*) Calculate the work done during the cooling process.

3·61 Determine by means of the Keenan, Keyes, Hill, and Moore (Reference 3·5) compressed liquid table the enthalpy of water at 40 000 kPa, 370°C. Can this value be found in the superheated vapor table? Is water at 40 000 kPa, 370°C, a compressed liquid or a superheated vapor? Is water at 40 000 kPa, 380°C, a liquid or a vapor?

3·62 Determine by means of the Keenan, Keyes, Hill, and Moore (Reference 3·5) compressed liquid table the enthalpy of water at 5500 psia, 700°F. Can this value be found in the superheated vapor table? Is water at 5500 psia, 700°F, a compressed liquid or a superheated vapor? Is water at 5500 psia, 750°F, a liquid or a vapor?

3·63 Sketch a *PT* diagram for liquid water, showing constant-enthalpy lines of h = 2000, 2700, and 3000 kJ/kg.

3·64 Sketch a *PT* diagram for liquid water, showing constant-enthalpy lines of h = 400, 1100, and 1300 B/lbm.

3·65 Ice at 0°C on the outside of a tube is melted to liquid at 0°C by a transfer of heat from water flowing inside the tube. Water enters the tube at 40°C and leaves at 25°C. Latent heat of fusion of water at 0°C is 333.4 kJ/kg. (*a*) How much warm water is needed per kilogram of ice? (*b*) Show how the latent heat of fusion given above can be verified from data given in the steam tables.

3·66 Ice at 32°F on the outside of a tube is melted to liquid at 32°F by a transfer of heat from water flowing inside the tube. Water enters the tube at 100°F and leaves at 80°F. Latent heat of fusion of water at 32°F is 143.3 B/lbm. (*a*) How much warm water is needed per pound of ice? (*b*) Show how the latent heat of fusion given above can be verified from data given in Reference 3·5 (including Table 6).

3·67 Calculate $(\partial u/\partial v)_T$ for steam at 10 MPa, 400°C.

3·68 Calculate $(\partial u/\partial v)_T$ for steam at 100 psia, 350°F.

3·69 For steam at 2700 kPa, 250°C, determine from the steam tables approximate values of $(\partial h/\partial T)_P$, $(\partial h/\partial T)_v$, and $(\partial P/\partial T)_v$. Why are the values you obtain approximate? Are these point functions or path functions? Is dh/dT a point or path function?

3·70 For steam at 400 psia, 520°F, determine from the steam tables approximate values of

$(\partial h/\partial T)_P$, $(\partial h/\partial T)_v$, and $(\partial P/\partial T)_v$. Why are the values you obtain approximate? Are these point functions or path functions? Is dh/dT a point or path function?

3·71 A method of obtaining a low pressure in a sealed vessel is to fill the vessel with steam before sealing, seal, and then condense the steam. A vessel having a volume of 0.05 m^3 is filled with dry saturated steam at 101 kPa. The vessel is sealed and then chilled to 30°C. (*a*) What is the final pressure? (*b*) Sketch *Pv* and *PT* diagrams of the process, showing saturation lines on each diagram. (*c*) Calculate the amount of heat added to or taken from the steam.

3·72 The water level in a storage tank 9 ft high is regulated by a float-controlled valve. The water level is indicated by a simple gage glass 8 ft long on the side of the tank. When storing cold water, the gage glass shows that the water level is maintained 8 ft above the tank bottom by the float-controlled valve, but when hot (200°F) water is in the tank, the gage shows that the same control device maintains the water level 3 in. lower. How do you account for this?

3·73 Two kilograms of steam at 14 bar, 90 percent quality, is heated in a closed-system frictionless process until the temperature is 260°C. Calculate the amount of heat transferred if the process is at (*a*) constant pressure and (*b*) constant volume.

3·74 Steam at 1 bar, 150°C, is contained in a closed rigid vessel which has a volume of 0.03 m^3. How much heat must be removed in order to lower the steam pressure to 0.7 bar?

3·75 One-tenth kilogram of steam initially dry and saturated at 200 kPa is heated in a closed system to a final condition of 300 kPa, 200°C. Work done on the steam during the process amounts to 8.0 kJ/kg. Calculate the amount of heat transferred per kilogram of steam.

3·76 One-tenth kilogram of steam initially dry and saturated at 2 bar is heated in a closed system to a final condition of 3 bar, 200°C. Work done on the steam during the process amounts to 760 J. Calculate the amount of heat transferred per kilogram of steam.

3·77 In a closed system, steam initially at 460 kPa, 80 percent quality, expands to 300 kPa, 200°C. During the expansion, heat is added to the steam in the amount of 540 kJ/kg. Calculate the work.

3·78 One-half kilogram of dry saturated steam at 140 kPa is heated in a closed, rigid container until its temperature is 220°C. Determine the amount of heat added to the steam.

3·79 One-fifth kilogram of dry saturated steam at 1.50 bar is heated in a closed, rigid container until its temperature is 220°C. Determine the amount of heat added to the steam.

3·80 One-tenth kilogram of steam at 95 kPa, 80 percent quality, is contained in a rigid, thermally insulated vessel. A paddle wheel inside the vessel is turned by an external motor until the steam is at 140 kPa. Determine the amount of work done on the steam.

3·81 One-tenth kilogram of steam at 1 bar, 80 percent quality, is contained in a rigid, thermally insulated vessel. A paddle wheel inside the vessel is turned by an external motor until the steam is at 1.50 bar. Determine the amount of work done on the steam.

3·82 One-tenth kilogram of dry saturated steam at 120 kPa is contained in closed, rigid tank. Work in the amount of 2.1 kJ is done on the steam by a paddle wheel while heat is transferred to or from the steam. The final pressure of the steam is 180 kPa. Sketch a *Pv* diagram of the process, and calculate the heat transferred.

3·83 In a closed system, dry saturated steam at 690 kPa is heated in a constant-volume process until its pressure is 1200 kPa. It is then expanded adiabatically to 690 kPa, 370°C, and later cooled to the saturation temperature at constant pressure. Calculate the net work done. State any assumptions made.

3·84 Ninety grams of steam initially dry and saturated at 70 kPa (condition 1) is heated at constant pressure until its volume is 0.25 m^3 (condition 2). It is then heated at constant volume until it is at 95 kPa (condition 3). Calculate the total heat added per kilogram of steam between conditions 1 and 3.

3·85 (*a*) Solve Problem 3·84 for a frictionless steady-flow system (v_2 = 2.8 m^3/kg and process 2-3 is at constant specific volume). (*b*) Is the amount of work done the same in Problems 3·84 and 3·85?

3·86 Three pounds of steam at 200 psia, 90 percent quality, is heated in a closed-system frictionless process until the temperature is 500°F. Calculate the amount of heat transferred if the process is at (*a*) constant pressure and (*b*) constant volume.

3·87 Steam at 14.7 psia, 300°F, is contained in a closed, rigid vessel which has a volume of 1.0 cu ft. How much heat must be removed in order to lower the steam pressure to 10 psia?

3·88 One-tenth pound of steam initially dry and saturated at 30 psia is heated in a closed system to a final condition of 40 psia, 400°F. Work done on the steam during the process amounts to 560 ft·lbf. Calculate the amount of heat transferred per pound of steam.

3·89 In a closed system, steam initially at 67 psia, 80 percent quality, expands to 40 psia, 300°F. During the expansion, heat is added to the steam in the amount of 231.0 B/lbm. Calculate the work.

3·90 One-half pound of dry saturated steam at 20 psia is heated in a closed, rigid container until its temperature is 440°F. Determine the amount of heat added to the steam.

3·91 One-tenth pound of steam at 14.0 psia, 80 percent quality, is contained in a rigid, thermally insulated vessel. A paddle wheel inside the vessel is turned by an external motor until the steam is at 20 psia. Determine the amount of work done on the steam.

3·92 One-tenth pound of dry saturated steam at 18.0 psia is contained in a closed, rigid tank. Work in the amount of 2.0 B is done on the steam by a paddle wheel while heat is transferred to or from the steam. The final pressure of the steam is 27.0 psia. Sketch a Pv diagram of the process, and calculate the heat transferred.

3·93 In a closed system, dry saturated steam at 95 psia is heated in a constant-volume process until its pressure is 175 psia. It is then expanded adiabatically to 95 psia, 780°F, and later cooled to the saturation temperature at constant pressure. Calculate the net work done. State any assumptions made.

3·94 Two-tenths pound of steam initially dry and saturated at 10 psia (condition 1) is heated at constant pressure until its volume is 9.0 cu ft (condition 2). It is then heated at constant volume until it is at 14.0 psia (condition 3). Calculate the total heat added per pound of steam between conditions 1 and 3.

3·95 (*a*) Solve Problem 3·94 for a frictionless steady-flow system (v_2 = 44.99 cu ft/lbm and process 2-3 is at constant specific volume). (*b*) Is the amount of work done the same in Problems 3·94 and 3·95?

3·96 Steam enters a turbine with negligible velocity at 7000 kPa, 550°C, and leaves at 3 kPa, 5 percent moisture, with a velocity of 200 m/s. The flow rate is 25 000 kg/h and the power output is 7500 kW. The cross-sectional area of the exhaust opening is 1.51 m^2. Determine the heat loss (in kJ/kg) from the steam passing through the turbine.

3·97 Steam enters a turbine with negligible velocity at 70 bar, 540°C, and leaves at 0.1 bar, 10 percent moisture, with a velocity of 200 m/s. The flow rate is 23 000 kg/h, and the power output is 7000 kW. The cross-sectional area of the exhaust opening is 0.55 m^2. Determine the heat loss (in J/g) from the steam passing through the turbine.

3·98 Steam flows steadily through a turbine at a rate of 900 kg/h, entering at 700 kPa, 200°C, with negligible velocity and leaving at 7 kPa, 92 percent quality, through an opening of 0.031 m^2 cross-sectional area. The turbine shaft speed is 3000 rpm, and the power output is 125 kW. Calculate the heat transfer in kJ/kg.

3·99 Steam enters a turbine at 280 kPa, 150°C, with negligible velocity and is exhausted at 8 kPa with a velocity of 150 m/s. The flow rate is 18 000 kg/h, and the turbine power output is 1500 kW. The flow is adiabatic. Determine the quality (if wet) or temperature (if superheated) of the exhaust steam.

3·100 Steam enters a turbine at 2.50 bar, 150°C, with negligible velocity and is exhausted at 0.1 bar with a velocity of 150 m/s. The flow rate is 18 000 kg/h, and the turbine power output is 1500 kW. The flow is adiabatic. Determine the quality (if wet) or temperature (if superheated) of the exhaust steam.

3·101 Calculate the throttle temperature required for a steam turbine that is to develop 10 000 kW from a flow rate of 45 000 kg/h if the steam enters at 4100 kPa and leaves at 30 mm of mercury absolute containing 10 percent moisture. Assume that there is negligible heat transfer and negligible change in kinetic energy.

3·102 Steam enters a turbine at 2800 kPa, 310°C, and leaves at 60 mm of mercury (h = 2550 kJ/kg). The flow rate is 35 000 kg/h. If the inlet opening has a cross-sectional area of 0.02 m^2, what exhaust opening area is required in order to have the change in kinetic energy between inlet and exhaust not more than 1 kJ/kg?

3·103 Saturated liquid water at 14 bar is throttled at a steady rate of 140 kg/h to a pressure of 1 bar. What fraction of the water vaporizes during this throttling process? What is the density of the H_2O leaving the throttle valve?

3·104 Steam flows at a rate of 1.54 kg/s through a tube that has a constant cross-sectional area of 45 cm^2. The steam enters at 480 kPa, 90 percent quality, with a velocity of 120 m/s, and leaves at 4600 kPa, 420°C. Calculate the amount of heat added per kilogram of steam.

3·105 A nozzle discharges steam at 1.75 bar, 170°C, through an exit area of 0.01 m^2. The flow rate is measured as 3.30 kg/s. The inlet pressure is 3 bar. Assuming that the flow is adiabatic and that the kinetic energy of the steam at inlet is negligibly small, determine the inlet temperature.

3·106 Steam at 1.25 bar containing 60 percent moisture enters a condenser through a flow area of 0.014 m^2 with a velocity of 200 m/s. The condensate leaves at 1.25 bar, 40°C. Calculate the heat transfer in kJ/h.

3·107 Steam enters a turbine with negligible velocity at 1000 psia, 1000°F, and leaves at 1 psia, 10 percent moisture, with a velocity of 700 fps. The flow rate is 50 000 lbm/h, and the power output is 9300 hp. The cross-sectional area of the exhaust opening is 5.96 sq ft. Determine the heat loss from the steam passing through the turbine in B/lbm.

3·108 Steam flows steadily through a turbine at a rate of 2000 lbm/h, entering at 100 psia, 400°F, with negligible velocity and leaving at 1 psia, 90 percent quality, through an opening of 0.333 sq ft cross-sectional area. The turbine shaft speed is 7000 rpm, and the power output is 150 hp. Calculate the heat transfer in B/lbm.

3·109 Steam enters a turbine at 40 psia, 300°F, with negligible velocity and is exhausted at 1.0 psia with a velocity of 500 fps. The flow rate is 40 000 lbm/h, and the turbine power output is 2000 hp. The flow is adiabatic. Determine the quality (if wet) or temperature (if superheated) of the exhaust steam.

3·110 Calculate the throttle temperature required for a steam turbine which is to develop 10 000 kW from a flow rate of 95 000 lbm/h if the steam enters at 600 psia and leaves at 1.0 in. of mercury absolute containing 10 percent moisture. Assume that there is negligible heat transfer and negligible change in kinetic energy.

3·111 Steam enters a turbine at 400 psia, 600°F, and leaves at 2.0 in. of mercury (h = 1110 B/lbm). The flow rate is 80 000 lbm/h. If the inlet opening has a cross-sectional area of 0.2 sq ft, what exhaust opening area is required in order to have the change in kinetic energy between inlet and exhaust not more than 1 B/lbm?

3·112 One hundred cubic feet per minute of steam at 20 psia and 95 percent quality enters a compressor and is discharged at 100 psia and 500°F. If the amount of heat transfer is negligible, what is the power consumption of the machine?

3·113 Saturated liquid water at 200 psia is throttled at a steady rate of 300 lbm/h to a pressure of 14.7 psia. What fraction of the water vaporizes during this throttling process? What is the density of the H_2O leaving the throttle valve?

3·114 Steam flows at a rate of 3.59 lbm/s through a tube which has a constant cross-sectional area of 0.05 sq ft. The steam enters at 70 psia, 90 percent quality, with a velocity of 400 fps, and leaves at 65 psia, 800°F. Calculate the amount of heat added per pound of steam.

3·115 A nozzle discharges steam at 25 psia, 340°F, through an exit area of 1.0 sq in. The flow rate is measured as 0.728 lbm/s. Inlet pressure is 60 psia. Assuming that the flow is adiabatic and that the kinetic energy of the steam at inlet is negligibly small, determine the inlet temperature.

3·116 Steam at 16 psia containing 60 percent moisture enters a condenser through a flow area of 0.15 sq ft with a velocity of 700 fps. The condensate leaves at 16 psia, 100°F. Calculate the heat transfer in B/h.

3·117 A footnote in Sec. 1·13 states that the Celsius scale is very nearly the same as the original centigrade scale. The difference is that now 1 kelvin is defined as the fraction 1/273.16 of the thermodynamic temperature of the triple point of water, and temperature on the Celsius scale is defined as $T - 273.15$, where T is the temperature in kelvins. Explain with the aid of a PT diagram the difference between the ice point (neglecting the effect of the air dissolved in the water) and the triple point, thereby explaining why the two numerical values mentioned above are different. As more precise *measurements* of the triple point and the ice point are made, which temperatures such as 273.15 K, 273.16 K, 0°C, and 0.01°C of these points will change and which will remain unchanged?

3·118 Ammonia enters a refrigerator compressor at 100 kPa, 95 percent quality, is compressed adiabatically, and leaves at 1000 kPa, 100°C. Power delivered to the ammonia by the compressor is 22.0 kW. Determine the ammonia flow rate.

3·119 In a refrigeration system ammonia is throttled from 1500 kPa, 30°C, to 200 kPa. Determine the quality of the ammonia leaving the throttling valve.

3·120 Ammonia enters a condenser at 1400 kPa, 40°C, and leaves at the same pressure and 30°C. The flow rate is 1.0 kg/s. Determine the rate of heat transfer from the ammonia.

3·121 Ammonia enters a refrigeration system evaporator at 150 kPa, 20 percent quality, and leaves at 150 kPa, −20°C. The flow rate is 0.5 kg/s. Determine the amount of heat transfer in kJ/kg.

3·122 Refrigerant 12 is compressed at a rate of 1.20 kg/min from 100 kPa, 90 percent quality, to

800 kPa. Power input is 1.05 kW, and heat is removed from the refrigerant during compression at a rate of 6.10 kJ/min. Determine the discharge temperature.

3·123 Refrigerant 12 is compressed adiabatically from 125 kPa, 90 percent quality, to 1000 kPa, 60°C. Power delivered to the refrigerant is 22.0 kW. Determine the refrigerant flow rate.

3·124 Refrigerant 12 is throttled from 1400 kPa, 40°C, to 400 kPa. Determine the quality of the refrigerant leaving the throttling valve.

3·125 Refrigerant 12 enters a condenser at 1200 kPa, 60°C, and leaves as a liquid at the same pressure and 45°C. The flow rate is 0.5 kg/s. Determine the rate of heat transfer.

3·126 Refrigerant 12 enters a refrigeration system evaporator at 200 kPa, 15 percent quality, and leaves at 200 kPa, 0°C. The flow rate is 0.8 kg/s. Determine the amount of heat transfer in kJ/kg.

3·127 Carbon dioxide enters a heat exchanger at −40°C, 30 percent quality, and leaves as dry saturated vapor at the same temperature. Calculate (*a*) the heat added in kJ/kg and (*b*) the change in internal energy in kJ/kg. (See the list of references at the end of Chapter 3 for sources of data.)

3·128 Carbon dioxide at −23°C containing 10 percent moisture enters a condenser through a flow area of 40 cm^2 with a velocity of 135 m/s. The condensate leaves with a low velocity at 1790 kPa, −33°C. Calculate the heat transfer in kJ/h.

3·129 One-fifth pound of ammonia initially dry and saturated at 70 psia is heated in a closed system to a final condition of 100 psia, 150°F. Work done on the ammonia during the process amounts to 5000 ft·lbf. Calculate the amount of heat transferred per pound of ammonia.

3·130 Ammonia at 20 psia, 90 percent quality, enters a compressor at a rate of 12 lbm/min. Power input to the compressor is 46.7 hp, and heat is removed from the ammonia during compression at a rate of 330 B/min. Discharge pressure is 200 psia. Determine the discharge temperature.

3·131 Carbon dioxide enters a heat exchanger at −40°F, 30 percent quality, and leaves as dry saturated vapor at the same temperature. Calculate (*a*) the heat added in B/lbm and (*b*) the change in internal energy in B/lbm. (See the list of references at the end of Chapter 3 for sources of data.)

3·132 Carbon dioxide at −5°F containing 10 percent moisture enters a condenser through a flow area of 0.05 sq ft with a velocity of 400 fps. The condensate leaves with a low velocity at 281 psia, −15°F. Calculate the heat transfer in B/h.

3·133 An insulated tank with a volume of 6 m^3 is completely evacuated except for a sealed vial of water. The liquid water in the vial is initially at 100°C. The vial is then opened. Determine the final pressure in the tank and the fraction of the water that evaporates for a vial volume of (*a*) 0.0600 m^3, (*b*) 0.0060 m^3.

3·134 In a steam power plant, an open feedwater heater is a vessel into which water from various places in the plant flows to be mixed with steam so that water at near saturation temperature leaves to go to the steam generator. An open feedwater heater operating at 110 kPa receives the following measured water flows: 30 000 kg/h at 40°C, 2500 kg/h at 80°C, and 800 kg/h at 20°C. Also, saturated liquid at 200 kPa is throttled into the heater at a rate of 2000 kg/h. Steam enters the heater at 110 kPa containing 2 percent moisture. Water leaving the heater is at a temperature 2 degrees Celsius below the saturation temperature, and its measured flow rate is 39 000 kg/h. The heater is insulated, but there is a question as to whether the insulation is fully effective. What conclusion can you draw?

3·135 In a refrigeration system, refrigerant 12 at 800 kPa, 40°C, enters a condenser and liquid at a temperature 2 degrees Celsius lower than the saturation temperature leaves. The refrigerant flow rate is 680 kg/h. The cooling medium is water entering at 13°C. The water exit temperature is not to exceed 22°C, and it is estimated that heat transfer from the condenser to the surrounding atmosphere amounts to 4 percent of the energy transferred from the refrigerant. Determine the required cooling water flow rate.

3·136 A vessel initially contains 4.0 kg of water at 100 kPa, 20°C. How much heat must be added to boil the water until only one-fourth of the initial amount is still liquid, if saturated vapor leaves at a rate to maintain constant pressure in the vessel?

3·137 A vessel has a volume of 1.22 m^3. Initially, 80 percent of its volume is filled with saturated liquid ammonia at -20°C and the rest is filled with saturated ammonia vapor. Liquid leaves through a small valve while heat is added to maintain the temperature constant. The final mass of liquid in the vessel is one-third of the initial mass of liquid. Determine the amount of heat that must be added.

3·138 A well-insulated vessel has a volume of 1.22 m^3. Initially, 80 percent of its volume is filled with saturated liquid ammonia at -20°C and the rest is filled with saturated vapor. Vapor is allowed to escape through a small valve until the liquid remaining in the vessel occupies only 76 percent of the vessel volume. Determine the final state of the ammonia remaining in the vessel.

3·139 A well-insulated vessel has a volume of 1.22 m^3. Initially, 80 percent of its volume is filled with saturated liquid ammonia at -20°C and the rest is filled with saturated vapor. Vapor is allowed to escape through a small valve. Plot pressure in the vessel as a function of the fraction of the vessel filled with liquid.

3·140 An insulated tank with a volume of 0.80 m^3 is initially filled with liquid water at 20°C. Saturated steam at 1 atm is then bled into the vessel to heat the water while a pressure relief valve allows liquid to leave to maintain the pressure constant at 1 atm. Determine the amount of steam which must be added to raise the water temperature to 90°C.

3·141 An insulated tank with a volume of 0.8 m^3 contains dry saturated steam at 100 kPa. It is connected to a large line in which steam at 800 kPa, 200°C, is flowing. Determine the final temperature of the steam in the tank if the filling process is adiabatic.

3·142 Refrigerant 12 vapor initially saturated at 10°C is held in an insulated tank having a volume of 0.30 m^3. An electric heating element in the tank adds heat at a constant rate of 0.20 kJ/s. Refrigerant is bled from the tank to hold the pressure constant. Assuming that conditions are uniform throughout the tank at any instant, determine the time required for the refrigerant to reach a temperature of 70°C. (A stepwise solution is needed for an accurate result. Using a mean value of outlet stream properties gives an error of less than 3 percent.)

3·143 A vessel of 5.6 m^3 volume contains steam initially saturated at 800 kPa. Heat is added at a constant rate of 5 kJ/s while an automatic valve allows steam to leave at a constant rate of 0.02 kg/s. What is the state of the steam remaining in the tank 5 min after the initial condition? (Use a mean value for any property of the vapor leaving.)

3·144 A tank of 10 cu ft volume is half filled with liquid water, and the remainder is filled by vapor. Heat is added until one-half of the liquid is evaporated while an automatic valve lets saturated vapor escape at such a rate that the pressure is held constant at 500 psia. Determine the heat transfer.

3·145 Solve Problem 3·144 for a pressure of 3000 psia.

3·146 A vessel has a volume of 4.20 ft^3. Initially, 80 percent of its volume is filled with saturated liquid refrigerant 12 at 60°F and the rest is filled with saturated R-12 vapor. Vapor leaves through a small valve while heat is added to maintain the temperature constant. The final mass of liquid in the vessel is one-third of the initial mass of liquid. Determine the amount of heat that must be added.

3·147 A well-insulated vessel has a volume of 18.4 ft^3. Initially, 80 percent of its volume is filled with saturated liquid water at 240°F and the rest is filled with saturated vapor. Vapor is allowed to escape through a small valve until the liquid remaining in the vessel occupies only 76 percent of the vessel volume. Determine the final state of the water remaining in the vessel.

3·148 A well-insulated vessel has a volume of 18.4 ft^3. Initially, 80 percent of its volume is filled with saturated liquid water at 240°F and the rest is filled with saturated vapor. Vapor is allowed to escape through a small valve. Plot pressure in the vessel as a function of the fraction of the vessel filled with liquid.

3·149 An insulated tank with a volume of 45.0 ft^3 is initially filled with liquid water at 60°F. Saturated steam at 1 atm is then bled into the vessel to heat the water while a pressure relief valve allows liquid to leave to maintain the pressure constant at 1 atm. Determine the amount of steam which must be added to raise the water temperature to 180°F.

3·150 An insulated tank with a volume of 30.0 ft^3 contains dry saturated ammonia vapor at 100 psia. It is connected to a large line in which ammonia at 200 psia, 140°F, is flowing. Determine the final temperature of the ammonia in the tank if the filling process is adiabatic.

3·151 Steam initially saturated at 212°F is held in an insulated tank having a volume of 8.30 ft^3. An electric heating element in the tank adds heat at a constant rate of 0.50 kW. Steam is bled from the tank to hold the pressure constant. Assuming that conditions are uniform throughout the tank at any instant, determine the time required for the steam to reach a temperature of 300°F. (A stepwise solution is needed for an accurate result. Using a mean value of outlet stream properties gives only a small error.)

3·152 A vessel of 24.6 ft^3 volume contains refrigerant 12 vapor initially saturated at 78 psia. Heat is added at a constant rate of 5 kW while an automatic valve allows R-12 to leave at a constant rate of 0.05 lbm/s. What is the state of the R-12 remaining in the tank 5 min after the initial condition? (Use a mean value for any property of the vapor leaving.)

CHAPTER FOUR

Ideal Gases

An equation that expresses the relationship among pressure, specific volume, and temperature of a substance is called an *equation of state*. The $Pv T$ relationship for most substances is quite complex, so that accurate equations of state for wide ranges of pressure and temperature have been developed for only a few substances. Experience shows that gases, particularly those of low molar mass at low pressure and relatively high temperature, can often be represented by a very simple equation of state, $Pv = RT$, where R is a constant for each gas. The relation $Pv = RT$ is known as the ideal-gas equation of state. This chapter discusses the ideal-gas equation of state, some of its implications regarding properties other than P, v, and T, and its application to real gases.

4·1 The Ideal Gas

An *ideal gas* is defined as one for which the equation of state is

$$Pv = RT \tag{4·1}$$

where R is a different constant for each gas. R is called the *gas constant* for each gas and has units such as kJ/kg·K, ft·lbf/lbm·R, or B/lbm·R. R is given by

$$R = \frac{R_u}{M}$$

where R_u is the *universal gas constant,* which has the same value for all gases, and M is the molar mass.* Therefore, alternative forms of the ideal gas equation of state are

$$Pv = RT \qquad PV = NR_uT$$

$$PV = mRT \qquad PV = \frac{mR_uT}{M}$$

$$P = \rho RT \qquad PV = NMRT$$

where N is the number of moles of the gas and the other symbols are as previously defined.

*Recall that *molar mass* is frequently called *molecular weight.* This latter term is, of course, a misnomer for this quantity, which is independent of gravitational effects. However, long usage has firmly established the name *molecular weight,* so it is widely used.

TABLE 4·1 *R* Values for Several Gases

Gas	Molar Mass	Gas Constant, R	
		kJ/kg·K	ft·lbf/lbm·R
Acetylene, C_2H_2	26.04	0.319	59.3
Air	28.97	0.287	53.3
Ammonia, NH_3	17.03	0.488	90.7
Argon, Ar	39.94	0.208	38.7
Butane, C_4H_{10}	58.12	0.143	26.6
Carbon dioxide, CO_2	44.01	0.189	35.1
Carbon monoxide, CO	28.01	0.297	55.2
Ethane, C_2H_6	30.07	0.276	51.4
Ethene, C_2H_4	28.05	0.296	55.1
Helium, He	4.003	2.077	386.0
Hydrogen, H_2	2.016	4.124	766.0
Methane, CH_4	16.04	0.518	96.3
Nitrogen, N_2	28.01	0.297	55.2
Octane, C_8H_{18}	114.23	0.0728	13.5
Oxygen, O_2	32.00	0.260	48.3
Propane, C_3H_8	44.10	0.189	35.0
Steam, H_2O	18.02	0.461	85.7

The value of R_u expressed in various units is

$$R_u = 8.314 \text{ kJ/kmol·K} = 1.986 \text{ B/lbmol·R} = 1545 \text{ ft·lbf/lbmol·R}$$

R values for several gases are given in Table 4·1.

An ideal gas is a hypothetical substance because we define it simply as a substance that follows the equation of state $Pv = RT$. Our definition is not based on any assumption that such a substance actually exists. Historically, the concept of an ideal gas resulted from the work of Boyle, Charles, and Gay-Lussac, but you should remember that they measured properties of real gases. Although both their work and other later work indicate that for real gases at low pressure and relatively high temperature the equation of state is very nearly $Pv = RT$, our use of $Pv = RT$ for an ideal gas does not depend on the accuracy of any measurements: We *define* an ideal gas as one that follows $Pv = RT$.

A PvT surface for an ideal gas is shown in Fig. 4·1.

Occasionally the term *perfect gas* is used for what we here call an *ideal gas*.

4·2 Real Gases and the Ideal-Gas Equation of State

How closely does $Pv = RT$ represent the characteristics of real gases? This is an important question. It is convenient to use such a simple equation of state, but we must know how much accuracy we sacrifice for the convenience. First we give a qualitative answer: Usually $Pv = RT$ tends to be more accurate for real gases

1. As molar mass decreases
2. As pressure decreases
3. As temperature increases

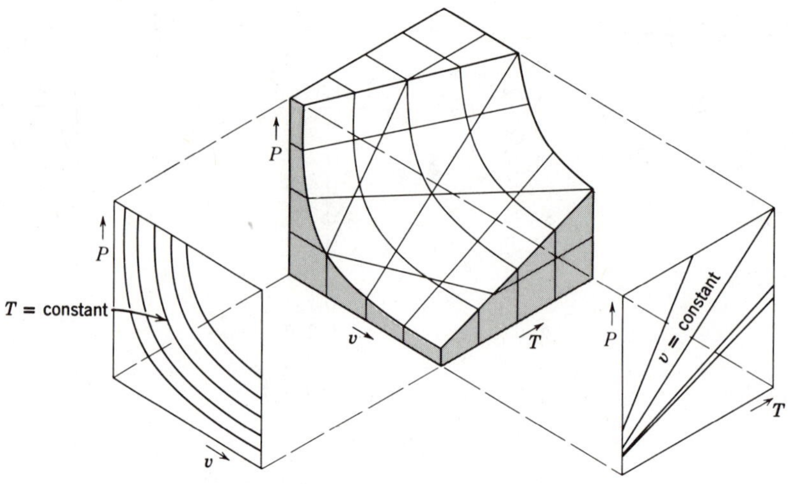

Figure 4·1. PvT surface for an ideal gas.

This is a statement of a general trend, but it cannot be relied on in all ranges of pressure and temperature. (Of course, if the pressure of a real gas is decreased at constant temperature to the extent that the molecules are so far apart that we cannot treat the gas as a continuous or homogeneous substance, then the state cannot be described in terms of properties such as P, v, and T, and a PvT relation is meaningless.) A complete quantitative answer to the question regarding the accuracy of $Pv = RT$ for real gases is given in Chapter 9, but at this point it is well to observe, for example, that, for air at room temperature, the error in $Pv = RT$ is less than 1 percent for pressures as high as 2.7 MPa (nearly 400 psia). For air at 1 atm, the error in $Pv = RT$ is less than 1 percent for temperatures as low as -130°C. For hydrogen at 1 atm, the error in $Pv = RT$ is less than 1 percent for temperatures as low as -220°C. For the present let it suffice to say that $Pv = RT$ represents the characteristics of many real gases accurately enough for many engineering calculations as long as the gases are at pressures well below their critical pressures and at temperatures well above their critical temperatures. In Chapter 9 we will investigate this point more thoroughly.

In using the ideal gas equation of state for real gases, be aware that uncertainty regarding the actual composition of gases involved in engineering calculations may limit accuracy. For example, the composition of atmospheric air varies with location and time even beyond expected changes in humidity (water vapor content) that are treated in Chapter 11. Commercially available gases which are nominally pure usually contain traces of other gases. Consequently, the number of significant digits you may find in tables for molar masses or gas constants may not be justified throughout many calculations.

Example 4·1. Compute the pressure of 3 kg of ethene at 20°C occupying a volume of 1.20 m^3.

Solution. Assume that ethene under these conditions behaves as an ideal gas. Then

$$P = \frac{mRT}{V} = \frac{mR_uT}{VM} = \frac{3[\text{kg}]8.314[\text{kJ/kmol·K}]293[\text{K}]}{1.20[\text{m}^3]28.05[\text{kg/kmol}]} = 217 \text{ kPa}$$

This pressure is slightly more than twice normal atmospheric pressure, so in view of the discussion in the paragraphs preceding this example, our assumption that $Pv = RT$ holds appears reasonable.

Example 4·2. Determine the pressure of hydrogen at 32°F which has a density of 0.00155 slug/cu ft.

Solution. Assume that the pressure is low enough so that the hydrogen follows the ideal-gas equation of state. Then

$$P = \frac{mRT}{V} = \rho RT = \frac{\rho R_u T}{M}$$

For consistency of units we must express either (1) the density ρ in lbm/cu ft and the molar mass M in lbm/lbmol or (2) ρ in slug/cu ft and M in slug/lbmol. (This is necessary because we will use a value of R_u involving lbmol, not slugmol.) We then have

$$\rho = 0.00155 \text{ slug/cu ft} = 0.00155(32.2) \text{ lbm/cu ft}$$

$$M = 2.016 \text{ lbm/lbmol} = \frac{2.016}{32.2} \text{ slug/lbmol}$$

Either way, substitution of numerical values gives

$$P = \frac{\rho R_u T}{M} = \frac{0.00155(32.2)1545(492)}{2.016} = 18\ 840 \text{ psfa} = 131 \text{ psia}$$

At 131 psia, 32°F, hydrogen does follow the ideal-gas equation of state so that our assumption is justified.

4·3 The Ideal-Gas Thermometer Scale

In Sec. 1·13 it was pointed out that different real gases used in constant-volume gas thermometers give results which agree with each other more and more closely as the pressure at any one temperature is reduced. Also, as the pressure of a real gas approaches zero, the gas behaves more and more like an ideal gas. Thus the gas thermometer scale defined by the extrapolation of real gas behavior to the condition of zero pressure is called the ideal-gas thermometer scale. The ideal-gas thermometer scale defined in this manner does not depend on the properties of any one gas but on the properties of gases in general. The defining equation then is

$$\frac{T_2}{T_1} = \left(\frac{P_2}{P_1}\right)_{v=\text{const},\ P_1\to 0} \qquad \text{or} \qquad \frac{T_2}{T_1} = \left(\frac{v_2}{v_1}\right)_{P=\text{const},\ P\to 0}$$

The ideal-gas thermometer scale can also be defined in terms of the behavior of an ideal gas without reference to a real-gas thermometer. Let an ideal gas be defined as a substance which behaves so that

$$\left(\frac{P_{T2}}{P_{T1}}\right)_{v=\text{const}} = \left(\frac{v_{T2}}{v_{T1}}\right)_{P=\text{const}} \tag{a}$$

where the subscripts $T1$ and $T2$ denote properties measured at temperatures T_1 and T_2. Notice that no scale of temperature nor any numerical value even for T_1 or T_2 is needed in order to define an ideal gas in this manner. Then we can *define* the ideal-gas thermometer scale by either

$$\frac{T_2}{T_1} \equiv \left(\frac{P_{T2}}{P_{T1}}\right)_{v=\text{const}} \qquad \text{or} \qquad \frac{T_2}{T_1} \equiv \left(\frac{v_{T2}}{v_{T1}}\right)_{P=\text{const}}$$

where the pressures or specific volumes must be those of a substance that obeys Eq. a. The temperature scale or thermometer scale so defined does not depend on the characteristics of any real substance but only on those of a hypothetical substance—an ideal gas. For purposes of practical thermometry, one must then search for a substance which behaves like the hypothetical ideal gas. Such a substance can be identified by its behavior in accordance with Eq. a.

The ideal-gas temperature scale depends either on the characteristics of gases in general as their pressure approaches zero or on the characteristics of a hypothetical ideal gas. At this point it still appears that the ideal-gas temperature scale, since it depends on physical properties of some substance or substances, is no better than one based on the expansion of mercury or on the change in electrical resistance of some material. It will be shown in Chapter 7, however, that the ideal-gas scale is equivalent to one that is entirely independent of physical properties.

4·4 Internal Energy and Enthalpy of Ideal Gases

From the first and second laws of thermodynamics it can be proved that, for any substance that follows the equation of state $Pv = RT$,

$$\left(\frac{\partial u}{\partial v}\right)_T = 0$$

(The proof is given in Example 9·1 which follows the introduction of the second law.) This means that the internal energy of an ideal gas is a function of temperature only, because, as long as T is constant, u is constant. This is an important fact. It is usually referred to as Joule's law because James Prescott Joule conducted a series of experiments that indicated at first that the internal energy of a *real* gas is a function of temperature only. Further work by Joule and others showed that only for an ideal gas is internal energy a function of temperature alone, but Joule's name is usually connected with this fact on account of his original experiments.

Since Joule's law can be proved from the first and second laws of thermodynamics, the following description is chiefly of historical interest. However, it does offer an example of valuable reasoning based on the first law.

If the internal energy of some substance is a function of temperature only, then a process with $q = 0$ and $w = 0$—since the first law would show $\Delta u = 0$—would be a process with $\Delta T = 0$. Thus the interdependence of u and T could be checked experimentally by means of an adiabatic system. However, perfect thermal insulation is not needed, because, if $\Delta T = 0$ and the surroundings are at the initial temperature of the system, there will be no heat transfer because there will be no temperature difference between the system and the surroundings. Thus, to check his hypothesis that $u = f(T)$, Joule submerged two tanks connected by a stopcock in a tank of water. One tank was filled with air at high pressure and the other was evacuated. The air, the tanks, and the surrounding water were allowed to come to the same temperature. Then the stopcock between the tanks was opened to let air pass from one tank to the other. There was of course no work done by the air. Joule observed no change in the temperature of the water bath. From this he concluded that no heat was transferred to or from the air, and hence—since there was also no work done—the internal energy of the air did not change. Since there was no heat exchange between the air and the water, the temperature of the air must have remained constant, even though the volume changed. Thus $(\partial u/\partial v)_T = 0$, and the internal energy of the gas is a function of temperature only.

If in an experiment like Joule's the temperature of the water changes, does this prove that the internal energy of the gas under study is not a function of temperature only? Yes, it does so conclusively. Suppose the water temperature rises. Then there must have been a heat transfer to the water from the gas, and the gas temperature must have risen above that of the water during the expansion process. Since heat was removed from the gas and no work was done, the internal energy of the gas must have decreased. This expansion process is shown as process 1-2 on the *uv* and *Tv* diagrams of Fig. 4·2. In order to restore the gas that finally fills both tanks to its initial temperature, heat must be removed. In this constant-volume process, shown as 2-3 in Fig. 4·2, no work is done as heat is removed, and so the internal energy of the gas decreases. Thus, when the gas has been restored to its initial temperature ($T_3 = T_1$), its internal energy is lower than its initial value ($u_3 < u_1$). Clearly, the internal energy of a gas which behaves in this manner is not a function of temperature only.

For real gases internal energy is not a function of temperature only, but Joule did not detect a temperature change of the water, because the mass (or, to state the matter completely, the product of mass and specific heat) of the water was so great compared to that of the air that the temperature rise of the water was too small to be detected by his instruments. Joule later ran other experiments with different equipment to show that for real gases internal energy is not a function of temperature alone. However, in the ranges of pressure and temperature where $Pv = RT$ is sufficiently accurate for real gases, Joule's law holds with the same order of accuracy.

From Joule's law it is easy to eastablish the relationship between u and T for ideal

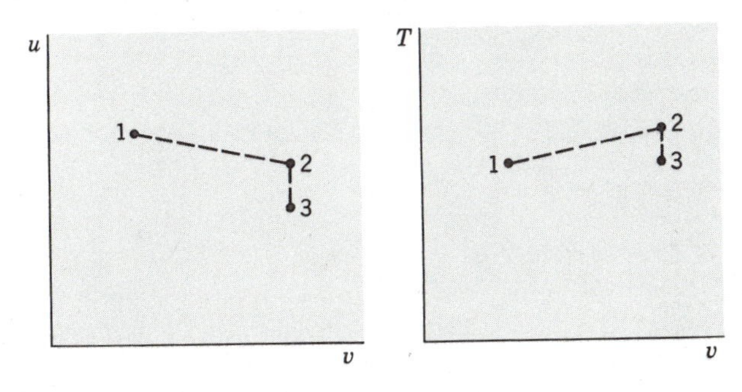

Figure 4·2. Analysis of Joule's experiment.

gases. For any pure substance in the absence of electric, magnetic, solid distortion, and surface tension effects, internal energy is usually a function of any two independent properties. (Recall that a *pure substance* is defined as a substance that is chemically homogeneous and fixed in chemical composition.) For a gas, T and v are certainly independent properties, so we may write

$$u = f(T, v)$$

Then

$$du = \left(\frac{\partial u}{\partial T}\right)_v dT + \left(\frac{\partial u}{\partial v}\right)_T dv$$

Recalling the definition of c_v, we can write

$$du = c_v\, dT + \left(\frac{\partial u}{\partial v}\right)_T dv$$

For an ideal gas, however, Joule's law states that $(\partial u/\partial v)_T = 0$, so that

$$du = c_v\, dT \tag{4·2}$$

for any process of an ideal gas. Since u is a function of T only, then c_v must be a function of T only. For any finite change of state,

$$\Delta u = \int c_v\, dT \tag{4·2}$$

To emphasize that this relationship holds for *all processes* of an ideal gas, let us look at the Pv diagram for an ideal gas shown in Fig. 4·3. In accordance with the equation of state $Pv = RT$, the constant-temperature lines are hyperbolas. In accordance with Joule's law, the constant-temperature lines are also lines of constant internal energy. Therefore, $u_a = u_b = u_c = u_d$, and

$$u_a - u_1 = u_b - u_1 = u_c - u_1 = u_d - u_1 = \int_1^2 c_v\, dT \tag{4·2}$$

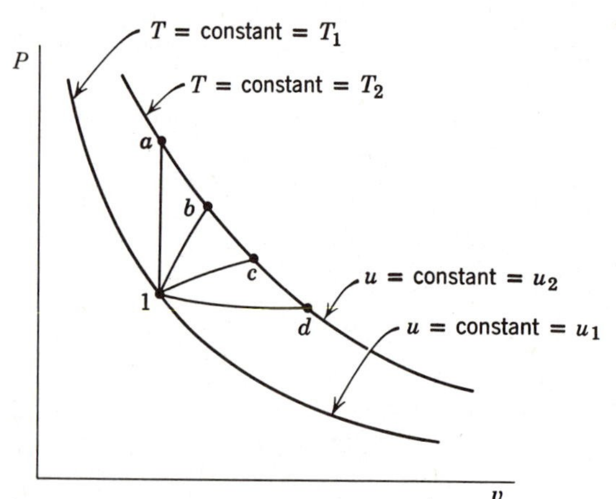

Figure 4·3. Joule's law: Pv diagram for an ideal gas.

No matter what path is followed by an ideal gas between a state 1 and any state at a temperature T_2, the change in internal energy per unit mass is given by $\int_1^2 c_v\, dT$. This point troubles many students because they see c_v, which we (unfortunately) call specific heat at constant volume, and then conclude *incorrectly* that Eq. 4·2 holds only for constant-volume processes. The relationship which is restricted to *constant-volume frictionless* processes is $\delta q = c_v\, dT$ (see Sec. 3·5). In Fig. 4·3, $q = \int_1^2 c_v\, dT$ only for a frictionless process along the constant volume path 1-*a*. For the other paths shown, q is different for each path. For the *change in internal energy of an ideal gas,* however, the relation $du = c_v\, dT$ applies to *all processes.*

As mentioned earlier, classical thermodynamics does not depend on any assumptions regarding the microscopic structure of matter. Molecules are not considered in classical thermodynamics because their existence has no bearing on the first or second law or on the deductions made from these laws. A consideration of the molecular picture of an ideal gas, however, helps in the understanding of Joule's law. Before discussing molecules, let us first consider a body that has a mass of 1 kg. At least at some locations on the surface of the earth this body weighs 9.8 N, and the work required to lift it 1 m is 9.8 N·m. Thus, as the body is moved 1 m farther from the earth, its potential energy increases by 9.8 N·m or 9.8 J. This same body when at an elevation of 100 km above the surface of the earth weighs only about 9.5 N, so only 9.5 N·m of work is required to move it 1 m farther from the earth. As the body moves farther from the earth, less work is required to move it through each meter of travel because the gravitational force continues to decrease. The condition is approached in which a negligible amount of work is required to move the body farther from the earth. Then the body can be moved through space with only negligible changes in its potential energy. (This does not mean that its potential energy is zero. If we call its potential energy zero when it is at the surface of the earth, its potential energy at any point above the earth is equal to the total amount of work done in lifting it to that point.) Now consider the molecules of an ideal gas. It is the sum of their potential and kinetic energies which comprises the internal energy of the gas. Like the body and the earth we have just discussed, the molecules of an ideal gas are so far from each other that the gravitational forces among them are negligibly small. The distances between molecules are so great that, even if the distances are halved or doubled, there is no appreciable change in the potential energy of the molecules. Thus the internal energy of an ideal gas is independent of the gas volume. When the gas temperature is changed, however, the velocity and the kinetic energy of the molecules change, so the internal energy of an ideal gas is a function of temperature. If the volume and pressure are changed at constant temperature, there is no change in internal energy because (1) molecular potential energy does not change as the spacing of the molecules varies, and (2) molecular kinetic energy does not change as long as the gas temperature is constant. In the light of such reasoning, an ideal gas is sometimes described microscopically as a gas in which there are no gravitational forces among the molecules.

The enthalpy of an ideal gas is also a function of temperature only. By definition, $h \equiv u + Pv$, and for an ideal gas this becomes $h = u + RT$. Both terms on the right-hand side of this equation depend solely on temperature, so h is a function of temperature only. In order to establish the relationship between h and T for an ideal gas, let us start by considering that for any pure substance (again in the absence of certain effects) as long as pressure and temperature are independent properties

$$h = f(P, T)$$

and
$$dh = \left(\frac{\partial h}{\partial T}\right)_P dT + \left(\frac{\partial h}{\partial P}\right)_T dP$$

Recalling the definition of c_p, we can write this as

$$dh = c_p\, dT + \left(\frac{\partial h}{\partial P}\right)_T dP$$

This equation holds for any process of any pure substance in the absence of electric, magnetic, solid distortion, and surface tension effects. For an ideal gas, however, h is a function of T only, so that $(\partial h/\partial P)_T = 0$, and the last expression for dh reduces to

$$dh = c_p\, dT \tag{4·3}$$

for any process of an ideal gas. Since h is a function of T only, then c_p must be a function of T only. For any finite change of state,

$$\Delta h = \int c_p\, dT \tag{4·3}$$

The quantities $\int c_v\, dT$ and $\int c_p\, dT$ which appear in Eqs. 4·2 and 4·3 can be evaluated by several methods. One method is to use c_vT and c_pT equations such as those presented in the next section and Table A·4 so that the integration can be performed analytically. A special case of this procedure is to use a mean value of c_v or c_p. A second method is to perform the integration graphically by means of c_vT or c_pT charts such as those presented in the next article. A third method is to use published tables of u and h versus temperature, a sample of which is given for air in Table A·5 of the appendix. Notice that this is a single-argument (T) table, because u, h, and the other properties tabulated there are functions of temperature only.

In closing this article, let us repeat for emphasis: *For an ideal gas*

$$du = c_v\, dT \tag{4·2}$$

and
$$dh = c_p\, dT \tag{4·3}$$

for all processes.

Example 4·3. Work in the amount of 85 kJ is required to compress 2 kg of a certain ideal gas (c_v = 1.25 kJ/kg·°C throughout the temperature range involved) in a closed system from an initial pressure of 90 kPa to a final pressure P_2. The temperature increases by 30°C during the compression. Compute the heat transfer.

Solution. For any process of an ideal gas, $\Delta u = \int c_v\, dT$. Since in this instance c_v is constant throughout the temperature range involved, we have

$$\Delta U = m\, \Delta u = m\int_1^2 c_v\, dT = mc_v(T_2 - T_1)$$

Application of the first law to this closed system gives

$$Q = \Delta U + W$$

Substituting the expression shown above for ΔU,

$$Q = mc_v(T_2 - T_1) + W = 2(1.25)30 - 85 = -10 \text{ kJ}$$

The minus sign indicates that heat was removed from the system.

4·5 Specific Heats of Ideal Gases

It has been pointed out that c_p and c_v of ideal gases are functions of temperature only. We should also observe that for ideal gases the difference $(c_p - c_v)$ is constant. This can be shown by first noting that the definition of enthalpy, $h \equiv u + Pv$, gives us for the case of an ideal gas

$$h = u + RT$$

and, since R is constant,

$$dh = du + R\,dT$$

For any process of an ideal gas $dh = c_p\,dT$ and $du = c_v\,dT$, so that

$$c_p\,dT = c_v\,dT + R\,dT$$

$$c_p = c_v + R$$

$$c_p - c_v = R \tag{4·4}$$

The ratio of the specific heats c_p/c_v is designated by k:

$$k \equiv \frac{c_p}{c_v}$$

Combining Eq. 4·4 and the definition of k gives

$$c_p = \frac{Rk}{k - 1} \qquad \text{and} \qquad c_v = \frac{R}{k - 1}$$

Specific heat data for real gases are usually obtained most accurately by spectroscopic methods that are not discussed here. As pressure decreases, the behavior of real gases approaches that of ideal gases, so the specific heats of real gases measured at very low pressures are called either the ideal-gas specific heats or the zero-pressure specific heats. The symbols c_{p0} and $c_{v\infty}$ are often used to indicate that the values concerned are those for zero pressure or for very large specific volume. When dealing only with ideal gases, we omit the subscripts 0 and ∞, but we must remember that the c_p and c_v values we use for an ideal gas are the c_{p0} and $c_{v\infty}$ values for the corresponding real gas.

Figures 4·4 and 4·5 present zero-pressure specific heat data for several gases. Before drawing hasty conclusions from these charts regarding the variation of specific heat with temperature, look carefully at the scales. Notice that a suppressed origin is used, making

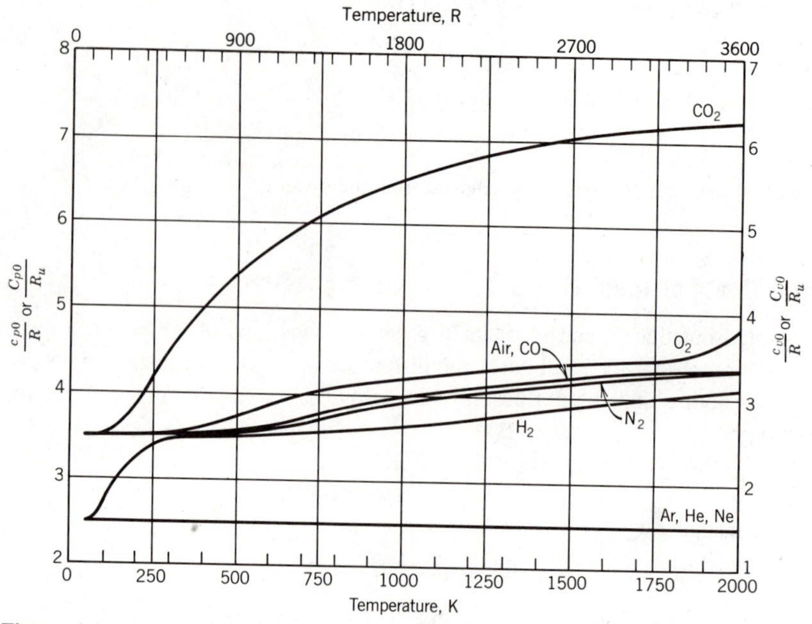

Figure 4·4. Zero-pressure molar specific heats of nine gases. Data from Reference 4·5.

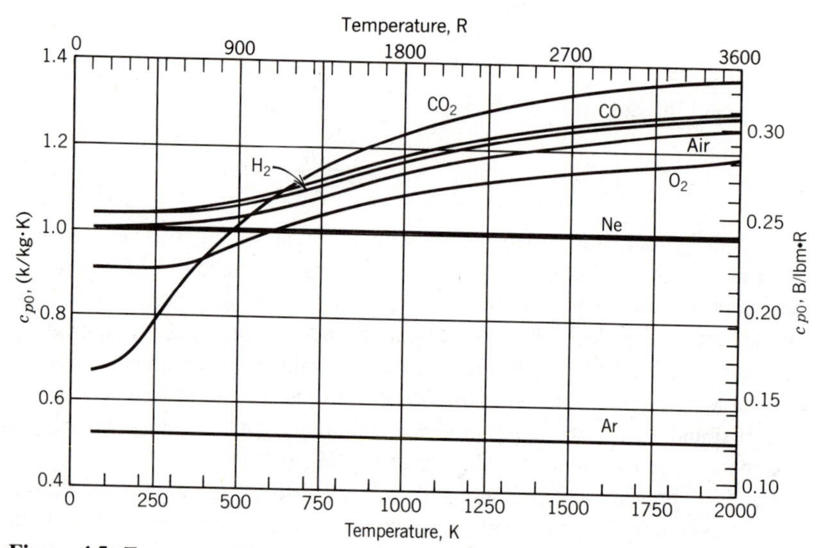

Figure 4·5. Zero-pressure specific heats of seven gases. Data from Reference 4·5.

the variation of specific heat at first glance appear larger than it is.* Specific heat data are also given in Table 4·2.

For ideal gases, changes in internal energy and in enthalpy are represented by areas on $c_v T$ and $c_p T$ diagrams, respectively. An area representing Δh between temperatures T_1 and T_2 is shown in Fig. 4·6. The mean specific heat for a given temperature interval can be obtained from specific-heat–temperature charts by the method illustrated in Fig. 4·7, if we recall from Sec. 3·5 that

$$\overline{c_p} \equiv \frac{\int_1^2 c_p \, dT}{T_2 - T_1} \qquad \text{and} \qquad \overline{c_v} \equiv \frac{\int_1^2 c_v \, dT}{T_2 - T_1}$$

For any process of an ideal gas

$$\Delta h = \overline{c_p} \, \Delta T \qquad \text{and} \qquad \Delta u = \overline{c_v} \, \Delta T$$

Remember that in general the value of $\overline{c_p}$ or $\overline{c_v}$ is different for each different temperature interval.

As mentioned in the preceding section, changes in enthalpy and internal energy of ideal gases can be determined as a function of temperature in various ways:

1. Use mean values of specific heats as given by Figs. 4·4 and 4·5 or by a table such as Table 4·2.

2. Integrate graphically by means of the $c_p T$ or $c_v T$ charts.

3. Use a $c_p T$ or $c_v T$ equation such as those given in Table A·4. Such equations are developed through curve-fitting of results obtained from experiment and the methods of statistical thermodynamics. Table A·4 gives for ten gases the coefficients in two temperature ranges for an equation of the form

$$\frac{c_p}{R} = \frac{C_p}{R_u} = a_1 + a_2 T + a_3 T^2 + a_4 T^3 + a_5 T^4$$

(The source referenced gives the coefficients for more than 400 substances.) Other forms of equations may be more accurate for some gases, but one advantage of using the same format for several gases is that the same computer program can be used for several gases with only the coefficients changed.

4. Use published tables such as those shown in abridged form in Tables A·5 and A·6. Such tables have been developed from equations similar to those given in Table A·4. (The use of gas tables for further purposes is discussed in Chapter 9 following the introduction of a property *entropy*. Entropy is defined on the basis of the second law of thermodynamics, which is introduced in the next chapter.)

*Graphs with suppressed origins make possible larger scales on a given-size sheet of paper so that more accurate reading is possible, but they do have the disadvantage of distortion, which is often misleading. Because they exaggerate the variation of one quantity with respect to another, they have been called "gee-whiz graphs." See *How to Lie With Statistics*, by Darrel Huff, W. W. Norton & Co., 1954, Chapter 5.

TABLE 4·2 (SI) Zero-Pressure Specific Heats for Various Gases in kJ/kg·°C

	Air			Carbon Dioxide, CO_2			Carbon Monoxide, CO			
Temp., °C	c_p	c_v	k	c_p	c_v	k	c_p	c_v	k	Temp., °C
0	1.004	0.717	1.401	0.817	0.628	1.301	1.040	0.743	1.400	0
50	1.006	0.719	1.399	0.869	0.680	1.278	1.041	0.745	1.399	50
100	1.010	0.723	1.397	0.916	0.727	1.260	1.045	0.748	1.397	100
150	1.016	0.729	1.394	0.958	0.769	1.246	1.050	0.754	1.394	150
200	1.024	0.737	1.389	0.995	0.806	1.234	1.074	0.777	1.382	200
400	1.068	0.781	1.367	1.113	0.924	1.204	1.106	0.809	1.367	400
600	1.115	0.828	1.347	1.195	1.006	1.188	1.157	0.860	1.345	600
800	1.154	0.867	1.331	1.253	1.064	1.178	1.199	0.902	1.329	800
1000	1.185	0.898	1.320	1.294	1.105	1.171	1.231	0.934	1.318	1000
1500	1.235	0.948	1.303	1.354	1.165	1.162	1.280	0.983	1.302	1500
2000	1.266	0.978	1.293	1.387	1.198	1.158	1.306	1.010	1.294	2000
2500	1.287	1.000	1.287	1.407	1.218	1.155	1.323	1.026	1.289	2500

	Hydrogen, H_2			Nitrogen, N_2			Oxygen, O_2			
	c_p	c_v	k	c_p	c_v	k	c_p	c_v	k	
0	14.19	10.07	1.410	1.039	0.742	1.400	0.915	0.655	1.397	0
50	14.37	10.25	1.402	1.040	0.743	1.399	0.922	0.663	1.392	50
100	14.46	10.33	1.399	1.042	0.745	1.398	0.934	0.674	1.386	100
150	14.49	10.37	1.398	1.046	0.749	1.396	0.948	0.688	1.378	150
200	14.51	10.38	1.397	1.052	0.755	1.393	0.963	0.703	1.369	200
400	14.59	10.46	1.394	1.091	0.795	1.374	1.024	0.764	1.340	400
600	14.79	10.66	1.387	1.139	0.842	1.352	1.069	0.809	1.321	600
800	15.12	10.99	1.375	1.181	0.885	1.335	1.100	0.840	1.309	800
1000	15.53	11.41	1.362	1.215	0.918	1.323	1.122	0.863	1.301	1000
1500	16.58	12.46	1.331	1.269	0.972	1.305	1.164	0.904	1.287	1500
2000	17.45	13.33	1.309	1.298	1.001	1.296	1.200	0.940	1.276	2000
2500	18.12	14.00	1.295	1.316	1.019	1.291	1.234	0.975	1.267	2500

Source: Tables of Thermal Properties of Gases, National Bureau of Standards Circular 564, Washington, D.C., 1955.

TABLE 4·2 (E) Zero-Pressure Specific Heats for Various Gases in B/lbm·°F

	Air			Carbon Dioxide, CO_2			Carbon Monoxide, CO			
Temp., °F	c_p	c_v	k	c_p	c_v	k	c_p	c_v	k	Temp., °F
32	0.240	0.171	1.401	0.195	0.150	1.300	0.248	0.177	1.400	32
100	0.240	0.172	1.400	0.205	0.160	1.283	0.249	0.178	1.399	100
200	0.241	0.173	1.397	0.217	0.172	1.262	0.249	0.179	1.397	200
300	0.243	0.174	1.394	0.229	0.184	1.246	0.251	0.180	1.394	300
500	0.248	0.179	1.383	0.247	0.202	1.223	0.256	0.185	1.384	500
750	0.255	0.187	1.368	0.266	0.221	1.204	0.264	0.193	1.367	750
1000	0.263	0.194	1.353	0.280	0.235	1.192	0.273	0.202	1.351	1000
1500	0.276	0.208	1.330	0.298	0.253	1.178	0.287	0.216	1.328	1500
2000	0.286	0.217	1.316	0.312	0.267	1.169	0.297	0.226	1.314	2000
3000	0.297	0.229	1.300	0.326	0.281	1.160	0.308	0.237	1.299	3000
4000	0.305	0.236	1.291	0.333	0.288	1.156	0.314	0.243	1.292	4000
5000				0.338	0.293	1.154	0.318	0.247	1.287	5000

	Hydrogen, H_2			Nitrogen, N_2			Oxygen, O_2			
	c_p	c_v	k	c_p	c_v	k	c_p	c_v	k	
32	3.391	2.406	1.409	0.248	0.177	1.400	0.219	0.156	1.397	32
100	3.426	2.441	1.404	0.248	0.178	1.399	0.220	0.158	1.394	100
200	3.451	2.466	1.399	0.249	0.178	1.398	0.223	0.161	1.387	200
300	3.461	2.476	1.398	0.250	0.179	1.396	0.226	0.164	1.378	300
500	3.469	2.484	1.397	0.254	0.183	1.388	0.235	0.173	1.360	500
750	3.483	2.498	1.394	0.261	0.190	1.374	0.245	0.182	1.340	750
1000	3.513	2.528	1.390	0.269	0.198	1.359	0.252	0.190	1.326	1000
1500	3.618	2.633	1.374	0.283	0.212	1.334	0.263	0.201	1.309	1500
2000	3.758	2.773	1.355	0.293	0.222	1.319	0.270	0.208	1.298	2000
3000	4.051	3.066	1.321	0.306	0.235	1.302	0.281	0.219	1.284	3000
4000	4.238	3.253	1.303	0.312	0.241	1.294	0.290	0.228	1.272	4000
5000	4.400	3.415	1.288	0.316	0.245	1.289	0.299	0.237	1.262	5000

Source: Tables of Thermal Properties of Gases, National Bureau of Standards Circular 564, Washington, D.C., 1955.

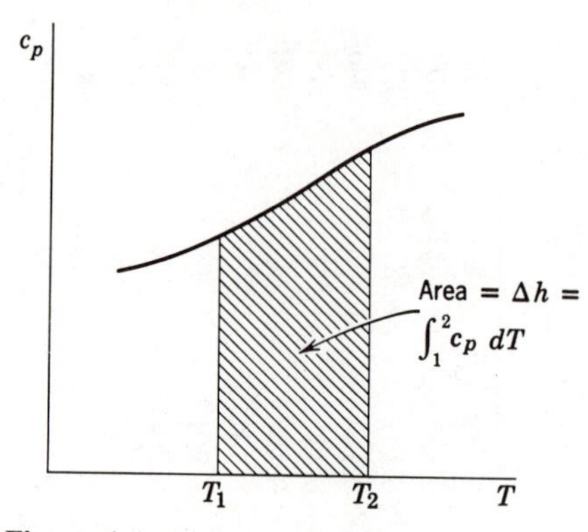

Figure 4·6. Δh for an ideal gas.

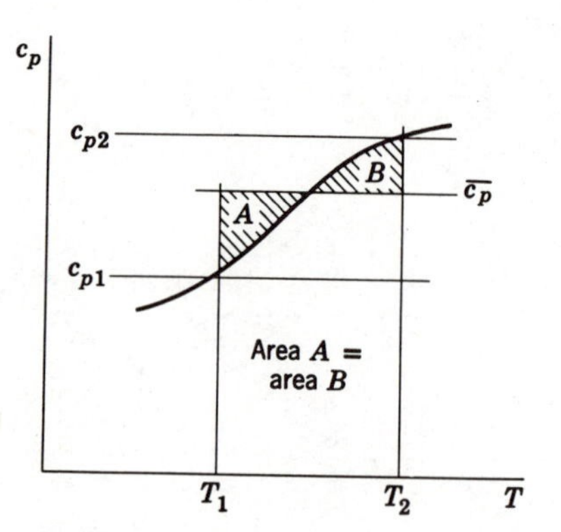

Figure 4·7. Determination of mean specific heat.

Any source of c_p values also provides c_v values, because

$$c_p - c_v = R \qquad \text{and} \qquad C_p - C_v = R_u \tag{4·4}$$

Since the difference $(c_p - c_v)$ is constant, the ratio c_p/c_v, which is designated by k, varies with temperature. The variation of k with temperature is shown in Fig. 4·8 for several gases.

Example 4·4. Compute the amount of heat required to raise the temperature of nitrogen from 25 to 600°C in a frictionless steady-flow process under a constant pressure of 1 atm.

Solution. For any steady-flow process

$$q = \Delta h + w + \Delta ke + \Delta pe$$

Since the flow is also frictionless, $w = -\int v\,dP - \Delta ke - \Delta pe$, and so

$$q = \Delta h - \int v\,dP - \cancel{\Delta ke} - \cancel{\Delta pe} + \cancel{\Delta ke} + \cancel{\Delta pe}$$

and, since the pressure is constant, this reduces to

$$q = \Delta h$$

Under the specified conditions of pressure and temperature, nitrogen may be treated as an ideal gas. For any process of an ideal gas, $\Delta h = \int c_p\,dT$. Using the c_pT relation from Table A·4 we have

$$q = \Delta h = \int_1^2 c_p dT = R\int_1^2 (c_p/R)dT = R\int_1^2 (a + bT + cT^2 + dT^3 + eT^4)dT$$

$$= R\left[aT + \frac{b}{2}T^2 + \frac{c}{3}T^3 + \frac{d}{4}T^4 + \frac{e}{5}T^5\right]_{T_1 = 298\text{ K}}^{T_2 = 873\text{ K}}$$

After this equation is entered into a computer or calculator, coefficient values are obtained

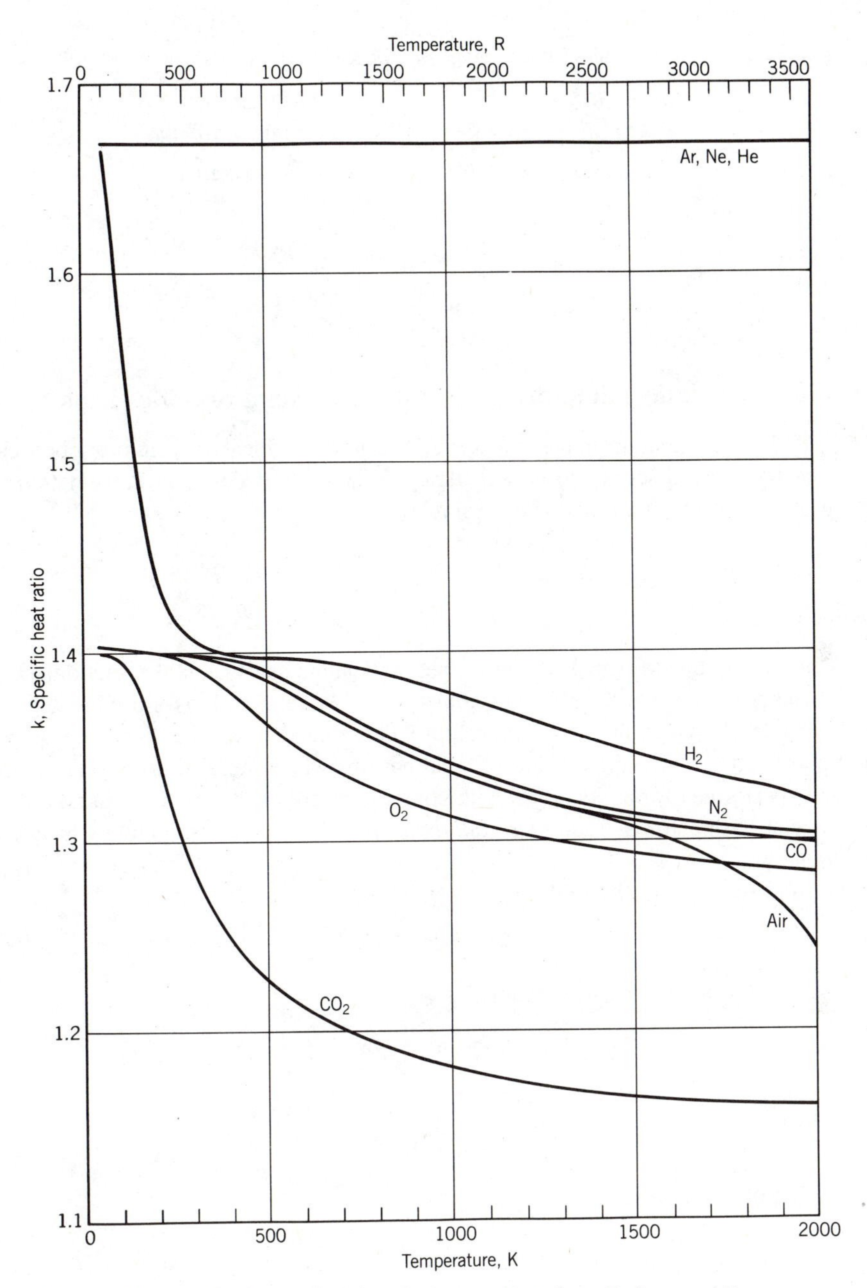

Figure 4·8. Specific-heat ratio k for nine gases. Data from Reference 4·5.

from Table A·4 (for the temperature range of 300 K to 1000 K) and entered as follows:

$$a = 3.675 \qquad d = 0.6322 \times 10^{-9}\ \text{K}^{-3}$$
$$b = 1.208 \times 10^{-3}\ \text{K}^{-1} \qquad e = 0.2258 \times 10^{-12}\ \text{K}^{-4}$$
$$c = 2.324 \times 10^{-6}\ \text{K}^{-2} \qquad R = 0.297\ \text{kJ/kg·K}$$

and the result is

$$q = 620\ \text{kJ/kg}$$

4·6 Special Relations for Ideal Gases with Constant Specific Heats

Within limited temperature ranges the specific heats of an ideal gas can often be treated as constant without a serious loss of accuracy. When this is done, several relationships among properties are simplified. For example,

$$\Delta u = \int c_v\, dT \qquad \text{and} \qquad \Delta h = \int c_p\, dT \qquad (4·2,3)$$

become

$$\Delta u = c_v\, \Delta T \qquad \text{and} \qquad \Delta h = c_p\, \Delta T$$

A process frequently encountered in thermodynamic analyses is the frictionless adiabatic. Therefore, it is worthwhile to establish for it a Pv relationship for use in integrating $\int P\, dv$ and $\int v\, dP$. Once the Pv relationship is determined, it can be combined with the equation of state to give the PT and vT relationships which are also useful. These steps are easily performed for an ideal gas with constant specific heats, so we will now derive the Pv relationship for a *frictionless adiabatic process of an ideal gas with constant specific heats.*

For a closed system the first law may be written

$$\delta q = du + \delta w$$

If the process is *adiabatic,* this becomes

$$0 = du + \delta w$$

and, if it is also *frictionless,* we have

$$0 = du + P\, dv$$

If the substance is an *ideal gas,*

$$0 = c_v\, dT + P\, dv$$

In order to get a Pv relation, we must eliminate T. We can do this by use of the ideal-gas equation of state which gives $T = Pv/R$ and $dT = (P\, dv + v\, dP)/R$. Thus we have

$$0 = \frac{c_v}{R}(P\, dv + v\, dP) + P\, dv$$

that we rearrange as follows:

$$0 = v\,dP + \left(\frac{R}{c_v} + 1\right)P\,dv = v\,dP + \left(\frac{R + c_v}{c_v}\right)P\,dv$$

$$0 = v\,dP + \frac{c_p}{c_v}P\,dv = v\,dP + kP\,dv$$

$$0 = \frac{dP}{P} + k\frac{dv}{v}$$

Up to this point we have not imposed the restriction of constant specific heats, but now we do so in order to integrate the last expression. Recall that the difference between c_p and c_v is constant (and equal to R), so that a constant k (which equals c_p/c_v) implies constant values of c_p and c_v. Integration gives

$$Pv^k = \text{constant} \tag{4·5a}$$

This equation holds for *frictionless adiabatic processes of ideal gases with constant specific heats*. For such a process between states 1 and 2

$$P_1 v_1^k = P_2 v_2^k \tag{4·5b}$$

Although the derivation above was based on a closed system, the Pv relation is the same if the frictionless adiabatic process occurs in an open system. Now we develop the corresponding PT and vT relationships. From the ideal-gas equation of state we have

$$\frac{P_1 v_1}{T_1} = \frac{P_2 v_2}{T_2}$$

which when combined with Eq. 4·5b gives

$$\frac{T_2}{T_1} = \left(\frac{P_2}{P_1}\right)^{(k-1)/k} = \left(\frac{v_1}{v_2}\right)^{k-1} \tag{4·5c}$$

Equations 4·5 are Pv, PT, and Tv relations that apply only to a particular type of process of an ideal gas with constant specific heats; they are not equations of state. They involve only two properties at a time. Since for a pure substance there are generally two *independent* properties, an equation of state must involve three properties.

For many, but not all, frictionless processes of an ideal gas the Pv relation is

$$Pv^n = \text{constant} \tag{4·6a}$$

where n is a constant for each process. Any process represented by such an equation is called a *polytropic* process and n is called the *polytropic exponent* for the process. n is *not* a property of the gas. Combining Eq. 4·6a with the equation of state gives

$$\frac{T_2}{T_1} = \left(\frac{P_2}{P_1}\right)^{(n-1)/n} = \left(\frac{v_1}{v_2}\right)^{n-1} \tag{4·6b}$$

Notice that the frictionless constant-pressure, constant-volume, constant-temperature, and adiabatic processes are special cases of polytropic processes for which the values of n are 0, ∞, 1, and k, respectively. (If the reason for $n = \infty$ in the constant-volume case is not apparent, consider that, when $n = \infty$, $P^{1/\infty}v$ = constant; hence v = constant.) Various polytropic processes are shown on a Pv diagram in Fig. 4·9. During a polytropic expansion for which $k > n > 1$, heat is added to the gas but its temperature decreases because the work done by the system exceeds the heat added.

Example 4·5. Two kilograms of argon in a closed system is compressed frictionlessly and adiabatically from 100 kPa, 20°C, to 400 kPa. Compute (*a*) the initial volume, (*b*) the final volume, (*c*) the final temperature, and (*d*) the work.

Solution. (***a***) The initial volume is computed from the ideal-gas equation of state.

$$V_1 = \frac{mRT_1}{P_1} = \frac{2(8.314)293}{39.94(100)} = 1.22 \text{ m}^3/\text{kg}$$

(***b***) For this frictionless adiabatic process,

$$V_2 = V_1 \left(\frac{P_1}{P_2}\right)^{1/k} = 1.22 \left(\frac{100}{400}\right)^{0.313/0.521} = 0.530 \text{ m}^3$$

(The value of k is $c_p/c_v = 0.521/0.313 = 1.66$, with specific heat values from Table 4·2. If we were to assume that the values of c_p and c_v as shown are accurate to two additional significant digits, the value of k would be 1.6645. To only one more significant digit than the data justify, the value would be 1.665. The reason for this comment follows.)

(***c***) The final temperature can be obtained as

$$T_2 = T_1 \left(\frac{P_2}{P_1}\right)^{(k-1)/k} = 293 \left(\frac{400}{100}\right)^{(1.665-1)/1.665} = 510 \text{ K} = 237°\text{C}$$

or, alternatively, from the ideal-gas equation of state

$$T_2 = T_1 \frac{P_2 V_2}{P_1 V_1} = 293 \left(\frac{400}{100}\right)\left(\frac{0.530}{1.22}\right) = 509 \text{ K}$$

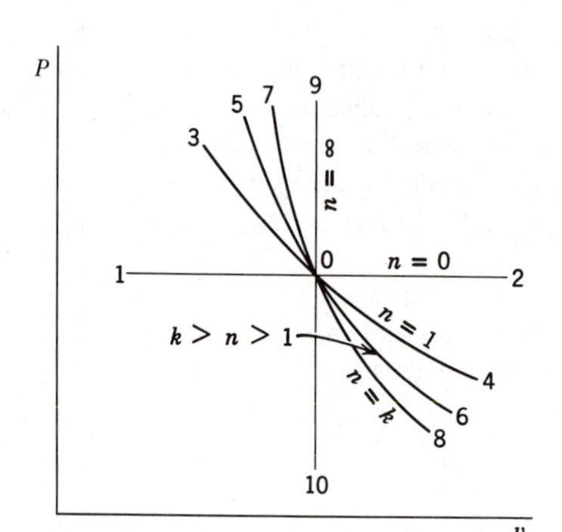

Figure 4·9. Polytropic processes.

(The difference between these two values of T_2 is the result of round-off, and you can readily confirm this by carrying through the calculations without rounding off to the number of significant digits apparently justified by the data. Notice that the accuracy of the result of part *b* would be overstated by writing it $V_2 = 0.5305\ m^3$, as would be allowed by a calculation assuming one additional significant digit in the data; yet the equation-of-state solution for T_2 using this value—or with no rounding off anywhere prior to the final step—would give 510 K. If $k = 1.66$ is used in the first calculation above for T_2, the result is 508 K, because of the accuracy lost in rounding off the value of k before subtracting 1 from it. A common, usually reliable rule in such calculations is to delay rounding off as far as possible, especially when differences between numbers are involved.)

(*d*) The work done on the gas is computed by means of the first law applied to the closed system and Joule's law for the evaluation of ΔU

$$W_{in} = \Delta U - Q = U_2 - U_1 - 0 = mc_v(T_2 - T_1)$$
$$= 2(0.313)(510 - 293) = 136\ \text{kJ}$$

The work can also be calculated as follows for this frictionless process in which $PV^k = C = P_1V_1^k = P_2V_2^k$:

$$W_{in} = -\int_1^2 P\,dV = -C\int_1^2 V^{-k}\,dV = \frac{-C}{1-k}[V^{1-k}]_1^2 = \frac{P_1V_1^k}{k-1}[V_2^{1-k} - V_1^{1-k}]$$

$$= \frac{P_1V_1}{k-1}\left[\left(\frac{V_2}{V_1}\right)^{1-k} - 1\right] = \frac{mRT_1}{k-1}\left[\frac{T_2}{T_1} - 1\right] = m\left(\frac{R}{k-1}\right)[T_2 - T_1]$$

$$= mc_v(T_2 - T_1)$$

Of course, the same result is obtained from $W = \int P\,dV$ as from the first law.

Example 4·6. In a closed system 0.070 kg of air is compressed polytropically from 100 kPa and a volume of 0.060 m³ to a volume of 0.030 m³ with a polytropic exponent of 1.3. Compute (*a*) the final temperature, (*b*) the work, (*c*) the change in internal energy, and (*d*) the heat transfer.

Solution. **(*a*)** The air is initially at a pressure of only 100 kPa, so unless its temperature is extremely low the ideal-gas equation of state can be used. Let us check the value of T_1 first of all by means of the ideal-gas equation of state.

$$T_1 = \frac{P_1V_1}{mR} = \frac{100(0.060)}{0.070(0.287)} = 299\ \text{K} = 26°\text{C}$$

At 100 kPa, 26°C, $Pv = RT$ is accurate for air, so we can proceed. The final temperature is

$$T_2 = T_1\left(\frac{V_1}{V_2}\right)^{n-1} = 299(2)^{0.3} = 368\ \text{K} = 95°\text{C}$$

(*b*) As a means of determining work we should first consider the first law, $W_{in} = \Delta U - Q$. Since we know T_1 and T_2, we can compute ΔU, but a method of computing Q is not apparent so it looks as though we cannot determine work by means of the first law. Let us then turn to the expression $W = \int P\,dV$ which holds for frictionless processes of a closed system. Since we know the PV relation for the process in question, we can evaluate $\int P\,dV$ as follows:

$$W_{in} = -\int P\,dV = \frac{-C}{1-n}[V^{1-n}]_1^2 = \frac{-P_1V_1^n}{1-n}[V_2^{1-n} - V_1^{1-n}]$$

$$= \frac{-P_1V_1}{1-n}\left[\left(\frac{V_2}{V_1}\right)^{1-n} - 1\right] = \frac{P_1V_1}{n-1}\left[\left(\frac{V_1}{V_2}\right)^{n-1} - 1\right]$$

$$= \frac{100(0.060)}{0.3}[(2)^{0.3} - 1] = 4.62\text{ kJ}$$

(*c*) Figure 4·4 shows that, in the temperature range of 26 to 95°C, c_v for air is nearly constant, so we have for the change in internal energy

$$U_2 - U_1 = m\int_1^2 c_v\,dT = mc_v(T_2 - T_1) = 0.070(0.717)(368 - 299)$$

$$= 3.46\text{ kJ}$$

(*d*) Applying the first law to this closed system,

$$Q = U_2 - U_1 - W_{in} = 3.46 - 4.62 = -1.16\text{ kJ}$$

The minus sign indicates that heat was removed from the air.

Example 4·7. A 60-cu-ft tank contains carbon monoxide initially at 20 psia, 100°F. A pump delivers carbon monoxide to the tank, raising the pressure from 20 to 40 psia. As the pump discharge pressure increases, so does the discharge temperature, according to $T = 80P^{0.25}$, where T is in Rankine and P is in psfa. Determine the heat transfer necessary to maintain the temperature of the carbon monoxide in the tank at 100°F.

Solution. First a sketch of the pump and tank is made, and then we define the following symbols:

Subscript i denotes properties of the CO initially in the tank.

Subscript f denotes properties of the CO finally in the tank.

Subscript 1 denotes properties of the CO entering the tank.

m, P, T, U, etc. = the mass, pressure, temperature, internal energy, etc. of CO in the tank at any instant.

m_1 = the mass of CO which has already entered the tank = $m - m_i$.

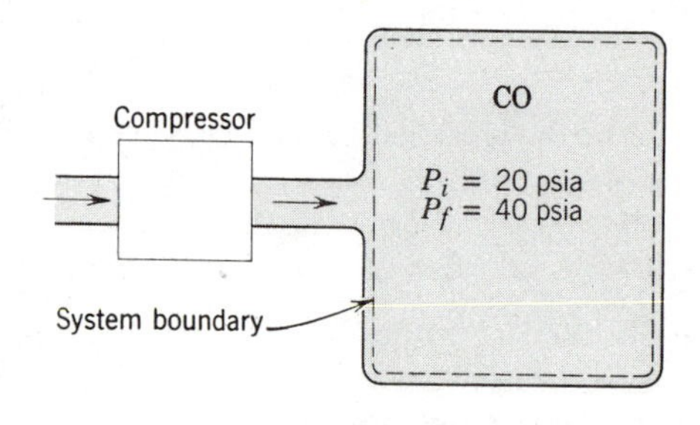

Example 4·7.

Consider the space within the tank as the system. This is an open system, and its boundary is the inner surface of the tank. There is no work done and no mass leaves the system; so an energy balance for this open system is

$$\begin{bmatrix}\text{heat added}\\ \text{to system}\end{bmatrix} + \begin{bmatrix}\text{stored energy and flow}\\ \text{work of entering mass}\end{bmatrix} = \begin{bmatrix}\text{increase in stored}\\ \text{energy of system}\end{bmatrix}$$

As an infinitesimal mass δm_1 enters the system, the energy balance formulation is

$$\delta Q + h_1\,\delta m_1 = dU$$
$$\delta Q = m\,du + u\,dm - h_1\,\delta m_1$$

Since the temperature of the CO in the tank is held constant, u is constant, and therefore $m\,du = 0$.

$$\delta Q = u\,dm - h_1\,\delta m_1$$

We must evaluate u and h. Let us see if the temperature range is narrow enough for specific heats to be constant. The pump-discharge (i.e., tank entrance) temperatures for the 20 psia and 40 psia pressure limits are

$$T_{1,\,20\text{ psia}} = 80P_1^{0.25} = 80[20(144)]^{0.25} = 586\text{ R} = 126°\text{F}$$
$$T_{1,\,40\text{ psia}} = 80P_1^{0.25} = 80[40(144)]^{0.25} = 697\text{ R} = 237°\text{F}$$

Therefore, the temperature range we are concerned with is that of 100 to 237°F. In this range we see from Fig. 4·4 or 4·5 that we can treat the specific heats as constant. (We will use $c_p = 0.250$ B/lbm·R and $c_v = 0.179$ B/lbm·R.) Thus we have $u - u_0 = c_v(T - T_0)$ and $h - h_0 = c_p(T - T_0)$. For this problem, let $u_0 = 0$ and $h_0 = 0$ when $T_0 = 0$, so that we have $u = c_vT$ and $h = c_pT$. The energy balance now becomes

$$\delta Q = c_vT\,dm - c_pT_1\delta m_1$$
$$= c_vT\,dm - c_p(80)P_1^{0.25}\,\delta m_1$$

But

$$P_1 = P = \frac{mRT}{V}$$

and

$$\delta m_1 = d(m - m_i) = dm = d\left(\frac{PV}{RT}\right) = \frac{V}{RT}\,dP$$

Making these two substitutions in the energy balance above

$$\delta Q = c_v\frac{V}{R}\,dP - 80c_p\frac{V}{RT}P^{0.25}\,dP$$

$$Q = \frac{V}{R}\left\{c_v\int_i^f dP - \frac{80c_p}{T}\int_i^f P^{0.25}\,dP\right\}$$

$$= \frac{V}{R}\left\{c_v(P_f - P_i) - \frac{80c_p}{1.25T}(P_f^{1.25} - P_i^{1.25})\right\}$$

$$= \frac{60}{55.2}\left\{0.179(144)(40 - 20) - \frac{80(0.25)}{1.25(560)}(144)^{1.25}[(40)^{1.25} - (20)^{1.25}]\right\}$$

$$= -343\text{ B}$$

The minus sign indicates that heat must be removed from the CO.

4·7 The Ideal-Gas Equation of State and Kinetic Theory

The ideal-gas equation of state can be derived from the kinetic theory of gases. Such a derivation has no bearing on classical thermodynamics, however, because classical thermodynamics neither depends on nor gives us any information on the microscopic structure of matter. This article and the next are included to give a glimpse of the kinetic theory as it explains ideal-gas behavior for two reasons. One is that most engineering students like to have a physical picture of the structure of gases even if it is unnecessary in the logical development of thermodynamics. The other reason is that even a superficial knowledge of kinetic theory helps one to understand why the behavior of real gases differs from that of ideal gases. The following presentation is not rigorous, but it gives an idea of the kinetic theory approach to the behavior of ideal gases.

Let us consider a monatomic gas and make the following assumptions:

1. The gas consists of many, many molecules. In a pure gas, all of the molecules are alike.
2. The molecules are small compared to the average distance between them. That is, the volume occupied by the molecules themselves is a small fraction of the volume of the container which the gas fills.
3. The molecules move about in all directions; all directions of motion are equally probable.
4. The molecules exert no force on each other except when they collide.
5. The laws of macroscopic mechanics (Newton's laws) apply to individual molecules.
6. Collisions of molecules with each other and with the walls of the container are perfectly elastic.

According to the kinetic theory, the pressure exerted by an ideal gas on the walls of a container results from the action of a great number of molecules striking and rebounding from the walls. In order to develop an expression for the pressure, we will consider first the behavior of a single gas molecule. Let this molecule be one of many identical molecules that comprise an ideal gas held in a container of dimensions A, B, and C as shown in Fig. 4·10. The velocity V of the molecule has components V_x, V_y, and V_z as shown in Fig. 4·10. When the molecule with this velocity strikes the container wall BC, it rebounds without change in V_y and V_z but with its x component of velocity changed from V_x to $-V_x$. The x component of its momentum changes from $m'V_x$ to $-m'V_x$, where m' is the mass of the molecule. The magnitude of the change in the x component of momentum is $2m'V_x$. The molecule travels the distance A in the time A/V_x if V_x is constant; so it could cross the container in the x direction and return in a time $2A/V_x$. The time between successive collisions with the wall BC is $2A/V_x$, and the number of collisions with wall BC per unit time is $V_x/2A$. The change in x momentum per unit time is the product of the change per collision and the number of collisions per unit time, $2m'V_x(V_x/2A) = m'V_x^2/A$. Thus the force on wall BC is

$$F'_x = \frac{d}{dt}(m'V_x) = \frac{m'V_x^2}{A}$$

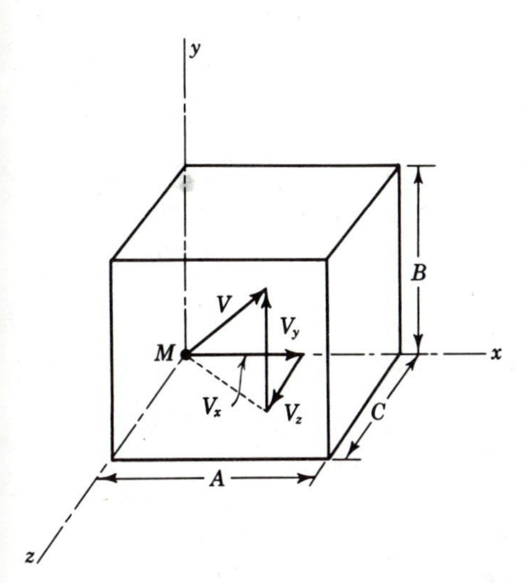

Figure 4·10. Molecular velocity components.

This is the force of just one molecule. The total force on wall BC is the sum of the forces of individual molecules,

$$F_x = \sum \frac{m' V_x^2}{A}$$

All the molecules have the same mass m', and A is constant; so

$$F_x = \frac{m'}{A} \sum V_x^2$$

Define the average V^2 for η molecules as

$$\overline{V_x^2} \equiv \frac{(V_{x1}^2 + V_{x2}^2 + V_{x3}^2 + \cdots)}{\eta} \tag{a}$$

Then

$$F_x = \frac{m' \eta \overline{V_x^2}}{A}$$

The pressure on wall BC is

$$P_x = \frac{F_x}{\text{area}} = \frac{m' \eta \overline{V_x^2}}{A(BC)} = \frac{m' \eta \overline{V_x^2}}{(\text{vol})} \tag{b}$$

(''vol'' is used for volume to avoid confusion between symbols for *volume* and for *velocity*.) If it is assumed that the pressure is the same in all directions and that all directions of velocity are equally probable,

$$P = \frac{m' \eta \overline{V_x^2}}{(\text{vol})} = \frac{m' \eta \overline{V_y^2}}{(\text{vol})} = \frac{m' \eta \overline{V_z^2}}{(\text{vol})} \tag{c}$$

and therefore

$$\overline{V_x^2} = \overline{V_y^2} = \overline{V_z^2} \tag{d}$$

Also, since V_x, V_y, and V_z are components of V,

$$V^2 = V_x^2 + V_y^2 + V_z^2$$

and it can be shown that

$$\overline{V^2} = \overline{V_x^2} + \overline{V_y^2} + \overline{V_z^2} \tag{e}$$

where $\overline{V^2}$ is the average of the squared velocity magnitudes for all molecules and the right-hand terms are defined as $\overline{V_x^2}$ is defined in Eq. a.

Combining Eqs. d and e shows that

$$\overline{V^2} = 3\overline{V_x^2} = 3\overline{V_y^2} = 3\overline{V_z^2}$$

so that Eq. c for the pressure becomes

$$P = \frac{m'\eta\overline{V^2}}{3(\text{vol})}$$

$$P(\text{vol}) = \tfrac{1}{3}\,\eta m'\overline{V^2} \tag{f}$$

If we assume the temperature of an ideal gas to be proportional to the kinetic energy of its molecules, we have

$$T = D\,\frac{m'\overline{V^2}}{2} \tag{g}$$

where D is the factor of proportionality. Combining Eqs. f and g gives

$$P(\text{vol}) = \frac{2}{3D}\,\eta T = \left(\frac{2\eta}{3DN}\right) NT \tag{h}$$

where η/N is Avogadro's number, the number of molecules per mole. In Eq. h, $2\eta/3DN$ is a constant. If we call it R_u, we have

$$P(\text{vol}) = NR_uT$$

which we recognize as the ideal-gas equation of state.

Following Eq. f, we assumed that temperature is proportional to molecular kinetic energy and showed that the ideal-gas equation of state results. Conversely, we could have compared Eq. f with the ideal-gas equation of state and showed that temperature is proportional to molecular kinetic energy because

$$P(\text{vol}) = \tfrac{1}{3}\,\eta m'\overline{V^2} \tag{f}$$

and

$$P(\text{vol}) = NR_uT$$

can be combined to give

$$T = \frac{1}{3}\frac{\eta}{NR_u}\,m'\overline{V^2} = \frac{2}{3}\frac{\eta}{NR_u}\frac{m'\overline{V^2}}{2} = \text{constant}\,\frac{m'\overline{V^2}}{2}$$

Using the important physical constant $\mathbf{k}$ which is defined as $\mathbf{k} \equiv NR_u/\eta$ and called

Boltzmann's constant, we can write the last equation in a form that is used in the next section:

$$\frac{3}{2}\,\mathbf{k}T = \frac{m'\overline{V^2}}{2} \tag{g}$$

Notice that the average kinetic energy is the same at a given temperature for molecules of any gas. Thus heavier gas molecules have lower mean speeds than lighter molecules at the same temperature.

4·8 Specific Heats of Ideal Gases and Kinetic Theory

The internal energy of an ideal gas is the sum of molecular kinetic and potential energies. In Sec. 4·4 we saw that the molecular potential energy of an ideal gas is constant. Therefore, it can be assigned the value of zero, so that in a monatomic ideal gas where the only molecular motion is translatory we can write

$$U = NMu = \frac{\eta m'\overline{V^2}}{2}$$

The product of molar mass and specific internal energy Mu is the internal energy per mole of gas. Using equation g of the preceding section, we can change this to

$$Mu = \frac{\eta}{N}\frac{3}{2}\,\mathbf{k}T = \frac{3}{2}R_uT$$

Then the molar specific heat at constant volume is

$$C_v = \left(\frac{\partial Mu}{\partial T}\right)_v = \frac{3}{2}R_u$$

The molar specific heat at constant pressure is then

$$C_p = C_v + R_u = \tfrac{3}{2}R_u + R_u = \tfrac{5}{2}R_u$$

and the ratio of specific heats for the monatomic ideal gas is

$$k = \frac{c_p}{c_v} = \frac{C_p}{C_v} = \frac{\frac{5}{2}}{\frac{3}{2}} = \frac{5}{3}$$

In order to extend this discussion to diatomic molecules, we must introduce the principle of equipartition of energy and the concept of degrees of freedom. By the number of degrees of freedom is meant the number of independent quantities that must be specified to determine the energy of a molecule. For example, a monatomic molecule constrained to move only in the x direction has one degree of freedom because if V_x is specified then the energy can be calculated as $m'V_x^2/2$. If the molecule is free to move in any direction, then V_x, V_y, and V_z must be specified, and the molecule has three degrees of freedom. If the moment of inertia of the molecule about the x axis is not zero and the molecule rotates about this axis, then there is rotational kinetic energy and the angular velocity must be specified, adding another degree of freedom. Rotation about other axes adds other degrees of freedom. An important principle in kinetic theory developments is the equipartition of energy principle that states that the total energy of molecules is divided equally among their degrees of freedom. Thus, since the kinetic energy per mole of monatomic molecules that involve translation only (three degrees of freedom) is $3R_uT/2$, the energy per degree of freedom is $\frac{1}{2}R_uT$. The equipartition of energy principle tells us that each additional degree of freedom involves additional energy in the amount $\frac{1}{2}R_uT$ per mole. For f degrees of freedom the internal energy is therefore given by

$$Mu = \frac{f}{2} R_u T$$

Thus

$$C_v = \left(\frac{\partial Mu}{\partial T}\right)_v = \frac{f}{2} R_u$$

$$C_p = C_v + R_u = \frac{f}{2} R_u + R_u = \frac{f+2}{2} R_u$$

and

$$k = \frac{C_p}{C_v} = \frac{f+2}{f}$$

Consider now a diatomic molecule comprised of two point masses. In addition to translatory kinetic energy, the diatomic molecule may possess rotational kinetic energy by virtue of rotation about the two mutually perpendicular axes that are perpendicular to the line joining the two atoms. Energy of rotation about the axis connecting the atoms is zero because the moment of inertia about this axis is zero for point particles. The atoms may also vibrate along the line connecting them. This involves two additional degrees of freedom because both potential and kinetic energy are involved. (The atoms within a molecule exert forces on each other even though the spacing between molecules is so great that intermolecular forces are negligible.) Both position and velocity of the atoms must be specified in order to determine the vibrational energy. Thus a diatomic molecule may have seven degrees of freedom: three associated with translation, two with rotation, and two with vibration. However, vibration occurs only at higher temperatures, so that at normal room temperature, for example, a diatomic gas usually has only five degrees of freedom. At very low temperatures rotation ceases; so $f = 3$ for a diatomic gas. Thus for a diatomic gas the kinetic theory as we have discussed it leads us to expect

At low temperatures:	$C_v = \frac{3}{2} R_u$	$C_p = \frac{5}{2} R_u$	$k = \frac{5}{3}$
At intermediate temperatures:	$C_v = \frac{5}{2} R_u$	$C_p = \frac{7}{2} R_u$	$k = \frac{7}{5}$
At high temperatures:	$C_v = \frac{7}{2} R_u$	$C_p = \frac{9}{2} R_u$	$k = \frac{9}{7}$

Such a variation of C_v is shown in Fig. 4·11. The reason for the gradual change shown in Fig. 4·11 is that, as the temperature increases, not all the molecules begin to rotate or

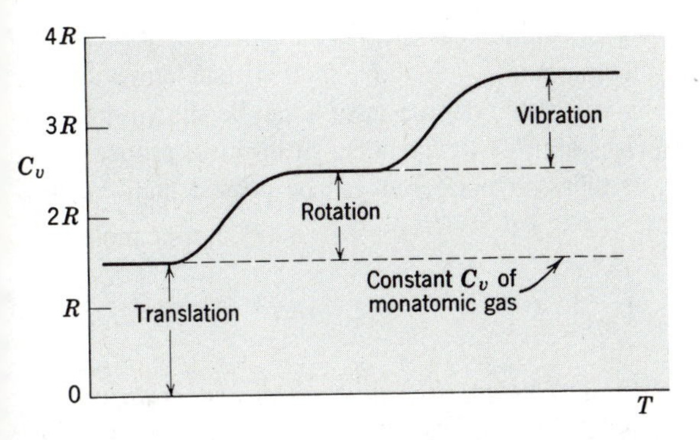

Figure 4·11. C_v variation based on kinetic theory.

vibrate at the same temperature, so that the cumulative effect as more molecules begin to rotate and then to vibrate is a gradual increase in specific heat.

The specific heat values predicted from kinetic theory should be compared with the values presented in Fig. 4·4 and in Table 4·2. As a sample comparison, see Table 4·3 which shows good agreement at 0°C for several gases.

The specific heats of gases with more complex molecules cannot be explained by a simplified approach like the one we have used here.

4·9 Summary

An ideal gas is defined as one for which the equation of state is

$$Pv = RT \tag{4·1}$$

where R is a different constant, called the *gas constant*, for each gas. R is given by

$$R = \frac{R_u}{M}$$

where R_u is the *universal gas constant* and M is molar mass. R_u has the same value for all gases, 8.134 kJ/kmol·K or 1545 ft·lbf/lbmol·R.

The ideal-gas equation of state represents the characteristics of many real gases

TABLE 4·3 Molar Specific Heats of Several Gases in kJ/kmol·K

	Experimental Values at 1 atm, 0°C		From Kinetic Theory (no vibration)	
Gas	C_v	C_p	C_v	C_p
Ar	12.5	20.8	$\frac{3}{2}R_u = 12.5$	$\frac{5}{2}R_u = 20.8$
He	12.7	21.0	$\frac{3}{2}R_u = 12.5$	$\frac{5}{2}R_u = 20.8$
Air	20.8	29.1	$\frac{5}{2}R_u = 20.8$	$\frac{7}{2}R_u = 29.1$
CO	20.8	29.1	$\frac{5}{2}R_u = 20.8$	$\frac{7}{2}R_u = 29.1$
N_2	20.8	29.1	$\frac{5}{2}R_u = 20.8$	$\frac{7}{2}R_u = 29.1$
O_2	21.0	29.3	$\frac{5}{2}R_u = 20.8$	$\frac{7}{2}R_u = 29.1$

accurately enough for many engineering calculations as long as the gases are at pressures well below their critical pressures and temperatures well above their critical temperatures.

The ideal-gas thermometer scale is a valuable one because it can be shown that it is equivalent to a temperature scale which is entirely independent of physical properties.

From the first and second laws of thermodynamics it can be proved that, for any substance which follows the equation of state $Pv = RT$,

$$\left(\frac{\partial u}{\partial v}\right)_T = 0$$

This means that *the internal energy of an ideal gas is a function of temperature only.* This important fact is known as Joule's law. It follows that *for any process of an ideal gas*

$$du = c_v\,dT \tag{4·2}$$

$$dh = c_p\,dT \tag{4·3}$$

The specific heats of an ideal gas are functions of temperature only and

$$c_p - c_v = R$$

under all conditions.

Whenever the temperature range is such that the specific heats can be considered as constant,

$$\Delta u = c_v\,\Delta T \qquad \text{and} \qquad \Delta h = c_p\,\Delta T$$

For a *frictionless adiabatic process of an ideal gas with constant specific heats,*

$$pv^k = \text{constant} \tag{4·5}$$

where $k \equiv c_p/c_v$. For a process from state 1 to state 2 under the same restrictions

$$\frac{T_2}{T_1} = \left(\frac{P_2}{P_1}\right)^{(k-1)/k} = \left(\frac{v_1}{v_2}\right)^{k-1} \tag{4·5}$$

It must be emphasized that Eqs. 4·5 apply *only* to *frictionless adiabatic processes of an ideal gas with constant specific heats.*

For many, but not all, frictionless processes of an ideal gas the Pv relation is

$$Pv^n = \text{constant} \tag{4·6}$$

A process which can be represented by such an equation is called a *polytropic* process, and n is called the *polytropic exponent* for the process.

At least for simple molecules, the ideal-gas equation of state and specific heat values can be developed from the kinetic theory of gases.

References

4·1 Van Wylen, Gordon J., and Richard E. Sonntag, *Fundamentals of Classical Thermodynamics,* 3rd ed., Wiley, 1985, Section 5.6.

4·2 Wales, Charles E., *Programmed Thermodynamics*, McGraw-Hill, New York, 1970, Chapters 1 and 4.

4·3 Zemansky, Mark W., Michael M. Abbott, and Hendrick C. Van Ness, *Basic Engineering Thermodynamics*, 2nd ed., McGraw-Hill, New York, 1975, Sections 5–7 to 5–9.

4·4 Keenan, Joseph H., Jing Chao, and Joseph Kaye, *Gas Tables*, 2nd ed., Wiley, New York, 1980 (English units), 1983 (SI units).

4·5 *Tables of Thermal Properties of Gases*, National Bureau of Standards Circular No. 564, Washington, D.C., 1955.

4·6 Gordon, Sanford, and B. J. McBride, "Computer Program for Calculation of Complex Chemical Equilibrium Compositions, Rocket Performance, Incident and Reflected Shocks, and Chapman-Jouget Detonations," NASA SP-273, 1971 (and interim revision, March 1976).

Problems

4·1 For an ideal gas, is x a property if $x = \int (c/T)dT + \int (P/T)dv$ and c is a constant?

4·2 A spherical balloon 3.2 m in diameter contains hydrogen at 25°C and 100 kPa absolute. Compute the mass of hydrogen in the balloon.

4·3 What volume must be provided in a methane tank to hold 10 kg of methane at 1500 kPa and 30°C? If the temperature drops to 0°C, what will be the pressure?

4·4 The density of a gas is 1.15 kg/m^3 at 0°C and 100 kPa. Using the ideal-gas equation of state, compute the molar mass.

4·5 Determine the volume required to store 25 kg of ethene at 320 kPa and 35°C. If the air temperature rises to 65°C, what will be the pressure in the storage tank?

4·6 A closed tank contains 50 kg of helium at 900 kPa and 50°C. Determine the mass of helium that leaks out if the final pressure is 400 kPa and the final temperature is 25°C.

4·7 Compute the volume occupied by 1 kmol of CO at 101.3 kPa and 80°C.

4·8 Hydrogen gas is to be stored in a cylindrical tank having an internal diameter of 0.2 m and a length of 0.6 m. If the maximum allowable pressure and temperature are 2 MPa and 60°C, how many moles of hydrogen can be stored at the maximum temperature and pressure?

4·9 Air at 25 kPa gage, 25°C, is contained in a tank which has a volume of 0.05 m^3. Barometric pressure is 90 kPa, and the local acceleration of gravity is 9.61 m/s^2. Calculate the weight of air in the tank.

4·10 Air at 200 mm of mercury vacuum, 25°C, is contained in a tank which has a volume of 0.040 m^3. Barometric pressure is 690 mm of mercury, and the local acceleration of gravity is 9.75 m/s^2. Calculate the mass and the weight of air in the tank.

4·11 An elevator shaft in a building is 100 m high. The lower end of the shaft is open so that the air pressure inside the shaft is the same as outside. The outdoor temperature is −5°C and the air inside the shaft is at 25°C. Barometric pressure is 101.35 kPa. What is the difference between the air pressure inside the shaft and outside the building at the top of the shaft? State clearly your assumptions.

4·12 A spherical balloon 48 ft in diameter contains helium at 80°F and 29.0 in. of mercury absolute. Compute the mass of helium in the balloon.

4·13 What volume must be provided in an air receiver to hold 20 lbm of air if the pressure is 250 psia and the temperature is 100°F? If the temperature drops to 32°F, what will be the pressure?

4·14 The density of a gas is 0.07704 lbm/cu ft at 32°F and 14.7 psia. Using the ideal-gas equation of state compute the molar mass.

4·15 Determine the volume required to store 50 lbm of air at 250 psia and 70°F. If the temperature rises to 150°F, what will be the pressure in the storage tank?

4·16 A closed tank contains 100 lbm of air at 140 psia and 120°F. What amount of air leaks out if the final pressure is 60 psia and the final temperature is 80°F?

4·17 Compute the volume occupied by 1 lbmol of CO at 14.7 psia and 190°F.

4·18 Hydrogen gas is to be stored in a cylindrical tank having an internal diameter of 8 in. and a length of 2 ft. If the maximum allowable pressure and temperature are 300 psia and 140°F, how many moles of hydrogen can be stored at the maximum temperature and pressure?

4·19 Two kilograms of a gas with c_v = 0.75 kJ/kg·K expands adiabatically in a closed system and performs 10 kJ of work. The final temperature is 90°C. Compute the initial temperature.

4·20 To compress 2 kg of an ideal gas with c_v = 0.85 kJ/kg·K requires 30 kJ of work while 25 kJ of heat is removed. Compute the temperature change.

4·21 To compress 3 kg of gas with c_v = 0.75 J/g·K requires 34 kJ of work while 26 kJ of heat is removed. Compute the temperature change.

4·22 A gas flows through a nozzle from a pressure of 1200 kPa to a final pressure of 120 kPa. The initial and final specific volume values are 1.186 and 6.289 m^3/kg respectively. If the initial and final temperatures are 345 K and 183 K, respectively, compute the final velocity. Neglect the initial velocity and the heat-loss terms. Assume that the specific heat at constant volume is 10.0 kJ/kg·K.

4·23 A cylinder fitted with a piston contains 0.10 kg of argon at 200 kPa, 5°C. The argon expands frictionlessly and isothermally until its pressure is 100 kPa. Calculate the heat transfer.

4·24 One kilogram of oxygen is compressed isothermally in a closed system from 100 kPa and 25°C to 300 kPa. Calculate (*a*) work, (*b*) heat transfer, and (*c*) change in internal energy.

4·25 A closed rigid tank contains 0.15 m^3 of helium initially at 100 kPa, 5°C. Work is done on the helium by means of a paddle wheel turned by an external motor, and 15 kJ of heat is added to the helium. The final pressure is then 200 kPa. Sketch a *PV* diagram of the process, and calculate the amount of work done on the helium during the process.

4·26 In a closed system, 0.09 kg of air initially at 200 kN/m^2, 4°C, expands frictionlessly and isothermally until its pressure is 100 kN/m^2. Calculate the heat transfer.

4·27 One kilogram of air is compressed isothermally in a closed system from 100 kN/m^2, 4°C, to 300 kN/m^2. Calculate (*a*) work, (*b*) heat transfer, and (*c*) change in internal energy.

4·28 A closed rigid tank contains 0.14 m^3 of air initially at 100 kN/m^2, 4°C. Work is done on the air by means of a paddle wheel turned by an external motor, and 5 kcal of heat is added to the air. The final pressure is 200 kN/m^2. Sketch a *PV* diagram of the process and calculate the work.

4·29 Air at 8.0 in. of mercury vacuum, 25°C, is contained in a tank which has a volume of 1.5 cu ft. Barometric pressure is 27.0 in. of mercury, and the local acceleration of gravity is 32.00 ft/s^2. Calculate the weight of air in the tank.

4·30 An elevator shaft in a building is 400 ft high. The lower end of the shaft is open so that the air pressure inside the shaft is the same as outside. The outdoor temperature is 0°F and the air inside the shaft is at 80°F. Barometric pressure is 14.7 psia. What is the difference between the air pressure inside the shaft and outside the building at the top of the shaft? Clearly state your assumptions.

4·31 For an ideal gas sketch a Pv diagram showing lines of constant T, a Tv diagram showing lines of constant P, and a PT diagram showing lines of constant v. Derive an expression for the slope of each line.

4·32 Prove that for an ideal gas

$$du = \frac{1}{k-1}\, d(Pv) \qquad \text{and} \qquad dh = \frac{k}{k-1}\, d(Pv)$$

4·33 Three pounds of a certain gas with $c_v = 0.18$ B/lbm·°F expands adiabatically in a closed system and in so doing performs 10 000 ft·lbf of work. The final temperature is 200°F. Compute the initial temperature.

4·34 To compress 5 lbm of an ideal gas with $c_v = 0.169$ B/lbm·°F requires 25 000 ft·lbf of work while 25 B of heat is removed. Compute the temperature change.

4·35 A gas flows through a nozzle from a pressure of 180 psia to a final pressure of 20 psia. The initial and final specific volume values are 1.277 and 6.09 cu ft/lbm, respectively. If the initial and final temperatures are 621 and 329 R respectively, compute the final velocity. Neglect the initial velocity and the heat-loss terms. Assume that the specific heat at constant volume is 0.169 B/lbm·°F.

4·36 A cylinder fitted with a piston contains 0.20 lbm of air at 30 psia, 40°F. The air expands frictionlessly and isothermally until its pressure is 15 psia. Calculate the heat transfer.

4·37 Two pounds of air is compressed isothermally in a closed system from 15 psia, 40°F, to 45 psia. Calculate (*a*) work, (*b*) heat transfer, and (*c*) change in internal energy.

4·38 A closed rigid tank contains 5 cu ft of air initially at 15 psia, 40°F. Work is done on the air by means of a paddle wheel turned by an external motor, and 20.0 B of heat is added to the air. The final pressure is 30 psia. Sketch a PV diagram of the process and calculate the work of the process.

4·39 An ideal gas is heated from state 1 to state 2 at constant pressure and is then further heated from state 2 to state 3 at constant volume. For this sequence of processes sketch three property diagrams: Pv, Tv, and PT, where in each case the first variable is to be the ordinate.

4·40 An ideal gas is heated from state 1 to state 2 at constant pressure and is then further heated from state 2 to state 3 at constant volume. Sketch for this sequence of processes three property diagrams: uv, Ph, and uT, where in each case the first variable is to be the ordinate.

4·41 An ideal gas is compressed frictionlessly and adiabatically from state 1 to state 2. It is then expanded frictionlessly and isothermally from state 2 to state 3, and $v_3 = v_1$. For this sequence of processes sketch three property diagrams: Pv, Tv, and PT, where in each case the first variable is to be the ordinate.

4·42 An ideal gas is compressed frictionlessly and adiabatically from state 1 to state 2. It is then expanded frictionlessly and isothermally from state 2 to state 3, and $v_3 = v_1$. Sketch for this sequence of processes three property diagrams: uv, Ph, and uT, where in each case the first variable is to be the ordinate.

4·43 Nitrogen in a closed system is heated at a constant pressure of 1 atm until its volume has increased by 30 percent. Sketch the following diagrams showing the process: Pv, PT, Pu, Ph, vT, vu, vh, Tu, Th, and uh. The first variable listed is to be plotted on the vertical axis.

4·44 Air is compressed isothermally from one atmosphere to 3 atm. Sketch the process on the following axes: *PV*, *PT*, *PU*, *PH*, *VT*, *VU*, *VH*, *TU*, *TH*, and *UH*.

4·45 In the discussion of Joule's experiment in the early part of Sec. 4·4, see the response to the question, "If in an experiment like Joule's the temperature of the water changes, does this prove that the internal energy of the gas under study is not a function of temperature only?" Answer this question by supposing that the water temperature decreases instead of rising as is supposed in Sec. 4·4.

4·46 A piece of electronic equipment fits into a case of dimensions 9 by 16 by 22 cm. Power consumption under steady conditions is 600 W. The output signals carry a negligible amount of energy. It is desired to maintain the average temperature of the equipment not higher than 60°C either by blowing atmospheric air through the unit or by circulating water through a heat exchanger inside the unit. Determine both the required mass flow rate (kg/s) and the required volume flow rate (m^3/s) if (*a*) the coolant is air entering at 1 atm, 20°C, with an allowable air temperature rise of 30 degrees C, (*b*) the coolant is water entering at 1 atm, 20°C, with an allowable water temperature rise of 40 degrees C.

4·47 A single-phase electric transformer operates steadily with a primary voltage of 2300 V and its secondary at 230 V. The rated power output is 750 kW, and the efficiency (the ratio of output power to input power) is 97.0 percent. If the transformer is cooled only by atmospheric air blown over it, what flow rate (kg/s) is required for an entry air temperature of 30°C and an air temperature rise not to exceed 20 Celsius degrees?

4·48 A Hilsch tube or Ranque tube is a short length of insulated pipe, open at one end and capped at the other, with a small opening at the center of the cap. A gas is injected tangentially into the pipe at some point along its length and, even though there are no moving parts, the gas streams leaving at the two ends are observed to be at different temperatures. If air enters such a device at 300 kPa, 40°C, at a rate of 0.055 kg/s and air leaves one end at 100 kPa, 50°C, at a rate of 0.035 kg/s, what is the temperature of the air leaving the other end?

4·49 In a closed system, a gas is expanded frictionlessly at constant pressure until its volume has increased by one-fifth. What fraction of the heat added would be converted into work if the gas were (*a*) air, (*b*) ethane, (*c*) helium?

4·50 Refer to Problem 2·97. What is the value of k for the gas?

4·51 A rigid tank having a volume of 0.2 m^3 contains air initially at 100 kPa, 10°C. An external motor connected to an impeller inside the tank delivers 0.15 kW to the air in the tank for a period of 3.0 min while heat is removed from the air in the amount of 10 kJ. Calculate the change in internal energy per kilogram of air in the tank.

4·52 One-tenth kilogram of an ideal gas having a molar mass of 40 is contained at 200 kPa, 60°C, in a closed, rigid vessel. By means of a paddle wheel 0.75 kJ of work is done on the gas while 1.5 kW of heat is added. During this process the temperature of the gas rises to 100°C. Calculate c_v of the gas.

4·53 One-tenth kilogram of an ideal gas with a molar mass of 40 expands frictionlessly at constant pressure in a closed system from 100 kPa, 60°C, to 160°C when 5 kJ of heat is added. Determine c_v.

4·54 A rigid tank having a volume of 0.5 m^3 is filled with a ideal gas at 0.150 MPa, 30°C. The addition of 14 kJ of heat raises the gas temperature to 60°C. The specific heat of the gas at constant pressure is 0.90 kJ/kg·K. Compute the value of k for the gas.

4·55 A rigid tank having a volume of 6.0 cu ft contains air initially at 15 psia, 40°F. An external motor connected to an impeller inside the tank delivers 0.20 hp to the air in the tank for a period of 3.0 min while heat is removed from the air in the amount of 9.0 B. Calculate the change in internal energy per pound of air in the tank.

4·56 One-half pound of an ideal gas having a molar mass of 40 is contained at 30 psia, 140°F, in a closed, rigid vessel. By means of a paddle wheel 560 ft·lbf of work is done on the gas while 3.0 B of heat is added. During this process the temperature of the gas rises to 240°F. Calculate c_v of the gas.

4·57 Two-tenths pound of an ideal gas with a molar mass of 40 expands frictionlessly at constant pressure in a closed system from 15 psia, 140°F, to 340°F when 4.9 B of heat is added. Determine c_v.

4·58 A rigid tank having a volume of 20.0 cu ft is filled with an ideal gas at 3000 psfa, 100°F. The addition of 13.7 B of heat raises the gas temperature to 140°F. The specific heat of the gas at constant pressure is 0.238 B/lbm·°F. Compute the value of k for the gas.

4·59 An ideal gas at 50 kPa, 10°C, fills a closed, rigid, thermally insulated container. An impeller inside the container is turned by an external motor until the pressure is 100 kPa. For this gas in the temperature range involved, $c_p = 0.84 + 0.00075\,T$, where T is in kelvins and c_p and c_v are in kJ/kg·K. Calculate in kJ/kg: (*a*) the enthalpy change of the gas, and (*b*) the work done on the gas.

4·60 The equation for C_p as a function of T displayed in Table A·4 has five terms. Comment on the suggestion that only the first two or three terms might be used for an approximation.

4·61 Over the temperature range 0 to 2500°C, compare C_p values for air as an ideal gas as given by Table 4·2, Table A·4, and Table 2 of Reference 4·4. Determine the greatest discrepancy among these tables in this temperature range.

4·62 Over the temperature range 0 to 2500°C, compare C_v values for CO_2 as an ideal gas as given by Table 4·2, Table A·4, and Table 18 of Reference 4·4. Determine the greatest discrepancy among these tables in this temperature range.

4·63 Compare the values of c_p for nitrogen from Tables 4·2 and A·4 in the temperature range of 0 to 2500°C. Display the results in a table with temperature values every 50 degrees to 200°C, then every 200 degrees to 1000°C, and then every 500 degrees to 2500°C.

4·64 Compute the amount of heat required to raise the temperature of 6 kg of CH_4 from 50 to 700°C at a constant pressure of 1 atm.

4·65 Ethene is heated under a constant pressure of 1 atm from 20 to 500°C. Compute the heat added.

4·66 Determine the amount of heat which must be added to heat 0.2 kg of carbon dioxide from 100 to 2000°C under a constant pressure of 1 atm.

4·67 Compute the amount of heat required to raise the temperature of 6 lbm of CH_4 from 500 to 900°F at a constant pressure of 1 atm. What assumption do you make?

4·68 Ethene is heated under a constant pressure of 1 atm from 100 to 500°F. Compute the heat added.

4·69 Determine the amount of heat which must be added to heat 0.2 lbm of carbon dioxide from 100 to 4000°F under a constant pressure of 14.7 psia.

4·70 Refer to the solution of Example 4·7. Occasionally someone objects to the assignment of $u_0 = 0$ and $h_0 = 0$ when $T_0 = 0$ on the basis that this affects the result. Carry out the solution with $u = c_v T + K$ and $h = c_p T + K'$, where K and K' are constants, to see how the result is affected. Also, what is the numerical value of $(K' - K)$?

4·71 An ideal gas at 10 psia, 40°F, fills a closed, rigid, thermally insulated container. An impeller inside the container is turned by an external motor until the pressure is 14 psia. For this gas in the temperature range involved, $c_p = 0.2 + 0.0001T$ and $c_v = 0.15 + 0.0001T$, where T is in Rankine and c_p and c_v are in B/lbm·R. Calculate, in B/lbm: (*a*) the enthalpy change of the gas and (*b*) the work done on the gas.

4·72 A gas which follows the ideal-gas equation of state and has a molar mass of 33.1 is heated in a closed, rigid tank having a volume of 6.0 cu ft. The gas is initially at 15 psia, 40°F. Calculate the amount of heat which must be added to raise the pressure to 21 psia if specific heats in B/lbm·R are given by

$$c_v = 0.200 - \frac{6}{T}$$

$$c_p = 0.260 - \frac{6}{T}$$

with T in Rankine, throughout the temperature range involved.

4·73 Using the data of Table A·4, plot curves of k versus temperature in the range of 300 to 1500 K for (*a*) propane, (*b*) ethene, and (*c*) methane.

4·74 In a closed system, 1.25 kg of ethane is heated at constant pressure from 200 kPa, 0°C, to 60°C. The work output is only 18.2 kJ. Calculate the amount of heat added to the gas.

4·75 One-hundredth kilogram of air at 300 kPa, 10°C, is trapped inside a vertical cylinder which is fitted at the top with a weighted piston so that the pressure of the air is held constant. There is no heat transfer. A paddle wheel in the cylinder is turned until the volume of the air has increased by 20 percent. Determine (*a*) the net amount of work done on the air and (*b*) the amount of work done on the air by the paddle wheel.

4·76 Two kilograms of carbon monoxide in a closed system expands from 200 kPa, 600°C, to 100 kPa, 550°C, while performing 150 kJ of work. Calculate the heat transfer.

4·77 One-tenth cubic meter of methane is contained in a cylinder at 100 kPa, 10°C. It is compressed to twice its initial pressure and one-half its initial volume. Calculate (*a*) the net work done on or by the gas, (*b*) the net heat added to or taken from the gas, and (*c*) the net change in internal energy if (1) the pressure is first doubled at constant volume and the volume is then halved at constant pressure, (2) the volume is first halved at constant pressure and the pressure is then doubled at constant volume.

4·78 Two one-hundredths cubic meter of nitrogen trapped in a cylinder at 100 kPa, 5°C, is compressed until its volume is 0.01 m^3 and its pressure is 175 kPa. Calculate the quantities listed below (if there is insufficient information for the calculation of any item, state what additional information is necessary for its determination): (*a*) total internal energy change of the air, (*b*) net heat added to the air during the process, and (*c*) net work done by the air during the process.

4·79 In a closed system, 2 lbm of air is heated at constant pressure from 30 psia, 40°F, to 140°F. Because of a frictional effect, the work output is only 10 B. Calculate the amount of heat added to the air.

4·80 One-tenth pound of air at 40 psia, 40°F, is trapped inside a vertical cylinder which is fitted at the top with a weighted piston so that the pressure of the air is held constant. There is no heat transfer. A paddle wheel in the cylinder is turned until the volume of the air has increased by 20 percent. Determine (*a*) the net amount of work done on the air and (*b*) the amount of work done on the air by the paddle wheel.

4·81 Four pounds of air in a closed system expands from 30 psia, 1240°F, to 15 psia, 1100°F, while performing 130 B of work. Calculate the heat transfer.

4·82 One-tenth pound of air is contained in a cylinder at 15 psia and 40°F. This air is to be compressed to twice its initial pressure and one-half its initial volume. Calculate (*a*) the net work done on or by the gas, (*b*) the net heat added to or taken from the gas, and (*c*) the net change in internal energy if (1) the pressure is first doubled at constant volume and the volume is then halved at constant pressure, (2) the volume is first halved at constant pressure and the pressure is then doubled at constant volume.

4·83 Air trapped in a cylinder expands frictionlessly against a piston in such a manner that PV = constant. Initially the air is at 400 kN/m^2, 4°C, and occupies a volume of 0.02 m^3. The local value of g is 9.51 m/s^2. (*a*) To what pressure must the air expand in order to perform 8100 J of work? (*b*) What is the mass of air in the system?

4·84 Twenty thousand cubic centimeters of argon at 35.0 kPa, 100°C, in a closed system expands frictionlessly until its volume is doubled and its temperature is 0°C. During the expansion, $PV^{1.5}$ = constant. Calculate the heat transfer.

4·85 Carbon dioxide is expanded frictionlessly in a closed system in such a manner that $PV^{1.5}$ = constant from initial conditions of 275 kPa, 170°C, V_1 = 0.06 m^3 to a final volume of V_2 = 0.12 m^3. Calculate (*a*) the work done and (*b*) the heat transferred.

4·86 Consider a rigid vessel 33.9 ft high (inside dimension) which is completely filled with water except for a small bubble of air which is held at the bottom. (The bubble might be held by means of an inverted cup, for example.) The pressure at the top of the water column is 1 atm, so the pressure of the bubble is 2 atm. The bubble is then released. After the bubble reaches the top of the vessel, what is the pressure of the water at the bottom? If two or more equal-size bubbles were initially at the bottom of the vessel, how would the pressure at the bottom vary if they were released one at a time?

4·87 One cubic foot of air trapped in a cylinder at 15 psia, 40°F, is compressed until its volume is 0.50 cu ft and its pressure is 25 psia. Calculate the quantities listed below (if there is insufficient information for the calculation of any item, state what additional information is necessary for its determination): (*a*) total internal energy change of the air, (*b*) net heat added to the air during the process, (*c*) net work done by the air during the process.

4·88 Air trapped in a cylinder expands frictionlessly against a piston in such a manner that PV = constant. Initially the air is at 60 psia, 40°F, and occupies a volume of 0.50 cu ft. The local value of g is 31.8 ft/s^2. (*a*) To what pressure must the air expand in order to perform 6000 ft·lbf of work? (*b*) What is the mass of air in the system?

4·89 One cubic foot of argon at 50 psia, 240°F, in a closed system expands frictionlessly until its volume is doubled and its temperature is 35°F. During the expansion, $PV^{1.5}$ = constant. Calculate the heat transfer.

4·90 Carbon dioxide is expanded frictionlessly in a closed system in such a manner that $PV^{1.30}$ = constant from initial conditions of 40 psia, 340°F, V_1 = 2.0 cu ft to a final volume of V_2 = 4.0 cu ft. Calculate (*a*) the work done and (*b*) the heat transferred.

4·91 Methane enters a machine at 95 kPa, 35°C, with a velocity of 9.0 m/s through a cross-sectional area of 0.040 m^2. It leaves the machine at 220 kPa, 90°C, through a cross-sectional area of 0.015

m^2. Heat removed from the methane flowing through the machine amounts to 60.4 kJ/kg. The flow is steady. Determine the outlet velocity.

4·92 Helium enters a gas turbine at 500 kPa, 200°C, and leaves at 100 kPa, 25°C. Heat removed from the helium passing through the turbine amounts to 35 kJ/kg. The flow rate is 6350 kg/h. Calculate the power output if the change in kinetic energy is negligible.

4·93 Propane enters a steady-flow system at 175 kPa, 200°C, with a velocity of 60 m/s through a cross-sectional area of 0.05 m^2. It leaves at 80 kPa, 150°C, at the same velocity. The system delivers 60 kW to the surroundings. Calculate the heat transfer in kJ/kg.

4·94 In a gas-turbine power plant, air is taken in at 101.35 kPa, 15°C, at a rate of 1150 m^3/min, compressed to 410 kPa, heated, and then expanded through a turbine and exhausted at 101.35 kPa, 260°C. If the net output of the plant is 2880 kW, calculate the net amount of heat added to the air in kJ/kg. Neglect changes in kinetic energy.

4·95 A compressor takes in 340 m^3/min of air at 95 kPa, 25°C, with negligible velocity and discharges air at 200 kPa, 120°C, through an opening that has a cross-sectional area of 0.025 m^2. Heat is removed from the air being compressed at a rate of 60 kJ/min. Determine the power input to the compressor.

4·96 Ethene is compressed isothermally from 95 kPa, 5°C, to 480 kPa at a rate of 0.90 kg/s by an ideal compressor. The kinetic energy of the air passing through the compressor increases by 11 kJ/kg. Heat removed from the air amounts to 128 kJ/kg. Determine the power input to the compressor.

4·97 Helium is drawn into a multistage centrifugal compressor at 95 kPa, 5°C, with negligible velocity and is discharged at 275 kPa, 150°C, with a velocity of 150 m/s through a cross-sectional area of 0.020 m^2. Power input to the compressor is 1000 kW. Determine the heat transfer in kJ/kg.

4·98 Methane enters a machine at 14.0 psia, 95°F, with a velocity of 30 fps through a cross-sectional area of 0.4 sq ft. It leaves the machine at 32.0 psia, 200°F, through a cross-sectional area of 0.15 sq ft. Heat removed from the methane flowing through the machine amounts to 26.1 B/lbm. The flow is steady. Determine the outlet velocity.

4·99 Air enters a gas turbine at 75 psia, 395°F, and leaves at 15 psia, 80°F. Heat removed from the air passing through the turbine amounts to 15 B/lbm. The flow rate is 14 000 lbm/h. Calculate the power output if the change in kinetic energy is neglected.

4·100 Air enters a steady-flow system at 25 psia, 400°F, with a velocity of 200 fps through a cross-sectional area of 0.50 sq ft. It leaves at 11.5 psia, 300°F, at the same velocity. The system delivers 20 hp to the surroundings. Calculate the heat transfer in B/lbm.

4·101 In a gas-turbine power plant, air is taken in at 14.7 psia, 60°F, at a rate of 40 000 cfm, compressed to 60 psia, heated, and then expanded through a turbine and exhausted at 14.7 psia, 500°F. If the net output of the plant is 3860 hp, calculate the net amount of heat added to the air in B/lbm. Neglect changes in kinetic energy.

4·102 A compressor takes in 12 000 cfm of air at 14.0 psia, 80°F, with negligible velocity and discharges air at 30.0 psia, 240°F, through an opening which has a cross-sectional area of 0.25 sq ft. Heat is removed from the air being compressed at a rate of 60 B/s. Determine the power input to the compressor.

4·103 Air is compressed isothermally from 14.0 psia, 40°F, to 70 psia at a rate of 2.0 lbm/s by an ideal compressor. The kinetic energy of the air passing through the compressor increases by 5.0 B/lbm. Heat removed from the air amounts to 55.2 B/lbm. Determine the power input to the compressor.

4·104 Air is drawn into a multistage centrifugal compressor at 14.0 psia, 40°F, with negligible velocity and is discharged at 40 psia, 300°F, with a velocity of 500 fps through a cross-sectional area of 0.20 sq ft. Power input to the compressor is 1500 hp. Determine the heat transfer in B/lbm.

4·105 Starting with the first law as applied to a steady-flow process, prove that Pv^k = constant for a frictionless adiabatic process of an ideal gas with constant specific heats.

4·106 Air enters a compressor at 100 kN/m^2, 4°C, with a velocity of 150 m/s through a cross-sectional area of 0.060 m^2. The air is compressed frictionlessly, steadily, and adiabatically to 200 kN/m^2. The discharge velocity is very low. Calculate the power input.

4·107 Three kilograms of nitrogen expands frictionlessly and adiabatically in a closed system from 2760 kN/m^2, 38°C, to 1380 kN/m^2. Compute the change in internal energy, the change in enthalpy, the work, and the heat transfer.

4·108 One kilogram of carbon monoxide is compressed frictionlessly and adiabatically in a closed system from 100 kPa, 20°C, to 200 kPa. Compute the change in internal energy, the change in enthalpy, and the work.

4·109 Determine the heat transfer during a polytropic process in which the temperature of octane in a closed system changes from 115 to 5°C and the octane does 45 kJ of work per kilogram.

4·110 One cubic meter of gas at 170°C expands polytropically in a cylinder until the temperature is 5°C and the volume is 4 m^3. Determine the value of n.

4·111 Nitrogen is compressed polytropically in a closed system from 70 kPa and a volume of 0.42 m^3 to 700 kPa and a volume of 0.06 m^3. Determine the value of n for the process.

4·112 Air enters a nozzle at 690 kN/m^2, 170°C, with negligible velocity. It expands adiabatically without friction as it flows through the nozzle and leaves at 345 kN/m^2. The nozzle exit area is 5.20 cm^2. Calculate the exit velocity.

4·113 Oxygen in a closed system expands frictionlessly and adiabatically from 200 kN/m^2, 170°C, to 100 kN/m^2. Calculate the work done per kilogram of oxygen.

4·114 Air at 275 kPa, 60°C, enters a nozzle with negligible velocity and expands frictionlessly and adiabatically to 140 kPa. The outlet area of the nozzle is 16.4 cm^2. Calculate (*a*) the outlet velocity and (*b*) the value of $\int_1^2 v\, dP$ for this process.

4·115 Calculate the power required to compress air frictionlessly and adiabatically from 83 kN/m^2, 4°C, to 166 kN/m^2, discharging it at an average velocity of 180 m/s through a cross-sectional area of 0.050 m^2. Inlet velocity is negligibly small.

4·116 An ideal centrifugal compressor compresses air polytropically, with n = 1.35, at a rate of 2.3 kg/s from 100 kPa, 5°C, to 300 kPa. Inlet velocity is negligibly small while the discharge velocity is 180 m/s. Heat is removed from the air being compressed at a rate of 20 kJ/s. Calculate the power input.

4·117 An ideal reciprocating compressor draws in 5.7 m^3/min of air at 100 kPa, 5°C, and compresses it polytropically with n = 1.35 to 500 kPa. Cooling water which removes heat from the air flows at a rate of 4.63 kg/min and undergoes a temperature rise of 5 degrees Celsius. Calculate the compressor power requirement.

4·118 An ideal centrifugal compressor takes in oxygen at 100 kPa, 5°C, with negligible velocity. The flow rate is 500 kg/h. It compresses the air polytropically with n = 1.35 and discharges it at 200 kPa with a velocity of 180 m/s. Calculate the power input.

4·119 Air is compressed in a frictionless, steady-flow process from 70 kN/m^2, 15°C, to 100 kN/

m^2 in such a manner that $P(v + 0.2) =$ constant, where v is in m^3/kg. Inlet velocity is negligibly small and discharge velocity is 100 m/s. Calculate the heat transfer.

4·120 Air flows steadily and frictionlessly through a compressor at a rate of 1.5 kg/s. At inlet the air is at 70 kPa, 5°C, and at outlet the air is at 140 kPa, 120°C. Between inlet and outlet of the system, the air is compressed in such a manner that $Pv^2 =$ constant. The inlet area is very large and the outlet area is 45 cm^2. Sketch the process on Pv and Tv coordinates, and calculate the heat transfer in kJ/kg.

4·121 A kilogram of argon which under a pressure of 1.4 MPa occupies a volume of 0.06 m^3 expands at constant temperature until the volume is doubled. It is then compressed at constant pressure to a volume of 0.06 m^3, after which it has its pressure raised to 1.4 MPa at constant volume. Draw approximately to scale the Pv diagram for the processes of this closed system. Determine the net change in internal energy for the processes involved. Calculate the net work done and the net amount of heat added or abstracted during these processes.

4·122 One kilogram of octane is compressed frictionlessly and adiabatically from 100 kPa and 0.40 m^3 to 0.20 m^3, then expanded at constant pressure to 0.4 m^3, and finally cooled at constant volume until the initial pressure is reached. Sketch the processes on a Pv diagram approximately to scale. Shade the area representing the work for the constant-pressure process. Determine the work done, the heat added, and the change in internal energy for the adiabatic process.

4·123 A kilogram of methane occupying 0.028 m^3 at 1.4 MPa undergoes a change to 0.056 m^3 at constant pressure, a constant-temperature process to 0.085 m^3, an adiabatic change to 0.056 m^3, and finally a constant-volume process to the original pressure. Draw a Pv diagram to approximate scale showing the processes. Compute the work done during the isothermal process and show the work area by crosshatching on the diagram.

4·124 Air enters a compressor at 15 psia, 40°F, with a velocity of 500 fps through a cross-sectional area of 0.60 sq ft. The air is compressed frictionlessly, steadily, and adiabatically to 30 psia. The discharge velocity is very low. Calculate the power input.

4·125 Six pounds of nitrogen expands frictionlessly and adiabatically in a closed system from 400 psia, 100°F, to 200 psia. Compute the change in internal energy, the change in enthalpy, the work, and the heat transfer.

4·126 Two pounds of air is compressed frictionlessly and adiabatically in a closed system from 15 psia, 70°F, to 150 psia. Compute the change in internal energy, the change in enthalpy, and the work.

4·127 Determine the heat transfer during a polytropic process in which the temperature of air in a closed system changes from 240 to 40°F and the air does 15 560 ft·lbf of work per pound.

4·128 Five cubic feet of gas at 340°F expands polytropically in a cylinder until the temperature is 40°F and the volume is 20 cu ft. Determine the value of n.

4·129 Nitrogen is compressed polytropically in a closed system from 10 psia and a volume of 15 cu ft to 100 psia and a volume of 2 cu ft. Determine the value of n for the process.

4·130 Air enters a nozzle at 100 psia, 340°F, with negligible velocity. It expands adiabatically without friction as it flows through the nozzle and leaves at 50 psia. The nozzle exit area is 0.80 sq in. Calculate the exit velocity.

4·131 Oxygen in a closed system expands frictionlessly and adiabatically from 30 psia, 340°F, to 15 psia. Calculate the work per pound.

4·132 Air at 40 psia, 140°F, enters a nozzle with negligible velocity and expands frictionlessly and adiabatically to 20 psia. The outlet area of the nozzle is 2.54 sq in. Calculate (*a*) the outlet velocity and (*b*) the value of $\int_1^2 v\,dP$ for this process.

4·133 Calculate the power required to compress air frictionlessly and adiabatically from 12 psia, 40°F, to 24 psia, discharging it at an average velocity of 600 fps through a cross-sectional area of 0.5 sq ft. Inlet velocity is negligibly small.

4·134 An ideal centrifugal compressor compresses air polytropically, with $n = 1.35$, at a rate of 5 lbm/s from 15 psia, 40°F, to 45 psia. Inlet velocity is negligibly small while the discharge velocity is 600 fps. Heat is removed from the air being compressed at a rate of 20 B/s. Calculate the power input.

4·135 An ideal reciprocating compressor draws in 200 cfm of air at 15 psia, 40°F, and compresses it polytropically with $n = 1.35$ to 75 psia. Cooling water that removes heat from the air flows at a rate of 10.2 lbm/min and undergoes a temperature rise of 10 degrees Fahrenheit. Calculate the compressor power requirement.

4·136 An ideal centrifugal air compressor takes in 10 000 cfm of air at 15 psia, 40°F, with negligible velocity. It compresses the air polytropically with $n = 1.35$ and discharges it at 30 psia with a velocity of 600 fps. Calculate the power input.

4·137 Air is compressed in a frictionless steady-flow process from 10 psia, 60°F, to 15 psia in such a manner that $P(v + 5) =$ constant, where v is in cu ft/lbm. Inlet velocity is negligibly small and discharge velocity is 350 fps. Calculate the heat transfer.

4·138 Air flows steadily and frictionlessly through a compressor at a rate of 3.2 lbm/s. At inlet the air is at 10.0 psia, 40°F, and at outlet the air is at 20.0 psia, 247°F. Between inlet and outlet of the system, the air is compressed in such a manner that $Pv^2 =$ constant. The inlet area is very large and the outlet area is 0.050 sq ft. Sketch the process on Pv and Tv coordinates, and calculate the heat transfer in B/lbm.

4·139 A cylinder has an inside cross-sectional area of 100 sq ft and a length of 30 ft. Its top is closed and its bottom open. It is partially submerged in a large body of water with its axis vertical until 2500 cu ft of air is trapped in the cylinder and the water level inside the cylinder is 15 ft below that outside. The air is at 150°F. Neglecting any effects of water vapor in the air, determine the amount of heat which must be removed from the air in the cylinder in order to lower its temperature to 100°F. The water level outside the cylinder remains 10 ft below the top of the cylinder.

4·140 A pound of air at 200 psia occupies a volume of 2 cu ft and expands at constant temperature until the volume is doubled. It is then compressed at constant pressure to a volume of 2 cu ft, after which it has its pressure raised to 200 psia at constant volume. Draw approximately to scale the Pv diagram for the processes of this closed system. Determine the net change in internal energy, the net work done, and the net heat transfer.

4·141 One pound of air is compressed adiabatically from 15 psia and 14 cu ft to 7 cu ft, then expanded at constant pressure to 14 cu ft, and finally cooled at constant volume until the initial pressure is reached. Sketch the processes on a Pv diagram approximately to scale. Shade the area representing the work for the constant-pressure process. Determine the work done, the heat added, and the change in internal energy for the adiabatic process.

4·142 A pound of air occupying 1 cu ft at 200 psia undergoes a change to 2 cu ft at constant pressure, a constant-temperature process to 3 cu ft, an adiabatic change to 2 cu ft, and finally a constant volume process to the original pressure. Draw a Pv diagram to approximate scale showing the processes. Compute the work done during the isothermal process, and show the work area by crosshatching on the diagram.

4·143 An ideal gas is to be compressed frictionlessly and adiabatically from initial atmospheric conditions of P_1 and T_1 to a higher pressure P_2. Plot curves of T_2 versus P_2/P_1 for various constant values of k of common gases for $T_1 = 25°C$.

4·144 An ideal gas is to be expanded frictionlessly and adiabatically from initial atmospheric conditions of P_1 and T_1 to a lower pressure P_2. Plot curves of T_2 versus P_2/P_1 for various constant values of k of common gases for $T_1 = 25°C$.

4·145 In a space heater, steam enters coils at 120 kPa, 96 percent quality, and condensate at 70°C leaves the coils. A fan blows air over the coils. If the entering air is at 16°C and the leaving air is to be at 28°C, what must be the ratio of air mass flow rate to steam mass flow rate?

4·146 Refrigerant 12 enters the condensing coils of a refrigerator at 35°C, 93 percent quality, and leaves as saturated liquid at a rate of 0.50 kg/min. Determine the volume flow rate of air (in m^3/s) required to condense the refrigerant if the air at 1 atm blown over the condenser is supplied at 20°C and leaves at 32°C.

4·147 Atmospheric air is to be cooled at a rate of 0.020 kg/s from 17 to −5°C by being blown over the evaporator coils of a refrigerator. Refrigerant 12 is supplied to the evaporator as saturated liquid at −10°C. Determine the flow rate of refrigerant needed if it is to leave the evaporator at a temperature not higher than 10°C.

4·148 Water at 110°C leaving a solar collector at a rate of 0.105 kg/s is partially vaporized. The quality of the mixture is 12 percent. Determine the flow rate of atmospheric air at 15°C required to cool the water to 90°C at constant pressure if the temperature of the air leaving the heat exchanger is not to exceed 40°C.

4·149 In a space heater, steam enters coils at 16 psia, 96 percent quality, and condensate at 160°F leaves the coils. A fan blows air over the coils. If the entering air is at 60°F and the leaving air is to be at 90°F, what must be the ratio of air mass flow rate to steam mass flow rate?

4·150 Refrigerant 12 enters the condensing coils of a refrigerator at 100°F, 91 percent quality, and leaves as saturated liquid at a rate of 0.90 lbm/min. Determine the volume flow rate of air (in cfm) required to condense the refrigerant if the air at 1 atm blown over the condenser is supplied at 70°F and leaves at 82°F.

4·151 A rigid vessel with a volume of 0.182 m^3 is divided into two equal volumes by a partition. Initially, hydrogen at 70.00 kPa, 20°C, is held on one side of the partition and the other side of the tank is evacuated. After 6 h it is discovered that some hydrogen has passed through the partition so that the pressure on the first side has dropped to 69.90 kPa with no change in temperature. Determine the average number of molecules per second passing through the partition.

4·152 A vessel having a volume of 0.32 m^3 containing air at 120.0 kPa, 12°C, was sealed 4200 years ago and placed in a tomb in Egypt. If, since that time enough air has leaked from the vessel to cause the pressure to drop to 119.9 kPa with no change in temperature, determine the average rate (molecules per second) at which air has left the vessel.

4·153 An ideal gas escapes from an insulated tank at a constant low rate of $\dot{m}$. Derive expressions for the rate of change of pressure and of temperature of the gas in the tank with respect to time in terms of $\dot{m}$ and the properties of the gas at any instant.

4·154 A tank of volume V contains an ideal gas initially at P_i and T_i. The gas leaks out of the tank through a small opening until the pressure drops to P_f. Heat is added to the gas to keep the temperature constant. Neglect kinetic energy changes. Determine the amount of heat which must be added. Express your answer in terms of quantities included in the following list: V, P_i, P_f, T_i, m_i, m_f.

4·155 An ideal gas in a well-insulated tank leaks out to the atmosphere through a porous plug. Write the first law as it applies to this system in terms of nothing more than the pressure, temperature, and mass of gas in the tank and specific heats and gas constant of the gas.

4·156 Air initially at P_i and T_i is held in an insulated tank. The pressure P_i is less than the pressure of the surrounding atmosphere, P_0. A small leak develops in the tank. Determine the temperature in the tank when the air pressure in the tank has reached P_0.

4·157 A gas accumulator is a device that stores varying amounts of gas to reduce flow rate variations in parts of a gas distribution system. A vertical cylinder with a weight-loaded piston supported by the gas in the cylinder is a constant-pressure accumulator. Consider such an accumulator that receives an ideal gas at a constant rate $\dot{m}_1$ and at constant pressure and temperature P_1 and T_1 and has an outward flow rate given by $\dot{m}_2 = \dot{m}_0 \sin \omega\tau$, where ω is a constant and τ is time. The accumulator is well insulated, and it operates continuously under the given conditions. Assume that the gas is well mixed in the accumulator so that the properties of the gas leaving are essentially the same as those of the gas in the accumulator. Determine the outlet temperature as a function of the specified conditions and time.

4·158 Air is drawn from a large test chamber by a vacuum pump at a variable mass rate of flow while a small amount of atmospheric air leaks into the chamber at a constant rate. The velocities of both entering and leaving air are very low. The surroundings are at a higher temperature than the air in the chamber. Write an energy balance equation for the system composed of the air in the test chamber, and simplify the equation as far as possible.

4·159 A cylindrical tank open at the bottom is mostly submerged in water, open end down, with the axis of the cylinder vertical. The pressure inside the cylinder is greater than atmospheric pressure P_0. A small valve in the top of the tank is opened, allowing air to escape and the water level inside the tank to rise. It is desired to determine the temperature of the air in the tank at any instant, assuming that the process is adiabatic. Write the differential equation for this system in terms of physical constants and P, V, T, and m, where these symbols stand for the pressure, volume, temperature, and mass of the air inside the tank at any instant.

4·160 Air in a rigid, well-insulated tank at any instant during a certain time interval is at a pressure P and a temperature T. Air at pressure P and a constant temperature T_1 enters the tank at a mass flow rate given at any instant by $\dot{m}_1 = B/P$, where B is a constant. Air at P and T leaves the tank at a different mass flow rate given at any instant by $\dot{m}_2 = DP$, where D is a constant. For this system write the first law and reduce it to a differential equation in terms of nothing other than variables P, T, and τ (time) and constants such as c_p, c_v, k, V (volume of tank), T_1, B, and D.

4·161 Derive an expression for the amount of heat which must be added to an ideal gas in a tank of volume V in order to increase its temperature from T_i to T_f while gas is bled from the tank to hold the pressure constant.

4·162 A cylinder, with its axis vertical and the cylinder head at the top, is fitted with a piston. Initially the piston is at the top of the cylinder with negligible clearance volume. The piston is held in position by the atmospheric pressure acting behind it. A small hole in the cylinder head is opened to let air from the atmosphere flow into the cylinder, allowing the piston to drop slowly while the pressure within the cylinder remains constant. The piston and cylinder are made of thermally insulating material. When the piston comes to rest against stops, the final temperature and volume of air in the cylinder are designated by T_f and V_f. Atmospheric temperature and pressure are T_0 and P_0. Determine T_f.

4·163 Solve Problem 4·162 with an initial clearance volume V_i containing a mass of air m_i at $P_i < P_0$ and $T_i = T_0$.

4·164 An arresting device is a large cylinder fitted with a piston and initially containing air at atmospheric pressure and temperature, P_0 and T_0. When the piston is struck, it immediately has a velocity V_i and then decelerates at a rate $dV/d\tau = AP$ while air escapes to the atmosphere through a small opening at a rate $\dot{m} = BP^{0.5}$, where A and B are constants. Neglecting heat transfer, derive an expression for the temperature of the air in the cylinder as a function of time, V_i, A, B, and the initial properties of the air. Sketch curves of the air pressure and the piston velocity versus time.

4·165 An empty one-room building has an internal volume of 6000 m³. The building is very well insulated. A small vent maintains the pressure inside the building equal to the outside atmospheric pressure of 101 kPa. How much heat must be added by an electric heater in the building to increase the inside air temperature from 7 to 20°C?

4·166 Carbon dioxide at 700 kPa, 5°C, is held in a tank of 0.50 m³ volume. Heat is added until the gas remaining in the tank is at 115°C while gas is bled from the tank to hold the pressure constant at 700 kPa. Determine the heat transfer.

4·167 Acetylene initially at 95 kPa, 40°C, is held in an insulated tank having a volume of 0.30 m³. An electric heating element within the tank adds heat at a constant rate of 0.10 kJ/s. Acetylene is bled from the tank to hold the pressure constant. Assuming that conditions are uniform throughout the tank at any instant, determine the time required for the acetylene to reach a temperature of 150°C.

4·168 A tank with a volume of 1.2 m³ is initially evacuated. Atmospheric air seeps into the tank through a porous plug so slowly that there is ample time for heat transfer to keep the temperature inside the tank equal to the atmospheric temperature of 20°C. Finally the pressure inside the tank is equal to the atmospheric pressure of 95 kPa. Determine the amount of heat added to or removed from the air in the tank.

4·169 An insulated tank with a volume of 0.80 m³ contains oxygen at 100 kPa, 20°C. It is connected to a large line in which oxygen at 800 kPa, 50°C, is flowing. Determine the final temperature of oxygen in the tank if the filling process is adiabatic.

4·170 A tank of 2.8 m³ volume contains air initially at 700 kPa, 40°C. Heat is added at a constant rate of 5 kJ/s while an automatic valve allows air to leave the tank at a constant rate of 0.02 kg/s. What is the temperature of the air in the tank 5 min after the initial condition?

4·171 Refer to Problem 4·170. Starting from the initial conditions, how long will it take for the air in the tank to reach 190°C?

4·172 Refer to Problem 4·170. Starting from the initial conditions, what is the pressure of the air in the tank after 10 min?

4·173 An insulated tank initially contains 10 kg of air at 600 kPa, 100°C. Air is to be bled from the tank to the atmosphere, causing both the pressure and the temperature of the air remaining in the tank to decrease. The bleed valve is to be regulated, however, so that the temperature drops at the constant rate of 0.10 degrees Celsius per second. Assuming that the process is adiabatic and that kinetic energy effects are negligible, determine the mass rate of flow from the tank as the bleeding of air *begins*.

4·174 Air at 100 psia, 40°F, is held in a tank of 20 cu ft volume. Heat is added until the air remaining in the tank is at 240°F while air is bled from the tank to hold the pressure constant at 100 psia. Determine the heat transfer.

4·175 Air initially at 14.0 psia, 100°F, is held in an insulated tank having a volume of 10 cu ft. An electric heating element within the tank adds heat at a constant rate of 0.10 B/s. Air is bled from the tank to hold the pressure constant. Assuming that conditions are uniform throughout the tank at any instant, determine the time required for the air to reach a temperature of 300°F.

4·176 A tank with a volume of 12 cu ft is initially evacuated. Atmospheric air seeps into the tank through a porous plug so slowly that there is ample time for heat transfer to keep the temperature of the air inside the tank equal to the atmospheric temperature of 60°F. Finally the pressure inside the tank is equal to the atmospheric pressure of 14.0 psia. Determine the amount of heat added to or removed from the air in the tank.

4·177 An insulated tank with a volume of 10 cu ft contains oxygen at 14.6 psia, 60°F. It is connected to a large line in which oxygen at 100 psia, 120°F is flowing. Determine the final temperature of oxygen in the tank if the filling process is adiabatic.

4·178 A tank of 100 cu ft volume contains air initially at 100 psia, 100°F. Heat is added at a constant rate of 5 B/s while an automatic valve allows air to leave the tank at a constant rate of 0.05 lbm/s. What is the temperature of the air in the tank 5 min after the initial condition?

4·179 Refer to Problem 4·178. Starting from the initial conditions, how long will it take for the air in the tank to reach 380°F?

4·180 Refer to Problem 4·178. Starting from the initial conditions, what is the pressure of the air in the tank after 10 min?

4·181 An insulated tank initially contains 10 lbm of air at 100 psia, 240°F. Air is to be bled from the tank to the atmosphere, causing both the pressure and the temperature of the air remaining in the tank to decrease. The bleed valve is to be regulated, however, so that the temperature drops at the constant rate of 0.10 degrees Fahrenheit per second. Assuming that the process is adiabatic and that kinetic energy effects are negligible, determine the mass rate of flow from the tank as the bleeding of air *begins*.

CHAPTER FIVE

The Second Law

5·1 Limitations of the First Law

The first law of thermodynamics expresses a relationship between heat and work and makes possible the definition of stored energy. When extended to give a simple relationship among work, heat, and the change in stored energy of a system, the first law is known as the law of conservation of energy. The first law does not limit the extent of any energy conversion nor does it indicate whether any particular process is possible or not. Yet experience shows that at least one type of desirable energy conversion—heat to work—cannot be carried out completely. Furthermore, certain processes, that would in no way violate the first law, cannot occur. Let us examine these two lessons of experience.

Consider a gasoline engine. Energy stored in the fuel and in the combustion air is delivered to the engine. Energy leaves the engine as work via the drive shaft, as heat, and as stored energy in the exhaust gases. There is also an energy transfer as flow work of the fluids entering and leaving. For economical operation the work output should be as great as possible for a given amount of energy input. Thus the energy leaving the engine in the exhaust gases and as heat should be reduced to a minimum. As far as the first law is concerned, these energy losses could be reduced to zero, and the work output would equal the energy input. All attempts to obtain such performance from an engine have failed. No matter what ingenious accessories have been used, the complete conversion of fuel energy into work by an engine has not been accomplished. Some energy has always been rejected either in the exhaust gases or as heat. It is reasonable to ask "Is this necessary?" or "Is there some law of nature that accounts for the limitation on this energy conversion?"

A steam power plant affords another example of a limited conversion of energy stored in fuel into work. Even the best steam power plants require for the production of 100 kJ of work an energy input of about 250 kJ. Thus, for every 100 kJ of work produced, about 150 kJ is rejected to the surroundings as some form of energy other than work. Again, the question arises: "Is this necessary?"

Consider now two blocks of steel at different temperatures that are enclosed together in a thermally insulated box. Of course, energy will be transferred from the higher-temperature to the lower-temperature block as heat. If there is nothing inside the box besides the two steel blocks, the amount of energy lost by the higher-temperature block will be equal to the amount of energy gained by the lower-temperature block. This is in accordance with the first law. The first law, however, would be satisfied also by a process whereby energy was transferred from the lower-temperature block to the higher-temperature block. This latter process never occurs, however, so satisfaction of the first law alone does not insure that a process can occur.

As another example, a spinning flywheel mounted on a shaft between two bearings will come to rest as a result of friction in the bearings. In this process the kinetic energy of the flywheel and shaft is reduced to zero while the internal energy of the bearings, the lubricant, and part of the shaft is increased by the same magnitude. Energy is conserved. Energy would also be conserved in the reverse process that would cool the bearings, lubricant, and parts of the shaft, and accelerate the flywheel and shaft until their kinetic energy equals the decrease in internal energy of those parts that were cooled. It is common knowledge that this process does not occur. Although we recognize this intuitively in this simple case, we naturally wonder if there is some law of nature which will support our intuition in this case and guide us in more complex cases where intuition alone will not tell us whether a certain process is possible or not.

These illustrations of the limitations of the first law of thermodynamics show us the need for another general principle of widespread application: the second law of thermodynamics.

5·2 The Second Law of Thermodynamics

Studies of questions similar to those in the preceding section and much experience, including many deliberate experiments, have led to the second law of thermodynamics. This law has been stated in many different forms, some of which appear at first to bear no relation whatsoever to the other forms. If any one of the statements of the second law is accepted as a postulate, all the other statements can then be proved from this starting point. The statement which is taken as the starting point, however, cannot be derived from any other law of nature. Two of the well-known statements of the second law of thermodynamics are known as the Clausius statement and the Kelvin–Planck* statement.

The Clausius statement may be given as follows: *It is impossible for any device to*

*Rudolph Julius Emmanuel Clausius (1822–1888) was a German mathematical physicist. After a study of the work of Sadi Carnot (see Secs. 6·6 and 7·1) he presented in 1850 a clear general statement of the second law. He applied the second law and showed the value of the property entropy (see Chapter 8) in an exhaustive treatise on steam engines. Although he made significant contributions in the areas of optics, kinetic theory of gases, and electrolysis, he is most famous for his work related to the second law. William Thomson (Lord Kelvin) (1824–1907), one of the outstanding physicists of all time, was for 53 years professor of natural philosophy at the University of Glasgow. In 1851 he presented a paper in which the first and second laws were combined for the first time. In addition to his work in helping to firmly establish the first law and in formulating the second law of thermodynamics, he published papers on geophysics, electricity, magnetism, telegraphy, navigation, and many other branches of science. He invented many instruments for scientific and engineering work. The encouragement which he gave his students and other scientists played a large part in scientific advances made by others. Max Planck (1858–1947), professor of physics at the University of Berlin, clarified several concepts in thermodynamics but is best known for his work on radiation in which he laid the foundation of the quantum theory.

operate in such a manner that it produces no effect other than the transfer of heat from one body to another body at a higher temperature. This statement warrants close investigation. It does not say that it is impossible to transfer heat from a lower-temperature body to a higher-temperature body. Indeed, this is exactly what a refrigerator does. A refrigerator does not operate, though, unless it receives an energy input, usually in the form of work. This transfer of energy from the surroundings constitutes an effect other than the transfer of heat from the lower-temperature body to the higher-temperature body. The "no effect" mentioned in the Clausius statement of the second law includes effects within the refrigerating device itself. It is possible to build and operate a device which will absorb heat from a lower-temperature body, reject heat to a higher-temperature body, and produce no other effect in the surroundings. However, this device will itself experience an effect and be in a different state at the conclusion of the process from the one it was in at the beginning. Thus it has produced some effects other than the transfer of heat from one body to another body at a higher temperature and therefore does not violate the Clausius statement. Questions about effects within the device are avoided by having the device operate cyclically so that it always returns to its initial state.

The Clausius statement is sometimes given as "Heat cannot of itself pass from a cold to a hot body." This statement requires interpretation. In the words of Planck, "As Clausius repeatedly and expressly pointed out, this principle does not merely say that heat does not flow directly from a cold to a hot body—that is self-evident, and is a condition of the definition of temperature—but it expressly states that heat can in no way and by no process be transported from a colder to a warmer body without leaving further changes." The form of the Clausius statement given in the preceding paragraph makes this interpretation clear. One advantage of the Clausius statement over many other statements of the second law is that many people find it easy to accept intuitively.

A statement that pertains more directly to heat engines is the Kelvin–Planck statement. The Kelvin–Planck statement of the second law is as follows: *It is impossible for any device to operate in a cycle and produce work while exchanging heat only with bodies at a single fixed temperature.* Notice that the value of the temperature of the bodies in the surroundings is not mentioned. If a device can operate cyclically and produce work while exchanging heat only with bodies at 500°C, it can be made to do the same thing while exchanging heat only with bodies at, say, 0°C. Application of the first law to a device which violates the Kelvin–Planck statement of the second law (recalling that $\oint dE = 0$) shows that the net work produced would be equal to the net amount of heat received from bodies in the surroundings which are all at the same temperature. Such a device is called a perpetual-motion machine of the second kind. Therefore, the Kelvin–Planck statement is sometimes paraphrased as "A perpetual-motion machine of the second kind is impossible."

A common version of the Kelvin–Planck statement is: It is impossible for any device operating in a cycle to absorb heat from a single reservoir and produce an equivalent amount of work. The term *reservoir* refers to an *energy reservoir,* which is defined as a body or system that a finite amount of energy can be drawn from or added to without causing any appreciable change in its temperature. In practice, the atmosphere or the water of a river or lake can serve as an energy reservoir. Thousands of kilojoules can be discharged to river water by a power plant without appreciably raising the temperature of the river. A furnace atmosphere that is maintained at a constant temperature by the combustion of fuel serves as an energy reservoir in a steam power plant.

The Clausius and Kelvin–Planck statements of the second law are entirely equivalent to each other in their consequences. This equivalence can be demonstrated by showing that the violation of either statement can result in a violation of the other one. Each case will now be shown.

Referring to Fig. 5·1*a*, the device marked "Clausius violator" causes heat Q to be transferred from the energy reservoir at T_L to the one at the higher temperature T_H without causing any other effects. This kind of device is impossible according to the Clausius statement. To show that the Kelvin–Planck statement can be violated any time the Clausius statement is violated, let a heat engine operate with heat input Q from the reservoir at T_H and with an amount of heat Q_L rejected to the reservoir at T_L. Experience shows that actual heat engines can operate in this manner. Application of the first law shows that the amount of heat rejected Q_L is equal to $(Q - W)$, where W is the work output of the heat engine. Since the Clausius violator rejects an amount of heat at T_H equal to that absorbed at the same temperature by the heat engine, the operation of these two devices, bounded by a broken line in Fig. 5·1*a*, produces no change in the reservoir at T_H. In fact, the reservoir could be eliminated by having the Clausius violator transfer heat directly to the heat engine. Thus the system comprised of the two devices absorbs a net amount of heat $(Q - Q_L)$ from a single reservoir (at T_L), produces work, and produces no other effects. This is a violation of the Kelvin–Planck statement. Thus it is shown that a violation of the Clausius statement results in a violation of the Kelvin–Planck statement.

The device labeled "Kelvin–Planck violator" in Fig. 5·1*b* absorbs an amount of heat Q from the reservoir at T_H, produces an equivalent amount of work W, and produces no other effects. This kind of device violates the Kelvin–Planck statement. To show that the operation of such a device can result in a violation of the Clausius statement, let a refrigerator be used to transfer heat from the reservoir at T_L to the reservoir at T_H. Such a refrigerator can certainly be operated, but it requires some work input. Let the refrigerator be driven by the Kelvin–Planck violator. Then see what the system composed of both devices is doing. It is operating cyclically, causing heat to be transferred from one reservoir to another one at a higher temperature, and producing no other effects! Thus a violation of the Kelvin–Planck statement results in a violation of the Clausius statement. This

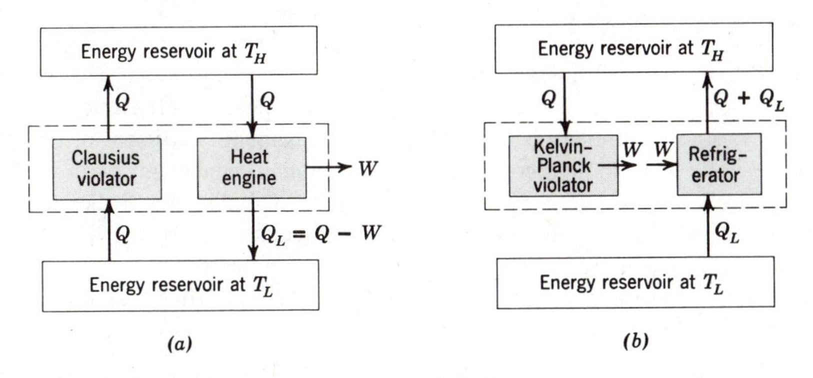

Figure 5·1. Equivalence of the Clausius and Kelvin–Planck statements of the second law.

paragraph and the preceding one therefore demonstrate the equivalence of these two statements of the second law of thermodynamics.

We will deduce several corollaries from the second law in the following chapter. To disprove the second law it is necessary only to disprove any one of its corollaries. As yet, all attempts to do this have failed.

5·3 Perpetual-Motion Machines

Any perpetual-motion machine may be classified as one of three kinds. A *perpetual-motion machine of the first kind* violates the first law by operating in a cycle and producing a greater net work output than the net amount of heat put into the machine. Many such machines have been proposed, and several of them have been patented, but none of these has actually operated as a perpetual-motion machine of the first kind.

As mentioned earlier, a *perpetual-motion machine of the second kind* violates the second law by producing work while operating cyclically and exchanging heat only with bodies at a single fixed temperature. No violation of the first law is involved. No energy is created, but such a device is just as valuable as a perpetual-motion machine of the first kind, because a virtually limitless supply of energy is available in the atmosphere or the oceans for input to such a machine.

Occasionally the name *perpetual-motion machine of the third kind* is applied to devices that, once set in motion, continue in motion for an indefinitely long time without slowing down. A spinning top on a frictionless pivot or a spinning flywheel mounted in frictionless bearings is an example of this type of perpetual-motion machine. It violates neither the first nor the second law of thermodynamics but requires only the elimination of friction. Although no one has succeeded in completely eliminating friction from such devices, the extent to which friction can be reduced, short of complete elimination, depends only on the time and money available. Notice that a perpetual-motion machine of the third kind produces no work and would therefore not be as valuable as one of the first or second kind.

5·4 The Value of the Second Law

Why is the second law valuable? In later chapters we will see that the second law and its corollaries provide means for (1) determining the maximum possible efficiency of a heat engine under various conditions, (2) determining the maximum coefficient of performance of a refrigerator under various conditions, (3) determining whether any particular process we may conceive is possible or not, (4) predicting in which direction a chemical reaction or any other process will proceed, (5) defining a temperature scale which is independent of physical properties, and (6) correlating physical properties.

For example, a few specific questions we will be able to answer after studying the applications of the second law are:

1. In a steam power plant, the maximum furnace temperature is 1400°C, and cooling

water is available at 15°C. What is the maximum possible thermal efficiency of the steam power plant, no matter how many refinements are included in its design?

2. A refrigerator located in a room where the air temperature is 20°C freezes 5 kg of ice per hour from water initially at 10°C. The refrigerator rejects heat to the air in the room. What is the absolute minimum power requirement of the refrigerator?
3. Air expands adiabatically from 200 kPa, 40°C, to 100 kPa. What is the lowest possible final temperature?
4. Is it possible to compress air adiabatically from 100 kPa, 15°C, to (*a*) 200 kPa, 30°C? (*b*) 200 kPa, 105°C?
5. If carbon dioxide is expanded adiabatically through a nozzle from zero velocity at 40 psia, 100°F, to 15 psia, what is the maximum velocity it can reach?
6. Sixteen kilograms of oxygen and 28 kg of carbon monoxide fill a tank. The mixture is heated until it is at 250 kPa, 500°C. To what extent does the reaction $CO + \frac{1}{2}O_2 \rightarrow CO_2$ occur?
7. Is there a limit as to how low a temperature can be attained?
8. Steel expands when heated. When it is stretched adiabatically within the elastic limit, does its temperature increase or decrease?

The answers to questions such as these can be obtained from the second law through a process of deductive reasoning. So that we do not have to reason all the way from the second law itself each time we answer such questions, we carry some of the reasoning through once and establish certain useful definitions and corollaries of the second law from which we can then begin our reasoning for particular applications. The next chapter answers none of the questions listed above, but it does establish some definitions which will help us in answering all these questions in later chapters.

5·5 Conclusion

The second law of thermodynamics is a far-reaching principle of nature that has been stated in many forms. For engineers, one of the following two forms is usually the most valuable:

The Clausius statement: *It is impossible for any device to operate in such a manner that it produces no effect other than the transfer of heat from one body to another body at a higher temperature.*

The Kelvin–Planck statement: *It is impossible for any device to operate in a cycle and produce work while exchanging heat only with bodies at a single fixed temperature.*

These two statements of the second law and many other statements are entirely equivalent in their consequences. If any one of them is taken as a starting point, all of the others can be deduced.

CHAPTER SIX

Reversible and Irreversible Processes and Cycles

The preceding chapter mentioned that certain processes that can be conceived do not occur. The second law is useful in distinguishing such impossible processes. Among the possible processes, some bring about changes in a system and its surroundings that can be completely undone: that is, both the system and the surroundings can be returned to their initial states. Other processes that occur cause changes such that the system and all of its surroundings can never *both* be returned to their initial states! These processes are called reversible and irreversible processes, respectively. Engineers are interested in reversible processes because for devices that produce work, such as engines and turbines, reversible processes deliver more work than corresponding irreversible processes. Also, refrigerators, compressors, fans, and pumps require less power input when reversible processes are used in place of the corresponding irreversible ones.

This chapter discusses reversible and irreversible processes and also some engine and refrigerator cycles comprised of reversible processes.

6·1 Reversible and Irreversible Processes: Definitions

A process is *reversible* if, after it has occurred, both the system and all the surroundings can by any means whatsoever be returned to the states they were in before the process occurred. Any other process which occurs is *irreversible*.

To determine whether the system and surroundings can be returned to their initial states after a process has occurred, it is necessary to apply the second law. Use of the second law for this purpose is illustrated in the following section.

6·2 Reversible and Irreversible Processes: Characteristics; Illustrations

We will examine irreversible processes before reversible processes for two reasons. First, it is often easier to show the impossibility of meeting the test for reversibility than it is to show a method of meeting it. Second, once certain irreversible occurrences are identified, we often recognize reversible processes simply by the absence of these irreversible events.

If a process is reversible, then the reverse process—the one which restores both the system and the surroundings to their initial states—is possible. If a process is irreversible, the reverse process is impossible. Determining whether a process is reversible or irreversible is then a matter of determining whether the reverse process is possible or impossible.

One way to prove that a process is impossible is as follows: (1) Assume that the process is possible. (2) Combine this process with other processes, known from experience to be possible, to form a cycle which violates the second law. If such a cycle can be devised, then the assumption of step 1 is false and the process in question is impossible.

Consider the process in which a gas in a closed, rigid, thermally insulated tank is stirred by a paddle wheel. The system is the gas within the tank. (Assume there is never more than a negligible amount of heat transfer between the gas and the paddle wheel itself.) Let the paddle wheel be turned by the action of a falling weight that turns a pulley on the shaft (see Fig. 6·1). The motion of the paddle wheel is resisted by shearing forces in the gas, and thus work is done on the gas by the paddle wheel. The gas changes from state *A* to state *B*. Application of the first law shows that the internal energy of the gas is increased. The temperature of the gas increases.

The question is now: Is this process reversible? If it is reversible, then it is possible to restore both the system and the surroundings to their initial states after the process has occurred. That is, there must be some process that results in the weight's being lifted to its initial position while the internal energy of the gas (and hence its temperature) decreases. The gas changes from state *B* to state *A*. No change other than the lowering of the weight was made in the surroundings during the original process. Therefore, no change in the surroundings other than the lifting of the weight can be made during the reverse process if it is to result in both the system and the surroundings being returned to their initial states. If this reverse process is possible, then the stirring process is reversible; if this reverse process is impossible, then the stirring process is irreversible. Thus a question equivalent to *Is the stirring process reversible?* is: *Is the reverse process possible?* Let us answer this by first *assuming* that the reverse process *is possible*. Then consider a cycle composed of two processes:

Process 1. The process described above in which the weight is raised as the temperature and

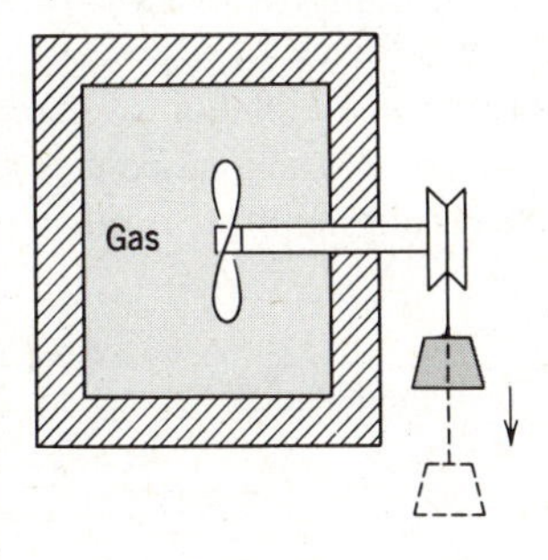

Figure 6·1. Stirring process. System changes from state *A* to state *B* as the weight drops.

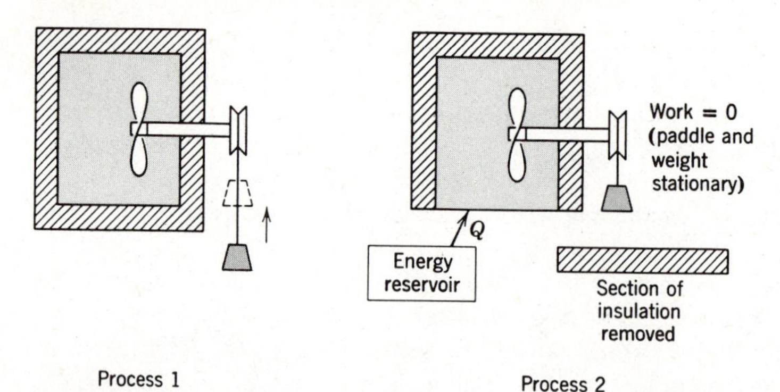

Figure 6·2. Process 1: System is changed from state B to state A as the weight is lifted (assumed to be possible). Process 2: System is changed from state A to state B.

internal energy of the gas decrease. (This is the process we have assumed to be possible.) The system changes from an initial state B to state A (see Fig. 6·2). $T_A < T_B$.

Process 2. A process in which heat is transferred from some constant-temperature energy reservoir in the surroundings to the gas while the paddle wheel is stationary. (Part of the thermal insulation of the tank must be removed during this process.) This process continues until the gas is brought to its initial temperature T_B. (The energy reservoir must therefore be at a temperature higher than T_B.)

The net results of this cycle are: (1) The system which has executed a cycle is restored to its initial state, and (2) two changes have occurred in the surroundings: (*a*) The weight is at a higher level than it was initially, and (*b*) the amount of energy stored in the energy reservoir has been decreased. Application of the first law shows that the energy decrease of the reservoir equals the energy increase of the weight. Thus the system is a device that operates in a cycle, exchanges heat with a single reservoir, and does work. This is precisely the kind of device that the second law declares to be impossible. Now it is a matter of experience that process 2 is possible; therefore if the cycle is impossible, it must be so because process 1 is impossible. Thus our assumption that process 1 is possible is false: Process 1 is impossible. Since process 1 is impossible, the original stirring process is irreversible. The only alternative is that the second law is false, and a tremendous amount of experience argues against this alternative.

Summarizing the reasoning above, the process considered is the stirring of a gas in a rigid, thermally insulated tank. The question is: Is this process reversible? It was shown that the assumption that the process is reversible permits the occurrence of a process which, when combined with a process known to be possible, leads to a violation of the second law. Thus we conclude that the assumption is false and the stirring process is irreversible. The following examples show the same line of reasoning applied to other processes.

Example 6·1. A block initially at rest on an inclined plane slides down the plane and comes to rest at a lower elevation. Is this process reversible?

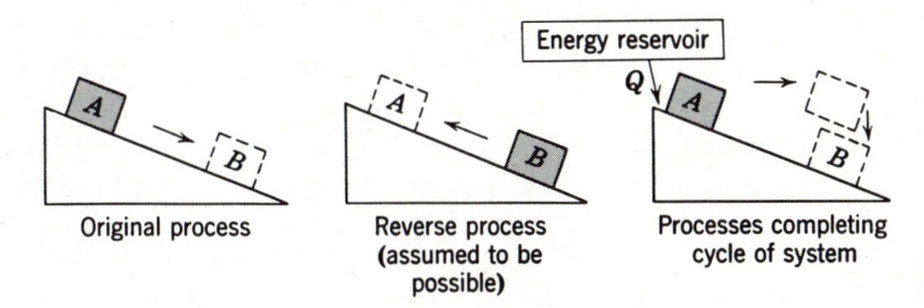

Example 6·1.

Solution. The system is comprised of the block and the plane. During this process, the internal energy of the block and plane increases by an amount equal to the decrease in potential energy of the block. The temperatures of at least parts of the block and plane increase. There is no change in the surroundings. If this process is reversible, the reverse process is possible. The *reverse* process would lift the block to its initial position while the internal energy of the system (block and plane) decreases by an amount equal to the increase in potential energy of the block. The temperatures of the block and plane would decrease. Since there was no change in the surroundings during the original process, there must be none during the reverse process.

Assume that the reverse process is possible. It can be incorporated into a cycle that violates the second law in the following manner: First, let the reverse process occur, lifting the block from position B to position A while decreasing the internal energy (and temperatures) of the block and the plane. Next, add heat to the system from some energy reservoir in the surroundings to increase the internal energy of the system by the amount it was just decreased. Move the block horizontally off the plane (doing no work) until it is directly above position B. Then lower the block to position B, allowing it to do work equal to its decrease in potential energy. (For example, a weight in the surroundings can be lifted by means of a pulley arrangement.) The system has thus executed a cycle. The net result of this cycle has been to take heat from a single energy reservoir, perform an equivalent amount of work, and produce no other effects.

In accordance with the second law, this cycle is impossible. The heating of the system is possible, and the lowering of the block as it does work is possible; therefore it must be the process in which the block moved from B to A while the system internal energy decreased that is impossible. This is the reverse of the original process in which the block slides down the plane; hence this original process must be irreversible.

Example 6·2. Demonstrate that the transfer of heat across a finite temperature difference is irreversible.

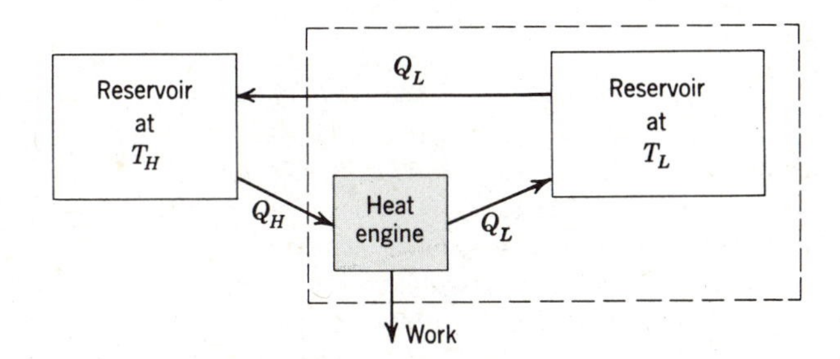

Example 6·2.

Solution. *If* the transfer of heat from a body at a temperature T_H to a body at a lower temperature T_L is reversible, then the transfer of heat from the body at T_L to the one at T_H without any other effects is possible. Assume that this transfer of heat from T_L to T_H is possible. Then construct a cycle as follows: (1) Let an amount of heat Q_H be transferred from a reservoir at T_H to a heat engine which operates cyclically, producing work and rejecting an amount of heat Q_L to a reservoir at a lower temperature T_L. It is a matter of experience that this can be done. (2) Then let an amount of heat Q_L be transferred from the reservoir at T_L to the one at T_H in accordance with the assumption that such a process is possible. Since Q_L was added to the low-temperature reservoir in the first process and the same amount of heat was withdrawn from it during the second process, this reservoir has executed a cycle. The heat engine also executed a cycle. Therefore the system enclosed by the broken line on the diagram (the heat engine and the reservoir at T_L) has executed a cycle. Notice that during this cycle this composite system has produced work while exchanging heat with a single reservoir. Such a cycle violates the second law. Checking back to see which process of the proposed cycle might actually be impossible and thus prevent the execution of such a cycle, we see that the first process is proved by experience to be possible, so the second process, which we *assumed* to be possible, must actually be impossible. If this process—the transfer of heat from T_L to T_H—is impossible, then the transfer of heat across the finite temperature difference from T_H to T_L must be irreversible.

Example 6·3. Demonstrate that the free expansion of a gas is an irreversible process. An example of a free or unrestrained expansion of a gas is the following: An insulated tank is separated into two parts by a partition. A gas is held in the tank on one side of the partition, and on the other side of the partition the tank is evacuated. An opening is then made in the partition, and the gas expands to fill the entire tank. No work is done.

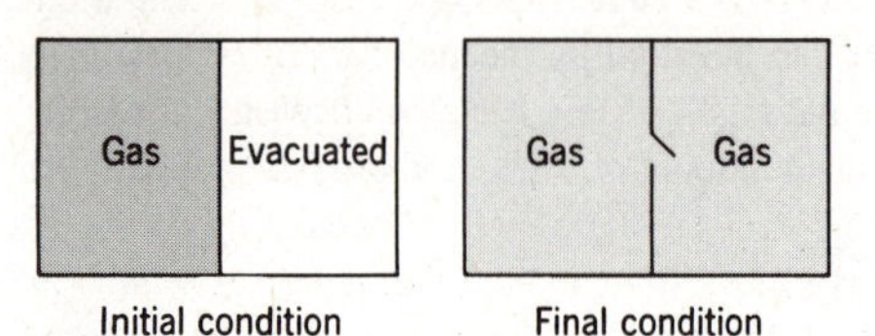

Example 6·3. Statement.

Solution. In order to show that a free expansion is irreversible, we will show that the reverse process is impossible. First, assume that the reverse process is possible. Consider the gas as the system. This reverse process would begin with the gas occupying the entire tank and would result in all the gas on one side of the partition passing through the partition to the other side against the increasing pressure of the gas. This would occur with no interaction with the surroundings since the free expansion process occurred with no interaction with the surroundings. Consider a cycle made up of three processes: (1) Starting with the gas all on one side of the partition, let part of it expand through an engine and into the other part of the tank until the pressure is the same on the two sides of the partition. In expanding through the engine the gas does work so that its internal energy is decreased. (2) Remove part of the tank insulation, and add heat from an external reservoir to the gas until its internal energy is restored to its initial value. (Notice that for a free expansion process $\Delta U = 0$, since during such a process work $= 0$ and $Q = 0$.) (3) Starting with the gas that is now in the condition it would have been in following a free expansion from its initial condition, let the reverse of a free expansion occur to restore the system to its initial state. Thus a cycle is completed.

Inspection of this *cycle* reveals that it results in a production of work while heat is absorbed from a single reservoir. This cycle thereby violates the second law. Reinspection of the cycle reveals that processes 1 and 2 are shown by experience to be possible whereas process 3 is

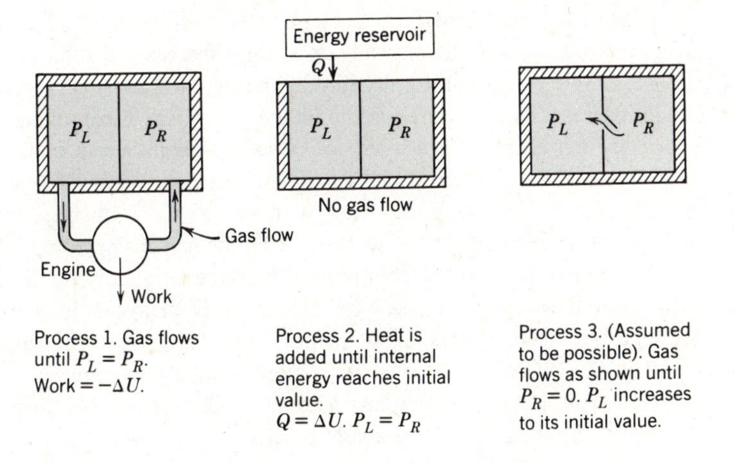
Energy reservoir
Q
P_L
P_R
P_L
P_R
P_L
P_R
No gas flow
Gas flow
Engine
Work
Process 1. Gas flows until $P_L = P_R$. Work $= -\Delta U$.
Process 2. Heat is added until internal energy reaches initial value. $Q = \Delta U$. $P_L = P_R$
Process 3. (Assumed to be possible). Gas flows as shown until $P_R = 0$. P_L increases to its initial value.

Example 6·3. Solution.

assumed to be possible. If the second law is accepted, then the assumption regarding process 3 is false. The reverse of a free expansion is therefore impossible and a free expansion must be irreversible.

By reasoning similar to that used in the preceding examples it can be shown that processes involving mixing,* inelastic deformation of a substance, and certain other effects are also irreversible. *A reversible process must therefore involve no*

a. Friction
b. Heat transfer across a finite temperature difference
c. Free expansion
d. Mixing
e. Inelastic deformation

Various other effects (such as an electric current flow through a resistance, to name one) are also irreversible but are not listed here. In all cases the test for reversibility involves the application of the second law of thermodynamics.

At this point you may infer from the three examples above that we are using a second law formalism to reach conclusions that are obvious from common experience. Indeed, it is well that these simple examples do confirm common experience. In one sense, doing so makes them good examples. The reasoning that may appear unnecessary here, however, is valuable because it applies to cases where common experience does not provide an obvious conclusion. Some of the questions listed in Sec. 5·4, in fact, refer to such cases, and we will encounter others in subsequent chapters.

Let us now examine some reversible processes and then draw conclusions as to some features that are common to all reversible processes. Consider first a system comprised of a gas trapped in a cylinder fitted with a frictionless gas-tight piston. Let the cylinder

*Under very special conditions mixing may be done reversibly by using semipermeable membranes, that is, partitions that are permeable to one or more substances and impermeable to all others.

and piston be made of a material that is a perfect heat insulator. If the piston is slowly pushed into the cylinder, the pressure and temperature of the gas increase. If the piston moves very slowly, the pressure increases uniformly throughout the gas, and so does the temperature. A very small decrease in the external force on the piston will permit the gas to expand, and, if the expansion is allowed to proceed very slowly, the pressure decreases uniformly throughout the system. For each position of the piston, the pressure of the gas during the expansion is the same as it was during the compression. Consequently, the work done by the gas during expansion equals the work done on the gas during compression. When the gas has expanded to its initial volume, all the work originally done on the system has been returned to the surroundings as work. There has been no heat transfer. The surroundings have therefore been returned to their initial state. The system has also been returned to its initial state. Thus, after a very slow frictionless adiabatic compression is completed, it is possible to restore both the system and the surroundings to their initial states. Consequently the very slow frictionless adiabatic process is reversible.

If the adiabatic compression is performed by a rapid inward motion of the piston, the process is not reversible. The pressure near the piston face is higher than that elsewhere in the cylinder. A pressure wave is thus initiated, and it travels through the gas until the pressure is again uniform. Then, even if the gas expanded slowly to its initial volume, for each position of the piston the pressure near the piston face is lower than it was during the compression process. The work done during the expansion is therefore less than that done on the gas during compression. Say that the work input to the system during the rapid compression amounts to 10 kJ and the work output obtained by expanding the fluid to its initial volume is 8 kJ. At the end of the expansion process, the stored energy of the system is 2 kJ greater than it was initially; but, since the system volume equals its initial value, the 2 kJ of excess stored energy cannot be removed *as work* while the system is restored to its initial state. Let the stored energy be reduced by a transfer of 2 kJ of heat from the system to the surroundings while the piston is stationary. The system has now been returned to its initial state. Turning our attention now to the surroundings, we see that 10 kJ of work was taken from the surroundings in order to compress the gas. Perhaps the work was done by the lowering of a weight or the unwinding of a coil spring in the surroundings. Then the system performed 8 kJ of work on the surroundings to raise the weight part of the way to its initial position or to rewind the spring partially. Then 2 kJ of heat was transferred to the surroundings. In order for the surroundings to be returned to their initial state, this 2 kJ of heat must be converted completely into work to raise the weight or wind the spring without causing any other effects. Any device which could perform this conversion would violate the second law; hence we conclude that the system and the surroundings cannot *both* be restored to their initial states. Therefore, the adiabatic compression of the gas during which the pressure is not uniform throughout the gas is irreversible.

A reversible process may be approximated by the elongation or compression of a spring. If a very small load is slowly applied to a tension spring, the spring will elongate a minute distance. After elongation, if the load is reduced so that the spring is allowed to contract to its original position, the work performed by the spring as it contracts will be approximately equal to that required to stretch the spring. It is important to note that this process approaches a reversible process only if an infinitesimal force is applied and reduced gradually; otherwise vibrations and other effects will occur which would make the process irreversible.

Another example of a reversible process is the frictionless isothermal process. Consider a gas in a cylinder having a gas-tight frictionless piston and cylinder walls that are perfect conductors of heat. If the temperature of the surroundings is slightly greater than the temperature of the gas, the gas will receive heat from the surroundings and expand, and work will be performed. It is assumed that the process takes place very slowly. If the temperature of the surroundings is reduced to a value slightly less than that of the gas, heat will flow from the gas, and work will be done in compressing it. In this process, minute temperature differentials are to be considered, and the time required for the expansion or compression to occur is extremely long. This type of process fulfills the requirements of a reversible process as the temperature difference between the system and the surroundings approaches zero. Notice that the temperature must be uniform throughout the gas so that only an infinitesimal change in the temperature of the surroundings will cause the surroundings to be at a higher or lower temperature than the entire system.

Next consider a cylinder fitted with a frictionless piston and containing a small quantity of water. The cylinder is assumed to be immersed in a *constant-temperature* bath. If the piston is loaded by a weight equal to the product of the area of the piston and the saturation pressure corresponding to the temperature of the bath, the system is in a condition of equilibrium. The force exerted by the vapor pressure of the water in the cylinder is equal to the weight applied to the piston. If the weight applied to the piston is decreased an infinitesimal amount, the water will start to evaporate, thus absorbing heat from the bath. At any instant, if the load is increased to a value infinitesimally greater than the product of the area and saturation pressure at the temperature of the bath, the vapor will start to condense, and heat will flow into the surrounding bath. Thus, by slightly altering the load on the piston, the process may be made to proceed in either direction. As the load differential approaches zero, the process becomes reversible. It is not possible to perform the experiment in the laboratory because of the frictional forces that would be present.

Can the transfer of heat to or from a system be reversible if the temperature of the system varies with time? The answer is yes, provided the temperature of the surroundings also varies with time so that the difference in temperature between the system and the surroundings is never more than an infinitesimal amount.

If heat is transferred from one body to another, the greater the temperature difference between the two bodies, the greater the rate of heat transfer will be. This is true whether the heat is transferred by conduction, convection, or radiation. As the temperature difference between the two bodies is made smaller, the time required to transfer a given quantity of heat increases. As the temperature difference approaches an infinitesimal value, the time required to transfer any finite amount of heat grows toward an infinite value. Obviously, in the design of heat transfer devices and processes, energy transfer rates approaching zero are unacceptable. However, it is only when the temperature difference is infinitesimal (so that the direction of the heat transfer can be changed by an infinitesimal change in the temperature of one of the bodies) that the process can be considered reversible. Thus reversible heat transfer is the limiting case of heat transfer as the temperature difference between two bodies approaches zero.

Another example of a reversible process is the steady frictionless adiabatic flow of a fluid through a nozzle. Application of the first law to such a system shows that, as kinetic energy increases in the direction of flow, the enthalpy decreases. If the nozzle is followed by a frictionless diffuser, as shown in Fig. 6·3, the fluid undergoes an increase in enthalpy and a decrease in kinetic energy between sections 2 and 3, and it can be discharged in a state 3 which is identical with state 1. Thus after the nozzle process has occurred, it is possible to restore the fluid flowing and all parts of the surroundings to their initial states.

Study of various reversible and irreversible processes such as those just described leads to certain conclusions regarding reversible processes:

1. A reversible process must be such that, after it has occurred, the system and the surroundings can be made to traverse in the reverse order the states they passed through during the original process. All energy transformations of the original process would be reversed in direction but unchanged in form or magnitude.

2. The direction of a reversible process can be changed by making infinitesimal changes in the conditions that control it.

3. During a reversible process, the system and the surroundings must each at all times be in states of equilibrium or infinitesimally close to states of equilibrium; that is, the process must be quasistatic.

4. A reversible process must involve no friction, unrestrained expansion, mixing, heat transfer across a finite temperature difference, or inelastic deformation.

A reversible process must meet each of the conditions listed. If any one of the conditions listed is not met by a process, the process is irreversible. Thus we can apply either these conditions or the definition of a reversible process to test any particular process for reversibility.

A test for reversibility which is occasionally useful is the following: If in any process work is used to accomplish some effect that could have been accomplished wholly or in part by heat, the process is irreversible. Consider the stirring of a gas in a closed tank by means of a paddle wheel. Since there is no decrease in volume as energy is added to the system as work, the same effect could be accomplished by the addition of heat to the system. The stirring process is irreversible.

Work can always be completely converted into heat, but the extent to which heat can be converted into work is limited. Therefore, work is the more valuable form of energy in transition. An irreversible process that uses work to produce an effect which could be

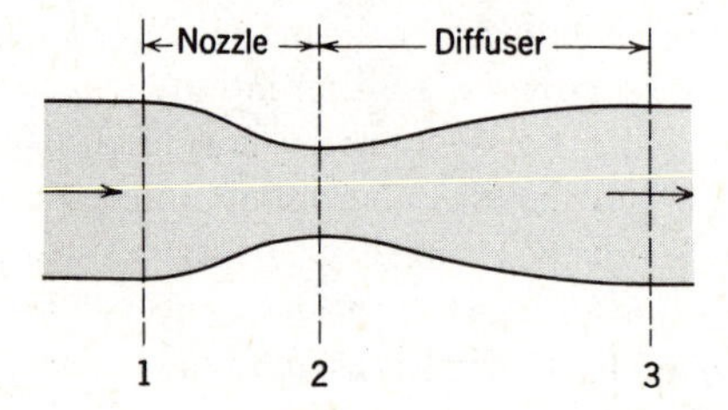

Figure 6·3. Nozzle and diffuser.

produced by the less valuable form of energy, heat, is consequently an undesirable process in this respect.

Reversible processes are occasionally spoken of as "maximum work" processes. The reason is that if a system can pass through a given series of states reversibly and also irreversibly, the amount of work done by the system will be greater for the reversible process than for any irreversible process following the same path. If the series of states requires that work be done on the system (as in the compression of a gas), the reversible process will require a smaller work input than the irreversible one. It is still correct to call the reversible process a *maximum* work process as long as the term work is taken to mean work_{out}. Say that the work input for a reversible compression is 8 kJ and for an irreversible compression is 10 kJ. Then

$$\text{work}_{\text{out,rev}} = -8 \text{ kJ} > \text{work}_{\text{out,irrev}} = -10 \text{ kJ*}$$

As an illustration of a system that can be made to pass through the same series of states reversibly and irreversibly, consider a gas which expands reversibly against a piston in a cylinder (see Fig. 6·4*a*) while heat is added at the same rate at which work is done. The internal energy of the gas consequently remains constant. On the other hand, the same mass of gas at the same initial pressure and temperature might be held in part of a vessel as shown in Fig. 6·4*b*. The vessel is thermally insulated and fitted with many very thin sliding partitions that can be withdrawn in sequence to let the gas expand. Initially all the gas is to the left of the extreme left-hand partition. The rest of the vessel is evacuated. No work is done as the partitions are withdrawn, and no heat is transferred. Consequently the gas expands at constant internal energy. As the number of partitions is increased, the number of equilibrium states through which the system passes increases, and the irreversible path approaches more and more closely the reversible path.† Even in the limit of an infinite number of partitions, however, no work is done in the irreversible process; but the work of the reversible process, given by $\int P\,dV$, is greater than zero.

All actual processes that involve finite energy transformations are irreversible. Reversible processes do not occur. One reason they do not occur is that frictional effects cannot be completely eliminated. Nevertheless, reversible processes are extremely useful in the analysis and design of actual processes and devices for several reasons. Reversible processes are useful as standards of comparison or guides to perfection. Reversible processes are often the limiting cases of actual processes as the irreversibilities present in any actual process are reduced further and further. Since many actual processes are difficult to analyze completely, an engineer frequently bases an analysis or design on reversible

*As a homely example to clarify this point, suppose that the statistics for a football game show

	Team *A*	Team *B*
yards gained, running	−5	−20

Although it may be nothing to be proud of, the running attack of team *A* was more effective than that of team *B*. In other words, as far as yards *gained* are concerned, $-5 > -20$.

†Strictly speaking, even after we locate an infinite number of points on the *PV* diagram of Fig. 6·4*b*, the path is not continuous because there are an infinite number of intervals between these points in which the path is not defined!

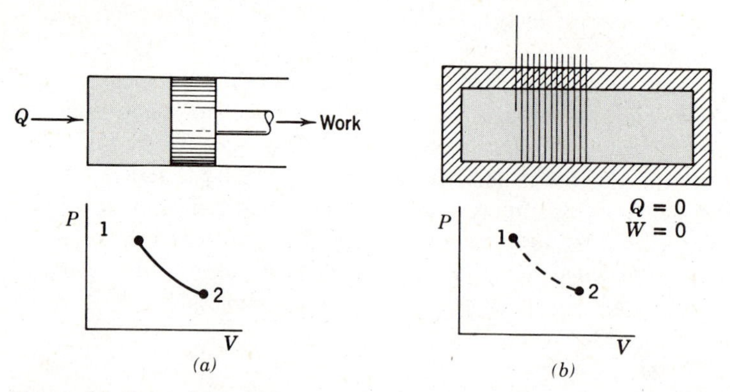

Figure 6·4. Reversible and irreversible processes through the same equilibrium states.

processes, and then adjusts the results before applying them to actual processes. This is similar to the use of ideal systems in the analysis or design of actual systems. (See Sec. 1·3 if you do not remember the discussion of this point.) Engineers use reversible processes in the same way that they use point masses, frictionless pulleys, weightless cords, and homogeneous beams: They use these idealizations to simplify the analysis of actual systems and processes.

The concept of reversible processes is important also because it permits the definition of a very useful property called entropy. Entropy is introduced in Chapter 8.

6·3 Internal and External Reversibility

A process is irreversible if it involves heat transfer across a finite temperature difference between the system and the surroundings. However, the system behaves during this irreversible process just as though the heat were being transferred to or from it reversibly across an infinitesimal temperature difference. Such a process is said to be *internally reversible* because nothing occurs within the system to make it irreversible, but it is *externally irreversible.*

Frictionless adiabatic and frictionless isothermal processes described in the preceding article are *internally reversible* and *externally reversible.*

A process that involves friction or some other irreversibility within the system and also heat exchange with surroundings at a different temperature is *internally irreversible* and *externally irreversible.*

Calculations of a system's behavior are the same for an internally reversible process, whether the process is externally reversible or irreversible. Consequently, from here on in this book, whenever a principle or relation is said to be valid for a reversible process, the process need be only internally reversible. In the few cases where it is required that a process be externally reversible, this additional restriction will be explicitly stated. It is to be understood that any process designated as externally reversible is also internally reversible. (A process *can* be externally reversible and internally irreversible, but no principles or general relationships are formulated for such processes. Therefore, we may always consider external reversibility to be an *additional* restriction instead of an *alternative* restriction.)

6·4 $\int P\,dv$ and $\int v\,dP$ in Irreversible Processes

In Secs. 1·15 and 1·17 two equations were developed from the principles of mechanics for the work of frictionless processes. Now that the concept of reversibility has been introduced, the restrictions on these equations should be changed from "for frictionless processes" to "for reversible processes" (meaning internally reversible in accordance with the convention established in the preceding section). Thus we have

$$\text{work} = \int P\,dV \tag{1·7}$$

for reversible closed system processes and

$$\text{work} = -\int v\,dP - \Delta ke - \Delta pe \tag{1·10}$$

for reversible steady-flow processes.

Equations 1·7 and 1·10 do not apply to irreversible processes. For many irreversible processes $\int P\,dV$ or $\int v\,dP$ cannot be evaluated because the system is not in equilibrium and consequently no single value of P or v is a property of the system as a whole at any instant. Even for irreversible processes in which $\int P\,dV$ or $\int v\,dP$ can be evaluated, Eqs. 1·7 and 1·10 do not apply. Two illustrations of this point follow.

Consider a system comprised of a gas in a closed, rigid vessel. The gas is stirred by a paddle wheel driven externally (see Fig. 6·5). During the stirring process, there is at any instant a variation of pressure from point to point within the vessel. However, the volume of the system is constant so that undoubtedly $\int P\,dV = 0$. Work is delivered to the system by means of the paddle wheel; so work $\neq 0$. Consequently, work $\neq \int P\,dV$, and Eq. 1·7 does not apply in this case.

As a second illustration, consider the flow of an incompressible liquid through a section of horizontal pipe with a constant cross-sectional area. $\Delta pe = 0$, and $\Delta ke = 0$. No work is done: work $= 0$. If there is a pressure drop caused by friction, $\int v\,dP \neq 0$. Consequently, work $\neq -\int v\,dP - \Delta ke - \Delta pe$, and Eq. 1·10 does not apply in this case.

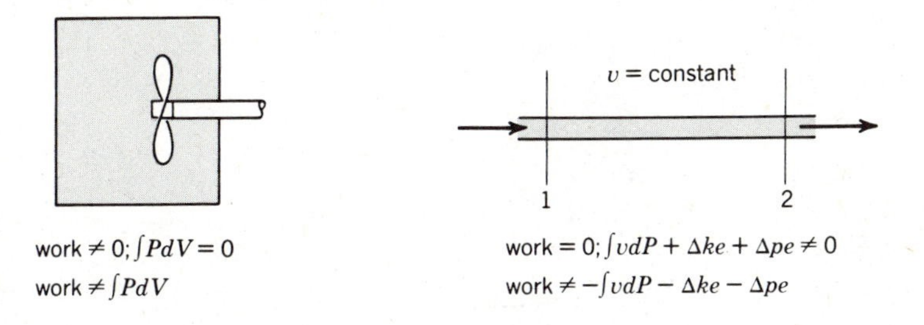

Figure 6·5. Irreversible processes in closed and steady-flow systems.

You may ask, "Since $\int P\,dV \neq$ work for an irreversible process of a closed system, what significance does $\int P\,dV$ have for such a process?" The answer to this question is that $\int P\,dV$ has no significance for an irreversible process except that it is equal to the work that could be done by the system if it passed reversibly through the same series of states. As pointed out in Sec. 6·2, the work done by a reversible process is the maximum work for any particular path.

The same answer applies to the similar question which can be asked regarding $-\int v\,dP - \Delta ke - \Delta pe$ in irreversible steady-flow processes, although a dynamic analysis of pipe flow does show some further physical significance of $-\int v\,dP - \Delta ke - \Delta pe$ in this particular application.

Repeating for emphasis,

$$\text{work} = \int P\,dV \tag{1·7}$$

holds only for reversible closed-system processes, and

$$\text{work} = -\int v\,dP - \Delta ke - \Delta pe \tag{1·10}$$

holds only for reversible steady-flow processes.

6·5 Reversible and Externally Reversible Cycles

A cycle composed entirely of reversible processes is called a reversible cycle. If all of the processes are externally reversible, the cycle is called an externally reversible cycle.

As an example of a reversible cycle, consider an ideal gas trapped in a cylinder behind a piston. Let the gas expand reversibly at constant pressure. During this process heat is added to the gas, and the gas does work on the surroundings. Then let heat be removed from the gas while the piston is stationary until the gas reaches its initial temperature. Then let the gas be compressed reversibly and isothermally to its initial state. During the isothermal compression, work is done on the gas, and heat is removed from it. This cycle is shown on a Pv diagram in Fig. 6·6*a*. Since the pressure during the expansion is greater than that during the compression, there is a net work output from this cycle. This work output is represented by the area within the cycle diagram of Fig. 6·6*a*. Each process of this cycle is reversible. (Recall from Sec. 6·3 that this means

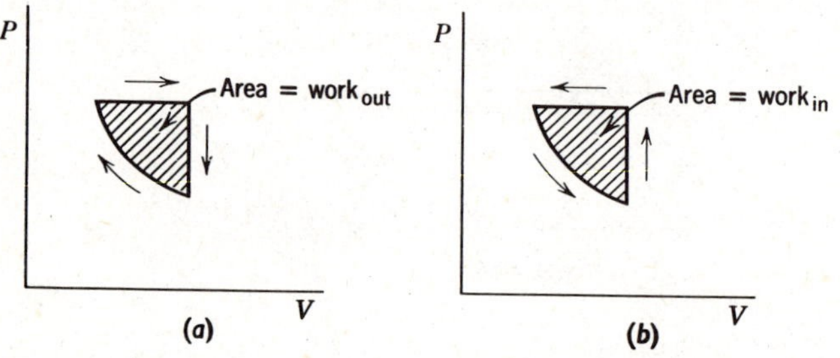

Figure 6·6. Reversible cycle.

internally reversible.) The cycle is therefore reversible. In the reversed operation of this cycle, the gas is first expanded reversibly and isothermally, doing work on the surroundings and receiving heat. Then, with the piston stationary, heat is added until the pressure reaches its initial value. The gas is then cooled reversibly at constant pressure to its initial state. Work is done on the gas during this constant-pressure cooling. This cycle is shown in Fig. 6·6*b*. The net work input to the cycle of Fig. 6·6*b* equals in magnitude the net work output of the cycle of Fig. 6·6*a*.

Example 6·4. Air in a closed system undergoes a cycle composed of the following three reversible processes: (1) a constant-pressure expansion from 80 kPa, 10°C, to 90°C, (2) a constant-volume cooling to 10°C, and (3) an isothermal compression to 80 kPa. Determine (*a*) the energy transfer per kilogram of air for each process, (*b*) the thermal efficiency of the cycle, and (*c*) the energy transfers for each process when air undergoes this cycle in reverse.

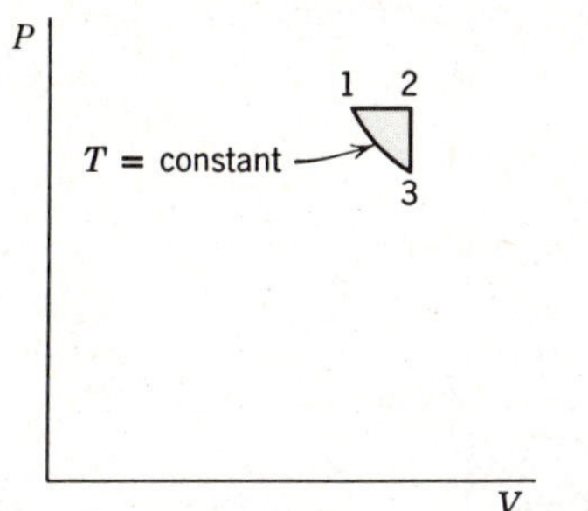

Example 6·4.

Solution. (***a***) Air under these conditions behaves as an ideal gas. For the reversible constant-pressure expansion,

$$\text{work} = \int_1^2 P\,dv = P(v_2 - v_1) = P_2v_2 - P_1v_1 = R(T_2 - T_1)$$

$$= 0.287(363 - 283) = 23.0 \text{ kJ/kg}$$

Using a constant value of c_v for the limited temperature range involved,

$$u_2 - u_1 = c_v(T_2 - T_1) = 0.7165(363 - 283) = 57.3 \text{ kJ/kg}$$

Applying the first law,

$$q = u_2 - u_1 + w = 57.3 + 23.0 = 80.3 \text{ kJ/kg}$$

For the constant-volume cooling, work $= 0$ and

$$q_{\text{out}} = u_2 - u_3 = c_v(T_2 - T_3) = 0.7165(363 - 283) = 57.3 \text{ kJ/kg}$$

For the isothermal compression,

$$\text{work} = \int_3^1 P\,dv = \int_3^1 \frac{RT}{v}\,dv = RT_1 \ln\frac{v_1}{v_3} = RT_1 \ln\frac{P_3}{P_1} = RT_1 \ln\frac{P_3}{P_2} = RT_1 \ln\frac{T_3}{T_2}$$

$$= 0.287(283) \ln\frac{283}{363} = -20.2 \text{ kJ/kg}$$

Applying the first law,

$$q_{out} = w_{in} + u_3 - u_1 = 20.2 + 0 = 20.2 \text{ kJ/kg}$$

Presenting the results in tabular form,

Process	q, kJ/kg	w, kJ/kg
1-2	80.3	23.0
2-3	−57.3	0
3-1	−20.2	−20.2
Cycle	+2.8	+2.8

(*b*)

$$\eta = \frac{\oint \delta w}{q_{in}} = \frac{2.8}{80.3} = 3.5 \text{ percent}$$

(*c*) For the reversed cycle the energy transfers for each process are equal in magnitude but opposite in sign to those of the original cycle.

Process	q, kJ/kg	w, kJ/kg
1-3	20.2	20.2
3-2	57.3	0
2-1	−80.3	−23.0
Cycle	−2.8	−2.8

In order for the cycle described in Example 6·4 and the preceding paragraph to be *externally* reversible, heat must be transferred only across infinitesimal temperature differences. Therefore, the temperature of part of the surroundings would have to vary during the constant-pressure and constant-volume processes. If the parts of the surroundings that exchange heat with the system are at constant temperatures, then the cycle must be externally irreversible. The calculations of Example 6·4 are the same, however, whether the processes are externally reversible or externally irreversible. They depend only on the condition that the processes are internally reversible.

Let us look now at a cycle that is externally reversible but exchanges heat with parts of the surroundings at only two fixed temperatures. If heat is exchanged reversibly with bodies at two fixed temperatures T_H and T_L, then heat must be transferred only when the system is at a constant temperature only infinitesimally higher or lower than T_H or T_L. In other words, a system that executes an externally reversible cycle while exchanging heat with only two constant-temperature regions cannot exchange heat with them during any processes except isothermal ones. This means that the processes that are not isothermal must be adiabatic.

As an example of an externally reversible cycle operating between two energy reservoirs at temperatures T_H and T_L, refer to Fig. 6·7, and consider the following cycle executed by a gas in either a closed or a steady-flow system:

Process 1-2. The gas which is at a temperature of $(T_H - dT)$ expands reversibly and isothermally from a state 1 to a state 2. Work is done by the gas, and heat is transferred from the energy reservoir at T_H to the system at $(T_H - dT)$. Since heat is being transferred across an infinitesimal temperature difference, this process is externally reversible.

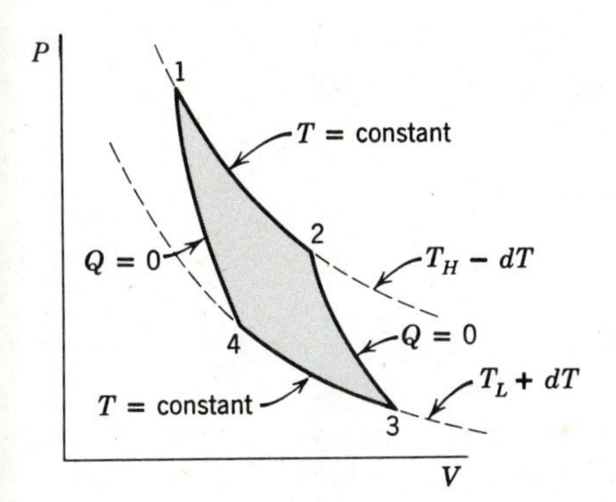

Figure 6·7. An externally reversible cycle operating between energy reservoirs at T_H and T_L.

Process 2-3. The gas expands reversibly and adiabatically from state 2 to state 3, doing work. During this process the temperature of the system drops from $(T_H - dT)$ to $(T_L + dT)$.

Process 3-4. The gas is compressed isothermally, rejecting heat to the energy reservoir at T_L. This process, like the heat addition process 1-2, is externally reversible because heat is transferred across only an infinitesimal temperature difference. The isothermal compression progresses until the gas reaches state 4, from which it can be compressed reversibly and adiabatically to state 1 to complete the cycle.

Process 4-1. The gas is compressed reversibly and adiabatically to the initial state. The temperature of the system increases from $(T_L + dT)$ to $(T_H - dT)$. The area within the Pv diagram of Fig. 6·7 represents the net work of this externally reversible cycle. Since the system executes a cycle, the net change in its stored energy is zero ($\oint dE = 0$). The net effects of this cycle are: (1) Heat is removed from the energy reservoir at T_H, (2) heat is added to the energy reservoir at T_L, and (3) work is produced. Application of the first law shows that the work produced is equal to the difference between the heat added to the system at T_H or $(T_H - dT)$ and the heat rejected at T_L or $(T_L + dT)$.

Since the cycle is externally reversible, it can be carried out in the opposite direction with the system passing through a sequence of states 1-4-3-2-1, absorbing heat from the reservoir at T_L and rejecting heat to the reservoir at T_H. During the reversed cycle, work must be done on the system.

The externally reversible cycle just described (which can also be executed by substances other than gases) is called the Carnot (ˈkär-ˈnō) cycle in honor of Sadi Carnot,* the French engineer who in 1824 first described it. The Carnot cycle is discussed further

*Nicholas Leonard Sadi Carnot (1796–1832), member of an illustrious family, studied at the Ećole Polytechnique and was an officer in the French Army Engineers. He had extremely broad interests and was an accomplished athlete. The only paper he published during his lifetime, *Reflections on the Motive Power of Heat,* is one of the milestones of scientific thought. In this paper he originated the use of cycles in thermodynamic analysis and laid the foundations for the second law by describing and analyzing the Carnot cycle and stating the Carnot principle which is discussed in Sec. 7·1. He employed the caloric theory in his reasoning, but his conclusions are correct because the second law is a principle that is independent of the first law or of any other "theory of heat." When his paper was first published, many people thought that his analysis and conclusions *depended* on the caloric theory. Consequently, the caloric theory, which had suffered at the hands of Benjamin Thompson and Humphry Davy, was given a new, but short, lease on life. In 1851 Rudolph Clausius and Lord Kelvin showed that there is no dependence of the second law on the caloric theory.

in the following two sections. The Carnot cycle is important because no cycle can be devised that is more efficient than a Carnot cycle operating between the same temperature limits. This fact will be demonstrated in Sec. 7·1.

6·6 The Carnot Cycle

A Carnot cycle can be executed by many different types of systems. The system can be a liquid, a gas, an electric cell, a soap film, a steel wire, or a rubber band, to name just a few. In any case, the operation of a Carnot cycle involves (1) a system, (2) an energy reservoir at some temperature T_H, (3) an energy reservoir at some lower temperature T_L, (4) some means of periodically insulating the system from one or both of the reservoirs, and (5) a part of the surroundings that can absorb work and periodically do work on the system.

In order to have a specific example of a Carnot cycle for study, consider a system comprised of a gas (not necessarily an ideal gas—see Example 6·5 for this special case) held in an insulated cylinder fitted with an insulating piston (Fig. 6·8). The insulation I of the cylinder head can be removed so that the cylinder can periodically be placed in intimate contact with the energy reservoir at T_H or the one at T_L.

Let the cycle begin with the gas in a state 1 as shown on the *PV* diagram labeled "Gas" in Fig. 6·8. The temperature of the gas is T_H. Then the insulation I is removed from the cylinder, and the cylinder head is placed in contact with the energy reservoir at T_H. The gas expands very slowly, doing work on the surroundings. The temperature of the gas *tends* to decrease, but the flow of the heat from the energy reservoir maintains the gas temperature constant at $(T_H - dT)$. Notice that the transfer of heat is reversible only as long as the temperature difference between the reservoir and the gas is infinitesimal. If the temperature of the gas were to fall lower than $(T_H - dT)$, the process would be externally irreversible. This reversible isothermal process continues until the piston reaches a position 2. The piston is then stopped, holding the gas in state 2, while the cylinder is

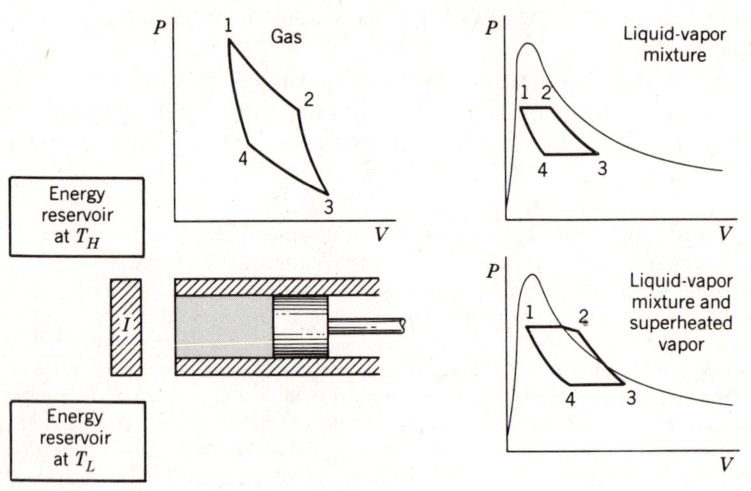

Figure 6·8. Closed-system Carnot cycle.

removed from the reservoir at T_H, and the insulation is put back on the cylinder head. The gas then pushes the piston farther outward as it expands reversibly and adiabatically. Work is done by the gas, and there is no heat input; so the temperature drops. The piston is allowed to move until the gas temperature becomes T_L. The gas is then in state 3. Notice that any heat transfer between the gas and one of the reservoirs while the gas was at a temperature between T_H and T_L would have made the process externally irreversible.

Now, if the piston were to be pushed inward while the cylinder is still completely insulated, the gas would be compressed adiabatically, and its temperature would again rise. If the compression were reversible and adiabatic, the gas would retrace the path between states 3 and 2. Obviously this will not help us obtain a net work output from the cycle. Therefore, while the piston is in position 3, the cylinder-head insulation I is again removed, and the cylinder is placed in contact with the low-temperature reservoir. Now, as the piston is pushed inward, the gas temperature *tends* to rise, but heat is transferred from the gas to the cold reservoir at such a rate that the gas temperature remains constant at $(T_L + dT)$. Since the heat is transferred across an infinitesimal temperature difference, the process is externally reversible. This reversible isothermal compression of the gas is continued until a state 4 is reached. State 4 is such that, if the insulation is put back on the cylinder head and the gas is compressed reversibly and adiabatically, its temperature and pressure will increase, and the gas will be returned to state 1. Thus the cycle is completed.

If the working fluid in the cycle just described were a liquid–vapor mixture of some pure substance instead of a gas, the isothermal processes 1-2 and 3-4 would also be constant-pressure processes. During process 1-2, evaporation would occur, and condensation would occur during process 3-4. If all of the liquid were evaporated and the vapor became superheated during the isothermal heat addition, only that part of process 1-2 which occurred with both liquid and vapor present would be a constant-pressure process. *PV* diagrams for these cases are shown in Fig. 6·8. Remember that the heat-addition and heat-rejection processes of the Carnot cycle are always isothermal. For a particular substance, an isothermal process may also be a constant-pressure process or a constant internal energy process, but this characteristic of the substance is merely incidental to the execution of a Carnot cycle. The Carnot cycle requires that the heat-addition and -rejection processes be only reversible and isothermal—nothing else.

Figure 6·9 shows a flow diagram and a *Pv* diagram for a steady-flow Carnot cycle

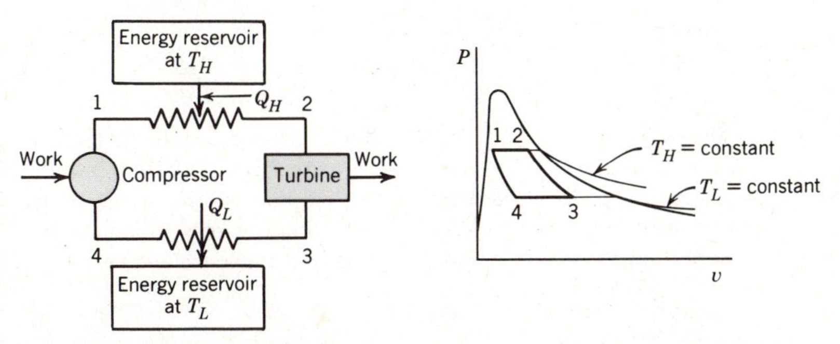

Figure 6·9. Steady-flow Carnot cycle using a wet vapor as working substance.

that uses a liquid–vapor mixture (often called a "wet vapor") as working substance. The reversible isothermal heat-addition process 1-2 is an evaporation which occurs in a heat exchanger where no work is done. The reversible adiabatic expansion, during which the temperature of the wet vapor drops from T_H to T_L, occurs in a turbine. Heat is rejected from the working fluid in a condenser. No work is done in the condenser. The fluid which leaves the condenser in state 4 is compressed reversibly and adiabatically to its initial state 1 to complete the cycle.

Example 6·5. Derive an expression in terms of the reservoir temperatures T_H and T_L for the thermal efficiency of a Carnot cycle using an ideal gas as working fluid. (This problem statement *implies* that the efficiency of this cycle is a function of the temperature limits only, but this fact has not yet been proved. It is proved in Sec. 7·1.)

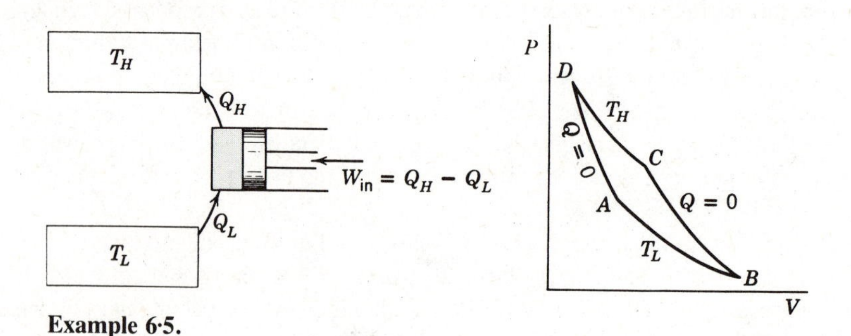

Example 6·5.

Solution. A schematic diagram of the engine and the reservoirs and a *PV* diagram of the ideal-gas working fluid are made first.

By the definition of thermal efficiency,

$$\eta = \frac{\oint \delta\text{work}}{Q_{\text{in}}} = \frac{W}{Q_{\text{in}}}$$

Applying the first law to this closed system which executes a cycle,

$$\eta = \frac{\oint \delta\text{work}}{Q_{\text{in}}} = \frac{\oint \delta Q}{Q_{\text{in}}} = \frac{Q_{\text{in}} - Q_{\text{out}}}{Q_{\text{in}}} = 1 - \frac{Q_{\text{out}}}{Q_{\text{in}}} = 1 - \frac{Q_L}{Q_H}$$

where Q_{in} or Q_H and Q_{out} or Q_L stand for the gross or total heat transferred from the reservoir at T_H to the engine and from the engine to the reservoir at T_L, respectively. All the heat added to the working fluid during the cycle is added during the isothermal expansion 1-2, during which the temperature of the working fluid is T_H (or lower than this by only an infinitesimal amount). Applying the first law to the system during the reversible isothermal expansion 1-2,

$$Q_H = U_2 - U_1 + W$$

For an ideal gas, internal energy is a function of temperature only; therefore $\Delta U = 0$ for an isothermal process. Also, for a reversible process of a closed system, $W = \int P\,dV$, so that

$$Q_H = 0 + \int_1^2 P\,dV$$

Substituting for P from the ideal-gas equation of state, and noting that m, R, and T are constant for this process,

$$Q_H = 0 + \int_1^2 \frac{mRT}{V}\,dV = mRT \int_1^2 \frac{dV}{V} = mRT_H \ln \frac{V_2}{V_1}$$

By the same reasoning,

$$Q_L = U_3 - U_4 - W = 0 - \int_3^4 P\,dV = -mRT_L \ln \frac{V_4}{V_3} = mRT_L \ln \frac{V_3}{V_4}$$

Substituting these values for Q_H and Q_L into the expression obtained above for thermal efficiency,

$$\eta = 1 - \frac{Q_L}{Q_H} = 1 - \frac{mRT_L \ln (V_3/V_4)}{mRT_H \ln (V_2/V_1)} = 1 - \frac{T_L \ln (V_3/V_4)}{T_H \ln (V_2/V_1)}$$

For the reversible adiabatic processes 2-3 and 4-1 of the ideal-gas working substance, at least if the ratio of specific heats is constant,

$$\frac{V_3}{V_2} = \left(\frac{T_2}{T_3}\right)^{1/(k-1)} = \left(\frac{T_H}{T_L}\right)^{1/(k-1)} \quad \text{and} \quad \frac{V_4}{V_1} = \left(\frac{T_1}{T_4}\right)^{1/(k-1)} = \left(\frac{T_H}{T_L}\right)^{1/(k-1)}$$

Thus

$$\frac{V_3}{V_2} = \frac{V_4}{V_1}$$

and

$$\frac{V_3}{V_4} = \frac{V_2}{V_1}$$

so that the expression for thermal efficiency reduces to

$$\eta = 1 - \frac{T_L}{T_H}$$

It is worth noting that the thermal efficiency of a Carnot cycle, using an ideal gas as a working fluid, increases as the ratio of T_H to T_L increases. Writing the efficiency expression as

$$\eta = \frac{T_H - T_L}{T_H}$$

shows that, for a given temperature of either reservoir, the efficiency is increased by increasing the temperature difference between the reservoirs.

6·7 The Reversed Carnot Cycle

A Carnot cycle operated in reverse is not a heat engine that produces work but is a refrigerator or heat pump. It absorbs heat from a low-temperature reservoir and rejects

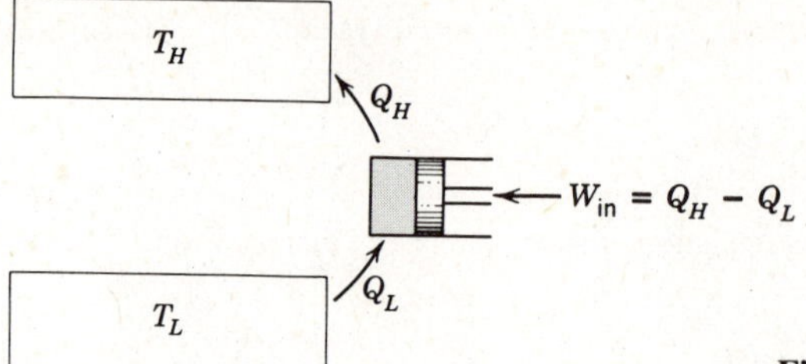

Figure 6·10. Refrigerator or heat pump.

heat to a high-temperature reservoir. In order to do this, it requires work from the surroundings. (In a household refrigerator, heat is absorbed from the refrigerator compartment, heat is rejected to the surrounding air, and an electric motor provides the work input.)

The working substance of a Carnot refrigerator absorbs heat from Q_L during an isothermal process at T_L (or $T_L - dT$, since the temperature of the working substance must be infinitesimally lower than T_L in order for heat to be transferred from the energy reservoir at T_L). Its temperature is then increased adiabatically to T_H. During an isothermal process at T_H, heat Q_H is rejected to the higher-temperature reservoir. See Fig. 6·10. The cycle is then completed by an adiabatic process that lowers the temperature of the working substance and returns it to its initial state. All processes are of course externally reversible. Application of the first law shows that

$$\text{Net work}_{\text{in}} = Q_H - Q_L$$

Example 6·6. Derive an expression, in terms of the reservoir temperatures T_L and T_H, for the coefficient of performance of a Carnot refrigerator using an ideal gas as working fluid.

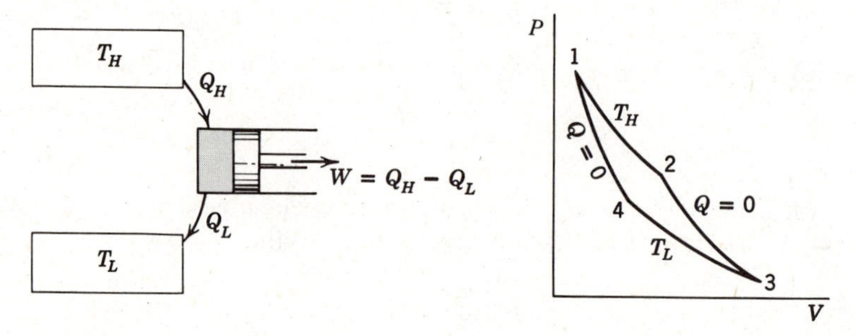

Example 6·6.

Solution. A schematic diagram of the Carnot refrigerator and the reservoirs and a *PV* diagram of the ideal-gas working fluid are made first.

By the definition of coefficient of performance for a refrigerator (Sec. 2·10) and application of the first law,

$$\beta_R = \frac{Q_L}{W_{\text{in}}} = \frac{Q_L}{Q_H - Q_L}$$

Q_L and Q_H can be evaluated by the first law, Joule's law, work $= \int P\,dV$, and $PV = mRT$ as in Example 6·5 to give

$$Q_L = Q_{A-B} = mRT_L \ln \frac{V_B}{V_A}$$

$$Q_H = Q_{\text{out},C\text{-}D} = mRT_H \ln \frac{V_C}{V_D}$$

It can also be shown, as in Example 6·5, that

$$\frac{V_B}{V_A} = \frac{V_C}{V_D}$$

so that substitutions in the expression for coefficient of performance give

$$\beta_R = \frac{mRT_L \ln (V_C/V_D)}{mRT_H \ln (V_C/V_D) - mRT_L \ln (V_C/V_D)} = \frac{T_L}{T_H - T_L}$$

Notice that this is *not* the reciprocal of the efficiency expression obtained in Example 6·5. Finally, notice some numerical values of β_R for various reservoir temperatures (recalling that absolute temperatures are used in the expression derived):

T_L, °C	T_H, °C	β_R
5	30	11.1
−5	20	10.7
−20	10	8.4
−50	20	3.2
−150	20	0.72
−250	20	0.085

6·8 Other Externally Reversible Cycles

The Carnot cycle is not the only externally reversible heat-engine cycle. Two others are known as the Stirling* cycle and the Ericsson† cycle. These warrant consideration here because they both employ the principle of *regeneration* that is used in many modern steam and gas-turbine power plants and elsewhere.

In order for a cycle to be externally reversible and to exchange heat with energy reservoirs at two fixed temperatures, heat can be transferred between the system and the reservoirs only during isothermal processes. However, during nonisothermal processes, it is possible for heat to be transferred between the system and some regenerative energy-

*The Reverend Robert Stirling (1790–1878) was the first person to propose the use of regeneration in heat-engine cycles.

†John Ericsson (1803–1889), Swedish-American engineer and inventor, built a steam locomotive, the *Novelty,* which competed with Stephenson's *Rocket* in 1829. Among his inventions were the revolving naval gun turret, the marine screw propeller, and the steam fire engine. He is well known as the designer and builder of the ironclad *Monitor* used by the United States in answer to the Confederate *Merrimac* during the Civil War. Under Ericsson's supervision the *Monitor* was built in only 126 days.

storage device which absorbs heat from the system during part of the cycle and returns the same amount of heat to the system during another part of the cycle. Of course, in order for the operation of the regenerator to be reversible, heat must always be transferred between the system and the regenerator across only an infinitesimal temperature difference.

Figure 6·11 shows a schematic diagram of a Stirling engine and a *PV* diagram for an ideal-gas working substance. (The engine shown here is not the same physical arrangement as that proposed by Stirling, but the thermodynamic cycle is the same. Also see Problem 6·26.) The engine consists of a cylinder with a piston at each end. In the middle of the cylinder, between the pistons, is the regenerator. This can be a plug of wire gauze or a porous plug made by holding small metal shot between two wire screens. Assume that the regenerator as a whole is a poor conductor of heat, so that even though a temperature gradient exists across it, a negligible amount of heat is conducted in the direction of the cylinder axis. The cylinder is completely insulated except for a contact with the hot reservoir at one end and a contact with the cold reservoir at the other end.

Starting with state 1, the cycle proceeds as follows, with each process being externally reversible:

Process 1-2. Heat is added to the gas at T_H (or, strictly speaking, at $T_H - dT$) from the reservoir at T_H. During this reversible isothermal process, the left piston moves outward, doing work as the system volume increases and the pressure falls.

Process 2-3. Both pistons are moved to the right at the same rate to keep the system volume constant. There is no heat transfer with either reservoir. As the gas passes through the regenerator, heat is transferred from the gas to the regenerator, causing the gas temperature to fall to T_L by the time the gas leaves the right end of the regenerator. In order for this heat transfer to be reversible, the temperature of the regenerator at each point must equal the gas temperature at that point. Therefore, there is a temperature gradient through the regenerator from T_H at the left end to T_L at the right end. No work is done during this process.

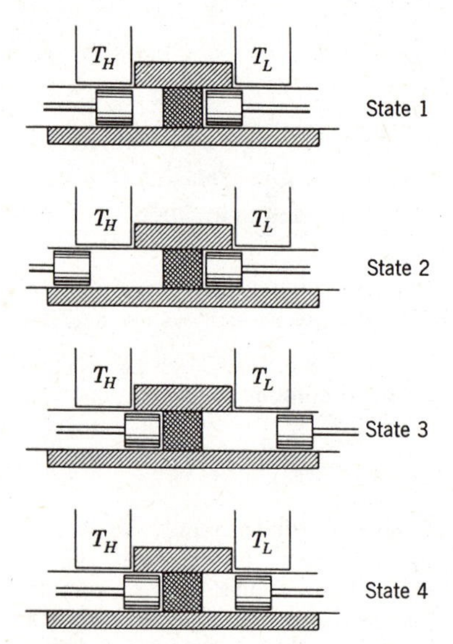

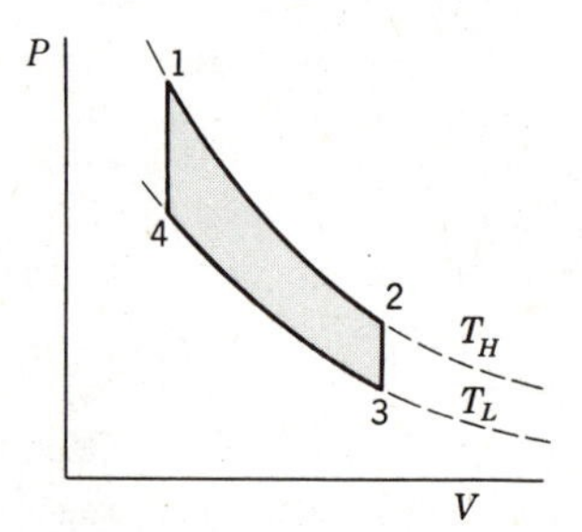

Process

1-2 Q in; T = constant = T_H.
2-3 V = constant; T decreases.
3-4 Q out; T = constant = T_L.
4-1 V = constant; T increases.

Figure 6·11. Stirling engine and cycle.

Process 3-4. Heat is removed from the gas at T_L (or $T_L + dT$) to the reservoir at T_L. In order to hold the gas temperature constant, the right piston is moved inward, doing work on the gas, and the pressure increases.

Process 4-1. Both pistons are moved to the left at the same rate to keep the system volume constant. (Notice that the pistons are closer together during this process than they were during process 2-3, because $V_4 = V_1 < V_2 = V_3$.) There is no heat transfer with either reservoir. As the gas passes back through the regenerator, the energy stored in the regenerator during process 2-3 is returned to the gas, so that it emerges from the left end of the regenerator at the temperature T_H. No work is done during this process because the system volume is constant.

The externally reversible cycle is thus completed. Notice that the system has exchanged a net amount of heat with only the two energy reservoirs at T_H and T_L.

Figure 6·12 shows a schematic diagram of a steady-flow power plant operating on the Ericsson cycle and a Pv diagram for an ideal-gas working fluid. (Ericsson's original engine, of which several models have been built, was a nonflow or closed system engine, but the ideal thermodynamic cycle is the same for the steady-flow engine. The steady-flow engine is described here in order to introduce another type of regenerator.) The steady-flow Ericsson engine involves (1) a turbine through which the gas expands isothermally, doing work and absorbing heat from an energy reservoir at T_H, (2) a compressor that compresses the gas isothermally while heat is rejected from the gas at T_L (or $T_L + dT$) to the energy reservoir at T_L, and (3) a counterflow heat exchanger which is used as a regenerator. In the regenerator, gas from the compressor enters at a temperature of T_L (or $T_L + dT$) and leaves at a temperature T_H (or $T_H - dT$). This is process 4-1. The gas from the turbine then flows through the regenerator in the opposite direction. This is process 2-3. Its temperature drops from T_H to T_L as it rejects heat to the gas that is undergoing process 4-1. Notice in Fig. 6·12 that the gas entering the left end of the regenerator (state 4) is at T_L, as is the gas that leaves the left end (state 3). Also, both the gas leaving the right end (state 1) and the gas entering the right end (state 2) are at T_H. Throughout the regenerator there is no more than an infinitesimal temperature difference between the two gas streams at any one section, so that the operation of the regenerator is reversible. (Of course, no actual regenerator would be designed for an infinitesimal temperature difference because then a finite amount of heat could not be transferred during a finite period of time. Remember that reversible heat transfer is the *limiting* case as the temperature difference is made smaller and smaller.) All processes

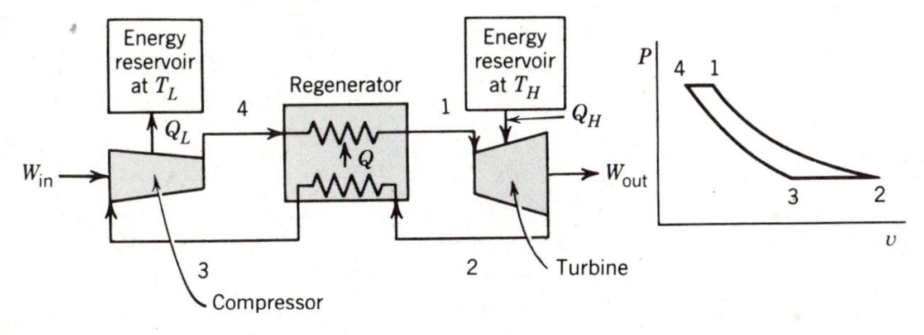

Figure 6·12. Ericsson engine and cycle.

of the Ericsson cycle are externally reversible. The only parts of the surroundings involved in heat transfer are the two energy reservoirs at T_H and T_L.

The Carnot, Stirling, and Ericsson cycles have been introduced here to show how an externally reversible cycle can be executed. The next chapter introduces some far-reaching consequences of the second law which involve the performance of externally reversible engines.

6·9 Irreversible Processes and Molecular Disorder

The principles of thermodynamics do not depend on any assumptions regarding the existence or behavior of molecules. However, a molecular picture occasionally supplements our understanding of some effect. As an illustration, we consider briefly a conclusion that can be deduced from the principles of statistical thermodynamics: Irreversible processes always result in an increase of molecular disorder.

Consider the sliding of a block along a horizontal plane. Friction between the block and plane causes the block to stop. Energy is transformed from kinetic energy of the block to internal energy of the block and plane. Near their rubbing surfaces, the block and plane experience an increase in temperature. This temperature increase shows up on the molecular scale as an increase in molecular velocities, and thus as an increase in kinetic energy of the molecules. The initial kinetic energy of the block can also be considered as kinetic energy of molecules, since all the molecules in the block are initially moving with the block; they have a common translatory motion superimposed on their individual oscillations or vibrations. Thus the energy which we speak of as being "transformed" is in the form of kinetic energy of molecules both before and after the transformation occurs. The essential difference is that initially the energy is related to an *ordered* motion of the molecules, whereas finally it is related to a *random* or *disordered* motion. Is the reverse transformation possible? The first law, which requires only conservation of energy, does not prohibit it; but the second law, as discussed earlier in this chapter, leads us to classify the original process as irreversible, so that the reverse transformation is impossible. The transformation that we declare to be impossible would proceed toward a more highly ordered molecular state. The irreversible process proceeds toward a less ordered (or more disordered) state.

As another example, a paddle wheel stirring a gas imposes an ordered motion on the gas molecules, but this ordered motion is rapidly dissipated into a random motion that is observed as an increase in gas temperature. Our discussion of the second law leads us to believe that the reverse process is impossible: the random molecular motions will not spontaneously become ordered in such a manner as to turn the paddle wheel. But, from the molecular point of view, is such an occurrence absolutely impossible? If the molecules are all in random motion, might there be *some chance* that at some time a sufficient number will be moving in such ways as to turn the paddle wheel? Statistical thermodynamics tells us that such an event *is* possible, but that the probability of its occurrence is *extremely* small—so small that for any calculations or predictions we make we can safely consider the occurrence to be impossible. Thus we can say that the ordered molecular state is a state of extremely low probability and that the disordered molecular state is more probable. We can apply similar reasoning from statistical thermodynamics to any irreversible process, and we reach the same conclusion in each instance: In an irreversible process, any isolated system proceeds toward more probable states. More probable states are ones of greater molecular disorder.

Remember that this conclusion from statistical thermodynamics cannot be reached from, and has no bearing on, the principles of thermodynamics as presented in this book, because this book deals with classical thermodynamics or thermodynamics from the macroscopic point of view. For purposes of engineering analysis and design, classical thermodynamics is at present more useful than statistical thermodynamics. However, statistical thermodynamics is becoming more valuable in explaining certain phenomena and physical properties, and its importance to engineers is increasing.

6·10 Summary

A *reversible process* is a process such that, after it has occurred, both the system and all the surroundings can be returned to the states they were in before the process occurred. Any other process that occurs is *irreversible*.

One way to prove that a process is irreversible is as follows: (1) Assume that the process is reversible so that the reverse process is possible. (2) Combine this reverse process with other processes, known from experience to be possible, to form a cycle that violates the second law. If such a cycle can be devised, then the assumption of step (1) is false, and the process originally considered is irreversible.

Any process is irreversible if it involves friction, heat transfer across a finite temperature difference, free expansion, mixing, or inelastic deformation.

Reversible processes must be executed very slowly to avoid friction and can involve heat transfer only across infinitesimal temperature differences. Also, a reversible process must be such that, after it has occurred, the system and the surroundings can be made to traverse in the reverse order the states they passed through during the original process. All energy transformations of the original process must be reversed in direction but remain unchanged in form or magnitude. During a reversible process, the system and the surroundings must each at all times be in states of equilibrium or infinitesimally close to states of equilibrium. The direction of a reversible process can be changed by making infinitesimal changes in the conditions that control it.

Reversible processes actually do not occur in nature. Neither do point masses, frictionless pulleys, weightless cords, and homogeneous beams; but engineers find reversible processes to be just as useful as these other idealizations in analyzing and designing actual systems and processes.

A process is *internally reversible* if all irreversible effects occur outside the system boundary. An *externally reversible* process is internally reversible and also involves no heat transfer across a finite temperature difference.

The relationship derived from the principles of mechanics

$$\text{work} = \int P\,dV \qquad (1\cdot7)$$

holds only for reversible closed-system processes. The analogous relation for steady flow,

$$\text{work} = -\int v\, dP - \Delta ke - \Delta pe \qquad (1\cdot10)$$

holds only for reversible steady-flow processes.

A cycle composed entirely of reversible processes is called a reversible cycle. If all the processes are externally reversible, the cycle is an externally reversible cycle. An example of an externally reversible cycle that exchanges heat with energy reservoirs at only two fixed temperatures is the Carnot cycle. A Carnot engine absorbs heat from an energy reservoir at one temperature and rejects heat to an energy reservoir at a lower temperature. Work is produced. The reversed Carnot engine is a refrigerator or heat pump which absorbs heat from the lower-temperature reservoir and discharges heat to the higher-temperature reservoir. Work must be supplied to drive the reversed Carnot engine. The Carnot cycle consists of two isothermal and two adiabatic processes, with all four processes being externally reversible.

Two other externally reversible cycles that operate between two constant-temperature reservoirs are the Stirling cycle and the Ericsson cycle.

Suggested Reading

6·1 Cravalho, Ernest G., and Joseph L. Smith, Jr., *Engineering Thermodynamics,* Pitman Publishing, Belmont, Calif., 1981, Chapter 5.

6·2 Dixon, J. R., *Thermodynamics I,* Prentice-Hall, Englewood Cliffs, N.J., 1975, Section 9-1.

Problems

6·1 One occasionally hears someone say, when evaluating an apparent perpetual motion machine of the first kind, that the machine would work if it were not for friction. Explain why any device that would operate as a perpetual motion machine of the first kind in the absence of friction would also work as such a machine despite frictional effects.

6·2 Comment on the following statement: If in any process work is used to produce an effect that could have been produced by heat, that process is irreversible. (In commenting on such a statement, consider questions such as the following: Is it true, false, meaningless, trivial, important, general?)

6·3 Comment on the following statement: A reversible process, when undone, leaves no history.

6·4 An electric motor doing work receives current from a storage battery. The battery and motor exchange heat with only the atmosphere. Is this a violation of the second law? Explain.

6·5 Person *A* states that in a reversible isothermal expansion of an ideal gas $\Delta U = 0$ so that the heat added equals the work done. Person *B* states that such a process violates the second law in that it amounts to a system absorbing heat from a constant-temperature energy reservoir and producing an equivalent amount of work. Resolve this conflict.

6·6 The solution of Example 6·2 is carried out using the Kelvin–Planck statement of the second law. Carry it out using the Clausius statement.

6·7 In Sec. 6·3 it is stated that the frictionless adiabatic and frictionless isothermal processes described in Sec. 6·2 are *internally* and *externally reversible*. Could either of them be *internally reversible* and *externally irreversible*? Explain.

6·8 Generate a list of several processes and classify them as reversible or irreversible. Then further classify the irreversible processes as ones that can be closely approximated by reversible processes and those that cannot.

6·9 Three kilograms of oxygen expands isothermally in a closed system from 300 kPa, 60°C, to 135 kPa. Determine the heat transfer if (a) the process is reversible, and (b) the process is irreversible so that the work is 85 percent of that produced by the reversible process.

6·10 Two kilograms of oxygen is to be compressed isothermally in a closed system from 95 kPa, 20°C, to 300 kPa. Determine the heat transfer if (a) the process is reversible, and (b) the process is irreversible so that 20 percent more work is required than is the case with the reversible process.

6·11 Nitrogen in a closed system is compressed isothermally from 10 N/cm^2, 50°C, to 20 N/cm^2. Calculate the heat transfer in J/g if (*a*) the process is reversible, (*b*) irreversibilities increase the amount of work required to 40 percent more than is required in the reversible process.

6·12 Air expands reversibly in a well-insulated cylinder. Determine the final temperature if the initial conditions are 180 kPa, 75°C, and the final pressure is 140 kPa.

6·13 Air enters a turbine at 40 kN/m^2, 900°C, and leaves at 10 kN/m^2. The flow is reversible and adiabatic. Inlet velocity is quite low, and outlet velocity is 150 m/s through a cross-sectional area of 0.20 m^2. Calculate the power output in kW.

6·14 Two pounds of air in a closed system expands isothermally from 30 psia, 140°F, to 15 psia. Calculate the amount of heat added if (*a*) the process is reversible and (*b*) irreversibilities are present which reduce the work to 80 percent of that produced by the reversible process.

6·15 Nitrogen in a closed system is compressed isothermally from 14 psia, 100°F, to 35 psia. Calculate the heat transfer per pound if (*a*) the process is reversible and (*b*) irreversibilities are present which increase the amount of work required by 50 percent.

6·16 In a Carnot engine using methane as the working fluid, the methane is at 500 kPa, 90°C, and occupies a volume of 0.05 m^3 at the beginning of the isothermal expansion. The volume at the end of the isothermal expansion is 0.10 m^3. The temperature at the end of the adiabatic expansion is 0°C. Compute (*a*) the heat added and (*b*) the heat rejected.

6·17 A reversed Carnot cycle using octane as the working fluid operates between temperature limits of 20 and 200°C. During the isothermal compression, the volume is halved, and the minimum specific volume during the cycle is 0.15 m^3/kg. Determine the coefficient of performance and the amount of heat absorbed from the low-temperature region per kilogram of octane.

6·18 Carbon dioxide expands isothermally in a Carnot engine from $V_1 = 0.03$ m^3 to $V_2 = 0.09$ m^3. During the adiabatic expansion process, the enthalpy of the air decreases from 200 to 100 kJ. Assuming the enthalpy and internal energy of the air to be zero at 0 K, and constant values of $c_p = 0.80$ kJ/kg·K and $c_v = 0.63$ kJ/kg·K, compute (*a*) the efficiency of the engine, (*b*) the internal energy at the end of the adiabatic expansion, and (*c*) the pressure at the end of the adiabatic expansion.

6·19 Consider a Carnot engine using air as the working substance. At the beginning of the isothermal expansion, the air is at 80 psia and occupies a volume of 2 cu ft. The pressure and volume at the end of the adiabatic expansion are 20 psia and 6 cu ft. Determine the efficiency of the cycle.

6·20 In a Carnot engine using air as the working fluid, the air is at 75 psia, 200°F, and occupies a volume of 2 cu ft at the beginning of the isothermal expansion. The volume at the end of the isothermal expansion is 4 cu ft. The temperature at the end of the adiabatic expansion is 30°F. Compute (*a*) the heat added and (*b*) the heat rejected.

6·21 A reversed Carnot cycle using air as the working fluid operates between temperature limits of 70 and 400°F. During the isothermal compression, the volume is halved, and the minimum specific volume during the cycle is 2 cu ft/lbm. Determine the coefficient of performance and the amount of heat absorbed from the low-temperature region per pound of air.

6·22 Air expands isothermally in a Carnot engine from $V_1 = 1$ cu ft to $V_2 = 3$ cu ft. During the adiabatic expansion process, the enthalpy of the air decreases from 200 to 100 B. Assuming the enthalpy and internal energy of the air to be zero at 0 R, and constant values of $c_p = 0.24$ B/lbm·R and $c_v = 0.171$ B/lbm·R compute (*a*) the efficiency of the engine, (*b*) the internal energy at the end of the adiabatic expansion, and (*c*) the pressure at the end of the adiabatic expansion.

6·23 Consider a Carnot engine using air as a working fluid and having an overall volume ratio of 9. During the isothermal heat rejection which begins with the air at 14.0 psia, 40°F, the volume is decreased to one third of its maximum value. Determine the efficiency of this cycle.

6·24 For a Carnot cycle, is the work of the adiabatic compression equal in magnitude to that of the adiabatic expansion if the working fluid is an ideal gas?

6·25 Derive an expression for the coefficient of performance (β_{HP}) of a Carnot heat pump, using an ideal gas as working fluid, in terms of the reservoir temperatures T_L and T_H. Make a table of β_{HP} values for the same reservoir temperatures as used in the table of Example 6·6.

6·26 The Stirling engine physical arrangement shown in Fig. 6·11 has serious drawbacks as a practical engine, so other physical arrangements are used. For one of the common ones, make a diagram like Fig. 6·11 showing the physical configuration at each of the four states of the working fluid, the *PV* diagram, and a description of each of the processes. Explain any terms such as "power piston" or "displacer piston" that you use, and state the chief advantages of this arrangement over the one shown in Fig. 6·11.

6·27 The Ericsson engine shown in Fig. 6·12 involves steady-flow equipment, but the Ericsson cycle can also be carried out in a piston-cylinder arrangement. Describe one such physical arrangement and explain its operation with the aid of a sketch and a *PV* diagram.

6·28 A closed system Stirling cycle using helium as the working fluid operates with maximum pressure and temperature of 1200 kPa, 1000°C. The cycle has an overall volume ratio of 4.2 and minimum pressure and temperature of 75 kPa, 61°C. Determine the thermal efficiency of the cycle.

6·29 Determine the thermal efficiency of the Ericsson cycle shown in Fig. 6·12 if $P_1 = 500$ kPa, $T_1 = 600$°C, $P_3 = 100$ kPa, $T_3 = 60$°C, $\dot{m} = 0.220$ kg/s, and the working fluid is argon.

6·30 A closed system Stirling cycle using helium as working fluid has an overall volume ratio of 3.8 and an overall temperature ratio of 3.0. Maximum pressure and temperature in the cycle are 300 psia and 1100°F. Determine the power output of the cycle for a heat input rate of 55 000 B/h.

6·31 A closed-system Ericsson cycle has an overall temperature ratio of 3.0 and an overall volume ratio of 4.8. The maximum temperature in the cycle is 1200°F and the maximum pressure is 500 psia. Determine the cycle thermal efficiency.

CHAPTER SEVEN

Some Consequences of the Second Law

Three valuable deductions from the second law of thermodynamics are discussed in this chapter. The first two pertain to the efficiency of reversible engines and are known as the two points of the *Carnot principle*. The third, which follows from the Carnot principle, deals with a temperature scale that is independent of the physical properties of any substance.

7·1 The Carnot Principle

The following two deductions make up the Carnot principle:*

I. *No engine can be more efficient than an externally reversible engine operating between the same temperature limits.* The "temperature limits" of a cycle are the temperatures of the two energy reservoirs with which the system exchanges heat.

II. *All externally reversible engines operating between the same temperature limits have the same efficiency.*

We shall now prove these two statements by showing that a violation of either one results in a violation of the Kelvin–Planck statement of the second law.

Referring to Fig. 7·1*a*, some engine designated as engine X and an externally reversible engine R operate between the same temperature limits. As a first step in proving that no engine can be more efficient than an externally reversible engine operating between the same temperature limits, let us *assume* that engine X has a higher efficiency than the externally reversible engine. Then for the same amount of heat Q_H supplied to each engine, $W_X > W_R$ and $Q_{LX} < Q_{LR}$. (For this discussion we drop our sign convention and let Q and W stand for absolute values.) Now let the externally reversible engine be reversed to operate as a refrigerator as shown in Fig. 7·1*b*. The reversed engine rejects heat Q_H to the energy reservoir at T_H and requires a work input of W_R. Since the magnitude of W_R is less than the work output W_X of engine X, engine X can drive the reversed engine and still deliver work in the amount $(W_X - W_R)$ to other parts of the surroundings. The reversed engine is rejecting heat in the amount Q_H to the reservoir at T_H, and engine X is absorbing

*The Carnot principle is also called the *Carnot theorem* or the *Carnot theorem and corollary*.

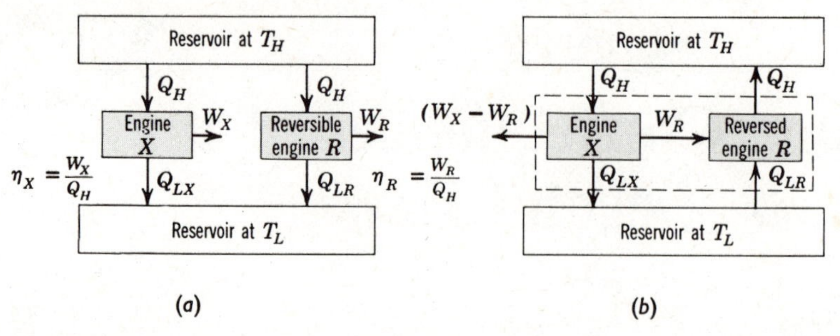

Figure 7·1. Proof of the Carnot principle.

the same amount of heat from this reservoir. Therefore, there is zero net exchange of heat with the reservoir, and the reservoir could in fact be eliminated by having the reversed engine discharge heat directly into engine X. Now look at the system that is comprised of engine X and the reversed engine together (enclosed by the broken line in Fig. 7·1*b*). It is operating cyclically, exchanging heat with a single reservoir (the one at T_L), and producing work.* This is precisely what the Kelvin–Planck statement of the second law declares to be impossible. Consequently our assumption that $\eta_X > \eta_R$ must be false. *Conclusion:* No engine can be more efficient than an externally reversible engine operating between the same temperature limits.

To prove the second point of the Carnot principle, let both of the engines in Fig. 7·1*a* be externally reversible engines. Assume that their efficiencies are different so that their work outputs are different for the same amount of heat input. Then reverse the less efficient engine. The more efficient engine can drive the reversed engine and have some work left over, even though a net amount of heat is drawn from only one reservoir. This is the same violation of the second law that we reached earlier when we made a false assumption. We must conclude that our assumption that two externally reversible engines operating between the same temperature limits can have different efficiencies is false. *Conclusion:* All externally reversible engines operating between the same temperature limits have the same efficiency.

7·2 The Efficiency of Reversible Engines

The second point of the Carnot principle means that the efficiency of any externally reversible engine depends only on the temperatures of the reservoirs with which it exchanges heat. The efficiency does not depend on the working substance. For given temperature limits, an externally reversible engine operating on air has the same efficiency as one operating on steam or any other substance. This means that if we can determine the efficiency of an externally reversible engine operating on any particular substance as

*Notice that the single reservoir could be the atmosphere, the ocean, or the water of a river or lake. The engine and reversed engine together comprise a device which could draw energy only from one of these sources and convert it into work continuously while producing no other effects. What a marvelous device this would be! Do you believe that such a device is possible? On what is your belief based?

a function of its temperature limits, then this same equation for efficiency in terms of temperature limits applies to all externally reversible engines.

Example 6·5 shows that for a Carnot engine using an ideal gas as working substance and operating between reservoir temperatures of T_H and T_L

$$\eta = 1 - \frac{T_L}{T_H}$$

where T_H and T_L are temperatures on the ideal-gas absolute temperature scale. In accordance with the second point of the Carnot principle, then, the efficiency of *any* externally reversible engine operating between the temperature limits T_H and T_L is given by

$$\eta = 1 - \frac{T_L}{T_H} \tag{7·1}$$

where T_H and T_L are temperatures on the ideal-gas absolute temperature scale. It is shown in the next article that these are the same as temperatures on a *thermodynamic temperature scale*.

7·3 The Thermodynamic Temperature Scale

In Sec. 1·13 the shortcoming of any temperature scale defined in terms of the physical properties of a substance was mentioned.* Now we know that (1) the efficiency of externally reversible engines is a function of temperature only, and (2) efficiency involves only heat and work and can therefore be measured independently of properties. In view of these two facts, perhaps a temperature scale can be defined by the relationship

$$\eta_{\text{externally reversible}} \equiv \frac{W}{Q_{\text{in}}} = f(T_H, T_L)$$

As a matter of fact, this can be done, and the procedure is discussed in the following paragraphs. A temperature scale that is independent of the properties of substances is called a thermodynamic temperature scale.

First, let us make clear just what we mean by "defining a temperature scale" and why we must do it. We are not trying to define temperature, because we accept it—along

*As shown in Sec. 4·3, the ideal-gas temperature scale does not depend on the properties of any one gas, but it does depend either on the characteristics of gases in general or on the characteristics of a hypothetical substance—an ideal gas. Which of the two it depends on is a matter of point of view. Still another point of view—that there is no ideal-gas temperature scale but that T in $Pv = RT$ is by definition an absolute thermodynamic temperature—is mentioned in Sec. 9·13. These three points of view are not in conflict in the sense that one is right and the other two are wrong. Each one is part of a different logical structure or a different philosophy of thermodynamics. If you go into more advanced study of thermodynamics, you may come to believe, as some people do, that one point of view is much to be preferred over the others. If you are interested chiefly in the applications and results of thermodynamic analysis, you will be happy to know that all three points of view lead to the same results. After all, there are many recipes for delicious chocolate cake. Some people are interested in recipes and prefer this one or that one over all the rest; other people just enjoy the result.

with mass, length, and time—as being verbally undefined. When we define a temperature scale, we specify a method for assigning numerical values to various temperatures.

From childhood you are used to reading thermometers—usually mercury-in-glass or alcohol-in-glass—to get numbers for different temperature levels; so, when part way through a thermodynamics course you are told that we must define a temperature scale, your reaction may well be: "What's wrong with the one we've been using?" Let us answer this question partly by referring to the thermometer calibration difficulty mentioned in Sec. 1·13. Even thermometers of the same kind calibrated at two temperatures often disagree with each other at other temperatures. You may say, "Well, why not arbitrarily establish one thermometer as a standard and thereby define a temperature scale?" This is a reasonable suggestion (provided other people would adopt the same thermometer as a standard), but if we were to use such a scale in scientific work we would be annoyed by some features of it. There may be a question as to whether this scale should be used outside the temperature range in which the standard thermometer itself operates. Also, we will find that certain quantities (such as the efficiency of externally reversible engines) are functions of temperature only, but the functional relationship is not quite the same for different temperature ranges. Certain other difficulties will also arise. Thus the thermometer scale we use in everyday life will not suffice. It is suitable for indicating when we should wear an overcoat or when an automobile engine is too hot; but as a scientific tool we need some other means of assigning numerical values to different temperatures. That is what we mean by saying that we must "define a temperature scale."

Consider three externally reversible engines operating as shown in Fig. 7·2. Engines A and C each absorb heat in the amount Q_1 from the reservoir at T_1. Engine C rejects heat Q_3 to the reservoir at T_3. Engine A rejects heat Q_2 at a constant temperature T_2 to

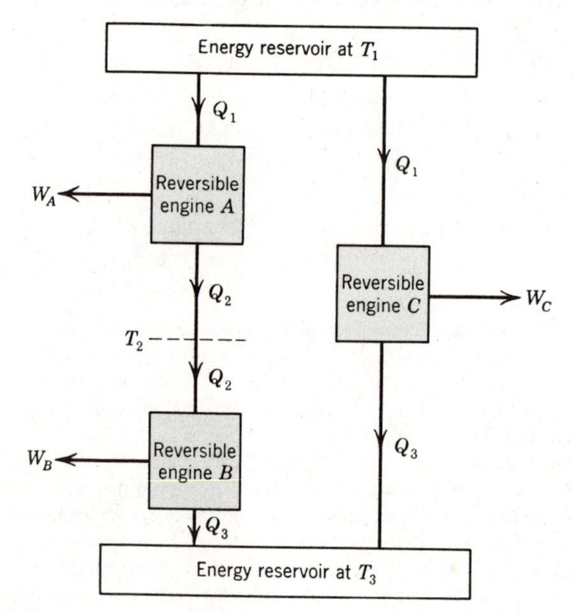

Figure 7·2. Schematic arrangement of externally reversible engines.

engine B. Engine B rejects heat to the reservoir at T_3, and the amount rejected must be Q_3, the same amount that engine C rejects. This must be true because engines A and B taken together constitute an externally reversible heat engine operating between the same temperature limits as engine C; so engines A and B taken together must have the same efficiency as engine C. Since the heat input to engines A and B combined is the same as the heat input to engine C, the heat rejected must be the same.

By the Carnot principle, the efficiency of an externally reversible engine is a function of its operating temperature limits only

$$\eta = \frac{W}{Q_{\text{in}}} = \frac{Q_{\text{in}} - Q_{\text{out}}}{Q_{\text{in}}} = 1 - \frac{Q_{\text{out}}}{Q_{\text{in}}} = 1 - \frac{Q_L}{Q_H} = \phi(T_H, T_L)$$

Therefore,

$$\frac{Q_H}{Q_L} = \psi(T_H, T_L)$$

For engines A, B, and C this is

$$\frac{Q_1}{Q_2} = \psi(T_1, T_2) \tag{a}$$

$$\frac{Q_2}{Q_3} = \psi(T_2, T_3) \tag{b}$$

$$\frac{Q_1}{Q_3} = \psi(T_1, T_3) \tag{c}$$

The product of Eqs. a and b is

$$\frac{Q_1}{Q_3} = \frac{Q_1}{Q_2}\frac{Q_2}{Q_3} = \psi(T_1, T_2) \cdot \psi(T_2, T_3) \tag{d}$$

Combining Eqs. c and d

$$\psi(T_1, T_3) = \psi(T_1, T_2) \cdot \psi(T_2, T_3) \tag{e}$$

Look closely at Eq. e. The left-hand side is a function of T_1 and T_3 only; therefore the right-hand side must be a function of T_1 and T_3 only. The value of T_2 does not affect the value of the product on the right-hand side. This tells us something about the form of the function ψ. In order to satisfy this condition, the function ψ must have the form

$$\psi(T_1, T_2) = \frac{f(T_1)}{f(T_2)}$$

$$\psi(T_2, T_3) = \frac{f(T_2)}{f(T_3)}$$

where f is another function. Substituting into Eq. e gives

$$\psi(T_1, T_3) = \psi(T_1, T_2) \cdot \psi(T_2, T_3) = \frac{f(T_1)}{f(T_2)} \cdot \frac{f(T_2)}{f(T_3)} = \frac{f(T_1)}{f(T_3)}$$

Substituting now into Eq. c gives

$$\frac{Q_1}{Q_3} = \frac{f(T_1)}{f(T_3)}$$

which tells us much more than Eq. c does regarding the dependence of Q_1/Q_3 on the reservoir temperatures. This is as far as deductive reasoning will take us. The second law thus requires *only* that

$$\frac{Q_H}{Q_L} = \frac{f(T_H)}{f(T_L)} \qquad (7\cdot2)$$

for any externally reversible engine. The function $f(T)$ can be chosen arbitrarily to define a temperature scale that is independent of the physical properties of any substance. Such a temperature scale is called a thermodynamic temperature scale. As proposed by Lord Kelvin, we can let $f(T) = T$ so that a thermodynamic temperature scale is *defined* by

$$\frac{Q_H}{Q_L} = \frac{T_H}{T_L} \qquad (7\cdot3)$$

On this scale, which is often called the Kelvin scale, the ratio of two temperatures is equal to the ratio of the amounts of heat transferred between an externally reversible heat engine and two reservoirs at these temperatures.

A characteristic of the Kelvin or thermodynamic temperature scale is that negative temperatures are impossible. Recall that in our discussion of externally reversible engines and in Eq. 7·3, Q_H and Q_L stand respectively for heat added to the engine at T_H and rejected from the engine at T_L. Therefore, if an externally reversible heat engine operates with Q_H absorbed from a reservoir at T_H and if T_L is negative, then we see from

$$\frac{Q_H}{Q_L} = \frac{T_H}{T_L} \qquad (7\cdot3)$$

that Q_L must be negative; that is, heat must be absorbed by the engine from the reservoir at T_L. Since the energy removed from the reservoir at T_L can always be replaced just by transferring heat from the reservoir at T_H to the one at T_L, the reservoir at T_L could operate cyclically, and it and the engine together would constitute a device that absorbed heat from the reservoir at T_H and produced an equivalent amount of work while operating cyclically. Thus the existence of a negative absolute thermodynamic temperature would make possible a violation of the second law.

In order to define zero on the Kelvin scale or absolute zero temperature, consider a series of reversible engines as shown in Fig. 7·3. One engine absorbs heat Q_1 from the energy reservoir at T_1. It rejects heat Q_2 at a temperature T_2 to another reversible engine that in turn rejects heat Q_3 at a temperature T_3 to another reversible engine and so forth. Let the temperatures be selected so that each engine does the same amount of work. Thus

$$W_1 = W_2 = W_3 = \ldots$$

and

$$Q_1 - Q_2 = Q_2 - Q_3 = Q_3 - Q_4 = \ldots$$

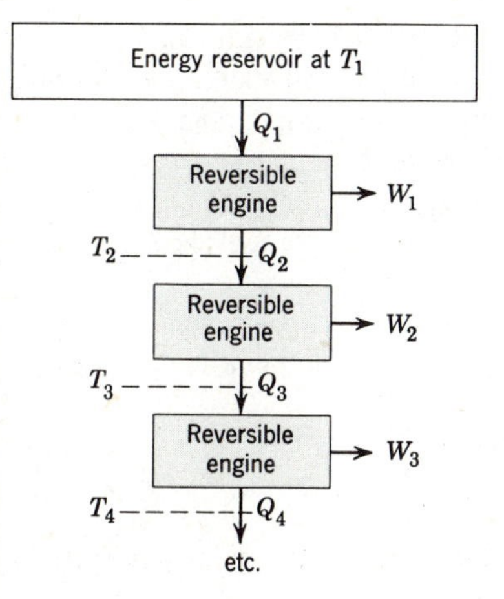

Figure 7·3. Schematic arrangement of externally reversible engines used in discussing the Kelvin temperature scale.

and application of the defining equation of the Kelvin scale (Eq. 7·3) shows that

$$T_1 - T_2 = T_2 - T_3 = T_3 - T_4 = \ldots$$

That is, the engines in the series do equal work when the temperature differences across them are equal.

As more engines are added to the series, the total work output increases. By the first law, the total work output cannot exceed Q_1. In the limiting case where there are enough engines in the series to make the total work output ($W_1 + W_2 + W_3 + \ldots$) equal Q_1, the last engine will reject zero heat. The last engine in the series (or the whole series of engines taken together) is then violating the Kelvin–Planck statement of the second law by operating cyclically and producing work while exchanging heat with only one reservoir. Thus the operation of the series of engines with zero heat rejection from the last engine cannot be accomplished, although it can be approached as a limiting case. As the number of engines in the series is increased and the heat rejection of the last engine approaches zero, the temperature at which this heat is rejected also approaches zero on the Kelvin scale in accordance with the defining equation

$$\frac{Q_H}{Q_L} = \frac{T_H}{T_L}$$

Absolute zero temperature can be defined as follows: *If an externally reversible heat engine operates between two energy reservoirs, absorbing a constant heat input from the hotter reservoir, and the temperature of the colder reservoir is successively lowered, the amount of heat rejected decreases. As the amount of heat rejected approaches zero, the temperature of the colder reservoir approaches absolute zero.* Notice that this definition makes no mention whatsoever of the physical properties of any substance either at absolute zero or at any other temperature. Also, the second law alone does not lead to the conclusion that it is impossible for the temperature of any system to be absolute zero. (This conclusion is in the realm of the third law of thermodynamics and is not discussed in this book.)

The Kelvin or thermodynamic temperature scale is not completely defined until the "size" of the degree is fixed. This was done by the international General Conference on Weights and Measures in 1954 and 1967: "The kelvin, unit of thermodynamic temperature, is the fraction 1/273.16 of the thermodynamic temperature of the triple point of water." Thus a numerical value in kelvins for any other temperature is given by

$$T = T_{\text{TP}} \frac{Q}{Q_{\text{TP}}} = 273.16 \frac{Q}{Q_{\text{TP}}}$$

where Q and Q_{TP} are the amounts of heat transferred by an externally reversible engine operating between a reservoir at T and a reservoir at T_{TP}, the triple-point temperature of water.

The Celsius temperature scale is then defined by

$$T[°\text{C}] = T[\text{K}] - 273.15$$

Notice that 0°C is assigned to the *ice-point* temperature which is approximately 0.01 K lower than the triple-point temperature. (The ice point temperature is the temperature of a mixture of ice and air-saturated water in equilibrium at 1 atm.) The size of the kelvin was selected to agree very closely with the size of a Celsius degree, but the definition of the kelvin is now independent of any other temperature scales. Other scales are defined in terms of the Kelvin or thermodynamic temperature scale.

Even though the thermodynamic temperature scale is defined in terms of the performance of externally reversible engines, it is unnecessary to operate an externally reversible engine in order to determine numerical values on the thermodynamic temperature scale or in order to establish the relationship between this scale and various other ones. Several relations between absolute thermodynamic temperatures and other properties of substances are now known. One of these we should observe now. From the definition of thermal efficiency and the first law, we can write for any heat engine

$$\eta = \frac{W}{Q_{\text{in}}} = \frac{Q_{\text{in}} - Q_{\text{out}}}{Q_{\text{in}}} = 1 - \frac{Q_{\text{out}}}{Q_{\text{in}}}$$

From the definition of the thermodynamic temperature scale, which depends on the second law, we can write for any externally reversible heat engine operating between reservoirs at T_H and T_L

$$\frac{Q_{\text{out}}}{Q_{\text{in}}} = \frac{Q_L}{Q_H} = \frac{T_L}{T_H}$$

Therefore, the efficiency of any externally reversible heat engine operating between reservoirs at T_H and T_L is

$$\eta = 1 - \frac{T_L}{T_H} \tag{7·1}$$

where T_H and T_L are temperatures on the absolute thermodynamic scale. But this same equation was given for this efficiency in the preceding section, and the temperatures were specified as being on the ideal-gas absolute temperature scale. The conclusion is that the ideal-gas absolute temperature scale agrees identically with the thermodynamic temperature scale.

In practice, highly accurate temperature measurements use the International Practical Temperature Scale (IPTS), an accurately reproducible scale for laboratory use. The IPTS is defined by (1) assigning values to certain accurately reproducible temperatures such as boiling and melting points, (2) specifying the type of thermometer to be used in each range of the scale, and (3) specifying the interpolation formula to be used for each thermometer between the assigned values. The IPTS was established to agree with the thermodynamic scale and it does so very closely.

Example 7·1. Determine the heat input to a Carnot engine which operates between 400 and 20°C and produces 100 kJ of work.

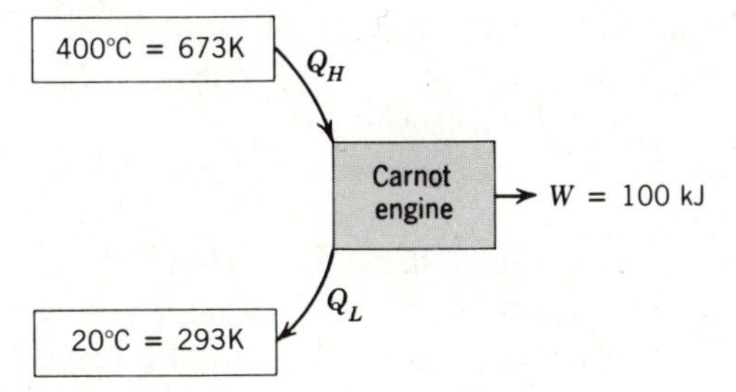

Example 7·1.

Solution. Applying the first law and the definition of the thermodynamic temperature scale to this engine, we have two simultaneous equations:

$$Q_H - Q_L = W$$

$$\frac{Q_H}{Q_L} = \frac{T_H}{T_L}$$

Solving these for Q_H,

$$Q_H - Q_H \frac{T_L}{T_H} = W$$

$$Q_H = \frac{W}{1 - T_L/T_H} = \frac{100}{1 - 293/673} = 177 \text{ kJ}$$

(Notice that the last equation is the same as $Q_H = W/\eta$.)

Example 7·2. Compute the power required to drive a reversed Carnot engine if 100 kJ/min is absorbed from the cold region and the isothermal processes occur at 200 and 10°C.

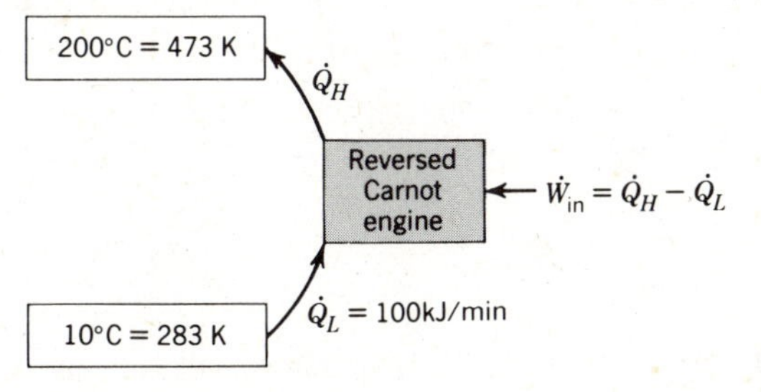

Example 7·2.

Solution. The power input is the difference between the rate at which heat is rejected and the rate at which heat is absorbed. The rate of heat rejection can be found by

$$\dot{Q}_H = \dot{Q}_L \frac{T_H}{T_L} = 100(473/283) = 167 \text{ kJ/min}$$

and then

$$\dot{W}_{\text{in}} = \dot{Q}_H - \dot{Q}_L = 167 - 100 = 67 \text{ kJ/min} = 1.1 \text{ kW}$$

7·4 Summary

No engine can be more efficient than an externally reversible engine operating between the same temperature limits. The "temperature limits" of a cycle are the temperatures of the two energy reservoirs with which the system exchanges heat.

All externally reversible engines operating between the same temperature limits have the same efficiency.

These two italicized statements are the two points of the *Carnot principle.*

A temperature scale that is entirely independent of the physical properties of any substance can be defined. Such a scale is called a *thermodynamic temperature scale.* The most commonly used thermodynamic temperature scale is one proposed by Kelvin and defined by the relationship

$$\frac{T_H}{T_L} = \frac{Q_H}{Q_L} \tag{7·3}$$

where T_H and T_L are the temperatures of the energy reservoirs between which an externally reversible engine operates when it absorbs heat Q_H from the hotter reservoir and rejects heat Q_L to the colder reservoir. The ideal-gas absolute temperature scale agrees identically with the Kelvin thermodynamic temperature scale.

Negative temperatures are impossible on the Kelvin scale. The question of whether the temperature of any system can be lowered to zero on the Kelvin scale cannot be answered by the second law alone.

It must be remembered that the relationship

$$\frac{Q_H}{Q_L} = \frac{T_H}{T_L} \tag{7·3}$$

applies only to externally reversible engines operating between reservoirs at T_H and T_L. *The efficiency of any externally reversible engine operating between temperatures of T_H and T_L is given by*

$$\eta = 1 - \frac{T_L}{T_H} = \frac{T_H - T_L}{T_H}$$

Suggested Reading

7·1 Cravalho, Ernest G., and Joseph L. Smith, Jr., *Engineering Thermodynamics,* Pitman Publishing, Belmont, Calif., 1981, Chapter 6.

7·2 Hatsopoulos, George N., and Joseph H. Keenan, *Principles of General Thermodynamics*, Wiley, New York, 1965, Chapter 15.

7·3 Van Wylen, Gordon J., and Richard E. Sonntag, *Fundamentals of Classical Thermodynamics*, 3rd ed., Wiley, New York, 1985, Sections 6.6, 7.

Problems

7·1 Refer to Fig. 7·1. Demonstrate the Carnot principle by letting one of the initial assumptions be that $Q_{LX} = Q_{LR}$.

7·2 Prove that no refrigerator can have a higher coefficient of performance than an externally reversible refrigerator operating between the same temperature limits.

7·3 Prove that all externally reversible refrigerators operating between the same temperature limits have the same coefficient of performance.

7·4 It is possible for an actual engine to be more efficient than a Carnot engine. Explain this.

7·5 Prove from the Clausius statement or the Kelvin–Planck statement the following, which is known as the Carathéodory statement of the second law: In the vicinity of any state of a substance there are some states that cannot be reached (from the state first mentioned) by adiabatic processes alone.

7·6 Lowering the temperature at which a heat engine rejects heat increases its efficiency. A steam power plant usually rejects its heat to water from a river, lake, or ocean; so would it be advisable to cool the water first by means of a refrigeration system driven by the power plant? Explain.

7·7 Write in terms of only the reservoir temperatures, T_H and T_L, expressions for the coefficient of performance of (*a*) a Carnot refrigerator and (*b*) a Carnot heat pump.

7·8 Answer questions 1 through 5 in Sec. 5·4.

7·9 Sketch a curve of Carnot cycle efficiency versus heat source temperature for a fixed heat-rejection temperature.

7·10 Sketch, on a common set of axes, curves of Carnot refrigerator coefficient of performance and Carnot heat pump coefficient of performance versus low temperature for a fixed heat-rejection temperature.

7·11 Referring to Fig. 7·2, demonstrate that $Q_1/Q_3 = f(T_1)/f(T_3)$ without making reference to any heat engines except *A* and *B*.

7·12 Someone has paraphrased the first law of thermodynamics as "You can't get something for nothing" and the second law as "You can't get as much as you thought you could." Someone else has suggested "You can't win" and "You can't even break even." Do these appear to you to be apt paraphrases? Explain.

7·13 Comment on the validity of the following as a statement of the second law: Heat cannot be converted into work unless there is a temperature difference.

7·14 Prove whether the following statement is true or false: A closed system cannot execute a cycle while receiving heat from a single energy reservoir.

7·15 Refer to the second footnote of Sec. 7·1. To what extent can it be proved that the device as described is impossible?

7·16 When Sir Isaac Newton established the first numerical temperature scale, would it have been possible to assign numbers so that temperatures of colder bodies were represented by larger numbers? (This means that a man in England stepping outdoors in January might turn up his coat collar and say, "Brrr, the temperature is really getting up there today.") If impossible, state why. If possible, do you see any objections?

7·17 Upon learning that a temperature of 0.0014 K has been reached in a cryogenic laboratory, someone states that the scientists "practically reached absolute zero." Comment on this statement.

7·18 A Carnot engine produces 3.5 kW while operating between temperatures of 200 and 35°C. Find the heat absorbed from the source by the medium per minute.

7·19 A Carnot engine receives 108 000 kJ/h from a source at 280°C and rejects 79 900 kJ/h to a cold body. At what temperature is the cold body?

7·20 A Carnot engine works between a hot body at 530°C and a cold body at 25°C. For each 100 kJ absorbed from the source, compute (*a*) the heat rejected and (*b*) the work developed by the engine.

7·21 A Carnot engine receives 8.5 kJ from a source at 260°C and during the cycle performs 4.7 kJ of work. What is the temperature of the cold body?

7·22 A heat engine operating on the Carnot cycle produces 13.5 kJ of work while operating between temperature limits of 260 and 5°C. Determine the efficiency.

7·23 A reversed Carnot engine is used as an ice machine; it operates between 0 and 30°C. How many kilograms of ice will it produce per hour if the power input is 1.5 kW? (Latent heat of fusion of ice is 334 kJ/kg.)

7·24 A reversed Carnot engine operates between 35 and 5°C. If 425 kJ/min is to be removed from the cold body, calculate the power required.

7·25 A reversed Carnot engine operating between −10 and 35°C rejects 42 000 kJ/h to the receiver at 35°C. Compute the power required to operate the machine.

7·26 A Carnot refrigerator is used to make ice. Freezing water at 0°C is the cold body, and heat is rejected from the system to a river at 20°C. What is the work required to freeze 45 kg of ice? (Latent heat of fusion of ice is 334 kJ/kg.)

7·27 A reversed Carnot engine absorbs 425 kJ/min from a source at a temperature of −25°C. If heat is rejected at 32°C, compute the coefficient of performance.

7·28 A reversed Carnot engine operating between 5 and 35°C delivers 105 000 kJ/h to the hot body. Compute the coefficient of performance.

7·29 A Carnot engine using wet steam as the working fluid operates between temperature limits of 150 and 35°C. Sketch a Pv diagram of the cycle. For a work output of 100 kJ, calculate the amount of heat rejected from the working fluid at the lower temperature.

7·30 A Carnot refrigerator using Refrigerant-12 as a working fluid removes 10 000 kJ/h from a region at 5°C. The highest pressure reached by the working fluid is 700 kPa. Sketch a Pv diagram of the cycle. Calculate the power input to the refrigerator.

7·31 A heat pump which picks up heat from the atmosphere at 100 kPa, −10°C, is used to maintain a building at 29°C. Doing so requires 125 000 kJ/h of heat input to the building. Determine the minimum possible power input to the heat pump, and state why the value you have calculated is the minimum value.

7·32 Maintaining a building at 29°C requires a heat input of 85 000 kJ/h. A heat pump that draws energy from the atmosphere at −5°C is used to heat the building. If the heat pump is a Carnot heat pump, how much power input is required?

7·33 A Carnot refrigerator removes heat at a rate of 40.0 kJ/s from a cold storage room at −20°C.

The refrigerator is driven by a Carnot engine that takes heat from a reservoir at 400°C and discharges heat to the atmosphere at 25°C. Determine (*a*) the power required to drive the refrigerator and (*b*) the rate at which heat is added to the Carnot engine from the high-temperature reservoir.

7·34 Power plants that operate on the difference in temperature between water at great depths in the ocean and near the surface have been proposed. In a location where the surface temperature is 35°C and the temperature at a great depth is 5°C, what is the maximum efficiency of such a power plant?

7·35 A Carnot engine develops 10 hp and at the same time rejects 40 000 B/h at 70°F. Compute the temperature of the source or hot body.

7·36 A Carnot engine operates between a source at 1200°F and a receiver at 70°F. If the output of the engine is 100 hp, compute the heat supplied, the heat rejected, and the efficiency of the engine.

7·37 A Carnot engine operates between a source at 800°F and a receiver at 100°F. If 200 B are rejected each minute to the receiver, compute the power output.

7·38 The efficiency of a Carnot engine discharging heat to a cooling pond at 80°F is 30 percent. If the cooling pond receives 800 B/min, what is the power output of the engine? What is the temperature of the high-temperature source?

7·39 A Carnot engine receives 15 B/s from a source at 900°F and delivers 6000 ft·lbf/s of power. Determine (*a*) the efficiency and (*b*) the temperature of the receiver.

7·40 A Carnot engine produces 5 hp while operating between temperatures of 400 and 100°F. Find the heat absorbed from the source by the medium per minute.

7·41 A Carnot engine receives 102 000 B/h from a source at 540°F and rejects 75 700 B/h to a cold body. At what temperature is the cold body?

7·42 A Carnot engine works between a hot body at 1000°F and a cold body at 80°F. For each 100 B absorbed from the source, compute (*a*) the heat rejected and (*b*) the work developed by the engine.

7·43 A Carnot engine receives 8 B from a source at 500°F and during the cycle performs 3500 ft·lbf of work. What is the temperature of the cold body?

7·44 A Carnot engine containing 8 lbm of air has at the beginning of the expansion stroke a volume of 10 cu ft and a pressure of 220 psia. The exhaust temperature is 40°F. If 8 B of heat is added during the cycle, find (*a*) the efficiency of the engine and (*b*) the work of the cycle.

7·45 A heat engine operating on the Carnot cycle produces 10 000 ft·lbf of work while operating between temperature limits of 500 and 40°F. Determine the efficiency.

7·46 If 100 B represents the net output for a Carnot engine operating between 100 and 300°F, how much heat must be absorbed from the source and how much rejected to the receiver?

7·47 A reversed Carnot engine is used as an ice machine; it operates between 32 and 84°F. How many pounds of ice will it produce per hour if the power input is 2 hp? (Latent heat of fusion of ice is 143 B/lbm.)

7·48 A reversed Carnot engine is used as a refrigerating machine for removing 6000 B/min from a cold storage room at −5°F. Heat is discharged to a hot body at 70°F. Compute the power required to operate the reversed engine.

7·49 A Carnot heat pump is used for heating a building. The outside air at 22°F is the cold body, the building at 72°F is the hot body, and 200 000 B/h is required for heating. Find (*a*) heat taken from the outside per hour and (*b*) power required.

7·50 A reversed Carnot engine operates between 100 and 40°F. If 400 B/min is to be removed from the cold body, calculate the power required.

7·51 A reversed Carnot engine operating between 0 and 100°F rejects 40 000 B/h to the receiver at 100°F. Compute the power required to operate the machine.

7·52 A Carnot refrigerator is used to make ice. Freezing water at 32°F is the cold body, and heat is rejected from the system to a river at 70°F. What is the work required to freeze 100 lbm of ice? (Latent heat of fusion of ice is 143 B/lbm.)

7·53 A reversed Carnot engine absorbs 400 B/min from a source at a temperature of −10°F. If heat is rejected at 90°F, compute the coefficient of performance.

7·54 A reversed Carnot engine operating between 40 and 100°F delivers 100 000 B/h to the hot body. Compute the coefficient of performance.

7·55 A Carnot engine using wet steam as the working fluid operates between temperature limits of 300 and 100°F. Sketch a Pv diagram of the cycle. For a work output of 100 B, calculate the amount of heat rejected from the working fluid at the lower temperature.

7·56 A Carnot refrigerator using wet steam as a working fluid removes 10 000 B/h from a region at 40°F. The highest pressure reached by the working fluid is 10 psia. Sketch a Pv diagram of the cycle. Calculate the power input to the refrigerator.

7·57 A Carnot refrigerator is to be used to remove 400 B/h from a region at −60°F and discharge heat to the atmosphere at 40°F. The Carnot refrigerator is to be driven by a Carnot engine operating between an energy reservoir at 1040°F and the atmosphere at 40°F. How much heat, in B/h, must be supplied to the Carnot engine at 1040°F?

7·58 A refrigeration system removes 200 B/h from a region at a constant temperature of −60°F while it receives heat at a rate of 50 B/h from an energy reservoir at 540°F and also receives a power input. It rejects heat to the atmosphere at 70°F. The system operates in a cyclic and externally reversible manner, but it does not operate on a Carnot cycle. Calculate (*a*) the rate of heat rejection to the atmosphere and (*b*) the power input.

7·59 It is desired to operate a refrigerator to remove heat at a rate of 40 000 B/h from a region at −60°F, discharging heat to the atmosphere at 60°F. The refrigerator is to be driven by a heat engine that receives heat at 1100°F and rejects heat to the atmosphere. Calculate the minimum rate of heat transfer into the engine (in B/h) and explain why this is the minimum amount.

7·60 Solve Problem 7·57 with the Carnot engine rejecting heat to the region at −60°F instead of to the atmosphere. The refrigerator must then absorb 400 B/h plus the heat rejected by the engine.

7·61 A Carnot cycle uses 0.2 lbm of steam as a working fluid. At the beginning of the isothermal expansion the working fluid is at 200 psia, 20 percent quality. At the end of the isothermal expansion, the steam is dry and saturated. During the isothermal compression the steam is at 10 psia. Calculate (*a*) the efficiency of the cycle and (*b*) the amount of work done per cycle.

7·62 A reversed Carnot cycle uses 0.2 lbm of steam as working fluid. At the beginning of the isothermal expansion, the steam is at 0.5 psia, 30 percent quality; at the end of the isothermal expansion, the steam is dry and saturated. Heat is rejected at 240°F. Determine (*a*) the change in internal energy of the steam during the isothermal expansion and (*b*) the work input per cycle.

7·63 A Carnot cycle uses 0.1 kg of steam as a working fluid. At the beginning of the isothermal expansion the working fluid is at 1.4 MPa, 20 percent quality. At the end of the isothermal expansion, the steam is dry and saturated. During the isothermal compression the steam is at 70 kPa. Calculate (*a*) the efficiency of the cycle and (*b*) the amount of work done per cycle.

7·64 A reversed Carnot cycle uses 0.1 kg of ammonia as a working fluid. At the beginning of the isothermal expansion, the ammonia is at 80 kPa, 30 percent quality; at the end of the isothermal expansion, the ammonia is dry and saturated. Heat is rejected at 20°C. Determine (*a*) the change in internal energy of the ammonia during the isothermal expansion and (*b*) the work input per cycle.

7·65 Describe precisely the steps involved in executing a Carnot cycle using a rubber band as the working substance and water at the ice point and water at the 1 atm boiling point as the energy reservoirs. Sketch force–length and Ts diagrams.

CHAPTER EIGHT

Entropy

The first law of thermodynamics leads to the definition of a very useful property: stored energy, E. The term *energy* is used in everyday conversation, often in the same sense as in engineering and science. Consequently, you were familiar with the term before meeting it in the study of thermodynamics, and this familiarity may even have helped you to gain an understanding of the nature of stored energy as rigorously defined by the first law.

The second law also leads to the definition of a very useful property: entropy, S. Unlike energy, the term *entropy* is not used in everyday conversation. Most high school physics courses and many college introductory physics courses do not mention it. Consequently, entropy is probably an unfamiliar term. Furthermore, it is defined in terms of a calculus operation, and no direct physical picture of it can be given. For these reasons, you may find at first that the concept of entropy is somewhat nebulous. To gain an understanding of entropy, you should study its uses and keep asking the questions, "What is it used for?" and "How is it used?" If you are looking for a physical description as an answer, the question "What *is* entropy?" is fruitless. (Recall that the development of the first law was impeded by the question, "What *is* heat?" Only when attention was turned to the question, "How is heat related to work and other effects?" was progress made toward the understanding of heat.) Consequently, this chapter not only introduces entropy but also discusses several of its uses. You will see that every subsequent chapter in this book except one makes use of the property entropy.

8·1 The Property Entropy

One way to prove that some quantity is a property is to show that the cyclic integral of the quantity is always zero. Physically, this means that if a system executes a cycle, the quantity is always returned to its initial value when the system is returned to its initial state. The value of the quantity thus depends only on the state of the system. This is true for all properties and for properties only. We will now prove that $\oint_{\text{rev}} (\delta Q/T) = 0$ and therefore that $\int_{\text{rev}} (\delta Q/T)$ is a property. This is the property called entropy, S. The definition of entropy (or entropy change) is therefore

$$\Delta S \equiv \int_{\text{rev}} \frac{\delta Q}{T} \tag{8·1a}$$

As a first step in the proof we show that any reversible process can be approximated by a series of reversible adiabatic and isothermal processes. For example, the reversible process represented by line 1-2 in Fig. 8·1 may be approximated by a series of reversible adiabatic and isothermal steps as shown by 1-*b*, *b*-*c*, and *c*-2, provided that the individual steps are so chosen that the area under curve 1-2 is equal to that under 1-*b*-*c*-2. Of course, the greater the number of adiabatic and isothermal steps, the closer the series of lines will approach the original reversible process. For purposes of discussion, the two adiabatic lines and the one isothermal line will suffice. Consider a closed system. Since the areas under the two curves are made equal, the work terms $\int P\,dv$ will be the same; hence

$$W_{1\text{-}2} = W_{1\text{-}b\text{-}c\text{-}2}$$

Applying the first law,

$$Q_{1\text{-}2} = U_1 - U_1 + W_{1\text{-}2}$$

and

$$Q_{1\text{-}b\text{-}c\text{-}2} = U_2 - U_1 + W_{1\text{-}b\text{-}c\text{-}2}$$

Since the initial and final internal energy (property) values are the same regardless of the path and the steps are selected so that the work terms are equal,

$$Q_{1\text{-}2} = Q_{1\text{-}b\text{-}c\text{-}2}$$

In other words, the heat transferred during the reversible process 1-2 is the same as that transferred during the isothermal change *b*-*c*, since no heat is transferred during the adiabatic steps 1-*b* and 2-*c*. This is a significant fact, because it is now possible to replace any reversible process by a series of reversible adiabatic and isothermal processes so that the internal energy change, the heat transferred, and the work performed are the same.

A system (for example, a gas) at state 1 as shown in Fig. 8·2*a* is assumed first to undergo a reversible process 1-2-3 and then to proceed along the reversible path 3-4-1 to the initial state 1. These processes form a cycle. The first step in this proof is to replace the original processes by a series of reversible adiabatic and isothermal lines as shown in

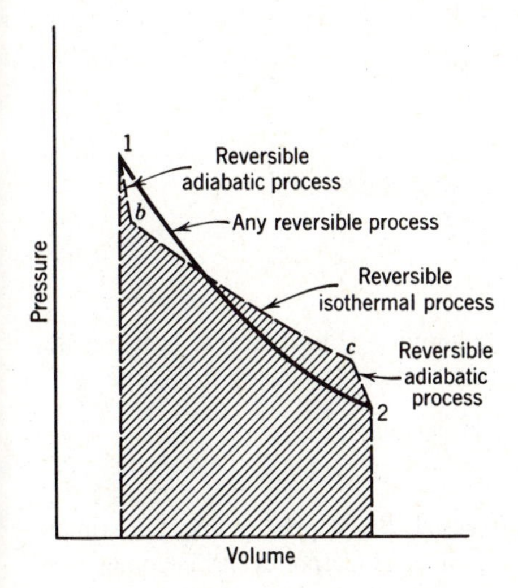

Figure 8·1. Simulating any reversible process by a series of reversible adiabatics and reversible isothermals.

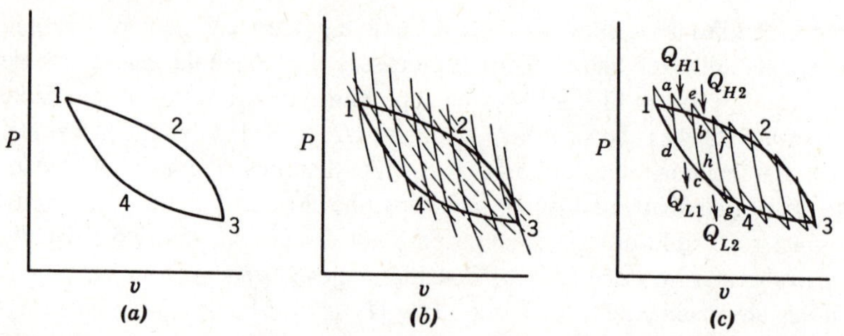

Figure 8·2. Simulating a reversible cycle by a series of reversible adiabatics and reversible isothermals.

Fig. 8·2*b* in a manner similar to that previously described. These various adiabatic (solid) and isothermal (dashed) lines may then be connected so as to represent a number of Carnot cycles, as shown in Fig. 8·2*c*. For the Carnot cycles *a-b-c-d* and *e-f-g-h*, the second law leads to

$$\frac{Q_{H1}}{T_{H1}} = \frac{Q_{L1}}{T_{L1}}$$

$$\frac{Q_{H2}}{T_{H2}} = \frac{Q_{L2}}{T_{L2}}$$

where the temperatures are the temperatures of the system during the isothermal processes. To simplify previous developments, both Q_H and Q_L were considered positive (or it could be said that only absolute values were used). Actually the heat rejected, Q_L, should be opposite in sign to the heat added, Q_H; hence, by using the sign convention of positive for heat absorbed and negative for heat rejected, the preceding equations become

$$\frac{Q_{H1}}{T_{H1}} + \frac{Q_{L1}}{T_{L1}} = 0$$

$$\frac{Q_{H2}}{T_{H2}} + \frac{Q_{L2}}{T_{L2}} = 0$$

From these relations it is apparent that the summation of all the ratios of heat transfer to absolute temperature for *all* the Carnot cycles equals zero, or

$$\left(\frac{Q_{H1}}{T_{H1}} + \frac{Q_{L1}}{T_{L1}}\right) + \left(\frac{Q_{H2}}{T_{H2}} + \frac{Q_{L2}}{T_{L2}}\right) + \cdots = 0$$

The summation expression can be simplified by employing Q to represent heat transfer and T the absolute temperature at which the heat is transferred. Hence for the cycle,

$$\sum \frac{Q}{T} = 0$$

If the number of Carnot cycles is greatly increased, the stepped paths approximate more closely the actual processes. Finally, as the number of Carnot cycles becomes very large, the summation of the Q/T terms becomes equal to the integral of $\delta Q/T$; hence the preceding equation becomes

$$\oint_{\text{rev}} \frac{\delta Q}{T} = 0$$

From this result alone *we conclude that* $\int_{\text{rev}} (\delta Q/T)$ is a property. However, it may be worthwhile to demonstrate once again* that $\oint_{\text{rev}} (\delta Q/T) = 0$ means that the value of $\int_{\text{rev}} (\delta Q/T)$ for any process depends only on the end states and not on the path followed in going from one state to another.

Consider two equilibrium states 0 and 1 for a system as shown in Fig. 8·3. The path 0-a-1 is a reversible path from state 0 to state 1. Paths 1-b-0 and 1-c-0 are any other two reversible paths between state 1 and state 0. Two cycles may be considered: 0-a-1-b-0 and 0-a-1-c-0, for which the cyclic integral is zero. In each cycle the cyclic integral may be represented as two integrals.

$$\text{Cycle 0-}a\text{-1-}b\text{-0:} \qquad \oint \frac{\delta Q}{T} = \int_{0\text{-}a}^{1} \frac{\delta Q}{T} + \int_{1\text{-}b}^{0} \frac{\delta Q}{T} = 0$$

and

$$\text{Cycle 0-}a\text{-1-}c\text{-0:} \qquad \oint \frac{\delta Q}{T} = \int_{0\text{-}a}^{1} \frac{\delta Q}{T} + \int_{1\text{-}c}^{0} \frac{\delta Q}{T} = 0$$

Subtracting these two expressions gives

$$\int_{1\text{-}b}^{0} \frac{\delta Q}{T} - \int_{1\text{-}c}^{0} \frac{\delta Q}{T} = 0$$

Transposing and rearranging results in

$$\int_{0}^{1\text{-}b} \frac{\delta Q}{T} = \int_{0}^{1\text{-}c} \frac{\delta Q}{T}$$

Since b and c are *any* two reversible paths between states 0 and 1, it follows that the

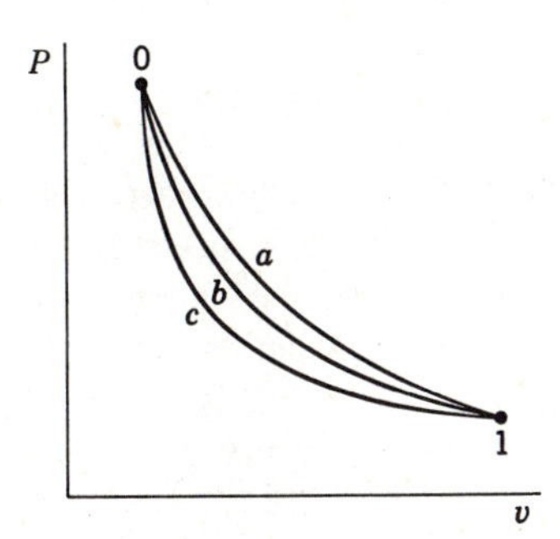

Figure 8·3. Reversible paths.

*It was demonstrated in Sec. 2·3 that $\oint(\delta Q - \delta W) = 0$ leads to the fact that the value of $\int(\delta Q - \delta W)$ depends only on the end states of a process.

value of $\int(\delta Q/T)$ is the same for *all* reversible paths between the two states. In other words, the value of $\int_{\text{rev}} (\delta Q/T)$ depends only on the end states of any process. The notation "rev" on the integral sign indicates that the integration must be carried out along some *reversible* path connecting the two states.

The convention that *reversible* without a modifier means *internally reversible* (see Sec. 6·3.) applies to the subscript "rev." Since Carnot cycles, which are externally reversible, were used in the development above, let us examine why only the restriction of *internal* reversibility is needed. Since a Carnot cycle is an externally reversible cycle, during the isothermal processes heat is transferred across only infinitesimal temperature differences between the system and the energy reservoirs in the surroundings. Therefore, as the number of Carnot cycles in Fig. 8·2 increases, the number of reservoirs needed at different temperatures increases. If instead of utilizing this large number of reservoirs the system executes the same processes of the Carnot cycles while exchanging heat with reservoirs only at the two fixed temperatures, T_1 and T_2, the processes would be externally irreversible. Notice, though, that the system could be passing through exactly the same states as it would while exchanging heat across infinitesimal temperature differences with the very large number of reservoirs. Therefore, if Q is the heat transfer of the system and T is the temperature of the system, the value of the integral $\oint_{\text{rev}} (\delta Q/T)$ is zero for the system whether the process is externally reversible or only internally reversible. Therefore, in evaluating the integral it is necessary only that the process be internally reversible.

The property $\int_{\text{rev}} (\delta Q/T)$ is called entropy and is denoted by the symbol S. Thus

$$\Delta S \equiv \int_{\text{rev}} \frac{\delta Q}{T} \tag{8·1a}$$

This is an operational definition. It tells us how to obtain numbers for ΔS, even though the operations prescribed are "pencil and paper" operations. Notice that we actually define the *change in entropy* ΔS instead of entropy S, just as we earlier defined ΔE instead of E. In engineering work it is usually only the *change* in S which is important; so a value of $S = 0$ can be assigned to any particular state of a system arbitrarily. Once this is done, the entropy at any other state x is given by

$$S_x = S_0 + \int_{0,\ \text{rev}}^{x} \frac{\delta Q}{T}$$

where state 0 is the one for which $S = S_0 = 0$.

Because S is a property, ΔS between two states is the same, no matter what path, reversible or irreversible, is followed as a system changes from one state to the other. Equation 8·1a states, however, that the numerical value for ΔS must be obtained by integrating $\int(\delta Q/T)$ along *some reversible* path. Examples of this calculation are given in the next section.*

The definition $\Delta S \equiv \int_{\text{rev}} (\delta Q/T)$ holds for any closed system or any fixed quantity

*If $\int (\delta Q/T)$ is integrated along *irreversible* paths between two states, it is found that in general a different numerical value is obtained for each path; that is, $\int_{\text{irrev}} (\delta Q/T)$ is not a property. In fact, it can be shown that $\oint_{\text{irrev}} (\delta Q/T) < 0$. The general statement, $\oint (\delta Q/T) \leq 0$, where the equality holds for reversible cycles and the inequality for irreversible ones is useful in some thermodynamic analyses and is known as the *inequality of Clausius*.

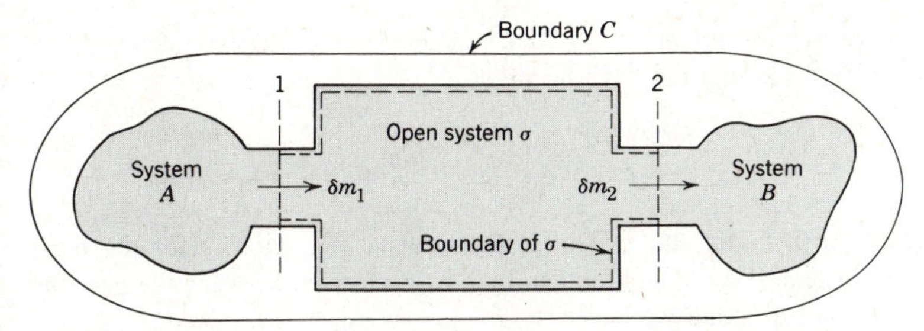

Figure 8·4. Entropy change of an open system.

of matter, just as the definition $\Delta E \equiv \int(\delta Q - \delta W)$ does. To determine a general expression for ΔS of an open system, consider any open system σ as in Fig. 8·4 (where for convenience only one inlet and one outlet are shown). Let the mass δm_1 entering the open system have a specific entropy s_1. Let this mass come from an open system A which might be in the particular form shown in Fig. 8·5. System A exchanges no heat with the surroundings, and the only work done on system A by the surroundings is equal to the flow work delivered to the open system as mass δm_1 crosses the boundary. (Notice that this restriction on system A does not restrict the generality of the open system's behavior at all.)

As mass δm_1 is transferred from system A to the open system σ, the entropy change of A is $-s_1\delta m_1$. Let the mass leaving the open system go to a system B which behaves in the same manner as system A. The entropy change of B is then $s_2\delta m_2$. Since there is no heat transfer to systems A and B, the heat transfer to closed system C, as defined in Fig. 8·4, is the same as that to open system σ. Thus for the closed system enclosed in boundary C,

$$dS_C = \left(\frac{\delta Q}{T}\right)_{\text{rev},C} = \left(\frac{\delta Q}{T}\right)_{\text{rev},\sigma}$$

where T is the temperature of that part of the system to which heat is transferred. (Recall that the temperature throughout an open system is generally not uniform.) Then, since entropy is an extensive property,

$$dS_C = dS_\sigma + dS_A + dS_B$$

and

$$dS_\sigma = dS_C - dS_A - dS_B$$

$$dS_\sigma = \left(\frac{\delta Q}{T}\right)_{\text{rev},\sigma} + s_1\,\delta m_1 - s_2\,\delta m_2 \qquad (8\cdot1b)$$

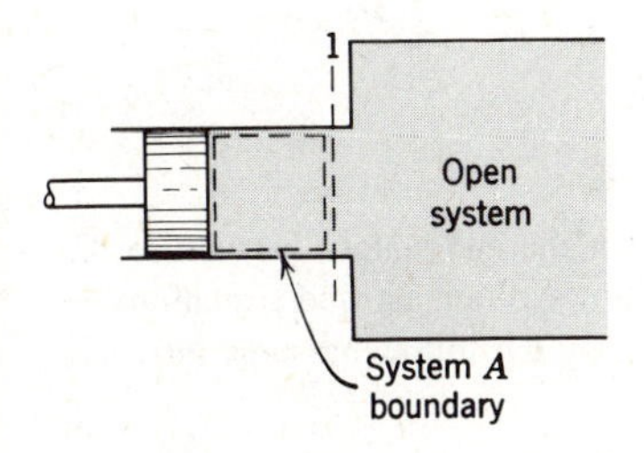

Figure 8·5. Possible form of system A in Fig. 8·4.

This is the general expression for dS of an open system. Notice that, for steady flow, $dS_\sigma = 0$ and $\delta m_1 = \delta m_2$, so that Eq. 8·1b becomes

$$0 = \left(\frac{\delta Q}{T}\right)_{\text{rev},\sigma} - (s_2 - s_1)\delta m$$

For steady flow, $\int_{\text{rev}} (\delta Q/T)$ for the system is the same as $\int_{\text{rev}} (\delta Q/T)$ for the mass flowing through the system. The integration can then be performed between the entering and leaving states of the fluid, and we have

$$s_2 - s_1 = \int_{1 \atop \text{rev}}^{2} \frac{\delta q}{T}$$

which agrees with the definition of Δs that holds for any fixed mass, including a mass which is moving through an open system.

8·2 Calculation of Entropy Changes

In order to calculate ΔS between any two states of a substance, select any reversible path connecting the two states, and integrate along that path. Some examples of this procedure follow.

Example 8·1. Calculate the change in entropy of 1 kg of helium which is heated reversibly at constant pressure from 140 kPa, 20°C, to 90°C in a closed system. c_p is constant at 5.24 kJ/kg·K.

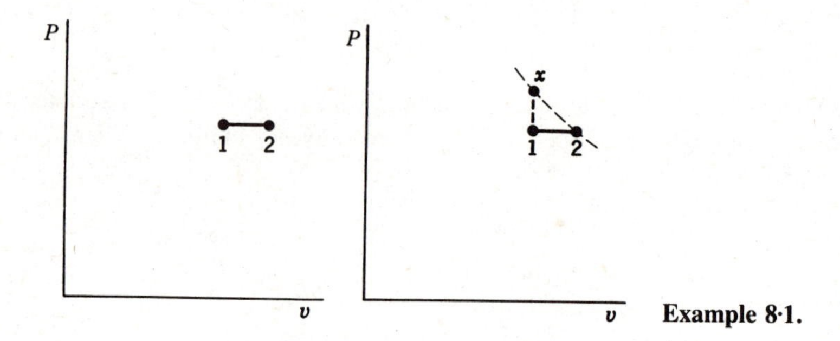

Example 8·1.

Solution A. ΔS can be calculated from the defining equation

$$\Delta S \equiv \int_{1 \atop \text{rev}}^{2} \frac{\delta Q}{T} \tag{8·1a}$$

by evaluating the integral along any reversible path between the end states. Let us use the reversible constant-pressure path that the system actually follows. We must find a relationship between δQ and dT in order to integrate. This can be done by the following steps in which each step is explained, or the restriction imposed is stated, by the words in parentheses:

$$\begin{aligned} \delta Q &= dU + \delta W && \text{(first law, closed system)} \\ &= dU + P\,dV && \text{(reversible process)} \\ &= dU + d(PV) && \text{(constant pressure)} \\ &= d(U + PV) = dH && \text{(definition of } H\text{)} \\ &= mc_p\,dT && \text{(ideal gas)} \end{aligned}$$

(Of course, it might have been recalled directly that for a *reversible constant-pressure process* $\delta Q = mc_p\,dT$.) Making this substitution for δQ,

$$\Delta S = \int_1^2 \frac{mc_p\,dT}{T}$$

$$\Delta S = mc_p \ln \frac{T_2}{T_1} = 1(5.24) \ln \frac{363}{293} = 1.12 \text{ kJ/kg·K}$$

(Notice that the units of c_p may be either kJ/kg·°C or kJ/kg·K, as explained in Example 3·1, but the units of entropy change should be kJ/K and not kJ/°C because the definition of ΔS involves *absolute* temperature—not a temperature change—in the denominator.)

Solution B. Let us evaluate

$$\Delta S = \int_{1\ \text{rev}}^2 \frac{\delta Q}{T}$$

over some reversible path other than the constant-pressure path. (There is no reason to choose any other path in this case except to illustrate that the same result will be obtained.) Let us choose a path as shown on the diagram that consists first of a reversible constant-volume heating from the initial state 1 to the final temperature. Call this state for which $V = V_1$ and $T = T_2$ state x. The second part of the reversible path is a reversible isothermal process from state x to state 2. Since ΔS is the same for all paths between states 1 and 2, we have

$$\Delta S = \int_{1\ \text{rev}}^2 \frac{\delta Q}{T} = \int_{1\ \text{rev}}^x \frac{\delta Q}{T} + \int_{x\ \text{rev}}^2 \frac{\delta Q}{T}$$

Applying the first law and noting that 1-x and x-2 are reversible processes,

$$\Delta S = \int_1^x \frac{dU}{T} + \int_x^2 \frac{dU + P\,dV}{T}$$

Since the system is comprised of an ideal gas,

$$\Delta S = \int_1^x \frac{mc_v\,dT}{T} + \int_x^2 \frac{0 + P\,dV}{T} = \int_1^x \frac{mc_v\,dT}{T} + \int_x^2 \frac{mR\,dV}{V}$$

Assuming c_v and R to be constant,

$$\Delta S = mc_v \ln \frac{T_x}{T_1} + mR \ln \frac{V_2}{V_x}$$

Noting that $T_x = T_2$, $V_x = V_1$, and $V_2/V_1 = T_2/T_1$,

$$\Delta S = mc_v \ln \frac{T_2}{T_1} + mR \ln \frac{T_2}{T_1} = m(c_v + R) \ln \frac{T_2}{T_1}$$

$$= mc_p \ln \frac{T_2}{T_1}$$

This, as expected, is the same expression that was obtained in solution A by integrating $\int(\delta Q/T)$ along the reversible constant-pressure path.

Example 8·2. Calculate the change in entropy per kilogram of helium heated reversibly at constant pressure from 140 kPa, 20°C, to 90°C in a steady-flow system.

Solution. Entropy is a property. For a pure substance, entropy is a function of any other two properties. Since here the initial pressure and temperature and the final pressure and temperature are the same as in Example 8·1, the entropy change per unit of mass must be the same. The fact that the helium in one case is in a closed system and in the other case is in a steady-flow system makes no difference. If we had not realized this, we might have calculated Δs as follows.

The first law applied to a steady-flow system gives an energy balance which, in differential form, is

$$\delta q = dh + dke + dpe + \delta w$$

For a reversible steady flow process

$$\delta w = -v\, dP - dke - dpe$$

so that

$$\delta q = dh + dke + dpe - v\, dP - dke - dpe$$

and for a constant-pressure process, $v\, dP = 0$ so that

$$\delta q = dh$$

For a steady-flow system, $dS_\sigma = 0$ and $\delta m_1 = \delta m_2$; so the expression for dS of an open system,

$$dS_\sigma = \left(\frac{\delta Q}{T}\right)_{\text{rev},\sigma} + s_1\, \delta m_1 - s_2\, \delta m_2 \tag{8·1b}$$

becomes

$$dS_\sigma = 0 = \left(\frac{\delta Q}{T}\right)_{\text{rev},\sigma} + (s_1 - s_2)\, \delta m$$

$$s_2 - s_1 = \int_{1\ \text{rev}}^{2} \frac{\delta q}{T} = \int_1^2 \frac{dh}{T} = \int_1^2 \frac{c_p\, dT}{T}$$

and, for a constant c_p of 5.24 kJ/kg·°C (or 5.24 kJ/kg·K)

$$\Delta s = c_p \ln \frac{T_2}{T_1} = 5.24 \ln \frac{363}{293} = 1.12 \text{ kJ/kg·K}$$

(Notice that this is the entropy change of the helium flowing through the system, not that of the system itself. From the definition of steady flow, $\Delta S_{\text{system}} = 0$.)

Example 8·3. Air expands irreversibly from 200 kPa, 167°C, to 100 kPa, 112°C. Calculate Δs, assuming that c_p and c_v are constant over this range of temperature with values of 1.015 kJ/kg·°C and 0.728 kJ/kg·°C, respectively.

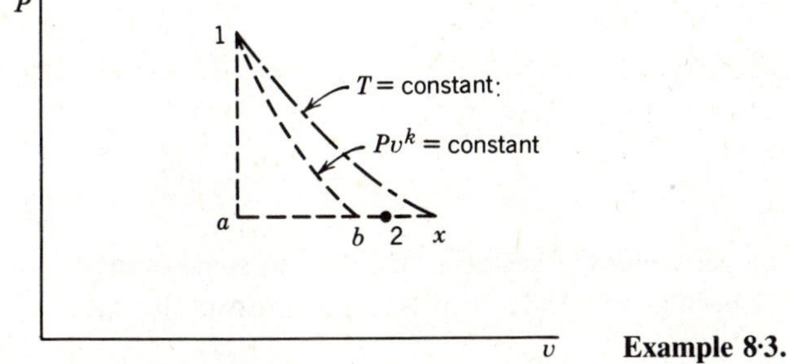

Example 8·3.

Solution. First we notice that no indication is given as to whether the air is in a closed system or is flowing into, from, or through an open system; but this lack of information is of no consequence because entropy is a property and is determined for a pure substance such as air by any two independent properties, and P and T are known at each of the end states. We also notice that the path between the two end states is not specified. Again this lack of information is of no consequence because entropy is a property: The change in entropy depends only on the end states and not on the path connecting them.

Calculate Δs for any process, reversible or irreversible, by evaluating $\int(\delta q/T)$ along *any reversible* path connecting the same end states. Three of the possible reversible paths between the end states 1 and 2 are those shown passing through states a, b, and x on the Pv diagram. [In order to sketch the Pv diagram approximately to scale, we note that $v_2/v_1 = P_1T_2/P_2T_1 = 200(385)/100(440) = 1.75$ and $v_b/v_1 = (P_1/P_2)^{1/k} = (2)^{1/1.394} = 1.644$.] The evaluation of $\int(\delta q/T)$ for each of these reversible paths is illustrated below, and of course we expect to obtain the same value in each case.

Path 1-a-2: A reversible constant-volume process followed by a reversible constant-pressure process.

$$\Delta s = \int_{\substack{1\text{-}a\\ \text{rev}}}^{2} \frac{\delta q}{T} = \int_{\substack{1\\ \text{rev}}}^{a} \frac{\delta q}{T} + \int_{\substack{a\\ \text{rev}}}^{2} \frac{\delta q}{T} = \int_1^a \frac{c_v\, dT}{T} + \int_a^2 \frac{c_p\, dT}{T}$$

$$= c_v \ln \frac{T_a}{T_1} + c_p \ln \frac{T_2}{T_a} = c_v \ln \frac{P_2}{P_1} + c_p \ln \frac{v_2}{v_1}$$

$$= 0.728 \ln \frac{100}{200} + 1.015 \ln 1.75 = 0.0634 \text{ kJ/kg·K}$$

Path 1-b-2: A reversible adiabatic process followed by a reversible constant-pressure process.

$$\Delta s = \int_{\substack{1\text{-}b\\ \text{rev}}}^{2} \frac{\delta q}{T} = \int_{\substack{1\\ \text{rev}}}^{b} \frac{\delta q}{T} + \int_{\substack{b\\ \text{rev}}}^{2} \frac{\delta q}{T} = 0 + \int_b^2 \frac{c_p\, dT}{T} = c_p \ln \frac{T_2}{T_b}$$

$$= c_p \ln \frac{v_2}{v_b} = c_p \ln \frac{v_2/v_1}{v_b/v_1} = 1.015 \ln \frac{1.75}{1.644} = 0.0634 \text{ kJ/kg·K}$$

Path 1-x-2: A reversible isothermal process followed by a reversible constant-pressure process.

$$\Delta s = \int_{\substack{1\text{-}x \\ \text{rev}}}^{2} \frac{\delta q}{T} = \int_{\substack{1 \\ \text{rev}}}^{x} \frac{\delta q}{T} + \int_{\substack{x \\ \text{rev}}}^{2} \frac{\delta q}{T} = \frac{1}{T_1} \int_{\substack{1 \\ \text{rev}}}^{x} \delta w + \int_{x}^{2} \frac{c_p\, dT}{T} = \frac{1}{T_1} \int_{1}^{x} P\, dV + c_p \ln \frac{T_2}{T_x}$$

$$= \frac{RT_1}{T_1} \int_{1}^{x} \frac{dv}{v} + c_p \ln \frac{T_2}{T_1} = R \ln \frac{v_x}{v_1} + c_p \ln \frac{T_2}{T_1}$$

$$= R \ln \frac{P_1}{P_2} + c_p \ln \frac{T_2}{T_1} = 0.287 \ln \frac{200}{100} + 1.015 \ln \frac{385}{440} = 0.0634 \text{ kJ/kg·K}$$

So far in this chapter the property entropy has been defined and some examples of entropy change calculations have been given. The next few sections answer the question, "Of what value is entropy?"

8·3 Entropy as a Coordinate

The definition of entropy

$$\Delta S \equiv \int_{\text{rev}} \frac{\delta Q}{T} \qquad \text{or} \qquad dS = \left(\frac{\delta Q}{T}\right)_{\text{rev}} \tag{8·1a}$$

can be rearranged as

$$Q_{\text{rev}} = \int T\, dS \qquad \text{or} \qquad \delta Q_{\text{rev}} = T\, dS \tag{8·2}$$

to show that the heat transfer of a reversible process is represented by an area on a diagram that uses absolute temperature and entropy as coordinates. Figure 8·6 shows the path on a *TS* diagram of a reversible process 1-2. The area beneath the path is $\int_1^2 T\, dS$, so it represents the heat transferred to the system during the reversible process. If the process

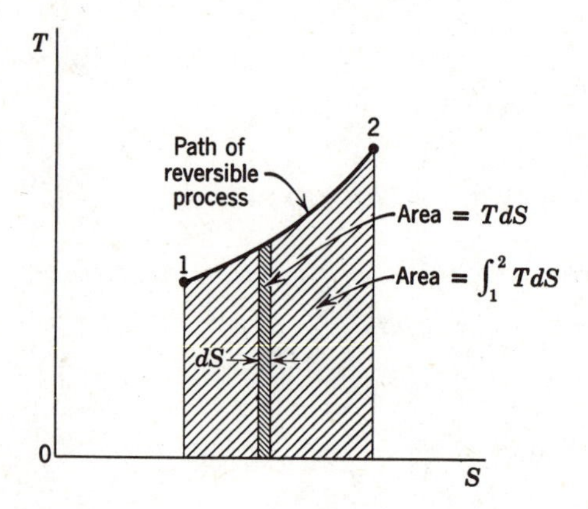

Figure 8·6. *TS* diagram of a reversible process.

is reversed so as to proceed from state 2 to state 1, then the heat transferred to the system is

$$Q = \int_2^1 T\, dS$$

For the same path, $\int_1^2 T\, dS$ and $\int_2^1 T\, dS$ are equal in magnitude but opposite in sign; likewise the area beneath the curve is the same for both directions of the process, but in one case the area represents a negative quantity.

The relation

$$q_{rev} = \int T\, ds \tag{8·2}$$

is in some respects analogous to the relations

$$w_{rev} = \int P\, dv \tag{1·7}$$

and

$$w_{rev} + \Delta ke + \Delta pe = -\int v\, dP \tag{1·10}$$

Each of these three equations relates heat or work to an area on a property diagram. Equation 1·7 applies only to a closed system and Eq. 1·10 applies only to a steady-flow system; but Eq. 8·2 applies to either type of system.

Temperature–entropy diagrams are frequently used in the analysis of processes and cycles because for reversible processes areas on the diagrams represent heat transfer. Another commonly used diagram is the enthalpy–entropy diagram. Areas on the *hs* diagram have no significance, but entropy is still a useful coordinate because the ideal process that we often try to approximate in actual machines is the reversible adiabatic process. Since a reversible adiabatic process must be a constant-entropy or *isentropic* process,* such a process is represented by a vertical line on any diagram which has entropy on the abscissa. Enthalpy is a convenient coordinate because for many processes in steady-flow systems enthalpy change is equal to the work done, the heat transferred, or the change in kinetic energy.

Temperature–entropy diagrams on which two and three phases of a substance are represented appear in Chapter 9. At present, let us look at a *Ts* diagram that shows only the liquid and vapor phases of some substance, Fig. 8·7.

The saturation line, which represents saturated liquid states to the left of the critical state and saturated vapor states to the right of the critical state, can be plotted by taking the s_f and s_g values from a table of properties. Of course, in the two-phase-mixture region or "wet-vapor" region of a pure substance, constant-pressure lines coincide with constant-temperature lines. It can be shown that at any point in a single-phase region a constant-volume line is steeper than a constant-pressure line through the same point.

Temperature–entropy diagrams for air and water are in the appendix. (*Caution*: When

*If you do not see why a reversible adiabatic process is a constant-entropy process, refer to the definition of entropy.

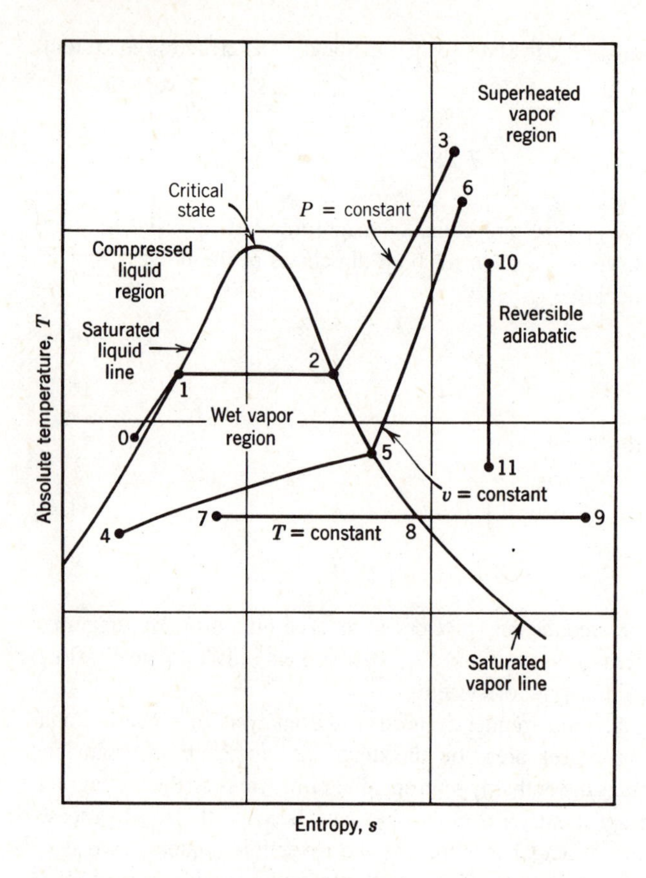

Figure 8·7. *TS* diagram for liquid and vapor.

air is condensed at constant pressure, the composition of the liquid differs from that of the vapor so that air is not a pure substance. Certain features of the *Ts* diagram for air are therefore different from those of a pure substance *Ts* diagram.)

A Carnot cycle is comprised of two reversible isothermal processes and two reversible adiabatic or isentropic processes. Therefore, it is always represented by a rectangle on a *TS* diagram. In Fig. 8·8, process *a-b* is the reversible isothermal heat addition at T_H. The area *a-b-f-e-a* beneath line *a-b* represents the heat added Q_H. Since process *b-c* is a reversible adiabatic process, the area beneath line *b-c* must be zero. Process *c-d* is the reversible isothermal rejection of heat Q_L to the energy reservoir at T_L. Q_L is represented by the area *c-d-e-f-c*. The cycle is completed by means of the reversible adiabatic process *d-a* that returns the system to its initial state *a*.

The area *a-b-c-d-a* is the difference between area *a-b-f-e-a* (representing Q_H) and area *c-d-e-f-c* (representing Q_L), so it represents $(Q_H - Q_L)$ or the work done by the cycle. Areas on a *TS* diagram represent heat transfer for all reversible processes of a substance, but they represent work only for reversible cycles, because for any cycle of a given quantity of substance

$$\oint \delta W = \oint \delta Q$$

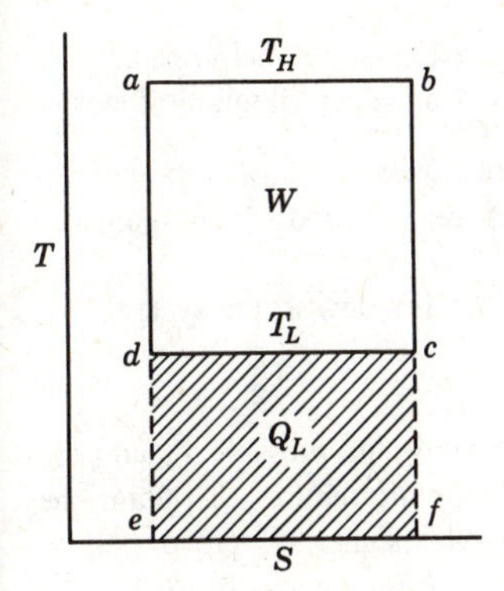

Figure 8·8. *TS* diagram of a Carnot cycle.

The area beneath the path of an irreversible process on a *TS* diagram has no significance. Certainly it does not represent heat transfer because

$$Q_{\text{irrev}} \neq \int T\, dS^*$$

As a reminder that the area beneath an irreversible process path has no significance, such paths are often shown as broken lines. Also, the exact path of an irreversible process between two states is frequently unknown or cannot be represented on a property diagram because the system may not be in an equilibrium state at all times. This is another reason for showing irreversible processes as broken lines on *TS* diagrams as illustrated in Fig. 8·9.

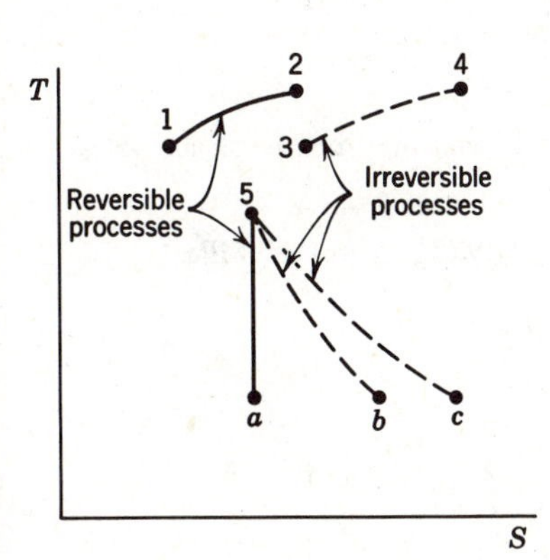

Figure 8·9. Reversible and irreversible processes on a *TS* diagram.

*It can be shown rigorously that $Q_{\text{irrev}} < \int T\, dS$. As *an example* that $Q_{\text{irrev}} \neq \int T\, dS$, consider a gas stirred by a paddle wheel inside a closed rigid thermally insulated vessel. The gas goes from state 1 to state 2. A reversible process between the same two end states would be a constant-volume heating, so that $S_2 > S_1$ and $\int_1^2 T\, dS > 0$. Yet the paddle-wheel process occurs in an insulated vessel so that $Q = 0$. Therefore, for this irreversible process

$$Q < \int T\, dS$$

Example 8·4. Dry saturated steam at 1.40 MPa undergoes a reversible isothermal expansion in a closed system until its pressure is 0.80 MPa. Calculate the work done per kilogram of steam.

Analysis. Make a *Ts* diagram first. As we sketch the diagram, we note that (1) the process is reversible so that the area beneath the path represents heat transfer, and (2) the path is a simple one on the *Ts* diagram: a horizontal line.

Since we are looking for the value of work, let us apply the first law to the system

$$w = u_1 - u_2 + q$$

State 1 is specified in the problem statement. For state 2 we know the pressure P_2 and the temperature, $T_2 = T_1 = T_{\text{sat, 1.40 MPa}}$. Therefore we have enough information to obtain the two internal energies from the steam tables. For this reversible isothermal process, $q = \int_1^2 T\,ds = T(s_2 - s_1)$ and since both end states are specified, this expression can be evaluated.

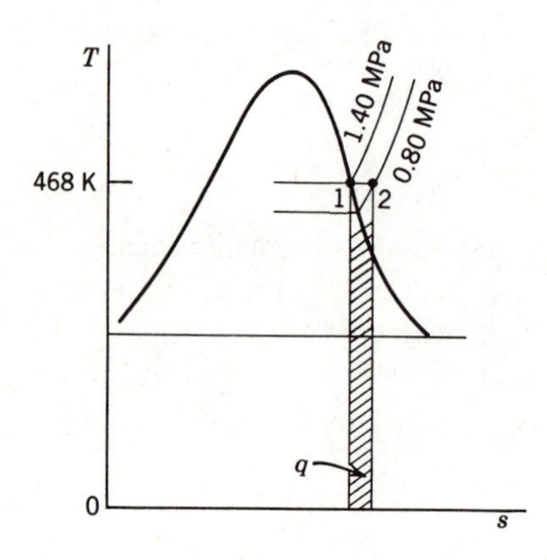

Example 8·4.

Solution. In accordance with the preceding analysis, the following are from the steam tables:

State 1	State 2
$T_1 = 195.07°\text{C} \approx 195°\text{C}$	$T_2 = T_1 \approx 195°\text{C}$; $P_2 = 0.80$ MPa
$u_1 = u_g = 2592.8$ kJ/kg	$u_2 = 2621.8$ kJ/kg
$s_1 = s_g = 6.4693$ kJ/kg·K	$s_2 = 6.7914$ kJ/kg·K

Applying the first law to the system,

$$w = u_1 - u_2 + q$$

and, for the reversible isothermal process,

$$\begin{aligned} w &= u_1 - u_2 + \int_1^2 T\,ds = u_1 - u_2 + T(s_2 - s_1) \\ &= 2592.8 - 2621.8 + 468(6.7914 - 6.4693) \\ &= 121.7 \text{ kJ/kg} \end{aligned}$$

Example 8·5. Ammonia is compressed reversibly and adiabatically at a steady rate of 5 lbm/min from 40 psia, 80 percent quality, to 200 psia. Determine the power input to the ammonia.

Analysis. For a steady-flow process, $\dot{W} = \dot{m}w$. Work can be found from the first law or steady-flow energy balance

$$w_{in} = h_2 - h_1 + \Delta ke + \Delta pe - q$$

which for an adiabatic process and no change in kinetic or potential energy becomes

$$w_{in} = h_2 - h_1$$

h_1 can be found from $h_1 = h_{f1} + x_1 h_{fg1}$. In addition to the pressure at the discharge, we know that $s_2 = s_1$ for the *reversible adiabatic* process. The Ts diagram helps us to see that, if s_2 is greater than s_g at 200 psia, the ammonia discharged is superheated, but, if s_2 is less than s_g at 200 psia, the ammonia discharged is wet. If the discharge is superheated, h_2 can be obtained from the tables by using P_2 and s_2; if the discharge is wet,

$$h_2 = h_{f2} + x_2 h_{fg2} \qquad \text{and} \qquad x_2 = \frac{s_2 - s_f}{s_{fg}}$$

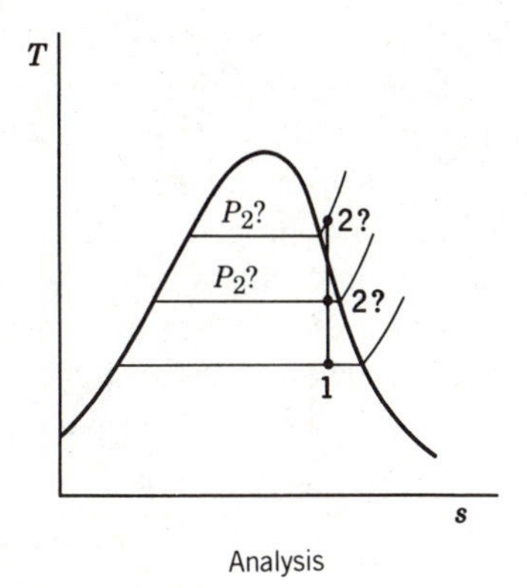

Analysis

Solution

Example 8·5.

Solution. Referring to the ammonia tables in the appendix,

$$h_1 = h_{f1} + x_1 h_{fg1} = 55.6 + 0.80(559.8) = 503.4 \text{ B/lbm}$$

$$s_2 = s_1 = s_{f1} + x_1 s_{fg1} = 0.1246 + 0.80(1.1879) = 1.0749 \text{ B/lbm·R}$$

The value of s_2 is less than s_{g2}, so the ammonia is wet at discharge.

$$x_2 = \frac{s_2 - s_f}{s_{fg}} = \frac{1.0749 - 0.3090}{0.8666} = 0.884$$

$$h_2 = h_{f2} + x_2 h_{fg2} = 150.9 + 0.884(481.8) = 576.8 \text{ B/lbm}$$

Applying the first law, recalling that $q = 0$, and assuming that $\Delta ke + \Delta pe = 0$,

$$w_{in} = h_2 - h_1 = 576.8 - 503.4 = 73.4 \text{ B/lbm}$$

and the power input is

$$\dot{W}_{in} = \dot{m}w_{in} = \frac{5(73.4)}{42.4} = 8.66 \text{ hp}$$

8·4 A Useful Relationship among Properties

Entropy is defined by

$$\Delta s \equiv \int_{\text{rev}} \frac{\delta q}{T} \tag{8·1a}$$

From the first law, for any closed system which passes through equilibrium states and for which the only type of work that can be done reversibly is evaluated as $\int P\,dv$, $\delta q_{\text{rev}} = du + P\,dv$.* Therefore,

$$\Delta s = \int \frac{du + P\,dv}{T}$$

or

$$ds = \frac{du + P\,dv}{T}$$

If the relationship among P, v, T, and u is known, this expression can be integrated to give the change in entropy between any two equilibrium states. Since entropy is a property (and thus a point function), Δs must be the same for any process, reversible or irreversible, between two given states. Also, the relation holds for a fluid flowing through an open system as well as for a closed system because entropy is a function of properties such as P, v, T, and u and is independent of the position or state of motion of the system.

It is well to repeat that the integration of the right-hand side of Eq. 8·1a must be performed for some reversible process between the end states. However, for an infinitesimal section of any path, $(du + P\,dv)$ is equal to δq for a *reversible* process along that section of the path. Therefore,

$$ds = \frac{du + P\,dv}{T}$$

holds for any process, provided that the path of the process is one which a reversible process can follow. In essence, this means that the path must connect equilibrium states. Also, such an equation implies that there is a continuous functional relationship among the properties. For our purposes we will meet this implied condition by restricting our use of the equation to pure substances. (If a chemical reaction occurs, entropy can change even though internal energy and volume are held constant; so obviously the equation does not apply. This point is treated in Chapter 13.)

The fact that this equation for ds applies to irreversible state changes as well as to reversible ones is an important but elusive point, so it is worth restating in different words

*The relation $\delta q_{\text{rev}} = du + P\,dv$ holds also for a steady-flow system as shown by

$$\delta q = du + d(Pv) + d(ke) + d(pe) + \delta w$$

which for a reversible process is

$$\begin{aligned}\delta q_{\text{rev}} &= du + P\,dv + v\,dP + d(ke) + d(pe) - v\,dP - d(ke) - d(pe)\\ &= du + P\,dv\end{aligned}$$

Notice, however, that for the steady-flow system $P\,dv \neq \delta w$.

for emphasis. Consider two states of a pure substance infinitesimally close to each other. The substance may go from one of these states to the other by many different processes, some reversible and some irreversible. Entropy is a property, however; so its change must be the same for all processes between the same two states. Since the change in entropy for (at least) any reversible process is given by

$$ds = \frac{du + P\,dv}{T}$$

it follows that this equation gives the change in entropy between the two states for any and all processes.

This equation can be rearranged as

$$T\,ds = du + P\,dv = dh - v\,dP \qquad (8\cdot3)$$

and it will be seen that this is a convenient form. *Conclusion: The relations*

$$T\,ds = du + P\,dv \qquad (8\cdot3)$$

$$T\,ds = dh - v\,dP \qquad (8\cdot3)$$

are valid for any process of a pure substance in the absence of electricity, magnetism, solid distortion effects, and surface tension. The only restriction is that the integration can be performed only between equilibrium states. Equations 8·3 are useful relationships among the properties of pure substances and are not restricted to any particular process. They are often referred to as the "$T\,ds$ equations."

Notice the differences among the restrictions that apply to the following four equations.

(a) $\delta q = du + \delta w$ — This is a statement of the first law, applicable to any closed system.

(b) $\delta q = du + P\,dv$ — This is a statement of the first law, restricted to reversible processes of a closed system.* By comparison with Eq. a, $P\,dv = \delta w$.

(c) $T\,ds = du + \delta w$ — This is a statement combining the first and second laws and is restricted to reversible processes of a closed system. By comparison with Eq. a, $T\,ds = \delta q$.

(d) $T\,ds = du + P\,dv$ — This is a relationship among properties (P, v, T, s, u) *valid for all processes between equilibrium states*. It is based on the first and second laws, but in itself it is a statement of neither. In general, $T\,ds \neq \delta q$ and $P\,dv \neq \delta w$. (By comparison with Eq. a it is seen, in fact, that for a closed system, $T\,ds = \delta q$ *only* when $P\,dv = \delta w$, and vice versa.)

To illustrate the difference between Eqs. a and d, consider the system comprised of a gas trapped in a closed, rigid, thermally insulated vessel fitted with a paddle wheel so

*This equation can also be applied to frictionless processes of a fluid flowing through an open system, but then $P\,dv \neq \delta w$.

that work can be done on the gas. Let the paddle wheel be turned by an external motor. Notice that $P\,dv = 0$ but $\delta w \neq 0$, and $\delta q = 0$ but $T\,ds \neq 0$. Equation a, the first law, reduces to

$$0 = du + \delta w$$

Equation d, the $T\,ds$ equation, reduces to

$$T\,ds = du + 0$$

Equations b and c do not apply because they are restricted to reversible processes.

You should learn Eqs. 8·3 because they are useful in a number of ways. For one thing, they provide a means of calculating the change in entropy of a system if the relationship among P, v, T, and u is known. For example, Eqs. 8·3 can be easily integrated for an ideal gas so that it is unnecessary to learn any special equations for the entropy change of an ideal gas. Many other useful results can be derived from these equations.

Example 8·6. Air expands irreversibly from 200 kPa, 167°C, to 100 kPa, 112°C. Calculate Δs, assuming that c_p and c_v are constant over this range of temperature with values of 1.015 kJ/kg·°C and 0.728 kJ/kg·°C, respectively.

Solution. Assuming that air under the stated conditions behaves as an ideal gas,

$$T\,ds = du + P\,dv \tag{8·3}$$

becomes

$$T\,ds = c_v\,dT + P\,dv$$

and

$$\Delta s = \int_1^2 ds = \int_1^2 \frac{c_v\,dT}{T} + \int_1^2 \frac{P\,dv}{T} = \int_l^2 \frac{c_v\,dT}{T} + \int_1^2 \frac{R\,dv}{v}$$

Integrating, for constant values of c_v and R,

$$\Delta s = c_v \ln \frac{T_2}{T_1} + R \ln \frac{v_2}{v_1} = c_v \ln \frac{T_2}{T_1} + R \ln \frac{P_1 T_2}{P_2 T_1}$$

$$= 0.728 \ln \frac{385}{440} + 0.287 \ln \frac{200(385)}{100(440)} = 0.0634 \text{ kJ/kg·K}$$

Alternative Solution. Using in similar fashion the relation

$$T\,ds = dh - v\,dP \tag{8·3}$$

we have

$$T\,ds = c_p\,dT - v\,dP$$

and

$$\Delta s = \int_1^2 ds = \int_1^2 \frac{c_p\,dT}{T} - \int_1^2 \frac{v\,dP}{T} = c_p \ln \frac{T_2}{T_1} - R \ln \frac{P_2}{P_1}$$

$$= 1.015 \ln \frac{385}{440} - 0.287 \ln \frac{100}{200} = 0.0634 \text{ kJ/kg·K}$$

See how much simpler the calculation of Δs is when the "$T\,ds$ equations" are used than when Δs is calculated from its definition as in Example 8·3.

Example 8·7. Determine s_{fg} for ammonia at 20°F if h_{fg} = 553.1 B/lbm.

Solution. For the calculation of an entropy change at constant pressure, which is what s_{fg} is, the relation

$$T\,ds = dh - v\,dP \tag{8·3}$$

simplifies to

$$T\,ds = dh$$

Since s_{fg} is also an entropy change at constant temperature, this equation can be integrated as

$$\Delta s = \frac{\Delta h}{T}$$

$$s_{fg} = \frac{h_{fg}}{T} = \frac{553.1}{479.7} = 1.153 \text{ B/lbm·R}$$

Example 8·8. Determine for an ideal gas the slope of (*a*) a constant-pressure line and (*b*) a constant-volume line on a *Ts* diagram.

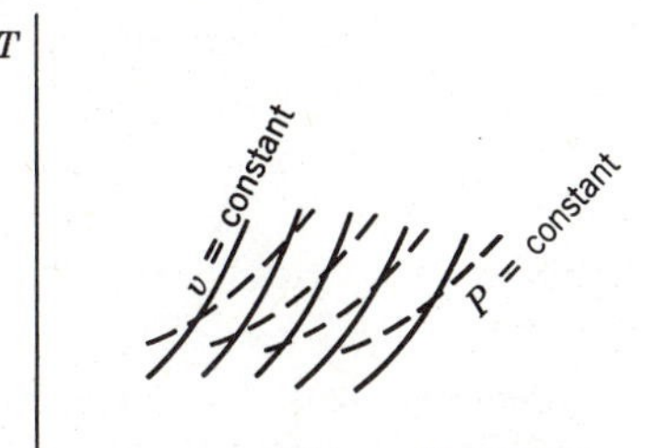

Example 8·8.

Solution. For an ideal gas, the $T\,ds$ equations

$$T\,ds = dh - v\,dP \tag{8·3}$$

$$T\,ds = du + P\,dv \tag{8·3}$$

become

$$T\,ds = c_p\,dT - v\,dP$$

$$T\,ds = c_v\,dT + P\,dv$$

For a constant-pressure process, the last term of the first equation is zero. For a constant-volume process, the last term of the second equation is zero. The desired slopes can consequently be obtained readily as

$$\text{Slope of constant-pressure line} = \left(\frac{\partial T}{\partial s}\right)_P = \frac{T}{c_p}$$

$$\text{Slope of constant-volume line} = \left(\frac{\partial T}{\partial s}\right)_v = \frac{T}{c_v}$$

Since $c_p > c_v$, the slope of a constant-pressure line is less than that of a constant-volume line *at the same temperature*. You can see now that the general form of these lines is as shown in the figure.

8·5 The Increase of Entropy Principle

The property entropy often provides a means of determining if a process is reversible, irreversible, or even possible. This application of entropy is based on the *principle of the increase of entropy,* which states that *the entropy of an isolated system always increases or, in the limiting case of a reversible process, remains constant with respect to time.* In mathematical form,

$$\left(\frac{dS}{d\tau}\right)_{\text{isolated system}} \geq 0$$

or, with the understanding that time is the independent variable, this statement is usually written

$$\Delta S_{\text{isolated system}} \geq 0 \tag{8·4}$$

The inequality holds for irreversible processes; the equality holds for reversible processes.

Since an isolated system is one that in no way interacts with its surroundings, this principle may at first appear to be severely restricted in application. However, you will see that a judicious selection of boundaries often makes it possible to work with isolated systems. A proof of the increase of entropy principle follows.

A system can always be made to go from one state to another by a series of reversible adiabatic and reversible isothermal processes. (See Sec. 8·1.) In fact, two states of a system can always be connected by one reversible adiabatic process and one reversible isothermal process. Let any closed system undergo an adiabatic process from a state 1 to a state 2 as shown in Fig. 8·10. Now the system can be restored to state 1 by means of

Process 2-*a*. A reversible adiabatic process which restores the system to its initial temperature, followed by

Process *a*-1. A reversible isothermal process which restores the system to its initial state.

The system has now executed a cycle. During the cycle it could exchange heat with its surroundings only during the reversible isothermal process *a*-1. This heat exchange could have been with an energy reservoir at constant temperature. Therefore, in accordance

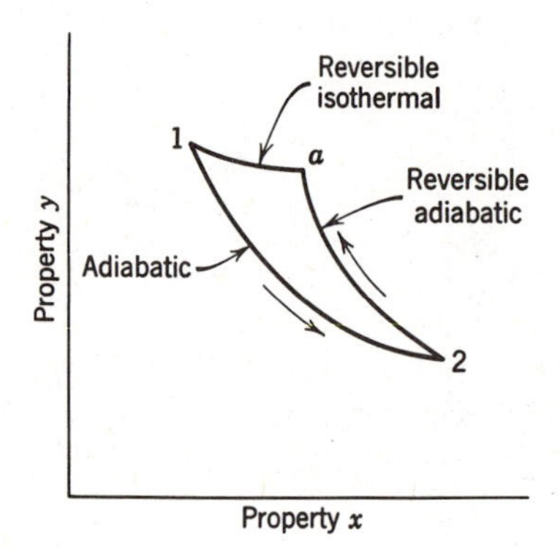

Figure 8·10. Proof of the increase of entropy principle.

with the second law, there can be no net work output from the system during the cycle; that is, $\oint \delta W \leq 0$. Then applying the first law,

$$\oint \delta Q = \oint \delta W$$

we see that $\oint \delta Q \leq 0$: Either (1) heat was removed from the system during process a-1, or (2) the cycle was completely adiabatic so that state a was identical with state 1. Consequently,

$$S_1 - S_a \leq 0 \tag{a}$$

where the equality holds if the reversible adiabatic process 2-a alone returned the system to state 1, state a and state 1 being identical. This could be the case only if the original adiabatic process 1-2 had been reversible. For process 2-a,

$$S_a - S_2 = 0 \tag{b}$$

For the cycle,

$$\Delta S = (S_2 - S_1) + (S_a - S_2) + (S_1 - S_a) = 0 \tag{c}$$

Comparison of Eqs. a, b, and c shows that

$$S_2 - S_1 \geq 0$$

where the equality holds only if the adiabatic process 1-2 is reversible. Recalling that process 1-2 was an adiabatic process of any closed system, we have shown that, in general,

$$\Delta S_{\text{adiabatic closed system}} \geq 0$$

This is one statement of the principle of the increase of entropy: *The entropy of an adiabatic closed system always increases or, in the limiting case of a reversible process, remains constant.* The independent variable is understood to be time.

An isolated system has already been defined as a system which in no way interacts with its surroundings. Therefore, every isolated system is (at least) an adiabatic closed system, and we conclude that

$$\Delta S_{\text{isolated system}} \geq 0 \tag{8·4}$$

Thus another statement of the principle of the increase of entropy is: *The entropy of an isolated system always increases or, in the limiting case of a reversible process, remains constant.* The independent variable is understood to be time. This statement is actually less general than the one that refers to an adiabatic closed system, because an isolated system is a special case (work = 0) of an adiabatic closed system. Nevertheless, this is a convenient and widely used form of the increase of entropy principle.

The increase of entropy principle provides another criterion of reversibility in addition to those listed in Sec. 6·2. For any process of an isolated system, if the entropy of the system remains constant, the process is reversible; if the entropy of the system increases, the process is irreversible. The entropy of an isolated system cannot decrease.

Consider the heating of a block of steel from a temperature T_1 to a temperature T_2. Heat is supplied from a constant-temperature reservoir (perhaps steam condensing at

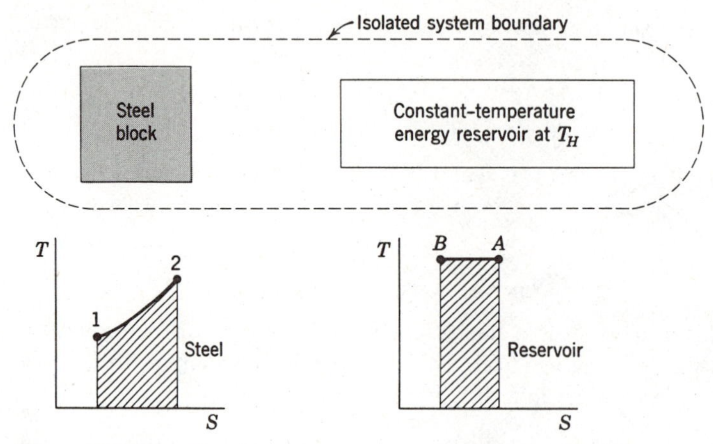

Figure 8·11. Transfer of heat across a finite temperature difference.

constant pressure) at T_H. T_H is greater than T_2. Neither the block nor the reservoir is an isolated system, but the block and the reservoir together do constitute an isolated system. Since the transfer of heat from the reservoir to the block across a finite temperature difference is irreversible, the entropy of the isolated system increases. The entropy increase of the block is greater in magnitude than the entropy decrease of the reservoir. This can also be seen from the *TS* diagrams of Fig. 8·11. Since the heating of the block and the cooling of the reservoir are *internally* reversible, the amount of heat transferred is represented by the area beneath the path on each *TS* diagram. The two areas must be equal; so the lower temperature of the block requires that $|\Delta S_{\text{block}}| > |\Delta S_{\text{reservoir}}|$ and thus $\Delta S_{\text{isolated system}} > 0$.

A Carnot cycle that is modified by having an irreversible adiabatic expansion is shown on a *TS* diagram in Fig. 8·12*a*. A Carnot cycle modified by having an irreversible adiabatic compression is shown in Fig. 8·12*b*. In each case the entropy of the system during the irreversible adiabatic process must increase. In each case the heat supplied is represented by the area beneath path 1-2, and the heat rejected is represented by the area beneath path 3-4. Remember that an area bounded in part by the path of an irreversible process has

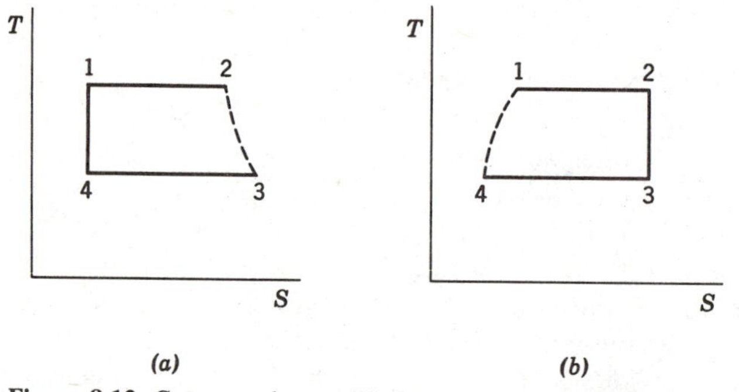

Figure 8·12. Carnot cycles modified by irreversible adiabatic processes.

no significance. Observe that in each case the entropy *increase* of the lower temperature reservoir is greater in magnitude than the entropy *decrease* of the higher temperature reservoir. (Remember that the *TS* diagrams of Fig. 8·12 are for the *systems,* and the entropy changes are of equal magnitude and opposite sign for the system and the reservoir during each reversible isothermal process.) Therefore, the entropy of an isolated system comprised of the engine and the two energy reservoirs increases.

An isolated system can always be formed by including any system and its surroundings within a single boundary. Sometimes the original system, which is then only a part of the isolated system, is called a *subsystem.* Since a system and its surroundings include, by definition, everything which is affected by the process, the combination is sometimes called the *universe,* so that the increase of entropy principle is stated as

$$\Delta S_{\text{universe}} \geq 0$$

where

$$\Delta S_{\text{universe}} = \Delta S_{\text{system}} + \Delta S_{\text{surr}}$$

Recall that we define the surroundings as everything outside the system boundary but usually restrict the term to things outside the system that in some way affect the behavior of the system. Consequently, the term *universe* generally refers to everything that is involved in a process and need not include things that have no effect on the process.

Notice that no conclusion has been drawn regarding ΔS of systems in general. The entropy of a *system* may increase, decrease, or remain constant. It is only the entropy of the universe or of an isolated system that must increase or, in the case of an externally reversible process, remain constant.

Now we must show how the increase of entropy principle can be applied to an open system. An isolated system can always be formed by including within one boundary an open system and its surroundings. Denoting the properties of the open system by the subscript σ, we then have

$$dS_{\text{isolated system}} = dS_\sigma + dS_{\text{surr}} \geq 0 \tag{a}$$

As shown in Sec. 8·1

$$dS_\sigma = \left(\frac{\delta Q}{T}\right)_{\text{rev},\sigma} + s_1\,\delta m_1 - s_2\,\delta m_2 \tag{b}$$

where δm_1 is the mass entering the open system and δm_2 is the mass leaving it. Likewise, for the surroundings of the open system (which constitute another open system),

$$dS_{\text{surr}} = \left(\frac{\delta Q}{T}\right)_{\text{rev, surr}} + s_2\,\delta m_2 - s_1\,\delta m_1$$

The surroundings may be divided into two parts: the parts that exchange mass with the system and those that do not. Usually, those parts of the surroundings that exchange mass with the system do not exchange heat or work with it; so we may refer to the two parts as (1) the parts that exchange mass with the system and (2) the parts that exchange work and/or heat with the system. There are exceptions to this, but for simplicity let us consider this to be the case. Then $(\delta Q/T)_{\text{rev,surr}}$ is dS for those parts of the surroundings that do

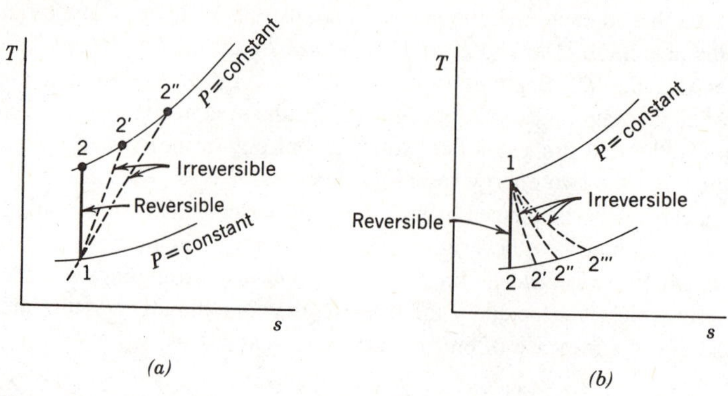

Figure 8·13. Reversible and irreversible adiabatic processes.

not exchange mass with the open system, so that the last equation becomes

$$dS_{\text{surr}} = dS_{\text{surr that exchange no mass with system}} + s_2\,\delta m_2 - s_1\,\delta m_1 \qquad \text{(c)}$$

The surroundings that exchange no mass with the system may exchange work and heat with it. Any part of the surroundings that exchanges only work with the system *can,* as far as the behavior of the system is concerned, operate reversibly and adiabatically so that its entropy change will be zero. Therefore, the entropy change of the surroundings that exchange no mass with the system is simply the entropy change of the surroundings that exchange heat with the system, and Eq. c can be written

$$dS_{surr} = dS_{\text{surr that exchange heat with system}} + s_2\,\delta m_2 - s_1\,\delta m_1 \qquad \text{(d)}$$

Making these substitutions from Eqs. b and c into Eq. a, we have *for any open system,*

$$\left(\frac{\delta Q}{T}\right)_{\text{rev},\sigma} + dS_{\text{surr that exchange heat with system}} \geq 0$$

For *steady flow,* $dS_\sigma = 0$, so that Eq. a becomes

$$dS_{\text{surr}} \geq 0$$

or

$$dS_{\text{surr that exchange heat with system}} + s_2\,\delta m - s_1\,\delta m \geq 0 \qquad (8\cdot5)$$

A frequently encountered special case in engineering is the adiabatic steady-flow process. The increase of entropy principle shows that

$$\Delta s_{\text{matter passing through adiabatic steady-flow system}} \geq 0$$

(Remember that ΔS for a steady-flow system itself, which is a region in space in which the properties at each point remain constant, is zero for all processes.) Some possible adiabatic compression and expansion paths for steadily flowing fluids are shown on *Ts* diagrams in Fig. 8·13.

Example 8·9. Air enters a turbine at 300 kPa, 155°C, and leaves at 100 kPa, 40°C. Heat removed from the air passing through the turbine amounts to 40 kJ/kg. The flow rate is 5000 kg/h. Is the process reversible?

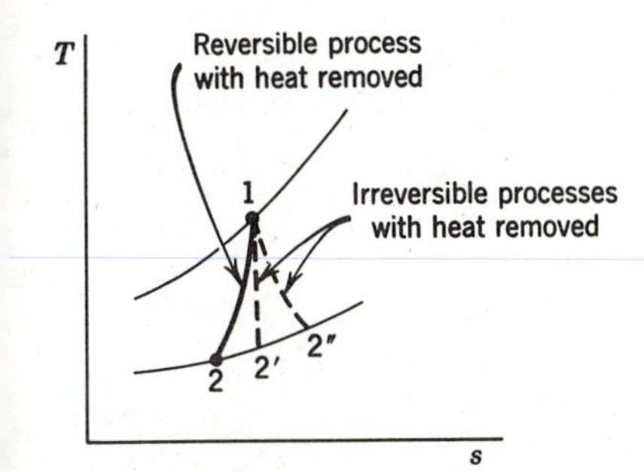

Example 8·9.

Solution. Consider the steady-flow system to be the region bounded by the turbine casing. Then,

$$\Delta s_{\text{air passing through system}} + \Delta s_{\text{surr that exchange heat with system}} \geq 0 \tag{8·5}$$

If the equality holds, the process is reversible; if the inequality holds, the process is irreversible. For the air,

$$\Delta s = \int_1^2 ds = \int_1^2 \frac{dh}{T} - \int_1^2 \frac{v\, dP}{T} = \int_1^2 \frac{c_p\, dT}{T} - \int_1^2 \frac{R\, dP}{P}$$

Assuming c_p to be constant for this temperature range, and using a value from Table 4·2,

$$\Delta s = c_p \ln \frac{T_2}{T_1} - R \ln \frac{P_2}{P_1} = 1.01 \ln \frac{313}{428} - 0.287 \ln \frac{100}{300} = 0.00 \text{ kJ/kg·K}$$

Heat is removed from the air passing through the turbine, so heat is added to the surroundings; therefore, the entropy of the surroundings increases. Thus

$$\Delta s_{\text{air passing through system}} = 0.00 \text{ kJ/kg·K}$$

and

$$\Delta s_{\text{surr that exchange heat with system}} > 0$$

so that the sum of these two quantities is greater than zero, and the process consequently must be irreversible. (Notice that, even though no information is available on the temperature of the surroundings, it is reasonable to assume that the temperature of the air in the turbine is never greater than 155°C so that the temperature of the surroundings which receive heat must be less than 155°C. Thus the extreme minimum Δs for the surroundings can be calculated and turns out to be much greater than 0.00 kJ/kg·K, in case anyone fears that the next significant digit might upset our conclusion.)

Example 8·10. What is the minimum final temperature for the adiabatic compression of ammonia from 30 psia, 90 percent quality, to 200 psia?

Solution. For an adiabatic compression, the entropy of the ammonia must increase if the process is irreversible and remain constant if the process is reversible. Sketching reversible and irreversible processes on a *Ts* diagram shows that the minimum final temperature is reached by means of the reversible process. It is therefore only necessary to determine the final temperature for a reversible adiabatic or isentropic process.

$$s_2 = s_1 = s_f + x s_{fg} = 0.0962 + 0.90(1.2402) = 1.2124 \text{ B/lbm·R}$$

Since s_2 is greater than s_g at 200 psia, the final condition (2) must be one of superheated ammonia. Referring to the superheated ammonia table, we see that for P = 200 psia and

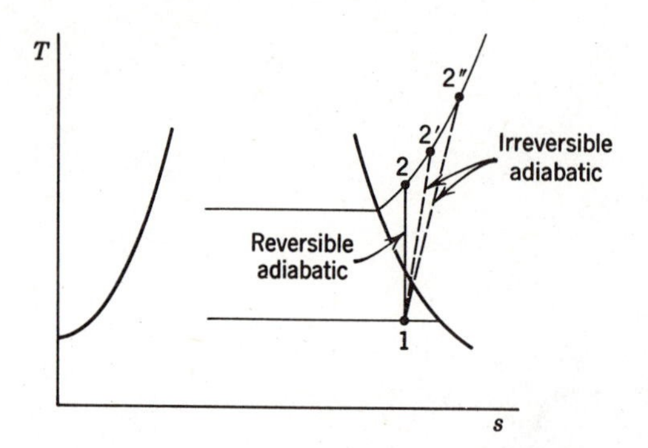

Example 8·10.

$s = 1.2124$ B/lbm·R, the temperature is 124°F. This is the minimum possible final temperature for an adiabatic compression.

Example 8·11. Is an adiabatic expansion of air from 25 psia, 140°F, to 15 psia, 40°F, possible?

Solution. For all possible adiabatic expansions, the entropy of the air must increase or, in the limiting case of a reversible adiabatic expansion, remain constant. For the end states specified,

$$\Delta s = \int_1^2 ds = \int_1^2 \frac{dh}{T} - \int_1^2 \frac{v\,dP}{T} = \int_1^2 \frac{c_p\,dT}{T} - \int_1^2 \frac{R\,dP}{P}$$

Assuming that the specific heats are constant,

$$\Delta s = c_p \ln \frac{T_2}{T_1} - R \ln \frac{P_2}{P_1} = 0.24 \ln \frac{500}{600} - \frac{53.3}{778} \ln \frac{15}{25} = -0.0088 \text{ B/lbm·R}$$

Since the adiabatic process considered would result in a decrease in entropy, it is impossible.

One of the most valuable uses of the increase of entropy principle is in connection with predicting the direction in which a chemical reaction will proceed. A reaction can never proceed in such a direction as to cause the entropy of the universe to decrease. This point is discussed in Chapter 12 in connection with combustion reactions.

8·6 Available and Unavailable Parts of Heat Transfer

As an introduction to the definition and discussion of the available and unavailable parts of heat transfer, let us consider two questions and their answers. First, suppose an energy reservoir at 600 K gives up 1000 kJ of heat. If the surrounding atmosphere, which is the coldest energy reservoir in the surroundings, is at 300 K, how much of this heat can be converted into work by devices which are not permanently changed, that is, that operate cyclically?* We answer this question by noting that the maximum work can be obtained by the use of an externally reversible engine (such as a Carnot, Stirling, or Ericsson engine) that receives the 1000 kJ at 600 K and rejects heat at the temperature of the atmosphere, 300 K. The efficiency of such an engine is 50 percent, so the answer to the

*We confine our attention to devices that operate cyclically in order to make certain that no part of the work produced results from the depletion of the stored energy of the devices.

question is that 500 kJ of the 1000 kJ given up by the reservoir can be converted into work, and this maximum amount of work can be obtained by the use of an externally reversible engine as shown in Fig. 8·14*a*. For the second question, suppose that the 1000 kJ is transferred from the reservoir at 600 K to a reservoir at 500 K. Then, after this transfer of heat has been made, and with the atmosphere still at 300 K, how much of the 1000 kJ can be converted into work by cyclically operating devices? In answering this question we first note that the 1000 kJ stored in the reservoir at 500 K can be supplied to an externally reversible engine as in Fig. 8·14*b* at 500 K but at no higher temperature. The efficiency of such an engine is $1 - T_{\text{low}}/T_{\text{high}} = 1 - 300/500 = 40.0$ percent; so the answer to this second question is that a maximum of 400 kJ of the 1000 kJ originally obtained from the reservoir at 600 K and then transferred to the 500 K reservoir can be converted into work. Yet this same 1000 kJ as removed from the reservoir at 600 K could have been converted to the extent of 500 kJ into work if it had not first been transferred to the reservoir at 500 K.

Such questions and their answers are important in engineering. For example, suppose that 1000 kJ of heat is added to an engine and a certain amount of work is actually performed. In evaluating the performance of the engine, it is helpful to know how much of the heat added could have been converted into work by an externally reversible engine (or how much of it could *not* have been converted into work *even* by an externally reversible engine). Or we may wish to know how much of the heat rejected by an actual engine could be converted into work by an externally reversible engine.

For another example of the type of question we want to answer, consider a steam power plant. For each 1000 kJ of energy input to the plant, suppose that 100 kJ is converted into work, 750 kJ is rejected as heat to the cooling water, and 150 kJ is carried away by the stack gases. Which loss is more serious—the 750 kJ to the cooling water or the 150 kJ in the stack gases? Obviously, one is five times bigger than the other in terms of energy quantities, but is it five times as valuable? It certainly is not, because usually only a very small fraction of the heat rejected to cooling water could be converted into work even by an externally reversible engine, whereas perhaps a third of the energy carried away by the stack gases could be converted into work. So in the evaluation of energy rejected by a system the question is not simply how much is rejected but how much of that rejected could be converted into work? This question can always be answered by considerations

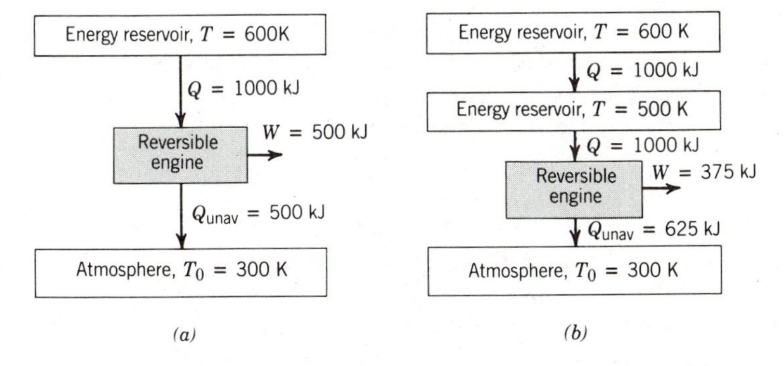

Figure 8·14. Different amounts of work obtainable from the same amount of heat.

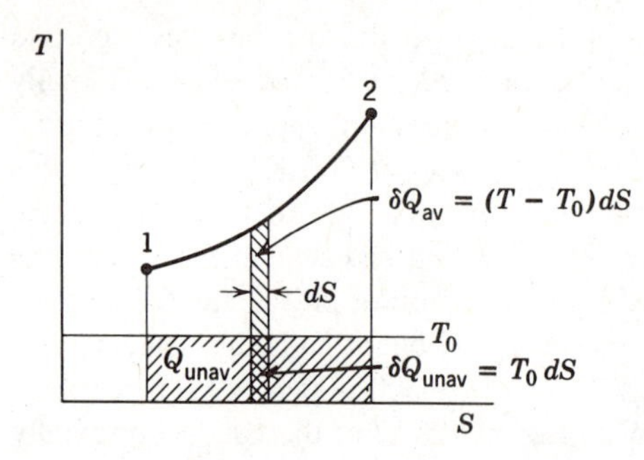

Figure 8·15. Available and unavailable parts of heat transferred.

of available energy (or exergy) and irreversibility which will be introduced in Chapter 10, but in many cases it can be answered by means of similar concepts—the available and unavailable parts of heat—that we now define.

The available part of the heat added to or taken from a system is that part that could be converted into work by externally reversible engines. (This is the maximum amount of work that could be obtained by any means whatsoever, because for specified temperature limits no engine can be more efficient than externally reversible engines.) *That part of heat added to or taken from a system that could not be converted into work even by externally reversible engines is called the unavailable part.* Using the symbols Q_{av} and Q_{unav} for the available and unavailable parts and Q for the total amount of heat, $Q = Q_{av} + Q_{unav}$. By using these definitions we could determine operationally the available and unavailable parts of the heat added to or removed from a system by transferring the heat to one or more externally reversible engines that reject heat to the surrounding atmosphere at T_0 and then determining the amount of work done by these engines or the amount of heat rejected to the atmosphere by them. (If part of the surroundings were cooled to a temperature lower than T_0, additional work could be obtained from these engines. However, the additional work would be offset or more than offset by the work input required to cool part of the surroundings to a temperature lower than T_0.)

For a Carnot engine that receives heat Q_H at T_H and rejects heat at T_0, the lowest temperature in the surroundings, the available and unavailable parts of Q_H are the same as the work done and the heat rejected.

T_0, the temperature of the atmosphere, is often called the *sink temperature* because an energy reservoir such as the atmosphere or the water of a river or lake that is used chiefly for absorbing heat from other systems is called a sink. T_0 is also referred to as the *lowest temperature in the surroundings.* This actually means the lowest temperature of an energy reservoir to which appreciable quantities of heat may be rejected.

If heat Q is added to a system in a reversible process such as 1-2 in Fig. 8·15, we can determine the available and unavailable parts by imagining that each small fraction of the heat δQ is added instead to a Carnot (or other externally reversible) engine at some temperature T. T varies from T_1 to T_2, but each Carnot engine receives so little heat that it operates between essentially constant temperature limits of T and T_0. The available and unavailable portions of δQ are represented by the areas $(T - T_0)\, dS$ and $T_0\, dS$ in Fig. 8·15 for any one of these engines. In order to add all of the heat Q to Carnot engines at the same temperatures at which the original system absorbs heat, an infinite number of

Carnot engines is required because only an infinitesimal amount of heat can be added to each engine at constant temperature. The available and unavailable parts of the heat Q are respectively the sums of δQ_{av} and of δQ_{unav} for all the engines. Thus

$$Q_{\text{unav},1\text{-}2} = \int_1^2 T_0\,dS = T_0(S_2 - S_1) = T_0\,\Delta S \tag{8·6}$$

$$Q_{\text{av},1\text{-}2} = \int_1^2 (T - T_0)\,dS = \int_1^2 T\,dS - T_0(S_2 - S_1) = \int_1^2 T\,dS - Q_{\text{unav}}$$

The *available* part of heat transferred can be represented by an area on a *TS* diagram only for reversible processes. Under some conditions the *unavailable* part can be represented by an area even for irreversible processes. However, we do not discuss these cases here because in Chapter 10 we introduce a quantity called *irreversibility, I,* that is simple to calculate and of much broader utility than the concept of the unavailable part of the energy transferred as heat.

A numerical illustration may be helpful at this point. Consider the transfer of 1000 kJ of heat from a reservoir at 1200 K to 5 kg of a gas initially at 200 kPa, 600 K, in a closed tank. For the gas, let $c_v = 0.800$ kJ/kg·K throughout the temperature range involved. The lowest temperature in the surroundings is 300 K. Let us determine how much of the heat removed from the reservoir is available and unavailable and how much of that absorbed by the gas is available and unavailable. In Fig. 8·16 the *TS* diagrams for the reservoir and for the gas are shown together. Since for each system (reservoir and gas) the process can be internally reversible, the area beneath the *TS* diagram path 1_r-2_r equals the area beneath 1_g-2_g because in each case the area represents the 1000 kJ of heat transferred. After finding $T_{g2} = 850$ K from $Q = \Delta U_g = mc_v(T_{g2} - T_{g1})$, we can find ΔS_g and ΔS_r by

$$\Delta S_g = \int_{1\ \text{rev}}^2 \frac{\delta Q}{T} = \int_1^2 \frac{dU}{T} = \int_1^2 \frac{mc_v\,dT}{T} = mc_v \ln \frac{T_2}{T_1}$$

$$= 5(0.800) \ln \frac{850}{600} = 1.39 \text{ kJ/K}$$

$$\Delta S_r = \int_{1\ \text{rev}}^2 \frac{\delta Q}{T} = \frac{Q}{T} = \frac{-1000}{1200} = -0.833 \text{ kJ/K}$$

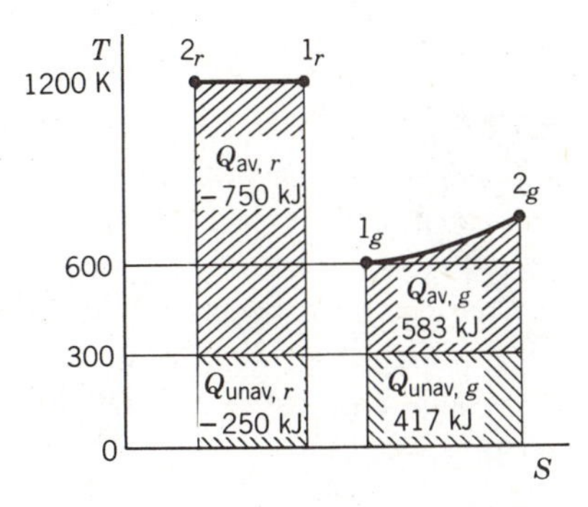

Figure 8·16. Heat transfer across a finite temperature difference.

Then

$$Q_{\text{unav},g} = T_0\,\Delta S_g = 300(1.39) = 417 \text{ kJ}$$

$$Q_{\text{av},g} = Q - Q_{\text{unav},g} = 1000 - 417 = 583 \text{ kJ}$$

$$Q_{\text{unav},r} = T_0\,\Delta S_r = 300(-0.833) = -250 \text{ kJ}$$

$$Q_{\text{av},r} = Q - Q_{\text{unav},r} = -1000 - (-250) = -750 \text{ kJ}$$

The last two values are negative because they are fractions of the heat *removed from* the reservoir. Notice that, as the 1000 kJ of heat was given up by the reservoir at 1200 K, 750 kJ of it could have been converted into work by an externally reversible engine; but, as the heat was absorbed by the gas at a temperature varying from 600 to 850 K, only 583 kJ of it could have been converted into work. Thus 167 kJ (= 750 − 583) has been made unavailable for conversion into work by the irreversible transfer of heat across a finite temperature difference.

The first law is the basis for ''energy accounting'' procedures in which heat, work, and stored energy are compared and balanced. The second law makes possible the division of heat into available and unavailable parts and thus adds valuable information to an energy balance. It shows that something more than just the quantity of energy is important. As an illustration, consider the same amount of heat, $Q_C = Q_D = 1000$ kJ, rejected reversibly by two vapors, C and D, that condense at constant temperature. Vapor C condenses at 260°C and vapor D condenses at 100°C. The lowest temperature in the surroundings is 15°C. The amount of heat rejected is the same in each case; but the available part of Q_C is

$$Q_{\text{av},C} = Q_C - Q_{\text{unav},C} = Q_C - T_0\,\Delta S_C = Q_C - T_0 \frac{Q_C}{T_C}$$

$$= -1000 - 288\frac{-1000}{533} = -540 \text{ kJ}$$

and the available part of Q_D is

$$Q_{\text{av},D} = Q_D - Q_{\text{unav},D} = Q_D - T_0\,\Delta S_D = Q_D - T_0 \frac{Q_D}{T_D}$$

$$= -1000 - 288\frac{-1000}{373} = -228 \text{ kJ}$$

More than twice as much (540/228 = 2.37) work can be obtained from the 1000 kJ rejected by vapor C as can be obtained from the 1000 kJ rejected by vapor D. Clearly, the equal amounts of energy rejected by C and D are not equal in all respects.

In this section we have discussed the available and unavailable parts of heat added to or removed from a system. In Chapter 10 we will consider the broader and more important matter of the fraction of other forms of energy available for doing work and the maximum amount of work that can be obtained as a system passes from one state to another.

Example 8·12. Dry saturated steam at 1.40 MPa undergoes a reversible isothermal expansion in a closed system until its pressure is 0.80 MPa. (This is the same process that was considered in Example 8·4.) The lowest temperature in the surroundings is -15°C. Determine the available part of the heat added.

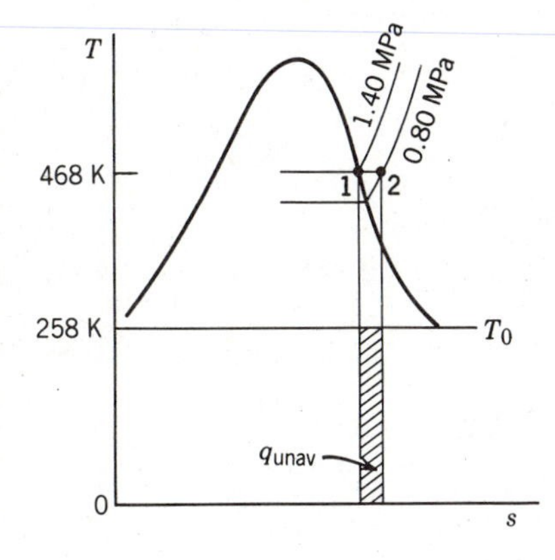

Example 8·12.

Solution. Since the process is reversible,

$$q = \int_1^2 T\,ds$$

and, since it is isothermal, integration gives

$$q = \int_1^2 T\,ds = T(s_2 - s_1) = 468(6.7914 - 6.4693) = 150.7\ \text{kJ/kg}$$

The unavailable part of this heat is given, as always, by $q_{unav} = T_0\,\Delta s$:

$$q_{unav} = T_0(s_2 - s_1) = 258(6.7914 - 6.4693) = 83.1\ \text{kJ/kg}$$

The available part is then

$$q_{av} = q - q_{unav} = 150.7 - 83.1 = 67.6\ \text{kJ/kg}$$

Example 8·13. A closed, rigid, thermally insulated tank contains 1.20 kg of air initially at 140 kPa, 7°C (state 1). The lowest temperature in the surroundings is 4°C. An impeller in the tank is driven by an external motor and stirs the air until its temperature becomes 63°C (state 2). After this process is completed, how much of the energy which was added can be reconverted into work?

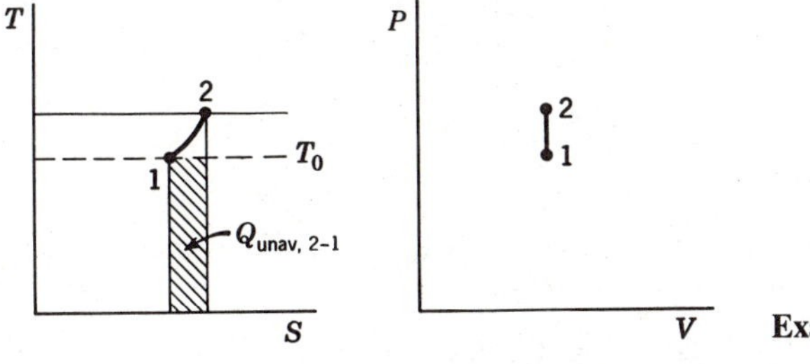

Example 8·13.

Solution. In order to reconvert to work any of the energy that was added to the system by means of the impeller, the energy must be removed from the system. All the energy added by the impeller will be removed if the system is returned to its initial state of 140 kPa, 7°C. This can be done by removing the insulation from the tank and removing heat. The fractions of this heat which can and cannot be converted into work by externally reversible engines are given by

$$Q_{av} = Q - Q_{unav}$$

$$Q_{unav} = T_0 \Delta S \tag{8·6}$$

Notice that the *cooling* process proceeds from state 2 to state 1 so that $\Delta S = S_1 - S_2$ and $\Delta U = U_1 - U_2$. Applying the first law and assuming that the air behaves as an ideal gas with constant $c_v = 0.718$ kJ/kg·K (see Table 4·2.),

$$Q = U_1 - U_2 = mc_v(T_1 - T_2) = 1.20(0.718)\,(7 - 63) = -48.2 \text{ kJ}$$

Integrating the *Tds* equation,

$$\Delta S = S_1 - S_2 = \int_2^1 \frac{dU}{T} + \int_2^1 \frac{P\,dV}{T} = mc_v \ln\frac{T_1}{T_2} + 0$$

$$= 1.20(0.718) \ln\frac{280}{336} = -0.157 \text{ kJ/K}$$

$$Q_{unav} = T_0 \Delta S = 277(-0.157) = -43.5 \text{ kJ}$$

$$Q_{av} = Q - Q_{unav} = -48.2 - (-43.5) = -4.7 \text{ kJ}$$

Of the 48.2 kJ added to the system as work by the impeller, 43.5 kJ cannot and 4.7 kJ can be reconverted into work by means of externally reversible engines. (The values of Q_{av} and Q_{unav} obtained are negative because during the cooling process each is a fraction of the heat *removed* from the system.)

Example 8·14. One-fourth kilogram of air in a closed system is carried through a cycle composed of four reversible processes: process 1-2 is an isentropic compression of the air from 100 kPa, 0°C, to one-fifth its initial volume; process 2-3 is a constant volume heating until $T_3 = 400$°C; process 3-4 is an isentropic expansion to the initial volume; and process 4-1 is a constant-volume cooling to the initial state. The lowest temperature in the surroundings is 0°C. Assume that the specific heats are constant at their 0°C values. For the cycle as a whole, make an accounting of the energy transfers and of the available parts of the energy transfers.

Solution. First a sketch of the system and *PV* and *TS* diagrams are made. Analysis of the problem shows that the temperatures at the various states must be known. The two that are not given above can be readily found by use of the ideal-gas equation of state and the *Pv*, *vT*, or *PT* relationship for an isentropic process of an ideal gas with constant specific heats:

$$T_2 = T_1\left(\frac{V_1}{V_2}\right)^{k-1} = 273(5)^{1.40-1} = 520 \text{ K}$$

$$T_4 = T_3\left(\frac{V_3}{V_4}\right)^{k-1} = T_3\left(\frac{V_2}{V_1}\right)^{k-1} = 673\left(\frac{1}{5}\right)^{0.40} = 354 \text{ K}$$

We determine the gross heat added and the gross heat rejected by applying the first law to processes 2-3 and 4-1 of the closed system:

$$Q_{2\text{-}3} = U_3 - U_2 + W = mc_v(T_3 - T_2) + 0 = 0.25(0.717)\,(673 - 520) = 27.4 \text{ kJ}$$

$$Q_{4\text{-}1} = U_1 - U_4 + W = mc_v(T_1 - T_4) + 0 = 0.25(0.717)\,(273 - 354) = -14.5 \text{ kJ}$$

Then the net work of the cycle is given by the first law as

$$W = \oint \delta W = \oint \delta Q = Q_{2\text{-}3} + Q_{4\text{-}1} = 27.4 - 14.5 = 12.9 \text{ kJ}$$

Of the heat added in process 2-3, the unavailable and available parts are

$$Q_{\text{unav},2\text{-}3} = T_0(S_3 - S_2) = T_0 \int_{2\ \text{rev}}^{3} \frac{\delta Q}{T} = T_0 \int_{2\ \text{rev}}^{3} \frac{mc_v\, dT}{T} = T_0 mc_v \ln \frac{T_3}{T_2}$$

$$= 273(0.25)0.717 \ln \frac{673}{520} = 12.6 \text{ kJ}$$

$$Q_{\text{av},2\text{-}3} = Q_{2\text{-}3} - Q_{\text{unav},2\text{-}3} = 27.4 - 12.6 = 14.8 \text{ kJ}$$

Notice that $(S_4 - S_1) = (S_3 - S_2)$, so that the unavailable part of $Q_{2\text{-}3}$ is equal in magnitude to the unavailable part of $Q_{4\text{-}1}$. Thus

$$Q_{\text{av},4\text{-}1} = Q_{4\text{-}1} - Q_{\text{unav},4\text{-}1} = -14.5 - (-12.6) = -1.9 \text{ kJ}$$

Now that all pertinent quantities have been calculated, the accounting can be tabulated as shown in the accompanying table. The work is, of course, counted as a 100 percent available part of the energy transfers.

Accounting of Energy Transfers (First Law)			Accounting of the Available Parts of Energy Transfers (Second Law)		
	kJ	% of $Q_{2\text{-}3}$		kJ	% of $Q_{\text{av},2\text{-}3}$
Energy in:			Available part of energy in:		
$Q_{2\text{-}3}$	27.4	100	$Q_{\text{av},2\text{-}3}$	14.8	100
Energy out:			Available part of energy out:		
W	12.9	47.1	W	12.9	87.2
$Q_{4\text{-}1}$	14.5	52.9	$Q_{\text{av},4\text{-}1}$	1.9	12.8

Only the *net* work is tabulated because the exchange of work is probably made with only one part of the surroundings, whereas the $Q_{2\text{-}3}$ and $Q_{4\text{-}1}$ are tabulated separately because they involve different parts of the surroundings. The results are shown graphically in the accompanying figure.

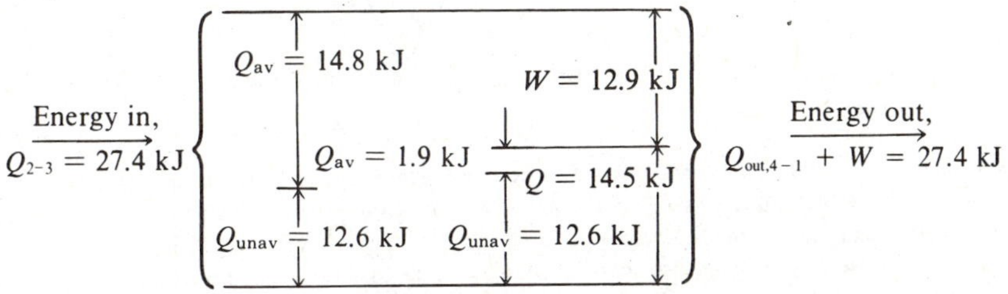

Energy in, $Q_{2\text{-}3} = 27.4$ kJ

$Q_{\text{av}} = 14.8$ kJ

$W = 12.9$ kJ

$Q_{\text{av}} = 1.9$ kJ

$Q = 14.5$ kJ

$Q_{\text{unav}} = 12.6$ kJ

$Q_{\text{unav}} = 12.6$ kJ

Energy out, $Q_{\text{out},4-1} + W = 27.4$ kJ

Discussion. It is clear that the second-law analysis gives information that changes greatly the picture obtained from the first-law analysis alone. The first-law analysis or energy accounting shows that 47.1 percent of the gross heat input to the cycle is converted into work. From the first-law analysis *alone,* one might conclude that it would be possible through various improvements to double the work output obtainable from a cycle with a heat-input process like 2-3. The second-law analysis, however, shows that 87.2 percent of the gross available energy added in process 2-3 is converted into work by this cycle; so that, no matter what improvements are made in the cycle (as long as the heat-addition process 2-3 is unchanged), the work output can be increased by only about 15 percent [(100 − 87.2)/87.2) = 14.7 percent]. The first law shows that 52.9 percent of the gross heat input is rejected by the cycle as heat, but the second law shows that, of the available energy added, only 12.8 percent is wasted. Of the heat rejected by the cycle, only 1.9/14.5 = 13.1 percent is available energy; the remaining 86.9 percent could not be converted into work even by externally reversible engines rejecting heat only at T_0.

In this cycle 12.6 kJ of the heat input is unavailable, and 12.6 kJ of the heat rejected is unavailable. There is no increase because all processes are reversible. If one or more of the processes were irreversible, the unavailable part of the energy rejected by the cycle would be greater than that of the energy added.

8·7 Helmholtz (*A*) and Gibbs (*G*) Functions

Enthalpy was introduced as a derived property, being defined as the sum of U and PV. Many other derived properties could be defined by such arbitrary combinations of other properties, but doing so is justified only if such derived properties are useful. Two derived properties that are quite useful in some areas of thermodynamics are the Helmholtz function, A, and the Gibbs* function, G, which are defined as

$$A \equiv U - TS \qquad \text{or} \qquad a \equiv u - Ts \tag{8·7a}$$

$$G \equiv H - TS \qquad \text{or} \qquad g \equiv h - Ts \tag{8·8a}$$

The differential of the Helmholtz function is

$$dA = dU - T\,dS - S\,dT$$

which becomes

$$dA = -P\,dV - S\,dT \tag{8·7b}$$

upon substitution from $T\,dS = dU + P\,dV$. For a constant-temperature process,

$$dA = -P\,dV \qquad \text{or} \qquad A_1 - A_2 = \int_1^2 P\,dV$$

Thus the decrease in A equals the amount of work done by a closed system during a reversible isothermal process.

*Josiah Willard Gibbs (1839–1903) received from Yale University in 1863 the first Ph.D. in engineering awarded in America. After further study in Europe he became professor of mathematical physics at Yale and held that position until his death. He made significant contributions in several fields, but a single paper, "On the Equilibrium of Heterogeneous Substances," places him in the first rank among scientists. Because the value of this contribution can be appreciated only by people with knowledge of thermodynamics, the name of this man who was one of the outstanding scientists of all time is virtually unknown to the general public.

We move now to a more widely useful conclusion. Consider a closed system that is initially and finally at the temperature T_0 of the atmosphere around it and exchanges heat only with the atmosphere. (During the process the system temperature may differ from T_0. For example, the system might be compressed or expanded adiabatically so that its temperature changes and then exchange heat with the atmosphere until its temperature is again T_0.) It can be shown (Problem 8·115) that the maximum amount of heat that can be transferred to the system is $T_0(S_2 - S_1)$, so that the maximum work output of the system is given by

$$\begin{aligned} W_{\text{max}} &= U_1 - U_1 + Q_{\text{max}} = U_1 - U_2 + T_0(S_2 - S_1) \\ &= U_1 - T_0S_1 - (U_2 - T_0S_2) = A_1 - A_2 \end{aligned}$$

Thus a more general conclusion than the one in the last paragraph is that, *if a closed system passes from one state to another state at the same temperature while exchanging heat only with the surrounding atmosphere at that temperature* ($T_2 = T_1 = T_0$), *the maximum work which can be produced during the process is equal to the decrease in A of the system*. This conclusion is not restricted to systems comprised of pure substances, so it may be applied, for example, to a process involving a chemical reaction.

Part of the work done by a closed system may be done in pushing back the atmosphere if the system expands. (See Sec. 1·15.) This part of the work is $P_0(V_2 - V_1)$, where P_0 is atmospheric pressure and V_1 and V_2 are the initial and final volumes of the system. If *useful work* means the work done in addition to pushing back the atmosphere,

$$W_{\text{useful}} = W - P_0(V_2 - V_1)$$

and the maximum useful work if the system exchanges heat only with the surroundings at T_0 is

$$\begin{aligned} W_{\text{max useful}} &= U_1 - U_2 + T_0(S_2 - S_1) - P_0(V_2 - V_1) \\ &= U_1 + P_0V_1 - T_0S_1 - (U_2 + P_0V_2 - T_0S_2) \end{aligned}$$

and, if $P_2 = P_1 = P_0$ and $T_2 = T_1 = T_0$,

$$W_{\text{max useful}} = H_1 - T_0S_1 - (H_2 - T_0S_2) = G_1 - G_2$$

Conclusion: If a closed system passes between two states, in each of which it is at the pressure and temperature of the surrounding atmosphere, and exchanges heat only with the atmosphere, the maximum useful work (that is, work done on systems other than the atmosphere) that can be produced during the process is equal to the decrease in G of the system. The conditions specified are met by a system undergoing a chemical reaction in an open vessel.

The differential of the Gibbs function is

$$dG = dH - T\,dS - S\,dT = V\,dP - S\,dT \tag{8·8b}$$

For a constant-temperature process,

$$dG = V\,dP \qquad \text{or} \qquad G_1 - G_2 = -\int_1^2 V\,dP$$

Thus the decrease in G equals the amount of work done by a steady-flow system in a reversible isothermal process in which $\Delta ke = \Delta pe = 0$. A more general conclusion can be reached by considering a steady-flow process for which $\Delta ke = \Delta pe = 0$, $T_1 = T_2 = T_0$, and the system exchanges heat with only the surrounding atmosphere at T_0. The maximum work output is

$$W_{max} = H_1 - H_2 + Q_{max} = H_1 - H_2 + T_0(S_2 - S_1)$$
$$= H_1 - T_0 S_1 - (H_2 - T_0 S_2) = G_1 - G_2$$

Conclusion: If a substance enters and leaves a steady-flow system at the temperature of the surrounding atmosphere and exchanges heat with only the atmosphere, the maximum work that can be produced is equal to the decrease in G of the substance.

As an example of the use of the Gibbs function, suppose we determine the G value for the air and fuel entering a gasoline engine and for the products of combustion after they are cooled to atmospheric temperature. The difference between these values is the maximum amount of work that can conceivably be obtained from this particular flow of air, fuel, and products. Fuel cells are another useful application.

The Helmholtz function, A, and the Gibbs function, G, have both been called free energy, and consequently some confusion between them exists in the literature. The symbols A, Ψ, and F have been used for A; and G, Z, and F (again!) have been used for G. Consequently, you must be careful when you encounter the name free energy or the symbol F in the literature. Neither is used in this book.

A and G have been introduced here simply as examples of useful properties that, like entropy, can be defined as a consequence of the second law. We will use them in Chapters 9, 13, and 20.

8·8 Uses of Entropy

The introduction to this chapter pointed out that searching for a general physical picture of entropy is fruitless. Instead of trying to answer the question "What *is* entropy?," you should concentrate on the questions "What is it used for?" and "How is it used?" Let us review some of the uses of entropy which have already been discussed.

1. *Entropy is used as a convenient coordinate.* The ideal process in many machines and flow passages is a constant-entropy process. Also, on a temperature–entropy diagram the area beneath the path of a reversible process represents heat transfer. By leafing through the remaining pages of this book you will see how frequently diagrams with entropy as a coordinate are used.

2. *Entropy is used in the calculation of heat transfer in reversible processes,* $Q_{rev} = \int T\,dS$. This is of limited practical value because, in most cases where $\int T\,dS$ can be integrated, sufficient data are available for the calculation of Q by means of the first law.

3. *Entropy is used to tell whether a process is reversible, irreversible, or impossible.* This use is based on the increase of entropy principle, $\Delta S_{\text{isolated system}} \geq 0$, where the equality holds only for reversible processes.

4. *Entropy is used in the correlation of physical property data.* This use is based on the relations $T\,ds = du + P\,dv = dh - v\,dP$ and on the definitions of a and g. Further illustrations appear in later chapters.

5. *Entropy is used in the calculation of the unavailable part of heat transfer,* $Q_{unav} = T_0\,\Delta S$. A much more general use in this connection is presented in Chapter 10.

8·9 Entropy and Probability; Entropy and Information Theory

We saw in Sec. 6·9 that a conclusion from statistical thermodynamics is that in an irreversible process an isolated system proceeds toward more probable states. In this chapter we have seen that a conclusion of classical thermodynamics is that during an irreversible process an isolated system proceeds toward states of higher entropy. A natural question is whether there is any direct relationship between probability and entropy. The answer to this question is yes. Before looking at this relationship, let us see what is meant by the probability of a state.

The most frequently used illustration begins with the consideration of a box which is divided into two equal parts by a half-partition. See Fig. 8·17. A marble is in the box, but the box has been shaken so that the marble is as likely to be on one side of the partition as on the other. There are two possible configurations or arrangements. Then the probability that the marble is in side X of the box is $\frac{1}{2}$. The probability that the marble is in side Y of the box is also $\frac{1}{2}$. (Since it is a certainty that the marble is somewhere in the box, the probability that it is in either side X or side Y is 1.) Now consider the case of two marbles, called A and B, in the box. There are four possible, and equally probable, arrangements as follows:

Arrangement	Marbles in X	Marbles in Y
1	A and B	None
2	A	B
3	B	A
4	None	A and B

The probability of each arrangement is $\frac{1}{4}$ because there are four possible arrangements. If the marbles cannot be distinguished from each other, arrangements 2 and 3 are identical. We say that arrangements 2 and 3 produce the same state: the state in which there is one marble in each side of the box. Therefore, only three states are possible:

State	Number of Marbles in X	Number of Marbles in Y	Probability
a	2	0	1/4
b	1	1	1/2
c	0	2	1/4

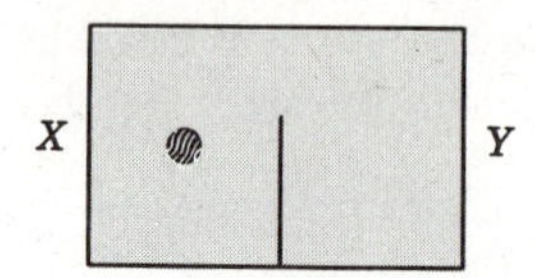

Figure 8·17. System that can exist in either one of two states.

State b has a probability of $\frac{1}{2}$ because two out of the four possible arrangements result in state b. The probability of state a is $\frac{1}{4}$ because only one out of the four possible arrangements results in this state. The same is true of state c. Since there are no states other than a, b, or c possible, the sum of the probabilities of these three states is 1. For four marbles in the box there are 16 possible arrangements which can produce five different states, and their probabilities can be calculated as shown below.

Number of Marbles in X	Number of Marbles in Y	Probability
4	0	1/16
3	1	1/4
2	2	3/8
1	3	1/4
0	4	1/16

If we go now to n marbles, we see that probability of having all n marbles in X is $(\frac{1}{2})^n$. In other words, the state in which all marbles are in one side of the box is a state of low probability. The probability of having r marbles out of n in side X (that is, the number of combinations of n things taken r at a time, divided by the total number of arrangements) is

$$\frac{n(n-1)(n-2)\cdots(n-r+1)}{r!(2)^n}$$

Let us now extend our consideration from a few marbles to many molecules of a gas that are free to move from one side of the box to the other. The opening in the partition can be very small in comparison with the size of the box and still permit enormous numbers of molecules to pass back and forth. It is assumed that any molecule is equally likely to be in side X or side Y. The probabilities of uniform and nearly uniform distributions of n molecules are shown in Table 8·1. It is seen that even for as few as 1000 molecules the probability of an appreciable departure from an equal distribution is very small. (Note that one cubic millimeter of air at normal room conditions contains more than 2×10^{16} molecules.) Any state in which the molecules are unequally distributed is a state of lower probability than one in which they are more nearly equally distributed. Having thus looked into the meaning of the probability of a state, we still face the question, "How is this related to entropy?"

In seeking the relationship between entropy and probability, let us first recall that entropy is an extensive property, so that, if systems A and B are combined to form system C,

$$S_C = S_A + S_B$$

TABLE 8·1 Probabilities of some distributions of molecules

		Number of Combinations in which Number of Molecules (r) in X Is			Fraction of Combinations in which Number of Molecules in X Is		
Number of Molecules, n	Number of Combinations of Molecules, 2^n	$0.5n$	$0.49n$ to $0.51n$	$0.45n$ to $0.55n$	$0.5n$	$0.49n$ to $0.51n$	$0.45n$ to $0.55n$
2	4	2			0.500		
4	16	6			0.375		
6	64	20			0.313		
8	256	70			0.273		
10	1024	252			0.246		
20	1.049×10^{6}	0.1848×10^{6}		0.5207×10^{6}	0.176		0.496
40	1.100×10^{12}	0.1378×10^{12}		0.6272×10^{12}	0.125		0.570
100	1.2676×10^{30}	0.1009×10^{30}	0.2987×10^{30}	0.9238×10^{30}	0.0796	0.236	0.729
200	1.6069×10^{60}	0.0905×10^{60}	0.4439×10^{60}	1.3862×10^{60}	0.0563	0.276	0.863
1000	1.0715×10^{301}	0.0270×10^{301}	0.5282×10^{301}	1.0700×10^{301}	0.0252	0.493	0.9986

On the other hand, the probability of the state of the combined system is

$$\mathcal{P}_C = \mathcal{P}_A \cdot \mathcal{P}_B^*$$

These conditions are satisfied by letting

$$S = C \ln \mathcal{P}$$

because then

$$S_C = S_A + S_B = C(\ln \mathcal{P}_A + \ln \mathcal{P}_B) = C \ln \mathcal{P}_A\mathcal{P}_B = C \ln \mathcal{P}_C$$

Thus the change in entropy of a system which passes from a state 1 to a state 2 is

$$S_2 - S_1 = C \ln \frac{\mathcal{P}_2}{\mathcal{P}_1}$$

where $\mathcal{P}_2$ and $\mathcal{P}_1$ are the probabilities of the final and initial states, respectively, and C is a constant.

As an example, consider an ideal gas which undergoes a free expansion in a closed system until its volume is doubled, $V_2 = 2V_1$ (see Fig. 8·18). From the reasoning above, the probability that all n molecules of the gas would be in the half of the enclosure called V_1 just after the partition is opened is $(\frac{1}{2})^n$. This is the probability of the initial state.

$$\mathcal{P}_1 = \left(\frac{1}{2}\right)^n$$

For a final equilibrium state, the distribution of molecules throughout the total volume

*For example, consider two semipartitioned boxes, each with ends labeled X and Y and each containing one marble. What is the probability that the Y ends of both boxes are empty? The probability that the Y end of one box is empty is $\frac{1}{2}$; the probability that the Y end of the other box is empty is $\frac{1}{2}$; the probability that the Y ends of both boxes are empty is $(\frac{1}{2}) \cdot (\frac{1}{2}) = \frac{1}{4}$. (This is the same as determining the probability of flipping a coin twice and getting a specified result on each flip.)

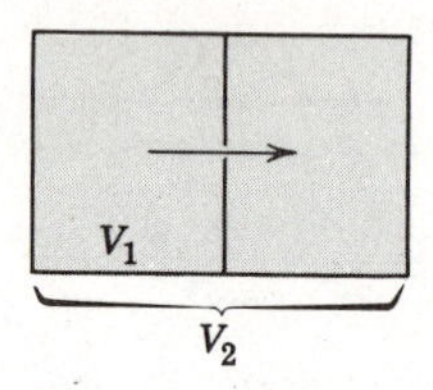

Figure 8·18. Free expansion of a gas.

V_2 is very nearly uniform so that the probability of the final state is very nearly 1:

$$\mathcal{P}_2 = (1)^n = 1$$

The constant C for this case is NR_u/n, where N is the number of moles of gas, R_u is the universal gas constant, and n is the number of molecules in the system. Then

$$S_2 - S_1 = \frac{NR_u}{n} \ln \frac{1}{(1/2)^n} = \frac{NR_u}{n} \ln (2)^n = NR_u \ln 2 = NR_u \ln \frac{V_2}{V_1}$$

and this is the same expression for ΔS that we obtain from classical thermodynamics.

When a solid is melted or a liquid is vaporized, there is a change toward a less ordered molecular state or an increase in the "randomness" of molecular motions and configurations. Such processes serve as additional illustrations of the relationship between entropy increase and the increase in molecular disorder.

In addition to this relationship between entropy and disorder, there are relationships between entropy and other quantities that instead of being conserved tend to increase or decrease. Indeed, the term *entropy,* which was first coined in classical thermodynamics, has been introduced into several fields.

One such field is information theory. One might intuitively expect some kind of inverse relationship between entropy and information, because if a message with a certain amount of information is put into a communication system, the message that comes out of the system can carry at most the same amount of information, but all events that occur in any real communication system tend to cause a loss in information.

Going beyond the intuitive connection requires that *information* be operationally defined and quantified. Elementary discussions often start by considering information on simple systems such as the partitioned box mentioned above containing four marbles. If we already know that four marbles are contained, the message, "There are four marbles in the box" gives us no additional information. Then we might consider the different amounts of information provided by the following three messages:

Message 1: There are more marbles inside X than inside Y.
Message 2: There is an even number of marbles inside X.
Message 3: There are two marbles inside X.

Another kind of example to test a quantitative definition of information is that of the information needed to avoid a collision of two aircraft known to be in proximity to each other. Obviously, knowing only their positions is of little value. Even knowing both positions and velocities would not give one complete confidence that a collision will be avoided. Would information on accelerations make our knowledge complete or sufficient? Would the maneuvering capabilities of the aircraft be pertinent? Does a message giving the positions, velocities, and accelerations of the aircraft at one instant have the same information value at that instant and at a later time?

Other terms must also be quantified. Some terms in use in addition to *information* are *uncertainty, expectance, lack of information, ignorance, message, measurement,* and *communication system.* Definitions well established in other fields may need revision. For example, if *isolated system* means that there is no interaction between the system and its surroundings, is it possible to make measurements on the system or gain any information about it? In the same way that great care is used in classical thermodynamics to establish operational, consistent, and useful definitions, such care must be used in information theory and other fields. Doing so is especially important where the terms used are ones which are used in everyday conversation.

Because information theory is a broad field itself, we discuss it no further here except to mention that it does lead to equations similar to those for entropy in classical thermodynamics, so the concept of entropy is important in areas outside the scope of classical thermodynamics.

The conclusion that concepts can be transferred from one field to another is important. Even more important, however, is the conclusion that the use of entropy in classical thermodynamics, statistical mechanics, information theory, and other fields does not generate any conflict among the fields but rather strengthens and promotes progress in all of them.

8·10 Summary

From the second law it can be shown that

$$\Delta S \equiv \int_{\text{rev}} \frac{\delta Q}{T} \tag{8·1a}$$

is a property. This property is called entropy. It is actually the *change* in entropy that is defined by Eq. 8·1a, but in engineering work we are usually concerned only with entropy changes rather than with absolute values.

ΔS between any two states of a system can be calculated by integrating $\int(\delta Q/T)$ along any reversible path between the two states. It is usually easier, however, to calculate ΔS from the equations

$$T\,dS = dU + P\,dV \tag{8·3}$$

$$T\,dS = dH - V\,dP \tag{8·3}$$

which can be developed from the first and second laws and which apply to any pure substance changing from one equilibrium state to another in the absence of electrical, magnetic, solid deformation, and surface tension effects. These equations are also useful in correlating the physical properties of substances.

The heat transfer of any reversible process is given by

$$Q_{\text{rev}} = \int T\,dS \tag{8·2}$$

Consequently, the heat transfer of any reversible process is represented as an area on a *TS* diagram. Entropy is also useful as a coordinate in conjunction with other properties besides temperature.

Whether a process is reversible, irreversible, or even possible can be determined by the *increase of entropy prinicple*:

$$\Delta S_{\text{isolated system}} \geq 0 \tag{8·4}$$

in which the equality applies to reversible processes. Application of this principle to a steady-flow system results in

$$\sum_{\text{out}} ms - \sum_{\text{in}} ms + \Delta S_{\text{surr that exchange heat with steady-flow system}} \geq 0 \tag{8·5}$$

The available part of the heat added to or taken from a system is that part that could be converted into work by externally reversible engines rejecting heat at the lowest temperature in the surroundings. That part of the heat that could not be converted into work by such engines is called the unavailable part. Thus

$$Q = Q_{\text{av}} + Q_{\text{unav}}$$

It can be shown that for any reversible process the unavailable part of the heat transferred is

$$Q_{\text{unav}} = T_0 \, \Delta S \tag{8·6}$$

where T_0 is the lowest temperature in the surroundings and ΔS is the entropy change of the system.

Two properties that are quite useful, particularly in processes involving chemical reactions, are the Helmholtz function, A, defined as

$$A \equiv U - TS \tag{8·7}$$

and the Gibbs function, G, defined as

$$G \equiv H - TS \tag{8·8}$$

For a closed system passing from one state to another at the same temperature while exchanging heat with only the surroundings at that temperature, the decrease in A of the system is equal to the maximum work that can be produced during the process. For a closed system passing between two states, in each of which it is at the pressure and temperature of the surrounding atmosphere, while exchanging heat with only the atmosphere, the decrease in G of the system is equal to the maximum *useful work* that can be produced during the process. *Useful work* is defined as the total work minus the work done on the surrounding atmosphere. The decrease in G of a substance which enters and leaves a steady-flow system at the temperature of the surrounding atmosphere and exchanges heat with only the atmosphere is equal to the maximum work that can be produced during the process.

In addition to its uses in classical thermodynamics which are listed here, the concept of entropy has proved to be valuable in several other fields.

Suggested Reading

8·1 Dixon, J. R., *Thermodynamics I,* Prentice-Hall, Englewood Cliffs, N.J., 1975, Chapter 10.

8·2 Hatsopoulos, George N., and Joseph H. Keenan, *Principles of General Thermodynamics,* Wiley, New York, 1965, Chapter 16.

8·3 Van Wylen, Gordon J., and Richard E. Sonntag, *Fundamentals of Classical Thermodynamics,* 3rd ed., Wiley, New York, 1985, Chapter 7.

8·4 Wark, K., *Thermodynamics,* 4th ed., McGraw-Hill, New York, 1983, Chapter 7.

Problems

8·1 A closed system can go from a state A to a state B by means of several different adiabatic paths. Is the work the same for all paths? Explain.

8·2 Is $\int_{rev} (\delta w/P)$ a property?

8·3 Prove that the minimum work input to a refrigerator that cools a rigid body from atmospheric temperature T_O to a lower temperature T_L is $W_{in} = T_0(S_0 - S_L) + U_L - U_0$, where S and U are properties of the body being cooled.

8·4 Prove that for any process of a closed system $\Delta S \geq \int(\delta q/T)$.

8·5 Five kilograms of carbon monoxide in a closed system expands reversibly with constant entropy from 250 kPa, 50°C, to 125 kPa. Calculate the work done per kilogram of carbon monoxide.

8·6 Nitrogen flowing through a compressor at a rate of 5 kg/s is compressed isentropically from 100 kPa, 5°C, to 200 kPa. The kinetic energy of the nitrogen is increased by 7.0 kJ/kg. Calculate the power input to the nitrogen.

8·7 Steam flows isentropically through a turbine from 1.4 MPa, 250°C, to an exhaust pressure of 50 mm Hg abs. The flow rate is 4500 kg/h. Calculate the power output, assuming $\Delta ke = 0$.

8·8 Calculate the work required per kilogram of Refrigerant 12 which is compressed reversibly and adiabatically in a steady-flow system from 700 kPa, 50°C, until its density is doubled. Kinetic energy changes are negligible.

8·9 Steam is supplied to a turbine at 1.4 MPa, 300°C, and exhausted at 101.3 kPa. Assume reversible adiabatic flow. To what pressure must the incoming steam be throttled in order to reduce the work per kilogram to two-thirds of that obtained without throttling? Assume that the flow through the turbine is still reversible adiabatic and that the exhaust pressure is unchanged.

8·10 In a closed system, steam expands reversibly from a dry saturated condition at 700 kPa to a dry saturated condition at 70 kPa in such a manner that the process is represented by a straight line on a *Ts* diagram. Determine the work per kilogram of steam.

8·11 In a reversible isothermal process, 250 kJ/kg of heat is added to a closed system comprised of ammonia that is initially saturated liquid at 20°C. Determine the work done during this process.

8·12 Dry saturated steam at 800 kPa enters a steady-flow system and is expanded reversibly and isothermally to 500 kPa. There is no change in kinetic energy. The flow rate is 1.0 kg/s. Sketch the process on a *Ts* diagram, and calculate the amount of work done per kilogram of steam.

8·13 In a reversible steady-flow process, steam initially at 275 kPa, 170°C, undergoes an isothermal expansion to 240 kPa. Sketch a *Ts* diagram for the process. If no work is done, calculate the change in kinetic energy of the steam in kJ/kg.

8·14 In a closed system, dry saturated steam at 700 kPa is heated in a constant-volume process until its pressure is 1.2 MPa. It is then expanded isothermally to 700 kPa and finally cooled to the saturation temperature at constant pressure. All processes are reversible. Calculate the net work done.

8·15 In a reversible process of a closed system, the heat added is given by $Q = aT + bT^2$, where $a = 200$ J/K and $b = 3.20$ J/K^2. Calculate the entropy change of the system as the system temperature changes from 50 to 200°C.

8·16 Dry saturated steam initially at 100 kPa, 99.6°C, is compressed reversibly in a cylinder in such a manner that $T = 151.7(s - 5.79)^2$ where T is in kelvins and s is in J/g·K. The final state is 400 kPa, 400°C. Calculate (*a*) the heat transfer in J/g and (*b*) the work in J/g.

8·17 A system undergoes a reversible process such that the process on a *TS* diagram is a straight line. Some property values are $T_1 = 60$°C, $U_1 = 170$ kJ, $H_1 = 220$ kJ, $S_1 = 0.2300$ kJ/K, $T_2 = 170$°C, $U_2 = 190$ kJ, $H_2 = 247$ kJ, and $S_2 = 0.3000$ kJ/K. Calculate the work of the process.

8·18 Ten pounds of air in a closed system expands reversibly with constant entropy from 40 psia, 140°F, to 20 psia. Calculate the work done per pound of air.

8·19 Air flowing through a compressor at a rate of 12 lbm/s is compressed isentropically from 15 psia, 40°F, to 30 psia. The kinetic energy of the air is increased by 3.0 B/lbm. Calculate the power input to the air.

8·20 Steam flows isentropically through a turbine from 200 psia, 500°F, to an exhaust pressure of 2 in. Hg abs. The flow rate is 10 000 lbm/h. Calculate the power output, assuming $\Delta ke = 0$.

8·21 Calculate the work required per pound of steam that is compressed reversibly and adiabatically in a steady-flow system from 100 psia, 330°F, until its density is doubled. Kinetic energy changes are negligible.

8·22 Steam is supplied to a turbine at 200 psia, 600°F, and exhausted at 14.7 psia. Assume reversible adiabatic flow. To what pressure must the incoming steam be throttled in order to reduce the work per pound to two-thirds of that obtained without throttling? Assume that the flow through the turbine is still reversible adiabatic and that the exhaust pressure is unchanged.

8·23 In a closed system, steam expands reversibly from a dry saturated condition at 100 psia to a dry saturated condition at 10 psia in such a manner that the process is represented by a straight line on a *Ts* diagram. Determine the work per pound of steam.

8·24 In a reversible isothermal process, 100 B/lbm of heat is added to a closed system comprised of water which is initially saturated liquid at 400°F. Determine the work done during this process.

8·25 Dry saturated steam at 118 psia enters a steady-flow system and is expanded reversibly and isothermally to 74 psia. There is no change in kinetic energy. The flow rate is 2.0 lbm/s. Sketch the process on a *Ts* diagram, and calculate the amount of work done per pound of steam.

8·26 In a reversible steady-flow process, steam initially at 40 psia, 340°F, undergoes an isothermal expansion to 35 psia. Sketch a *Ts* diagram for the process. If no work is done, calculate the change in kinetic energy of the steam in B/lbm.

8·27 In a closed system, dry saturated steam at 100 psia is heated in a constant-volume process until its pressure is 200 psia. It is then expanded isothermally to 100 psia, and finally cooled to the saturation temperature at constant pressure. All processes are reversible. Calculate the net work done.

8·28 A closed system undergoes a reversible process such that the heat added is given by the relationship $Q = aT + bT^2$, where $a = 2.0$ B/R and $b = 0.0010$ B/R^2. Calculate the entropy change of the system as the system temperature changes from 100 to 300°F.

8·29 An ideal gas is heated at constant volume from state 1 to state 2, expanded isothermally to state 3, expanded adiabatically to state 4 which is at the same pressure as state 1, and then restored to state 1 by a constant-pressure process. All four processes are reversible. (*a*) Sketch *Pv* and *Ts* diagrams of the cycle. (*b*) State whether each of the following quantities is greater than zero, less than zero, equal to zero, or of indeterminate sign: $\oint \delta q$, $\oint \delta w$, $\oint du$, $\oint ds$.

8·30 A vapor in a closed system passes through a cycle comprised of three processes: isentropic expansion 1-2, constant-pressure cooling 2-3, and constant-volume heating 3-1. Sketch *Pv* and *Ts* diagrams for the cycle if state 1 is superheated vapor and (*a*) states 2 and 3 are also superheated, (*b*) states 2 and 3 are "wet vapor" states.

8·31 An ideal gas in a closed system undergoes a cycle comprised of three processes: Process 1-2 is a reversible constant-volume process starting at P_1 and T_1 and going to $P_2 = 1.5P_1$; process 2-3 is an irreversible adiabatic process with an end state of $P_3 = P_1$ and $T_3 = 1.2T_1$; and process 3-1 is a reversible constant pressure process returning the system to state 1. Sketch to scale *Pv*, *Ts*, *vs*, and *uv* diagrams of the cycle.

8·32 A fluid passing through a steady-flow system can be heated reversibly from state 1 to state 2 in either of two ways. In one case, $T = a + bs$; in the other case, $T = c + es^2$. a, b, c, and e are constants. In which case is the heat transfer greater?

8·33 Sketch *Ts*, *Pv*, *Tv*, *Ph*, *hs*, and *hv* diagrams of a Carnot cycle using as a working fluid (*a*) an ideal gas, (*b*) a liquid-vapor mixture.

8·34 A Carnot engine operates between a source at 430°C and a receiver at 35°C. If 200 kJ is transferred to the receiver each minute, compute the changes in entropy per unit time for the system and the universe for the isothermal expansion.

8·35 A reversed Carnot engine operates between temperature limits of 20 and 50°C. Sketch the cycle on a *Ts* diagram, and indicate the areas that represent the heat transferred from the reservoir at 20°C and the work input. If 100 kJ is absorbed from the reservoir at 20°C, compute the change in entropy of the system for the isothermal compression process.

8·36 An engine operates on a slightly modified Carnot cycle between constant temperature limits of 250 and 30°C. During the isothermal expansion the entropy of the system changes 1.3 kJ/K for 2 kg of the working substance. Fluid friction causes an increase in entropy of 0.04 kJ/K during the adiabatic expansion. Sketch the cycle on a *Ts* diagram. Compute heat added, heat rejected, work performed, and efficiency of the cycle. Also, determine heat added, heat rejected, work performed, and efficiency on the basis that the adiabatic expansion is reversible.

8·37 An engine operates in accordance with a Carnot cycle except that, because of fluid friction, the entropy increases slightly during the adiabatic expansion. The constant operating temperature limits are 115 and 36°C. During the isothermal expansion, 50 kJ/min is added. The increase in entropy for the system during the adiabatic expansion is 5 percent of the entropy change for the isothermal expansion process. Sketch the cycle on a *Ts* diagram. Compute the power output of the engine.

8·38 A refrigerating machine patterned after a reversed Carnot cycle operates between −20 and 30°C; 10 500 kJ is to be removed each hour from the cold room at −20°C. One expansion process is an irreversible adiabatic process that causes an entropy increase for the system of 10 percent of the total entropy change for the reversible isothermal expansion. Draw a *Ts* diagram, and show by area the heat absorbed at −20°C. Compute the power required.

8·39 A reversed Carnot cycle operating on steam serves as a heat pump, discharging heat to a region at 25°C and picking up heat from the atmosphere at 5°C. During the heat rejection process, dry saturated vapor is condensed to saturated liquid. During one complete cycle, the heat pump delivers

50 kJ to the region being heated. (*a*) Sketch the cycle on *Pv* and *Ts* coordinates. (*b*) Calculate the work input to the heat pump per cycle. (*c*) Calculate the change in entropy of the atmosphere during one cycle.

8·40 A Carnot engine operates between temperature limits of 280°C and 5°C. It receives 100 kJ of heat from the high-temperature reservoir. Calculate (*a*) work of the engine, (*b*) the entropy change of the reservoir at 280°C, and (*c*) the entropy change of the reservoir at 5°C.

8·41 A Carnot engine operates between a source at 800°F and a receiver at 100°F. If 200 B is transferred to the receiver each minute, compute the changes in entropy per unit time for the system and the universe for the isothermal expansion.

8·42 A reversed Carnot engine operates between temperature limits of 70 and 120°F. Sketch the cycle on a *Ts* diagram, and indicate the areas that represent the heat transferred from the reservoir at 70°F and the work input. If 100 B is absorbed from the reservoir at 70°F, compute the change in entropy of the system for the isothermal compression process.

8·43 An engine operates on a slightly modified Carnot cycle between constant temperature limits of 500 and 100°F. During the isothermal expansion the entropy of the system changes 0.3 B/R for 2 lbm of the working substance. Because of fluid friction, an increase in entropy of 0.01 B/R occurs during the adiabatic expansion. Sketch the cycle on a *Ts* diagram. Compute heat added, heat rejected, and work performed. Also determine the efficiency of the cycle. Determine heat added, heat rejected, and work performed, and the efficiency on the basis that the adiabatic expansion is reversible.

8·44 An engine operates in accordance with a Carnot cycle except that, owing to fluid friction, the entropy increases slightly during the adiabatic expansion. The constant operating temperature limits are 240 and 100°F. During the isothermal expansion 50 B/min is added. The increase in entropy for the system during the adiabatic expansion is 5 percent of the entropy change for the isothermal expansion process. Sketch the cycle on a *Ts* diagram. Compute the power output of the engine.

8·45 A refrigerating machine patterned after a reversed Carnot cycle operates between −10 and 90°F; 10 000 B is to be removed each hour from the cold room at −10°F. One expansion process is an irreversible adiabatic process that causes an entropy increase for the system of 10 percent of the total entropy change for the reversible isothermal expansion. Draw a sketch on a *Ts* diagram, and show by area the heat absorbed at −10°F. Compute the power required.

8·46 A heat engine utilizing air as the working fluid completes its cycle in three steps as follows: process 1-2: adiabatic compression from T_1 = 100°F to T_2 = 300°F; process 2-3: isothermal expansion at 300°F; process 3-1: dT/ds = constant. Assuming each step in the cycle to be reversible, compute the efficiency of the engine.

8·47 Fifty pounds of water at 40°F and 80 lbm of water at 180°F are mixed in an insulated vessel under a constant pressure of 1 atm. Determine the entropy change of the system comprised of the 130 lbm of water.

8·48 State the conditions under which each of the following equations is true:

$$\oint T\,ds = -\oint v\,dP$$

$$\oint T\,ds = \oint P\,dv$$

8·49 Five kilograms of nitrogen is cooled in a closed tank from 250 to 40°C. The initial pressure is 2.8 MPa. Compute the changes in entropy, internal energy, and enthalpy.

8·50 One cubic meter of air at 25°C and 95 kPa is compressed isothermally to 0.3 m^3. Is the entropy of the air increased or diminished, and by how much?

8·51 The entropy of 2 kg of ethane decreases 1.8 kJ/K during a constant-temperature process at 130°C. For an initial pressure of 800 kPa, compute (*a*) the work done and (*b*) the final volume.

8·52 One-tenth kilogram of octane is heated at constant volume from 60°C and 300 kPa to 280°C, compressed reversibly and adiabatically to 1.4 MPa, and then cooled under constant-volume conditions to 110°C. Determine the change in entropy of the octane for the individual steps and for the complete series of steps.

8·53 A kilogram of carbon dioxide under a pressure of 1400 kPa and occupying a volume of 0.05 m^3 undergoes a constant-pressure change until the volume is doubled. The pressure is reduced to 700 kPa under constant-volume conditions, after which the gas is compressed isothermally to its original volume. Compute the total change in entropy of the system for the three steps.

8·54 One-tenth kilogram of propane at 100 kPa, 60°C, is contained in a closed, rigid, thermally insulated vessel. A paddle wheel inside the vessel is turned by an external motor until the pressure is 130 kPa. Sketch Pv and Ts diagrams for this process, and calculate the change in entropy of the propane.

8·55 Methane initially at 170 kPa, 5°C, has its volume doubled while its temperature rises to 90°C. Calculate, if possible, the change in enthalpy, the change in internal energy, the change in entropy, the work done, and the heat transferred.

8·56 One-twentieth kilogram of acetylene in a closed system is compressed irreversibly from 100 kPa, 5°C, to 200 kPa. During the process 1.8 kJ of heat is removed from the acetylene, and the work done on the acetylene amounts to 2.7 kJ. Determine the entropy change of the acetylene.

8·57 Air in a closed system is heated reversibly from 100 kPa, 7°C, to 119.7 kPa, 119°C, in such a manner that dT/ds is a constant. (*a*) Calculate the work done per kilogram of air. (*b*) Sketch a large, accurate Ts diagram for the process, showing lines of constant pressure for 100 and 119.7 kPa. Show also some constant-volume lines. (*c*) Sketch a large, accurate Pv diagram for the process. (You may wish to support your sketches by some calculations proving that you have shown the correct shapes or slopes.)

8·58 Air expands irreversibly in a cylinder from $P_1 = 280$ kPa, $T_1 = 60$°C, to $P_2 = 140$ kPa. The air does work in the amount of 30 kJ/kg and 14 kJ/kg of heat is removed from the air during the expansion. The initial volume is 0.008 78 m^3. Calculate the entropy change of the air per kilogram.

8·59 Air initially at 400 kPa, 60°C, expands reversibly in a cylinder against a piston. It expands isothermally from $P_1 = 400$ kPa to $P_2 = 200$ kPa and then expands adiabatically from $P_2 = 200$ kPa to $P_3 = 100$ kPa. Sketch the processes on Pv and Ts coordinates and calculate for the entire expansion from state 1 to state 3: (*a*) q, (*b*) $u_3 - u_1$, (*c*) $s_3 - s_1$.

8·60 Heat is transferred from a very large mass of water at 90°C to 0.1 kg of air that expands irreversibly in a cylinder fitted with a piston from 400 kPa, 60°C, to 150 kPa, 35°C, doing 4.00 kJ of work. Calculate (*a*) the heat transfer to the air and (*b*) the entropy change of the universe.

8·61 Ten pounds of nitrogen is cooled in a closed tank from 500 to 100°F. The initial pressure is 400 psia. Compute the changes in entropy, internal energy, and enthalpy.

8·62 Fifty cubic feet of air at 14 psia, 80°F, is compressed isothermally to 10 cu ft. Is the entropy of the air increased or diminished, and by how much?

8·63 The entropy of 4 lbm of air decreases 0.4 B/R during a constant-temperature process at 270°F. If the original pressure was 120 psia, compute (*a*) the work done and (*b*) the final volume.

8·64 One pound of air is heated at constant volume from 140°F and 50 psia to 540°F, compressed reversibly and adiabatically to 200 psia, and then cooled under constant-volume conditions to 240°F. Determine the change in entropy of the air for the individual steps and for the complete series of steps.

8·65 A pound of air that is under a pressure of 200 psia occupying a volume of 2 cu ft undergoes a constant-pressure change until the volume is doubled. The pressure is reduced to 100 psia under constant-volume conditions, after which it is compressed isothermally to its original volume. Compute the total change in entropy of the system for the three steps.

8·66 One-half pound of air at 15 psia, 140°F, is contained in a closed rigid thermally insulated vessel. A paddle wheel inside the vessel is turned by an external motor until the presure is 20 psia. Sketch *Pv* and *Ts* diagrams for this process, and calculate the change in entropy of the air.

8·67 Air initially at 25 psia, 40°F, has its volume doubled while its temperature rises to 200°F. Calculate, if possible, the change in enthalpy, the change in internal energy, the change in entropy, the work done, and the heat transferred.

8·68 One-tenth pound of air in a closed system is compressed irreversibly from 15 psia, 40°F, to 30 psia. During the process 1.7 B of heat is removed from the air, and the work done on the air amounts to 2.6 B. Determine the entropy change of the air.

8·69 Air in a closed system is heated reversibly from 15.0 psia, 40°F, to 17.92 psia, 240°F, in such a manner that dT/ds = constant. (*a*) Calculate the work done per pound of air. (*b*) Sketch a large accurate *Ts* diagram for the process, showing lines of constant pressure for 15 and 17.92 psia. Show also some constant-volume lines. (*c*) Sketch a large accurate *Pv* diagram for the process. (You may wish to support your sketches by some calculations proving that you have shown the correct shapes or slopes.)

8·70 Heat is transferred from a very large mass of water at 200°F to 0.20 lbm of air that expands irreversibly in a cylinder fitted with a piston from 60 psia, 140°F, to 20 psia, 100°F, doing 4.00 B of work. Calculate (*a*) the heat transfer to the air and (*b*) the entropy change of the universe.

8·71 Prove that the constant-pressure lines in the wet region of an *hs* diagram for steam are straight. Also prove that they are or are not parallel.

8·72 In Example 8·8 it is shown that for an ideal gas $(\partial T/\partial s)_P = T/c_p$ and $(\partial T/\partial s)_v = T/c_v$. Show that these relations are true for *any* pure substance.

8·73 Which one of the *Ts* diagrams shown here for an ideal gas properly shows two constant-pressure lines? Support your answer, and show on the same diagram two constant-volume lines.

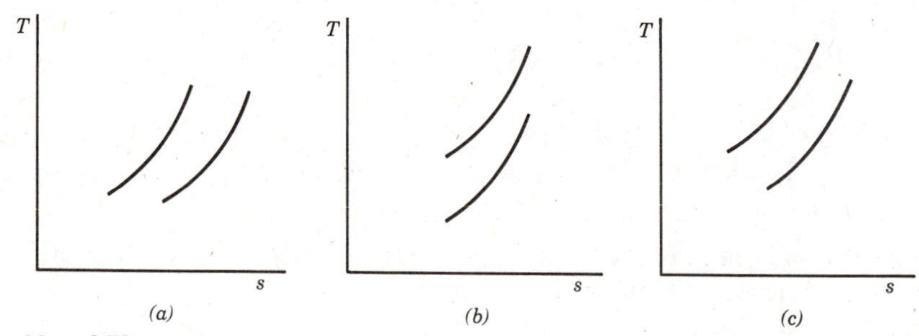

Problem 8·73.

8·74 Derive the following expressions for an ideal gas with constant specific heats:

$$s_2 - s_1 = c_p \ln \frac{T_2/T_1}{(P_2/P_1)^{(k-1)/k}}$$

$$s_2 - s_1 = c_v \ln \frac{P_2}{P_1} + c_p \ln \frac{v_2}{v_1}$$

8·75 (*a*) In a closed system, a gas undergoes a cycle made up of the following processes: 1-2, reversible isothermal compression; 2-3, reversible constant-volume heating; 3-4, reversible constant-pressure expansion; 4-1, reversible adiabatic expansion. Sketch *pv* and *Ts* diagrams, and state whether each of the following quantities is positive, zero, negative, or indeterminate in sign: $\oint \delta W$, $\oint \delta Q$, $\oint dS$, $\oint dU$, $\oint dH$. (*b*) Repeat part *a* with process 4-1 changed to an irreversible adiabatic expansion.

8·76 Is an adiabatic expansion of steam from 400 kPa, 150°C, to 101.3 kPa, 94 percent quality, reversible, irreversible, or impossible? State the quality limits for reversible, for irreversible, and for impossible adiabatic expansions between 400 kPa, 150°C, and 101.3 kPa.

8·77 Dry saturated sulfur dioxide at −20°C is compressed adiabatically to 400 kPa in a steady-flow process. What is the minimum final temperature? Is it the same for a closed-system adiabatic compression?

8·78 Is an adiabatic expansion of oxygen from 250 kPa, 60°C, to 125 kPa, 25°C, possible? Prove your answer. What is the lowest possible final temperature for an adiabatic expansion of air from 250 kPa, 60°C, to 125 kPa? What is the maximum possible final temperature?

8·79 A turbine operating on nitrogen has inlet conditions of 200 kPa, 90°C, and an exhaust pressure of 100 kPa. If the expansion is adiabatic and $\Delta ke = 0$, is an exhaust temperature of 5°C possible? Is it possible if $\Delta ke \neq 0$?

8·80 One-fourth kilogram of carbon monoxide in a closed system expands from 250 kPa, 120°C, to 125 kPa, 25°C, producing 8.0 kJ of work. During this process, the entropy of the surroundings increases by 0.0324 kJ/K. (*a*) Calculate the heat transfer. (*b*) Is the process reversible? Prove your answer.

8·81 One kilogram of air in a closed system expands from 400 kPa, 550°C, to 100 kPa. During the process, the entropy change of the surroundings is 0.135 kJ/K. Determine the minimum temperature possible for the air in the system.

8·82 Air enters a turbine at 300 kPa, 120°C, and leaves at 100 kPa, 50°C. The flow rate is 1.4 kg/s, and the power output is 80 kW. Some heat is transferred to the surrounding atmosphere at 5°C. Calculate, per kilogram of air passing through the turbine: (*a*) the heat transfer, (*b*) the entropy change of the universe.

8·83 Air expands steadily through a turbine from 200 kPa, 60°C, to 95 kPa, 15°C. The entropy of the surroundings decreases by 0.04 kJ/K for each kilogram of air passing through the turbine. Show whether the expansion as described is reversible, irreversible, or impossible.

8·84 A cylinder and piston contains 0.05 kg of air at 400 kPa and 60°C. The air expands isothermally until the volume is doubled while heat is supplied from a reservoir at 120°C. The air expands slowly and the process is frictionless. (*a*) How much work is done by the gas? (*b*) Is the process reversible?

8·85 Can a system undergo an isentropic process that is not reversible and adiabatic? Explain. Can an isolated system undergo an isentropic process that is not reversible and adiabatic?

8·86 Air is compressed irreversibly from 15 psia, 40°F, to 30 psia, 240°F. (*a*) Calculate Δs of the air. (*b*) Give all the information you can on the numerical value of the entropy change of the surroundings per pound of air.

8·87 Is an adiabatic expansion of steam from 60 psia, 300°F, to 14.7 psia, 94 percent quality, reversible, irreversible, or impossible? State the quality limits for reversible, for irreversible, and for impossible adiabatic expansions between 60 psia, 300°F, and 14.7 psia.

8·88 Dry saturated sulfur dioxide at −10°F is compressed adiabatically to 60 psia in a steady-flow process. What is the minimum final temperature? Is it the same for a closed-system adiabatic compression?

8·89 Is an adiabatic expansion of air from 40 psia, 140°F, to 20 psia, 80°F, possible? Prove your answer. What is the lowest possible final temperature for an adiabatic expansion of air from 40 psia, 140°F, to 20 psia? What is the maximum possible final temperature?

8·90 A turbine operating on air has inlet conditions of 30 psia, 200°F, and an exhaust pressure of 15 psia. If the expansion is adiabatic and $\Delta ke = 0$, is an exhaust temperature of 40°F possible? Is it possible if $\Delta ke \neq 0$?

8·91 One-half pound of air in a closed system expands from 40 psia, 240°F, to 20 psia, 80°F, producing 8.0 B of work. During this process, the entropy of the surroundings increases by 0.0114 B/R. (*a*) Calculate the heat transfer. (*b*) Is the process reversible? Prove your answer.

8·92 Air is compressed from 14 psia, 40°F, to 42 psia, 200°F, in a closed system. The mass of air is 3.0 lbm. During the process, the entropy of the surroundings increases by 0.100 B/R. Make the necessary calculations and prove whether the process is reversible, irreversible, or impossible.

8·93 Air expands reversibly and adiabatically through a turbine from 30 psia, 240°F, to 10 psia. Inlet and exhaust velocities are both 200 fps. The inlet cross-sectional area is 0.040 sq ft. (*a*) Sketch *Pv* and *Ts* diagrams of the process, indicating any areas or distances which represent heat or work. (*b*) Calculate the power output in hp.

8·94 Two pounds of steam in a closed system expands reversibly from a dry saturated condition at 500 psia to 20 psia, 300°F, in such a manner that the process is a straight line on a *Ts* diagram. The process is externally reversible. The lowest temperature in the surroundings is 40°F. Sketch the *Ts* diagram and calculate (*a*) the work done on or by the steam, (*b*) the entropy change of the surroundings, and (*c*) the amount of heat transferred that is available energy.

8·95 Air flows steadily through a compressor at a rate of 10 lbm/s, entering from the atmosphere at 12 psia, 40°F, and leaving at 24 psia, 140°F. Kinetic energy changes are negligible. During the compression, heat is transferred from the air to a coolant, and for *each pound of air compressed,* the entropy *of the coolant* increases by 0.010 B/R. Show whether or not the compressor operates reversibly.

8·96 Air is to be compressed adiabatically from 90 kPa, 20°C, to 400 kPa at a rate of 0.26 kg/min. Give all the information you can regarding the outlet temperature.

8·97 Heat is transferred from a furnace at a constant temperature of 1400°C to water that is evaporating at a constant temperature of 250°C. The lowest temperature in the surroundings is 30°C. For each 1000 kJ transferred, how much of the energy would be available if it were absorbed by a fluid at 1400°C? How much of the energy is available as it is received by the water at 250°C? Show these quantities as areas on a *TS* diagram.

8·98 Consider the transfer of heat from an energy reservoir at 250°C to 2.5 lbm of air initially at 100 kPa, 60°C, trapped in a closed, rigid tank. Heat is transferred until the temperature of the air is 170°C. The temperature of the surroundings is 5°C. (*a*) How much heat is transferred? (*b*) How much of the energy removed from the reservoir is available energy? How much is unavailable? (*c*) How much of the energy added to the air in the tank is available energy? How much is unavailable?

8·99 A closed, rigid container holds dry saturated Refrigerant 12 at 1.5 MPa. Heat is added until the pressure becomes 2.0 MPa. How much of the heat added is available energy if the sink temperature is 0°C?

8·100 Three kilograms of oxygen (the system) is cooled at a constant pressure of 140 kPa from 150 to 35°C. Sink (surroundings) temperature is 35°C. All the heat removed is dissipated to the surroundings. How much of the heat removed from the oxygen is available energy? Calculate the entropy change of the system and of the universe. By how much does the total energy of the universe change?

8·101 Ethene flows reversibly and at constant pressure through a steady-flow system at a rate of 0.7 kg/s. It enters at 140 kPa, 35°C, and leaves at 150°C. The lowest temperature in the surroundings (the sink temperature) is 5°C. Calculate, per kilogram of ethene, (*a*) the heat added to the ethene, (*b*) the work done on or by the ethene, and (*c*) the amount of heat added to the ethene that is available energy.

8·102 Air passing through a heating coil is heated from 95 kPa, 5°C, to 95 kPa, 115°C. Calculate the fraction of the heat added to the air that is unavailable energy if the atmospheric temperature is 5°C.

8·103 Air is compressed in a cylinder from initial conditions of 100 kPa, 24°C, to final conditions of 800 kPa, 36°C. The mass of air is 0.0235 kg, and the volume is initially 0.020 m^3. (*a*) Calculate the entropy change for the air (J/K). (*b*) If the entropy change of the surroundings is 22.10 J/K would such a process be reversible, irreversible, or impossible? Why?

8·104 Heat is transferred from a furnace at a constant temperature of 2540°F to water which is evaporating at a constant temperature of 540°F. The lowest temperature in the surroundings is 90°F. For each 1000 B transferred, how much of the energy would be available if it were absorbed by a fluid at 2540°F? How much of the energy is available as it is received by the water at 540°F? Show these quantities as areas on a *TS* diagram.

8·105 Consider the transfer of heat from an energy reservoir at 540°F to 5 lbm of air initially at 15 psia, 140°F, trapped in a closed, rigid tank. Heat is transferred until the temperature of the air is 340°F. The temperature of the surroundings is 40°F. (*a*) How much heat is transferred? (*b*) How much of the energy removed from the reservoir is available energy? How much is unavailable? (*c*) How much of the energy added to the air in the tank is available energy? How much is unavailable?

8·106 For each 1000 B supplied to a power plant, 600 B is rejected to the atmosphere at 80°F from the working fluid that is condensed at a constant temperature of 90°F, and 200 B of heat is rejected to the atmosphere from working fluid at 800°F. The work output is 200 B. Determine the maximum amount of work which might be obtained from (*a*) the heat rejected from the fluid at 90°F and (*b*) the heat rejected from the fluid at 800°F.

8·107 A closed, rigid container holds dry saturated steam at 17.0 psia. Heat is added until the pressure becomes 32.0 psia. How much of the heat added is available energy if the sink temperature is 32°F?

8·108 Ten pounds of air (the system) is cooled at a constant pressure of 20 psia from 300 to 100°F. Sink (surroundings) temperature is 100°F. All the heat removed is dissipated to the surroundings. How much of the heat removed from the air is available energy? Calculate the entropy change of the system and of the universe. By how much does the total energy of the universe change?

8·109 Air flows reversibly and at constant pressure through a steady-flow system at a rate of 1.4 lbm/s. It enters at 20 psia, 100°F, and leaves at 300°F. The lowest temperature in the surroundings (the sink temperature) is 40°F. Calculate, per pound of air, (*a*) the heat added to the air, (*b*) the work done on or by the air, and (*c*) the amount of heat added to the air that is available energy.

8·110 In a steady-flow reversible process with no change in potential or kinetic energy, a gas is cooled at a constant pressure of 20 psia from 340 to 140°F. The sink temperature is 40°F. For this gas, in the temperature range involved, $c_p = 0.2 + 0.0001T$ and $c_v = 0.15 + 0.0001T$, where T is in Rankine and c_p and c_v are in B/lbm·R. Sketch a Ts diagram and calculate, in B/lbm, (*a*) the amount of heat removed from the gas, (*b*) the entropy change of the gas, and (*c*) the amount of heat removed from the gas that is unavailable energy.

8·111 Nitrogen flowing steadily over the tubes of a heat exchanger enters at 15 psia, 400°F, and is cooled to 300°F. Nitrogen flows through the tubes at the same rate, entering at 60 psia, 150°F, and absorbing all the heat given up by the hotter nitrogen. Neglect all pressure drops in the heat exchanger. The temperature of the surrounding atmosphere is 70°F. Calculate (*a*) the fraction of the heat given up by the hotter stream that is available energy, and (*b*) the fraction of the heat absorbed by the cooler stream that is available energy.

8·112 Air enters a turbine at 30 psia, 240°F, and is exhausted at 15 psia, 160°F. The flow rate is 0.18 lbm/s, and the flow is adiabatic. (*a*) Calculate Δs. (*b*) Sketch the process on Pv and Ts coordinates. (*c*) Is the process reversible?

8·113 Refrigerant 12, initially saturated at 25°C, is held in an insulated vessel with a volume of 0.55 m^3. A valve is opened to allow the refrigerant to escape to the atmosphere. Determine the mass and the state of the refrigerant remaining in the vessel when the pressure inside the vessel has dropped to the pressure of the surrounding atmosphere, 95 kPa.

8·114 (*a*) An ideal gas undergoes a cycle made up of three processes: Process 1-2 is a constant-pressure cooling; process 2-3 is a constant volume heating until $T_3 = T_1$; and process 3-1 is shown as a straight line on a Pv diagram. Sketch a Ts diagram, numbering the states and showing clearly the shapes of the process lines. (*b*) An ideal gas undergoes a cycle made up of three processes: Process 1-2 is an isentropic expansion; process 2-3 is an isothermal compression until $P_3 = P_1$; and process 3-1 is shown as a straight line on a Ts diagram. Sketch a Pv diagram, numbering the states and showing clearly the shapes of the process lines.

8·115 Consider a closed system which is initially and finally at the temperature of the surrounding atmosphere, T_0, and exchanges heat with only the atmosphere. Show that the maximum amount of heat that can be transferred to the system is $T_0(S_2 - S_1)$, where S_2 and S_1 are the final and initial entropies of the system.

8·116 For any two phases of a pure substance existing together in equilibrium, the Gibbs functions must be the same. Verify that ammonia table values for a few cases of liquid–vapor equilibrium meet this condition.

8·117 Two students come to you, telling you they are having difficulty with the concept of entropy. You ask them to explain the difficulty, and the response is. "We do not know what entropy *is*." What would you tell these students to help them with their difficulty?

8·118 Write a statement explaining why "What *is* entropy?" is not a helpful question in the study of thermodynamics just as "What *is* momentum?" is not a helpful question in the study of mechanics.

8·119 A statement of the second law known as the Carathéodory statement is: In the vicinity of any equilibrium state of a closed system there are some states that cannot be reached by adiabatic processes. Demonstrate that this statement is equivalent to the Kelvin–Planck or Clausius statement.

8·120 Consider a closed system which is initially and finally at the temperature T_0 of the atmosphere around it and exchanges heat with only the atmosphere. During the process the system temperature may differ from T_0. For example, the system might be compressed or expanded adiabatically so that its temperature changes and then exchange heat with the atmosphere until its temperature is again T_0. Prove that the maximum amount of heat that can be transferred to the system is $T_0(S_2 - S_1)$.

CHAPTER NINE

Physical Properties

In Chapters 3 and 4, physical properties were discussed in terms of relationships among P, v, T, u, and h—properties which are either defined independently of the laws of thermodynamics or are defined by means of the first law. Now that the second law and the resulting properties s, a, and g have been introduced, we can greatly extend our study of physical property relationships.

Engineers are interested in physical property relationships for several reasons. One reason pertains to the selection of materials or substances to perform various functions. For example, just as the materials used for the various parts of an aircraft are selected on the basis of their strength, hardness, ductility, weight, wear properties, cost, and so forth, the fluids used in refrigerators, heat pumps, nuclear reactor cooling circuits, air-conditioning units, and other systems are selected on the basis of other properties.

Often the data that an engineer needs on a particular property of a substance are not available. In such cases a knowledge of general property relationships such as those discussed in this chapter may permit the calculation of needed properties from other properties. Occasionally an engineer finds that for some substance, data are needed outside the range of available data. Extrapolations are most reliably made on the basis of certain characteristics of substances in general. If the paucity of data on some substance is so restrictive that some experimental measurements must be made, a knowledge of general property relationships is essential in organizing the experimental program to give the needed information with the required accuracy and with the least expenditure of time and money. Indeed, the cost of preparing the extensive tables of properties that are now available for many substances would be prohibitive if it were not possible to gain much information from a relatively small amount of data.

Chapter 4 dealt with the equation of state and other relations for ideal gases. Now we consider real gases that deviate from $Pv = RT$ and for which there are often no extensive tables of properties. When $Pv = RT$ is not sufficiently accurate for real gases, we either modify it by means of correction factors or we use other, more complicated, equations of state. Also, if $Pv \neq RT$, specific heats are usually functions of both pressure and temperature so that relations for u and h are not as simple as they are for ideal gases. This chapter presents some PvT relations and also some specific-heat data for real gases. This chapter deals with properties of pure substances. Mixtures of variable composition are treated in Chapters 11, 13, and 19.

9·1 The Maxwell Equations

The following four equations which were introduced earlier are valid, at least for pure substances in the absence of electricity, magnetism, solid distortion effects, and surface tension, for any process between equilibrium states:

$$du = T\,ds - P\,dv \tag{8·3}$$

$$dh = T\,ds + v\,dP \tag{8·3}$$

$$da = -P\,dv - s\,dT \tag{8·7b}$$

$$dg = v\,dP - s\,dT \tag{8·8b}$$

Since u, h, a, and g are properties, du, dh, da, and dg are exact differentials. Therefore,

$$\left(\frac{\partial T}{\partial v}\right)_s = -\left(\frac{\partial P}{\partial s}\right)_v \tag{9·1}$$

$$\left(\frac{\partial T}{\partial P}\right)_s = \left(\frac{\partial v}{\partial s}\right)_P \tag{9·2}$$

$$\left(\frac{\partial P}{\partial T}\right)_v = \left(\frac{\partial s}{\partial v}\right)_T \tag{9·3}$$

$$\left(\frac{\partial v}{\partial T}\right)_P = -\left(\frac{\partial s}{\partial P}\right)_T \tag{9·4}$$

These four equations are called the *Maxwell equations.** They are important in the correlation of properties of pure substances because *they relate entropy to the directly measurable properties pressure, specific volume, and temperature.*

Many other relations among properties can be derived from Eqs. 8·3, 8·7b, and 8·8b. For example, taking u as a function of s and v, we have

$$du = \left(\frac{\partial u}{\partial s}\right)_v ds + \left(\frac{\partial u}{\partial v}\right)_s dv$$

Comparing this with Eq. 8·3

$$du = T\,ds - P\,dv \tag{8·3}$$

and noting that ds and dv are independent, we can equate the coefficients of ds and dv to give

$$\left(\frac{\partial u}{\partial s}\right)_v = T \qquad \left(\frac{\partial u}{\partial v}\right)_s = -P \tag{9·5}$$

*James Clerk Maxwell (1831–1879), a member of a wealthy Scottish family, received his formal education at the University of Edinburgh and Cambridge University. He was professor of physics and astronomy at King's College, London (1860–68), and in 1871 became the first professor of experimental physics at Cambridge University. He made scientific contributions in many areas, including mechanics, optics, electricity and magnetism, kinetic theory of gases, and thermodynamics.

Treating the expressions for dh, da, and dg similarly, we obtain

$$\left(\frac{\partial h}{\partial s}\right)_P = T \qquad \left(\frac{\partial h}{\partial P}\right)_s = v \tag{9·6}$$

$$\left(\frac{\partial a}{\partial v}\right)_T = -P \qquad \left(\frac{\partial a}{\partial T}\right)_v = -s \tag{9·7}$$

$$\left(\frac{\partial g}{\partial P}\right)_T = v \qquad \left(\frac{\partial g}{\partial T}\right)_P = -s \tag{9·8}$$

You do not have to memorize Eqs. 9·1 through 9·8 because they are so easy to derive from Eqs. 8·3, 8·7b, and 8·8b, which you should know or be able to derive quickly. Many other relations among the eight variables P, v, T, s, h, u, a, and g can be derived from Eqs. 9·1 through 9·8.* Some of these will be used later in this chapter.

Some interesting conclusions can be drawn directly from the Maxwell equations and Eqs. 9·5 to 9·8. For example, from the second of Eqs. 9·6, that is

$$\left(\frac{\partial h}{\partial P}\right)_s = v \tag{9·6}$$

we conclude that, since v is always positive, in any isentropic process the enthalpy of a substance decreases if the pressure decreases. Notice also that for isentropic processes of any system,

$$\Delta h = \int v\,dP \qquad (s = \text{constant, only})$$

An important fact is that this conclusion in no way depends on any equation of state.

Let us look at another example of a conclusion drawn from the Maxwell equations. The density of water under a pressure of 1 atm is a maximum at about 4°C. In other words, the specific volume is a minimum at this point, so that $(\partial v/\partial T)_P = 0$. The fourth Maxwell equation, Eq. 9·4, then shows that $(\partial s/\partial P)_T = 0$. Thus, at the temperature of maximum density, entropy is independent of pressure. This means that constant-pressure lines in the liquid region of a *Ts* diagram for water in the vicinity of 1 atm coincide with each other and cross the saturation line at around 4°C as shown in Fig. 9·1. Point *M* in Fig. 9·1 can represent either compressed liquid at 90 kPa or a low-quality mixture of liquid and vapor at some pressure lower than the saturation pressure for 4°C.

The first of Eqs. 9·5,

$$\left(\frac{\partial u}{\partial s}\right)_v = T \tag{9·5}$$

*See P. W. Bridgman, "A Complete Collection of Thermodynamic Formulas," *Physical Review*, vol. 3, 1914, for a systematic compilation that is ingeniously arranged so that millions (!) of relations can be obtained directly from fewer than one hundred tabulated relations.

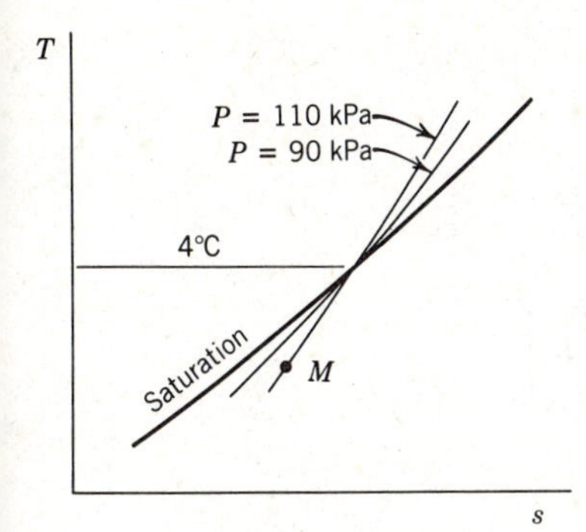

Figure 9·1. Constant-pressure lines for compressed liquid water.

can be written as

$$\left(\frac{\partial u}{\partial T}\right)_v \left(\frac{\partial T}{\partial s}\right)_v = T$$

Recalling the definition of c_v, we have

$$c_v \left(\frac{\partial T}{\partial s}\right)_v = T$$

or

$$\left(\frac{\partial T}{\partial s}\right)_v = \frac{T}{c_v} \tag{9·9}$$

Thus the slope of a constant-volume line on a Ts diagram is equal to T/c_v for any substance. This was shown in Example 8·8 to be true for an ideal gas, but here we see that the conclusion is independent of any equation of state.

The Maxwell equations provide a simple proof that the specific heats of an ideal gas are functions of temperature only. The third Maxwell equation is

$$\left(\frac{\partial s}{\partial v}\right)_T = \left(\frac{\partial P}{\partial T}\right)_v \tag{9·3}$$

and we have just shown that

$$\left(\frac{\partial s}{\partial T}\right)_v = \frac{c_v}{T} \tag{9·9}$$

As long as s is a continuous function of T and v, and the derivatives are also continuous, the order of differentiation is immaterial and the "mixed" second-order partial derivatives are equal,

$$\left[\frac{\partial}{\partial v}\left(\frac{\partial s}{\partial T}\right)_v\right]_T = \left[\frac{\partial}{\partial T}\left(\frac{\partial s}{\partial v}\right)_T\right]_v \tag{a}$$

This is usually written

$$\frac{\partial^2 s}{\partial v\, \partial T} = \frac{\partial^2 s}{\partial T\, \partial v}$$

Substituting from Eqs. 9·3 and 9·9 into Eq. a,

$$\left[\frac{\partial}{\partial v}\left(\frac{c_v}{T}\right)\right]_T = \left[\frac{\partial}{\partial T}\left(\frac{\partial P}{\partial T}\right)_v\right]_v$$

$$\left(\frac{\partial c_v}{\partial v}\right)_T = T\left(\frac{\partial^2 P}{\partial T^2}\right)_v$$

For an ideal gas, $Pv = RT$, and therefore $(\partial^2 P/\partial T^2)_v = 0$. Thus

$$\left(\frac{\partial c_v}{\partial v}\right)_T = 0$$

which indicates that c_v is a function of temperature only. In a similar manner it can be shown that c_p of an ideal gas is a function of temperature only, although this is unnecessary because it is even easier to show that, if either c_p or c_v of an ideal gas is a function of temperature only, then so is the other.

Let us now see how the Maxwell equations help in the evaluation of entropy from data on the directly measurable properties pressure, specific volume, temperature, and specific heats. For a pure substance entropy is a function of any two of the properties: P and v, P and T, or v and T. Two partial derivatives involving s are associated with each of the three pairs of measurable properties, making a total of six:

$$\left(\frac{\partial s}{\partial P}\right)_v \quad \left(\frac{\partial s}{\partial v}\right)_P \quad \left(\frac{\partial s}{\partial P}\right)_T \quad \left(\frac{\partial s}{\partial T}\right)_P \quad \left(\frac{\partial s}{\partial v}\right)_T \quad \left(\frac{\partial s}{\partial T}\right)_v$$

If we can express each of these derivatives in terms of P, v, T, and other readily measurable properties, we will have complete information on the entropy of any substance for which we have data on the other properties. The third and fifth derivatives can be evaluated from PvT data by means of two Maxwell equations alone:

$$\left(\frac{\partial s}{\partial P}\right)_T = -\left(\frac{\partial v}{\partial T}\right)_P \qquad \left(\frac{\partial s}{\partial v}\right)_T = \left(\frac{\partial P}{\partial T}\right)_v$$

The last derivative can be evaluated as

$$\left(\frac{\partial s}{\partial T}\right)_v = \frac{c_v}{T} \tag{9·9}$$

as shown above; and it can be shown in similar fashion (starting from the first of Eqs. 9·6) that the fourth derivative can be evaluated as

$$\left(\frac{\partial s}{\partial T}\right)_P = \frac{c_p}{T} \tag{9·10}$$

The two remaining derivatives can be expressed as

$$\left(\frac{\partial s}{\partial P}\right)_v = \left(\frac{\partial s}{\partial T}\right)_v\left(\frac{\partial T}{\partial P}\right)_v = \frac{c_v}{T}\left(\frac{\partial T}{\partial P}\right)_v$$

and

$$\left(\frac{\partial s}{\partial v}\right)_P = \left(\frac{\partial s}{\partial T}\right)_P\left(\frac{\partial T}{\partial v}\right)_P = \frac{c_P}{T}\left(\frac{\partial T}{\partial v}\right)_P$$

Thus we have obtained from the Maxwell equations a complete description of the entropy variation of a pure substance in terms of data on P, v, T, c_p, and c_v.

Notice that only *derivatives* of entropy have been expressed in terms of PvT and specific-heat data, so that entropy values can be obtained only by integration. The integration introduces arbitrary functions or "constants of integration" that cannot be evaluated from PvT and specific-heat data alone. The same difficulty is experienced in trying to evaluate u, h, a, or g from PvT and specific-heat data alone. On the other hand, inspection of Eqs. 9·5 through 9·8 shows that at least some properties can be completely evaluated by differentiation alone, and thus no arbitrary functions will be involved if we start with data on certain groups of properties. For example, if we have complete uvs data, we determine P and T by

$$P = -\left(\frac{\partial u}{\partial v}\right)_s \qquad T = \left(\frac{\partial u}{\partial s}\right)_v \tag{9·5}$$

Then

$$h \equiv u + Pv = u - \left(\frac{\partial u}{\partial v}\right)_s v$$

$$a \equiv u - Ts = u - \left(\frac{\partial u}{\partial s}\right)_v s$$

$$g \equiv h - Ts = u - \left(\frac{\partial u}{\partial v}\right)_s v - \left(\frac{\partial u}{\partial s}\right)_v s$$

Thus we can evaluate P, T, h, a, and g completely from uvs data alone. The uvs relation for a substance,

$$f(u, v, s) = 0$$

is called a *characteristic function*. A characteristic function is one from which all properties of a substance can be obtained by differentiation alone, so that no arbitrary functions that require supplementary data for their evaluation are introduced. Other characteristic functions are

$$f(h, P, s) = 0$$
$$f(a, v, T) = 0$$
$$f(g, P, T) = 0$$

As noted earlier, $f(P, v, T)$ is not a characteristic function. Unfortunately, the formulation of the properties of a substance cannot be started from a characteristic function because in none of them are all three properties directly measurable. The usefulness of the characteristic functions is that once any one of them is formulated for a substance, all properties can then be determined without the use of additional data.

Example 9·1. Joule's law states that for an ideal gas $(\partial u/\partial v)_T = 0$. Prove this.

Solution. We must express $(\partial u/\partial v)_T$ in terms of P, v, and T and their derivatives so that we can evaluate it from $Pv = RT$. Notice that Eq. 9·5 is an expression for $(\partial u/\partial v)_s$. $(\partial u/\partial v)_T$ can be expressed in terms of $(\partial u/\partial v)_s$ and other quantities by (see Appendix B for derivation)

$$\left(\frac{\partial u}{\partial v}\right)_T = \left(\frac{\partial u}{\partial v}\right)_s + \left(\frac{\partial u}{\partial s}\right)_v \left(\frac{\partial s}{\partial v}\right)_T$$

The three derivatives on the right-hand side can be expressed as

$$\left(\frac{\partial u}{\partial v}\right)_s = -P \tag{9·5}$$

$$\left(\frac{\partial u}{\partial s}\right)_v = T \tag{9·5}$$

$$\left(\frac{\partial s}{\partial v}\right)_T = \left(\frac{\partial P}{\partial T}\right)_v \tag{9·3}$$

so that

$$\left(\frac{\partial u}{\partial v}\right)_T = -P + T\left(\frac{\partial P}{\partial T}\right)_v$$

For an ideal gas, $Pv = RT$ and $(\partial P/\partial T)_v = R/v$.

$$\left(\frac{\partial u}{\partial v}\right)_T = -P + T\frac{R}{v} = -P + P = 0$$

9·2 The Clapeyron Equation

As an example of the use of the Maxwell equations in correlating physical properties, we will now derive the Clapeyron equation which indicates clearly certain property relationships during phase changes of a pure substance.

For two phases of a pure substance existing together in equilibrium, pressure and temperature are dependent only on each other. Therefore $(\partial P/\partial v)_T = 0$, so that the general expression

$$\frac{dP}{dT} = \left(\frac{\partial P}{\partial T}\right)_v + \left(\frac{\partial P}{\partial v}\right)_T \frac{dv}{dT}$$

reduces to

$$\frac{dP}{dT} = \left(\frac{\partial P}{\partial T}\right)_v$$

Also, if s' and v' represent the properties of one phase and s'' and v'' those of the other phase,

$$\left(\frac{\partial s}{\partial v}\right)_T = \frac{s'' - s'}{v'' - v'}$$

Thus the third Maxwell equation

$$\left(\frac{\partial P}{\partial T}\right)_v = \left(\frac{\partial s}{\partial v}\right)_T \tag{9·3}$$

becomes, for the case of two phases of a pure substance coexisting in equilibrium,

$$\frac{dP}{dT} = \frac{s'' - s'}{v'' - v'} \tag{a}$$

The equation

$$T\,ds = dh - v\,dP \tag{8·3}$$

for the case of a phase change at constant pressure (and, consequently, at constant temperature) becomes

$$T(s'' - s') = (h'' - h') \tag{b}$$

The quantity $(h'' - h')$ is the latent heat of the phase transformation. Combining Eqs. a and b,

$$\frac{dP}{dT} = \frac{h'' - h'}{T(v'' - v')} \tag{9·11}$$

This equation is called the Clapeyron equation and is useful because it relates three readily measurable properties: the slope of the saturation pressure–temperature line as shown on Fig. 9·2, the latent heat, and the change in volume during a phase transformation. Thus it can be used in checking the consistency of measurements or for obtaining information on one of these properties from data on the other two.

We have noticed earlier (Sec. 3·2) that the freezing temperature of water decreases as the pressure increases. We have also noted the unusual behavior of water in expanding as it freezes. Now we see that these two characteristics are related by the Clapeyron equation as it applies to the liquid–solid transformation:

$$\frac{dP}{dT} = \frac{h_f - h_i}{T(v_f - v_i)} = \frac{h_{if}}{Tv_{if}}$$

Since $h_{if} > 0$ and for water $v_{if} < 0$, the slope of the fusion line on a phase (PT) diagram must be negative. We already knew this about water from independent observations; but the Clapeyron equation now permits us to generalize: The freezing temperature of *any* substance that expands on freezing is lowered as the pressure is increased.

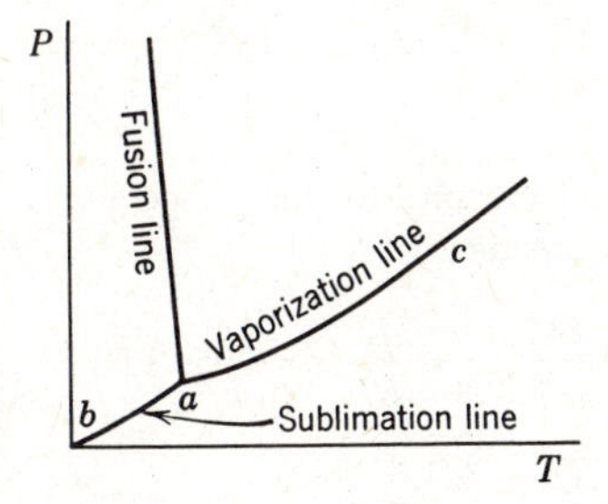

Figure 9·2. Phase diagram for water.

Because the latent heats h_{fg} and h_{ig} and the volume changes v_{fg} and v_{ig} are all positive for all substances, the sublimation and vaporization curves have positive slopes on the phase diagrams of all substances.

Example 9·2. Show that on a phase diagram for water the sublimation line and the vaporization line do not have the same slope at the triple point (i.e., that in Fig. 9·2 line *a-c* is not simply a continuation of line *b-a*).

Solution. We must prove that dP/dT has different values at the triple point for lines *b-a* and *a-c*. The difference between the slopes can be evaluated by means of the Clapeyron equation as follows:

$$\left(\frac{dP}{dT}\right)_{b\text{-}a} - \left(\frac{dP}{dT}\right)_{a\text{-}c} = \frac{s_{ig}}{v_{ig}} - \frac{s_{fg}}{v_{fg}}$$

At the triple state, $s_{ig} = s_{if} + s_{fg}$ and $v_{ig} = v_{if} + v_{fg}$, so that the difference in slopes is

$$\left(\frac{dP}{dT}\right)_{b\text{-}a} - \left(\frac{dP}{dT}\right)_{a\text{-}c} = \frac{(s_{if} + s_{fg})v_{fg} - s_{fg}(v_{if} + v_{fg})}{v_{fg}(v_{if} + v_{fg})}$$

$$= \frac{s_{if}v_{fg} - s_{fg}v_{if}}{v_{fg}(v_{if} + v_{fg})}$$

Since v_{if} is very small compared with v_{fg} while s_{if} is smaller than s_{fg} to a lesser degree, this reduces to approximately

$$\left(\frac{dP}{dT}\right)_{b\text{-}a} - \left(\frac{dP}{dT}\right)_{a\text{-}c} \approx \frac{s_{if}}{v_{fg}}$$

Both quantities included in the right-hand side are positive, so the slope of the sublimation line is greater than that of the vaporization line at the triple point.

Example 9·3. The following data are obtained from the steam tables.

Saturation Temperature, °C	Saturation Pressure, kPa	Specific Volume Liquid v_f, m³/kg	Specific Volume Vapor v_g, m³/kg
169	772.7	0.0011130	0.2485
170	791.7	0.0011143	0.2428
171	811.0	0.0011155	0.2373

Compute the approximate latent heat value at 170°C.

Solution. Assuming a linear variation between temperature and pressure for the temperatures specified, the latent heat may be found from the Clapeyron equation:

$$h_{fg} = v_{fg}T\frac{dp}{dT} \approx v_{fg}T\frac{\Delta P}{\Delta T} = (0.2428 - 0.0011143)\,443\,\frac{811.0 - 772.7}{171 - 169} = 2050 \text{ kJ/kg}$$

The steam-table value is 2049.5 kJ/kg.

9·3 The Joule–Thomson Coefficient

A useful property of substances that, like specific heats, is defined as a partial derivative is the Joule–Thomson coefficient μ:

$$\mu \equiv \left(\frac{\partial T}{\partial P}\right)_h$$

It is important partly because it can be measured more accurately and more readily than can certain other useful properties that are directly related to it.

In order to measure the Joule–Thomson coefficient of a liquid or a gas, the fluid is allowed to expand steadily through a porous plug as shown schematically in Fig. 9·3. (The plug may be made of steel wool, for example.) The plug and adjacent piping are thermally insulated, so $q = 0$. No work is done, and the changes in potential and kinetic energy can be made negligibly small. The first law as applied to any steady-flow system,

$$q + u_1 + \frac{V_1^2}{2} + gz_1 = u_2 + \frac{V_2^2}{2} + gz_2 + w + P_2v_2 - P_1v_1 \tag{2·9}$$

becomes for this system

$$u_1 + P_1v_1 = u_2 + P_2v_2$$

or,

$$h_1 = h_2$$

The inlet pressure and temperature (P_1 and T_1) are held constant. The downstream pressure is held at several different values successively, and at each of these T_2 is measured. Then T_2 is plotted against P_2 as in Fig. 9·4*a*. Each plotted point represents a state for which the enthalpy is equal to h_1. If enough measurements are made, we can pass a curve through the points and with reasonable confidence label the curve a constant-enthalpy line, (Fig. 9·4*b*). The slope of this line at any point is $(\partial T/\partial P)_h$, the Joule–Thomson coefficient. In order to obtain the Joule–Thomson coefficient at various pressures and temperatures it is necessary to use several combinations of initial pressure and temperature, each combination providing a different constant h line as shown in Fig. 9·5. Values of the Joule–Thomson coefficient for five gases are given in Fig. 9·6.

The broken line in Fig. 9·5 passes through the maximum temperature points of the constant-enthalpy lines. It is called the *inversion line*. In the region to the left of this line, the Joule–Thomson coefficient is positive; in the region to the right of the inversion line, the Joule–Thomson coefficient is negative. Expansions that occur to the left of the inversion line between states of equal enthalpy (such as from m to b) result in a decrease in

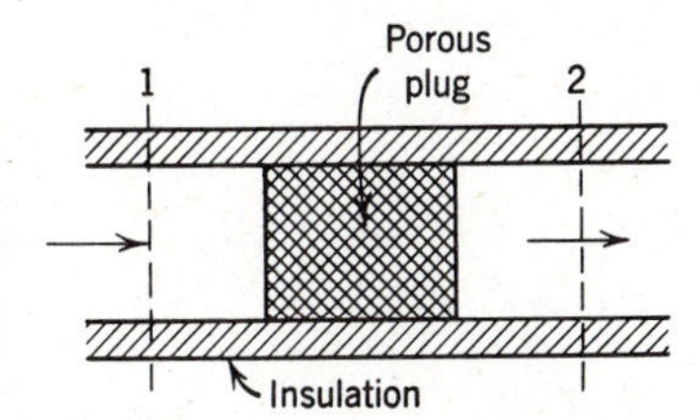

Figure 9·3. Schematic diagram of apparatus for measuring Joule–Thomson coefficient.

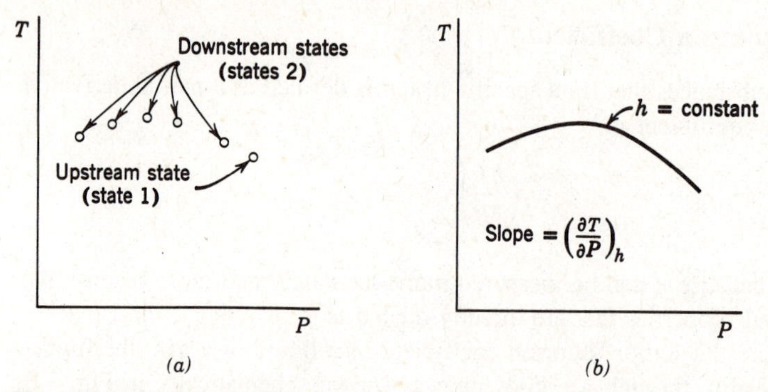

Figure 9·4. (*a*) Data from Joule–Thomson expansions, and (*b*) the resulting constant-enthalpy line.

temperature, while expansions (such as *a* to *m*) occurring to the right of the inversion line result in a temperature rise. Occasionally the Joule-Thomson expansion is used for refrigeration. Figure 9·5 shows that this is possible only when the initial temperature is lower than the maximum inversion temperature or the temperature at which the upper part of the inversion line cuts the axis. For example, starting from point *a*, a gas will experience an overall drop in temperature if its pressure is lowered sufficiently—say to point *c*. A gas initially at point *r* or *s* cannot be cooled by a Joule–Thomson expansion (i.e., throttling), no matter how low the downstream pressure is. The maximum inversion temperatures for several gases are given in Table 9·1.

The Joule–Thomson coefficient of a substance can be calculated from the PvT relation and data on c_p. Since entropy is a property and for a pure substance can be a function of P and T, its differential is exact and can be written as

$$ds = \left(\frac{\partial s}{\partial T}\right)_P dT + \left(\frac{\partial s}{\partial p}\right)_T dP$$

Substituting this value for ds into Eq. 8·3,

$$dh = T\,ds + v\,dP \tag{8·3}$$

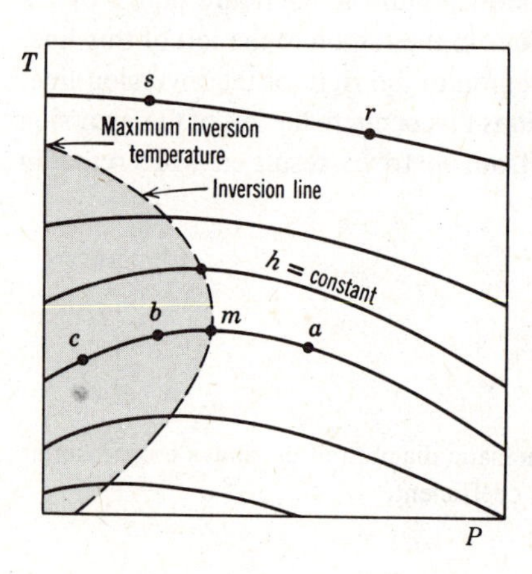

Figure 9·5. Constant-enthalpy lines and the inversion line for a substance.

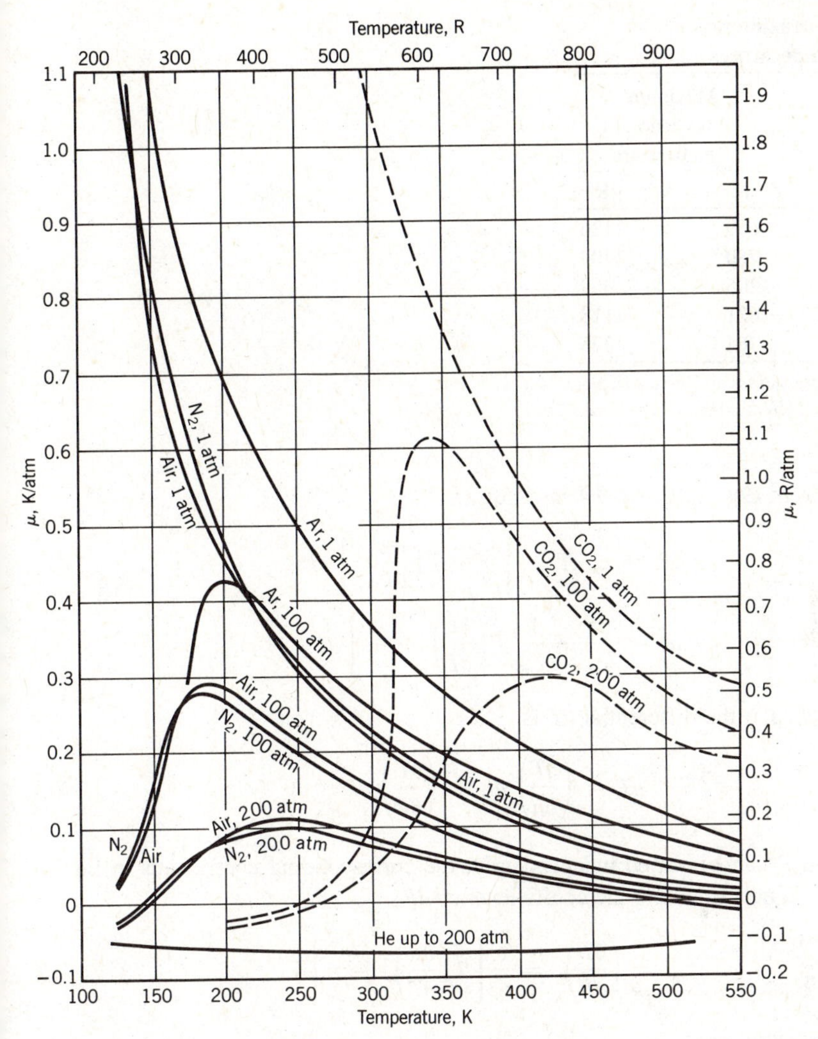

Figure 9·6. Joule–Thomson coefficients. Data from *Smithsonian Physical Tables*, 9th revised ed., The Smithsonian Institution, Washington, D.C., 1954.

we have

$$dh = T\left(\frac{\partial s}{\partial T}\right)_P dT + \left[T\left(\frac{\partial s}{\partial P}\right)_T + v\right] dP \qquad \text{(a)}$$

We noted in Sec. 9·1 that

$$\left(\frac{\partial s}{\partial T}\right)_P = \frac{c_p}{T} \qquad (9\cdot10)$$

and the fourth Maxwell equation is

$$\left(\frac{\partial s}{\partial P}\right)_T = -\left(\frac{\partial v}{\partial T}\right)_P \qquad (9\cdot4)$$

TABLE 9·1 Maximum Inversion Temperatures

	Maximum Inversion Temperature	
Gas	K	R
Air	659	1186
Argon	780	1404
Hydrogen	205	369
Nitrogen	621	1118
Oxygen	764	1375

Source: *American Institute of Physics Handbook,* 3rd. ed., McGraw-Hill, New York, 1972.

Substituting from Eqs. 9·10 and 9·4 into Eq. a,

$$dh = c_p\, dT + \left[v - T\left(\frac{\partial v}{\partial T}\right)_P\right] dP$$

or

$$dT = \frac{dh}{c_p} + \frac{1}{c_p}\left[T\left(\frac{\partial v}{\partial T}\right)_P - v\right] dP \tag{b}$$

Since $T = f(h, P)$, the differential dT is

$$dT = \left(\frac{\partial T}{\partial h}\right)_P dh + \left(\frac{\partial T}{\partial P}\right)_h dP \tag{c}$$

Because h and P can be varied independently, the corresponding coefficients in Eqs. b and c must be equal. Equating the coefficients of dP,

$$\mu = \left(\frac{\partial T}{dP}\right)_h = \frac{1}{c_p}\left[T\left(\frac{\partial v}{\partial T}\right)_P - v\right] \tag{d}$$

By inspection this can be written also as

$$\mu = \frac{T^2}{c_p}\left[\frac{\partial(v/T)}{\partial T}\right]_P \tag{e}$$

Equations d and e give μ in terms of the equation of state (PvT relation) and c_p.

For an ideal gas,

$$\mu = \left(\frac{\partial T}{\partial P}\right)_h = \frac{T^2}{c_p}\left[\frac{\partial(v/T)}{\partial T}\right]_P = \frac{T^2}{c_p}\left[\frac{\partial(R/P)}{\partial T}\right]_P = 0$$

The Joule–Thomson coefficient for a substance that follows the equation of state $Pv = RT$ is zero, and such a substance experiences no temperature change when it is throttled (that is, when $\Delta h = 0$).

Another coefficient which is useful and fairly easy to measure is the constant-temperature coefficient c defined by

$$c \equiv \left(\frac{\partial h}{\partial P}\right)_T$$

For its measurement, the fluid is expanded slowly through a porous plug and heat is added or removed (by means of an electric heating coil or a flow of coolant) to maintain the inlet and outlet temperatures equal. The amount of heat required to hold the temperature constant is measured so that enthalpy changes can be computed from the first law.

The general relationship among the partial derivatives of x, y, and z when the three are functionally related is

$$\left(\frac{\partial x}{\partial y}\right)_z \left(\frac{\partial y}{\partial z}\right)_x \left(\frac{\partial z}{\partial x}\right)_y = -1$$

Applying this to T, P, and h, we see the relationship among μ, c, and c_p:

$$\left(\frac{\partial T}{\partial P}\right)_h \left(\frac{\partial P}{\partial h}\right)_T \left(\frac{\partial h}{\partial T}\right)_P = -1$$

$$\frac{\mu c_p}{c} = -1$$

$$\mu = -\frac{c}{c_p} \tag{f}$$

This is a convenient relationship for checking the consistency of experimental data on these properties.

Equation f shows that for an ideal gas, since $\mu = 0$ and c_p is finite,

$$c = \left(\frac{\partial h}{\partial P}\right)_T = 0$$

which shows clearly that the enthalpy of an ideal gas is a function of temperature only.

We shall consider here one other application of the Joule–Thomson effect: its use in relating other temperature scales to the thermodynamic temperature scale. The thermodynamic temperature scale has been defined in terms of the performance of reversible engines, but of course reversible engines cannot be used in practice as thermometers. We have also seen that the ideal-gas temperature scale agrees with the thermodynamic scale; but an ideal gas does not exist and therefore cannot be used in a thermometer, even though the behavior of real gases can be extrapolated to zero pressure at which real gases behave in accordance with the ideal-gas equation of state. The problem is to find a means of relating temperatures on actual, practical thermometers to those on the thermodynamic scale.

The temperature T in Eq. d,

$$\left(\frac{\partial T}{\partial P}\right)_h = \frac{1}{c_p}\left[T\left(\frac{\partial v}{\partial T}\right)_P - v\right] \tag{d}$$

is the thermodynamic temperature. Let θ be the temperature indicated by some thermometer such as a real-gas thermometer or a mercury-in-glass thermometer. T and θ are functionally related; for each value of one there is a unique value of the other. In order to make the derivatives in Eq. d involve θ instead of T, we note that

$$\left(\frac{\partial T}{\partial P}\right)_h = \left(\frac{\partial \theta}{\partial P}\right)_h \frac{dT}{d\theta} \qquad \frac{1}{c_p} = \left(\frac{\partial T}{\partial h}\right)_P = \left(\frac{\partial \theta}{\partial h}\right)_P \frac{dT}{d\theta} \qquad \left(\frac{\partial v}{\partial T}\right)_P = \left(\frac{\partial v}{\partial \theta}\right)_P \frac{d\theta}{dT}$$

Making these substitutions in Eq. d,

$$\left(\frac{\partial \theta}{\partial P}\right)_h \frac{dT}{d\theta} = \left(\frac{\partial \theta}{\partial h}\right)_P \frac{dT}{d\theta}\left[T\left(\frac{\partial v}{\partial \theta}\right)_P \frac{d\theta}{dT} - v\right]$$

and rearranging gives

$$\frac{dT}{T} = \frac{(\partial v/\partial \theta)_P \, d\theta}{(\partial \theta/\partial P)_h(\partial h/\partial \theta)_P + v}$$

Letting T_0 and θ_0 be corresponding values of T and θ,

$$\int_{T_0}^{T} \frac{dT}{T} = \ln \frac{T}{T_0} = \int_{\theta_0}^{\theta} \frac{(\partial v/\partial \theta)_P \, d\theta}{(\partial \theta/\partial P)_h(\partial h/\partial \theta)_P + v}$$

All terms in the integrand of the right-hand side of this equation can be measured, and the integral is a function of T (and hence θ) only. Therefore, the right-hand side can be integrated numerically or graphically to provide a relationship between θ and T, and this is just what we are looking for. Notice that we placed no restrictions on the substance used for the measurement of the quantities in the integrand. Also notice that the three partial derivatives, $(\partial v/\partial \theta)_P$, $(\partial v/\partial P)_h$, and $(\partial h/\partial \theta)_P$, are, respectively, the coefficient of thermal expansion times v, the Joule–Thomson coefficient, and the specific heat at constant pressure, all defined in terms of θ instead of T. By measuring these three properties of a substance over a range of temperature, we are able to relate our laboratory thermometer readings to thermodynamic temperature scale values.

9·4 General Equations for Changes in Entropy, Internal Energy, and Enthalpy in Terms of P, v, T, and Specific Heats

In this section we restrict our attention to pure substances in the absence of any effects of electricity, magnetism, anisotropic stress, or surface tension. The absence of these effects defines a *simple compressible substance*. Throughout the literature of thermodynamics, *pure substance* often means *pure simple compressible substance*. This sense is used throughout this chapter, although we occasionally add the reminder, "in the absence of the effects of electricity, magnetism, etc." Notice that liquids and solids, which are much less compressible than gases, can also be simple compressible substances.

The entropy of a pure substance may be expressed as a function of T and v so that

$$ds = \left(\frac{\partial s}{\partial T}\right)_v dT + \left(\frac{\partial s}{\partial v}\right)_T dv$$

From Sec. 9·1,

$$\left(\frac{\partial s}{\partial T}\right)_v = \frac{c_v}{T} \tag{9·9}$$

and the third Maxwell equation is

$$\left(\frac{\partial s}{\partial v}\right)_T = \left(\frac{\partial P}{\partial T}\right)_v \tag{9·3}$$

Making these two substitutions in the expression for ds gives

$$ds = c_v \frac{dT}{T} + \left(\frac{\partial P}{\partial T}\right)_v dv \tag{9·12}$$

By starting with $s = f(P, T)$, it can be shown in a similar manner that

$$ds = c_p \frac{dT}{T} - \left(\frac{\partial v}{\partial T}\right)_P dP \tag{9·13}$$

Equations 9·12 and 9·13 are general equations for the entropy change of a pure substance. Notice that they can be integrated if specific-heat–temperature and PvT data are available. They are useful in obtaining general expressions for du and dh, as will now be shown.

For any process between equilibrium states of a pure substance,

$$du = T\,ds - P\,dv \tag{8·3}$$

Substituting into this equation the value of ds from Eq. 9·12,

$$du = c_v\,dT + \left[T\left(\frac{\partial P}{\partial T}\right)_v - P\right] dv \tag{9·14}$$

Substituting into

$$dh = T\,ds + v\,dP \tag{8·3}$$

the value of ds from Eq. 9·13,

$$dh = c_p\,dT - \left[T\left(\frac{\partial v}{\partial T}\right)_P - v\right] dP \tag{9·15}$$

Equations 9·14 and 9·15 are general equations for du and dh of a pure substance. They too can be integrated if specific-heat–temperature and PvT data are available. Notice that integration of Eqs. 9·12, 9·13, 9·14, and 9·15 can give only *changes* in s, u, and h. Absolute values cannot be obtained from PvT and specific-heat data alone.

These general equations make possible the generation of extensive and accurate property tables from relatively few measurements. As an illustration of this point, consider Fig. 9·7 which is a skeleton Ts diagram for some substance. All properties (h, u, s, g, etc.) on line 1-2-3-4-5 can be determined from measurements of the following quantities:

1. c_p of the liquid between states 1 and 2 where 1 is a base state at which values of h and s are assigned arbitrarily

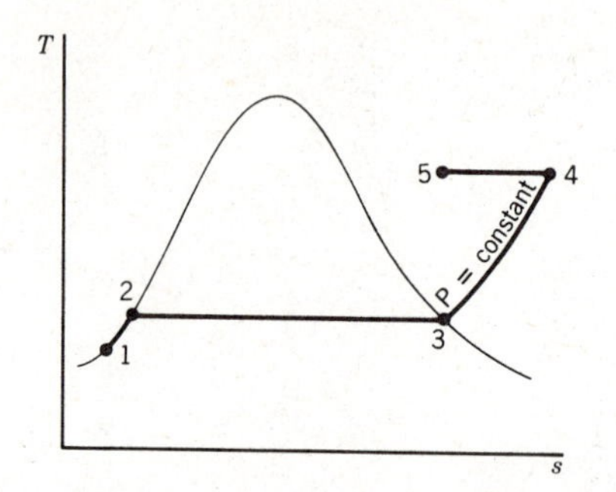

2. the saturation temperature and h_{fg} for the pressure of states 2, 3, and 4
3. c_p of the vapor along the constant pressure line from 3 to 4
4. PvT data in the region around the constant temperature line between 4 and 5

If the base state 1 is a saturated liquid state, the process 1-2 can be taken as an isentropic process from P_1 to P_2 followed by a constant-pressure process to state 2. For all liquids at low pressures, Δh and ΔT for the isentropic process are very small. For the constant-pressure part of the process, Eqs. 9·15 and 9·13 reduce to

$$\Delta h = \int_{T_1}^{T_2} c_p \, dT$$

$$\Delta s = \int_{T_1}^{T_2} c_p \frac{dT}{T}$$

From the measured values of the saturation temperature and h_{fg} at P_2, s_{fg} can be determined from a Tds equation such as Eq. 8·3.

For the constant-pressure process 3-4, Δh and Δs can again be obtained from Eqs. 9·15 and 9·13, because the c_p data along this constant-pressure line make numerical integration possible.

For the constant-temperature process 4-5, Eqs. 9·15 and 9·13 simplify to

$$\Delta h = \int_{P_4}^{P_5} \left[T \left(\frac{\partial v}{\partial T} \right)_P - v \right] dP$$

$$\Delta s = - \int_{P_4}^{P_5} \left(\frac{\partial v}{\partial T} \right)_P dP$$

The partial derivative in these equations can be evaluated numerically or graphically from the PvT data in the vicinity of process 4-5.

The result of these calculations is that all properties at a state 5 can be determined from properties at a state 1 and a small amount of experimental data. The approach outlined here is a simplified one. Much more extensive calculations would be made to ensure accuracy and consistency in published tables. For example, several paths might be used between states 1 and 5. These additional paths would require measurements of saturation temperature and h_{fg} at various pressures. If v_{fg} is also measured, the Clapeyron equation provides a check on the consistency of these measurements.

Published tables of properties usually include a listing of data sources and an explanation of the methods used in generating the table. Although these explanations are often at a level beyond that of an introductory course in thermodynamics, a perusal of them shows the key role of the general equations as presented in this chapter.

Example 9·4. Develop an equation for the enthalpy change of a gas which follows the ideal-gas equation of state.

Solution. Differentiating the ideal-gas equation of state,

$$\left(\frac{\partial v}{\partial T}\right)_P = \frac{R}{P}$$

Substituting this value into the general expression for dh,

$$dh = c_p\,dT - \left[T\left(\frac{\partial v}{\partial T}\right)_P - v\right] dP \tag{9·15}$$

gives

$$dh = c_p\,dT - \left[\frac{RT}{P} - v\right] dP = c_p\,dT - 0$$

Integration gives

$$\Delta h = \int c_p\,dT$$

which can be evaluated if the c_pT relation is known.

Example 9·5. Develop an expression for the change in internal energy of a gas which follows the equation of state

$$P = \frac{RT}{v - b} - \frac{a}{v^2}$$

Solution. Differentiating the equation of state,

$$\left(\frac{\partial P}{\partial T}\right)_v = \frac{R}{v - b}$$

Substituting this value into the general expression for du,

$$du = c_v\,dT + \left[T\left(\frac{\partial P}{\partial T}\right)_v - P\right] dv \tag{9·14}$$

gives

$$du = c_v\,dT + \left[\frac{RT}{v - b} - P\right] dv$$

Substituting $[RT/(v - b) - a/v^2]$ for P gives

$$du = c_v\,dT + \frac{a}{v^2}\,dv$$

Integrating,

$$u_2 - u_1 = \int_1^2 c_v\,dT - a\left(\frac{1}{v_2} - \frac{1}{v_1}\right)$$

Knowledge of the c_vT relation is needed for the integration of $\int c_v\,dT$.

9·5 Specific Heat Relations

General equations for the difference between specific heats ($c_p - c_v$) and for the ratio of specific heats c_p/c_v will now be derived.

Equating the two general equations for ds (Eqs. 9·12 and 9·13):

$$c_v \frac{dT}{T} + \left(\frac{\partial P}{\partial T}\right)_v dv = c_p \frac{dT}{T} - \left(\frac{\partial v}{\partial T}\right)_P dP$$

Solving for dT,

$$dT = \frac{T(\partial P/\partial T)_v}{c_p - c_v} dv + \frac{T(\partial v/\partial T)_P}{c_p - c_v} dP$$

The temperature may be expressed as a function of v and P. Hence,

$$dT = \left(\frac{\partial T}{\partial v}\right)_P dv + \left(\frac{\partial T}{\partial P}\right)_v dP$$

Equating the coefficients in the last two equations results in the two identical relations

$$\left(\frac{\partial T}{\partial v}\right)_P = \frac{T(\partial P/\partial T)_v}{c_p - c_v} \quad \text{and} \quad \left(\frac{\partial T}{\partial P}\right)_v = \frac{T(\partial v/\partial T)_P}{c_p - c_v}$$

Thus

$$c_p - c_v = T\left(\frac{\partial P}{\partial T}\right)_v \left(\frac{\partial v}{\partial T}\right)_P \tag{9·16}$$

The difference between c_p and c_v can thus be determined for any substance if the equation of state (PvT relation) is known or if the two partial derivatives can be measured. In the case of solids, it is difficult to measure $(\partial P/\partial T)_v$, so a different form of Eq. 9·16 is desired.

The general relation

$$\left(\frac{\partial x}{\partial y}\right)_z \left(\frac{\partial y}{\partial z}\right)_x \left(\frac{\partial z}{\partial x}\right)_y = -1$$

when applied to $f(P, v, T)$ is

$$\left(\frac{\partial P}{\partial T}\right)_v \left(\frac{\partial T}{\partial v}\right)_P \left(\frac{\partial v}{\partial P}\right)_T = -1$$

or

$$\left(\frac{\partial P}{\partial T}\right)_v = -\left(\frac{\partial P}{\partial v}\right)_T \left(\frac{\partial v}{\partial T}\right)_P$$

Making this substitution in Eq. 9·16,

$$c_p - c_v = -T\left(\frac{\partial P}{\partial v}\right)_T \left(\frac{\partial v}{\partial T}\right)_P^2 \tag{9·17}$$

The coefficient of volume expansion β and the isothermal compressibility κ_T are defined as

$$\beta \equiv \frac{1}{v}\left(\frac{\partial v}{\partial T}\right)_P \qquad \text{and} \qquad \kappa_T \equiv -\frac{1}{v}\left(\frac{\partial v}{\partial P}\right)_T$$

Making these substitutions in Eq. 9·17 gives

$$c_p - c_v = \frac{Tv\beta^2}{\kappa_T} \tag{9·17}$$

T, v, and κ_T are always positive, and β^2 is either positive or zero; so c_p can never be less than c_v. When $\beta = 0$ (as for water at 1 atm and about 4°C), $c_p = c_v$. For solids, c_v is difficult to measure, so Eq. 9·17 is used to calculate c_v from c_p and the other properties.

In order to obtain a general expression for the ratio of the specific heats c_p/c_v, let us combine

$$\frac{c_p}{T} = \left(\frac{\partial s}{\partial T}\right)_P \tag{9·10}$$

and

$$\frac{c_v}{T} = \left(\frac{\partial s}{\partial T}\right)_v \tag{9·9}$$

to give

$$k = \frac{c_p}{c_v} = \frac{(\partial s/\partial T)_P}{(\partial s/\partial T)_v}$$

In order to obtain derivatives that appear in the Maxwell equations, expand both the numerator and the denominator to give

$$k = \frac{c_p}{c_v} = \frac{(\partial s/\partial v)_P(\partial v/\partial T)_P}{(\partial s/\partial P)_v(\partial P/\partial T)_v}$$

Each of the four partial derivatives here appears in one of the Maxwell equations. Substituting for each one from the proper Maxwell equation gives

$$k = \frac{c_p}{c_v} = \frac{-(\partial P/\partial T)_s(\partial s/\partial P)_T}{-(\partial v/\partial T)_s(\partial s/\partial v)_T}$$

$$= \left(\frac{\partial P}{\partial v}\right)_s\left(\frac{\partial v}{\partial P}\right)_T \tag{9·18}$$

If we define isentropic compressibility κ_s by

$$\kappa_s \equiv -\frac{1}{v}\left(\frac{\partial v}{\partial P}\right)_s$$

we have

$$k = \frac{c_p}{c_v} = \frac{\kappa_T}{\kappa_s} \tag{9·18}$$

This relationship holds for gases, liquids, and solids.

Many methods have been developed for obtaining approximate values of the specific heats of substances from other data. A knowledge of some of these is often helpful. In Sec. 4·8 it was shown that for monatomic and diatomic gases which can be treated as ideal gases the C_p values are approximately $5R_u/2$ and $7R_u/2$, respectively, and the C_v values are approximately $3R_u/2$ and $5R_u/2$, respectively.

Figure 9·8 shows C_v versus temperature for several solid substances. Two things are readily noticed: (1) As temperature increases, C_v approaches a value of about 25 kJ/kmol·K (6 B/lbmol·R) for all the substances, and (2) the curves are approximately the same for all substances except for a temperature scale factor.

The first of these observations was reported in 1819 by Dulong and Petit and is sometimes referred to as *the law of Dulong and Petit: For elements in the solid phase at high temperatures, the molar specific heat at constant volume is equal to approximately 25 kJ/kmol·K (6 B/lbmol·R).* There is a question as to what is meant by "high temperature." Figure 9·8 shows that for several solids the Dulong and Petit value is accurate at normal room temperature but that for diamond it is far too high at this temperature. This point was cleared up by Peter Debye, who by use of quantum statistical mechanics showed that for all isotropic solids C_v is given very nearly by

$$C_v = 3\,R_u\,f\left(\frac{T}{\Theta}\right)$$

where R_u is the universal gas constant (Yes, R_u is used in connection with solids!), Θ is a constant—called the *Debye temperature*—for each substance, and the function $f(T/\Theta)$ is the same for all substances. A plot of Debye's equation is shown in Fig. 9·9, and values of Θ for several substances are given in Table 9·2. Notice that, as T/Θ increases, C_v

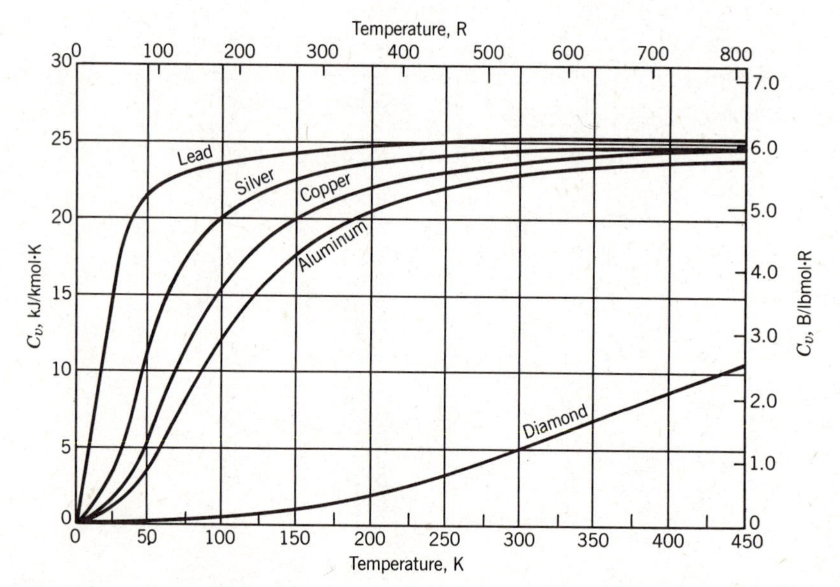

Figure 9·8. C_v for several solids.

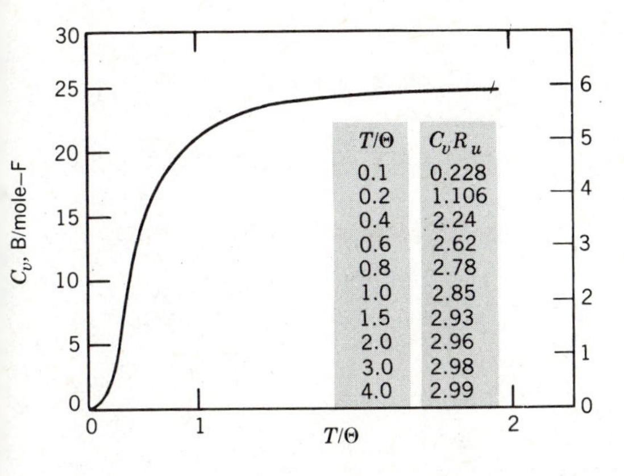

Figure 9·9. Debye's function.

approaches the Dulong and Petit value of approximately 25 kJ/kmol·K, or, more precisely, $3R_u$. As T/Θ approaches zero, Debye's equation becomes

$$C_v \propto T^3$$

and this is often called the *Debye T^3 law*. For T/Θ less than 0.1, Debye's T^3 law is accurate to within about 1 percent or less for isotropic nonmetals. For metals at low temperatures a correction must be made. There is some evidence that the T^3 law applies to all substances at extremely low temperatures.

At very high values of T/Θ, C_v of solids is higher than the $3R_u$ given by Debye's equation and by the law of Dulong and Petit. Some isotropic solids have C_v values well over $3R_u$ even at fairly low temperatures. Examples are ice with $C_v = 4.5R_u$ at 0°C and the element barium with $C_v = 4.7R_u$ at 20°C. Thus it is apparent that Debye's equation (as well as the law of Dulong and Petit which is the limiting case) must be applied with caution. Nevertheless, it is a convenient means of estimating C_v values, and it is an important step in the development of a complete theory of the solid phase.

TABLE 9·2 Debye Temperatures, Θ

Substance	Θ, K	Θ, R
Aluminum	428	770
Cadmium	209	376
Calcium	230	414
Copper	343	617
Diamond	2230	4014
Graphite	420	756
Iron	467	841
Lead	105	189
Mercury	72	130
Silver	225	405
Sodium	158	284
Zinc	327	589

Source: American Institute of Physics Handbook, 3rd. ed., McGraw-Hill, New York, 1972, pp. 4–115, 116.

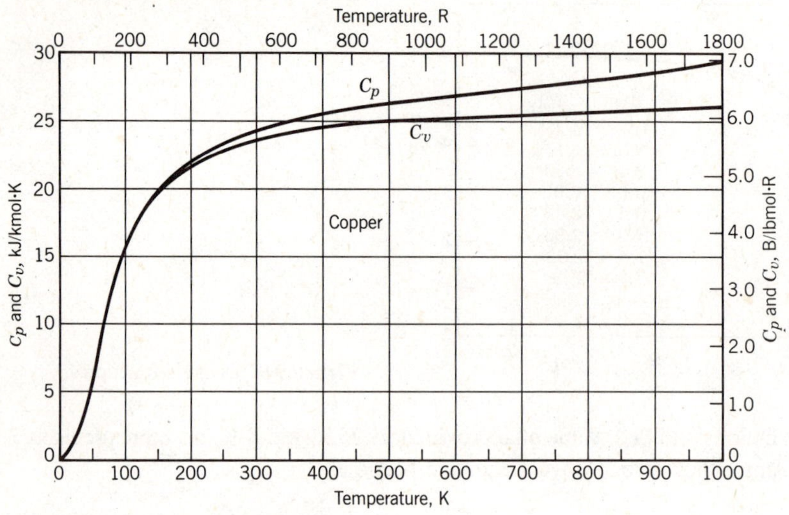

Figure 9·10. C_p and C_v for copper.

Remember that Debye's equation pertains to C_v, not C_p. For solids and liquids, data referred to simply as "specific-heat" values are usually C_p (or c_p) values. The difference between C_p and C_v increases as temperature increases, as shown by Eq. 9·17,

$$C_p - C_v = \frac{T v_N \beta^2}{\kappa_T} \qquad \text{(9·17, molar basis)}$$

Figure 9·10 shows the magnitude of this difference over a wide range of temperature for one solid.

9·6 Ideal-Gas Property Tables

In many engineering applications the behavior of real gases can be adequately represented by the ideal-gas equation of state, $Pv = RT$. For any gas which follows this equation of state, changes in internal energy, enthalpy, and entropy for all processes can be calculated from

$$du = c_v \, dT \qquad (4·2)$$

$$dh = c_p \, dT \qquad (4·3)$$

$$ds = \frac{dh}{T} - \frac{v \, dP}{T} = \frac{c_p \, dT}{T} - R \frac{dP}{P} \qquad (8·3, 4·3)$$

and c_p and c_v are functions of temperature only. The specific-heat–temperature functions for wide ranges of temperature are generally not simple, however, so in order to avoid laborious computations we often use tabulated values of temperature-dependent properties. One valuable collection of ideal-gas properties is the Keenan, Chao, and Kaye *Gas Tables* for air and certain other gases and gas mixtures. A greatly abridged version of the air table is given in the appendix as Table A·5, and a sample of the original table is shown below as Table 9·3. The first two columns give temperatures in kelvins and degrees Celsius. The third and fifth columns are values of h and u, based arbitrarily on $h = 0$ and $u = 0$ at 0 K. Notice that this is a single-argument table, unlike the superheated steam table, for example, in which both pressure and temperature are arguments or independent variables.

Because temperature is the single argument of this gas table, only properties that are functions of temperature alone (such as u and h) can be tabulated. Entropy is not a function of temperature alone, so it cannot be tabulated; but for an ideal gas

$$s = f(P, T)$$

can be put in the form

$$s = \phi(T) + f(P)$$

Then at least the first term on the right-hand side can be tabulated against temperature. We have seen that for a pure simple compressible substance a valuable relationship among properties is

$$T\,ds = dh - v\,dP \tag{8·3}$$

For an ideal gas we can replace dh by $c_p\,dT$ and v by RT/P. Thus

$$s_2 - s_1 = \int_1^2 ds = \int_1^2 \frac{c_p\,dT}{T} - R\int_1^2 \frac{dP}{P}$$

The first term on the right-hand side can be tabulated as a function of temperature only, and the second one can be evaluated from the pressure limits of a process. The symbol ϕ is used for the temperature function that is defined by

$$\phi_2 - \phi_1 \equiv \int_1^2 \frac{c_p\,dT}{T}$$

TABLE 9·3 Properties of Air at Low Pressure

T	t	h	p_r	u	v_r	ϕ
1200	926.85	1277.73	236.69	933.29	1.4552	7.1783
1201		1278.90	237.50	934.18	1.4515	7.1792
1202		1280.08	238.31	935.07	1.4478	7.1802
1203		1281.25	239.12	935.95	1.4441	7.1812
1204		1282.43	239.93	936.84	1.4403	7.1822
1205	931.85	1283.60	240.75	937.73	1.4367	7.1831
1206		1284.78	241.57	938.62	1.4330	7.1841
1207		1285.95	242.39	939.50	1.4293	7.1851
1208		1287.13	243.21	940.39	1.4256	7.1861
1209		1288.30	244.04	941.28	1.4220	7.1870

Source: An excerpt from *Gas Tables* (SI Units), by J. H. Keenan, J. Chao, and J. Kaye, 2nd. ed., Wiley, New York, 1983.

so that the change in entropy between two states is given by

$$s_2 - s_1 = \phi_2 - \phi_1 - R \ln \frac{P_2}{P_1} \tag{9·19}$$

Relationships (Eqs. 4·5) between P and v, between v and T, and between p and T were developed in Sec. 4·6 for a reversible adiabatic or isentropic process of an ideal gas with constant specific heats. For variable specific heats the relationships are not so simple; so tabulated functions are again used. For an *isentropic process* of an ideal gas we have from Eq. 9·19

$$s_2 - s_1 = 0 = \phi_2 - \phi_1 - R \ln \frac{P_2}{P_1}$$

Rearranging, we have, for an *isentropic process*,

$$\frac{P_2}{P_1} = \exp\left(\frac{\phi_2 - \phi_1}{R}\right) = \frac{\exp(\phi_2/R)}{\exp(\phi_1/R)} = \frac{f(T_2)}{f(T_1)}$$

This new function of temperature is called *relative pressure** p_r. By definition,

$$p_r \equiv \exp(\phi/R)$$

Then for an *isentropic process* we have

$$\frac{P_2}{P_1} = \frac{p_{r2}}{p_{r1}}$$

Thus a table of p_r versus temperature gives us the pressure–temperature relationship for isentropic processes.

For any two states at the same entropy the ratio of specific volumes can also be expressed as a ratio of temperature functions. For any two states 1 and 2 of an ideal gas,

$$\frac{v_2}{v_1} = \frac{P_1 T_2}{P_2 T_1}$$

If the states are at the same entropy,

$$\frac{v_2}{v_1} = \frac{P_1 T_2}{P_2 T_1} = \frac{p_{r1}}{p_{r2}}\frac{T_2}{T_1} = \frac{T_2/p_{r2}}{T_1/p_{r1}} = \frac{\text{function of } T_2}{\text{function of } T_1}$$

We now define the *relative specific volume*, $v_r \equiv T/p_r$, so that, for an *isentropic process*,

$$\frac{v_2}{v_1} = \frac{v_{r2}}{v_{r1}}$$

We have noted that the selection of a temperature at which the enthalpy or internal energy is zero is entirely arbitrary. Also, in the tabulation of ϕ, p_r, and v_r, various additive terms or multipliers may be introduced that have no effect on the use of the table. They

*Notice that p_r is dimensionless and is not a pressure. The term *relative pressure* is consequently somewhat misleading. Also, be careful not to confuse *relative pressure* p_r with *reduced pressure* P_R.

must be considered, however, in verifying tabular values by means of defining equations such as $p_r \equiv \exp(\phi/R)$. These terms or factors should be defined in the documentation that accompanies each table.

The following examples illustrate the use of the Keenan, Chao, and Kaye *Gas Tables*.

Example 9·6. Air in a closed, rigid tank is initially at 100 kPa, 5°C. It is heated until its pressure is 400 kPa. The lowest temperature in the surroundings is 5°C. Determine the available part of the heat added.

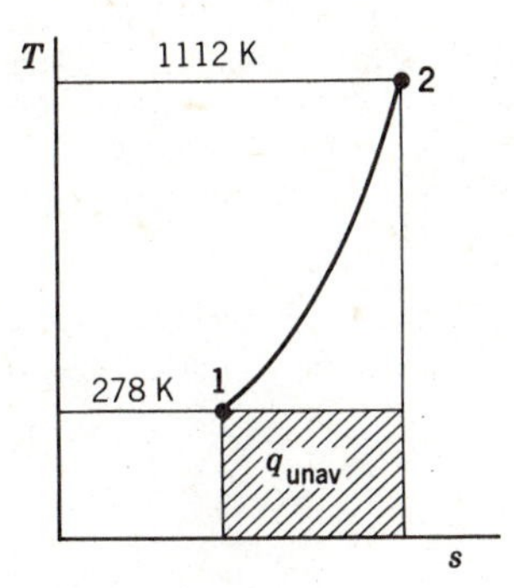

Example 9·6.

Solution. The ideal-gas equation of state shows that for a constant-volume process the pressure and absolute temperature are directly proportional to each other. The final temperature is then

$$T = T_1 \frac{P_2}{P_1} = 278 \frac{400}{100} = 1112 \text{ K}$$

The system under consideration is comprised of the air trapped in the tank. Application of the first law to this closed system and substitution of internal energy values from the Keenan, Chao, and Kaye table for air gives

$$q = u_2 - u_1 + w = 855.84 - 198.55 = 657.29 \text{ kJ/kg}$$

The change in entropy is

$$s_2 - s_1 = \phi_2 - \phi_1 - R \ln \frac{P_2}{P_1} = 7.0894 - 5.6251 - 0.287 \ln \frac{400}{100}$$

$$= 1.066 \text{ kJ/kg·K}$$

and the available part of the heat added is

$$q_{av} = q - q_{unav} = q - T_0(s_2 - s_1) = 657.29 - 278(1.066) = 361 \text{ kJ/kg}$$

Example 9·7. Air undergoes an isentropic process during which the temperature increases from 300 to 600 K. The initial pressure is 90 kPa. Compute (*a*) the final specific volume and (*b*) the final pressure.

Solution. **(*a*)** From the ideal-gas equation of state we can compute the initial specific volume as

$$v_1 = \frac{RT_1}{P_1} = \frac{0.287(300)}{90} = 0.957 \text{ m}^3/\text{kg}$$

Then, for this isentropic process we have

$$v_2 = v_1 \frac{v_{r2}}{v_{r1}} = 0.957 \frac{10.623}{62.393} = 0.163 \text{ m}^3/\text{kg}$$

(*b*) The final pressure for this isentropic is

$$P_2 = P_1 \frac{p_{r2}}{p_{r1}} = 90 \frac{16.212}{1.3801} = 1060 \text{ kPa}$$

(Of course, after either P_2 or v_2 is found, the other can be most quickly calculated by means of the ideal-gas equation of state directly. Here we have solved for these quantities in a manner that illustrates the use of both p_r and v_r from the gas tables.)

Example 9·8. A turbine expands air isentropically from 5 atm, 1660 R, to a final pressure of 1 atm. The flow rate is 40 lbm/s. Determine the power output.

Solution. This is a steady-flow system. In order to apply the first law, we must find the enthalpy values at inlet and outlet. The inlet temperature is known; so we can find h_1 immediately from the air table. We find h_2 by first finding p_{r2}:

$$p_{r2} = p_{r1} \frac{P_2}{P_1} = 82.44 \left(\frac{1}{5}\right) = 16.49$$

Corresponding to this value of p_{r2}, we find $h_2 = 262.33$ B/lbm. If we neglect changes in kinetic energy, the first law as applied to the steady-flow system for this adiabatic process is

$$\text{Work} = h_1 - h_2 = 411.89 - 262.33 = 149.6 \text{ B/lbm}$$

The power output is then

$$\dot{W} = (\text{work})\dot{m} = \frac{149.6(40)}{0.707} = 8460 \text{ hp}$$

(The conversion factor in the last step is 0.707 B/s = 1 hp.)

9·7 Compressibility Factors

The gas tables discussed in the preceding section account for variations of specific heat with temperature, but they are based on the equation of state $Pv = RT$. For many gases, especially at higher pressures and lower temperatures, $Pv = RT$ is not sufficiently accurate for many engineering applications. This section and the following three discuss some of the many other equations of state that are in use for gases.

The simplest modification of the ideal-gas equation of state to fit real-gas behavior is the introduction of a compressibility factor Z, which is defined as

$$Z \equiv \frac{Pv}{RT}$$

For an ideal gas, obviously, $Z = 1$. For a real gas, Z is a function of pressure and temperature and is usually determined empirically, although principles of both classical thermodynamics and statistical thermodynamics are used in correlating Z with other properties. The value of Z for any state is a direct indication of the error involved in using $Pv = RT$ for that state.

Compressibility factor data in the usual form, Z versus P for various values of T, are given for nitrogen in Fig. 9·11. (The general trend mentioned in Sec. 4·2 of increasing accuracy of $Pv = RT$ as pressure decreases and temperature increases does not hold for all ranges, and Fig. 9·11 shows this clearly. For example, at 0°C there is less error in $Pv = RT$ at 14 MPa than at 7 MPa; and at a constant pressure of 10 MPa the error in $Pv = RT$ increases as the temperature increases!) See also Fig. 9·12.

A few compressibility factor values for four other gases are given in Table 9·4 as an indication of the error involved in using $Pv = RT$ without correction.

Example 9·9. Determine the pressure in a steel vessel having a volume of 0.0150 m³ and containing 3.40 kg of nitrogen at 400°C.

Solution. As a first approximation, assume that the ideal-gas equation of state holds; hence the compressibility factor equals 1.

$$P = \frac{ZmRT}{V} = \frac{ZmR_uT}{MV} = \frac{1(3.40)8.314(673)}{28(0.0150)1000} = 45.3\text{ MPa}$$

For this pressure and temperature, the value from Fig. 9·11 for Z is 1.22. Assuming a new value for Z of 1.30 gives

$$P = \frac{1.30(3.40)8.314(673)}{28(0.0150)1000} = 58.9\text{ MPa}$$

For this pressure and temperature, the value for Z is 1.29. It is evident that the pressure

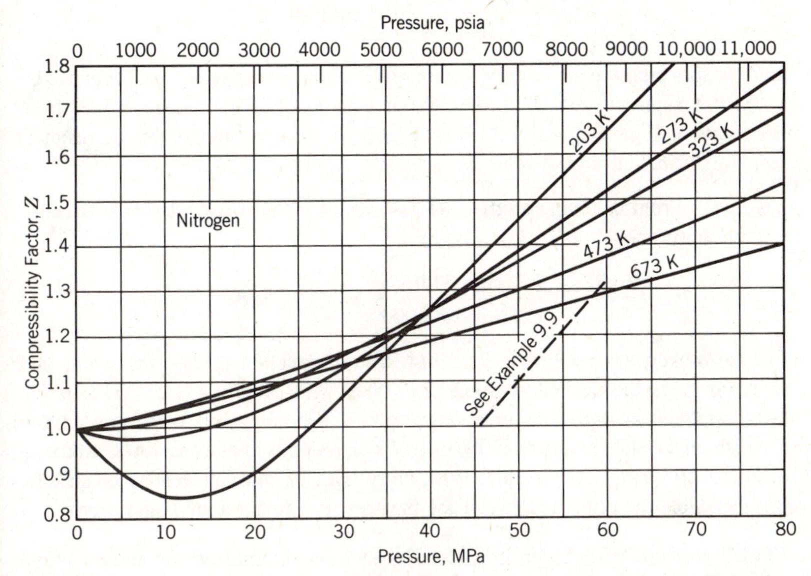

Figure 9·11. Approximate compressibility factors for nitrogen. From B. F. Dodge, *Chemical Engineering Thermodynamics*, McGraw-Hill, New York, 1944.

TABLE 9·4 Compressibility Factors for Four Gases

	Pressure			Temperature				
Gas	atm	MPa	psia	200 K 360 R	270 K 486 R	350 K 630 R	700 K 1260 R	1400 K 2520 R
Argon	1	0.101	14.7	0.997	0.999	1.000	1.000	1.000
	10	1.01	147	0.970	0.990	0.998	1.003	1.002
	40	4.05	588	0.878	0.962	0.992	1.011	1.009
	100	10.13	1470	0.692	0.921	0.988	1.029	1.021
Carbon	1	0.101	14.7		0.993	0.997	1.000	1.000
dioxide	10	1.01	147			0.970	1.000	1.003
	40	4.05	588			0.874	1.001	1.010
	100	10.13	1470			0.651	1.007	1.026
Carbon	1	0.101	14.7	0.997	0.999	1.000	1.000	1.000
monoxide	10	1.01	147	0.973	0.993	1.001	1.005	1.003
	40	4.05	588		0.977	1.004	1.018	1.012
	100	10.13	1470			1.022	1.047	1.030
Hydrogen	1	0.101	14.7	1.001	1.001	1.001		
	10	1.01	147	1.007	1.006	1.005		
	40	4.05	588	1.028	1.026	1.021		
	100	10.13	1470	1.076	1.065	1.054		

Source: Data from *Tables of Thermal Properties of Gases,* National Bureau of Standards Circular 564, 1955.

is between 45.3 and 58.9 MPa, and very close to 58.9 MPa; therefore assume that Z equals 1.28. The pressure equals

$$P = \frac{1.28(3.40)8.314(673)}{28(0.0150)1000} = 58.0 \text{ MPa}$$

The value for Z for this pressure and a temperature of 400°C is approximately 1.28; hence by trial the value for P is approximately 58.0 MPa. (Notice that reading the Z value as 1.29 from the small-scale chart of Fig. 9·11 gives $P = 58.4$ MPa, so the uncertainty of the result is greater than one might infer from the value written as 58.0 MPa.

Alternative solution. From the data specified we can write the following relation between P and Z for the state in question:

$$P = \frac{ZmRT}{V} = \frac{Z(3.40)8.314(673)}{28(0.0150)1000} = 45.3Z \text{ MPa}$$

This represents the broken straight line in Fig. 9·11. The intersection of the broken line and the 400°C isotherm gives the desired solution: $P = 58.0$ MPa and $Z = 1.28$. This is the solution of two simultaneous equations involving P and Z. One equation is the one written above from the data on the state in question; the other PZ equation is that of the 400°C isotherm on Fig. 9·11. Some ZP charts, such as Fig. 9·12, carry lines of constant density or specific volume, making possible the direct reading of the pressure as a function of T and ρ or v.

Example 9·10. Two kilograms of nitrogen in a closed system is compressed frictionlessly and isothermally from 28.0 MPa, 0°C, to 56.0 MPa. Determine the amount of work required.

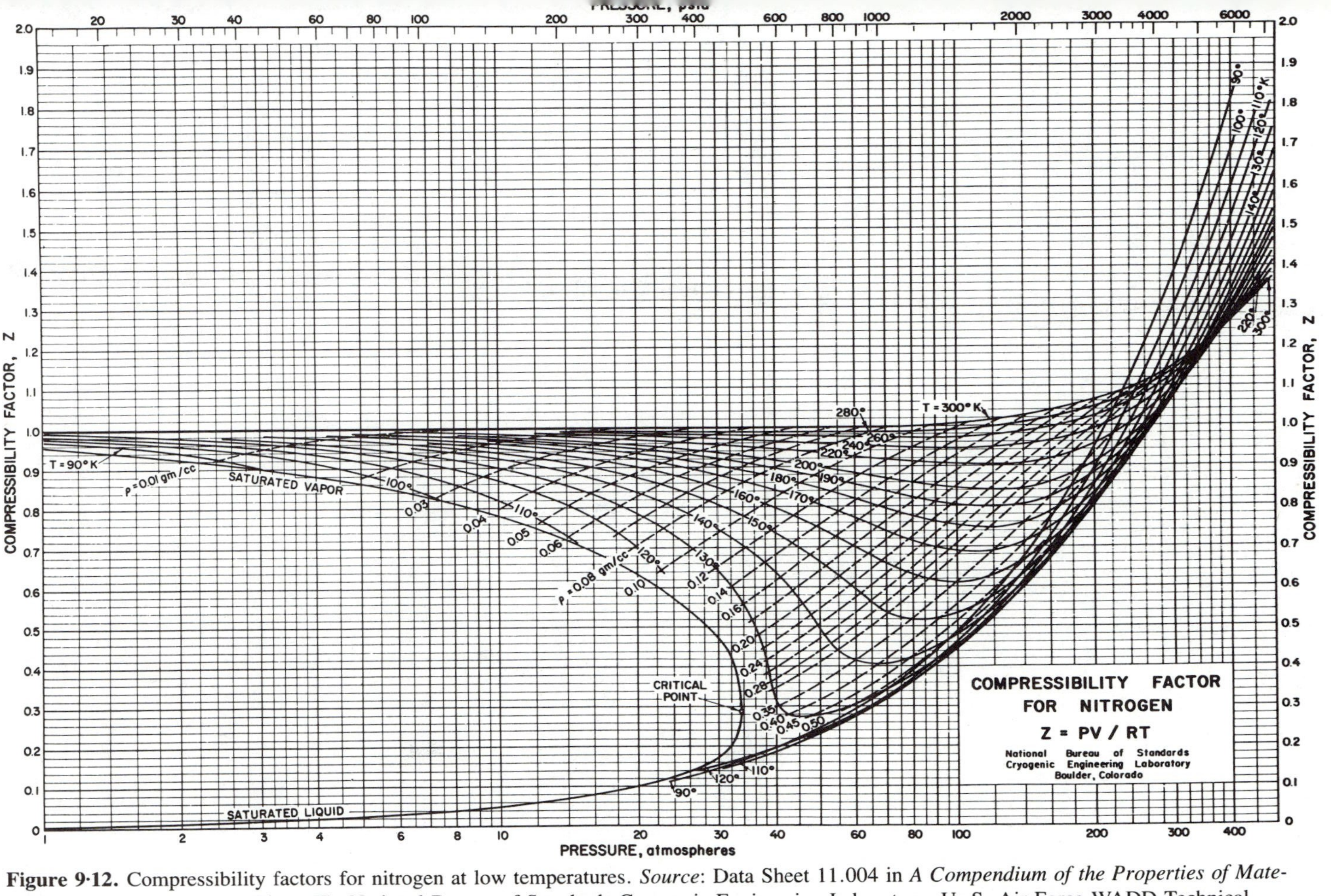

Figure 9·12. Compressibility factors for nitrogen at low temperatures. *Source*: Data Sheet 11.004 in *A Compendium of the Properties of Materials at Low Temperature,* (Phase II), National Bureau of Standards Cryogenic Engineering Laboratory, U. S. Air Force WADD Technical Report 60-56, 17 November 1960.

Solution. For a closed-system frictionless process,

$$W = \int P\, dV$$

In order to integrate this expression, we must have the relationship between P and V at constant temperature. The pressures involved are too high for the ideal-gas equation to be used accurately, but perhaps we can use $PV = ZmRT$. Z is usually a function of pressure and temperature, but for the constant-temperature path of this process we have simply $Z = f(P)$. Reference to Fig. 9·11 shows that between 28.0 MPa and 56.0 MPa the 0°C isotherm is a straight line. Let $Z = a + bP$, and a and b can be determined from Fig. 9·11. Since we know the pressure limits instead of the volume limits of the compression, let us replace V in the expression for work in terms of P. Thus

$$V = \frac{ZmRT}{P} = \frac{(a + bP)mRT}{P}$$

$$dV = -amRT\frac{dP}{P^2} + 0$$

$$W = \int_1^2 P\, dV = -\int_1^2 \frac{PamRT\, dP}{P^2} = -amRT \ln\frac{P_2}{P_1}$$

a is the $P = 0$ intercept of the 0°C isotherm between 28.0 and 56.0 MPa as it is extended. A measurement on Fig. 9·11 gives $a = 0.78$. (It turns out that we do not have to calculate b.) Thus

$$W = -amRT \ln\frac{P_2}{P_1} = -0.78(2)\frac{8.314}{28}(273)\ln\frac{56.0}{28.0} = -87.6 \text{ kJ}$$

The minus sign indicates that work is done on the system.

Example 9·11. A closed tank contains 12 kg of nitrogen at 40 MPa and 200°C. Compute the pressure that results if the temperature is lowered to −75°C.

Solution. The mass of nitrogen and the R value are constant; so we can write

$$\frac{P_1V_1}{Z_1T_1} = mR = \frac{P_2V_2}{Z_2T_2}$$

or

$$P_2 = \frac{P_1Z_2T_2}{Z_1T_1}$$

From Fig. 9·11 the value of Z_1 (at 40 MPa, 200°C) is approximately 1.23. Since the final temperature is lower than any shown on Fig. 9·11, we must use Fig. 9·12 for state 2. Since the final pressure is unknown, we make a first approximation on the basis of the ideal gas equation of state at state 2 so that $Z_2 = 1$ and

$$P_2 \approx \frac{P_1Z_2T_2}{Z_1T_1} = \frac{40(1)198}{1.23(473)} = 13.6 \text{ MPa}$$

For this pressure and 198 K, Fig. 9·12 gives $Z_2 = 0.84$. Therefore, a better approximation is

$$P_2 \approx \frac{P_1Z_2T_2}{Z_1T_1} = \frac{40(0.84)198}{1.23(473)} = 11.4 \text{ MPa}$$

At 11.4 MPa, 198 K, Fig. 9·12 shows that the compressibility factor remains 0.84 (the value used in the preceding step), so the final pressure is 11.4 MPa.

(Using data from a single source such as one *ZP* chart is likely to be more accurate, but in this case we do not have that choice because we use only the data presented in this book.)

9·8 Reduced Coordinates

The disadvantage of compressibility factor charts such as Figs. 9·11 and 9·12 is that a separate chart is needed for each gas. It would be convenient if one chart could be used for several gases. Two methods of doing this are often used. One is by means of the approximation sometimes called the law of corresponding states: if any two gases have equal values for the ratio of pressure to critical pressure and equal values for the ratio of temperature to critical temperature, then the ratio of specific volume to critical specific volume is the same for the two gases. The ratios of pressure, temperature, and specific volume to the corresponding critical values are called *reduced coordinates* or *reduced properties*. Reduced pressure, reduced temperature, and reduced specific volume are defined by

$$P_R \equiv \frac{P}{P_c} \qquad T_R \equiv \frac{T}{T_c} \qquad v_R \equiv \frac{v}{v_c}$$

where the subscript R denotes a reduced property and the subscript c denotes a property at the critical state. Values of P_c and T_c are given in Table A·4. In terms of these symbols, the approximation sometimes called the law of corresponding states says that for all gases

$$v_R = f(P_R, T_R) \tag{a}$$

and the function is the same for all gases.

If this approximation were accurate, a diagram of P_R versus T_R with v_R as a parameter plotted from data on any gas would apply to all others. Actually, the law of corresponding states is not sufficiently accurate over a wide range for much engineering work.

To illustrate the inaccuracy of the law of corresponding states, notice that the left-hand side of Eq. a can be written as

$$v_R = \frac{v}{v_c} = \frac{ZRTP_c}{PZ_cRT_c} = \frac{Z}{Z_c}\left(\frac{T_R}{P_R}\right)$$

so that

$$\frac{Z}{Z_c}\frac{T_R}{P_R} = f(P_R, T_R)$$

and

$$\frac{Z}{Z_c} = \phi(P_R, T_R)$$

Thus it follows that at the same values of P_R and T_R, all gases have the same value of Z/Z_c. However, at very low values of P_R, $Z = 1$ for all gases; so Z_c must be the same for all gases if the approximation stated by Eq. a holds. Experiments show that this is not the case. For example, Z_c is 0.230 for water, 0.275 for carbon dioxide, and 0.305 for hydrogen. Consequently, this approximation is inaccurate at least at low pressures. Further investigation shows that it is most useful in the vicinity of the critical state.

Another approach is based on the observation that at the same reduced pressure and reduced temperature all gases have approximately the same compressibility factor, except near the critical state. Thus

$$Z = f(P_R, T_R) \tag{b}$$

where the function is the same for all gases. (This is also sometimes called the law of corresponding states, but Eqs. a and b are not the same.) It is apparent that this does not hold for the critical state, where $P_R = 1$ and $T_R = 1$, because we have seen above that Z_c is not the same for all gases. For states well removed from the critical state, however, this *generalized compressibility factor* gives more accurate results than Eq. a. Chart A·4 is a generalized compressibility factor chart.

Better results can be obtained for some gases by using, in place of P_R and T_R, *pseudoreduced* pressure and temperature such as

$$P_R \equiv \frac{P}{P_c + A} \qquad \text{and} \qquad T_R \equiv \frac{T}{T_c + B}$$

where A and B are constants. For some gases pseudoreduced coordinates improve the accuracy of both Eqs. a and b.

The property *pseudoreduced specific volume,* defined as

$$v_R' \equiv \frac{vP_c}{RT_c}$$

is also used (and sometimes called the reduced specific volume, so you must be alert.) If Eq. a is written in terms of v_R' instead of v_R, then Eq. a becomes the same as Eq. b, and this is why the two approximations are often considered to be the same and even called by the same name.

In the interest of accuracy, generalized compressibility factor data should be resorted to only when compressibility factor data for the particular gas in question are unavailable.

9·9 Other Equations of State

Many equations of state have been developed for gases. Where extensive PvT data are already available, standard procedures for fitting equations to data might be used, with increased accuracy provided by increasing the number of constants or parameters. Usually, however, available data are not sufficiently extensive and accurate for the determination in this manner of equations of state covering extensive property ranges. Therefore, it is advantageous to introduce some theoretical considerations in order to formulate extensive equations of state from limited data. Also, a theoretical basis for the form of an equation of state makes it more likely that the same form of equation will be satisfactory for many gases.

Four examples of the many published equations of state for gases are presented briefly in this article. The van der Waals equation of state, aside from its historical interest, is of interest for the physical reasoning on which it is based. The more recent Redlich–Kwong equation is much more accurate over a wide range of property values than the van der Waals equation. Both of these involve two empirical constants for each gas. The Beattie–Bridgeman equation of state is an example using five constants for each gas. Several published equations of state for gases use many more constants. Finally, the virial form of equations of state is presented.

In 1873, J. D. van der Waals presented an equation of state that compensated for two assumptions made in developing the ideal-gas equation of state from kinetic theory: (1) The volume of the molecules themselves is negligible in comparison with the volume occupied by the gas and (2) the attractive forces between molecules are negligible. An equation of state that fits gases at high densities cannot be developed on the basis of these assumptions, so van der Waals modified the ideal-gas equation of state to take into account the finite size of the molecules and the attractive forces between them. He reasoned that $Pv = RT$ would be improved by replacing v by the volume (per unit mass of the gas) of the space between the molecules. If the volume of the molecules themselves is b, then the volume of the space between the molecules per unit mass of gas is $(v - b)$. Since b is the volume occupied by the molecules if they were all jammed together, it is of the same order of magnitude as the specific volume of the liquid. This modification of the ideal-gas equation of state gives us $P(v - b) = RT$. Van der Waals further reasoned that the pressure of a real gas is less than that of an ideal gas at the same temperature and density because of the attractive forces between molecules, and that this reduction in pressure is proportional to the square of the density or to $(1/v)^2$. Thus the van der Waals equation of state in molar form is

$$P = \frac{R_u T}{v_N - b} - \frac{a}{v_N^2} \tag{9·20}$$

or

$$\left(P + \frac{a}{v_N^2}\right)(v_N - b) = R_u T$$

Notice that the effect of each modification is greatest for states of small v_N.

The constants a and b can be evaluated from experimental data. Slightly different values fit the data well in various ranges of pressure and temperature, so the van der Waals equation is still only an approximate equation of state for real gases. Values of a and b for use in the van der Waals equation are given in Table A·4. Care must be exercised to ensure homogeneous units.

The van der Waals equation gives isotherms on a Pv plot which show maxima and minima, unlike the behavior of any real gas. As temperature increases, the v_N values at these maxima and minima of temperature approach each other, and the isotherm on which they coincide has neither a maximum nor a minimum but an inflection point with a tangent that is a constant-pressure line. This is a characteristic of the isotherm that passes through the critical state of a substance, so the critical state properties can in this way be related to the van der Waals constants as

$$v_{Nc} = 3b \qquad P_c = \frac{a}{27b^2} \qquad T_c = \frac{8a}{27R_u b}$$

These equations show that once the critical pressure and temperature of a substance are known, the van der Waals equation can be used as an approximate equation of state. Values of a and b obtained at the critical state may not be accurate for states far removed from P_c and T_c. Therefore, in order to make the equation more accurate over a wide range of properties, published values of the van der Waals constants may be slightly different from those obtained at the critical state.

In 1949, Redlich and Kwong presented a two-constant equation of state which is more accurate than the van der Waals equation over a wide range:

$$P = \frac{R_u T}{(v_N - b)} - \frac{a}{v_N(v_N + b)T^{1/2}} \tag{9·21}$$

(The a and b in the Redlich–Kwong equation are not the same as those in the van der Waals equation.) Values of the two empirical constants can be obtained from critical-state data as

$$a = 0.427 \frac{R_u^2 T_c^{2.5}}{P_c} \qquad \text{and} \qquad b = 0.0866 \frac{R_u T_c}{P_c}$$

The numerical coefficients in these expressions are dimensionless, so they can be used with any consistent set of units.

Another widely used equation of state is known as the Beattie–Bridgeman equation:

$$P = \frac{R_u T(1 - \varepsilon)}{v_N^2}(v_N + B) - \frac{A}{v_N^2} \tag{9·22}$$

where

$$A = A_0\left(1 - \frac{a}{v_N}\right)$$

$$B = B_0\left(1 - \frac{b}{v_N}\right)$$

$$\varepsilon = \frac{c}{v_N T^3}$$

This is a five-constant equation. A_0, a, B_0, b, and c for 10 gases are given in Table 9·5. ε is a function of molar volume and temperature as shown. The Beattie–Bridgeman equation has been used for formulating properties of the vapor close to the liquid phase at pressures less than the critical pressure. In general, the equation is accurate when the specific volumes involved are greater than twice the critical specific volume.

For the purpose of preparing extensive tables of properties and for computational purposes, it is often convenient to have an equation of state in the form

$$\frac{Pv}{RT} = A_0 + A_1 P + A_2 P^2 + A_3 P^3 + \cdots$$

or

$$\frac{Pv}{RT} = B_0 + \frac{B_1}{v} + \frac{B_2}{v^2} + \frac{B_3}{v^3} + \cdots$$

where the A's and B's are functions of temperature only. These equations are called *virial equations,* and the A's and B's are called *virial coefficients*. Several other forms of virial

TABLE 9·5 Beattie–Bridgeman Constants

	A_0		a		B_0		b		$c(10^{-4})$	
Gas	$\frac{atm \cdot m^6}{kmol^2}$	$\frac{atm \cdot ft^6}{lbmol^2}$	$\frac{m^3}{kmol}$	$\frac{ft^3}{lbmol}$	$\frac{m^3}{kmol}$	$\frac{ft^3}{lbmol}$	$\frac{m^3}{kmol}$	$\frac{ft^3}{lbmol}$	$\frac{m^3 \cdot K^3}{kmol}$	$\frac{ft^3 \cdot R^3}{lbmol}$
Air	1.302	334	0.0193	0.309	0.0461	0.738	−0.00110	−0.0176	4.35	406
Ar	1.294	332	0.0233	0.373	0.0393	0.629	0	0	5.99	560
CO	1.345	345	0.0262	0.419	0.0504	0.808	−0.00693	0.111	4.21	393
CO_2	5.008	1285	0.0712	1.14	0.1055	1.69	0.0724	1.16	66.05	6170
H_2	0.198	50.7	−0.0051	−0.081	0.0210	0.336	−0.0436	−0.698	0.050	4.7
He	0.022	5.6	0.0598	0.958	0.0140	0.224	0	0	0.040	0.37
N_2	1.345	345	0.0262	0.419	0.0504	0.808	−0.00693	−0.111	4.21	393
Ne	0.213	54.6	0.0220	0.352	0.0206	0.33	0	0	0.101	9.4
N_2O	5.008	1285	0.0712	1.14	0.1049	1.68	0.0724	1.16	66.05	6170
O_2	1.493	383	0.0257	0.411	0.0463	0.741	0.00421	0.0674	4.81	449

Note: For A_0 units of atm·m⁶/kmol², R_u = 0.0820 atm·m³/kmol·K. For A_0 units of atm·ft⁶/lbmol², R_u = 0.730 atm·ft³/lbmol·R.

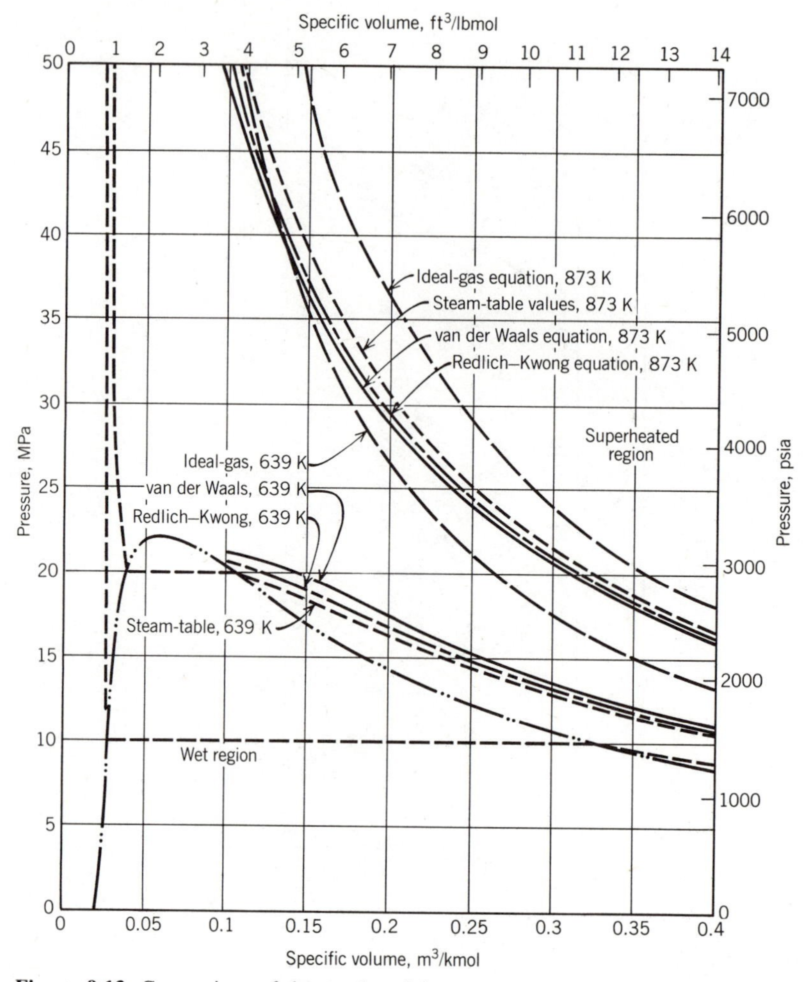

Figure 9·13. Comparison of the results of three equations of state with steam table data.

equations are also used. One of the advantages of virial equations of state is that it is relatively easy to determine the virial coefficients from experimental PvT data.

The accuracies of various equations of state vary significantly from one substance to another and from one property range to another. A sample comparison of the results of some equations of state with the accurate and extensive data from the steam tables is shown in Fig. 9·13. Do not conclude, however, that the relative accuracies of these equations of state are the same for all substances.

Example 9·12. Determine the pressure in a steel vessel having a volume of 0.0150 m³ and containing 3.40 kg of nitrogen at 400°C by means of (a) the van der Waals equation, (b) the Redlich–Kwong equation, and (c) the Beattie–Bridgeman equation.

Solution. In each case, the calculation is readily made using a calculator or computer with input values of $R_u = 8.314$ kJ/kmol·K, $T = 673$ K, and $v_N = V/N = VM/m = [0.0150(28)/3.40]$ m^3/kmol. Additional input values for each calculation are shown below.

(a) The van der Waals equation

$$P = \frac{R_u T}{(v_N - b)} - \frac{a}{v_N^2} \tag{9·20}$$

with $a = 136.6$ kPa·m^6/kmol^2 and $b = 0.0386$ m^3/kmol from Table A·4(SI) gives $P = 56.9$ MPa.

(b) The Redlich–Kwong equation is

$$P = \frac{R_u T}{(v_N - b)} - \frac{a}{v_N(v_N + b)T^{1/2}} \tag{9·21}$$

with $a = 0.427\, R_u^2 T_c^{2.5}/P_c$ and $b = 0.0866\, R_u T_c/P_c$. Input values of $P_c = 3.39$ MPa and $T_c = 126.2$ K give $P = 54.6$ MPa.

(c) The Beattie–Bridgeman equation is

$$P = \frac{R_u T(1 - \varepsilon)}{v_N^2}(v_N + B) - \frac{A}{v_N^2}$$

with $A = A_0(1 - a/v_N)$, $B = B_0(1 - b/v_N)$, and $\varepsilon = c/v_N T^3$. Input values of $A_0 = 136$ kPa·m^6/kmol^2, $a = 0.0262$ m^3/kmol, $B_0 = 0.0504$ m^3/kmol, $b = -0.00693$ m^3/kmol, and $c = 4.21$ m^3·K^3/kmol from Table 9·5 give $P = 57.8$ MPa.

(These results should be compared with the more accurate result of Example 9·9. Conclusions regarding the relative accuracy of the various equations of state over a wide range of conditions should not be drawn from this single example, however.)

9·10 Specific Heats of Real Gases

Chapter 4 pointed out that the specific heats of ideal gases are functions of temperature only. The specific heats of real gases are functions of both pressure and temperature. In order to learn when the influence of pressure is greatest, consider first the hT diagram of Fig. 9·14. This diagram is plotted from data on steam, but diagrams for other substances are similar. Recalling the definition of c_p

$$c_p \equiv \left(\frac{\partial h}{\partial T}\right)_P$$

we see that c_p is equal to the slope of a constant-pressure line on an hT diagram. Notice in Fig. 9·14 the steep and changing slope of the constant-pressure lines in the vicinity of the critical state. In fact, the constant-pressure line that passes through the critical state is vertical at that point, so c_p is infinite at the critical state. Notice also that c_p is discontinuous at the saturation line because the slope of a constant-pressure line on an hT diagram changes abruptly there. The plot of c_p versus T in Fig. 9·15 shows the marked increase in c_p near the critical state. This characteristic can be explained qualitatively in terms of the increase in molecular potential energy required during the expansion of a dense vapor

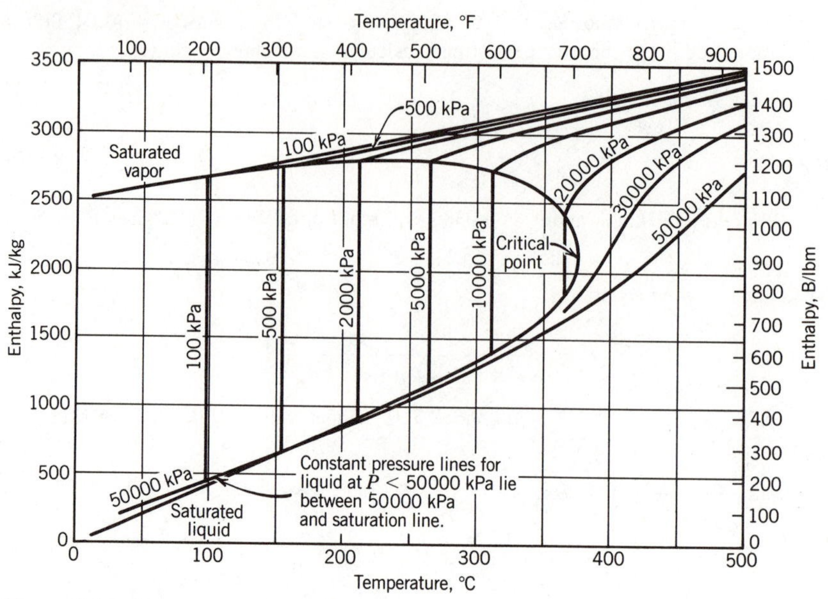

Figure 9·14. *hT* diagram for water.

as the temperature is increased at constant pressure. In a less dense vapor, expansion is accompanied by a smaller increase in molecular potential energy because the forces between molecules are smaller. In a liquid the intermolecular forces are high, but there is little increase in molecular potential energy when the temperature is increased at constant pressure because the increase in volume is usually small.

The increase in c_p near the critical state and the increase in c_p with temperature, explained by kinetic theory for lower densities, accounts for the general shape of c_p versus T curves of real gases at various pressures. Samples of such curves for air are shown in Fig. 9·16. Notice the similarity between Fig. 9·15 for water vapor and Fig. 9·16 for air. For a real gas, pressure has less influence on the value of c_v than it does on the value of c_p. The quantity $(c_p - c_v)$ is not a constant as it is for ideal gases.

9·11 Enthalpy and Entropy of Real Gases

The simplest equation of state for a gas is the ideal-gas equation, $Pv = RT$, and the more complex equations of state developed for other gases are often extensions of this simple one. We have seen that any substance that follows the ideal-gas equation of state also has a simple relationship of enthalpy to other properties: enthalpy is a function of temperature only. For other gases, the relationships involving enthalpy are more complex, but again it is convenient to formulate these as extensions to the relationships for ideal gases.

Consider now the determination of the enthalpy change for a process of a real gas from state 1 to state 2 as shown on the *Ts* diagram of Fig. 9·17*a*. At state 1, $P_R > 1$. At

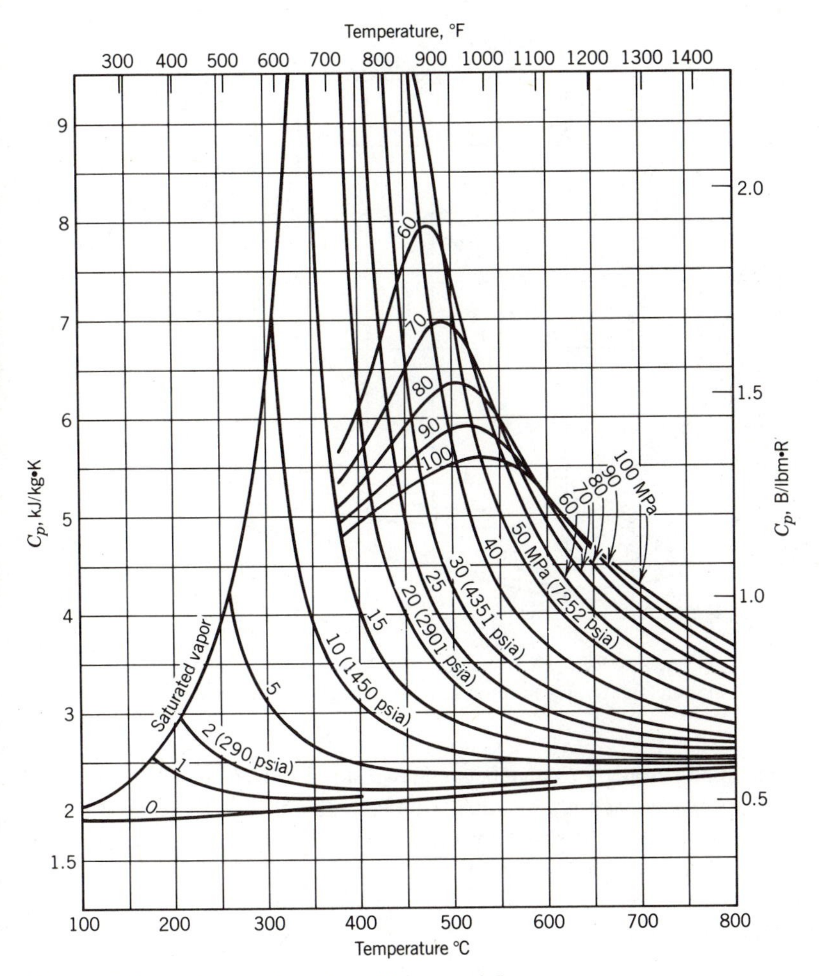

Figure 9·15. c_p of water. Data from Reference 3·5.

state 2, $T_r < 1$ and P_R is less than 1 but not very low. Therefore, it is unlikely that the ideal-gas equation of state would be accurate for these states.

For any gas, the change in enthalpy is given by

$$h_2 - h_1 = \int_1^2 c_p\, dT - \int_1^2 \left[T \left(\frac{\partial v}{\partial T} \right)_P - v \right] dP \tag{9·15}$$

The second term on the right-hand side can be evaluated from PvT data; the first one requires a knowledge of the c_pT relationship along the process path from state 1 to state 2. For many gases, the PvT data needed might be approximated by means of a generalized compressibility factor relationship using reduced coordinates, but data on c_pT relationships for gases at high pressures are scarce.

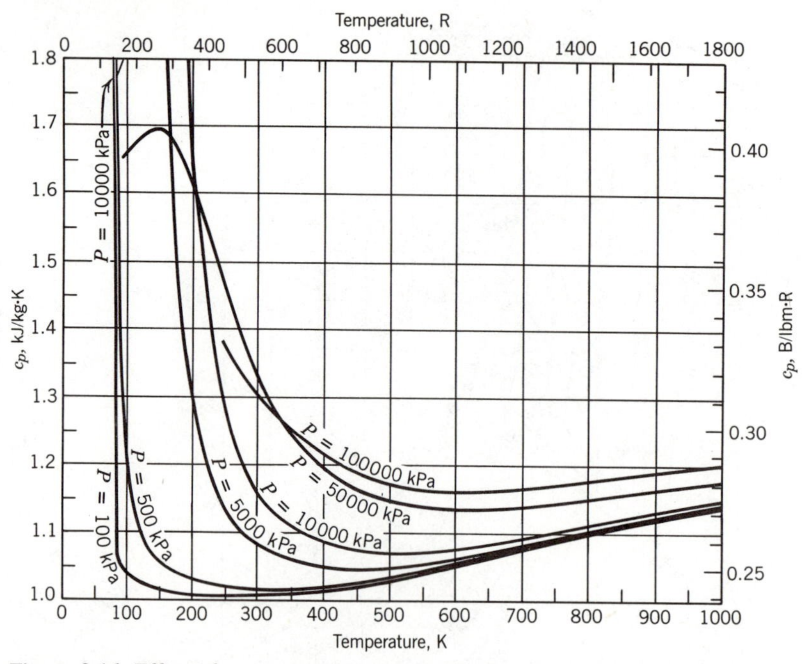

Figure 9·16. Effect of pressure and temperature on c_p of air. Data from Reference 9·5.

Since the value of any property change between two states is the same for any process between them, a fruitful approach is to evaluate the enthalpy change by using the path 1-1*-2*-2 shown in Fig. 9·17*b*. This path consists of a constant temperature process from state 1 to a state 1* at a low pressure, a constant pressure process from state 1* to a state 2*, and a constant temperature process from state 2* to state 2. States 1* and 2* are at such a low pressure that the ideal-gas equation of state applies. Thus we evaluate the

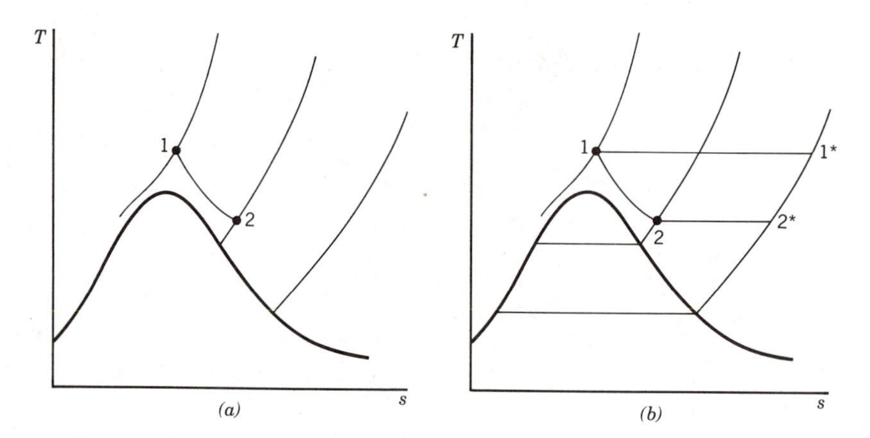

Figure 9·17. Process of real gas from state 1 to state 2.

enthalpy change between states 1 and 2 as

$$h_2 - h_1 = (h_2 - h^*_2) + (h^*_2 - h^*_1) + (h^*_1 - h_1)$$

Each of the three terms on the right-hand side can be evaluated by means of Eq. 9·15. For process 1*-2*, the pressure is constant at a low value, so Eq. 9·15 gives

$$h^*_2 - h^*_1 = \int_{1^*}^{2^*} c_p \, dT$$

which can be evaluated from the ideal-gas c_pT relation.

Process 1-1* is a constant temperature process, so Eq. 9·15 gives

$$h^*_1 - h_1 = \int_1^{1^*} \left[v - T \left(\frac{\partial v}{\partial T} \right)_P \right] dP$$

This expression can be evaluated from a PvT relationship. In a like manner, the enthalpy change between states 2* and 2 can be calculated. Consequently, for evaluating the enthalpy change between any two states of a real gas we can use the same procedure illustrated here: Join the two states by two constant temperature processes, each connecting one of the end states with a state at the same temperature and a pressure low enough for the ideal-gas equation to be valid, and a constant low-pressure process during which the total change in temperature occurs. This procedure is used so commonly that we assign a symbol to the properties of the ideal-gas states corresponding to the original end states. For any state x of a real gas, for which the enthalpy is h_x, the state at the same temperature and very low pressure (that is, low enough for the ideal gas equation of state to be valid) has an enthalpy value of h^*_x. The same relationship exists between u_x and u^*_x.

The difference between h and h^* (sometimes called the *enthalpy departure* or *enthalpy residual*) is obtained by means of Eq. 9·15 which, for the constant temperature process connecting these two states is

$$h^* - h = \int_P^0 \left[v - T \left(\frac{\partial v}{\partial T} \right)_P \right] dP$$

In the absence of any other PvT relationship for the gas at hand, we use a compressibility factor so that in this equation we make the substitution $v = ZRT/P$ to obtain

$$h^* - h = -R \int_P^0 \frac{T^2}{P} \left(\frac{\partial Z}{\partial T} \right)_P dP$$

If ZPT data are available for the gas, this expression can be evaluated graphically or numerically. If ZPT data are not available, we can use the approximation of a generalized compressibility factor or ZP_RT_R chart. Substituting $P = P_RP_c$ and $T = T_RT_c$ into the equation results in

$$\frac{h^* - h}{R_uT_c} = \int_0^{P_R} T_R^2 \left(\frac{\partial Z}{\partial T_R} \right)_{P_R} d \ln P_R \qquad (9\cdot16)$$

where the h's are molar specific enthalpies to permit the introduction of R_u instead of R. Again, the integral can be evaluated numerically or graphically from a table or chart of generalized compressibility factors. A resulting plot of $(h^* - h)/R_uT_c$ in terms of P_R and T_R, called a generalized enthalpy chart, is given as Chart A·5. This chart is based on a generalized compressibility factor chart with a value of $Z = 0.27$. Remember that it is an approximation based on properties of many gases.

A generalized entropy chart can be developed in a similar manner, starting with the general equation for entropy change

$$s_2 - s_1 = \int_1^2 c_p \frac{dT}{T} - \left(\frac{\partial v}{\partial T}\right)_P dP \tag{9·13}$$

One difference between the development of the generalized entropy chart and the development of the generalized enthalpy chart is that for an ideal gas, entropy is not a function of temperature only as enthalpy is. Furthermore, for an ideal gas $s \to \infty$ as $P \to 0$ at constant T. (Sketch a few constant-pressure lines on a Ts diagram to illustrate this.) The difficulty this introduces can be avoided by using some arbitrary low pressure, P_{ar}, as a standard; then it is seen that the arbitrary low pressure can be eliminated from the calculations. Applying Eq. 9·13 to a constant temperature process between the state of the real gas (at P,T) and a state (at P_{ar},T) at a low enough pressure for the gas to behave as an ideal gas gives

$$s_{P,T} - s^*_{P_{ar},T} = -\int_{P_{ar}}^{P} \left(\frac{\partial v}{\partial T}\right)_P dP \tag{a}$$

where the asterisk indicates an ideal gas property.

If the gas could exist as an ideal gas at state P,T the change in entropy between the state at P_{ar},T and that at P,T would be

$$s^*_{P,T} - s^*_{P_{ar},T} = -\int_{P_{ar}}^{P} \left(\frac{\partial v}{\partial T}\right)_P dP = -R\int_{P_{ar}}^{P} \frac{dP}{P} \tag{b}$$

Subtracting Eq. a from Eq. b gives

$$s^*_{P,T} - s_{P,T} = -\int_{P_{ar}}^{P} \left[\frac{R}{P} - \left(\frac{\partial v}{\partial T}\right)_P\right] dP$$

which is the difference in entropy between the hypothetical state of the gas as an ideal gas at P and T and the real gas state at the same pressure and temperature. (State P_{ar},T has been eliminated.) This entropy difference depends only on the PvT relationship, so by using a generalized compressibility factor and letting P_{ar} approach 0 it can be expressed as

$$\frac{s^*_P - s_P}{R_u} = -\int_0^{P_R} \frac{1 - Z}{P_R} dP_R + T_R\int_0^{P_R} \left(\frac{\partial Z}{\partial T_R}\right)_{P_R} d\ln P_R \tag{9·17}$$

where the s's are molar specific entropies to permit the use of R_u instead of R. This expression can be evaluated by numerical or graphical integration using ZP_RT_R data to produce a generalized entropy chart similar to the generalized enthalpy chart that was developed from Eq. 9·16. Chart A·6 is a generalized entropy chart. Again, it must be remembered that the accuracy of results based on such a chart is limited by the approximations involved in the generalized compressibility factor chart.

The calculation of entropy change of a real gas between states 1 and 2 is carried out by

$$s_2 - s_1 = (s_2 - s^*_2) + (s^*_2 - s^*_1) + (s^*_1 - s_1) \tag{c}$$

The first term on the right-hand side is the entropy difference between the real gas and the hypothetical ideal gas, both at P_2 and T_2. The second term is the entropy difference of the hypothetical ideal gas between states 1 and 2. The third term is the entropy difference between the real gas and the hypothetical ideal gas, both at P_1 and T_1.

Notice that s^*_x is defined as the entropy of a gas at state x *if it were an ideal gas*. h^*_x was defined as the enthalpy of the real gas at T_x and a very low pressure, but since the enthalpy of an ideal gas is independent of pressure, an almost identical definition would be: h^*_x is the enthalpy of a gas at state x *if it were an ideal gas*. The difference between the two definitions is that in the first case of enthalpy, states such as 1* and 2* can be shown on a property diagram for the real gas as in Fig. 9·17, since there is a region on the diagram where the real gas behaves at least very nearly as an ideal gas. If states 1* and 2* are defined as hypothetical ideal-gas states, as in the definition of s^*, they cannot be shown on a single property diagram of the real gas. For example, no diagram similar to Fig. 9·17*b* can be created to accompany Eq. c.

Example 9·13. Methane expands adiabatically through a turbine at a rate of 0.32 kg/s from 10 MPa, 27°C, to 2.0 MPa, −45°C. Kinetic energy changes are negligible. Determine the turbine power output and the change in entropy of the methane during the expansion.

Analysis. The first law must be applied to determine the power, and it will require enthalpy values for the methane. At the conditions specified, methane is unlikely to behave as an ideal gas. As a check on this point, we obtain values of P_c = 4.60 MPa and T_c = 190.6 K for methane (from Table A·4) and then calculate P_R and T_R for each state. Then we obtain values of Z from Chart A·4. The results are:

$$P_{R1} = 2.17 \qquad T_{R1} = 1.57 \qquad Z_1 = 0.87$$
$$P_{R2} = 0.43 \qquad T_{R2} = 1.20 \qquad Z_2 = 0.92$$

This much departure from the ideal-gas value of $Z = 1$ confirms that we cannot treat the methane as an ideal gas. Therefore, we will use the generalized enthalpy and generalized entropy charts in the calculations.

Solution. From the first law,

$$w = h_1 - h_2$$

and we evaluate the enthalpy change by means of the generalized enthalpy chart, Chart A·5, so that

$$w = h_1 - h_2 = h^*_1 - h^*_2 - (h^*_1 - h_1) + (h^*_2 - h_2)$$

Using a mean specific heat for the term $h^*_1 - h^*_2$ and arranging the other two terms on the right-hand side in a form suitable for using the generalized enthalpy chart

$$w = c_p(T_1 - T_2) - \frac{R_u T_c}{M}\left(\frac{h^*_1 - h_1}{R_u T_c} - \frac{h^*_2 - h_2}{R_u T_c}\right)$$

By means of the equation in the heading of Table A·4 and the coefficients for methane as an ideal gas given in the table, we obtain $c_{p1} = 2.22$ kJ/kg·K and $c_{p2} = 2.04$ kJ/kg·K. Since these values are relatively close together, we use as a mean specific heat for the temperature range between states 1 and 2 the value of 2.13 kJ/kg·K. Therefore, substitution of numerical values, including Chart A·5 values corresponding to the P_R and T_R values calculated in the analysis above, gives

$$w = 2.13(300 - 228) - \frac{8.314(190.6)}{16.04}(0.92 - 0.32) = 94 \text{ kJ/kg}$$

$$\dot{W} = \dot{m}w = 0.32(94) = 30 \text{ kW}$$

In a similar manner, the entropy change is

$$s_2 - s_1 = s^*_2 - s^*_1 - (s^*_2 - s_2) + (s^*_1 - s_1)$$

The first term on the right-hand side is the entropy change of methane as an ideal gas between states 1 and 2, and each of the other two terms is the difference between the entropy of an ideal gas and that of a real gas at the same pressure and temperature. For the ideal gas term, we integrate the Tds equation, $Tds = dh - v dP$, for an ideal gas with a constant c_p (the mean value determined earlier). The other two terms we arrange for convenient use of the generalized entropy chart, Chart A·6, which we enter with the P_R and T_R values already determined. Thus

$$\begin{aligned} s_2 - s_1 &= c_p \ln\frac{T_2}{T_1} - R\ln\frac{P_2}{P_1} - \frac{R_u}{M}\left(\frac{s^*_2 - s_2}{R_u} - \frac{s^*_1 - s_1}{R_u}\right) \\ &= 2.13 \ln\frac{228}{300} - \frac{8.314}{16.04}\ln\frac{2}{10} - \frac{8.314}{16.04}(0.20 - 0.47) \\ &= 0.39 \text{ kJ/kg·K} \end{aligned}$$

(An alternative solution method would be the use of tables or charts of methane properties as presented, for example, in Reference 9·3.)

A word of caution is in order. From these sections on real gases it must not be inferred that all the gas property information an engineer ever needs has been collected and organized so that it can be found simply by searching the literature. All too often, important problems that engineers face are made more difficult by a dearth of information on some substance, on some particular property of even a common substance, or on the properties of a substance in an extreme pressure or temperature range.

9·12 Property Diagrams

Various property diagrams are used extensively in thermodynamics as aids in analysis and design. Several of these have already been discussed in this book, and you should be familiar with their characteristics. The characteristics of others you should be able to deduce from your knowledge of general physical property relationships. Let us review some of the frequently used property diagrams.

The *Ts* Diagram

The outstanding characteristic of the *Ts* diagram is that for *reversible* processes heat transfer is represented by an area, since $q_{rev} = \int T\, ds$. Also, the unavailable part of the heat transfer of any reversible process can be represented by an area on a *Ts* diagram, since $q_{unav} = T_0\, \Delta s$. For purposes of analysis, any diagram using entropy as one of its coordinates offers some convenience because for many actual compressions and expansions the corresponding ideal process is an isentropic one. *Ts* diagrams for water and carbon dioxide are shown in Fig. 9·18 (see also Charts A·1 and A·2 in the appendix). A disadvantage of the *Ts* diagram is that a convenient scale for showing two- or three-phase regions causes the slopes of constant-pressure and constant-volume lines in the superheat region to be nearly the same.

The *hs* or Mollier Diagram

The *hs* diagram is used in connection with steam and gas turbines because (1) the ideal process is usually an isentropic one and is therefore easily traced, and (2) Δh values that are used extensively in first-law analyses of turbines are easily scaled or read from this diagram. On an *hs* diagram for a vapor the constant-pressure lines in the wet region are straight and diverge with increasing quality so that a diagram of moderate size can be read quite accurately in the high quality and superheat regions (see Chart A·3 in the Appendix).

The *hs* diagram is convenient for the determination of the quality of a wet vapor from measurements made with a throttling calorimeter as shown in Fig. 9·19. The quality of the wet vapor flowing in the line is to be determined. Measuring the pressure and temperature of the vapor in the line will not suffice because they are dependent on each other and independent of the quality. If a small amount of the vapor is throttled through an orifice to a pressure sufficiently low, it becomes superheated. Then *P* and *T* measurements suffice to determine the state of the vapor in the calorimeter. If the process is adiabatic

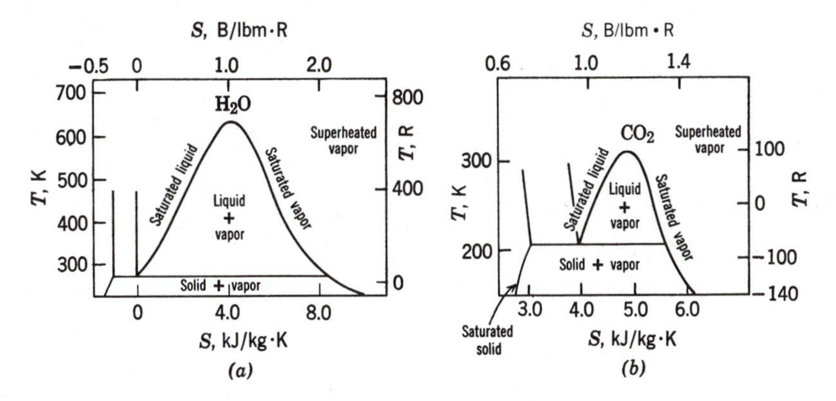

Figure 9·18. *Ts* diagrams showing solid, liquid, and vapor states.

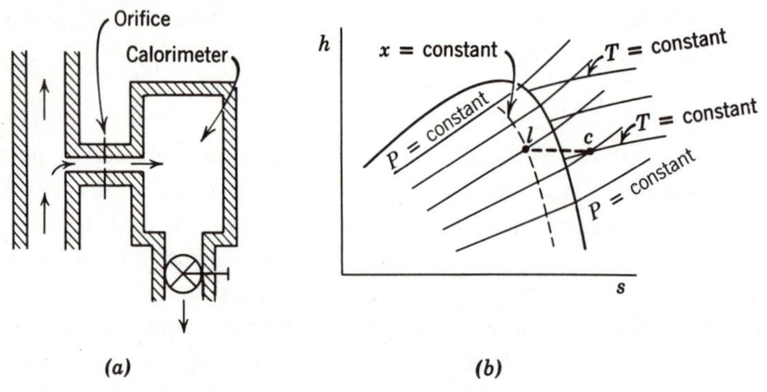

Figure 9·19. Operation of a throttling calorimeter.

and the change in kinetic energy is zero, the first law applied to the open system formed by the calorimeter and its connecting line gives

$$h_{\text{line}} = h_{\text{calorimeter}}$$

Then from the pressure and enthalpy of the vapor in the line its quality can be determined. This use of the *hs* diagram is shown in Fig. 9·19*b*. From the calorimeter pressure and temperature, point *c* is located. Extending a line horizontally (constant *h*) to its intersection with the line of constant pressure P_l establishes point *l* and the quality x_l that is sought. Notice that this method cannot be used unless the unknown quality is high enough so that the throttling process carries the vapor into the superheat region.

The *Ph* Diagram

The *Ph* diagram is also called a Mollier diagram and is used extensively in refrigeration. Values of Δh can be read or scaled directly from it. Vapor-compression refrigerating machines, such as used in most household refrigerators and air conditioners, operate on a cycle which includes two constant-pressure processes and one throttling process. The simplicity of such a cycle on a *Ph* diagram is shown in Fig. 9·20. A *Ph* diagram for ammonia is presented in the appendix (Chart A·8).

The *Pv* Diagram

The *Pv* diagram is useful in certain analyses of reversible processes because the quantities $\int P\,dv$ and $\int v\,dP$ are represented by areas on it. It is not a convenient diagram for presenting the properties of both liquid (or solid) and vapor phases of a substance, because the difference between v_f and v_g (except near the critical state) is so much greater than the range of variation of *v* in the compressed-liquid region. Another disadvantage of the *pv* diagram is that the slopes of constant-entropy and constant-temperature lines on it in the superheat region are so close together that reading the diagram is difficult.

The *hv* Diagram

The *hv* diagram is convenient for presenting vapor properties because over a wide range the constant-pressure, constant-temperature, and constant-entropy lines on it make good

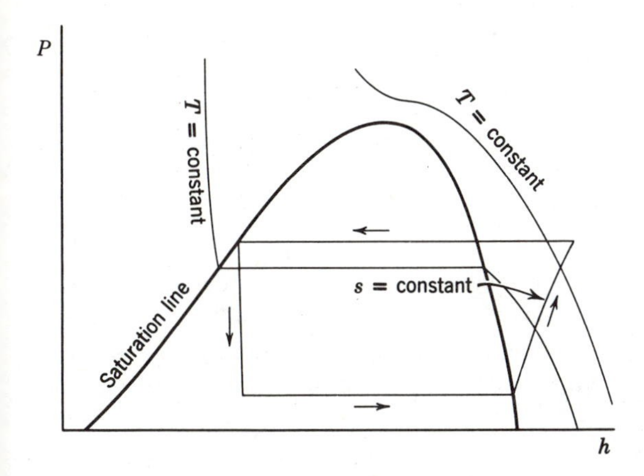

Figure 9·20. *Ph* diagram showing a typical vapor compression refrigeration cycle.

intersections; that is, their slopes are quite different (see Reference 3·6). Also, as pointed out in the comments on *hs* and *Ph* diagrams, *h* is a convenient coordinate for the purpose of first-law steady-flow analyses. On the *hv* (or *uv*) diagram the triple state is represented by an area.

The Stress–Strain Diagram

The stress–strain diagram used in studying the elastic behavior of solids is generally not thought of as a *thermodynamic* diagram, but one as shown in Fig. 9·21 is the logical one to use in certain thermodynamic analyses. Notice that there are at least two moduli of elasticity: the isothermal modulus and the isentropic modulus.

The above list of diagrams is by no means exhaustive. For one thing, it includes only two-dimensional diagrams, and we have seen in Chapter 3 that three-dimensional ones are occasionally convenient or necessary. Also, the diagrams listed above are for pure substances, but engineers frequently use diagrams (such as the psychrometric chart used in air-conditioning work, for example) that are for mixtures of variable composition. The important thing to remember when selecting diagrams for use in any particular thermodynamic analysis is to use those that are most helpful in that specific instance. If the most helpful ones happen to be unconventional, use them anyway.

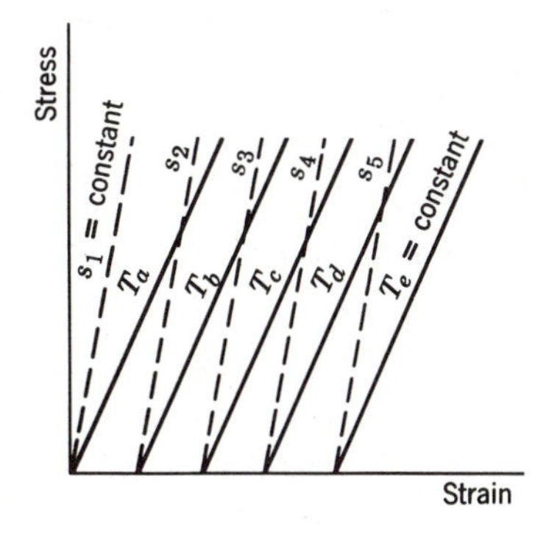

Figure 9·21. Tensile stress–strain diagram for an elastic substance.

9·13 Other Approaches to Thermodynamics

So far in this book topics have been treated in the following order:

1. Introduction: Undefined terms; definitions, including heat and work (see Chapter 1 for details).
2. The first law: The definition of energy; applications of the first law to various systems (Chapter 2).
3. Physical properties: Phase relationships, tables of properties, the ideal-gas equation of state and specific-heat relationships (Chapters 3 and 4).
4. The second law: Reversibility, the Carnot principle, the thermodynamic temperature scale, the definition of entropy, $T\,ds = du + P\,dv = dh - v\,dP$, the uses of entropy (Chapters 5, 6, 7, and 8).
5. Physical property relationships involving the second law as well as the first law (Chapter 9).

This order was selected for ease of learning. For example, the two chapters on physical properties follow the first-law chapter so that you could practice applying the first law and thereby strengthen your understanding of it before taking up the new material on the second law. Many property relationships depend on the second law, however, so it is necessary to return to the discussion of physical properties after the chapters on the second law.

People who have already studied thermodynamics usually prefer to develop this material in a more direct and perhaps more logical manner, covering all the basic principles which are established by induction (the first and second laws) first before considering the applications which are based on deductions from the basic principles. We have reached a point in this book where you should undertake a thorough review; therefore, because it may help you in making such a review, let us outline one of these logical approaches to thermodynamics:

1. The undefined terms: Length, mass, temperature, etc.
2. Definitions: System, equilibrium, state, property, process, heat, work.
3. The first law—by induction from experience: Definition of stored energy, proof that stored energy is a property.
4. The second law—by induction from experience: Reversibility, the Carnot principle, definition of thermodynamic temperature scale, definition of entropy, proof that entropy is a property.
5. Physical properties: (*a*) PvT relations: These are independent of the first and second laws except for the definition of the temperature scale. (*b*) Property relations involving u (or e) and s, and therefore depending on the first and second laws. (*c*) The ideal gas, a hypothetical substance for which it is postulated that $Pv = RT$, where T is the thermodynamic temperature.

Many variations of this basic approach are used. For example, one similar approach differs by defining, in step 2 above, work and an adiabatic system. Then the first law defines stored energy in terms of the work of an adiabatic system. Heat is then defined in terms of work and stored energy changes.

Some other approaches reach the first and second laws not by induction from experience but start with them as postulates. Some statements of the second law appear upon first reading to be quite unrelated to others, but from any statement one should be able to deduce any of the others. One approach (Reference 9·6) even uses a single postulate instead of the first and second laws.

Although classical thermodynamics—a macroscopic science—and statistical or microscopic thermodynamics are logically independent, their uses are complementary, and some approaches mix the two.

You must remember that there are many different logical approaches, depending on which terms are selected as undefined, the sequence of definitions, and the manner in which the basic inductions or the postulates are formulated. It is desirable to organize a science so that it depends on the minimum number of inductive conclusions or basic principles and to state these principles in such a manner that the maximum number of useful deductions can be readily made from them. In addition to this goal of logical economy, there are goals of rigor, simplicity, plausibility, convenience, utility, and ease of learning. Various approaches differ in the extent to which they meet these various goals. No single approach is the best in all respects.

9·14 Summary

The Maxwell equations

$$\left(\frac{\partial T}{\partial v}\right)_s = -\left(\frac{\partial P}{\partial s}\right)_v \tag{9·1}$$

$$\left(\frac{\partial T}{\partial P}\right)_s = \left(\frac{\partial v}{\partial s}\right)_P \tag{9·2}$$

$$\left(\frac{\partial P}{\partial T}\right)_v = \left(\frac{\partial s}{\partial v}\right)_T \tag{9·3}$$

$$\left(\frac{\partial v}{\partial T}\right)_P = -\left(\frac{\partial s}{\partial P}\right)_T \tag{9·4}$$

are useful in the correlation of properties of pure substances because they relate entropy to the directly measurable properties pressure, specific volume, and temperature. Many other useful relations can be derived from the four equations

$$du = T\,ds - P\,dv \tag{8·3}$$

$$dh = T\,ds + v\,dP \tag{8·3}$$

$$da = -P\,dv - s\,dT \tag{8·7b}$$

$$dg = v\,dP - s\,dT \tag{8·8b}$$

One example of a relation derived from the Maxwell equations is the Clapeyron equation,

$$\frac{dP}{dT} = \frac{h'' - h'}{T(v'' - v')} \tag{9·11}$$

which is useful in the study of phase transformations. (The single and double primes denote different phases.)

The importance of the Joule–Thomson coefficient,

$$\mu \equiv \left(\frac{\partial T}{\partial P}\right)_h$$

stems partly from the fact that it can be measured more accurately and more readily than certain other useful properties that are directly related to it.

Another coefficient that is useful and fairly easy to measure is the constant-temperature coefficient that is defined as

$$c \equiv \left(\frac{\partial h}{\partial P}\right)_T$$

The simple relationship among c_p, c, and μ is

$$\mu = -\frac{c}{c_p}$$

General expressions for entropy, internal energy, and enthalpy in terms of PvT and specific-heat data can be derived as

$$ds = c_v \frac{dT}{T} + \left(\frac{\partial P}{\partial T}\right)_v dv = c_p \frac{dT}{T} - \left(\frac{\partial v}{\partial T}\right)_P dP \tag{9·12,13}$$

$$du = c_v\, dT + \left[T\left(\frac{\partial P}{\partial T}\right)_v - P\right] dv \tag{9·14}$$

$$dh = c_p\, dT - \left[T\left(\frac{\partial v}{\partial T}\right)_P - v\right] dP \tag{9·15}$$

General relations involving c_p and c_v are

$$c_p - c_v = T\left(\frac{\partial P}{\partial T}\right)_v \left(\frac{\partial v}{\partial T}\right)_P \tag{9·16}$$

$$c_p - c_v = -T\left(\frac{\partial P}{\partial v}\right)_T \left(\frac{\partial v}{\partial T}\right)_P^2 = \frac{Tv\beta^2}{\kappa_T} \tag{9·17}$$

$$k = \frac{c_p}{c_v} = \left(\frac{\partial P}{\partial v}\right)_s \left(\frac{\partial v}{\partial P}\right)_T = \frac{\kappa_T}{\kappa_s} \tag{9·18}$$

The laws of Debye and of Dulong and Petit are useful in estimating the specific heats of many solids.

In the tabulation of ideal-gas properties as a function of temperature, three quantities

ϕ, p_r (called relative pressure), and v_r (called relative specific volume) are used. ϕ is defined by

$$\phi_2 - \phi_1 \equiv \int_1^2 \frac{c_p\, dT}{T}$$

so that, for any two states 1 and 2 of an ideal gas

$$s_2 - s_1 = \phi_2 - \phi_1 - R \ln \frac{P_2}{P_1} \tag{9·19}$$

p_r and v_r are so defined that for *isentropic processes only* of an ideal gas

$$\frac{p_{r2}}{p_{r1}} = \frac{P_2}{P_1} \quad \text{and} \quad \frac{v_{r2}}{v_{r1}} = \frac{v_2}{v_1}$$

For real gases in many applications the ideal-gas equation of state is sufficiently accurate. When it is unsatisfactory, we modify it by the use of *compressibility factors* or use other equations of state. A compressibility factor Z is defined as

$$Z \equiv \frac{Pv}{RT}$$

For an ideal gas, $Z = 1$. For a real gas, Z is a function of pressure and temperature. The value of Z for any state is a direct indication of the error involved in using $Pv = RT$ for that state.

An approximation that is most accurate in the vicinity of the critical state is that if any two gases have equal values for the ratio of pressure to critical pressure and equal values for the ratio of temperature to critical temperature, then the ratio of specific volume to critical specific volume is the same for the two gases. The ratios of pressure, temperature, and specific volume to the corresponding critical values are called *reduced coordinates* or *reduced properties*. By definition,

$$P_R \equiv \frac{P}{P_c} \qquad T_R \equiv \frac{T}{T_c} \qquad v_R \equiv \frac{v}{v_c}$$

where the subscript R denotes a reduced property, and the subscript c denotes a property at the critical state.

A better approximation, except near the critical state, is that all gases at the same reduced pressure and reduced temperature have approximately the same compressibility factor. This makes it possible to use a single *generalized compressibility factor* chart of Z versus P_R at various T_R for all gases.

For some gases these approximations give more accurate results if empirically determined *pseudoreduced* pressures and temperatures are used instead of the reduced pressures and temperatures as defined above.

Many equations of state have been developed for real gases. Some samples have been presented in this chapter.

The specific heats of a real gas are functions of both pressure and temperature. The influence of pressure on specific heats is greater at lower temperatures. In the absence of specific data on real gases, properties such as h, u, and s can be approximated by means of generalized property charts based on generalized compressibility factors.

Suggested Reading and References

9·1 Holman, J. P., *Thermodynamics,* 2nd ed., McGraw-Hill, New York, 1974, Chapter 7.

9·2 Kestin, Joseph, *A Course in Thermodynamics,* revised printing, McGraw-Hill, New York, 1979, Chapter 20.

9·3 Reynolds, William C., and Henry C. Perkins, *Engineering Thermodynamics,* 2nd ed., McGraw-Hill, New York, 1977, Chapter 8.

9·4 Smith, J. M., and H. C. Van Ness, *Introduction to Chemical Engineering Thermodynamics,* 2nd ed., McGraw-Hill, New York, 1975, Chapter 6.

9·5 Vargaftik, N. B., *Tables on the Thermophysical Properties of Liquids and Gases in Normal and Dissociated States,* 2nd ed., Wiley, New York, 1975.

9·6 Hatsopoulos, George N., and Joseph H. Keenan, *Principles of General Thermodynamics,* Wiley, New York, 1965.

Problems

9·1 For a single phase of a pure substance, sketch lines of constant entropy and constant temperature on Pv coordinates and prove that the relative slopes you show are correct.

9·2 Determine the numerical value of the slope of a constant-pressure line on an hs diagram for air at 2 atm, 35°C (95°F). Would the slope be greater than, less than, or the same as this for liquid water at the same pressure and temperature? Explain your reasoning.

9·3 Determine the value of $(\partial s/\partial v)_T$ for ethene at 150 kPa, 20°C.

9·4 Determine $(\partial P/\partial T)_s$ at 100 kPa, 200°C, (*a*) for nitrogen, and (*b*) approximately for Refrigerant 12.

9·5 Determine the value of $(\partial s/\partial v)_T$ for air at 10 psia, 40°F.

9·6 Determine $(\partial P/\partial T)_s$ at 11 psia, 480°F, (*a*) for air, and (*b*) approximately for steam.

9·7 From the data below on superheated ammonia, determine for ammonia at 90 psia, 70°F, (*a*) c_p, and (*b*) the value of Y in the following approximate expression for enthalpy change: $\Delta h = c_p \Delta T + Y \Delta P$.

Pressure, psia		Temperature, °F		
		60	70	80
80	h	634.3	640.6	646.7 B/lbm
	v	3.812	3.909	4.005 cu ft/lbm
90	h	631.8	638.3	644.7
	v	3.353	3.442	3.529
100	h	629.3	636.0	642.6
	v	2.985	3.068	3.149

9·8 Show that the latent heat of vaporization may be expressed as

$$h_{fg} = T \int_{v_f}^{v_g} \left(\frac{\partial P}{\partial T}\right)_v dv$$

9·9 From the data given below, compute the value of h_{fg} for steam at 0.70 MPa, and check the results with the steam-table value.

P Saturation, MPa	T Saturation, °C	Specific Volume, m³/kg	
		$v_f \times 10^3$	$v_g \times 10^3$
0.60	158.85	1.1006	315.7
0.70	164.97	1.1080	272.9
0.80	170.43	1.1148	240.4

9·10 From the following data determine the value for the latent heat of refrigerant 13 at 240 K.

T Saturation, K	P Saturation, MPa	Specific Volume, m³/kg	
		v_f	v_g
250	1.046	0.000792	0.01501
240	0.7647	0.000761	0.02071
230	0.5431	0.000734	0.02912

9·11 From the data of Fig. 3·3 and the compressed-liquid and saturated-solid data of the Keenan, Keyes, Hill, and Moore *Steam Tables* (Reference 3·5), estimate the latent heat of fusion of water at 140 MPa. State clearly any assumptions or approximations that you make.

9·12 By means of the Clapeyron equation and the data of Table A·1.5, determine the saturation pressure for water vapor at −55°C.

9·13 Develop the following relation on the basis of the Clapeyron equation, if the vapor in equilibrium with a liquid at a given temperature and pressure behaves as an ideal gas.

$$\frac{d(\ln P)}{dT} = \frac{h_{fg}}{RT^2}$$

9·14 Determine the change in atmospheric pressure with altitude near sea level. Then using this value and values of T_{sat}, h_{fg}, and v_{fg} for water at a single pressure, determine the change in the water boiling temperature with altitude near sea level.

9·15 From the data given below, compute the value of h_{fg} for steam at 100 psia, and check the results with the steam-table value.

P Saturation, psia	T Saturation, °F	Specific Volume, cu ft/lbm	
		v_f	v_g
90	320.31	0.01766	4.898
100	327.86	0.01774	4.434
110	334.82	0.01781	4.051

9·16 From the following data determine the value for the latent heat of Refrigerant 113 at −20°F.

T Saturation, °F	P Saturation, psia	Specific Volume, cu ft/lbm	
		v_f	v_g
−10	0.6046	0.00959	42.48
−20	0.4288	0.00953	58.61
−30	0.2987	0.00947	82.86

9·17 Compute the amount that the melting-point temperature of ice is lowered for an increase in pressure of 1 atm. The specific volumes of ice and water at 32°F are 0.01747 and 0.01602 cu ft/lbm, respectively.

9·18 Ice and water at 0.0886 psia and 32.02°F have the following properties.

Phase	Enthalpy, B/lbm	Specific Volume, cu ft/lbm
Liquid	0.01	0.01602
Solid	−143.34	0.01747

Assuming that the enthalpy and specific volume are independent of the pressure, compute the melting temperature or equilibrium temperature at 1000 atm.

9·19 By means of the Clapeyron equation and the data of Table A·2.1 or A·2.2, determine the saturation pressure for ammonia at −80°F.

9·20 Show that an approximate expression for the variation of saturation pressure with temperature, $P = Ae^{-b/T}$, where A and b are constants, can be derived from the Clapeyron equation by making certain assumptions (such as the independence of h_{fg} and pressure).

9·21 Determine the Joule–Thomson coefficient of Refrigerant 12 at (*a*) 0.02 MPa, −20°C, and (*b*) 2.0 MPa, 90°C.

9·22 Determine the Joule–Thomson coefficient of water at (*a*) 70 kPa, 150°C, and (*b*) 20.5 MPa, 400°C.

9·23 Determine the Joule–Thomson coefficient of water at (*a*) 10 psia, 300°F, and (*b*) 3000 psia, 750°F.

9·24 Determine the Joule–Thomson coefficient of a van der Waals gas in terms of a, b, R_u, T, v_N, and C_p.

9·25 Prove that for a van der Waals gas the inversion temperature is given by

$$T = \frac{2a}{bR_u}\left(1 - \frac{b}{v_N}\right)^2$$

9·26 The velocity of sound c in a medium is given by

$$c = \sqrt{\left(\frac{\partial P}{\partial \rho}\right)_s}$$

Determine the velocity of sound in terms of quantities such as P, v, T, R, k, and so forth for (*a*) an ideal gas and (*b*) an incompressible liquid.

9·27 For an ideal gas show that the coefficient of volume expansion is equal to the reciprocal of the absolute temperature.

9·28 For an ideal gas show that the isothermal bulk modulus, $-v(\partial P/\partial v)_T$, is equal to the pressure of the gas.

9·29 Derive an expression for the change in enthalpy of a gas that follows the equation of state $P(v - b) = RT$.

9·30 Establish a relation for the change in enthalpy of a gas that follows the van der Waals equation of state.

9·31 Determine an expression for the change in entropy of a van der Waals gas.

9·32 From Eq. 9·16 determine the value of $(c_p - c_v)$ for an ideal gas.

9·33 Show that for a reversible adiabatic process of a gas which follows the equation of state $P(v - b) = RT$

$$T(v - b)^{R/c_v} = \text{constant}$$

9·34 Derive the following expression from fundamental relations.

$$\left(\frac{\partial u}{\partial v}\right)_T = T\left(\frac{\partial P}{\partial T}\right)_v - P$$

9·35 Develop the following relation for the Helmholtz function, a.

$$a = u + T\left(\frac{\partial a}{\partial T}\right)_v$$

9·36 Derive the following from fundamental relations.

$$\left(\frac{\partial h}{\partial P}\right)_T = v - T\left(\frac{\partial v}{\partial T}\right)_P$$

9·37 Prove the following equality:

$$\left(\frac{\partial c_v}{\partial v}\right)_T = T\left(\frac{\partial^2 P}{\partial T^2}\right)_v$$

9·38 Derive the relation

$$\left(\frac{\partial^2 a}{\partial T^2}\right)_v = \frac{c_v}{T}$$

9·39 Determine the coefficient of volume expansion at 27°C for a gas which obeys the equation of state $P(v_N - b) = R_uT$. Under the conditions stated, v equals 1500 cm³/mole and b equals 20 cm³/mole.

9·40 Determine from steam-table data the value of β for water at (*a*) 70 kPa, 150°C, (*b*) 20.5 MPa, 400°C, and (*c*) 20.5 MPa, 150°C.

9·41 Determine from steam-table data the value of β for water at (*a*) 10 psia, 300°F, (*b*) 3000 psia, 750°F, and (*c*) 3000 psia, 300°F.

9·42 Starting with the air table values at 300 K and using the specific-heat relation given in Table A·4, calculate the values of ϕ and h at 1000 K. Compare your results with the tabular values.

9·43 During a constant-pressure heating process, 230 kJ/kg of heat is added to air initially at 1 atm, 60°C. Determine the final temperature.

9·44 During a constant-pressure heating process, 230 kJ/kg of heat is added to hydrogen at 1 atm, 60°C. Determine the final temperature.

9·45 During a constant-pressure heating process, 230 kJ/kg of heat is added to propane at 1 atm, 60°C. Determine the final temperature.

9·46 One-half kilogram of dry air expands isentropically until the final specific volume is six times the initial value. If the temperature at the start of the process is 1500 K, compute (*a*) the final temperature, (*b*) the final enthalpy, and (*c*) the entropy change.

9·47 Air at 700 kPa, 650°C, enters a nozzle with negligible velocity and expands isentropically to 350 kPa. Calculate the exit velocity.

9·48 Air enters a turbine at 350 kPa, 650°C, at a rate of 4.5 kg/s and expands isentropically to 100 kPa. Calculate the work delivered per kilogram of air. State your assumptions.

9·49 Air enters a gas turbine at 725 kPa, 650°C, and expands adiabatically to 100 kPa, 315°C. Neglecting changes in kinetic energy, calculate the work done per kilogram of air. Is the process reversible?

9·50 In a closed system, air is compressed isentropically from 95 kPa, 150°C, to 400 kPa and is then heated reversibly at constant volume until its pressure is 550 kPa. Sink temperature is 5°C. Using air tables, calculate, per kilogram of air, (*a*) the heat added and (*b*) the amount of this heat that is unavailable energy.

9·51 Air enters a well-insulated turbine at 500 kPa, 900°C, and leaves at 100 kPa. The flow is steady at a rate of 3.4 kg/s. State either the minimum possible exhaust temperature or the additional information required in order to determine it.

9·52 Starting with the air table values at 600 R and using the specific-heat relation given in Table A·4, calculate the values of ϕ and h at 2000 R. Compare your results with the tabular values.

9·53 An insulated tank having a volume of 5000 cu ft contains dry air initially at 60 psia, 1000 R. Air flows out through a small nozzle. Determine the state of the air in the tank after one-half the original mass has left. State any assumptions which you make.

9·54 During a constant-pressure heating process, 100 B/lbm of heat is added to air initially at 140°F. Determine the final temperature.

9·55 One pound of dry air expands isentropically until the final specific volume is 6 times the initial value. If the temperature at the start of the process is 3000 R, compute (*a*) the final temperature, (*b*) the final enthalpy, and (*c*) the entropy change.

9·56 Air at 75 psia, 1420°F, enters a nozzle with negligible velocity and expands isentropically to 15 psia. Calculate the exit velocity.

9·57 Air enters a turbine at 50 psia, 1200°F, at a rate of 10 lbm/s and expands isentropically to 14.5 psia. Calculate the work delivered per pound of air. State your assumptions.

9·58 Air enters a gas turbine at 105 psia, 1200°F, and expands adiabatically to 15 psia, 600°F. Neglecting changes in kinetic energy, calculate the work done per pound of air. Is the process reversible?

9·59 In a closed system, air is compressed isentropically from 14.0 psia, 300°F, to 56.5 psia and is then heated reversibly at constant volume until its pressure is 80.6 psia. Sink temperature is 40°F. Using air tables, calculate, per pound of air, (*a*) the heat added, and (*b*) the amount of this heat that is unavailable energy.

9·60 A closed, rigid tank with a volume of 3.3 cu ft contains air initially at 20 psia, 140°F. Heat is added reversibly until the pressure becomes 40 psia. Atmospheric temperature is 60°F. Determine the fraction of the heat added that is unavailable energy.

9·61 Air enters a well-insulated turbine at 70 psia, 1600°F, and exhausts at 14 psia. The flow is steady at a rate of 5.8 lbm/s. State either the minimum possible exhaust temperature or the additional information required in order to determine it.

9·62 Compute the pressure in a tank having a volume of 0.009 m^3 and containing 4.5 kg of nitrogen at 273 K. Use Fig. 9·11.

9·63 Determine the specific volume of nitrogen under a pressure of 62 MPa and at a temperature of 50°C.

9·64 Compute the compressibility factor for steam at 14 MPa and 500°C. The specific volume under these conditions is 0.02252 m^3/kg.

9·65 A closed, rigid tank that has a volume of 0.085 m^3 contains nitrogen at 28 MPa, 200°C. Calculate the nitrogen pressure that results when the gas in the tank is cooled to 0°C.

9·66 Estimate the pressure of nitrogen at a temperature of 50°C if the specific volume is 0.00246 m^3/kg.

9·67 Compare the specific volume of steam at 30 MPa, 500°C, obtained by use of the generalized compressibility factor chart with the steam-table value.

9·68 Determine the density of helium at 3.0 MPa, −220°C.

9·69 Compressibility factor has been defined as the ratio of the specific volume of a gas to the specific volume predicted by the ideal-gas equation of state. Does this agree with the definition presented in Sec. 9·7?

9·70 Compute the pressure in a tank having a volume of 0.34 cu ft and containing 10 lbm of nitrogen at 32°F.

9·71 Determine the specific volume of nitrogen under a pressure of 9000 psia and at a temperature of 122°F.

9·72 Compute the compressibility factor for steam at 2000 psia, 900°F. The specific volume under these conditions is 0.3532 cu ft/lbm.

9·73 A closed, rigid tank that has a volume of 3.0 cu ft contains nitrogen at 4000 psia, 392°F. Calculate the nitrogen pressure that results when the gas in the tank is cooled to 32°F.

9·74 If the values for the reduced pressure and compressibility factor for ethene are 20 and 1.25, respectively, compute the temperature.

9·75 If the compressibility factor and reduced pressure are 0.7 and 1.2, respectively, for isopentane, compute the temperature.

9·76 Estimate the pressure of nitrogen at a temperature of 122°F if the specific volume is 0.0394 cu ft/lbm.

9·77 Compare the specific volume of steam at 5000 psia, 800°F, obtained by use of the generalized compressibility factor chart with that in the steam tables.

9·78 Determine the density of helium at 500 psia, −300°F.

9·79 Section 9·9 states that the constant b in the van der Waals' equation of state is of the same order of magnitude as the specific volume of the liquid phase of a substance. For any four of the substances of Table A·4, report on the validity of this statement.

9·80 Develop the following form of van der Waals' equation in terms of the reduced coordinates.

$$\left(P_R + \frac{3}{v_R^2}\right)(v_R - \tfrac{1}{3}) = \tfrac{8}{3} T_R$$

9·81 By means of van der Waals' equation of state, plot for nitrogen on Pv coordinates two isotherms: one for the ice-point temperature and one for the steam-point temperature, both in a pressure range extending to 300 atm. Show the points at 100, 200, and 300 atm that would be obtained by the ideal-gas equation of state and by means of compressibility factors.

9·82 Derive an expression in terms of the van der Waals constants, other properties of the gas, and the pressure limits, for the work of a reversible isothermal expansion of a van der Waals gas for (*a*) a closed system and (*b*) a steady-flow system with no changes in kinetic or potential energy.

9·83 Derive an expression in terms of the Redlich–Kwong constants, other properties of the gas, and the pressure limits, for the work of a reversible isothermal expansion of a Redlich–Kwong gas for (*a*) a closed system and (*b*) a steady-flow system with no changes in kinetic or potential energy.

9·84 Determine $(\partial c_p/\partial P)_T$ for a van der Waals gas.

9·85 Compute the specific volume of steam at 1.4 MPa, 400°C by means of (*a*) the ideal-gas equation of state, (*b*) a compressibility factor, (*c*) the Redlich–Kwong equation of state.

9·86 Compute the specific volume of propane at 4.25 MPa, 244°C by means of (*a*) a generalized compressibility factor, (*b*) the Redlich–Kwong equation.

9·87 Compute the specific volume of nitrogen at 500 atm, 200°C, by means of (*a*) the ideal-gas equation of state, (*b*) van der Waals' equation, (*c*) the Redlich–Kwong equation, and (*d*) reduced coordinates.

9·88 By means of the ideal-gas equation and van der Waals' relation, compute the pressure of 3 moles of sulfur dioxide at 65°C when the total volume is 3.14 m^3.

9·89 Compute the pressure of argon for a temperature of 425°C and a specific volume of 0.0145 m^3/kg by means of the Redlich–Kwong equation.

9·90 One kilogram of acetylene at 30°C occupies 2.5 m^3. Compute the pressure of the gas according to the Redlich–Kwong equation.

9·91 Compute the pressure of neon for a temperature of 35°C and a specific volume of 0.0095 m^3/kg by means of the Beattie–Bridgeman equation.

9·92 Using the Beattie–Bridgeman equation, compute the pressure of a quantity of air compressed to a volume of 0.0002 m^3 at a temperature of 5°C. Before compression, the air occupied a volume of 0.03 m^3 at 20°C and 101.3 kPa.

9·93 Compute the specific volume for nitrogen at 70 MPa, 90°C, by means of (*a*) the Beattie–Bridgeman equation of state and (*b*) a compressibility factor.

9·94 Compute the specific volume of nitrogen at 1000 atm, 400°F, by means of (*a*) the ideal-gas equation of state, (*b*) van der Waals' equation, (*c*) the Redlich–Kwong equation, and (*d*) reduced coordinates.

9·95 By means of the ideal-gas equation and van der Waals' relation, compute the pressure of 3 lbmol of sulfur dioxide at 150°F when the total volume is 111 cu ft.

9·96 Compute the pressure of argon for a temperature of 800°F and a specific volume of 0.2 cu ft/lbm by means of van der Waals' equation.

9·97 Two pounds of nitrogen at 800°F occupies 90 cu ft. Compute the pressure of the gas according to van der Waals' equation.

9·98 Compute the pressure of neon for a temperature of 100°F and a specific volume of 0.15 cu ft/lbm by means of the Beattie–Bridgeman equation.

9·99 Using the Beattie–Bridgeman equation, compute the pressure of a quantity of air compressed to a volume of 0.008 cu ft at a temperature of 40°F. Before compression, the air occupied a volume of 1 cu ft at 70°F and 14.7 psia.

9·100 Compute the specific volume for nitrogen at 10 000 psia, 200°F, by means of the Beattie–Bridgeman equation of state.

9·101 Acetylene in a closed cylinder is initially at 50°C with a density of 60 kg/m^3. It expands isothermally until its volume is doubled. The lowest temperature in the surroundings is 60°C. Determine the fraction (percent) of the heat added that is unavailable energy.

9·102 Propane in a closed cylinder with an initial specific volume of 0.30 ft^3/lbm and a temperature of 300°F expands isothermally until its volume is doubled. The lowest temperature in the surroundings is 60°F. Determine the fraction (percent) of the heat added that is unavailable energy.

9·103 Sketch, and label corresponding points and processes on, force–length and temperature–entropy diagrams for a Carnot engine using as a working material (*a*) a steel wire, (*b*) a rubber band. (Suggestion: Some simple experiments with a rubber band show that rubber and steel are quite different in regard to thermal expansion. Further interesting conclusions result from applying the equations of Appendix B as we have done in this chapter to relate properties.)

9·104 Determine the power required to compress ethene steadily, reversibly, and isothermally at a rate of 0.31 kg/s from 10 MPa, 40°C, to 40 MPa. Determine also the rate of heat removal and the size of the discharge line if the mean velocity at outlet is not to exceed 20 m/s.

9·105 Determine the power required to compress ethane steadily, reversibly, and isothermally at a rate of 0.45 lbm/s from 100 atm, 200°F, to 400 atm. Determine also the rate of heat removal and the size of the discharge line if the mean velocity at outlet is not to exceed 50 fps.

9·106 Oxygen in a closed system expands reversibly and adiabatically from 50 MPa, 100°C, to 15 MPa. Calculate the work done per kilogram of oxygen.

9·107 Oxygen in a closed system expands reversibly and adiabatically from 4500 psia, 240°F, to 1500 psia. Calculate the work done per pound of oxygen.

9·108 Air at 30 MPa is to be cooled at constant pressure at a rate of 0.020 kg/s from 17 to -5°C by flowing over the evaporator coils of a refrigerator. Refrigerant 12 is supplied to the evaporator as saturated liquid at -10°C. Determine the flow rate of refrigerant needed if it is to leave the evaporator at a temperature not higher than 10°C.

9·109 Air enters a compressor at 3.0 MPa, 40°C, with a velocity of 100 m/s through a cross-sectional area of 4.2 cm^2. The air is compressed steadily and adiabatically to 9.0 MPa. The discharge velocity is very low. The power input is 30 percent greater than that required for reversible adiabatic compression. Determine the power input and the discharge temperature.

9·110 Oxygen enters a compressor at 800 psia, 140°F, with a velocity of 200 ft/s through a cross-sectional area of 2.12 sq in. The oxygen is compressed steadily and adiabatically to 3000 psia. The discharge velocity is very low. The power input is 30 percent greater than that required for reversible adiabatic compression. Determine the power input and the discharge temperature.

9·111 Nitrogen expands reversibly and isothermally in a closed system from 2000 psia, 100°F, to 1000 psia. Determine the heat transfer per pound by using (*a*) van der Waals' equation of state and the general equation for a property change, (*b*) generalized property charts.

9·112 Air in a vessel of 0.50 m^3 volume is initially at 25 MPa, 20°C. Air is bled from the vessel to hold the pressure constant while heat is added to raise the temperature of the air remaining in the vessel to 120°C. Determine the heat transfer.

9·113 Carbon dioxide in a vessel of 0.50 m^3 volume is initially at 50 MPa, 40°C. Gas is bled from the vessel to hold the pressure constant while heat is added to raise the temperature of the gas remaining in the vessel to 130°C. Determine the heat transfer.

9·114 An insulated vessel with a volume of 0.15 m^3 initially contains air at atmospheric pressure, 20°C. It is connected to a line through which air at 30 MPa, 40°C, flows. Determine the final temperature of the air in the vessel if the filling process is adiabatic.

9·115 Nitrogen expands reversibly and isothermally in a closed system from 14 MPa, 35°C, to 7 MPa. Determine the heat transfer per kilogram by using (*a*) van der Waals' equation of state and the general equation for change in a property, (*b*) generalized property charts.

9·116 Compute the heat required to raise the temperature of 0.5 kg of air from 300 to 500°C at a constant pressure of 70 MPa.

9·117 Compute the heat required to increase the temperature of 1 mole of air from 150 to 250°C at a constant pressure of 25 MPa.

9·118 Compute the heat required to raise the temperature of 1 lbm of air from 600 to 1000°F at a constant pressure of 10 000 psia.

9·119 Compute the heat required to increase the temperature of 1 lbmol of air from 300 to 700°F at a constant pressure of 5000 psia.

9·120 It was pointed out in Sec. 1·9 that properties may be classified as three types: (1) those which are directly observable, (2) those which can be defined by means of the laws of thermodynamics, and (3) those which are defined as combinations of other properties. Categorize each property in the following list of quantities accordingly: P, T, u, v, h, s, Q, R, A, U, W, m, S, c_p, V, G.

9·121 State the conditions under which each of the following relations is valid:

(*a*) $\text{work} = \int P\, dV$

(*b*) $h_2 - h_1 = c_p(T_2 - T_1)$

(*c*) $q_{\text{in}} + \text{work}_{\text{in}} = \Delta h + \Delta ke$

(*d*) $c_p - c_v = R$

(*e*) $\rho_1 A_1 V_1 = \rho_2 A_2 V_2$

(*f*) $\text{work}_{\text{in}} = \int v\, dP + \Delta ke + \Delta pe$

(*g*) $du = c_v\, dT$

(*h*) $h = u + Pv$

(*i*) $Q = U_2 - U_1 + \text{work}$

(*j*) $Pv^k = \text{constant}$

9·122 State the conditions under which each of the following relations is valid:

(*a*) $du = T\, ds - P\, dv$

(*b*) $\Delta S_{\text{system}} = 0$

(*c*) $\Delta S_{\text{isolated system}} = 0$

(*d*) $\Delta S_{\text{universe}} = 0$

(*e*) $\dfrac{Q_1}{Q_2} = \dfrac{T_1}{T_2}$

(*f*) $\left(\dfrac{\partial h}{\partial s}\right)_P = T$

(*g*) $Tds = dh - vdP$

(*h*) $\Delta S = \int \dfrac{\delta Q}{T}$

(*i*) $\left(\dfrac{\partial T}{\partial s}\right)_v = \dfrac{T}{c_v}$

(*j*) $\oint dx = 0$

9·123 State in not more than two pages which parts of Chapters 1 through 9 you believe you have a sound understanding of and which parts you are weakest in, giving as well as you can the reasons why any parts have caused difficulty.

CHAPTER TEN

Availability and Irreversibility

The second law limits the extent to which heat can be converted into work by any continuously operating device. Work can always be converted completely and continuously into heat. Therefore, work is a more valuable form of energy than heat. Because work is a valuable form of energy, an engineer often strives either to increase the work output of systems (such as engines, turbines, or entire power plants) or to decrease the work input to other systems (such as pumps, compressors, and refrigerators). An important question is, therefore, "What is the maximum amount of work that can be obtained when a system passes from one state to another?" The first law tells us how much change in stored energy results when a system goes from one state to another, but the second law is needed to tell us how much work can be obtained.

This chapter deals with the calculation of the maximum amount of work obtainable and the evaluation of the effects of irreversibility.

10·1 Maximum Work

In this chapter we will frequently refer to the *atmosphere,* so let us establish the meaning of this term. We have defined a system as any region in space within prescribed boundaries. We have defined the surroundings as everything outside the system boundary. For almost all thermodynamic systems one part of the surroundings is an atmosphere of uniform pressure and temperature. This atmosphere is so large in comparison with the system that its pressure and temperature are not changed by any process of the system; but the atmosphere does influence the behavior of the system. The atmosphere is only part of the surroundings. In Fig. 10·1, for example, the system is the gas in the cylinder, and important parts of the surroundings in addition to the atmosphere are the energy reservoir at T_R and the coil spring. The system exchanges work with the spring and heat with the energy reservoir. Whenever the system changes volume it either does work on the atmosphere (as well as on the spring) or the atmosphere does work on it. Therefore, the total work done by the system is equal to that delivered to systems other than the atmosphere plus that done on the atmosphere.

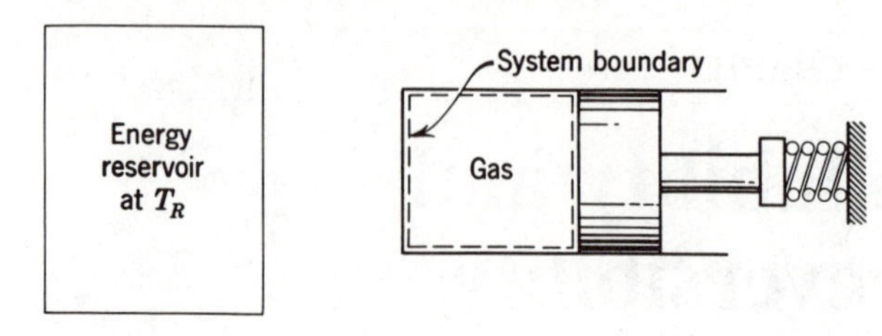

Figure 10·1. System and surroundings.

Let us determine the maximum amount of work that can be done by a system that goes from one specified state to another while exchanging heat with only the atmosphere. To be general, let us consider an open system, as shown in Fig. 10·2, that for some infinitesimal change has a mass δm_1 entering and a mass δm_2 leaving. The special cases of a closed system and of a steady-flow system can be developed later from this general approach. The first law applied to this system for an infinitesimal change is

$$\delta Q - \delta W + (e_1 + P_1 v_1)\,\delta m_1 - (e_2 + P_2 v_2)\,\delta m_2 = dE \tag{2·8c}$$

For specified end states and specified flows into and from the system, the first law can give us *only the difference* $(Q - W)$, and many different values of Q and W are possible. We must use the second law to answer the question: For specified end states, specified flows into and from the system, and heat exchange with only the atmosphere, what is the maximum amount of work obtainable? We will permit the use of auxiliary devices in producing this maximum work only if they operate cyclically so that they experience no permanent change. We find as we study this problem that the maximum work can be done by means of an externally reversible process, as shown by the following reasoning: Consider a system as shown in Fig. 10·2 that exchanges heat with only the atmosphere, has specified amounts of mass in specified states crossing its boundaries, and passes from a state i to a state f. *Suppose* that under these conditions some process A can occur that results in more work output W_A than some externally reversible process R between the same end states. (This is our hypothesis.) Let this process A occur. Then return the system to state i by the reverse of the reversible process R. During this reverse process those parts of the surroundings that exchanged mass with the system during process A are also returned to their initial states. The amount of work done on the system during the reverse process is W_R which (by our hypothesis) is smaller in magnitude than W_A. Thus for the cycle of the system from state i to state f and back to state i there is a net work output of $(W_A - W_R)$, heat has been exchanged with only the atmosphere, and there have been no other effects in the surroundings. This is a violation of the second law. Therefore, our hypothesis is false. *Conclusion*: For a specified change in state and specified mass exchange

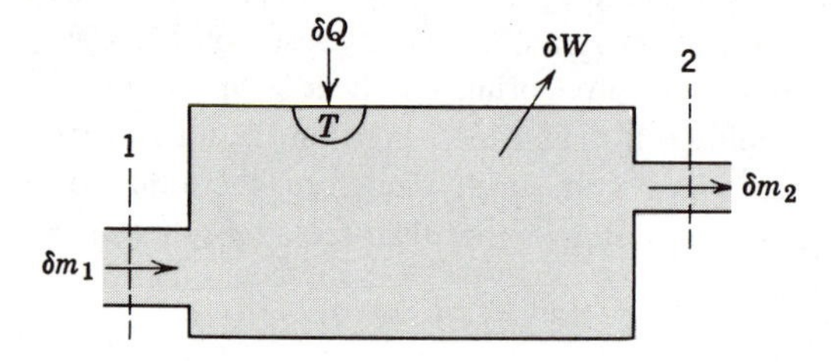

Figure 10·2. Open system.

with the surroundings, a system exchanging heat with only the atmosphere produces the maximum work by means of an externally reversible process. We can also conclude that all externally reversible processes conducted under the same conditions result in the same work output.

For the system shown in Fig. 10·2, let T be the temperature of that part of the system where heat enters or leaves. (Remember that for an open system in general T may vary from point to point throughout the system.) T_0 is the temperature of the atmosphere. Heat cannot be transferred reversibly directly between the system and the atmosphere because in general there is a finite difference between T and T_0, but a transfer of heat can be accomplished reversibly by means of an externally reversible engine (such as a Carnot engine) interposed between the system and the atmosphere. This arrangement is shown in Fig. 10·3. Let δQ be the heat added to the system. The heat entering the engine from the atmosphere is δQ_0. For the externally reversible engine

$$\frac{\delta Q}{T} = \frac{\delta Q_0}{T_0}$$

and

$$\delta W_{\text{eng}} = \delta Q_0 - \delta Q = \left(\frac{T_0}{T} - 1\right) \delta Q$$

(Notice that δW_{eng} is positive whether $T < T_0$ or $T > T_0$. When $T < T_0$, heat enters the system so that δQ is positive. Then both δQ and $(T_0/T - 1)$ are positive, so that their product δW_{eng} is positive. When $T > T_0$, heat leaves the system so that δQ is negative. $(T_0/T) - 1$ is also negative, so the product δW_{eng} is positive.) The maximum amount of work that can be obtained then is

$$\begin{aligned}\delta W_{\text{max}} &= \delta W_{\text{system}} + \delta W_{\text{eng}} \\ &= \delta W_{\text{system}} + \left(\frac{T_0}{T} - 1\right) \delta Q \\ &= (\delta W_{\text{system}} - \delta Q) + T_0 \frac{\delta Q}{T}\end{aligned}$$

Remember that δQ is transferred reversibly because for maximum work the system must undergo an externally reversible process. By the first law, for the open system

$$\delta W - \delta Q = -dE_\sigma + (e_1 + P_1 v_1)\, \delta m_1 - (e_2 + P_2 v_2)\, \delta m_2 \qquad (2\text{·}8c)$$

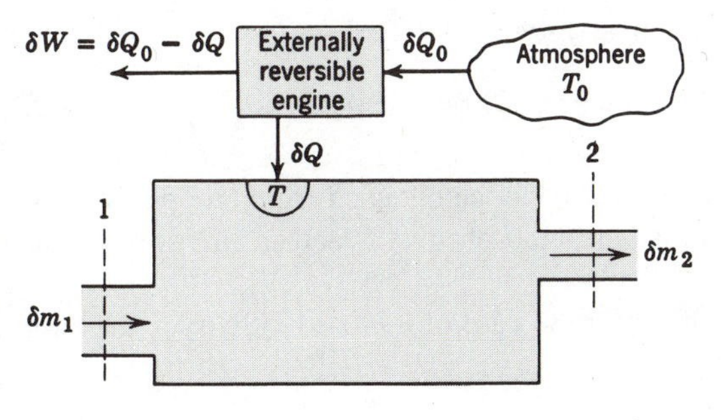

Figure 10·3. Reversible heat transfer between system and atmosphere.

Also, for an open system we have shown that

$$\left(\frac{\delta Q}{T}\right)_{\text{rev}} = dS_\sigma - s_1\,\delta m_1 + s_2\,\delta m_2 \tag{8·1b}$$

Making these two substitutions from Eqs. 2·8c and 8·1b into the expression for $\delta W_{\max}$ and collecting terms, we have

$$\delta W_{\max} = -d(E - T_0 S)_\sigma + (e_1 + P_1 v_1 - T_0 s_1)\,\delta m_1 - (e_2 + P_2 v_2 - T_0 s_2)\,\delta m_2 \tag{10·1}$$

Now consider the case in which the system exchanges heat with some reservoir at a temperature T_R as well as with the atmosphere. Let the reservoir exchange no mass with the surroundings. Applying Eq. 10·1 to the combination of the open system and the reservoir, and denoting properties of the reservoir by the subscript R,

$$\delta W_{\max} = -d(E - T_0 S)_\sigma - d(E - T_0 S)_R + (e_1 + P_1 v_1 - T_0 s_1)\,\delta m_1 - (e_2 + P_2 v_2 - T_0 s_2)\,\delta m_2 \tag{a}$$

An energy reservoir is usually considered to undergo only internally reversible changes and to do no work. Thus the first law applied to the reservoir, with δQ_R being the heat transferred *to* the system, is

$$\delta Q_R = -dE_R \tag{b}$$

Then we have for the second term on the right-hand side of Eq. a

$$-dE_R + T_0 dS_R = \delta Q_R + T_0 dS_R \tag{c}$$

But for a reversible process of the reservoir

$$dS_R = -\frac{\delta Q_R}{T_R}$$

(The minus sign is required because δQ_R is heat transferred *to* the system *from* the reservoir.) Then Eq. c becomes

$$-dE_R + T_0 dS_R = \delta Q_R - T_0\frac{\delta Q_R}{T_R} = \delta Q_R\left(1 - \frac{T_0}{T_R}\right)$$

and substitution of this into Eq. a followed by rearrangement gives

$$\delta W_{\max} = -d(E - T_0 S)_\sigma + (e_1 + P_1 v_1 - T_0 s_1)\,\delta m_1 - (e_2 + P_2 v_2 - T_0 s_2)\,\delta m_2 + \delta Q_R\left(1 - \frac{T_0}{T_R}\right) \tag{10·2}$$

This is the expression for the maximum work that can be obtained from an open system which exchanges heat with nothing other than the atmosphere at T_0 and an energy reservoir at T_R.

Equation 10·2 can be readily simplified for the case of a closed system or the case of steady flow. For a *closed system,* $\delta m_1 = \delta m_2 = 0$, and

$$\delta W_{max} = -d(E - T_0 S)_\sigma + \delta Q_R \left(1 - \frac{T_0}{T_R}\right) \tag{10·3a}$$

For a change from state i to state f,

$$W_{max} = (E_i - T_0 S_i) - (E_f - T_0 S_f) + Q_R \left(1 - \frac{T_0}{T_R}\right) \tag{10·3b}$$

Under *steady-flow* conditions, $dE_\sigma = 0$, $dS_\sigma = 0$, and $\delta m_1 = \delta m_2 = \delta m$. Then

$$\delta W_{max} = [(e_1 + P_1 v_1 - T_0 s_1) - (e_2 + P_2 v_2 - T_0 s_2)]\, \delta m + \delta Q_R \left(1 - \frac{T_0}{T_R}\right) \tag{10·4a}$$

Per unit mass of fluid entering at section 1 and leaving at section 2,

$$w_{max} = (e_1 + P_1 v_1 - T_0 s_1) - (e_2 + P_2 v_2 - T_0 s_2) + q_R \left(1 - \frac{T_0}{T_R}\right) \tag{10·4b}$$

Recall that in the absence of electrical, magnetic, solid distortion, and surface tension effects

$$e = u + \frac{V^2}{2} + gz$$

Example 10·1. A tank with a volume of 0.85 m³ is evacuated. Atmospheric air outside the tank is at 100 kPa, 20°C. What is the maximum possible amount of work that can be done by allowing atmospheric air to enter the tank until the tank contents come to equilibrium with the atmosphere?

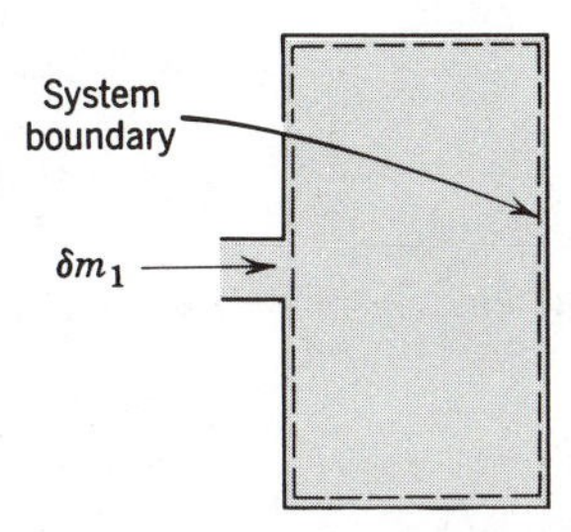

Example 10·1.

Solution. For any open system which exchanges heat with only the atmosphere,

$$\delta W_{max} = -d(E - T_0 S)_\sigma + (e_1 + P_1 v_1 - T_0 s_1)\, \delta m_1 - (e_2 + P_2 v_2 - T_0 s_2)\, \delta m_2 \tag{10·1}$$

In this case, $\delta m_2 = 0$. All the air which enters the tank crosses the open-system boundary at T_0 and P_0 with negligible kinetic energy, so that the properties of the mass δm_1 are constant and equal to those of the atmosphere. (If the kinetic energy of the mass crossing the system boundary is appreciable, then the pressure and temperature of it are lower than P_0 and T_0. Application of the first law to fluid flowing between some stagnant part of the atmosphere and the system boundary then shows that $h_0 = h_1 + V_1^2/2$, so that any kinetic energy possessed by the fluid at section 1 is balanced by a decrease in h_1 below h_0.) Thus the equation above can be readily integrated to give

$$W_{\max} = (E_i - T_0 S_i) - (E_f - T_0 S_f) + (e_0 + P_0 v_0 - T_0 s_0)(m_f - m_i)$$

where the subscripts i and f denote initial and final conditions in the tank. E_i, S_i, and m_i are each zero because there is initially nothing in the tank; so we can simplify the last equation to

$$\begin{aligned} W_{\max} &= 0 - (E_f - T_0 S_f) + (e_0 + P_0 v_0 - T_0 s_0) m_f \\ &= -e_f m_f + T_0 s_f m_f + e_0 m_f + P_0 v_0 m_f - T_0 s_0 m_f \end{aligned}$$

The gas finally in the tank comes to equilibrium with the atmosphere and is at rest; so $P_f = P_0$, $T_f = T_0$ and therefore $e_f = e_0$, $s_f = s_0$, and $v_f = v_0$.

$$W_{\max} = P_0 v_0 m_f = P_0 V_{\text{tank}} = 100(0.85) = 85 \text{ kJ}$$

(Notice that the mechanism for producing this work is entirely unspecified. No matter what devices are used, however, more work than 85 kJ cannot be obtained from the interaction of this system and the atmosphere.)

Example 10·2. Steam enters a steady-flow system at 800 kPa, 250°C, with negligible velocity and leaves at 100 kPa, 120°C, with a velocity of 160 m/s. The flow rate is 6600 kg/h. Heat is exchanged only with the surrounding atmosphere at 15°C. Determine the maximum possible power output.

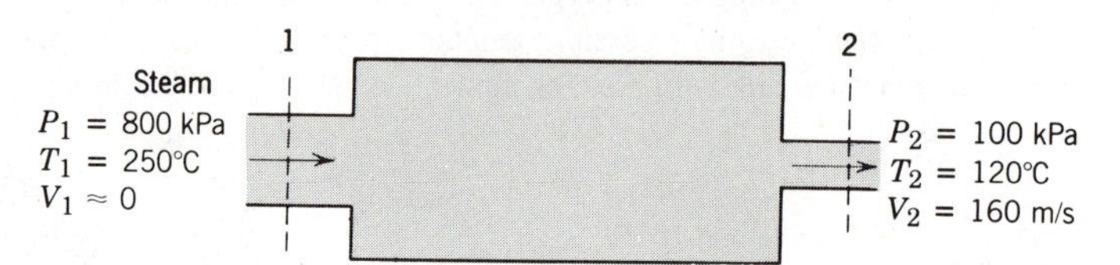

Example 10·2.

Solution. For a steady-flow system exchanging heat with only the atmosphere, the maximum work is

$$w_{\max} = (e_1 + P_1 v_1 - T_0 s_1) - (e_2 + P_2 v_2 - T_0 s_2) \qquad \text{(10·4, simplified)}$$

In this case, changes in potential energy can be neglected. Initial kinetic energy is negligible; so $e_1 = u_1$ and $e_1 + P_1 v_1 = u_1 + P_1 v_1 = h_1$. At the outlet, $e_2 = u_2 + V_2^2/2$, so that $e_2 + P_2 v_2 = h_2 + V_2^2/2$. The equation above becomes

$$w_{\max} = (h_1 - T_0 s_1) - \left(h_2 + \frac{V_2^2}{2} - T_0 s_2\right) = h_1 - h_2 - T_0(s_1 - s_2) - \frac{V_2^2}{2}$$

Obtaining property values from the steam tables

$$w_{\max} = 2950.0 - 2716.6 - 288(7.0384 - 7.4668) - \frac{(160)^2}{2(1000)} = 370 \text{ kJ/kg}$$

The maximum possible power output is then

$$\dot{W}_{max} = \dot{m}w_{max} = \frac{6600(370)}{3600} = 678 \text{ kW}$$

(Notice again that the mechanism for producing this much power is unspecified, but, no matter what devices are used, this power output cannot be exceeded. An actual machine operating with the specified end conditions will produce less power because of irreversibilities.)

Example 10·3 Air is compressed steadily from 14.0 psia, 60°F (conditions of the atmosphere) to 70.0 psia, 240°F, by a compressor that is cooled only by atmospheric air. Neglect kinetic energy changes, and determine the minimum work required per pound of air.

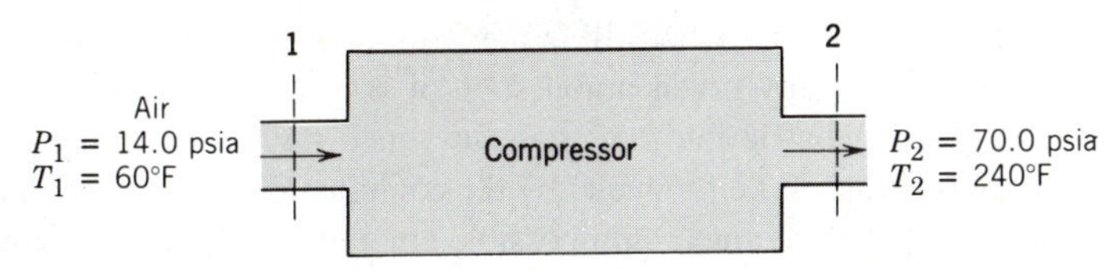

Example 10·3.

Solution. For this compression process, $w_{max} < 0$ because work must be done *on* the air. Then $-w_{max}$ = minimum w_{in}. For example, if we find that $w_{max} = -10$ units, this means that $w = -11$ units is possible but $w = -9$ units is impossible and that the minimum work input is 10 units. We calculate w_{max} for this steady-flow process (with $\Delta ke = \Delta pe = 0$) by

$$\begin{aligned} w_{max} &= (h_1 - T_0 s_1) - (h_2 - T_0 s_2) \qquad \text{(10·4, simplified)} \\ &= h_1 - h_2 - T_0(s_1 - s_2) \end{aligned}$$

Treating air as an ideal gas with constant specific heats (see Table 4·2) and using the *Tds* equation, $Tds = dh - vdP$,

$$\begin{aligned} w_{max} &= c_p(T_1 - T_2) - T_0 \int_2^1 ds = c_p(T_1 - T_2) - T_0 \left[\int_2^1 \frac{dh}{T} - \int_2^1 \frac{vdP}{T} \right] \\ &= c_p(T_1 - T_2) - T_0 \left[\int_2^1 c_p \frac{dT}{T} - R \int_2^1 \frac{dP}{P} \right] \\ &= c_p(T_1 - T_2) - T_0 \left[c_p \ln \frac{T_1}{T_2} - R \ln \frac{P_1}{P_2} \right] \\ &= 0.241(60 - 240) - 520 \left[0.241 \ln \frac{520}{700} - \frac{53.3}{778} \ln \frac{14}{70} \right] = -63.5 \text{ B/lbm} \end{aligned}$$

Since $w_{max} = -63.5$ B/lbm, $w \leq -63.5$ B/lbm or $w_{in} \geq 63.5$ B/lbm. Thus the minimum work input is 63.5 B/lbm.

10·2 Availability

In the preceding section we have seen how to calculate the maximum work obtainable from a system that goes from one specified state to another while exchanging heat with only the atmosphere and other constant-temperature energy reservoirs. As noted earlier, some of the work done by a system may be done on the atmosphere and may consequently serve no useful purpose. Also, the atmosphere does work on any system that decreases in volume.

Useful work is defined as the work done by a system exclusive of that done on the atmosphere. If P_0 is the uniform and constant pressure of the atmosphere and V_i and V_f are the initial and final volumes of a system,

$$W_{\text{useful}} = W - P_0(V_f - V_i)$$

Notice that if $V_f < V_i$, then $W_{\text{useful}} > W$, where W is the work done by the system.

An important question is: For any given initial state of a system which exchanges heat with only the atmosphere, what is the final state to which the system must go in order for the maximum useful work to be produced by the combination of the system and the atmosphere? The answer is that the maximum possible amount of useful work can be done if the system goes to a state in which its pressure and temperature equal those of the atmosphere. This answer is supported by the following reasoning: Let the atmosphere be at a pressure P_0 and a temperature T_0. If the system temperature T is different from T_0, work can always be produced by transferring heat from either the system or the atmosphere, whichever is at the higher temperature, to a heat engine that operates cyclically, produces work, and rejects heat to the lower-temperature region, either the atmosphere or the system. The cases of $T > T_0$ and $T < T_0$ are both shown schematically in Fig. 10·4. In both cases, the temperature of the system approaches T_0 as work is done. When $T = T_0$, no work can be obtained in this manner. However, even when $T = T_0$, work can be obtained if $P \neq P_0$ by letting the system change in volume until $P = P_0$. When $T = T_0$ and $P = P_0$, no work can be produced by interaction of the system and the atmosphere. They are in equilibrium with each other, and the system is said to be in the *dead state*. (The stored energy of the system in the dead state is designated as E_0 and is not necessarily zero.) Since (1) work can always be produced by the interaction of a

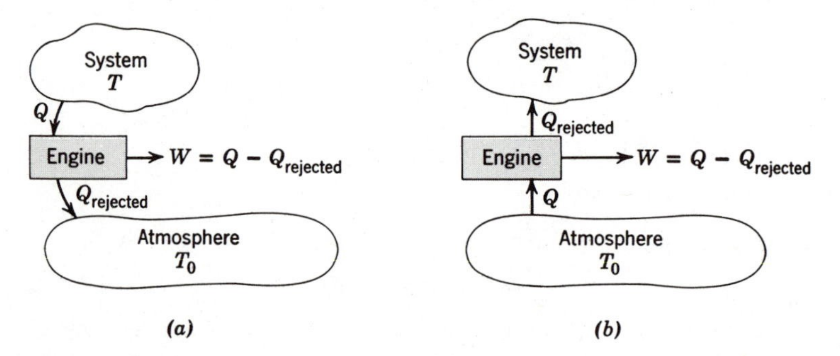

Figure 10·4. Obtaining work by transfer of heat between system and atmosphere, whether (a) $T > T_0$ or (b) $T < T_0$.

system and the atmosphere if the system is initially not in the dead state, but (2) no work can be produced if the system is initially in the dead state, the *maximum useful work can be produced by means of a process that carries the system to the dead state*. This fact leads to the definition of *availability*.

The *availability* of a system in a given state is defined as the *maximum useful work that can be obtained from the system-atmosphere combination as the system goes from that state to the dead state while exchanging heat with only the atmosphere*. Notice that availability is not a property of the system alone; its value depends on P_0 and T_0 as well as on the properties of the system. It would be more precise to speak of the availability of a system-atmosphere combination, but common usage is to refer simply to the availability of the system.

Since we have seen in the preceding article how to calculate maximum work for *any* two end states and heat exchange with only the atmosphere, we can readily calculate availability by letting the final state be the dead state and subtracting the work done by the system on the atmosphere.

The availability of a *closed system* is given the symbol Φ (and $\phi \equiv \Phi/m$) and is given by

$$\begin{aligned}\Phi_1 &\equiv W_{\text{max useful, } 1-0} = W_{\text{max, } 1-0} - P_0(V_0 - V_1) \\ &= (E_1 - T_0S_1) - (E_0 - T_0S_0) - P_0(V_0 - V_1) \\ &= (E_1 + P_0V_1 - T_0S_1) - (E_0 + P_0V_0 - T_0S_0)\end{aligned} \tag{10·5a}$$

where the subscript 0 denotes properties of the system when in the dead state. Since useful work can always be obtained if a closed system is in a state other than the dead state, $\Phi \geq 0$ *for all states*. Notice that the maximum useful work obtainable from a closed system which exchanges heat with only the atmosphere as it passes from state 1 to state 2 is given by the decrease in Φ:

$$\begin{aligned}W_{\text{max useful}} &= -\Delta\Phi = \Phi_1 - \Phi_2 \\ &= [(E_1 + P_0V_1 - T_0S_1) - (E_0 + P_0V_0 - T_0S_0)] \\ &\quad -[(E_2 + P_0V_2 - T_0S_2) - (E_0 + P_0V_0 - T_0S_0)] \\ &= (E_1 + P_0V_1 - T_0S_1) - (E_2 + P_0V_2 - T_0S_2)\end{aligned} \tag{10·5b}$$

If the system also receives heat Q_R from an energy reservoir at T_R, we have, from Eq. 10·3,

$$W_{\text{max useful}} = -\Delta\Phi + Q_R\left(1 - \frac{T_0}{T_R}\right) \tag{10·5c}$$

It can easily be shown that the availability Φ of a closed system can be increased only by doing work on the system or by transferring heat to it from a body at a temperature other than T_0.

We turn now to the availability of a *steadily flowing* fluid. Since the volume of a steady-flow system does not change, no work is done on or by the atmosphere (except flow work as described in Sec. 2·7), and there is no distinction between work and useful work. We call the availability of the steadily flowing fluid *stream availability* Y (and

$y = Y/m$), to distinguish it from the availability of a closed system Φ (or ϕ). *Stream availability* is defined as the maximum work that can be obtained as the fluid goes reversibly to the dead state while exchanging heat with only the atmosphere. Thus from Eq. 10·4, in the dead state the fluid is at P_0 and T_0 and is also at rest, $V_0 = 0$.

$$y_1 \equiv w_{\max,\ 1-0} = (e_1 + P_1 v_1 - T_0 s_1) - (e_0 + P_0 v_0 - T_0 s_0) \qquad (10\cdot 6a)$$

In the absence of electrical, magnetic, solid distortion, and surface tension effects,

$$e = u + \frac{V^2}{2} + gz \qquad \text{and} \qquad e + Pv = h + \frac{V^2}{2} + gz \qquad (10\cdot 6b)$$

Thus

$$y_1 = \left(h_1 + \frac{V_1^2}{2} + gz_1 - T_0 s_1\right) - \left(h_0 + gz_0 - T_0 s_0\right)$$

In availability calculations it is often convenient to use the *Darrieus function* which is defined as $b \equiv h - T_0 s$. For any given temperature of the atmosphere, b is a property of a substance. Thus

$$y_1 = \left(b_1 + \frac{V_1^2}{2} + gz_1\right) - \left(b_0 + gz_0\right) \qquad (10\cdot 6c)$$

and, for a change from state 1 to state 2

$$w_{\max} = -\Delta y = y_1 - y_2 = \left(b_1 + \frac{V_1^2}{2} + gz_1\right) - \left(b_2 + \frac{V_2^2}{2} + gz_2\right) \qquad (10\cdot 6d)$$

Stream availability y is often called simply the availability of a flowing fluid, but remember that for a fluid that is part of a flowing stream the maximum work which can be obtained as the fluid goes reversibly to the dead state while exchanging heat with only the atmosphere is y and not ϕ.

For a steadily flowing fluid which goes from one state to another while exchanging heat with only the atmosphere, the maximum work obtainable per unit mass is $-\Delta y$. If the fluid also receives heat in the amount q_R per unit mass from an energy reservoir at T_R, the maximum work which can be obtained per unit mass is

$$w_{\max} = -\Delta y + \left(1 - \frac{T_0}{T_R}\right) q_R \qquad (10\cdot 6e)$$

It can easily be shown that the stream availability y of a flowing fluid can be increased only by doing work on the fluid or transferring heat to it from a body at a temperature other than T_0.

Example 10·4 A turbine exhausts 38 000 kg of steam per hour at 30 mm Hg absolute, 94 percent quality, with a velocity of 140 m/s. Atmospheric pressure and temperature are 720 mm Hg, 15°C. How much power can be obtained from the steam?

Solution. The maximum work which can be obtained per kilogram of the exhaust steam is its stream availability,

$$w_{\max} = y = \left(b + \frac{V^2}{2} + gz\right) - \left(b_0 + gz_0\right) \tag{10·6c}$$

Noting that gravitational effects are negligible and that $b \equiv h - T_0 s$

$$w_{\max} = y = h - T_0 s + \frac{V^2}{2} - (h_0 - T_0 s_0) \tag{a}$$

Water in the dead state of 720 mm Hg, 15°C, is a compressed liquid. For such a low pressure, the enthalpy and entropy of compressed liquid are very nearly equal to those of saturated liquid at the same temperature. The exhaust pressure is

$$P = \gamma h = \rho_{Hg} g h_{Hg} = 1000(13.6)9.80\left(\frac{30}{1000}\right) = 4000 \text{ Pa} = 0.0040 \text{ MPa}$$

The pertinent properties are obtained from the steam tables as

$$h = h_f + x h_{fg} = 121.46 + 0.94(2432.9) = 2408.4 \text{ kJ/kg}$$

$$s = s_f + x s_{fg} = 0.4226 + 0.94(8.0520) = 7.9915 \text{ kJ/kg·K}$$

$$h_0 = h_{f,60F} = 62.99 \text{ kJ/kg}$$

$$s_0 = s_{f,60F} = 0.2245 \text{ kJ/kg·K}$$

Making these substitutions in Eq. a

$$w_{\max} = y = 2408.4 - 288(7.9915) + \frac{(140)^2}{2(1000)} - [62.99 - 288(0.2245)]$$

$$= 118.3 \text{ kJ/kg}$$

$$\dot{W}_{\max} = \dot{m} w_{\max} = \frac{38\,000(118.3)}{3600} = 1250 \text{ kW}$$

(Notice that this figure is obtained without reference to any specific means for obtaining power from the exhaust steam.)

Example 10·5 Suppose that steam enters the turbine of Example 10·4 at 700 kPa, 260°C, with negligible velocity and that the turbine is well insulated. From the specified end conditions, determine the maximum power, and compare it with the actual power output.

Solution. The maximum work per kilogram is equal to the decrease in stream availability y between inlet (section 1) and outlet (section 2). y_2 has been determined in Example 10·4. Obtaining property values from the steam tables, we calculate y_1:

$$y_1 = h_1 - T_0 s_1 - (h_0 - T_0 s_0)$$

$$= 2974.9 - 288(7.1455) - [62.99 - 288(0.2245)] = 918.7 \text{ kJ/kg}$$

The maximum work for the specified end conditions and heat exchange with only the atmosphere is

$$w_{\max} = y_1 - y_2 = 918.7 - 118.3 = 800.4 \text{ kJ/kg}$$

and

$$\dot{W}_{\max} = \dot{m} w_{\max} = \frac{38\,000(800.4)}{3600} = 8450 \text{ kW}$$

The actual work output can be obtained by application of the first law to the steady-flow system, assuming adiabatic flow:

$$w = h_1 - h_2 - \frac{V_2^2}{2} + q = 2974.9 - 2408.4 - \frac{(140)^2}{2(1000)} + 0$$

$$= 556.7 \text{ kJ/kg}$$

and $$\dot{W} = \dot{m}w = \frac{38\,000(556.7)}{3600} = 5880 \text{ kW}$$

(Section 10·4 discusses the significance of the difference between the maximum work or power and the actual work or power.)

10·3 Available and Unavailable Energy

The second law limits the conversion of energy from one form to another: Work can always be converted completely into heat, but heat cannot be converted completely into work.

In Sec. 8·6, the available and unavailable parts of heat transfer were defined as the fractions of heat which could or could not be converted into work. In Secs. 10·1 and 10·2 the maximum work obtainable from a system in any specified state was discussed, and availability and stream availability were defined. We now introduce a more general concept of available energy that covers both stored energy and energy in transition.

Available energy is energy that can by some means be converted into work. This applies to both energy in transition and all forms of stored energy. Kinetic energy, potential energy, and work itself are all available energy. Internal energy and heat may be partially available energy, but they cannot be entirely available energy. Available energy is also called *exergy. Unavailable energy is energy that can by no means be converted into work.* Some energy is totally unavailable. For example, the energy of the atmosphere is huge in amount, but no fraction of it can be converted into work by any means.

The earlier discussions of maximum work, availability, and the available and unavailable parts of heat transfer have shown that the maximum conversion of energy into work is achieved through reversible processes. Consequently, unavailable energy is sometimes defined as energy that cannot be converted into work *even by reversible processes.*

All energy, stored or transitional, can be classified as one of the following:

1. Energy that can be completely converted into work: available energy
2. Energy that can be partially converted into work: a combination of available and unavailable energy
3. Energy that cannot be converted into work: unavailable energy

In terms of available and unavailable energy, the first law might be stated as follows:

First law: Energy is comprised of available energy and unavailable energy, and the sum of the available energy and the unavailable energy of the universe remains constant.

The second law might be stated in any of the following forms, among others (see Reference 10·1):

> ***Second law:*** It is impossible to convert unavailable energy into available energy.
> ***Second law:*** The unavailable energy of the universe continually increases or, in the case of reversible processes only, remains constant.
> ***Second law:*** The available energy of the universe continually decreases or, in the case of reversible processes only, remains constant.

Each of these three statements can be deduced from either of the others or from any of the other second law statements. Notice that each of them limits the direction or extent of the energy conversion which the first law states is possible.

In nonscientific literature, the terms *energy waste* and *energy loss* are occasionally used. In many such cases it would be more accurate to refer to the waste or loss of *available energy*.

10·4 Irreversibility

A reversible process is defined as a process such that, after it has occurred, both the system and all the surroundings can be returned to their initial states. An irreversible process always produces some effects that cannot be undone, so it is impossible to restore *both* the system and all the surroundings to their initial states after an irreversible process has occurred. The total energy of the system and its surroundings remains constant (in accordance with the first law), but one result of an irreversible process is a decrease in the amount of energy that can be converted into work. A quantitative evaluation of irreversibility can be made in terms of this decrease in the amount of energy that can be converted into work. Consider a system that exchanges heat with only the surrounding atmosphere. For a process in which no work is done, the decrease in the amount of useful work which can be done is simply the decrease in availability. If, however, some work is done during the process, then some of the decrease in availability is compensated for by the work actually done. With this line of reasoning in mind, let us define the irreversibility I of *any* process as

$$I \equiv W_{max} - W \tag{10·7a}$$

W in this equation is the work actually produced. During the process, heat may be exchanged with various energy reservoirs as well as with the atmosphere. W_{max} is the maximum work for a process that (1) is between the same end states as the actual process, (2) involves the same amounts of heat exchange with the various energy reservoirs other than the atmosphere, and (3) involves heat exchange with the atmosphere also, if necessary. Since the work done on the atmosphere, $P_0(V_f - V_i)$, depends only on the end states of a process, it is the same for a maximum work process and for any other process between the same states. If we subtract this quantity from each term on the right-hand side of Eq. 10·7a, we have

$$I = W_{max\ useful} - W_{useful} \tag{10·7b}$$

Notice that the definition of irreversibility I given above is *not* restricted to systems that exchange heat with only the atmosphere. Following the usual convention, the symbol i is used for irreversibility per unit mass, I/m.

Irreversibility I is not a property. It is not a characteristic of a system. For any process, however, it can be evaluated from end states of the system *and the surroundings* as we shall now demonstrate.

Consider an open system as shown in Fig. 10·5. The properties of the system are denoted by the subscript σ; δm_1 is the mass entering the system, δm_2 is the mass leaving the system, and δQ_R is the amount of heat transferred to the system from an energy reservoir at a temperature T_R. δQ_R can be either positive or negative. For such a system, an expression for $\delta W_{\max}$ is

$$\delta W_{\max} = -d(E - T_0 S)_\sigma + (e_1 + P_1 v_1 - T_0 s_1)\,\delta m_1 - (e_2 + P_2 v_2 - T_0 s_2)\,\delta m_2 + \delta Q_R\left(1 - \frac{T_0}{T_R}\right) \tag{10·2}$$

and application of the first law gives the following expression for δW:

$$\delta W = \delta Q - dE + (e_1 + P_1 v_1)\,\delta m_1 - (e_2 + P_2 v_2)\,\delta m_2 \tag{2·8c}$$

Making these two substitutions in the expression above for I (Eq. 10·7) gives

$$\delta I = \delta W_{\max} - \delta W = T_0 dS_\sigma - T_0 s_1\,\delta m_1 + T_0 s_2\,\delta m_2 + \delta Q_R - T_0\frac{\delta Q_R}{T_R} - \delta Q \tag{a}$$

Since δQ is the total heat transferred to the system and δQ_R is the heat transferred from the energy reservoir to the system,

$$\delta Q - \delta Q_R = \delta Q_0$$

where δQ_0 is the heat transferred from the atmosphere to the system. As long as the atmosphere experiences no internal irreversibilities (mixing, friction, turbulence, etc.), $-\delta Q_0 = T_0 dS_{\text{atm}}$. (The minus sign is needed because δQ is heat transferred *from* the atmosphere *to* the system.) In the same manner, $-\delta Q_R/T_R = dS_R$. Making these substitutions in Eq. a gives

$$\delta I = T_0(dS_\sigma - s_1\,\delta m_1 + s_2\,\delta m_2 + dS_R + dS_{\text{atm}}) \tag{b}$$

This equation can be readily simplified for the cases of closed systems ($\delta m_1 = \delta m_2 = 0$)

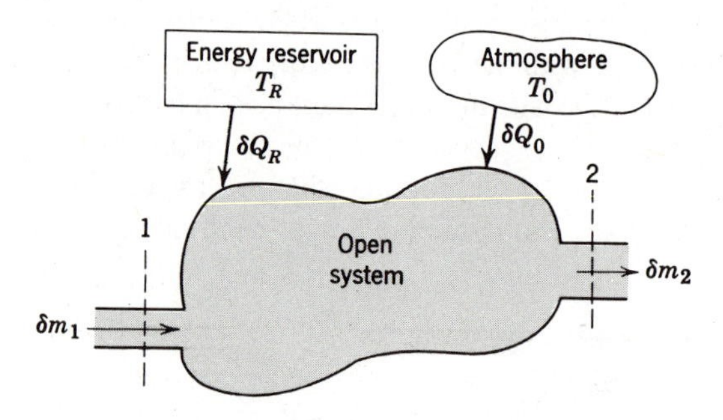

Figure 10·5. Open system.

and steady-flow systems ($dS_\sigma = 0$; $\delta m_1 = \delta m_2$). The quantity ($s_2\,\delta m_2 - s_1\,\delta m_1$) is the entropy change of those parts of the surroundings which exchange mass with the open system. Therefore, the last four terms in the parentheses of Eq. b comprise dS of the surroundings. (Those parts of the surroundings that exchange work with the system can operate reversibly and adiabatically, and hence for them $dS = 0$. Of course, those parts of the surroundings *may* operate irreversibly, but such irreversibility in no way influences the behavior of the system, atmosphere, or energy reservoir.) Thus Eq. b can be rewritten as

$$\delta I = T_0(dS_\sigma + dS_{\text{surr}}) \qquad (10\text{·}8a)$$

and, since T_0 = constant,

$$I = T_0(\Delta S_\sigma + \Delta S_{\text{surr}}) \qquad (10\text{·}8a)$$

$$= T_0\,\Delta S_{\text{isolated system}} \qquad (10\text{·}8b)$$

This last form is quite general because an isolated system can always be formed by including within a single boundary everything that affects a system during a process. The designation $\Delta S_{\text{universe}}$ is also used, with *universe* referring to the system plus all parts of the surroundings which have any interaction with the system.

Equation 10·8 is usually much easier to apply than Eq. 10·7, and it is one of the most useful of all thermodynamic equations. One reason for its usefulness is that it requires the knowledge of only end states and the heat transfer with the atmosphere. Also, unlike Eq. 10·7, it does not call for the devising of a reversible process between the actual end states for the calculation of W_{max}.

From Eq. 10·8b and the increase of entropy principle, we conclude that, for all processes

$$I \geq 0$$

where the equality holds only for reversible processes. For a *reversible* process, $W = W_{\text{max}}$, and both the system and the surroundings can be returned to their initial states. Therefore, there is no decrease in the amount of energy that can be converted into work, and it is reasonable to have $I = 0$.

For any *irreversible process,* $W < W_{\text{max}}$; so even without reference to Eq. 10·8b and the increase of entropy principle we see that $I > 0$. An irreversible process cannot be completely undone; it always results in some permanent change in the system or the surroundings. Before an irreversible process occurs, a system and its surroundings are in states such that a certain amount of their energy can be converted into work. After the irreversible process occurs, the system and surroundings are in states such that a smaller fraction of their energy can be converted into work; and this decrease in the amount of energy (stored in the system and surroundings) that can be converted into work is equal to I.

Consider a closed system that changes from state 1 to state 2 while exchanging heat with only the atmosphere at T_0. The maximum useful work is given by

$$W_{\text{max useful}} = \Phi_1 - \Phi_2 \qquad (10\text{·}5c)$$

so that

$$I = W_{\text{max useful}} - W_{\text{useful}} = \Phi_1 - \Phi_2 - W_{\text{useful}}$$

By definition Φ_1 is the maximum amount of useful work that can be obtained by the interaction of the system initially at state 1 and the atmosphere. Φ_2 has the same significance for state 2. If useful work is done during the process, it could lift a body or compress a spring so that energy in the amount W_{useful} is stored and can be reconverted into work. Thus, after the process has occurred, the total amount of useful work obtainable is $(\Phi_2 + W_{\text{useful}})$. Thus

$$\begin{aligned} I &= \Phi_1 - (\Phi_2 + W_{\text{useful}}) \\ &= \begin{bmatrix} \text{maximum useful work} \\ \text{initially obtainable} \end{bmatrix} - \begin{bmatrix} \text{maximum useful work} \\ \text{finally obtainable} \end{bmatrix} \\ &= \text{decrease in maximum useful work obtainable} \end{aligned}$$

It can readily be shown that this interpretation of I is valid for both open and closed systems and for cases of heat exchange with energy reservoirs other than the atmosphere.

I has several other valuable interpretations. It is sometimes called "energy made unavailable" because it is the increase in unavailable energy caused by a process. It is also the decrease in available energy or the decrease in exergy, sometimes called the *exergy loss*. I has also been called the *degradation* (Reference 10·2), a highly descriptive name, because it is the amount of energy degraded from the valuable available form to the less valuable unavailable form.

Example 10·6. One-half kilogram of acetylene initially at 200 kPa, 20°C is in an insulated tank. An impeller inside the tank is turned by an external motor until the pressure is 230 kPa. Determine the irreversibility of this process if the surrounding atmosphere is at 95 kPa, 20°C.

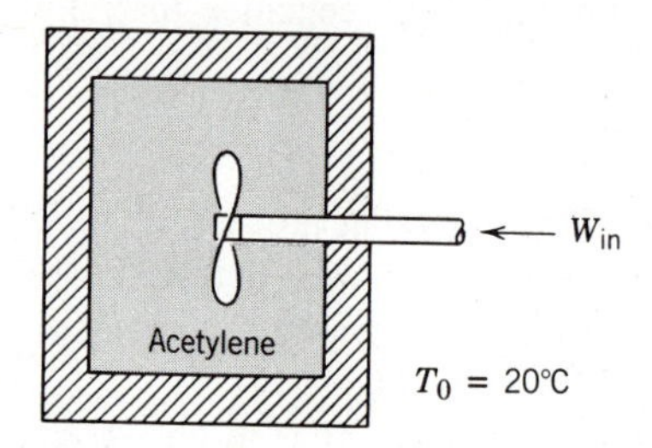

Example 10·6.

Solution.

$$I = T_0\, \Delta S_{\text{isolated system}} = T_0(\Delta S_{\text{system}} + \Delta S_{\text{surr}})$$

Since there is no change in entropy *of the surroundings* for this adiabatic process of a closed system

$$I = T_0\, \Delta S_{\text{system}} = T_0 m \int_1^2 ds = T_0 m \int_1^2 \left(\frac{du}{T} + P\,\frac{dv}{T} \right)$$

Under the conditions specified, acetylene (C_2H_2, molar mass = 26.04 kg/kmol) follows the ideal-gas equation of state. We must check to see if using a constant specific heat value is satisfactory. The final temperature for this constant volume process is

$$T_2 = T_1 \frac{P_2}{P_1} = 293 \left(\frac{345}{300} \right) = 337 \text{ K}$$

Calculating the c_p values for T_1 and T_2 from the equation and coefficients given in Table A·4,

$$c_p = R(a + bT + cT^2 + dT^3 + eT^4)$$
$$= 0.319[1.410 + 19.06(10^{-3})T - 24.50(10^{-6})T^2 + 16.39(10^{-9})T^3 - 4.135(10^{-12})T^4]$$
$$c_{p1} = 1.682 \text{ kJ/kg·K} \quad \text{at } T_1 = 293 \text{ K}$$
$$c_{p2} = 1.794 \text{ kJ/kg·K} \quad \text{at } T_2 = 337 \text{ K}$$

This small variation over the temperature range of the process indicates that the use of a mean value is satisfactory.

$$\bar{c}_p = \frac{c_{p1} + c_{p2}}{2} = \frac{1.682 + 1.794}{2} = 1.74 \text{ kJ/kg·K}$$
$$\bar{c}_v = \bar{c}_p - R = 1.74 - 0.319 = 1.42 \text{ kJ/kg·K}$$

Therefore,

$$I = T_0 m \int_1^2 \frac{c_v dT}{T} + 0 = T_0 m \bar{c}_v \ln \frac{T_2}{T_1} = T_0 m \bar{c}_v \ln \frac{P_2}{P_1}$$
$$= 293(0.5)1.42 \ln \frac{230}{200} = 29.1 \text{ kJ/K}$$

Alternative Solution.

$$I = W_{max} - W$$

For this adiabatic process of a closed system, $W_{max} = (E_1 - T_0S_1) - (E_2 - T_0S_2)$, and application of the first law gives $W = E_1 - E_2$; so

$$I = (E_1 - T_0S_1) - (E_2 - T_0S_2) - (E_1 - E_2) = T_0(S_2 - S_1)$$

From this point the solution is the same as the one presented above.

Discussion. The problem was solved without calculating $\Delta\Phi$ or W_{useful}. For discussion purposes, let us calculate these.

$$W_{useful} = W - P_0(V_2 - V_1) = E_1 - E_2 + Q - P_0(V_2 - V_1)$$
$$= U_1 - U_2 + 0 - 0 = mc_v(T_1 - T_2) = 0.5(1.42)(293 - 337)$$
$$= -31.2 \text{ kJ}$$
$$\Delta\Phi = \Phi_2 - \Phi_1 = (E_2 + P_0V_2 - T_0S_2) - (E_1 + P_0V_1 - T_0S_1)$$
$$= E_2 - E_1 - T_0(S_2 - S_1)$$
$$= 31.2 - 29.1 = 2.1 \text{ kJ}$$

Now we see that by means of the impeller 31.2 kJ of work was done on the system but the availability of the system (and atmosphere) increased by only 2.1 kJ. That is, only 2.1 kJ more useful work can be obtained from a process that starts at state 2 than can be obtained from one that starts at state 1. The surroundings can now (after the process) supply 31.2 kJ less work than before; so the net *decrease* in the amount of work obtainable is (31.2 − 2.1 = 29.1 kJ), and this is the irreversibility of the process.

Example 10·7. Air enters a counterflow heat exchanger at 75 psia, 90°F, and is heated to 220°F. Call this air stream the cold air. The cold air flow rate is 0.8 lbm/s. Heat is transferred to the cold air from an air stream that enters at 14.2 psia, 400°F, and leaves at 200°F. Call this latter stream the hot air. Both streams pass through the heat exchanger with negligible change in pressure and in kinetic energy. There are no stray heat losses. The temperature of the surrounding atmosphere is 65°F. Determine the irreversibility per pound of cold air.

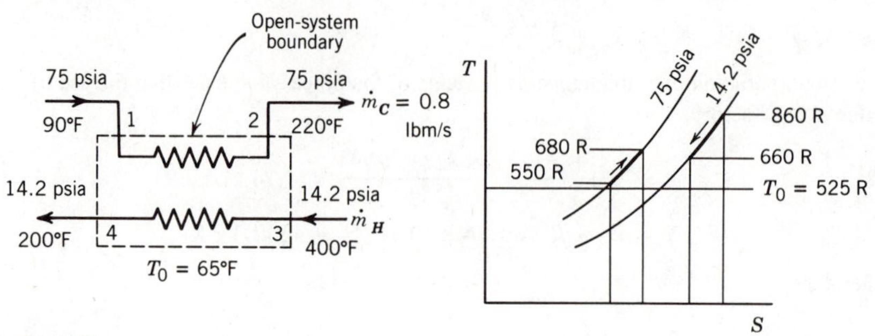

Example 10·7.

Solution. For any process of any system, $I = T_0(\Delta S_\sigma + \Delta S_{\text{surr}})$. Consider the entire heat exchanger as the system. (See the sketch.) Since it is a steady-flow system, $\Delta S_\sigma = 0$. In order to evaluate ΔS_{surr} per pound of cold air, we must determine the mass ratio of hot air to cold air. The system operates adiabatically (remember that we are taking the entire heat exchanger as the system), there is no work done, and there is no change in kinetic energy; so the first law gives us

$$\dot{m}_C(h_2 - h_1) = \dot{m}_H(h_3 - h_4)$$

$$\frac{\dot{m}_H}{\dot{m}_C} = \frac{h_2 - h_1}{h_3 - h_4}$$

For the small temperature ranges involved, c_p is nearly constant so that $\Delta h = c_p\Delta T$. From Table 4·2 we have mean values of c_p of 0.241^- B/lbm·°F between 90 and 220°F and 0.243 B/lbm·°F between 200 and 400°F. As a reasonable approximation, let us use $c_p = 0.242$ B/lbm·°F for both streams, so that

$$\frac{\dot{m}_H}{\dot{m}_C} = \frac{h_2 - h_1}{h_3 - h_4} = \frac{c_p(T_2 - T_1)}{c_p(T_3 - T_4)} = \frac{220 - 90}{400 - 200} = 0.65\ \frac{\text{lbm hot air}}{\text{lbm cold air}}$$

Then
$$\dot{I} = T_0\Delta\dot{S}_{\text{surr}} = T_0[\dot{m}_C(s_2 - s_1) + \dot{m}_H(s_4 - s_3)] \tag{10·8}$$

Per pound of cold air this is

$$i = \frac{\dot{I}}{\dot{m}_C} = \frac{T_0\Delta\dot{S}_{\text{surr}}}{\dot{m}_C} = T_0\left[(s_2 - s_1) + \frac{\dot{m}_H}{\dot{m}_C}(s_4 - s_3)\right]$$

For a constant-pressure process, the important relationship among properties $Tds = dh - vdP$ simplifies to $Tds = dh$, so that $ds = dh/T$. For an ideal gas, $dh = c_pdT$, so that $ds = c_pdT/T$, and we have

$$i = T_0\left(\int_1^2 \frac{c_pdT}{T} + \frac{\dot{m}_H}{\dot{m}_C}\int_3^4 \frac{c_pdT}{T}\right)$$

Again assuming that the c_p values are constant and the same,

$$i = T_0 c_p \left(\ln \frac{T_2}{T_1} + \frac{\dot{m}_H}{\dot{m}_C} \ln \frac{T_4}{T_3} \right) = 525(0.242) \left(\ln \frac{680}{550} + 0.65 \ln \frac{660}{860} \right)$$

$$= 5.1 \text{ B/lbm cold air}$$

(It is well to compare the value of i with the heat transferred per pound of cold air, $h_2 - h_1 = c_p(T_2 - T_1) = 0.24\ (220 - 90) = 31.5$ B/lbm). (Investigation will show that the maximum work that can be done by the two streams of air passing between the same end states and exchanging heat with only the atmosphere is 5.1 B/lbm of cold air. No work was done in the heat exchanger, so after the heat exchange process is completed, 5.1 B/lbm of cold air less work can be produced by the interaction of the streams with the atmosphere than could have been produced beforehand. This is one meaning of irreversibility.)

Example 10·8. A steam turbine operates adiabatically with a flow rate of 38 000 kg/h entering at 700 kPa, 260°C, with negligible velocity and leaving at 3.0 cm Hg absolute, 94 percent quality, with a velocity of 140 m/s. The surrounding atmosphere is at 720 mm Hg, 15°C. Determine the time rate of irreversibility.

Solution.

$$i = T_0(\Delta s_\sigma + \Delta s_{\text{surr}}) \qquad (10\cdot8)$$

$$= T_0(0 + s_2 - s_1)$$

Taking the entropy values from the steam tables (as in Examples 10·4 and 10·5),

$$i = 288(7.9915 - 7.1455) = 243.6 \text{ kJ/kg}$$

$$\dot{I} = \dot{m}i = \frac{38\ 000(243.6)}{3600} = 2570 \text{ kW}$$

It is seen that the time rate of irreversibility calculated here is the same as the difference between $\dot{W}_{\text{max}}$ and $\dot{W}$ calculated in Example 10·5: $(8450 - 5880) = 2570$ kW.

Example 10·9. Saturated liquid ammonia enters an expansion valve at 1000 kPa and leaves at 200 kPa. The inlet and outlet areas of the valve are such that there is no change in kinetic energy. While passing through the valve, the ammonia receives 40 kJ/kg heat from an energy reservoir at 0°C. The atmosphere is at 20°C. Determine the irreversibility of the process.

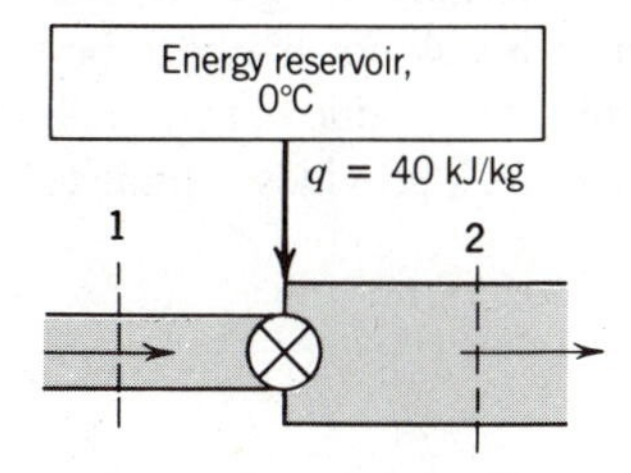

Example 10·9.

Solution.

$$i = \frac{I}{m} = \frac{T_0}{m}(\Delta S_\sigma + \Delta S_{\text{surr}})$$

For this steady-flow system, $\Delta S_\sigma = 0$. The entropy of the surroundings changes by an amount $m(s_2 - s_1)$ because of the mass transfer between the system and the surroundings and also by an amount $-Q_R/T_R$, where Q_R is heat transfer *to* the system *from* the reservoir at T_R. Thus

$$i = \frac{T_0}{m}\left[0 + m(s_2 - s_1) - \frac{Q_R}{T_R}\right] = T_0\left(s_2 - s_1 - \frac{q_R}{T_R}\right)$$

s_1 can be read directly from ammonia tables as s_f at 1000 kPa. We know the pressure at the outlet, but we must know one other independent property in order to determine the state and the value of s_2. (Since at the outlet the ammonia is a liquid-vapor mixture, we can find its temperature immediately from the saturated ammonia table, but we must still find another property because P and T are not independent for two-phase mixtures.) By application of the first law we can find h_2 because the first law applied to a steady-flow system,

$$h_2 = h_1 + q - w - \Delta ke - \Delta pe \tag{2·9}$$

in this case reduces to

$$h_2 = h_1 + q = 298.3 + 40 = 338.3 \text{ kJ/kg}$$

Now we can find the quality at the outlet,

$$x_2 = \frac{h_2 - h_f}{h_{fg}} = \frac{338.3 - 94.8}{1325.7} = 0.184$$

and
$$s_2 = s_f + x_2 s_{fg} = 0.3383 + 0.184(5.2131) = 1.2975 \text{ kJ/kg·K}$$

Now we can calculate i by substituting numerical values:

$$i = T_0\left(s_2 - s_1 - \frac{q_R}{T_R}\right) = 293\left(1.2975 - 1.1219 - \frac{40}{273}\right) = 8.5 \text{ kJ/kg}$$

Continually estimating or predicting results of engineering calculations is always good practice. Doing so can facilitate the calculations and also reveal calculation errors. However, estimating numerical values for entropy changes may be difficult for anyone without considerable experience in such calculations, so it should be noticed that the general equation $I = T_0 \Delta S_{\text{universe}}$ is useful for this purpose. As examples, in a compression process, the irreversibility cannot (except under highly unusual circumstances) be greater than the work input; in a heat exchanger process the irreversibility is usually not greater than the total heat transfer; and in a turbine or engine the irreversibility is not likely to be of a greater order of magnitude than the work output. Other limits of the value of irreversibility can be reasoned for various other processes. Then the ratio I/T_0 provides a limit to the possible entropy change of the universe for the process.

10·5 Availability Accounting

The first law is the basis for energy balances or energy accounting. The second law makes possible availability accounting which gives information that cannot be obtained from the first law. Let us look at an illustration of the fact that comparisons of energy quantities alone may be quite inconclusive if not actually misleading. Consider two streams of water, one (designated A) at 20 psia, 200°F, flowing at a rate of 1000 lbm/h, and the other (designated B) at 20 psia, 88°F, flowing at a rate of 3000 lbm/h. Let the kinetic energy in each case be negligibly small. The lowest temperature in the surroundings is 32°F. These streams flowing into a system would deliver equal amounts of energy because $\dot{m}_A h_A = 1000(167.99) = 167\,990$ B/h $\approx$ 168 000 B/h, and $\dot{m}_B h_B = 3000(56.0) =$ 168 000 B/h. The stream availabilities are quite different, however, because

$$\dot{m}_A b_A = \dot{m}_A(h_A - T_0 s_A) = 1000[167.99 - 492(0.2938)] = 23\,400 \text{ B/h}$$

and

$$\dot{m}_B b_B = \dot{m}_B(h_B - T_0 s_B) = 3000[56.0 - 492(0.1079)] = 8730 \text{ B/h}$$

Therefore, even though the streams deliver equal energy to a system, much more power can be obtained from stream A than from stream B.

Availability accounting involves the determination of availability changes for each of a series of processes. Comparison of the work done with the availability change gives a measure of the degree to which each process approaches the ideal. It thereby indicates where efforts spent in improving performance are likely to be most fruitful. A slight variation of *availability accounting* is *available energy* (or *exergy*) accounting. Both are used in subsequent chapters. The following example illustrates the general procedure.

Example 10·10. A flow diagram and a *Ts* diagram are shown for a steady-flow power plant using air as a working medium. Both the compressor and the turbine operate adiabatically. Changes in kinetic energy are negligible. Pressures and temperatures are shown on the diagram. The heat-addition process is externally reversible. Atmospheric temperature T_0 is 15°C. Make an energy accounting and an availability accounting for the air flowing through the plant.

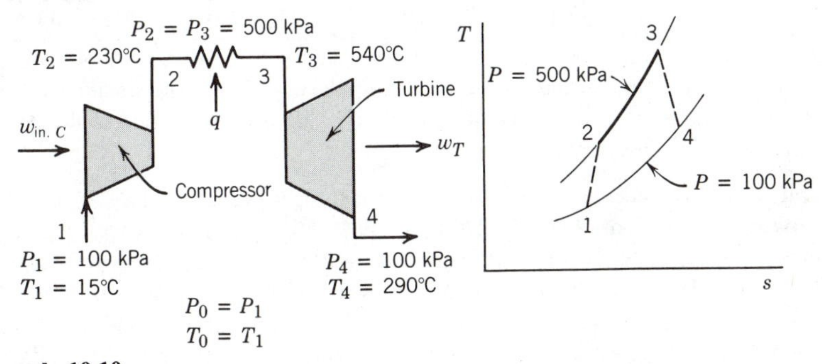

Example 10·10.

Solution. For the temperature range involved, the variation in c_p is significant (see Table 4·2), so we use for the determination of enthalpy and entropy values *Gas Tables* (Reference 4·4, abridged in Table A·5.) (Alternatively, the c_pT relation of Table A·4 or one of the available computer programs for gas properties could be used.)

Application of the first law to each of the three pieces of equipment gives

$$\text{Work}_{\text{in},C} = h_2 - h_1 = 506.39 - 288.38 = 218.01 \text{ kJ/kg}$$
$$q = h_3 - h_2 = 836.45 - 506.39 = 330.06 \text{ kJ/kg}$$
$$\text{Work}_T = h_3 - h_4 = 836.45 - 568.54 = 267.91 \text{ kJ/kg}$$

The energy balance on the plant is then as follows:

Energy in		Energy out	
With entering fluid	$h_1 = 288.38$ kJ/kg	With leaving fluid	$h_4 = 568.54$ kJ/kg
Compressor work$_{\text{in}}$	$w_{\text{in}} = 218.01$	Turbine work	$w = 267.91$
Heat	$q = 330.06$		
Total	$= 836.45$ kJ/kg	Total	836.45 kJ/kg

Since changes in kinetic and potential energy are being neglected, the changes in stream availability equal the changes in the Darrieus function, or $\Delta y = \Delta b = \Delta(h - T_0 s) = \Delta h - T_0 \Delta s$. The entropy changes for the three processes that occur within the plant are

$$s_2 - s_1 = \phi_2 - \phi_1 - R \ln \frac{P_2}{P_1} = 6.2254 - 5.6606 - 0.287 \ln \frac{500}{100} = 0.1029 \text{ kJ/kg·K}$$

$$s_3 - s_2 = \phi_3 - \phi_2 - R \ln \frac{P_3}{P_2} = 6.7352 - 6.2254 - 0 = 0.5098 \text{ kJ/kg·K}$$

$$s_4 - s_3 = \phi_4 - \phi_3 - R \ln \frac{P_4}{P_3} = 6.3421 - 6.7352 - 0.287 \ln \frac{100}{500} = 0.0688 \text{ kJ/kg·K}$$

An availability accounting can then be tabulated as follows, with all quantities in kJ/kg.

Process	q	w	$\Delta y = \Delta b$	$i = T_0 \Delta s_{\text{universe}}$
Compressor, 1-2	0	−218.0	188.4	29.6
Heat exchanger, 2-3	330.1	0	183.3	0
Turbine, 3-4	0	267.9	−287.7	19.8
Summation, 1-4	330.1	49.9	84.0	49.4

Since state 1 is the state of the surrounding atmosphere, $y_1 = 0$. The stream availability of the air leaving the plant at state 4 is therefore 84.0 kJ/kg, the summation of the Δy's for the three processes between states 1 and 4.

Conclusions. From the energy balance or first-law analysis we conclude that 330.1 kJ/kg of heat was added, a net work of $(267.9 - 218.0) = 49.9$ kJ/kg was produced, and the exhaust gas removed from the system (as $u + Pv$, or h) 280.2 kJ/kg more energy than the entering stream brought in. The extent of possible improvements in the cycle cannot be determined directly from the energy balance. From the availability accounting or second-law analysis, we see that the 49.9 kJ/kg net work output was produced from an increase of stream availability in the heat exchanger of 183.3 kJ/kg. The second-law analysis also shows why more work was not obtained: The air left the plant with a stream availability of 84.0 kJ/kg more than the incoming fluid, and the irreversibilities in the compressor and turbine cost 29.6 and 19.8 kJ/kg of stream availability, respectively. Thus the 183.3 kJ/kg increase in availability in the heat exchanger is disposed of as follows:

Work	49.9 kJ/kg
Leaving stream availability	84.0
Compressor irreversibility	29.6
Turbine irreversibility	19.8
Total	183.3 kJ/kg

10·6 Summary

The behavior of most systems is influenced by an *atmosphere* of uniform and constant pressure and temperature that is one part of the surroundings. The pressure and temperature of the atmosphere are designated as P_0 and T_0. T_0, the temperature of the atmosphere, is also called the *sink temperature* or the *lowest temperature in the surroundings* because there is usually no energy reservoir available at a lower temperature for supplying or receiving large quantities of heat.

For a general open system which exchanges heat with nothing but the atmosphere at T_0 and another energy reservoir at T_R, the maximum work from the combination of system and surroundings for any infinitesimal change in the system is

$$\delta W_{max} = -d(E - T_0S)_\sigma + (e_1 + P_1v_1 - T_0s_1)\,\delta m_1 - (e_2 + P_2v_2 - T_0s_2)\,\delta m_2 + \delta Q_R\left(1 - \frac{T_0}{T_R}\right) \qquad (10·2)$$

where δm_1 is the mass entering, δm_2 is the mass leaving, δQ_R is the heat transferred to the system from the reservoir at T_R, and the subscript σ designates properties of the system. This equation can be readily simplified for a *closed system* as follows:

$$\delta W_{max} = -d(E - T_0S) + \delta Q_R\left(1 - \frac{T_0}{T_R}\right) \qquad (10·3a)$$

and for a *steady-flow system* as follows:

$$\delta W_{max} = [(e_1 + P_1v_1 - T_0s_1) - (e_2 + P_2v_2 - T_0s_2)]\,\delta m + \delta Q_R\left(1 - \frac{T_0}{T_R}\right) \qquad (10·4a)$$

Useful work is defined as the work done by a system exclusive of that done on the atmosphere. If P_0 is the pressure of the atmosphere and V_i and V_f are the initial and final volumes of a system,

$$W_{useful} = W - P_0(V_f - V_i)$$

The greatest possible amount of work can be obtained from a system that exchanges heat with only the atmosphere if the system goes to a final state in which it is in equilibrium with the atmosphere. The pressure and temperature of the system will then be P_0 and T_0, and the system is said to be in the *dead state.*

The *availability* of a system in any state is defined as the maximum useful work that can be obtained as the system goes from that state to the dead state while exchanging heat with only the atmosphere. Availability is not a property of the system alone; its value depends on P_0 and T_0 as well as on the properties of the system.

The availability of a *closed system,* Φ (and $\phi = \Phi/m$), is given by

$$\Phi_1 \equiv W_{max\ useful,\,1\text{-}0} = (E_1 + P_0V_1 - T_0S_1) - (E_0 + P_0V_0 - T_0S_0) \qquad (10·5a)$$

where the subscript 0 denotes properties of the system when in the dead state. The maximum useful work obtainable from a closed system that exchanges heat with only the atmosphere as it passes from state 1 to state 2 is given by the decrease in Φ:

$$W_{\text{max useful}} = -\Delta\Phi = \Phi_1 - \Phi_2$$
$$= (E_1 + P_0V_1 - T_0S_1) - (E_2 + P_0V_2 - T_0S_2) \qquad (10\text{·}5b)$$

Since useful work can always be obtained if a closed system is in a state other than the dead state, $\Phi \geq 0$ for all states. If the system also receives heat Q_R from an energy reservoir at T_R,

$$W_{\text{max useful}} = -\Delta\Phi + Q_R\left(1 - \frac{T_0}{T_R}\right) \qquad (10\text{·}5c)$$

The availability of a *steadily flowing* fluid is called *stream availability* and is denoted by the symbol Y (and $y = Y/m$) to distinguish it from the availability of a closed system Φ (or ϕ). It is defined as the maximum work that can be obtained as the fluid goes to the dead state while exchanging heat with only the atmosphere. In the dead state the fluid is at P_0 and T_0 and is also at rest, $V_0 = 0$. For a steadily flowing fluid in a state 1,

$$y_1 \equiv w_{\text{max},1\text{-}0} = (e_1 + P_1v_1 - T_0s_1) - (e_0 + P_0v_0 - T_0s_0) \qquad (10\text{·}6a)$$

In the absence of electrical, magnetic, solid distortion, and surface tension effects, this becomes

$$y_1 = \left(h_1 + \frac{V_1^2}{2} + gz - T_0s_1\right) - \left(h_0 + gz_0 - T_0s_0\right) \qquad (10\text{·}6b)$$

The *Darrieus function* is defined as $b \equiv h - T_0s$. For any given temperature of the atmosphere, b is a property of a substance. The use of b shortens the preceding equation to

$$y_1 = \left(b_1 + \frac{V_1^2}{2} + gz_1\right) - \left(b_0 + gz_0\right) \qquad (10\text{·}6c)$$

and, for a change from state 1 to state 2,

$$w_{\text{max}} = -\Delta y = y_1 - y_2 = \left(b_1 + \frac{V_1^2}{2} + gz_1\right) - \left(b_2 + \frac{V_2^2}{2} + gz_2\right) \qquad (10\text{·}6d)$$

Stream availability y is often called simply the availability of a flowing fluid, but remember that for a fluid which is part of a flowing stream the pertinent quantity is y and not ϕ.

For a steadily flowing fluid that goes from one state to another while exchanging heat with only the atmosphere, the maximum work obtainable per unit mass is $-\Delta y$. If the fluid also receives heat q_R per unit mass from an energy reservoir at T_R, the maximum work that can be obtained per unit mass is

$$w_{\text{max}} = -\Delta y + \left(1 - \frac{T_0}{T_R}\right)q_R \qquad (10\text{·}6e)$$

All energy, whether stored or transitional, can be divided into two parts: *available energy* (also called *exergy*) and *unavailable energy*. Available energy is energy that can by some means be converted into work. Unavailable energy is energy that cannot be converted into work, even by reversible engines. Unavailable energy cannot be converted into available energy. An equivalent to the increase of entropy principle is: The unavailable energy of the universe continually increases or, in the limit of a reversible process, remains constant.

The *irreversibility* I of any process is defined as

$$I \equiv W_{\text{max}} - W = W_{\text{max useful}} - W_{\text{useful}} \tag{10·7}$$

W in this equation is the work actually produced. During the process, heat may be exchanged with various energy reservoirs as well as with the atmosphere. W_{max} is the maximum work for a process which (1) is between the same end states as the actual process, (2) involves the same amounts of heat exchange with the various energy reservoirs other than the atmosphere, and (3) involves heat exchange with the atmosphere also if necessary. For any process,

$$I = T_0(\Delta S_\sigma + \Delta S_{\text{surr}}) = T_0 \, \Delta S_{\text{isolated system}} = T_0 \, \Delta S_{\text{universe}} \tag{10·8}$$

For all processes, $I \geq 0$, with the equality holding for reversible processes. I can be interpreted as the decrease in available energy caused by a process.

Availability accounting provides information that cannot be obtained from energy accounting alone. It involves the determination of availability changes for each of a series of processes. Comparison of the work done with the availability change gives a measure of the degree to which each process approaches the ideal. Available energy accounting is similar. Both indicate where efforts spent in improving performance are likely to be most fruitful.

References and Suggested Reading

10·1 Baehr, Hans Dieter, *Thermodynamik,* 5th ed., Springer-Verlag, Berlin, 1984.

10·2 Dixon, J. R., *Thermodynamics I,* Prentice-Hall, Englewood Cliffs, N.J., 1975, Chapter 11.

10·3 Hatsopoulos, George N., and Joseph H. Keenan, *Principles of General Thermodynamics,* Wiley, New York, 1965, Chapter 17.

10·4 Keenan, J. H., *Thermodynamics,* Wiley, New York, 1941, Chapter XVII.

10·5 Keenan, J. H., "Availability and Irreversibility in Thermodynamics," *British Journal of Applied Thermodynamics,* vol. 2, July 1951, pp. 183–192.

10·6 Moran, M. J., *Availability Analysis: A Guide to Efficient Energy Use,* Prentice-Hall, Englewood Cliffs, N.J., 1982

10·7 Sussman, M. V., *Availability (Exergy) Analysis, A Self Instruction Manual,* published by M. V. Sussman, Tufts University, Medford, Mass., 1980.

10·8 Van Wylen, Gordon J., and Richard E. Sonntag, *Fundamentals of Classical Thermodynamics,* 3rd ed., Wiley, New York, 1985, Chapter 8.

10·9 Wark, K., *Thermodynamics,* 4th ed., McGråw-Hill, New York, 1983, Chapter 8.

Problems

10·1 One-twentieth kilogram of acetylene in a closed system expands from 200 kPa, 60°C, to 100 kPa, 35°C, while receiving 1.0 kJ of heat from a reservoir at 120°C. The surrounding atmosphere is at 95 kPa, 25°C. Determine the maximum work. How much of this work would be done on the atmosphere?

10·2 Saturated liquid Refrigerant 12 at 1.2 MPa is throttled to 200 kPa. Determine the maximum work if the sink temperature (see the first paragraph of Sec. 10·6) is 17°C.

10·3 Calculate the availability of a 20-kg block of ice at 0°C if the surrounding atmosphere is at 95 kPa, 15°C.

10·4 A tank with a volume of 0.3 m^3 contains air at 700 kPa, 20°C. The surrounding atmosphere is at 95 kPa, 20°C. Determine the availability of the air in the tank. Is this the same as the maximum work which can be obtained from expansion of the air into the surrounding atmosphere?

10·5 A tank with a volume of 0.3 m^3 contains air at 95 kPa, 260°C. The surrounding atmosphere is at 95 kPa, 20°C. Determine the availability of the air in the tank.

10·6 Solve Problem 10·5 with the air in the tank initially at 14 kPa, 20°C.

10·7 Solve Problem 10·5 with the air in the tank initially at 95 kPa, −70°C.

10·8 One-twentieth kilogram of ethane in a closed system expands from 200 kPa, 60°C, to 100 kPa, 35°C, while receiving heat from a reservoir at 120°C. Determine the maximum amount of heat that can be transferred from the reservoir to the ethane under these conditions.

10·9 Solve Example 10·1 if the tank initially contains air at 35 kPa, 20°C.

10·10 Solve Example 10·1 if the tank initially contains air at 35 kPa, −20°C.

10·11 In Sec. 10·2 the following statement appears: It can easily be shown that the availability Φ of a closed system can be increased only by doing work on the system or by transferring heat to it from a body at a temperature other than T_0. Prove this.

10·12 One-tenth pound of air in a closed system expands from 30 psia, 140°F, to 15 psia, 100°F, while receiving 1.0 B of heat from a reservoir at 250°F. The surrounding atmosphere is at 14.0 psia, 80°F. Determine the maximum work. How much of this work would be done on the atmosphere?

10·13 Steam enters a radiant superheater at 1000 psia, 700°F, with a velocity of 110 fps and leaves at 950 psia, 900°F, with a velocity of 145 fps. Heat is transferred to the steam from a furnace at 2540°F. The atmosphere is at 14.7 psia, 60°F. Determine the maximum work for this process. Devise a means for producing this amount of work without changing the process end states.

10·14 Saturated liquid ammonia at 170 psia is throttled to 30 psia. Determine the maximum work if the sink temperature (see the first paragraph of Sec. 10·6) is 65°F.

10·15 Calculate the availability of a 50-lbm block of ice at 32°F if the surrounding atmosphere is at 14.0 psia, 60°F.

10·16 A tank with a volume of 10 cu ft contains air at 100 psia, 70°F. The surrounding atmosphere is at 14.0 psia, 70°F. Determine the availability of the air in the tank. Is this the same as the maximum work that can be obtained from expansion of the air into the surrounding atmosphere?

10·17 A tank with a volume of 10 cu ft contains air at 14.0 psia, 500°F. The surrounding atmosphere is at 14.0 psia, 70°F. Determine the availability of the air in the tank.

10·18 Solve Problem 10·17 with the air in the tank initially at 2 psia, 70°F.

10·19 Solve Problem 10·17 with the air in the tank initially at 14.0 psia, −100°F.

10·20 One-tenth pound of air in a closed system expands from 30 psia, 140°F, to 15 psia, 100°F, while receiving heat from a reservoir at 250°F. Determine the maximum amount of heat which can be transferred from the reservoir to the air under these conditions.

10·21 Air initially at atmospheric conditions of 100 kPa, 20°C, is held in a rigid, well-insulated tank which has a volume of 2.80 m^3. A paddle wheel in the tank is driven by an external motor until the temperature of the air in the tank is 313°C. (*a*) Calculate the work input per kilogram of air. (*b*) After the process has occurred, what fraction of the energy put in as work during the process could possibly be reconverted to work?

10·22 Air enters a compressor at the atmospheric conditions of 100 kPa, 5°C, and is compressed to 200 kPa, 75°C. The flow rate is 6.0 kg/s. Determine, per kilogram of air, the stream availability of the entering air, the stream availability of the leaving air, and the minimum work input. What is the physical meaning of the minimum work input? Does it refer to an adiabatic compressor? To a water-cooled compressor?

10·23 Helium enters an actual turbine at 300 kPa, 200°C, and expands to 100 kPa, 150°C. Heat transfer to the atmosphere at 101.3 kPa, 25°C, amounts to 7.0 kJ/kg of helium passing through the turbine. Calculate the entering stream availability, the leaving stream availability, and the maximum work. What is the physical meaning of the maximum work? Does it refer to an adiabatic turbine? To one with a heat loss of 7.0 kJ/kg?

10·24 Heat is transferred from an energy reservoir at 840°C to steam flowing at a rate of 900 kg/h. Steam enters the heat exchanger dry and saturated at 5.2 MPa, and leaves at 5.1 MPa, 360°C. The lowest temperature in the surroundings is 5°C. Determinc the amount of energy made unavailable by this process in kJ/h.

10·25 Steam enters a turbine at 700 kPa, 200°C, and leaves at 7 kPa. The expansion is adiabatic. The flow rate is 9000 kg/h. The power output is 3.7 W. Condenser cooling water is available from a large river at 5°C. Per kilogram of steam, how much energy is made unavailable by the operation of the turbine?

10·26 In a steam power plant, the heat rejected to the condenser cooling water may be four times as great as the energy lost up the stack. Why is it that the stack loss is of greater concern; that is, why are greater efforts made to reduce it rather than the condenser loss?

10·27 For a closed system, the first law gives $\oint \delta Q = \oint \delta W$. Comment on the following equation for a closed system which is proposed as an analogous relation given by the second law:

$$\oint \delta Q_{\text{av}} = \oint \delta W + T_0 \,\Delta S_{\text{surr}}$$

10·28 Are the following true? (*a*) All reversible processes between the same two end states produce the same amount of work. (*b*) All reversible adiabatic processes between the same two end states produce the same amount of work. (*c*) All reversible processes between the same two end states produce the same amount of work if the system exchanges heat with only a constant-temperature reservoir.

10·29 Is it possible for the availability of a closed system to be negative? Explain your answer, giving reasons or examples.

10·30 Dry saturated steam at 300°F is held in a rigid, well-insulated tank that has a volume of 12.0 ft^3. A paddle wheel in the tank is driven by an external motor until the steam is at 200 psia, 1000°F. Atmospheric conditions are 14.0 psia, 80°F. (*a*) Calculate the work input in Btu/lbm. (*b*) After the process has occurred, how much of the energy put in as work during the process could possibly be reconverted to work?

10·31 Air enters a compressor at the atmospheric conditions of 15 psia, 40°F, and is compressed to 30 psia, 165°F. The flow rate is 13.0 lbm/s. Determine, per pound of air, the stream availability of the entering air, the stream availability of the leaving air, and the minimum work input. What is the physical meaning of the minimum work input? Does it refer to an adiabatic compressor? To a water-cooled compressor?

10·32 Air enters an actual turbine at 45 psia, 400°F, and expands to 15 psia, 300°F. Heat transfer to the atmosphere at 14.7 psia, 80°F, amounts to 3 B/lbm of air passing through the turbine. Calculate the entering stream availability, the leaving stream availability, and the maximum work. What is the physical meaning of the maximum work? Does it refer to an adiabatic turbine? To one with a heat loss of 3 B/lbm?

10·33 Heat is transferred from an energy reservoir at 1540°F to steam flowing at a rate of 2000 lbm/h. Steam enters the heat exchanger dry and saturated at 750 psia and leaves at 740 psia, 680°F. The lowest temperature in the surroundings is 40°F. Determine the amount of energy made unavailable by this process in B/h.

10·34 Steam enters a turbine at 100 psia, 400°F, and leaves at 1 psia. The expansion is adiabatic. The flow rate is 20 000 lbm/h. The power output is 1500 hp. Condenser cooling water is available from a large river at 40°F. Per pound of steam, how much energy is made unavailable by the operation of the turbine?

10·35 Air expands through a turbine from 400 kPa, 500°C, to 100 kPa, 340°C, while heat in the amount of 11 kJ/kg is lost to the atmosphere at 101.3 kPa, 20°C. The net change in kinetic energy is negligibly small. Calculate for this process (*a*) the decrease in stream availability, (*b*) the maximum work, and (*c*) the irreversibility.

10·36 One-fourth kilogram of air, initially at 400 kPa, 60°C, is expanded adiabatically in a closed system until its volume is doubled and its temperature equals that of the surrounding atmosphere, 5°C. Calculate for this process (*a*) the maximum work, (*b*) the change in availability, and (*c*) the irreversibility.

10·37 Steam enters a condenser at 7 kPa, 90 percent quality, (state 1) and is condensed to a saturated liquid (state 2) by cooling water which enters at 15°C (state *A*) and leaves at 20°C (state *B*). The steam flow rate is 90 kg/min. The sink temperature is 15°C. Sketch a *Ts* diagram for each of the fluids. Calculate (*a*) the maximum work and (*b*) the irreversibility.

10·38 A closed, rigid, thermally insulated tank contains 2 kg of methane at 100 kPa, 60°C. A paddle wheel inside the tank is turned by an external motor until the methane is at 170°C. The sink temperature is 15°C. Calculate the irreversibility of the process.

10·39 In an air-to-air heat exchanger, the "hot side" air enters at 260°C, 100 kPa, and the "cold side" air enters at 60°C, 1.7 MPa, and leaves at 115°C. Frictional pressure losses are negligible. Flow rate of air is 4.5 kg/s on each side. Calculate (*a*) the irreversibility of the process per kilogram of "hot side" air for a sink temperature of 15°C, and (*b*) the entropy change of the universe resulting from operation of this process for 1 h.

10·40 Steam at 700 kPa, 270°C, flowing at a rate of 9000 kg/h is desuperheated to dry saturated steam at 700 kPa by having water at 90°C sprayed into it. The spray chamber in which this occurs is called a desuperheater. The sink temperature is 15°C. Calculate, per kilogram of steam entering the desuperheater, (*a*) the amount of water needed, and (*b*) the irreversibility of the desuperheating process. Devise a means of carrying out this process reversibly.

10·41 Steam which is dry and saturated at 2.0 MPa is throttled through a well-insulated valve to 600 kPa. There is no change in kinetic energy. The flow rate is 360 kg/h. The sink temperature is 5°C. Sketch a *Ts* diagram and calculate, in kJ/kg, the irreversibility of the process.

10·42 Air is compressed adiabatically from 100 kPa, 5°C, to 300 kPa at a rate of 1.26 kg/s. Kinetic energy changes are negligible. Work input is 140 kJ/kg. (*a*) Calculate the irreversibility of the process per kilogram. (*b*) Explain the physical significance of the answer to *a*.

10·43 Steam trapped in a cylinder behind a piston is compressed reversibly and isothermally from 1000 kPa, 230°C, to 2800 kPa, 90 percent quality. The initial volume of the steam is 1.20 m^3. Although the compression is internally reversible, any heat removed goes directly to the surrounding atmosphere which is at 100 kPa, 15°C, so the process is externally irreversible. Calculate (*a*) the work of the process per kilogram of steam and (*b*) the irreversibility of the process per kilogram of steam.

10·44 In a steam-heated water heater, water at 95 kPa flowing at a rate of 2.3 kg/min is heated from 15°C to 70°C by means of steam which enters the heater at 95 kPa, 80 percent quality, and leaves as saturated liquid. The surrounding atmosphere is at 95 kPa, 15°C. Calculate, per kilogram of water heated, the irreversibility of the process.

10·45 Air expands through a turbine from 60 psia, 940°F, to 15 psia, 640°F, while heat in the amount of 5 B/lbm is lost to the atmosphere at 14.7 psia, 70°F. The net change in kinetic energy is negligibly small. Calculate for this process (*a*) the decrease in stream availability, (*b*) the maximum work, and (*c*) the irreversibility.

10·46 One-half pound of air, initially at 60 psia, 140°F, is expanded adiabatically in a closed system until its volume is doubled and its temperature equals that of the surrounding atmosphere, 40°F. Calculate for this process (*a*) the maximum work, (*b*) the change in availability, and (*c*) the irreversibility.

10·47 Steam enters a condenser at 1 psia, 90 percent quality, (state 1) and is condensed to a saturated liquid (state 2) by cooling water which enters at 60°F (state *A*) and leaves at 70°F (state *B*). The steam flow rate is 200 lbm/min. The sink temperature is 60°F. Sketch a *Ts* diagram for each of the fluids. Calculate (*a*) the maximum work and (*b*) the irreversibility.

10·48 A closed, rigid, thermally insulated tank contains 5 lbm of air initially at 15 psia, 140°F. A paddle wheel inside the tank is turned by an external motor until the air is at 340°F. The sink temperature is 60°F. Calculate the irreversibility of the process.

10·49 In an air-to-air heat exchanger, the "hot side" air enters at 500°F, 15 psia, and the "cold side" air enters at 140°F, 250 psia, and leaves at 240°F. Frictional pressure losses are negligible. Flow rate of air is 10 lbm/s on each side. Calculate (*a*) the irreversibility of the process per pound of "hot side" air for a sink temperature of 60°F, and (*b*) the entropy change of the universe resulting from operation of this process for 1 h.

10·50 Steam at 100 psia, 520°F, flowing at a rate of 20 000 lbm/h is desuperheated to dry saturated steam at 100 psia by having water at 200°F sprayed into it. The spray chamber in which this occurs is called a desuperheater. The sink temperature is 60°F. Calculate per pound of steam entering the desuperheater, (*a*) the amount of water needed and (*b*) the irreversibility of the desuperheating process. Devise a means of carrying out this process reversibly.

10·51 Steam which is dry and saturated at 300 psia is throttled through a well-insulated valve to 87 psia. There is no change in kinetic energy. The flow rate is 800 lbm/h. The sink temperature is 40°F. Sketch a *Ts* diagram, and calculate, in B/lbm, the irreversibility of the process.

10·52 Air is compressed adiabatically from 15 psia, 40°F, to 45 psia at a rate of 2.78 lbm/s. Kinetic energy changes are negligible. Work input is 60 B/lbm. (*a*) Calculate the irreversibility of the process per pound. (*b*) Explain the physical significance of the answer to *a*.

10·53 A compressor draws in air from the surrounding atmosphere at 14.0 psia, 40°F, and discharges it at 56.0 psia. The flow rate is 5.0 lbm/s, power input to the compressor is 600 hp, and during compression heat is removed from the air being compressed to the surrounding atmosphere in the amount of 10.0 B/lbm. Determine the irreversibility of the compression process per pound of air being compressed.

10·54 Measurements on a compressed air turbine operating steadily show that air enters at 20 psia, 140°F, with a velocity of 100 fps through a cross-sectional area of 0.020 sq ft and leaves at 10 psia, 80°F, with a velocity of 100 fps. The power output is 3.16 hp. The turbine casing is uninsulated. The surrounding atmosphere is at 10 psia, 40°F. Calculate the irreversibility of the process per pound of air.

10·55 Under what conditions can the irreversibility of a process be negative? Give reason(s) for your answer.

10·56 A household refrigerator has two compartments, one which is maintained at −15°C and the other which is maintained at 5°C. A container of frozen orange juice concentrate containing 350 grams of concentrate has been stored in the lower-temperature compartment. The ambient room temperature is 20°C. Making reasonable assumptions regarding physical properties and the mass of the container, calculate (*a*) the amount of energy made unavailable by taking the sealed container from the lower-temperature compartment into the surrounding room and allowing it to come to room temperature; and (*b*) the amount of energy made unavailable by first moving the sealed container into the 5°C compartment and letting it come to equilibrium and then removing it to the surrounding room and allowing it to come to equilibrium there.

10·57 Refer to Problem 10·56. Explain why there would or would not be a saving of energy required to operate the household refrigerator if procedure *a* or *b* were used regularly instead of the other procedure.

10·58 Make the investigation that is mentioned in the comment at the end of Example 10·7.

CHAPTER ELEVEN

Gas and Gas–Vapor Mixtures

A pure substance has been defined as a substance that is homogeneous and unchanging in chemical composition. Homogeneous mixtures of gases that do not react with each other are therefore pure substances, and the properties of such mixtures can be determined, correlated, and tabulated or fitted by equations just like the properties of any other pure substance. This has been done for common mixtures such as air and certain combustion products, but, since an unlimited number of mixtures is possible, properties of all of them cannot be determined experimentally and tabulated. It is therefore important to be able to calculate the properties of any mixture from the properties of its constituents. This chapter discusses such calculations, first for gas mixtures and then for gas–vapor mixtures.

11·1 Mass Fraction; Mole Fraction

The mass analysis of a gas mixture is based on the fact that the mass of a mixture is equal to the sum of the masses of its constituents:

$$m_m = m_A + m_B + m_C + \cdots \tag{a}$$

where the subscript m refers to the mixture and the subscripts A, B, and C refer to individual constituents of the mixture. The ratio m_A/m_m is called the mass fraction of constituent A. (A mass analysis is sometimes called a gravimetric analysis.)

The total number of moles* in a mixture is defined as the sum of the number of

*The General Conference of Weights and Measures in 1971 adopted, as part of the basis for the international system of units, the following definition of the mole: "The mole is the amount of substance of a system which contains as many elementary entities as there are atoms in 0.012 kilogram of carbon 12." By extension, the kilomole (kmol), pound mole (lbmol), and slug mole (slugmol) are related to carbon 12 amounts of 12 kg, 12 lbm, and 12 slug. From the relative atomic masses of various substances, the mass of a mole of any substance can then be determined and is commonly called the molar mass (or molecular weight, an unfortunate term since it has no connection with gravity). For example, one kilomole of water has a mass of 18.02 kg whether the water is a solid, a liquid, or a gas. This point is mentioned because the fact that one mole of an ideal gas occupies the same volume as one mole of any other ideal gas *at the same pressure and temperature* often misleads people into regarding the mole as a volume unit. It can be treated as a volume unit only for ideal gases at the same pressure and temperature.

moles of its constituents:

$$N_m \equiv N_A + N_B + N_C + \cdots$$

The mole fraction x is defined as

$$x \equiv \frac{N}{N_m}, \qquad x_A \equiv \frac{N_A}{N_m}, \qquad \text{etc.}$$

The number of moles N, the mass m, and the molar mass M of a substance are related by

$$m = NM \tag{b}$$

Substituting from Eq. b into Eq. a

$$m_m = N_m M_m = N_A M_A + N_B M_B + N_C M_C + \cdots$$

and

$$M_m = x_A M_A + x_B M_B + x_C M_C + \cdots$$

where M_m is called the *apparent* (or *average*) *molar mass* of the mixture.

11·2 Partial Pressure; Partial Volume

The *partial pressure* P_i of a constituent i in a gas mixture is defined as

$$P_i \equiv x_i P_m, \qquad P_A \equiv x_A P_m, \qquad \text{etc.}$$

where x is the mole fraction. From this definition, the sum of the partial pressures of the constituents of a gas mixture equals the mixture pressure

$$P_m = P_A + P_B + \cdots$$

This applies to any gas mixture, whether it is an ideal gas or not. (Sometimes partial pressure is defined in a slightly different manner. This is often the case when it is to be used for only ideal-gas mixtures. As we have defined it here, it is also useful in connection with real-gas mixtures.)

The partial volume V_i' of a constituent i in a gas mixture is defined as

$$V_i' \equiv x_i V_m, \qquad V_A' \equiv x_A V_m, \qquad \text{etc.}$$

The sum of the partial volumes of the constituents of a gas mixture equals the volume of the mixture. The partial volume is of course not an actual volume of a constituent as it exists in the mixture because each constituent fills the entire volume of the vessel that holds the mixture. The symbol V_i' instead of V_i is used as a reminder of this fact. We will see that for ideal-gas mixtures partial volume has a physical significance, but for gas mixtures in general no physical picture of partial volume can be given.

11·3 Dalton's Law or the Law of Additive Pressures

If a mixture of ideal gases is also an ideal gas, then for a mixture of ideal gases A, B, C, etc.,

$$P_m = \frac{N_m R_u T_m}{V_m} = \frac{(N_A + N_B + \cdots)R_u T_m}{V_m} = \frac{N_A R_u T_m}{V_m} + \frac{N_B R_u T_m}{V_m} + \cdots$$
$$= P_A(T_m, V_m) + P_B(T_m, V_m) + \cdots$$

where $P_A(T_m, V_m)$ is the pressure of constituent A existing at the temperature T_m and the volume V_m. This relationship is known as Dalton's law, or the law of additive pressures. Restated, *the pressure of a mixture of ideal gases equals the sum of the pressures of its constituents if each existed alone at the temperature and volume of the mixture*. Dalton's law is strictly true only for ideal-gas mixtures. It holds approximately for real-gas mixtures even in some ranges of pressure and temperature where $Pv = RT$ is quite inaccurate.

It follows from Dalton's law and the definition of partial pressure that in a mixture of *ideal gases* the partial pressure of each constituent equals the pressure which that constituent would exert if it existed alone at the temperature and volume of the mixture. This is shown as follows:

$$P_A \equiv x_A P_m = \frac{N_A}{N_m} P_m = \frac{N_A R_u T_m}{V_m} = P_A(T_m, V_m)$$

Another statement of Dalton's law is that *in a mixture of ideal gases each constituent behaves in all respects as though it existed alone at the temperature and volume of the mixture*. The internal energy and entropy of an ideal-gas mixture are equal respectively to the sums of the internal energies and entropies of the constituents if each existed alone at the temperature and volume of the mixture.

11·4 Amagat's Law, Leduc's Law, or the Law of Additive Volumes

Starting again from the premise that a mixture of ideal gases is itself an ideal gas, we can write

$$V_m = \frac{N_m R_u T_m}{P_m} = \frac{(N_A + N_B + \cdots)R_u T_m}{P_m} = \frac{N_A R_u T_m}{P_m} + \frac{N_B R_u T_m}{P_m} + \cdots$$
$$= V_A(P_m, T_m) + V_B(P_m, T_m) + \cdots$$

where $V_A(T_m, P_m)$ is the volume of constituent A when it exists at the temperature T_m and the pressure P_m. Thus *the volume of a mixture of ideal gases equals the sum of the volumes of its constituents if each existed alone at the temperature and pressure of the mixture*. This is known as Amagat's law, Leduc's law, or the law of additive volumes. Like Dalton's law it is strictly true only for ideal gases but holds approximately for real-gas mixtures even in some ranges of pressure and temperature where $Pv = RT$ is inaccurate. When the temperature of a real-gas mixture is well above the critical temperatures of all its constituents, the additive volume law is usually more accurate than the additive pressure law.

For ideal-gas mixtures, volumetric analyses are frequently used. The volume fraction is defined as

$$\text{volume fraction of } A \equiv \frac{V_A(P_m, T_m)}{V_m} = \frac{\text{volume of } A \text{ existing alone at } P_m, T_m}{\text{volume of the mixture at } P_m, T_m}$$

Notice that in a gas mixture each constituent occupies the total volume, and so volume fraction is *not* defined as the ratio of a constituent volume to the mixture volume because this ratio is always unity. Notice also that we define volume fraction or volumetric analysis only for mixtures of *ideal gases* because only for ideal gases does the law of additive volumes hold strictly. The volume fraction of a constituent in an ideal-gas mixture equals its mole fraction, as can be shown by

$$\frac{V_A(P_m, T_m)}{V_m} = \frac{N_A R_u T_m P_m}{P_m N_m R_u T_m} = \frac{N_A}{N_m} = x_A$$

and the volume of an ideal-gas mixture constituent if it existed alone at P_m and T_m equals the partial volume of the constituent in the mixture

$$V_A(P_m, T_m) = x_A V_m = V_A'$$

The equality of volume fraction and mole fraction in an ideal-gas mixture enables us to write the units of volume fraction as moles of constituent per mole of mixture, and doing so simplifies the conversion between volumetric and mass analyses. Such conversions must be made because gas mixtures are often analyzed on a volumetric basis, but a mass analysis is generally more useful in relating properties of a mixture to the properties of its constituents. Conversion from one basis to the other is illustrated in the two examples that follow. Notice that the pressure and temperature of the mixture have no bearing on the conversion. An ideal-gas mixture can be heated, cooled, compressed, or expanded, and its volumetric analysis remains constant as long as its mass analysis does. Two suggestions on making the conversions are (1) use a tabular form if there are more than two constituents, and (2) write down the units at the head of each column, and observe them carefully.

The gas mixture model based on Dalton's law is more widely used than that based on Amagat's law, but each has advantages, so they are used together often. For example, volumetric analyses, which are based on the Amagat model, are often used in connection with calculations based largely on the Dalton model.

Example 11·1. A blast-furnace gas has the following volumetric analysis in percentages: H_2, 9; CO, 24; CH_4, 2; CO_2, 6; O_2, 3; and N_2, 56. Determine the mass analysis.

Solution. In the table below, the given data are in columns *a* and *b*. The approximate molar masses are listed in column *c*. The values in column *d* are the products of those in columns *b* and *c*. The summation of column *d* is the mass of 100 moles of mixture or 100 times the apparent molar mass of the mixture. Column *e* values are obtained by dividing the column *d* values by the column *d* total and multiplying by 100.

a Constituent	*b* Volumetric Analysis, kmol/100 kmol of Mixture	*c* Molar Mass, kg/kmol	*d* kg/100 kmol of Mixture	*e* Gravimetric Analysis, kg/100 kg of Mixture
H_2	9	2	18	0.7
CO	24	28	672	25.4
CH_4	2	16	32	1.2
CO_2	6	44	264	10.0
O_2	3	32	96	3.6
N_2	56	28	1568	59.1
	100		2650	100.0

Example 11·2. A gas mixture has the following mass analysis in percentages: H_2, 10; CO, 60; and CO_2, 30. Determine the volumetric analysis.

Solution.

Constituent	Mass Analysis, kg/100 kg of Mixture	Molar Mass, kg/kmol	kmol/100 kg of Mixture	Volumetric Analysis, kmol/100 kmol of Mixture
H_2	10	2	5.00	63.9
CO	60	28	2.14	27.4
CO_2	30	44	0.68	8.7
	100		7.82	100.0

11·5 Properties of Ideal-Gas Mixtures

For the discussion of the properties of ideal-gas mixtures, consider a mixture of three ideal gases, *A*, *B*, and *C*. (The results can readily be generalized to a mixture of any number of constituents.) The properties of such a mixture in terms of the properties of its constituents are discussed here.

Temperature

For any uniform mixture the temperature is the same for each constituent and for the mixture.

$$T_m = T_A = T_B = T_C.$$

Mass, Number of Moles, and Apparent Molar Mass

The mass of a mixture, the number of moles of mixture, and its apparent molar mass are shown in Sec. 11·1 to be given by

$$m_m = m_A + m_B + m_C$$
$$N_m = N_A + N_B + N_C$$
$$M_m = x_A M_A + x_B M_B + x_C M_C$$

(These relationships hold for all mixtures, not just for ideal gases.)

Pressure

According to Dalton's law, the pressure of an ideal-gas mixture equals the sum of the pressures of the constituents if each existed alone at the temperature and volume of the mixture.

$$P_m = P_A(T_m, V_m) + P_B(T_m, V_m) + P_C(T_m, V_m)$$

For ideal-gas mixtures the pressure of any constituent if it existed alone at the temperature and volume of the mixture equals its partial pressure,

$$P_A(T_m, V_m) = P_A = x_A P_m$$

For ideal-gas mixtures the partial pressure of a constituent or its pressure if it existed alone at the temperature and volume of the mixture (these two pressures are equal only for ideal gases) can be thought of as the pressure exerted by that constituent *as it exists in the mixture*. This concept is in keeping with the molecular picture that shows the pressure of an ideal gas to be caused by the bombardment of the vessel walls by the gas molecules. From this point of view it is easy to separate the pressure of a mixture into parts, each attributable to the bombardment of the vessel walls by the molecules of one constituent. It is impossible to measure directly the pressure of just one constituent of a mixture, but nevertheless it is often convenient to treat the partial pressure of a constituent in an ideal-gas mixture as the pressure exerted by that constituent *as it exists in the mixture*.

Volume

The volume of each constituent of a gas mixture is the same as the volume of the mixture because the molecules of each constituent are free to move throughout the entire space occupied by the mixture.

$$V_m = V_A = V_B = V_C$$

The law of additive volumes equates the volume of an ideal-gas mixture to the sum of the volumes of its constituents if each existed alone at the pressure and temperature of the mixture,

$$V_m = V_A(P_m, T_m) + V_B(P_m, T_m) + V_C(P_m, T_m)$$

For ideal gases, the volume of a constituent existing alone at the pressure and temperature of the mixture is equal to its partial volume,

$$V_A(P_m, T_m) = V'_A = x_A V_m = x_A V_A(P_A, T_m)$$

Internal Energy, Enthalpy, Entropy, and so on

The Dalton's law statement that in a mixture of ideal gases each constituent behaves in all respects as though it existed alone at the temperature and volume of the mixture leads to

$$
\begin{aligned}
U_m &= U_A(V_m, T_m) + U_B(V_m, T_m) + U_C(V_m, T_m)\\
H_m &= H_A(V_m, T_m) + H_B(V_m, T_m) + H_C(V_m, T_m)\\
S_m &= S_A(V_m, T_m) + S_B(V_m, T_m) + S_C(V_m, T_m)
\end{aligned}
$$

and to

$$
\begin{aligned}
u_m &\equiv \frac{U_m}{m_m} = \frac{U_A + U_B + U_C}{m_m} = \frac{m_A u_A + m_B u_B + m_C u_C}{m_m}\\
h_m &\equiv \frac{H_m}{m_m} = \frac{m_A h_A + m_B h_B + m_C h_C}{m_m}\\
s_m &= \frac{S_m}{m_m} = \frac{m_A s_A + m_B s_B + m_C s_C}{m_m}
\end{aligned}
$$

and similar expressions for other properties such as a and g. In these equations, the constituent properties must be evaluated as though each existed alone at the temperature and volume of the mixture. The internal energy and enthalpy of an ideal gas are functions of temperature only, and the only temperature we use in evaluating properties of a mixture or its constituents is the mixture temperature T_m; so $U_A(T_m, V_m) = U_A(T_m) = U_A$, and $H_A(T_m, V_m) = H_A(T_m) = H_A$. However, the entropy of an ideal gas is a function of two properties, so the constituent entropies must be evaluated at the temperature and volume of the mixture or at the mixture temperature and the constituent partial pressures. The entropy of any constituent at the volume and temperature of the mixture (and hence at its partial pressure) is greater than its entropy when it exists alone at the pressure and temperature of the mixture (and hence at its partial volume). One simple way to prove this is by means of the increase of entropy principle.

Specific Heats, Gas Constant

Since

$$
u_m = \frac{m_A}{m_m} u_A + \frac{m_B}{m_m} u_B + \frac{m_C}{m_m} u_C
$$

the c_v of a mixture is given by

$$
\begin{aligned}
c_{vm} &\equiv \left(\frac{\partial u_m}{\partial T}\right)_v = \frac{m_A}{m_m}\left(\frac{\partial u_A}{\partial T}\right)_v + \frac{m_B}{m_m}\left(\frac{\partial u_B}{\partial T}\right)_v + \frac{m_C}{m_m}\left(\frac{\partial u_C}{\partial T}\right)_v\\
&= \frac{m_A c_{vA} + m_B c_{vB} + m_C c_{vC}}{m_m} = \frac{m_A}{m_m} c_{vA} + \frac{m_B}{m_m} c_{vB} + \frac{m_C}{m_m} c_{vC}
\end{aligned}
$$

In a similar manner,

$$c_{pm} = \frac{m_A c_{pA} + m_B c_{pB} + m_C c_{pC}}{m_m}$$

and

$$R_m = \frac{m_A R_A + m_B R_B + m_C R_C}{m_m}$$

The gas constant of the mixture can also be obtained by

$$R_m = \frac{R_u}{M_m}$$

Example 11·3. The mass analysis of a gas mixture is 10 percent hydrogen, 30 percent nitrogen, 40 percent carbon monoxide, and 20 percent carbon dioxide. For the mixture at 100 kPa, 20°C, compute (*a*) the partial pressures of the constituents, (*b*) the constant-volume specific heat, (*c*) the internal energy, and (*d*) the enthalpy. Assign $u_m = 0$ and $h_m = 0$ at $T = 0$ K. Approximate specific heats of the constituents are given in the following table.

Gas	c_p, kJ/kg·K	c_v, kJ/kg·K
Hydrogen	14.3	10.1
Nitrogen	1.04	0.742
Carbon monoxide	1.04	0.743
Carbon dioxide	0.848	0.649

Solution. **(*a*)** In order to determine the partial pressures we first convert the mass analysis to a volumetric analysis.

Constituent	Mass Analysis, kg/100 kg of Mixture	Molar Mass, kg/kmol	kmol/100 kg of Mixture	Volumetric Analysis, kmol/100 kmol of Mixture
H_2	10	2	5.00	62.8
N_2	30	28	1.07	13.5
CO	40	28	1.43	18.0
CO_2	20	44	0.45	5.7
			7.95	100.0

The partial pressures could conveniently be tabulated in an additional column, but to make clear the calculation we list them separately as

$$P_{H_2} = x_{H_2} P_m = 0.628(100) = 62.8 \text{ kPa}$$

$$P_{N_2} = x_{N_2} P_m = 0.135(100) = 13.5 \text{ kPa}$$

$$P_{CO} = x_{CO} P_m = 0.180(100) = 18.0 \text{ kPa}$$

$$P_{CO_2} = x_{CO_2} P_m = 0.057(100) = 5.7 \text{ kPa}$$

$$P_m = 100.0 \text{ kPa}$$

(*b*) The constant-volume specific heat of the mixture is the weighted average of the constituent c_v's.

$$c_{vm} = \sum \frac{m}{m_m} c_v = 0.10(10.1) + 0.30(0.742) + 0.40(0.743) + 0.20(0.649)$$
$$= 1.66 \text{ kJ/kg·K}$$

(*c*) For an ideal-gas mixture

$$u_m - u_{m0} = c_{vm}(T - T_0)$$

where u_{m0} is the internal energy of the mixture at some reference temperature T_0. Here we assign $u_{m0} = 0$ when $T_0 = 0$ K, so we have

$$u_m = c_{vm}T = 1.66(293) = 486 \text{ kJ/kg}$$

(*d*) The enthalpy could be determined by first determining c_{pm} and following the procedure used for finding u_m. After noting from part *a* that the mixture molar mass is 100/7.95, it is quicker to write

$$h_m = u_m + P_m v_m = u_m + R_m T = u_m + \frac{R_u}{M_m} T$$
$$= 486 + \frac{8.314(7.95)}{100} 293 = 680 \text{ kJ/kg}$$

Example 11·4. A mixture which has a volumetric analysis of 30 percent argon and 70 percent nitrogen is compressed reversibly and adiabatically in a closed system from 14 psia, 60°F, to 50 psia. Assume that the specific heats are constant at the following values: Argon, $c_p = 0.124$ B/lbm·R and $c_v = 0.0743$ B/lbm·R; nitrogen, $c_p = 0.249$ B/lbm·R and $c_v = 0.178$ B/lbm·R. Determine (*a*) the final temperature, (*b*) the work done per pound of mixture, and (*c*) the entropy change of each constituent per pound of mixture.

Solution. (*a*) In order to determine the specific heats of the mixture, we first determine the mass analysis.

Constituent	Volumetric Analysis, lbmol/100 lbmol of Mixture	Molar Mass, lbm/lbmol	lbm/100 lbmol of Mixture	Mass Analysis, lbm/100 lbm of Mixture
A	30	40	1200	38.0
N_2	70	28	1960	62.0
			3160	

Then the mixture specific heats are given by

$$c_{pm} = \frac{m_A}{m_m} c_{pA} + \frac{m_{N_2}}{m_m} c_{pN_2} = 0.38(0.124) + 0.62(0.249) = 0.2015 \text{ B/lbm·R}$$

$$c_{vm} = \frac{m_A}{m_m} c_{vA} + \frac{m_{N_2}}{m_m} c_{vN_2} = 0.38(0.0743) + 0.62(0.178) = 0.1386 \text{ B/lbm·R}$$

and $$k_m = \frac{c_{pm}}{c_{vm}} = \frac{0.2015}{0.1386} = 1.454$$

Since we are assuming that the specific heats are constant, the final temperature for the reversible adiabatic process is

$$T_2 = T_1 \left(\frac{P_2}{P_1}\right)^{(k_m - 1)/k_m} = 520 \left(\frac{50}{14}\right)^{0.454/1.45} = 774.6 \text{ R} = 775 \text{ R}$$

(*b*) Applying the first law to the closed system for this adiabatic process, we have

$$\begin{aligned} w_{\text{in}} &= u_{m2} - u_{m1} - q = c_{vm}(T_2 - T_1) - 0 \\ &= 0.1386(775 - 520) = 35.3 \text{ B/lbm} \end{aligned}$$

(*c*) For this reversible adiabatic process the entropy of the mixture must remain constant. The entropy of each constituent may change, but the sum of the entropy changes of the two constituents must be zero. For any process of an ideal gas,

$$\Delta s = \int_1^2 ds = \int_1^2 \frac{dh}{T} - \int_1^2 \frac{v\,dP}{T} = \int_1^2 \frac{c_p\,dT}{T} - R\int_1^2 \frac{dP}{P}$$

and, if c_p is constant,

$$\Delta s = c_p \ln \frac{T_2}{T_1} - R \ln \frac{P_2}{P_1}$$

In applying this equation to each constituent in the mixture, the pressures to be used are the partial pressures, but note that for each gas,

$$\frac{P_2}{P_1} = \frac{x_2 P_{m2}}{x_1 P_{m1}} = \frac{P_{m2}}{P_{m1}}$$

Thus, applying the equation for Δs to the argon, we have

$$\Delta s_{\text{A}} = 0.124 \ln \frac{774.6}{520} - \frac{38.7}{778} \ln \frac{50}{14} = -0.0139 \text{ B/lbm·R}$$

Per pound of mixture,

$$\frac{\Delta S_{\text{A}}}{m_m} = \frac{m_{\text{A}} \Delta s_{\text{A}}}{m_m} = 0.38(-0.0139) = -0.0053 \text{ B/lbm mixture·R}$$

For the nitrogen,

$$\Delta s_{\text{N}_2} = 0.249 \ln \frac{774.6}{520} - \frac{55.2}{778} \ln \frac{50}{14} = 0.0091 \text{ B/lbm·R}$$

$$\frac{\Delta S_{\text{N}_2}}{m_m} = \frac{m_{\text{N}_2} \Delta s_{\text{N}_2}}{m_m} = 0.62(0.0091) = 0.0055 \text{ B/lbm mixture·R}$$

Within the limits of accuracy of the data, these two Δs values are equal in magnitude (although opposite in sign). It is actually unnecessary to calculate both Δs values in this manner, except as a check on the computations, because we know that their sum is zero.

11·6 Mixing of Ideal Gases Initially at Different Pressures and Temperatures

Thus far in this chapter we have dealt with mixtures that have already been formed. Now let us look at the *formation* of an ideal-gas mixture or the *mixing* of ideal gases initially at different pressures and temperatures. The general problem is to determine the properties of a mixture that results from mixing constituents of known properties. No new principles are involved. We simply apply the first law and the conservation of mass principle to a conveniently selected system to relate the mixture properties to the properties of the constituents before mixing. For example, consider the adiabatic mixing of three gases, A, B, and C, in a closed system of fixed volume. The gases might be initially in three tanks connected by piping, or they might be in three parts of a tank separated by partitions as in Fig. 11·1. If the partitions are ruptured or removed, the three gases will form a mixture that has a mass and a volume given by

$$m_m = m_A + m_B + m_C$$

$$V_m = V_A + V_B + V_C$$

where V_A, V_B, and V_C are the volumes of the constituents *before mixing*. The mixing process was specified as adiabatic and there is no work done; so the internal energy of the system remains constant and

$$U_m = U_A + U_B + U_C$$

where U_A, U_B, and U_C are the internal energies of the constituents before mixing. The internal energy of the mixture is also equal to the sum of the internal energies of the constituents after mixing, but the internal energy of each constituent is generally not the same before and after mixing. Since the internal energy of the entire system remains constant, the sum of the internal energy changes of the constituents is zero:

$$\Delta U = \Delta U_A + \Delta U_B + \Delta U_C = 0$$

If c_v for each constituent is constant in the temperature range between the constituent's initial temperature and the mixture temperature, an equation such as

$$\Delta U_A = m_A c_{vA}(T_m - T_A)$$

can be written for each constituent. Then

$$\Delta U = m_A c_{vA}(T_m - T_A) + m_B c_{vB}(T_m - T_B) + m_C c_{vC}(T_m - T_C) = 0$$

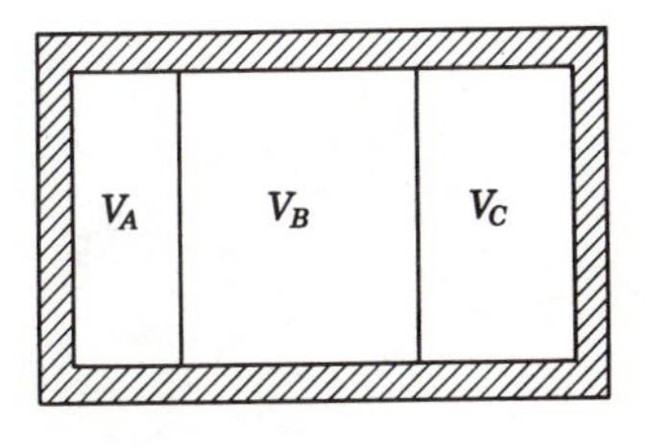

Figure 11·1. Three gases in an adiabatic system before mixing.

Solving for the mixture temperature,

$$T_m = \frac{m_A c_{vA} T_A + m_B c_{vB} T_B + m_C c_{vC} T_C}{m_A c_{vA} + m_B c_{vB} + m_C c_{vC}}$$

The denominator is equal to $m_m c_{vm}$. Notice that in deriving this expression for the mixture temperature no assumption was made regarding a temperature at which $U = 0$, nor was it stipulated that $U = 0$ at the same temperature for all constituents.

After the temperature of the mixture has been determined, the pressure can be calculated from

$$P_m = \frac{m_m R_m T_m}{V_m}$$

R_m can be determined from the mass analysis of the mixture.

Since the mixing process we are considering is irreversible and adiabatic, the entropy of the system must increase. The entropy of the mixture, while equal to the sum of the entropies of the constituents as they exist in the mixture, is greater than the sum of the entropies of the constituents before mixing. The entropy change of the entire system is

$$\Delta S = \Delta S_A + \Delta S_B + \Delta S_C > 0$$

and the entropy change for each constituent can be calculated as though each constituent existed alone and expanded from its initial conditions to the mixture temperature and volume, its final pressure being therefore its partial pressure in the mixture.

We have illustrated here that no new principles or techniques are involved in the determination of the properties of an ideal-gas mixture formed by mixing constituents of known properties in a closed, rigid, adiabatic system. Nonadiabatic mixing and mixing in open systems, both steady flow and transient flow, can also be analyzed by means of the principles that have already been introduced.

Example 11·5. In a steady-flow system, hydrogen initially at 20°C and methane initially at 65°C are to be mixed adiabatically to form a mixture which is 50 percent hydrogen by volume. Each gas enters the mixing chamber at a pressure of 1 atm, and the mixing occurs under a constant total pressure of 1 atm. The hydrogen mass flow rate is 0.012 kg/s. The lowest temperature in the surroundings is 20°C. Assuming that kinetic-energy changes are negligible, determine (*a*) the temperature of the mixture, and (*b*) the irreversibility rate in kJ/s.

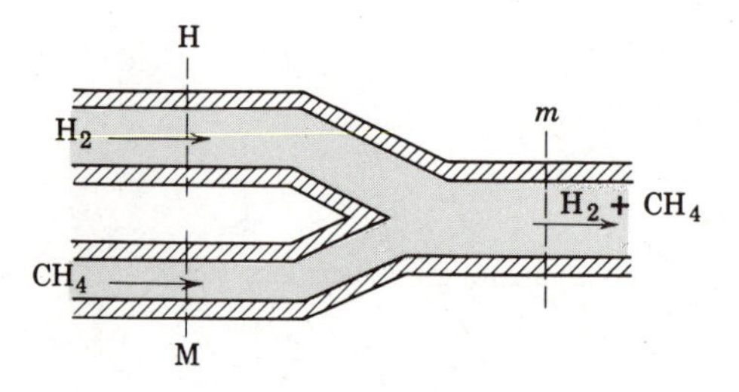

Example 11·5.

Solution. (*a*) The mass analysis (or gravimetric analysis) of the mixture is first found.

Constituent	Volumetric Analysis, kmol/100 kmol of Mixture	Molar Mass, kg/kmol	kg/100 kmol of Mixture	Mass Analysis, kg/100 kg of Mixture
H_2	50	2	100	11.11
CH_4	50	16	800	88.89
			900	

The methane mass flow rate is thus 0.096 kg/s.

The mixing is adiabatic, the kinetic-energy change is negligible, and there is no work done; so application of the first law to the steady-flow system in which the mixing occurs gives (with the subscripts H and M referring to hydrogen and methane, respectively, and the subscript m referring to the mixture)

$$(\dot{m}_{\rm H}h_{\rm H} + \dot{m}_{\rm M}h_{\rm M})_{\text{before mixing}} = (\dot{m}_{\rm H} + \dot{m}_{\rm M})h_m = (\dot{m}_{\rm H}h_{\rm H} + \dot{m}_{\rm M}h_{\rm M})_{\text{after mixing}}$$

where each $\dot{m}$ stands for a mass rate of flow. If we now let $h_{\rm H}$ and $h_{\rm M}$ represent the enthalpies of hydrogen and methane at their initial temperatures $T_{\rm H}$ and $T_{\rm M}$, and let the enthalpies of the constituents in the mixture be represented by $h_{{\rm H},T_m}$ and $h_{{\rm M},T_m}$, the energy balance above can be written as

$$\dot{m}_{\rm H}h_{\rm H} + \dot{m}_{\rm M}h_{\rm M} = \dot{m}_{\rm H}h_{{\rm M},T_m} + \dot{m}_{\rm M}h_{{\rm M},T_m}$$

Assuming that the c_p values are constant,

$$\dot{m}_{\rm H}c_{p\rm H}(T_{\rm H} - T_m) + \dot{m}_{\rm M}c_{p\rm M}(T_{\rm M} - T_m) = 0$$

$$T_m = \frac{\dot{m}_{\rm H}c_{p\rm H}T_{\rm H} + \dot{m}_{\rm M}c_{p\rm M}T_{\rm M}}{\dot{m}_{\rm H}c_{p\rm H} + \dot{m}_{\rm M}c_{p\rm M}}$$

Table A·4 shows that the c_p of hydrogen is about seven times that of methane. The mass of methane in this mixture is eight times the mass of hydrogen. Therefore the product mc_p (or $\dot{m}c_p$) is approximately the same for the two constituents so that the mixture temperature will be approximately midway between $T_{\rm H}$ and $T_{\rm M}$ and a little closer to $T_{\rm M}$. Let us estimate it as 45°C. This estimate of T_m enables us to select reasonable values for the mean c_p between $T_{\rm H}$ and T_m for hydrogen and between $T_{\rm M}$ and T_m for methane. From Table 4·2 we get a c_p value of approximately 14.25 kJ/kg·K for hydrogen and from the Table A·4 equation a value of approximately 2.28 kJ/kg·K for methane. Substituting these values into the last equation and noting that either Celsius or Kelvin temperatures can be used,

$$T_m = \frac{0.012(14.25)20 + 0.096(2.28)65}{0.012(14.25) + 0.096(2.28)} = 45°\text{C}$$

(This happens to be the estimated value used in selecting c_p values; so no correction is needed.)

(*b*) The irreversibility is given by

$$I = T_0 \Delta S = T_0[m_{\rm H}\,\Delta s_{\rm H} + m_{\rm M}\,\Delta s_{\rm M}]$$

Per unit time, this is

$$\frac{I}{\tau} = \dot{I} = T_0[\dot{m}_{\rm H}\,\Delta s_{\rm H} + \dot{m}_{\rm M}\Delta s_{\rm M}]$$

For each constituent, assuming that c_p values are constant,

$$\Delta s = \int ds = \int \frac{dh}{T} - \int \frac{v\,dP}{T} = \int \frac{c_p\,dT}{T} - R\int \frac{dP}{P} = c_p \ln \frac{T_m}{T_{\text{initial}}} - R \ln \frac{P}{P_{\text{initial}}}$$

The volume fraction of each constituent is 0.5; so the partial pressure of each constituent is one half of the mixture pressure. The mixture pressure equals the initial pressure of each constituent; so $P/P_{\text{initial}} = 0.5$.

$$\begin{aligned} I &= T_0\left\{\dot m_{\text{H}}\left[c_{p\text{H}} \ln \frac{T_m}{T_{\text{H}}} - R_{\text{H}} \ln \frac{P_{\text{H},m}}{P_{\text{H}}}\right] + \dot m_{\text{M}}\left[c_{p\text{M}} \ln \frac{T_m}{T_{\text{M}}} - R_{\text{M}} \ln \frac{P_{\text{M},m}}{P_{\text{M}}}\right]\right\} \\ &= 293\left\{0.012\left[14.25 \ln \frac{318}{293} - 4.124 \ln 0.5\right]\right. \\ &\quad \left. + 0.096\left[2.28 \ln \frac{318}{338} - 0.518 \ln 0.5\right]\right\} \\ &= 20.3 \text{ kJ/s} \end{aligned}$$

11·7 Mixtures of Real Gases

We have observed that the behavior of real gases can often be adequately represented by the ideal-gas equation of state. In such cases, mixtures of these gases can in turn be treated as ideal gases. When the constituents of a gas mixture deviate considerably from ideal-gas behavior, however, the simple relations that apply to ideal-gas mixtures do not hold, and we must resort to certain approximations in determining the properties of a mixture from the properties of its constituents. Three approximation procedures will be considered here: the use of the additive pressure law in conjunction with equations of state of the constituents, the use of the additive volume law in a similar manner, and the use of constituent compressibility factors.

Law of Additive Pressures

As pointed out in Sec. 11·3, the law of additive pressures (Dalton's law) holds approximately for real-gas mixtures even in some ranges of pressure and temperature where $Pv = RT$ is quite inaccurate. If we assume that it does hold for a real-gas mixture, then we write

$$P_m = P_A(T_m, V_m) + P_B(T_m, V_m) + \cdots$$

In this equation $P_A(T_m, V_m)$, $P_B(T_m, V_m)$, etc., denote the pressures that would be exerted by individual constituents if they existed alone at the temperature and volume of the mixture. These are not partial pressures, because partial pressure is defined by $P_i \equiv x_i P_m$, and $P_i = P_i(T_m, V_m)$ only for ideal-gas mixtures. By selecting a suitable equation of state it is possible to determine the pressure of each constituent in a mixture if it existed alone at T_m and V_m. For example, if we assume that each gas follows the van der Waals equation of state, the pressure of each gas in the mixture will be represented by an equation such as

$$P_A(T_m, V_m) = \frac{R_u T}{v_{NA} - b_A} - \frac{a_A}{(v_{NA})^2}$$

Other equations of state may also be used. Following through with the use of the van der Waals equation, we relate the molar specific volume of each constituent at T_m and V_m to that of the mixture by

$$v_{NA} \equiv \frac{V_A}{N_A} = \frac{V_m}{N_A} = \frac{V_m}{x_A N_m} = \frac{v_{Nm}}{x_A}$$

and make this substitution in the van der Waals equation to get

$$P_A(T_m, V_m) = \frac{x_A R_u T}{v_{Nm} - x_A b_A} - \frac{x_A^2 a_A}{v_{Nm}^2}$$

Substituting this type of relation for the pressure of each constituent in the additive pressure law equation and collecting terms gives

$$P_m = R_u T \left[\frac{x_A}{v_{Nm} - x_A b_A} + \frac{x_B}{v_{Nm} - x_B b_B} + \cdots \right] - \frac{1}{v_{Nm}^2} \left[a_A x_A^2 + a_B x_B^2 + \cdots \right]$$

Example 11·6. A gas mixture that is 55 and 45 percent by mass of nitrogen and carbon dioxide, respectively, occupies 150 cu ft at 14.7 psia, 70°F. It is compressed to a final volume of 0.5 cu ft. If the final temperature of the mixture is 100°F, compute the approximate pressure if the gases are assumed to follow the van der Waals equation of state. Values for the constants in the equation of state follow.

Gas	a, atm·ft⁶/(lbmol)²	b, cu ft/lbmol
Nitrogen	346	0.618
Carbon dioxide	926	0.686

The universal gas constant 0.730 atm·cu ft/lbmol·R is used in connection with these constants.

Solution. The mole fractions are first determined.

Constituent	Mass Fraction, lbm/100 lbm of Mixture	Molar Mass, lbm/lbmol	Mol/100 lbm of Mixture	Mole Fraction, lbmol/100 lbmol of Mixture
N_2	55	28	1.965	65.8
CO_2	45	44	1.023	34.2
			2.988	

The next step is to determine the number of moles in the mixture. If it is assumed that the mixture behaves as an ideal gas under the initial low-pressure condition, the number of moles equals

$$N = \frac{PV_m}{R_u T} = \frac{1(150)}{0.730(530)} = 0.388 \text{ mole}$$

After compression the molar specific volume for the mixture is

$$v_{Nm} = \frac{0.5}{0.388} = 1.289 \text{ cu ft/mole}$$

When the mole fractions, molar specific volume for the mixture, and the various constants are known, it is possible to obtain the pressure by substitution into the last equation derived before the statement of this example:

$$P_m = 0.730(560)\left[\frac{0.658}{1.289 - 0.658(0.618)} + \frac{0.342}{1.289 - 0.342(0.686)}\right] - \frac{1}{(1.289)^2}[346(0.658)^2 + 926(0.324)^2]$$
$$= 437 - 155 = 282 \text{ atm}$$

Law of Additive Volumes

As pointed out in Sec. 11·4, the law of additive volumes also holds approximately for real-gas mixtures even in some ranges of pressure and temperature where $Pv = RT$ is quite inaccurate. If we assume that it does hold for a real-gas mixture, then we write

$$V_m = V_A(P_m, T_m) + V_B(P_m, T_m) + \cdots$$

In this equation $V_A(P_m, T_m)$, $V_B(P_m, T_m)$, etc., denote the volumes of the individual constituents if they existed alone at the pressure and temperature of the mixture. These are not partial volumes, because partial volume is defined by $V_i' \equiv x_i V_m$, and $V_i' = V_i(P_m, T_m)$ only for ideal-gas mixtures. To get the additive volume formulation in terms of molar specific volumes, divide the form given above by the number of moles in the mixture and use the definition of mole fraction, $x \equiv N/N_m$, to give

$$v_{Nm} = \frac{V_m}{N_m} = \frac{V_A(P_m, T_m)}{N_m} + \frac{V_B(P_m, T_m)}{N_m} + \cdots$$
$$= \frac{x_A V_A(P_m, T_m)}{N_A} + \frac{x_B V_B(P_m, T_m)}{N_B} + \cdots$$
$$= x_A v_{NA}(P_m, T_m) + x_B v_{NB}(P_m, T_m) + \cdots$$

By selecting a suitable equation of state it is possible to determine the molar specific volume of each constituent if it existed alone at P_m and T_m, but a trial-and-error method may be required if the equations of state used are not in an explicit volume form. (For ideal gases, the molar specific volume is the same for all constituents if they existed alone at the same pressure and temperature, but remember that this is not true for real gases.)

Example 11·7. Solve Example 11·6 by using the law of additive volumes and the van der Waals equation of state.

Solution. Substituting the mixture molar specific volume and the mole fractions determined in the solution of Example 11·6 into the additive volume formulation

$$v_{Nm} = x_A v_{NA}(P_m, T_m) + x_B v_{NB}(P_m, T_m)$$

gives

$$v_{Nm} = 1.289 = 0.658 v_{NA}(P_m, T_m) + 0.342 v_{NB}(P_m, T_m)$$

where the subscript A refers to nitrogen and the subscript B refers to carbon dioxide. Solving for v_{NA} and dropping the notation which reminds us that the molar specific volumes are to be evaluated at the pressure and temperature of the mixture gives

$$v_{NA} = 1.959 - 0.520 v_{NB}$$

The remainder of the solution consists of solving this equation and the van der Waals equation for the two constituents simultaneously by iteration. Convergence gives $P_A = P_B = 334$ atm.

Compressibility Factors

If compressibility factors for various gas mixtures were available, it would be possible to employ the relation

$$P_m V_m = Z_m N_m R_u T_m$$

Values of the compressibility factor Z_m for real-gas mixtures are difficult to obtain; hence it is convenient to employ an approximate relation of the type

$$Z_m = x_A Z_A + x_B Z_B + \cdots$$

x_A, x_B, etc., are the mole fractions of the constituents, and Z_A, Z_B, etc., are their compressibility factors. If these compressibility factors are evaluated at the pressure and temperature of the mixture, this equation reduces to the law of additive volumes. This is a convenient procedure because charts of Z as a function of P_R and T_R are available. If the compressibility factors are evaluated at the volume and temperature of the mixture, the equation above reduces to the law of additive pressures. The use of the additive pressure law with compressibility factors is complicated by the fact that charts of Z as a function of v_R and T_R are required, and these are not readily available.

Example 11·8. Solve Example 11·6 by using a compressibility factor for the mixture.

Solution. We will use the constituent compressibility factors at the pressure and temperature of the mixture. Thus we will be using the additive volume law. The first step is to determine the reduced pressure and temperature values for the nitrogen and carbon dioxide. From Table A·4, the following values are obtained for the critical temperatures and pressures of the gases.

Gas	Critical Temperature, R	Critical Pressure, atm
Nitrogen	227	33.5
Carbon dioxide	548	72.9

The reduced values are

Nitrogen: $$T_R = \frac{T}{T_c} = \frac{560}{22} = 2.46$$

$$P_R = \frac{P}{P_c} = \frac{P}{33.5}$$

Carbon dioxide: $$T_R = \frac{T}{T_c} = \frac{560}{548} = 1.02$$

$$P_R = \frac{P}{P_c} = \frac{P}{72.9}$$

Since the pressure is unknown, a trial value of 300 atm will be assumed, for which the reduced pressures are

Nitrogen: $P_R = 8.96$
Carbon dioxide: $P_R = 4.12$

By using the reduced values and Table A·4, the following compressibility factor values are found:

Nitrogen: $Z_A = 1.15$
Carbon dioxide: $Z_B = 0.57$

The composite compressibility factor for the mixture may be found from

$$Z_m = x_A Z_A + x_B Z_B = 0.658(1.15) + 0.342(0.57) = 0.952$$

Then

$$P = \frac{Z_m R_u T}{v_{Nm}} = \frac{0.952(0.730)560}{1.289} = 301 \text{ atm}$$

If this calculated pressure value were not close to the assumed pressure, it would be necessary to make other trial solutions.

Few experimental data for real-gas mixtures are available from which it would be possible to draw conclusions relative to the accuracy of the three methods used to solve the same problem in Examples 11·6, 7, and 8. This is one illustration of the fact that the general problem of determining the properties of real-gas mixtures has not been fully solved.

11·8 Mixtures of Ideal Gases and Vapors

For the purpose of the following discussion, let us call a gas which exists at a temperature lower than its critical temperature a vapor. Thus a vapor can be liquefied by an increase in pressure at constant temperature. The analysis of problems dealing with gas–vapor mixtures is, in general, similar to that for a gas mixture. However, an important additional fact must be considered; that is, the maximum pressure of a vapor in a mixture depends on the temperature of the mixture. To illustrate this, consider a mixture of nitrogen and oxygen at a mixture pressure of 100 kPa and a temperature of 40°C. The mole fraction of either constituent can have any value from 0 to 1.0, and the partial pressure can have a corresponding value from 0 to 100 kPa. Under the stated conditions these gases follow very closely the ideal-gas equation of state, so a partial pressure equals the pressure of the constituent existing alone at the temperature and volume of the mixture. Now, in contrast, consider a mixture of nitrogen and water vapor at 40°C and a mixture pressure of 100 kPa. The steam tables show that at 40°C the maximum pressure under which water vapor can exist is 7.384 kPa, so the mole fraction and partial pressure of water vapor in this mixture are strictly limited. Furthermore, the pressure and temperature of the nitrogen-oxygen mixture can be varied over wide ranges without affecting the composition of the mixture; but increasing the pressure or decreasing the temperature of the nitrogen–water-vapor mixture even slightly may cause some of the water vapor to condense, thereby changing the composition of the gas–vapor mixture. Several examples will show the application of the principles involved.

Example 11·9. A mixture of 0.10 kg of saturated water vapor and 0.50 kg of air is contained in a tank at a temperature of 120°C. Compute the total pressure of the mixture and the volume of the tank.

Solution. From the tables of the properties of steam, the saturation pressure and saturation specific volume for water vapor at 120°C are

$$P_v = 198.53 \text{ kPa}$$

$$v_g = 0.8919 \text{ m}^3/\text{kg}$$

The actual volume occupied by the water vapor or the volume of the tank is

$$V = m_v v_g = 0.10(0.8919) = 0.0892 \text{ m}^3$$

This is also the volume occupied by the air. If we assume in accordance with Dalton's law that the air behaves as though it existed alone at the temperature and volume of the mixture, the pressure of the air may be determined from the ideal-gas equation of state.

$$P_a = \frac{m_a R_a T}{V} = \frac{0.50(0.287)393}{0.0892} = 632.2 \text{ kPa}$$

Again assuming that Dalton's law holds, the total pressure of the mixture is equal to the sum of the pressures exerted by the water vapor and the air.

$$P_m = P_v + P_a = 198.53 + 632.2 = 831 \text{ kPa}$$

Example 11·10. A vessel contains 0.10 lbm of wet steam having a quality of 4 percent and 0.20 lbm of nitrogen at a temperature of 100°F. Compute the total pressure of the mixture and the mass of nitrogen per pound of liquid water.

Solution. The data taken from the steam tables for water at 100°F follow:

$$P_v = 0.9503 \text{ psia}$$

$$v_f = 0.01613 \text{ cu ft/lbm}$$

$$v_g = 350.0 \text{ cu ft/lbm} = v_{fg} \text{ approximately}$$

The specific volume of the wet vapor (liquid–vapor mixture) equals $v_x = v_f + xv_{fg} = 0.01613 +$

0.04(350.0) = 14.016 cu ft/lbm or, using the approximate relation for v_x which actually gives the specific volume of the vapor only

$$v_x = xv_g = 0.04(350.0) = 14.0 \text{ cu ft/lbm}$$

The volume of the vapor is therefore

$$V = mv_x = 0.10(14.0) = 1.40 \text{ cu ft}$$

This is the volume occupied by the 0.2 lbm of nitrogen as well as by the water vapor. The pressure exerted by the nitrogen can be computed by use of the ideal-gas equation of state.

$$P_{N_2} = \frac{mR_uT}{VM} = \frac{0.20(1545)560}{1.4(28)144} = 30.65 \text{ psia}$$

The total pressure of the mixture is

$$P_m = P_v + P_{N_2} = 0.950 + 30.65 = 31.6 \text{ psia}$$

Since the quality is 4 percent, 96 percent of the water is liquid; therefore the mass of liquid present is

$$m_f = 0.96(0.10) = 0.096 \text{ lbm}$$

The mass of nitrogen per pound of liquid water is

$$\frac{m_{N_2}}{m_f} = \frac{0.20}{0.096} = 2.08 \text{ lbm } N_2\text{/lbm liquid water}$$

Example 11·11. One kilogram of water vapor and 0.10 kg of air are contained in a tank having a volume of 0.563 m^3. If the temperature of the mixture is 200°C, compute the pressure of the mixture.

Solution. From the superheated steam table, the pressure of steam at a temperature of 200°C and a specific volume of 0.563 m^3/kg is 380 kPa. The air pressure may be determined from the ideal-gas equation of state:

$$P_a = \frac{m_aR_aT}{V} = \frac{0.10(0.287)473}{0.563} = 24.1 \text{ kPa}$$

The total pressure of the mixture is, therefore, equal to

$$P_m = P_a + P_v = 24.1 + 380 = 404 \text{ kPa}$$

11·9 Atmospheric Air

Many different gas–vapor mixtures are encountered in engineering, but the one that receives the most attention by far is the mixture of air and water vapor commonly referred to as *atmospheric air*. In most applications dealing with atmospheric air the temperatures—and consequently the maximum vapor partial pressures—are low enough so that the water vapor behaves as an ideal gas. That is, the water vapor follows $Pv = RT$, its enthalpy is a function of temperature only, and its partial pressure in the mixture equals the pressure it would exert if it existed alone at the temperature and volume of the mixture.

For support of the statement that water vapor in atmospheric air follows $Pv = RT$, you can calculate from steam-table data some compressibility factor values for steam at low pressures. For example, even at a partial pressure as high as 20 kPa (2.9 psia), which cannot be reached unless the mixture is at a temperature of over 60°C (140°F), the compressibility factor is 0.996^+. For vapor pressures and temperatures usually encountered in atmospheric air, the compressibility factor is even closer to unity.

The fact that the enthalpy of water at low pressure depends only on temperature can be verified by means of a *Ts* diagram for steam such as Chart A·2 in the appendix. For superheated steam at low pressures the constant-enthalpy lines coincide with constant-temperature lines. Thus no matter what the pressure of water vapor in atmospheric air is, its enthalpy can be read from the superheated vapor tables at the lowest pressure entry in the table and the atmospheric air temperature. Some tables include a separate table for low pressures. For temperatures lower than the lowest temperature entry in the superheated steam or low-pressure steam table, the enthalpy of water vapor at any pressure equals very nearly the enthalpy of saturated vapor at the same temperature. Repeating this important fact for emphasis, *in atmospheric air the enthalpy of the water vapor equals (very nearly) the enthalpy of saturated vapor at the same temperature.*

For an ideal-gas mixture the partial pressure of each constituent equals the pressure that that constituent would exert if it existed alone at the temperature and volume of the mixture. For this reason we will henceforth often refer to the partial pressure of water vapor in atmospheric air simply as *the pressure of the vapor* in the air, since the vapor behaves in all respects as though it were alone at this pressure and the mixture temperature. (The presence of air actually has a slight effect on the equilibrium temperature of liquid and vapor water at a given pressure, making it different from that given in the steam tables for water as a pure substance, but this effect is negligible with atmospheric air.)

The air constituent of atmospheric air is often referred to as the *dry air* to distinguish it from the mixture. Thus atmospheric air is composed of dry air plus water vapor. In situations involving atmospheric air the temperature range is nearly always so limited that the c_p of dry air can be considered constant at 1.005 kJ/kg·K (0.240 B/lbm·R).

11·10 Relative Humidity and Humidity Ratio

Two terms frequently used in dealing with mixtures of air and water vapor are *relative humidity* and *humidity ratio*. It is important to learn the definitions of these terms and the relationship between them.

Relative humidity ϕ is defined as the ratio of the pressure (that is, partial pressure) of the vapor in the mixture to the saturation pressure of the vapor at the temperature of the mixture. Let the pressure of the vapor in the mixture be designated by P_v, and let the vapor saturation pressure at the mixture temperature be designated by P_g. (The subscript g has been adopted because it is used in the steam tables to refer to saturated vapor.) Then the relative humidity is defined as

$$\phi \equiv \frac{P_v}{P_g} \tag{11·1}$$

If the water vapor in the mixture and saturated water vapor at the same temperature follow the ideal-gas equation of state, as they do to a high degree of accuracy at the temperatures normally encountered with atmospheric air, relative humidity can be expressed in other forms by substituting for each pressure its equivalent RT/v. Thus

$$\phi \equiv \frac{P_v}{P_g} = \frac{RTv_g}{v_v RT} = \frac{v_g}{v_v} = \frac{\rho_v}{\rho_g} \tag{11·2}$$

The specific volume of the vapor in atmospheric air can thus be determined from the mixture temperature and the relative humidity,

$$v_v = \frac{v_g}{\phi} \tag{11·2}$$

Notice that relative humidity pertains only to the *vapor* in atmospheric air. It is independent of the pressure and density of the dry air in the mixture. It is independent of the barometric pressure.

Atmospheric air that contains saturated water vapor has a relative humidity of 1.0 or 100 percent and is called *saturated air*, although it is only the water vapor in the air which is in a saturation state.

A *Ts* diagram for water vapor in atmospheric air is shown in Fig. 11·2. If point 1 represents the state of the water vapor as it exists in the mixture, relative humidity is the ratio of P_1 to P_2, where state 2 is a saturation state at the mixture temperature.

Humidity ratio ω is defined as the ratio of the mass of vapor in atmospheric air to the mass of dry air. Notice that it is not the same as the mass fraction of water vapor in the mixture. Using the subscript v for vapor and a for dry air,

$$\omega \equiv \frac{m_v}{m_a} \tag{11·3}$$

and, since the volume is the same for both constituents of the mixture,

$$\omega \equiv \frac{m_v}{m_a} = \frac{\rho_v}{\rho_a} = \frac{v_a}{v_v} \tag{11·3}$$

Assuming that both the air and the water vapor behave as ideal gases, we can express

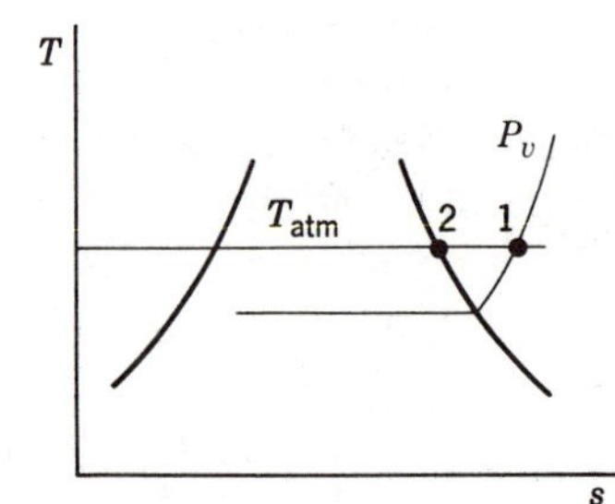

Figure 11·2. *Ts* diagram of water vapor in atmospheric air.

the humidity ratio as

$$\omega \equiv \frac{m_v}{m_a} = \frac{P_v V R_a T}{R_v T P_a V} = \frac{P_v R_a}{P_a R_v} = \frac{P_v R_a}{(P_m - P_v) R_v}$$

Solving for P_v gives

$$P_v = \frac{\omega R_v P_m}{\omega R_v + R_a}$$

This relation is also quite useful in dealing with combustion products at low pressures, R_a being taken as the gas constant of the dry products of combustion.

Humidity ratio is also called specific humidity or absolute humidity. The relationship between humidity ratio and relative humidity is obtained by combining Eq. 11·2 and 11·3:

$$\omega = \frac{v_a}{v_g} \phi \tag{11·4}$$

Example 11·12. Atmospheric air at 100 kPa, 30°C has a relative humidity of 70 percent. Compute the humidity ratio.

Solution. From the steam tables, the saturation pressure for water vapor at 30°C is 4.246 kPa. The partial pressure of the water vapor in the mixture is then

$$P_v = \phi(P_g) = 0.70(4.246) = 2.97 \text{ kPa}$$

Since the total atmospheric pressure is equal to the sum of the partial pressures, the air pressure equals

$$P_a = P_m - P_v = 100 - 2.97 = 97.0 \text{ kPa}$$

The specific volume of the air may now be found from the ideal-gas equation

$$v_a = \frac{RT}{P} = \frac{0.287(303)}{97.0} = 0.897 \text{ m}^3/\text{kg}$$

The value for v_g, taken from the steam tables, is 32.89 m^3/kg. The humidity ratio is then

$$\omega = \frac{v_a \phi}{v_g} = \frac{0.897(0.70)}{32.89} = 0.0191 \frac{\text{kg water vapor}}{\text{kg dry air}}$$

11·11 Temperatures Used in the Determination of the Properties of Atmospheric Air

Four temperatures referred to in the determination of the properties of atmospheric air are dry-bulb temperature, dew-point temperature, adiabatic saturation temperature or thermodynamic wet-bulb temperature, and wet-bulb temperature.

Dry-Bulb Temperature

The dry-bulb temperature is simply the temperature of the mixture as it would be measured by any of several types of ordinary thermometers placed in the mixture. Care must be exercised in measuring the temperature of atmospheric air to avoid errors caused by radiant heat transfer between the thermometer and the walls of the containing vessel. Where temperature differences exist, shielded thermometer elements should be used. The term *dry-bulb* temperature is used to distinguish the temperature of the mixture from the temperature reading obtained from a thermometer that has its temperature-sensitive element wrapped in gauze that is soaked in water.

Dew-Point Temperature

The dew-point temperature of an air-vapor mixture is defined as the saturation temperature of the vapor corresponding to its partial pressure in the mixture. It is thus the temperature at which condensation begins if the mixture is cooled at constant pressure.

A simple laboratory determination of dew-point temperature consists of partially filling a metal cup with water, adding ice, and then stirring while observing the water temperature as it is lowered. The temperature at which moisture begins to collect on the outside of the cup is approximately the dew-point temperature of the room air. One of the errors in this method results from the fact that the water temperature and the temperature of the air in contact with the cup are not exactly the same.

If point 1 in Fig. 11·3 represents the water vapor in atmospheric air, the dew point is the temperature at point 2 or the saturation temperature corresponding to the vapor pressure. States 1, 2, and x as well as any other states with the same vapor pressure have the same dew point. For air at 100 percent relative humidity (saturated air) the dew-point temperature equals the dry-bulb temperature.

Example 11·13. Determine the dew point of atmospheric air at 100 kPa, 30°C, 70 percent relative humidity:

Solution. From the steam tables, the saturation pressure for water vapor at 30°C is 4.246 kPa. The partial pressure of the water vapor in the mixture is then

$$P_v = \phi(P_g) = 0.70(0.6982) = 0.70(4.246) = 2.97 \text{ kPa}$$

The dew point is the saturation temperature corresponding to this pressure. This is found from the steam tables to be approximately 24°C.

Adiabatic Saturation Temperature

In atmospheric air with a relative humidity of less than 100 percent, the water vapor is at a pressure lower than its saturation pressure. Therefore, if this air is placed in contact

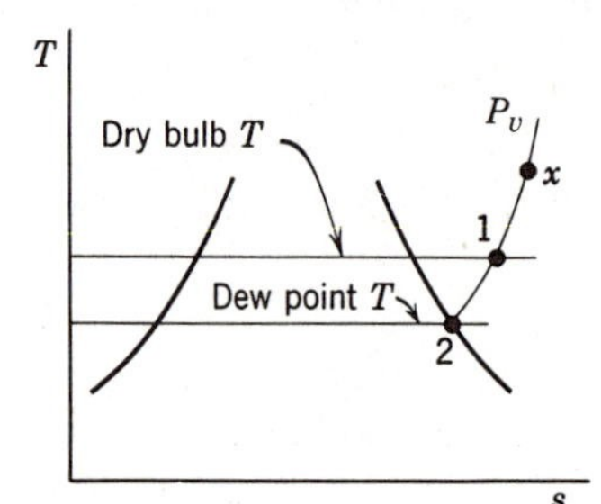

Figure 11·3. T_2 is the dew point of states 1, 2, and x.

with liquid water, some of the water will evaporate into the air. The humidity ratio of the air will increase. If the evaporation occurs in a thermally insulated container, the temperature of the air will decrease because at least part of the latent heat of vaporization of the water that evaporates will come from the air. The lower the initial humidity ratio is, the greater the amount of evaporation, and the greater the temperature decrease will be; so we have here the basis of an indirect measurement of humidity ratio.

The adiabatic saturation temperature of atmospheric air is defined as the temperature which results from adiabatically adding water to the atmospheric air in steady flow until it becomes saturated, the water being supplied at the final temperature of the mixture. At first it appears that this definition is circular, because to determine the adiabatic saturation temperature we must supply water which is at that temperature. Actually the definition is operational and sufficient, and the adiabatic saturation temperature can be found by the following operations: (1) Add water at any temperature adiabatically to steadily flowing atmospheric air until it becomes saturated. (2) Measure the temperature of the saturated air. (3) Change the temperature of the water being added to equal that of the saturated air as measured in step 2. (4) Repeat steps 2 and 3 until the temperature of the saturated air equals that of the water being added. This is the adiabatic saturation temperature of the atmospheric air.

To see how a measurement of the adiabatic saturation temperature can be used to determine the humidity ratio of atmospheric air, consider the steady-flow system shown in Fig. 11·4. Air of unknown humidity ratio ω_1 enters at section 1. The air leaving at section 2 is saturated, and, since the water added is at the same temperature, this is the adiabatic saturation temperature. The total or mixture pressure is constant throughout the system. The mixture pressure and the temperatures at sections 1 and 2 can be measured.

Noting that there is no work done, the process is adiabatic, and changes in kinetic and potential energy are negligible, we can apply the first law to this steady-flow system to get

$$m_a h_{a1} + m_{v1} h_{v1} + (m_{v2} - m_{v1}) h_{f2} = m_a h_{a2} + m_{v2} h_{v2}$$

In this equation, $(m_{v2} - m_{v1})$ is the amount of water added, and the enthalpy of the water is written as h_{f2} because the water is introduced at the temperature T_2. Dividing by the mass of air m_a, and noting that $h_{v1} = h_{g1}$ very nearly, as explained in Sec. 11·9, and

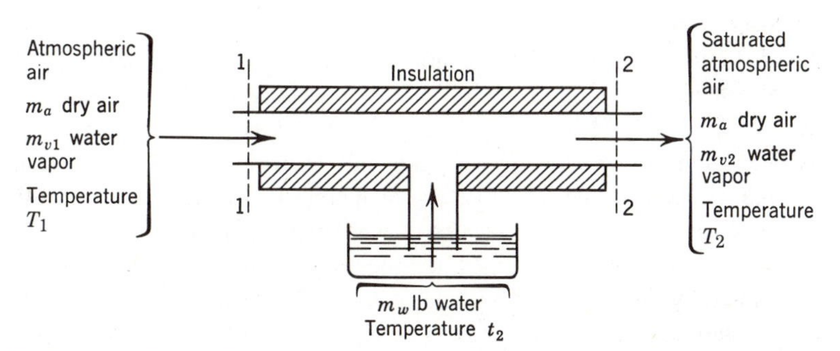

Figure 11·4. Steady-flow system for determining adiabatic saturation temperature.

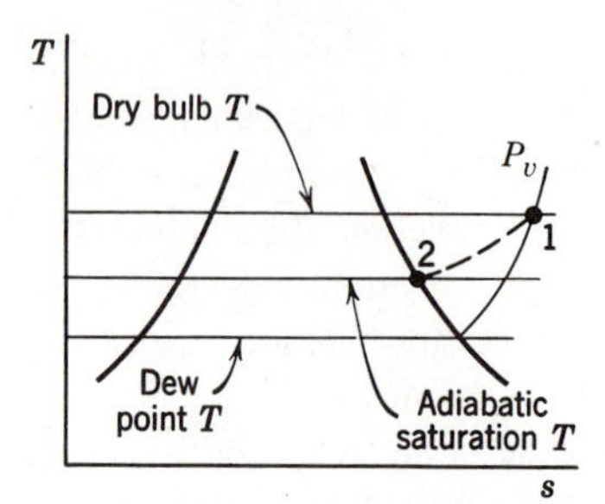

Figure 11·5. Adiabatic saturation.

$h_{v2} = h_{g2}$ exactly because the vapor at section 2 is saturated, we have

$$h_{a1} + \omega_1 h_{g1} + (\omega_2 - \omega_1) h_{f2} = h_{a2} + \omega_2 h_{g2}$$

$$\omega_1 = \frac{h_{a2} - h_{a1} + \omega_2 (h_{g2} - h_{f2})}{h_{g1} - h_{f2}}$$

$$= \frac{c_{pa}(T_2 - T_1) + \omega_2 h_{fg2}}{h_{g1} - h_{f2}}$$

This expression can be evaluated if T_1, T_2, and P_m are measured. Since $\phi_2 = 1.0$,

$$\omega_2 = \frac{v_{a2}}{v_{g2}} = \frac{R_a T_2}{(P_m - P_{g2}) v_{g2}}$$

so that ω_2 is a function of only P_m and T_2.

The states of the water vapor in the mixture during the adiabatic saturation process are shown on a *Ts* diagram in Fig. 11·5. During the process the vapor pressure increases and the temperature decreases; so the adiabatic saturation temperature is higher than the dew-point and lower than the dry-bulb temperature. For the limiting case of a saturated mixture, the dry-bulb, dew-point, and adiabatic saturation temperatures are the same.

The procedure and analysis just outlined provide a method of determining the humidity ratio of atmospheric air from temperature and barometric-pressure measurements. Actually it is inconvenient to saturate atmospheric air by the procedure described, so other methods are used.

Wet-Bulb Temperature

To avoid the difficulty of adiabatically saturating a sample of atmospheric air, the wet-bulb temperature analysis has been devised. This procedure involves the passage of an unsaturated air–vapor mixture over a wetted surface until a condition of dynamic equilibrium has been attained. When this condition has been reached, the heat transferred from the air–vapor stream to the liquid film to evaporate part of it is equal to the energy carried from the liquid film to the air–vapor stream by the diffusing vapor.

The equilibrium condition is obtained and the temperature of the resulting air–vapor mixture is measured by means of a thermometer, the bulb of which is covered with gauze soaked in clean water. A schematic diagram is shown as Fig. 11·6. The flow of atmospheric air is provided either by a fan or by mounting the thermometer in a holder with a swivel

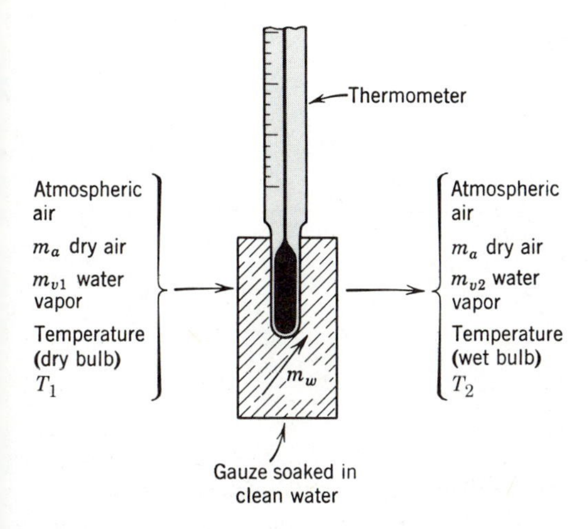

Figure 11·6. Wet-bulb thermometer.

handle so that it can be rotated or whirled through the air. The thermometer reading is called the wet-bulb temperature.

The relationship between the wet-bulb temperature and the adiabatic saturation temperature for any gas–vapor mixture depends on the heat-transfer and diffusion characteristics of the mixture. It happens that for air–water-vapor mixtures in the normal pressure and temperature range of atmospheric air the wet-bulb temperature measured by the usual type of instrument is very nearly equal to the adiabatic saturation temperature. This is simply fortuitous, and the equality does not hold for most gas–vapor mixtures. For example, in the air–vapor mixtures in oil-storage tanks and in alcohol–air mixtures, the difference between the wet-bulb temperature and the adiabatic saturation temperature may be quite large, and serious errors would follow from the assumption that they are equal.

Example 11·14. Determine the humidity ratio of air that has a dry-bulb temperature of 30°C, a wet-bulb temperature of 20°C, and a barometric pressure of 95.0 kPa.

Solution. The solution involves first assuming that the wet-bulb temperature is equal to the adiabatic saturation temperature and then applying the first law to an adiabatic saturation process that proceeds from the specified state to the saturated state at 20°C. Call these two states 1 and 2, respectively.

The relative humidity at state 2 is 100 percent, since complete saturation is assumed. For this condition the vapor pressure equals the saturation pressure at 20°C, which is found from the steam tables to be 2.339 kPa. The pressure of the dry air is then

$$P_{a2} = P_m - P_{v2} = 95.0 - 2.339 = 92.66 \text{ kPa}$$

The specific volume of dry air is

$$v_{a2} = \frac{R_a T_2}{P_{a2}} = \frac{0.287(293)}{92.66} = 0.908 \text{ m}^3/\text{kg}$$

and the humidity ratio at state 2 is

$$\omega_2 = \frac{v_{a2}}{v_{g2}} \phi = \frac{0.908(1.0)}{57.791} = 0.0157 \text{ kg/kg dry air}$$

Application of the first law to the adiabatic saturation process leads (as is shown earlier in this section) to

$$\omega_1 = \frac{c_{pa}(T_2 - T_1) + \omega_2 h_{fg2}}{h_{g1} - h_{g2}}$$

$$\omega_1 = \frac{1.005(20 - 30) + 0.0157(2454.1)}{(2556.3 - 83.96)}$$

$$= 0.0115 \text{ kg/kg dry air}$$

11·12 Psychrometric Charts

A psychrometric chart for atmospheric air is a plot of the properties of air–water-vapor mixtures at a fixed total pressure of the mixture. The chart is usually based on a pressure of one standard atmosphere, but charts for other barometric pressures are available. The specific properties are expressed not per kilogram (or pound) of mixture but per kilogram (or pound) of dry air in the mixture.

A skeleton outline of a psychrometric chart is shown in Fig. 11·7. (Complete charts are included in the appendix.) The horizontal and vertical axes are selected to represent dry-bulb temperature and humidity ratio, respectively. (Charts using English units often express humidity ratio in grains of water vapor per pound of dry air, where 7000 grains = 1 lbm.)

The line *A-B* is the saturation line and represents states for which the relative humidity is 100 percent. This line and other constant ϕ lines can be readily plotted because for states of constant relative humidity the humidity ratio is a function of only the mixture

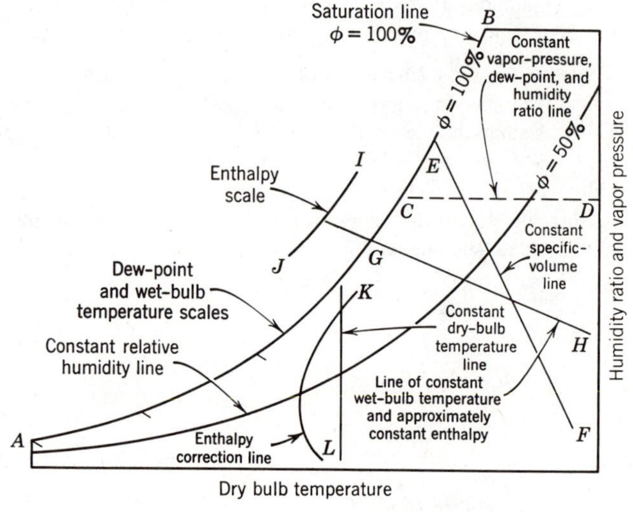

Figure 11·7. Skeleton outline of psychrometric chart.

pressure and the dry-bulb temperature. This fact is shown by

$$\omega = \frac{v_a}{v_g}\phi = \frac{R_a T}{P_a v_g}\phi = \frac{R_a T}{(P_m - P_v)v_g}\phi = \frac{R_a T}{(P_m - \phi P_g)v_g}\phi$$

As long as the air and water vapor both behave as ideal gases and the mixture pressure is constant, there is a unique relationship between the vapor pressure and the humidity ratio as can be seen from

$$\omega \equiv \frac{m_v}{m_a} = \frac{P_v V R_a T}{P_a V R_v T} = \frac{P_v R_a}{P_a R_v} = \frac{P_v R_a}{(P_m - P_v)R_v}$$

Thus a vapor-pressure scale can be constructed along with the humidity ratio scale on the vertical axis. The humidity ratio scale is usually selected as a linear scale, so the vapor-pressure scale is not linear.

The dew point of atmospheric air is a function of the vapor pressure only; so a dew-point temperature scale can also be constructed along the vertical axis. To facilitate the reading of the chart, the dew-point temperature scale is laid out along the saturation line *A-B* instead of as another scale at one side of the chart. For each state on line *A-B* (ϕ = 100 percent) the dew-point equals the dry-bulb temperature; so the dew-point scale values can be plotted by simply following vertical lines up from the various dry-bulb temperature scale values.

Horizontal lines such as *C-D* therefore are lines of constant humidity ratio, vapor pressure, and dew point.

The chart also has lines of constant specific volume, given as volume of mixture (or, identically, of dry air or of vapor) per kilogram (or pound) of dry air. One of these lines is shown as line *E-F*.

Consider again the adiabatic saturation process that was discussed in the preceding section. By definition, every state that air passes through during the adiabatic saturation process has the same adiabatic saturation temperature. Thus the path of an adiabatic saturation process, along which the vapor pressure increases and the dry-bulb temperature decreases, is a line of constant adiabatic saturation temperature and for the special case of air–water-vapor mixtures it is a constant wet-bulb temperature line. Such a line is shown as *G-H*.

Lines of constant wet-bulb temperature are approximately lines of constant mixture enthalpy. To demonstrate this, we first write an energy balance for an adiabatic saturation process,

$$h_{a1} + \omega_1 h_{v1} + (\omega_2 - \omega_1)h_{f2} = h_{a2} + \omega_2 h_{v2}$$

where the subscript 1 refers to any section in the adiabatic saturation flow path and the subscript 2 refers to the fully saturated state. If we now define the mixture enthalpy as $h_m \equiv h_a + \omega h_v$, noting that this is the enthalpy *of the mixture* expressed *per unit mass of dry air,* the energy balance becomes

$$h_{m1} - \omega_1 h_{f2} = h_{m2} - \omega_2 h_{f2}$$

Since subscript 1 refers to *any* section in the adiabatic saturation flow path, this last equation can be written as

$$h_m - \omega h_{f2} = \text{constant}$$

Since ωh_f evaluated at the adiabatic saturation temperature is always very small in comparison with h_m, we have

$$h_m \approx \text{constant}$$

for the adiabatic saturation process. Thus, a constant wet-bulb temperature line (which for atmospheric air is very nearly a constant adiabatic saturation temperature line) such as *G-H* is approximately a constant-enthalpy line, so the psychrometric chart carries an enthalpy scale that is read against the constant wet-bulb temperature lines. For accurate determination of the enthalpy of an unsaturated air–vapor mixture, a correction must be applied to the scale value. If we do not neglect the ωh_{f2} term, the enthalpy of the mixture at any state is

$$h_m = h_{m2} - (\omega_2 - \omega)h_{f2} = h_{m2} + \text{correction}$$

where the subscript 2 refers to the adiabatic saturation end state. The correction term, $-(\omega_2 - \omega)h_{f2}$, which literally corrects for the assumption that the mixture enthalpy is constant along a constant wet-bulb temperature line, is a function of specific humidity and wet-bulb temperature only, and so it can be plotted on the chart. *K-L* is such a line of constant enthalpy correction. The enthalpy of any mixture then is the enthalpy at the corresponding adiabatic saturation state plus the correction read from a curve such as *K-L* passing through the mixture state on the chart. (Several other methods are used to show the correction on psychrometric charts, and the correction is usually so small that many charts do not show it at all.)

Example 11·15. If the dry-bulb and wet-bulb temperatures of atmospheric air are 90 and 70°F, respectively, determine by use of the psychrometric chart (*a*) the humidity ratio, (*b*) the relative humidity, (*c*) the dew-point temperature, (*d*) the pressure of the water vapor, (*e*) the volume per pound of dry air, and (*f*) the enthalpy per pound of dry air.

Solution. The particular point on the chart which designates the condition of the atmospheric air is located by finding the intersection of the vertical 90°F dry-bulb temperature line and the sloping 70°F wet-bulb temperature line.

(***a***) The humidity ratio is found by moving along a horizontal line to the proper scale on the diagram. The approximate value is 78 grains of water vapor per pound of dry air.

(***b***) The relative humidity is estimated by observing the location of the point between two of the curved relative humidity lines. The approximate value is 37 percent.

(***c***) The dew-point temperature is found by moving along a horizontal line through the point until the dew-point temperature scale is reached. The approximate value is 60.2°F.

(***d***) The partial pressure of the water vapor is found by following a horizontal line until the vapor-pressure scale is reached. The value is 0.26 psia.

(***e***) The volume per pound of dry air is estimated by observing the location of the point between two of the constant-volume lines. The approximate value is 14.1 cu ft/lbm dry air.

(***f***) The enthalpy may be found by first following the sloping 70°F wet-bulb temperature line until it intersects the enthalpy scale. The enthalpy value is 34.1 B/lbm for saturated air at 70°F wet-bulb temperature. For 90°F dry-bulb and 70°F wet-bulb temperatures, the enthalpy correction is approximately −0.2 B/lbm, so the mixture enthalpy is 34.1 − 0.2 = 33.9 B/lbm.

11·13 Processes of Air–Vapor Mixtures under Constant Total Pressure

Many processes of air–vapor mixtures occur under conditions of constant total pressure. Among these are the heating, cooling, humidification, and dehumidification that occur in air-conditioning systems, evaporative cooling, and drying. The application of the principles discussed in this chapter to some of these processes is illustrated in the following three examples. Many more applications are treated in the problems at the end of this chapter.

Example 11·16. Atmospheric air has a relative humidity of 80 percent at a dry-bulb temperature of 20°C. If the air is heated to 30°C, determine the amount of heat that must be transferred per kilogram of dry air under steady-flow conditions. Atmospheric pressure equals 101.3 kPa.

Solution. Under the specified condition there will be no change in the quantity of vapor present for each kilogram of dry air, since moisture is neither added nor removed during the process.

The first step is to determine the value of the initial humidity ratio. The saturation pressure and specific volume values for the water vapor corresponding to 20°C, taken from the steam tables, are 2.339 kPa and 57.791 m^3/kg, respectively. From the relation for relative humidity, the vapor pressure is

$$P_v = \phi(p_g) = 0.80(2.339) = 1.87 \text{ kPa}$$

The air pressure equals

$$P_a = P_m - P_v = 101.3 - 1.87 = 99.4 \text{ kPa}$$

From the ideal-gas equation of state the specific volume of the air is

$$v_a = \frac{R_a T}{P_a} = \frac{0.287(293)}{99.4} = 0.846 \text{ m}^3/\text{kg}$$

The humidity ratio is then

$$\omega = \frac{v_a}{v_g}\phi = \frac{0.846(0.80)}{57.791} = 0.0117 \text{ kg/kg dry air}$$

Application of the first law to this steady-flow process in which work = 0 and $\Delta ke = 0$ gives

$$\begin{aligned} q &= h_{a2} + \omega h_{v2} - (h_{a1} + \omega h_{v1}) \\ &= h_{a2} - h_{a1} + \omega(h_{v2} - h_{v1}) = c_p(T_2 - T_1) + \omega(h_{g2} - h_{g1}) \\ &= 1.005(30 - 20) + 0.0117(2556.3 - 2538.1) \\ &= 10.3 \text{ kJ/kg dry air} \end{aligned}$$

(The energy balance can also be written as

$$q = h_{m2} - h_{m1}$$

and the mixture enthalpy values can be obtained from a psychrometric chart.)

The next example deals with the dehumidification of air in an air-conditioning system. See Fig. 11·8. The air is first cooled to a temperature lower than its initial dew point to cause some of the water vapor to condense. The air leaving the cooler is saturated at a low temperature and is unsuitable for use in rooms occupied by people because it is clammy. It must be heated or mixed with warmer air to produce a condition that is normally regarded as comfortable. The dehumidification and reheating processes are shown on a psychrometric chart in Fig. 11·8. Notice that the temperature to which the mixture must be cooled is the dew point corresponding to the desired final condition.

Example 11·17. Atmospheric air at 80°F and 60 percent relative humidity is to be brought to 75°F and 40 percent relative humidity by means of a system comprised of a cooler, with provision for removing condensate, followed by a heater. Barometric pressure is 14.696 psia. The mass rate of flow of atmospheric air entering the system is 202.6 lbm/h. Determine (*a*) the mass of water removed per hour, (*b*) the heat removed in the cooler per hour, and (*c*) the heat added in the heater per hour.

Solution. Refer to Fig. 11·8 for a sketch of the system. The first step in the solution is to determine the humidity ratios for the initial and final conditions.

Initial Condition	Final Condition
$P_v = \phi P_g = 0.60(0.5073) = 0.304$ psia	$P_v = 0.40(0.4300) = 0.172$ psia
$P_a = P_m - P_v = 14.696 - 0.304 = 14.392$ psia	$P_a = 14.696 - 0.172 = 14.524$ psia
$v_a = \dfrac{R_a T}{p_a} = \dfrac{53.3(540)}{14.392(144)} = 13.89$ cu ft/lbm	$v_a = \dfrac{53.3(535)}{14.524(144)} = 13.63$ cu ft/lbm
$\omega = \dfrac{v_a}{v_g}\phi = \dfrac{13.89(0.60)}{632.8} = 0.01317$ lbm/lbm	$\omega = \dfrac{13.64(0.40)}{739.7} = 0.00737$ lbm/lbm

We now find the mass flow rate of dry air from

$$\dot{m}_{\text{atm air},1} = \dot{m}_{\text{da}} + \dot{m}_{v1} = \dot{m}_{\text{da}} + \omega_1 \dot{m}_{\text{da}} = \dot{m}_{\text{da}}(1 + \omega_1)$$

$$\dot{m}_{\text{da}} = \frac{\dot{m}_{\text{atm air},1}}{1 + \omega_1} = \frac{202.6}{1.01317} = 200 \text{ lbm/h}$$

(***a***) The mass flow rate of water removed is

$$\dot{m}_w = \dot{m}_{\text{da}}(\omega_1 - \omega_2) = 200(0.01317 - 0.00737) = 1.16 \text{ lbm/h}$$

(***b***) The mixture leaving the cooler will be saturated and must have the same humidity ratio as the mixture in the desired final condition. Thus the temperature at the cooler outlet must be the dew point corresponding to the final condition. This is the saturation temperature corresponding to the final vapor pressure of 0.172 psia and is approximately 49°F. The cooler must, therefore, cool the air from 80 to 49°F. During the latter part of this cooling process (while the mixture is at temperatures lower than the dew point of the initial state) water is removed. Per pound of dry air the amount of water removed is $(\omega_1 - \omega_2) =$

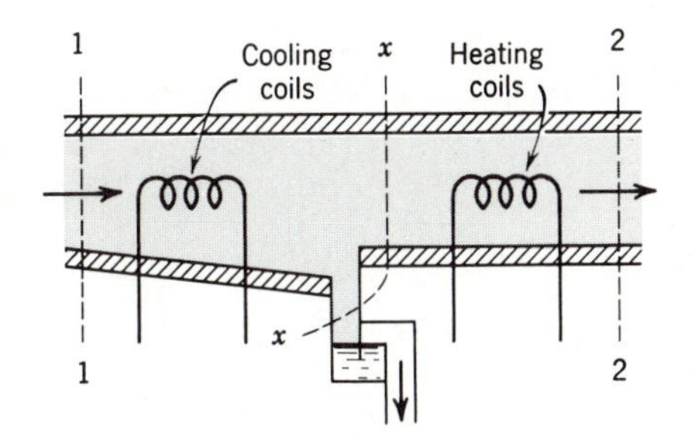

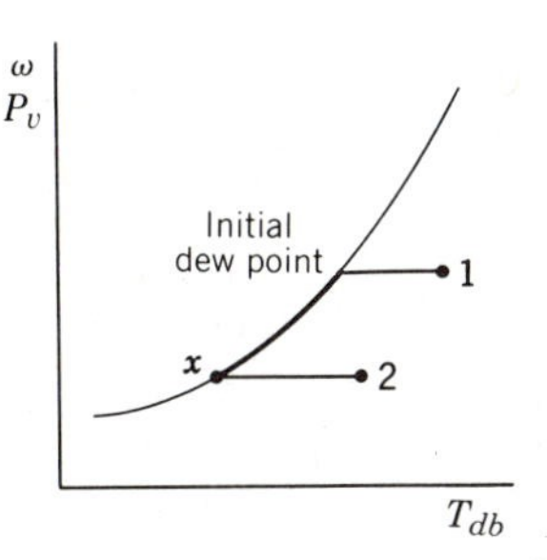

Figure 11·8. Dehumidification and reheating.

0.01317 − 0.00737 = 0.00580 lbm/lbm dry air. Application of the first law to the cooler in which work = 0, Δke = 0, and the flow is steady gives

$$q_{\text{out}} = h_{a1} + \omega_1 h_{v1} - h_{ax} - \omega_x h_{vx} - (\omega_1 - \omega_2) h_{fx}$$

We assume that all the condensate leaves at the temperature T_x. Noting that $\omega_x = \omega_2$, $h_v = h_g$ at the same temperature, and $\Delta h_a = c_{pa}\,\Delta T$, we have then

$$\begin{aligned} q_{\text{out}} &= c_{pa}(T_1 - T_x) + \omega_1 h_{g1} - \omega_2 h_{gx} - (\omega_1 - \omega_2) h_{fx} \\ &= 0.24(80 - 49) + 0.01317(1096.4) - 0.00737(1082.9) - 0.00580(17.1) \\ &= 13.8 \text{ B/lbm dry air} \end{aligned}$$

The rate of heat removal is

$$\dot{Q} = \dot{m}_{da} q_{\text{out}} = 200(13.8) = 2760 \text{ B/h}$$

(*c*) Application of the first law in similar fashion to the heater gives

$$\begin{aligned} q &= h_{a2} + \omega_2 h_{v2} - h_{ax} - \omega_x h_{vx} \\ &= c_p(T_2 - T_x) + \omega_2(h_{g2} - h_{gx}) \\ &= 0.24(75 - 49) + 0.00737(1094.2 - 1082.9) \\ &= 6.3 \text{ B/lbm dry air} \end{aligned}$$

The rate of heat addition is

$$\dot{Q} = \dot{m}_{da} q = 200(6.3) = 1260 \text{ B/h}$$

(This problem can also be solved quickly and with only a slight loss in accuracy by means of the psychrometric chart.)

In locations where sufficient cooling water cannot be obtained from a river, lake, or ocean, it is necessary to cool water by means of a cooling tower or spray pond so that it can be used over and over again to remove heat from buildings, power plants, refrigerators, and other equipment. A schematic sketch of an induced draft cooling tower is shown in Fig. 11·9. The warm water that is to be cooled is introduced at the top of the tower through distributing troughs or spray nozzles and falls through a series of trays that are arranged to keep the falling water broken up in fine drops that have a large surface area from which evaporation can occur. Fans located at the top of the tower draw in atmospheric air at the bottom of the tower and cause it to flow upward through the falling water. The water

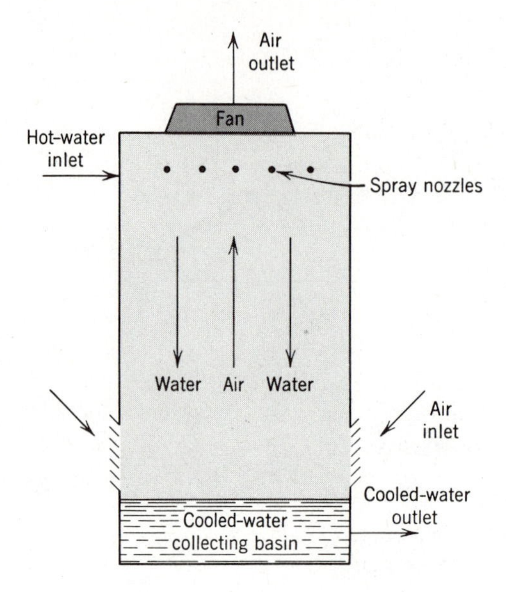

Figure 11·9. Schematic diagram of induced-draft cooling tower.

falling through the tower may be cooled somewhat by a transfer of heat from it to the air, but the cooling results chiefly from the evaporation of some of the water, because the water that evaporates must be supplied with its latent heat of vaporization. This is obtained chiefly from the water that does not evaporate.

In making an energy balance on a cooling tower, we can usually neglect the heat transfer to the fluids within the tower from the surrounding atmosphere. Also, the work of the fans is negligible in comparison with the other energy quantities involved. The following example illustrates the application of the first law (energy balance) and the law of conservation of mass (mass balance) to a cooling tower.

Example 11·18. A cooling tower is used to cool water from 45 to 25°C. Water enters the tower at a rate of 110 000 kg/h. The air entering the tower is at 20°C with a relative humidity of 55 percent; the air leaving is at 40°C with a relative humidity of 95 percent. The air enters at

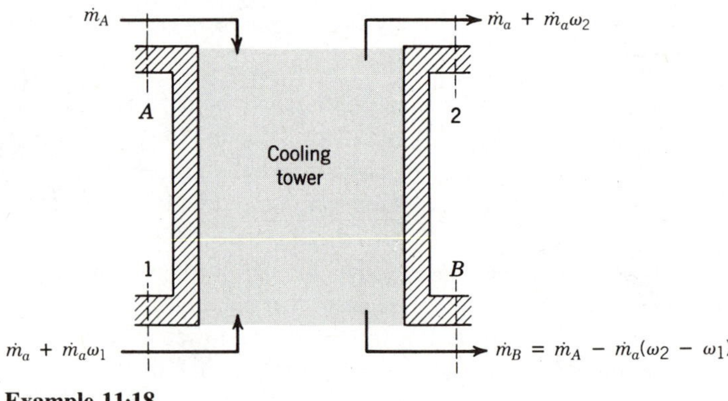

Example 11·18.

the bottom of the tower and flows upward. Barometric pressure is 92.0 kPa. Compute (*a*) the required flow rate of atmospheric air in kg/h and (*b*) the amount of water lost by evaporation per hour.

Solution. First a sketch of the steady-flow system comprised of the cooling tower is made. Designate the air inlet as section 1, the air outlet as section 2, the water inlet as section A, and the water outlet as section B. The properties of the entering and leaving air streams are calculated as in Example 11·17 with the following results:

Entering Air	Leaving Air
$P_{v1} = 1.286$ kPa	$P_{v2} = 7.015$ kPa
$P_{a1} = 90.7$ kPa	$P_{a2} = 85.0$ kPa
$\omega_1 = 0.00822$ kg/kg dry air	$\omega_2 = 0.0541$ kg/kg dry air

(***a***) Let $\dot{m}_a$ be the mass flow rate of dry air required. $\dot{m}_{a1} = \dot{m}_{a2} = \dot{m}_a$. Applying the first law to the cooling tower, assuming that $q = 0$, $w = 0$, and $\Delta ke = 0$,

$$\dot{m}_a h_{a1} + \dot{m}_{v1} h_{v1} + \dot{m}_A h_A = \dot{m}_a h_{a2} + \dot{m}_{v2} h_{v2} + \dot{m}_B h_B$$

However,

$$\dot{m}_{v1} = \omega_1 \dot{m}_a$$

$$\dot{m}_{v2} = \omega_2 \dot{m}_a$$

$$\dot{m}_B = \dot{m}_A - \dot{m}_a(\omega_2 - \omega_1)$$

so that the energy balance becomes

$$\dot{m}_a h_{a1} + \dot{m}_a \omega_1 h_{v1} + \dot{m}_A h_A = \dot{m}_a h_{a2} + \dot{m}_a \omega_2 h_{v2} + [\dot{m}_A - \dot{m}_a(\omega_2 - \omega_1)]h_B$$

$$\dot{m}_a = \frac{\dot{m}_A(h_A - h_B)}{h_{a2} - h_{a1} + \omega_2 h_{v2} - \omega_1 h_{v1} - (\omega_2 - \omega_1)h_B}$$

$$= \frac{\dot{m}_A(h_A - h_B)}{c_{pa}(T_2 - T_1) + \omega_2 h_{g2} - \omega_1 h_{g1} - (\omega_2 - \omega_1)h_B}$$

$$= \frac{110\,000(188.45 - 104.89)}{1.005(40-20) + 0.0541(2574.3) - 0.00822(2538.1) - (0.0541 - 0.00822)104.89}$$

$$= 68\,750 \text{ kg dry air/h}$$

(***b***) The amount of water evaporated is

$$\dot{m}_{\text{evaporated}} = \dot{m}_a \omega_2 - \dot{m}_a \omega_1 = 68\,750(0.0541 - 0.00822) = 3150 \text{ kg/h}$$

11·14 Summary

The mass of a mixture is equal to the sum of the masses of its constituents:

$$m_m = m_A + m_B + m_C + \cdots$$

The total number of moles of a mixture is defined as the sum of the number of moles of its constituents:

$$N_m \equiv N_A + N_B + N_C + \cdots$$

The mole fraction is defined as

$$x \equiv \frac{N}{N_m}, \qquad x_A \equiv \frac{N_A}{N_m}, \qquad \text{etc.}$$

The partial pressure P_i of a constituent i in a gas mixture is defined as

$$P_i \equiv x_i P_m, \qquad P_A \equiv x_A P_m, \qquad \text{etc.}$$

The partial volume V_i' of a constituent i in a gas mixture is defined as

$$V_i' \equiv x_i V_m, \qquad V_A' \equiv x_A V_m, \qquad \text{etc.}$$

A mixture of ideal gases is also an ideal gas.

Dalton's law or the law of additive pressures states that the pressure of a mixture of ideal gases equals the sum of the pressures of its constituents if each existed alone at the temperature and volume of the mixture. Another statement of this law is that, in a mixture of ideal gases, each constituent behaves in all respects as though it existed alone at the temperature and volume of the mixture. It follows that in a mixture of ideal gases the partial pressure of each constituent is equal to the pressure that that constituent would exert if it existed alone at the temperature and volume of the mixture.

Amagat's law, Leduc's law, or the law of additive volumes states that the volume of a mixture of ideal gases equals the sum of the volumes of its constituents if each existed alone at the temperature and pressure of the mixture. It follows that in a mixture of ideal gases the partial volume of each constituent is equal to the volume that that constituent would occupy if it existed alone at the pressure and temperature of the mixture.

The law of additive pressures and the law of additive volumes hold strictly only for ideal-gas mixtures, but they hold approximately for real-gas mixtures even in some ranges of pressure and temperature where $Pv = RT$ is inaccurate.

Properties of ideal-gas mixtures can be obtained accurately and properties of real-gas mixtures can be obtained approximately in terms of constituent properties by application of the principles stated above.

In many gas–vapor mixtures all constituents can be treated as ideal gases, but an additional fact must be considered: The maximum pressure of a vapor in a mixture depends on the temperature, since it cannot be higher than the saturation pressure corresponding to the mixture temperature.

In *atmospheric air* the water vapor can be treated as an ideal gas; so it follows the equation of state $Pv = RT$, and its enthalpy is a function of temperature only. Thus the enthalpy of water vapor in atmospheric air can be read from the steam tables by reading the enthalpy of saturated vapor at the same temperature.

Relative humidity ϕ is defined as the ratio of the pressure (i.e., partial pressure) of the vapor in atmospheric air to the saturation pressure of the vapor at the temperature of the mixture:

$$\phi \equiv \frac{P_v}{P_g} \tag{11·1}$$

Humidity ratio ω is defined as the ratio of the mass of vapor in atmospheric air to the mass of dry air:

$$\omega \equiv \frac{m_v}{m_a} \tag{11·3}$$

The relationship between relative humidity and humidity ratio for gas–vapor mixtures in which each constituent behaves as an ideal gas is

$$\omega = \frac{v_a}{v_g}\,\phi \tag{11·4}$$

The *dew-point temperature* of an air–vapor mixture is defined as the saturation temperature corresponding to the partial pressure of the vapor in the mixture.

The *adiabatic saturation temperature* of an air–vapor mixture is defined as the temperature that results from adiabatically adding water to the mixture in steady flow until it becomes saturated, the water being supplied at the final temperature of the mixture.

The *wet-bulb temperature* is the temperature measured by a thermometer wrapped in gauze soaked in clean water and placed in a stream of atmospheric air. For atmospheric air it happens that the wet-bulb temperature is very close to the adiabatic saturation temperature. This approximation does not hold for other air–vapor mixtures.

Charts of atmospheric air properties called *psychrometric charts* are useful in reducing the time required for various calculations.

Suggested Reading

11·1 Holman, J. P., *Thermodynamics,* 2nd ed., McGraw-Hill, New York, 1974, Chapter 10.

11·2 McQuiston, Faye C., and Jerald D. Parker, *Heating, Ventilating, and Air Conditioning,* 2nd ed., Wiley, New York, 1982, Chapter 2.

11·3 Wark, K., *Thermodynamics,* 4th ed., McGraw-Hill, New York, 1983, Chapters 11 and 12.

Problems

11·1 Prove that the density of an ideal-gas mixture equals the sum of the densities of its constituents.

11·2 A mixture at 20°C (68°F) consists of 2 parts by mass carbon dioxide, 4 parts hydrogen, and 6 parts nitrogen. Compute the apparent molar mass and the gas constant for the mixture.

11·3 A mixture having a volumetric analysis as follows is cooled under constant-volume conditions from 1 atm, 85°C (185°F) to a final temperature of 10°C (50°F): oxygen 40 percent, nitrogen 60 percent. Compute the partial pressures of the constituents at 10°C and the volumetric analysis at 10°C.

11·4 The mass analysis of a gas mixture shows that it consists of 60 percent nitrogen, 30 percent carbon dioxide, and 10 percent carbon monoxide. For a mixture pressure of 1 atm, compute (*a*) the partial pressures of the constituents, (*b*) the apparent molar mass, and (*c*) the apparent gas constant.

11·5 The volumetric analysis for a gas mixture shows that it consists of 50 percent nitrogen, 30 percent carbon dioxide, and 20 percent carbon monoxide. The mixture is at 7 atm, 30°C (86°F). Compute (*a*) the partial pressures, (*b*) the apparent molar mass, and (*c*) the apparent gas constant.

11·6 A mixture having the following volumetric analysis is contained in a closed tank: water vapor 11 percent, oxygen 12 percent, nitrogen 77 percent. Compute the mass analysis and the apparent molar mass of the mixture.

11·7 A sample of blast-furnace gas had the following volumetric analysis: hydrogen 3 percent, carbon monoxide 27 percent, carbon dioxide 10 percent, nitrogen 60 percent. Compute (*a*) the mass analysis, (*b*) the partial pressures, and (*c*) the specific heats for the mixture at 20°C (68°F). The total pressure is 1 atm.

11·8 A mixture composed of the following gases is at 1 atm, 0°C (32°F): 3 parts oxygen, 2 parts carbon monoxide, 0.5 parts hydrogen, 1 part helium, 3 parts nitrogen, and 0.5 parts argon, by mass. Compute the specific heat values and the partial pressures.

11·9 A tank having a volume of 0.20 m^3 contains oxygen at 72 kPa, 35°C. Carbon monoxide is forced into the tank until the mixture is at 200 kPa, 35°C. Determine for the mixture (*a*) the volumetric analysis and (*b*) the mass analysis.

11·10 Calculate the specific volume of a mixture of 30 percent CH_4, 20 percent C_2H_6, and 50 percent N_2 by mass at 95 kPa, 35°C.

11·11 The mass fraction of each constituent in a mixture of N_2 and CO_2 is 50 percent. The mixture is compressed from 100 kPa, 5°C, to 500 kPa, 170°C. Calculate the entropy change of the CO_2.

11·12 A gas mixture is made up of 60 percent air and 40 percent argon on a volumetric basis. The mixture is compressed from 100 kPa, 20°C, to 200 kPa, 120°C. Determine the entropy change of the gas in kJ/kg·K.

11·13 A tank containing 0.4 m^3 of helium at 700 kPa and 10°C is connected by means of a pipe and valve system to a 0.3-m^3 tank containing nitrogen at 350 kPa and 65°C. If no heat losses occur, compute the resulting pressure and temperature after the valve has been opened and mixing occurs.

11·14 Eight one-hundredths of a cubic meter of methane at 300 psia, 100°F, is mixed with 0.20 m^3 of oxygen at 700 kPa, 5°C, by opening a valve between the two tanks. Calculate the heat transfer if the final temperature is 35°C.

11·15 Three kilograms of nitrogen at 1.4 MPa, 35°C, and 2 kg of oxygen at 200 kPa, 90°C, are mixed adiabatically with no change in the total volume. Determine the mixture pressure and temperature.

11·16 A tank containing 1 kg of methane at 140 kPa, 10°C, and a tank holding 2 kg of oxygen at 700 kPa, −5°C, are connected through a valve. The valve is opened and the gases mix adiabatically. Atmospheric temperature is 10°C. Determine (*a*) the mixture pressure, (*b*) the mixture temperature, (*c*) the volumetric analysis of the mixture, and (*d*) the irreversibility of the process.

11·17 A system consisting of three tanks, all interconnected by pipes and valves, contains nitrogen, oxygen, and argon. One tank contains 5 kg of nitrogen at 700 kPa, 30°C; another 2.5 kg of argon at 500 kPa, 50°C; and the third 7.5 kg of oxygen at 350 kPa, 80°C. After the valves are open, compute (*a*) the temperature of the mixture, (*b*) the apparent molar mass, (*c*) the constant-pressure specific heat of the mixture, (*d*) the gas constant, (*e*) the pressure of the mixture, and (*f*) the partial pressure of each constituent.

11·18 Determine the minimum work input required per kilogram of air to separate air at 1 atm, 15°C, into nitrogen and oxygen, each at 1 atm, 15°C.

11·19 One tank contains 3.76 moles of nitrogen at 140 kPa, 35°C, and another tank contains 1 mole of oxygen at 140 kPa, 30°C. A valve connecting the tanks is opened, and the gases mix adiabatically. The lowest temperature in the surroundings is 5°C. Determine the irreversibility of the process. Explain the physical significance of this value.

11·20 Propane at 700 kPa, 35°C, flows steadily through a pipe line. A tank of 0.06 m^3 volume that contains air at 101.3 kPa, 15°C, is connected to the pipe line, and propane is allowed to flow into the tank until the total pressure in the tank is 350 kPa. Assuming that no air leaves the tank and that the filling is adiabatic, determine the mass of propane that flows into the tank.

11·21 Propane at 700 kPa, 35°C, flows steadily through a pipe line. A tank of 0.06 m^3 volume that contains air at 101.3 kPa, 15°C, is connected to the pipe line, and propane is allowed to flow into the tank until the total pressure in the tank is 350 kPa. Assuming that no air leaves the tank, determine the heat transfer that is required to maintain the contents of the tank at 15°C.

11·22 Helium at 350 kPa, 50°C, flows through an insulated pipe line. A well-insulated tank 0.06 m^3 in volume containing air at 101.3 kPa, 15°C, is connected to the helium line through a small valve that is opened until the pressure in the tank becomes 200 kPa. The valve is then closed. Assuming that no air flows from the tank and that the process is adiabatic, determine the final temperature of the mixture in the tank.

11·23 Prove that for the adiabatic mixing of ideal gases initially at the same pressure and temperature in a rigid tank the entropy change depends only on the number of moles of constituents and not on what the constituents are.

11·24 Is it true that adiabatic mixing in a constant-volume system of two ideal gases with different k values results in a change in enthalpy but no change in internal energy? Prove your answer.

11·25 A tank having a volume of 6.0 cu ft contains oxygen at 10.5 psia, 100°F. Carbon monoxide is forced into the tank until the mixture is at 30 psia, 100°F. Determine for the mixture (*a*) the volumetric analysis and (*b*) the mass analysis.

11·26 Calculate the specific volume of a mixture of 30 percent CO_2, 20 percent CO, and 50 percent N_2 by mass at 14.0 psia, 100°F.

11·27 The mass fraction of each constituent in a mixture of N_2 and CO_2 is 50 percent. The mixture is compressed from 15 psia, 40°F, to 75 psia, 340°F. Calculate the entropy change of the CO_2.

11·28 Hydrogen and oxygen, each at 14.0 psia, 60°F, flow into an adiabatic mixing chamber and leave as a mixture of stoichiometric proportions at 14.0 psia. The flow is steady. The temperature of the surrounding atmosphere is 60°F. Calculate (*a*) the partial pressure of hydrogen in the mixture and (*b*) the irreversibility of the mixing process per pound of hydrogen.

11·29 A tank containing 15 cu ft of helium at 100 psia and 50°F is connected by means of a pipe and valve system to a 10-cu-ft tank containing nitrogen at 50 psia and 150°F. If no heat losses occur, compute the resulting pressure and temperature after the valve has been opened and mixing occurs.

11·30 Three cubic feet of methane at 300 psia, 100°F, is mixed with 7 cu ft of oxygen at 100 psia, 40°F, by opening a valve between the two tanks. Calculate the heat transfer if the final temperature is 100°F.

11·31 Six pounds of nitrogen at 200 psia, 100°F, and 4 lbm of oxygen at 30 psia, 200°F, are mixed adiabatically with no change in the total volume. Determine the mixture pressure and temperature.

11·32 A tank containing 2 lbm of methane at 20 psia, 50°F, and a tank holding 4 lbm of oxygen at 100 psia, 20°F, are connected through a valve. The valve is opened and the gases mix adiabatically.

Atmospheric temperature is 50°F. Determine (*a*) the mixture pressure, (*b*) the mixture temperature, (*c*) the volumetric analysis of the mixture, and (*d*) the irreversibility of the process.

11·33 A system consisting of three tanks, all interconnected by pipes and valves, contains nitrogen, oxygen, and argon. One tank contains 10 lbm of nitrogen at 100 psia, 90°F; another 5 lbm of argon at 75 psia, 120°F; and the third 15 lbm of oxygen at 50 psia, 180°F. After the valves are open, compute (*a*) the temperature of the mixture, (*b*) the apparent molar mass, (*c*) the constant-pressure specific heat of the mixture, (*d*) the gas constant, (*e*) the pressure of the mixture, and (*f*) the partial pressure of each constituent.

11·34 Determine the minimum work input required per pound of air to separate air at 1 atm, 60°F, into nitrogen and oxygen, each at 1 atm, 60°F.

11·35 One tank contains 3.76 lbmol of nitrogen at 20 psia, 100°F, and another tank contains 1 lbmol of oxygen at 20 psia, 100°F. A valve connecting the tanks is opened, and the gases mix adiabatically. The lowest temperature in the surroundings is 40°F. Determine the irreversibility of the process. Explain the physical significance of this value.

11·36 Argon at 100 psia, 100°F, flows steadily through a pipe line. A tank of 2 cu ft volume that contains air at 14.7 psia, 60°F, is connected to the pipe line, and argon is allowed to flow into the tank until the total pressure in the tank is 50 psia. Assuming that no air leaves the tank and that the filling is adiabatic, determine the mass of argon that flows into the tank.

11·37 Argon at 100 psia, 100°F, flows steadily through a pipe line. A tank of 2 cu ft volume that contains air at 14.7 psia, 60°F, is connected to the pipe line, and argon is allowed to flow into the tank until the total pressure in the tank is 50 psia. Assuming that no air leaves the tank, determine the heat transfer that is required to maintain the contents of the tank at 60°F.

11·38 Helium at 50 psia, 120°F, flows through an insulated pipe line. A well-insulated tank 2 cu ft in volume containing air at 14.7 psia, 70°F, is connected to the helium line through a small valve that is opened until the pressure in the tank becomes 30 psia. The valve is then closed. Assuming that no air flows from the tank and that the process is adiabatic, determine the final temperature of the mixture in the tank.

11·39 One breathing mixture used for divers contains 20 percent oxygen by volume, 15 percent nitrogen, and the balance helium. Another contains the same fractions of oxygen and nitrogen, but the balance is a mixture of helium and neon in the same volumetric ratio they have in air (3.47 mol Ne/mol He). Compare the c_p values and the densities of these breathing mixtures with those of air.

11·40 A mixture consisting of 0.1 kg of dry saturated steam and 0.005 kg of air is contained in a tank at a temperature of 90°C. Compute the total pressure in the tank and the total volume occupied by the mixture.

11·41 A mixture consists of 0.10 kg of dry saturated water vapor, 0.15 kg of nitrogen, and 0.025 kg of oxygen. If the temperature of the mixture is 90°C, compute the total and partial pressures for the mixture.

11·42 One-half kilogram of nitrogen and 0.1 kg of water vapor occupy a volume of 0.12 m^3. Compute the total pressure for the mixture if the temperature is 250°C.

11·43 For wet-bulb and dry-bulb temperatures for atmospheric air of 20°C and 25°C, respectively, determine (*a*) the partial pressure of the water vapor, (*b*) the relative humidity, (*c*) the mass of water vapor per kilogram of dry air, (*d*) the specific volume in cubic meters per kilogram of dry air, and (*e*) the enthalpy per kilogram of dry air. Use the psychrometric chart. Assume barometric pressure of 1 atm.

11·44 For a relative humidity of 30 percent and a dry-bulb temperature of 35°C, determine by means of the psychrometric chart the humidity ratio and the pressure of the water vapor. Atmospheric pressure equals 101.325 kPa.

11·45 If the mass of water vapor per kilogram of dry air is 12 grams and the relative humidity is 60 percent, determine from the psychrometric chart the dry-bulb temperature and the pressure of the water vapor.

11·46 Atmospheric air at 25°C and a total pressure of 101 kPa has a relative humidity of 50 percent. What are the partial pressures of the constituents? What is the dew-point temperature?

11·47 The partial pressure of water vapor in moist air at a total pressure of 101 kPa and temperature of 30°C is 2.8 kPa. Determine the relative humidity and dew-point temperature.

11·48 Determine the humidity ratio of moist air at 100 kPa, 30°C, for a relative humidity of 60 percent.

11·49 Determine the humidity ratio for moist air at a total pressure of 85 kPa and a temperature of 30°C. The relative humidity is 60 percent.

11·50 Determine the mass of water vapor per kilogram of dry air for moist air at 35°C and 50 percent relative humidity. The total pressure is 101 kPa. What is the dew-point temperature?

11·51 An insulated tank contains dry air and a covered vessel of water at a uniform temperature. The cover of the vessel is removed so that water evaporates into the air. Finally, all parts of the system are again at a uniform temperature. State whether each of the following properties of the entire system has increased, decreased, remained constant, or varied indeterminately: (*a*) temperature, (*b*) internal energy, (*c*) entropy.

11·52 A room of 250 m^3 volume contains dry air at 1 atm, 25°C. How many tumblers of water (250 ml each) must be evaporated to bring the relative humidity to 0.50? If the room is sealed and insulated, what changes must occur in the pressure and temperature of the atmosphere?

11·53 A mixture consisting of 0.2 lbm of dry saturated steam and 0.01 lbm of air is contained in a tank at a temperature of 200°F. Compute the total pressure in the tank and the total volume occupied by the mixture.

11·54 A mixture consists of 0.2 lbm of dry saturated water vapor, 0.3 lbm of nitrogen, and 0.05 lbm of oxygen. If the temperature of the mixture is 200°F, compute the total and partial pressures for the mixture.

11·55 One pound of nitrogen and 0.2 lbm of water vapor occupy a volume of 14.12 cu ft. Compute the total pressure for the mixture if the temperature is 480°F.

11·56 If the wet-bulb and dry-bulb temperatures for atmospheric air are 70 and 80°F, respectively, determine (*a*) the partial pressure of the water vapor, (*b*) the relative humidity, (*c*) the mass of water vapor per pound of dry air, (*d*) the specific volume in cubic feet per pound of dry air, and (*e*) the enthalpy per pound of dry air. Use the psychrometric chart. Assume that the atmospheric pressure is 14.696 psia.

11·57 If the relative humidity is 30 percent and the dry-bulb temperature is 100°F, determine by means of the psychrometric chart the humidity ratio and the pressure of the water vapor. Atmospheric pressure equals 14.696 psia.

11·58 If the mass of water vapor per pound of dry air is 90 grains and the relative humidity is 60 percent, determine from the psychrometric chart the dry-bulb temperature and the pressure of the water vapor.

11·59 Atmospheric air at 80°F and a total pressure of 14.7 psia has a relative humidity of 50 percent. What are the partial pressures of the constituents? What is the dew-point temperature?

11·60 The partial pressure of water vapor in moist air at a total pressure of 14.7 psia and temperature of 90°F is 0.4 psia. Determine the relative humidity and dew-point temperature.

11·61 Determine the humidity ratio of moist air at 14.5 psia, 90°F for a relative humidity of 60 percent.

11·62 Determine the humidity ratio for moist air at 12.5 psia and 90°F. The relative humidity is 60 percent.

11·63 Determine the mass of water vapor per pound of dry air for moist air at 100°F and 50 percent relative humidity. The total pressure is 14.7 psia. What is the dew-point temperature?

11·64 A rigid tank contains 10.0 cu ft of oxygen at 14.0 psia, 80°F, and no other gas or vapor. A small covered jar inside the tank contains liquid water at 80°F. The jar is uncovered and some of the liquid evaporates until equilibrium is reached at 80°F, with some heat exchange taking place with the surroundings to maintain the temperature of the system (oxygen, liquid water, and vapor water) at 80°F. (*a*) What is the final mixture pressure in the tank? (*b*) Calculate the mass of water evaporated. (*c*) Indicate whether each of the following quantities increases, decreases, remains constant, or varies indeterminately: (1) entropy of the oxygen, (2) entropy of the H_2O in the system, (3) entropy of the surroundings.

11·65 How many kilograms of moisture are removed from moist air at 25°C and 95 kPa when cooled to 5°C? The initial relative humidity is 60 percent. Consider this as a constant-total-pressure cooling process.

11·66 Air at 10°C and relative humidity of 10 percent is heated under constant pressure of 101 kPa to a final temperature of 30°C. Determine the humidity ratio at 30°C.

11·67 Air at 30°C has a relative humidity of 100 percent. What is the dew-point temperature? What mass of water per kilogram of dry air will be removed if the mixture is cooled to 5°C at a constant pressure of 85 kPa?

11·68 Three hundred kilograms per hour of moist air at 30°C and relative humidity of 40 percent is to be cooled to 10°C at constant atmospheric pressure. Compute the heat removed per hour. The barometric pressure is 101.325 kPa.

11·69 Atmospheric air at 650 mm Hg pressure, 25°C, and relative humidity of 50 percent is heated at constant pressure to a temperature of 35°C. Compute the final relative humidity.

11·70 Air at 10°C, 90 percent relative humidity, and air at 30°C, 90 percent relative humidity, are mixed steadily and adiabatically. The mass flow rate of the colder stream is twice that of the other stream. What is the resulting state?

11·71 A glass tank with a volume of 0.35 m^3 contains an air–water-vapor mixture. When the temperature of the mixture is 35°C, its pressure is 140 kPa. If the mixture cools to 15°C, condensation on the glass begins. Determine the mass of water in the tank.

11·72 An air–water-vapor mixture initially at 140 ka, 60°C, 100 percent relative humidity, undergoes a reversible adiabatic expansion to 100 kPa. Give all the information you can about the entropy change of (*a*) the air, (*b*) the water, (*c*) the mixture, and (*d*) the surroundings.

11·73 How many pounds of moisture are removed from moist air at 80°F and 14 psia when cooled to 40°F? The initial relative humidity is 60 percent. Consider this as a constant-total-pressure cooling process.

11·74 Air at 50°F and relative humidity of 10 percent is heated under constant pressure of 14.7 psia to a final temperature of 90°F. Determine the humidity ratio at 90°F.

11·75 Air at 90°F has a relative humidity of 100 percent. What is the dew-point temperature? What mass of water per pound of dry air will be removed if the mixture is cooled to 40°F at a constant pressure of 12.5 psia?

11·76 Six hundred pounds per hour of moist air at 90°F and relative humidity of 40 percent is to be cooled to 50°F at constant atmospheric pressure. Compute the heat removed per hour. The barometric pressure is 14.696 psia.

11·77 Atmospheric air at 25 in. mercury pressure, 80°F, and relative humidity of 50 percent is heated at constant pressure to a temperature of 100°F. Compute the final relative humidity.

11·78 The percentage composition by mass for a gas–vapor mixture is nitrogen 70, carbon monoxide 4, carbon dioxide 22, water 4. Compute the dew point of the mixture for a total pressure of 1 atm.

11·79 Air at 50°F, 90 percent relative humidity, and air at 90°F, 90 percent relative humidity, are mixed steadily and adiabatically. The mass flow rate of the colder stream is twice that of the other stream. What is the resulting state?

11·80 A glass tank with a volume of 12 cu ft contains an air–water-vapor mixture. When the temperature of the mixture is 100°F, its pressure is 20 psia. If the mixture cools to 60°F, condensation on the glass begins. Determine the mass of water in the tank.

11·81 An air–water-vapor mixture initially at 20 psia, 140°F, 100 percent relative humidity undergoes a reversible adiabatic expansion to 15 psia. Givc all the information you can about the entropy change of (*a*) the air, (*b*) the water, (*c*) the mixture, and (*d*) the surroundings.

11·82 Which has the greater density, dry air or air with a high humidity ratio? Does this explain why fog collects in valleys and low places?

11·83 Under what conditions does a water pipe or a glass of water ''sweat''? Under what conditions may water from the atmosphere contaminate the gasoline in an automobile tank?

11·84 Why do building and automobile windows ''fog'' or ''frost''? If a double layer of glass is used, should the air space between them be vented to the inside or the outside?

11·85 What is dew? Under what conditions does it form? Under what conditions does frost form?

11·86 Under what conditions is a person's breath visible?

11·87 A slice of toast is taken from the toaster and placed on a dry plate. A minute later the toast is removed from the plate, and the plate is seen to be wet where the toast was lying. Explain this.

11·88 An engineer speaks of the ''moisture in the atmosphere'' and the ''moisture in a steam-turbine exhaust.'' Compare the two meanings of the term *moisture*.

11·89 Air at 35°C, 40 percent relative humidity, is cooled to 25°C by spraying water at 15°C into it. Mixture pressure remains constant at 101.3 kPa. Assuming that all of the water evaporates and that the mixing occurs in an insulated pipe, calculate the mass of water added per kilogram of air.

11·90 A gas-vapor mixture having a composition by mass of nitrogen 70 percent, carbon monoxide 4 percent, carbon dioxide 22 percent, water 4 percent, is heated under a constant total pressure of 101.3 kPa from an initial temperature of 10°C to a final temperature of 35°C. Compute the heat required per kilogram of mixture.

11·91 Atmospheric air at 101.3 kPa, 20°C, 60 percent relative humidity, is to be used to cool a transformer at a rate of 210 000 kJ/h. To increase the cooling effect per kilogram of air and thereby decrease the amount of air needed, the air is first adiabatically saturated with water and then passed over the transformer. If the temperature of the air leaving the transformer should not exceed 30°C, determine the required flow rate in kilograms of dry air per hour.

11·92 Air at 25°C and relative humidity of 100 percent is heated at constant pressure to a final temperature of 35°C. Determine (*a*) the dew-point temperature at 25°C, (*b*) the initial partial pressure of the water vapor, (*c*) the initial humidity ratio, and (*d*) the final humidity ratio.

11·93 Outside air having a temperature and relative humidity of 30°C and 45 percent, respectively, is to be conditioned so that the final temperature and relative humidity are 20°C and 30 percent. If the flow process occurs under constant-pressure conditions, compute (*a*) the quantity of water removed per kilogram of dry air, (*b*) the heat removed in the initial cooling process per kilogram of dry air, and (*c*) the heat added per kilogram of dry air. Assume that the atmospheric pressure equals 101.3 kPa.

11·94 An air-conditioning unit consists of a cooler (with provision for removing condensate) followed by a heater. Air at 101.3 kPa enters at 30°C, 80 percent relative humidity, and leaves at 20°C, 49 percent relative humidity. The flow rate is 500 kg dry air per hour. Calculate (*a*) the amount of water removed per hour and (*b*) the heat removed (in kJ/kg dry air) in the cooler.

11·95 Air at 15°C, 100 percent relative humidity, enters a dehumidifier and passes over the cold coils of a refrigerating unit. The moisture condensed is drained away and then the air passes over the motor and warm coils of the refrigerating unit. All heat rejected by the refrigerating unit thus goes into reheating the air which emerges at 30°C, 30 percent relative humidity. The flow rate is 5 kg of dry air per minute. Barometric pressure is 95 kPa. Determine the power input to the refrigerating unit.

11·96 A cooling tower is used to cool 7500 kg of water per minute from an initial temperature of 40°C to a final temperature of 31°C. Air enters the cooling tower at 25°C and 40 percent relative humidity and leaves at a temperature and relative humidity of 38°C and 100 percent. Compute the mass of air required per minute, and the mass of water lost per minute. Assume that 7500 kg enters the tower per minute. Atmospheric pressure equals 101.3 kPa.

11·97 Water enters a cooling tower at 50°C and leaves at 30°C. The dry-bulb and wet-bulb temperatures of the entering air are 27 and 25°C, respectively. The air leaving the tower is completely saturated and has a temperature of 40°C. If 50 000 kg of water enters the tower each hour, compute the mass of air required and the water lost by evaporation. Atmospheric pressure equals 101.325 kPa.

11·98 Determine the dew point of the products of the following reaction if the products are at 100 kPa, 260°C:

$$CH_4 + 2O_2 + 7.52N_2 \rightarrow CO_2 + 2H_2O + 7.52N_2$$

11·99 Dry air at 5°C enters a heating and humidifying unit, and air at 25°C, 80 percent relative humidity, leaves the unit. Spray water for humidification is supplied to the unit at 15°C. Barometric pressure is 98 kPa. Determine the heat transfer per kilogram of dry air.

11·100 Atmospheric air at 1 atm, 35°C, ϕ = 80 percent, enters an air-conditioning system at a rate of 8.20 m^3/s and is cooled to 30°C. (*a*) Determine the mass rate of flow of atmospheric air and the mass rate of flow of dry air within the atmospheric air. (*b*) Determine the rate (kJ/s) of heat removal required. (*c*) Compare this with the heat transfer rate required if dry air at 1 atm, 35°C, enters at a rate of 8.20 m^3/s. (*d*) Compare the heat transfer rate obtained in *b* also against the heat transfer rate required if dry air at 1 atm, 35°C, enters at the same mass flow rate as that of dry air in the 8.20 m^3/s of atmospheric air.

11·101 Methane is heated at constant pressure as it passes through a duct at a rate of 0.56 kg/s. The gas enters at 110 kPa, 35°C, and a control system maintains the outlet temperature at 300°C. Changes in velocity are small. (*a*) Determine the rate of heat transfer in kJ/s. (*b*) Determine the rate of heat transfer for the case of an entering gas that is saturated with water vapor at a mixture pressure of 110 kPa, still with a flow rate of 0.56 kg of methane per second.

11·102 Compute the amount of heat required, in kilojoules per kilogram of nitrogen, to raise the temperature of water-saturated nitrogen (i.e., $\phi = 1.0$) from 25 to 600°C in a frictionless steady-flow process under a constant pressure of 1 atm. Compare the result with that of Example 4·4.

11·103 A mixture of helium and water vapor at 1.30 atm, 30°C (86°F), has a relative humidity of 80 percent. (*a*) Calculate the humidity ratio. (*b*) If the pressure is held constant, to what temperature must the mixture be cooled in order to initiate condensation? (*c*) If the temperature is held constant, to what pressure must the mixture be brought in order to initiate condensation?

11·104 Air at 100°F, 40 percent relative humidity, is cooled to 80°F by spraying water at 60°F into it. Mixture pressure remains constant at 14.7 psia. Assuming that all of the water evaporates and that the mixing occurs in an insulated pipe, calculate the mass of water added per pound of air.

11·105 Atmospheric air at 29.6 in. mercury, 50 percent relative humidity, is compressed isothermally to 3 psig. State whether each of the following quantities increases, decreases, remains constant, or varies indeterminately: (*a*) humidity ratio, (*b*) relative humidity, and (*c*) dew point.

11·106 A gas-vapor mixture having a composition by mass of nitrogen 70 percent, carbon monoxide 4 percent, carbon dioxide 22 percent, water 4 percent, is heated under a constant total pressure of 14.7 psia from an initial temperature of 50°F to a final temperature of 100°F. Compute the heat required per pound of mixture.

11·107 Atmospheric air at 14.7 psia, 70°F, 60 percent relative humidity, is to be used to cool a transformer at a rate of 200 000 B/h. To increase the cooling effect per pound of air and thereby decrease the amount of air needed, the air is first adiabatically saturated with water and then passed over the transformer. If the temperature of the air leaving the transformer should not exceed 90°F, determine the required flow rate in pounds of dry air per hour.

11·108 Air at 80°F and 100 percent relative humidity is heated at constant pressure to a final temperature of 100°F. Determine (*a*) the dew-point temperature at 80°F, (*b*) the initial partial pressure of the water vapor, (*c*) the initial humidity ratio, and (*d*) the final humidity ratio.

11·109 Outside air having a temperature and relative humidity of 90°F and 45 percent, respectively, is to be conditioned so that the final temperature and relative humidity are 70°F and 30 percent. If the flow process occurs under constant-pressure conditions, compute (*a*) the quantity of water removed per pound of dry air, (*b*) the heat removed in the initial cooling process per pound of dry air, and (*c*) the heat added per pound of dry air. Assume that the atmospheric pressure equals 1 atm.

11·110 An air-conditioning unit consists of a cooler (with provision for removing condensate) followed by a heater. Air at 14.696 psia enters at 90°F, 80 percent relative humidity, and leaves at 70°F, 49 percent relative humidity. The flow rate is 1000 lbm of dry air per hour. Calculate (*a*) the amount of water removed per hour, and (*b*) the heat removed (in B/lbm dry air) in the cooler.

11·111 Air at 60°F, 100 percent relative humidity, enters a dehumidifier and passes over the cold coils of a refrigerating unit. The moisture condensed is drained away and then the air passes over the motor and warm coils of the refrigerating unit. All heat rejected by the refrigerating unit thus goes into reheating the air which emerges at 85°F, 30 percent relative humidity. The flow rate is 10 lbm of dry air per minute. Barometric pressure is 14.0 psia. Determine the power input to the refrigerating unit.

11·112 A cooling tower is used to cool 2000 gal of water per minute from an initial temperature of 110°F to a final temperature of 88°F. Air enters the cooling tower at 80°F and 40 percent relative humidity and leaves at a temperature and relative humidity of 105°F and 100 percent. Compute the mass of air required per minute and the mass of water lost per minute. Assume that 2000 gal enter the tower per minute. Atmospheric pressure equals 14.69 psia.

11·113 Water enters a cooling tower at 120°F and leaves at 90°F. The dry-bulb and wet-bulb temperatures of the entering air are 85 and 70°F, respectively. The air leaving the tower is completely saturated and has a temperature of 105°F. If 100 000 lbm of water enters the tower each hour, compute the mass of air required and the water lost by evaporation. Atmospheric pressure equals 14.696 psia.

11·114 Determine the dew point of the products of the following reaction if the products are at 15.0 psia, 500°F:

$$H_2 + O_2 + 3.76N_2 \rightarrow H_2O + \tfrac{1}{2}O_2 + 3.76N_2$$

11·115 Dry air at 40°F enters a heating and humidifying unit, and air at 80°F, 80 percent relative humidity, leaves the unit. Spray water for humidification is supplied to the unit at 60°F. Barometric pressure is 14.2 psia. Determine the heat transfer per pound of dry air.

11·116 Atmospheric air at 1 atm, 90°F, ϕ = 70 percent, enters a heating system at a rate of 7000 cfm and is cooled to 80°F. (*a*) Determine the mass rate of flow of atmospheric air and the mass rate of flow of dry air within the atmospheric air. (*b*) Determine the rate (B/s) of heat removal required. (*c*) Compare this with the heat transfer rate required if dry air at 1 atm, 90°F, enters at a rate of 7000 cfm. (*d*) Compare the heat transfer rate obtained in *b* also against the heat transfer rate required if dry air at 1 atm, 90°F, enters at the same mass flow rate as that of dry air in the 7000 cfm of atmospheric air.

11·117 Air is compressed steadily from atmospheric conditions of 14.0 psia, 60°F, 100 percent relative humidity, to 70.0 psia, 240°F by a compressor that is cooled by only atmospheric air. Neglect kinetic energy changes and determine the minimum work required per pound of dry air. Compare the result with that of Example 10·3.

11·118 Explain how it is possible to dehumidify air by passing it through water sprays.

11·119 Consider the design of a cooling tower. What are the relative advantages and disadvantages of a high air velocity? Is it desirable to have the air leaving the tower carry liquid drops?

11·120 Sketch a *Ts* diagram for the water vapor in atmospheric air, showing a few lines of constant relative humidity (i.e., loci of vapor states for which the relative humidity is the same).

CHAPTER TWELVE

Chemical Reactions: Combustion

The preceding chapters have treated chiefly pure substances and other nonreacting systems. In this chapter we apply thermodynamic principles to chemical reactions. For simplicity, we deal with a particular type of chemical reaction: combustion. Combustion reactions are chosen because of their importance in engineering, but you should keep in mind that the analysis presented here of combustion reactions may be applied to other chemical reactions.

Three aspects of chemical reactions are considered in this book: (1) the mass balance by which we determine the products formed by known reactants or the reactants required to form known products, (2) the energy balance or application of the first law by means of which we determine the energy transfers and conversions accompanying a reaction, and (3) the application of the second law to determine the extent to which a reaction will proceed and to determine the irreversibility of a reaction. Mass and energy balances are treated in this chapter. Second-law considerations, except for a brief mention in this chapter, are taken up in Chapter 13. (A fourth important aspect of chemical reactions—the rate at which reactions occur—is not treated in this book.)

12·1 The Basic Combustion Reactions

The combustible constituents of fuels are usually carbon, hydrogen, and sulfur and their compounds. The basic reactions for complete combustion of these three elements are

$$C + O_2 \rightarrow CO_2$$

$$H_2 + \tfrac{1}{2}O_2 \rightarrow H_2O$$

$$S + O_2 \rightarrow SO_2$$

The substances present before a reaction occurs are called the reactants, and those present after the reaction has occurred are called the products.

In a reaction such as

$$C + \tfrac{1}{2}O_2 \rightarrow CO$$

combustion is said to be incomplete because the products are not completely oxidized.

The complete combustion of hydrocarbon compounds results in the formation of carbon dioxide and water. Thus

$$C_8H_{18} + 12\tfrac{1}{2}O_2 \rightarrow 8CO_2 + 9H_2O$$

This equation may be interpreted as follows:

$$1 \text{ mole } C_8H_{18} + 12\tfrac{1}{2} \text{ moles } O_2 = 8 \text{ moles } CO_2 + 9 \text{ moles } H_2O$$

where the general term *mole* could be gmol, kmol, lbmol, slugmol, etc.,

or $$114 \text{ kg } C_8H_{18} + 400 \text{ kg } O_2 = 352 \text{ kg } CO_2 + 162 \text{ kg } H_2O$$

where any unit of mass could be used. Notice that the total mass of matter is the same on both sides of the equation. Also the total mass of each chemical element is the same on each side of the equation. The total number of moles on each side of the equation may not be the same.

The masses in the last equation are based on approximate molar masses of 12 for carbon and 2 for hydrogen. The more accurate values of 12.011 for the naturally occurring mixture of carbon isotopes and 2.016 for hydrogen should be used where greater precision is required. As an illustration of this point, the use of the approximate molar masses indicates that the complete burning of 1 kg of hydrogen produces 9 kg of water, but more precise molar masses give a value of 8.94 kg of water per kilogram of hydrogen.

12·2 The Composition of Dry Air

The oxygen for most combustion reactions comes from air. The composition of dry air is given by the following mole fractions: $0.7809N_2$, $0.2095O_2$, 0.0093A (argon), and $0.0003CO_2$. The molar mass of this mixture is 28.967 kg/kmol. For nearly all combustion calculations we can treat the argon and carbon dioxide as nitrogen and use the approximate composition of 0.79 mole N_2 per mole of air and 0.21 mole O_2 per mole of air, corresponding to mass fractions of $0.768N_2$ and $0.232O_2$. Convenient forms in which to remember the approximate composition of air are

$$1 \text{ mole } O_2 + 3.76 \text{ moles } N_2 = 4.76 \text{ moles air}$$

$$1 \text{ kg } O_2 + 3.31 \text{ kg } N_2 = 4.31 \text{ kg air}$$

The nitrogen, being inert, does not enter into the combustion reaction and merely appears in the products of combustion as a diluent. It is therefore often omitted in writing the combustion reactions.

Stoichiometric air is the quantity of air required to burn one unit mass or unit volume of fuel completely with no oxygen appearing in the products of combustion. The products of combustion would then consist of carbon dioxide, water, sulfur dioxide, and the nitrogen that accompanied the oxygen in the air as well as any nitrogen from the fuel.

Excess air is air supplied in excess of that necessary to burn the fuel completely and appears in the products of combustion unchanged, that is, as oxygen and nitrogen. The amount of excess air is normally expressed as a percentage of the stoichiometric amount required for complete combustion of the fuel.

12·3 Ideal Combustion

It is often necessary to determine for a fuel of known composition the amount of air (or other oxidizer) required to burn the fuel and to determine the analysis of the resulting products, assuming that combustion is complete. This calculation is an application of elementary chemistry and is illustrated in the following two examples. The first example pertains to a gaseous fuel for which the analysis is given on a molar or (for ideal gases) volumetric basis. The second example pertains to solid fuel that, like a liquid fuel, is analyzed on a mass basis. Two forms of analysis of a solid fuel such as coal are used by engineers. The *proximate analysis* comprises an analysis of the coal into arbitrary constituents that are designated as moisture, volatile matter, fixed carbon, and ash. The *ultimate analysis* shows the composition of coal in terms of chemical elements, except for ash which is reported as such and consists of various oxides.

No matter what type of fuel or which type of analysis—mass or molar—we have to start with, it is always possible to convert from the mass basis to the molar basis or vice versa in order to handle all combustion calculations in the same way. This is unnecessary, however, and in each of the following examples the calculations are made directly from the fuel analysis as given.

Example 12·1. A blast-furnace gas has the following volumetric analysis: $H_2$9 percent, CO 24, CH_4 2, CO_2 6, O_2 3, and N_2 56. For the burning of this gas at 1 atm with 50 percent excess air, determine the following:

(*a*) The volume of air in cubic meters required per cubic meter of fuel (both measured at the same pressure and temperature).
(*b*) The volumetric analysis of the dry products of combustion.
(*c*) The mass of the total products of combustion per kilogram of fuel.
(*d*) The mass of the dry products of combustion per kilogram of fuel.
(*e*) The mass of air supplied per kilogram of fuel.
(*f*) The dew point of the products.

Solution. This problem will be solved by dealing with 100 kmol of fuel gas. Necessary preliminary computations are presented in the table below. In this table the combustion reaction for CH_4 indicates that 1 kmol of CH_4 requires 2 kmol of O_2 and produces 1 kmol of CO_2 and 2 kmol of water vapor. Therefore 2 kmol of CH_4 requires 4 kmol of oxygen and produces 2 kmol of CO_2 and 4 kmol of water vapor. The products of combustion of H_2 and CO are determined similarly from the reactions and tabulated. The CO_2 and N_2 in the fuel are inert. The oxygen in the fuel reduces the oxygen required from the air to 17.5 kmol per 100 kmol of fuel. The 17.5 kmol of O_2 is accompanied by 66.0 kmol of N_2. For 50 percent excess air 8.75 kmol of additional O_2 is supplied, which appears unchanged in the products of combustion. The oxygen is accompanied by 33.0 kmol of inert nitrogen. The total products of combustion without and with 50 percent excess air are obtained by addition.

Constituents	Number of Kilomoles Based on 100 kmol	Combustion Reaction	Number of Kilomoles of Oxygen Supplied by Air	Number of Kilomoles in Products of Combustion			
				CO_2	H_2O	N_2	O_2
H_2	9	$H_2 + \frac{1}{2}O_2 = H_2O$	4.5	..	9	.....	....
CO	24	$CO + \frac{1}{2}O_2 = CO_2$	12.0	24	..	.....	....
CH_4	2	$CH_4 + 2O_2 = CO_2 + 2H_2O$	4.0	2	4	.....	....
						→ 65.8	....
CO_2	6	....	......	6	..	.....	....
O_2	3	....	(−3.0)	..	..	.....	....
N_2	56	....	......	..	..	56	....
Total	100	No excess air	17.5	32	13	121.8	....
		Excess O_2 and N_2	8.75	..	..	33.0	8.75
		Total for 50% excess air	26.25	32	13	154.8	8.75

(***a***) The volume of air required per cubic meter of fuel burned without excess air is

$$V_a = \frac{17.5 \text{ kmol } O_2}{100 \text{ kmol fuel}} \times \frac{4.76 \text{ kmol air}}{\text{kmol } O_2} = \frac{0.833 \text{ kmol air}}{\text{kmol fuel}}$$

Since 1 kmol of air and 1 kmol of fuel at the same pressure and temperature occupy the same volume, the result, 0.833 kmol of air per kmol of fuel, also represents 0.833 m^3 of air per cubic meter of fuel.

For 50 percent excess air, the air supplied equals 1.50(0.833) or 1.250 m^3/m^3 of fuel.

(***b***) The volumetric analysis of the dry products of combustion may be obtained from the tabular data as follows. In the determination of a dry analysis, the water vapor formed is not included.

Products Constituent	Kilomoles of Constituent per 100 kmol of Fuel	Volumetric Analysis
CO_2	32.0	16.4
O_2	8.75	4.5
N_2	154.8	79.1
	195.6	100.0

(***c***) The following table shows the computation of the total mass of products per 100 kmol of fuel

Products Constituent	Kilomoles of Constituent per 100 kmol of Fuel	Mass of Constituent per 100 kmol of Fuel
CO_2	32.0	32(44) = 1408
H_2O	13.0	13(18) = 234
O_2	8.75	8.75(32) = 280
N_2	154.8	154.8(28) = 4334
	208.6	6256

We now determine the mass of the fuel per 100 kmol of fuel (or 100 times the molar mass of the fuel) by using the molar analysis, with x = mole fraction,

$$100M_{\text{fuel}} = \sum 100xM = 9(2) + 24(28) + 2(16) + 6(44) + 3(32) + 56(28)$$
$$= 2650 \text{ kg/100 kmol}$$

Then the mass of products per kilogram of fuel is given by

$$\frac{m_p}{m_f} = \frac{6256}{2650} = 2.36 \text{ kg products/kg fuel}$$

(d) The mass of dry products equals the mass of total products minus the mass of water vapor formed:

$$\frac{m_{dp}}{m_f} = \frac{m_p}{m_f} - \frac{m_v}{m_f} = 2.36 - \frac{234}{2650} = 2.27 \text{ kg/kg fuel}$$

(e) The mass of air supplied per kilogram of fuel equals the mass of products per kilogram of fuel minus the mass of fuel per kilogram of fuel (which is identically 1.0):

$$\frac{m_a}{m_f} = \frac{m_p}{m_f} - \frac{m_f}{m_f} = 2.36 - 1.00 = 1.36 \text{ kg air/kg fuel}$$

(f) The dew point of the products is the saturation temperature corresponding to the partial pressure of the water vapor in the products. (We are assuming that the products behave as an ideal gas.) The water-vapor partial pressure is

$$P_v = x_v P_m = \frac{N_v}{N_m} P_m = \frac{13}{208.8}(101.325) = 6.31 \text{ kPa}$$

The corresponding saturation temperature, and thus the dew point of the products, is approximately 37°C.

Example 12·2. A coal has the following ultimate analysis in percentages: carbon 70, hydrogen 5, sulfur 1, oxygen 12, nitrogen 2, and ash 10. For complete combustion with 20 percent excess air at 14.7 psia, determine

(a) The mass of air required per pound of fuel.
(b) The volumetric analysis of the dry products of combustion.
(c) The dew point of the products.

Solution. The solution for *a*, the mass of air required, and for the mass of the various products is shown in the following table.

Ultimate Analysis	Reactions	Mass of O_2 Required per lbm of Coal	Mass of Products per lbm of Coal
C = 0.70	$C + O_2 = CO_2$ $12 + 32 = 44$ lbm	$\frac{32 \text{ lbm } O_2}{12 \text{ lbm C}}(0.70)\frac{\text{lbm C}}{\text{lbm coal}} = 1.87$ lbm	$\frac{44}{12}(0.70) = 2.57$ lbm CO_2
H = 0.05	$2H_2 + O_2 = 2H_2O$ $4 + 32 = 36$	$\frac{32 \text{ lbm } O_2}{4 \text{ lbm H}}(0.05)\frac{\text{lbm H}}{\text{lbm coal}} = 0.40$	$\frac{36}{4}(0.05) = 0.45$ lbm H_2O
S = 0.01	$S + O_2 = SO_2$ $32 + 32 = 64$ lbm	$\frac{32 \text{ lbm } O_2}{32 \text{ lbm S}}(0.01)\frac{\text{lbm S}}{\text{lbm coal}} = 0.01$	$\frac{64}{32}(0.01) = 0.02$ lbm SO_2
O = 0.12	Less O_2 supplied by coal	−0.12	. . . .
N = 0.02		. . . .	0.02 lbm N_2
Ash = 0.10		. . . .	0.10 lbm ash
	Total O_2 required, no excess	2.16	. . . .
	Additional O_2 for 20% excess	0.43	0.43 lbm O_2
	Total O_2 required, 20% excess	2.59	. . . .
	N_2 supplied with air = 2.59(3.31)	= 8.57	8.57 lbm N_2
	Total air required, 20% excess	11.16	. . . .
	Plus 1 lbm of fuel	1.00	. . . .
Total mass per lbm of coal		In: 12.16 lbm	Out: 12.16 lbm

(*b*) The dry products of combustion consist of all of the gaseous products except water vapor. From the masses of these products determined in the table above we can obtain the volumetric analysis as follows:

Con-stituent	Mass of Constituent per lbm of Coal	Molar Mass	Moles of Constituent per lbm of Coal	Volumetric Analysis
CO_2	2.57	44	0.0584	0.154
SO_2	0.02	64	0.0003	0.001
O_2	0.43	32	0.0134	0.035
N_2	8.59	28	0.307	0.810
Total	11.61		0.379	1.000

(*c*) Assuming that all the water formed is in the vapor phase, the number of moles of water vapor per pound of coal is $0.45/18 = 0.025$, so the total number of moles of gaseous products is $0.379 + 0.025 = 0.40$ lbmol/lbm of coal. The mole fraction of water vapor in the gaseous products is then $0.025/0.404 = 0.0619$. The partial pressure of the water vapor is then

$$P_v = x_v P_m = 0.0619(14.7) = 0.91 \text{ psia}$$

and the dew point is the saturation temperature corresponding to this pressure, approximately 99°F.

12·4 Actual Combustion Mass Balance

In the preceding section we calculated the amount of air that should be supplied to burn a fuel and calculated the products analysis that results from the complete burning of a fuel with a specified amount of air. This is a calculation that can be made from the fuel analysis alone; no measured data on an actual combustion process are involved. Another problem engineers encounter is determining how much air was actually supplied in a combustion process. Often it is difficult to measure the air flow into a combustion chamber—whether it is a jet engine, a boiler furnace, or an internal-combustion engine—so an inferential method is used which is based on the analyses of the fuel, the solid refuse (if any), and the dry gaseous products.

A commonly used apparatus for analyzing the gaseous products of combustion is the Orsat gas analyzer. In this apparatus a sample of gas of known volume is first trapped at atmospheric pressure in a water-jacketed measuring burette. The purpose of the water jacket is to maintain the sample at constant temperature. The sample is then passed sequentially into vessels containing reagents that absorb CO_2, O_2, and CO. After the gas sample is passed through each vessel, it is returned to the burette where its volume is measured at the original pressure and temperature. The reduction in volume during each absorption divided by the original sample volume is the fraction by volume of the constituent that was absorbed. After CO_2, O_2, and CO are measured, the remainder of the sample is usually assumed to be nitrogen. Sulfur dioxide is absorbed as carbon dioxide, introducing a small error that can be corrected for by adding three eighths (12/32) of the mass of sulfur in the fuel to the mass of carbon, thus treating it as additional carbon. In applications where appreciable amounts of hydrogen and hydrocarbon gases are expected in the products, several refinements are made in the usual Orsat apparatus to measure these constituents. The Orsat analysis is on a dry basis; that is, no water vapor is considered. The gas sample in the Orsat apparatus does contain saturated water vapor, but since the sample temperature is constant, the partial pressure of the saturated water vapor remains constant. From this fact it can easily be shown that the analysis obtained is the same that would be obtained if no water vapor were present. Even when other analysis instruments are used, the results are sometimes called an Orsat analysis.

In analyzing products of combustion, care must be taken to insure that the sample drawn into the gas analyzer is a representative sample of the total flow of gaseous products.

A mass balance for a steady-flow combustion process is

$$\text{mass entering} = \text{mass leaving}$$

If the fuel is burned in air and the total products of combustion contain some solid material, the mass balance can be written as

$$m_f + m_a = m_{dg} + m_v + m_{\text{solid}}$$

where the subscripts f, a, dg, and v stand for fuel, air, dry gaseous products, and water vapor, respectively. m_f and m_{solid} can usually be measured accurately. If we assume that all the hydrogen in the fuel combines with oxygen, then m_v is nine (or, more accurately, 8.94) times the amount of hydrogen in the fuel. We still cannot solve for m_a because there is another unknown, m_{dg}, in the equation. Therefore, something more than a total mass balance is needed.

A mass balance can also be written for each element as well as for all the matter passing through the system. The distribution of the various elements is shown in Fig. 12·1. Remembering that we are trying to determine the mass of air supplied per unit mass of fuel and that this can be determined if we find the mass of dry gaseous products per unit mass of fuel, we first look for an element that appears in both the fuel and the dry gaseous products and only in those two places. This would solve the problem because, if the same mass of such an element amounts to x percent (by mass) of the fuel and y percent (by mass) of the dry gaseous products, then the mass ratio of dry gaseous products to fuel is x/y. Sulfur is distributed in the manner desired, but a sulfur balance is unsatisfactory, because (1) the fraction of sulfur in the fuel is usually so low that a sulfur balance would be inaccurate and (2) SO_2 (and therefore S) is usually not measured in exhaust gases. An Orsat gas analyzer measures SO_2 as part of the CO_2 content.

The next choice of an element on which to base a mass balance is carbon, which appears in the fuel, in the dry gaseous products, and in the solid refuse. In each of these three places the mass fraction of carbon can be readily determined. The amount of solid refuse per unit mass of fuel can be measured, or it can be determined by assuming that all the ash in the fuel becomes part of the solid refuse; so a carbon balance, such as

$$\begin{bmatrix} \text{mass of carbon} \\ \text{in dry gases} \end{bmatrix} = \begin{bmatrix} \text{mass of carbon} \\ \text{in fuel} \end{bmatrix} - \begin{bmatrix} \text{mass of carbon} \\ \text{in solid refuse} \end{bmatrix}$$

can be used in the form

$$\begin{bmatrix} \text{mass of carbon} \\ \text{in dry gases} \\ \text{per kg of fuel} \end{bmatrix} = \begin{bmatrix} \text{mass of carbon} \\ \text{in fuel per} \\ \text{kg of fuel} \end{bmatrix} - \begin{bmatrix} \text{mass of carbon} \\ \text{in solid refuse} \\ \text{per kg of fuel} \end{bmatrix}$$

Thus from the fuel ultimate analysis and the solid refuse analysis that together give us the terms on the right-hand side of this equation we can determine the mass of carbon in the dry gaseous products *per unit mass of fuel*. If this quantity is called x and the mass fraction of carbon in the dry gaseous products as determined from their analysis is y, then x/y is the mass of dry gases per unit mass of fuel. Then from a total mass balance (or a nitrogen or oxygen mass balance) we can determine the amount of air supplied.

The following example problem solution illustrates the mass balance analysis of an

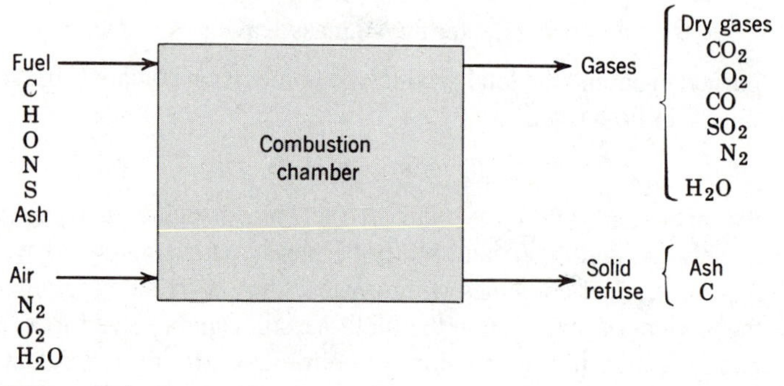

Figure 12·1. Combustion mass balance.

actual combustion process. The first solution is based on a step-by-step application of mass balance reasoning. The second solution is shorter because it is based on a single equation which expresses the mass balances for all elements.

Example 12·3. A coal (the same one referred to in Example 12·2) has an ultimate analysis as follows in percentages: carbon 70, hydrogen 5, sulfur 1, oxygen 12, nitrogen 2, and ash 10. It is burned in a furnace, and the solid refuse is found to contain 33 percent carbon. The Orsat analysis of the dry products of combustion is as follows: CO_2 14.3 percent, O_2 4.0, CO 1.2. Barometric pressure is 14.7 psia. Determine (*a*) the amount of dry air supplied per pound of fuel, (*b*) the percent excess air, and (*c*) the dew point of the products.

Solution. (***a***) The first solution will be obtained by determining the mass of dry gaseous products per pound of fuel and then using a total mass balance to find the amount of air supplied. The mass of dry gas will be found by means of a carbon balance, so the first step is to calculate the mass of carbon in the dry gas (abbreviated *dg*) per pound of fuel.

$$\frac{\text{lbm C in } dg}{\text{lbm fuel}} = \frac{\text{lbm C in fuel}}{\text{lbm fuel}} - \frac{\text{lbm C in refuse}}{\text{lbm fuel}}$$

$$= \frac{\text{lbm C in fuel}}{\text{lbm fuel}} - \left(\frac{\text{lbm C in refuse}}{\text{lbm ash}}\right)\frac{\text{lbm ash}}{\text{lbm fuel}}$$

$$= 0.70 - \left(\frac{0.33}{0.67}\right) 0.10 = 0.65 \frac{\text{lbm C in } dg}{\text{lbm fuel}}$$

Now from the Orsat analysis we obtain the mass of carbon in the dry gas per pound of dry gas.

Constituent	Mole Fraction, lbmol/100 lbmol *dg*	Molar Mass, lbm/lbmol	lbm/100 lbmol *dg*	lbm C/100 lbmol *dg*
CO_2	14.3	44	629	171.6
O_2	4.0	32	128	
CO	1.2	28	33.6	14.4
N_2	80.5	28	2254	
			3045	186.0

Thus we have (186/3045) lbm C in *dg*/lbm *dg* and

$$\frac{m_{dg}}{m_f} = 0.65 \frac{\text{lbm C in } dg}{\text{lbm fuel}} \frac{3045}{186} \frac{\text{lbm } dg}{\text{lbm C in } dg} = 10.64 \frac{\text{lbm } dg}{\text{lbm fuel}}$$

A total mass balance then gives the air-fuel ratio:

$$\frac{m_a}{m_f} = \frac{m_{dg}}{m_f} + \frac{m_v}{m_f} + \frac{m_{\text{ref}}}{m_f} - \frac{m_f}{m_f}$$

$$= 10.64 + 9(0.05) + 0.10/0.67 - 1.0 = 10.24 \text{ lbm air/lbm fuel}$$

(In this equation the mass of vapor is nine times the mass of hydrogen in the fuel, and the mass of refuse is 0.10 lbm of ash per pound of fuel divided by 0.67 lbm of ash per pound of refuse.)

(***b***) In Example 12·2 we found that the stoichiometric amount of oxygen to burn this coal is 2.16 lbm of O_2 per pound of coal. Thus the stoichiometric amount of air is 2.16(4.31) = 9.30 lbm of air per pound of coal and the percent excess air is

$$\text{Excess air} = \frac{m_a - m_{a,\text{stoichiometric}}}{m_{a,\text{stoichiometric}}} = \frac{10.24 - 9.30}{9.30} = 10.1 \text{ percent}$$

(*c*) In order to find the dew point of the gaseous products, we must determine the partial pressure of the water vapor, and this depends on the mole fraction of the vapor in the gases. We have already determined in part *a* the mass of dry gas and the mass of water vapor per pound of coal. From the Orsat analysis we also have the molar mass of the dry gases, 30.45 lbm/lbmol. Thus

$$x_v = \frac{N_v}{N_{dg} + N_v} = \frac{m_v/M_v}{m_{dg}/M_{dg} + m_v/M_v} = \frac{9(0.05)/18}{10.64/30.45 + 9(0.05)/18} = 0.0668$$

$$P_v = x_v P_m = 0.0668(14.7) = 0.982 \text{ psia}$$

$$T_{dp} = T_{\text{sat.,0.982 psia}} = 101^+ \text{ °F}$$

Alternative Solution. (***a***) The amount of dry air supplied can be determined more directly by writing the chemical equation for the actual combustion process with unknown coefficients where necessary and then solving for these coefficients by means of mass balances for the various elements. Let there be *a* moles of O_2 supplied per pound of coal and *x* moles of dry gas formed per pound of coal. As in the initial step of the first solution we find that per pound of fuel only 0.65 lbm of carbon is burned. The remainder of the carbon and the ash undergo no change during the combustion process, so they can be omitted from the chemical equation. Then we have, per pound of fuel

$$\frac{0.65}{12}\text{C} + \frac{0.05}{2}\text{H}_2 + \frac{0.01}{32}\text{S} + \frac{0.12}{32}\text{O}_2 + \frac{0.02}{28}\text{N}_2 + a\text{O}_2 + 3.76a\text{N}_2$$
$$\rightarrow 0.143x\text{CO}_2 + 0.04x\text{O}_2 + 0.012x\text{CO} + 0.805x\text{N}_2 + 0.025\text{H}_2\text{O}$$

The coefficient of H_2O on the right-hand side of this equation was established from the fact that hydrogen appears in only one term on each side of the equation; that is, a hydrogen balance was used to establish 0.025 as the number of moles of hydrogen (in H_2O) in the products. *x* and *a* can be determined by means of carbon and oxygen balances as follows:

C balance: $$\frac{0.65}{12} = 0.143x + 0.012x$$

$$x = 0.349 \text{ lbmol dry gas/lbm coal}$$

O_2 balance: $$\frac{0.12}{32} + a = x\left(0.143 + 0.04 + \frac{0.012}{2}\right) + \frac{0.025}{2}$$

$$a = 0.0748 \text{ lbmol O}_2\text{/lbm coal}$$

As a check on the oxygen balance, a nitrogen balance can be used:

$$\frac{0.02}{28} + 3.76a = 0.805x$$

$$a = 0.0745 \text{ lbmol O}_2\text{/lbm coal}$$

This is as close a check as can be expected. The mass of air supplied is then

$$\frac{m_a}{m_f} = 0.0748(32)4.31 = 10.3 \text{ lbm air/lbm fuel}$$

using the a value from the O_2 balance, and

$$\frac{m_a}{m_f} = 0.0744(32)4.31 = 10.3 \text{ lbm air/lbm fuel}$$

using the a value from the N_2 balance.

(c) From the values found in part a we can determine the mole fraction of water vapor in the products as

$$x_v = \frac{N_v}{N_{dg} + N_v} = \frac{0.025}{x + 0.025} = \frac{0.025}{0.349 + 0.025} = 0.0668$$

and the dew point can then be found as in the first solution.

Example 12·4. A fuel gas has the following volumetric analysis: H_2 40 percent, CH_4 30, C_2H_6 20, N_2 10. It is burned with air supplied at 100 kPa, 35°C, 90 percent relative humidity, and the Orsat analysis of the products is as follows: CO_2 8.2 percent, O_2 4.1, CO 0.6. Determine (a) the air–fuel (mass) ratio and (b) the dew point of the products.

Analysis. To determine the air–fuel ratio, the chemical equation for this actual combustion reaction can be written without regard to the atmospheric moisture because the moisture undergoes no chemical change during the combustion process. Then, for the determination of the products dew point, the atmospheric moisture, calculated separately, can be added to the water vapor formed by the combustion of hydrogen. An alternative method is to include the atmospheric moisture in the chemical equation so that the coefficient of H_2O on the products side can be used directly to determine the mole fraction of H_2O in the products. We shall use the former method and write the chemical equation without regard to the atmospheric moisture.

Solution. Letting a be the number of moles of oxygen supplied per mole of fuel and b be the number of moles of dry gaseous products per mole of fuel, we have per mole of fuel

$$0.4H_2 + 0.3CH_4 + 0.2C_2H_6 + 0.1N_2 + aO_2 + 3.76aN_2$$
$$= 0.082bCO_2 + 0.006bCO + 0.041bO_2 + 0.871bN_2 + (0.4 + 0.6 + 0.6)H_2O$$

The coefficient of H_2O was obtained by means of a hydrogen balance. b and a are determined by means of carbon and oxygen balances:

C balance: $$0.3 + 0.2(2) = 0.082b + 0.006b$$
$$b = 7.95 \text{ kmol dry gas/kmol fuel}$$

$$a = 0.082b + \frac{0.006}{2}b + 0.041b + \frac{1.6}{2}$$

O_2 balance: $$= (0.082 + 0.003 + 0.041)7.95 + 0.8$$
$$= 1.80 \text{ kmol } O_2\text{/kmol fuel}$$

As a check, a nitrogen balance gives

$$0.1 + 3.76a = 0.871b$$
$$a = 1.82 \text{ kmol } O_2\text{/kmol fuel}$$

In order to convert from a molar to a mass basis, we determine the apparent molar mass of the fuel

$$M_f = \sum xM = 0.4(2) + 0.3(16) + 0.2(30) + 0.10(28) = 14.4 \text{ kg/kmol}$$

Then the air-fuel (mass) ratio is

$$\frac{m_a}{m_f} = \frac{N_a M_a}{N_f M_f} = \frac{1.80(4.76)28.966}{14.4} = 17.2 \frac{\text{kg air}}{\text{kg fuel}}$$

(*b*) To find the dew point of the products, we must know the mole fraction of water vapor in the products. The amount of water formed by the combustion of hydrogen was determined for balancing the chemical equation in part *a*. The amount of water introduced as atmospheric moisture can be determined if we first calculate the humidity ratio of the air.

$$P_a = P_m - P_v = P_m - \phi P_g = 100 - 0.9(5.628) = 94.9 \text{ kPa}$$

$$v_a = \frac{RT}{P_a} = \frac{0.287(308)}{94.9} = 0.931 \text{ m}^3/\text{kg}$$

$$\omega = \frac{v_a}{v_g}\phi = \frac{0.931(0.90)}{25.216} = 0.0332 \text{ kg vapor/kg dry air}$$

The number of moles of water vapor carried in with the air per mole of fuel is

$$\frac{N_{v,\text{with air}}}{N_f} = 4.76a \left[\frac{\text{kmol air}}{\text{kmol fuel}}\right] \omega \frac{M_a}{M_v} \left[\frac{\text{kmol vapor}}{\text{kmol air}}\right]$$

$$= 4.76(1.80)0.0332 \frac{28.966}{18} = 0.46 \frac{\text{kmol vapor}}{\text{kmol fuel}}$$

The mole fraction of water vapor in the products is

$$x_v = \frac{N_v}{N_{dg} + N_v} = \frac{N_v/N_f}{b + N_v/N_f} = \frac{1.60 + 0.46}{7.95 + 1.60 + 0.46} = 0.206$$

Then

$$P_v = x_v P_m = 0.206(100) = 20.6 \text{ kPa}$$

$$T_{dp} = T_{\text{sat.,20.6 kPa}} = 60.7°\text{C}$$

12·5 Energy Balance for a Chemical Reaction

The first law of thermodynamics as applied to any system has the same form whether a chemical reaction occurs within the system or not. For example, for any (stationary) closed system we have

$$Q = U_2 - U_1 + W \tag{2·7}$$

where the subscripts 1 and 2 refer to the initial and final states of the system. The chemical composition of the system may be different in states 1 and 2. This fact does not alter Eq. 2·7 at all, but it may complicate the calculation of ΔU of the system from other properties. We have discussed already some means of evaluating ΔU as a function of P, v, T, c_v, etc., for pure substances; now we must investigate the determination of ΔU (and ΔH and ΔPV) for systems that vary in chemical composition. (Remember that a pure substance is defined as one that is chemically homogeneous and also chemically *invariant* with respect to time. If a chemical reaction occurs, we are not dealing with a pure substance.)

Consider the application of the first law to a closed system comprised initially of 1 mole of CO and $\frac{1}{2}$ mole of O_2 that undergo the reaction

$$CO + \tfrac{1}{2}O_2 \rightarrow CO_2$$

so that the system is comprised finally of 1 mole of CO_2. Suppose that this process occurs in a rigid thermally insulated tank so that $Q = 0$ and $W = 0$. Then the first law,

$$Q = U_2 - U_1 + W$$

shows that

$$U_2 = U_1$$

We know from experience that the final temperature of this system would be appreciably greater than the initial temperature. Both the reactants (CO and O_2) and the product (CO_2) at low pressures can be treated as ideal gases so that their internal energies are functions of temperature only, but notice that 1 mole of CO_2 at a temperature T_2 has the same internal energy as 1 mole of CO plus $\frac{1}{2}$ mole of O_2 at T_1 when $T_2 > T_1$. To examine this point from another angle, suppose that the reaction occurs in a rigid tank but a sufficient amount of heat is removed to bring the product to the initial temperature of the reactants, $T_2 = T_1$. Then application of the first law shows that

$$Q_{out} = U_1 - U_2$$

From experience we know that in order to bring the CO_2 to the same temperature that the CO and O_2 started at we must remove heat. Thus $Q_{out} > 0$ and $U_1 > U_2$ if $T_2 = T_1$. Thus a mixture of 1 mole of CO plus $\frac{1}{2}$ mole of O_2 has a higher internal energy than 1 mole of CO_2 at the same temperature. Of course we can calculate ΔU of an ideal gas by $\Delta U = \int mc_v\, dT$ (as discussed in Sec. 4·4), but how do we determine the difference between the internal energies of two different substances such as the products and reactants of a chemical reaction? This is the question to which we now turn our attention.

If we know the specific heats, we can plot curves of U versus T for the reactants $CO + \frac{1}{2}O_2$ and for the product CO_2 as shown in Fig. 12·2*a* and *b*, respectively. In each of these plots the temperature at which $U = 0$ is chosen arbitrarily. In part *c* of Fig. 12·2, the U versus T curves for the reactants and for the product are superposed with $U = 0$ at the same temperature for both. By the use of such a plot or the plots of Figs. 12·2*a* and *b* we can determine ΔU of $CO + \frac{1}{2}O_2$ and ΔU of CO_2 for any ΔT, but we cannot determine the difference between $U_{CO+\frac{1}{2}O_2}$ and U_{CO_2} at any temperature. Suppose now that we determine experimentally $U_{CO_2} - U_{CO+\frac{1}{2}O_2}$ at any one temperature by allowing the reaction to occur in a closed rigid vessel and measuring the amount of heat transfer required to bring the product to the initial temperature of the reactants. Let us call the ΔU between reactants and products *at the same temperature* ΔU_R, the subscript R denoting that the internal energy change is associated with a chemical reaction and not with a temperature change. Then we can shift one of the lines of Fig. 12·2*c* so that the vertical distance between the two lines at this one temperature equals the measured value of ΔU_R. Thus we have Fig. 12·2*d* that enables us to calculate ΔU for any process involving the reaction $CO + \frac{1}{2}O_2 \rightarrow CO_2$, regardless of the temperatures of the reactants and the prod-

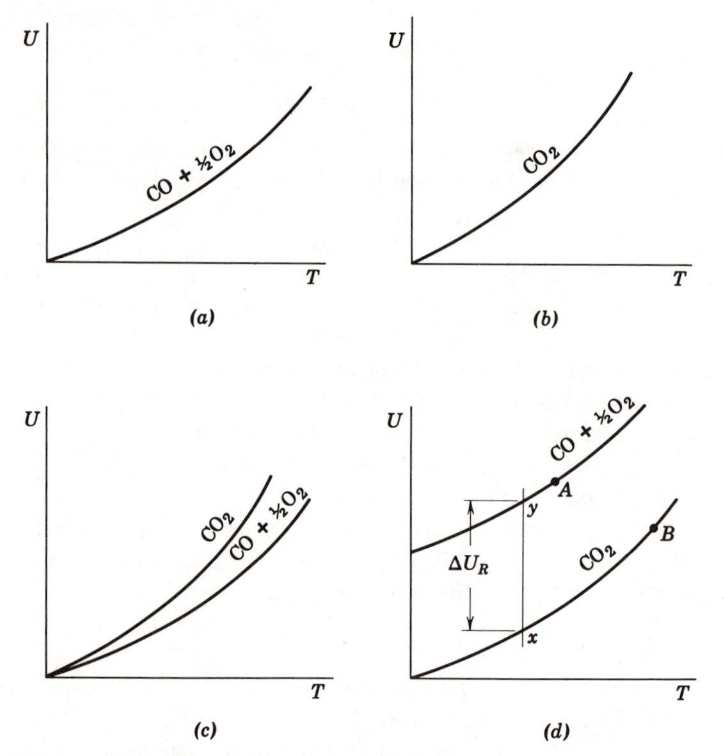

Figure 12·2. U versus T for reactants and products.

uct. For example, referring to Fig. 12·2*d*, ΔU for a process that begins with the reactants at state A and ends with the product at state B is

$$\Delta U = U_B - U_A = (U_B - U_x) + (U_x - U_y) + (U_y - U_A)$$
$$= U_B - U_x + \Delta U_R + U_y - U_A$$

Treating both the reactants and the products as ideal gases,

$$\Delta U = U_B - U_A = \int_{T_x}^{T_B} (mc_v)_{CO_2}\, dT + \Delta U_R + \int_{T_A}^{T_x} (mc_v)_{CO+\frac{1}{2}O_2}\, dT$$

where T_x is the temperature at which ΔU_R is measured. The two integrals on the right-hand side of this equation can be evaluated from data on the individual gases, but ΔU_R must be evaluated from data on the chemical reaction.

In a similar manner for a steady-flow process between states A and B in which $\Delta KE = \Delta PE = 0$, the first law states that

$$Q = H_B - H_A + W \tag{a}$$

and this equation is the same whether a chemical reaction occurs or not. If a chemical reaction does occur, though, we must have in addition to enthalpy data on the reactants and products separately a ΔH_R value—the enthalpy change for the reaction without a

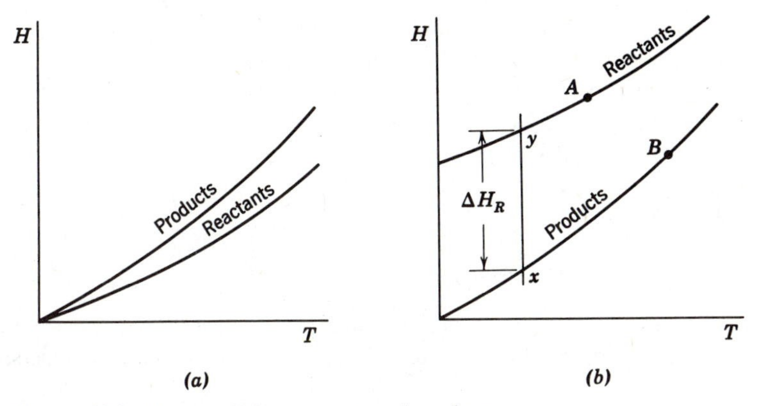

Figure 12·3. H versus T for reactants and products.

change in temperature. If the reactants and products are ideal gases, the HT data for the reactants and products separately can be shown as in Fig. 12·3a, but for the proper relationship between these data a plot like Fig. 12·3b must be used. Thus

$$H_B - H_A = (H_B - H_x) + \Delta H_R + (H_y - H_A)$$

and the first-law formulation becomes

$$Q = (H_B - H_x) + \Delta H_R + (H_y - H_A) + W \tag{b}$$

Notice that Eq. b is the same as Eq. a except that the term $H_B - H_A$ has been expanded in order to use ΔH_R data for relating the enthalpy of the products to that of the reactants.

The purpose of this section is to show that the first law is applied in the same form to chemically reacting systems as to chemically inert systems, and the only complication with a chemical reaction is to evaluate ΔU or ΔH for use in the first law. In the following two sections we take up some useful definitions and discuss in more general terms the evaluation of ΔU and ΔH for processes that involve chemical reactions.

12·6 Enthalpy of Reaction

Consider a process in which some substances called the reactants react chemically to form other substances called the products. Let us consider only cases in which the enthalpies of the reactants and products depend only on temperature. At a given temperature, the enthalpy of the reactants is fixed, and so is the enthalpy of the products, but measurements of Q and W for a steady-flow reaction with $\Delta KE = \Delta PE = 0$ and with the same initial and final temperature show that *the enthalpy of the reactants and the enthalpy of the products at the same temperature* are not equal. The difference between the enthalpy of the products H_p and the enthalpy of the reactants H_r, both at the same temperature, is called the *enthalpy of reaction* ΔH_R. (Enthalpy *change* of reaction would be a more accurate description, but this terminology is not widely used.) Thus

$$\Delta H_R \equiv (H_p - H_r)_{T_p=T_r}$$

ΔH_R might be determined by measuring Q and W (with $\Delta KE = \Delta PE = 0$) for a steady-flow reaction which is carried out so that the products leave at the temperature of the entering reactants. Then, if the variation of enthalpy with temperature both for the products and for the reactants is known, a plot like Fig. 12·4 can be made. Clearly, ΔH_R varies with temperature. Then the enthalpy change between reactants at state 1 and products at state 2 is, referring to Fig. 12·5,

$$H_2 - H_1 = (H_2 - H_{pa}) + (H_{pa} - H_{ra}) + (H_{ra} - H_1) \tag{a}$$

$$= (H_2 - H_{pa}) + \Delta H_{R,a} - (H_1 - H_{ra}) \tag{b}$$

where the subscript a refers to the temperature T_a at which ΔH_R is measured. If the reactants mixture is comprised of substances j and k, the enthalpy of the reactants H_1 is given by

$$H_1 = N_j h_j + N_k h_k \tag{c}$$

where the h's are on a molar basis. A similar expression can be written for the enthalpy of the products in terms of the numbers of moles and specific molar enthalpies of the various constituents of the products mixture. Substituting such expressions for the H values on the right-hand side of Eq. b gives the following relation for the enthalpy change for a process involving a chemical reaction:

$$H_2 - H_1 = \sum_{\text{prod}} N(h_2 - h_a) + \Delta H_{R,a} - \sum_{\text{reac}} N(h_1 - h_a) \tag{12·1}$$

where h_a is in each case the enthalpy of a constituent at the temperature at which ΔH_R is measured. Notice that in Eq. 12·1 the specific enthalpies of the various constituents in the mixtures do not have to be on the same scale because for each constituent only an enthalpy *difference* is required. For example, the enthalpies of some constituents might be taken from Table A·6 that is based on $h = 0$ at $T = 0$, the enthalpies of other constituents might be taken from other tables with $h = 0$ at 0°C or at 25°C, and the enthalpy differences of still other constituents might be calculated by $\int_{T_a}^{T} c_p\, dT$ or $\overline{c_p}(T - T_a)$ as long as they are ideal gases. This is fortunate, because often data on all constituents involved in a reaction cannot be found in the same form or in the same source. Several sources of enthalpy data on ideal gases are listed among the references at the end of this chapter.

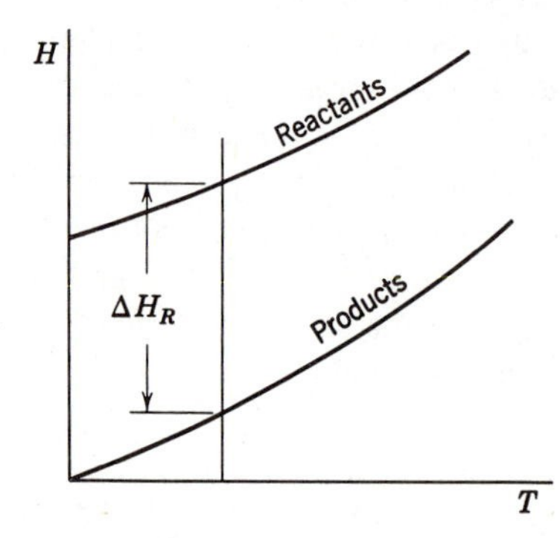

Figure 12·4. Definition of ΔH_R.

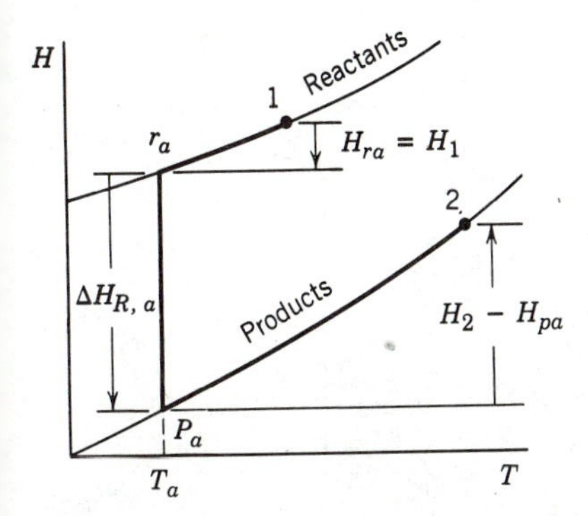

Figure 12·5. Evaluation of $H_2 - H_1$.

For each chemical reaction, ΔH_R might be determined by calorimetric measurements, either by having the reactants and products at the reference temperature or by having them at other temperatures and applying corrections to the measurements. This would be called a first-law method. In practice, better methods are used. One, called the second-law method, is based on a relationship involving ΔH_R and the equilibrium constant that is introduced in Sec. 13·2. A third-law method, based on the third law of thermodynamics that is mentioned in Sec. 12·10, has further advantages. Usually, published values of ΔH_R and related quantities are based on two or more methods to ensure accuracy and consistency.

Enthalpy of reaction is usually expressed per mole or per unit mass of one of the reactants, but, contrary to convention, the symbol ΔH_R instead of Δh_R is widely used. The most common reference state is 1 atm, 25°C (77°F), and for this state we use the symbol Δh_R°. h° designates an enthalpy at this standard reference temperature. The difference between enthalpies of reaction measured at temperatures within a few degrees of each other can usually be neglected. Pressure increases of a few atmospheres also have a negligible effect on enthalpies of reaction.

The enthalpy of reaction for any chemical reaction depends on the phases in which the reactants and products appear. Therefore, in Table 12·1, which lists Δh_R° values for a few reactions, the phase of each constituent is indicated by an *s*, *l*, or *g* (for solid, liquid, or gas). Such notation should always be used where there might be doubt as to the phase.

For a steady-flow process with no change in kinetic or potential energy, the first law gives

$$Q - W = H_2 - H_1$$

Expanding the right-hand side in accordance with Eq. 12·1, we obtain

$$Q - W = \sum_{\text{prod}} N(h_2 - h^\circ) + \Delta H_R^\circ - \sum_{\text{reac}} N(h_1 - h^\circ)$$

If there is no work done and the reactants and products are both at the reference temperature, this first-law expression reduces to

$$Q_{\text{out}} = -Q = -\Delta H_R$$

For this reason, $-\Delta H_R^\circ$ is sometimes called the *heat of reaction*. Since $-\Delta H_R$ is equal

TABLE 12·1 Enthalpies and Internal Energies of Reaction at 25°C (77°F), Low Pressure

Reaction	Δh_R (kJ/kmol)	Δu_R (kJ/kmol)	Δh_R (B/lbmol)	Δu_R (B/lbmol)
$H_2(g) + \frac{1}{2}O_2(g) \rightarrow H_2O(g)$	−241 820	−240 581	−103 964	−103 431
$H_2(g) + \frac{1}{2}O_2(g) \rightarrow H_2O(l)$	−285 830	−282 114	−122 885	−121 285
$C(s) + \frac{1}{2}O_2(g) \rightarrow CO(g)$	−110 530	−111 769	−47 519	−48 052
$C(s) + O_2(g) \rightarrow CO_2(g)$	−393 510	−393 510	−169 179	−169 179
$CO(g) + \frac{1}{2}O_2(g) \rightarrow CO_2(g)$	−282 980	−281 741	−121 660	−121 127
$S(s) + O_2(g) \rightarrow SO_2(g)$	−296 842	−296 842	−127 619	−127 619
$H_2O(g) + CO(g) \rightarrow H_2(g) + CO_2(g)$	−41 160	−41 160	−17 696	−17 696
$H_2(g) + CO_2(g) \rightarrow H_2O(g) + CO(g)$	+41 160	+41 160	+17 696	+17 696
$CH_4(g) + 2O_2(g) \rightarrow CO_2(g) + 2H_2O(g)$	−802 300	−802 300	−344 927	−344 927
$CH_4(g) + 2O_2(g) \rightarrow CO_2(g) + 2H_2O(l)$	−890 320	−885 365	−382 769	−380 636
$C_2H_6(g) + 3\frac{1}{2}O_2(g) \rightarrow 2CO_2(g) + 3H_2O(g)$	−1 428 630	−1 429 869	−614 200	−614 733
$C_2H_6(g) + 3\frac{1}{2}O_2(g) \rightarrow 2CO_2(g) + 3H_2O(l)$	−1 560 660	−1 554 466	−670 963	−668 297
$C_3H_8(g) + 5O_2(g) \rightarrow 3CO_2(g) + 4H_2O(g)$	−2 043 130	−2 045 608	−878 388	−879 454
$C_3H_8(g) + 5O_2(g) \rightarrow 3CO_2(g) + 4H_2O(l)$	−2 219 170	−2 211 737	−954 071	−950 872
$C_3H_8(l) + 5O_2(g) \rightarrow 3CO_2(g) + 4H_2O(g)$	−2 028 310	−2 033 265	−872 016	−872 146
$C_3H_8(l) + 5O_2(g) \rightarrow 3CO_2(g) + 4H_2O(l)$	−2 204 350	−2 199 395	−947 700	−945 567

Note: Calculated from data in Table A·8 and Reference 12·7.

to a quantity of heat only under certain conditions, the term heat of reaction is somewhat misleading, but it is widely used. Heat of reaction is usually defined as $-\Delta H_R$ instead of $+\Delta H_R$ because a convention was established many years ago that the heat of reaction for an exothermic reaction should be a positive number. In some publications, however, $+\Delta H_R$ is called heat of reaction, so you must be careful about signs in reading tables headed "heats of reaction." If the tabulated heats of reaction for exothermic reactions are positive, then the table is based on the usual definition of heat of reaction as $-\Delta H_R$.

If a reaction occurs in a closed system at constant pressure and the only work done is that involved in changing the volume of the system,

$$Q_{\text{out}} = -\Delta U - W = -\Delta U - \int P\,dV = -\Delta U - P\,\Delta V$$
$$= -\Delta U - \Delta PV = -\Delta H$$

and, if the reactants and products are at the reference temperature

$$Q_{\text{out}} = -\Delta H_R$$

Consequently, $-\Delta H_R$ is also sometimes called the *heat of reaction at constant pressure*.

Example 12·5. Determine the amount of heat transfer per kilogram of fuel during the complete combustion of methane, CH_4, in an open steady-flow burner without excess air if the methane enters at 50°C, the air enters at 5°C, and the products leave at 450°C.

Solution. The reaction equation is

$$CH_4 + 2O_2 + 2(3.76)N_2 \rightarrow CO_2 + 2H_2O + 2(3.76)N_2$$

and the energy balance, if we neglect changes in kinetic energy, is

$$Q_{\text{out}} = -Q = H_1 - H_2 = \sum_{\text{reac}} N(h_1 - h_a) - \Delta H_{R,a} - \sum_{\text{prod}} N(h_2 - h_a)$$
$$= N_{CH_4}(h_1 - h_a)_{CH_4} + N_{O_2}(h_1 - h_a)_{O_2} + N_{N_2}(h_1 - h_a)_{N_2} - \Delta H_{R,a}$$
$$- N_{CO_2}(h_2 - h_a)_{CO_2} - N_{H_2O}(h_2 - h_a)_{H_2O} - N_{N_2}(h_2 - h_a)_{N_2}$$

Since the water leaving as a product at 450°C will be a vapor, the $-\Delta H_R$ value to be used from Table 12·1 is 802 300 kJ/kmol CH_4. Obtaining enthalpy values from Tables A·6 and A·7 for substitution into the energy balance

$$Q_{\text{out}} = 1(10\,965 - 10\,025) + 2(8092 - 8679)$$
$$+ 7.52(8083 - 8666) - (-802\,300)$$
$$- 1(28\,266 - 9360) - 2(24\,961 - 9899)$$
$$- 7.52(21\,316 - 8666)$$
$$= 653\,500 \text{ kJ}$$

This calculation has been made for one kilomole of CH_4, so the result can be stated more concisely as

$$Q_{\text{out}} = 653\,500 \text{ kJ/kmol } CH_4 = 40\,730 \text{ kJ/kg } CH_4$$

(In the final step, a molar mass of 16.043 kg/kmol is used for CH_4.)

Example 12·6. Determine the enthalpy of reaction for $CO + \frac{1}{2}O_2 \rightarrow CO_2$ at 600 K.

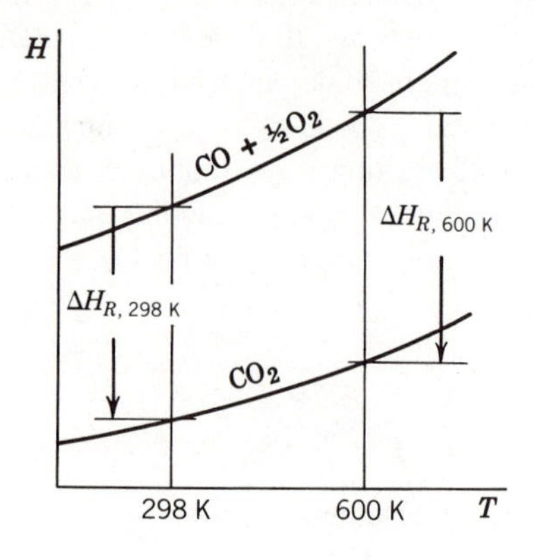

Example 12·6.

Solution. An *HT* diagram is first made. It shows clearly the relationship

$$\begin{aligned}\Delta H_{R,600\text{K}} &= H_{p,600\text{K}} - H_{r,600\text{K}} = (H_{p,600\text{K}} - H_{p,298\text{K}}) + (H_{p,298\text{K}} \\ &\quad - H_{r,298\text{K}}) + (H_{r,298\text{K}} - H_{r,600\text{K}}) \\ &= (H_{p,600\text{K}} - H_p^\circ) + \Delta H_R^\circ - (H_{r,600\text{K}} - H_r^\circ) \\ &= N_{CO_2}(h_{600\text{K}} - h^\circ) + \Delta H_R^\circ - N_{CO}(h_{600\text{K}} - h^\circ)_{CO} - N_{O_2}(h_{600\text{K}} - h^\circ)O_2\end{aligned}$$

Substituting values from Tables A·6 and 12·1, and dividing by the number of moles of CO

$$\begin{aligned}\Delta h_{R,600\text{K}} &= (22\,273 - 9360) + (-282\,981) - (17\,612 - 8667) \\ &\quad - \frac{1}{2}(17\,927 - 8679) \\ &= -283\,600 \text{ kJ/kmol CO}\end{aligned}$$

Enthalpies of reaction for various reactions are frequently given more specific names such as *enthalpy of combustion, enthalpy of formation, enthalpy of hydration*, and so forth. Similarly, the terms *heat of formation*, etc., are widely used. *Enthalpy of combustion* or heating value of a fuel is defined as the enthalpy of reaction for the complete combustion of the fuel. *Enthalpy of formation* is defined as the enthalpy of reaction for the formation of a compound from its elements and is expressed per mole or per unit mass of the compound. (The enthalpy of formation of an element is zero.) The symbol $-\Delta H_f$ is often used for enthalpy of formation and $-\Delta H_c$ for enthalpy of combustion.

The value of the enthalpy of combustion of any fuel containing hydrogen is called the higher enthalpy of combustion or higher heating value (*hhv*) if the H_2O in the products is a liquid. It is called the lower enthalpy of combustion or lower heating value (*lhv*) if the H_2O formed is a gas.

Example 12·7. Verify the lower enthalpy of combustion of ethane (C_2H_6) at 77°F as given in Table A·8 using the tabulated higher enthalpy of combustion as the starting point.

Solution A. The lower and higher enthalpies of combustion are given respectively by

$$-\Delta H_{lhv} = H_r - H_{p,\mathrm{H_2O}(g)} = m_\mathrm{E}h_\mathrm{E} + m_\mathrm{O}h_\mathrm{O} - m_\mathrm{C}h_\mathrm{C} - m_\mathrm{W}h_\mathrm{W}(g)$$

$$-\Delta H_{hhv} = H_r - H_{p,\mathrm{H_2O}(l)} = m_\mathrm{E}h_\mathrm{E} + m_\mathrm{O}h_\mathrm{O} - m_\mathrm{C}h_\mathrm{C} - m_\mathrm{W}h_\mathrm{W}(l)$$

where E, O, C, and W stand for ethane, oxygen, carbon dioxide, and water, respectively. Combining these two equations

$$-\Delta H_{lhv} = -\Delta H_{hhv} - m_\mathrm{W}[h_\mathrm{W}(g) - h_\mathrm{W}(l)] = -\Delta H_{hhv} - m_\mathrm{W}(h_g - h_f)$$
$$= -\Delta H_{hhv} - m_\mathrm{W}h_{fg}$$

Three moles of water are formed for each mole of ethane; so per pound of ethane the mass of water formed is 3(18.016)/1(30.070) = 1.7974 lbm. The value of h_{fg} at 77°F is obtained from the steam tables so that

$$-\Delta h_{lhv} = -\Delta h_{hhv} - \frac{m_\mathrm{W}}{m_\mathrm{E}}h_{fg} = 22\,314 - 1.7974(1050.0) = 20\,427\ \mathrm{B/lbm\ C_2H_6}$$

Solution B. In the steady-flow complete combustion of ethane with no work done and no change in kinetic or potential energy and the reactants and products both at 77°F, the heat removed is 22 314 B/lbm of ethane if all the water formed is condensed to a liquid. If all the water leaves as a vapor, the heat removed is less by the latent heat of vaporization of the water. Thus, by this direct reasoning the difference between the higher and lower enthalpies of combustion is

$$-\Delta h_{lhv} = -\Delta h_{hhv} - \frac{m_\mathrm{W}}{m_\mathrm{E}}h_{fg} = 22\,314 - \frac{3(18.016)}{1(30.070)}(1050.0) = 20\,427\ \mathrm{B/lbm\ ethane}$$

Example 12·8. Starting from the higher enthalpy of combustion of gaseous benzene as given in Table A·8 determine the higher enthalpy of combustion of liquid benzene at the same temperature, 77°F.

Solution. For liquid benzene

$$-\Delta H_{hhv,\mathrm{C_6H_6}(l)} = m_\mathrm{B}h_\mathrm{B}(l) + m_\mathrm{O}h_\mathrm{O} - m_\mathrm{C}h_\mathrm{C} - m_\mathrm{W}h_\mathrm{W}$$

and, for gaseous benzene

$$-\Delta H_{hhv,\mathrm{C_6H_6}(g)} = m_\mathrm{B}h_\mathrm{B}(g) + m_\mathrm{O}h_\mathrm{O} - m_\mathrm{C}h_\mathrm{C} - m_\mathrm{W}h_\mathrm{W}$$

where B, O, C, and W stand for benzene, oxygen, carbon dioxide, and water, respectively. Combining these two equations

$$-\Delta h_{hhv,\mathrm{C_6H_6}(l)} = -\Delta h_{hhv,\mathrm{C_6H_6}(g)} - h_\mathrm{B}(g) + h_\mathrm{B}(l)$$
$$= -\Delta h_{hhv,\mathrm{C_6H_6}(g)} - (h_g - h_f)_\mathrm{B}$$
$$= -\Delta h_{hhv,\mathrm{C_6H_6}(g)} - h_{fg}$$
$$= 18\,171 - \left(\frac{14\,552}{78.11}\right) = 17\,985\ \mathrm{B/lbm\ C_6H_6}$$

This calculation can also be carried out by more direct physical reasoning as was done in solution B of Example 12·7.

Example 12·9. Liquid hydrazine, N_2H_4, which is used as a rocket fuel, is burned with one fourth of the stoichiometric amount of liquid oxygen in a steady-flow process. The hydrazine enters at 25°C (h_{fg} = 44 727 kJ/kg), and the liquid oxygen enters at −173°C (h_f = −3700 kJ/kmol on the same scale as Table A·6; h_{fg} = 6482 kJ/kmol). The products leave at 727°C. The pressure is low enough so that the vapors can be treated as ideal gases. Assuming that the products consist of only N_2H_4, N_2, and H_2O, determine the amount of heat removed per pound of hydrazine.

Solution. The reaction is

$$N_2H_4(l) + \tfrac{1}{4}O_2(l) \rightarrow \tfrac{3}{4}N_2H_4(g) + \tfrac{1}{4}N_2(g) + \tfrac{1}{2}H_2O(g) \qquad \text{(a)}$$

but in order to use the data in Table A·8, we consider the reaction

$$N_2H_4(l) + \tfrac{1}{4}O_2(g) \rightarrow \tfrac{3}{4}N_2H_4(l) + \tfrac{1}{4}N_2(g) + \tfrac{1}{2}H_2O(g) \qquad \text{(b)}$$

for which the enthalpy of reaction is one-fourth the lower enthalpy of combustion ($-lhv$) of hydrazine as listed in Table A·8 because only one fourth of the hydrazine burns and the rest is unchanged. When using ΔH_R for reaction *b* in an energy balance for reaction a, care must be taken to account for the change in phase of hydrazine and oxygen. The energy balance for reaction a is

$$Q = \Delta H_R^\circ + \sum_{\text{prod}} N(h_2 - h^\circ) - \sum_{\text{reac}} N(h_1 - h^\circ)$$

The h° terms must be evaluated for the same phases as those connected with ΔH_R°; so, if we use ΔH_R° for reaction b in the equation above, and divide by the number of moles of N_2H_4 entering

$$q = \Delta h_R^\circ + \tfrac{3}{4}[h_2(g) - h^\circ(l)]_{N_2H_4} + \tfrac{1}{4}[h_2(g) - h^\circ(g)]_{N_2} + \tfrac{1}{2}[h_2(g) - h^\circ(g)]_{H_2O} - [h_1(l) - h^\circ(l)]_{N_2H_4} - \tfrac{1}{4}[h_1(l) - h^\circ(g)]_{O_2}$$

For hydrazine h° is for the liquid and for oxygen h° is for the gas because these are the phases for which we have a value of ΔH_R°. For hydrazine we evaluate $h^\circ(l)$ by subtracting h_{fg} from the $h^\circ(g)$ value from Table A·7. Thus, substituting values from Tables A·6, A·7, and A·8

$$\begin{aligned} q &= \tfrac{1}{4}(-534\,280) + \tfrac{3}{4}[64\,584 - (11\,510 - 44\,727)] + \tfrac{1}{4}[30\,132.4 - 8665.8] \\ &\quad + \tfrac{1}{2}[35904.1 - 9899.0] - [0] - \tfrac{1}{4}[-3700 - 8676.9] \\ &= -38\,760 \text{ kJ/kmol } N_2H_4 \end{aligned}$$

Enthalpy of formation is defined as the enthalpy of reaction for the formation of a compound from its elements. It is a very useful property because it reduces the amount of data needed on enthalpies of reaction. For an illustration of the calculation of ΔH_R from Δh_f data, consider the reaction

$$CH_4 + 2O_2 \rightarrow CO_2 + 2H_2O$$

For this reaction the enthalpy of reaction is given by

$$\Delta H_R = H_p - H_r = h_{CO_2} + 2h_{H_2O} - h_{CH_4} - 2h_{O_2} \qquad \text{(a)}$$

where the enthalpies of the individual constituents are on a molar basis, are all at the same temperature, and are all evaluated on the same scale. Now suppose that we know the Δh_f values of the constituents CH_4, CO_2, and H_2O. ($\Delta h_f = 0$ for O_2 as for all elements.)

That is, for the reactions

$$C + O_2 \rightarrow CO_2$$
$$H_2 + \tfrac{1}{2}O_2 \rightarrow H_2O$$
$$C + 2H_2 \rightarrow CH_4$$

we know the ΔH_R values which are, by definition, Δh_f values:

$$\Delta h_{f,CO_2} = h_{CO_2} - h_C - h_{O_2} \tag{b}$$
$$\Delta h_{f,H_2O} = h_{H_2O} - h_{H_2} - \tfrac{1}{2}h_{O_2} \tag{c}$$
$$\Delta h_{f,CH_4} = h_{CH_4} - h_C - 2h_{H_2} \tag{d}$$

Rearranging these last three equations, we have

$$h_{CO_2} = \Delta h_{f,CO_2} + h_C + h_{O_2} \tag{b}$$
$$h_{H_2O} = \Delta h_{f,H_2O} + h_{H_2} + \tfrac{1}{2}h_{O_2} \tag{c}$$
$$h_{CH_4} = \Delta h_{f,CH_4} + h_C + 2h_{H_2} \tag{d}$$

Now, if all these enthalpies are on a common scale—as those of Eq. a must be—we can substitute from these last three equations into Eq. a to get

$$\Delta H_R = \Delta h_{f,CO_2} + h_C + h_{O_2} + 2\Delta h_{f,H_2O} + 2h_{H_2} + h_{O_2} - \Delta h_{f,CH_4} - h_C - 2h_{H_2} - 2h_{O_2}$$

Collecting terms,

$$\Delta H_R = \Delta h_{f,CO_2} + 2\Delta h_{f,H_2O} - \Delta h_{f,CH_4}$$

Thus we have expressed ΔH_R solely in terms of the Δh_f values of the various constituents involved in the reaction. The procedure illustrated can be generalized so that for any reaction

$$\Delta H_R = \sum_{\text{prod}} N\Delta h_f - \sum_{\text{reac}} N\Delta h_f \tag{12·2}$$

In using this relationship remember that the enthalpy of formation of an element is zero. All the Δh_f values must of course be for the same temperature.

Equation 12·2 makes possible the calculation of ΔH_R for complex reactions from data on the individual constituents. Another advantage of the formulation of ΔH_R in terms of Δh_f values is that it applies to arbitrary mixtures of reactants and to incomplete reactions for which ΔH_R would not be tabulated. For example, if for some reason the burning of CH_4 is incomplete, we might have

$$CH_4 + 2O_2 \rightarrow 0.1CH_4 + 0.7CO_2 + 0.2CO + 1.8H_2O + 0.3O_2$$

For this reaction the enthalpy of reaction is not likely to be found tabulated, but it can be calculated by

$$\Delta H_R = \sum_{\text{prod}} N\Delta h_f - \sum_{\text{reac}} N\Delta h_f = 0.1\Delta h_{f,CH_4} + 0.7\Delta h_{f,CO_2}$$
$$+ 0.2\Delta h_{f,CO} + 1.8\Delta h_{f,H_2O} + 0.3(0) - \Delta h_{f,CH_4} - 2(0)$$

Combining Eqs. 12·1 and 12·2, with Δh_f values taken at the standard reference state, gives a useful equation,

$$H_2 - H_1 = \sum_{\text{prod}} N(h_2 - h^\circ + \Delta h_f^\circ) - \sum_{\text{reac}} N(h_1 - h^\circ + \Delta h_f^\circ) \qquad (12\cdot3)$$

Table A·8 gives the enthalpies of formation of several compounds.

By means of Δh_f data it is possible to build tables of the enthalpies of different substances on a common scale. Tables A·6 and A·7 list h versus T for several substances with $h = 0$ at $T = 0$ for each substance; that is, a common enthalpy scale is not employed. This is apparent because the enthalpy of a compound is much different from the enthalpy of its elements at the same temperature. This difference is known for many compounds and is of course Δh_f. This means that, by arbitrarily assigning $h = 0$ at $T = 0$ (or some other temperature) *for elements* and adding a constant dependent on Δh_f to all h values for each compound, we would have a single table giving for several substances h values that could be directly compared, added, and subtracted. Let us designate such h values by h^{abs} for purposes of this discussion. If h^{abs} versus T tables were available, first-law analyses would be simplified. Also, Δh_R and Δh_f values could be obtained from them and the enthalpy information of Tables A·6, A·7, A·8, and 12·1 could be presented in a single table. However, few data have been published in this form, partly because for certain calculations data in other forms are more useful. One disadvantage of h^{abs} versus T data is that they depend on Δh_f values that are more subject to refinement than the h versus T data of a substance. Thus the data of Tables A·6 and A·7 can be used in conjunction with any Δh_R or Δh_f data, but refinements in the Δh_f data on which an h^{abs} versus T table is based would call for additive corrections to the whole table.

Even though presenting data in the form of h^{abs} versus T or giving Δh_f values at 0 K instead of at 25°C (77°F) would simplify matters in this textbook, the data have been presented in more conventional forms because you must understand these in order to use effectively the standard sources of thermochemical data, several of which are listed in the references at the end of this chapter.

Some of the standard sources of thermochemical property data can be imposing at first. Novices are often surprised by the variety of presentation formats, by the large numbers of significant digits, and by small discrepancies among references. Among the reasons for these small discrepancies are differences in molar masses (atomic weights) of constituents and differences in temperature scales. (Does 298 K mean exactly 298 or 298.15 or 298.16?) Unit conversions are occasional sources of discrepancies. The difference between the thermochemical calorie (= 4.184 kJ) and the international steam table (or IT) calorie (which is equal to 4.1868 kJ) must be observed.

Sometimes an entire table is converted from one set of units to another by using the original equations to recalculate values in the new units. In other cases, unit conversions are made on individual table values, and this may involve interpolation either before or after the conversion is made. The interpolation may be linear or it may involve several neighboring values.

One table may carry a later date than another but include some properties taken from tables published much earlier.

Ideally, documentation covering these various points should accompany every published compilation of property values. However, one may find the documentation to be overwhelmingly copious, difficult for a novice to use, sparse, or missing altogether. Nevertheless, engineering practice requires one to find and use data from many sources.

12·7 Internal Energy of Reaction

For any chemical reaction the difference between the internal energy of the products U_p and the internal energy of the reactants U_r, both at the same temperature, is called the *internal energy of* reaction ΔU_R. Thus

$$\Delta U_R \equiv (U_p - U_r)_{T_p=T_r}$$

Calculations involving ΔU_R are made in the same manner as those involving ΔH_R. For example, we have for the internal energy change between reactants at state 1 and products at state 2

$$U_2 - U_1 = \sum_{\text{prod}} N(u_2 - u_0) + \Delta U_R - \sum_{\text{reac}} N(u_1 - u_0)$$

This relation is similar to Eq. 12·1 for the change in enthalpy.

For a process of a closed system, whether a chemical reaction occurs or not, the first law gives

$$Q - W = U_2 - U_1$$

If the process does involve a chemical reaction and *the reactants and products are at the reference temperature*, this becomes

$$Q - W = \Delta U_R$$

If the process occurs in a constant-volume container so that no work is done and $T_p = T_r$ (as in a bomb calorimeter), then

$$Q_{\text{out}} = -Q = -\Delta U_R$$

For this reason, $-\Delta U_R$ is often called the *heat of reaction at constant volume*. The preceding section pointed out that $-\Delta H_R$ for a similar reason is often called the *heat of reaction at constant pressure*, although more general usage is to refer to $-\Delta H_R$ as simply the *heat of reaction*.

The relationship between ΔU_R and ΔH_R follows from the definition of enthalpy and is

$$\Delta H_R = \Delta U_R + \Delta(PV)_R$$

where $\Delta(PV)_R = (PV)_{\text{prod}} - (PV)_{\text{reac}}$ with the products and reactants at the same temperature. If the reactants and products may be treated as ideal gases, $\Delta(PV) = \Delta(mRT) = \Delta(NR_uT) = R_uT\,\Delta N$. This relationship may also be used if some of the reactants or products are solid or liquid and the rest are ideal gases if ΔN for only the gaseous constituents is used. This may be done because the molar volume of a solid or liquid is usually negligibly small compared to that of a gas at the same pressure, so the volume change of the entire system is very nearly equal to that of the gaseous constituents.

Since $\Delta(PV)$ may be greater than, less than, or equal to zero, the enthalpy of reaction may be greater than, less than, or equal to the internal energy of reaction. For example, for $H_2 + \frac{1}{2}O_2 \rightarrow H_2O(g)$, $\Delta(PV) < 0$ and $\Delta H_R < \Delta U_R$. The values, on a molar basis, are $\Delta h_R^\circ = -241\ 820$ kJ/kmol of H_2 and $\Delta u_R = -240\ 581$ kJ/kmol of H_2. Inequalities between negative numbers occasionally confuse people, because the inequality is different for the numbers and for their magnitudes only. For exothermic reactions, therefore, it is often convenient to make comparisons between the positive numbers $(-\Delta H_R)$ and $(-\Delta U_R)$. Thus

$$(-\Delta H_R) = (-\Delta U_R) - \Delta(PV)$$

For example, for $H_2 + \frac{1}{2}O_2 \rightarrow H_2O(g)$, $\Delta(PV) < 0$ and $(-\Delta H_R) > (-\Delta U_R)$; for $CH_4 + 2O_2 \rightarrow CO_2 + 2H_2O(g)$, $\Delta(PV) = 0$ and $(-\Delta H_R) = (-\Delta U_R)$; and, for $C_2H_6 + 3\frac{1}{2}O_2 \rightarrow 2CO_2 + 3H_2O(g)$, $\Delta(PV) > 0$ and $(-\Delta H_R) < (-\Delta U_R)$. To make sure that you know how to calculate ΔH_R from ΔU_R and vice versa for any reaction, you can test your reasoning by verifying the differences between the numerical values given in Table 12·1. One method of physical reasoning is to compare the amount of heat transferred out of a closed system during a constant-pressure process and during a constant-volume process between the same temperatures, the difference being the work of the constant-pressure process which is $\int P\,dV = P\,\Delta V = \Delta(PV)$.

12·8 Maximum Adiabatic Combustion Temperature

Sometimes we need to know the temperature of the products of a combustion reaction which occurs adiabatically. For example, the combustion in a rocket motor or in a gas-turbine combustion chamber occurs nearly adiabatically. If no work or heat is removed from the system during the process, the final temperature is a maximum for the reaction which occurs. For a steady-flow adiabatic combustion reaction in which work $= 0$, $\Delta ke = 0$, and $\Delta pe = 0$, application of the first law shows that

$$H_{\text{prod}} = H_{\text{reac}}$$

so that the states 1 and 2 of the reactants and products are as shown in Fig. 12·6. Inspection of Figs. 12·6 and 12·5 shows that

$$0 = H_2 - H_1$$

can be expressed in terms of ΔH_R° and the enthalpies of the constituents as

$$0 = \sum_{\text{reac}} N(h - h^\circ) - \sum_{\text{prod}} N(h - h^\circ) - \Delta H_R^\circ$$

Consider a combustion reaction for which ΔH_R and the initial properties of the reactants are known. The energy balance can be solved for $\Sigma_{\text{prod}}\, N(h - h^\circ)$. The number of moles of the products can be obtained from stoichiometric calculations. The enthalpy change of the products between the standard reference temperature and the end state of the adiabatic combustion process can then be determined. From the enthalpy at the end state, the

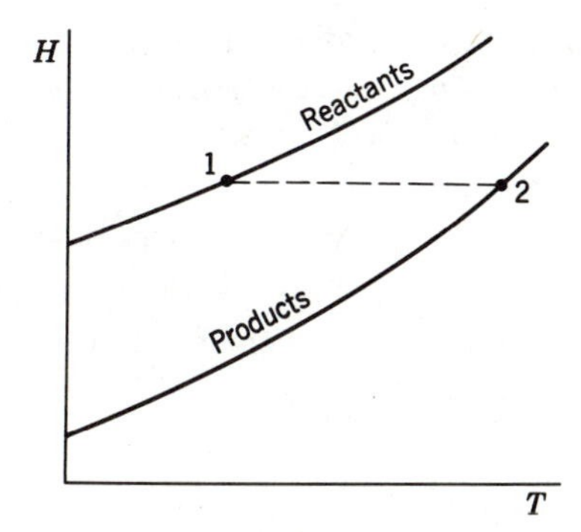

Figure 12·6. Steady-flow adiabatic combustion.

temperature can be determined. If there are two or more products and enthalpy tables are being used, a trial-and-error or graphical solution is necessary. This is illustrated in Example 12·10. If the hT equations of the products are known or equations are fitted to the tabular data in the range of interest, a direct solution is possible. One chooses among various methods on the basis of the computational tools at hand and the number of similar problems to be solved.

Example 12·10. Calculate the maximum adiabatic combustion temperature for the steady-flow burning of methane, CH_4, at 77°F with 100 percent excess air at 240°F.

Solution. Since the *maximum* temperature is desired, complete combustion will be considered. Water in the products will be a vapor. The reaction equation is

$$CH_4 + 2(2)O_2 + 2(2)3.76N_2 \rightarrow CO_2 + 2H_2O(g) + 1(2)O_2 + 2(2)3.76N_2$$

The energy balance is

$$Q = 0 = N_{CH_4}(h_1 - h°)_{CH_4} + N_{O_2}(h_1 - h°)_{O_2} + N_{N_2}(h_1 - h°)_{N_2} - N_{CO_2}(h_2 - h°)_{CO_2} - N_{H_2O}(h_2 - h°)_{H_2O} - N_{O_2}(h_2 - h°)_{O_2} - N_{N_2}(h_2 - h°)_{N_2} - \Delta H_{lhv}$$

Substituting enthalpy values from Table A·6 and the enthalpy of reaction value from Table 12·1 or A·8, and dividing by the number of moles of CH_4

$$0 = 0 + 4(4890.3 - 3735.4) + 15.04(4865.2 - 3729.8) - 1(h_{2,CO_2} - 4029.1) - 2(h_{2,H_2O} - 4260.3) - 2(h_{2,O_2} - 3735.4) - 15.04(h_{2,N_2} - 3729.8) + 344\,940$$

Collecting terms and rearranging

$$h_{2,CO_2} + 2h_{2,H_2O} + 2h_{2,O_2} + 15.04h_{2,N_2} = 442\,753 \text{ B/lbmol } CH_4$$

Now it is necessary to find the temperature at which the sum of these enthalpies is 442 753 B/lbmol of CH_4. One method is to calculate and tabulate the sum for enough temperature values so that the temperature value sought can be obtained by interpolation.

T,R	h_{CO_2}	$2h_{H_2O}$	$2h_{O_2}$	$15.04h_{N_2}$	Total, kJ/kmol CH_4
2400	26 407	43 294	37 169	267 127	373 997
2600	29 165	47 682	40 631	291 894	409 372
2800	31 954	52 180	44 122	316 921	445 177

By interpolation the temperature sought is approximately 2790 R or 2330°F. If hT equations are available for all of the products or if equations can be fit to hT data for the limited temperature range of interest, then these equations and the one above for the sum of constituent enthalpies can be solved simultaneously for the products temperature.

Actually the maximum adiabatic combustion temperature as calculated in Example 12·10 will not be reached, even though the process is adiabatic and the reactants are well mixed, for a reason that is explained qualitatively in the following section and is treated in more detail in Chapter 13.

12·9 Chemical Equilibrium

In actual combustion, the maximum adiabatic temperature as calculated in Example 12·10 above is not attained for several reasons. One of these is that the reaction may not reach completion because the system reaches an equilibrium state in which both reactants and products are present instead of products alone. To illustrate this point, let us consider a reaction which occurs in a closed system instead of an open one. The qualitative conclusion is the same in either case, and the closed system affords a simpler brief explanation.

Consider a rigid thermally insulated tank that contains 12 kg of carbon and 32 kg of oxygen. All the carbon is combined with some oxygen so that the composition of the tank contents may vary from 44 kg of pure CO_2 to 28 kg of CO mixed with 16 kg of O_2. The entropy of the system will in general be different for each different composition. A plot of the entropy of the system versus composition is shown in Fig. 12·7. Since the tank is rigid and thermally insulated, there is no energy transfer across the system boundary as either work or heat, and the system is therefore an *isolated* system; its internal energy must be constant. For this reason the entropy-composition curve is labeled as a constant U curve. The value of the internal energy depends on the temperature and composition of the gas when the tank was filled or when the insulation was put on the tank, so several constant U curves are possible.

For any process of an *isolated* system,

$$\Delta S_{\text{isolated system}} \geq 0 \tag{8·4}$$

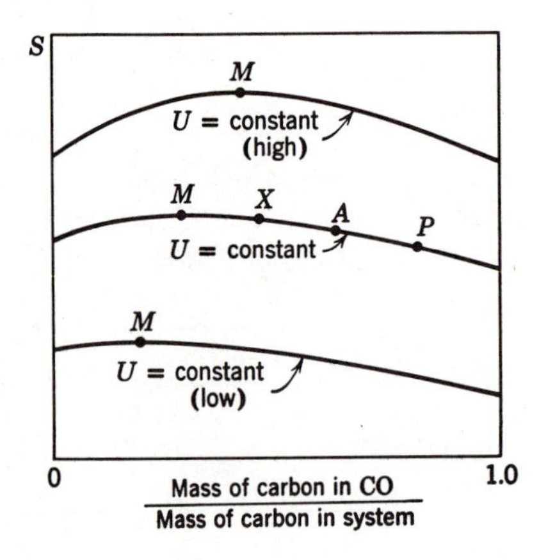

Figure 12·7. Entropy of an isolated system.

Therefore, if the gas mixture in the closed thermally insulated tank is represented by a point A on the figure, there can be no possible process of the isolated system that will carry the system to state P because such a process would violate the increase of entropy principle. A process that carries the system from point A to point X is possible (and irreversible). Further reasoning along this line tells us that, once the system exists in one of the states marked M (in each case the state of maximum entropy for a given internal energy), its state must remain fixed as long as the system is isolated. Furthermore, in any isolated system existing at states other than M states, there is always a tendency for the system to change toward the M state for its particular value of internal energy, although certain factors may prevent the system from undergoing the change toward the M state.

Now it is seen why the maximum adiabatic combustion temperature as calculated in Example 12·10 will not be attained: A state of equilibrium (state of maximum entropy if we are dealing with an isolated system) will be reached before the reaction is completed, and thus not all the reactants will react to form products, as was assumed in the calculation of the maximum adiabatic combustion temperature. The determination of the equilibrium composition and the temperature that is actually attained is treated in Chapter 13.

12·10 Second-Law Analysis of Chemical Reactions

The preceding sections applied the first law to chemically reacting systems, but it was necessary to use new methods for evaluating ΔU and ΔH for use in first-law analyses because all our previous study of property relationships pertained to pure substances. In applying the second law to chemically reacting systems, we use the property entropy, and we realize immediately that up to this point we have usually calculated entropy changes for only pure substances. Now we consider the calculation of ΔS for processes involving chemical reactions.

First, let us review the calculation of ΔS for pure substances. We can of course make use of the definition

$$\Delta S \equiv \int_{\text{rev}} \frac{\delta Q}{T} \tag{8·1}$$

for any fixed quantity of matter whether a chemical reaction is involved or not; but the various relationships we have considered (such as the Maxwell equations) that relate entropy to other properties stem from the equation

$$T\,dS = dU + P\,dV \tag{8·3}$$

which holds only for pure substances. As an illustration that Eq. 8·3 holds only for pure substances, consider a system consisting initially of a mixture of several substances in a closed, rigid, thermally insulated vessel. Let this mixture be in mechanical and thermal equilibrium but not in chemical equilibrium; that is, the pressure and temperature are uniform throughout the system, and there is no relative motion among parts of the system, but the constituents are chemically reactive. The reaction may be occurring very slowly, or a catalyst may be needed to initiate it. (Remember that, if a catalyst is employed, it is entirely unaffected by the process.) In either case, allow the reaction to occur so that the system does reach a state of chemical equilibrium as well as mechanical and thermal equilibrium. This process in which the system goes from a nonequilibrium to an equilibrium state is irreversible, and it occurs in an isolated system; therefore

$$dS > 0 \qquad \text{and consequently} \qquad T\,dS > 0$$

Since the system is isolated, however, U and V are constant so that

$$dU = 0 \qquad \text{and} \qquad P\,dV = 0$$

Therefore

$$T\,dS > dU + P\,dV$$

or, in other words, Eq. 8·3 does not hold for a process involving a chemical reaction.*

A reasonable suggestion for a means of determining ΔS for a chemical reaction is to make measurements of $\int_{\text{rev}} \delta Q/T$ during a reaction to determine ΔS_R in the same way that ΔH_R might be determined from measurements of Q under specified conditions; but finding and using a reversible path between end states involving different substances presents formidable difficulties. To avoid these difficulties we employ instead the *third law of thermodynamics*, an independent principle that cannot be deduced from the first and second laws or from any other principles of nature.

The third law of thermodynamics, like the second, can be stated in several forms that at first appear to be quite unrelated. The third law is mentioned here solely as the basis for determining entropies of various substances on a common scale. For our purpose a suitable statement of the third law is the following: As the temperature of a pure substance approaches zero on the Kelvin scale, the entropy of the substance approaches zero in most cases. There are some important restrictions on the third law as stated here. For example, just which cases the statement does apply to must be pointed out. However, the result that is important to us at this point is that the third law makes possible the determination of absolute entropies based on $S = 0$ at $T = 0$. Table A·8 gives the absolute entropies of several substances, each in a standard state† of 1 atm and 25°C (77°F). There are on a common scale and can be directly added or compared. The symbol for absolute entropy in the standard state is $S°$ (or $s°$ on a per unit mass or per mole basis). The absolute entropy of a substance in any other state x is given by

$$S_x = S° + \int_{\circ}^{x} dS$$

*Sometimes the range of application of Eq. 8·3 is defined by saying that the relationship holds only between states in which a system is in complete (chemical, mechanical, thermal) equilibrium. Another manner of stating the restriction on Eq. 8·3 is to limit it to states of systems in which $S = f(U, V)$. In the process described above, $S \neq f(U, V)$ because the entropy of the system changed during the process while U and V remained constant. The relative merits of these various statements are of little concern in this introductory textbook.

†The term standard state is given various meanings in the thermodynamics literature. Often it refers not to a specific state but only to a specified pressure without regard to temperature so that one encounters references to the standard-state entropy as a function of temperature, and tables of standard-state entropy versus temperature are published. This use of the term standard state to refer to only a standard pressure is confusing, but it is widespread.

where the integral is evaluated by the methods already discussed for pure substances. For ideal gases

$$s_x = s^\circ + \int_\circ^x \frac{dh - v\,dP}{T} = s^\circ + \int_\circ^x \frac{c_p\,dT}{T} - R\int_\circ^x \frac{dP}{P}$$

$$= s^\circ + \phi_x - \phi^\circ - R \ln \frac{P_x}{P^\circ}$$

The standard state used for solids, liquids, and *ideal* gases is often (always in this book) 1 atm and 25°C (77°F). For gases, this is sometimes a hypothetical state in which the substance actually cannot exist as a gas. For example, s° for $H_2O(g)$ is listed in Table A·8. At 25°C, however, the highest pressure under which water can exist as a gas is 0.03128 atm. The s° value in Table A·8 is therefore equal to the absolute entropy of saturated water vapor at 25°C plus the entropy change of an isothermal compression of water vapor *as an ideal gas* from 0.03128 atm, 25°C, to 1 atm.

In the Keenan, Chao, and Kaye *Gas Tables* (except for air) and in Table A·6 the ϕ values at 25°C are equal to the absolute entropy values at 25°C and 1 atm. Thus *in connection with these tables* the equation above for the entropy of an ideal gas in a state x is given by

$$s_x = \phi_x - R \ln \frac{P_x}{P^\circ}$$

where $P^\circ = 1$ atm. Thus for ideal gases at 1 atm the tabulated ϕ values may be used directly as absolute entropy values. (Be careful: This applies only to ideal gases at 1 atm in connection with tables that have $\phi^\circ = s^\circ$.)

Being able to calculate ΔS for processes involving chemical reactions makes it possible for us to apply the increase of entropy principle to determine whether a particular reaction is possible. Furthermore, we can now calculate the irreversibility of a chemical reaction.

Example 12·11. Consider the adiabatic steady-flow burning of carbon monoxide with the stoichiometric amount of air. The air and the carbon monoxide are initially at 25°C, and the total pressure remains constant at 1 atm. *Assuming* that it is possible for 90 percent of the carbon monoxide to be burned, is it possible for all of it to be burned under the conditions specified? In other words, with reference to the figure, is the process from state 2 to state 3 possible?

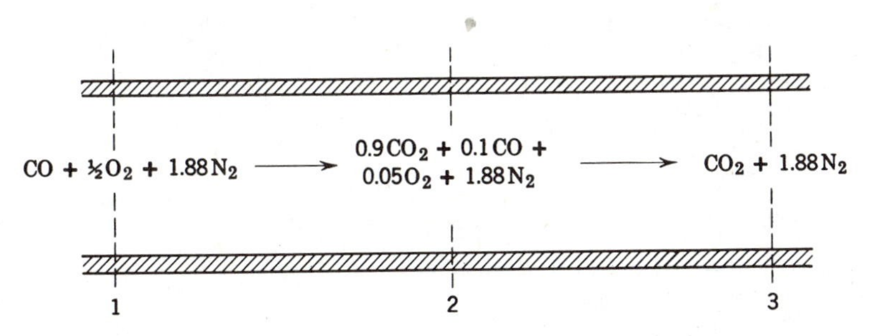

Example 12·11.

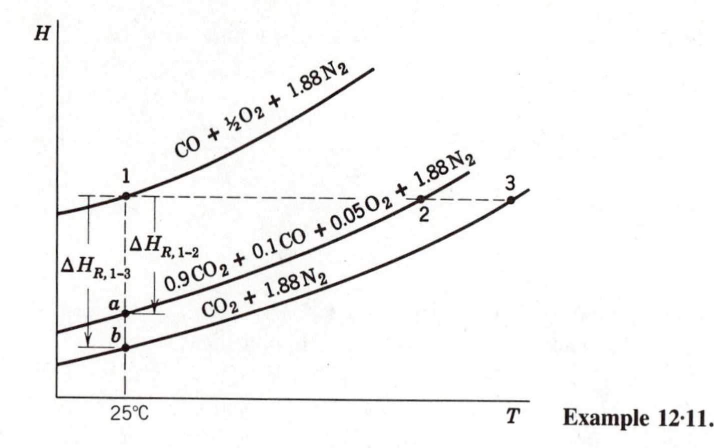

Example 12·11.

Solution. An adiabatic steady-flow process is possible only if the entropy of the flowing material increases or remains constant; so we can answer the question by determining the entropy change of the gas mixture between states 2 and 3. To do this we must first determine T_2 and T_3. This can be done by means of the first law. For the adiabatic process 1-2-3 with no work done* and $\Delta KE = 0$, $H_1 = H_2 = H_3$. An *HT* diagram of the process is made to help in writing energy balances involving ΔH_R. For process 1-2 we have (recalling that Δh_f° for an element is zero)

$$\Delta H^\circ_{R,1\text{-}2} = \sum_{\text{prod}} N\,\Delta h_f - \sum_{\text{reac}} N\,\Delta h_f = 0.9(-393\,522) + 0.1(-110\,541) - 1(-110\,541)$$

$$= -254\,683 \text{ kJ for each kilomole of CO entering}$$

The energy balance is

$$H_2 = H_1 = H_1 - H_a + H_a = -\Delta H_{R,1\text{-}2} + H_a$$

Expressing H_2 and H_a in terms of the molar enthalpies of the individual constituents

$$0.9h_{2,CO_2} + 0.1h_{2,CO} + 0.05h_{2,O_2} + 1.88h_{2,N_2}$$

$$= -\Delta H_{R,1\text{-}2} + 0.9h_{a,CO_2} + 0.1h_{a,CO} + 0.05h_{a,O_2} + 1.88h_{a,N_2}$$

$$= 254\,683 + 0.9(9359.7) + 0.1(8666.7) + 0.05(8676.9) + 1.88(8665.8)$$

$$= 280\,699 \text{ kJ for each kilomole of CO entering}$$

By one of the methods at the end of the solution of Example 12·10, the temperature at which this equation is satisfied is $T_2 = 2453$ K.

For process 1-3 the corresponding calculations are as follows:

$$\Delta H_{R,1\text{-}3} = -282\,981 \text{ kJ/kmol CO (from Table 12·1)}$$

$$H_3 = H_1 = H_1 - H_b + H_b = -\Delta H_{R,1\text{-}3} + H_b$$

$$h_{3,CO_2} + 1.88h_{3,N_2} = -\Delta H_R + h_{b,CO_2} + 1.88h_{b,N_2} = 282\,981 + 9359.7 + 1.88(8665.8)$$

$$= 308\,632 \text{ kJ for each kilomole of CO entering}$$

*Of course there is flow work, but recall the convention established in Sec. 2·7 that the term *work* without modifiers stands for all other forms of work besides flow work.

This equation is satisfied at $T_3 = 2664$ K.

Now to see if the process from state 2 to state 3 is possible we calculate ΔS.

$$S_3 - S_2 = \sum_3 N\left(s° + \phi - \phi° - R_u \ln \frac{P}{P°}\right) - \sum_2 N\left(s° + \phi - \phi° - R_u \ln \frac{P}{P°}\right)$$

where $s°$ is the absolute entropy of each constituent at 1 atm 25°C; P is the partial pressure of each constituent; and $P°$ is the standard-state pressure of 1 atm. Since the mixture pressure is also 1 atm, $P/P° = P/P_m = x$, where x is the mole fraction. R_u is used in place of R because all specific properties are on a molar basis. If we use the Keenan, Chao, and Kaye *Gas Tables* or Table A·6, $\phi° = s°$ for each constituent. Thus the expression for ΔS is simplified to

$$\begin{aligned} S_3 - S_2 &= \sum_3 N(\phi - R_u \ln x) - \sum_2 N(\phi - R_u \ln x) \\ &= [(\phi - R_u \ln x)_{CO_2} + 1.88(\phi - R_u \ln x)_{N_2}]_3 - [0.9(\phi - R_u \ln x)_{CO_2} \\ &\quad + 0.1(\phi - R_u \ln x)_{CO} + 0.05(\phi - R_u \ln x)_{O_2} + 1.88(\phi - R_u \ln x)_{N_2}]_2 \\ &= \left[326.691 - 8.314 \ln \frac{1}{2.88} + 1.88\left(262.355 - 8.314 \ln \frac{1.88}{2.88}\right)\right] \\ &\quad - \left[0.9\left(321.612 - 8.314 \ln \frac{0.9}{2.93}\right) + 0.1\left(266.017 - 8.314 \ln \frac{0.1}{2.93}\right)\right. \\ &\quad \left. + 0.05\left(276.381 - 8.314 \ln \frac{0.05}{2.93}\right) + 1.88\left(259.332 - 8.314 \ln \frac{1.88}{2.93}\right)\right] \\ &= -2.30 \text{ kJ/K for each kilomole of CO at section 1} \end{aligned}$$

Process 2-3 would result in a *decrease* in the entropy of material flowing adiabatically through a steady-flow system; therefore in accordance with the increase of entropy principle, *process 2-3 is impossible*. The reaction $CO + \frac{1}{2}O_2 + 1.88N_2 \rightarrow CO_2 + 1.88N_2$ will not go to completion in a steady-flow adiabatic process at 1 atm with the reactants entering at 25°C.

Example 12·11 illustrates a method for determining if a process is possible and is another indication of the usefulness and scope of thermodynamics as a tool of the engineer and scientist. Some of the examples of irreversible and impossible processes discussed in earlier chapters were quite obvious: You could draw the correct conclusions intuitively, and the support of your conclusions by the second law seemed hardly necessary. But now in Example 12·11 we have a case in which the second law clearly leads us to a conclusion that cannot be reached intuitively or by deductions from other principles.

The extent to which a reaction will proceed could be determined by repeated calculations like those of Example 12·11, but a more direct method is presented in Chapter 13.

Example 12·12 Assume that the reaction $CO + \frac{1}{2}O_2 + 1.88N_2 \rightarrow 0.9CO_2 + 0.1CO + 0.05O_2 + 1.88N_2$ occurs adiabatically at 1 atm in a steady-flow system which the reactants enter at 25°C. (This is process 1-2 of Example 12·11.) The lowest temperature in the surroundings is 7°C. (This means that the combustion air has been preheated if the atmosphere is at 7°C, or it could mean that a large body of water is available as an energy reservoir at 7°C.) Determine the irreversibility of the process and the stream availability of the products, both per mole of CO entering.

Solution. The lowest temperature in the surroundings is $T_0 = 7°C$ and $I = T_0 \Delta S_{\text{isolated system}}$.

For an adiabatic steady-flow process the total change in entropy of the universe (or of an isolated system defined as the steady-flow system plus all parts of the surroundings which interact with it) is $S_2 - S_1$, where S_2 and S_1 are respectively the entropies of the material leaving and the material entering the system.

$$S_2 - S_1 = \sum_2 N\left(s^\circ + \phi - \phi^\circ - R_u \ln \frac{P}{P^\circ}\right) - \sum_1 N\left(s^\circ + \phi - \phi^\circ - R_u \ln \frac{P}{P^\circ}\right)$$

As noted in the solution of Example 12·11, this equation can be simplified to

$$S_2 - S_1 = \sum_2 N(\phi - R_u \ln x) - \sum_1 N(\phi - R_u \ln x)$$

Using $T_2 = 2453$ K as determined in Example 12·11, we have

$$\begin{aligned} S_2 - S_1 = &\left[0.9\left(321.612 - 8.314 \ln \frac{0.9}{2.93}\right) + 0.1\left(266.017 - 8.314 \ln \frac{0.1}{2.93}\right)\right. \\ &\left.+ 0.05\left(276.381 - 8.314 \ln \frac{0.05}{2.93}\right) + 1.88\left(259.332 - 8.314 \ln \frac{1.88}{2.93}\right)\right] \\ &- \left[1\left(197.499 - 8.314 \ln \frac{1}{3.38}\right) + 0.5\left(204.975 - 8.314 \ln \frac{0.5}{3.38}\right)\right. \\ &\left.+ 1.88\left(191.448 - 8.314 \ln \frac{1.88}{3.38}\right)\right] \\ = &\ 150.5 \text{ kJ/K for each kilomole of CO entering} \end{aligned}$$

$$I = T_0 \Delta S_{\text{isolated system}} = 280(150.5) = 42\,140 \text{ kJ for each kilomole of CO entering}$$

If we neglect kinetic energy, the stream availability of the products is

$$Y_2 = (H_2 - T_0 S_2) - (H_0 - T_0 S_2) = H_2 - H_0 - T_0(S_2 - S_0)$$

where H_0 and S_0 are the enthalpy and entropy *of the products* (not of the reactants) at the dead state. (Recall that the stream availability is the maximum work that could be obtained as the fluid goes reversibly to the dead state while exchanging heat with only the atmosphere.) For the substitution of tabular values we rearrange the equation for stream availability

$$Y_2 = \Sigma N(h_2 - h_0) - T_0 \,\Sigma N\left(\phi_2 - \phi_0 - R_u \ln \frac{P_2}{P_0}\right)$$

Since there is no change in the total pressure or the composition of the products between state 2 and the dead state, the partial pressure of each constituent remains constant, $P_2 = P_0$. Thus the equation for stream availability is

$$\begin{aligned} Y_2 &= \Sigma N(h_2 - h_0) - T_0 \Sigma N(\phi_2 - \phi_0) \\ &= 0.9(128\,395.7 - 8698.6) + 0.1(81\,938.7 - 8142.3) + 0.05(85\,150.1 - 8150.8) \\ &\quad + 1.88(81\,233.9 - 8141.6) - 280[0.9(321.612 - 211.373) \\ &\quad + 0.1(266.017 - 195.687) + 0.05(276.381 - 203.150) \\ &\quad + 1.88(259.332 - 189.637)] \\ &= 188\,910 \text{ kJ for each kilomole of CO entering} \end{aligned}$$

This is the maximum amount of work that can be obtained by any process that takes the products to the dead state while exchanging heat with only the energy reservoir at 7°C. The stream availability of the products is less than the enthalpy of combustion for the reaction; so it is impossible, following the adiabatic combustion process, to produce work in an amount equal to the enthalpy of combustion. Notice that this example problem is based on the *assumption* that the process 1-2 is possible.

12·11 Summary

In this chapter mass balances and energy balances for processes involving chemical reactions have been discussed, and the second-law analysis of such processes has been considered briefly.

The oxygen for most combustion reactions comes from air. For nearly all combustion calculations the approximate composition of air can be taken as

$$1 \text{ mole } O_2 + 3.76 \text{ mole } N_2 = 4.76 \text{ moles air}$$

$$1 \text{ kg } O_2 + 3.31 \text{ kg } N_2 = 4.31 \text{ kg air}$$

Stoichiometric air is the quantity of air required to burn a unit quantity of fuel completely with no oxygen appearing in the products of combustion.

Two combustion mass balance problems are encountered in engineering: One is the determination of the amount of air or other oxidizer required to burn a fuel completely and the composition of the resulting products; the other is the determination, from an analysis of the products of an actual combustion process, of the amount of air actually supplied. An instrument frequently used for analyzing combustion products is the Orsat gas analyzer, which measures the *volumetric* fractions of CO_2, O_2, and CO on a dry basis.

The first law of thermodynamics as applied to any system has the same form whether a chemical reaction occurs within the system or not, the only complication with a chemical reaction being to evaluate ΔU or ΔH for use in the first law. To assist in the evaluation of ΔH and ΔU we define the *enthalpy of reaction* ΔH_R as the difference between the enthalpy of the products and the enthalpy of the reactants at the same temperature (as long as the enthalpy of each is a function of temperature only); and we define *the internal energy of reaction* ΔU_R as the difference between the internal energy of the products and of the reactants at the same temperature (again assuming that the internal energy of each is a function of temperature only). The change in enthalpy for a process between the reactants at a state 1 and the products at a state 2 can be written as

$$H_2 - H_1 = \sum_{\text{prod}} N(h_2 - h_a) + \Delta H_{R,a} - \sum_{\text{reac}} N(h_1 - h_a) \tag{12·1}$$

where h_a is in each case the enthalpy of a constituent at the temperature at which Δh_R is measured. A similar expression can be written for $U_2 - U_1$. $-\Delta h_R$ is often called the *heat of reaction at constant pressure*, and $-\Delta u_R$ is often called the *heat of reaction at constant volume*.

Enthalpies of reaction for various reactions are often given more specific names such as enthalpy of combustion, enthalpy of formation (Δh_f), enthalpy of hydration, and so forth. The enthalpy of formation is a valuable property because ΔH_R for any reaction can be expressed in terms of the Δh_f values of the various constituents involved in the reaction in accordance with the relationship

$$\Delta H_R = \sum_{\text{prod}} N\Delta h_f - \sum_{\text{reac}} N\Delta h_f \tag{12·2}$$

Combining Eqs. 12·1 and 12·2, with Δh_f values taken at the standard reference state, gives

$$H_2 - H_1 = \sum_{\text{prod}} N(h_2 - h^\circ + \Delta h_f^\circ) - \sum_{\text{reac}} N(h_1 - h^\circ + \Delta h_f^\circ) \tag{12·3}$$

The maximum adiabatic combustion temperature for any fuel-oxidizer combination can be calculated by means of the first law and data on ΔH_R and on the hT relationships of the constituents involved in the reaction. A reaction that appears possible from the standpoint of stoichiometric and first-law considerations may actually be impossible, however. This can be shown by means of the second law. Application of the second law to tell whether a given process is possible or to calculate the irreversibility of a process involves the determination of ΔS for the process. The entropy change for a process involving a chemical reaction can be determined by use of absolute entropy values that can be found on the basis of the third law of thermodynamics which, for our purposes here, can be stated as follows: As the temperature of a pure substance approaches zero on the Kelvin scale, the entropy of the substance approaches zero in most cases. Absolute entropies of several substances at a specified standard state are given in Table A·8 and are designated as s°. The absolute entropy of a substance in any other state x is then

$$s_x = s^\circ + \int_\circ^x ds$$

where the integral is evaluated by any of several methods that apply to entropy changes of pure substances.

References

12·1 Wark, K., *Thermodynamics*, 4th ed., McGraw-Hill, New York, 1983, Chapter 14.

12·2 Dixon, J. R., *Thermodynamics I*, Prentice-Hall, Englewood Cliffs, N.J., 1975, Sections 15-1 to 15-3.

12·3 Reynolds, William C., and Henry C. Perkins, *Engineering Thermodynamics*, 2nd ed., McGraw-Hill, New York, Chapter 11.

12·4 Van Wylen, Gordon J., and Richard E. Sonntag, *Fundamentals of Classical Thermodynamics*, 3rd ed., Wiley, New York, 1985, Chapter 12.

Brief Reviews of the Third Law

12·5 Hatsopoulos, George N., and Joseph H. Keenan, *Principles of General Thermodynamics*, Wiley, New York, 1965, pp. xxix–xxxii.

12·6 Potter, J. H., "The Third Law of Thermodynamics: A Half-Century Appraisal of the Nernst Heat Theorem," *Transactions ASME*, Vol. 80, 1958, pp. 895–903.

For Thermodynamic and Thermochemical Property Data

12·7 *JANAF Thermochemical Tables*, 2nd ed., National Bureau of Standards, NSRDS-NBS 37, U.S. Department of Commerce, Washington, D.C., 1971, and revisions published in *Journal of Physical and Chemical Data*, 1974, 1975, 1978, 1982.

12·8 *Selected Values of Properties of Hydrocarbons and Related Compounds*, American Petroleum Institute Project 44, Thermodynamics Research Center, Texas A & M University, College Station, TX 77843. (Data presented here from loose-leaf data sheets, extant 1984.)

12·9 *Selected Values of Chemical Thermodynamic Properties*, National Bureau of Standards Technical Note 270-3, 1968.

12·10 Keenan, Joseph H., Jing Chao, and Joseph Kaye, *Gas Tables*, 2nd ed., Wiley, New York, 1980 (English Units), 1983 (SI Units).

12·11 Hilsenrath, Joseph, et al., *Tables of Thermodynamic and Transport Properties*, Pergamon Press, New York, 1960.

12·12 Gallant, R. W., "Physical Properties of Hydrocarbons," *Hydrocarbon Processing*, Vol. 45, No. 10, October 1966.

12·13 *Selected Values of Chemical Thermodynamic Properties*, National Bureau of Standards Circular 500, 1952.

12·14 *Selected Values of Properties of Hydrocarbons*, National Bureau of Standards Circular C461, 1947.

12·15 *Tables of Thermal Properties of Gases*, National Bureau of Standards Circular 564, 1955.

12·16 *Handbook of Chemistry and Physics*, CRC Press, Boca Raton, Fla., annual editions.

12·17 *Marks' Mechanical Engineers' Handbook*, 8th ed., McGraw-Hill, New York, 1978.

Problems

12·1 Calculate (*a*) the mass of water formed by burning completely 1000 m^3 of ethane, measured at 95 kPa, 10°C, with 20 percent excess air, and (*b*) the dew point of the products which are at 95 kPa, 210°C.

12·2 Calculate (*a*) the mass of water formed by burning completely 1000 cu ft of ethane, measured at 14.0 psia, 40°F, with 20 percent excess air, and (*b*) the dew point of the products which are at 14.0 psia, 515°F.

12·3 A gaseous mixture at 1 atm, 85°C (185°F), has a mass analysis of 80 percent CH_4, 14 percent N_2, and 6 percent H_2O. Determine (*a*) the dew point of this mixture and (*b*) the minimum amount of air required to burn this mixture completely.

12·4 A gasoline with an ultimate analysis of 85 percent carbon and 15 percent hydrogen has a specific gravity of 0.72 and a higher heating value of 48 150 kJ/kg (20 700 B/lbm). For the complete combustion of this gasoline with 10 percent excess air, determine (*a*) the air–fuel ratio and (*b*) the number of liters of water formed per liter of gasoline burned.

12·5 A gaseous fuel has the following volumetric analysis: CH_4 60 percent, CO 20, O_2 10, N_2 10. Compute (*a*) the mass analysis of this fuel by chemical compounds, (*b*) the ultimate analysis, (*c*) the specific volume at 1 atm, 40°C (104°F), (*d*) the volume of dry air at 1 atm, 40°C (104°F), required to burn one unit volume of this fuel with 30 percent excess air, (*e*) the volumetric analysis of the dry products of complete combustion with 30 percent excess air, (*f*) the mass of the total products per unit mass of fuel with 30 percent excess air, (*g*) the mass of water vapor in the products per unit mass of fuel, (*h*) the mass of air supplied per unit mass of fuel, (*i*) the dew point of the products if dry air is supplied for combustion, and (*j*) the dew point of the products if the air supplied for combustion is at 1 atm, 40°C (104°F), 70 percent relative humidity.

12·6 A gaseous fuel has the following volumetric analysis: CO 10 percent, H_2 10, CH_4 20, C_2H_6 10, O_2 10, CO_2 10, N_2 30. Compute the same items as in Problem 12·5.

12·7 Coke-oven gas has the following volumetric analysis: CO 6 percent, H_2 42, CH_4 34, C_2H_4 2, C_2H_6 3, O_2 1, CO_2 2, N_2 10. Compute the same items as in Problem 12·5.

12·8 A Mond producer gas has the following volumetric analysis: CO_2 12 percent, CO 15, CH_4 3, H_2 26, N_2 44. Compute the same items as in Problem 12·5.

12·9 A coal has the following mass analysis: C 68 percent, H 5, O 16, N 2, S 2, and ash 7. Calculate (*a*) the amount of air required to burn this coal completely with 30 percent excess air, (*b*) the volumetric analysis of the dry products of complete combustion with 30 percent excess air, (*c*) the mass of the total products per unit mass of fuel with 30 percent excess air, (*d*) the mass of water vapor in the products per unit mass of fuel, (*e*) the mass of air supplied, (*f*) the dew point of the products if dry air is supplied for combustion, and (*g*) the dew point of the products if air at 1 atm, 40°C (104°F), 70 percent relative humidity is supplied for combustion.

12·10 A hydrocarbon fuel is completely burned in air. How does the dew point of the products vary with (*a*) the fraction of hydrogen in the fuel? (*b*) the humidity ratio of the combustion air? (*c*) the barometric pressure?

12·11 The molar mass of air is 28.967. Explain why this value is not obtained by calculating the molar mass from the composition

$$1 \text{ mole } O_2 + 3.764 \text{ moles } N_2 = 4.764 \text{ moles air}$$

12·12 A liquid fuel has an ultimate analysis of 84 percent carbon, 13 percent hydrogen, and 3 percent oxygen. When this fuel was burned, the Orsat analysis of the gaseous products was 11.5 percent CO_2, 4.1 percent O_2, and 0.9 percent CO. The products were analyzed at 0.93 atm. Determine the amount of air supplied per unit mass of fuel.

12·13 Ethyl alcohol, C_2H_5OH, is burned in air, and a volumetric analysis of the dry products is given by CO_2 9.8 percent, O_2 6.5, CO 0.6, HCHO 0.6, N_2 82.5. Air for combustion is supplied at 0.92 atm, 25°C (77 K). Determine per unit mass of ethyl alcohol (*a*) the mass of air supplied and (*b*) the mass of water formed.

12·14 A powdered substance has a mass analysis of 60 percent carbon, 30 percent sulfur, 10 percent ash. It is burned in air, and a modified Orsat analyzer which measures SO_2 as well as the usual constituents of the gaseous products gives the following: CO_2 11.9 percent, O_2 4.5, CO 3.0, SO_2 2.8. Assuming that all of the carbon and sulfur is burned, determine the amount of air supplied per unit mass of substance.

12·15 A gasoline with an ultimate analysis of 85 percent carbon and 15 percent hydrogen has a specific gravity of 0.72. This gasoline was burned in an automobile engine, and the products of combustion gave the following Orsat readings: CO_2 12.1 percent, O_2 2.6, CO 1.1. Barometric pressure was 0.95 atm. The air supplied for combustion had a humidity ratio of 0.030 kg vapor/kg dry air (lbm vapor/lbm of dry air). Determine per unit mass of fuel (*a*) the amount of air supplied and (*b*) the amount of water in the products of combustion.

12·16 A coal has the following mass analysis: C 76 percent, H 5, O 6, S 1, N 1, ash 5, free moisture 6. Volumetric analysis of the dry products of combustion by an Orsat shows: CO_2 12.7 percent, CO 0.4, O_2 6.1. Assume that all ash is discharged to the ashpit and that the fraction of carbon in the refuse is 25 percent. Calculate (*a*) the mass of air supplied per unit mass of fuel, (*b*) the percent excess air, (*c*) the mass of carbon burned to CO per unit mass of fuel, (*d*) the dew point of the products if dry air at 1 atm, 40°C (104°F), is supplied, and (*e*) the dew point of the products if the air supplied for combustion is at 1 atm, 40°C (104°F), 70 percent relative humidity.

12·17 Blast-furnace gas has the following volumetric analysis: CO_2 11 percent, CO 29, H_2 3, CH_4 2, N_2 55. Its higher heating value is 3763 kJ/m^3 (101 B/cu ft) at 1 atm, 15°C (59°F). The volumetric analysis of the products of combustion is: CO_2 19.4 percent, CO 1.0, O_2 4.5, N_2 75.1. Compute the same items as in Problem 12·16.

12·18 Determine from the following test data the amount of water in the gaseous products. Coal ultimate analysis: C 70 percent, H 5, O 10, N 4, S 3, ash 8. Higher heating value: 12 500 B/lbm. Orsat analysis: CO_2 13.6 percent, O_2 4.5, CO 1.1. Amount of coal fired during test: 5000 lbm. Refuse collected: 500 lbm with 20 percent carbon. Combustion air conditions: 29.2 in. Hg, 90°F, humidity ratio of 0.030 lbm/lbm dry air. Products temperature: 510°F.

12·19 Prove that an Orsat analysis is a dry analysis, even though the gas sample contains saturated water vapor. Is there any change in the amount of water vapor in the sample each time it is returned to the measuring burette?

12·20 The heat of combustion of carbon is determined by means of a bomb calorimeter submerged in a water bath. In addition to a small amount of carbon, a considerable excess of air is placed in the bomb. How does the amount of air, assuming that in all cases there is enough for complete combustion, affect the value obtained for the heat of combustion?

12·21 The heat of combustion of a liquid hydrocarbon is to be measured by means of a bomb calorimeter. It is proposed to place a drop of liquid water in the bomb so that the amount of heat removed to return the bomb and its contents to their initial temperature will be the higher heat of combustion at constant volume, the reasoning being that part of the water drop will evaporate and make the air–vapor atmosphere in the bomb saturated so that all of the water formed by combustion will condense. Comment on this proposal.

12·22 Nitrogen is generally considered to be inert in combustion reactions. Still, oxides of nitrogen do exist. Therefore, why do we not burn the nitrogen in the air and thereby obtain some heating value from the air as well as from the fuel being burned?

12·23 Refer to Table 12·1. The sum of ΔH_R for $C + \frac{1}{2}O_2 \rightarrow CO$ and ΔH_R for $CO + \frac{1}{2}O_2 \rightarrow CO_2$ equals ΔH_R for $C + O_2 \rightarrow CO_2$. Is this a coincidence or is it a consequence of the first or second law? Prove your answer.

12·24 Determine the lower heat of combustion of hydrogen at 560°C (680°F).

12·25 Hydrogen is supplied to a burner at 35°C (95°F) and is completely burned with 20 percent excess air which is also supplied at 35°C (95°F). The products leave at 800 K (1440 R). Determine the amount of heat released per unit mass of hydrogen.

12·26 Methane at 25°C (77°F) is burned with 20 percent excess air supplied at 25°C (77°F) in a steady-flow process. The products leave at 410°C (770°F). Calculate (*a*) the amount of heat transfer per unit mass of methane, and (*b*) the dew point of the products.

12·27 Determine the amount of heat released by the complete combustion of methane with 30 percent excess air if the methane is supplied at 35°C (95°F), air is supplied at 110°C (230°F), and the products leave at 700 K (1260 R).

12·28 Liquid benzene, C_6H_6, at 25°C (77°F) is completely burned in an open burner with the stoichiometric amount of oxygen supplied at 500 K (900 R). Products leave the burner at 700 K (1260 R). Determine the amount of heat released during the combustion.

12·29 Gaseous propane at 25°C (77°F) is to be mixed with air at 25°C (77°F) and burned adiabatically in a steady-flow system. Determine the air–fuel ratio required for a products temperature of 800 K (1440 R).

12·30 Verify the enthalpy of formation of butane, C_4H_{10}, as given in Table A·8, from other data in that table.

12·31 Verify the higher heat of combustion given in Table A·8 for ethene gas at 25°C (77°F) from the enthalpy of formation given in that table.

12·32 On the basis of data in Tables A·6 and 12·1 construct a skeleton table of h^{abs} versus T for CO_2, H_2, H_2O, and O_2, showing values at 0, 298, and 500 K (0, 537, and 900 R). Assign $h^{abs} = 0$ at $T = 0$ for elements.

12·33 Solve Problem 12·32 but assign $h^{abs} = 0$ at $T = 0$ for CO_2 and H_2O.

12·34 State whether the heat of combustion at constant pressure is greater than, less than, or equal to the heat of combustion at constant volume for each of the following fuels: C, CO, C_3H_8, C_2H_2, C_2H_4.

12·35 An exothermic reaction occurs in a closed, rigid, thermally insulated vessel. The number of moles of products equals the number of moles of reactants. The specific heats are constant for both reactants and products, and they are higher for the products than for the reactants. Sketch a complete *UT* diagram for the reactants, the products, and the process.

12·36 Determine ΔU_R at 500 K (900 R) for the complete combustion of acetylene.

12·37 Write an equation in terms of properties tabulated in Tables A·6 and A·8 for the higher heating value at constant volume of a hydrocarbon at 1 atm and a temperature higher than 25°C (77°F).

12·38 Carbon monoxide is burned completely with 100 percent excess air in a steady-flow process at 1 atm. The carbon monoxide and the air are supplied at 25°C (77°F). Calculate the maximum adiabatic combustion temperature for this reaction, assuming that the reaction goes to completion.

12·39 Determine the maximum adiabatic combustion temperature for the combustion of hydrogen supplied at 25°C (77°F) with 100 percent excess air also supplied at 25°C (77°F). Combustion occurs in a steady-flow system under a constant pressure of 1 atm. Assume that the reaction goes to completion.

12·40 Determine the maximum adiabatic combustion temperature for the steady-flow burning of propane, C_3H_8, with 100 percent excess air with propane and air supplied at 1 atm, 25°C (77°F). Assume that the reaction goes to completion.

12·41 Refer to Problem 12·26. All the heat released by the combustion process is absorbed by water that flows into the heat exchanger at 25 atm, 110°C (230°F), and leaves at 24 atm, 310°C (590°F). The lowest temperature in the surroundings is 25°C (77°F). Determine, per kilogram of methane, the irreversibility.

12·42 Determine the irreversibility of the process of Problem 12·29.

12·43 Plot a curve of the adiabatic combustion temperature, for the combustion of hydrogen supplied at 25°C (77°F), against percent excess air [also supplied at 25°C (77°F)]. Let the percent excess air range from 0 to 100 percent. Combustion occurs in a steady-flow system under a constant pressure of 1 atm. Assume that the reaction goes to completion.

12·44 Plot a curve of the adiabatic combustion temperature against the percent excess air for the steady-flow burning of propane, C_3H_8, with both propane and air supplied at 1 atm, 25°C (77°F). Assume that the reaction goes to completion.

CHAPTER THIRTEEN

Chemical Equilibrium in Ideal-Gas Reactions

In Chapter 12 it was demonstrated that the extent to which a chemical reaction can proceed under specified conditions is limited. In this chapter we take up the conventional manner of determining the extent to which a reaction can proceed by introducing a valuable parameter known as the *equilibrium constant*. You will see that all considerations of equilibrium are based on the second law. For simplicity, we confine our attention to chemical reactions in which all the constituents may be treated as ideal gases.

13·1 Criteria of Equilibrium

A system is said to be in equilibrium (or in an equilibrium state) if no changes can occur in the state of the system without the aid of an external stimulus. A test to see if a system is in equilibrium is to isolate the system and observe whether any changes in its state occur. The pressure and temperature must be the same throughout a system in equilibrium, and there must be no velocity or concentration gradients within the system, because otherwise spontaneous changes would occur. These conditions alone ensure that a system is in thermal and mechanical equilibrium, but there is still the possibility that the system is not in chemical equilibrium and a chemical reaction may occur. If the state of the system is such that no chemical reaction can occur without an external stimulus, then the system is in complete (mechanical, thermal, and chemical) equilibrium. When in such a state, the system is chemically homogeneous and invariant; so *it is a pure substance* and the relationship

$$T\,dS = dU + P\,dV \qquad \text{(pure substance only)}$$

applies.

For any isolated system

$$dU = 0 \qquad \text{and} \qquad P\,dV = 0$$

but if a spontaneous chemical reaction occurs it is an irreversible process so that

$$dS_{\text{isolated system}} > 0$$

Therefore

$$T\,dS > dU + P\,dV \qquad \text{(irreversible chemical reactions)}$$

for a process involving an irreversible chemical reaction, at least in an isolated system. Assuming that this relationship holds also for reactions in nonisolated systems (and this can be proved rigorously), we have the general relationship for all possible processes

$$T\,dS \geq dU + P\,dV \tag{13·1}$$

where the equality holds if no chemical reaction is possible and the inequality holds for chemical reaction processes, the only exception being the limiting case of a *reversible* chemical reaction for which the equality holds. For a system of constant U and constant V it is apparent that $T\,dS \geq 0$, or, using a brief notation,

$$(dS)_{U,V} \geq 0 \tag{13·2a}$$

For all processes at constant U and V, the entropy increases or remains constant; so, when a system is in complete equilibrium (that is, no process is possible) at constant U and V, its entropy must be a maximum. This is one criterion of equilibrium.

Other criteria of equilibrium for different conditions can be obtained from Eq. 13·1 by rewriting it in terms of H, G, and A as we did in deriving the Maxwell equations (Sec. 9·1). For example, if we would like to establish a criterion of equilibrium under the constraints of *constant pressure* and *constant temperature*, we notice that $T\,dS = d(TS)$ and $P\,dV = d(PV)$, so that Eq. 13·1 becomes

$$d(TS) \geq dU + d(PV) \qquad (P \text{ and } T \text{ constant})$$

$$d(U + PV - TS)_{P,T} \leq 0$$

Recalling the definition of the Gibbs function, we write this as

$$(dG)_{P,T} \leq 0 \tag{13·2b}$$

where the equality holds for the condition of complete equilibrium. This conclusion can also be reached by writing Eq. 13·1 as

$$T\,dS \geq dU + P\,dV = dH - V\,dP \tag{13·1}$$

and combining this with

$$dG = d(H - TS) = dH - T\,dS - S\,dT$$

to give

$$dG \leq V\,dP - S\,dT$$

From this,

$$(dG)_{P,T} \leq 0 \tag{13·2b}$$

Thus the Gibbs function of any system in complete equilibrium must be a minimum with regard to all states at the same pressure and temperature. Otherwise there would be some process for which $(dG)_{P,T} < 0$, and this process could occur, indicating that the system had not been in a state of complete equilibrium. This is a valuable criterion of equilibrium not only because many chemical reactions take place at constant pressure and temperature but also because it can be shown that if a system is in such a state that no spontaneous process can occur at constant pressure and temperature, then no spontaneous process at all can occur. To illustrate this point, suppose that a system initially at the pressure and temperature of its surroundings is in such a state that some spontaneous process can occur that changes its pressure or temperature. This can certainly be followed by a spontaneous process that restores the system to the pressure and temperature of the surroundings. (The system can be allowed to expand or contract until its pressure again equals that of the surroundings, and heat can be transferred until there is temperature equality.) These two processes together constitute a spontaneous process between two states at the same pressure and temperature, so, *if no change to another state at the same pressure and temperature is possible, then no spontaneous process at all is possible.*

The $(dG)_{P,T} \leq 0$ criterion can be used to show that when two or more phases of a pure substance coexist in equilibrium, the (specific) Gibbs function of each phase must be the same. Consider a mixture of liquid and vapor of a pure substance. The masses of the liquid and vapor are m_f and m_g, respectively, and g_f and g_g are the corresponding specific Gibbs functions. The Gibbs function of the system is

$$G = m_f g_f + m_g g_g$$

Now suppose that a mass dm evaporates at constant pressure so that the mass of the vapor changes by dm_g and the mass of the liquid changes by $dm_f = -dm_g$. The change in the Gibbs function of the system is

$$dG = m_f\, dg_f + m_g\, dg_g + g_f\, dm_f + g_g\, dm_g$$

g_f and g_g are each constant because the pressure and temperature of the system are constant; hence

$$dG = 0 + 0 - g_f\, dm_g + g_g\, dm_g = (g_g - g_f)\, dm$$

But for this process at constant pressure and temperature in which the system is in equilibrium, $dG = 0$, hence

$$g_g = g_f$$

For any two phases (not just liquid and vapor) in equilibrium, the specific Gibbs functions are the same. At a triple state all three phases have the same g values.

The Clapeyron equation which was derived in Sec. 9·2 from the Maxwell equations can be derived also from the fact that two phases of a pure substance existing together in equilibrium must have equal g values. Consider for the purpose of illustration a mixture of liquid and vapor. If the mixture is in equilibrium, then $g_f = g_g$. Let the temperature (and consequently the pressure) of the mixture change by an infinitesimal amount. The g values will change by dg_f and dg_g as the system goes to the new equilibrium state infinitesimally close to the original state, but, since the g values of the two phases must remain equal to each other, $dg_f = dg_g$. For each phase we can write $dg = v\, dP - s\, dT$; so, noting that dP and dT are the same for the two phases, we have

$$v_f\, dP - s_f\, dT = v_g\, dP - s_g\, dT$$

Rearranging,

$$\frac{dP}{dT} = \frac{s_g - s_f}{v_g - v_f} = \frac{s_{fg}}{v_{fg}}$$

From $T\,ds = dh - v\,dP$ we have $Ts_{fg} = h_{fg}$ which we substitute into the equation above to give the Clapeyron equation

$$\frac{dP}{dT} = \frac{h_{fg}}{Tv_{fg}} \tag{9·11}$$

13·2 The Equilibrium Constant

We will now apply the equilibrium criterion $(dG)_{P,T} = 0$ to ideal-gas chemical reactions in order to determine the equilibrium state for any group of such constituents. The first step is to derive an expression for the Gibbs function of an ideal-gas mixture. Then we apply the equilibrium criterion, and finally we rearrange the resulting equation to show that for any reaction the relationship among the partial pressures of the constituents is a function of temperature only. We call this temperature function the equilibrium constant.

First step: The Gibbs Function of an Ideal-Gas Mixture

If we assume that the expressions for the enthalpy and the entropy of a mixture of inert ideal gases apply also to a reactive mixture, we have

$$H_m = \sum Nh \tag{a}$$

and

$$S_m = \sum Ns(T, V_m) = \sum Ns(P_i, T) \tag{b}$$

where, as indicated, the entropy of each constituent is that of the constituent existing alone at the temperature and volume of the mixture. P_i is the partial pressure of a constituent. The entropy of each constituent in terms of the *mixture* pressure and temperature can be obtained by first writing

$$s(P_i, T) = s^\circ + \phi - \phi^\circ - R_u \ln \frac{P_i}{P^\circ}$$

where s° is at the standard state (See Sec. 12·10.) and $\phi - \phi^\circ = \int_{T^\circ}^{T} c_p\,dT/T$ as introduced in Sec. 9·6. If we then add and subtract $R_u \ln x$ on the right side, we have

$$s(P_i, T) = s^\circ + \phi - \phi^\circ - R_u \ln \frac{P_i/x}{P^\circ} - R_u \ln x$$

But $P_i/x = P_m$, where P_m is the mixture pressure,* so

$$s(P_i, T) = s^\circ + \phi - \phi^\circ - R_u \ln \frac{P_m}{P^\circ} - R_u \ln x$$

$$= s(P_m, T) - R_u \ln x \tag{c}$$

Then we combine Eqs. a, b, and c with the defining equation for Gibbs function to get

$$G_m = H_m - TS_m = \sum Nh - T\sum Ns(P_m, T) + T\sum NR_u \ln x$$

$$= \sum N[g(P_m, T) + R_u T \ln x] \tag{13·3}$$

*The subscript m of P_m must be retained to avoid confusion between the mixture pressure and the constituent partial pressures, but no such subscript is needed on T to denote the mixture temperature because all constituents are at the same temperature and we are concerned with no other temperature.

Equation 13·3 gives the Gibbs function of the mixture in terms of the Gibbs functions of the constituents at the pressure and temperature *of the mixture* instead of in terms of the constituent partial pressures. The reason for deriving the equation in this form is that we will use it under the constraints of constant *mixture* pressure and constant temperature.

Second step: Application of the Equilibrium Criterion to an Ideal-Gas Mixture
Now we use Eq. 13·3 in the application of the equilibrium criterion $(dG)_{P,T} = 0$ to a mixture of ideal gases, so that

$$(dG_m)_{P_m,T} = \sum N\, d[g(P_m, T) + R_u T \ln x] + \sum [g(P_m, T) + R_u T \ln x] dN = 0 \quad \text{(d)}$$

Under the conditions of constant mixture pressure and temperature, however, the first summation term of Eq. d is zero because (1) for each constituent $g(P_m, T)$ is constant and hence $dg(P_m, T) = 0$ and (2)

$$\begin{aligned} \sum N d \ln x &= \sum N \frac{dx}{x} = \sum N \frac{dx}{N/N_m} = \sum N_m\, dx = N_m \sum dx \\ &= N_m(dx_1 + dx_2 + \cdots) \\ &= 0 \end{aligned}$$

because the sum of x_1, x_2, etc., is unity by definition. Thus Eq. d becomes

$$(dG_m)_{P_m,T} = \sum [g(P_m, T) + R_u T \ln x]\, dN = 0 \quad \text{(e)}$$

Let us now apply the equilibrium criterion in this form to an ideal-gas chemical reaction

$$\nu_1 A_1 + \nu_2 A_2 \rightarrow \nu_3 A_3 + \nu_4 A_4 \quad \text{(f)}$$

where ν_1 moles of ideal gas A_1 react with ν_2 moles of ideal gas A_2 to form ν_3 moles of ideal gas A_3 and ν_4 moles of ideal gas A_4. The ν's are the *mole numbers or stoichiometric coefficients which satisfy the reaction equation and are independent of the amounts (moles) of constituents actually present* at any time which are represented by N_1, N_2, etc. In a mixture of reacting gases the changes in the number of moles of the various constituents present (dN_1, dN_2, etc.) are not independent of each other. They are proportional to the corresponding stoichiometric coefficients so that

$$-\frac{dN_1}{\nu_1} = -\frac{dN_2}{\nu_2} = \frac{dN_3}{\nu_3} = \frac{dN_4}{\nu_4} \quad \text{(g)}$$

The two terms in Eq. g pertaining to the left-hand side of the reaction Eq. f carry minus signs because all of the ν's are positive, and, as the reaction proceeds in the direction shown, N_1 and N_2 decrease while N_3 and N_4 increase. Equation e can then be written as

$$\sum_{3,4} [g(P_m, T) + R_u T \ln x]\nu - \sum_{1,2} [g(P_m, T) + R_u T \ln x]\nu = 0 \quad \text{(h)}$$

Third step: Definition of the Equilibrium Constant

The quantities within brackets in Eq. h can be expanded and rearranged as sums of partial pressure functions and temperature functions. For each constituent,

$$
\begin{aligned}
[g(P_m, T) + R_u T \ln x] &= h - Ts(P_m, T) + R_u T \ln x \\
&= h - T\left(s^\circ + \phi - \phi^\circ - R_u \ln \frac{P_m}{P^\circ}\right) + R_u T \ln x \\
&= h - Ts^\circ - T\phi + T\phi^\circ + R_u T \ln \frac{xP_m}{P^\circ} \\
&= f(T) + R_u T \ln \frac{P_i}{P^\circ}
\end{aligned} \tag{i}
$$

Equation h becomes

$$
\left[f_3(T) + R_u T \ln \frac{P_3}{P^\circ}\right]\nu_3 + \left[f_4(T) + R_u T \ln \frac{P_4}{P^\circ}\right]\nu_4 - \left[f_1(T) + R_u T \ln \frac{P_1}{P^\circ}\right]\nu_1 - \left[f_2(T) + R_u T \ln \frac{P_2}{P^\circ}\right]\nu_2 = 0
$$

Rearranging

$$
\begin{aligned}
\nu_3 \ln \frac{P_3}{P^\circ} + \nu_4 \ln \frac{P_4}{P^\circ} - \nu_1 \ln \frac{P_1}{P^\circ} - \nu_2 \ln \frac{P_2}{P^\circ} \\
= \frac{1}{R_u T}[\nu_1 f_1(T) + \nu_2 f_2(T) - \nu_3 f_3(T) - \nu_4 f_4(T)]
\end{aligned} \tag{j}
$$

$$
\ln \frac{(P_3/P^\circ)^{\nu 3}\,(P_4/P^\circ)^{\nu 4}}{(P_1/P^\circ)^{\nu 1}\,(P_2/P^\circ)^{\nu 2}} = F(\nu_1, \nu_2, \nu_3, \nu_4, T)
$$

For any particular ideal-gas reaction (that is, for fixed values of the ν's), we see then that

$$
\frac{(P_3/P^\circ)^{\nu 3}\,(P_4/P^\circ)^{\nu 4}}{(P_1/P^\circ)^{\nu 1}\,(P_2/P^\circ)^{\nu 4}} = \text{function of } T
$$

For a given group of ideal gases, the relationship among their partial pressures at equilibrium depends only on the temperature. For convenience in handling this temperature function, we define the equilibrium constant K_P *for ideal gases only** by

$$
K_P \equiv \frac{(P_3/P^\circ)^{\nu 3}\,(P_4/P^\circ)^{\nu 4}}{(P_1/P^\circ)^{\nu 1}\,(P_2/P^\circ)^{\nu 2}}
$$

*A more general definition of equilibrium constant is possible and is mentioned in Sec. 13·4. So far in this book we have used broad definitions in place of restricted ones (e.g., the definitions of specific heat in Sec. 3·5 and of partial pressure in Sec. 11·2); so why do we here introduce a definition that is restricted to ideal gases when a broader one is possible? The answer is that the purpose of this chapter is to give you some familiarity with chemical equilibrium calculations, and this can be done in a brief chapter if we consider the special case of an ideal gas. A more general approach to this particular subject, including the development of a more general definition of equilibrium constant, requires much more space and does not appreciably increase the understanding of chemical equilibrium calculations as an application of the second law.

For more than two reactants and two products,

$$K_P \equiv \frac{\text{product of } (P/P^\circ)^\nu \text{ for all products}}{\text{product of } (P/P^\circ)^\nu \text{ for all reactants}}$$

Conclusion

Let us repeat for emphasis the conclusion reached in the preceding paragraph: If a mixture of ideal gases A_1, A_2, A_3, and A_4 which can undergo the reaction

$$\nu_1 A_1 + \nu_2 A_2 \rightarrow \nu_3 A_3 + \nu_4 A_4$$

is in equilibrium, the quantity

$$\frac{(P_3/P^\circ)^{\nu 3}\,(P_4/P^\circ)^{\nu 4}}{(P_1/P^\circ)^{\nu 1}\,(P_2/P^\circ)^{\nu 2}} = K_P$$

(where the P's are partial pressures) that we call the equilibrium constant is a function of temperature only. It does *not* depend on the amount of the various constituents initially present. Therefore, a knowledge of equilibrium constants for an ideal-gas reaction provides us with one relationship among the partial pressures of the constituents.

For any equilibrium state, the value of K_P does not depend on any past state of the mixture, although some earlier state may be referred to in establishing the amount of each element present (see Examples 13·2, 4, and 5). Also, the value of K_P is not affected by the presence of constituents other than those that take part in the reaction used to define K_P (see Example 13·3).

K_P for a given reaction can be measured at various temperatures by analyzing the equilibrium gas mixture. For an ideal-gas reaction in which the number of moles of products is different from the number of moles of reactants, the extent to which the reaction has proceeded when equilibrium is reached can be determined by measuring the total volume of the constituents at equilibrium. Several different methods of determining equilibrium constants experimentally and from other thermochemical data are used.

K_P data for eight ideal-gas reactions are given in Table A·9. Figure 13·1 presents K_P data graphically for some of these reactions to show typical variations of K_P with temperature. Often in equilibrium calculations several relations, including one or more K_PT relations, must be solved simultaneously. It is convenient to have the relationships in the form of equations instead of charts or tables, and Figure 13·1 indicates that a curve-fitting technique can be readily used to obtain K_PT equations over limited temperature ranges. (Section 13·3 gives a suggestion to help in the curve-fitting process.)

Standard data sources for K_P, including some listed in the references at the end of this chaper, often give K_P values for substances rather than for reactions. In such a case, K_p for a substance is defined as the K_P value for the reaction forming that substance from its elements. It can be called the equilibrium constant of formation, and sometimes the symbol K_f is used. Therefore, obtaining the K_P value for a reaction from published K_P values of the constituents involved in the reaction is analogous to determining the Δh_R value for a reaction from Δh_f values of the reaction constituents as mentioned in Sec.

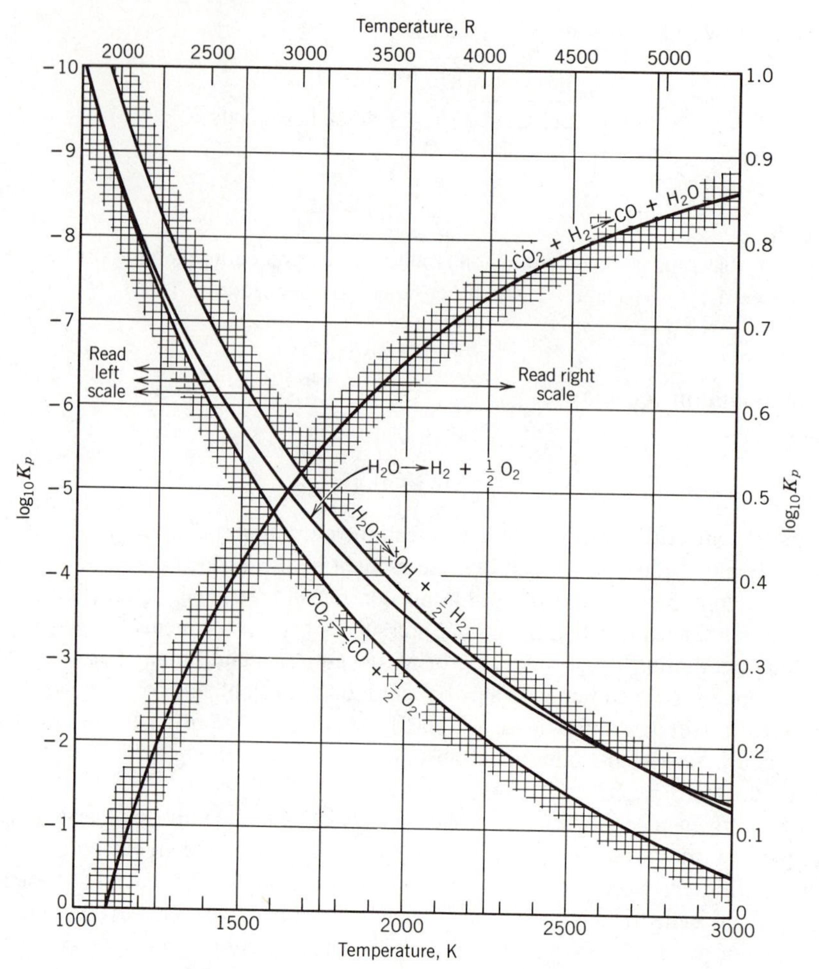

Figure 13·1. Equilibrium constants for four reactions. Data from Table A·9.

12·6. The K_P value for any element in its reference state is zero. (The reference state for nitrogen, for example, is the ideal gas N_2; therefore, the monatomic gas N has a nonzero value of K_P.)

Equilibrium calculations based on classical thermodynamics provide no information on the rate at which a reaction occurs. Calculations may show that a system is not in a state of equilibrium, but the indicated reaction may occur so slowly that measurements made on the system even over a period of years will reveal no change in state. Such a condition is sometimes referred to as *frozen equilibrium*. A familiar example is that hydrogen and oxygen can exist together at room pressure and temperature for an indefinitely long period with no measurable reaction occurring, even though the equilibrium constant for the reaction $H_2 + \frac{1}{2}O_2 \rightarrow H_2O$ at room conditions shows that the equilibrium mixture is comprised almost entirely of H_2O. The reaction rate for the combining of hydrogen and oxygen at room conditions is so low that it can be taken as zero for virtually all purposes. The presence of a catalyst or a stimulus such as an electric spark speeds the reaction, however, and the reaction can be observed to proceed always in the direction indicated by equilibrium calculations.

We now take up several examples illustrating the use of the equilibrium constant. *Remember that we are treating only the special case of ideal-gas reactions.*

Example 13·1. For the reaction $\frac{1}{2}O_2 + \frac{1}{2}N_2 \rightarrow NO$, K_P at 3300 K (or 5940 R) is 0.1694. Determine K_P at this temperature for (*a*) the reaction $O_2 + N_2 \rightarrow 2NO$ and (*b*) the reaction $NO \rightarrow \frac{1}{2}O_2 + \frac{1}{2}N_2$.

Solution. The equilibrium constant given is

$$K_{P1} = \frac{P_{NO}/P^\circ}{(P_{O_2}/P^\circ)^{1/2}(P_{N_2}/P^\circ)^{1/2}} = 0.1694$$

(***a***) For the reaction $O_2 + N_2 \rightarrow 2NO$, call the equilibrium constant K_{P2}:

$$K_{P2} = \frac{(P_{NO}/P^\circ)^2}{(P_{O_2}/P^\circ)(P_{N_2}/P^\circ)} = K_{P1}^2 = (0.1694)^2 = 0.0287$$

(***b***) For the reaction $NO \rightarrow \frac{1}{2}O_2 + \frac{1}{2}N_2$, the equilibrium constant K_{P3} is

$$K_{P3} = \frac{(P_{O_2}/P^\circ)^{1/2}(P_{N_2}/P^\circ)^{1/2}}{P_{NO}/P^\circ} = \frac{1}{K_{P1}} = \frac{1}{0.1694} = 5.90$$

Example 13·2. At 2500 K (or 4500 R), $K_P = 27.5$ for the reaction $CO + \frac{1}{2}O_2 \rightarrow CO_2$. For an initial mixture of $CO + \frac{1}{2}O_2$, determine the composition of the equilibrium mixture at 2500 K (or 4500 R) and (*a*) 1 atm, (*b*) 5 atm.

Solution. Let y be the fraction of the carbon in the system which is in CO under equilibrium conditions. Then the equilibrium mixture is

$$yCO + (1 - y)CO_2 + \tfrac{1}{2}yO_2$$

This can also be expressed by saying that the actual reaction is

$$CO + \tfrac{1}{2}O_2 \rightarrow yCO + (1 - y)CO_2 + \tfrac{1}{2}yO_2$$

but remember that the initial state has no bearing on the calculation except to specify the amounts of the various elements present. The total number of moles of mixture is $y + 1 - y + \frac{1}{2}y = 1 + \frac{1}{2}y$. The partial pressures of the three constituents are

$$P_{CO} = x_{CO}P_m = \frac{y}{1 + \frac{1}{2}y}P_m = \frac{2y}{2 + y}P_m$$

$$P_{CO_2} = x_{CO_2}P_m = \frac{1 - y}{1 + \frac{1}{2}y}P_m = \frac{2 - 2y}{2 + y}P_m$$

$$P_{O_2} = x_{O_2}P_m = \frac{\frac{1}{2}y}{1 + \frac{1}{2}y}P_m = \frac{y}{2 + y}P_m$$

Substituting these values into the defining equation for K_P

$$K_P = \frac{P_{CO_2}/P^\circ}{(P_{CO}/P^\circ)(P_{O_2}/P^\circ)^{1/2}} = \frac{2 - 2y}{2y\left(\dfrac{y}{2+y}\right)^{1/2}(P_m/P^\circ)^{1/2}}$$

$$K_P^2 = \frac{(1-y)^2}{y^2\left(\dfrac{y}{2+y}\right)(P_m/P^\circ)} = \frac{(1-y)^2(2+y)}{y^3(P_m/P^\circ)} = (27.5)^2$$

For $P_m = P^\circ = 1$ atm, the solution of this equation is $y = 0.129$; and, for $P_m = 5$ atm, $y = 0.0776$.

(*a*) For $P_m = 1$ atm, $y = 0.129$, and the mole fractions are

$$x_{CO} = \frac{2y}{2+y} = 0.121$$

$$x_{CO_2} = \frac{2-2y}{2+y} = 0.818$$

$$x_{O_2} = \frac{y}{2+y} = 0.061$$

(*b*) For $P = 5$ atm, $y = 0.0776$, and the mole fractions are

$$x_{CO} = \frac{2y}{2+y} = 0.075$$

$$x_{CO_2} = \frac{2-2y}{2+y} = 0.888$$

$$x_{O_2} = \frac{y}{2+y} = 0.037$$

A comparison of parts *a* and *b* shows that increasing the pressure causes the reaction $CO + \frac{1}{2}O_2 \rightarrow CO_2$ at this temperature to proceed further to the right, which is the direction of decreasing volume.

Example 13·3. Using the K_P data of Example 13·2, determine the equilibrium mixture at 1 atm and 2500 K (or 4500 R) for CO and the stoichiometric amount of air.

Solution. The complete reaction equation is

$$CO + \tfrac{1}{2}O_2 + \tfrac{1}{2}(3.76)N_2 \rightarrow CO_2 + \tfrac{1}{2}(3.76)N_2$$

If y stands for the fraction of the carbon which is in CO under equilibrium conditions, then the equilibrium mixture is

$$yCO + (1-y)CO_2 + \tfrac{1}{2}yO_2 + \frac{3.76}{2}N_2$$

The total number of moles of mixture is $2.88 + \frac{1}{2}y$. The partial pressures of the four constituents are

$$P_{CO} = x_{CO}P_m = \frac{y}{2.88 + \frac{1}{2}y}P_m = \frac{2y}{5.76 + y}P_m$$

$$P_{CO_2} = x_{CO_2}P_m = \frac{1 - y}{2.88 + \frac{1}{2}y}P_m = \frac{2 - 2y}{5.76 + y}P_m$$

$$P_{O_2} = x_{O_2}P_m = \frac{\frac{1}{2}y}{2.88 + \frac{1}{2}y}P_m = \frac{y}{5.76 + y}P_m$$

$$P_{N_2} = x_{N_2}P_m = \frac{1.88}{2.88 + \frac{1}{2}y}P_m = \frac{3.76}{5.76 + y}P_m$$

Substituting these values into the defining equation for K_P

$$K_P \equiv \frac{P_{CO_2}/P^\circ}{(P_{CO}/P^\circ)(P_{O_2}/P^\circ)^{1/2}} = \frac{2 - 2y}{2y\left(\dfrac{y}{5.76 + y}\right)^{1/2}(P_m/P^\circ)^{1/2}}$$

$$K_P^2 = \frac{(1 - y)^2(5.76 + y)}{y^3(P_m/P^\circ)} = (27.5)^2$$

For $P_m = P^\circ = 1$ atm, the solution of this equation is $y = 0.175$, so the mole fractions are

$$x_{CO} = \frac{2y}{5.76 + y} = 0.059$$

$$x_{CO_2} = \frac{2 - 2y}{5.76 + y} = 0.278$$

$$x_{O_2} = \frac{y}{5.76 + y} = 0.029$$

$$x_{N_2} = \frac{3.76}{5.76 + y} = 0.634$$

(Notice that the presence of nitrogen affects the extent to which the reaction proceeds at this temperature, but it does not change the equilibrium constant. The presence of excess oxygen would cause the reaction to go further toward completion at this temperature, but again this result would follow from the same equilibrium constant value that is independent of the amounts of the various constituents actually present.)

The three examples above have dealt with only the conditions of equilibrium at given states. The following examples involve equilibrium calculations as part of the application of the first and second laws to combustion processes. In one kind of application, the temperatures and pressures of the end states are specified. Equilibrium calculations determine mixture compositions that must be known to evaluate mixture properties for first and second law applications. (See Example 13·4.) The relationships involved in determining mixture compositions at a fixed temperature are

1. Mass balances for individual elements in the reacting constituents, and
2. The defining relationships between K_P values at a given temperature and the mole fractions or partial pressures of the constituents.

Depending on the mixture composition, more than one reaction may be involved. Solving for the mixture composition requires a number of equations equal to the number of reacting constituents. One equation is provided by the mass balance for each element involved in a reaction. One K_P equation is provided for each reaction considered. Therefore, in general, the number of reactions that must be considered is equal to the number of reacting constituents in the mixture minus the number of elements in those constituents.

Another kind of application determines an end state resulting from specified energy transfers. An example is the determination of an adiabatic flame temperature where the specified energy transfers are $Q = 0$ and $W = 0$. For an exothermic reaction carried out adiabatically, the temperature of the mixture increases as the reaction proceeds, the energy for the increased enthalpy of the constituents being supplied by the reaction. However, equilibrium calculations show that for exothermic reactions, the higher the temperature of the products, the less complete the reaction can be. The less complete the reaction is, the less energy there is available from the enthalpy of reaction for raising the products temperature, so the equilibrium temperature is lower than the (nonequilibrium) adiabatic flame temperature calculated on the basis of complete combustion.

The calculation of adiabatic flame temperature involves several simultaneous relationships:

1. Mass balances for the individual elements in the reacting constituents
2. Relationships between mixture composition and K_P as given by K_P definitions
3. $K_P T$ relationships that are often given in tabular or graphical form but which usually can be readily fitted by equations, at least for limited temperature ranges
4. hT relationships, which may be in tables, charts, or equations
5. The first law, which for the special case of steady flow through an open system is

$$Q = 0 = \sum_{\text{prod}} N(h - h^\circ + \Delta h_f^\circ) - \sum_{\text{reac}} N(h - h^\circ + \Delta h_f^\circ)$$

The number of reactions that must be considered is still equal to the number of reacting constituents in the mixture minus the number of elements in those constituents. However, relationships 1 and 2 cannot be solved without involving relationships 3, because K_P values are functions of the unknown temperature. The temperature is involved also in relationships 4, which are coupled to relationship 5 through the enthalpy values. Therefore, all five of the relationships listed must be solved simultaneously. (See Examples 13·5 and 6.)

The calculation procedures one uses depend on (1) the data at hand, including the accuracy and form of the data; (2) the computational tools at hand, including equipment, software, and one's skill in using them; (3) the number of similar problems to be solved; and (4) the trade-off between the accuracy desired and the computational and data-gathering effort.

Example 13·4. Methane is burned with 80 percent of stoichiometric air in a steady-flow process at 1 atm. Methane and air are both supplied at 25°C (77°F), and the products leave at 1700 K (3060 R). Assuming that no CH_4, OH, NO, or free oxygen appears in the products, determine the amount of heat transferred per kilogram of methane.

Solution. The reaction is

$$CH_4 + 0.8(2)O_2 + 0.8(2)3.76N_2 \rightarrow yCO_2 + (1 - y)CO + zH_2O + (2 - z)H_2 + 0.8(2)3.76N_2$$

y and z must be determined before an energy balance can be made. One relation between y and z is given by a mass balance for oxygen

$$1.6 = y + \tfrac{1}{2}(1 - y) + \tfrac{1}{2}z$$

$$2.2 = y + z \tag{A}$$

A second relationship between y and z can be found by means of the equilibrium constant for the reaction $CO_2 + H_2 \rightarrow CO + H_2O$, which is known as the *water gas reaction*

$$K_P = \frac{(P_{CO}/P^\circ)(P_{H_2O}/P^\circ)}{(P_{CO_2}/P^\circ)(P_{H_2}/P^\circ)}$$

The partial pressures for substitution in this expression are obtained from the equilibrium mixture

$$yCO_2 + (1 - y)CO + zH_2O + (2 - z)H_2 + 6.02N_2$$

as

$$P_{CO} = \frac{(1 - y)P_m}{y + (1 - y) + z + (2 - z) + 6.02} = \frac{1 - y}{9.02} P_m$$

$$P_{H_2O} = \frac{z}{9.02} P_m$$

$$P_{CO_2} = \frac{y}{9.02} P_m$$

$$P_{H_2} = \frac{2 - z}{9.02} P_m$$

Thus, for $K_P = 3.39$ at 1700 K (or 3060 R) (from Table A·9 as $\log_{10} K_P = 0.530$),

$$K_P = \frac{(1 - y)z}{y(2 - z)} = 3.39 \tag{B}$$

Solving Eqs. A and B gives

$$y = 0.567 \qquad z = 1.633$$

Thus the actual reaction equation is

$$CH_4 + 1.6O_2 + 6.02N_2 \rightarrow 0.567CO_2 + 0.433CO + 1.633H_2O + 0.367H_2 + 6.02N_2$$

The enthalpy of reaction at 25°C for this particular reaction is

$$\Delta H_R^\circ = N_{CH_4}\,\Delta h_R^\circ = \sum_{\text{prod}} N\Delta h_f^\circ - \sum_{\text{reac}} N\Delta h_f^\circ \tag{13·2}$$

$$\begin{aligned}\Delta h_R^\circ &= 0.567\Delta h_{f,CO_2}^\circ + 0.433\Delta h_{f,CO}^\circ + 1.633\Delta h_{f,H_2O}^\circ \\ &\quad + 0 + 0 - \Delta h_{f,CH_4}^\circ - 0 - 0 \\ &= 0.567(-393\,522) + 0.433(-110\,541) \\ &\quad + 1.633(-241\,826) - 1(-74\,873) \\ &= -591\,020 \text{ kJ/kmol } CH_4\end{aligned}$$

where the Δh_f° values are taken from Table A·8.

The first law applied to this steady-flow reaction is

$$Q = \Delta H_R^\circ + \sum_{\text{prod}} N(h_2 - h^\circ) - \sum_{\text{reac}} N(h_1 - h^\circ)$$

Since the reactants enter at 25°C, the last term on the right-hand side is zero. Then, dividing by the number of kilomoles of CH_4

$$\begin{aligned} q &= h_R^\circ + 0.567(h_2 - h^\circ)_{CO_2} + 0.433(h_2 - h^\circ)_{CO} + 1.633(h_2 - h^\circ)_{H_2O} \\ &\quad + 0.367(h_2 - h^\circ)_{H_2} + 6.02(h_2 - h^\circ)_{N_2} \\ &= -591\ 020 + 0.567(82\ 847.5 - 9359.7) + 0.433(54\ 605.8 - 8666.7) \\ &\quad + 1.633(67\ 662.0 - 9899.0) + 0.367(51\ 303.3 - 8463.1) \\ &\quad + 6.02(54\ 096.3 - 8665.8) \\ &= -145\ 920 \text{ kJ/kmol CH}_4 = -9096 \text{ kJ/kg CH}_4 \end{aligned}$$

Example 13·5. Determine the equilibrium adiabatic flame temperature for the steady flow constant pressure burning of a stoichiometric mixture of carbon monoxide and dry air with inlet conditions of 1 atm, 25°C. Determine also the irreversibility of the process.

Analysis. Since we are looking for the temperature of the products mixture, we do not know the mixture composition because it in turn depends on the temperature. In addition to not knowing mole fractions to use in the first law, we do not have enthalpy values because they depend on the unknown temperature. Therefore, we will write the expression for K_P in terms of a composition variable. For several values of product temperature, we will solve this equation for the composition. Then using for each temperature the corresponding composition and the enthalpy values obtained from hT tables, we will calculate the heat transfer by means of the first law. In this manner we will determine the temperature for which $Q = 0$. After the adiabatic flame temperature and composition are determined, we will calculate the irreversibility by $i = T_0 \Delta s_{\text{universe}}$.

Solution. The reaction equation is

$$CO + \tfrac{1}{2}O_2 + \frac{3.76}{2}N_2 \rightarrow (1 - y)CO_2 + yCO + \frac{y}{2}O_2 + 1.88\ N_2$$

The total number of moles of products is $1 - y + y + y/2 + 1.88 = 2.88 + y/2$. The partial pressures are

$$P_{CO_2} = \frac{1 - y}{2.88 + \dfrac{y}{2}} P_m \qquad P_{CO} = \frac{y}{2.88 + \dfrac{y}{2}} P_m$$

$$P_{O_2} = \frac{y/2}{2.88 + \dfrac{y}{2}} P_m \qquad P_{N_2} = \frac{1.88}{2.88 + \dfrac{y}{2}} P_m$$

For the reaction $CO_2 \rightarrow CO + \frac{1}{2}O_2$

$$K_P = \frac{(P_{CO}/P^\circ)(P_{O_2}/P^\circ)^{1/2}}{P_{CO_2}/P^\circ} = \frac{x_{CO}x_{O_2}^{1/2}}{x_{CO_2}}\left(\frac{P_m}{P^\circ}\right)^{1/2} = \frac{y}{1 - y}\left(\frac{y}{5.76 + y}\right)^{1/2}(1)^{1/2} \qquad \text{(A)}$$

The first law for the steady flow system is

$$Q = H_{\text{prod}} - H_{\text{reac}} = \sum_{\text{prod}} N(h - h^\circ + \Delta h_f^\circ) - \sum_{\text{reac}} N(h - h^\circ + \Delta h_f^\circ)$$

where each Δh_f° is on a per mole of constituent basis. Applying this equation to the case at hand, and dividing by the number of moles of CO entering

$$q = (1 - y)(h - h^\circ + \Delta h_f^\circ)_{CO_2} + y(h - h^\circ + \Delta h_f^\circ)_{CO} + \frac{y}{2}(h^\circ - \Delta h_f^\circ)_{O_2}$$
$$+ 1.88(h - h^\circ + \Delta_f^\circ)_{N_2} - (h - h^\circ + \Delta h_f^\circ)_{CO,r} - \tfrac{1}{2}(h - h^\circ + \Delta h_f^\circ)_{O_2,r}$$
$$- 1.88(h - h^\circ + \Delta h_f^\circ)_{N_2,r}$$

where the subscript r refers to the incoming mixture of $r$eactants. Since $T^\circ = T_r = 25°C$ and Δh_f° of elements is zero, the first law simplifies to

$$q = (1 - y)(h - h^\circ + \Delta h_f^\circ)_{CO_2} + y(h - h^\circ + \Delta h_f^\circ)_{CO}$$
$$+ \frac{y}{2}(h - h^\circ)_{O_2} + 1.88(h - h^\circ)_{N_2} - (\Delta h_f^\circ)_{CO}$$

Substituting from Tables A·6 and A·8 the numerical values that are independent of the products temperature

$$q = (1 - y)(h - 9360 - 393\,522)_{CO_2} + y(h - 8667 - 110\,541)_{CO} + \frac{y}{2}(h - 8677)_{O_2}$$
$$+ 1.88(h - 8666)_{N_2} - (-110\,541) \quad \text{(B)}$$

Equations A and B are now entered into a computer equation solver and solved for several values of T and the corresponding log K_P values obtained from Table A·9 as shown in the following table. The resulting values of y and q are also shown in the table. Fortuitously, the $q = 0$ condition is seen to be met by a temperature close to 2400 K, one of the table values, so no closer search is needed. (If hT equations were entered instead of discrete T and h values, or if an interpolation procedure is available to use with the hT data, the calculation can be made directly under the constraint of $q = 0$.)

T, K	Log K_P	K_P	y	q, kJ/kmol CO
2100	−2.539	0.002 89	0.0356	−63 333
2200	−2.226	0.005 94	0.0568	−44 627
2300	−1.94	0.011 48	0.0863	−23 581
2400	−1.679	0.020 94	0.1254	104
2500	−1.44	0.036 31	0.1747	26 564
2600	−1.219	0.060 39	0.2341	55 798
2700	−1.015	0.096 61	0.3021	87 347
2800	−0.825	0.149 62	0.3766	120 639

For the steady-flow system per mole of CO entering

$$i = T_0 \Delta s_{\text{universe}} = T_0(\Delta s_{\text{sys}} + \Delta s_{\text{surr}}) = T_0(0 + s_2 - s_1) = T_0(s_p - s_r)$$
$$= T_0\left[\sum_p N\left(s^\circ + \phi - \phi^\circ - R_u \ln \frac{P}{P^\circ}\right) - \sum_r N\left(s^\circ + \phi - \phi^\circ - R_u \ln \frac{P}{P^\circ}\right)\right]$$

where N is the number of moles of each constituent per mole of CO entering, s° is the absolute entropy of each constituent at 1 atm, 25°C; the P's are partial pressures, and P° is the standard-state pressure of 1 atm. Because the mixture pressure is also 1 atm, $P/P^\circ = P/P_m = x$, where x is the mole fraction. In Table A·6 and in the Keenan, Chao, and Kaye *Gas Tables*, $\phi^\circ = s^\circ$ for each constituent. Therefore, the last equation simplifies to

$$i = T_0[(1 - y)(\phi - R_u \ln x)_{CO_2,p} + y(\phi - R_u \ln x)_{CO,p} + \frac{y}{2}(\phi - R_u \ln x)_{O_2,p}$$
$$+ 1.88(\phi - R_u \ln x)_{N_2,p} - (\phi - R_u \ln x)_{CO,r} - \tfrac{1}{2}(\phi - R_u \ln x)_{O_2,r}$$
$$- 1.88(\phi - R_u \ln x)_{N_2,r}]$$

For substituting mole fractions, note that per mole of CO entering there are 3.38 moles of reactants and $(2.88 + y/2) = 2.94$ moles of products.

$$i = 298\left[(1 - 0.125)\left(320.271 - 8.314 \ln \frac{1 - 0.125}{2.94}\right) + 0.125\left(265.215\right.\right.$$
$$\left. - 8.314 \ln \frac{0.125}{2.94}\right) + \frac{0.125}{2}\left(275.538 - 8.314 \ln \frac{0.125}{2(2.94)}\right)$$
$$+ 1.88\left(258.534 - 8.314 \ln \frac{1.88}{2.94}\right) - \left(197.499 - 8.314 \ln \frac{1}{3.38}\right)$$
$$\left. - \tfrac{1}{2}\left(204.975 - 8.314 \ln \frac{0.5}{3.38}\right) - 1.88\left(191.448 - 8.314 \ln \frac{1.88}{3.38}\right)\right]$$
$$= 44\,878 \text{ kJ/kmol CO entering}$$

This much energy has been made unavailable by the combustion process. Thus the maximum work obtainable from this adiabatic steady flow burning of CO is less than the enthalpy of combustion of CO (−282 981 kJ/kmol) by at least 44 878 kJ/kmol.

Example 13·6. Methane supplied at 1 atm, 25°C (77°F), is burned adiabatically in a steady-flow burner with the stoichiometric amount of air supplied at the same conditions. Assuming that the products contain no CH_4 or NO, determine the temperature of the products.

Solution. The reaction equation is

$$CH_4 + 2O_2 + 7.52N_2 \rightarrow yCO_2 + (1 - y)CO + zH_2O$$
$$+ \tfrac{1}{2}(1 - z + y + 2x)H_2 + (3 - y - 2x - z)OH + xO_2 + 7.52N_2$$

The molar coefficients of the products in this equation were determined by C, O, and H mass balances. The products temperature and the values of x, y, and z must be such that they satisfy an energy balance and also satisfy the equilibrium conditions readily expressed by means of equilibrium constants. In the products are six reacting constituents involving three elements, so we must consider three independent reactions for determining the equilibrium composition. We then have an equal number of relationships and variables.

Before carrying out numerical calculations, we set up in convenient forms the relations to be used. The total number of moles of products is $y + (1 - y) + z + \tfrac{1}{2}(1 - z + y + 2x) + (3 - y - 2x - z) + x + 7.52$ or $(12.02 - \tfrac{1}{2}y - \tfrac{1}{2}z)$. The partial pressures of the products constituents are

$$P_{CO_2} = \frac{y}{12.02 - \frac{1}{2}y - \frac{1}{2}z} P_m$$

$$P_{CO} = \frac{1 - y}{12.02 - \frac{1}{2}y - \frac{1}{2}z} P_m$$

$$P_{H_2O} = \frac{z}{12.02 - \frac{1}{2}y - \frac{1}{2}z} P_m$$

$$P_{H_2} = \frac{\frac{1}{2}(1 - z + y + 2x)}{12.02 - \frac{1}{2}y - \frac{1}{2}z} P_m$$

$$P_{OH} = \frac{3 - y - 2x - z}{12.02 - \frac{1}{2}y - \frac{1}{2}z} P_m$$

$$P_{O_2} = \frac{x}{12.02 - \frac{1}{2}y - \frac{1}{2}z} P_m$$

$$P_{N_2} = \frac{7.54}{12.02 - \frac{1}{2}y - \frac{1}{2}z} P_m$$

We can use equilibrium constant data for any three reactions that involve constituents of the products mixture. Three such reactions for which equilibrium constant data are available from Table A·9 and Fig. 13·1 are

$$CO + \tfrac{1}{2}O_2 \rightarrow CO_2 \tag{1}$$

$$OH + \tfrac{1}{2}H_2 \rightarrow H_2O \tag{2}$$

$$H_2 + \tfrac{1}{2}O_2 \rightarrow H_2O \tag{3}$$

The equilibrium constants for these reactions are

$$K_{P1} = \frac{P_{CO_2}/P^\circ}{(P_{CO}/P^\circ)(P_{O_2}/P^\circ)^{1/2}} \qquad K_{P2} = \frac{P_{H_2O}/P^\circ}{(P_{OH}/P^\circ)(P_{H_2}/P^\circ)^{1/2}} \qquad K_{P3} = \frac{P_{H_2O}/P^\circ}{(P_{H_2}/P^\circ)(P_{O_2}/P^\circ)^{1/2}}$$

Substituting the partial pressure values from above into these expressions, we have

$$K_{P1} = \frac{y(12.02 - \frac{1}{2}y - \frac{1}{2}z)^{1/2}}{(1 - y)x^{1/2}\,(P_m/P^\circ)^{1/2}} \tag{A}$$

$$K_{P2} = \frac{z(12.02 - \frac{1}{2}y - \frac{1}{2}z)^{1/2}}{(3 - y - 2x - z)(\frac{1}{2} - \frac{z}{2} + \frac{y}{2} + x)^{1/2}\,(P_m/P^\circ)^{1/2}} \tag{B}$$

$$K_{P3} = \frac{z(12.02 - \frac{1}{2}y - \frac{1}{2}z)^{1/2}}{(\frac{1}{2} - \frac{z}{2} + \frac{y}{2} + x)x^{1/2}\,(P_m/P^\circ)^{1/2}} \tag{C}$$

The first law for this adiabatic steady-flow process is

$$Q = 0 = H_{prod} - H_{reac} = \sum_{prod} N(h - h^\circ + \Delta h_f^\circ) - \sum_{reac} N(h - h^\circ + \Delta h_f^\circ) \tag{12·3}$$

For the reaction under consideration, recalling that $\Delta h_f^\circ = 0$ for elements and that the reactants enter at the reference temperature, we expand the last equation, on a per mole of CH_4 basis, to

$$\begin{aligned} q = 0 &= y(h - h^\circ + \Delta h_f^\circ)_{CO_2} + (1 - y)(h - h^\circ + \Delta h_f^\circ)_{CO} + z(h - h^\circ + \Delta h_f^\circ)_{H_2O} \\ &+ \tfrac{1}{2}(1 - z + y + 2x)(h - h^\circ)_{H_2} + (3 - y - 2x - z)(h - h^\circ)_{OH} + x(h - h^\circ)_{O_2} \\ &+ 7.52(h - h^\circ)_{N_2} - \Delta h_{f,CH_4}^\circ \end{aligned}$$

Substitution of h° and Δh_f° values from Tables A·6 and A·8 gives

$$\begin{aligned} 0 &= y(h_{CO_2} - 402\,882) + (1 - y)(h_{CO} - 119\,208) + z(h_{H_2O} - 252\,486) \\ &+ \tfrac{1}{2}(1 - z + y + 2x)(h_{H_2} - 8463) + (3 - y - 2x - z)(h_{OH} + 30\,292) \\ &+ x(h_{O_2} - 8677) + 7.52(h_{N_2} - 8666) + 74\,850 \end{aligned} \tag{D}$$

Now we are ready to determine the final temperature as that value which satisfies Eqs. A, B, C, and D. The unknowns in these equations and supporting relationships are K_{P1}, K_{P2}, K_{P3}, x, y, z, T, and the h's of the products constituents. The supporting relationships available are the K_PT relations in Table A·9 and the various hT relations in Table A·6.

Several approaches are now possible. One that is suitable to use even with limited computer capability is to select several T values in a likely range of product temperatures and then obtain from the appropriate tables the corresponding values of K_{P1}, K_{P2}, K_{P3}, and enthalpies. Then by a suitable computer program for solving simultaneous algebraic equations, x, y, z, and q can be calculated for each T value. This will show the temperature for which $q = 0$. The result of this procedure is an adiabatic combustion temperature of approximately 2240 K.

(So that you might check this calculation, the values for $T = 2200$ K are given here: $x = 0.0493$, $y = 0.9199$, $z = 1.953$; and Eq. D becomes

$$\begin{aligned} q &= 0.9199(112\,929 - 402\,882) + 0.0801(72\,671 - 119\,208) + 1.953(93\,054 \\ &\quad - 252\,486) + 0.0328(68\,344 - 8463) + 0.0285(69\,923 + 30\,292) \\ &\quad + 0.0493(75\,432 - 8677) + 7.52(72\,023 - 8666) + 74\,850 \\ q &= -22\,400 \text{ kJ/kmol CH}_4 \text{ entering} \end{aligned}$$

This means that in order for the products to be at 2200 K, heat must be removed in the amount of 22 400 kJ/kmol. The adiabatic combustion temperature must therefore be higher than 2200 K.)

13·3 The Relationship between K_P and Δh_R

Section 13·2 mentioned methods of determining K_P values from other thermochemical data. An important relationship for this purpose, as well as for others, is the equation involving K_P and Δh_R which we now derive.

If we combine Eq. j of the preceding section and the definition of K_P, we have

$$\ln K_P = \frac{1}{R_u T}[\nu_1 f_1(T) + \nu_2 f_2(T) - \nu_3 f_3(T) - \nu_4 f_4(T)] \tag{k}$$

for a reaction of the form

$$\nu_1 A_1 + \nu_2 A_2 \rightarrow \nu_3 A_3 + \nu_4 A_4$$

If we differentiate Eq. j or Eq. k with respect to temperature, we will have some terms $d/dT\ (f(T)/T)$; so let us examine these terms. From Eq. i

$$f(T) = h - T(\phi - \phi^\circ) - Ts^\circ$$

and

$$\frac{f(T)}{T} = \frac{h}{T} - \phi + \phi^\circ - s^\circ$$

Then

$$\frac{d}{dT}\left(\frac{f(T)}{T}\right) = \frac{T(dh/dT) - h}{T^2} - \frac{d}{dT}\left[\int \frac{c_p dT}{T}\right]$$

$$= \frac{1}{T}\frac{dh}{dT} - \frac{h}{T^2} - \frac{c_P}{T} = -\frac{h}{T^2}$$

where the last simplification is possible because *for an ideal gas h* is a function of T only so that $c_p = dh/dT$. (In general, of course, $c_p = (\partial h/\partial T)_P$.) Differentiation of Eq. k thus gives

$$\frac{d \ln K_P}{dT} = \frac{1}{R_u T^2}[-\nu_1 h_1 - \nu_2 h_2 + \nu_3 h_3 + \nu_4 h_4]$$

The term within the brackets is recognized as Δh_R; so we have

$$\frac{d \ln K_P}{dT} = \frac{\Delta h_R}{R_u T^2} \qquad (13\cdot4)$$

This equation is valuable because, if the $K_P T$ variation is known, Δh_R can be calculated. Notice that for exothermic reactions ($\Delta h_R < 0$), K_P decreases as the temperature increases, and for endothermic reations ($\Delta h_R > 0$), K_P increases with increasing temperature. Therefore, the higher the temperature, the more an exothermic reaction falls short of going to completion. Equation 13·4 can also be written as

$$\frac{d \ln K_P}{d(1/T)} = -\frac{\Delta h_R}{R_u}$$

This form shows that if Δh_R varies only slightly with temperature, a plot of ln K_P or $\log_{10}$ K_P versus $1/T$ is nearly a straight line. (Caution: Data are usually given in terms of $\log_{10}$ K_P instead of in terms of ln K_P.)

13·4 The Relationship Between K_P and Δg_R

An equation which is even more useful then Eq. 13·4 relates ln K_P (not its derivative) for a reaction to the Gibbs function (change) of reaction Δg_R at a standard pressure of 1 atm. The definition of Δg_R is analogous to that of Δh_R and Δu_R. The derivation of this valuable equation can start with Eq. k of the preceding section:

$$\ln K_P = \frac{1}{R_u T}[\nu_1 f_1(T) + \nu_2 f_2(T) - \nu_3 f_3(T) - \nu_4 f_4(T)] \qquad (k)$$

Referring back to Eq. i in Sec. 13·2, we see that $f(T)$ is defined as

$$f(T) \equiv h - Ts^\circ - T\phi + T\phi^\circ$$
$$= (h - Ts)_{P=1\,\text{atm}} = g_{1\,\text{atm}}$$

It must not be inferred that Eq. k or the function $f(T)$ can be used for only a pressure of 1 atm. The preceding steps show only the $f(T)$ for each constituent *at any pressure* turns out to be the Gibbs function of the constituent at the same temperature and at 1 atm. Thus Eq. k becomes

$$\ln K_P = \frac{1}{R_u T}[\nu_1 g_1 + \nu_2 g_2 - \nu_3 g_3 - \nu_4 g_4]_{1\,\text{atm}}$$

The quantity within the brackets is $-\Delta g_R$ at 1 atm and temperature T; so we have

$$\ln K_P = -\frac{\Delta g_{R,1\,\text{atm}}}{R_u T} \tag{13·5}$$

$\Delta g_{R,1\,\text{atm}}$ is often represented by the symbol Δg_R° and is sometimes called the *standard free-energy* change, although the use of the name free energy for the Gibbs function is inadvisable because the same name has been used for the Helmholtz function as was pointed out in Sec. 8·7. Δg_R° can be calculated from Δh_R° and absolute entropy values, and it can also be found in tables of thermochemical data. Some data sources provide values of Δg_R° for substances instead of for reactions, in the same manner that K_P values are sometimes presented for substances as mentioned in Sec. 13·2. These values can be designated by Δg_f°. (Caution in using various sources of thermochemical data: Usually, Δg_R° or Δg° refers only to a specified standard pressure; so that for any reaction, tables of Δg_R° or Δg° versus temperature are possible; but sometimes it refers to a specified pressure *and* a specified temperature so that there is but a single value for each reaction.)

This chapter is concerned with only ideal-gas reactions, but Eq. 13·5 actually applies to other reactions and in fact is widely used as a general definition of the equilibrium constant.

13·5 Summary

A more general form of the relation

$$T\,dS = dU + P\,dV \qquad \text{(pure substances)} \tag{8·3}$$

which holds only for pure substances is

$$T\,dS \geq dU + P\,dV \tag{13·1}$$

which holds even for systems which undergo chemical reactions. The equality holds for systems in complete (mechanical, thermal, and chemical) equilibrium, and the inequality holds for all other cases including those of irreversible chemical reactions.

From Eq. 13·1 we obtain equilibrium criteria such as

$$(dS)_{U,V} \geq 0 \qquad \text{and} \qquad (dG)_{P,T} \leq 0 \tag{13·2}$$

Referring to the second of these criteria, one interpretation is that, when a system is in complete equilibrium (i.e., no spontaneous process is possible), its Gibbs function must be a minimum with regard to all states at the same pressure and temperature. This is a valuable criterion of equilibrium, not only because many chemical reactions occur at constant pressure and temperature, but also because it can be shown that if a system is in such a state that no spontaneous process can occur at constant pressure and temperature, then no spontaneous process at all can occur.

Application of one of the equilibrium criteria shows that for a mixture of ideal gases A_1, A_2, A_3, and A_4 which can undergo a reaction

$$\nu_1 A_1 + \nu_2 A_2 \rightarrow \nu_3 A_3 + \nu_4 A_4$$

the equilibrium mixture is such that

$$\frac{(P_3/P^\circ)^{\nu_3}(P_4/P^\circ)^{\nu_4}}{(P_1/P^\circ)^{\nu_1}(P_2/P^\circ)^{\nu_2}} = K_P$$

where the P's are partial pressures and K_P, called the equilibrium constant, is a function of temperature only. K_P does not depend on the amount of the various constituents initially present. Therefore, a knowledge of equilibrium constants for an ideal-gas reaction provides us with one relationship among the partial pressures of the constituents. Stoichiometric considerations give us other relationships; so we are able to determine the extent to which a reaction can proceed at a given temperature. Classical thermodynamics gives no information on reaction rates, however; so we cannot predict on this basis whether a reaction will proceed to the indicated extent within a given time interval.

Two valuable relationships between K_P and other thermochemical quantities are

$$\frac{d \ln K_P}{dT} = \frac{\Delta h_R}{R_u T^2} \tag{13·4}$$

and

$$\ln K_P = -\frac{\Delta g_R^\circ}{R_u T} \tag{13·5}$$

Suggested Reading and References

13·1 Dixon, J. R., *Thermodynamics I*, Prentice-Hall, Englewood Cliffs, N.J., 1975, Section 15-4.

13·2 Holman, J. P., *Thermodynamics*, 2nd ed., McGraw-Hill, New York, 1974, Chapter 11.

13·3 Reynolds, William C., and Henry C. Perkins, *Engineering Thermodynamics*, 2nd ed., McGraw-Hill, New York, 1977, Chapter 12.

13·4 Van Wylen, Gordon J., and Richard E. Sonntag, *Fundamentals of Classical Thermodynamics*, 3rd ed., Wiley, New York, 1985, Chapter 13.

13·5 Wark, K., *Thermodynamics*, 4th ed., McGraw-Hill, New York, 1983, Chapter 15.

13·6 Zemansky, Mark W., Michael M. Abbott, and Hendrick C. Van Ness, *Basic Engineering Thermodynamics*, 2nd ed., McGraw-Hill, New York, 1975, Sections 11-6 to 11-11.

13·7 *JANAF Thermochemical Tables*, 2nd ed., National Bureau of Standards, NSRDS-NBS 37, U.S. Department of Commerce, Washington, D.C., 1971, and revisions published in *Journal of Physical and Chemical Data*, 1974, 1975, 1978, 1982.

13·8 *Selected Values of Properties of Hydrocarbons and Related Compounds*, American Petroleum Institute Project 44, Thermodynamics Research Center, Texas A & M University, College Station, Texas 77843. (Data presented here from loose-leaf data sheets, extant 1984.)

13·9 *Selected Values of Chemical Thermodynamic Properties*, National Bureau of Standards Technical Note 270-3, 1968.

13·10 Gallant, R. W., "Physical Properties of Hydrocarbons," *Hydrocarbon Processing*, Vol. 45, No. 10, October 1966.

Problems

13·1 Demonstrate that each of the following is a valid criterion of equilibrium:

$$(dU)_{S,V} \leq 0 \qquad (dS)_{H,P} \geq 0 \qquad (dA)_{T,V} \leq 0$$

13·2 Refer to Eq. h in Sec. 13·2. For a mixture of reactants 1 and 2 and products 3 and 4 *not* in equilibrium, the equality sign would be replaced by an inequality sign. State a rule relating the inequality to the direction (toward more products or toward more reactants) in which the reaction would proceed.

13·3 For the reaction $\frac{1}{2}Br_2 \rightarrow Br$, $K_P = 0.04099$ at 1200 K. Determine K_P at 1200 K for (*a*) the reaction $Br_2 \rightarrow 2\,Br$ at 5 atm and (*b*) the reaction $2\,Br \rightarrow Br_2$ at 5 atm.

13·4 Reference 13·7 lists K_P values for substances instead of reactions. For hydrazine, N_2H_4, $\log K_P = -16.709$ at 1000 K. Determine at this temperature K_P for the reaction $N_2H_4 + O_2 \rightarrow 2H_2O + N_2$.

13·5 Reference 13·7 lists K_P values for substances instead of reactions. For ammonia, $\log K_P = -2.523$ at 800 K. Determine the extent to which ammonia dissociates into diatomic hydrogen and nitrogen at 800 K and (*a*) 1 atm and (*b*) 5 atm.

13·6 For an initial mixture of 1 mole of CO and 1 mole of O_2, determine the composition of the equilibrium mixture at 3000 K (5400 R) and (*a*) 1 atm, (*b*) 5 atm.

13·7 Solve Example 13·2 for a temperature of 3000 K (5400 R), using an appropriate K_P value from Table A·9.

13·8 Determine the equilibrium mixture at 2500 K (or 4500 R) for the steady-flow burning of CO and 1.5 times the stoichiometric amount of oxygen at one atmosphere. Compare the results with those of Example 13·2.

13.9 Solve Example 13·3 for pressures of (*a*) 5 atm and (*b*) 10 atm.

13·10 Determine the equilibrium mixture at 2500 K (or 4500 R) for the steady-flow burning of CO and 1.5 times the stoichiometric amount of air at 1 atm. Compare the results with those of Example 13·3.

13·11 A system comprised initially of H_2O is heated to 3000 K (5400 R). Determine the composition of the equilibrium mixture at (*a*) 1 atm, (*b*) 10 atm.

13·12 Determine the equilibrium composition at 1 atm, 2200 K (3500°F), for an initial mixture of one mole of CO_2 and one mole of H_2O, considering CO, OH, and H_2 as possible dissociation products.

13·13 Solve Problem 13·12 for an initial mixture of one mole of CO_2, one mole of H_2, and one mole of N_2.

13·14 Solve Problem 13·12 for an initial mixture of one mole of CO_2 and two moles of H_2.

13·15 At what temperature is 10 percent of CO_2 dissociated at 1 atm?

13·16 Solve Problem 13·15 if the CO_2 is initially mixed with an equal number of moles of air.

13·17 In a steady-flow constant-pressure process, CO_2 that enters at 1 atm, 25°C (77°F), is heated to 3000 K (5400 R). Determine the heat transfer.

13·18 In a steady-flow constant-pressure process, CO_2 that enters at 10 atm, 25°C (77°F), is heated to 3000 K (5400 R). Determine the heat transfer.

13·19 In a steady-flow constant-pressure process, H_2O that enters at 1 atm, 400 K (720 R), is heated to 3000 K (5400 R). Determine the heat transfer.

13·20 Acetylene is burned with 20 percent excess air in a steady-flow process at 1 atm with the reactants entering at 25°C (77°F) and the products leaving at 2400 K (4320 R). Considering as dissociation products CO, OH, and H_2, determine the heat transfer per unit mass of acetylene. What does this result indicate regarding the temperature that would result from adiabatic burning of this mixture?

13·21 Ethene is burned with 20 percent excess air in a steady-flow process at 1 atm with the reactants entering at 25°C (77°F) and the products leaving at 2400 K (4320 R). Considering as dissociation products CO, OH, and H_2, determine the heat transfer per unit mass of ethene. What does this result indicate regarding the temperature that would result from adiabatic burning of this mixture?

13·22 Ethane is burned with 20 percent excess air in a steady-flow process at 1 atm with the reactants entering at 25°C (77°F) and the products leaving at 2400 K (4320 R). Considering as dissociation products CO, OH, and H_2, determine the heat transfer per unit mass of ethane. What does this result indicate regarding the temperature that would result from adiabatic burning of this mixture?

13·23 Solve Problem 13·20 for pressures of 5 and 15 atm.

13·24 Solve Problem 13·21 for pressures of 5 and 15 atm.

13·25 Solve Problem 13·22 for pressures of 5 and 15 atm.

13·26 Solve Problem 13·20 for air at 100 percent relative humidity instead of for dry air.

13·27 Solve Problem 13·20 with the air and acetylene entering at 35°C and with an air relative humidity of 100 percent.

13·28 Determine the adiabatic flame temperature of a stoichiometric mixture of carbon monoxide and air reacting in a steady-flow system under a pressure of 2 atm if the initial mixture temperature is 25°C (77°F).

13·29 Solve Problem 13·28 for a pressure of 5 atm.

13·30 Solve Problem 13·28 for the mixture $CO + O_2$ instead of CO and the stoichiometric amount of air.

13·31 Solve Problem 13·28 for a mixture of carbon monoxide and 20 percent excess air.

13·32 Solve Problem 13·28 for a mixture of carbon monoxide and 80 percent of the stoichiometric amount of air.

13·33 Determine the adiabatic flame temperature of a stoichiometric mixture of hydrogen and air, initially at 25°C (77°F), reacting in a steady-flow system at 1 atm.

13·34 Solve Problem 13·33 for a mixture of hydrogen and 100 percent excess air.

13·35 Solve Problem 13·33 for a mixture of hydrogen and 20 percent excess air.

13·36 Solve Problem 13·33 for a mixture of hydrogen and 80 percent of the stoichiometric amount of air.

13·37 Determine the adiabatic flame temperature of a stoichiometric mixture of methane and air, initially at 25°C (77°F), burning in a steady-flow system at 1 atm. Assume that the products contain no constituents other than CO, CO_2, O_2, N_2, and H_2O.

13·38 Determine the adiabatic flame temperature of a stoichiometric mixture of ethane and air, initially at 25°C (77°F), burning in a steady-flow system at 1 atm. Assume that the products contain no constituents other than CO, CO_2, O_2, N_2, and H_2O.

13·39 Determine the adiabatic flame temperature of a stoichiometric mixture of air and a fuel with a volumetric analysis of 60 percent methane and 40 percent ethane. Combustion occurs at 1 atm in a steady-flow system, and the reactants enter at 25°C (77°F). Assume that the products contain no constituents other than CO, CO_2, O_2, N_2, and H_2O.

13·40 Solve Problem 13·37 with consideration of additional products OH and H_2.

13·41 Solve Problem 13·38 with consideration of additional products OH and H_2.

13·42 Solve Problem 13·39 with consideration of additional products OH and H_2.

13·43 A natural gas is comprised of 83 percent methane and 17 percent ethane by volume. Determine the adiabatic flame temperature for constant pressure burning at 1 atm with the reactants and air entering at 25°C (77°F) and 30 percent excess air.

13·44 Determine the adiabatic flame temperature of a stoichiometric mixture of propane and air, initially at 25°C (77°F), burning in a steady-flow system at 1 atm.

13·45 A stoichiometric mixture of methane and air, initially at 25°C (77°F), is to be burned adiabatically in a steady-flow system at 1 atm. To limit the temperature to 2000 K (3600 R), liquid water is to be injected into the combustion chamber. Assuming that the products contain no constituents other than CO, CO_2, O_2, N_2, and H_2O, determine the mass of water to be added per unit mass of methane.

13·46 From the data of Fig. 13·1, determine Δh_R at 2000 K (3600 R) for the reaction $CO + \frac{1}{2}O_2 \rightarrow CO_2$.

13·47 From the data of Fig. 13·1, determine Δh_R at 2000 K (3600 R) for the reaction $CO_2 + H_2 \rightarrow CO + H_2O$.

13·48 From the data of Fig. 13·1, determine Δh_R at 2000 K (3600 R) for the reaction $H_2 + \frac{1}{2}O_2 \rightarrow H_2O$.

13·49 For the reaction $\frac{1}{2}H_2 + \frac{1}{2}I_2 \rightarrow HI$, the $K_P T$ relation in the vicinity of 1000 K is given by $\log K_P = 0.385 + 346.6/T$. From this relationship determine for this reaction at 1000 K (*a*) the enthalpy of reaction and (*b*) the Gibbs function of reaction.

13·50 For ideal gas compounds containing carbon, the values of K_P in Reference 13·7 are based on the constituent element C as a solid in the standard state. Therefore, $K_{P,CH_4} \equiv (P_{CH_4}/P^\circ)/(P_{H_2}/P^\circ)^2$ and $K_{P,CO} \equiv (P_{CO}/P^\circ)/(P_{O_2}/P^\circ)^{1/2}$; that is, the sum of the partial pressures of only the gaseous constituents equals the mixture pressure. Determine K_P at 1000 K for the reaction $CH_4 + H_2O \rightarrow 3H_2 + CO$ (*a*) using values at 1000 K of $\log K_{P,CH_4} = -1.011$ and $\log K_{P,CO} = 10.459$ and (*b*) using data from Tables A·4 and A·8.

13·51 For the reaction $O_2 \rightarrow 2O$ at 4000 K, $\Delta g_R^\circ = -26\ 100$ kJ/kmol O_2. At what pressure at 4000 K would one-third of the O_2 be dissociated?

CHAPTER FOURTEEN

Thermodynamic Aspects of Fluid Flow

In the study of compressible fluid flow, the principles of both fluid mechanics and thermodynamics must be applied, and there is no sharp demarcation between the areas covered by these two engineering sciences. This chapter is devoted to some of the thermodynamic aspects of the flow of fluids, and in order to treat these we must touch briefly upon some points that are usually considered to be in the realm of fluid mechanics. This chapter shows that the application to flowing fluids of the first law, the second law, the principles of mechanics, and physical property relationships leads to many interesting and useful results.

14·1 One-, Two-, and Three-dimensional Steady Flow

One-dimensional steady flow is flow in which the fluid properties (pressure, temperature, velocity, etc.) depend on only one space coordinate. In two-dimensional steady flow the properties depend on two space coordinates, and in three-dimensional steady flow they depend on three.

One-dimensional flow exists if the velocity, temperature, etc. are uniform across each cross section that is normal to the direction of flow. (See Fig. 14·1*a*.) Properties may change from one cross section to another. Also, the flow direction may change. Still, for each value of the coordinate defined as distance in the direction of flow there is a single value of velocity, a single value of temperature, etc.; so the flow is one dimensional. Most of the flows treated in earlier chapters of this book have been assumed to be one dimensional.

In the actual flow of a fluid in a pipe, shear forces of the pipe wall on the fluid cause a variation in velocity across any cross section, as shown in Fig. 14·1*b*. If the pipe is circular and the flow is axially symmetric, the flow is two dimensional because the velocity (and any other property) distribution can be completely described in terms of two space coordinates: distance along the pipe and radius from the pipe centerline. If the flow is unsymmetrical, it is three dimensional.

As a further illustration of the difference between two- and three-dimensional flow, consider the flow across a long airplane wing as shown in Fig. 14·2. Over wing sections

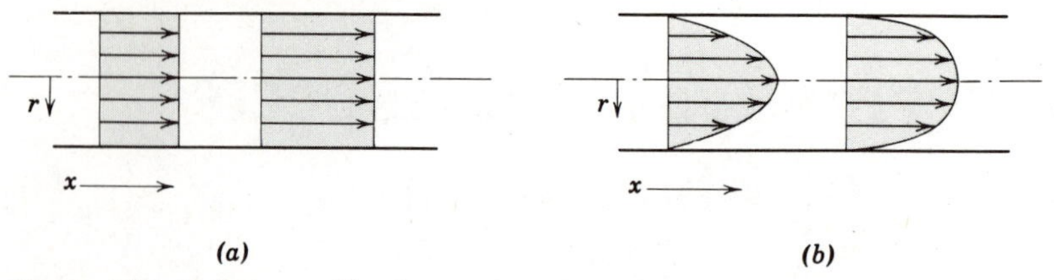

Figure 14·1. Velocity profiles for one- and two-dimensional flow in a circular pipe.

AA, *BB*, and *CC* in the uniform part of the wing, far from the root and the tip, the flow patterns are the same. They are independent of the distance l along the wing span; so this flow is two dimensional. Across a tapered section of the wing and near the root or tip the flow is three dimensional.

To simplify analyses we frequently approximate two-dimensional flow by one-dimensional flow in which the uniform velocity across a cross section is given by

$$V = \frac{\dot{m}v}{A}$$

which is often called the average velocity at the section. Notice that the kinetic energy of a one-dimensional stream is not the same as that of a two-dimensional stream with the same average velocity. (See Problem 14·1.) In many cases the difference can be neglected, but you should recognize that an approximation is being made because in cases where other energy changes are small compared to this difference, serious errors can result from the approximation.

14·2 Total Enthalpy, Total Temperature, and Total Pressure

In the application of the first law of thermodynamics to flowing fluids the group $h + V^2/2$ appears frequently. This group has been given the name of total enthalpy h_t, so that

$$h_t \equiv h + \frac{V^2}{2}$$

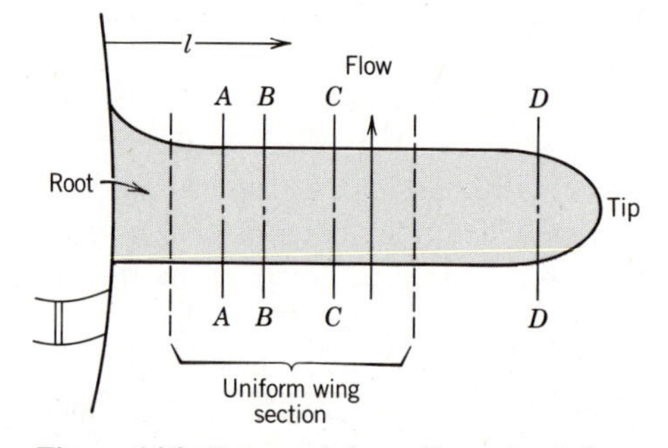

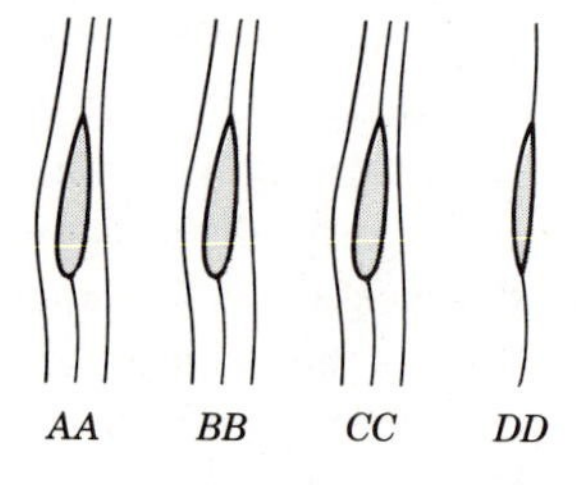

Figure 14·2. Two- and three-dimensional flow.

This definition involves no assumption or restriction regarding the type of fluid or the type of process. Total enthalpy is useful because it simplifies energy balances. For example, the steady-flow energy balance or first law applied to the one-dimensional adiabatic flow through a turbine,

$$w = h_1 - h_2 + \frac{V_1^2 - V_2^2}{2}$$

becomes with the total enthalpy

$$w = h_{t1} - h_{t2}$$

The first law for one-dimensional adiabatic steady flow through a nozzle,

$$h_1 + \frac{V_1^2}{2} = h_2 + \frac{V_2^2}{2}$$

becomes

$$h_{t1} = h_{t2}$$

The steady-flow energy balance applied to flow through a heat exchanger can be written in either of two forms:

$$q = h_2 - h_1 + \frac{V_2^2 - V_1^2}{2}$$

$$q = h_{t2} - h_{t1}$$

Property diagrams of h_t versus s are often more valuable than hs diagrams. For example, in the case of an adiabatic turbine, ordinate distances on an hs diagram represent work only if there is no change in kinetic energy; but on an h_ts diagram ordinate distances represent work even if the kinetic energy change is appreciable.

Total enthalpy is also called *stagnation enthalpy* because the total enthalpy of a flowing fluid is equal to the enthalpy of the fluid if it were brought to rest adiabatically. The process of bringing the fluid to rest need not be reversible. A *stagnation point* is defined as a point in a fluid-flow field where the velocity is zero. Writing an energy balance between some point 1 in a fluid stream and a stagnation point along the same streamline gives

$$h_1 + \frac{V_1^2}{2} = h_{\text{stagnation point}} = h_t$$

which is the same for any adiabatic no-work process, reversible or irreversible.

The *total temperature* or *stagnation temperature* of a flowing fluid is defined as the temperature that would result if the fluid were brought to rest *reversibly* and adiabatically or, in other words, isentropically. (For an ideal gas, total temperature could be defined as the temperature corresponding to the total enthalpy, but such a definition cannot be used for a substance for which temperature and enthalpy are not uniquely related.)

The pressure which results when a flowing fluid is brought to rest reversibly and adiabatically (isentropically) is called the *total pressure* or the *stagnation pressure*. (In some publications there is a slight difference in definition between *total pressure* and *stagnation pressure*—and between *total temperature* and *stagnation temperature*—but here the two terms are synonymous. Also, in some publications what we call *stagnation pressure* is called *isentropic stagnation pressure*.)

In principle, the total pressure of a flowing fluid can be measured by means of a pressure probe that opens directly upstream so that the flowing fluid is brought to rest isentropically at the opening. Static pressure can be measured by means of a pressure probe that moves with the fluid, or where the fluid flows along a straight wall the static pressure can be measured by means of a small opening in the wall. It is usually assumed that static pressure is constant along a line normal to such a wall.

In principle, total temperature can be measured by a probe placed so that the temperature-sensitive element is in contact with only the fluid that has been brought to rest isentropically. Static temperature is measured by a thermometer that moves with the fluid. (In both cases the usual precautions must be taken in regard to radiation and other sources of error.) A thermometer element placed in a wall along which a fluid flows does not measure the static temperature.

The temperature of the fluid at the wall and its relationship to the fluid temperature far from the wall require some explanation. When a fluid flows along a wall, shear forces reduce the fluid velocity to zero right at the wall, establishing a velocity gradient between the wall and the region where the flow is unaffected by the shear forces. The pressure is usually considered to be constant along a line normal to a straight wall, but the deceleration of the fluid layers near the wall causes an increase in enthalpy of those layers, resulting in a temperature gradient that causes heat transfer from one layer to the next. For a given velocity distribution, the temperature distribution depends on the interrelationship of the momentum-transfer and heat-transfer characteristics of the flow. The fluid temperature at the wall is called the *adiabatic wall temperature* if there is no heat transfer between the fluid and the wall. (This condition requires that the temperature gradient at the wall be zero.) For many gases, the relationship of properties (including c_p, dynamic viscosity, and thermal conductivity) is such that for adiabatic flow the total enthalpy is nearly constant from layer to layer across the stream. Let us call the adiabatic wall temperature that is calculated from the assumption of constant total enthalpy across the stream the *approximate adiabatic wall temperature*. This temperature corresponds to the total enthalpy and the static pressure of the main stream.

Figure 14·3 shows the difference between the approximate adiabatic wall temperature and the total or stagnation temperature. Part *a* of Fig. 14·3 is for any gas and part *b* is for the special case of an ideal gas. In both cases the state of a flowing fluid is represented by point *f*. The corresponding stagnation state is represented by point *t*. Process *f-t* is an isentropic deceleration to zero velocity. The approximate adiabatic wall state *w* is also the result of an adiabatic deceleration from V_f to zero velocity; so the enthalpy increases $h_w - h_f$ and $h_t - h_f$ are equal; therefore, $h_w = h_t$. however, process *f-w* is *irreversible* so that states *w* and *t* are not the same, the pressure at state *w* being equal to that at state *f*. Therefore, $T_w \neq T_t$ as a general case. For the case of an ideal gas, however, states of equal enthalpy are states of equal temperature (regardless of the pressure); so the approximate adiabatic wall temperature equals the stagnation temperature as shown in Fig. 14·3*b*.

Example 14·1. Determine the total pressure and the total temperature of air at 100 kPa, 20°C, flowing with a velocity of 300 m/s.

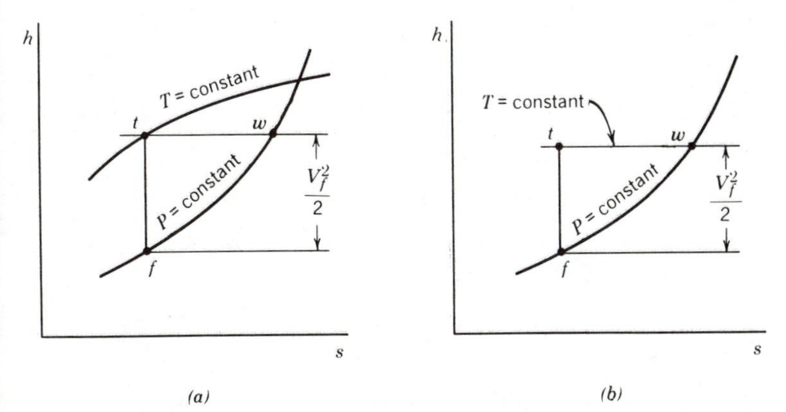

Figure 14·3. Total or stagnation temperatures and approximate adiabatic wall temperature. (*a*) Most gases. (*b*) Ideal gases.

Solution. The total pressure and total temperature are the pressure and temperature that correspond to the total enthalpy and the same entropy as the flowing fluid. From the definition of total enthalpy,

$$h_t - h = \frac{V^2}{2}$$

For an ideal gas enthalpy is a function of temperature only; so we can determine T_t without reference to the value of entropy or any other property of the gas besides h_t. Applying Joule's law (i.e., h is a function of T only for an ideal gas) and assuming that c_p is constant in the temperature range involved,

$$c_p(T_t - T) = \frac{V^2}{2}$$

$$T_t = T + \frac{V^2}{2c_p} = 20.0 + \frac{(300)^2}{2(1.005)1000} = 64.8°\text{C}$$

(The temperature range is thus small enough to justify our assumption of a constant c_p.) For a reversible adiabatic process of an ideal gas with constant specific heats we know the relationship between P and T; so we have for the total pressure

$$P_t = P\left(\frac{T_t}{T}\right)^{k/(k-1)} = 100\left(\frac{337.8}{293}\right)^{1.4/0.4} = 165 \text{ kPa}$$

Example 14·2. Determine the total pressure, the total temperature, and the approximate adiabatic wall temperature of steam at 60 psia, 300°F, flowing at 1070 fps.

Solution. The total pressure and total temperature are the pressure and temperature of the state defined by

$$h_t = h + \frac{V^2}{2} = 1181.9 + \frac{(1070)^2}{2(32.2)778} = 1204.8 \text{ B/lbm}$$

$$s_t = s = 1.6496 \text{ B/lbm·R}$$

Interpolation on a Mollier chart or in the steam tables or the use of a steam properties computer program gives for this stagnation state

$$P_t = 79 \text{ psia} \qquad \text{and} \qquad T_t = 350°\text{F}$$

The approximate adiabatic wall temperature T_w is the temperature of the state defined by

$$h_w = h_t = 1204.8 \text{ B/lbm}$$
$$P_w = P = 60 \text{ psia}$$

From the chart, tables, or program we obtain

$$T_w = 343^{+}°\text{F}$$

14·3 The Dynamic Equation for Steady One-Dimensional Flow

Effective study of the thermodynamics of fluid flow requires attention to the mechanics or dynamics of flow. We now derive the basic dynamic equation for steady one-dimensional flow by starting with Newton's second law of motion,

$$F = \frac{d}{d\tau}(mV) \tag{1·1b}$$

where F is the resultant of all forces acting on a body which at any instant has a mass m and a velocity V. Equation 1·1 is a *vector* equation: F must be in the same direction as the change in momentum.* Applying it to any body of fixed mass m,

$$F = V\frac{dm}{d\tau} + m\frac{dV}{d\tau} = 0 + m\frac{dV}{d\tau}$$

If dL is the distance traveled by the body in time $d\tau$, then $dV/d\tau = (dL/d\tau)(dV/dL) = V(dV/dL)$, and we have

$$F = mV\left(\frac{dV}{dL}\right)$$

Thus the resultant force on a mass m at any instant is proportional to the rate of change of velocity with distance. Let us consider now a fluid stream, directing our attention to an element of mass δm which has a length dL in the direction of flow. (See Fig. 14·4.)

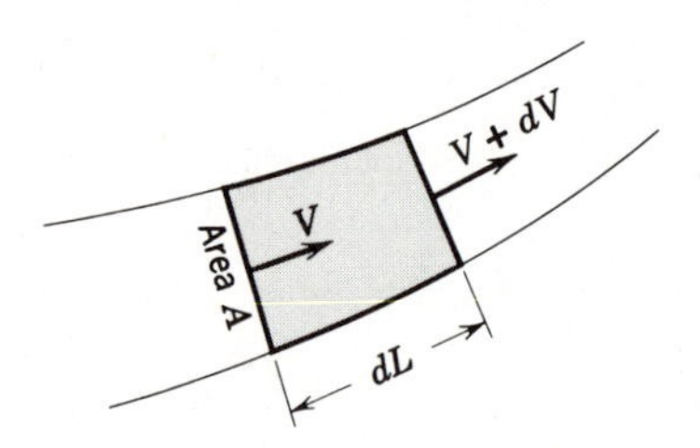

Figure 14·4. An element of a fluid stream.

*Vector quantities are often designated by boldface type or overscores, **F** and **V** or $\bar{F}$ and $\bar{V}$. For our limited use of vectors this is unnecessary.

The mass and the length of the element are related by $\delta m = \rho A\, dL$, where A is the cross-sectional area of the element normal to the flow direction. An infinitesimal force dF causes this infinitesimal element to change velocity by dV as it moves a distance dL,

$$dF = (\delta m)V\left(\frac{dV}{dL}\right) = (\rho A\, dL)V\frac{dV}{dL} = \rho AV\, dV = \dot{m}\, dV$$

For steady flow, $\dot{m} = \rho AV =$ constant; so we can integrate to get the resultant force on a stream which changes velocity by ΔV:

$$F = \dot{m}\,\Delta V \tag{14·1}$$

This is the basic dynamic equation for steady one-dimensional flow. The restriction to one-dimensional flow means that the velocity is uniform across any cross section of the flow. Remember that Eq. 14·1 is a vector equation: the resultant force is in the same direction as ΔV, as shown in Fig. 14·5. The equation holds also for corresponding components of F and ΔV.

A convenient way of writing Eq. 14·1 for the resultant force on a fluid flowing steadily from a section 1 to a section 2 is

$$F = \dot{m}V_2 - \dot{m}V_1$$

The vector quantity $\dot{m}V$ is called the *momentum flux* at a section. For a steady-flow system with more than one stream entering or leaving, the resultant force is the vector difference between the sum of the momentum fluxes of streams leaving and the sum of the momentum fluxes of streams entering.

In applying the dynamic equation (Eq. 14·1) it is advisable to sketch the steady-flow system under consideration, showing all of the forces *on the fluid.** The sum of these forces is then equal to $\dot{m}\,\Delta V$.

Example 14·3. Air at 200 kPa, 50°C, enters a duct with a velocity of 150 m/s. The duct entrance area is 0.200 m². The air leaves the duct at 150 kPa with a velocity of 250 m/s through a

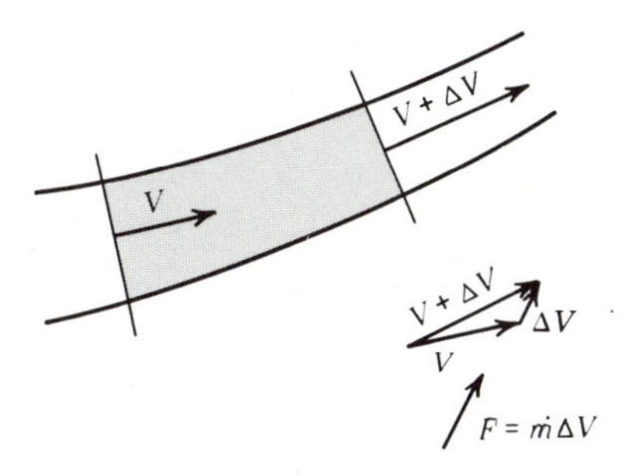

Figure 14·5. The basic dynamic equation for steady one-dimensional flow F is the resultant force and is in the same direction as ΔV.

*Notice that such a diagram, although often called a "free-body diagram," is not the same as the free-body diagram made for a body of fixed mass where the sum of the forces acting on the body equals the time rate of change of momentum of the body. The time rate of change of momentum of a steady-flow system itself is always zero in accordance with the definition of steady flow.

cross-sectional area of 0.157 m² and in a direction 30 degrees different from the entrance direction. Determine the resultant force of the air on the duct.

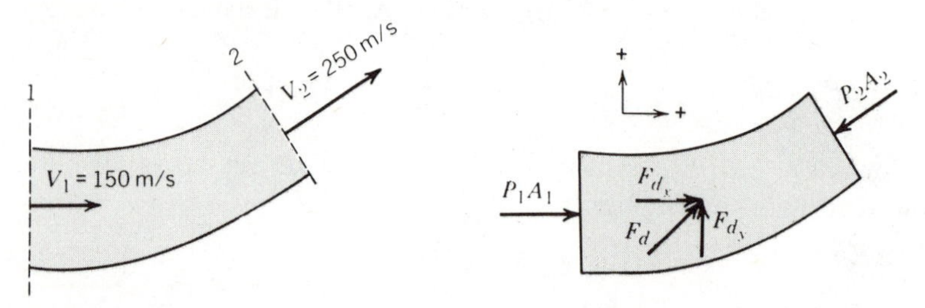

Example 14·3.

Solution. First a sketch of the open system and a diagram showing the forces on the fluid within the duct are made. The only external forces on the fluid are the force of the adjacent fluid at inlet P_1A_1 and outlet P_2A_2 and the force of the duct on the fluid F_d. We assign the positive directions of velocity and force as upward and to the right in the sketch. We do not know at first the direction of F_d; so we show it on the diagram in the positive directions. Then, if we observe the sign convention we have established, the signs of the components F_{d_x} and F_{d_y} that we solve for will indicate the actual direction of the force *on the fluid*.

Application of the dynamic equation (Eq. 14·1) requires a knowledge of the mass rate of flow; so we use the ideal-gas equation of state to determine the inlet air density,

$$\rho_1 = \frac{P_1}{RT_1} = \frac{200}{0.287(323)} = 2.16 \text{ kg/m}^3$$

and then

$$\dot{m} = \rho_1 A_1 V_1 = 2.16(0.200)150 = 64.8 \text{ kg/s}$$

Equating the sum of the x components of the forces to the x component of $\dot{m}\Delta V$ we have

$$P_1A_1 + F_{d_x} - P_2A_2 \cos 30° = \dot{m}(V_{2x} - V_{1x})$$

$$\begin{aligned} F_{d_x} &= \dot{m}(V_2 \cos 30° - V_1) - P_1A_1 + P_2A_2 \cos 30° \\ &= 64.8(250 \cos 30° - 150) - 200(1000)\,0.200 \\ &\quad + 150(1000)0.157 \cos 30° \\ &= -15\,300 \text{ N} \end{aligned}$$

Thus the x component of the force of the duct *on the fluid* is 15 300 N to the left.

Applying the dynamic equation now in the y direction,

$$F_{d_y} - P_2A_2 \sin 30° = \dot{m}(V_{2y} - V_{1y})$$

$$\begin{aligned} F_{d_y} &= \dot{m}(V_2 \sin 30° - 0) + P_2A_2 \sin 30° \\ &= 64.8(250 \sin 30°) + 150(1000)0.157 \sin 30° \\ &= 19\,900 \text{ N} \end{aligned}$$

Adding F_{d_x} and F_{d_y} vectorially gives the resultant force of the duct *on the fluid* as 25 100 N in a direction (referring to directions in the sketch) upward and to the left, 37.6 degrees from the vertical. The resultant force of the fluid *on the duct* is of course opposite in direction.

Example 14·4 Air with a density of 0.0020 slug/cu ft enters a turbojet engine with a velocity of 500 fps relative to the engine. Fuel is burned at one thirtieth of the mass rate of air flow.

Products of combustion leaving the engine have a density of 0.000 80 slug/cu ft. Inlet and outlet have equal cross-sectional areas of 2.0 sq ft, and the pressure is atmospheric at both inlet and outlet. Determine the thrust developed by the engine.

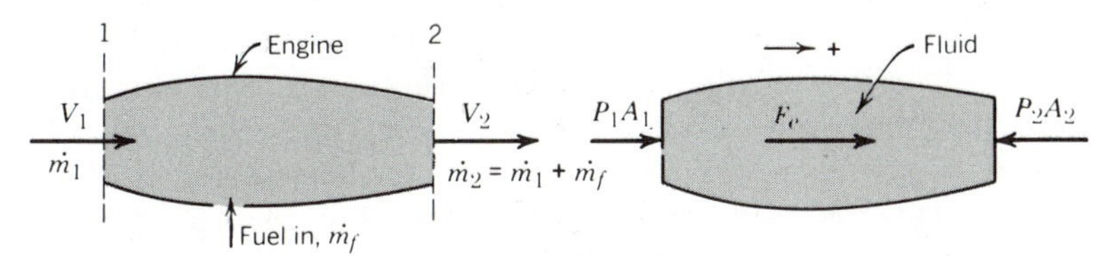

Example 14·4.

Solution. A diagram of the engine is made first. It is unnecessary to show any internal features of the engine in the sketch because application of the dynamic equation involves only the *resultant* force on the fluid and conditions at the boundary of the open system. Thrust is the net force of the fluid on the engine; so we find the equal-in-magnitude but opposite-in-direction force of the engine on the fluid by applying the dynamic equation in the axial direction, with the positive direction as shown above the sketch of the fluid.

$$F = P_1A_1 - P_2A_2 + F_e = (\dot{m}_2V_2 - \dot{m}_1V_1 - \dot{m}_fV_f)$$

where F is the resultant force on the fluid in the engine and F_e is the net force of the engine on the fluid. Since $P_1A_1 = P_2A_2$, the resultant force happens to be equal to F_e. Since the fuel is carried with the craft, its initial velocity has a zero component in the axial direction; so the last term on the right-hand side of the equation above is zero. Also, $\dot{m}_2 = (\dot{m}_1 + \dot{m}_f) = (\dot{m}_1 + \dot{m}_1/30) = 31\dot{m}_1/30$. Thus the dynamic equation becomes

$$F = F_e = (\tfrac{31}{30}\dot{m}_1V_2 - \dot{m}_1V_1) = \dot{m}_1(\tfrac{31}{30}V_2 - V_1)$$

We calculate $\dot{m}_1$ and V_2 for use in this equation by

$$\dot{m}_1 = \rho_1A_1V_1 = 0.0020(2)500 = 2 \text{ slug/s}$$

$$V_2 = \frac{\dot{m}_2}{\rho_2A_2} = \frac{\frac{31}{30}\dot{m}_1}{\rho_2A_2} = \left(\frac{31}{30}\right)\frac{2}{0.000\,80(2)} = 1291 \text{ fps}$$

Substitution of these values into the dynamic equation gives

$$F_e = \dot{m}_1(\tfrac{31}{30}V_2 - V_1) = 2[\tfrac{31}{30}(1291) - 500] = 1670 \text{ lbf}$$

This is the force of the engine on the fluid. It is positive; so, in accordance with the sign convention established at the beginning of this solution, this force is to the right for the diagrams as drawn. The force of the fluid on the engine, the thrust, is therefore to the left and has a magnitude of 1670 lbf.

14·4 Sonic Velocity; Mach Number

An important property in the study of gas flow is the velocity of sound through the gas. This is called also the *sonic velocity* or the *acoustic velocity*. We therefore derive now an expression for the sonic velocity in a gas.

The plan of the derivation is to consider a pressure wave traveling through a gas and, from the viewpoint of an observer moving with the wave, to apply the continuity equation and the dynamic equation to determine the relative velocity between the wave and the fluid ahead of it. We do this first for a pressure wave of any size (that is, the pressure behind the wave many differ appreciably from that ahead of it) and then apply the restriction that the pressure difference across the wave is very small, as it is for a sound wave.

For a physical picture of the generation and propagation of a pressure wave in a gas, refer to Fig. 14·6. A gas fills a long tube of constant cross-sectional area. A piston in one end of the tube is moved as shown with a velocity u. The gas in a region near the piston face also moves with the velocity u. The boundary of this region in which the gas velocity is u is called the wave. Behind (that is, to the left of) the wave the pressure is greater than it is ahead of it. Ahead of the wave the gas is unaffected by the motion of the piston or the approaching wave. The wave moves with a velocity a that is equal to or greater than u.

Our problem is to determine the wave velocity a in terms of properties of the fluid ahead of the wave. To simplify the derivation, however, let us use a frame of reference in which the wave is fixed, or, in other words, let us adopt the viewpoint of an observer moving with the wave. The resulting picture is shown in Fig. 14·6*b*, which is the same as Fig. 14·6*a* with a uniform velocity a to the left imposed on the entire system. Thus the stationary wave is approached from the right by fluid with a velocity $V_1 = a$, and fluid leaves the wave to the left with a velocity $V_2 = a - u$. The flow is steady. Since the cross-sectional area is constant, the continuity equation $\rho_1 A_1 V_1 = \rho_2 A_2 V_2$ becomes

$$\rho_1 V_1 = \rho_2 V_2$$

If we assume the thickness of the wave is so small that shearing forces at the wall are negligible, the dynamic equation becomes

$$P_1 A_1 - P_2 A_2 = \dot{m} \Delta V = \rho_1 A_1 V_1 (V_2 - V_1)$$

Substituting V_2 from the continuity equation into this equation and solving for V_1 (which

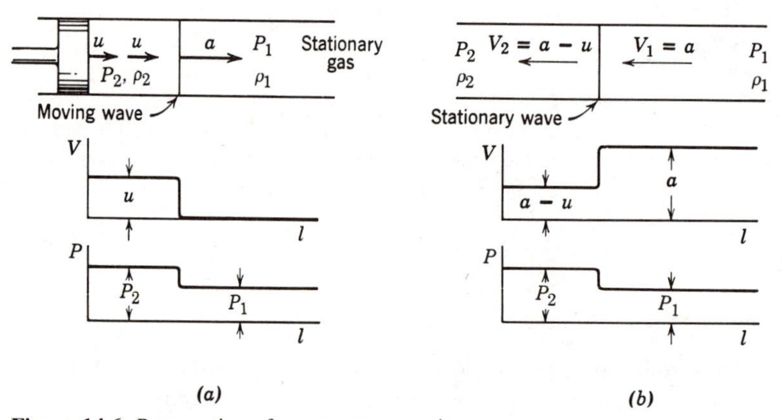

Figure 14·6. Propagation of a pressure wave in a gas.

equals a) gives

$$a = V_1 = \sqrt{\frac{\rho_2}{\rho_1}\left(\frac{P_1 - P_2}{\rho_1 - \rho_2}\right)}$$

This is the velocity of the gas approaching the wave in Fig. 14·6*b* or the velocity of the wave in Fig. 14·6*a*. Such a wave is called a *shock wave,* and its velocity is higher for higher values of P_2/P_1. The sonic velocity or velocity of sound is the velocity of a pressure wave of very small amplitude. As P_2 approaches P_1 and ρ_2 approaches ρ_1, the quantity beneath the radical sign in the equation above becomes $dP/d\rho$. Of course, the value of $dP/d\rho$ depends on the process because P and ρ can be varied independently. However, if we assume that the propagation of a sound wave is reversible and adiabatic, and hence isentropic, we restrict $dP/d\rho$ to $(\partial P/\partial \rho)_s$ which has a single value at any state. Thus the sonic velocity, which is represented by the symbol c, is

$$c = \sqrt{\left(\frac{\partial P}{\partial \rho}\right)_s} \tag{14·2}$$

This expression has been derived for a plane wave, but it holds also for cylindrical and spherical waves. c can be readily expressed in terms of the isentropic compressibility (defined in Sec. 9·5) or the bulk modulus of the fluid. For ideal gases with constant specific heats, the $P\rho$ relationship for isentropic processes is P/ρ^k = constant, so that $(\partial P/\partial \rho)_s = kP/\rho$, and the *velocity of sound in an ideal gas* is

$$c = \sqrt{\frac{kP}{\rho}} = \sqrt{kPv} = \sqrt{kRT}$$

In an ideal gas the sonic velocity, which is a property of the gas, is a function of temperature only. In the derivation above we have assumed that the propagation of a sound wave is isentropic, but we have placed no restriction on the type of process, if any, that the bulk of the gas may be undergoing.

Although the equation above for the sonic velocity in an ideal gas is simple, the consistency of units demands special attention whether one is working in SI or English units.

In the study of gas flow a useful parameter is the ratio of the velocity of the gas at any point to the sonic velocity at the same point. This ratio is called the Mach* number and is designated by the symbol M:

$$M \equiv \frac{V}{c}$$

Flow with $M < 1$ is called *subsonic*, and flow with $M > 1$ is called *supersonic*. The term *hypersonic* is used in connection with Mach numbers much greater than one, and the term *transonic* refers to flows in which the Mach number is close to 1. Just how close to 1

*Ernst Mach (1838–1916) was an Austrian physicist and psychologist who is probably best known for his contributions to the philosophy of science.

the Mach number must be in order for a flow to be called transonic is not specified, and also the boundary between supersonic and hypersonic is not sharply defined; but notice that the value of $M = 1$ denotes clearly the boundary between subsonic and supersonic flow.

14·5 The Basic Relations

For the remainder of this chapter we will be concerned with the steady one-dimensional flow of fluids in which no work is done. In the analysis of such flow we have available five powerful tools:

1. The first law, which under the conditions stated can be expressed as

$$q = \Delta h + \Delta ke + \Delta pe$$

   In connection with gases and vapors, Δpe is nearly always negligible.
2. The continuity equation

$$\dot{m} = \rho_1 A_1 V_1 = \rho_2 A_2 V_2$$

3. The dynamic equation

$$F = \dot{m}\Delta V$$

4. The second law and its corollaries that indicate whether a conceivable process is possible.
5. Physical property relationships, either in the form of tabular or graphical data or in the form of equations of state and other property equations resulting from the first and second laws.

In analyzing fluid flow in which no work is done, these are the basic tools that you should use. Often you will find it convenient or even necessary to combine and restrict some of the basic relations to form equations that apply only to special cases. Always remember the restrictions on these special case equations and work as much as possible from the basic relations that are more general. In engineering practice, to be sure, it is the special cases that demand solution; but new special cases can be solved only by the application of the basic principles, and the great value of the basic principles is precisely that they apply not to just some special cases but to all special cases, including ones that are as yet unthought of.

14·6 Area Variation for the Isentropic Flow of Any Fluid

We now develop an important conclusion regarding the variation of the cross-sectional flow area with respect to velocity change and pressure change for the steady, one-dimensional isentropic flow of any fluid.

For adiabatic flow with no work done and no change in potential energy, the first law can be expressed as

$$\Delta h + \Delta\left(\frac{V^2}{2}\right) = 0$$

or in differential form

$$dh + V\,dV = 0 \tag{a}$$

For an isentropic process, Eq. 8·3 can be written

$$T\,ds = 0 = dh - \frac{dP}{\rho} \tag{b}$$

Combining these two equations gives one that can also be developed from the dynamic equation. It is

$$dP = -\rho V\,dV \tag{c}$$

Differentiating the continuity equation, ρAV = constant, gives

$$\frac{d\rho}{\rho} + \frac{dA}{A} + \frac{dV}{V} = 0 \tag{d}$$

Substituting from Eq. c into Eq. d gives

$$\frac{dA}{A} = \frac{dP}{\rho}\left[\frac{1}{V^2} - \frac{d\rho}{dP}\right] \tag{e}$$

Since the process is isentropic, $dP/d\rho = (\partial P/\partial\rho)_s$. Thus the second term within the brackets in Eq. e is $1/c^2$ so that we have

$$\frac{dA}{A} = \frac{dP}{\rho V^2}[1 - M^2] \tag{f}$$

Substituting from Eq. c into Eq. f gives

$$\frac{dA}{A} = -\frac{dV}{V}[1 - M^2] \tag{g}$$

Inspection of Eqs. f and g leads to the following conclusions:

1. When $M < 1$, $dA/dP > 0$ and $dA/dV < 0$
2. When $M > 1$, $dA/dP < 0$ and $dA/dV > 0$
3. When $M = 1$, $dA/dP = 0$ and $dA/dV = 0$

These results are shown graphically in Fig. 14·7. A flow channel we call a nozzle accelerates a fluid and a channel called a diffuser is used to decelerate a fluid. The conclusions reached above and illustrated in Fig. 14·7 mean that in order to accelerate a fluid that is flowing subsonically, a converging nozzle must be used. Once the speed of sound is reached by the fluid, however, further acceleration can occur only in a diverging section. Conversely, a diverging section is used to decelerate a fluid that is flowing subsonically, but if the fluid were initially flowing supersonically, the diverging diffuser would have to be preceded by a converging diffuser for decelerating the fluid to the sonic velocity. This behavior of a supersonically flowing fluid seems strange at first because all our everyday experience is with subsonic flow. For a fluid flowing isentropically, the density and the velocity vary oppositely; and we see from the above and from the continuity

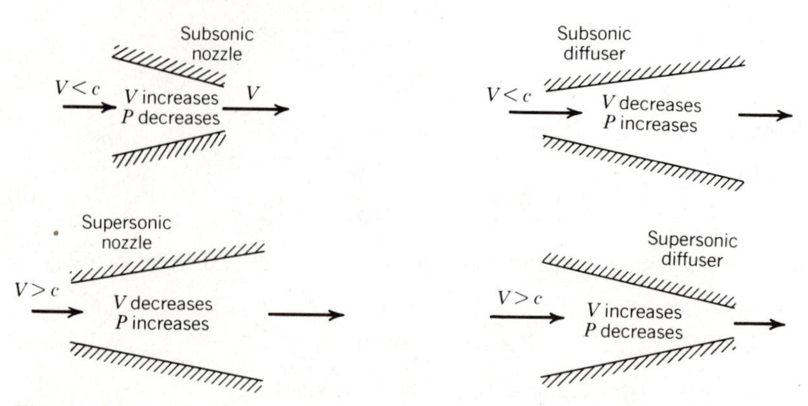

Figure 14·7. Area–velocity–pressure variations for isentropic flow.

equation ρAV = constant that for supersonic flow the density must vary more rapidly than velocity does with respect to area. This point warrants some reflection.

Example 14·5. Helium flows isentropically through a channel of varying cross-sectional area. At one section where the area is 0.175 sq ft, the helium is at 6.0 psia, 90°F, and has a velocity of 5000 fps. Determine the velocity, Mach number, and cross-sectional area at a section where the pressure is 12.0 psia.

Analysis. Designate as section 1 that at which the pressure is 6.0 psia, and designate as section 2 that at which the pressure is 12.0 psia. Since the flow is isentropic, we can immediately calculate T_2 from $T_2/T_1 = (P_2/P_1)^{(k-1)/k}$ if we assume that the specific heats are constant. Then V_2 can be determined from an energy balance and the fact that $\Delta h = c_p \Delta T$. We determine the sonic velocity at section 2 by $c_2 = \sqrt{kRT_2}$ so that M_2 can be calculated. A_2 can be calculated from the continuity equation.

Solution. Following the procedure outlined in the analysis

$$T_2 = T_1\left(\frac{P_2}{P_1}\right)^{(k-1)/k} = 550\left(\frac{12.0}{6.0}\right)^{(1.67-1)/1.67} = 726 \text{ R}$$

$$\frac{V_2^2 - V_1^2}{2} = h_1 - h_2 = c_p(T_1 - T_2)$$

$$V_2 = \sqrt{V_1^2 + 2c_p(T_1 - T_2)} = \sqrt{(5000)^2 + 2(32.2)778(1.25)(550 - 726)}$$

$$= 3740 \text{ fps}$$

$$c_2 = \sqrt{kRT_2} = \sqrt{1.67(32.2)\left[\frac{\text{lbm}\cdot\text{ft}}{\text{lbf}\cdot\text{s}^2}\right]386\left[\frac{\text{ft}\cdot\text{lbf}}{\text{lbm}\cdot\text{R}}\right]726[\text{R}]} = 3880 \text{ fps}$$

$$M_2 = \frac{V_2}{c_2} = \frac{3740}{3880} = 0.964$$

$$A_2 = A_1\frac{V_1}{V_2}\frac{\rho_1}{\rho_2} = A_1\frac{V_1}{V_2}\frac{P_1RT_2}{RT_1P_2} = 0.175\,\frac{5000(6.0)726}{3740(550)12.0} = 0.154 \text{ sq ft}$$

(Since $M_1 > 1$ and $M_2 < 1$, there must be a throat or minimum area section between sections 1 and 2.)

14·7 Adiabatic Flow of Ideal Gases with Constant Specific Heats

For ideal gases, the simple equation of state and the resulting simple hT relation make the application of the basic principles to adiabatic flow quite simple. *We will consider in this article only steady-flow systems in which there is no work done and the change in potential energy is negligible.*

For the adiabatic flow of an ideal gas with no work done, the total enthalpy and consequently the total temperature also are constant. If we add the restriction of constant specific heats, we can derive a simple and useful expression relating the temperature at any section where $M = 1$ to the total temperature of the gas. Let us denote the properties at a section where $M = 1$ by symbols such as h^* and T^*. Then from the definition of total enthalpy, recalling that the total enthalpy is constant in adiabatic flow, we have

$$h_t = h_t^* = h^* + \frac{V^{*2}}{2}$$

Rearranging, substituting for V^* its equivalent $\sqrt{kRT^*}$, and applying Joule's law, we have

$$c_p(T_t - T^*) = \frac{kRT^*}{2}$$

Use of the relationship $c_p = kR/(k - 1)$ and rearrangement gives

$$\frac{T^*}{T_t} = \frac{2}{k + 1} \tag{14·3}$$

This equation holds for the steady adiabatic (reversible or irreversible) flow of an ideal gas with constant specific heats and with no work done and no change in potential energy.

One important conclusion from Eq. 14·3 is reached by first noting that at any section of the flow path where the velocity is negligibly small, the temperature is the total temperature. Therefore, accelerating an ideal gas with constant specific heats from rest to a sonic velocity by adiabatic expansion requires that the expansion proceed until the temperature has decreased in the ratio of $2/(k + 1)$. This holds whether the expansion is reversible or irreversible. Since sonic velocity is proportional to the square root of the temperature and since T^* is less than T_t, the sonic velocity at T^* is less than that in the gas at rest.

If the flow is reversible as well as adiabatic, and hence isentropic, the total pressure is also constant along the flow, and

$$\frac{P^*}{P_t} = \left(\frac{T^*}{T_t}\right)^{k/(k-1)} = \left(\frac{2}{k + 1}\right)^{k/(k-1)} \tag{h}$$

Section 14·6 shows that if a fluid enters a converging passage subsonically, it cannot reach a Mach number greater than 1 in the converging section. Therefore, for isentropic flow the pressure in a converging passage cannot be lower than a certain fraction of the total pressure as indicated by Eq. h. For gases with $k = 1.4$, this fraction is 0.53. Let us investigate this point further.

Consider a converging nozzle as shown in Fig. 14·8. The upstream pressure P_1 and the temperature T_1, at a section where the velocity is negligibly small, are held constant. The pressure P_b (*b*ackpressure) in the downstream region can be varied. Designate the pressure in the minimum area section or throat as P_{th}. If $P_b = P_1$, there is no flow through the nozzle. If P_b is lowered slightly, flow occurs with $P_{th} = P_b$ and $V_{th} < c_{th}$. If P_b is lowered further, the pressure at the throat decreases to remain equal to P_b, and the mass rate of flow through the nozzle increases as shown in Fig. 14·8*b* until P_b and P_{th} reach the value of P^*. The velocity at the throat is then sonic, $V_{th} = c_{th} = V^*$. Under these conditions the nozzle is said to be *choked* or to have reached *limiting flow* or *critical flow. Further reduction of* P_b *then has no effect on* P_{th}, *the velocity at the throat, or the mass rate of flow through the nozzle.* In a converging passage with isentropic flow that is initially subsonic, the Mach number cannot exceed 1. Also, the pressure cannot be less than P^* as given by Eq. h.

Whenever $P_b < P^*$, the pressure at the throat remains equal to P^*, and the drop in pressure from P^* to P_b occurs outside the nozzle just beyond the throat. This sudden expansion outside the nozzle is not one dimensional, and it is irreversible; it is only the flow within the nozzle that can be isentropic. Thus we can write $P_1 v_1^k = P_{th} v_{th}^k$, but $P_1 v_1^k \neq P_b v_b^k$.

Converging nozzles with choked flow are sometimes used as flow-measuring devices, because as long as the backpressure applied is known to be sufficiently low that $P_b/P_1 < P^*/P_1$, the flow rate through a nozzle of known throat area can be determined from measurements of only P_1 and T_1. Notice that the minimum pressure at the throat is determined by only the inlet pressure and the k value of the gas, not by the shape of the nozzle. The actual flow through a well-designed smooth converging nozzle can be quite close to isentropic, so that the relations derived for isentropic flow apply accurately.

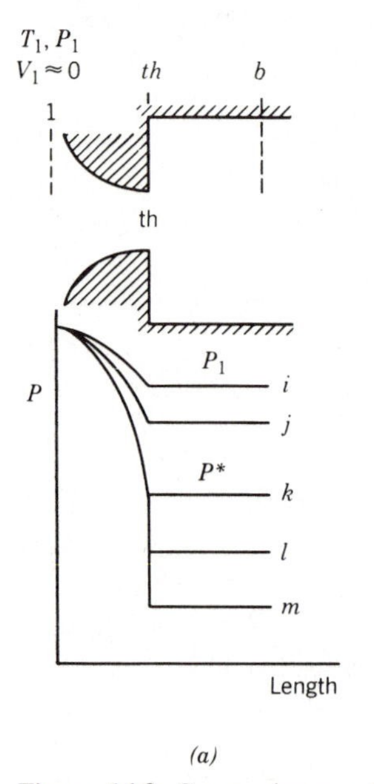

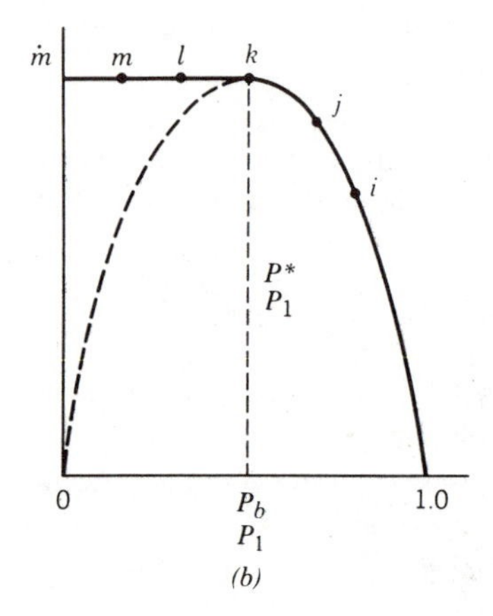

Figure 14·8. Converging nozzle flow ($M_1 < 1$).

Converging nozzles are used in turbines and other pieces of equipment to produce jets of subsonic or sonic velocity.

Example 14·6. Air at 500 kPa, 57°C, enters a converging nozzle with negligible velocity. The nozzle discharges into a receiver where the pressure is 120 kPa. Assuming isentropic flow, calculate the velocity at the nozzle exit.

Solution. We have seen that a fluid that enters subsonically cannot reach a Mach number greater than 1 in a converging section if the flow is isentropic. Therefore, the lowest temperature that can be reached by the air flowing through this converging nozzle is that corresponding to $M = 1$. This is of course the value we designate T^* and is given by

$$T^* = T_t\left(\frac{2}{k+1}\right) = 330\left(\frac{2}{2.4}\right) = 275 \text{ K} \qquad (14.2)$$

in which we have made use of the fact that $T_t = T_1$ since the inlet velocity is negligibly small. The lowest pressure that can exist in the converging nozzle is then

$$P^* = P_t\left(\frac{T^*}{T_t}\right)^{k/(k-1)} = 500\left(\frac{275}{330}\right)^{3.5} = 264 \text{ kPa}$$

Thus the receiver pressure is lower than P^*; so the pressure at the exit or throat of the converging nozzle is P^*, not the receiver pressure, and the velocity there is sonic:

$$V_2 = c_2 = \sqrt{kRT_2} = \sqrt{kRT^*} = \sqrt{1.4(1)\left[\frac{\text{kg}\cdot\text{m}}{\text{N}\cdot\text{s}^2}\right]1000\left[\frac{\text{N}}{\text{kN}}\right]0.287\left[\frac{\text{kJ}}{\text{kg}\cdot\text{K}}\right]275[\text{K}]}$$

$$= 332 \text{ m/s}$$

(The same value of V_2 can be obtained by writing an energy balance between the inlet and the exit or throat.)

To accelerate a fluid from a subsonic to a supersonic velocity, a converging–diverging nozzle, as shown in Fig. 14·9, is used. We will now examine the various flow regimes in such a nozzle. Which regime or kind of flow occurs is determined by the ratio P_b/P_1 applied; that is, by the backpressure applied for given inlet conditions.

Consider a converging–diverging nozzle as shown in Fig. 14·9 with P_1 and T_1 fixed and with V_1 very low. When the backpressure P_b is equal to P_1, as indicated by point c on the pressure–length plot, there is no flow through the nozzle. As the backpressure is lowered, several different regimes of flow occur.

$P_c > P_b > P_f$: When the backpressure is just slightly below P_1 (point d), there is some flow through the nozzle. The maximum velocity and minimum pressure occur at the throat. As the backpressure is lowered from c to f, the flow rate increases. For backpressures between c and f, the flow throughout the nozzle is subsonic. The diverging section of the nozzle acts as a diffuser, with the pressure rising and the velocity decreasing in the direction of flow. In this regime the converging–diverging nozzle can serve as a venturi meter, a device used for flow rate measurement.

$P_b = P_f$: As the backpressure is reduced, the throat pressure decreases and the throat velocity increases. The backpressure for which the throat velocity becomes sonic and the throat pressure becomes P^* is P_f. At this condition, the maximum velocity still occurs at the throat, so the diverging section is still a subsonic diffuser. However, the maximum flow rate through the nozzle has been reached, because the throat velocity is sonic. As

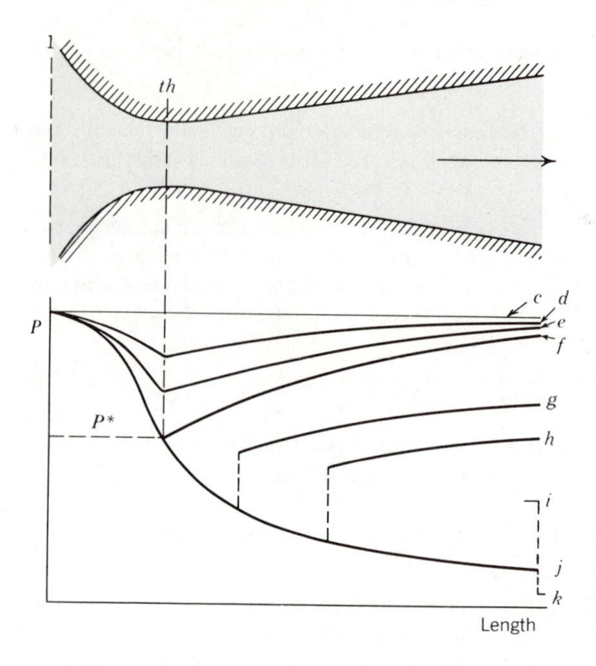

Figure 14·9. Converging-diverging nozzle flow ($M_1 < 1$). Solid lines are for isentropic processes; broken lines are for irreversible processes.

concluded earlier, flow through a converging passage with a subsonic entrance velocity can never result in a velocity higher than sonic or a pressure lower than P^*.

$P_b < P_f$: As P_b is reduced below P_f, the flow in the *converging* portion of the nozzle is completely unaffected; the pressure at the throat remains P^*, the velocity at the throat remains sonic, and the flow rate does not change. As P_b is decreased, the character of the flow in the *diverging* portion of the nozzle does change, however, and four distinct flow regimes occur.

$P_f > P_b > P_i$: As P_b is reduced to g and h, the fluid passing through the throat continues to expand and accelerate in the diverging section of the nozzle. Its velocity just downstream of the throat is supersonic. However, at a section downstream of the throat there is an abrupt irreversible increase in pressure accompanied by a deceleration from supersonic to subsonic velocity. This discontinuity in the flow is called a *normal shock*. (The modifier *normal* indicates that the plane of the discontinuity is normal to the flow direction.) Flow through the shock is steady and adiabatic but irreversible, so it is not isentropic. Downstream of the shock the gas is decelerated further isentropically as the diverging passage acts as a subsonic diffuser. As P_b is decreased, the shock moves downstream, approaching the nozzle exit plane as P_b approaches the value represented by point i.

$P_b = P_i$: When the backpressure is at the value represented by point i, the normal shock stands in the exit plane of the nozzle. The flow within the nozzle is isentropic: subsonic in the converging portion, sonic at the throat, and supersonic in the diverging portion. The jet leaving the nozzle is subsonic, however, because the fluid passes through the shock as it leaves the nozzle.

$P_i > P_b > P_j$: When the backpressure is below P_i but above the value P_j that is discussed below, the fluid expands to P_j at the nozzle exit plane, and no normal shock forms within or outside the nozzle. Downstream of the exit plane the pressure increases irreversibly from P_j to the backpressure through some discontinuities called oblique shocks that destroy the one-dimensionality of the flow. Flow through oblique shocks is not discussed in this textbook.

$P_b = P_j$: When the backpressure is maintained equal to the exit-plane pressure that causes isentropic expansion throughout the nozzle, no shocks occur within or outside the nozzle and a one-dimensional supersonic jet leaves the nozzle.

$P_b < P_j$: For any backpressure lower than P_j, the flow within and at the exit plane of the nozzle is the same as for $P_b = P_j$. Just downstream of the nozzle exit, the flow loses its one-dimensional character, and irreversible expansion and mixing occur. No matter how far the backpressure is reduced, the pressure within the nozzle never drops below P_j, and the flow rate and exit velocity do not change.

A review of the flow regimes shows that isentropic subsonic flows throughout the converging–diverging nozzle are possible for flow rates from zero to the maximum. The maximum flow rate corresponds to the attainment of sonic velocity at the throat. For the maximum flow rate, only two backpressures (P_f and P_j) provide isentropic one-dimensional flow throughout the entire nozzle and the region downstream of the nozzle exit.*

One of these (P_f) causes subsonic diffuser flow and the other (P_j) causes supersonic nozzle flow in the diverging portion. (A nozzle operating with a backpressure below P_j is said to be *underexpanded*; and operating with a backpressure greater than P_j but less than P_g is said to be *overexpanded*.)

The precise shape of a converging nozzle or the converging portion of a converging–diverging nozzle is not crucial in nozzle design, as long as the convergence is sufficiently gradual, but the geometry of a supersonic nozzle must be determined carefully if it is to produce essentially one-dimensional isentropic flow. Design methods to give the variation of area with length of supersonic nozzles are outside the scope of this textbook.

If the adiabatic flow through a nozzle is irreversible, as it always is to some extent in an actual nozzle because of friction, the entropy of the fluid must increase. Reversible adiabatic (isentropic) and irreversible adiabatic expansions from the same initial state and to the same final pressure are shown in Fig. 14·10. Point 2_i is the *i*sentropic (or *i*deal) exhaust state and point 2_a is the *a*ctual one. (Notice that on an *hs* diagram only lines of static pressure are uniquely determined because there exists only an *hPs* relationship and not a unique hP_ts relationship. In a like manner, a single point on an h_ts diagram can represent several static pressures but only one total pressure; so total pressure lines instead of static pressure lines are shown on h_ts diagrams.) The total or stagnation enthalpy is constant for any adiabatic expansion in a nozzle; the total or stagnation pressure is constant only for an isentropic expansion and decreases in an irreversible adiabatic expansion.

*That two different pressures at each cross section of the diverging section are possible for isentropic flow can be seen by (1) combining the energy balance, the ideal-gas equation of state, Pv^k = constant, the continuity equation, and the definition of Mach number to give

$$\frac{A}{A^*} = \frac{1}{M}\left[\left(\frac{2}{k+1}\right)\left(1 + \frac{k-1}{2}M^2\right)\right]^{(k+1)/2(k-1)}$$

where A^* is the throat area, and (2) noting that two different values of M, one greater than unity and the other less than unity, satisfy this equation for any value of A greater than A^*. For the subsonic Mach number, $P > P^*$, and for the supersonic Mach number, $P < P^*$.

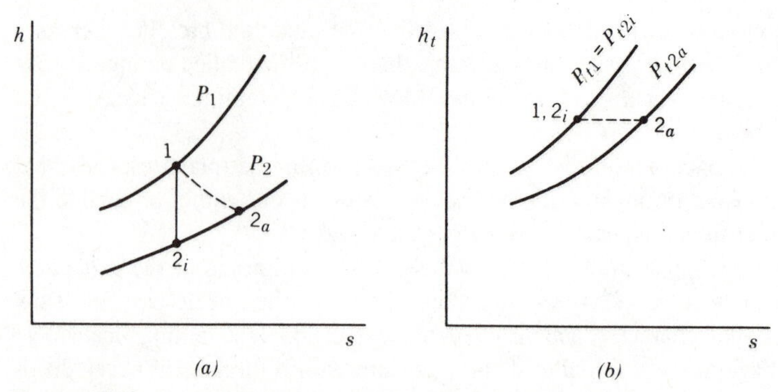

Figure 14·10. Reversible and irreversible adiabatic expansions.

Nozzle efficiency η_N is defined as the ratio of the kinetic energy at the nozzle outlet to the kinetic energy that would result at that section if the flow through the nozzle were isentropic between the same initial conditions and the same final pressure. Writing the definition and then extending it by means of the first law, we have

$$\eta_N \equiv \frac{V_{2a}^2/2}{V_{2i}^2/2} = \frac{h_{t1} - h_{2a}}{h_{t1} - h_{2i}}$$

Nozzle efficiencies of 0.95 and higher are easily obtained with converging nozzles. Similar efficiencies with converging–diverging nozzles can be obtained only by careful design. As nozzle sizes increase, fluid friction has relatively less effect on the flow, so nozzle efficiencies increase.

For various specialized studies of fluid flow and for repetitive calculations, it is often convenient to combine some of the basic equations to form equations that may be quite complex but that are suitable for the special purpose at hand. Some of the problems at the end of this chapter involve the derivation of some of these. Many elements of such equations are tabulated in the *Gas Tables* by Keenan, Chao, and Kaye (See Tables A·10 and A·11.) and elsewhere. They can be readily programmed for computers or programmable calculators, and such programs have been published. For an introductory study of compressible fluid flow and for illustrating the application of basic principles, you should avoid complete dependence on such equations and tables. Notice that the following example problems are solved by the direct application of basic principles.

Example 14·7. Air at 110 kPa, 90°C, with a velocity of 180 m/s is to be expanded isentropically until its Mach number is 1.5. The flow rate is 0.15 kg/s. Determine the final pressure and cross-sectional area.

Analysis. Since the initial Mach number is obviously less than one and the final velocity is to be supersonic, a converging–diverging nozzle must be used. The final area can be found from the continuity equation, $A_2 = Mv_2/V_2$ if the final velocity and specific volume can be found. We know that $V_2 = 1.5c_2 = 1.5\sqrt{kRT_2}$, and an energy balance will give us another relationship between V_2 and T_2; so we can solve for these two quantities. The final specific volume can be found from $v_2 = v_1(T_1/T_2)^{1/(k-1)}$ for the isentropic process, but, since the final pressure is also sought anyway, we might as well calculate $P_2 = P_1(T_2/T_1)^{k/(k-1)}$ and then determine v_2 by means of the ideal-gas equation of state.

Alternatively, we can put the basic equations into a computer program for solving simultaneous algebraic equations. We will use this approach for this problem.

Solution. The pertinent basic equations, assuming that the temperature range is small enough so that a constant value of c_p can be used, are

$$\frac{V_2^2 - V_1^2}{2} = h_1 - h_2 = c_p(T_1 - T_2)$$

$$\dot{m} = \rho_2 A_2 V_2$$

$$\rho_2 = \frac{P_2}{RT_2}$$

$$M_2 = \frac{V_2}{c_2}$$

$$c_2 = \sqrt{kRT_2}$$

$$\frac{P_2}{P_1} = \left(\frac{T_2}{T_1}\right)^{k/(k-1)}$$

Input values in addition to those listed in the problem statement are $R = 0.287$ kJ/kg·K, $c_p = 1.005$ kJ/kg·K, and $k = 1.4$. Output values are: $P_2 = 34.9$ kPa, $A_2 = 0.663 \times 10^{-3}$ m^2 (and also $T_2 = 262$ K, $V_2 = 486$ m/s, and $\rho_2 = 0.465$ kg/m^3). The temperature range is such that our assumption of constant specific heats is justified. (See Figs. 4·4 and 4·8.)

14·8 Adiabatic Flow of Vapors

The basic principles listed in Sec. 14·5 and the isentropic flow relationships derived in Sec. 14·6 apply of course to the flow of vapors or nonideal gases. The method of analysis for vapor flow is similar to that for ideal-gas flow except that for vapors no simple equation of state or *hT* relation is available. Therefore tabulated or graphical data or a computer program for properties must be used. The *hs* or Mollier diagram which was discussed in Sec. 9·12 is useful in this respect for nozzle flow calculations. For steam, convenient *hv* charts are available (Reference 14·3).

For limited ranges of pressure and temperature, the c_p and k values of vapors may sometimes be considered as constant so that relationships derived for ideal gases, such as Eqs. 14·2 and h of the preceding section, can be used for vapors. For example, the mean value of k for slightly superheated steam at pressures less than about 1400 kPa (about 200 psia) is around 1.3, so that Eq. h of the preceding section gives $P^*/P_t = 0.55$, and this value is frequently used in predicting whether a converging or a converging–diverging nozzle should be used for a given expansion of steam. The same value of P^*/P_t is used for slightly wet steam because it has been observed experimentally that steam expanding isentropically (or nearly so) in a nozzle can expand for some distance into the wet region on an *hs* diagram before condensation actually begins. This delay of condensation is known as *supersaturation.* Steam that is represented by a point within the wet region, as far as its enthalpy and entropy are concerned, and yet contains no liquid is called *supersaturated steam.* It is said to be in a metastable state, and we do not treat such states in this textbook.

Nozzle efficiency as defined in the preceding section is used in connection with both vapor and ideal-gas nozzles. For vapor nozzles the term *reheat fraction* is also used and may be defined as $(1 - \eta_N)$. The frictional effect in a converging–diverging nozzle occurs principally between the throat and exit of the nozzle; hence the flow from the entrance to the throat may be safely assumed to be isentropic even in actual nozzles.

Example 14·8. A nozzle expands steam at a rate of 2.0 lbm/s from 200 psia, 600°F, to 40 psia. The initial velocity is negligible. Assuming that the flow is isentropic, compute the velocity and the area at sections along the nozzle where the pressure is 150, 100, 80, and 40 psia.

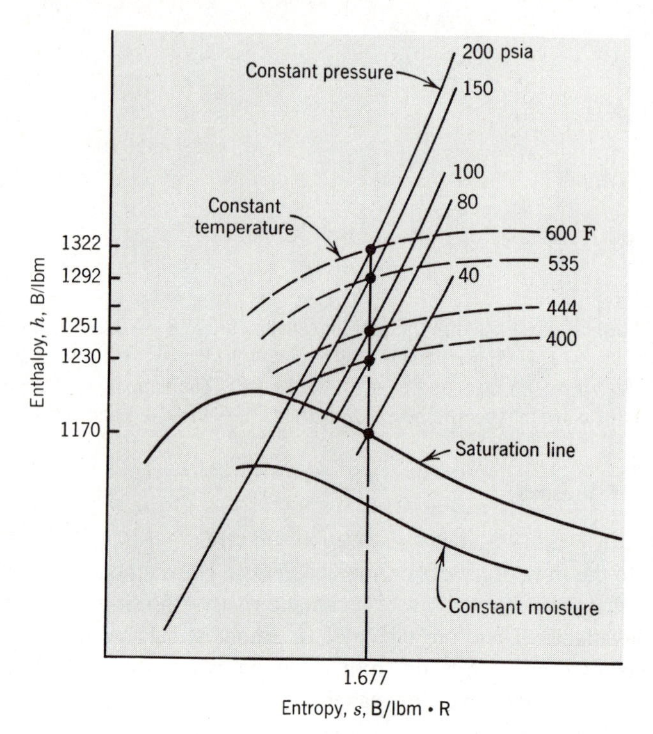

Example 14·8.

Solution. Since an isentropic process is assumed, the four state points we are interested in can be found at the intersections of the constant-entropy line from the initial state with the proper pressure lines on a Mollier chart. A skeleton sketch of the chart is shown.

Since the initial kinetic energy is negligible, application of the first law shows that the kinetic energy at any section equals the decrease in enthalpy between the inlet and that section. Thus the velocities are given by

$$V = \sqrt{2(h_1 - h)} = \sqrt{2(32.2)778(h_1 - h)} = 223.8\sqrt{1322 - h}$$

with h in B/lbm. The specific volumes are found from the steam tables, $v = v(P,s)$; and the areas are found by use of the continuity equation, $A = \dot{m}v/V$. The results are tabulated as

P, psia	h, B/lbm	V, fps	v, ft³/lbm	A, in.²
150	1292	1226	3.83	0.90
100	1251	1886	5.31	0.81
80	1230	2147	6.21	0.83
40	1170	2759	10.50	1.10

These figures show that the nozzle area is a minimum at some section between the 150-psia section and the 80-psia section.

Example 14·9. Determine the flow rate through a properly designed nozzle which steam enters at 2.0 MPa, 300°C, with negligible velocity and leaves at 0.2 MPa if the throat area is 190 mm^2.

Solution. The ratio P_2/P_1, where P_2 is the outlet pressure, is so low that undoubtedly the exit velocity is supersonic, and therefore limiting flow exists. Using the value of 0.55 for P^*/P_t and noting that $P_1 = P_t$ because the entrance velocity is negligible, we have at the throat

$$P_{th} = P^* = 0.55P_1 = 0.55(2.0) = 1.10 \text{ MPa}$$

At this pressure and $s_{th} = s_1 = 6.7664$ kJ/kg·K, the steam is superheated; so from a Mollier chart or steam tables we have $h_{th} = 2883$ kJ/kg and from the tables we have $v_{th} = 0.199$ m^3/kg. Applying the first law and then the continuity equation gives

$$V_{th} = \sqrt{2(h_1 - h_{th})} = \sqrt{2(1000)(3023.5 - 2883)} = 530 \text{ m/s}$$

$$\dot{m} = \frac{A_{th}V_{th}}{v_{th}} = \frac{190(10)^{-6}\, 530}{0.199} = 0.506 \text{ kg/s}$$

Example 14·10. A nozzle is to be designed to expand steam at a rate of 0.10 kg/s from 500 kPa, 210°C, to 100 kPa. Inlet velocity is to be very low. For a nozzle efficiency of 0.90, determine the exit area.

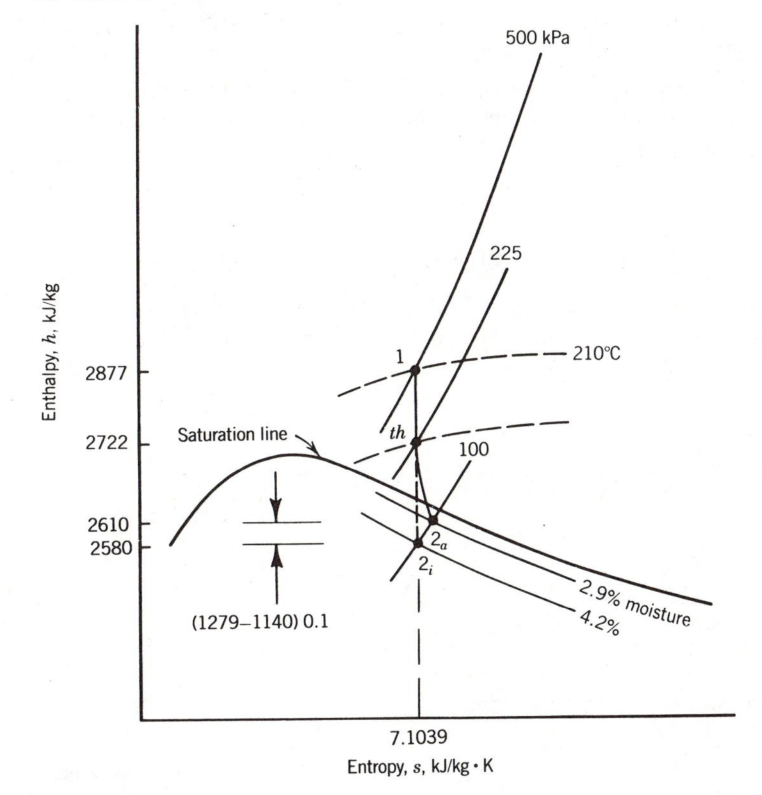

Example 14·10.

Solution. A skeleton *hs* diagram is made first, and values read from a Mollier chart are placed on it. Point 2_a, the actual end state, is located by first finding $(h_1 - h_{2i}) = 2877 - 2580 = 297$ kJ/kg and then applying the fact that the actual enthalpy decrease is 0.90 of the ideal enthalpy decrease. $h_{2a} = 2610$ kJ/kg. The actual exit velocity is

$$V_{2a} = \sqrt{2(h_1 - h_{2a})} = \sqrt{2(h_1 - h_{2i})\eta_N}$$
$$= \sqrt{2(1000)(2877 - 2580)0.90} = 730 \text{ m/s}$$

From the steam tables at 100 kPa and 2610 kJ/kg ($x_{2a} = 0.971$), $v_{2a} = 1.645$ m³/kg. (Notice that this is larger than v_{2i}.) The area is then found from the continuity equation

$$A_2 = \frac{\dot{m}v_{2a}}{V_{2a}} = \frac{0.10(1.645)}{730} = 0.225 \times 10^{-3} \text{ m}^2$$

14·9 Flow through Orifices

Up to this point in our discussion of nozzle and diffuser flow we have treated only one-dimensional flow in which the fluid completely fills the flow channel. In the flow through an orifice as shown in Fig. 14·11 the fluid stream continues to contract after passing through the orifice until a minimum area section called the *vena contracta* is reached at which the pressure is also a minimum. Simple analysis shows that neither the pressure nor the velocity is uniform across any section, even of the jet itself, between the orifice and the vena contracta. The size and location of the vena contracta can be predicted by means discussed in books on fluid mechanics and fluid metering. Mention of orifice flow is made here only as a reminder that some common flows cannot be modeled successfully on a one-dimensional basis.

14·10 The Fanno Line; Flow in Pipes

Consider the adiabatic steady flow of a fluid in a constant-area passage with no work done. The flow in an insulated pipe meets such conditions. Application of the first law

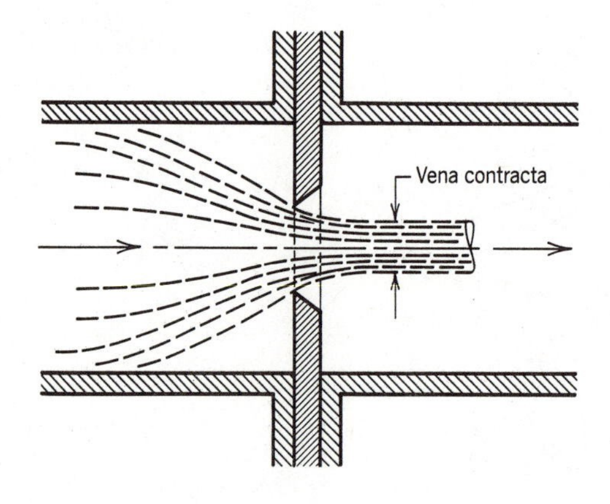

Figure 14·11. Flow through an orifice.

and the continuity equation gives

$$h_t = h + \frac{V^2}{2} = h + \left(\frac{\dot{m}}{A}\right)^2 \frac{v^2}{2} = \text{constant} \tag{14·4}$$

Thus the relationship between h and v (and consequently between any two properties if we are dealing with a pure substance) is fixed; so the states that satisfy Eq. 14·4 can be plotted on an hs diagram. Figure 14·12 shows three such plots for the same total enthalpy but different values of $\dot{m}/A$ (lines A, B, and C) and one plot for a higher total enthalpy (line D). These lines are called *Fanno lines** and each one is the locus of states through which a fluid passes in adiabatic pipe flow for given entrance conditions. A limiting case Fanno line is the constant-enthalpy line E that inspection of Eq. 14·4 shows to be the limiting case as $\dot{m}/A$ approaches zero or the case for incompressible flow (v = constant).

We know from the increase of entropy principle that for adiabatic flow the entropy of the fluid cannot decrease; therefore, the process represented by the Fanno lines in Fig. 14·12 can proceed only in the directions shown by the arrows on the lines. Let us investigate the conditions at a maximum entropy point on a Fanno line such as point a in Fig. 14·12. The energy balance in differential form is

$$dh + V\,dV = 0$$

From the continuity equation we have ρV = constant so that $dV = -V\,d\rho/\rho$, and making this substitution in the energy balance gives

$$dh - \frac{V^2\,d\rho}{\rho} = 0$$

Then substituting for dh from the $T\,ds$ equation,

$$T\,ds = dh - v\,dP \tag{8·3}$$

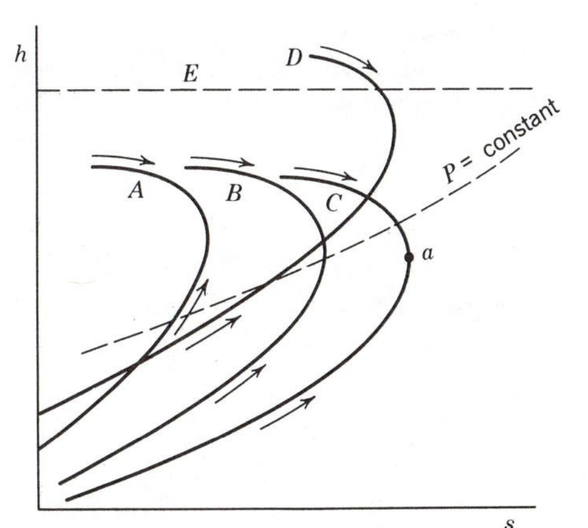

Figure 14·12. Fanno lines.

*Gino Fanno (1882–1962) described these lines in his diploma thesis submitted to the Swiss Federal Institute of Technology, Zürich, in 1904.

gives

$$T\,ds + v\,dP - \frac{V^2\,d\rho}{\rho} = 0$$

$$T\,ds + \frac{dP}{\rho} - \frac{V^2\,d\rho}{\rho} = 0$$

At a maximum entropy point, $ds = 0$, so that this last equation can be simplified and solved for V as

$$V = \sqrt{\left(\frac{\partial P}{\partial \rho}\right)_s} = c \tag{14·2}$$

Thus the limiting state such as a on Fig. 14·12 is one for which the velocity equals the sonic velocity. Further consideration of Fig. 14·12 shows that the upper part of each Fanno line (that is, the part above the maximum entropy point) represents states of subsonic flow, and the lower part corresponds to supersonic flow. Therefore, for subsonic adiabatic flow in a pipe, the pressure and enthalpy decrease and the velocity increases in the direction of flow, but the velocity cannot exceed the velocity of sound. For supersonic adiabatic flow in a pipe, the pressure and enthalpy increase and the velocity decreases in the direction of flow, the limiting velocity again being the velocity of sound that in this case is approached from above. In both cases, of course, the entropy of the fluid must increase in the direction of flow. If a pipe carrying a fluid at subsonic velocities discharges into a region where the pressure is lower than that corresponding to the point of maximum entropy on the Fanno line P_a, then the expansion from P_a to the receiver pressure must occur outside the end of the pipe, because inside the pipe only states on the Fanno line that can be reached without a decrease in entropy can be attained. If the flow in the pipe is initially supersonic and the receiver pressure is higher than P_a, then a normal shock must occur inside the pipe, causing the flow to change from supersonic to subsonic. The flow is represented first by states on the lower part of the Fanno line and then, downstream of the shock, by states on the upper part. (See Sec. 14·12.) (*Caution*: The limiting pressure in a pipe where the flow is irreversible is not the same as the limiting pressure in a converging nozzle under isentropic flow conditions. In both cases, however, the total enthalpy is constant and the limiting velocity is sonic.)

The variation of pressure, enthalpy, velocity, etc., of the fluid as a function of pipe length can be predicted only if information is available on friction factors or fluid shearing stresses. These matters are treated in books on fluid mechanics.

14·11 The Rayleigh Line

Consider steady frictionless flow (that is, no fluid shearing forces) in a constant-area passage with no work done. The dynamic equation applied to an element of fluid between two sections 1 and 2 is

$$P_1 A - P_2 A = \dot{m}(V_2 - V_1) \tag{14·1, expanded}$$

Dividing both sides of this equation by the area and substituting from the continuity equation $V = \dot{m}v/A$, we have

$$P_1 - P_2 = \left(\frac{\dot{m}}{A}\right)^2 (v_2 - v_1)$$

or

$$P + \left(\frac{\dot{m}}{A}\right)^2 v = \text{constant}$$

This is the equation of a *Rayleigh* line* which holds for steady frictionless flow in a constant-area passage with no work done. It is based on only the dynamic and continuity equations. If property relationships for the fluid are known, the equation can be plotted on various coordinates. Figure 14·13 shows a Rayleigh line on *hs* coordinates and, for comparison, a Fanno line for the same value of $\dot{m}/A$. The Fanno line is a line of constant total enthalpy, so obviously the total enthalpy varies along a Rayleigh line. This means that a fluid can follow a Rayleigh line *only* if there is heat transfer. As in the case of the Fanno line, it can be shown that the maximum entropy point on a Rayleigh line corresponds to $M = 1$. The lower branch of the line is for supersonic flow, and the upper branch is for subsonic flow. The process is not adiabatic, so of course there is no restriction on the sign of entropy change of the fluid alone. The direction in which a process proceeds along a Rayleigh line depends on the direction of heat transfer as indicated in Fig. 14·13. If heat is added to the fluid in this frictionless process, its entropy must increase; if heat is removed, the entropy of the fluid must decrease.

A number of interesting conclusions can be drawn from inspection of the Rayleigh line. For subsonic flow, adding heat causes the pressure to drop; for supersonic flow, adding heat causes the pressure to rise. Along the Rayleigh line between the maximum enthalpy point and the maximum entropy point, adding heat causes the enthalpy to decrease.

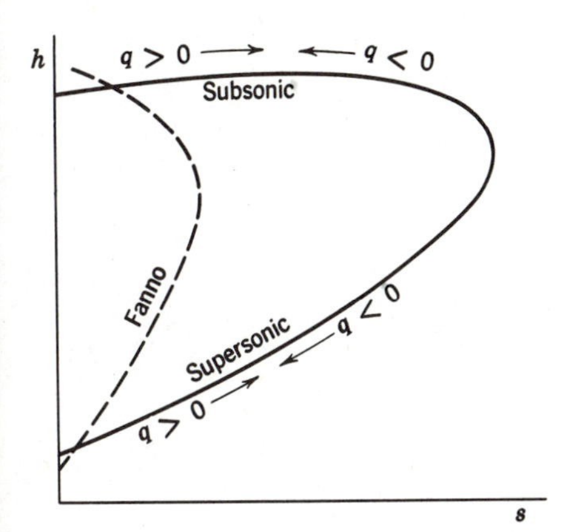

Figure 14·13. Rayleigh line.

*Lord Rayleigh (1842–1919) was an English physicist who is perhaps now best known for his work in acoustics, but he also made notable contributions in the fields of optics, hydrodynamics, electromagnetic phenomena, and gas properties. He received a Nobel prize in 1904.

14·12 Flow across a Normal Shock

A combination of Fanno and Rayleigh lines is useful in determining the change in state that occurs across a normal shock, which was mentioned in Sec. 14·7 as a flow discontinuity in which the pressure rises suddenly and the velocity changes from supersonic to subsonic. The application of Fanno and Rayleigh lines to the same flow requires some explanation, since one is for frictional adiabatic flow and the other is for frictionless diabatic flow.

Referring to Fig. 14·14, fluid in state 1 approaches the shock at a supersonic velocity. The thickness of the shock is very small so that between section 1 just upstream of the shock and section 2 just downstream of it the area is virtually the same, even though the shock may occur in a diverging passage. If the process is also adiabatic, then states 1 and 2 must lie *on the same Fanno line*. Again because the shock is so thin, wall friction forces can be neglected (that is, they are very small compared to the difference in pressure forces acting on a fluid element that contains the shock). Therefore, states 1 and 2 are connected by a frictionless (but not necessarily reversible) constant-area flow and hence lie *on the same Rayleigh line*. The only point on the *hs* diagram of Fig. 14·14 that lies on both the same Fanno line and the same Rayleigh line as point 1 is the point 2 at the other intersection of the two lines. Thus the conditions just downstream of a shock can be determined by plotting the Fanno and Rayleigh lines through the state of the fluid just upstream of it. For an ideal gas the equation of state is simple enough so that the Fanno and Rayleigh line equations can be solved directly for the Mach number, pressure, and temperature ratios across a shock. For vapors, an empirical *hPv* relation can often be formulated which makes possible the direct solution of the two equations.

Figure 14·14 shows $s_2 > s_1$ as expected for an irreversible adiabatic process. It is possible to show that for all flows that satisfy both the Fanno and Rayleigh line conditions the entropy of the subsonic state is always higher than that of the supersonic state. Therefore, since the shock is an adiabatic process, it can occur in only one direction: from a supersonic state to a subsonic state. This is another illustration of the value of the

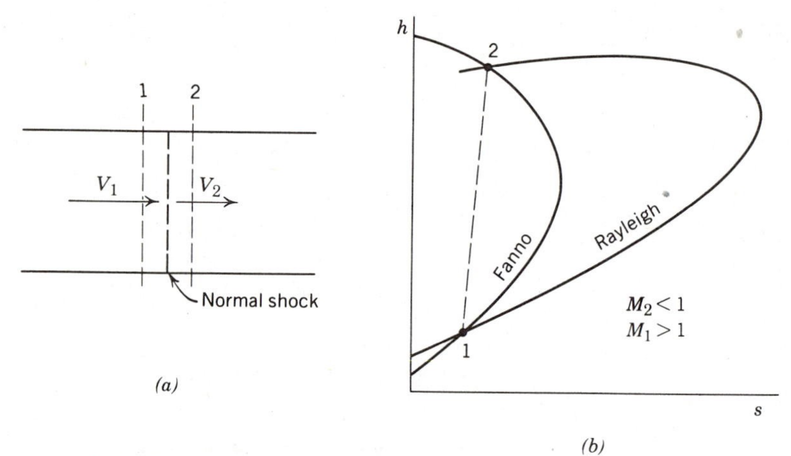

Figure 14·14. State change across a normal shock.

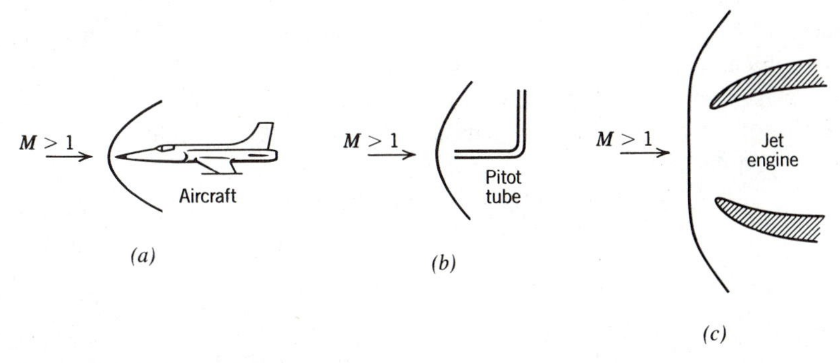

Figure 14·15. Normal shocks.

second law in showing whether a conceivable process is possible. From inspection of the Fanno and Rayleigh lines we conclude that two states satisfy the requirements of the first law, the dynamic equation, and the continuity equation, but the second law is needed to tell us in which direction a process between these two states can occur.

Shocks, which can occur only in supersonic flow, affect the flow significantly. Their occurrence in supersonic nozzles has already been mentioned, and is the subject of Example 14·11. They occur also in the flow ahead of a craft moving at a supersonic speed as illustrated in Fig. 14·15*a*. The normal shock exists only directly ahead of the nose of the craft where the velocity is normal to the wave shown. The other parts of the wave structure shown in the diagram are related to the oblique shocks mentioned earlier and are not discussed in this textbook. Since a shock occurs upstream of a probe inserted into a supersonic stream as shown in Fig. 14·15*b*, the probe clearly does not sense directly the supersonic stream; instead, the flow impinging on the probe is subsonic. The flow entering the engine of a supersonic craft may pass through a normal shock as shown in Fig. 14·15*c*, and the engine design may utilize the pressure rise across the shock as part of the total compression process of the engine.

14·13 Steady Adiabatic Flow of Ideal Gases with Constant Specific Heats and No Work: Recapitulation

The analysis of steady adiabatic flow of ideal gases with constant specific heats and no work or change in potential energy requires only the basic relations reviewed in Sec. 14·5. For computational purposes many special equations can be derived from these basic relations. For example, the first law and the definition of Mach number lead to:

$$T\left[1 + \frac{k-1}{2}M^2\right] = \text{constant} \tag{i}$$

Notice that Eq. 14·3 is a special case of this equation. Equations are often written to involve either total properties or properties at a section where the velocity is sonic, because the respective Mach number values of 0 and 1 simplify them. The use of properties at the section where $M = 1$ can involve the cross-sectional area, while this is not the case when using total properties.

If the flow is also reversible, we can derive from the same basic relations and the added relationship Pv^k = constant or s = constant other relationships such as

$$P\left[1 + \frac{k-1}{2}M^2\right]^{(k-1)/k} = \text{constant} \tag{j}$$

and

$$\frac{A}{A^*} = \frac{1}{M}\left[\frac{2}{k+1}\left(1 + \frac{k-1}{2}M^2\right)\right]^{(k+1)/2(k-1)} \tag{k}$$

The flow across a normal shock is adiabatic but irreversible, so Eq. i holds but Eqs. j and k do not. The dynamic equation and the continuity equation along with Eq. i lead to equations such as

$$M_y^2 = \frac{M_x^2 + \left(\dfrac{2}{k-1}\right)}{\dfrac{2k}{k-1}M_x^2 - 1} \tag{l}$$

and

$$\frac{P_y}{P_x} = \frac{2k}{k+1}M_x^2 + \frac{k-1}{k+1} \tag{m}$$

where sections x and y are just upstream and just downstream of the shock, respectively.

All of the equations above can be quickly derived, and they are presented here only as samples of those that might be useful in some applications. Remember that they contain no information beyond that in the basic equations, and each carries certain restrictions. Obviously, any problem that can be solved by the use of these special equations can be solved by using the basic equations, and the basic equations can be applied to many more situations. Clearly, though, computational labor can sometimes be reduced by using the appropriate special-case equations. Some quantities that cannot be solved for directly can be quickly obtained through iterative procedures of a computer program.

Tabulations of these equations and many others are published in the Keenan, Chao, and Kaye *Gas Tables* and elsewhere. Greatly abridged versions of two of these tabulations are given as Tables A·10 and A·11. These equations can also be readily entered into a simultaneous equation solver program for computer solution. This is the best way to approach numerous problems. After developing facility in applying the basic relations, one can quickly learn to use the special equations or tables with confidence.

Example 14·11. A nozzle is to be used to expand air steadily and isentropically from an initial state of 100 kPa, 200°C, with negligible velocity, to a Mach number of 2.00 where the cross-sectional area is 0.00300 m². Determine (*a*) the mass rate of flow, (*b*) the discharge pressure for the design conditions, (*c*) the discharge pressure for the maximum flow rate and isentropic compression in the diverging portion of the nozzle, and (*d*) the backpressure to cause a normal shock to stand in the exit plane.

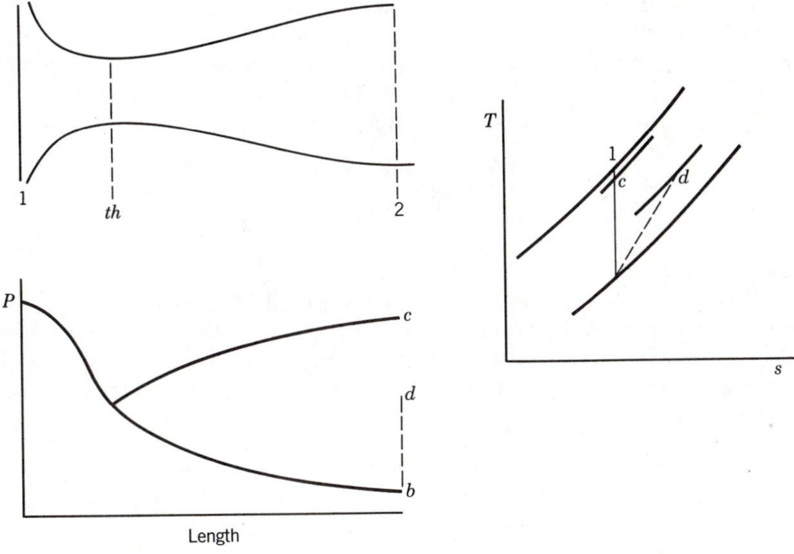

Example 14·11.

Analysis. Several solution methods are possible for the various parts of the problem. To illustrate different methods we will solve the problem first from basic equations and then again using published tables. As a first step for either method, we sketch pressure–length and *Ts* diagrams, labeling the different exit states to correspond to lettered sections of the problem statement.

Solution Using Basic Equations (a, b) We can obtain the mass rate of flow from the continuity equation

$$\dot{m} = \rho_b A_2 V_b$$

We must also use the ideal-gas equation of state

$$\rho_b = \frac{P_b}{RT_b}$$

and the first law for steady adiabatic flow of an ideal gas with constant specific heats, zero work, and zero change in potential energy

$$\frac{V_b^2}{2} = h_1 - h_b = c_P (T_1 - T_b) \qquad (T_1 = T_{t1})$$

We will also use the relation for an isentropic process between states 1 (or t) and b

$$P_b = P_1 \left(\frac{T_b}{T_1}\right)^{k/(k-1)}$$

and the defining equation for Mach number combined with the expression for the sonic velocity of an ideal gas

$$V_b = M_b c_b = M_b\sqrt{kRT_b}$$

These five equations (excluding the intermediate parts involving h_1, h_b, and c_b) involve five unknowns ($\dot{m}$, ρ_b, V_b, P_b, T_b). Inputs to a simultaneous algebraic equation solver program are the data included in the problem statement and $R = 0.287$ kJ/kg·K, $c_p = 1.01$ kJ/kg·K, and $k = 1.397$. (The c_p and k values are taken from Table 4·2 on the assumption that the mean temperature will be around 100°C. k, c_p, and R are related by $c_p = kR/(k - 1)$.) The results are

$$\dot{m} = 0.330 \text{ kg/s} \qquad P_b = 12.79 \text{ kPa}$$
$$V_b = 530.3 \text{ m/s} \qquad T_b = 263.7 \text{ K}$$
$$\rho_b = 0.169 \text{ kg/m}^3$$

We see that our assumption of a mean temperature around 100°C is justified.

(*c*) To determine the outlet pressure, P_c, for isentropic flow at the same rate (and therefore with $M_{\text{th}} = 1$, still) but with subsonic flow in the diverging portion of the nozzle, we have the same equations we used for parts *a* and *b*, except that we do not need the one involving Mach number. Thus

$$\dot{m} = \rho_c A_2 V_c$$

$$\rho_c = \frac{P_c}{RT_c}$$

$$\frac{V_c^2}{2} = c_p(T_1 - T_c)$$

$$P_c = P_1\left(\frac{T_c}{T_1}\right)^{k/(k-1)}$$

The unknowns in these four equations are ρ_c, V_c, P_c, and T_c. Obviously, the results of parts *a* and *b* satisfy these equations, but we now seek a set of values involving a lower value of V_c and higher values of P_c, T_c, and ρ_c. Our simultaneous algebraic equation solver program gives

$$V_c = 159.9 \text{ m/s} \qquad T_c = 460.3 \text{ K}$$
$$P_c = 90.9 \text{ kPa} \qquad \rho_c = 0.688 \text{ kg/m}^3$$

(*d*) For determining the pressure rise across a normal shock standing in the exit plane, we use for the properties of the fluid entering the shock those listed above for isentropic expansion to the exit plane (state *b*). The dynamic equation across a shock (inlet *b*, outlet *d*) is

$$(P_d - P_b)A = \dot{m}(V_b - V_d)$$

The continuity equation for $A_b = A_d$ is

$$\rho_b V_b = \rho_d V_d$$

The ideal-gas equation of state is

$$\rho_d = \frac{P_d}{RT_d}$$

and the first law for the adiabatic flow of an ideal gas with constant specific heats is

$$\frac{V_b^2 - V_d^2}{2} = c_p(T_d - T_b)$$

Introducing these four equations (with unknowns P_d, V_d, ρ_d, and T_d) into our simultaneous algebraic equation solver computer program gives $P_d = 57.6$ kPa.

Solution Using Tables. Using Tables A·10 and A·11 provides a quick solution to this problem. However, we must use $k = 1.40$ because we do not have a table for $k = 1.397$.

(*a*, *b*) For $M_b = 2.0$, Table A·10 shows $\rho_b/\rho_t = 0.23005$, $P_b/P_t = 0.12780$, and $T_b/T_t = 0.55556$. Since $P_t = 100$ kPa, we immediately have the desired result for part *b*, $P_b = 12.78$ kPa. For use in the continuity equation, we calculate

$$\dot{m} = \rho_b A_b V_b = \rho_b A_b M_b c_b = \frac{\rho_b}{\rho_t}\rho_t A_b M_b \sqrt{kRT_b}$$

$$= \frac{\rho_b}{\rho_t}\frac{P_t}{RT_t} A_b M_b \sqrt{kR\frac{T_b}{T_t}T_t}$$

$$= 0.230\,05 \frac{100}{0.287(473)} 0.00300(2)\sqrt{1.4(1000)0.287(0.555\,56)473}$$

$$= 0.330 \text{ kg/s}$$

(*c*) Table A·10 shows that for producing $M_b = 2.0$, the nozzle has $A_2/A^* = 1.6875$. It also shows that this same value of the ratio of outlet area to throat area can result in $M_c = 0.372$ and $P_c = (P_c/P_t)P_t = 0.9087(100) = 90.9$ kPa.

(*d*) For $M_b = 2.0$ entering a normal shock, Table A·11 shows $P_d = (P_d/P_b)P_b = 4.500(12.78) = 57.5$ kPa.

Comment. In these two solutions we have illustrated both approaches: (1) using basic relations and (2) using published tables of compressible flow functions. Each approach has advantages.

14·14 Summary

This chapter deals with the thermodynamic aspects of the steady one-dimensional flow of fluids. A useful property in such studies is the total enthalpy that is defined as

$$h_t \equiv h + \frac{V^2}{2}$$

The total temperature and total pressure of a flowing fluid are defined as the temperature and pressure that would result if the fluid were brought to rest isentropically. They are also called the stagnation temperature and stagnation pressure.

The basic dynamic equation for steady one-dimensional flow is

$$F = \dot{m}\Delta V \tag{14·1}$$

where F is the resultant force on a stream that undergoes a velocity change ΔV and that has a flow rate $\dot{m}$. This is a vector equation: F and ΔV are in the same direction.

The velocity of sound in a fluid is given by

$$c = \sqrt{\left(\frac{\partial P}{\partial \rho}\right)_s} \tag{14·2}$$

which for an ideal gas becomes

$$c = \sqrt{\frac{kP}{\rho}} = \sqrt{kPv} = \sqrt{kRT}$$

A useful parameter in the study of gas flow is the Mach number M, defined by

$$M \equiv \frac{V}{c}$$

Application of the first and second laws, the continuity equation, and the dynamic equation shows that for the isentropic flow of any fluid the pressure, velocity, and cross-sectional area variations depend on the Mach number as follows:

1. When $M < 1$, $dA/dP > 0$ and $dA/dV < 0$
2. When $M > 1$, $dA/dP < 0$ and $dA/dV > 0$
3. When $M = 1$, $dA/dP = 0$ and $dA/dV = 0$

Further application of the basic relations reveals that if a fluid enters a converging nozzle subsonically the Mach number at the outlet cannot exceed unity, regardless of how far the receiver pressure is lowered. A converging–diverging nozzle must be used to accelerate a fluid from a subsonic to a supersonic velocity and a converging–diverging diffuser must be used to decelerate a fluid isentropically from a supersonic to a subsonic velocity. When $M = 1$ at the throat of a converging–diverging passage, there are, for isentropic flow, only two possible pressures for the fluid at each other section of the passage, one for subsonic and the other for supersonic flow.

The Fanno line is the locus of states of a fluid that flows adiabatically through a constant-area channel with no work done. The Rayleigh line is the locus of states of a fluid that flows frictionlessly through a constant-area passage with no work done but with heat transfer allowed. The Fanno and Rayleigh lines are useful in several types of analyses, one of which is the determination of the change of state across a normal shock. A normal shock is a flow discontinuity that occurs only if the velocity is initially supersonic and that results in a deceleration to a subsonic velocity and an abrupt pressure rise.

References and Suggested Reading

14·1 Fox, Robert W., and Alan T. McDonald, *Introduction to Fluid Mechanics,* Wiley, New York, 1973, Chapters 9 and 10.

14·2 Keenan, Joseph H., Jing Chao, and Joseph Kaye, *Gas Tables,* 2nd ed., Wiley, New York, 1980 (English Units), 1983 (SI Units).

14·3 Potter, J. H., *Steam Charts,* American Society of Mechanical Engineers, 1976.

14·4 Reynolds, William C., and Henry C. Perkins, *Engineering Thermodynamics,* 2nd ed., McGraw-Hill, New York, 1977, Chapter 13.

14·5 Roberson, J. A., and C. T. Crowe, *Engineering Fluid Mechanics,* 2nd ed., Houghton Mifflin, Boston, 1980, Chapter 12.

14·6 Shames, Irving H., *Mechanics of Fluids*, 2nd ed., McGraw-Hill, New York, 1982, Chapter 11.

14·7 Shapiro, A. H., *The Dynamics and Thermodynamics of Compressible Flow*, Ronald Press, New York, 1953, Vol. I, Chapters 3, 4, and 5.

Problems

14·1 A liquid flows steadily through a circular pipe with a velocity distribution given by $u = u_{max}[1 - (r/r_0)^2]$, where u is the velocity at any radius r, u_{max} is the velocity at the pipe axis, and r_0 is the pipe radius. Determine the ratio of the kinetic energy per unit mass of this stream to that of a one-dimensional flow in the same pipe with the same flow rate.

14·2 Consider a liquid in a circular pipe of radius r_0. At one instant the velocity is a maximum u_{max} at the pipe axis and zero at the pipe wall. Determine the ratio V/u_{max}, where V is the average velocity at one section, if the velocity u at any radius r is given by $u = u_{max}(1 - r/r_0)^2$.

14·3 Consider a fluid flowing steadily in a circular pipe of radius r_0. Determine the ratio V/u_{max}, where V is the average velocity across the pipe and u_{max} is the maximum velocity, if the velocity u at any radius is given by $u = u_{max}[1 - (r/r_0)^2]$.

14·4 An aircraft flies steadily at 400 m/s through air at 60 kPa, 15°C. Calculate the stagnation temperature and pressure on the craft.

14·5 Determine the stagnation pressure and temperature on a craft moving at 1000 m/s through air at 12 kPa, −20°C.

14·6 Determine the stagnation temperature and the approximate adiabatic wall temperature of steam at 3.60 MPa, 250°C, flowing with a velocity of 300 m/s.

14·7 An aircraft flies steadily at 1450 fps through air at 10 psia, 50°F. Determine the stagnation temperature and pressure on the craft.

14·8 You can readily derive the following expression for the velocity of a gas stream, as measured by a pressure probe, when the velocity is sufficiently low that the air can be considered incompressible.

$$V = \sqrt{\frac{2\,(P_t - P)}{\rho}}$$

Such a velocity measurement was made and reported as 26.4 m/s for air at 25°C under a barometric pressure of 755 mm Hg. If the experimenter obtained this velocity assuming the air was dry but the wet bulb temperature was actually 22°C, how much error did this assumption introduce into the velocity? How much error would be involved in the kinetic energy?

14·9 Determine the stagnation temperature and the stagnation pressure for air at 5 psia, 0°F, moving at 1000 fps.

14·10 Determine the stagnation pressure and the stagnation temperature on a craft moving at 6000 fps through air at 2 psia, −40°F.

14·11 Determine the stagnation temperature and the approximate adiabatic wall temperature of steam at 600 psia, 500°F, flowing with a velocity of 1000 fps.

14·12 Air at 120 kPa, 40°C, flows in a stream with a cross-sectional area of 0.40 m^2 with a velocity of 150 m/s into a system of vanes that turns the stream through an angle of 30 degrees without changing the magnitude of the velocity or the pressure. Determine the force of the stream on the vane system.

14·13 A jet of water having a cross-sectional area of 0.0320 m^2 strikes a flat plate that is normal to the jet. The force on the plate is 400 N. If the jet velocity is doubled, what will be the magnitude of the force on the plate? What is the magnitude of the higher velocity?

14·14 Ethane (C_2H_6) flows at a rate of 20 000 kg/h through a long straight section of pipe that has an inside diameter of 10 cm. At inlet the ethane is at 800 kPa, 30°C; at outlet it is at 200 kPa, 10°C. Determine the axial force of the ethane on the pipe.

14·15 An aircraft turbojet engine is to provide a thrust of 60 kN when operating at a speed of 800 km/h through air at 40 kPa, −35°C. The inlet velocity relative to the engine will be the same as the craft speed. It is estimated that the ratio of air mass to fuel mass will be 40, the exhaust gas leaving at 40 kPa will have a density of 0.15 kg/m^3, and the exhaust velocity relative to the craft will be 420 m/s. Determine the required fuel flow rate and the diameter of the air inlet, which has a circular cross-section.

14·16 Air at 20 psia, 100°F, flows in a stream of 4 sq ft cross-sectional area with a velocity of 500 fps into a system of vanes which turns the stream through an angle of 30 degrees without changing the magnitude of the velocity or the pressure. Determine the force of the stream on the vane system.

14·17 A jet of water having a cross-sectional area of 0.1 sq ft strikes a flat plate that is normal to the jet. The force on the plate is 100 lbf. If the jet velocity is doubled, what will be the magnitude of the force on the plate? What is the magnitude of the higher velocity?

14·18 Air flows through a long straight section of 4-in.-diameter pipe at a rate of 600 lbm/min, entering at 100 psia, 80°F, and leaving at 18.4 psia, −7°F. Determine the force of the air on the pipe.

14·19 Determine the ratio of the sonic velocity in air at 2000 K (3140°F) to that in air at 300 K (80°F).

14·20 On a certain day when sea-level atmospheric temperature is 14°C, the temperature decreases with altitude at a rate of 9 degrees C per 1000 m. How long does it take for a sound wave to travel from sea level to an altitude of 5000 m?

14·21 Determine the Mach number of the steam flow of problem 14·6.

14·22 Helium at 200 kPa, 200°C, has a velocity of 600 m/s through a cross-sectional area of 0.0680 m^2. Calculate: stagnation pressure, stagnation temperature, Mach number, and mass rate of flow.

14·23 On a certain day when sea-level atmospheric temperature is 60°F, the temperature decreases with altitude at a rate of 5 degrees F per 1000 ft. How long does it take for a sound wave to travel from sea level to an altitude of 10 000 ft?

14·24 Determine the Mach number of the steam flow of problem 14·11.

14·25 Air at 50 psia, 540°F, has a velocity of 1000 fps through a cross-sectional area of 10.0 sq in. Calculate the stagnation pressure, stagnation temperature, Mach number, and mass rate of flow.

14·26 If the sonic velocity of a breathing mixture used for divers is greatly different from that of air, voice intelligibility is degraded. One breathing mixture contains 20 percent oxygen by volume, 15 percent nitrogen, and the balance helium. Another contains the same fractions of oxygen and nitrogen, but the balance is a mixture of helium and neon in the same volumetric ratio they have in air (3.47 mol Ne/mol He). Compare the sonic velocities of these breathing mixtures with that of air.

14·27 A vertical pipe contains water. A pressure gage on the pipe reads 10 psi (68.9 kPa) higher than another gage located 30 ft (9.1 m) above the first. Is there upward flow, downward flow, no flow, or flow of indeterminate direction in the pipe?

14·28 Steam flowing through a pipe enters a pressure-reducing valve at 2.00 MPa, 250°C, with a velocity of 100 m/s. The pressure on the downstream side of the valve is 400 kPa. The pipe inside diameter is 10.0 cm on both sides of the valve. The flow is steady and adiabatic. Determine the exit velocity and the exit temperature.

14·29 Is the flow of problem 14·14 adiabatic?

14·30 Air flows through a long tapered tube that is well insulated. At section 1, where the cross-sectional area is 0.165 m^2, the air is at 140 kPa, 60°C, and has a velocity of 150 m/s. Determine the cross-sectional area at a section 2 where the air is at 70 kPa, 15°C.

14·31 Helium flows steadily through an insulated passage. At one section, where the cross-sectional area is 0.080 m^2, the helium is at 70 kPa, 25°C, with a velocity of 1800 m/s. At another section the helium is at 350 kPa, 330°C. Calculate the Mach number and the cross-sectional area at this other section. Prove which one is the upstream section and sketch the variation in cross-sectional area between the two sections.

14·32 At section 1 of a well-insulated wind tunnel through which air is flowing, P_1 = 70 kPa, T_1 = 5°C, V_1 = 100 m/s, and A_1 = 0.140 m^2. At a section 2, T_2 = 280°C. Determine the Mach number at section 2.

14·33 Nitrogen flows steadily and isentropically through a passage. At section 1, where the cross-sectional area is 0.186 m^2 the nitrogen is at 100 kPa, 5°C, and has a velocity of 600 m/s. At section 2, the velocity is 360 m/s. Calculate (a) the cross-sectional area of section 2 and (b) the Mach number at section 2.

14·34 At section 1 of a duct where the cross-sectional area is 0.0450 m^2, air at 50 kPa, 5°C, flows with a Mach number of 2.0. Further downstream at section 2 where the cross-sectional area is 0.0360 m^2, the temperature is 59°C. The flow is steady and adiabatic. Determine (*a*) the pressure at section 2 and (*b*) whether the process between 1 and 2 is reversible or irreversible, showing the calculations on which you base your conclusion.

14·35 An airplane flies through air at 88.3 kPa, 15°C, at a velocity of 200 m/s. Determine the Mach number at a point on the plane where the air velocity relative to the plane is 100 m/s.

14·36 Calculate the flight Mach number that causes the stagnation density on an aircraft to differ by 5 percent from the density of the surrounding air.

14·37 In a throttling process, the final enthalpy equals the initial enthalpy. The flow through a pressure-reducing valve is often considered to be a throttling process, even though there is actually a change in velocity of the fluid if the inlet and outline lines are the same size. Determine the change in enthalpy for the adiabatic flow of steam through a pressure-reducing valve in a 6-in.-diameter line if steam enters at 600 psia, 600°F, with a velocity of 200 fps and leaves at 100 psia.

14·38 Is the flow of problem 14·18 adiabatic?

14·39 Air flows through a long tapered tube that is well insulated. At section 1, where the cross-sectional area is 2.0 sq ft, the air is at 20 psia, 140°F, and has a velocity of 500 fps. Determine the cross-sectional area at section 2 where the air is at 10 psia, 60°F.

14·40 Air flows steadily through an insulated passage. At one section, where the cross-sectional area is 0.500 sq ft, the air is at 10.0 psia, 40°F, with a velocity of 2000 fps. At another section the air is at 50 psia, 360°F. Calculate the Mach number and the cross-sectional area at this other section. Prove which one is the upstream section.

14·41 At section 1 of a well-insulated wind tunnel through which air is flowing, $P_1 = 10$ psia, $T_1 = 40°C$, $V_1 = 300$ fps, and $A_1 = 1.48$ sq ft. At section 2, $T_2 = 540°F$. Determine the Mach number at section 2.

14·42 Nitrogen flows steadily and isentropically through a passage. At section 1, where the cross-sectional area is 2.0 sq ft, the nitrogen is at 10 psia, 40°F, and has a velocity of 2000 fps. At section 2, the velocity is 1200 fps. Calculate (*a*) the cross-sectional area of section 2 and (*b*) the Mach number at section 2.

14·43 At section 1 of a duct where the cross-sectional area is 0.50 sq ft, air at 8.0 psia, 40°F flows with a Mach number of 2.0. Farther downstream at section 2, where the cross-sectional area is 0.40 sq ft, the temperature is 137°F. The flow is steady and adiabatic. Determine (*a*) the pressure at section 2 and (*b*) whether the process between 1 and 2 is reversible or irreversible, showing the calculations on which you base your conclusion.

14·44 An airplane flies through air at 12.8 psia, 59°F, at a velocity of 600 fps. Determine the Mach number at a point on the plane where the air velocity relative to the plane is 300 fps.

14·45 Derive the following equation for the stagnation pressure of an ideal gas with constant specific heats:

$$P_t = P\left(1 + \frac{k-1}{2} M^2\right)^{k/(k-1)}$$

14·46 Air flows steadily and reversibly through an insulated passage. At one section, where the cross-sectional area is 0.0420 m^3, the air is at 70 kPa, 5°C, with a velocity of 600 m/s. At another section, farther downstream, the air is at 350 kPa. (*a*) Calculate the Mach number and the cross-sectional area at this second section. (*b*) Sketch the shape of the passage.

14·47 Air in a large tank is maintained by means of vacuum pumps and water cooling at 10 kPa, 15°C. The surrounding atmosphere is at 100 kPa, 25°C, and atmospheric air leaks into the tank through a converging nozzle, which has a throat or discharge cross-sectional area of 2.0×10^{-6} m^2. Calculate the maximum flow rate of air through the nozzle and into the tank.

14·48 Ethene (C_2H_4) escapes from a huge tank where the pressure and temperature are maintained at 400 kPa and 30°C into a furnace through a converging nozzle. The exit area of the nozzle is 10^{-3} m^2. The atmosphere is at 100 kPa, 25°C. List the assumptions you make and calculate the maximum flow rate through the nozzle.

14·49 Air at 850 kPa, 80°C, with negligible velocity enters a converging nozzle that discharges to a region where the pressure can be varied from 850 kPa down to 100 kPa. The flow is adiabatic. The nozzle throat has a cross-sectional area of 4 cm^2. Tabulate mass rate of flow versus backpressure, with 50 kPa increments of backpressure.

14·50 Air at 200 kPa, 93°C, enters a horizontal converging tube with a velocity of 60 m/s. The tube tapers from a cross-sectional area of 0.050 to 0.030 m^2 and is insulated. The pressure at the outlet is 70 kPa. Calculate the temperature and velocity of the air at the outlet.

14·51 Air at 120 psia, 180°F, with negligible velocity enters a converging nozzle that discharges to a region where the pressure is 14.6 psia. The flow is adiabatic. The nozzle throat has a cross-sectional area of 0.60 sq in. Determine the flow rate.

14·52 Air at 30 psia, 200°F, enters a horizontal converging tube with a velocity of 200 fps. The tube tapers from a cross-sectional area of 0.50 to 0.30 sq ft and is insulated. The pressure at the outlet is 10.0 psia. Calculate the temperature and velocity of the air at the outlet.

14·53 Acetylene (C_2H_2) at 100 psia and 140°F flows through a converging tube into a receiver in which the pressure is 20 psia. Assuming reversible adiabatic flow and negligible upstream velocity, calculate the velocity at the throat.

14·54 Ethene (C_2H_4) flows isentropically through a converging nozzle having a throat area of 0.1 sq ft. Ethene enters with negligible velocity at 60 psia, 200°F. The receiver pressure is 15 psia. Calculate the mass rate of flow.

14·55 Ethene (C_2H_4) flows through a converging nozzle that has a throat or outlet area of 0.050 sq ft. The ethene upstream is at 14.0 psia, 80°F, with negligible velocity, and the Mach number at the throat is 0.50. Calculate the pressure at the throat.

14·56 Air at 95 kPa, 15°C, flows through a tube at 550 m/s. If the flow is reversible and adiabatic, does the velocity increase, decrease, remain constant, or vary indeterminately in a diverging section following the tube?

14·57 Air is to be expanded isentropically through a nozzle and discharged at 101.3 kPa, 35°C, with a Mach number of 2.0 through a cross-sectional area of 35.0 cm^2. Determine the initial pressure and temperature if the initial velocity is very low.

14·58 Air flows through the test section of a wind tunnel at 70 kPa, 5°C, with a Mach number of 1.15. At the outlet of the diffuser that follows the test section, the air is at 130 kPa, 70°C. The flow is adiabatic. Determine the velocity leaving the diffuser. Is the flow reversible?

14·59 Carbon dioxide at 275 kPa, 115°C, enters a nozzle with a velocity of 180 m/s and expands reversibly and adiabatically to 135 kPa. The nozzle discharge area is 9.29 cm^2. Determine the flow rate.

14·60 Determine the inlet and throat areas of a nozzle that is to discharge hydrogen at 35 kPa, 5°C, at a Mach number of 2.0 through an exit area of 0.05 m^2. The inlet velocity is to be 150 m/s. Assume that the flow is isentropic.

14·61 Helium at 2 MPa, 200°C, expands isentropically in a nozzle to 600 kPa. If the discharge rate is 0.15 kg/s, compute the areas at sections where the pressures are 1.9, 1.7, 1.5, 1.3, 1.1, 0.9 and 0.7 MPa. Disregard the entrance velocity.

14·62 Nitrogen at 1000 kPa, 150°C, enters a nozzle with negligible velocity and expands isentropically to 140 kPa. For a flow rate of 0.20 kg/s, determine the cross-sectional area and the Mach number at sections in the nozzle where the pressures are 900, 800, 700, 600, 500, 400, 300, and 200 kPa.

14·63 Helium is to be expanded isentropically through a nozzle and discharged at 100 kPa, 40°C, with a Mach number of 2.0 through a cross-sectional area of 0.00400 m^2. Sketch the shape of the nozzle and calculate the cross-sectional area at the section of minimum area.

14·64 Air at 100 kPa, 15°C, is to leave a nozzle at a Mach number of 2.0. Flow through the nozzle is to be steady and adiabatic at a rate of 0.550 kg/s. For a nozzle efficiency of 0.900, determine the required pressure and temperature of the air that enters the nozzle at low velocity.

14·65 Air flows steadily, reversibly, and adiabatically through a passage. At one section, the cross-sectional area is 0.0620 m^2, the pressure is 20 kPa, the temperature is 20°C, and the Mach number is 3.00. At a section downstream, the pressure is 30 kPa. (*a*) Determine the velocity at the second section. (*b*) Is the cross-sectional area at the second section greater than, equal to, or smaller than that of the first section? (This can be determined without calculating the area or other quantities at the second section.)

14·66 Solve problem 14·57 for adiabatic flow with a nozzle efficiency of 90 percent. Also calculate the irreversibility of the process if the sink temperature is 15°C.

14·67 Air flows steadily and reversibly through an insulated passage. At one section, where the cross-sectional area is 0.50 sq ft, the air is at 10 psia, 40°F, with a velocity of 2000 fps. At another section, farther downstream, the air is at 50 psia. (*a*) Calculate the Mach number and the cross-sectional area at this second section. (*b*) Sketch the shape of the passage.

14·68 Air at 300 psia, 400°F, expands isentropically in a nozzle to 100 psia. If the discharge rate is 2 lbm/s, compute the areas at sections where the pressures are 250, 200, 150, and 100 psia. Disregard the entrance velocity.

14·69 Air at 150 psia, 300°F, enters a nozzle with negligible velocity and expands isentropically to 20 psia. For a flow rate of 0.50 lbm/s, plot the cross-sectional area and the Mach number through the nozzle as a function of pressure.

14·70 Air is to be expanded isentropically through a nozzle and discharged at 14.7 psia, 100°F, with a Mach number of 2.0 through a cross-sectional area of 5.42 sq in. Determine the initial pressure and temperature if the initial velocity is very low.

14·71 Air flows through the test section of a wind tunnel at 10 psia, 40°F, with a Mach number of 1.15. At the outlet of the diffuser which follows the test section, the air is at 19 psia, 160°F. The flow is adiabatic. Determine the velocity leaving the diffuser. Is the flow reversible?

14·72 Determine the inlet and throat areas of a nozzle which is to discharge air at 5.0 psia, 40°F, at a Mach number of 2.0 through an exit area of 0.5 sq ft. The inlet velocity is to be 500 fps. Assume that the flow is isentropic.

14·73 Air at 14.0 psia, 60°F, flows through a tube at 1800 fps. If the flow is reversible and adiabatic, does the velocity increase, decrease, remain constant, or vary indeterminately in a diverging section following the tube?

14·74 Air at 40 psia, 240°F, enters a nozzle with a velocity of 600 fps and expands reversibly and adiabatically to 20 psia. The nozzle discharge area is 1.44 sq in. Determine the flow rate.

14·75 Air flows steadily and reversibly through an insulated passage. At one section, where the cross-sectional area is 0.50 sq ft, the air is at 10 psia, 40°F, with a velocity of 2000 fps. At another section, farther downstream, the air is at 50 psia. Calculate the Mach number and the cross-sectional area at the second section.

14·76 Air flows steadily and isentropically through a passage. At one section, where the cross-sectional area is 2.0 sq ft, the air is at 5.0 psia, 0°F, with a velocity of 2520 fps. At a section downstream, which has the same cross-sectional area, the pressure is 70.0 psia. The flow rate is 148 lbm/s. (*a*) Determine the stagnation pressure and the stagnation temperature at section 1. (*b*) Determine the velocity at the downstream section. (*c*) Determine the net force (magnitude and direction) of the air on the passage between the two sections. (*d*) Sketch a possible shape of the flow passage between the two sections.

14·77 Air enters a supersonic diffuser at 5.0 psia, 40°F, with a Mach number of 3.0. The diffuser inlet cross-sectional area is 2.0 sq ft. Sketch the diffuser and determine the temperature at the section where the Mach number is 1.0, assuming isentropic flow.

14·78 A converging–diverging nozzle has a very large inlet area, a throat area of 0.020 sq ft, and a discharge area of 0.048 sq ft. For inlet conditions of air at 200 psia, 400°F, determine the proper backpressures for isentropic flow of air at the maximum flow rate.

14·79 For the flow conditions of problem 14·78, what is the area at a section where the pressure is 50 psia?

14·80 The nozzle described in problem 14·78 has the same inlet conditions applied but has a backpressure of 196 psia applied. Calculate the Mach number at the throat.

14·81 Propane (C_3H_8) with a stagnation pressure of 150 psia and stagnation temperature of 340°F enters a converging nozzle having an exit area of 1.8 sq. in. and an efficiency of 94 percent. The exit pressure is 100 psia. Atmospheric temperature is 70°F. (*a*) Calculate the mass rate of flow through the nozzle. (*b*) Calculate the irreversibility per pound of propane.

14·82 Air at a stagnation pressure of 80 psia and a stagnation temperature of 160°F enters a diffuser at a Mach number of 1.0. The diffuser inlet cross-sectional area is 2.40 sq ft. The flow is adiabatic. The temperature of the surrounding atmosphere is 60°F. At the diffuser outlet, the stagnation pressure is 74 psia. Determine (*a*) the irreversibility of the diffuser flow per pound of air and (*b*) the mass rate of flow.

14·83 Air flows steadily through an insulated nozzle, entering with low velocity at 150 psia, 500°F, and leaving at 15 psia, 70°F. The mass rate of flow is 10.0 lbm/s. Atmospheric conditions are 14.7 psia, 40°F. Calculate (*a*) the exit cross-sectional area, (*b*) the exit Mach number, (*c*) the irreversibility (in B/lbm).

14·84 Solve problem 14·70 for adiabatic flow with a nozzle efficiency of 90 percent. Also calculate the irreversibility of the process if the sink temperature is 60°F.

14·85 Derive the following equation for the velocity at any section of a channel through which an ideal gas with constant specific heats flows isentropically. The subscript 1 refers to any other section of the channel.

$$V = \sqrt{\frac{2\,kRT_1}{k-1}\left[1 - \left(\frac{P}{P_1}\right)^{(k-1)/k}\right] + V_1^2}$$

14·86 Derive the following equation for the mass rate of flow of an ideal gas with constant specific heats flowing isentropically from an initial condition 1 of negligible velocity:

$$\dot{m} = A\sqrt{\frac{2\,k}{k-1}\,\frac{P_1}{v_1}\left[\left(\frac{P}{P_1}\right)^{2/k} - \left(\frac{P}{P_1}\right)^{(k+1)/k}\right]}$$

14·87 Derive the following equation for P^* of an ideal gas with constant specific heats flowing isentropically:

$$\frac{P^*}{P} = \left[\frac{2}{k+1} + M^2\left(\frac{k-1}{k+1}\right)\right]^{k/(k-1)}$$

14·88 Produce a table showing, for the flow of an ideal gas with constant specific heats, A/A^*, P/P_t, ρ/ρ_t, and T/T_t as a function of M with argument values from 0.25 to 3.00, in steps of 0.25, for (*a*) $k = 1.24$ and (*b*) $k = 1.25$.

14·89 A craft flies through air at 0.5 atm, $-20°C$ ($-4°F$). Plot on one set of axes the stagnation temperature on the craft versus the craft Mach number for a range of Mach numbers from 1.0 to 4.0 for (*a*) specific heats constant at their ambient values and (*b*) variable specific heats.

14·90 A wind tunnel is to provide a steady supersonic stream of air of circular cross section 14.0 cm in diameter at 100 kPa, 15°C. Plot for isentropic flow the required upstream (where velocity is negligible) pressure and temperature versus stream Mach number for a range of Mach numbers from 1.0 to 2.4, using constant specific heats. For comparison, plot at $M = 2.4$ the required pressure and temperature if the variation of specific heats is considered.

14·91 Plot the curve called for in problem 14·90 if the gas is a mixture having a volumetric analysis of 90 percent helium and 10 percent nitrogen.

14·92 Consider a large tank that contains air initially at 30 psig. A small converging nozzle allows the air to escape to the atmosphere when it is unplugged. A certain amount of time is required for the pressure to drop to 5 psig after the plug is removed. It has been suggested that this time can be reduced by adding a diverging section to the nozzle. Write a clear statement (not longer than 600 words) explaining the effect of adding the diverging section. No equations need be used, but a sketch is likely to help.

14·93 A fable tells of a traveler who was turned out of a peasant's house because he blew on his hands to warm them and blew on his soup to cool it. The peasant would not allow any person enchanted enough to be able to blow hot and cold from the same mouth to stay in his house. Explain how it is possible to "blow either hot or cold."

14·94 A nozzle is to expand steam adiabatically from 700 to 550 kPa. The steam enters the nozzle with negligible velocity in a dry saturated condition. Should this be a converging nozzle, a converging–diverging nozzle, either type, or neither type?

14·95 Steam flows isentropically through a nozzle from 2000 to 170 kPa. If the initial temperature of the steam is 270°C and the flow rate is 0.20 kg/s, compute the areas at the throat and outlet. Disregard the entrance velocity.

14·96 Determine the areas at the throat and outlet of a nozzle that discharges 1900 kg of steam per hour. The steam is initially at 800 kPa, 260°C, and is discharged to a vacuum of 355 mm Hg. The barometric pressure is 760 mm Hg. The entrance velocity is 115 m/s. Assume reversible adiabatic flow conditions.

14·97 Determine the throat area of a nozzle that is to discharge dry saturated steam at 700 kPa with a velocity of 785 m/s at a rate of 7.35 kg/s. Assume that the flow is isentropic.

14·98 A nozzle with an efficiency of 90 percent is used to expand steam adiabatically from 2000 kPa, 280°C, to 400 kPa at a rate of 900 kg/h. The initial velocity is negligible. The lowest temperature in the surroundings is 20°C. Compute the nozzle discharge area and the irreversibility of the process per kilogram of steam.

14·99 A nozzle is to expand steam adiabatically from 100 to 80 psia. The steam enters the nozzle with negligible velocity in a dry saturated condition. Should this be a converging nozzle, a converging-diverging nozzle, either type, or neither type?

14·100 Steam flows isentropically through a nozzle from 300 to 25 psia. If the initial temperature of the steam is 520°F and the flow rate is 0.5 lbm/s, compute the areas at the throat and outlet. Neglect the entrance velocity.

14·101 Determine the areas at the throat and outlet of a nozzle which discharges 4200 lbm of steam per hour. The steam is initially at 85 psia, 500°F, and is discharged to a vacuum of 14 in. of mercury. The barometric pressure is 30 in. of mercury. The entrance velocity is 375 fps. Assume reversible adiabatic flow conditions.

14·102 Determine the throat area of a nozzle which is to discharge dry saturated steam at 100 psia with a velocity of 2580 fps at a rate of 16.2 lbm/s. Assume that the flow is isentropic.

14·103 A nozzle with an efficiency of 90 percent is used to expand steam adiabatically from 300 psia, 550°F, to 60 psia at a rate of 2000 lbm/h. The initial velocity is negligible. The lowest temperature in the surroundings is 70°F. Compute the nozzle discharge area and the irreversibility of the process per pound of steam.

14·104 Shown in the figure is a sketch of a steam accumulator. Accumulators are used to accommodate varying steam flow demands of systems that periodically require more steam flow than can be provided by the steam generators supplying them. In the figure, steam comes from the steam generators at A, and the system requiring steam is connected to B. When the demand of the system is less than the capacity of the steam generators, the pressure near the accumulator rises, the surplus of the steam flow at A over that at B goes into the accumulator and partially condenses. When the steam flow demand at B exceeds the capacity of the steam generators, the pressure near the accumulator drops slightly, and some of the saturated water in the accumulator flashes into steam to provide the additional steam flow required at B. It has been found that a pipe rupture between A and B causes the accumulator to be driven downward. It has been proposed to solve this problem by inserting a converging nozzle within the pipe near the accumulator, as shown at C in the figure. Some people have been discussing this proposal and have not fully agreed on (1) why the accumulators are driven downward when the steam line ruptures and (2) whether the converging section will eliminate the problem without causing any undesirable effects. Write a clear statement (not more than 600 words long) that clarifies these points for this group. Equations should be used only sparingly if at all, but sketches may be helpful.

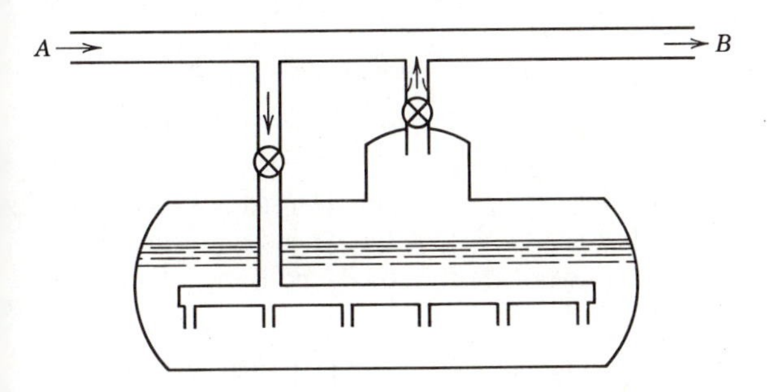

Problem 14·104.

14·105 Sketch hs and $h_t s$ diagrams for an ideal gas flowing adiabatically through (*a*) a long pipe line, (*b*) a nozzle, (*c*) a diffuser.

14·106 For the steady flow of an ideal gas with constant specific heats in a long insulated pipe, sketch the variation of the following quantities versus distance along the pipe: P, T, P_t, T_t, T^*, V, ρ, s.

14·107 Air at 380 kPa, 128°C, enters a 20-cm-diameter pipe at 30 m/s and leaves at 118°C. The flow is adiabatic. Calculate the Mach number at the outlet.

14·108 At one section in a long, well-insulated pipeline, 5 cm in inside diameter, methane at 400 kPa, 50°C, has a velocity of 100 m/s. Determine the minimum pressure that can exist in the pipe.

14·109 For an atmospheric condition of 101.3 kPa, 20°C, determine the change in stream availability between the two specified sections in problem 14·107.

14·110 Ethane flows through a well-insulated 10-cm-diameter pipe at a rate of 275 kg/min. At one section the ethane is at 700 kPa, 25°C. Determine the minimum pressure and the maximum velocity that can occur in the pipe.

14·111 Plot the Fanno line for the flow of problem 14·110.

14·112 At one section in an insulated pipe, steam at 2.7 MPa, 260°C, has a velocity of 120 m/s. Plot the Fanno line for this flow, and determine the minimum pressure and the maximum velocity that can occur in the pipe.

14·113 Consider the steady flow of air in a well-insulated pipe. The pipe diameter is 10 cm. At one section, called section *a*, P_a = 120 kPa, T_a = 5°C, and V_a = 500 m/s. At another section, called section *b*, V_b = 400 m/s. Determine, showing all work clearly, (*a*) P_b and (*b*) the direction of flow, from *a* to *b* or from *b* to *a*.

14·114 Air at 55 psia, 263°F, enters an 8-in.-diameter pipe at 100 fps and leaves at 243°F. The flow is adiabatic. Calculate the Mach number at the outlet.

14·115 At one section in a long, well-insulated pipeline 2 in. in inside diameter air at 60 psia, 120°F, has a velocity of 350 fps. Determine the minimum pressure that can exist in the pipe.

14·116 For an atmospheric condition of 14.7 psia, 70°F, determine the change in stream availability between the two specified sections in problem 14·110.

14·117 Air flows through a well-insulated 4-in.-diameter pipe at a rate of 600 lbm/min. At one section the air is at 100 psia, 80°F. Determine the minimum pressure and the maximum velocity that can occur in the pipe.

14·118 Plot the Fanno line for the flow of problem 14·113.

14·119 At one section in an insulated pipe, steam at 400 psia, 500°F, has a velocity of 400 fps. Plot the Fanno line for this flow, and determine the minimum pressure and the maximum velocity that can occur in the pipe.

14·120 Consider the steady flow of air in a well-insulated pipe. The pipe diameter is 3.0 in. At one section, called section *a*, P_a = 10 psia, T_a = 40°F, and V_a = 1500 fps. At another section, called section *b*, V_b = 1200 fps. Determine, showing all work clearly, (*a*) P_b and (*b*) the direction of flow, from *a* to *b* or from *b* to *a*.

14·121 Methane is heated as it passes through a constant-area duct at a rate of 0.56 kg/s. The duct cross-sectional area is 0.0080 m^2. The gas enters at 110 kPa, 35°C; and a control system maintains the outlet temperature at 300°C. Assume that the flow is frictionless and one dimensional. Determine the rate of heat transfer.

14·122 State whether each of the following increases, decreases, remains constant, or varies indeterminately across a normal shock: pressure, temperature, density, velocity, Mach number, entropy, stream availability.

14·123 A gas flows steadily through a converging–diverging nozzle, entering at section *i* where the velocity is low, passing through the throat *th*, and discharging at section *e*. A normal shock occurs in the nozzle between sections *x* and *y*. The flow is adiabatic and is reversible except at the shock. (*a*) Sketch the pressure–length and *Ts* diagrams. (*b*) Circle the correct symbol (>, =, <, or I) in each of the following statements below (where I means "is indeterminate with respect to")

T_e	$>$	$=$	$<$	I	T_i
s_e	$>$	$=$	$<$	I	s_i
s_e	$>$	$=$	$<$	I	s_{th}
V_e	$>$	$=$	$<$	I	V_{th}
ρ_e	$>$	$=$	$<$	I	ρ_{th}
M_e	$>$	$=$	$<$	I	M_y
c_y	$>$	$=$	$<$	I	c_x
s_x	$>$	$=$	$<$	I	s_{th}
V_y	$>$	$=$	$<$	I	V_{th}
M_e	$>$	$=$	$<$	I	M_{th}
T_y	$>$	$=$	$<$	I	T_x
P_{ty}	$>$	$=$	$<$	I	P_{tx}

14·124 Air flows through a long insulated pipe. At inlet, the velocity is supersonic. Inside the pipe a normal shock occurs. Sketch for this flow *hs, hl, Pl,* and *sl* diagrams where l is distance measured along the pipe.

14·125 Air at 95 kPa, 15°C, with a Mach number of 2.0, enters a normal shock. Determine the downstream pressure, temperature, and velocity and the irreversibility of the process if the sink temperature is taken as 15°C.

14·126 Air flows steadily through an insulated pipe that has a diameter of 0.100 m. At section 1, $M_1 = 2.00$, $V_1 = 680.3$ m/s, $P_1 = 100$ kPa, $T_1 = 15$°C. A normal shock occurs just downstream from section 1, and at section 2, which is just downstream of the shock, $V_2 = 255.1$ m/s. Determine (*a*) the pressure at section 2, P_2, and (*b*) the irreversibility of the flow between sections 1 and 2.

14·127 Consider the nozzle of Example 14·7. Air flows through the nozzle with the same inlet conditions but with a backpressure of 95 kPa applied. Determine the outlet velocity and the maximum velocity reached in the nozzle.

14·128 Refer to Example 14·11. For the same nozzle and the same inlet conditions, determine the pressure at which a normal shock occurs if a backpressure of 69 kPa is applied.

14·129 A nozzle is properly designed for isentropic flow to discharge nitrogen at 40 kPa, 20°C, with a Mach number of 1.80 through a cross-sectional area of 0.0160 m^2. Assuming that design inlet conditions are maintained, determine (*a*) the discharge pressure that causes a normal shock to stand in the exit plane of the nozzle and (*b*) the pressure at which the shock occurs if the backpressure is 175 kPa.

14·130 For the purpose of making some measurements in a normal shock, it is desired to have a flow of air such that a normal shock stands at the exit plane of a nozzle that is to have a discharge cross-sectional area of 0.0120 m^2 (18.6 in^2). The Mach number entering the shock is to be 2.00, and the conditions of the air as it leaves the shock are to be 100 kPa (14.5 psi), 25°C (77°F). Determine the throat area of the nozzle and the pressure and temperature at which air at low velocity must be supplied to the nozzle.

14·131 Air at 14.0 psia, 60°F, with a Mach number of 2.0 enters a normal shock. Determine the downstream pressure, temperature, and velocity and the irreversibility of the process, using a sink temperature of 60°F.

14·132 A nozzle is properly designed for isentropic flow to discharge helium at 20 psia, 50°F, with a Mach number of 1.50 through a cross-sectional area of 1.400 sq in. Assuming that design inlet conditions are maintained, determine (*a*) the discharge pressure that causes a normal shock to stand in the exit plane of the nozzle and (*b*) the pressure at which the shock occurs if the backpressure is 65 kPa.

14·133 Under some conditions, as a fluid flows in a pipe, friction causes the fluid velocity to increase and the fluid temperature to decrease. Also, under some conditions adding heat to a flowing fluid causes its temperature to decrease. These phenomena are not what one is likely to expect, so write a convincing physical explanation that would resolve any reasonable doubts on these matters.

CHAPTER FIFTEEN

Compression and Expansion Processes: Fluid Machines

Machines that compress flowing fluids or that produce work from the expansion of flowing fluids are components of many engineering systems. These machines may generally be classified either as turbo- or dynamic machines or as positive displacement machines. In turbomachines work is done on or by a fluid by means of a rotor that exerts a torque on the fluid, causing a change in angular momentum of the fluid as it passes through the rotor. Examples of turbomachines are centrifugal and axial-flow pumps, fans, blowers, compressors, and turbines. The designation *pump* is generally used in connection with liquids. *Fans* move gases against such small pressure differences that density changes are negligible. Machines that move gases against somewhat greater pressure differences are called *blowers*, and *compressors* are used for still greater pressure differences. There are no sharply defined lines of demarcation among these designations. Examples of positive-displacement machines are reciprocating and rotary engines, pumps, blowers, and compressors. Gear pumps and vane pumps are rotary machines.

In this chapter we first review the application of the first and second laws to all types of compression and expansion processes and define some performance parameters. Then with the help of the principles of mechanics we consider the energy transfer between a fluid and a rotor for various types of turbomachinery. In conclusion, we look briefly at some performance features of reciprocating machines.

15·1 Steady-Flow Compression Processes

To any steady-flow system we can apply (1) the first law for determining energy quantities, (2) the relationship $w_{in} = \int v\, dP + \Delta ke + \Delta pe$ for determining work if the process is reversible, and (3) the second law and its corollaries to determine the possibility of a process and its irreversibility. These are the tools we now use in analyzing steady-flow compression and expansion processes.

For given end states it is generally desirable to reduce to a minimum the work input to a compressor. This of course means that the actual process should approximate as closely as possible a reversible process between the two specified states. Often, however, the purpose of a compressor is to bring a fluid not to a specified final state but only to a specified final pressure. This is usually the case when the compressed fluid is to be stored before it is used further. Therefore, we should first investigate the various types of reversible processes between an initial state and some given final pressure to see which involves the least work. Figure 15·1 shows on *Pv* and *Ts* diagrams for a gas three different reversible compression paths between a state 1 and the same final pressure P_2. Process 1-*a* is a reversible adiabatic compression, process 1-*b* is a reversible isothermal compression, and process 1-*c* is a reversible compression with some heat removed from the gas but not enough to hold the temperature constant. Process 1-*c* might be a polytropic process: one for which Pv^n = constant. If it is, then $k > n > 1$ because the cases of $n = k$ and $n = 1$ are the reversible adiabatic (process 1-*a*) and reversible isothermal (process 1-*b*) cases, respectively.

For a reversible steady-flow process

$$w_{in} = \int v\,dP + \Delta ke + \Delta pe \tag{1·10}$$

and the integral term is represented by an area on a *Pv* diagram. Inspection of the *Pv* diagram of Fig. 15·1 shows that, for equal changes in kinetic and potential energy, the work required to compress a gas reversibly from a state 1 to a final pressure P_2 is reduced as the compression path approaches the isothermal. (The question might be raised as to the advisability of cooling the gas during compression to a temperature lower than T_1. If a coolant at a sufficiently low temperature is available for doing this, it would be better, as shown by the solution of Problem 15·9, to cool the gas before it enters the compressor; therefore we generally consider the compressor inlet temperature to be the minimum temperature attainable by means of heat transfer to a coolant.)

In an actual machine, the amount of heat that can be transferred during the compression process is limited by both the small surface area available for heat transfer and the short length of time required for the gas to pass through the machine. Consequently, the ideal process that may be simulated in actual cooled compressor is a polytropic process with *n* closer to *k* than to 1.

Sometimes no cooling is desired during a compression process, even though the lack of cooling increases the work input for given pressure limits. This is the case in a simple

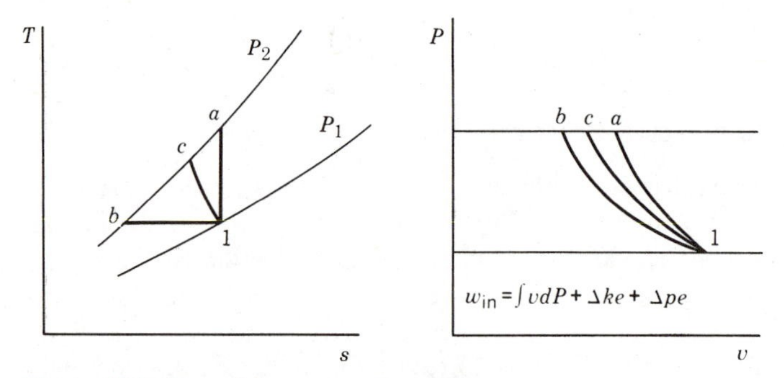

Figure 15·1. Reversible compression paths.

gas-turbine power plant where immediately after compression the gas is to be heated by means of fuel. For this application a reversible adiabatic compression is the most desirable from the standpoint of overall plant efficiency. (See Chapter 16.)

Comparisons between actual and ideal compressor performance are expressed by compressor efficiencies. We have seen above that the ideal compression process may be either reversible isothermal, polytropic, or isentropic, so care must be taken to use the proper ideal process as a basis of comparison in any particular case. We will introduce here only one compressor efficiency, the *adiabatic compressor efficiency,* which is defined by

$$\eta_C \equiv \frac{\text{work of } \textit{reversible adiabatic} \text{ compression from state 1 to } P_2}{\text{work of } \textit{actual adiabatic} \text{ compression from state 1 to } P_2}$$

Slightly different definitions are also in use. For the ideal compression the kinetic energy change is often considered to be negligible. Referring to Fig. 15·2 and applying the first law, we have

$$\eta_C = \frac{\text{input work}_i}{\text{input work}_a} = \frac{h_{2i} - h_1}{h_{2a} - h_1 + \dfrac{V_{2a}^2 - V_1^2}{2g_c}} = \frac{h_{2i} - h_1}{h_{t2a} - h_{t1}}$$

If there is no change in kinetic energy, then

$$\eta_C = \frac{h_{2i} - h_1}{h_{2a} - h_1}$$

Following a practice which is rather general, especially in the gas-turbine literature, we shall in this book use the term *compressor efficiency,* without modifiers, to mean *adiabatic compressor efficiency* as defined above. Remember that *this parameter is useful only in connection with adiabatic compressors.*

Multistage compressors consist of two or more compressors, each of which is called a *stage,* in series. If each stage operates adiabatically, and there is no heat transfer from the gas as it passes from one stage to another, the entire compressor operates adiabatically. Then we can apply the definition of (adiabatic) compressor efficiency to individual stages and to the entire compressor. It should be noticed, however, that, if the efficiency of each

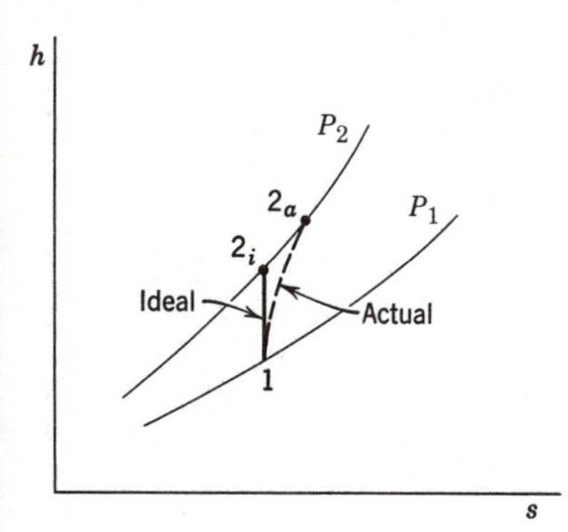

Figure 15·2. Reversible and irreversible adiabatic compression.

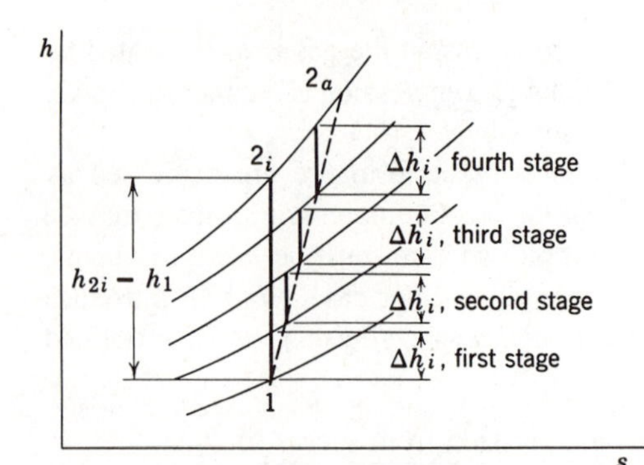

Figure 15·3. Multistage compression.

stage is the same (and less than 100 percent), the efficiency of the entire compressor is less than that of the stages. This can be seen by reference to Fig. 15·3 which shows the actual adiabatic compression path through a four-stage compressor. Let $\Delta ke = 0$ for each stage, and therefore for the compressor. If each stage has the same efficiency, the ratio $\Delta h_i/\Delta h_a$ is the same for each stage. For the entire compressor the efficiency is $(h_{2i} - h_1)/(h_{2a} - h_1)$. Notice that $h_{2a} - h_1 = \Sigma\, \Delta h_a$, but, because of the divergence of constant-pressure lines on the diagram in the direction of increasing entropy, $h_{2i} - h_1 < \Sigma\, \Delta h_i$; therefore, the efficiency of the entire compressor is less than that of the individual stages.

Multistaging of compressors is sometimes used to allow for cooling between the stages to reduce the total work input. Gas can be cooled more effectively in heat exchangers called intercoolers between stages than during the compression process. Figure 15·4 shows a polytropic compression process 1-*a*. If $\Delta ke = 0$, the work done on the gas is represented by the area 1-*a*-*j*-*k*-1 on the *Pv* diagram. A constant-temperature line is shown as 1-*x*. If the polytropic compression from state 1 to P_2 is divided into two parts, 1-*c* and *d*-*e*, with constant pressure cooling to $T_d = T_1$ between them, the work done is represented by area 1-*c*-*d*-*e*-*j*-*k*-1. The area *c*-*a*-*e*-*d*-*c* represents the work saved by means of the two-stage compression with intercooling to the initial temperature.

Inspection of Fig. 15·4 indicates that for specified values of P_1 and P_2 there is an optimum pressure for intercooling. (This conclusion is suggested by reflection on extreme cases. If the process proceeds for a very short distance along the polytropic path before intercooling, or if it proceeds almost to *a* before intercooling, the work saved in either

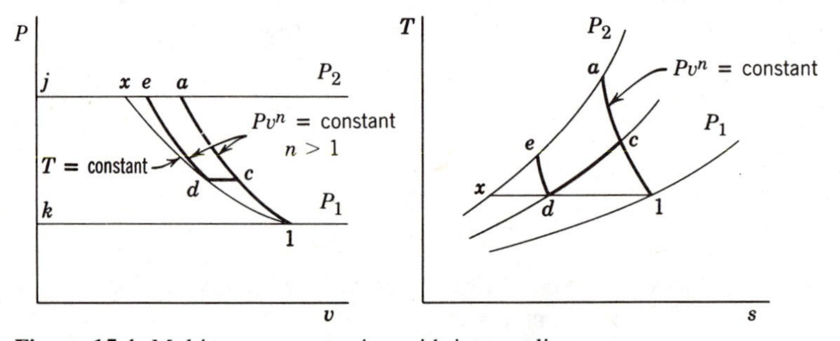

Figure 15·4. Multistage compression with intercooling.

case is small compared to that saved with an interstage pressure such as shown in Fig. 15·4.) Call the optimum interstage pressure P_i, so that, for two stages with $\Delta ke = 0$

$$w_{\text{in}} = \int_1^{P_i} v\, dP + \int_{P_i}^{P_2} v\, dP$$

If we expand this equation for the case of Pv^n = constant and then set $dw/dP_i = 0$, we find the conditions for minimum work. The result is that if the intercooling brings the gas to its initial temperature, the minimum work input is required when $P_2/P_i = P_i/P_1$, or when the work is the same for the two stages. Further analysis shows that no matter how many stages there are, minimum work input is required when the total work is divided equally among the stages. The pressure ratio is then the same for each stage.

Another means of reducing the work required to compress a gas is to inject a liquid into the gas being compressed. As it evaporates, the liquid takes energy from the gas and thereby reduces the gas temperature rise.

Example 15·1. A compressor operates with inlet air at 92 kPa, 15°C, and a pressure ratio (i.e., ratio of discharge pressure to inlet pressure) of 3. Kinetic energy changes are negligible. The compressor efficiency is 70 percent. Determine the irreversibility per kilogram of air.

Analysis. We will calculate the irreversibility as $i = T_0 \Delta s$. Since no information to the contrary is given, we will assume that the temperature of the inlet air is the temperature of the surrounding atmosphere which can be considered the sink temperature. In order to calculate Δs, we must determine the end states of the process. The initial state is specified. The final state can be determined from the initial state and the work done, since the process is adiabatic, so how is the work determined? The work done can be obtained from the ideal (isentropic) work and the compressor efficiency. The ideal work can be obtained by means of the first law and the ideal-gas enthalpy-temperature relationship after the ideal end state temperature is calculated. For the small temperature range likely to be encountered we will use constant values of $c_p = 1.01$ kJ/kg·K and $k = 1.4$. (The calculations can also be made using air tables, but for the limited temperature range involved there is but a slight gain in accuracy.)

Solution. Following the procedure outlined in the analysis above

$$T_{2i} = T_1\left(\frac{P_2}{P_1}\right)^{(k-1)/k} = 288(3)^{0.286} = 394 \text{ K}$$

$$\text{Input work}_i = h_{2i} - h_1 = c_p(T_{2i} - T_1) = 1.01(394 - 288) = 107 \text{ kJ/kg}$$

$$\text{Input work}_a = \frac{\text{input work}_i}{\eta_C} = \frac{107}{0.7} = 153 \text{ kJ/kg}$$

$$T_{2a} - T_1 = \frac{h_{2a} - h_1}{c_p} = \frac{\text{input work}_a}{c_p} = \frac{153}{1.01} = 151 \text{ degrees C}$$

$$T_{2a} = T_1 + 151 = 288 + 151 = 439 \text{ K}$$

$$\Delta s = s_{2a} - s_1 = \int_1^{2a} ds = \int_1^{2a} \frac{dh}{T} - \int_1^{2a} \frac{v\, dP}{T} = c_p \ln \frac{T_{2a}}{T_1} - R \ln \frac{P_2}{P_1}$$

$$= 1.01 \ln \frac{439}{288} - 0.287 \ln 3 = 0.110 \text{ kJ/kg·K}$$

$$i = T_0 \Delta s = 288(0.110) = 31.7 \text{ kJ/kg}$$

This result means that, of the 153 kJ/kg of energy added to the air as work, 31.7 kJ/kg can never be reconverted into work. For the reversible process used as a standard of comparison and involved in the definition of compressor efficiency, only 107 kJ/kg of work is required, and of course it can all be reconverted into work by reversing the process. The difference between the actual work and the ideal work is greater than the irreversibility because some of the additional work required by the actual process can be reconverted into work after the actual process has occurred.

15·2 Steady-Flow Expansion Processes

The process in a turbine or expansion engine is usually adiabatic because the heat addition process in a steady-flow power plant is usually separate from the expansion process. Furthermore, the expansion machine is insulated to prevent heat loss to the surroundings that would reduce the work output. The ideal process used as a basis of comparison is therefore the reversible adiabatic or isentropic process between the same initial state and the same final total pressure.

Adiabatic turbine efficiency, which we will call simply *turbine efficiency* or the efficiency of a turbine, is defined as

$$\eta_T \equiv \frac{\text{work of } \textit{actual adiabatic} \text{ expansion}}{\text{work of } \textit{reversible adiabatic} \text{ expansion between same initial state and same final total pressure}}$$

This term is also used in connection with positive-displacement expansion machines such as steam engines and compressed-air motors. It has also been called engine efficiency, and slightly different definitions are occasionally found. Applying the first law, we have

$$\eta_T = \frac{\text{work}_a}{\text{work}_i} = \frac{h_1 - h_{2a} + \dfrac{V_1^2 - V_{2a}^2}{2}}{h_1 - h_{2i} + \dfrac{V_1^2 - V_{2i}^2}{2}} = \frac{h_{t1} - h_{t2a}}{h_{t1} - h_{t2i}}$$

where state 2_i is one for which the entropy is the same as at state 1 and the total pressure

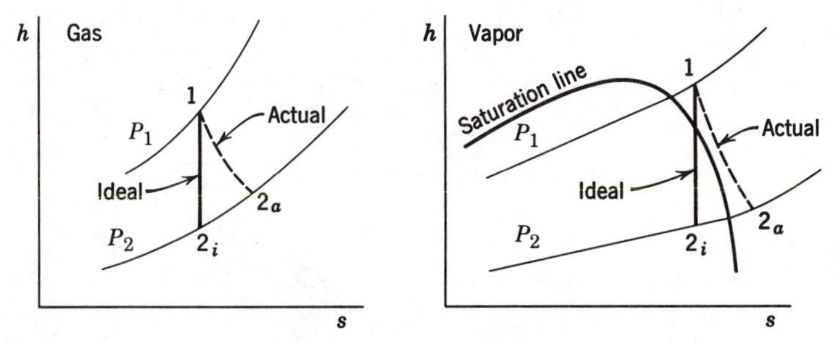

Figure 15·5. Reversible and irreversible adiabatic expansion.

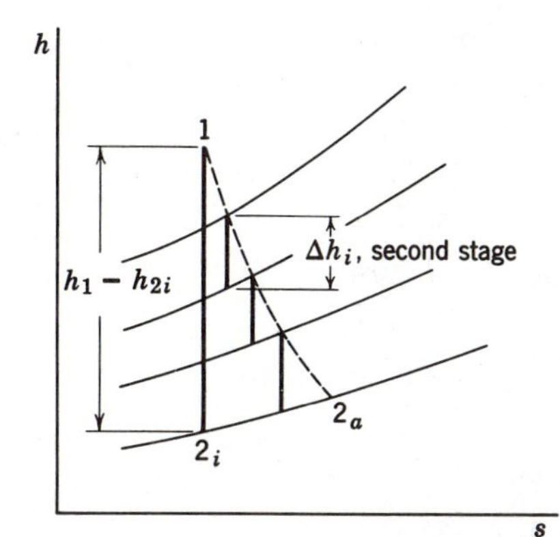

Figure 15·6. Multistage expansion.

is the same as at the actual exhaust condition. If there is no change in kinetic energy, then

$$\eta_T = \frac{h_1 - h_{2a}}{h_1 - h_{2i}}$$

Actual and ideal expansion paths are shown in Fig. 15·5 for an ideal gas and for a vapor.

The efficiency of a multistage turbine can be higher than the efficiency of any of its stages. For an illustration of this point, refer to Fig. 15·6 which shows the adiabatic expansion path through a four-stage turbine. Let $\Delta ke = 0$ for each stage and therefore for the entire turbine. If each stage has the same efficiency, the ratio $\Delta h_a/\Delta h_i$ is the same for each stage. For the entire turbine, $h_1 - h_{2a} = \Sigma\Delta h_a$, but, because of the divergence of constant-pressure lines on the diagram in the direction of increasing entropy, $|h_1 - h_{2i}| < \Sigma|\Delta h_i|$; consequently, the efficiency of the entire turbine is higher than that of the individual stages. This paradoxical behavior is sometimes referred to as the *reheat effect,* although this term is a misnomer because the entire process under consideration is adiabatic.

Adiabatic turbine effectiveness, which is often called just *turbine effectiveness,* is defined as

$$\text{turbine effectiveness} \equiv \frac{\text{work}}{-\Delta b}$$

where b is the Darrieus function, $h - T_0 s$. Turbine effectiveness, like turbine efficiency, can be applied to individual stages or groups of stages in a turbine as well as to the entire machine. Effectiveness gives a clearer picture of the overall effects of losses. For example, if two stages have the same efficiency, the stage at the higher pressure has a higher effectiveness and consequently a lower irreversibility.

Example 15·2. Determine the exhaust temperature of an air turbine that operates adiabatically between 70 psia, 1540°F, and 15 psia with an efficiency of 80 percent and no change in kinetic energy.

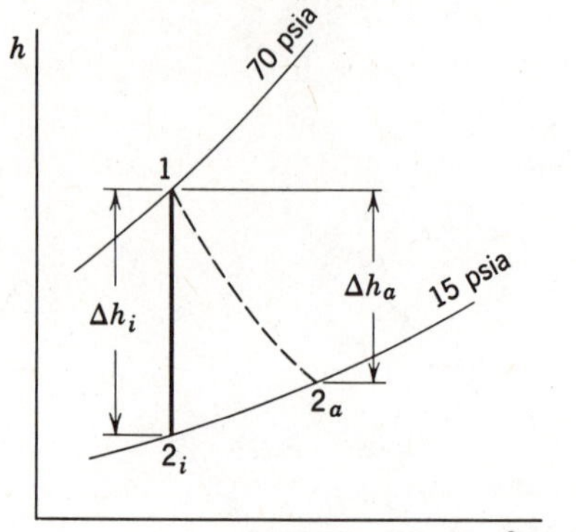

Example 15·2.

Solution. An *hs* diagram showing the actual process and the corresponding ideal process is made first. We will use the Keenan, Chao, and Kaye *Gas Tables* (or Table A·5). To determine the ideal exhaust condition,

$$p_{r2i} = p_{r1}\frac{P_2}{P_1} = 173.10\,\frac{15}{70} = 37.09$$

$$h_{2i} = f(p_{r2i}) = 329.7\text{ B/lbm}$$

Then from the definition of turbine efficiency,

$$h_1 - h_{2a} = \eta_T(h_1 - h_{2i})$$

$$h_{2a} = h_1 - \eta_T(h_1 - h_{2i})$$

$$= 504.7 - 0.8(504.7 - 329.7) = 364.7\text{ B/lbm}$$

From the tables,

$$T_{2a} = f(h_{2a}) = 1483\text{ R} = 1023°\text{F}$$

15·3 Incompressible Flow through Machines

Consider the steady flow of an incompressible (ρ = constant) fluid through a machine. The first law becomes

$$q - w = u_2 - u_1 + P_2v - P_1v + \Delta ke + \Delta pe \qquad \text{(a)}$$

If the process is also reversible,

$$-w = \int v\,dP + \Delta ke + \Delta pe \qquad (1\cdot 10)$$

$$= (P_2 - P_1)v + \Delta ke + \Delta pe \qquad \text{(b)}$$

Comparison of Eqs. a and b shows that for reversible incompressible steady flow $q - \Delta u = 0$. If the process is irreversible, then $q - \Delta u \neq 0$, but the heat transfer and internal energy change are usually small and difficult to measure in comparison with the other energy quantities in Eqs. a and b; therefore the quantity $(q - \Delta u)$ in incompressible flow is occasionally given a name such as "friction loss" or "head loss due to friction" and is measured indirectly by measuring all the other quantities in an energy balance such as Eq. a.

Incompressible flow analyses are not restricted to liquids. A gas or vapor can be treated as incompressible if its density changes very little during a process.* For example, atmospheric air flowing through a fan and undergoing an overall pressure change of a few inches of water can be treated as incompressible because its density change is so small. The greater the density change is, the greater are the errors that result from treating the flow as incompressible.

It must also be remembered that liquids are not entirely incompressible: Their densities are affected by pressure, and for large pressure differences isentropic pumping may cause appreciable changes in temperature. This can be seen from compressed liquid tables such as Table A·1.4. (Reference 14·3 presents *hv* diagrams for compressed liquid that are convenient for determining enthalpy and temperature changes in isentropic processes.)

Pump and fan efficiencies for incompressible flow are defined just as compressor efficiency is except that for incompressible flow the kinetic energy change is always the same for the actual and the ideal processes. Thus

$$\eta_P \equiv \frac{\text{input work}_i}{\text{input work}_a} = \frac{v\,\Delta P + \Delta ke + \Delta pe}{v\,\Delta P + \Delta ke + \Delta pe + \Delta u - q}$$

For an actual pump (or fan) the numerator is evaluated by measurements of pressure, velocity, elevation, etc., and the denominator is obtained by power and flow rate measurements from which the actual work input can be calculated. (The shaft input work to a pump exceeds the work done on the fluid at least by the work required to overcome bearing and shaft seal friction.)

The term *head* that is frequently used in connection with pumps, fans, and hydraulic turbines refers to an energy quantity per unit mass or per unit weight. When energy is expressed on a unit weight basis, units such as ft·lbf/lbf can be reduced to a unit of length alone, so expressions like "a velocity head of 40 ft" (meaning kinetic energy of 40 ft·lbf/lbf weight) are common. For most engineering analyses, treating the weight of a given mass of substance as constant is satisfactory, but in a varying gravitational field (as in a high-altitude vehicle) it is much better to use a mass basis instead of a weight basis.

An important phenomenon in the flow of liquids is *cavitation*. Changes in the velocity of a liquid flowing through a passage are accompanied by pressure changes. If the pressure is lowered to the saturation pressure corresponding to the liquid temperature, vapor bubbles form. If these bubbles are then carried by the stream to a region where the pressure is higher than the vapor pressure, they collapse. This alternate formation and collapse of vapor bubbles in a flowing liquid is called *cavitation*. When the bubbles collapse on a solid wall of a flow passage such as a pump rotor, very high impact pressures are exerted on minute areas of the wall and serious damage to the wall may result. Also, the vapor bubbles may disrupt the liquid flow to the detriment of the overall performance of the machine.

*It has been suggested that such flows be referred to as *uncompressed* rather than *incompressible*. This terminology has not been widely adopted, but it is mentioned for its descriptive value.

15·4 Energy Transfer between a Fluid and a Rotor

The performance of turbomachines depends on the mechanism of energy transfer between a fluid and a rotor. A working knowledge of this subject cannot be obtained from a discussion as limited in scope as the one in this section, but we derive the general equations for the torque and the work transfer between a fluid and a rotor for steady one-dimensional flow in order to give you a picture of the general mechanism. Flow through a rotor is usually two or three dimensional, but for simplicity we restrict our attention here to one-dimensional flow, which in many cases is a satisfactory approximation.

Consider the flow of a fluid through a rotating rotor as shown in the general diagram of Fig. 15·7. Some particular types of rotors are shown in Fig 15·8 to illustrate that the general diagram applies to various rotor types which differ markedly from each other. (In none of these rotors is the flow one-dimensional except as an approximation.) The linear velocity of any point on the rotor is u. The radius to any point is r, and the angular velocity of the rotor is ω, so that $u = r\omega$. The fluid velocity V at any point has three mutually perpendicular components:

V_{rad}, the radical component.
V_{ax}, the axial component (parallel to the rotor axis of rotation).
V_u, the tangential component.

The torque of the fluid on the rotor is equal and opposite in sign to the torque of the rotor on the fluid. The net torque on the fluid is equal to the time rate of change of angular momentum of the fluid. The angular momentum with respect to the axis of rotation of a mass m of fluid entering is mr_1V_{u1}, and that of a mass m of fluid leaving is mr_2V_{u2}. For this mass m the change in angular momentum is $m(r_2V_{u2} - r_1V_{u1})$, and for steady flow the rate of change of angular momentum is $\dot{m}(r_2V_{u2} - r_1V_{u1})$. Thus, for the steady flow

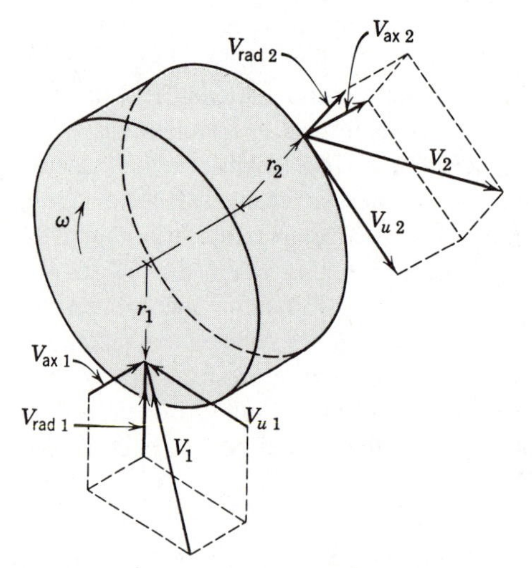

Figure 15·7. Flow entering and leaving a rotor. General diagram.

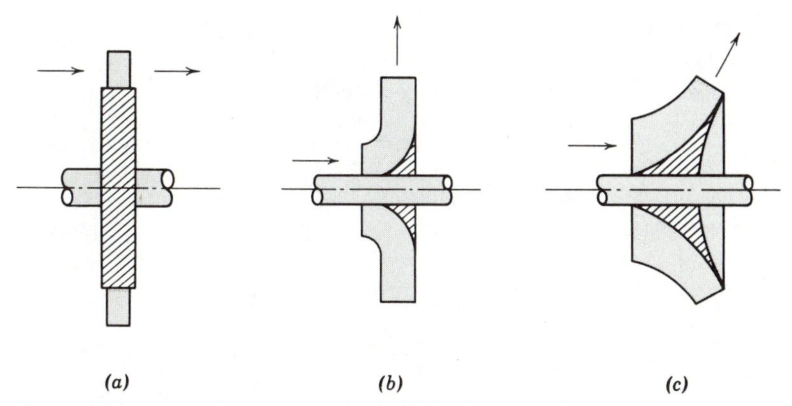

Figure 15·8. Types of rotors. (*a*) Axial. (*b*) Radial. (*c*) Mixed.

of a fluid the basic relation of mechanics,

$$\text{torque} = \frac{d}{d\tau}(mrV_u)$$

becomes

$$\text{torque on fluid} = \dot{m}(r_2V_{u2} - r_1V_{u1})$$

This is the basic dynamic equation for the steady one-dimensional flow through a rotor. (Remember that one-dimensional means that properties are uniform in any cross section normal to the direction of flow; it does not restrict the flow to a constant direction.) The fluid pressure does not appear in the torque equation because the pressure does not vary circumferentially. No matter what the path inside the rotor is, the net torque depends only on the flow rate, the inlet and outlet radii, and the tangential components of the fluid velocity at inlet and outlet. The inlet and outlet velocities depend on the flow rate, the rotor speed, and the rotor geometry, but this dependence is not a simple one because the velocities at inlet and outlet are not necessarily tangent to the rotor vanes or blades.

The power delivered to the fluid by a rotor is

$$\dot{W}_{\text{in}} = (\text{torque on fluid})\omega = \dot{m}\omega(r_2V_{u2} - r_1V_{u1})$$

and the work done on the fluid is

$$w_{\text{in}} = \frac{\dot{W}_{\text{in}}}{\dot{m}} = \omega(r_2V_{u2} - r_1V_{u1}) = (u_2V_{u2} - u_1V_{u1}) \tag{15·1}$$

The product uV_u is sometimes called the "whirl" or the "whirl velocity," so the work equation can be written as w_{in} = change in whirl. No matter what the flow path is through the rotor, work is done only if there is a change in whirl.

Equation 15·1 applies only to the flow through a rotor. For example, sections 1 and 2 in Eq. 15·1 cannot be taken as the inlet of one rotor and the outlet of a subsequent rotor in a multistage machine, because the torque exerted on the fluid by a stationary part of the machine does no work on the fluid. The change in the product uV_u between the inlet and outlet of an entire machine, in fact, is often zero.

In addition to the above equation for work, which was derived from dynamic considerations, the first law may be used to determine the work done on a fluid by a rotor. If the process is reversible, we can also use the expression

$$w_{\text{in}} = \int v\,dP + \Delta ke + \Delta pe \tag{1·10}$$

This equation and the first law can also be applied to the flow through a rotor with the frame of reference fixed to the rotor. In such a frame of reference, or from the viewpoint of an observer moving with the rotor, there is no work done, and of course the velocities used must be velocities relative to the rotor. The first law and Eq. 1·10 become

$$q = h_2 - h_1 + \frac{V_{r2}^2 - V_{r1}^2}{2} + \Delta pe$$

and

$$0 = \int v\,dP + \frac{V_{r2}^2 - V_{r1}^2}{2} + \Delta pe$$

where V_r designates a velocity relative to the rotor. The flow through a rotor is usually adiabatic, and the change in potential energy across the rotor is zero or negligible except in the case of some large hydraulic turbines.

The continuity equation can also be applied to the flow through a rotor if relative velocities are used in connection with flow areas in the rotor. Thus the basic relations (in addition to physical property relationships) that are useful in analyzing the one-dimensional steady flow through turbomachine rotors and in designing such rotors are (1) the torque and work equations derived in this article, (2) the first law, (3) Eq. 1·10, and (4) the continuity equation. From these basic relations many special forms useful for particular phases of analysis and design can be derived. These are treated in books on pumps, compressors, turbines, and turbomachines in general.

15·5 Dynamic Compressors

A fundamental difference between the flow through turbines and the flow through pumps and compressors must be recognized. In all flows of a fluid through a passage, shearing forces at the wall retard the fluid near the wall in a region called the boundary layer. In turbines, the pressure generally decreases in the direction of flow and thus the pressure gradient, acting oppositely to the wall shearing forces, tends to prevent the retardation of fluid in the boundary layer. In pumps and compressors, however, there is an adverse pressure gradient (that is, pressure increases in the direction of flow) that acts with the wall shearing forces to reduce the fluid velocity near the wall. These two actions together may reduce the velocity to zero near the wall and then the adverse pressure gradient may cause the flow to reverse direction near the wall. This phenomenon is called *separation* and is highly detrimental to the performance of fluid machines. The design of pumps and compressors is therefore greatly influenced by the necessity to limit the magnitude of adverse pressure gradients, which tend to cause separation.

Both centrifugal and axial-flow compressors are comprised of two parts: a rotor that does work on the fluid, increasing its kinetic energy and its pressure, and a stator that at

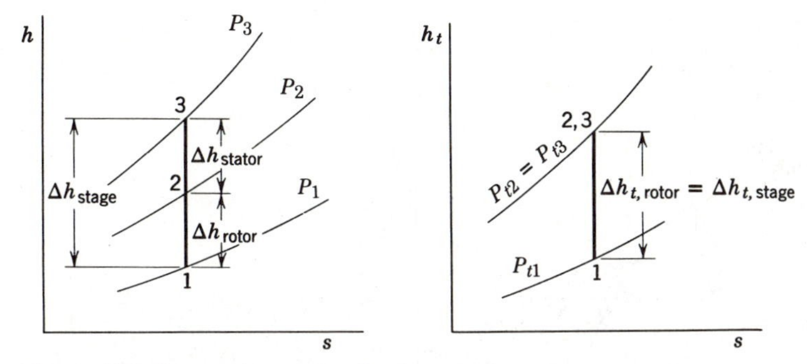

Figure 15·9. Isentropic compression in a single stage.

least serves as a diffuser to reduce the velocity of the fluid and increase its pressure. The stator may also serve to redirect the fluid into following rotors or into a collector that leads to the machine outlet. Work is done on a fluid passing through a compressor only by the rotor, but in the stator or diffuser there is a decrease in kinetic energy, so the enthalpy increases in both the rotor and the stator. The fraction of the overall increase in enthalpy that occurs in the rotor is called the *degree of reaction.*

Figure 15·9 shows *hs* and $h_t s$ diagrams for one stage, including rotor and stator, of either a centrifugal or axial-flow compressor operating isentropically. Process 1-2 occurs in the rotor, and process 2-3 occurs in the stator or stationary diffuser. In a centrifugal compresor the stationary diffuser may have vanes that form diverging passages or it may be a vaneless annular passage through which the fluid spirals outward with decreasing velocity and increasing pressure. In an axial-flow compressor the stator of each stage is comprised of a row of vanes that form diverging flow passages. Details can be found in the specialized literature.

To illustrate the application of the basic dynamic equation to an axial-flow machine, a developed view of an axial-flow compressor stage is shown in Fig. 15·10. The cylindrical

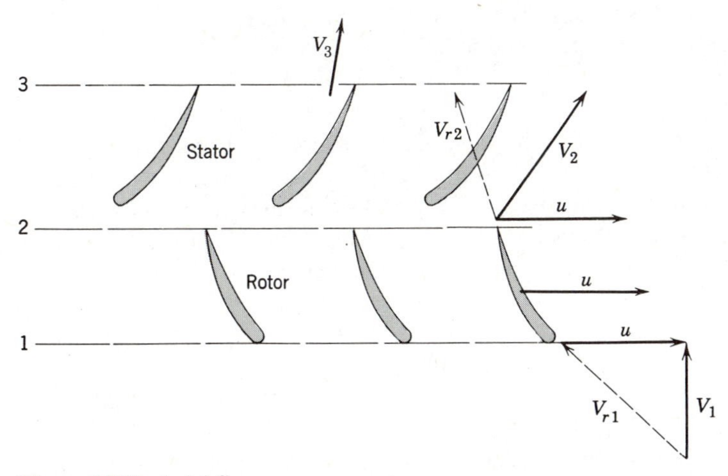

Figure 15·10. Axial-flow compressor stage.

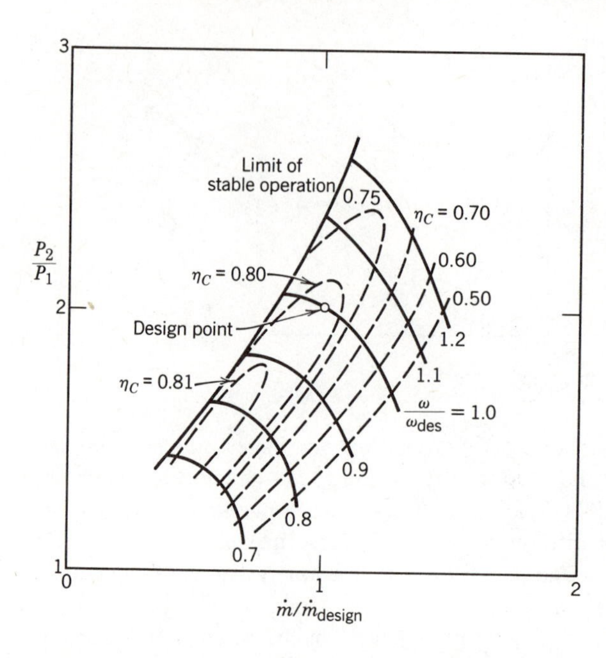

Figure 15·11. Typical performance characteristics of centrifugal compressor.

surfaces to which the rotor and stator blades are attached are flattened into the plane of this figure. Velocity vectors relative to the moving blades are shown as broken lines. Since the inlet and outlet of the rotor blade passages are at the same distance from the rotor axis of rotation, $u_2 = u_1$. The work done by the rotor on the fluid is then, from Eq. 15·1,

$$w_{in} = (u_2 V_{u2} - u_1 V_{u1}) = u(V_{u2} - V_{u1})$$

The work per stage is proportional to the change in the tangential component of velocity across the rotor. The more the fluid stream is turned by the rotor blades, the greater the work and the pressure ratio per stage. The amount of turning that can be accomplished is limited by the occurrence of separation, so axial-flow compressors for high pressure ratios need many stages.

Refer again to Fig. 15·10. The velocity V_1 is proportional to the flow rate through the machine, and u is proportional to the shaft speed. Obviously, a proper relationship between these parameters is required for satisfactory flow geometry and machine performance. This is a key factor in the low efficiency of dynamic machines when operating under "off-design" conditions. Notice in Fig. 15·11, which shows typical performance characteristics of a centrifugal compressor, how rapidly compressor efficiency changes as the flow rate varies.

15·6 Turbines

The flow through turbines is usually adiabatic or very nearly so. A stage consists of at least one stationary row of blades or nozzles and at least one row of moving blades, which are a part of the rotor. The *degree of reaction* is defined as the fraction of the total decrease in enthalpy that occurs across the rotor. A turbine with a zero degree of reaction is called an impulse turbine or sometimes a pure impulse turbine. In such a turbine the fluid is expanded to the stage exhaust pressure in stationary nozzles; then the pressure remains constant and kinetic energy decreases as the fluid does work on the rotor blades. The blades on the rotor of an impulse turbine form flow passages of approximately constant cross-sectional area, since changes in area would involve changes in pressure.

Both axial-flow and radial-flow turbines are in use, the latter usually in the smaller sizes. Many different types of flow paths through turbines are used for various reasons, but the basic dynamic equation for turbomachines (Eq. 15·1) and the other basic relations listed in Sec. 15·4 apply in all cases.

To illustrate the application of the basic dynamic equation to an axial-flow turbine stage, Fig. 15·12 shows a developed view of such a stage. Fluid enters the stationary nozzles with a low velocity V_1, is expanded, and leaves with a higher velocity V_2. Relative to the moving blades, this nozzle exit velocity is V_{r2}. Within the moving blade passages, the fluid is further expanded so that its velocity relative to the rotor blades increases from V_{r2} to V_{r3}. The absolute velocity decreases across the rotor; that is, $V_3 < V_2$. The work done *by* the fluid is

$$w = (uV_{u2} - uV_{u3}) = u(V_{u2} - V_{u3}) = u(V_{ru2} - V_{ru3})$$

Figure 15·13 is a diagram of an impulse stage. If there are no frictional effects, the pressure and the relative velocity do not change across the rotor.

For given inlet conditions and exhaust pressure, maximum work is obtained from the fluid passing through a turbine stage if the kinetic energy (based on absolute velocity) of the fluid leaving the stage is a minimum. Inspection of the velocity vector diagrams in Figs. 15·12 and 15·13 shows that for a given blade shape (or a given degree of reaction) the leaving kinetic energy can be minimized by the proper selection of the ratio u/V_{u2} or $u/(V_2 \cos \alpha)$. For an impulse stage the optimum value of this ratio is 0.5. The optimum value increases as the degree of reaction increases. The blade speed u of a turbine is limited by centrifugal stresses in the rotor. Therefore, maintaining the optimum value of $u/V_2 \cos \alpha$ limits the pressure or enthalpy decrease across a stage. This is the reason for multistaging of turbines. Several ingenious arrangements of fixed nozzles and moving and stationary blades are in use for increasing the allowable enthalpy drop across a single

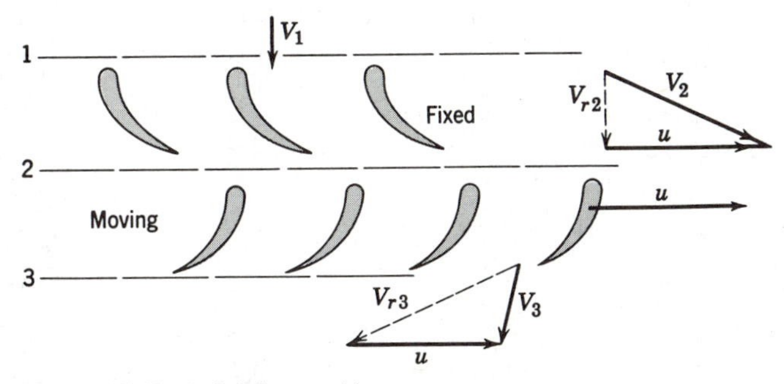

Figure 15·12. Axial-flow turbine stage.

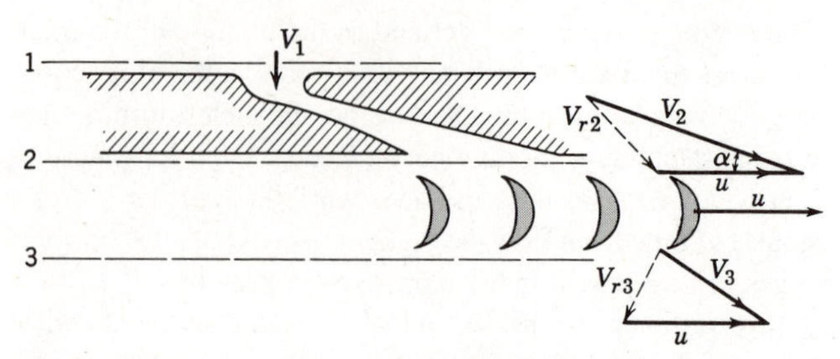

Figure 15·13. Axial-flow impulse turbine stage.

stage. Some of these, called *reentry stages* or *Curtis stages*, are described in most books on turbines.

15·7 Reciprocating Machines

In analyzing the processes that occur in the cylinder of a reciprocating machine, we deal with a closed system when the cylinder valves are closed and with transient flow in an open system the rest of the time. As pointed out in Sec. 1·16, the flow through a reciprocating machine can be treated as steady if the properties at various points within the system vary cyclically; but it is often difficult to arrive at average values of the properties. Furthermore, in designing a reciprocating machine it is necessary to analyze the processes which occur in a cylinder. A convenient diagram in such analyses is a *PV* diagram such as the one shown in Fig. 15·14 for an ideal reciprocating gas compressor. This compressor is ideal in that the pressure and temperature are uniform throughout the cylinder, the valves act instantaneously, and there is no pressure drop across the valves while gas is flowing through them. The abscissa is the volume of gas within the cylinder. The mass of gas in the cylinder varies, so *P* and *V* do not define the state of the gas. Therefore,

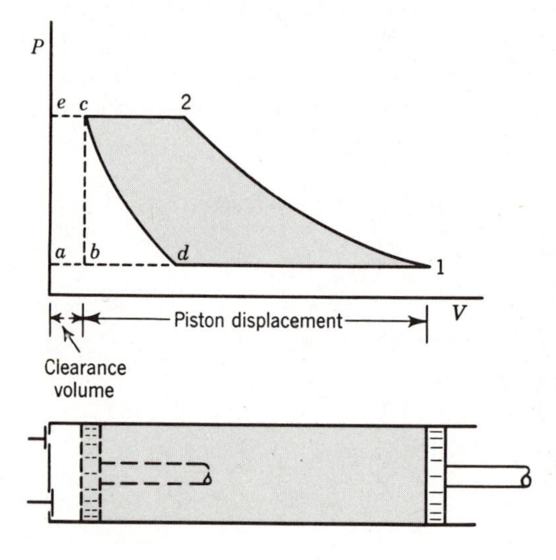

Figure 15·14. *PV* diagram for an ideal reciprocating gas compressor.

gas at a given pressure and temperature may be represented by many different points on the *PV* diagram.

For the compressor operation illustrated by Fig. 15·14, all valves are closed as the piston travels from point 1 to point 2, so path 1-2 is a closed-system process. When the piston reaches point 2, the pressure inside the cylinder is sufficient to open the discharge valve against the pressure of the gas in the discharge line. As the piston moves from point 2 to point *c*, gas is pushed out of the cylinder at constant pressure and constant temperature. Point *c* is the end of the piston stroke. The remaining volume in the cylinder V_c is called the *clearance volume*. The volume swept by the piston during its entire stroke $(V_1 - V_c)$ is called the *piston displacement*, and the volume V_1 is called the *cylinder volume*. The percent clearance is defined as the ratio of the clearance volume to the piston displacement. As the piston moves back to the right from point *c*, the gas that was trapped in the clearance volume at the discharge pressure expands. The intake valve does not open until the pressure inside the cylinder reaches the intake pressure. This occurs at point *d*. From point *d* to point 1 gas is being drawn into the cylinder.

The volume of gas, measured at P_1, which is drawn in is less than the piston displacement. Volumetric efficiency η_{vol} is defined as

$$\eta_{\text{vol}} \equiv \frac{\text{mass of gas drawn in per stroke}}{\text{mass of gas at intake line pressure and temperature required to fill piston displacement}}$$

If P_1 equals the intake line pressure, $\eta_{\text{vol}} = (V_1 - V_d)/(V_1 - V_c)$. In an actual compressor, P_1 is lower than the pressure in the intake line because of the throttling of the gas through the intake valve. (For an air compressor, the intake line pressure may be lower than the ambient pressure because of pressure drop through intake filters.)

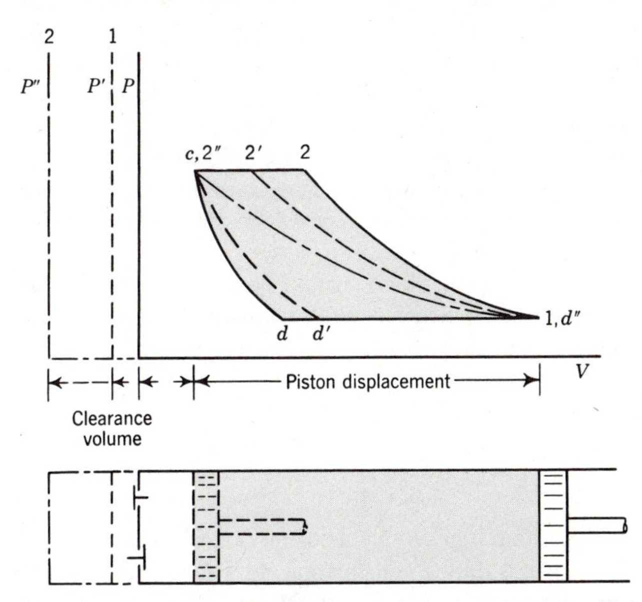

Figure 15·15. Influence of clearance volume on volumetric efficiency.

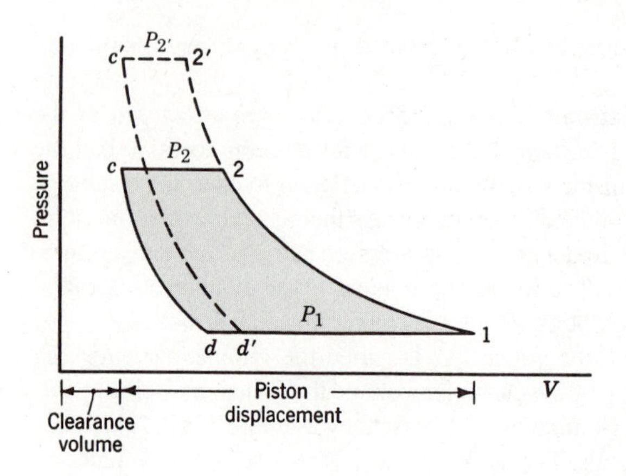

Figure 15·16. Influence of discharge pressure on volumetric efficiency.

If the compression 1-2 and the expansion *c*-*d* are both polytropic with the same value of n in Pv^n = constant, the amount of clearance has no effect on the amount of work done per unit mass of gas delivered.

Increasing the clearance decreases the volumetric efficiency of a reciprocating compressor. The compressor of Fig. 15·15 has a fixed stroke and piston displacement, but the cylinder can be extended to increase the clearance volume. As the clearance volume increases, the pressure axis of the *PV* diagram must be moved to the left. Three different values of clearance volume are illustrated. For the minimum clearance shown, the *PV* diagram is 1-2-*c*-*d*-1. If the clearance is increased, the pressure axis must be moved to the left to a position such as shown for the axis marked P'. The *PV* diagram is then 1-2′-*c*-*d*′-1. The volume of gas inducted per stroke has decreased from $(V_1 - V_d)$ to $(V_1 - V_{d'})$, so the volumetric efficiency has decreased. If the clearance is increased further, a value will be reached for which the pressure of the gas in the cylinder reaches the discharge pressure just at the end of the stroke. Thus no gas is delivered. If the processes are reversible, the gas expands along the same path during the return stroke of the piston and no gas is drawn in. The pressure axis for this case is marked P''. Points 2″ and *c* on the *PV* diagram coincide, and so do points *d*″ and 1. The volumetric efficiency is zero. The delivery of constant-speed compressors is sometimes regulated by changing the clearance. This is done by valves which open and close auxiliary chambers connected to the clearance volume of the cylinder.

Inspection of Fig. 15·16 shows that for constant suction pressure, increasing the discharge pressure decreases the volumetric efficiency. With discharge pressure P_2, gas is inducted along the line *d*-1; but, with the higher discharge pressure $P_{2'}$, the induction process is shortened to *d*′-1.

15·8 Summary

The work required to compress a gas from a given initial state to a given final pressure is reduced by removing heat from the gas during compression. In actual machines, the amount of heat which can be transferred during the compression process is limited; so the ideal process for simulating an actual compression may be a polytropic process with the polytropic exponent n closer to k than to 1. In simple gas-turbine plants and in other applications where immediately after compression the gas is to be heated by means of a fuel, adiabatic compression is most desirable from the standpoint of overall plant efficiency, even though the compression work is greater than for compression with cooling.

For an adiabatic compressor, the compressor efficiency is defined as

$$\eta_C \equiv \frac{\text{work of } \textit{reversible adiabatic} \text{ compression from state 1 to } P_2}{\text{work of } \textit{actual adiabatic} \text{ compression from state 1 to } P_2}$$

In a multistage adiabatic compressor, the efficiency of the entire machine is lower than that of the individual stages if they have equal efficiencies of less than 100 percent.

If isothermal compression is impossible or impractical, as it usually is, a reduction in the work required for given pressure limits can be achieved by cooling the gas at constant pressure between stages. For polytropic compression with the same value of n in each stage and intercooling to the initial temperature, minimum total work is required when the pressure ratio is the same for each stage.

Adiabatic turbine efficiency, which we call simply *turbine efficiency* or the efficiency of a turbine, is defined as

$$\eta_T \equiv \frac{\text{work of } \textit{actual adiabatic} \text{ expansion}}{\begin{array}{c}\text{work of } \textit{reversible adiabatic} \text{ expansion between}\\ \text{same initial state and same final total pressure}\end{array}}$$

The efficiency of a multistage turbine can be higher than the efficiency of any of its stages.

Adiabatic turbine effectiveness, often called simply *turbine effectiveness,* is defined as

$$\text{turbine effectiveness} \equiv \frac{\text{work}}{-\Delta b}$$

The basic dynamic equation for the steady, one-dimensional flow of a fluid through a turbomachine rotor is

$$\text{Torque on fluid} = \dot{m}(r_2 V_{u2} - r_1 V_{u1})$$

where r is the radial distance from the axis of rotation, V_u is the tangential component of the fluid velocity, and subscripts 1 and 2 refer to inlet and outlet of the rotor. The work done on the fluid is

$$w_{\text{in}} = (u_2 V_{u2} - u_1 V_{u1}) \tag{15·1}$$

where u is the linear velocity of a point on the rotor.

The degree of reaction of a turbomachine stage is defined as the fraction of the enthalpy change that occurs in the rotor.

In analyzing the performance of a reciprocating machine, it must be remembered that the piston and the cylinder walls enclose a closed system part of the time and form most of the boundary of an open transient-flow system at other times during a cycle.

The work done by an ideal reciprocating compressor is the same per unit mass of gas delivered regardless of the amount of clearance in the machine, but the amount of gas delivered per stroke is decreased as the clearance increases. Increasing the discharge pressure also decreases the amount of gas delivered per stroke.

Suggested Reading

15·1 Burghardt, M. David, *Engineering Thermodynamics with Applications,* 2nd ed., Harper & Row, New York, 1982, Chapter 15.

15·2 El-Wakil, M. M., *Powerplant Technology,* McGraw-Hill, New York, 1984, Chapter 5.

15·3 Fox, Robert W., and Alan T. McDonald, *Introduction to Fluid Mechanics,* 3rd ed., Wiley, New York, 1985, Section 4–7.

15·4 Logan, Earl, Jr., *Turbomachinery: Basic Theory and Applications,* Marcel Dekker, New York, 1981.

15·5 Moss, S. A., C. W. Smith, and W. R. Foote, ''Energy Transfer between a Fluid and a Rotor for Pump and Turbine Machinery,'' *Transactions ASME,* vol. 64, August 1942, pp. 567–597. (A classic paper.)

15·6 Sorensen, Harry A., *Energy Conversion Systems,* Wiley, New York, 1983, Chapters 3, 5, and 6.

15·7 Wood, Bernard, D., *Applications of Thermodynamics,* 2nd ed., Addison-Wesley, Reading, Mass. 1982, Chapter 3.

Problems

15·1 Air enters an axial-flow compressor at 30°C (86°F) and is compressed adiabatically through a pressure ratio of 3. Plot the air discharge temperature versus compressor efficiency for an efficiency range of 50 to 100 percent.

15·2 Solve problem 15·1 for air entering with a relative humidity of 100 percent.

15·3 Air at 1 atm, 30°C (86°F), $\phi = 0.50$, is compressed to 2 atm. Calculate the final relative humidity for (*a*) reversible adiabatic compression and (*b*) reversible isothermal compression.

15·4 Isothermal compression of a gas–vapor mixture can cause condensation; adiabatic compression does not. For processes that follow Pv^n = constant, determine the limiting value of n to avoid condensation if the entering gas is atmospheric air at 1 atm, 20°C (86°F), and $\phi_1 = 0.60$ and the pressure ratio is 3.

15·5 Solve problem 15·1 for the compression of natural gas having a volumetric analysis of 84 percent methane (CH_4) and 16 percent ethane (C_2H_6).

15·6 Solve problem 15·1 for the compression of coke-oven gas having a volumetric analysis of 46 percent hydrogen, 6 percent carbon monoxide, 32 percent methane, and 16 percent nitrogen.

15·7 A reversible compression process of an ideal gas might be approximated by a process equation, Pv^n = constant. For CO_2 being compressed from 1 atm, 20°C (68°F), to 3 atm at a rate of 0.260 kg/s, plot power input and discharge temperature versus n in the range $1 \leq n \leq k$.

15·8 It is desired to produce a steady jet of air at atmospheric pressure, 10°C, 6 cm in diameter, with a velocity of 60 m/s. This is to be produced by means of a compressor inducting atmospheric air and discharging through a heat exchanger into a nozzle. The cooling medium passing through the heat exchanger is water entering at 15°C and experiencing a temperature rise of no more than 10 degrees C. For adiabatic compression, plot compressor power input and the minimum mass flow rate of cooling water versus compressor efficiency in a range of 60 to 100 percent.

15·9 An ideal gas with constant specific heats is to be compressed from P_1 and T_1 to a state at P_2 and the same temperature. Consider two methods for doing this: (*a*) isentropic compression from state 1 to P_2 followed by constant-pressure cooling to the initial temperature, and (*b*) constant-pressure cooling from state 1 to a temperature T_c, isentropic compression to P_2, and then constant-pressure cooling to the initial temperature. Sketch the processes on a *Ts* diagram and calculate (in terms of P_1, P_2, T_1, T_c, and properties of the gas) the amount of work that can be saved by the proper choice of method. (Do not consider the power required for refrigerating the coolant.)

15·10 Air at 200 kPa, 150°C, is to be produced from air at 100 kPa, 20°C, by isentropic compression and constant-pressure heating. Calculate the work required per kilogram of air if (*a*) the isentropic compression precedes the heating, and (*b*) the isentropic compression follows the heating.

15·11 Solve problem 15·10 with the compression for each case being adiabatic with an efficiency of 70 percent.

15·12 Air at 100 kPa, 15°C, enters a compressor with negligible velocity and is discharged at 300 kPa, 115°C, through a cross-sectional area of 0.040 m^2. The flow rate through the compressor is 20 kg/s, and the power input is 2900 kW. Heat exchange is with only the surrounding atmosphere. Calculate the heat transfer per kilogram of air, stating clearly its direction.

15·13 A compressor draws in ambient air at 95 kPa, 15°C, at a rate of 0.80 lbm/min and compresses it to 440 kPa, 205°C. The compressor is uncooled except for possible heat loss to the surrounding atmosphere. Calculate the minimum possible power input to any compressor system working between these end states and exchanging heat with only the atmosphere.

15·14 For the wind tunnel of problem 14·90, plot a curve of power required for isentropic compression versus test section Mach number if the tunnel is operated in a once-through manner.

15·15 Ethene is compressed from 101.3 kPa, 20°C, to 350 kPa, 120°C, by a centrifugal compressor at a rate of 20 kg/min. The rise in temperature of the cooling water flowing through the jacket is 4 degrees C for a water flow rate of 20 kg/min. Compute the power required to operate the compressor. Disregard the change in kinetic energy.

15·16 An air compressor uses a lubricating oil having a flash point of 175°C. If air is taken in at 105 kPa, 15°C, and compressed polytropically with an exponent of 1.35, what is the maximum allowable discharge pressure if the maximum allowable temperature is 25°C below the flash point of the oil?

15·17 An ideal centrifugal compressor compresses air polytropically, with $n = 1.35$, at a rate of 2 kg/s from 100 kPa, 5°C, to 300 kPa. Inlet velocity is negligibly small, and the discharge velocity is 180 m/s. Heat is removed from the air being compressed at a rate of 21.8 kJ/s. Calculate the compressor power requirement.

15·18 An ideal centrifugal air compressor takes in 180 m^3/min of air at 100 kPa, 5°C, with negligible velocity. It compresses the air polytropically with $n = 1.35$ and discharges it at 200 kPa with a velocity of 180 m/s. Calculate the power input to the compressor.

15·19 Determine the total amount of work required by a compressor that draws in methane at 95 kPa, 25°C, and compresses it isentropically to raise the pressure to 300 kPa in a 0.5 m^3 insulated tank that initially contains methane at 95 kPa, 25°C.

15·20 Determine the power required to compress air adiabatically at a rate of 340 m^3/min (measured at the inlet) from 100 kPa, 24°C, to 240 kPa with an efficiency of 72 percent.

15·21 Carbon monoxide is compressed adiabatically from 100 kPa, 25°C, to 580 kPa, 260°C. Kinetic energy changes are negligible. Determine the compressor efficiency and the irreversibility.

15·22 Determine the irreversibility of the steady-flow adiabatic compression of air initially at 101.3 kPa, 25°C, through a pressure ratio of 5 if the compressor efficiency is 74 percent.

15·23 A two-stage compressor draws in 8.5 m^3/min of air at 98.5 kPa and compresses it polytropically with $n = 1.3$ to 985 kPa. The intercooler cools the air to its initial temperature before it enters the second stage. How much work does two-stage compression save in comparison with compression in a single stage with the same value of n?

15·24 Air is to be compressed from 1 atm, 25°C (77°F) to 5 atm. For two-stage ideal compression with intercooling to the initial temperature and reversible adiabatic compression in each stage, plot work per unit mass of air against first-stage pressure ratio. (Use a constant mean specific heat.)

15·25 Solve problem 15·24 for an adiabatic efficiency of 0.75 for each stage.

15·26 Solve problem 15·24 for the case in which some cooling is effected during compression so that for each stage $Pv^{1.35}$ = constant.

15·27 Ambient air enters an axial-flow compressor at 30°C (86°F) and is compressed adiabatically through a pressure ratio of 3. Plot the work required per unit mass and the irreversibility of the process per unit mass versus compressor efficiency for an efficiency range of 50 to 100 percent.

15·28 A natural gas compressor station receives natural gas at 50 psia (345 kPa), 59°F (15°C), and is to discharge it at 760 psia (5240 kPa). The gas has a volumetric analysis of 84 percent methane and 16 percent ethane. The flow rate is 88 000 lbm/h (39 900 kg/h). Two-stage and three-stage compression are to be compared. In each case, the stages are to have equal pressure ratios, and each stage but the last is to be followed by an intercooler in which the gas will be cooled at essentially constant pressure to within 20 degrees F (8.3 degrees C) of the temperature of available cooling water, 59°F. Adiabatic compressor efficiency is estimated to be 65 percent for each stage. Calculate for each of the cases (*a*) the required power input and (*b*) the irreversibility per unit mass of gas.

15·29 Air at 14.0 psia, 60°F, enters a compressor with negligible velocity and is discharged at 42 psia, 240°F, through a cross-sectional area of 0.40 sq ft. The flow rate through the compressor is 40 lbm/s, and the power input is 3600 hp. Heat exchange is with only the surrounding atmosphere. Calculate the heat transfer per pound of air, stating clearly its direction.

15·30 A compressor draws in ambient air at 14.0 psia, 60°F, at a rate of 0.80 lbm/min and compresses it to 64.0 psia, 400°F. The compressor is uncooled except for possible heat loss to the surrounding atmosphere. Calculate the minimum possible power input to any compressor system working between these end states and exchanging heat with only the atmosphere.

15·31 Air is compressed from 14.7 psia, 70°F, to 50 psia, 250°F, by a centrifugal compressor at a rate of 50 lbm/min. The rise in temperature of the cooling water flowing through the jacket is 7 degrees F for a water flow rate of 50 lbm/min. Compute the power required to operate the compressor. Disregard the change in kinetic energy.

15·32 An air compressor uses a lubricating oil having a flash point of 350°F. If air is taken in at 15 psia, 60°F, and compressed polytropically with an exponent of 1.35, what is the maximum allowable discharge pressure if the maximum allowable temperature is 50°F below the flash point of the oil?

15·33 An ideal centrifugal compressor compresses air polytropically, with $n = 1.35$, at a rate of 5 lbm/s from 15 psia, 40°F, to 45 psia. Inlet velocity is negligibly small while the discharge velocity is 600 fps. Heat is removed from the air being compressed at a rate of 20 B/s. Calculate the compressor power requirement.

15·34 An ideal centrifugal air compressor takes in 10 000 cfm of air at 15 psia, 40°F, with negligible velocity. It compresses the air polytropically with $n = 1.35$ and discharges it at 30 psia with a velocity of 600 fps. Calculate the power input to the compressor.

15·35 Determine the total amount of work required by a compressor that draws in air at 14.0 psia, 80°F, and compresses it isentropically to raise the pressure to 42 psia in a 20-cu-ft insulated tank that initially contains air at 14.0 psia, 80°F.

15·36 Determine the power required to compress air adiabatically at a rate of 12 000 cfm (measured at the inlet) from 14.5 psia, 75°F, to 35 psia with an efficiency of 72 percent.

15·37 Air is compressed adiabatically from 14.5 psia, 80°F, to 85 psia, 500°F. Kinetic energy changes are negligible. Determine the compressor efficiency.

15·38 Determine the irreversibility of the steady-flow adiabatic compression of air initially at 14.7 psia, 80°F, through a pressure ratio of 5 if the compressor efficiency is 74 percent.

15·39 A two-stage compressor draws in 300 cfm of air at 14.3 psia and compresses it polytropically with $n = 1.3$ to 143 psia. The intercooler cools the air to its initial temperature before it enters the second stage. How much work does two-stage compression save in comparison with compression in a single stage with the same value of n?

15·40 Is it true that for a turbine operating on an ideal gas with constant specific heats the work output for a given pressure ratio is proportional to the inlet gas absolute temperature? Prove your answer.

15·41 Helium enters a turbine at 572 kPa, 830°C, and expands adiabatically to 100 kPa, 504°C, with negligible change in kinetic energy. Determine the turbine efficiency.

15·42 Determine the mass rate of flow through a turbine that takes in air at 500 kPa, 870°C, and exhausts at 100 kPa if the flow is adiabatic with a turbine efficiency of 75 percent and the power output is 1050 kW.

15·43 A three-stage gas turbine operating on air is to produce equal work per stage. Determine the intermediate stage pressures if the inlet conditions are 620 kPa, 980°C; the exhaust pressure is 100 kPa; and the adiabatic efficiency for each stage is 75 percent. Calculate also the effectiveness of each stage.

15·44 Determine the irreversibility of an adiabatic expansion of air from 507 kPa, 870°C, through a pressure ratio of 5 if the turbine efficiency is 78.5 percent and the lowest temperature in the surroundings is 25°C.

15·45 Air flows through a well-insulated turbine at a rate of 1.22 kg/s, entering at 300 kPa, 200°C, and exhausting into the atmosphere. The atmosphere is at 100 kPa, 20°C. Plot the irreversibility of the turbine process in kJ/kg versus turbine efficiency for an efficiency range from 50 to 100 percent.

15·46 Air enters a turbine at 83.0 psia, 1540°F, and expands adiabatically to 14.6 psia, 940°F, with negligible change in kinetic energy. Determine the turbine efficiency and the turbine effectiveness.

15·47 Determine the mass rate of flow through a turbine that takes in air at 75 psia, 1600°F, and exhausts at 15 psia if the flow is adiabatic with a turbine efficiency of 75 percent and the power output is 1500 hp.

15·48 A three-stage gas turbine operating on air is to produce equal work per stage. Determine the intermediate stage pressures if the inlet conditions are 90 psia, 1800°F; the exhaust pressure is 15 psia; and the adiabatic efficiency for each stage is 75 percent. Calculate also the effectiveness of each stage.

15·49 Determine the irreversibility of an adiabatic expansion of air from 73.5 psia, 1600°F, through a pressure ratio of 5 if the turbine efficiency is 78.5 percent and the lowest temperature in the surroundings is 80°F.

15·50 Air enters a turbine at 6 atm, 1175°F (635°C). If the final pressure is 1 atm, plot the exhaust temperature against turbine efficiency for an efficiency range of 60 to 100 percent.

15·51 Solve problem 15·49, using constant specific heats, and compare your results with those of problem 15·49.

15·52 Two air streams are available for power production by expansion to 1 atm. One is at 2 atm, 500°F (260°C) and the other is at 2 atm, 1130°F (610°C). The flow rates are the same. For maximum power output, should the two streams be mixed and then expanded through a single turbine, or should each be expanded through a separate turbine without mixing? Prove your answer.

15·53 For an emergency power supply it is proposed to use a turbine driven by compressed air from a storage tank. The turbine will operate adiabatically between the decreasing pressure in the tank and atmospheric pressure. Disregard heat transfer to the air in the tank. Apply the first law to the tank to obtain a differential equation relating m (the mass of air in the tank at any instant), T (the temperature of the air in the tank at that instant) and such constant properties of the air as R, c_p, c_v, and k. Then solve the differential equation for T in terms of m, T_i (the initial temperature of the air in the tank), m_i (the initial mass of air in the tank), and constant properties of the air.

15·54 As an emergency power supply it is proposed to use a turbine driven by compressed air from a storage tank in which the air is initially at 7 atm, 25°C (77°F). The turbine will operate adiabatically between the decreasing pressure in the tank and atmospheric pressure of 1 atm with an expected efficiency of 55 percent until the tank pressure drops to 1.3 atm. Disregard heat transfer to the air in the tank. What size tank is needed in order for the turbine to deliver 2 kW for 10 min? Would doubling the initial pressure halve the required volume?

15·55 Solve problem 15·54 for the case of a turbine efficiency of 100 percent.

15·56 As a means of reducing the tank volume required for the power supply described in problem 15·54, it is suggested that a large number of short, hollow metal cylinders be placed in the tank to reduce the decrease in temperature of the air in the tank during the operation of the system. Readily available steel tubing has dimensions such that the mass of steel that can be packed into the tank is 38 times the mass of air intially in the tank. The steel has a specific heat of 0.50 kJ/kg·K (0.119 B/lbm·R). Assuming that the steel temperature follows very closely the air temperature, determine the size tank needed for the service specified in problem 15·54.

15·57 The efficiency of the turbine for the power supply described in problem 15·54 will actually depend on the pressure ratio and will consequently not be constant during the operation. To achieve a smaller tank volume, would it be better to design the turbine for higher efficiency at higher or lower pressure ratios?

15·58 Suppose the air is heated between the tank and the turbine of problem 15·54 to maintain a constant turbine inlet temperature of 485°C (905°F). What size tank would be needed? How much heat would be required during the 10 min?

15·59 Determine the efficiency and the effectiveness of a turbine that expands steam at a rate of 9000 kg/h from 1.4 MPa, 225°C, to 35 kPa, 90.5 percent quality, if the power output is 1150 kW.

15·60 Steam is supplied to a turbine at 1.4 MPa, 315°C, and is exhausted at 101.3 kPa. The flow is adiabatic, the power output of the turbine is 10 000 kW, and the turbine efficiency is 65 percent. The lowest temperature in the surroundings is 25°C. Determine (*a*) the steam rate in kg/kW·h, (*b*) the amount of energy made unavailable during the turbine expansion in kJ/kg, and (*c*) the turbine effectiveness.

15·61 An ideal steam turbine operates between pressures of 2.00 MPa and 10 kPa. The power output is 250 kW. Determine the inlet temperature needed to ensure that the exhaust steam contains not more than y percent moisture, where y = 12, 10, 8, 6, 4, 2 and 0.

15·62 Determine the efficiency and effectiveness of a turbine that expands steam at a rate of 20 000 lbm/h from 200 psia, 440°F, to 5 psia, 90.5 percent quality if the power output is 1560 hp.

15·63 Steam is supplied to a turbine at 200 psia, 600°F, and is exhausted at 14.7 psia. The flow is adiabatic, the power output of the turbine is 10 000 kW, and the turbine efficiency is 65 percent. The lowest temperature in the surroundings is 80°F. Determine (*a*) the steam rate in lbm/kW·h, (*b*) the amount of energy made unavailable during the turbine expansion in B/lbm, and (*c*) the turbine effectiveness.

15·64 Devise a single-stage turbine that can be made to rotate in either direction, the change in direction of rotation to be brought about by the movement of a single lever or valve. Explain its operation clearly by means of sketches or diagrams as well as words.

15·65 Prove that, if a turbine discharges wet steam at the sink temperature, the product of $(1 - \eta_T)$ and the isentropic enthalpy drop between the initial condition and the exhaust pressure is equal to the irreversibility.

15·66 The water flow rate through a centrifugal pump running at 1200 rpm is 0.06 m³/s. The inlet and discharge lines are 15 cm in diameter and are at the same elevation. The inlet pressure is 200 mm Hg vacuum, and the discharge pressure is 140 kPa. Power input to the pump is 15 kW. Barometric pressure is 740 mm Hg. Determine the pump efficiency.

15·67 A fan draws air at 95 kPa, 25°C, from a room at a rate of 17 m³/min and discharges it through a duct of 0.20 m² cross-sectional area at a static gage pressure of 3.0 in. of water. The fan efficiency is 60 percent. Determine the power input.

15·68 The water flow rate through a centrifugal pump running at 1200 rpm is 2 cfs. The inlet and discharge lines are 6 in. in diameter and are at the same elevation. The inlet pressure is 8 in. Hg vacuum, and the discharge pressure is 20 psig. Power input to the pump is 20 hp. Barometric pressure is 29.0 in. Hg. Determine the pump efficiency.

15·69 A fan draws air at 14.0 psia, 80°F, from a room at a rate of 6000 cfm and discharges it through a duct of 2.0-sq-ft cross-sectional area at a static gage pressure of 3.0 in. of water. The fan efficiency is 60 percent. Determine the power input.

15·70 Consider a steam-turbine impulse stage. Steam enters the nozzle at 400 kPa, 240°C, with a velocity of 150 m/s and expands adiabatically through the nozzle to 275 kPa, 200°C. The nozzle angle is 20 degrees. The blade velocity is 150 m/s, and the symmetrical blades are so shaped that the steam enters the blades tangentially to the blade surface. The steam flow rate is 4500 kg/h. Sketch a complete vector diagram, and, disregarding friction in the blade passage, calculate (*a*) the kinetic energy of the steam leaving the turbine and (*b*) the power output of the turbine.

15·71 Consider an ideal single-stage impulse turbine. Steam enters the nozzles at 200°C, 700 kPa, with a velocity of 150 m/s, and the exhaust pressure is 200 kPa. Flow is frictionless and adiabatic. The blades are symmetrical, and the blade speed is 300 m/s. The flow rate is 45 kg/min. Sketch a complete accurate vector diagram for the stage, and calculate (*a*) the power output and (*b*) the kinetic energy of the steam leaving the blades. What change, if any, should be made in the blade speed to increase the stage efficiency?

15·72 Consider a steam-turbine impulse stage. Steam enters the nozzle at 60 psia, 460°F, with a velocity of 500 fps and expands adiabatically through the nozzle to 40 psia, 400°F. The nozzle angle is 20 degrees. The blade velocity is 500 fps, and the symmetrical blades are so shaped that the steam enters the blades tangentially to the blade surface. The steam flow rate is 10 000 lbm/h. Sketch a complete vector diagram, and, disregarding friction in the blade passage, calculate (*a*) the kinetic energy of the steam leaving the turbine and (*b*) the power output of the turbine.

15·73 Consider an ideal single-stage impulse turbine. Steam enters the nozzles at 100 psia, 400°F, with a velocity of 500 fps, and the exhaust pressure is 30 psia. Flow is frictionless and adiabatic. The blades are symmetrical, and the blade speed is 1000 fps. Steam leaving the nozzles makes an angle of 20 degrees with the line of blade motion. The flow rate is 100 lbm/min. Sketch a complete accurate vector diagram for the stage, and calculate (*a*) the power output and (*b*) the kinetic energy of the steam leaving the blades. What change, if any, should be made in the blade speed to increase the stage efficiency?

15·74 An ideal reciprocating compressor draws in 2 m^3/min of air at 100 kPa, 5°C, and compresses it polytropically with $n = 1.35$ to 500 kPa. Cooling water that removes heat from the air flows at a rate of 4.63 kg/min and undergoes a temperature increase of 6 Celsius degrees. Calculate the compressor power input.

15·75 A single-acting reciprocating compressor compresses air reversibly and adiabatically. The piston displacement is 280 cm^3, and there is 4.0 percent clearance. The speed is 60 rpm. Suction pressure is 95 kPa; discharge pressure is 480 kPa. Ambient temperature is 15°C. Disregard pressure drop across the valves. Sketch a *PV* diagram and determine the volume of ambient air compressed per minute.

15·76 A single-stage air compressor having a clearance of 3 percent takes in air at 97.9 kPa, 15°C, and compresses it to 1.4 MPa. The mass of air compressed per minute is 2.2 kg, and n for both compression and expansion processes is 1.35. Compute (*a*) the power required, (*b*) the piston displacement if the compressor operates at 100 rpm, and (*c*) the heat transferred to the water jacket.

15·77 Compute the power and the cylinder volume required to compress 15 m^3/min of ambient air to 825 kPa. Atmospheric pressure is 101.3 kPa, and the suction pressure is 95 kPa. The clearance is 2.5 percent, and the value of n during compression and expansion is 1.3. The compressor is double-acting and operates at 50 rpm.

15·78 An ideal reciprocating compressor draws in 200 cfm of air at 15 psia, 40°F, and compresses it polytropically with $n = 1.35$ to 75 psia. Cooling water that removes heat from the air flows at a rate of 10.2 lbm/min and undergoes a temperature rise of 10 degrees Fahrenheit. Calculate the compressor power input.

15·79 A single-acting reciprocating compressor compresses air reversibly and adiabatically. The piston displacement is 1.0 cu ft, and there is 4.0 percent clearance. The speed is 60 rpm. Suction pressure is 14.0 psia; discharge pressure is 70.0 psia. Ambient temperature is 60°F. Disregard pressure drop across the valves. Sketch a *PV* diagram and determine the volume of ambient air compressed per minute.

15·80 A single-stage air compressor having a clearance of 3 percent takes in air at 14.2 psia, 60°F, and compresses it to 200 psia. The mass of air compressed per minute is 50 lbm, and n for both compression and expansion processes is 1.35. Compute (*a*) the power required, (*b*) the piston displacement if the compressor operates at 100 rpm, and (*c*) the heat transferred to the water jacket.

15·81 Compute the power and the cylinder volume required to compress 500 cfm of ambient air to 120 psia. Atmospheric pressure is 14.7 psia, and the suction pressure is 14 psia. The clearance is 2.5 percent, and the value of n during compression and expansion is 1.3. The compressor is double-acting and operates at 50 rpm.

15·82 Consider an ideal reciprocating gas compressor that operates with polytropic compression and with polytropic expansion of the clearance gas, but the polytropic exponents are not the same for the two processes. If the cylinder walls are generally cooler than the gas, for which process will the n be greater? In such a case, how does increasing the clearance affect the work done per pound of gas delivered?

CHAPTER SIXTEEN

Gas Power Cycles

Nearly all systems that convert heat or the stored energy of a fuel into work use fluid working substances. One way to classify such systems is according to whether the working fluid changes phase. This chapter treats power cycles that use a gas as the working substance, and the next chapter deals with power cycles that use a working substance that is alternately vaporized and condensed.

16·1 Air-Standard Analyses

An ideal system is simpler than the corresponding actual system in that it is described in terms of only a few characteristics instead of the many needed to describe the actual system fully. Depending on the nature and degree of simplification, several ideal systems can be devised to correspond to the same actual system. Likewise, an actual process or cycle can be simulated by several different ideal processes or cycles.

The processes that occur in a reciprocating internal-combustion engine and to a lesser degree those which occur in gas-turbine power plants are so complex that considerable simplification is required to make possible an elementary analysis. A common procedure is to use an *air-standard analysis.* In an air-standard analysis two idealizations are always made: (1) The working substance is air which behaves like an ideal gas, and (2) a simple heat-addition process replaces the combustion process. The second of these idealizations must accompany the first because the combustion of a fuel in the system would of course involve substances other than air. In most actual gas turbines and all reciprocating internal-combustion engines, air is drawn in at ambient conditions, and gases at relatively high temperature are discharged. The system is an open one, and the same working fluid is not used repeatedly. In an air-standard analysis, however, the cycle is usually completed by a heat-rejection process that restores the fluid (which is taken to be air alone) to its initial state. Thus the same fluid is used repeatedly, and there is no exchange of mass with the surroundings.

A *cold air-standard analysis* involves the further simplification that the specific heats of air are constant at their room-temperature values. The cold air-standard analysis leads to very simple relations for quantities such as thermal efficiency, and these relations are often valuable as qualitative indications of cycle characteristics. For other purposes, more accurate quantitative results can so readily be obtained by means of an air-standard analysis using variable specific heats that there is little reason for the use of the cold air-standard analysis.

Since sufficient data are published on the products of combustion of hydrocarbon fuels, it is not difficult to eliminate the use of air as the ideal-gas power-cycle working substance. In the interest of brevity in this chapter, however, we treat only air-standard power cycles, but some of the problems at the end of the chapter explore other working substances.

16·2 The Simple Gas-Turbine Cycle

In the simple gas-turbine power plant shown in Fig. 16·1, air is drawn into the compressor and compressed adiabatically. Fuel is mixed with the compressed air and burned in a combustion chamber. The air-fuel ratio is quite high in order to limit the temperature of the gas entering the turbine. The gases expand adiabatically through the turbine to the ambient pressure. Usually more than half the turbine power output is required to drive the compressor, and the rest of the turbine power output is the net power of the gas-turbine plant. An actual plant also requires auxiliary equipment such as starting motors, fuel pumps, lubricating oil pumps, ignition systems, and control and safety equipment.

The air-standard ideal-gas turbine cycle is called the Brayton* cycle or Joule cycle. Flow, Pv, and Ts diagrams are shown in Fig. 16·2. Notice that the compression and expansion processes are isentropic, and the heat addition and heat rejection occur in reversible constant-pressure processes. (In sketching Ts and hs diagrams of gas-turbine cycles, show clearly the divergence of constant-pressure lines with increasing entropy to

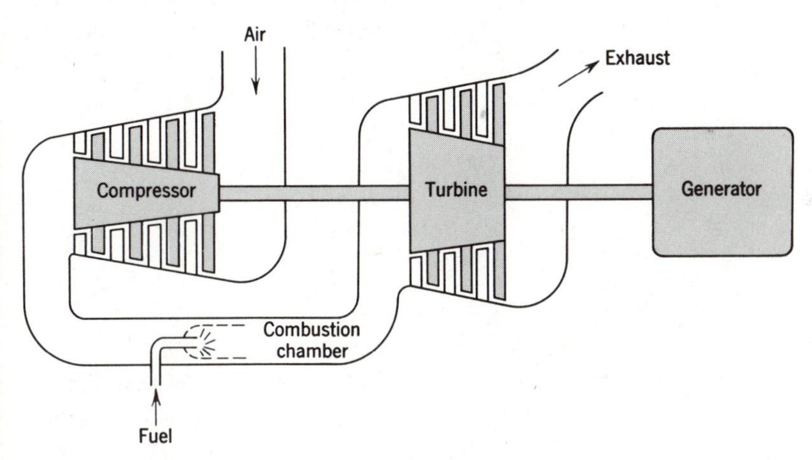

Figure 16·1. Simple gas-turbine power plant.

*George B. Brayton (1830–1892), American engineer, invented a breech-loading gun, a riveting machine, and a sectional steam generator in addition to the internal-combustion engine for which he is best remembered. The Brayton engine, developed around 1870, was a reciprocating oil-burning engine with fuel injection directly into the cylinder, and a compressor which was separate from the power cylinder. The Brayton cycle, which is now used only for gas turbines, was thus first used with reciprocating machines.

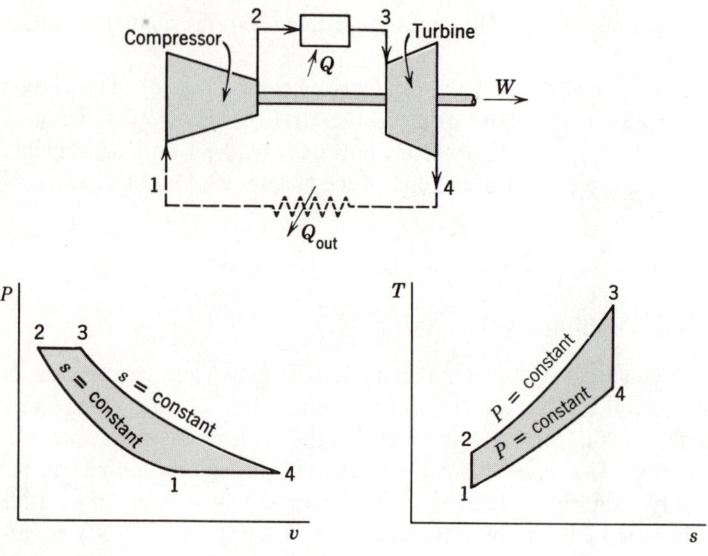

Figure 16·2. Air-standard gas-turbine cycle.

insure that the diagram indicates the Δh of the turbine expansion to be greater in magnitude than the Δh of compression. Otherwise, the turbine cannot drive the compressor and deliver work to the surroundings.)

Application of the first law to an air-standard gas-turbine cycle numbered as in Fig. 16·2 shows the turbine work, compressor work, net work, and cycle efficiency to be given by

$$w_T = h_3 - h_4 + \frac{V_3^2 - V_4^2}{2} = h_{t3} - h_{t4}$$

$$w_{\text{in},C} = h_2 - h_1 + \frac{V_2^2 - V_1^2}{2} = h_{t2} - h_{t1}$$

$$w_{\text{cycle}} = w_T - w_{\text{in},C} = h_{t3} - h_{t4} - (h_{t2} - h_{t1})$$

$$\eta = \frac{w_{\text{cycle}}}{q_{2\text{-}3}} = \frac{h_{t3} - h_{t4} - (h_{t2} - h_{t1})}{h_{t3} - h_{t2}} = 1 - \frac{h_{t4} - h_{t1}}{h_{t3} - h_{t2}}$$

For the *special case of constant specific heats* (that is, the cold air-standard analysis), the last equation can be reduced to show that the thermal efficiency of the cycle is a function of the pressure ratio P_2/P_1 only.

The *backwork ratio,* defined as

$$\text{backwork ratio} \equiv \frac{\text{compression work}_{\text{in}}}{\text{gross work of prime mover}} = \frac{w_{\text{in},C}}{w_T} = \frac{w_T - w_{\text{cycle}}}{w_T}$$

$$= 1 - \frac{w_{\text{cycle}}}{w_T}$$

is high for a gas-turbine cycle.* Consequently, the efficiency of the cycle is reduced appreciably by relatively small reductions in compressor and turbine efficiencies. For example, suppose that the gross power output of a turbine is 9000 kW and the compressor requires an input power of 6000 kW so that the net power produced by the plant is 3000 kW. A reduction of 10 percent in the compressor efficiency from 0.80 to 0.72 would increase the compressor power requirement to 6670 kW and would thereby reduce the plant power output by 22.3 percent. (The plant thermal efficiency would not be decreased by 22.3 percent because a decrease in compressor efficiency raises the compressor outlet temperature and thus decreases the amount of heat input required for the cycle.) The efficiencies of the turbine and compressor in a gas-turbine plant are consequently of great importance. The practical development of gas-turbine power plants was delayed many years by the low efficiencies of the available compressors and turbines.

Figure 16·3 shows *hs* and *Ts* diagrams of a gas-turbine cycle with irreversible adiabatic compression and expansion. For an adiabatic steady-flow process, the entropy change of the universe equals the entropy change of the matter flowing through the steady-flow system. Therefore, for the compression process 1-2_a,

$$i = T_0 \Delta s_{\text{universe}} = T_0(s_{2a} - s_1)$$

and this can be represented by an area on the *Ts* diagram. The irreversibility of the adiabatic turbine expansion can likewise be shown. Remember that the irreversibility of a process is the decrease, caused by the process, in the amount of energy of the universe that can be converted into work.

In an actual gas-turbine plant, frictional effects in the passages between components and in the combustion chamber also add to the irreversibility of the cycle.

Increasing the turbine inlet temperature helps to offset the effects of various irreversibilities in the cycle, but this temperture is limited by the loss of strength of turbine construction materials with increasing temperature. This temperature limitation is more stringent in a steady-flow machine like a turbine, where the gas temperature at each point

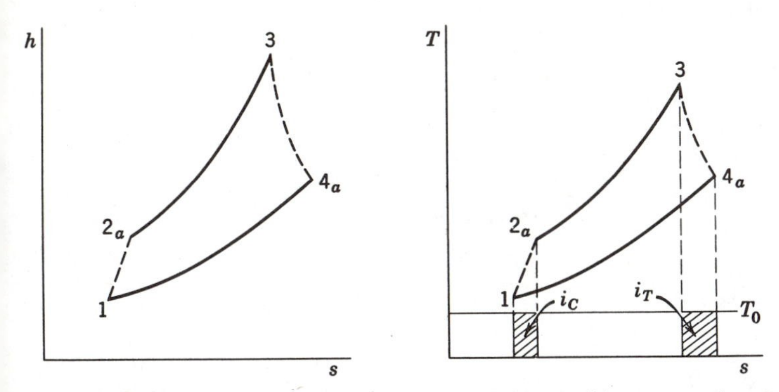

Figure 16·3. Air-standard gas-turbine cycle with irreversible compression and expansion.

*Backwork ratio is sometimes defined as $w_{in,C}/w_{cycle}$. As it is defined and used in this book, the limiting values are 0 and 1.

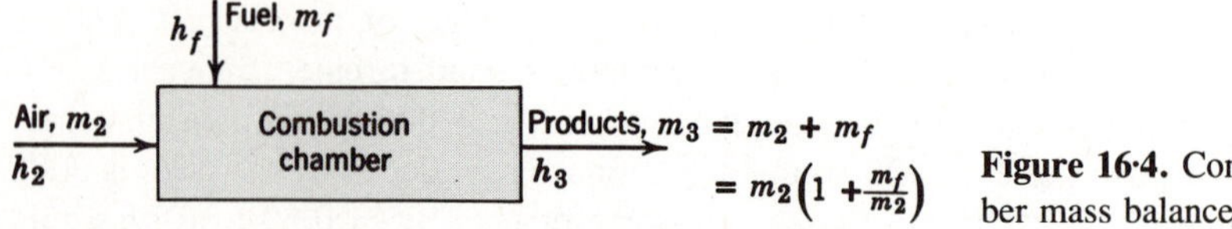

Figure 16·4. Combustion-chamber mass balance.

is constant, than it is in a reciprocating engine, where the metal parts that are at one instant in contact with gases at the maximum temperature are for much of the cycle in contact with relatively cool gases. For this reason, much higher maximum gas temperatures can be used in reciprocating engines than in gas turbines.

In order to limit the maximum temperature in a gas-turbine cycle, high air–fuel ratios are used. Therefore, the products of combustion can be treated as air with little loss of accuracy. Because the backwork ratio is so high, however, the small increase of mass flowing through the turbine resulting from the addition of fuel may increase the net work of the cycle by an appreciable amount. A cycle analysis that considers the mass of the fuel added but still treats the combustion products as air is called an *air-standard analysis with mass of fuel considered.* The air–fuel ratio required to produce a given temperature with a given fuel can be found by starting with an energy balance for adiabatic combustion in a steady-flow burner,

$$-Q = 0 = \sum_{\text{reac}} m(h - h^\circ) - \sum_{\text{prod}} m(h - h^\circ) - \Delta H_R^\circ \qquad (12\cdot1)$$

This energy balance is based on the assumption that the change in kinetic energy is negligible. Designating the air entering by 2, the fuel entering by f, and the products leaving by 3, as shown in Fig. 16·4,

$$0 = m_2(h_2 - h^\circ)_{\text{air}} + m_f(h_f - h^\circ)_{\text{fuel}} - m_3(h_3 - h^\circ)_{\text{prod}} - \Delta H_R^\circ$$

This can be rearranged to

$$\frac{m_2}{m_f} = \frac{\Delta H_R^\circ/m_f + (h_3 - h^\circ)_{\text{prod}} - (h_f - h^\circ)_{\text{fuel}}}{(h_2 - h^\circ)_{\text{air}} - (h_3 - h^\circ)_{\text{prod}}}$$

where $\Delta H_R^\circ/m_f$ is the enthalpy of combustion per unit mass of fuel. If the products are treated as air, the air–fuel ratio expression can be simplified to

$$\frac{m_2}{m_f} = \frac{\Delta H_R^\circ/m_f + h_3 - h^\circ - (h_f - h^\circ)_{\text{fuel}}}{h_2 - h_3}$$

where all of the h's except those marked for fuel can be obtained from air tables or from c_pT relations as in Table A·4.

Example 16·1. A Brayton cycle operates with air entering the compressor at 100 kPa and 25°C, a pressure ratio of 5, and a turbine inlet temperature of 1000°C. Changes in kinetic energy are negligible. Determine the compressor work, the turbine work, and the cycle thermal efficiency based on (*a*) an air-standard analysis using air tables, and (*b*) a cold air-standard analysis.

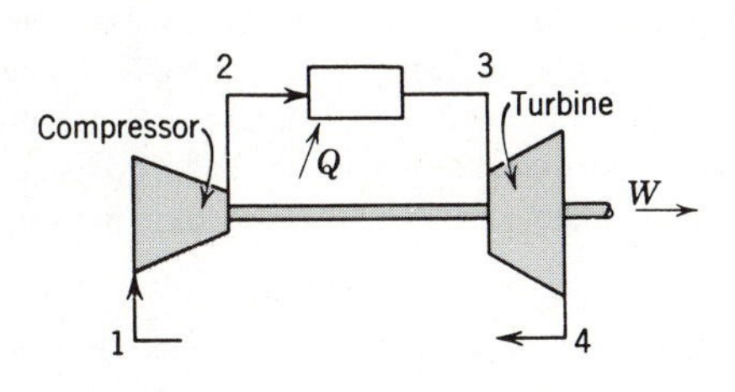

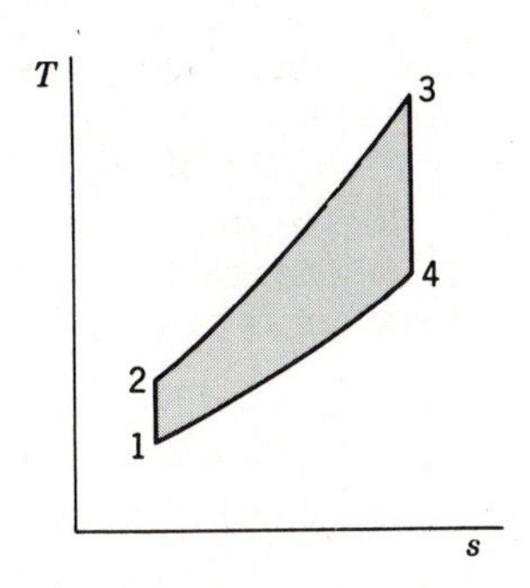

Example 16·1.

Solution. A sketch of the system and a Ts diagram are made first. For either analysis, the work and heat quantities can be found from the first law if the enthalpy changes across the compressor, turbine, and heat exchanger 2-3 are determined.

(*a*) Use of air tables:

$$p_{r2} = p_{r1}\frac{P_2}{P_1} = 1.3482(5) = 6.741 \qquad h_2 = 473.0 \text{ kJ/kg}$$

$$p_{r4} = p_{r3}\frac{P_4}{P_3} = 301.66\left(\frac{1}{5}\right) = 60.33 \qquad h_4 = 878.6 \text{ kJ/kg}$$

Applying the first law to the compressor, the turbine, and the heat exchanger 2-3, respectively,

$$w_{\text{in},C} = h_2 - h_1 = 473.0 - 298.4 = 174.6 \text{ kJ/kg}$$
$$w_T = h_3 - h_4 = 1363.8 - 878.6 = 485.2 \text{ kJ/kg}$$
$$q_{2\text{-}3} = h_3 - h_2 = 1363.8 - 473.0 = 890.8 \text{ kJ/kg}$$

Then the cycle thermal efficiency is

$$\eta = \frac{w_{\text{cycle}}}{q_{2\text{-}3}} = \frac{w_T - w_{\text{in},C}}{q_{2\text{-}3}} = \frac{485.2 - 174.6}{890.8} = 0.349$$

(*b*) Cold air-standard analysis:

$$T_2 = T_1\left(\frac{P_2}{P_1}\right)^{(k-1)/k} = 298(5)^{0.286} = 472 \text{ K}$$

$$w_{\text{in},C} = h_2 - h_1 = c_p(T_2 - T_1) = 1.005(472 - 298) = 174.9 \text{ kJ/kg}$$

$$T_4 = T_3\left(\frac{P_4}{P_3}\right)^{(k-1)/k} = 1273(1/5)^{0.286} = 803 \text{ K}$$

(As a matter of interest, notice that $T_2/T_1 = T_3/T_4$ in the cold air-standard cycle.)

$$w_T = h_3 - h_4 = c_p(T_3 - T_4) = 1.005(1273 - 803) = 472.4 \text{ kJ/kg}$$

$$q_{2\text{-}3} = h_3 - h_2 = c_p(T_3 - T_2) = 1.005(1273 - 472) = 805.0 \text{ kJ/kg}$$

$$\eta = \frac{\text{w}_{\text{cycle}}}{q_{2\text{-}3}} = \frac{w_T - w_{\text{in},C}}{q_{2\text{-}3}} = \frac{472.4 - 174.9}{805.0} = 0.370$$

Discussion. The compressor work is nearly the same in both analyses because the compressor operates in a temperature range where c_p is very nearly constant at 1.005 kJ/kg·K. Using the same value of c_p in the higher temperature range up to 1000°C introduces some inaccuracy, however. The cold air-standard analysis gives a work output that is too low and a cycle thermal efficiency that is too high in comparison with the more accurate results based on the variable specific heat data of the air tables.

Example 16·2. Solve Example 16·1*a* on the basis of an air-standard analysis with the mass of fuel considered. The fuel is added at 25°C and has a lower heating value of 44 000 kJ/kg at 25°C.

Solution. The compressor work is unchanged by the consideration of the mass of fuel added. We determine the air-fuel ratio by means of the expression developed from an energy balance just prior to Example 16·1, using a base temperature of 25°C

$$\frac{m_a}{m_f} = \frac{\Delta H_R^\circ/m_f + h_3 - h^\circ - (h_f - h^\circ)_{\text{fuel}}}{h_2 - h_3} = \frac{-44\,000 + 1363.8 - 298.4 - 0}{473.0 - 1363.8}$$
$$= 48.2 \text{ kg air/kg fuel}$$

From the solution of Example 16·1, the turbine work is 485.2 kJ/kg of air (i.e., combustion products treated as air) passing through the turbine. Per kilogram of air flowing through the compressor

$$w_T = 485.2\left(\frac{\text{kJ}}{\text{kg air to turbine}}\right)\frac{49.2}{48.2}\left(\frac{\text{kg air to turbine}}{\text{kg air to compressor}}\right)$$
$$= 495.3 \text{ kJ/kg air compressed}$$

Using a thermal efficiency based on the lower heating value

$$\eta = \frac{w_T - w_{\text{in},C}}{(-\Delta H_R^\circ/m_f)(m_f/m_a)} = \frac{495.3 - 174.9}{44\,000/48.2} = 0.351$$

16·3 The Regenerative Gas-Turbine Cycle

In a simple gas-turbine cycle, the turbine exhaust temperature is appreciably higher than the temperature of the air leaving the compressor and entering the combustion chamber. Obviously, the amount of fuel needed can be reduced by the use of a heat exchanger in which the hot turbine exhaust gas is used to preheat the air between the compressor and the combustion chamber. This heat exchanger is called a *regenerator*. In the ideal case the flows through the regenerator are at constant pressure.

An air-standard regenerative cycle is shown in Fig. 16·5. Application of the first law to the regenerator, assuming that kinetic energy changes are negligible and that no heat is lost to the surroundings, shows that $h_3 - h_2 = h_5 - h_6$. Since $q_{2-3} = -q_{5-6}$, the two crosshatched areas on the *Ts* diagram of Fig. 16·5 are equal in magnitude. If, as shown in the *Ts* diagram, heat is transferred across a finite temperature difference in the regenerator, then the increase of entropy principle shows that, for this irreversible steady-flow process, $s_3 - s_2 + s_6 - s_5 > 0$, or $|\Delta s_{2-3}| > |\Delta s_{5-6}|$. The irreversibility of the regenerator process is $i = T_0(s_3 - s_2 + s_6 - s_5)$.

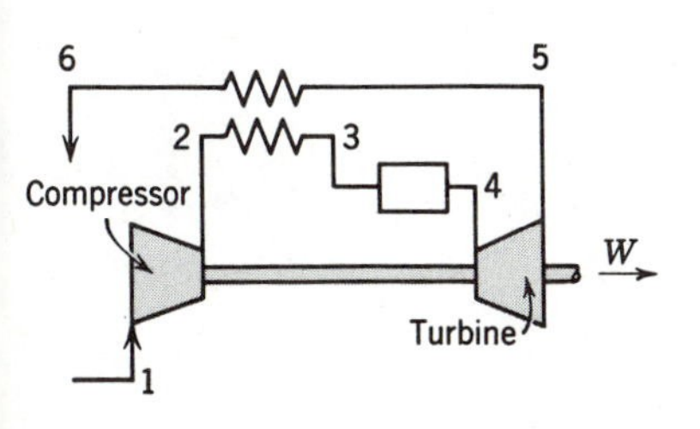

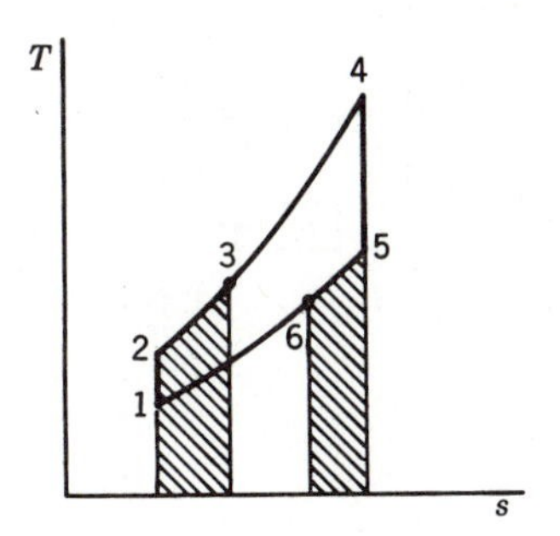

Figure 16·5. Air-standard regenerative cycle.

Another way to look at the irreversibility of the regenerator is this: The magnitude of the unavailable portion of the heat removed from the hot gas is $T_0(s_5 - s_6)$. This part of the heat removed from the hot gas could not be converted into work even by an externally reversible engine. As the heat is added to the cold gas in the regenerator, however, the unavailable part of it is $T_0(s_3 - s_2)$. The difference between $T_0(s_3 - s_2)$ and $T_0(s_5 - s_6)$ is therefore the amount of *energy made unavailable* by the irreversible transfer of heat in the regenerator. This is the same as the irreversibility of the process.

For maximum thermal efficiency of a regenerative cycle, T_3 in Fig. 16·5 should be as high as possible, its limiting value being T_5. The extent to which this limit is approached in any cycle is expressed by the regenerator *effectiveness* which, with reference to Fig. 16·5, is defined as

$$\text{Effectiveness} \equiv \frac{T_3 - T_2}{T_5 - T_2}$$

Increasing the effectiveness of a regenerator calls for more heat-transfer surface area which increases both cost and space requirements. Using regenerators of very high effectiveness cannot be justified economically because the fixed costs (that is, costs that are largely independent of plant output such as depreciation, interest, taxes, insurance, maintenance, etc.) exceed the savings in fuel cost that result from the higher plant thermal efficiency. Consequently, effectiveness values rarely exceed about 0.7 in actual plants.

Example 16·3. In a regenerative air-standard gas-turbine cycle, air enters the compressor at 14.7 psia, 80°F, and leaves it at 73.5 psia, 500°F. The temperature of air entering the combustion chamber is 800°F, entering the turbine is 1600°F, and leaving the turbine is 1060°F. The lowest temperature in the surroundings is 80°F. Neglect the mass of fuel. Assuming no pressure drop in the regenerator or "combustion chamber," and neglecting changes in kinetic energy, calculate (*a*) the compressor efficiency, (*b*) the turbine efficiency, (*c*) the cycle thermal efficiency, (*d*) the irreversibility of the compressor process, (*e*) the irreversibility of the turbine expansion, and (*f*) the irreversibility of the regenerator. (*g*) How much of the heat added to the cycle is available energy? (*h*) How much of the heat rejected from the cycle is available energy?

Solution. First a flow diagram and a *Ts* diagram are made. Points 2_i and 5_i are the *isentropic* end states of the compression and expansion processes which are needed in the calculation of compressor and turbine efficiencies. At various stages in the solution, property values will be

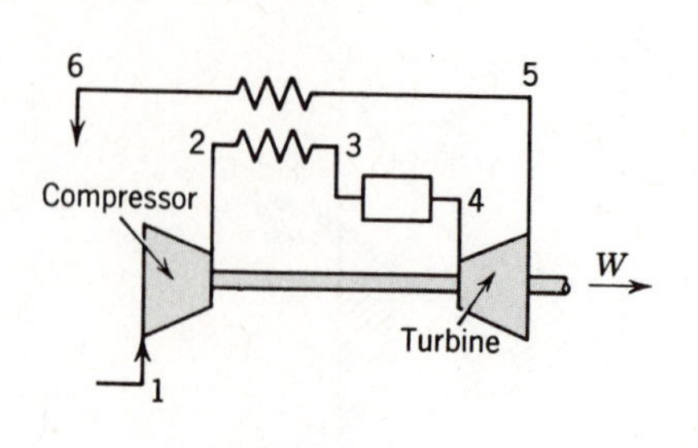

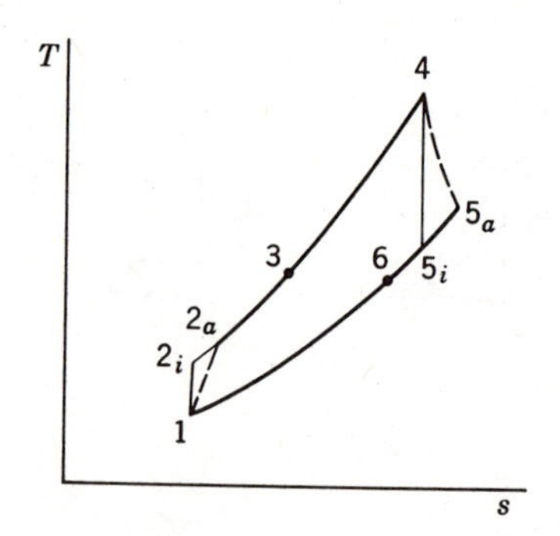

Example 16·3.

obtained; so for convenience a table of these values is made, showing both the specified values and the calculated values which are entered as soon as they are obtained. Air tables are used.

Point	P, psia	T, °F	h, B/lbm	p_r	ϕ
1	14.7	80	129.2	1.380	0.6007
2_i	73.5		204.6	6.90	
2_a		500	231.2		0.7402
3		800	306.7		0.8086
4		1600	521.4	195.1	0.9401
5_i	14.7		334.4	39.2^+	
5_a		1060	374.5		0.8575
6			299.0		0.8025

The solution of the first three parts involves the application of (1) the definitions of the three efficiencies, and (2) the first law.

$$(a)\ p_{r2i} = p_{r1}\frac{P_2}{P_1} = 1.380\frac{73.5}{14.7} = 6.90 \qquad h_{2i} = 204.7\text{ B/lbm}$$

$$\eta_C = \frac{w_{\text{in},i}}{w_{\text{in},a}} = \frac{h_{2i} - h_1}{h_{2a} - h_1} = \frac{204.7 - 129.2}{231.2 - 129.2} = 74.0\text{ percent}$$

$$(b)\ p_{r5i} = p_{r4}\frac{P_5}{P_4} = 195.1\frac{14.7}{73.5} = 39.02 \qquad h_{5i} = 334.4\text{ B/lbm}$$

$$\eta_T = \frac{w_a}{w_i} = \frac{h_4 - h_{5a}}{h_4 - h_{5i}} = \frac{521.4 - 374.5}{521.4 - 334.4} = 78.6\text{ percent}$$

$$(c)\ \eta = \frac{w_{\text{cycle}}}{q_{3\text{-}4}} = \frac{w_T - w_{\text{in},C}}{h_4 - h_3} = \frac{521.4 - 374.5 - (231.2 - 129.2)}{521.4 - 306.7}$$

$$= 20.9\text{ percent}$$

The compressor, the turbine, and the entire regenerator (including both fluid streams) operate steadily and adiabatically, so for each one the irreversibility per pound of fluid is given by $i = T_0\Delta s$, where Δs is the entropy change of the fluid flowing through the device. (See Eq. b of Sec. 10·4.)

(*d*) Compressor:

$$i = T_0(s_{2a} - s_1) = T_0\left[\phi_{2a} - \phi_1 - R \ln \frac{P_2}{P_1}\right]$$

$$= 540\left[0.7402 - 0.6007 - 0.0685 \ln \frac{73.5}{14.7}\right]$$

$$= 15.8 \text{ B/lbm}$$

(*e*) Turbine:

$$i = T_0(s_{5a} - s_4) = T_0\left[\phi_{5a} - \phi_4 - R \ln \frac{P_5}{P_4}\right]$$

$$= 540\left[0.8575 - 0.9401 - 0.0685 \ln \frac{14.7}{73.5}\right]$$

$$= 14.9 \text{ B/lbm}$$

(*f*) Regenerator:

$$h_6 = h_{5a} - (h_3 - h_{2a}) = 374.5 - (306.7 - 231.2) = 299.0 \text{ B/lbm}$$

$$i = T_0[\Delta s_{2\text{-}3} + \Delta s_{5\text{-}6}] = T_0\left[\phi_3 - \phi_2 - R \ln \frac{P_3}{P_2} + \phi_6 - \phi_5 - R \ln \frac{P_6}{P_5}\right]$$

$$= 540[0.8086 - 0.7402 - 0 + 0.8025 - 0.8575 - 0]$$

$$= 7.24 \text{ B/lbm}$$

(*g*) Of the heat added:

$$q_{\text{unav}} = T_0(s_4 - s_3) = T_0(\phi_4 - \phi_3 - R \ln 1) = 540(0.9401 - 0.8086)$$

$$= 71.0 \text{ B/lbm}$$

$$q_{\text{av}} = q - q_{\text{unav}} = 214.7 - 71.0 = 143.7 \text{ B/lbm}$$

(*h*) For the heat rejected:

$$q_{\text{out}} = h_6 - h_1 = 299.0 - 129.2 = 169.8 \text{ B/lbm}$$

$$q_{\text{unav}} = T_0(s_6 - s_1) = T_0(\phi_6 - \phi_1) = 540(0.8025 - 0.6007) = 109.0 \text{ B/lbm}$$

$$q_{\text{out,av}} = q_{\text{out}} - q_{\text{out,unav}} = 169.8 - 109.0 = 60.8 \text{ B/lbm}$$

The energy flow diagram shows graphically that the unavailable energy rejected from the

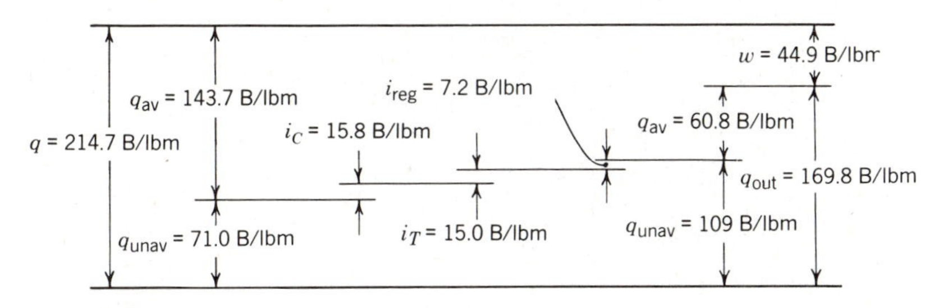

Example 16·3.

cycle exceeds the unavailable energy added to the cycle by the sum of the irreversibilities in the cycle.

16·4 Intercooling and Reheating in Gas-Turbine Cycles

The net work of a gas-turbine cycle is given by

$$w_{\text{cycle}} = w_T - w_{\text{in},C}$$

and can be increased either by decreasing the compressor work or by increasing the turbine work. These are the purposes of *intercooling* and *reheating,* respectively.

It was shown in Sec. 15·1 that intercooling between stages reduces the work input required to compress a gas from a given initial state to a specified final pressure. Therefore, if a simple gas-turbine cycle is modified by having the compression accomplished in two or more adiabatic processes with intercooling between them, the backwork ratio is reduced and consequently the net work of the cycle is increased with no change in the work of the turbine.

Several different lines of reasoning can be used to show that the thermal efficiency of an ideal simple gas-turbine cycle is lowered by the addition of an intercooler. One line of reasoning is based on a *Ts* diagram like Fig. 16·6 in which the ideal simple gas-turbine cycle is 1-2-3-4-1, and the cycle with the intercooler added is 1-*a*-*b*-*c*-2-3-4-1. Any reversible cycle can be simulated by a number of Carnot cycles. (See Sec. 8·1.) If the simple gas-turbine cycle 1-2-3-4-1 is divided into a number of cycles like *i*-*j*-*k*-*l*-*i* and *m*-*n*-*o*-*p*-*m*, these little cycles approach Carnot cycles as their number increases. Notice that, *if the specific heats are constant*

$$\frac{T_3}{T_4} = \frac{T_m}{T_p} = \frac{T_i}{T_l} = \frac{T_2}{T_1} = \left(\frac{P_2}{P_1}\right)^{(k-1)/k}$$

Thus all the Carnot cycles making up the simple gas-turbine cycle have the same efficiency. Likewise, all the Carnot cycles into which the cycle *a*-*b*-*c*-2-*a* might similarly be divided have a common value of efficiency that is *lower* than the Carnot cycles which comprise cycle 1-2-3-4-1. Thus the addition of an intercooler, which adds *a*-*b*-*c*-2-*a* to the simple cycle, lowers the efficiency of the ideal gas-turbine cycle. We have based our reasoning

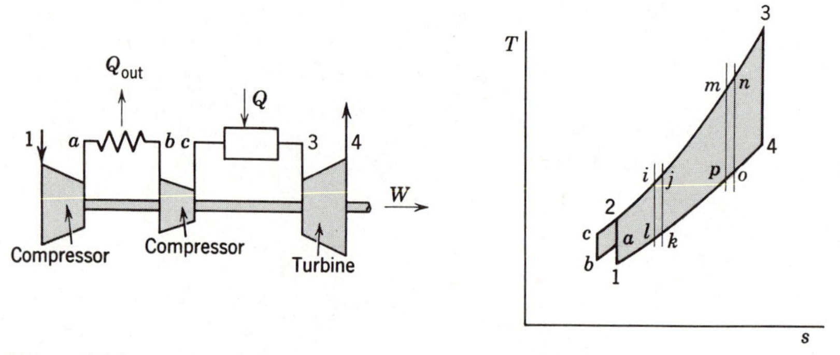

Figure 16·6. Air-standard gas-turbine cycle with intercooling.

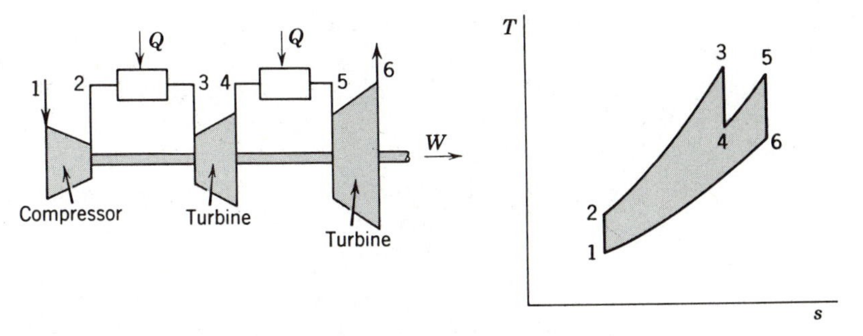

Figure 16·7. Air-standard gas-turbine cycle with reheat.

here on constant specific heats (the cold air-standard analysis), but the same conclusions can be reached for variable specific heats.

The addition of an intercooler to a regenerative gas-turbine cycle increases the cycle thermal efficiency because the heat required for the process *c*-2 in Fig. 16·6 can be obtained from the hot turbine exhaust gas passing through the regenerator instead of from burning additional fuel (or, in the case of the air-standard cycle, from adding more heat from outside the cycle). The use of an intercooler consequently calls for a larger regenerator.

The turbine work, and consequently the net work of the cycle, can be increased without changing the compressor work or the maximum temperature in the cycle by dividing the turbine expansion into two or more parts with constant-pressure heating (that is, combustion in an actual cycle, heat transfer in an air-standard cycle) before each expansion. This cycle modification is known as *reheating*. Flow and *Ts* diagrams of an ideal simple gas-turbine cycle modified by a single reheat are shown in Fig. 16·7. By

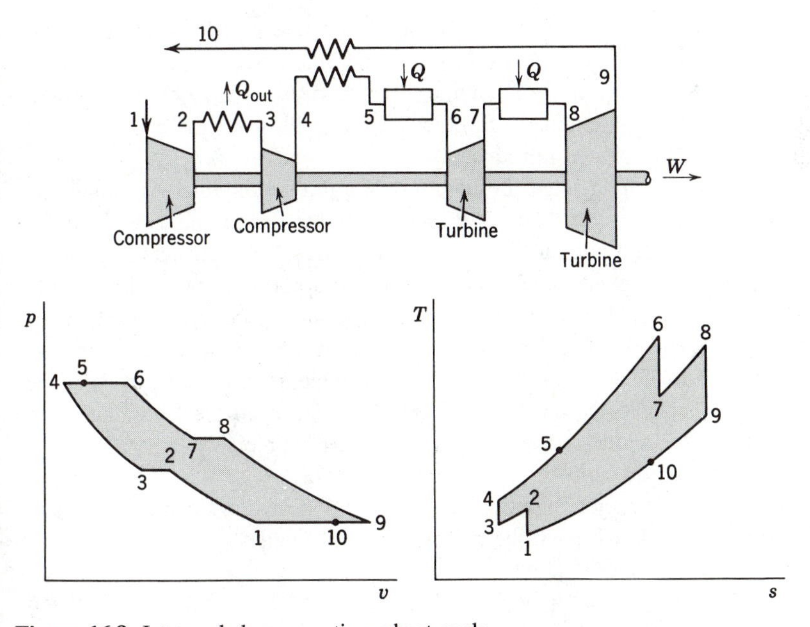

Figure 16·8. Intercooled-regenerative-reheat cycle.

reasoning similar to that used in connection with intercooling, one can show that the thermal efficiency of a simple cycle is lowered by the addition of reheat but that a combination of regeneration and reheating can increase the thermal efficiency.

Notice that reheating increases the turbine exhaust temperature so that regeneration can bring the temperature of the air entering the combustion chamber closer to the turbine inlet temperature than is possible without reheat. As the number of reheats is increased, the exhaust temperature increases, making regeneration even more valuable.

Diagrams of an intercooled–regenerative–reheat cycle are shown in Fig. 16·8. If the number of intercools, the number of reheats, and the regenerator effectiveness are increased, this cycle approaches the Ericsson cycle discussed in Sec. 6·8. With adiabatic compression and expansion, the costs of equipment for multiple intercools and reheats exceed the fuel-cost savings, but the advantages of approaching isothermal compression and isothermal turbine expansion (as by burning fuel in the turbine in an actual cycle) are apparent.

16·5 Gas-Turbine Jet Propulsion

Powered aircraft and water craft are propelled by accelerating fluid rearward by exerting a force on it. The equal-magnitude and opposite-direction force of the fluid on the craft drives the craft. A propeller accelerates slightly a large mass of fluid while an aircraft jet engine causes a relatively small mass of air to undergo a large change in velocity. In both cases the action is in accordance with the basic dynamic equation for steady flow which for the one-dimensional case is

$$F = \dot{m}\Delta V \tag{14·1}$$

where F is the resultant force on the fluid which flows at a rate $\dot{m}$. (See Example 14·4, but notice that a special case of $P_1 = P_2 = P_{atm}$ is treated there.)

From among the various jet-propulsion engines that are in use or have been proposed, we will select the gas-turbine jet engine or turbojet engine to illustrate the application of the principles that have already been introduced.

Figure 16·9 shows the flow, Pv, and Ts diagrams of a turbojet engine. Air enters the engine at state 1 and is compressed to state 2. It then flows into the combustion chamber, where it is mixed with fuel and combustion occurs, or where, in the air-standard cycle, heat is added. The gas then expands adiabatically through the turbine, which drives only the compressor, so that *the turbine work equals the compressor work input*. In order for the turbine to produce enough work to drive the compressor it is necessary for it to expand the gas only to a pressure P_4, which is higher than the ambient pressure $P_1(=P_5)$. The gas is then expanded isentropically (in the ideal case) through a nozzle from state 4 to state 5. The gas leaves the engine at a velocity V_5 which is appreciably higher than V_1, and consequently a thrust is developed.

If the flow through the turbojet engine is reversible and steady, we can apply to any process the relationship developed from the principles of mechanics,

$$\int v\,dP = w_{in} - \Delta ke - \Delta pe \tag{1·10}$$

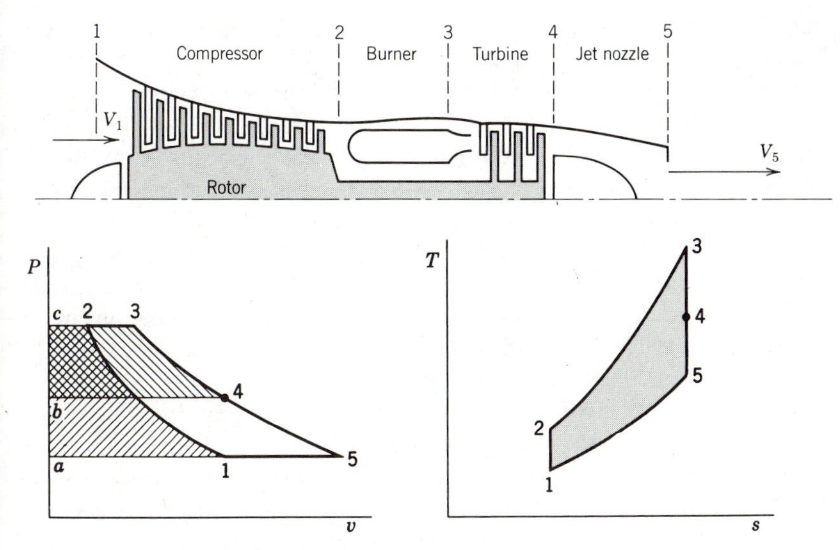

Figure 16·9. Turbojet engine.

For a gas flowing through a jet engine, Δpe is negligibly small. Equation 1·10 shows the physical significance of areas on a Pv diagram as in Fig. 16·9. For example,

$$w_{\text{in},1\text{-}2} = w_{3\text{-}4}$$

for an air-standard analysis. Therefore,

$$\int_1^2 v\,dP + \frac{V_2^2 - V_1^2}{2} = -\int_3^4 v\,dP + \frac{V_3^2 - V_4^2}{2}$$

and, if $V_2 = V_1$ and $V_3 = V_4$, then the two crosshatched areas 3-4-*b*-*c*-3 and 1-2-*c*-*a*-1 in Fig. 16·9 must be equal in magnitude. Also,

$$\frac{V_5^2 - V_4^2}{2} = -\int_4^5 v\,dP$$

so that the area 4-5-*a*-*b*-4 in Fig. 16·9 represents the increase in kinetic energy of the gas flowing through the turbojet nozzle.

A study of Fig. 16·9 shows that low compressor and turbine efficiencies reduce markedly the thrust obtainable from a turbojet engine.

Space and weight limitations prohibit the use of regenerators and intercoolers on aircraft engines, although the compressor work can be reduced by injecting water into the compressor inlet to provide "internal cooling" by its evaporation. The counterpart of reheating is *afterburning*. The air–fuel ratio in a jet engine is so high that the turbine exhaust gases are sufficiently rich in oxygen to support the combustion of more fuel. Such burning of fuel (or, in the air-standard cycle, addition of heat) raises the temperature of the gas before it expands in the turbojet nozzle, increasing the kinetic energy change in the nozzle and consequently increasing the thrust.

Several modifications of the simple turbojet engine are in use to meet special performance needs of various aircraft. Two examples of a modification known as the turbofan engine are shown schematically in Fig. 16·10. In each of these engines, some of the air that enters passes through the compressor, combustion chamber, and turbine just as in a simple turbojet engine. The rest passes through only a fan or the first stages of the compressor and is then either mixed with the exhaust jet or discharged separately at a velocity higher than the inlet velocity to produce part of the propulsive force. The ratio of the flow rates of the two air streams is controllable and has a major influence on engine performance. It is apparent that the ratio is quite different for the two engines shown in Fig. 16·10.

Many other arrangements involving various combinations of propellers, fans, and various numbers of compressors and turbines are also used.

This brief discussion of turbojet engines has not touched on the performance characteristics of the individual components—compressors, combustion chambers, and turbines—and the important problem of matching these characteristics over a wide operating range. Another major problem is the prediction of the system behavior under transient conditions. For example, a change in fuel flow changes the turbine inlet temperature, which changes the turbine power output and hence the shaft speed. A change in shaft speed changes the flow rate and the pressure ratio, which in turn cause further changes in the turbine inlet temperature. Such problems must be thoroughly analyzed in designing an engine. They are mentioned here as a reminder that the design of jet engines involves much more than a thermodynamic analysis of steady-flow cycles.

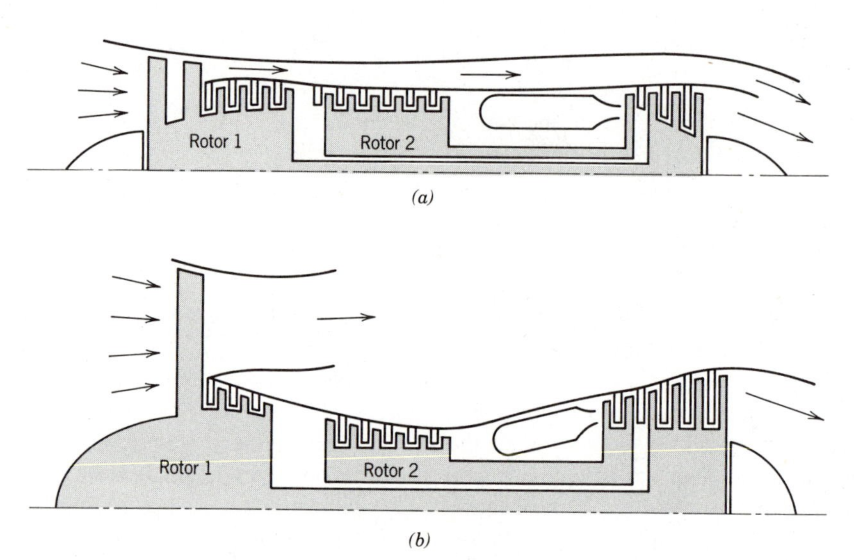

Figure 16·10. Two examples of turbofan gas-turbine aircraft engines.

16·6 The Air-Standard Otto Cycle

The air-standard Otto* cycle is comprised of four reversible processes of air in a closed system: adiabatic compression, constant-volume heat addition, adiabatic expansion, and constant-volume heat rejection. Pv and Ts diagrams of an Otto cycle are shown in Fig. 16·11. The *compression ratio* r of the cycle is defined by

$$r \equiv \frac{V_1}{V_2} = \frac{v_1}{v_2}$$

Compression ratio is thus a volume ratio that is fixed by the geometry of the engine.

A first-law analysis of this cycle is very simple because no work is done during the two heat-transfer processes (2-3 and 4-1) and both processes that involve work (1-2 and 3-4) are adiabatic; so for each process one term is zero in the first-law formulation

$$Q = \Delta U + W \tag{2·7}$$

A cold air-standard analysis is of course even simpler and yields results such as $T_2/T_1 = T_3/T_4$ and $\eta = 1 - r^{1-k}$. The latter result shows that the cold air-standard Otto-cycle thermal efficiency is a function of compression ratio only. The thermal efficiency of an air-standard (that is, variable specific heats) Otto cycle depends on the temperature limits as well as on the compression ratio.

The qualitative effects on efficiency and on work per cycle of varying the compression ratio, temperature limits, and pressure limits are similar for an air-standard Otto cycle and for an actual spark-ignition engine, and this is why the air-standard analysis is of value. The differences between the air-standard and the actual cycles are so great, however, that close quantitative agreement is not to be expected.

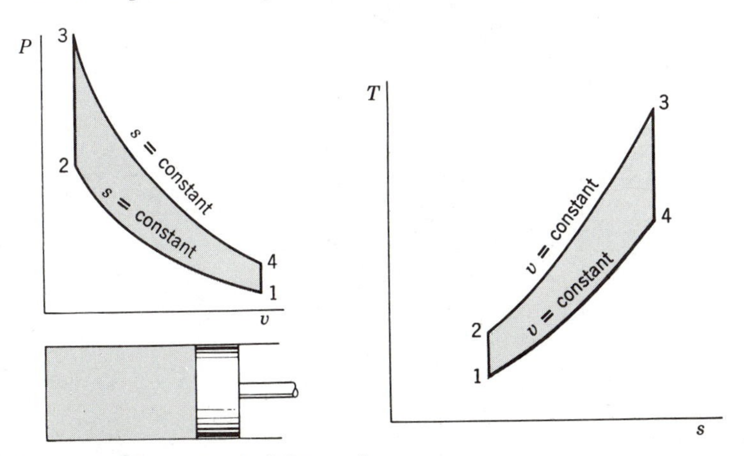

Figure 16·11. Air-standard Otto cycle.

*Nikolaus A. Otto (1832–1891) and his partner Eugen Langen built a gas engine in 1867 in Deutz, Germany, and began commercial manufacture of it. In 1876 Otto produced a successful four-stroke cycle engine that was far superior to any internal-combustion engine previously built. The four-stroke cycle had been worked out in principle in 1862 by a Frenchman, Alphonse Beau de Rochas.

16·7 The Air-Standard Diesel Cycle

Figure 16·12 shows *Pv* and *Ts* diagrams of an air-standard Diesel* cycle. All four processes are reversible:

1-2: adiabatic compression
2-3: constant-pressure heat addition
3-4: adiabatic expansion to the initial volume
4-1: constant-volume heat rejection

The *cutoff ratio* r_c is defined by

$$r_c \equiv \frac{V_3}{V_2} = \frac{v_3}{v_2}$$

and the *percent cutoff* is defined by

$$\text{Percent cutoff} \equiv \frac{V_3 - V_2}{V_1 - V_2} = \frac{v_3 - v_2}{v_1 - v_2}$$

that is, it is the fraction of the stroke during which heat is added.

The thermal efficiency of a cold air-standard Diesel cycle is a function of the compression ratio and cutoff ratio only, but other factors are pertinent with variable specific heats.

Actual Diesel engines are *compression–ignition* (as distinguished from *spark ignition*) engines. Only air is compressed in the cylinder, and, since there is no fuel in the cylinder, the compression ratio can be high enough to raise the air temperature above the fuel

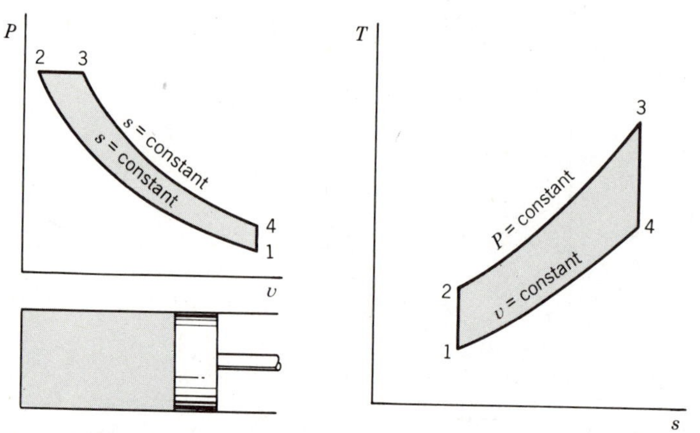

Figure 16·12. Air-standard Diesel cycle.

*Rudolph Diesel (1858–1913) was born in Paris of German parents and educated in Munich. In 1893 he published a book, *The Theory and Construction of a Rational Heat Motor*, and obtained a patent on a compression–ignition engine. By 1899 he had developed this new type of engine to the point where he was able to begin commercial production in his factory at Augsburg.

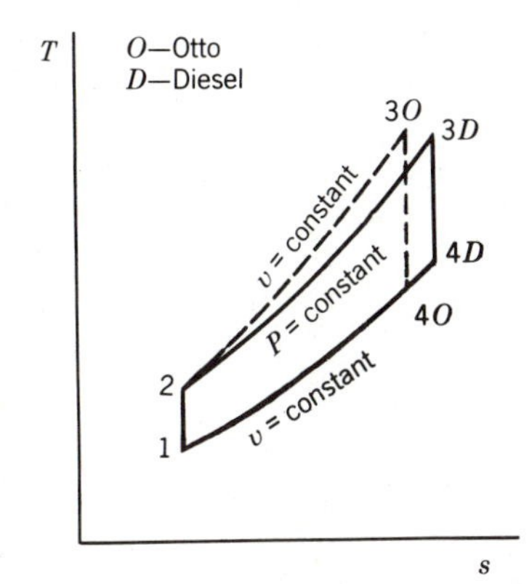

Figure 16·13. Comparison of Otto and Diesel cycles.

ignition temperature without causing ignition. Then, as the piston moves away from the cylinder head, fuel is injected and burned. The pressure variation during the first part of the expansion stroke depends on the rate at which fuel is injected and burned. In low-speed engines at least part of the fuel may be burned at approximately constant pressure as it is injected. In high-speed engines, ignition lag delays the start of combustion until most or all of the fuel is in the cylinder, so the ideal picture of the pressure being held constant by a regulated injection of fuel as the gas expands is unrealistic. Once again the differences between an actual cycle and a corresponding ideal cycle must be remembered so that conclusions based on one will not be blindly applied to the other.

Performance comparisons among various gas power cycles are readily made by using *Ts* diagrams to aid in reasoning that is based on the first and second laws and ideal-gas properties. As one example, let us determine which has the higher thermal efficiency, an air-standard Otto cycle or an air-standard Diesel cycle with the same compression ratio and the same heat input. Superimposed *Ts* diagrams of the two cycles are shown in Fig. 16·13. Since the two cycles have the same compression ratio, process 1-2 is common to them. After one of the cycle *Ts* diagrams is sketched, the relative locations of points 3*O* (Otto) and 3*D* are fixed, because equal heat input means equal areas under the constant-volume line 2-3*O* and the constant-pressure line 2-3*D*. Thus points 3*O* and 4*O* are at a lower entropy than points 3*D* and 4*D*. Since both cycles reject heat in a constant-volume process ending at state 1, the relative location of points 4*O* and 4*D* on the *Ts* diagram shows that less heat is rejected by the Otto cycle. The heat inputs are the same; so the work of the Otto exceeds the work of the Diesel cycle. Consequently the Otto cycle has the higher thermal efficiency. *Caution*: This conclusion applies to cycles which have the same compression ratio and the same heat input. Comparisons can be made in a similar manner for various other conditions.

Example 16·4. In an air-standard Diesel cycle, compression starts at 14.5 psia, 120°F. At the end of compression the pressure is 550 psia. The mass of air in the system is 0.15 lbm, and the amount of heat added per cycle is 42 B. Calculate for this cycle (*a*) the compression ratio, (*b*) the work, and (*c*) the amount of heat rejected which is available energy.

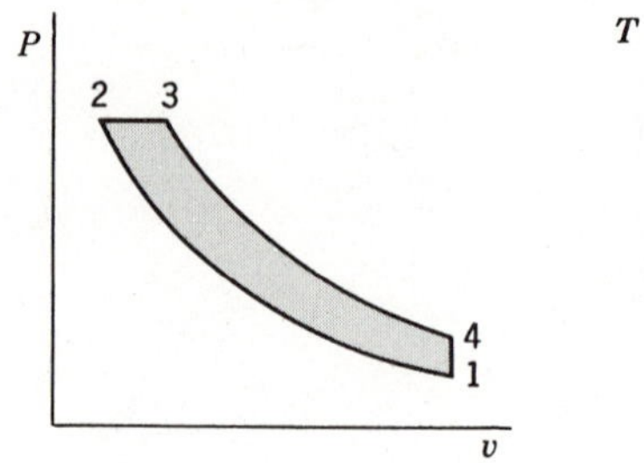

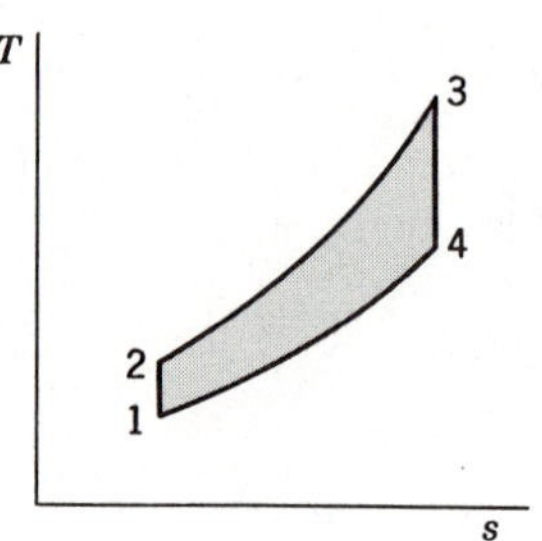

Example 16·4.

Solution. Pv and Ts diagrams of the cycle are made first. Since in order to solve part c it will be necessary to determine the amount of heat rejected and since part b can also be solved by use of the heat rejected, it appears advantageous to determine state 4 at which the heat-rejection process starts. Since this involves calculations "around the cycle," we make a table in which we write specified property values immediately and calculated values as soon as they are obtained:

Point	P, psia	T, R	p_r	v_r	u, B/lbm	h, B/lbm	ϕ
1	14.5	580	1.7725	121.23	99.00		
2	550	1575	67.2	8.684	281.2	389.2	0.86703
3	550	2582		1.9299		669.2	1.00402
4		1255		16.44	219.5		

The manner in which the calculated values are obtained for the table is shown in the following:

$$p_{r2} = p_{r1}\frac{P_2}{P_1} = 1.7725\left(\frac{550}{14.5}\right) = 67.2$$

(From p_{r2}, values of v_r, T, u, and h at state 2 can be determined from the air tables.)

(*a*)

$$\text{Compression ratio} = \frac{v_1}{v_2} = \frac{v_{r1}}{v_{r2}} = \frac{121.23}{8.684} = 13.96$$

(*b*) Applying the first law to process 2-3 and noting that for a reversible constant-pressure process of a closed system $q = \Delta u + w = \Delta u + \int P\,dv = \Delta u + \Delta Pv = \Delta h$,

$$h_3 = h_2 + \frac{Q}{m} = 389.2 + \frac{42}{0.15} = 389.2 + 280.0 = 669.2 \text{ B/lbm}$$

(Knowing h_3, we take from the air tables values of T and v_r at state 3.)

$$v_{r4} = v_{r3}\frac{v_4}{v_3} = v_{r3}\frac{v_1}{v_3} = v_{r3}\frac{T_1P_3}{T_3P_1} = 1.9299\left(\frac{580}{2582}\right)\frac{550}{14.5} = 16.44$$

(Knowing v_{r4}, we take from the air tables values of T, u, and ϕ for state 4.)

$$W = \oint \delta Q = Q_{2\text{-}3} - Q_{\text{out},4\text{-}1} = Q_{2\text{-}3} - m(u_4 - u_1) = 42 - 0.15(219.5 - 99.0)$$
$$= 42 - 18.1 = 23.9 \text{ B}$$

(c)
$$Q_{\text{av},4\text{-}1} = Q_{4\text{-}1} - Q_{\text{unav},4\text{-}1} = Q_{4\text{-}1} - T_0(S_1 - S_4) = Q_{4\text{-}1} - T_0(S_2 - S_3)$$
$$= Q_{4\text{-}1} - T_0 m\left(\phi_2 - \phi_3 - R \ln \frac{P_2}{P_3}\right)$$
$$= -18.1 - 580(0.15)(0.86703 - 1.00402 - 0) = -6.2 \text{ B}$$

We have used the lowest temperature in the cycle as T_0 since no other information was provided. The negative sign indicates that 6.2 B of available energy is *removed* from the system. The physical interpretation is that if the 18.1 B of heat rejected by the Diesel cycle were transferred into a series of Carnot engines (or other externally reversible engines) rejecting heat at 580 R, 6.2 B out of the 18.1 B could be converted into work by the Carnot engines. A shortcoming of a Diesel cycle in comparison with a Carnot cycle is that some of the heat rejected by the former *can* be converted into work while the Carnot cycle rejects only heat which *cannot* be converted into work.

16·8 The Dual Cycle and Others

Neither the air-standard Otto nor the air-standard Diesel cycle closely approximates the pressure–volume variation of the working substance in an actual engine. Some analyses call for an air-standard cycle in which the pressure–volume variation does simulate that in an actual engine. One such cycle, called the dual cycle, involves two heat-addition processes: one at constant volume and one at constant pressure as shown in Fig. 16·14. The relative amounts of heat added in the two processes can be adjusted to make the air-standard cycle more like an actual one.

A further refinement is to add heat first at constant volume, then at constant pressure, and then at constant temperature in an air-standard cycle. Such cycles can be analyzed and evaluated by the same fundamental methods used for the simpler ones. Remember, however, that the most highly refined air-standard cycle is still much different from what occurs in an actual internal-combustion engine. An ideal system can be more confusing

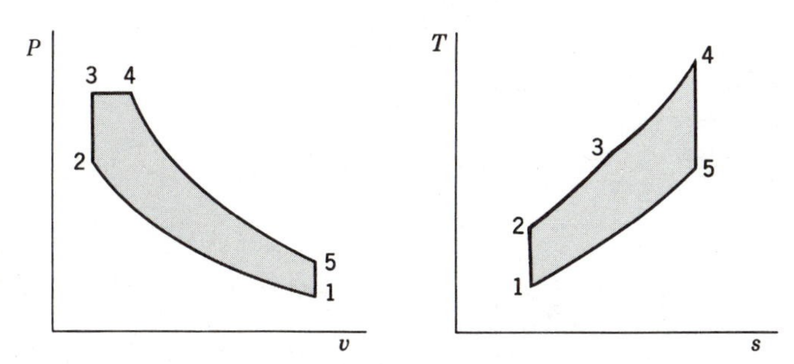

Figure 16·14. Air-standard dual cycle.

than helpful if the differences between it and the corresponding actual system are not kept in mind.

16·9 Compound Power Plants

As mentioned in Sec. 16·2, reciprocating engines can employ much higher maximum gas temperatures than turbines. Since adiabatic expansions are used in most gas power cycles, the higher temperatures are accompanied by higher pressures. On the other hand, reciprocating engines are poor in contrast with turbines for expanding gas to low pressures with resulting large volumes because the engine size increases more rapidly than the power output. Either method of increasing the mass flow rate through a reciprocating engine—increasing engine size or increasing speed—augments the design difficulties that stem from increased inertia loads on the reciprocating parts. In contrast, turbines are readily adaptable to handling large gas volumes. To capitalize on the advantages and minimize the drawbacks of each type of prime mover, *compound power plants* or *compound engines* are used in which the high-pressure and high-temperature processes occur in a reciprocating engine and the low-pressure expansion occurs in a turbine.

Three different arrangements of compound engines are shown in Fig. 16·15. In the first one, exhaust gases from the reciprocating engine drive a turbine that supplies part of the net power output of the plant. The compressor for supercharging the engine is driven from the engine shaft. Figure 16·15*b* shows the exhaust gas turbine driving only the compressor or supercharger and the reciprocating engine supplying all the plant power output. The turbine and compressor are mounted on a single shaft with no mechanical connection to the engine. In the arrangement of Fig. 16·15*c*, the reciprocating engine

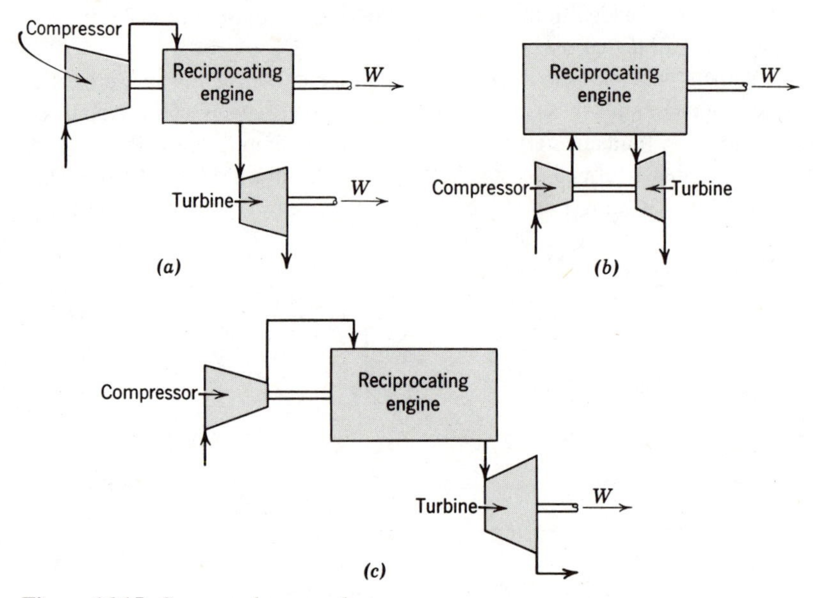

Figure 16·15. Compound power plants.

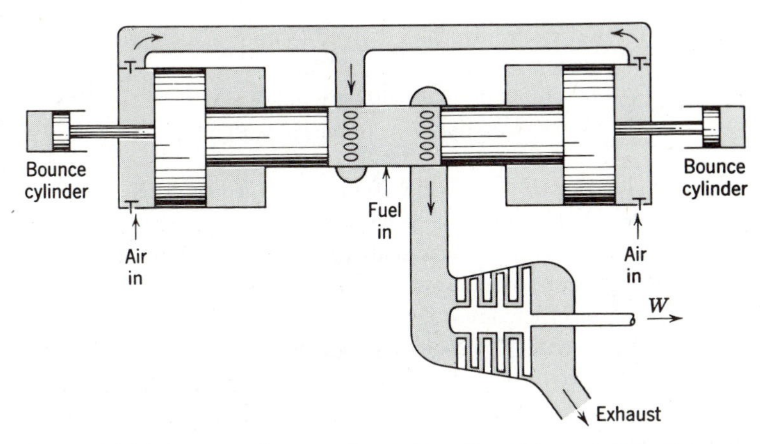

Figure 16·16. Compound power plant using a free-piston gas generator.

drives only the compressor, and the turbine provides the plant power output. If the air-fuel ratio of the reciprocating engine is sufficiently high, some fuel can be burned in a combustion chamber between the engine and the turbine to raise the turbine inlet temperature and thereby increase the plant power output. In addition to the thermodynamic advantage of a compound engine, there are some mechanical advantages of a turbine, instead of a reciprocating engine, driving the plant load.

In the compound power plant shown in Fig. 16·15*c*, the reciprocating engine and compressor comprise a gas generator, the sole function of which is to supply gas for driving the turbine. The gas generator can consist of a conventional engine geared to a centrifugal or axial-flow compressor, or, since there is no work output, it can be a free-piston engine–compressor. A schematic diagram of a compound power plant using a free-piston gas generator is shown in Fig. 16·16. The air in the bounce cylinders is compressed during the power stroke of the opposed pistons and then expands to move the pistons toward each other to compress the air in the combustion chamber and start a new cycle of operation.

In elementary analyses of compound power plants, the flow through reciprocating engines can often be treated as steady, using mean values of properties.

16·10 Summary

In an *air-standard* cycle analysis, the working substance is air that behaves like an ideal gas, and the combustion process of an actual power cycle is replaced by a heat-addition process. A *cold air-standard* analysis involves the further simplification that the specific heats of air are constant at their room-temperature values.

The air-standard ideal gas-turbine cycle is the *Brayton cycle* that consists of isentropic compression, reversible constant-pressure heating, isentropic expansion through a turbine, and reversible constant-pressure heat rejection to complete the cycle.

The *backwork ratio* of a power cycle, which is defined as

$$\text{backwork ratio} \equiv \frac{\text{compression work}_{\text{in}}}{\text{gross work of prime mover}}$$

is high for a gas-turbine cycle. Consequently, the efficiency of a gas-turbine cycle is reduced appreciably by relatively small reductions in compressor and turbine efficiencies.

The addition of a *regenerator* increases the thermal efficiency of a simple gas-turbine cycle. Intercooling and reheating increase the power output, and when used in conjunction with regeneration they increase the efficiency.

In a gas-turbine jet-propulsion engine (turbojet engine), the turbine drives the compressor, and the exhaust gas from the turbine expands through a nozzle to leave at a high velocity in a rearward direction. The change in momentum of the gas passing through the engine results in a forward thrust on the engine, and thus on the aircraft. In a jet engine, the counterpart of reheating is *afterburning,* which increases the thrust by increasing the enthalpy drop and hence the velocity change across the jet nozzle.

The air-standard *Otto cycle* is comprised of four reversible processes of air in a closed system: adiabatic compression, constant-volume heat addition, adiabatic expansion, and constant-volume heat rejection. The *compression ratio* of a reciprocating engine is defined as the ratio of the maximum to the minimum volume of the working substance in the cylinder.

The air-standard *Diesel cycle* is also comprised of four reversible processes of air in a closed system: adiabatic compression, constant-pressure heat addition, adiabatic expansion to the initial volume, and constant-volume heat rejection. For a Diesel cycle the *cutoff ratio* is defined as the ratio of the cylinder volume after heat addition to that before heat addition. The *percent cutoff* is the fraction of the stroke during which heat is added.

Other air-standard cycles such as the dual cycle, in which the heat addition is divided between a constant-volume and a constant-pressure process, are used in analyses where it is desired to simulate the pressure-volume variation of actual engine operation more closely than can be done with an air-standard Otto or Diesel cycle.

Compound power plants or compound engines are combinations of reciprocating engines and turbines designed to capitalize on the advantages and minimize the drawbacks of each type of prime mover.

Suggested Reading

16·1 Black, William Z., and James G. Hartley, *Thermodynamics,* Harper & Row, New York, 1985, Chapter 7.

16·2 El-Wakil, M. M., *Powerplant Technology,* McGraw-Hill, New York, 1984, Chapter 8.

16·3 Sorensen, Harry A., *Energy Conversion Systems,* Wiley, New York, 1983, Chapters 4 and 7.

16·4 Wood, Bernard D., *Applications of Thermodynamics,* 2nd ed., Addison-Wesley, Reading, Mass., 1982, Sections 1.1,3.

Problems

16·1 An air-standard gas turbine operates on the Brayton cycle between pressure limits of 100 and 500 kPa. The inlet air temperature to the compressor is 20°C, and the air entering the turbine is at 800°C. Air enters the compressor at a rate of 25.0 m^3/s. Compute the power output and the cycle thermal efficiency.

16·2 Solve problem 16·1 for turbine inlet temperatures of 500, 600, 700, 800, and 900°C and plot the results versus turbine inlet temperature. (The c_p equation given in Table A·4 for the lower temperature range is sufficiently accurate to 900°C.)

16·3 Solve problem 16·1 for upper pressure limits of 300, 400, 500, 600, 700, and 800 kPa and plot the results versus compressor pressure ratio.

16·4 Air enters the compressor of a simple gas-turbine cycle at 100 kPa, −10°C, and discharges at 700 kPa, 230°C. The temperatures of the gases entering and leaving the turbine are 870 and 480°C, respectively. If such data are possible, compute the net work per kilogram of air and the efficiency of the cycle. Assume that the compression and expansion are adiabatic.

16·5 A simple gas-turbine cycle operates at a pressure ratio of 5. The inlet temperature to the compressor is 293 K. If the turbine and compressor efficiencies are 0.85 and 0.84, respectively, compute the cycle efficiency for a turbine inlet temperature of 870 K, and sketch an energy flow diagram showing irreversibilities quantitatively. Assume that the compression and expansion are adiabatic.

16·6 For the pressure ratio and limiting temperatures as given in problem 16·5, plot a curve of cycle thermal efficiency versus compressor efficiency with turbine efficiency as a parameter. (Cover machine efficiencies from 40 to 100 percent, if possible.)

16·7 Solve problem 16·5 using a pressure ratio of 6 and an inlet turbine temperature of 500°C.

16·8 Consider a simple air-standard gas-turbine cycle operating with compressor inlet conditions of 90 kPa, 10°C. The pressure ratio is 5, and the maximum temperature is 800°C. What reduction in compressor efficiency would have the same effect on the cycle efficiency as a reduction of the turbine efficiency to 0.75?

16·9 For equal pressure and temperature limits, rank in order of decreasing efficiency cold air-standard simple gas-turbine cycles using as working fluids helium, air, and carbon dioxide.

16·10 The thermal efficiency of a cold air-standard Brayton cycle depends only on the pressure ratio and k of the gas. Make a plot of efficiency versus pressure ratio for various k values of common gases to show the influence of these parameters.

16·11 For an ideal simple gas-turbine cycle with compressor inlet conditions of 1 atm, 25°C, and a maximum temperature of 900°C, make a comparative plot of efficiency versus pressure ratio for a cold air-standard analysis and an air-standard analysis.

16·12 An air-standard gas turbine operates on the Brayton cycle between pressure limits of 14.7 and 70 psia. The inlet air temperature to the compressor is 60°F, and the air entering the turbine is at 1500°F. Air enters the compressor at a rate of 50 000 cfm. Compute the power output and the cycle thermal efficiency.

16·13 Solve problem 16·12 for turbine inlet temperatures ranging from 900 to 1500°F and plot the results versus turbine inlet temperature. (The c_p equation given in Table A·4 for the lower temperature range is sufficiently accurate to 1500°F.)

16·14 Solve problem 16·12 for upper pressure limits ranging from 40 to 120 psia and plot the results versus compressor pressure ratio.

16·15 Air enters the compressor of a simple gas-turbine cycle at 14.7 psia and 15°F, and discharges at 103 psia, 445°F. The temperatures of the gases entering and leaving the turbine are 1600 and 716°F, respectively. If such data are possible, compute the net work per pound of air and the efficiency of the cycle.

16·16 A simple gas-turbine cycle operates at a pressure ratio of 5. The inlet temperature to the compressor is 530 R. If the turbine and compressor efficiencies are 0.85 and 0.84, respectively, compute the cycle efficiency for a turbine inlet temperature of 1560 R, and sketch an energy flow diagram showing irreversibilities quantitatively. Assume that the compression and expansion are adiabatic.

16·17 For the pressure ratio and limiting temperatures as given in problem 16·16, plot a curve of cycle thermal efficiency versus turbine efficiency with compressor efficiency as a parameter. (Cover machine efficiencies from 40 to 100 percent, if possible.)

16·18 Solve problem 16·16 using a pressure ratio of 6 and an inlet turbine temperature of 1400 R.

16·19 Consider a simple air-standard gas-turbine cycle operating with compressor inlet conditions of 14.7 psia, 70°F. The pressure ratio is 5, and the maximum temperature is 1500°F. $\eta_C = 0.80$, and $\eta_T = 0.80$. What reduction in compressor efficiency would have the same effect on the cycle efficiency as a reduction of the turbine efficiency to 0.75?

16·20 Air enters the compressor of a simple gas turbine plant at 100 kPa, 20°C, at a rate of 2.20 kg/s. The compressor efficiency is 60 percent. Discharge pressure is 450 kPa. Calculate the amount of heat (kJ/kg) that must be added in order to provide a turbine inlet temperature of 650°C.

16·21 A gas turbine (Brayton cycle) mounted on a trailer is used to provide a supply of hot gas to help in snow and ice removal from an airport runway. Air enters the compressor at 95 kPa, −10°C. The turbine drives only the compressor. The compressor efficiency is 65 percent and the pressure ratio is 4. Gas at 165 kPa, 95°C, leaves the turbine at a rate of 20 kg/s. Assume adiabatic compression and expansion. Using an air-standard analysis, calculate (*a*) the temperature of the gas entering the turbine and (*b*) the turbine efficiency.

16·22 Derive an expression for the efficiency of a cold air-standard Brayton cycle in terms of the compressor pressure ratio and one property of the working substance.

16·23 A cold air-standard Brayton cycle has a minimum temperature of T_{min} and a maximum temperature of T_{max}. Show that the compressor pressure ratio for maximum cycle thermal efficiency equals $(T_{max}/T_{min})^{k/2(k-1)}$.

16·24 An air-standard simple gas turbine cycle operates with an inlet pressure of 1 atm, a pressure ratio of 5, and a turbine inlet temperature of 660°C (1220°F). The working fluid is air. Plot cycle thermal efficiency and backwork ratio against compressor inlet temperature for a range of −40°C (−40°F) to 40°C (104°F).

16·25 Solve problem 16·24 with turbine and compressor efficiencies of 75 percent. Compare the results.

16·26 An air-standard simple gas turbine cycle operates with an inlet pressure of 1 atm, a pressure ratio of 5, and a turbine inlet temperature of 660°C (1220°F). The working fluid is air. A filter is placed ahead of the compressor inlet, causing a reduction in compressor inlet pressure, but the pressure ratio across the compressor remains unchanged. Also, the turbine exhaust pressure remains unchanged. Plot the decrease in cycle thermal efficiency against pressure drop across the filter for a pressure drop range of 0 to 15 kPa (2.18 psi).

16·27 Solve problem 16·26 with turbine and compressor efficiencies of 75 percent. Compare the results.

16·28 How would the performance of an ideal simple gas turbine cycle be affected by injecting into the compressor inlet liquid water in an amount that would result in no liquid water leaving the compressor? (Consider efficiency and net power output for a fixed mass flow rate of dry air.) What would be the effect of injecting the same amount of liquid water at the turbine inlet? (Both of these procedures have been used in practice for certain purposes.)

16·29 For a power output of 1000 kW, determine the minimum flow rate for a simple air-standard gas turbine cycle with minimum and maximum temperatures of 15°C (59°F) and 660°C (1220°F) if compressor and turbine efficiencies are each 75 percent.

16·30 An air-standard gas turbine operates on a regenerative cycle with a regenerator effectiveness of 100 percent. Compression and expansion are isentropic. The inlet temperature to the compressor is 15°C (59°F). Compute the air-standard cycle efficiency for the following conditions: (*a*) inlet temperature to the turbine equals 635°C (1175°F), pressure ratio equals 5; (*b*) inlet temperature to the turbine equals 885°C (1625°F), pressure ratio equals 5; and (*c*) inlet temperature to the turbine equals 635°C (1175°F), pressure ratio equals 10.

16·31 The following data apply to a regenerative air-standard gas-turbine cycle with adiabatic compression and expansion. Atmospheric temperature is 80°F.

Location	Pressure, psia	Temperature, °F
Entering compressor	14.5	80
Leaving compressor	85.0	500
Entering combustion chamber	84.9	620
Entering turbine	83.0	1540
Leaving turbine	14.6	940
Leaving plant	14.5	

Determine (*a*) the effectiveness of the regenerator, (*b*) the turbine efficiency, (*c*) the turbine effectiveness, and (*d*) the irreversibility of the turbine expansion.

16·32 Solve problem 16·30*a* if the regenerator effectiveness is (*a*) 75 percent, (*b*) 50 percent.

16·33 Solve problem 16·30*a* if there is a leak in the high-temperature end of the regenerator that allows 5 percent of the entering air to flow into the exhaust line.

16·34 Solve problem 16·30*a* if there is a frictional pressure drop of 22 kPa (3.2 psi) in each gas stream passing through the regenerator.

16·35 Solve problem 16·30*a* for $\eta_C = \eta_T = 0.80$.

16·36 For fixed inlet temperatures, sketch a curve of regenerator irreversibility versus regenerator effectiveness. Does the maximum irreversibility correspond to the least desirable value of effectiveness from the standpoint of cycle thermal efficiency?

16·37 For an ideal regenerative gas-turbine cycle with compressor inlet conditions of 1 atm, 25°C, and a maximum temperature of 900°C, make a comparative plot of efficiency vs. regenerator effectiveness for a cold air-standard analysis and for an air-standard analysis.

16·38 Air enters a gas turbine plant at 95 kPa, 5°C. Compression is adiabatic with an efficiency of 0.70 and a pressure ratio of 5. The regenerator effectiveness is 0.60. Turbine inlet conditions are 475 kPa, 850°C. The turbine expansion is adiabatic with an efficiency of 0.70. The plant power output is 1500 kW. Calculate (*a*) the mass rate of flow of air through the plant and (*b*) the irreversibility (kJ/kg) of the turbine expansion.

16·39 A gas-turbine plant is used to supply compressed air by having the turbine drive only the compressor, and the air stream from the compressor is divided into two parts: that which goes to the combustion chamber and turbine, and that which is the compressed air delivered by the plant. Inlet conditions are 14.0 psia (96.5 kPa), 70°F (21°C), and the compressor discharge pressure is 70 psia (483 kPa). The fuel lower heating value is 17 300 B/lbm (40 240 kJ/kg). The turbine inlet temperature is 1500°F (816°C), and the exhaust products can be treated as air. All velocities are negligible except for the turbine exhaust velocity of 500 fps (152 m/s). Determine the mass ratio of compressed air delivered to air drawn in.

16·40 In the gas-turbine plant shown in the figure, the compressor is driven by one turbine, and the other turbine delivers 2000 kW as the power output of the plant. The compressor and the turbines operate adiabatically. The compressor efficiency is 80 percent; the efficiency of each turbine is 80 percent. Air drawn into the compressor is at 98.0 kPa, 15°C. Both turbines exhaust at 98.0 kPa. The inlet condition for each turbine is 410 kPa, 700°C. High-pressure air leaves the regenerator at 315°C. Disregard pressure drop in the heat exchangers and changes in kinetic energy. Treat the working fluid as air. Calculate the heat added in each combustion chamber in kJ/h.

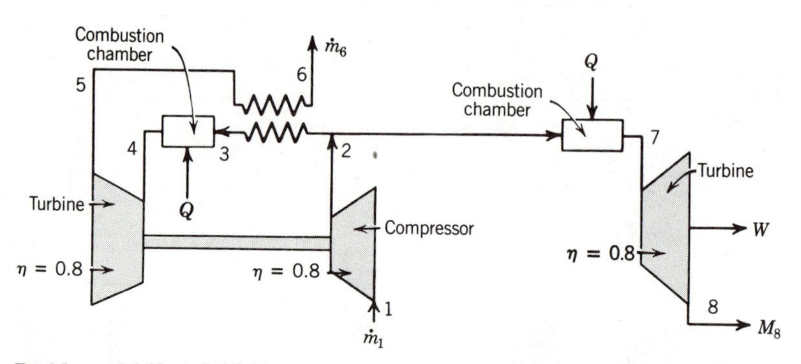

Problems 16·40 and 16·41.

16·41 In the gas-turbine plant shown in the figure, the compressor is driven by one turbine, and the other turbine delivers 1000 hp as the power output of the plant. The compressor and the turbines operate adiabatically. The compressor efficiency is 80 percent; the efficiency of each turbine is 80 percent. Air drawn into the compressor is at 14.7 psia, 60°F. Both turbines exhaust at 14.7 psia. The inlet condition for each turbine is 61 psia, 1280°F. High-pressure air leaves the regenerator at 600°F. Disregard pressure drop in the heat exchangers and changes in kinetic energy. Treat the working fluid as air. Calculate the heat added in each combustion chamber in B/h.

16·42 In the gas turbine plant shown in the figure, the entire net power output of 3750 kW is produced by turbine *A*, while turbine *B* drives only the compressor. Each turbine has an efficiency of 70 percent. The compressor efficiency is 80 percent. Disregard frictional pressure drop in the heat exchangers and piping. $P_1 = P_5 = P_6 = P_9 = P_{10} = 1$ atm; $P_2 = P_3 = P_4 = P_7 = P_8 =$ 5 atm; $T_1 = 5$°C (41°F); $T_4 = T_8 = 840$°C (1544°F); $T_{10} - T_2 = T_6 - T_2 = 25$ degrees C (45 degrees F). Calculate (*a*) the mass rates of flow $\dot{m}_6$ and $\dot{m}_{10}$ and (*b*) the amount of energy per unit mass made unavailable by friction in turbine *A*.

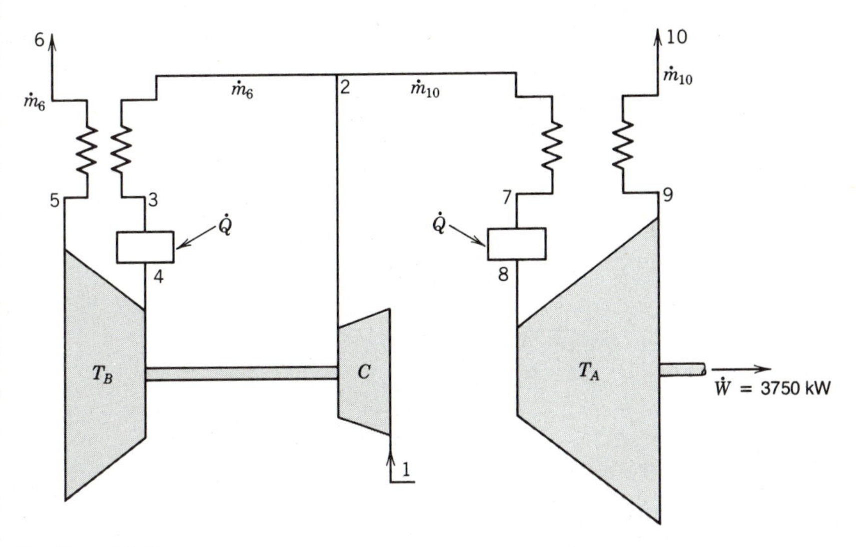

Problem 16·42.

16·43 The following data apply to an air-standard intercooled regenerative cycle: air temperatures entering and leaving the low-pressure stage of the compressor are 25 and 97°C; air temperatures entering and leaving the high-pressure stage of the compressor are 60 and 138°C; air temperature leaving the regenerator is 340°C; gas temperatures entering and leaving the turbine are 870 and 410°C. The initial pressure is 101.3 kPa, and the pressure ratio is 6. Cooling water enters the intercooler at 20°C and leaves at 25°C. Compute (*a*) the net work of the cycle per kilogram of air, (*b*) the thermal efficiency, (*c*) the amount of water required per kilogram of air in the intercooler, and (*d*) the irreversibility of the regenerator per kilogram of air.

16·44 The following data apply to an air-standard intercooled regenerative cycle: air temperatures entering and leaving the low-pressure stage of the compressor are 80 and 208°F; air temperatures entering and leaving the high-pressure stage of the compressor are 140 and 280°F; air temperature leaving the regenerator is 645°F; gas temperatures entering and leaving the turbine are 1600 and 770°F. The initial pressure is 14.7 psia, and the pressure ratio is 6. Cooling water enters the intercooler at 70°F and leaves at 80°F. Compute (*a*) the net work of the cycle per pound of air, (*b*) the thermal efficiency, (*c*) the amount of water required per pound of air in the intercooler, and (*d*) the irreversibility of the regenerator per pound of air.

16·45 Consider an air-standard reheat cycle with the high-pressure turbine driving the compressor and the low-pressure turbine supplying the net plant power output of 9000 kW. Air enters the plant at 1 atm, 15°C, and the compressor pressure ratio is 6. The inlet temperature to each turbine is 815°C. The compressor efficiency is 0.85, and each turbine efficiency is 0.82. Determine the mass rate of flow and the cycle thermal efficiency. Make a diagram showing available and unavailable parts of the heat added and rejected and the irreversibility of various processes.

16·46 Consider an air-standard reheat cycle with the high-pressure turbine driving the compressor and the low-pressure turbine supplying the net plant power output of 5000 hp. Air enters the plant at 14.7 psia, 60°F, and the compressor pressure ratio is 6. The inlet temperature to each turbine is 1500°F. The compressor efficiency is 0.85, and each turbine efficiency is 0.82. Determine the mass rate of flow and the cycle thermal efficiency. Make a diagram showing available and unavailable parts of the heat added and rejected and the irreversibility of various processes.

16·47 The flow diagram of a gas turbine power plant is shown in the figure. The high-pressure turbine (HPT) drives only the compressor, and the low-pressure turbine (LPT) supplies the net power output of the plant. $P_1 = P_8 = 1$ atm; $P_2 = P_3 = P_4 = 6$ atm; $T_1 = 5$°C (41°F); $T_4 = 760$°C (1400°F); regenerator effectiveness = 0.60; net power output = 570 kW.

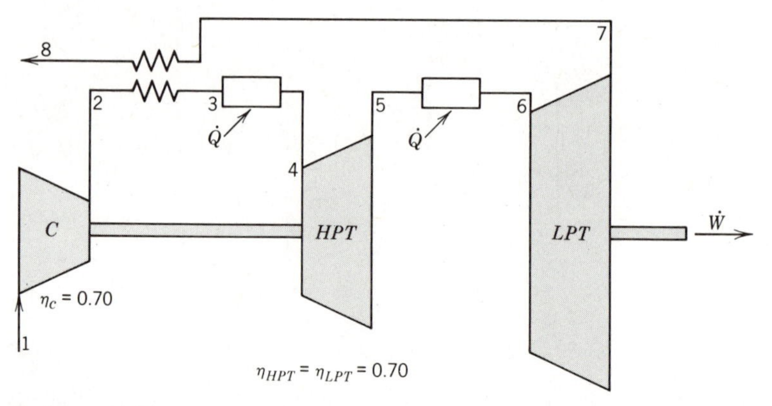

Problem 16·47.

16·48 The system shown in the figure is used to provide 10 lbm/s of air at 28.0 psia, 1000°F. Pertinent data are given on the flow diagram. Notice that the turbine drives only the compressor. Assuming isentropic compression and expansion, no frictional effects, and negligible changes in kinetic energy, calculate on the basis of an air-standard analysis the rate of heat transfer in processes 2-3 and 4-5.

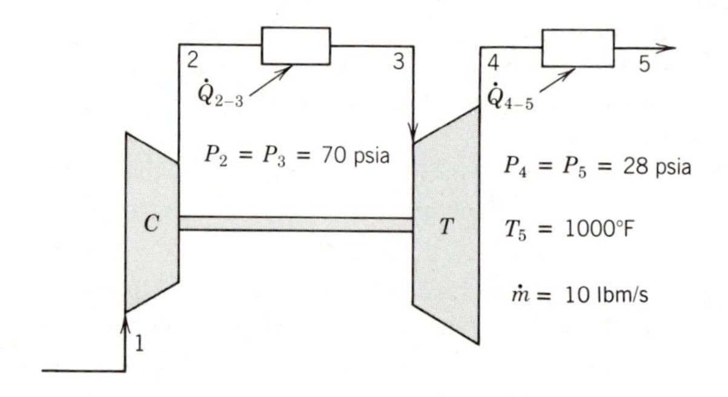

Problem 16·48.

16·49 The following temperatures are for the intercooled regenerative air-standard cycle with reheat as shown in Fig. 16·8: $T_1 = 30$°C, $T_2 = 120$°C, $T_3 = 75$°C, $T_4 = 180$°C, $T_5 = 375$°C, $T_6 = 875$°C, $T_7 = 585$°C (at 320 kPa), $T_8 = 875$°C, and $T_9 = 585$°C. The initial pressure is 1 atm and the overall pressure ratio is 10. Cooling water enters the intercooler at 20°C and leaves at 25°C. Compute per kilogram of air (*a*) heat transferred to the water in the intercooler, (*b*) mass of cooling air, (*c*) heat supplied in the first combustion chamber, (*d*) heat supplied in the second combustion chamber, (*e*) work required to compress the air, and (*f*) net work for the complete unit.

16·50 The following temperatures are for the intercooled regenerative air-standard cycle with reheat as shown in Fig. 16·8: T_1 = 80°F, T_2 = 245°F, T_3 = 160°F, T_4 = 349°F, T_5 = 700°F, T_6 = 1600°F, T_7 = 1082°F, (at 46.5 psia), T_8 = 1600°F, and T_9 = 1082°F. The initial pressure is 14.7 psia, and the overall pressure ratio is 10. Cooling water enters the intercooler at 70°F and leaves at 78°F. Compute per pound of air (*a*) heat transferred to the water in the intercooler, (*b*) mass of cooling water, (*c*) heat supplied in the first combustion chamber, (*d*) heat supplied in the second combustion chamber, (*e*) work required to compress the air, and (*f*) net work for the complete unit, as well as (*g*) thermal efficiency of the cycle.

16·51 The plant shown in the figure is a closed-circuit plant; that is, the air that passes through the system does not mix with fuel and is not exhausted to the atmosphere. The plant is used for both heating and cooling of parts of its surroundings. Between 3 and 4 the air is heated by external combustion. This plant provides heat for space heating and water heating at the heat exchangers 2-8 and 6-7. In the heat exchanger 9-10 the air in the system absorbs heat from one part of the surroundings at a rate of 500 B/s. The two turbines together drive the compressor and nothing else (i.e., there is no net power output from the plant). Disregard pressure drops in heat exchangers and connecting lines. Calculate the mass rates of flow $\dot{m}_1$ and $\dot{m}_4$.

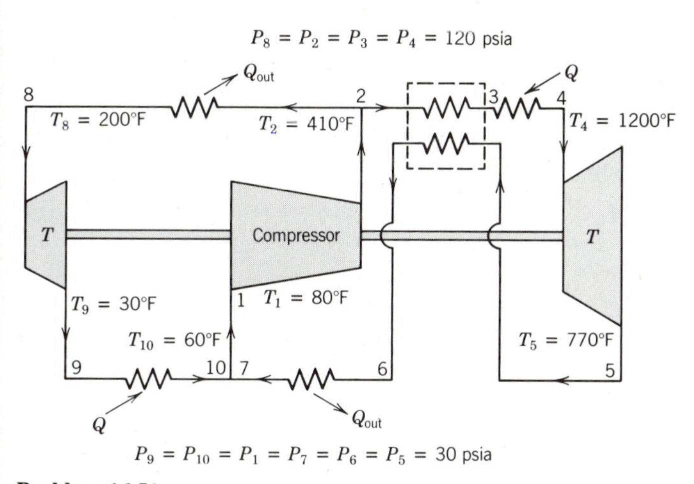

Problem 16·51.

16·52 An ideal turbojet engine draws in air at 95 kPa, 25°C, with a velocity of 150 m/s at a rate of 15 kg/s. After passing through the compressor and the combustion chamber, the air enters the turbine at 400 kPa, 870°C. The velocity is 150 m/s at the compressor outlet, turbine inlet, and turbine exhaust. Calculate the velocity of the exhaust jet, assuming isentropic flow through the exhaust nozzle to a pressure of 95 kPa.

16·53 Determine the maximum thrust obtainable from an air-standard turbojet engine that takes in air at 70 kPa, 5°C, with a velocity of 250 m/s (the craft speed) at a rate of 32.0 kg/s. The compressor total pressure ratio is 3, compressor outlet and turbine inlet and outlet velocities are 150 m/s, and the maximum temperature is 980°C. Assume that the total pressure is constant from the compressor outlet to the turbine inlet. The exhaust pressure is 70 kPa.

16·54 Solve problem 16·53 with an afterburner raising the temperature to 1800°F at the entrance to the jet nozzle.

16·55 In a prop-jet aircraft engine, the gas turbine drives a propeller and also the compressor, and the exhaust gases from the turbine are expanded through a nozzle to form an exhaust jet that provides additional thrust. Determine the required compressor pressure ratio for an air-standard prop-jet engine that provides 4.5 kN of thrust in addition to 750 kW to a propeller if air at 55 kPa, −20°C, is drawn in with a velocity of 180 m/s. The velocity is also 180 m/s at the compressor outlet, turbine inlet, and turbine exhaust. The maximum temperature is 870°C, and the flow rate is not to exceed 9.0 kg/s. Assume that all processes are reversible.

16·56 In a ramjet engine, air at 70 kPa, 5°C, enters the diffuser at 460 m/s (the craft speed) through a cross-sectional area of 0.075 m^2 and is decelerated to 95 m/s relative to the engine. Fuel at 0.02 times the air flow rate is burned to bring the temperature to 870°C, and the combustion products that can be treated as air are then expanded through a nozzle to the ambient pressure and leave at 760 m/s. Calculate the thrust.

16·57 An ideal turbojet engine draws in air at 14.0 psia, 80°F, with a velocity of 500 fps at a rate of 32 lbm/s. After passing through the compressor and the combustion chamber, the air enters the turbine at 60 psia, 1600°F. The velocity is 500 fps at the compressor outlet, turbine inlet, and turbine exhaust. Calculate the velocity of the exhaust jet, assuming isentropic flow through the exhaust nozzle to a pressure of 14.0 psia.

16·58 Determine the maximum thrust obtainable from an air-standard turbojet engine that takes in air at 10 psia, 40°F, with a velocity of 800 fps (the craft speed) at a rate of 70 lbm/s. The compressor total pressure ratio is 3, compressor outlet and turbine inlet and outlet velocities are 500 fps, and the maximum temperature is 1800°F. Assume that the total pressure is constant from the compressor outlet to the turbine inlet. The exhaust pressure is 10 psia.

16·59 Solve problem 16·58 with an afterburner raising the temperature to 1800°F at the entrance to the jet nozzle.

16·60 In a prop-jet aircraft engine, the gas turbine drives a propeller and also the compressor, and the exhaust gases from the turbine are expanded through a nozzle to form an exhaust jet that provides additional thrust. Determine the required compressor pressure ratio for an air-standard prop-jet engine that provides 1000 lbf of thrust in addition to 1000 hp to a propeller if air at 8 psia, 0°F, is drawn in with a velocity of 600 fps. The velocity is also 600 fps at the compressor outlet, turbine inlet, and turbine exhaust. The maximum temperature is 1600°F, and the flow rate is not to exceed 20 lbm/s. Assume that all processes are reversible.

16·61 In a ramjet engine, air at 10 psia, 40°F enters the diffuser at 1500 fps (the craft speed) through a cross-sectional area of 0.8 sq ft and is decelerated to 300 fps relative to the engine. Fuel at 0.02 times the air flow rate is burned to bring the temperature to 1600°F, and the combustion products that can be treated as air are then expanded through a nozzle to the ambient pressure and leave at 2500 fps. Calculate the thrust.

16·62 The following pressure and temperature measurements were made on an actual aircraft jet engine during flight: at compressor inlet, 14.2 psia, 59°F (98.0 kPa, 15°C); at compressor outlet, 161.0 psia, 660°F (1110 kPa, 349°C); at turbine inlet, 154 psia, 1570°F (1062 kPa, 855°C); and at turbine outlet 34.8 psia, 1000°F (240 kPa, 538°C). The fuel flow rate was 165 lbm/s, and a gas analysis of the exhaust showed an air–fuel ratio of 70. The thrust was 10 000 lbf. (*a*) Assuming that the jet exit-plane pressure equals the inlet-plane pressure, estimate the difference between the velocity of the exhaust jet and the incoming flow. (*b*) Estimate the compressor and turbine efficiencies, stating any assumptions you make.

16·63 Estimate the static takeoff thrust of a turbofan engine as shown in Fig. 16·10*b* if ambient conditions are 1 atm, 59°F (15°C), the total mass rate of flow is 1500 lbm/s (680 kg/s), one-sixth of the flow passes through the turbines, the turbine exhaust jet at 850°F (455°C) has a velocity of 1190 fps (363 m/s), the fan exhaust flow at 130°F (55°C) has a velocity of 885 fps (270 m/s), and the maximum temperature in the engine is 1970°F (1077°C).

16·64 A turbofan engine is of the configuration shown in Fig. 16·10*a*. At takeoff 42 percent of the total mass flow passes through the turbines. Estimate the static takeoff thrust under ambient conditions of 1 atm, 59°F, if the compressor discharge conditions are 200 psia, 715°F; turbine inlet conditions are 190 psia, 1600°F; low-pressure turbine exhaust pressure is 28 psia; the turbine exhaust jet has a velocity of 1560 fps, and the fan exhaust jet has a velocity of 990 fps.

16·65 What effects do you believe a heavy rain would have on the performance of a turbojet aircraft engine in flight?

16·66 In each case, *a* and *b*, one property diagram for an ideal gas is shown in the figure. Sketch the other property diagram in each case, numbering corresponding points and clearly showing the shapes of process lines you draw.

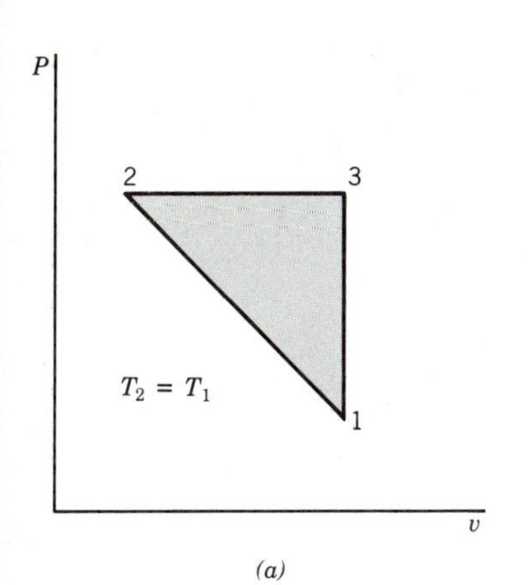

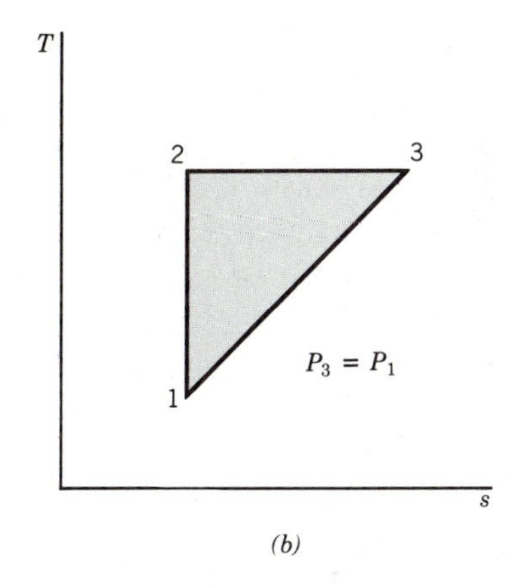

Problem 16·66.

16·67 Derive the following expression for the cold air-standard Otto-cycle thermal efficiency; $\eta = 1 - r^{1-k}$, where r is the compression ratio.

16·68 At the beginning of compression in a cold air-standard Otto cycle, the working substance is at 95 kPa, 30°C, and has a volume of 0.030 m³. At the end of compression, the pressure is 950 kPa, and 10 kJ is added during the constant-volume process. Calculate (*a*) the thermal efficiency, (*b*) the amount of heat added that is available energy, and (*c*) the amount of heat rejected that is available energy. (In the absence of other information, the sink temperature should be taken as equal to the lowest temperature in the cycle.)

16·69 Solve problem 16·68 on an air-standard basis.

16·70 For the conditions of problem 16·68, calculate (*a*) the percentage increase of stroke necessary to allow the adiabatic expansion to proceed until the initial pressure (95 kPa) is reached, causing the heat-rejection process to be at constant pressure instead of at constant volume, and (*b*) the thermal efficiency of the modified cycle described in *a*.

16·71 Solve problem 16·70 with the adiabatic expansion proceeding to the initial temperature (30°C) so that the heat-rejection process is isothermal.

16·72 An air-standard Otto cycle has a compression ratio of 8 and a maximum temperature of 1097°C. At the beginning of the compression stroke, the air is at 100 kPa, 25°C. Determine the maximum pressure in the cycle, the amount of heat added in kJ/kg, and the available fraction of the heat added.

16·73 Consider a cold air-standard Otto cycle having a compression ratio of 9. At the beginning of compression, the air is at 95 kPa, 5°C (13.8 psia, 41°F); during the heat-addition process, the pressure of the air is doubled. Calculate the efficiency and the backwork ratio (How do you think it should be defined?) of this cycle and the efficiency of a Carnot cycle operating between the same overall temperature limits. What would be the minimum overall volume ratio for a Carnot cycle with the same temperature limits and the same maximum specific volume?

16·74 In a cold air-standard Otto cycle with a compression ratio of 6, the working fluid is 0.045 kg (0.099 lbm) of air that is at 95 kPa, 5°C (13.8 psia, 41°F), at the beginning of the compression stroke. The maximum temperature in the cycle is 1740°F. Heat is received from a constant-temperature reservoir at 1740°F and is rejected to the atmosphere at 40°F. How much energy is made unavailable by the operation of one cycle?

16·75 At the beginning of compression in a cold air-standard Otto cycle, the working substance is at 14 psia, 90°F, and has a volume of 1 cu ft. At the end of compression, the pressure is 140 psia, and 10 B is added during the constant-volume process. Calculate (*a*) the thermal efficiency, (*b*) the amount of heat added that is available energy, and (*c*) the amount of heat rejected that is available energy. (In the absence of other information, the sink temperature should be taken as equal to the lowest temperature in the cycle.)

16·76 Solve problem 16·75 on an air-standard basis.

16·77 For the conditions of problem 16·76, calculate (*a*) the percentage increase of stroke necessary to allow the adiabatic expansion to proceed until the initial pressure (14 psia) is reached, causing the heat-rejection process to be at constant pressure instead of at constant volume, and (*b*) the thermal efficiency of the modified cycle described in *a*.

16·78 Solve problem 16·76 with the adiabatic expansion proceeding to the initial temperature (90°F) so that the heat-rejection process is isothermal.

16·79 In a cold air-standard Otto cycle, the cylinder volume is 0.10 cu ft and the compression ratio is 7. At the beginning of the compression stroke the working fluid is at 14.0 psia, 80°F. Heat added during one cycle amounts to 1.41 B, and the pressure at the end of the heat addition is 427 psia. The sink temperature is 80°F. Calculate the fraction of the heat rejected that is available energy.

16·80 Refer to Fig. 16·11. For the cold air-standard Diesel cycle, determine the compression ratio that results in $T_3 - T_4 = T_2 - T_1$.

16·81 A cold air-standard Diesel cycle has a compression ratio of 16 and a cutoff ratio of 2. The cylinder volume is 0.0140 m^3 (0.494 ft^3), and at the beginning of the compression stroke the air is at 95 kPa, 5°C (13.8 kPa, 41°F). The lowest temperature in the surroundings is 5°C (41°F). How much of the heat added is available energy?

16·82 Solve problem 16·81 on an air-standard basis.

16·83 Solve Example 16·4 on a cold air-standard basis, and compare the results of the two analyses.

16·84 State clearly the differences between a *Pv* diagram as used with air-standard cycles and an indicator diagram for an actual engine.

16·85 In the closed-system *Pv* diagram shown, paths 1-2 and *a-b-c-d* are reversible adiabatic paths. All processes shown are reversible. (*a*) Sketch the corresponding *Ts* diagram. (*b*) Four cycles are possible: 1-2-a-d-1; 1-2-a-c-1; 1-2-b-c-1; and 1-2-b-d-1. Rank these cycles in order of thermal efficiency.

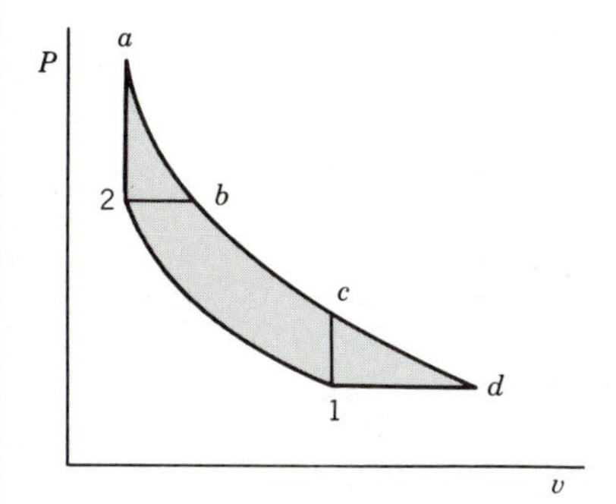

Problem 16·85.

16·86 For the same temperature limits and equal amounts of heat added, which air-standard cycle has the higher thermal efficiency, Otto or Diesel? Which has the higher compression ratio?

16·87 The compression ratio for an engine operating on the cold air-standard dual cycle is 7, the cylinder diameter is 25 cm, and the stroke is 30 cm. The air at the start of compression is at 101 kPa, 20°C. At the end of the constant-volume process, the pressure is 5600 kPa. If heat is added at constant pressure during 3 percent of the stroke, compute (*a*) the net work of the cycle, (*b*) the thermal efficiency, (*c*) the amount of heat added that is available energy, and (*d*) the amount of heat rejected that is available energy.

16·88 The compression ratio of an engine operating on the cold air-standard dual cycle is 7, the cylinder diameter is 10 in., and the stroke is 12 in. The air at the start of compression is at 101 kPa, 20°C. At the end of the constant-volume process, the pressure is 5600 kPa. If heat is added at constant pressure during 3 percent of the stroke, compute (*a*) the net work of the cycle, (*b*) the thermal efficiency, (*c*) the amount of heat added that is available energy, and (*d*) the amount of heat rejected that is available energy.

16·89 Compare the net work and the heat input of cold air-standard Ericsson and Stirling cycles that have identical isothermal heat rejection processes and identical temperature limits. Superpose the two cycles on Pv and Ts diagrams.

16·90 For Stirling cycles with the same temperature limits and the same heat addition, how does the work output vary with k of the working substance?

16·91 A compound power plant consists of a high-speed multicylinder Diesel engine that drives only a centrifugal compressor and a turbine that operates on the gases that leave the Diesel engine. Air enters the compressor at 14.7 psia, 70°F. The compressor operates adiabatically with an efficiency of 80 percent and a pressure ratio of 3. Air from the compressor is supplied to the Diesel engine. The air–fuel ratio is high because a high amount of excess air is used for scavenging and cooling. The pressure in the engine exhaust line is 43.5 psia. The brake specific fuel consumption of the Diesel engine alone when operating at 3 atm is 0.50 lbm/hp·h. The lower heating value of the fuel is 18 000 B/lbm. Twenty percent of the lower heating value of the fuel is transferred to cooling water, which is used to cool the hottest parts of the engine cylinders. Gases leaving the engine expand adiabatically to 14.7 psia through a turbine that has an efficiency of 80 percent. The turbine output is 1000 hp. Calculate the specific fuel consumption of the plant. (*Note:* In view of the high air–fuel ratio, treat the engine exhaust products as air. Assume that the flow through the entire plant is steady.)

16·92 For fixed inlet conditions and pressure ratio, how does the thermal efficiency of an air-standard Brayton cycle vary with the maximum temperature in the cycle? Explain your reasoning clearly and compare the result with that for a cold-air-standard analysis.

16·93 Compare the thermal efficiencies of air-standard and cold-air-standard Brayton cycles having the same inlet conditions, the same pressure ratio, and the same maximum temperature.

CHAPTER SEVENTEEN

Vapor Power Cycles

This chapter treats power cycles in which working fluids are alternately vaporized and condensed. The most common working fluid is water, and the cycles are called steam cycles, even though the water is in the liquid phase during part of the cycle.

Steam power plants are engineering systems of widespread importance because they generate a major and increasing fraction of the electric power produced in the world. Also, steam power generation is often combined with the use of steam for building heating or process heating such as sugar cooking or paper drying. The steam power cycle itself is essentially the same whether heat is supplied from the burning of a fuel in a furnace or from the fission process in a nuclear reactor.

The chief purpose of this chapter is to illustrate further the application of thermodynamic principles to engineering systems.

17·1 The Carnot Cycle Using Steam

Two corollaries of the second law are (1) the efficiency of a Carnot cycle depends only on the temperature limits and not on the working substance and (2) no cycle can have a higher efficiency than a Carnot cycle operating between the same temperature limits. (See Chapter 7.) Therefore, from the standpoint of efficiency, a steam power cycle might well be designed to simulate a Carnot cycle; however, for several reasons this is not done.

Ts diagrams of steady-flow steam Carnot cycles for two different temperature ranges are shown in Fig. 17·1. In the first one, the maximum cycle temperature is below the critical temperature, the heat-addition process converts liquid water completely to dry saturated steam, and the working fluid is always in a saturated state. In the second cycle shown, the maximum temperature is above the critical temperature, the heat-rejection process is a complete condensation of dry saturated vapor to saturated liquid, and the working fluid passes also through compressed liquid and superheated vapor states.

Referring to the Carnot cycle that is entirely within the wet region (Fig. 17·1*a*), process 1-2 is of course a constant-pressure as well as constant-temperature process that can be simulated very closely in a boiler. In fact, it is easier to control a constant-temperature process with a two-phase working substance by controlling the pressure than it is to maintain an isothermal process of a gas. The isentropic expansion process 2-3 can be simulated by a well-designed engine or turbine. Constant-temperature (and hence constant-

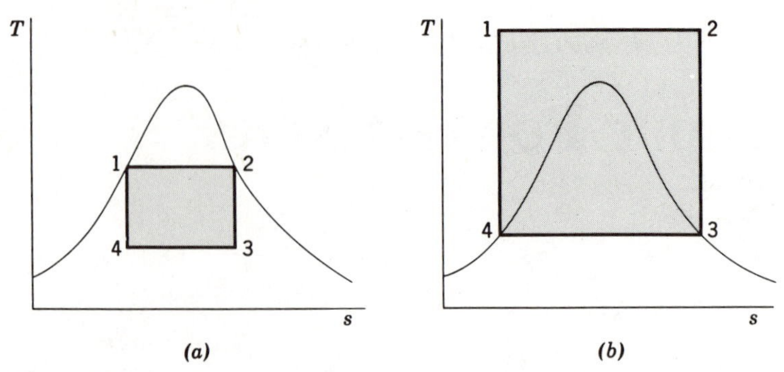

Figure 17·1. Carnot cycles using steam as the working fluid.

pressure) condensation as in process 3-4 can be accomplished in an actual condenser except that controlling the final quality is difficult. If the final quality is not correct, that is, if state 4 is not located correctly, an isentropic compression will not end at state 1. Also, it may be difficult to compress the two-phase mixture isentropically. Another difficulty in using this cycle is that heat cannot be added to an evaporating fluid at a constant temperature higher than the critical temperature, but for maximum thermal efficiency heat should be added at the highest possible temperature. For steam, the critical temperature of 374.1°C (705.4°F) is well below the temperature limit set by the strength characteristics of construction materials.

The Carnot cycle of Fig. 17·1*b* eliminates the difficulties of the first one but presents some new ones. The isentropic compression 4-1 involves extremely high pressures. Also, the isothermal heat-addition process 1-2 involves a pressure variation that introduces a control problem not encountered when the temperature can be held constant simply by maintaining a constant pressure as in heat addition to a two-phase mixture.

In summary, the Carnot cycle is not a suitable model for the design of steam power-plant cycles because of the difficulties in carrying out the required processes in actual machines, especially while utilizing the maximum possible temperature for heat addition.

17·2 The Rankine Cycle

Flow, *Pv*, and *Ts* diagrams of a Rankine cycle, which is the ideal simple steam power-plant steady-flow cycle, are shown in Fig. 17·2. Dry saturated steam enters the prime mover, which may be either an engine or a turbine, and expands isentropically to pressure P_2. The steam is then condensed at constant pressure and temperature to a saturated liquid, state 3. In the condenser, heat is transferred from the condensing steam to water, frequently from a lake or river, which is circulated through many tubes that provide a large heat-transfer surface.* The saturated liquid leaving the condenser is then pumped isentropically

*Your study of this chapter will be more rewarding if you refer to a handbook or book on steam power plants or applied thermodynamics for physical descriptions, including pictures, of the various cycle components mentioned.

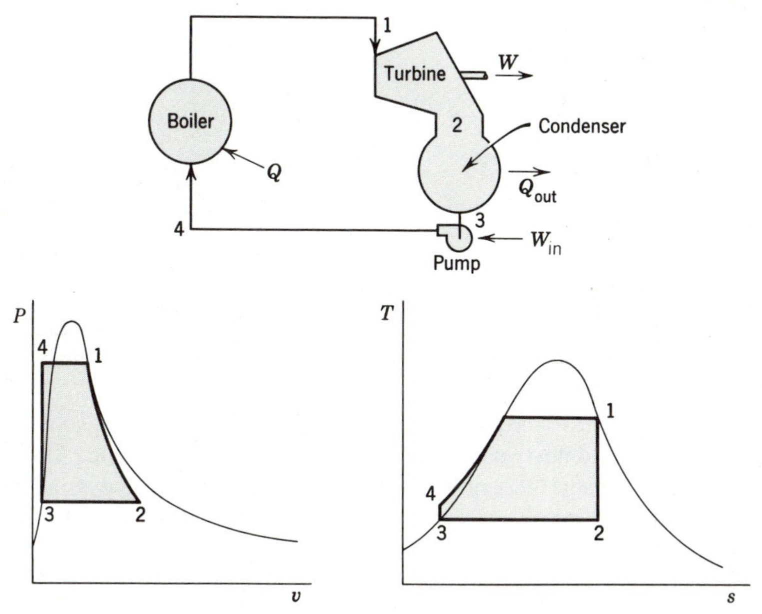

Figure 17·2. Rankine cycle.

into the boiler at pressure P_4 $(=P_1)$ where at constant pressure it is first heated to the saturation temperature and then evaporated to state 1 to complete the cycle. The temperature rise that results from isentropic compression of liquid water is very small, so the length of line 3-4 in the *Ts* diagram of Fig. 17·2 is greatly exaggerated.

In order to use higher temperatures without increasing the maximum pressure of the cycle, the steam after leaving the boiler is heated further at constant pressure in a superheater. The combination of boiler and superheater is called a steam generator. Flow and *Ts* diagrams of a Rankine cycle with superheat are shown in Fig. 17·3. Comparison of the *Ts* diagrams of Figs. 17·2 and 17·3 shows that for given pressure limits the thermal efficiency is increased by superheating.

If the steam-generator pressure is higher than the critical pressure, there can be no constant-temperature heat-addition process in a Rankine cycle. *Pv, Ts,* and *hs* diagrams for such a cycle are shown in Fig. 17·4.

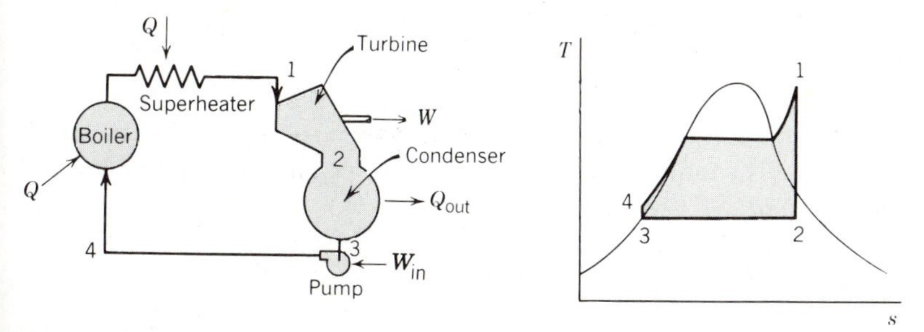

Figure 17·3. Rankine cycle with superheat.

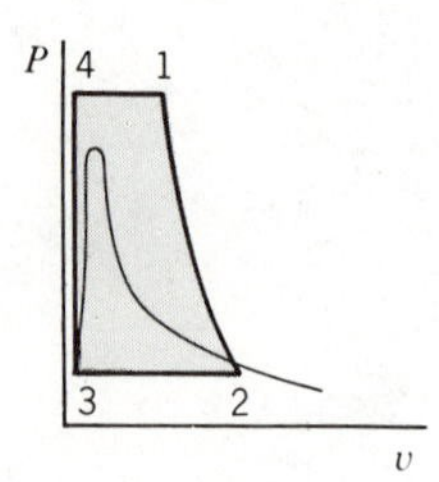

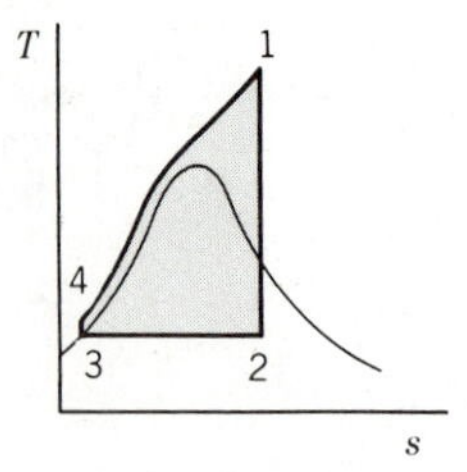

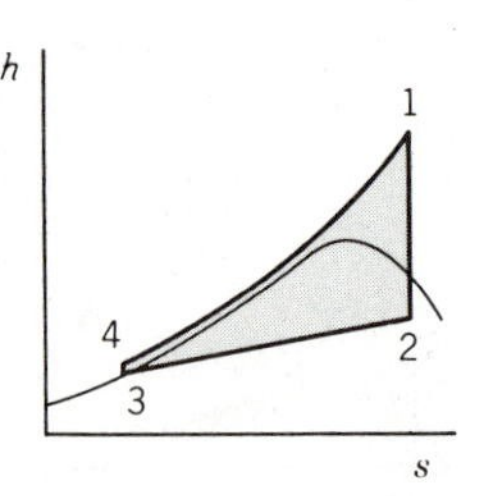

Figure 17·4. Supercritical-pressure Rankine cycle.

Inspection of *Ts* diagrams shows that, *for given temperature limits,* the Rankine cycle efficiency is always lower than the Carnot cycle efficiency because not all the heat is added at the highest temperature.

Performance calculations for a Rankine cycle are simple because no work is done in the two heat-transfer processes and the two processes involving work are adiabatic. Also, the changes in kinetic and potential energy across each cycle component are usually negligible. The first law applied to each piece of steady-flow equipment thus gives

Steam generator:	$q_{4-1} = h_1 - h_4$	$(P_1 = P_4)$
Turbine:	$w_T = h_1 - h_2$	$(s_2 = s_1)$
Condenser:	$q_{\text{out},2\text{-}3} = h_2 - h_3$	$(P_3 = P_2;\ h_3 = h_f)$
Pump:	$w_{\text{in},P} = h_4 - h_3$	$(s_4 = s_3)$

The cycle thermal efficiency is

$$\eta = \frac{w_{\text{cycle}}}{q_{4-1}} = \frac{w_T - w_{\text{in},P}}{q_{4-1}} = \frac{(h_1 - h_2) - (h_4 - h_3)}{(h_1 - h_4)}$$

The enthalpy change across the pump can be determined in various ways. Some steam property computer programs allow one to enter the value for P_4 and the value for s_4 $(=s_1)$ to obtain h_4. Another method is to make crossplots from Table A·1.4 (or from the more extensive source table of Reference 3·5). Published *hv* charts (Reference 3·6) can also be used. Another way to determine Δh across the pump is to use $T\,ds = dh - v\,dP$ which, *for an isentropic process,* reduces to

$$\Delta h = \int v\,dP$$

and then to make the approximation

$$\int v\,dP \approx v\,\Delta P$$

This approximation is quite accurate because the specific volume of liquid water is nearly independent of pressure. (The same expression for reversible pump work can be obtained from Eq. 1·10, $w_{\text{in}} = \int v\,dP + \Delta ke + \Delta pe$.)

The analysis of a steam power cycle can easily cover deviations from the Rankine cycle caused by pressure drops in piping, stray heat losses, and turbine and pump efficiencies of less than unity.

An advantage of the Rankine cycle over all other power cycles introduced so far is its low backwork ratio that is given by

$$\text{backwork ratio} = \frac{w_{\text{in},P}}{w_T}$$

where w_T and $w_{\text{in},P}$ are respectively the turbine work and the pump work input.

Example 17·1. Consider a Rankine cycle that operates with throttle (i.e., turbine inlet) conditions of 8000 kPa, 500°C, and a condenser pressure of 5 kPa. The lowest temperature of available cooling water is 20°C. Determine (*a*) the thermal efficiency, (*b*) the backwork ratio, and (*c*) the fraction of the heat added which is available energy.

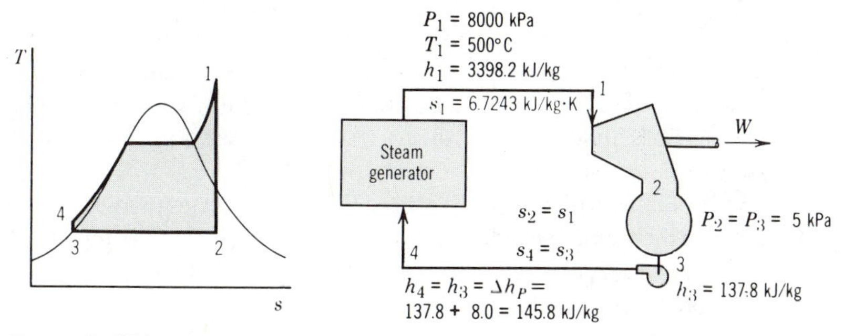

Example 17·1.

Solution. A *Ts* diagram and a flow diagram are made first, and then property values from the steam tables are entered on the flow diagram. The enthalpy change across the pump can be calculated by

$$\Delta h_P = h_4 - h_3 = \int_3^4 v\,dP \approx v(P_4 - P_3) = 0.001005(8000 - 5) = 8.0 \text{ kJ/kg}$$

$$h_4 = h_3 + \Delta h_P = 137.82 + 8.0 = 145.8 \text{ kJ/kg}$$

h_2 can be determined from $P_2 = 5$ kPa and $s_2 = s_1 = 6.7240$ kJ/kg·K as follows:

$$x_2 = \frac{s_2 - s_f}{s_{fg}} = \frac{6.7240 - 0.4764}{7.9187} = 0.789$$

$$h_2 = h_f + x_2 h_{fg} = 137.82 + 0.789(2423.7) = 2050.0 \text{ kJ/kg}$$

or this value can be determined from an *hs* (Mollier) chart or *hv* chart for steam. Having obtained all the *h* and *s* values needed, we can now complete the solution.

(*a*)
$$\eta = \frac{w_{\text{cycle}}}{q_{4-1}} = \frac{w_T - w_{\text{in},P}}{q_{4-1}} = \frac{h_1 - h_2 - \Delta h_P}{h_1 - h_4}$$

$$= \frac{3398.3 - 2050.0 - 8.0}{3398.3 - 145.8} = \frac{1348.3 - 8.0}{3252.5} = 0.412$$

(*b*)
$$\text{backwork ratio} = \frac{w_{\text{in},P}}{w_T} = \frac{\Delta h_P}{h_1 - h_2} = \frac{8.0}{1348.3} = 0.0059$$

(Notice how low this value is compared with those for gas-turbine cycles.)

(*c*) Of the heat added,

$$q_{av} = q - q_{unav} = q - T_0(s_1 - s_4) = 3252.5 - 293(6.7240 - 0.4764)$$
$$= 1422.0 \text{ kJ/kg}$$
$$\frac{q_{av}}{q} = \frac{1422.0}{3252.5} = 0.437$$

17·3 The Regenerative Steam Power Cycle

When we try to improve the Rankine cycle, we notice that some heat is added to the working fluid at a very low temperature. We know from our study of the second law that this impairs the efficiency of the cycle, so we look for a means of raising the temperature of the condensate leaving the pump without transferring heat to it from outside the cycle. Isentropic compression of liquid water as in the Carnot cycle of Fig. 17·1*b* is clearly infeasible because of the very high pressures that would be involved. A more promising method is regeneration similar to that of the Stirling and Ericsson cycles (Sec. 8·8) whereby heat is transferred from the vapor expanding in the turbine (or engine) to the liquid flowing between the pump and the steam generator. To make this heat transfer reversible, a counterflow arrangement is required, allowing heat to be transferred across only an infinitesimal temperature difference at each section of the heat exchanger which would be built into the turbine. Disadvantages of this scheme are the difficulties of building an effective heat exchanger into a turbine and the increased moisture content of the expanding steam that causes erosion of turbine blades.

A better solution is to extract or ''bleed'' steam from the turbine at various points in the expansion process and to use this bled steam to heat water flowing from the pump to the steam generator. This energy transfer occurs in *feedwater heaters.*

Flow and *Ts* diagrams of a steam power cycle using one *open feedwater heater* (also called a *direct-contact heater*) are shown in Fig. 17·5. In an open feedwater heater, the bled steam and the condensate pumped from the condenser are mixed, and, under optimum operating conditions, saturated liquid leaves the heater. The heater operates at the bleed-point pressure (or slightly below it in an actual plant because of frictional pressure drop in the bleed line), so another pump is needed to force the water into the steam generator.

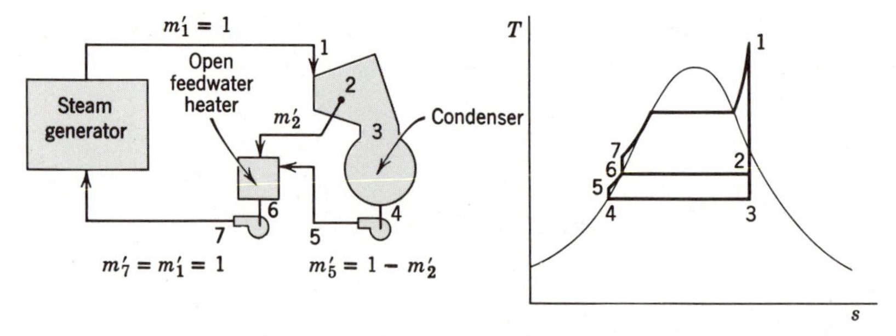

Figure 17·5. Regenerative cycle with one open feedwater heater.

For energy and mass balances on regenerative cycles it is convenient to define the mass ratio m' for any point a in the cycle as

$$m'_a \equiv \frac{\dot{m}_a}{\dot{m}_1}$$

where $\dot{m}_a$ is the mass rate of flow at point a, and $\dot{m}_1$ is the mass rate of flow leaving the steam generator. By definition, $m'_1 = 1$. Thus an energy balance on the open feedwater heater of Fig. 17·5 for which $w = 0$, $q = 0$, $\Delta ke = 0$, and $\Delta pe = 0$ is

$$\dot{m}_2 h_2 + \dot{m}_5 h_5 = \dot{m}_6 h_6$$
$$\dot{m}_2 h_2 + (\dot{m}_1 - \dot{m}_2) h_5 = \dot{m}_1 h_6$$

and this can also be written as

$$m'_2 h_2 + (1 - m'_2) h_5 = h_6$$

If the states of the fluid at points 2, 5, and 6 are known, this equation can be solved for m'_2, the fraction of the throttle flow that is bled to the open feedwater heater. Notice that on a *Ts* diagram for a regenerative cycle, as shown in Fig. 17·5, the state of the fluid at each point in the cycle is correctly shown but the mass of fluid is not the same at all points. Therefore, the area beneath the path of a reversible process represents heat transfer *per unit mass of fluid undergoing that process,* and care must be taken to account for variations in mass when comparing areas.

In addition to raising the temperature of feedwater to reduce the amount of heat that is added to the cycle at low temperature, an open feedwater heater in an actual cycle also serves as a *deaerator* to remove air and other noncondensable gases that would cause corrosion. This deaeration occurs because the solubility of gases in water decreases with increasing temperature. If the water is brought to its saturation temperature and provision is made for letting the noncondensable gases escape, they can be removed from the system.

Flow and *Ts* diagrams of a steam power cycle using one *closed feedwater heater* are shown in Fig. 17·6. The feedwater (that is, the condensate from the condenser) is pumped through many tubes in the feedwater heater, and the bled steam condenses (after first being desuperheated, if necessary) on the outside of the tubes. The two fluid streams do not mix in the closed feedwater heater. In the ideal case, the condensate leaving at 8 is saturated liquid, and the feedwater leaving at 6 is at a temperature only infinitesimally lower than T_8. In practice, in order to limit the amount of heat-transfer surface area needed, the difference $(T_8 - T_6)$, called the *terminal temperature difference,* is usually of the order of 5 degrees Celsius.

The solid lines in Fig. 17·6 show the heater drain being pumped into the feedwater line between the heater and the steam generator. The drain after being pumped to state 9 is mixed with the feedwater at state 6 to form feedwater at state 7. A magnified portion of the *Ts* diagram in Fig. 17·6 shows these processes. Another method of handling the heater drain is shown by the broken lines in Fig. 17·6. It is throttled through a trap or valve into the condenser. (A trap operates so that liquid that enters is throttled to the lower pressure on the discharge side, but vapor that enters the trap is not allowed to pass. The same function is performed by a valve in the drain line that is controlled by a float

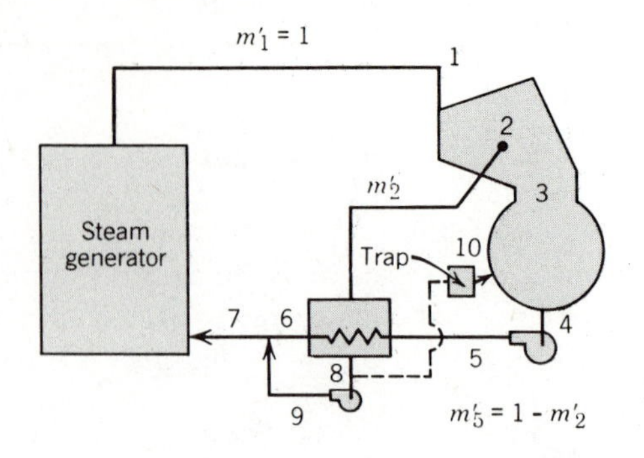

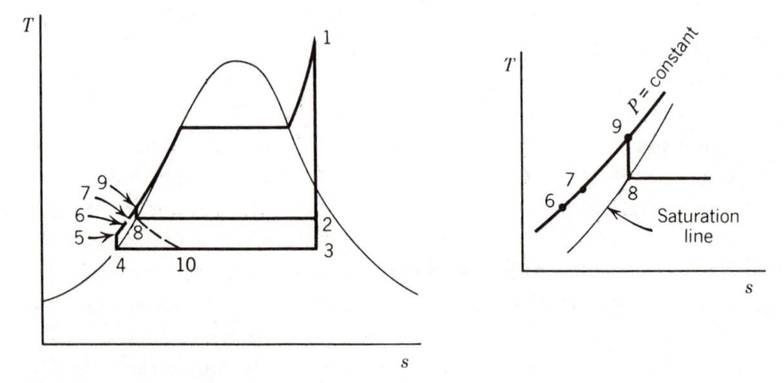

Figure 17·6. Regenerative cycle with one closed feedwater heater. An alternative method of handling the heater drain is shown by broken lines.

to maintain a constant liquid level in the lower part of a heater.) Application of the first law to the trap results in $h_8 = h_{10}$.

A mass and energy balance on the closed feedwater heaters gives

$$\dot{m}_2 h_2 + \dot{m}_5 h_5 = \dot{m}_2 h_8 + \dot{m}_5 h_6$$
$$\dot{m}_2 h_2 + (\dot{m}_1 - \dot{m}_2) h_5 = \dot{m}_2 h_8 + (\dot{m}_1 - \dot{m}_2) h_6$$
$$m'_2 h_2 + (1 - m'_2) h_5 = m'_2 h_8 + (1 - m'_2) h_6$$

A significant advantage of closed heaters is that they do not require for each heater a separate pump to handle the feedwater flow, but a drawback of closed heaters is that they do not bring the feedwater to the heater saturation temperature as open heaters do.

Example 17·2. Consider an ideal steam power cycle that operates with throttle conditions of 8000 kPa, 500°C, an open feedwater heater at 450 kPa, and a condenser pressure of 5 kPa. Determine the thermal efficiency.

Solution. Flow and *Ts* diagrams should be made first, and for this solution we will refer to those of Fig. 17·5. The following enthalpy values are obtained from the steam tables as in Example 17·1, or more quickly by following an isentropic line on a Mollier (*hs*) or *hv* chart:

$$h_1 = 3398.3 \text{ kJ/kg} \qquad h_2 = 2688.1 \text{ kJ/kg} \qquad h_3 = 2050.0 \text{ kJ/kg}$$

From the steam tables,

$$h_4 = h_{f,5\text{ kPa}} = 137.82 \text{ kJ/kg} \qquad h_6 = h_{f,450\text{ kPa}} = 623.25 \text{ kJ/kg}$$

The first law shows that h_5 and h_7 exceed h_4 and h_6, respectively, by the pump work in each case. Using the approximation $\int v\,dP \approx v\Delta P$ for the reversible pump work input,

$$h_5 = h_4 + v\Delta P = 137.82 + 0.001\,005(450 - 5) = 138.3 \text{ kJ/kg}$$
$$h_7 = h_6 + v\Delta P = 623.25 + 0.001\,088(8000 - 450) = 631.5 \text{ kJ/kg}$$

The amount of steam bled per kilogram of throttle steam m_2' can be found by making an energy balance on (i.e., by applying the first law to) the open feedwater heater. We assume that there is no heat transfer with the surroundings and that the changes in kinetic and potential energy are negligible. Then

$$m_2'h_2 + (1 - m_2')h_5 = h_6$$

$$m_2' = \frac{h_6 - h_5}{h_2 - h_5} = \frac{623.25 - 138.3}{2688.1 - 138.3} = 0.190 \text{ kg/kg throttle}$$

Applying the first law to the steady-flow turbine with $q = 0$ and $\Delta ke + \Delta pe = 0$, we have for the work *per kilogram of steam entering the turbine*

$$w_T = h_1 - m_2'h_2 - (1 - m_2')h_3 \tag{a}$$

This energy balance has been written from the general form $w = \Sigma_{\text{in}}m'h - \Sigma_{\text{out}}m'h$. Slightly different physical reasoning gives

$$w_T = h_1 - h_2 + (1 - m_2')(h_2 - h_3) \tag{b}$$

or

$$w_T = m_2'(h_1 - h_2) + (1 - m_2')(h_1 - h_3) \tag{c}$$

Any one of these three forms is satisfactory here, but for complex systems, with multiple inlets and outlets, form a is the simplest to apply. Substituting values already determined into Eq. a

$$w_T = 3398.3 - 0.190(2688.1) - 0.810(2050.0) = 1227.1 \text{ kJ/kg}$$

Then $$\eta = \frac{w_{\text{cycle}}}{q_{7\text{-}1}} = \frac{w_T - w_{\text{in},4\text{-}5} - w_{\text{in},6\text{-}7}}{h_1 - h_7} = \frac{1227.1 - (0.810)0.5 - 8.2}{3398.3 - 631.5} = 0.440$$

Discussion. Comparing this result with that of Example 17·1 for a Rankine cycle with the same throttle and exhaust conditions shows that the use of a single feedwater heater increases the efficiency by $(0.437 - 0.412)/0.412 = 6.1$ percent. Other things being equal, the feedwater heater is justified if its total annual cost is less than 6.1 percent of the annual fuel cost for the simple Rankine cycle.

High-pressure steam power plants use several stages of regenerative feedwater heating. A flow diagram with three closed heaters and one open heater is shown in Fig. 17·7. Some plants use twice as many. The gain in efficiency resulting from the addition of a heater drops as the number of heaters increases. The number of heaters to be used in a plant is determined by an economic study. Roughly speaking, the number of heaters is increased until the addition of one more heater would increase the fixed charges more than it would decrease fuel costs. Consequently, more heaters will be used in a plant that

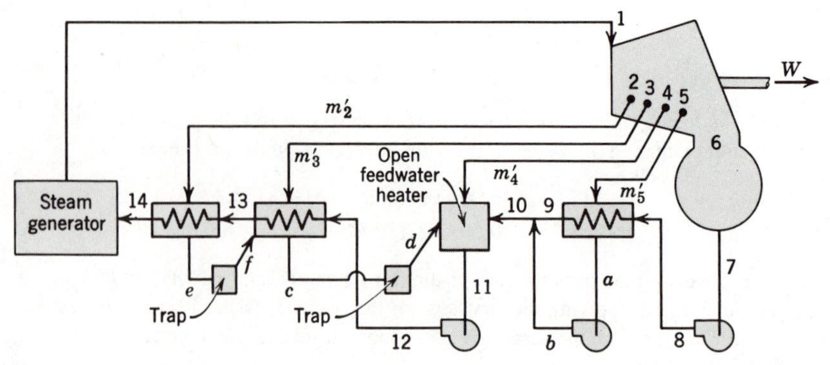

Figure 17·7. Cycle with four stages of regenerative feedwater heating.

operates near full capacity most of the time than in one of the same capacity that is lightly loaded much of the time.

The steam flow distribution required for several heaters can be determined by starting at the highest-pressure heater and making mass and energy balances on each one in turn.

In passing, it should be mentioned that in actual plants many features not discussed here affect the steam and energy distribution in the cycle. For example, steam may be used for soot blowing and for atomizing fuel oil; water is lost through blowdown, and make-up water must be added; steam may be used to drive auxiliaries; and there are numerous components such as shaft seals and oil coolers that have small but significant effects on the plant energy balance. For analyzing these various effects there are no more powerful tools than the first and second laws of thermodynamics, which on account of their generality, can be studied and learned in connection with some systems and then applied to quite different ones.

Example 17·3. Consider a regenerative steam power cycle as shown in the figure. The flow through the turbine is adiabatic but not isentropic. Assume that the pumps operate isentropically. Pressure, temperature, and enthalpy values are shown on the diagram. The lowest temperature in the surroundings is 20°C. Determine the efficiency of the cycle and the irreversibilities, per

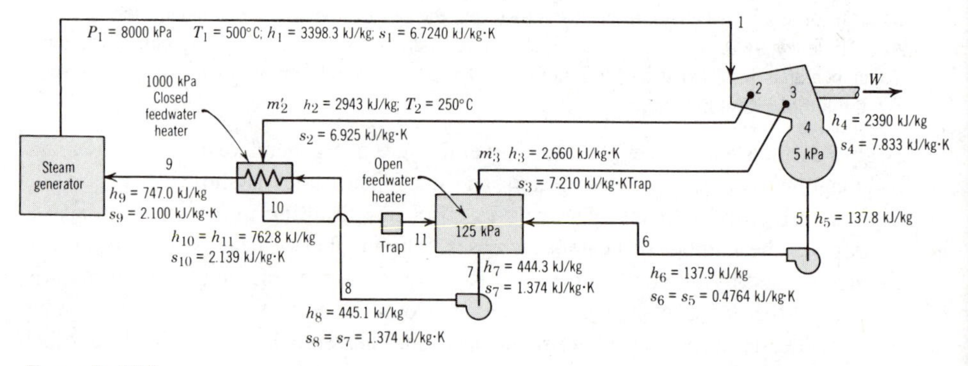

Example 17·3.

kilogram of throttle steam, of the processes in the turbine, closed feedwater heater (including trap), and open feedwater heater. Make an energy flow diagram showing the results.

Solution. Energy and mass balances on the closed heater and then on the open heater are made to determine m_2' and m_3'.

Closed heater: $$m_2'h_2 + h_8 = m_2'h_{10} + h_9$$

$$m_2' = \frac{h_9 - h_8}{h_2 - h_{10}} = \frac{747.0 - 445.1}{2943 - 762.8} = 0.138 \text{ kg/kg throttle}$$

Open heater: $$m_3'h_3 + (1 - m_2' - m_3')h_6 + m_2'h_{11} = h_7$$

$$m_3' = \frac{h_7 - h_6 - m_2'(h_{11} - h_6)}{h_3 - h_6} = \frac{444.3 - 137.9 - 0.138(762.8 - 137.9)}{2660 - 137.9}$$
$$= 0.0873 \text{ kg/kg throttle}$$

Applying the first law to the turbine, and noting that $m_4' = 1 - m_2' - m_3'$

$$w_T = h_1 - m_2'h_2 - m_3'h_3 - m_4'h_4$$
$$= 3398.3 - 0.138(2943) - 0.0873(2660) - 0.775(2390) = 908 \text{ kJ/kg throttle}$$

Then $$\eta = \frac{w_T - m_4'w_{\text{in},5\text{-}6} - w_{\text{in},7\text{-}8}}{q_{9\text{-}1}} = \frac{w_T - m_4'\Delta h_{5\text{-}6} - \Delta h_{7\text{-}8}}{h_1 - h_9}$$
$$= \frac{908 - 0.775(0.1) - 0.8}{3398.3 - 747.0} = 0.342$$

For each piece of adiabatic steady-flow equipment, $i = T_0\Delta s$, where $\Delta s = \Sigma_{\text{out}}m's - \Sigma_{\text{in}}m's$. For the turbine,

$$i = T_0[m_4's_4 + m_3's_3 + m_2's_2 - s_1]$$
$$= 293[0.775(7.833) + 0.0873(7.210) + 0.138(6.925) - 6.724] = 273 \text{ kJ/kg throttle}$$

To find s_{11}, we notice that $P_{11} = 125$ kPa and $h_{11} = h_{10} = 762.8$ kJ/kg because the flow through the trap is a throttling process. Then

$$x_{11} = \frac{h_{11} - h_f}{h_{fg}} = \frac{762.8 - 444.3}{2241.0} = 0.142$$

$$s_{11} = s_f + x_{11}s_{fg} = 1.374 + 0.142(5.9104) = 2.214 \text{ kJ/kg·K}$$

Then for the closed feedwater heater and its trap,

$$i = T_0[s_9 + m_2's_{11} - s_8 - m_2's_2] = T_0[s_9 - s_8 + m_2'(s_{11} - s_2)]$$
$$= 293[2.100 - 1.374 + 0.138(2.214 - 6.925)] = 22.2 \text{ kJ/kg throttle}$$

For the open feedwater heater,

$$i = T_0[s_7 - m_3's_3 - m_4's_6 - m_2's_{11}]$$
$$= 293[1.374 - 0.0873(7.210) - 0.775(0.4764) - 0.138(2.214)] = 20.5 \text{ kJ/kg throttle}$$

Other data needed for making the energy flow diagram are the unavailable part of the heat added,

$$q_{\text{unav},9\text{-}1} = T_0(s_1 - s_9) = 293(6.7240 - 2.100) = 1355 \text{ kJ/kg throttle}$$

the amount of heat rejected per kilogram of throttle steam,

$$q_{out,4\text{-}5} = m'_4(h_4 - h_5) = 0.775(2390 - 137.8) = 1745 \text{ kJ/kg throttle}$$

and the unavailable part of the heat rejected,

$$q_{out,unav,9\text{-}1} = m'_4 T_0(s_4 - s_5) = 0.775(293)(7.883 - 0.4764) = 1670 \text{ kJ/kg throttle}$$

The energy flow diagram shows that the difference between the unavailable energy rejected by the cycle and that added to the cycle is accounted for completely by the sum of the irreversibilities of the cycle components.

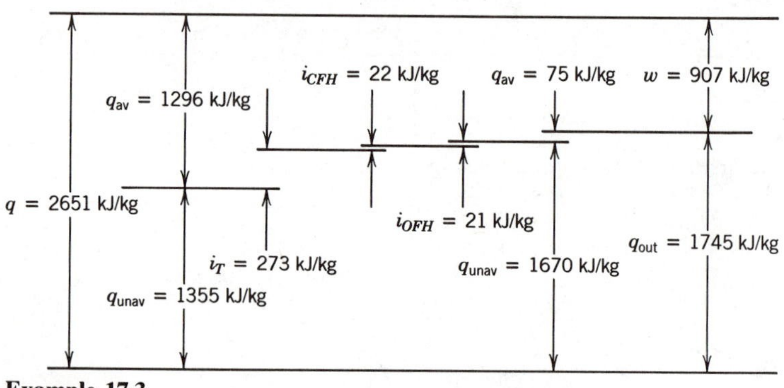

Example 17·3.

17·4 The Reheat Cycle

The thermal efficiency of a Rankine cycle is increased by increasing the steam-generator pressure or the maximum temperature. For a given maximum temperature, which is usually determined by the strength characteristics of construction materials, increasing the steam-generator pressure causes a decrease in the quality of the steam leaving the prime mover. This is undesirable if the prime mover is a turbine because moisture content of more than 10 to 12 percent causes serious erosion of turbine blades. (Notice that the exhaust quality in Examples 17·1 and 17·2 is only 0.789, far below the erosion limit of 0.88 to 0.9. Of course, an actual turbine with an efficiency less than 100 percent has a higher exhaust

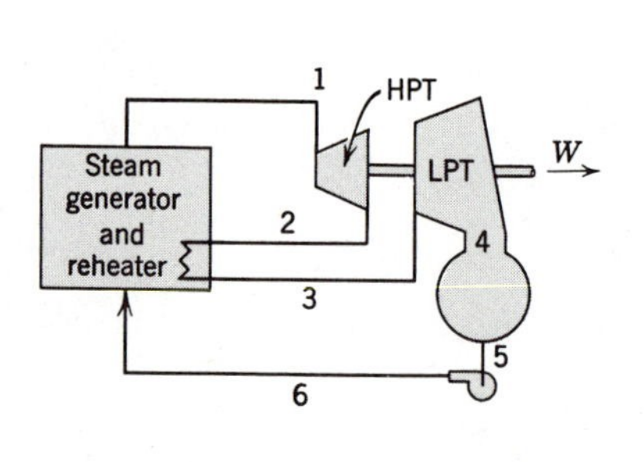

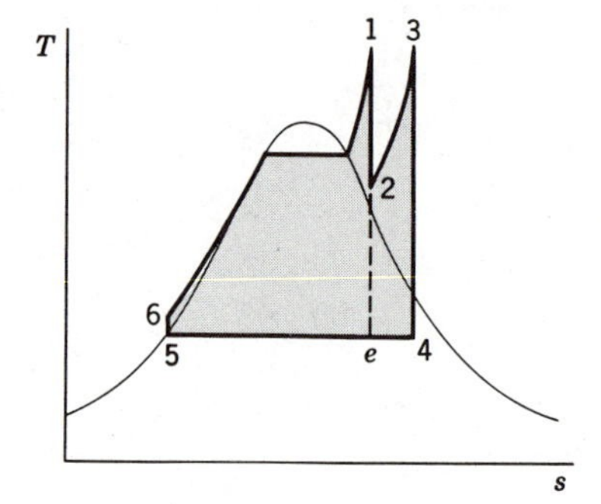

Figure 17·8. Reheat cycle.

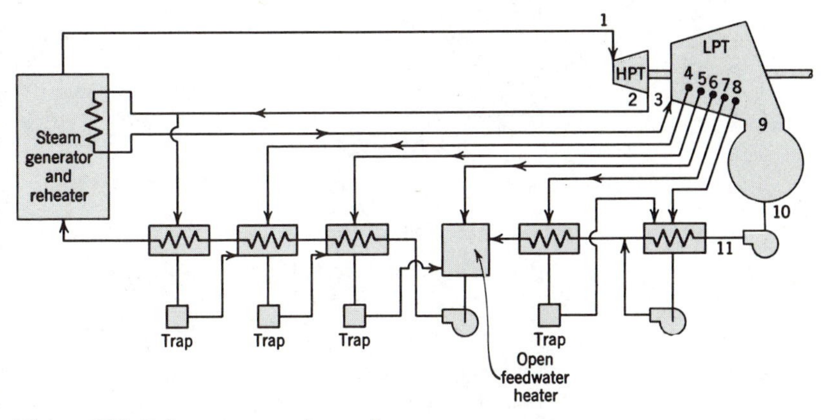

Figure 17·9. Reheat–regenerative cycle.

quality than the corresponding ideal turbine with isentropic expansion.) One way to prevent the exhaust moisture content from exceeding the limiting value is to remove the steam from the turbine after part of its expansion, add heat to it at constant pressure, and return it to the turbine for further expansion to the condenser pressure. This procedure characterizes the *reheat cycle*.

Flow and *Ts* diagrams of a reheat cycle are shown in Fig. 17·8. Expansion 1-2 occurs in the high-pressure turbine (HPT), process 2-3 is the constant-pressure reheating, and expansion 3-4 occurs in the low-pressure turbine (LPT), Point *e* is the exhaust state without reheating.

Whether adding reheat to a Rankine cycle increases the cycle thermal efficiency depends on the cycle conditions. Inspection of a reheat-cycle *Ts* diagram shows that, the higher the reheat pressure and the higher the final reheat temperature, the higher the cycle thermal efficiency is. For the cycle of Fig. 17·8, the thermal efficiency is

$$\eta = \frac{w_{\text{HPT}} + w_{\text{LPT}} - w_{\text{in},P}}{q_{6\text{-}1} + q_{2\text{-}3}}$$

In practice, reheat equipment is economically justifiable only in high-capacity plants; consequently, several stage of regeneration are used in any cycle that involves reheating, and a reheat cycle without regeneration, as shown in Fig. 17·8, is not used. A flow diagram of a reheat-regenerative cycle is shown in Fig. 17·9. For very high-pressure cycles, double reheating is used.

17·5 Cycles for Heating and Power

In all the cycles discussed so far in this chapter, the useful energy output is work or power, and the heat rejected is considered to be of no value, especially since it is largely unavailable energy. There are many instances, however, where energy in the form of heat is needed, and, since there is no intention of producing power from it, its value is not lessened by the fact that its available fraction is small. Where this heat can be supplied

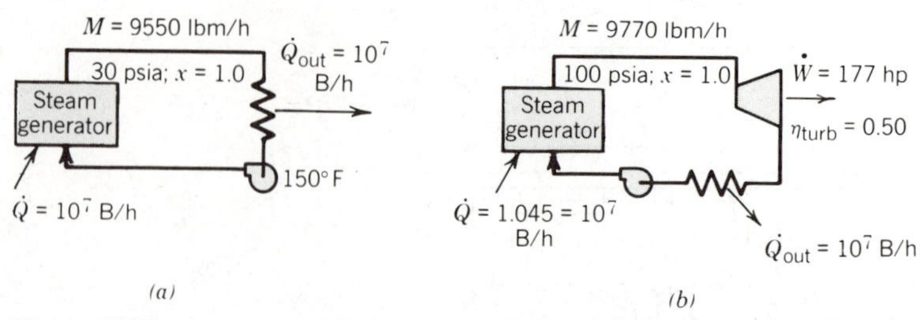

Figure 17·10. (*a*) Cycle for heating only. (*b*) Cycle for heating and byproduct power: cogeneration.

by steam, as in building heating and the heating required by many industrial processes, the functions of heating and power production can often be combined very effectively. This combination is often called *cogeneration*.

As an example, consider a plant that needs 10^7 B of heat from steam per hour at 30 psia with a minimum temperature in the heating system of 150°F. Figure 17·10*a* shows how this heating load can be met by 9550 lbm of dry saturated steam per hour at 30 psia with condensate returned to the boiler at 150°F. The pump is needed to overcome frictional pressure drop in the system. Its work input is negligibly small in comparison with 10^7 B/h. An alternative method of meeting this heating load is shown in Fig. 17·10*b* where dry saturated steam is generated at 100 psia, expanded through a turbine that has an efficiency of 50 percent, and then gives up 10^7 B/h as it condenses and is subcooled to 150°F at 30 psia. Analysis of this cycle shows that the same heating load is carried and also 177 hp or 450 000 B/h of power is produced with an additional heat input of only 450 000 B/h. Thus power is produced in an amount equivalent to the additional rate of heat input; so this *byproduct power* or *cogenerated power* is much cheaper than that from any cycle devised for power only.

The power output and heat output of the cycle in Fig. 17·10*b* are tied together because the same steam flow passes through the turbine and the heating system. Such a cycle is satisfactory where the steam flow can be controlled only by the heating needs, and whatever power produced can always be used, probably to supplement in varying amounts power that is obtained from other sources. Where varying heat and power loads must be carried by a single plant, a cycle such as shown in Fig. 17·11 is used. When the power load is zero, all the steam passes through the pressure-reducing valve (PRV), and none through the turbine. When the heating load is zero, all of the steam expands through the turbine and into the condenser. For meeting both heating and power demands, the steam flow

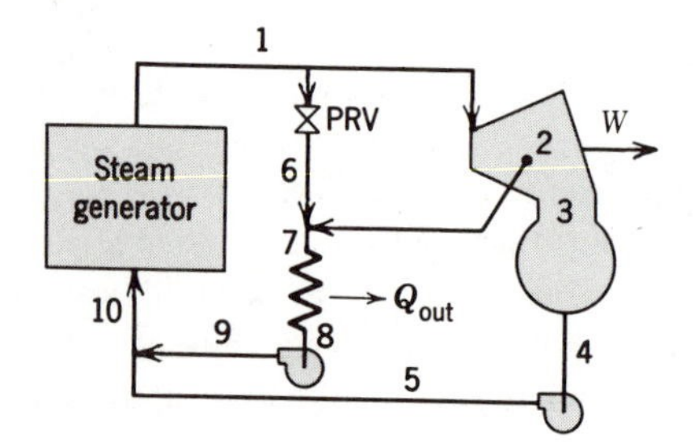

Figure 17·11. Cycle for heating and power: cogeneration.

distribution is determined as follows: Assuming that all steam to the heating system comes from the turbine extraction point* (state 2), the flow rate in the extraction line is determined from an energy balance on the heating system as

$$\dot{m}_2 = \frac{\dot{Q}_{\text{heating}}}{h_2 - h_8}$$

Then the power produced by this steam flowing through the turbine from the throttle to the extraction point is determined from an energy balance on that part of the turbine

$$\dot{W} = \dot{m}_2(h_1 - h_2)$$

If this power is less than the total required, the remainder is produced by additional steam that expands all the way through the turbine to the condenser. If this power obtained from the flow $\dot{m}_2$ is greater than the total required, then the flow $\dot{m}_2$ into the turbine and through the extraction line must be reduced to the value that produces the required amount of power, even though it is inadequate for the heating load. The deficit in energy to the heating system is then made up by steam passed through the pressure-reducing valve. Remember that $h_6 = h_1$ and $h_2 = h_1 - w$, so that $h_6 > h_2$.

In short, the steam distribution problem is to determine the values of $\dot{m}_2$, $\dot{m}_3$, and $\dot{m}_6$ such that their sum is a minimum and they satisfy the first-law equations

$$\dot{W} = \dot{m}_2(h_1 - h_2) + \dot{m}_3(h_1 - h_3)$$
$$\dot{Q} = \dot{m}_2(h_2 - h_8) + \dot{m}_6(h_6 - h_8)$$

where $\dot{W}$ and $\dot{Q}$ are the power required and the heating load, respectively. For the minimum total steam flow, $\dot{m}_3$ or $\dot{m}_6$ must be zero.†

When $\dot{m}_3$ is zero, all the heat added in the steam generator is being utilized; none is rejected to condenser cooling water. When $\dot{m}_6$ is zero, the maximum amount of byproduct power is being obtained for a given heating load. When $\dot{m}_3 = \dot{m}_6 = 0$, the heating and power loads are said to be balanced, and this is the most economical condition. Loads are usually variable, so a plant cannot be operated with balanced loads all the time; however, the plant designer must select operating conditions that give the closest approach to balanced conditions in the long run.

17·6 Binary Vapor Cycles

The only working fluid we have considered and the one which is used almost exclusively in practice for vapor power cycles is water. No better fluid has been found, although

*As used here, a turbine extraction point differs from a turbine bleed point in that the pressure at an extraction point is held constant under all flow conditions by an automatic valve arrangement in the turbine while a bleed point is simply an opening that takes steam from a section of the turbine where the pressure may vary slightly as the flow rate changes.

†In an actual turbine, changes in flow rate affect the efficiency so that h_2 and h_3 vary somewhat as the total flow rate and the flow distribution are changed. Also, it may not be advisable from an operating viewpoint to reduce $\dot{m}_3$ to zero.

water is quite undesirable in some respects. Let us list the desirable characteristics of a vapor-cycle working fluid.

1. Critical temperature well above the highest temperature that can be used in a cycle as fixed by construction material limitations. This would make it possible to vaporize the fluid, and thus add a considerable amount of heat to it, at the maximum temperature.
2. Saturation pressures at the maximum and minimum cycle temperatures within a range that involves neither very high pressures that introduce strength problems nor very low pressures that introduce problems of sealing against infiltration of the atmosphere.
3. A high ratio of h_{fg} to c_p of the liquid so that most of the heat added in a Rankine cycle is added at the maximum temperature. This reduces the need for regeneration.
4. Chemical inertness and stability throughout the cycle temperature range.
5. Triple-state temperature below the expected minimum ambient temperature. This ensures that the fluid will not solidify at any point in the cycle or while being handled outside the cycle.
6. Saturated vapor line (on a property diagram) that is close to a turbine expansion path. This would prevent excessive moisture in the turbine exhaust, thus eliminating the need for reheating, and still permit all or nearly all the heat rejection to occur at the minimum temperature.
7. Cheapness and ready availability.
8. Nontoxicity.

No fluid has all these desirable characteristics. Water is better than any other in an overall evaluation, but it is poor in regard to desirable characteristics 1 and 2. The critical temperture of water is 374.1°C (705.4°F), approximately 300 degreees Celsius below the temperature limit set by material strength properties. Also, the saturation pressure of water is quite high even at moderate vaporization temperatures (8.58 MPa at 300°C and 16.5 MPa at 350°C, for example). Thus water is especially poor at the high-temperature end of the operating range.

Since no single working fluid better than water has been found, searches have been made for a combination of fluids such that one is well suited to the high-temerature part and the other to the low-temperature part of the cycle. A successful combination that has been used is mercury and water.

Flow and *Ts* diagrams for a mercury-water *binary vapor cycle* are shown in Figs.

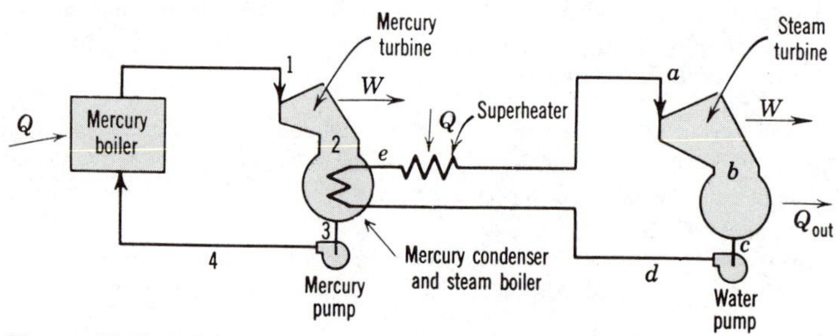

Figure 17·12. Mercury–water binary vapor cycle.

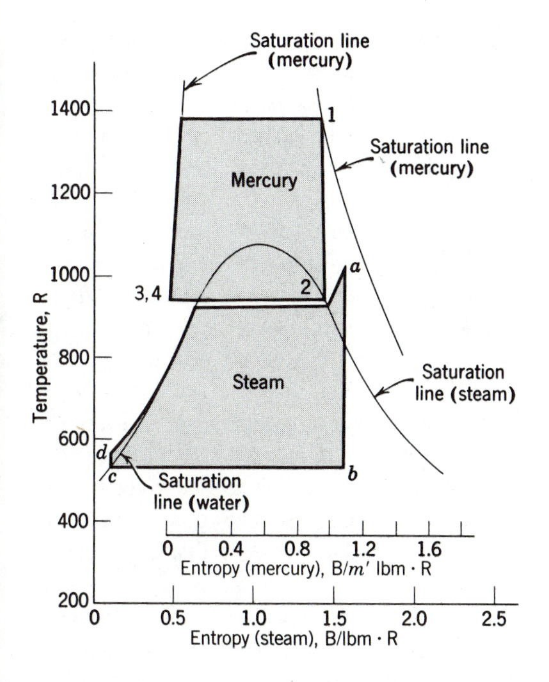

Figure 17·13. *Ts* diagram of mercury–water binary vapor cycle.

17·12 and 17·13. The mercury is vaporized in a boiler at, say, 100 psia for which the saturation temperature is 907°F. It is then expanded in a turbine to a pressure of 2 psia for which the saturation temperature is about 505°F. (Expanding the mercury to a condenser temperature of 100°F would require maintaining a condenser pressure of 0.005 mm Hg!) Heat removed from the mercury to condense it is used to boil water at 540 psia and 475°F. The steam is then superheated and expanded through a turbine to a pressure determined by the temperature of available cooling water. Thus the Rankine cycle using mercury as a working substance rejects heat to the Rankine cycle using water. (In practice, the water cycle would be a regenerative one.) The mercury cycle is called a ''topping'' cycle. Since h_{fg} for water at the steam-generation temperature is several times as large as h_{fg} for mercury, several pounds (m′ on Fig. 17·13) of mercury must circulate per pound of water in the binary vapor cycle.

Inspection of the *Ts* diagram of Fig. 17·13 shows that the binary vapor cycle approaches a Carnot cycle more closely than a steam cycle can for the same temperature limits. Consequently, higher thermal efficiency can be reached by the binary vapor cycle. The overriding drawback of the mercury-steam cycle in practice is that the savings in fuel costs are largely or entirely offset by the increased fixed costs of the more complicated equipment. Higher fuel costs would therefore tend to favor the use of such a cycle.

17·7 Summary

The Carnot cycle is unsuitable as a model for the design of steam power-plant cycles because of the difficulty of carrying out the required processes in actual machines. Actual cycles can, however, simulate the *Rankine cycle* which is comprised of four reversible processes: (1) constant-pressure heat addition that takes the fluid from a compressed liquid state to a saturated or superheated vapor state, (2) isentropic expansion in a prime mover, (3) constant-pressure heat rejection to condense the vapor to a saturated liquid, and (4) isentropic pumping to a compressed liquid state at the steam-generation pressure. The backwork ratio of a Rankine cycle is low, so cycle thermal efficiency suffers relatively little from low prime-mover or pump efficiency.

The efficiency of a steam power cycle can be increased by *regeneration,* which involves the *bleeding* of some vapor from the prime mover after it has expanded part of the way to the condenser pressure and using it in *feedwater heaters* to preheat the liquid going to the steam generator. In open feedwater heaters, the bled steam is mixed with the feedwater in the heater; in closed feedwater heaters, the bled steam condenses on the outside of tubes through which the feedwater flows.

In order to reduce the moisture content that causes erosion in the low-pressure stages of turbines, *reheating* is used. In a reheat cycle, steam is removed from the turbine after part of its expansion is completed, heated at constant pressure, and returned to the turbine for the remainder of the expansion to the condenser pressure. Reheating is used only in conjunction with regeneration because economic studies always show that regeneration should be added to a Rankine cycle before reheating.

Whenever power and heat at a relatively low temperature are both to be produced, a combination steam cycle, often using an *extraction turbine,* is much more economical than separate power and heating cycles.

Water has some serious shortcomings as a vapor-power-cycle working fluid, but no better fluid has been found. A *binary vapor cycle* uses two fluids: One with good high-temperature characteristics is used in a Rankine cycle which rejects heat into another Rankine cycle (or regenerative cycle) that uses a second fluid with good low-temperature characteristics.

It must be remembered that modifications which improve the efficiency of a power plant can be economically justified only if the saving in fuel costs exceeds the additional fixed costs of the modifications.

Suggested Reading

17·1 Black, William Z., and James G. Hartley, *Thermodynamics*, Harper & Row, New York, 1985, Chapter 8.

17·2 Burghardt, M. David, *Engineering Thermodynamics with Applications*, Harper & Row, New York, 1978, Chapter 11.

17·3 El-Wakil, M. M., *Powerplant Technology*, McGraw-Hill, New York, 1984, Chapter 2.

17·4 Li, Kam W., and A. Paul Priddy, *Power Plant System Design,* Wiley, New York, 1985.

17·5 Sorensen, Harry A., *Energy Conversion Systems*, Wiley, New York, 1983, Chapter 8.

17·6 Weisman, Joel, and Roy Eckart, *Modern Power Plant Engineering,* Prentice-Hall, Englewood Cliffs, N.J., 1985.

17·7 Wood, Bernard D., *Applications of Thermodynamics*, 2nd ed., Addison-Wesley, Reading, Massachusetts, 1982, Sections 1.1,2.

Problems

17·1 Determine the efficiency of a steam power plant operating on a Carnot cycle such that during the isothermal expansion the working fluid is changed from a saturated liquid to a saturated vapor at 2000 kPa. Heat is rejected from the steam at a pressure of 10 kPa.

17·2 Determine the efficiency of a Rankine cycle operating with throttle conditions of 3.0 MPa, 350°C, and a condenser pressure of (*a*) 100 kPa (*b*) 10 kPa.

17·3 Determine the efficiency and the backwork ratio of a Rankine cycle operating between 10.0 MPa and 4 kPa with a throttle temperature of 550°C.

17·4 Solve Example 17·1 with a turbine efficiency of 0.70 and a pump efficiency of 0.65.

17·5 Solve Example 17·1 with the same throttle pressure but a 200 kPa frictional pressure drop between the steam generator and the turbine, turbine efficiency of 0.70, pump efficiency of 0.60, and a pump discharge pressure of 8300 kPa to allow for frictional pressure drops in the piping, feedwater regulator valve, and steam generator. In addition,(*d*) make an availability accounting (as in Example 10·10) and (*e*) make an energy flow diagram (as in Example 16·3).

17·6 For a Rankine cycle with throttle conditions of 10.0 MPa, 500°C, plot cycle thermal efficiency as a function of exhaust pressure in the range of 1 atm down to 0.01 atm.

17·7 For a throttle temperature of 600°C, an exhaust pressure of 3.0 kPa, and a turbine efficiency of 0.74, determine the maximum throttle temperature if the moisture content of the exhaust steam is not to exceed 10 percent.

17·8 For a throttle pressure of 4.0 MPa and an exhaust pressure of 4 kPa, plot Rankine cycle efficiency against throttle temperature in the range from the saturation temperature to 600°C.

17·9 For a throttle temperature of 500°C and an exhaust pressure of 4 kPa, plot Rankine cycle efficiency against throttle pressure in the range of 100 kPa to 5.0 MPa. Indicate the part of the curve that involves superheated exhaust steam.

17·10 A steam turbine is to operate with a throttle temperature of 500°C. Cooling water is available to maintain an exhaust pressure of 3 kPa. Moisture fraction in the turbine exhaust is not to exceed 10 percent, but maximum power output for a given flow rate is desired. Determine the throttle pressure to use for turbine efficiencies of 70, 80, 90, and 100 percent.

17·11 Steam is supplied to a turbine at 1.50 MPa, 300°C, and exhausted at 100 kPa. Assume reversible adiabatic flow. To what pressure must the incoming steam be throttled in order to reduce the work per pound to two thirds of that obtained without throttling? Assume that the flow through the turbine is still reversible and adiabatic and that the exhaust pressure is unchanged.

17·12 Determine the efficiency of a steam power plant operating on a Carnot cycle if the fluid states at the beginning and end of the isothermal expansion are saturated liquid and dry saturated vapor, respectively, at a pressure of 200 psia. The pressure during the heat rejection is 10 psia.

17·13 Determine the efficiency of a Rankine cycle with steam entering the turbine at 400 psia, 700°F, and an exhaust pressure of (*a*) 1 psia and (*b*) 1 in. Hg absolute.

17·14 A Rankine cycle operates with throttle conditions of 1000 psia, 800°F, and a condenser pressure of 1 psia. Available cooling water is at 70°F. Determine (*a*) the thermal efficiency, (*b*) the backwork ratio, and (*c*) the fraction of the heat added that is available energy.

17·15 Solve problem 17·14 with a turbine efficiency of 0.75 and a pump efficiency of 0.60. Also, (*d*) make an availability accounting (as in Example 10·10) and (*e*) make an energy flow diagram (as in Example 16·3).

17·16 Solve problem 17·14, allowing for a 30-psi drop in pressure between the steam-generator outlet and the turbine (i.e., the steam-generator outlet pressure is 1030 psia), a turbine efficiency of 0.70, a pump efficiency of 0.65, and a pump discharge pressure of 1300 psia to allow for pressure drops in the piping and in the steam generator.

17·17 For a Rankine cycle with throttle conditions of 1000 psia, 900°F, plot cycle thermal efficiency as a function of exhaust pressure in the range of 1 atm down to 0.01 atm.

17·18 For a throttle temperature of 1150°F, an exhaust pressure of 0.5 psia, and a turbine efficiency of 0.76, what is the maximum throttle pressure that can be used if the moisture in the exhaust is not to exceed 10 percent?

17·19 For a throttle pressure of 600 psia and an exhaust pressure of 1 psia, plot Rankine cycle efficiency against throttle temperature in the range from the saturation temperature to 1000°F.

17·20 For a throttle temperature of 900°F and an exhaust pressure of 1 psia, plot Rankine cycle efficiency against throttle pressure in the range of 15 to 700 psia. Indicate the part of the curve that involves superheated exhaust steam.

17·21 A steam turbine is to operate with a throttle temperature of 900°F. Cooling water is available to maintain an exhaust pressure of 0.50 psia. Moisture fraction in the turbine exhaust is not to exceed 10 percent, but maximum power output for a given flow rate is desired. Determine the throttle pressure to use for turbine efficiencies of 70, 80, 90, and 100 percent.

17·22 Steam enters a turbine at 5000 psia, 1100°F, and expands to 1 psia. At the section of the turbine where the quality reaches 90 percent, 90 percent of the liquid present is separated mechanically from the steam flow and removed from the turbine. Calculate the amount by which the ideal turbine work is changed.

17·23 Steam is supplied to a turbine at 200 psia, 600°F, and exhausted at 14.7 psia. Assume reversible adiabatic flow. To what pressure must the incoming steam be throttled in order to reduce the work per pound to two thirds of that obtained without throttling? Assume that the flow through the turbine is still reversible and adiabatic and that the exhaust pressure is unchanged.

17·24 Consider any type of steam power cycle operating with fixed conditions at the steam-generator inlet and outlet and a fixed condenser pressure. The turbine exhaust is always wet (i.e., not superheated). Demonstrate that an increase in irreversibility i anywhere in the cycle always results in an increased heat rejection in the condenser.

17·25 Mechanical design limitations are similar for both gas and steam turbines. What reasons can you give for the fact that steam power plants are built in much larger capacities than gas-turbine power plants?

17·26 Prove that the constant-pressure lines in the wet region of an *hs* diagram for steam are straight. Also prove that they are or are not parallel.

17·27 Dry saturated steam at a pressure P_1 enters a turbine at a rate of $\dot{m}_1$ lbm/h. Partway through the turbine, $\dot{m}_2$ lbm/h of dry saturated steam at a pressure P_2 is added. All of the steam is exhausted at a pressure P_3. The expansion in the turbine is adiabatic. The sink temperature is T_3. A throttling calorimeter is used to measure the quality of the exhaust steam. Write equations in terms of flow rates and properties at 1, 2, and 3 for (*a*) the turbine power output, (*b*) the irreversibility of the turbine process, and (*c*) the minimum possible exhaust quality if no measurement of it had been made. (Of course, without the measurement of x_3 or h_3, parts *a* and *b* could not be solved.)

17·28 Liquid sodium leaving a nuclear reactor is at 20 psia, 700°F. It goes to a heat exchanger where it is cooled to 600°F before returning to the reactor. The sodium flow rate in this circuit (the *primary circuit*) is 100 000 lbm/h. In the heat exchanger, heat is transferred to liquid sodium in an *intermediate circuit*, which in turn transfers heat to a boiler that produces dry saturated steam at 1000 psia. The steam expands adiabatically through a turbine to 1 psia. The turbine efficiency is 0.60. Sketch a flow diagram showing all three (primary, intermediate, and steam) circuits. Assuming (1) negligible pressure drops due to friction, (2) negligible work required by sodium pumps, (3) negligible stray heat losses, and (4) the specific heat of sodium to be constant at 0.30 B/lbm·°F, determine (*a*) the net amount of power available from the plant, (*b*) the irreversibility, in B/h, of the process of transferring energy from the primary sodium circuit through the intermediate circuit to the water.

17·29 An ideal regenerative cycle operates with throttle conditions of 3.0 MPa, 400°C; a single open feedwater heater at 250 kPa; and a condenser pressure of 30 mm Hg absolute. Compare its efficiency with that of a simple Rankine cycle having the same throttle conditions and same exhaust pressure.

17·30 An ideal regenerative cycle operates with throttle conditions of 3.0 MPa, 400°C; one open feedwater heater at 140 kPa; one closed feedwater heater (with its drain trapped to the open heater) at 700 kPa; and a condenser pressure of 30 mm Hg absolute. The terminal temperature difference of the closed feedwater heater is 5 Celsius degrees. Determine the thermal efficiency of the cycle.

17·31 Solve problem 17·29, assuming that the turbine efficiency between the throttle and any point in the turbine is 0.65 and any pump efficiency is 0.70. Also make an energy flow diagram similar to that of Example 17·3.

17·32 An ideal regenerative steam cycle has throttle conditions of 6000 kPa, 400°C; one open feedwater heater at 700 kPa; and a condenser pressure of 4 kPa. Compare its cycle efficiency with that of a simple Rankine cycle having the same throttle conditions and exhaust pressure.

17·33 An ideal regenerative steam cycle has throttle conditions of 6000 kPa, 400°C; an open feedwater heater at 170 kPa; a closed feedwater heater at 1100 kPa with a terminal temperature difference of 5 degrees and its drain pumped into the feed line on the steam generator side of the heater; and a condenser pressure of 4 kPa. Determine the cycle efficiency and make an energy flow diagram showing the irreversibilities of various components and the fractions of the heat added and of the heat rejected that are unavailable.

17·34 An ideal regenerative steam cycle has throttle conditions of 6000 kPa, 400°C; an open feedwater heater at 105 kPa; two closed feedwater heaters, one at 400 kPa and one at 1100 kPa; and a condenser pressure of 4 kPa. Each closed heater has a terminal temperature difference of 5 degrees and its drain is pumped into the feed line on the steam generator side of the heater. Determine the cycle efficiency.

17·35 Solve problem 17·33, assuming that the turbine efficiency between the throttle and any point in the turbine is 0.75 and that the efficiency of each pump is 0.70.

17·36 From the standpoint of cycle efficiency, determine to within 100 kPa the optimum pressure for a single open feedwater heater in an ideal steam power plant cycle operating between throttle conditions of 6000 kPa, 400°C, and an exhaust pressure of 4 kPa.

17·37 Solve Example 17·2 if a closed heater is used and the condensate in the heater is trapped into the condenser. (*a*) Assume that the feedwater is heated to the saturation temperature of the heater. (*b*) Assume that the feedwater leaves the heater at a temperature of 10 Celsius degrees lower than the saturation temperature of the heater.

17·38 Calculate the irreversibility per kilogram of throttle steam of the operation of the feedwater heater of (*a*) Example 17·2, (*b*) problem 17·37.

17·39 A regenerative steam power-plant cycle uses one open and one closed feedwater heater. The closed heater operates at a higher pressure than the open one. There are three ways that the condensate drain from the closed heater can be handled: (*a*) It can be pumped into the feedwater line between the closed heater and the boiler. (*b*) It can be trapped into the open heater. (*c*) It can be trapped into the condenser. From the standpoint of cycle efficiency, which method is the best? Which is the worst? Explain your reasoning.

17·40 Most steam power plant equipment and piping are insulated to minimize heat losses to the surroundings. For a given plant operating steadily, does a heat loss from a high-pressure steam line (for example, between a turbine bleed point and the highest-pressure heater) have the same effect on cycle efficiency as an equal magnitude of heat loss from a low-pressure line (for example, the bleed line to the lowest-pressure heater)? Explain your reasoning convincingly.

17·41 An ideal regenerative cycle operates with throttle conditions of 400 psia, 700°F; a single open feedwater heater at 35 psia; and a condenser pressure of 1 in. Hg absolute. Compare its efficiency with that of a simple Rankine cycle having the same throttle conditions and exhaust pressure.

17·42 An ideal regenerative cycle operates with throttle conditions of 400 psia, 700°F; one open feedwater heater at 20 psia; one closed feedwater heater (with its drain trapped to the open heater) at 100 psia; and a condenser pressure of 1 in. Hg absolute. The terminal temperature difference of the closed feedwater heater is 8 Fahrenheit degrees. Determine the thermal efficiency of the cycle.

17·43 Solve problem 17·41, assuming that the turbine efficiency between the throttle and any point in the turbine is 0.65 and the boiler feed pump efficiency is 0.70. Also make an energy flow diagram similar to that of Example 17·3.

17·44 An ideal regenerative steam cycle has throttle conditions of 1000 psia, 800°F; one open feedwater heater at 100 psia; and a condenser pressure of 1 psia. Compare its cycle efficiency with that of a simple Rankine cycle having the same throttle conditions and exhaust pressure.

17·45 An ideal regenerative steam cycle has throttle conditions of 1000 psia, 800°F; an open feedwater heater at 25 psia; a closed feedwater heater at 150 psia with a terminal temperature difference of 10 degrees and its drain pumped into the feed line on the steam generator side of the heater; and a condenser pressure of 1 psia. Determine the cycle efficiency and make an energy flow diagram showing the irreversibilities of various components and the fractions of the heat added and of the heat rejected that are unavailable.

17·46 An ideal regenerative steam cycle has throttle conditions of 1000 psia, 800°F; an open feedwater heater at 15 psia; two closed feedwater heaters, one at 60 psia and one at 160 psia; and a condenser pressure of 1 psia. The low-pressure closed heater has a terminal temperature difference of 10 degrees and its drain is pumped into the feed line on the steam generator side of the heater. Feedwater leaving the high-pressure heater is at 360°F. Determine the cycle efficiency.

17·47 Solve problem 17·45, assuming that the turbine efficiency between the throttle and any point in the turbine is 0.75 and that the efficiency of each pump is 0.70.

17·48 From the standpoint of cycle efficiency, determine to within 20 psi the optimum pressure for a single open feedwater heater in an ideal steam power plant cycle operating between throttle conditions of 1000 psia, 800°F, and an exhaust pressure of 1 psia.

17·49 Consider a regenerative steam cycle with one open heater. If the turbine expansion is adiabatic but irreversible, the values of several flow and energy quantities in the cycle may be different from their values with isentropic turbine expansion. Explain clearly how and why each of the following values is affected: (*a*) the mass of steam bled, (*b*) the heat rejected in the condenser per unit mass of throttle steam, and (*c*) the heat rejected in the condenser per unit mass of steam entering the condenser.

17·50 An ideal reheat cycle has throttle conditions of 14.0 MPa, 450°C, and reheats at 3.8 MPa to 480°C. The exhaust pressure is 5 kPa. Compute the thermal efficiency of the cycle. Per kilogram of throttle steam, how much unavailable energy is rejected by the cycle if the sink temperature is 25°C?

17·51 Solve problem 17·50 for a high-pressure turbine efficiency of 0.80 and a low-pressure turbine efficiency of 0.85. Calculate also the effectiveness of each turbine.

17·52 Determine the efficiency and the required flow rate of an ideal reheat cycle that is to produce 150 000 kW at the turbine coupling if the throttle conditions are 15.0 MPa, 600°C; reheat is at 1.4 MPa to 600°C; and the condenser pressure is 5 kPa.

17·53 Solve problem 17·52 if a closed feedwater heater with a terminal temperature difference of 3 Celsius degrees is supplied with steam from the high-pressure turbine exhaust. The heater drain is pumped into the feed line.

17·54 Determine the efficiency and the required steam generator flow rate for an ideal reheat-regenerative cycle that is to produce 150 000 kW at the turbine coupling if the throttle conditions are 15.0 MPa, 600°C; reheat is at 1.4 MPa to 600°C; there are closed feedwater heaters at 1400 kPa, 300 kPa, and 50 kPa; there is an open feedwater heater at 150 kPa; and the condenser pressure is 7 kPa. The drain from each closed heater is trapped into the next lower heater except that the drain from the 50-kPa heater is pumped into the open heater.

17·55 Steam is to be supplied to a turbine at 35.0 MPa, 600°C, and exhausted to a condenser at 5 kPa. For isentropic expansion it is desired to have no moisture entering a reheater and not more than 10 percent moisture entering the condenser. These conditions call for a double reheat arrangement. If reheat temperatures are not to exceed 550°C and it is desired to have the same enthalpy drop across each turbine, determine the reheat pressures to be used.

17·56 A nuclear reactor provides saturated steam at 7000 kPa for a steam power cycle as shown in the figure. The separator mechanically removes moisture from the steam passing through it so that the steam leaving (state 6) contains only 1 percent moisture. The steam leaving the first reheater is at $T_7 = 250$°C, and that leaving the second reheater is at $T_8 = 280$°C. Saturated liquid leaves the separator and each of the reheaters and feedwater heaters. The terminal temperature difference of each of the closed feedwater heaters is 5 Celsius degrees. Immediately downstream from the bleed point (9) in the low-pressure turbine, a separator removes liquid water so that the quality of the steam starting the remainder of the expansion through the turbine is $x_{11} = 0.96$. Disregard frictional pressure drops in piping and heat exchangers and assume that turbine expansions are isentropic. Pressures are $P_1 = P_{27} = P_{16} = P_{21} = 7000$ kPa, $P_3 = P_{26} = P_{30} = 4200$ kPa, $P_4 = P_{24} = P_{25} = P_{28} = P_8 = 2500$ kPa, $P_9 = P_{10} = P_{11} = P_{15} = 85$ kPa, $P_{12} = P_{13} = 5.0$ kPa. For a net plant power output of 100 000 kW, determine the steam flow rate through the reactor and the cycle thermal efficiency.

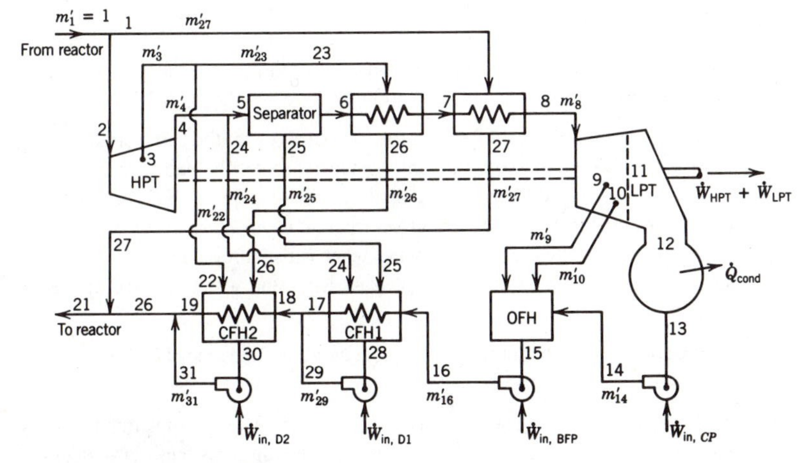

Problems 17·56 and 17·57.

17·57 A nuclear reactor provides saturated steam at 1000 psia for a steam power cycle as shown in the figure. The separator mechanically removes moisture from the steam passing through it so that the steam leaving (state 6) contains only 1 percent moisture. The steam leaving the first reheater is at $T_7 = 480°F$, and that leaving the second reheater is at $T_8 = 540°F$. Saturated liquid leaves the separator and each of the reheaters and feedwater heaters. The terminal temperature difference of each of the closed feedwater heaters is 8 Fahrenheit degrees. Immediately downstream from the bleed point (9) in the low-pressure turbine, a separator removes liquid water so that the quality of the steam starting the remainder of the expansion through the turbine is $x_{11} = 0.96$. Disregard frictional pressure drops in piping and heat exchangers and assume that turbine expansions are isentropic. Pressures are $P_1 = P_{27} = P_{16} = P_{21} = 1000$ psia, $P_3 = P_{26} = P_{30} = 600$ psia, $P_4 = P_{24} = P_{25} = P_{28} = P_8 = 350$ psia, $P_9 = P_{10} = P_{11} = P_{15} = 12$ psia, $P_{12} = P_{13} = 0.70$ psia. For a net plant power output of 100 000 kW, determine the steam flow rate through the reactor and the cycle thermal efficiency.

17·58 An ideal reheat cycle has throttle conditions of 2000 psia, 836°F, and reheats at 550 psia to 900°F. The exhaust pressure is 1 psia. Compute the thermal efficiency of the cycle. Per pound of throttle steam, how much unavailable energy is rejected by the cycle if the sink temperature is 70°F?

17·59 Solve problem 17·58 for a high-pressure turbine efficiency of 0.80 and a low-pressure turbine efficiency of 0.85. Calculate also the effectiveness of each turbine.

17·60 Determine the efficiency and the required flow rate of an ideal reheat cycle that is to produce 150 000 kW at the turbine coupling if the throttle conditions are 2400 psia, 1100°F; reheat is at 200 psia to 1050°F; and the condenser pressure is 1 psia.

17·61 Solve problem 17·60 if a closed feedwater heater with a terminal temperature difference of 5 Fahrenheit degrees is supplied with steam from the high-pressure turbine exhaust. The heater drain is pumped into the feed line.

17·62 Determine the efficiency and the required steam generator flow rate for an ideal reheat–regenerative cycle that is to produce 150 000 kW at the turbine coupling if the throttle conditions are 2400 psia, 1100°F; reheat is at 200 psia to 1050°F; there are closed feedwater heaters at 200 psia, 40 psia, and 8 psia; there is an open feedwater heater at 20 psia; and the condenser pressure is 1 psia. The drain from each closed heater is trapped into the next lower heater except that the drain from the 8-psia heater is pumped into the open heater.

17·63 Steam is to be supplied to a turbine at 5000 psia, 1100°F, and exhausted to a condenser at 1 psia. For isentropic expansion it is desired to have no moisture entering a reheater and not more than 10 percent moisture entering the condenser. These conditions call for a double reheat arrangement. If reheat temperatures are not to exceed 1000°F and it is desired to have the same enthalpy drop across each turbine, determine the reheat pressures to be used.

17·64 Consider a reheat cycle as shown in Fig. 17·8 with the two turbines on the same shaft driving a single generator. The load on the generator is suddenly reduced from full load to zero. Steam flow through the turbines must be stopped immediately or they will overspeed, but some flow of steam must be maintained through the superheater and reheater to keep them from overheating before the furnace or reactor cools. Make a sketch showing what valves are needed in the system to take care of this emergency, each one's normal position, and its action when the turbine load is suddenly dropped.

17·65 Comment on the advisability of extracting steam for regenerative feedwater heating from the line between a reheater and the following turbine.

17·66 A steam accumulator is an energy-storage device that consists of an insulated tank that normally contains both liquid water and steam. Steam is admitted below the liquid surface, and steam is withdrawn from the highest point in the tank. A certain accumulator contains 4.0 m^3 of vapor and 20 m^3 of liquid at 1200 kPa. For how long a period, with no steam entering, can steam at 300 kPa be withdrawn through a pressure-reducing valve at a steady rate of 2000 kg/h? As an approximation, a constant mean enthalpy value for the steam leaving can be used.

17·67 An accumulator (See problem 17·66) is filled with 5.0 m^3 of steam and 10.0 m^3 of liquid water at 300 kPa. The accumulator is charged from a line in which steam is at 1200 kPa, 250°C. If steam is supplied to the accumulator at a steady rate of 2000 kg/h, how long will it take to charge it to 1200 kPa?

17·68 If a steam accumulator is alternately charged with dry saturated steam at 1200 kPa and discharged to pressures as low as 300 kPa with only dry saturated vapor leaving, is it necessary to add or remove water periodically to keep the accumulator operating? Explain.

17·69 A steam accumulator is an energy-storage device that consists of an insulated tank which normally contains both liquid water and steam. Steam is admitted below the liquid surface, and steam is withdrawn from the highest point in the tank. A certain accumulator contains 100 cu ft of vapor and 500 cu ft of liquid at 200 psia. For how long a period, with no steam entering, can steam at 50 psia be withdrawn through a pressure-reducing valve at a steady rate of 4000 lbm/h? As an approximation, a constant mean enthalpy value for the steam leaving can be used.

17·70 An accumulator is filled with 200 cu ft of steam and 400 cu ft of liquid water at 50 psia. The accumulator is charged from a line in which steam is at 200 psia, 500°F. If steam is supplied to the accumulator at a steady rate of 4000 lbm/h, how long will it take to charge it to 200 psia?

17·71 Steam is supplied to a turbine at 1400 kPa, 300°C. The turbine, which has an efficiency of 65 percent, exhausts at 100 kPa into a heating system that supplies 500 000 kJ/h to heat buildings. Condensate leaves the heating system at 50°C. Determine the power of the turbine.

17·72 A large building is heated by having steam that is dry and saturated at 1.5 MPa throttled through a pressure-reducing valve into the heating system at 100 kPa, where it is condensed and cooled to 40°C. The heating system delivers 20 000 kJ/s to the air in the building at 20°C. It is proposed to replace the pressure-reducing valve with a turbine that has an efficiency of 60 percent in order to obtain some power. For the same heating system energy requirement and the same temperature of condensate leaving the heating system, determine the required mass rate of flow through the turbine in kg/s and the power output.

17·73 A steam turbine operates with throttle conditions of 1500 kPa, 250°C. At 250 kPa steam is extracted to provide 25 $\times$ 10^6 kJ/h for heating. Condensate from the heating system is at 200 kPa, 85°C, when it enters a trap that discharges into an open tank. The turbine exhausts into a condenser where the pressure is 5 kPa. Condensate, which is cooled to 30°C in the condenser, is pumped into the same open tank that collects the heating system condensate. From this tank, water is pumped into the steam generator. Turbine efficiency between the throttle and any point in the turbine is estimated to be 65 percent. Atmospheric conditions are 92 kPa, 20°C. The power output is 4000 kW. Determine the required throttle flow rate.

17·74 A plant is designed to supply a power load of 5000 kW and a heating load of 50 million kJ/h. Steam is generated at 4.5 MPa, 450°C, and is expanded in a turbine exhausting to a condenser at a pressure of 50 mm Hg absolute. Assume that expansion is isentropic and that the condensate leaves the condenser as saturated water. The heating load is supplied by steam at 150 kPa which is condensed and subcooled to 50°C and returned to the boiler at that temperature. Two types of plant can be used, (1) power produced by a condensing turbine and heating steam supplied by throttling high-pressure steam, (2) power produced by an automatic extraction turbine and heating steam supplied by bleeding 150 kPa steam from the turbine. Compute for each type (*a*) the amount of steam required per hour, (*b*) the heat input to the boiler, kJ/h, and (*c*) the heat rejected in the condenser, kJ/h.

17·75 Verify the numerical values given in the second paragraph of Sec. 17·5, which refers to Fig. 17·10.

17·76 Steam enters an extraction turbine at 200 psia, 500°F; extraction occurs at 20 psia, $h = 1140$ B/lbm; and steam entering the condenser is at 1 psia, $h = 1005$ B/lbm. Part of the extracted steam goes to an open feedwater heater, which operates at 20 psia, and the rest goes to a heating system that provides 30 million B/h and from which condensate at 152°F is pumped into the open heater. Saturated liquid leaves the open heater. The turbine power output is 3500 kW. Determine (*a*) the efficiency of the turbine between the throttle and the extraction point, (*b*) the irreversibility of the turbine expansion upstream of the extraction point, and (*c*) the flow rate of steam into the turbine.

17·77 A plant is designed to supply a power load of 5000 kW and a heating load of 45 million B/h. Steam is generated at 650 psia, 850°F, and is expanded in a turbine exhausting to a condenser at a pressure of 2 in. Hg absolute. Assume that expansion is isentropic and that the condensate leaves the condenser as saturated water. The heating load is supplied by steam at 22 psia, which is condensed and subcooled to 120°F and returned to the boiler at that temperature. Two types of plant can be used, (1) power produced by a condensing turbine and heating steam supplied by throttling high-pressure steam, (2) power produced by an automatic extraction turbine and heating steam supplied by bleeding 22 psia steam from the turbine. Compute for each type (*a*) amount of steam required per hour, (*b*) heat input to boiler, B/h, and (*c*) heat rejected in condenser, B/h.

17·78 In a manufacturing plant, steam for heating purposes is obtained at 20 psia by drawing steam through a pressure-reducing valve from a line where it is dry and saturated at 100 psia. The heating load is 20 million B/h. Condensate leaves the heating system at 20 psia, 180°F. It is proposed to use a turbine in place of the pressure-reducing valve in order to obtain cogenerated or "byproduct" power while still carrying the same heating load. Assuming a turbine efficiency of 60 percent, determine (*a*) the amount of power that can be obtained with the proposed arrangement, and (*b*) the change in steam flow rate, in lbm/h, which will be required if the proposal is adopted.

17·79 A steam plant is to produce 7500 hp at the turbine coupling and supply a heating load of 70 million B/h by means of steam extracted at 30 psia, which is condensed and subcooled to 200°F in the heating system. Condenser pressure is 1 psia and the condenser flow rate must not be less than 5000 lbm/h. Determine the optimum steam-generator pressure for balanced operation under these conditions, and calculate the amount of heat that must be added in B/h. A throttle pressure below 300 psia should be accompanied by a throttle temperature of 600°F; 300 to 600 psia, by 700°F; and over 600 psia, by 800°F.

CHAPTER EIGHTEEN

Refrigeration

The purpose of a refrigerating system is to remove heat continuously from a body to maintain its temperature lower than that of the surroundings. If the refrigerator operates cyclically (so that there is no net change in its stored energy), the energy transfers are as shown in Fig. 18·1*a*. A consequence of the first law is that $W_{in} = Q_H - Q_L$ and a consequence of the second law is that $W_{in} > 0$. The same cycles of operation and the same energy transfers are involved in the heat-pump application, where heat is removed from the atmosphere or some other part of the surroundings for the purpose of supplying heat to a building or other body at a higher temperature, as shown in Fig. 18·1*b*.

The purpose of this chapter is to introduce a few refrigeration or heat-pump systems as further illustrations of the use of the first law, the second law, and physical property relationships. (Your study of this chapter will be more rewarding if you refer to a handbook or other sources for physical descriptions, including pictures, of the various cycle components mentioned.)

18·1 The Reversed Carnot Cycle

The reversed Carnot cycle (or the Carnot refrigerator cycle) has been described in Sec. 6·7 and discussed further in Chapter 7. Figure 18·2 shows a *TS* diagram for a reversed Carnot cycle. The heat absorbed at temperature T_L during process 1-2 is represented by area 1-2-*b*-*a*-1; the heat rejected at T_H during process 3-4 is represented by area 3-4-*a*-*b*-3; and the net work input is therefore represented by the difference between these two areas which is area 1-2-3-4-1.

The Carnot cycle is used as a standard of comparison for heat-engine cycles because its efficiency is the maximum for given temperature limits. In a similar manner, the reversed Carnot cycle is used as a standard of comparison for refrigerators and heat pumps because for given temperature limits its coefficient of performance is a maximum. (See Problem 7·2.). The coefficient of performance β_R of any refrigerator is defined as

$$\beta_R \equiv \frac{Q_L}{W_{in}}$$

where Q_L is the heat absorbed from the low-temperature body. Application of the first

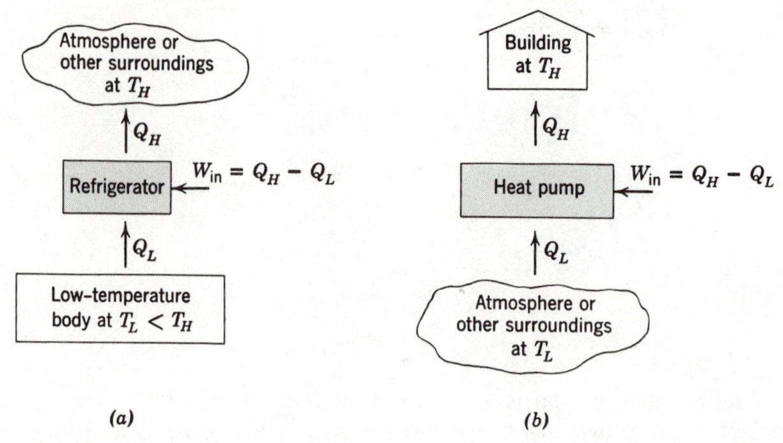

Figure 18·1. Refrigeration cycle application (*a*) as a refrigerator and (*b*) as a heat pump.

law and the important relationship

$$\frac{Q_H}{Q_L} = \frac{T_H}{T_L} \tag{7·3}$$

(which applies only to externally reversible cycles) shows that the coefficient of performance of the Carnot refrigerator is

$$\beta_{R,\text{Carnot}} = \frac{T_L}{T_H - T_L}$$

For a heat pump, the coefficient of performance is defined as

$$\beta_{\text{HP}} = \frac{Q_{\text{out},\ H}}{W_{\text{in}}}$$

where $Q_{\text{out},H}$ is the heat delivered to the high-temperature part of the surroundings. For a Carnot heat pump, one can readily show that

$$\beta_{\text{HP,Carnot}} = \frac{T_H}{T_H - T_L}$$

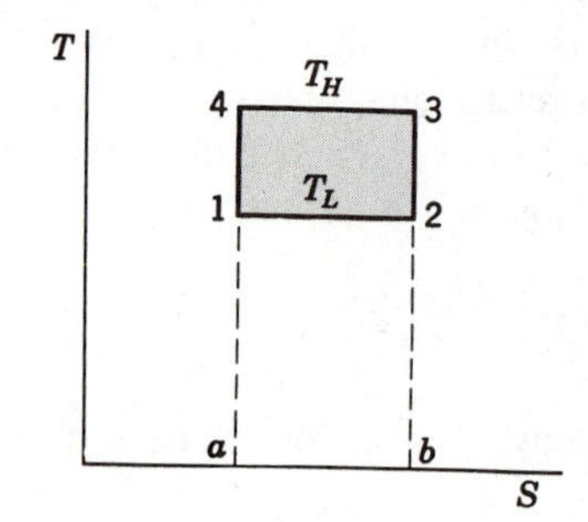

Figure 18·2. Reversed Carnot cycle.

18·2 The Reversed Brayton Cycle

Actual refrigeration cycles using gaseous working substances are not based on the reversed Carnot cycle, because of the difficulties involved in carrying out the isothermal heat-absorption and heat-rejection processes. An ideal cycle that can be more easily simulated in practice is the reversed Brayton cycle. Flow, *Pv,* and *Ts* diagrams of a reversed Brayton cycle are shown in Fig. 18·3. T_H and T_L are the temperatures of those parts of the surroundings to which heat can be rejected and from which heat is to be removed, respectively. The usual working fluid is air. The air is compressed isentropically to a temperature above T_H, and then it is cooled reversibly* at constant pressure until its temperature is T_H or slightly higher than T_H. The air is then expanded isentropically through an engine that supplies some of the power requirement of the compressor. The temperature of the air passing through the engine drops to a value lower than T_L. Therefore, heat can be absorbed from the surroundings at T_L as the air flows at constant pressure through a heat exchanger to state 1 to complete the cycle. Application of the first law to

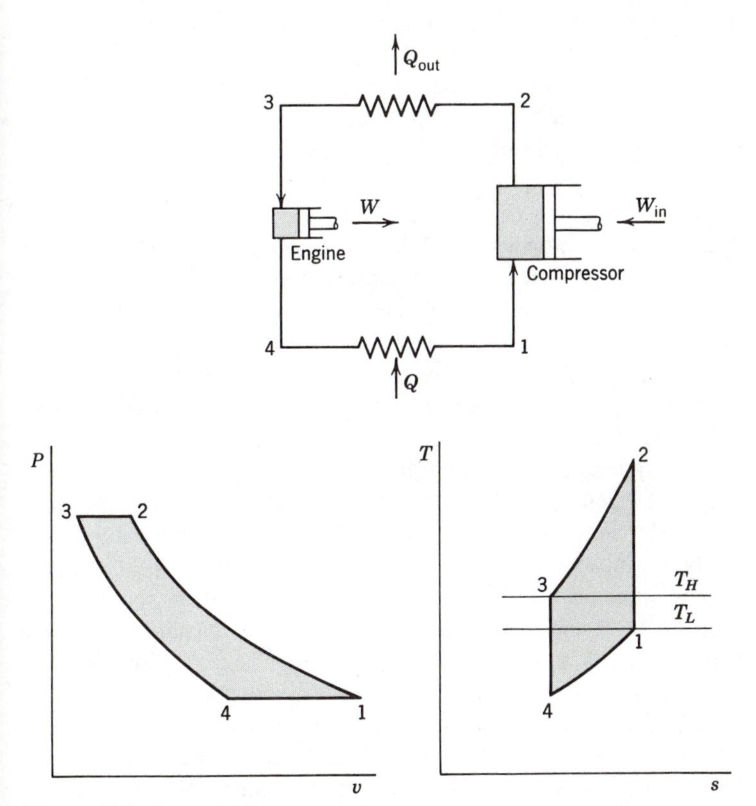

Figure 18·3. Reversed Brayton cycle.

*Remember that *reversible,* without a modifier, means *internally reversible.* (See Sec. 6·3.) The direct transfer of heat from the gas undergoing process 2-3 across a finite temperature difference to the surroundings at T_H is of course externally irreversible.

each of the four pieces of equipment in the steady-flow cycle, assuming that $\Delta ke + \Delta pe = 0$, gives

$$w_{in,C} = h_2 - h_1$$
$$q_{out,2-3} = h_2 - h_3$$
$$w_{eng} = h_3 - h_4$$
$$q_{4-1} = h_1 - h_4$$

There is a net work input to the cycle because the work input to the compressor is greater than the engine work. This can be seen from either the Pv or the Ts diagram of Fig. 18·3 and is of course in accordance with the second law.

The engine of the reversed Brayton cycle cannot be successfully replaced by a throttling valve if the air in the cycle behaves as an ideal gas, because for an ideal gas the Joule-Thomson coefficient, $(\partial T/\partial P)_h$, is zero. That is, there is no temperature change across a throttling valve, since there is no change in enthalpy. For information on how closely air at low temperatures approximates an ideal gas, refer to the Ts diagram for air in the appendix, Chart A·1. For an ideal gas, constant-enthalpy lines coincide with constant-temperature lines.

In the temperature ranges normally encountered in refrigeration cycles, the specific heats of air can be treated as constant with very little loss in accuracy.

Recall that for any process the irreversibility is given by

$$I = T_0 \, \Delta S_{\text{isolated system}}$$

where T_0 is the temperature of the atmosphere. If some other energy reservoir such as the water of a river or lake is used instead of the atmosphere as a source or absorber of heat,* then its temperature is used as T_0. In the case of a refrigeration cycle, it is not strictly correct to refer to T_0 as "the lowest temperature in the surroundings" unless it is understood that the temperature of the body being cooled is excluded from consideration because that body is certainly not an energy reservoir that can absorb or reject large amounts of heat without experiencing a temperature change. If it were, there would probably be no need for a refrigerating system to remove heat from it.

A drawback of an air refrigeration system is that the low density of air at moderate pressures calls for either very high pressures or very high volume flow rates in order to obtain moderate refrigerating capacity. The early solution to this problem was to use very high pressures in order to limit the physical size of the equipment. Consequently, the early systems were called "dense air refrigerating machines." Such machines have been completely displaced by vapor-compression systems, which are cheaper to build because of the lower pressures and smaller volumes involved, and cheaper to operate because their coefficients of performance are higher for any specified temperature range. The major application for air refrigerating systems is now aircraft cabin or cockpit cooling where cool air must be supplied and low system weight is desirable. Small high-speed, low-weight compressors and turbines that handle large volumes of air are used. An example

*In a power cycle, heat is rejected to the atmosphere or to the water of a river, lake, or ocean. A heat pump for heating a building, however, takes heat *from* these same energy reservoirs.

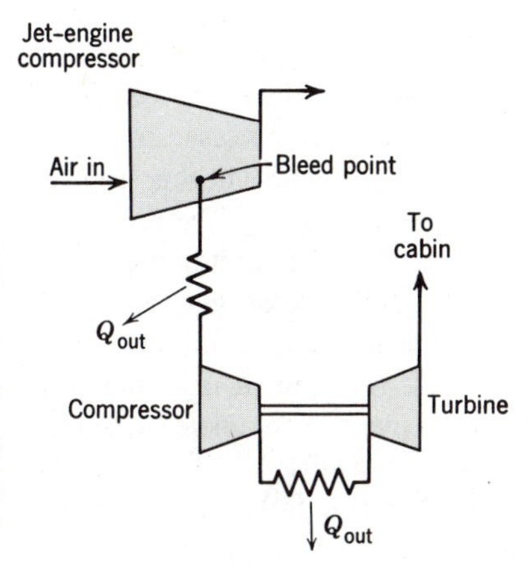

Figure 18·4. Aircraft cooling system.

of an aircraft cooling system is shown in Fig. 18·4. Some air is bled from the jet-engine compressor, cooled in a heat exchanger, and then compressed further by a small compressor that is driven by the turbine. The air is cooled in a heat exchanger between the compressor discharge and the turbine inlet. The cool air leaving the turbine is then ducted to the cabin. Many variations of this cycle are possible. A major problem with such systems in high-speed aircraft is cooling the air before it enters the turbine, because the ambient air experiences a temperature rise when it is brought into the craft and slowed down (relative to the craft).

The chief disadvantage of the reversed Brayton cycle from a thermodynamic standpoint is illustrated by a *Ts* diagram. In Fig. 18·5, 1-2-3-4-1 is a reversed Brayton cycle that absorbs the same amount of heat (the area beneath 1-2 equals the area beneath *a*-2) as the Carnot cycle *a*-2-*b*-*c*-*a*, which exchanges heat with the same energy reservoirs at T_H and T_L. The area within each diagram represents the work input, and it is apparent that the coefficient of performance of the reversed Brayton cycle is much lower than that of the reversed Carnot cycle. The constant-pressure heat-transfer processes result in an operating temperature range much greater than the minimum range established by the temperatures of those parts of the surroundings with which heat is exchanged.

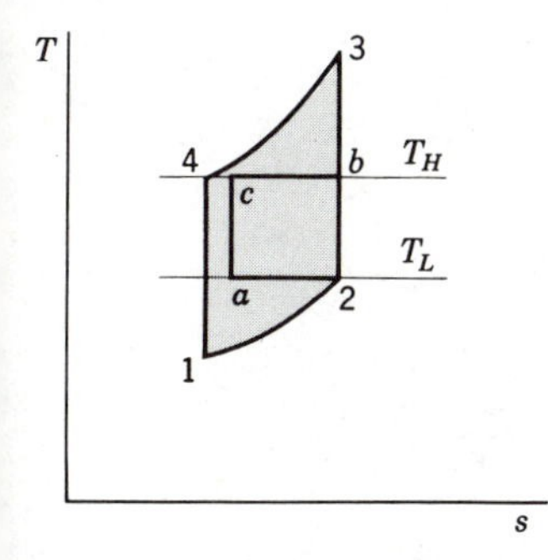

Figure 18·5. Comparison of reversed Brayton and reversed Carnot cycles.

18·3 Vapor-Compression Refrigeration

Flow, *Ts*, and *Ph* diagrams of a conventional vapor-compression refrigerating system are shown in Fig. 18·6. Wet vapor enters the compressor at 1 and is compressed reversibly and adiabatically to state 2. The vapor then flows into a condenser where heat is removed to condense the vapor to a saturated liquid at the same pressure, state 3. The refrigerant then expands through a valve or capillary tube to state 4. For the throttling process, $h_4 = h_3$. Some of the liquid is vaporized as it passes through the expansion valve, so a low-quality mixture enters the evaporator, and this mixture is at a temperature lower than that of the body being refrigerated. In the evaporator, heat is absorbed from the body being refrigerated to evaporate most of the liquid. Thus the cycle is completed when the refrigerant leaves the evaporator and enters the compressor in state 1.

Notice that three of the four processes in the ideal vapor-compression refrigeration cycle are reversible but that process 3-4 is irreversible. An engine is not used in place of the expansion valve in *actual* cycles because the work obtained by expanding a saturated liquid in an engine is much too small to justify the addition of the engine and its related equipment. Therefore, an engine is not used in the *ideal* cycle because doing so would increase the differences between the actual cycle and the corresponding ideal cycle.

The *Ph* diagram, which also is called a Mollier diagram,* is convenient in analyzing vapor-compression refrigeration cycles because (1) three of the four processes appear on it as straight lines, and (2) for the evaporator and condenser processes the heat transfer is proportional to the lengths of the process paths. Such a diagram for ammonia is in the

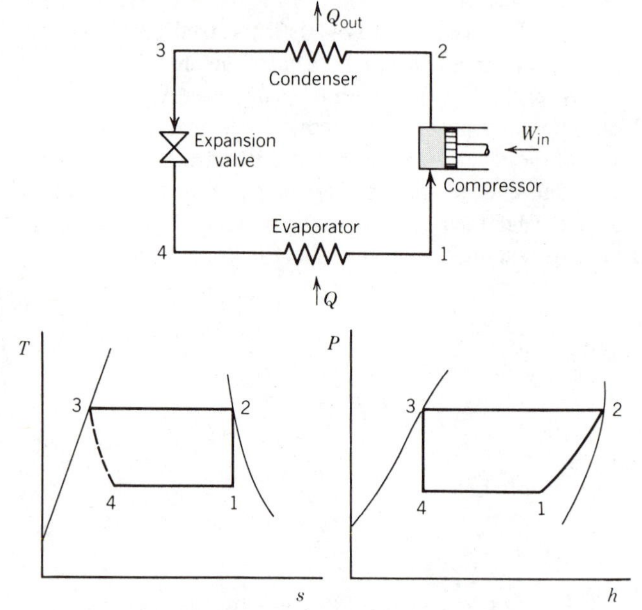

Figure 18·6. Vapor compression refrigeration cycle.

*This name is used for several different thermodynamic diagrams.

appendix, Chart A·8. On Chart A·8 the pressure scale is logarithmic. The *Ph* diagram of Fig. 18·6 has a linear pressure scale.

The cycle shown in Fig. 18·6 is called a *wet compression* cycle because it involves the compression of a liquid–vapor mixture. Notice on the *Ts* diagram how closely it approximates a reversed Carnot cycle. In fact, replacing the expansion valve throttling process by an isentropic expansion would make the two cycles identical.

If the vapor entering the compressor is dry and saturated or superheated, as in Fig. 18·7, the cycle is said to involve *dry compression*. Notice that dry compression, especially with superheated vapor entering the compressor, makes part of the cycle resemble a reversed Brayton cycle and usually reduces the coefficient of performance for given evaporation and condensation temperatures. Nevertheless, dry compression is usually favored over wet compression because it results in higher compressor efficiency, higher compressor volumetric efficiency, and less danger of damage to the compressor caused by slugs of liquid entering it.

In a dry compression cycle, the temperature of the vapor during part of the compression exceeds the condensation temperature, so it is possible to cool the compressor somewhat by the same coolant used in the condenser. Doing so decreases the work required by the compressor.

For a given condensation temperature, the refrigerating capacity of a vapor-compression system is increased by subcooling the condensate before it reaches the expansion valve. In actual systems the incoming (coldest) coolant is sometimes used for this purpose before it is used for condensing the refrigerant.

Many factors must be considered in the selection of a refrigerant for a vapor-compression system. One of the most important properties is the saturation pressure-temperature relationship that establishes the operating pressure range for any particular application. Saturation curves for several refrigerants are shown in Fig. 18·8. Also, for given temperature limits the coefficient of performance that can be obtained varies markedly from one refrigerant to another. Tables showing the relative performance obtainable with various refrigerants are published in virtually all engineering handbooks that include sections on refrigeration.

Many modifications of the basic vapor-compression cycle are possible for the purpose of increasing the coefficient of performance. In an actual plant, various refinements are

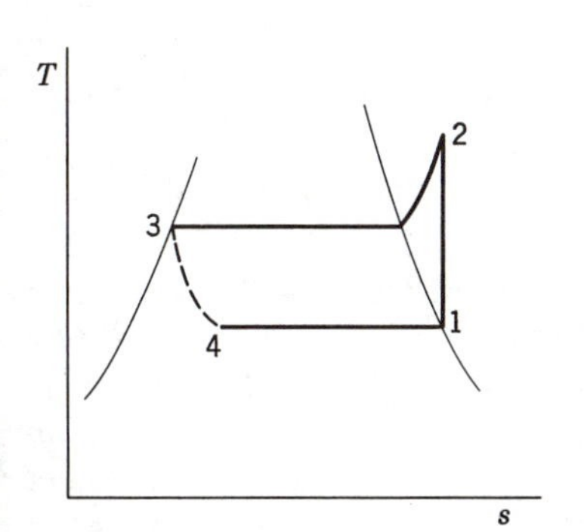

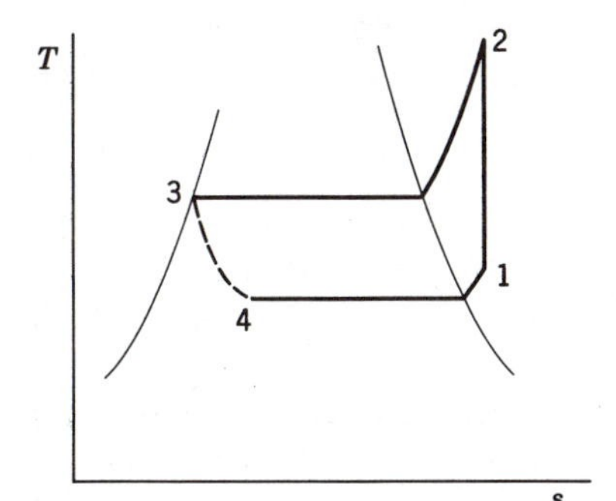

Figure 18·7. Dry compression cycles.

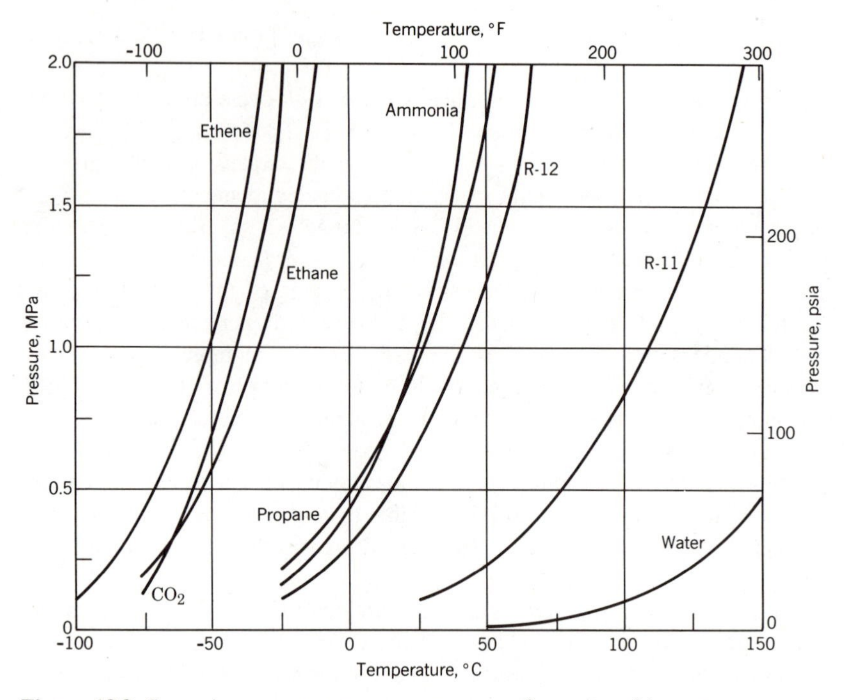

Figure 18·8. Saturation pressure–temperature curves for some refrigerants.

used only if the saving in power costs is greater than the additional costs incurred by adding the refinements. Details of these cycle variations can be found in the literature on refrigeration. Here, simply as an example of such a variation, we will describe the use of multistage compression with intercooling by means of a flash chamber.

For a fixed condensation temperature, lowering the evaporator temperature increases the compressor pressure ratio. For a reciprocating compressor, a high pressure ratio across a single stage means low volumetric efficiency. Also, with dry compression the high pressure ratio results in a high compressor discharge temperature that may damage the refrigerant. Multistage compression with intercooling is obviously called for, but effective intercooler temperatures may be well below the temperature of available cooling water that is used for the condenser. Therefore, several methods of using the refrigerant as an intercooling medium have been devised. One of these methods employs a flash chamber as shown in Fig. 18·9.

Liquid from the condenser in Fig. 18·9 expands through the first expansion valve into the flash chamber where the pressure is the same as the compressor interstage pressure. The refrigerant entering the flash chamber is a liquid-vapor mixture in state 6. The liquid fraction, which alone is in state 7 as shown on the *Ts* diagram, flows through the second expansion valve into the evaporator. The vapor from the flash chamber, which is in state 9, is mixed with the vapor (state 2) from the low-pressure compressor to form vapor in state 3 which enters the high-pressure compressor. For the same pressure ratio, a smaller work input to the compressor per unit mass of refrigerant is required with refrigerant entering in state 3 instead of in state 2. The ratio $\dot{m}_9/\dot{m}_7$ can be determined from an energy

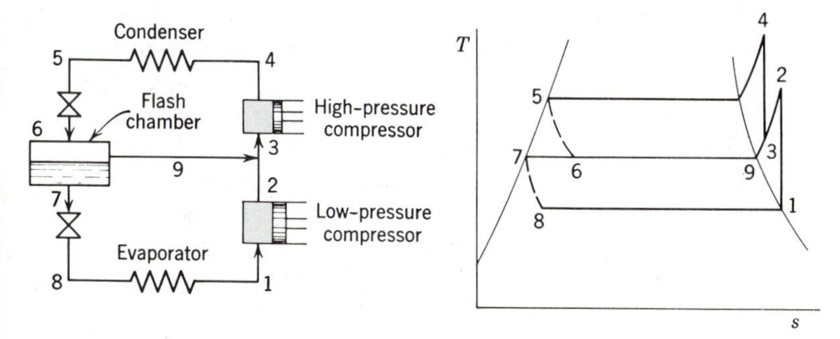

Figure 18·9. Two-stage vapor compression cycle with flash chamber intercooler.

balance on the flash chamber. State 3 can then be determined by an energy balance on the mixing point between the compressors.

For very wide temperature ranges, binary vapor cycles are used in which the condenser for the lower-temperature fluid is the evaporator for the higher-temperature fluid. Three- and four-fluid cycles have also been used.

Control systems for refrigerating units range from very complex ones used in large industrial installations to the simple on–off control of household refrigerators. One aspect of the control problem should be noted: The heat absorbed is (1) the product of the mass rate of flow through the evaporator and the specific enthalpy increase in the evaporator and also (2) proportional to the temperature difference between the evaporating refrigerant and the surroundings of the evaporator. Therefore, no one of these three variables ($\dot{m}$, Δh in the evaporator, and ΔT for heat transfer) can be changed without affecting the others.

One other refrigeration system that involves the compression of a vapor will be described even though it does not operate on the usual vapor-compression refrigeration cycle. This is the *steam-jet system* or *vacuum system*. For refrigeration at temperatures above 0°C, water is a satisfactory refrigerant. Its chief drawback is the high specific volume of water vapor at low temperatures which calls for large compressors. Centrifugal compressors have been used for this service, but steam-jet ejectors for the same flow rate and pressure ratio are cheaper and involve less maintenance expense, even though they use large quantities of steam. The application of a steam-jet ejector to a system for chilling water is shown in Fig. 18·10. Water to be chilled is sprayed into the flash chamber where

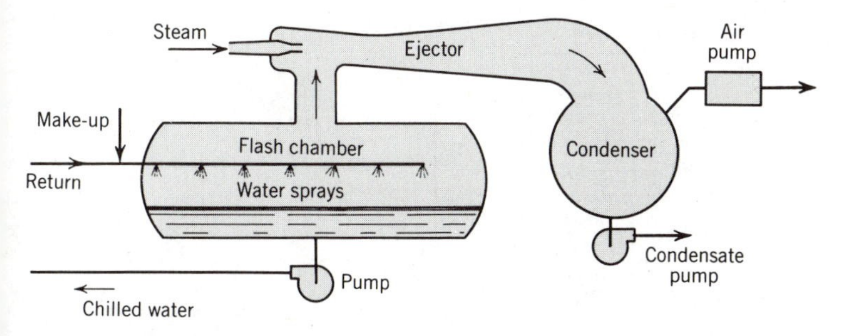

Figure 18·10. Steam-jet refrigeration system.

part of it evaporates. The pressure is kept low by the steam ejector which removes the vapor formed. The flash chamber is thermally insulated; so the latent heat of the spray water that evaporates is taken from the spray water that does not evaporate. Thus the liquid is chilled. Make-up water must be supplied continually because some of the spray water is carried away as vapor. The ejector discharges into a condenser where the pressure is determined by the temperature of available cooling water. This pressure is usually below atmospheric, so a pump or ejector must be used to remove air and other noncondensable gases from the condenser. The steam-jet or vacuum system is well-suited to air-conditioning applications where ample supplies of steam and condenser cooling water are available.

In connection with English units, a widely used unit of refrigeration capacity is the *ton of refrigeration,* which is defined as a heat absorption rate of 200 B/min. The name of this unit stems from the fact that this rate of heat removal will freeze approximately one ton of water at 32°F into ice at 32°F in 24 h. (It would be exactly one ton if the latent heat of fusion were 144 B/lbm.) Thus a 10-ton refrigerator can absorb 2000 B/min *when operating under design conditions.* This does not mean that the rate of heat absorption by this system at any instant actually is 2000 B/min, because many factors other than the system design affect the actual performance. (This is analogous to the fact that the power delivered by a 10-hp electric motor is not 10 hp just because the motor is running.)

18·4 Absorption Refrigeration

A vapor-compression refrigeration system requires a power input to compress the refrigerant vapor from the evaporator pressure to the condenser pressure. Pumping a liquid through the same pressure difference at the same mass flow rate requires less power, so it is worthwhile to seek means for having the refrigerant in a liquid phase as its pressure is raised. To condense the pure refrigerant at the evaporator pressure would defeat the purpose of the refrigeration system; however, a workable arrangement is the absorption refrigeration cycle in which the refrigerant is dissolved in a liquid at the evaporator pressure, the liquid is pumped to the condenser pressure, and the refrigerant is then separated at the condenser pressure from the liquid.

Figure 18·11 is a flow diagram of an elementary absorption cycle that uses ammonia

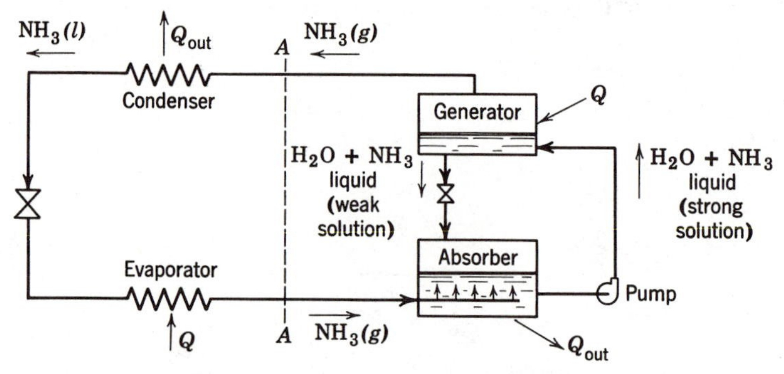

Figure 18·11. Elementary ammonia absorption refrigeration cycle.

as the refrigerant and water (or a water-ammonia mixture) as the liquid in which the refrigerant is carried from the low pressure to the high pressure. Ammonia enters the condenser as a vapor, is condensed, and enters the expansion valve as a liquid. It partially flashes to vapor when it expands through the valve, and it is further vaporized as it absorbs heat in the evaporator. Thus the part of the flow diagram to the left of line *A-A* in Fig. 18·11 is identical with that of a vapor-compression cycle. However, to the right of line *A-A* we see that in place of a compressor are an absorber, a pump, a generator, and a return line for liquid. Vapor from the evaporator is absorbed in water in the absorber. This dissolving process is exothermic, so heat must be removed from the absorber in order to keep its temperature constant. Also, at a given pressure the amount of ammonia that can be dissolved in water increases as the temperature is decreased; so the absorber temperature must be kept as low as possible by means of cooling water. The strong ammonia–water solution is then pumped to the generator, which is at the condenser pressure. Heat is added to the solution in the generator to drive some of the ammonia out of solution (an endothermic process). Ammonia vapor, or a vapor mixture that is very rich in ammonia, then goes to the condenser, and the weak ammonia–water solution left in the generator passes through a valve back to the absorber. The liquid that travels through the absorber–pump–generator–valve cycle is simply a transport medium for the refrigerant, carrying it from the evaporator pressure to the condenser pressure in the liquid phase.

An understanding of the operation of an absorption cycle requires some knowledge of the characteristics of binary mixtures such as ammonia and water. (See Chapter 19.) A skeleton *equilibrium diagram* for aqua ammonia, as the ammonia–water mixture is called, is shown in Fig. 18·12. At any pressure, the boiling temperature of aqua ammonia liquid depends on the concentration X which is the mass of ammonia per unit mass of mixture. When $X = 0$, the boiling temperature is the saturation temperature of water; when $X = 1.0$, it is the saturation temperature of ammonia at the specified pressure. An important fact is that, when liquid and vapor exist together in equilibrium, the concentration

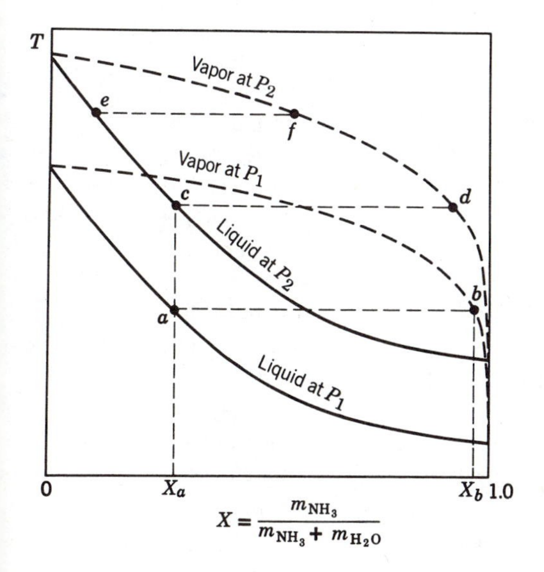

Figure 18·12. Aqua–ammonia equilibrium diagram.

is not the same in the two phases. In Fig. 18·12, the solid curves represent saturated liquid states at two pressures, P_1 and P_2; and the broken curves represent saturated vapor states at the same pressures. P_2 is greater than P_1. Points *a* and *b* represent states of saturated liquid and saturated vapor, respectively, which can exist together in equilibrium. Notice that the concentration of ammonia is higher in the vapor than in the liquid.

In an absorption cycle, saturated liquid in state *a* might leave the absorber and enter the pump. Leaving the pump, the concentration is the same, and the temperature is very nearly the same, but the pressure is P_2. In order to evaporate any of the liquid, its temperature must be raised to bring the liquid to state *c*. This is done in the generator. Then, as more heat is added, some of the liquid is evaporated to form vapor in state *d*. Since X_d is greater than X_c, the formation of the ammonia-rich vapor reduces the concentration of ammonia in the liquid. Evaporation then ceases if the pressure and temperature are held constant, because the liquid at a lower concentration is not at its boiling point. If more heat is added so that evaporation continues, the temperature of the two phases increases, and the states of the liquid and vapor change toward *e* and *f*, respectively. Liquid *e* flows from the generator back to the absorber, and vapor *f* flows to the condenser.

Equilibrium diagrams like Fig. 18·12 are adequate for mass balance analyses of absorption systems, but, for energy balance or first-law analyses, enthalpy data are also needed. These cannot be conveniently presented on a diagram like Fig. 18·12, so other property diagrams have been devised. One that relates P, T, h, X' (concentration in the liquid), and X'' (concentration in the vapor) for aqua ammonia is included in the appendix as Chart A·10.

Actual ammonia absorption systems always involve some modifications not shown in Fig. 18·11. The pump delivers cold liquid from the absorber to the generator where the temperature must be high, and hot liquid flows from the generator back to the absorber which must be kept cold; so a heat exchanger is always used to transfer heat from the weak solution to the strong solution. Also, carrying water vapor into the condenser must be avoided because it will freeze in the expansion valve and evaporator; therefore, a device called a *rectifier* is placed between the generator and the condenser. Its function is to remove traces of water from the refrigerant before it reaches the condenser.

Another pair of substances commonly used for absorption refrigeration systems is lithium bromide and water. An aqueous solution of lithium bromide is the carrier and water is the refrigerant; therefore, lithium bromide absorption systems are used only for applications where the minimum temperature is above 0°C.

The LiBr–H_2O absorption system is strikingly different from the aqua-ammonia system because LiBr has a very low vapor pressure; it is essentially nonvolatile. The water that is driven from the liquid solution of LiBr in water by heat addition in the generator is therefore nearly pure water, so its properties can be determined from the steam tables. Thus the property diagrams used for LiBr absorption systems (Charts A·11 and A·12) are simpler than for aqua-ammonia (Chart A·10). They give properties of only the liquid solution. The equilibrium chart (Chart A·11) does show the vapor pressure for pure water as a function of the refrigerant (water) temperature, and these values are directly from the steam tables for saturation states. The enthalpy of the liquid depends chiefly on temperature and concentration, so Chart A·12 can be used over a wide range of pressures. The crystallization line marks the maximum concentration that can be used at each temperature without causing crystallization of LiBr from the liquid solution.

The basic pieces of equipment—absorber, generator, condenser, and evaporator—and the expansion valve and liquid pump as shown in Fig. 18·11 are parts of any absorption cycle. Flow diagrams such as Fig. 18·11 are shown in elemental form, however, and the actual equipment may be arranged much differently. For example, most lithium bromide absorption refrigerators have all the major pieces of equipment contained within only one or two cylindrical shells. Such an arrangement has many advantages over connecting the equipment by external piping. Also, actual installations involve various pieces of auxiliary equipment that are essential to plant operation, although they have little effect on the thermodynamic analysis of the cycle. For example, absorption refrigerators usually involve two or more pumps instead of the one shown in elementary flow diagrams. You should refer to books on refrigeration for information on actual systems and their components.

Absorption cycles are generally used in industrial applications only where heat that would otherwise be wasted is available. The choice between a compression cycle and an absorption cycle depends on the relative costs of power and heat.

The advantage of an absorption cycle over a compression cycle is the large reduction in power input. An ingenious method of reducing to zero the required power input of an absorption cycle was devised by two undergraduates at the Royal Institute of Technology in Stockholm, Carl G. Munters and Baltzar von Platen. It is usually referred to as the Servel system. The basic feature is that the total or mixture pressure is constant throughout the system. Condensation of the refrigerant ammonia occurs where the ammonia exists alone under the system pressure; but evaporation occurs where the ammonia is mixed with hydrogen so that the ammonia behaves as though it existed alone at a pressure approximately (exactly, if it were an ideal gas) equal to its partial pressure in the mixture. Thus the ammonia is condensed at one temperature and evaporated at a lower one, even though the total pressure is the same in the condenser and the evaporator. Refrigeration is achieved without the use of a mechanical pump or compressor.

A highly simplified flow diagram of a Servel refrigerator is shown in Fig. 18·13. A water–ammonia solution flows down into the generator where heat is added to vaporize part of the mixture, the vapor formed being very rich in ammonia. The formation of vapor in the dome of the generator forces liquid up the tube toward the separator until the liquid level in the generator drops to the end of the tube. Then some vapor from the dome passes into the tube, the liquid level rises as more mixture enters the generator, and the process is repeated. The result is that alternate ammonia-rich bubbles and slugs of weak liquid solution flow up to the separator from which the ammonia passes to the condenser, and the water (or weak solution) flows by gravity to the absorber. In the condenser, heat is removed from the ammonia, which exists alone at the total pressure of the system, to condense it. The liquid drains from the condenser through a U-tube liquid seal or trap, which allows only liquid to pass, and into the evaporator. The total pressure is the same in both the condenser and the evaporator; so the liquid seal does not blow out. Hydrogen is present in the evaporator, however, so the ammonia passing through the liquid seal from the condenser evaporates at a low partial pressure and the corresponding low saturation temperature, thus absorbing heat from the region that is to be refrigerated. The

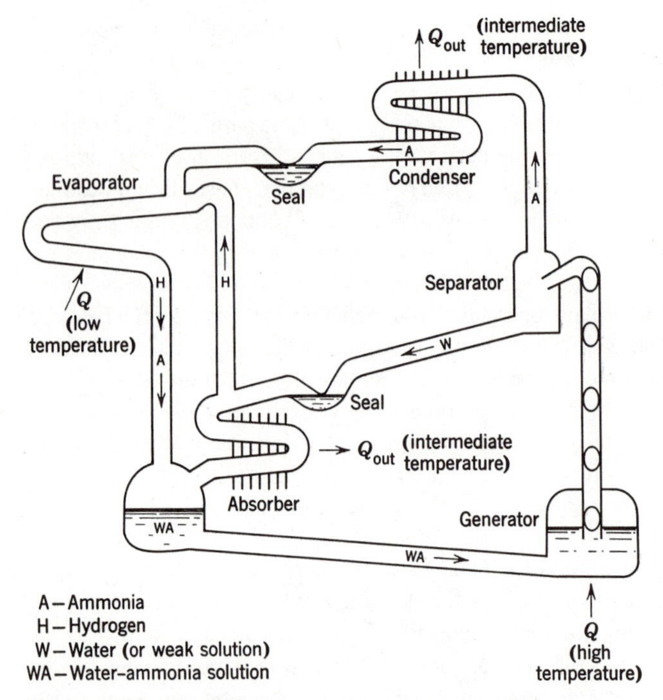

Figure 18·13. Simplified flow diagram of Servel refrigerator.

cold ammonia–hydrogen mixture flows down from the evaporator and back up through the absorber, where the ammonia is absorbed by water that is flowing from the separator through a liquid seal that prevents the entry of hydrogen into the separator and condenser. The ammonia is thereby carried back to the generator to complete its cycle and the hydrogen is left in the evaporator and absorber circuit. Heat must be removed from the absorber to keep its temperature constant, just as in the ammonia-absorption cycle discussed earlier.

In analyzing a Servel system, notice that there are three fluids, each of which flows in a different circuit. The refrigerant ammonia passes through the generator, separator, condenser, evaporator, and absorber, and back to the generator. The water is a transport medium that carries ammonia from the absorber, where the ammonia partial pressure is low, to the separator, where it is high. The function of the hydrogen is to circulate through the evaporator and absorber to hold down the ammonia partial pressure.

18·5 The Liquefaction of Gases

The liquefaction of a gas is an important step in most very low-temperature refrigeration systems, and also in the preparation of pure oxygen and nitrogen from air.

As pointed out in Sec. 9·3, a gas can be cooled by throttling if its Joule–Thomson coefficient, $(\partial T/\partial P)_h$, is positive, and this means that its initial temperature must be less than its maximum inversion temperature. Further information on the conditions necessary

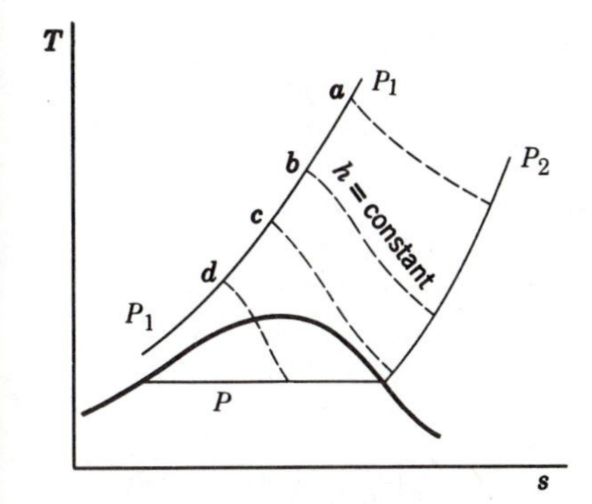

Figure 18·14. Throttling processes from gaseous states.

for gas liquefaction by throttling can be obtained from a *Ts* diagram such as Fig. 18·14. (A *Ts* diagram for air is included in the appendix as Chart A·1.) A throttling process between pressures P_1 and P_2 will not result in liquefaction if it begins at state *a, b,* or *c,* but it will if it starts from a state such as *d.*

Flow and *Ts* diagrams for an ideal Hampson–Linde gas liquefaction system are shown in Fig. 18·15. After multistage compression, the gas is cooled from state 2 to state 3 at constant pressure in an aftercooler, which is cooled either by water or by a refrigerating system. The gas is further cooled to state 4 in a regenerative heat exchanger, which is supplied with very cold gas from elsewhere in the cycle. After expansion through a throttle valve, the fluid is in the liquid-vapor mixture state 5 and is mechanically separated into liquid (state *a*) and vapor (state 6) parts. The liquid is drawn off as the desired product, and the vapor flows through the regenerative heat exchanger to cool high-pressure gas flowing toward the throttle valve. The area under line 6-7 on the *Ts* diagram equals that under 3-4 after adjustment for the smaller mass flow rate for process 6-7. The gas at state 7 is mixed with an amount of gas from outside equal to the amount of liquid removed, and this mixture in state 1 enters the compressor.

From a thermodynamic viewpoint, better performance of a gas liquefaction plant could be obtained by replacing the highly irreversible throttling process by expansion in an engine. The operation of an engine at very low temperatures presents difficulties,

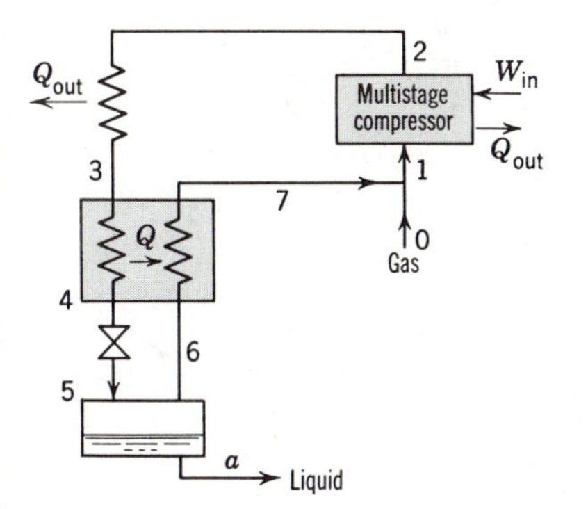

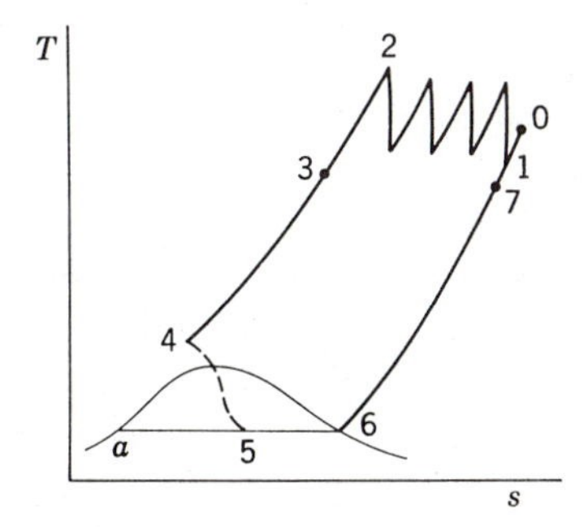

Figure 18·15. Hampson–Linde system for liquefying gases.

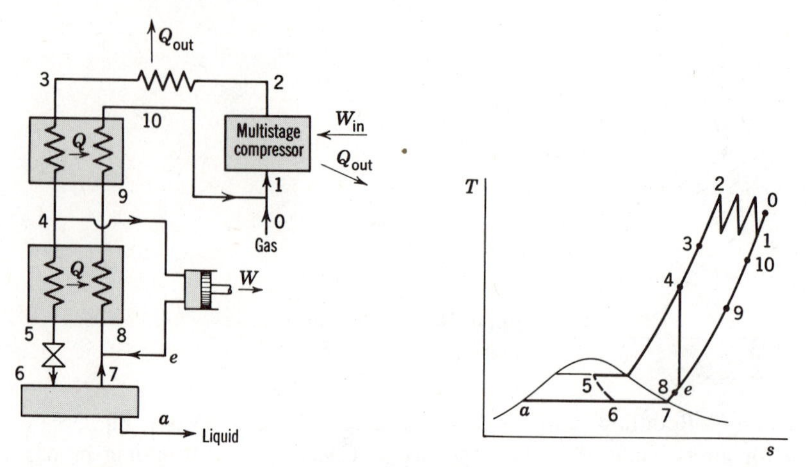

Figure 18·16. Claude system for liquefying gases.

however, especially if liquefaction occurs in the engine. A compromise solution is the Claude system for liquefying gases. Simplified flow and *Ts* diagrams are shown in Fig. 18·16. From state 1 to state 4 the Claude system processes are the same as those of the Hampson–Linde system. After the gas is cooled to state 4 by the compressor aftercooler followed by a regenerative heat exchanger, most of it is expanded through an engine and then is mixed with vapor from the separator and flows back toward the compressor through a heat exchanger, which precools appreciably the small fraction of the flow that is directed toward the throttle valve instead of the engine. Thus in the ideal Claude system the cooling of a gas by isentropic expansion is utilized but the presence of liquid in the engine is avoided. In studying the *Ts* diagram of Fig. 18·16, remember that the flow rates are different at points 3, 5, 7, and 9.

18·6 Summary

This chapter describes several refrigeration systems in sufficient detail so that you can analyze their performance by means of the principles introduced earlier.

The reversed Carnot cycle serves as a standard of performance for refrigerating cycles but it is unsuitable for actual systems.

The reversed Brayton cycle employs a gas as working fluid and is widely applied in aircraft cooling. Shortcomings of a gas as a refrigerant are its low specific heat, its low density at moderate pressures, and the difficulty of executing isothermal heat absorption and rejection processes that are necessary if the performance of any refrigerating system is to approach that of a reversed Carnot cycle.

Vapor-compression refrigeration systems employing many different refrigerants are widely used. In large commercial or industrial applications, the basic cycle is modified in various ways to improve performance. For air conditioning, water is sometimes used as the refrigerant in a variation of the basic vapor-compression cycle known as the steam-jet or vacuum system.

An absorption refrigeration system requires very little power input, but it takes large amounts of heat. Power input is saved by dissolving the refrigerant in a liquid at the evaporator pressure, pumping it in the liquid phase to the condenser pressure, and then driving the refrigerant out of solution. An ingenious method of reducing the work input of a small absorption system to zero is known as the Servel system. The total pressure throughout the system is constant, and the pressure of the refrigerant is varied by alternately mixing with it and separating from it an inert gas.

Several methods of producing very low temperatures or liquefying gases are based on the adiabatic expansion of a high-pressure gas either through a throttling valve or in an engine. In this chapter the Hampson-Linde and the Claude systems were described briefly.

Suggested Reading

18·1 *ASHRAE Handbook, 1985 Fundamentals,* American Society of Heating, Refrigerating, and Air-Conditioning Engineers, Atlanta, 1985, Chapters 1 and 17.

18·2 Black, William Z., and James G. Hartley, *Thermodynamics,* Harper & Row, New York, 1985, Chapter 8.

18·3 Burghardt, M. David, *Engineering Thermodynamics with Applications,* 2nd ed., Harper & Row, New York, 1982, Chapter 12.

18·4 Gosney, W. B., *Principles of Refrigeration,* Cambridge University Press, Cambridge, England, 1982.

18·5 McQuiston, Faye C., and Jerald D. Parker, *Heating, Ventilating, and Air Conditioning,* 2nd ed., Wiley, New York, Chapter 15.

18·6 Sorensen, Harry A., *Energy Conversion Systems,* Wiley, New York, 1983, Chapter 11.

18·7 Stoecker, W. F., and J. W. Jones, *Refrigeration and Air Conditioning,* 2nd ed., McGraw-Hill, New York, 1982.

18·8 Wood, Bernard D., *Applications of Thermodynamics,* 2nd ed., Addison-Wesley, Reading, Mass., 1982, Chapter 4.

Problems

18·1 Thirty-five kilojoules per second of refrigeration are to be provided by a reversed Carnot engine. The condenser temperature is 30°C, and the refrigeration temperature is −20°C. Compute the power required to operate the refrigeration machine and compute the coefficient of performance.

18·2 A dense air refrigerating machine operates between 350 and 1400 kPa. The air temperatures entering the compressor and expansion cylinder are 0 and 35°C, respectively. If the expansion is reversible adiabatic and the compression is polytropic with $n = 1.35$, determine (*a*) the power required for a capacity of 2000 kJ extracted from the cold room per minute, and (*b*) the coefficient of performance.

18·3 A reversed Brayton cycle operates between 700 and 1400 kPa. Air leaves the cold heat exchanger at 5°C and enters the air engine at 35°C. Refrigerating capacity is 3.0 kJ/s. Determine the coefficient of performance.

18·4 Air at 70 kPa, 5°C, is to be supplied to the cockpit of an aircraft at a rate of 0.1 kg/s by bleeding air at 350 kPa, 200°C, from the aircraft jet-engine compressor, cooling it in a heat exchanger, and expanding it isentropically through a turbine that exhausts into the cockpit. Disregarding friction and changes in kinetic energy, determine (*a*) the heat-transfer rate in the heat exchanger, and (*b*) the power output of the turbine.

18·5 For the cabin cooling system of an aircraft, air at 200 kPa, 115°C, is bled from the jet engine compressor at a rate of 0.80 kg/s, cooled in a heat exchanger to 60°C, and then expanded adiabatically through a turbine to 65 kPa. The turbine efficiency is 0.70. (*a*) Calculate the turbine power output. (*b*) Explain how effective the cooling system would be if the turbine were replaced by an expansion valve.

18·6 A house is to be supplied with 50 000 B/h by a heat-pump heating system that takes heat from the atmosphere at 20°F. The house temperature is to be 70°F. (*a*) What is the minimum power input to the heat pump? (*b*) Explain why this is the minimum power input.

18·7 Ten tons of refrigeration are to be supplied by a reversed Carnot engine. The condenser temperature is 85°F and the refrigeration temperature is −5°F. Compute the power required to operate the refrigeration machine and the coefficient of performance.

18·8 A dense air refrigerating machine operates between 50 and 200 psia. The air temperatures entering the compressor and expansion cylinder are 33 and 96°F, respectively. If the expansion is reversible and adiabatic and the compression is polytropic with $n = 1.35$, determine (*a*) the power required for a capacity of 2000 B extracted from the cold room per minute, and (*b*) the coefficient of performance.

18·9 A reversed Brayton cycle operates between 100 and 200 psia. Air leaves the cold heat exchanger at 40°F and enters the air engine at 100°F. Refrigerating capacity is one ton. Determine the coefficient of performance.

18·10 Air at 10 psia, 40°F, is to be supplied to the cockpit of an aircraft at a rate of 0.2 lbm/s by bleeding air at 50 psia, 400°F, from the aircraft jet-engine compressor, cooling it in a heat exchanger, and expanding it isentropically through a turbine which exhausts into the cockpit. Disregarding friction and changes in kinetic energy, determine (*a*) the heat-transfer rate in the heat exchanger, and (*b*) the power output of the turbine.

18·11 For an aircraft cabin cooling system, a turbine is to exhaust air at 10 psia, 40°F, at a rate of 4.0 lbm/s. The turbine drives a compressor that takes in air at 10 psia, 120°F. The air discharged from this compressor is mixed with air at the same state bled from the main jet-engine compressor. It is then cooled to 140°F before entering the cooling-system turbine. Assume that compression and expansion are reversible and adiabatic. Disregard pressure drops in the heat exchanger and ducting and also changes in kinetic energy. Determine the pressure and the flow rate of air to be bled from the main engine compressor.

18·12 Explain how a reversed Brayton cycle can be modified to approach a reversed Carnot cycle by means of multistage compression with intercooling and also multistage expansion. Sketch flow, Pv, and Ts diagrams for the modified cycle.

18·13 Comment on the advisability of using a reversed regenerative gas-turbine cycle for refrigeration.

18·14 An ideal vapor-compression refrigerating cycle operates between temperature limits of 5°C and 35°C. Dry saturated vapor leaves the compressor and saturated liquid enters the expansion valve. For a refrigerating capacity of 20 kJ/s, calculate the required power input if the refrigerant is (*a*) ammonia, (*b*) water, (*c*) refrigerant 12, (*d*) propane.

18·15 Solve problem 18·14 with an isentropic expansion in an engine replacing the throttling through the expansion valve.

18·16 In a vapor-compression refrigeration system the condenser temperature is 25°C and the evaporator temperature is −10°C. Saturated liquid enters the expansion valve, and dry saturated vapor enters the compressor which operates reversibly and adiabatically. For a refrigeration effect of 3.5 kJ/s, determine the flow rate and the power input if the refrigerant is (*a*) ammonia, (*b*) refrigerant 12.

18·17 A refrigerating plant uses 6450 kg/h of cooling water in the condenser. The average inlet and outlet water temperatures are 13 and 27°C, respectively. The power required to drive the compressor is 15 kW. Compression is adiabatic. Calculate the coefficient of performance and the capacity of the plant in kilojoules per second of refrigeration.

18·18 For a condensing temperature of 20°C and adiabatic dry compression, plot the minimum compressor discharge temperature against evaporator temperature for a range of −15 to −50°C of the latter for (*a*) ammonia, (*b*) refrigerant 12. What conclusions do you draw from these plots?

18·19 An ideal refrigerant-12 compression refrigerating cycle operates between 80.7 and 700 kPa. Dry saturated vapor enters the compressor, and R-12 at 700 kPa, 15°C, enters the expansion valve. Compression is adiabatic. The refrigerating capacity is 35 kJ/s. Determine the power input.

18·20 A 17-kJ/s refrigerant-12 compression refrigeration machine operates at rated capacity with pressures in the evaporator and condenser of 261 and 652 kPa, respectively. Dry saturated vapor enters the compressor, and vapor leaving thc compressor is at 35°C. Heat removed from the refrigerant 12 being compressed amounts to 2.0 kJ/kg. Liquid entering the expansion valve is at 20°C. Determine the power input and the irreversibility rate.

18·21 In an ammonia compression refrigeration cycle, the evaporator and condenser pressures are 140 and 1240 kPa, respectively. Dry saturated vapor enters the compressor, and vapor at 135°C leaves. Liquid at 25°C leaves the condenser. When the refrigerating effect is 70 kJ/s, cooling water passing through the compressor cylinder water jacket picks up heat at a rate of 270 kJ/min. (*a*) Calculate the power input. (*b*) Is the compression process reversible? Prove your answer.

18·22 In an ammonia compression refrigeration cycle, saturated liquid leaves the condenser at 1000 kPa and dry saturated vapor leaves the evaporator at 140 kPa. These fluids then pass through a heat exchanger in which the vapor is superheated to −5°C while the liquid is subcooled before entering the expansion valve. Compression is adiabatic, and the compressor discharge temperature is 160°C. The refrigerating effect is 175 kJ/s. Calculate the power input.

18·23 A portable emergency refrigerating unit consists of a bottle containing 9.0 kg of liquid ammonia and a coil of tubing that is fed from the bottle through a small valve that prevents the pressure in the tubing from exceeding 140 kPa. The volume of the tubing is negligible compared to the volume of the bottle. The tubing discharges to the atmosphere through a thermostatic valve that is closed only when the temperature of ammonia at the discharge end of the tubing is less than 0°C. Determine the maximum amount of heat that can be absorbed by completely discharging this unit if the ammonia in the bottle is initially at 10°C. (Some assumptions must be made. State them clearly.)

18·24 Refer to Fig. 18·9. Determine the power input to an ideal refrigerant-12 refrigerating system that operates on the cycle shown with condenser, flash chamber, and evaporator pressures of 700, 261, and 80.7 kPa, respectively. Saturated liquid enters each expansion valve, and the refrigerating effect is 70 kJ/s.

18·25 Solve problem 18·24 for the same capacity and the same condenser, flash chamber, and evaporator temperatures if the refrigerant is ammonia.

18·26 An ideal vapor-compression refrigerating cycle operates between temperature limits of 40 and 100°F. Dry saturated vapor leaves the compressor and saturated liquid enters the expansion valve. For a refrigerating capacity of 5 tons, calculate the required power input if the refrigerant is (*a*) ammonia, (*b*) water, (*c*) refrigerant 12, (*d*) propane.

18·27 Solve problem 18·26 with an isentropic expansion in an engine replacing the throttling through the expansion valve.

18·28 In a vapor-compression refrigeration system the condenser temperature is 80°F and the evaporator temperature is 10°F. Saturated liquid enters the expansion valve, and dry saturated vapor enters the compressor which operates reversibly and adiabatically. For a refrigeration effect of one ton, determine the flow rate and the power input if the refrigerant is (*a*) ammonia, (*b*) refrigerant 12.

18·29 A refrigerating plant uses 14 220 lbm/h of cooling water in the condenser. The average inlet and outlet water temperatures are 55 and 81°F, respectively. The power required to drive the compressor is 20 hp. Compression is adiabatic. Calculate the coefficient of performance and the capacity of the plant in tons of refrigeration as indicated by these figures.

18·30 For a condensing temperature of 70°F and adiabatic dry compression, plot the minimum compressor discharge temperature against evaporator temperature for a range of 10 to −60°F of the latter for (*a*) ammonia, (*b*) refrigerant 12. What conclusions do you draw from these plots?

18·31 An ideal refrigerant-12 compression refrigerating cycle operates between 12 and 100 psia. Dry saturated vapor enters the compressor, and R-12 at 100 psia, 60°F, enters the expansion valve. Compression is adiabatic. The refrigerating capacity is 10 tons. Determine the power input.

18·32 A 5-ton refrigerant-12 compression refrigeration machine operates at rated capacity with pressures in the evaporator and condenser of 35.7 and 90 psia, respectively. Dry saturated vapor enters the compressor, and vapor leaving the compressor is at 100°F. Heat removed from the R-12 being compressed amounts to 1.0 B/lbm. Liquid entering the expansion valve is at 70°F. Determine the power input in hp and the irreversibility rate in B/s.

18·33 In an ammonia compression refrigeration cycle, the evaporator and condenser pressures are 20 and 180 psia, respectively. Dry saturated vapor enters the compressor, and vapor at 280°F leaves. Liquid at 75°F leaves the condenser. When the refrigerating effect is 20 tons, cooling water passing through the compressor cylinder water jacket picks up heat at a rate of 250 B/min. (*a*) Calculate the power input. (*b*) Is the compression process reversible? Prove your answer.

18·34 In an ammonia compression refrigeration cycle, saturated liquid leaves the condenser at 140 psia, and dry saturated vapor leaves the evaporator at 20 psia. These fluids then pass through a heat exchanger in which the vapor is superheated to 20°F while the liquid is subcooled before entering the expansion valve. Compression is adiabatic, and the compressor discharge temperature is 320°F. The refrigerating effect is 50 tons. Calculate the power input.

18·35 A portable emergency refrigerating unit consists of a bottle containing 20 lbm of liquid ammonia and a coil of tubing that is fed from the bottle through a small valve that prevents the pressure in the tubing from exceeding 20 psia. The volume of the tubing is negligible compared to the volume of the bottle. The tubing discharges to the atmosphere through a thermostatic valve that is closed only when the temperature of ammonia at the discharge end of the tubing is less than 30°F. Determine the maximum amount of heat that can be absorbed by completely discharging this unit if the ammonia in the bottle is initially at 50°F. (Some assumptions must be made. State them clearly.)

18·36 Refer to Fig. 18·9. Determine the power input to an ideal refrigerant 12 refrigerating system that operates on the cycle shown with condenser, flash chamber, and evaporator pressures of 100,

35.7, and 12 psia, respectively. Saturated liquid enters each expansion valve, and the refrigerating effect is 20 tons.

18·37 Solve problem 18·36 for the same capacity and the same condenser, flash chamber, and evaporator *temperatures* if the refrigerant is ammonia.

18·38 Consider an ideal vapor compression refrigeration cycle with two-stage compression and interstage cooling by means of a flash chamber as shown in Fig. 18·9. The refrigerant is ammonia. Evaporator pressure is 10 psia, interstage pressure is 30 psia, and condenser pressure is 150 psia. Saturated liquid leaves the condenser and saturated vapor leaves the evaporator. Both compressions are isentropic. For a refrigerating capacity of 50 tons, determine the total power input to the cycle.

18·39 Derive the relationship between coefficient of performance and horsepower per ton of refrigeration. For refrigerating machines used in ice making and food preservation, a rule of thumb for the power required is 1 hp/ton. What is the corresponding coefficient of performance?

18·40 Calculate the slope of an isentropic line on a pressure-enthalpy diagram of a pure substance. Check your result qualitatively by the ammonia *Ph* chart in the appendix, Chart A·8.

18·41 A reciprocating ammonia compressor used in a refrigerating system operates normally between pressures of 10 and 180 psia (69 and 1240 kPa). It is found that, when the suction pressure rises to 20 psia (138 kPa) while the discharge pressure remains 180 psia (1240 kPa), the overload protection device on the electric motor driving the compressor disconnects the motor from the power line. How do you account for this overloading of the driving motor when the compressor pressure ratio has been reduced?

18·42 A household electric refrigerator is operated in a kitchen that is closed and thermally insulated. If it is operated continuously for 2 h, will the average temperature of the air in the kitchen increase, decrease, or remain constant if the refrigerator door is kept (*a*) open, (*b*) closed?

18·43 Determine the makeup water flow rate for a system such as the one shown in Fig. 18·10 if 23 000 kg of water per hour at 7°C is to be delivered, the return water is at 20°C, and the makeup water is at 13°C. Assume that the ejector removes only vapor.

18·44 Refer to problem 18·43 and Fig. 18·10. If the ejector is replaced by a centrifugal compressor and cooling water is available to maintain a condenser temperature of 18°C, determine the minimum power input to the compressor if it operates adiabatically.

18·45 Determine the makeup water flow rate for a system such as the one shown in Fig. 18·10 if 50 000 lbm of water per hour at 45°F is to be delivered, the return water is at 70°F, and the makeup water is at 55°F. Assume that the ejector removes only vapor.

18·46 Refer to problem 18·45 and Fig. 18·10. If the ejector is replaced by a centrifugal compresssor and cooling water is available to maintain a condenser temperature of 65°F, determine the minimum power input to the compressor if it operates adiabatically.

18·47 In Sec. 18·4 it is stated that pumping a liquid requires less power input than pumping a vapor through the same pressure ratio and at the same mass flow rate. Explain why this is true.

18·48 In Sec. 18·4 it is stated that at a given pressure the amount of ammonia that can be dissolved in water increases as the temperature decreases. Explain how this fact is related to (*a*) the observation that drinking water that has been boiled tastes "flat" and (*b*) what one observes if a glass of ice water is allowed to stand until it reaches room temperature.

18·49 The Servel refrigeration system causes heat to be transferred from one body to another one at a higher temperature with zero work input. Explain any possible conflict with the Clausius statement of the second law.

18·50 Aqua-ammonia liquid that is 50 percent ammonia by mass enters a generator of an absorption refrigeration system at 2.0 MPa, 20°C. Saturated liquid and saturated vapor, both at 110°C, leave. For steady flow, determine the mass of vapor leaving per kilogram of liquid entering and the amount of heat added per kilogram of liquid entering.

18·51 Refer to Fig. 18·11. For pressure limits of 1400 kPa and 400 kPa, a generator temperature of 100°C, and an absorber temperature of 20°C, determine (*a*) the amount of liquid that must be pumped per kilogram of ammonia entering the condenser and (*b*) the amount of heat added to the generator per kilogram of ammonia entering the condenser.

18·52 A simple ideal ammonia absorption refrigeration system has condenser and evaporator pressures of 1.40 and 0.30 MPa. Saturated liquid leaves the condenser and saturated vapor leaves the evaporator. Generator and absorber temperatures are maintained at 80 and 20°C. For a refrigerating effect of 12.0 MJ/h, determine (*a*) the rate of heat supply to the generator, (*b*) the rate of heat removal from the condenser, and (*c*) the rate of heat removal from the absorber.

18·53 Aqua-ammonia liquid that is 50 percent ammonia by mass enters a generator of an absorption refrigeration system at 300 psia, 80°F. Saturated liquid and saturated vapor, both at 210°F, leave. For steady flow, determine the mass of vapor leaving per pound of liquid entering and the amount of heat added per pound of liquid entering.

18·54 Refer to Fig. 18·11. For pressure limits of 215 and 55 psia, a generator temperature of 200°F, and an absorber temperature of 80°F, determine (*a*) the amount of liquid that must be pumped per pound of ammonia entering the condenser and (*b*) the amount of heat added to the generator per pound of ammonia entering the condenser.

18·55 A simple ideal ammonia absorption refrigeration system has condenser and evaporator pressures of 200 and 40 psia. Saturated liquid leaves the condenser and saturated vapor leaves the evaporator. Generator and absorber temperatures are maintained at 180 and 70°F. For a refrigerating effect of 80 tons, determine (*a*) the rate of heat supply to the generator, (*b*) the rate of heat removal from the condenser, and (*c*) the rate of heat removal from the absorber.

18·56 The flow diagram of a lithium bromide-water absorption refrigeration system is shown. Evaporator and condenser temperatures are 5 and 45°C, respectively. Saturated liquid leaves the condenser, and saturated vapor leaves the evaporator. The solution leaves the absorber at 45°C and enters the generator at 80°C. The solution leaving the generator is at 100°C. The refrigerating effect is 400 kJ/s. Dry saturated steam at 120 kPa enters the generator heating coils, and subcooled liquid water at 95°C leaves. Cooling water is supplied at 12°C and should undergo a temperature rise not greater than 15 Celsius degrees. (*a*) Sketch the states 1, 2, 3, 4, 5, and 6 and the processes connecting them on diagrams like Charts A·11 and A·12. (*b*) Determine the flow rate through the pump.

18·57 For the conditions of Problem 18·56 determine the required flow rate of (*a*) heating steam to the generator, (*b*) cooling water to the condenser, and (*c*) cooling water to the absorber.

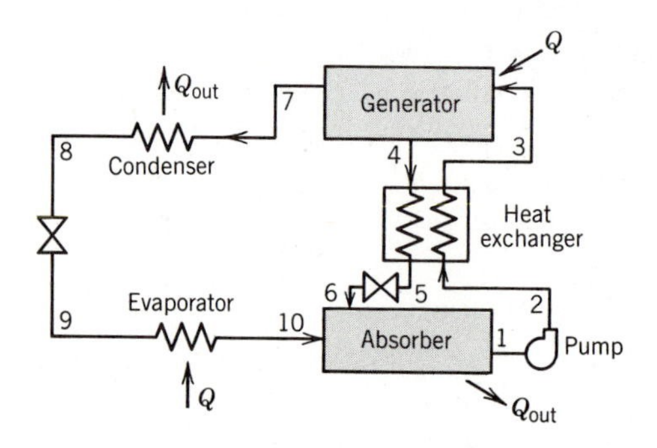

Problems 18·56 through 18·61.

18·58 In a lithium bromide-water absorption refrigeration system, the evaporator and condenser temperatures are 5 and 50°C, respectively. Condensate leaves the condenser at 45°C, and vapor leaves the evaporator at 8°C. The solution leaves the absorber at 30°C and enters the generator at 75°C. Solution leaving the generator is at 95°C. The refrigerating effect is 250 kJ/s. Dry saturated steam at 125 kPa enters the generator heating coils, and subcooled liquid water at 95°C leaves. Cooling water is supplied at 15°C and should undergo a temperature rise not greater than 10 Celsius degrees. Determine the required flow rate of (*a*) heating steam to the generator, (*b*) cooling water to the condenser, and (*c*) cooling water to the absorber.

18·59 The flow diagram of a lithium bromide–water absorption refrigeration system is shown. Evaporator and condenser temperatures are 40 and 115°F, respectively. Saturated liquid leaves the condenser, and saturated vapor leaves the evaporator. The solution leaves the absorber at 115°F and enters the generator at 175°F. The solution leaving the generator is at 210°F. The refrigerating effect is 100 tons. Dry saturated steam at 17 psia enters the generator heating coils, and subcooled liquid water at 200°F leaves. Cooling water is supplied at 45°F and should undergo a temperature rise not greater than 20 Fahrenheit degrees. (*a*) Sketch the states 1, 2, 3, 4, 5, and 6 and the processes connecting them on diagrams like Charts A·11 and A·12. (*b*) Determine the flow rate through the pump.

18·60 For the conditions of problem 18·59, determine the required flow rate of (*a*) heating steam to the generator, (*b*) cooling water to the condenser, and (*c*) cooling water to the absorber.

18·61 In a lithium bromide-water absorption refrigeration system, the evaporator and condenser temperatures are 40 and 125°F, respectively. Condensate leaves the condenser at 110°F, and vapor leaves the evaporator at 45°F. The solution leaves the absorber at 85°F and enters the generator at 165°F. Solution leaving the generator is at 200°F. The refrigerating effect is 60 tons. Dry saturated steam at 20 psia enters the generator heating coils, and subcooled liquid water is 210°F leaves. Cooling water is supplied at 55°F and should undergo a temperature rise not greater than 15 Fahrenheit degrees. Determine the required flow rate of (*a*) the heating steam to the generator, (*b*) cooling water to the condenser, and (*c*) cooling water to the absorber.

18·62 What are the principal advantages and disadvantages of a lithium-bromide absorption system in comparison with an ammonia absorption system? Why is no rectifier used in a lithium-bromide system?

18·63 Air is to be liquefied by the Hampson-Linde regenerative process (Fig. 18·15) operating with pressure limits of 200 and 1 atm. It is desired to obtain 1 mass unit of liquid from each 6 mass units of air flowing through the expansion valve. Assume that the regenerative heat exchanger is perfectly insulated from its surroundings and that the minimum temperature differential between the two streams of air passing through this heat exchanger is to be 10°C (18°F). Neglect pressure drops in heat exchangers. In order to maintain steady conditions, to what temperature must the air be cooled by the compressor aftercooler before it enters the regenerative heat exchanger?

CHAPTER NINETEEN

Binary Mixtures

Phase relationships for a pure substance were discussed in Chapter 3, and in Chapter 9 pure substances were treated further by the application of the first and second laws. Chapter 11 treated gas and gas-vapor mixtures, but such mixtures always form a single phase, and, as long as the chemical composition is constant, such mixtures are pure substances, regardless of the number of chemical species present. In the humidification or dehumidification of atmospheric air, the gaseous phase is not a pure substance, because the amount of water vapor in it changes; however, this is still a simple case because only two phases are present, and the liquid phase is a pure substance consisting of only the single substance water. Thus the only multisubstance phase relationships considered so far in this textbook have been for a rather elementary case.

This chapter extends Chapter 9 by treating phase relationships for binary mixtures—mixtures of two substances, where both may exist in more than one phase. Extensive applications of the first and second laws are not considered.

Before taking up the phase relationships for binary mixtures, we discuss briefly the *phase rule* which is a valuable generalization that pertains to many systems in addition to binary mixtures.

19·1 The Phase Rule

The phase rule, which was first published by J. Willard Gibbs in 1876, is a simple, powerful rule for determining the maximum number of independent intensive properties of various types of systems in equilibrium. This number of independent intensive properties is necessary and sufficient to specify the state of each phase in the system, but it is usually insufficient to specify the amount of each phase present. The number of independent intensive properties is also called the number of *degrees of freedom* of the system, so the symbol F is used for it. It is also called the *variance*.

Although the phase rule can be simply stated, its use requires a thorough understanding of some definitions and of elementary chemistry. Let us review the pertinent definitions.

Two or more properties are independent if any one can be varied while any other one is held constant. (For example, for an ideal gas T and P are independent properties, h and T are not. For a liquid–vapor mixture of a pure substance, T and P are not independent properties.) For a multiphase system, a property is an *intensive property of the system* if it is an intensive property of any phase of the system. (Examples are pressure, temperature, and the concentration of any one component in one phase.) As defined in Chapter 3, a phase is any homogeneous part of a system that is physically distinct and is separated from other parts of the system by distinct bounding surfaces. If all the intensive properties of one part of a system are identical with those of another part, the two parts comprise a single phase, even though they may be physically separated. For example, if many drops of water are present in condensing steam, all of the drops constitute a single liquid phase, even though they are separated from each other.

The *components* of a system are the *independently variable* substances (that is, chemical species) from which the system in any of its states can be prepared. The *number of components* is the least number of independently variable substances in terms of which the composition of any phase can be specified.

Determining the number of components in a given system requires careful analysis. The number of components is not in all cases equal to the number of different substances present. Several systems will be considered in order to demonstrate how the number of components may be ascertained.

A system comprised of ice, liquid water, and water vapor

For this system the number of components in equilibrium is only one; that is, the chemical substance *water*. Hydrogen and oxygen are not regarded as components since they are combined in definite proportions, and cannot be varied independently. A variation in the amount of hydrogen requires a definite variation in the quantity of oxygen.

A system composed of a saturated solution of water, NaCl, and excess NaCl

The only phases in this system are the saturated solutions of NaCl and water and solid NaCl. Each phase is completely described in terms of either one or both of the components H_2O and NaCl. There are therefore two components.

A system consisting of Glauber's salt, $Na_2SO_4 \cdot 10H_2O$, with solution and vapor

The composition of each phase may be expressed in terms of Na_2SO_4 and H_2O as follows:

Phase	Relations Specifying Each Phase in Terms of the Components H_2O and Na_2SO_4
$Na_2SO_4 \cdot 10H_2O$ (solid)	$Na_2SO_4 + 10H_2O$
Saturated solution	$Na_2SO_4 + xH_2O$
Vapor	$0Na_2SO_4 + H_2O$

Since two components completely specify the phases, the system may be classified as a two-component system.

A system in equilibrium consisting of calcium carbonate, $CaCO_3$; calcium oxide, CaO; and carbon dioxide, CO_2

Under equilibrium conditions three different substances are present; however, they are not all regarded as components for they are not mutually independent. The three substances are related by the following equilibrium reaction:

$$CaCO_3 \rightleftharpoons CaO + CO_2$$

If any two of the substances are specified, the third is fixed according to this relation. In order to be considered as components, the various substances must be capable of independent variation. The number of components of this system is the number of *independently variable* substances (molecular species) in the system. According to this chemical reaction, this number is two. Any two of the three substances may be selected as the components. Choosing CaO and CO_2 as the two components, the composition of each phase may be specified as follows:

Phase	Equation Specifying Each Phase in Terms of the Components CaO and CO_2
$CaCO_3$ (solid)	$CaCO_3 = CaO + CO_2$
CaO (solid)	$CaO = CaO + 0CO_2$
CO_2 (gas)	$CO_2 = 0CaO + CO_2$

From the two terms on the right side of the equations it is apparent that each phase may be expressed in terms of one or the other or both of the components CaO and CO_2.

These examples of determining the number of components are by no means exhaustive. For more complex systems, counting the components may be difficult.

For any system which is (1) homogeneous or comprised of a finite number of homogeneous parts in contact, (2) in equilibrium, and (3) free of electric, magnetic, gravitational, solid distortion, and surface tension effects, the *phase rule* states that the maximum number of independent intensive properties F is given by

$$F = 2 + C - P \tag{19·1}$$

where C is the number of components and P is the number of phases in the system.

Notice that we present the phase rule here without derivation. Plausible derivations can be based on little more than the principles treated in this textbook; a rigorous derivation requires a more extensive background.

A system for which there are no independent intensive properties is said to be *invariant*; one with one independent property is called *univariant*; one with two is called *divariant*; and so forth. Now we shall apply the phase rule to determine the *variance,* or the number of independent intensive properties, of several systems. In each case the system is in equilibrium and meets the other conditions necessary for systems to which the phase rule applies.

A system composed of ice, liquid water, and water vapor

In this system there are three phases and one component, H_2O. Application of the phase rule gives

$$F = 2 + C - P = 2 + 1 - 3 = 0$$

This result indicates that there are no independent variables; that is, neither the pressure nor the temperature can be varied as long as the three phases exist together in equilibrium. The system is at the triple state, which is the only state in which these three phases can exist together. This system is said to be invariant.

A system composed of liquid water and water vapor
There are two phases and one component, so the phase rule gives $F = 1$. The system is univariant, or, in other words, there is one independent intensive property. If either pressure or temperature is specified, then the state of the system is determined, and this is in accordance with the conclusions reached in Chapter 3 from experiments. The phase rule is not concerned with the amount of each phase present; so it might be said more precisely that the *intensive* state of this system is determined by a single intensive property.

A system composed of water vapor
This system is comprised of one phase and one component, so the phase rule indicates that it is divariant, or $F = 2$. This is in accordance with the observation in Chapter 3 that two properties must be specified to determine the state of superheated steam. Superheated vapor tables are therefore double-argument tables, while tables of saturation properties have a single argument.

A system composed of ammonia and water in a liquid solution and a vapor
There are two phases, liquid and vapor. There are two components, NH_3 and H_2O (or two components can also be taken as NH_4OH and H_2O in the liquid phase). The phase rule shows that the system is divariant, $F = 2$. This means that specifying the pressure and temperature of this system completely determines the state of the system and hence the composition of each phase. If the composition of one phase and the pressure are specified, then there is only one possible temperature under which the two phases can exist together, and the composition of the other phase is fixed.

A system composed of liquid mercury, liquid water, and vapor
The water and mercury are immiscible and so form two liquid phases. These plus the vapor make a total of three phases. There are two components, Hg and H_2O. The phase rule shows that $F = 1$, so the system is univariant. This means that, at a given pressure, there is only one temperature at which these three phases can coexist, and the composition of the vapor is fixed under those conditions. Another conclusion is that, in order to have a mercury vapor and water vapor mixture of specified composition coexist with liquid water and liquid mercury, there is no choice of either pressure or temperature of the system.

A system comprised of three gases: phosphorus pentachloride, PCl_5; phosphorus trichloride, PCl_3; and chlorine, Cl_2.
There is one phase because any gaseous mixture forms only one phase. There are two components because the equilibrium equation for these three gases

$$PCl_5 \rightleftharpoons PCl_3 + Cl_2$$

shows that the composition of the single phase can be expressed in terms of any two of the three substances. The phase rule gives $F = 3$, so pressure, temperature, and composition must all be specified to determine the state of this system.

19·2 Miscibility

If two liquids when mixed in any concentration form a single phase, they are said to be *miscible*. If when mixed they remain in two phases, each identical in composition with one of the liquids before mixing, they are *immiscible*. If the mixing results in a single phase for some concentrations and two phases for the others, the liquids are said to be *partially miscible*. The miscibility of two liquids may change markedly with temperature.

Figure 19·1 is a temperature-composition diagram or phase diagram for two liquids, A and B, that are miscible at temperatures above T_a and partially miscible at lower temperatures. The diagram is for a fixed pressure. Any point such as b, c, or d in the single-phase region represents a solution of A and B. For two components in a single phase, the phase rule gives us $F = 2 + C - P = 2 + 2 - 1 = 3$, so pressure, temperature, and composition must all be specified to define a state such as b, c, or d. A point such as k does not represent a single-phase solution but rather a mixture of two phases. Point k represents a mixture of a phase in state e and a phase in state f. The proportion of the two phases present is such that the overall composition of the system is represented by point k.

Pure B at temperature T_e is represented by point j in Fig. 19·1. If A is gradually added to B, the state point on the diagram moves to the right and represents a single-phase solution until state e is reached. Further addition of A causes the formation of a second phase that is in state f. As more of pure A is added to the system, the total mass in state f increases and that in state e decreases. The mixture of the two phases represented by point g is composed of about one part in state f and nine parts in state e. The mixture represented by point k is about six parts in state f and four parts in state e. When the state point, which has passed from j through e, g, and k reaches f, the last trace of the phase in state e disappears, and further addition of pure A increases the concentration of A in a single phase represented by state points between f and m. For mixtures represented by points such as g and k, notice that $F = 2 + C - P = 2 + 2 - 2 = 2$, so there are

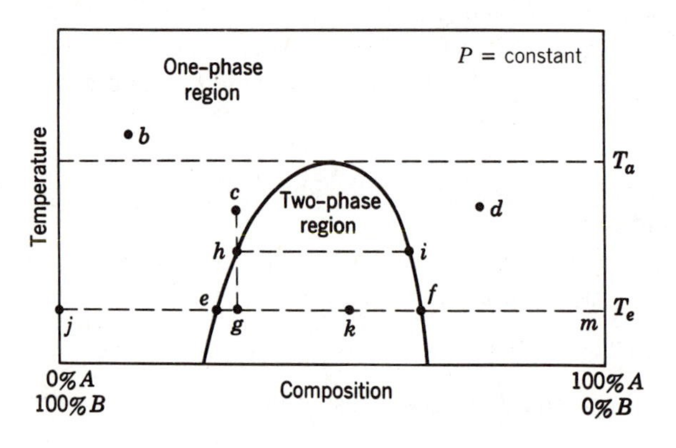

Figure 19·1. Phase diagram for two liquids that are miscible at temperatures higher than T_a and partially miscible at lower temperatures.

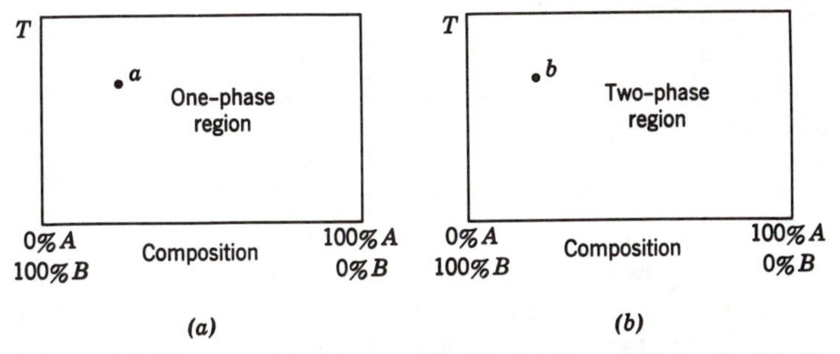

Figure 19·2. Phase diagrams for (*a*) miscible liquids and (*b*) immiscible liquids.

only two independent intensive properties. This is in accordance with the fact that, for the specified pressure of the diagram and temperature T_e, two phases existing together must be in states *e* and *f*, and there is consequently no choice of concentration in either phase.

If a solution in state *c* of Fig. 19·1 is cooled, only one phase is present until state *h* is reached, whereupon a separate phase, in state *i*, begins to form. As the temperature is lowered to T_e, the two phases approach states *e* and *f*. The relative amounts of the two phases must be such that the overall concentration remains constant.

Notice that in a two-phase region on a phase diagram a point always represents a mixture of the two phases represented by state points on the phase boundaries.

Figure 19·2*a* is a phase diagram for a system composed of two liquids that are miscible throughout the temperature range of the diagram. Figure 19·2*b* is for two liquids that are immiscible throughout the diagram temperature range. These two simple diagrams are shown to illustrate that point *a* in the single-phase region represents a solution, while point *b* in the two-phase region represents a mixture of two separate phases, pure *A* and pure *B*.

Solids, like liquids, show varying degrees of miscibility, and their characteristics can be shown in a similar manner on phase diagrams. Gases are always completely miscible.

19·3 Liquid–Vapor Equilibrium: Miscible Liquids

This section treats liquid–vapor equilibrium of binary mixtures that meet two conditions: (1) the two liquids are miscible, at least when they are in equilibrium with vapor, and (2) as the composition of the mixture is changed from one extreme to the other, the boiling and condensation temperatures change progressively from the saturation temperature of one pure component to that of the other. The second condition excludes any *azeotropic* mixture, one that has a minimum or maximum boiling or condensation temperature at some intermediate composition.

For a binary mixture in a single phase, the phase rule shows that there are three independent intensive properties. A complete property diagram for such a mixture would therefore be a three-dimensional diagram. This difficulty is avoided by using plane diagrams on which two of the properties are plotted for constant values of the third one, usually pressure. On such a diagram there are areas in which points representing single-phase states can be located, because for a fixed pressure the other two properties are independent. If two phases exist together in equilibrium, the phase rule indicates that there are only two independent properties. Since pressure is fixed for a given diagram, only one other property is independent. Therefore, any points representing states in which two phases can coexist in equilibrium must lie on *lines* that relate the two plotted properties.

Two such diagrams commonly used are *Txy* and *yx* diagrams, each with pressure constant. T is the temperature, x is the mole fraction of one of the substances in the liquid, and y is the mole fraction of the same substance in the vapor. Figures 19·3 and 19·4 are *Txy* and *yx* diagrams, respectively, for mixtures of toluene and benzene.

In Fig. 19·3, points A and C are at the boiling points of pure toluene and pure benzene, respectively. The curve A-B-C gives the boiling point for any mixture composition. A point such as i below the boiling-point line represents a liquid mixture. If the mixture of state i is heated, it will boil when it reaches state d at temperature T_2. The composition of the boiling liquid at T_2 is designated by x_2. The vapor in equilibrium with this liquid has a composition y_2 and is represented by state e. Since two phases exist, the phase rule shows that at a specified pressure, the temperature and composition cannot be independent; so an increase in temperature causes a change in composition of the two phases along the lines A-B-C and A-D-C. At temperature T_3 the mixture that was originally in state i is comprised of a liquid phase in state a with a composition x_3 and a vapor phase in state c with a composition y_3. *If the total amount of each component in the system has remained the same,* point b at the original composition represents the mixture state at T_3, but remember that this is actually a mixture of a liquid phase and a vapor phase, and no single phase with the composition of state b can exist at T_3. For the state represented by point b the relative proportions of the liquid and vapor are shown by the lengths of lines b-c

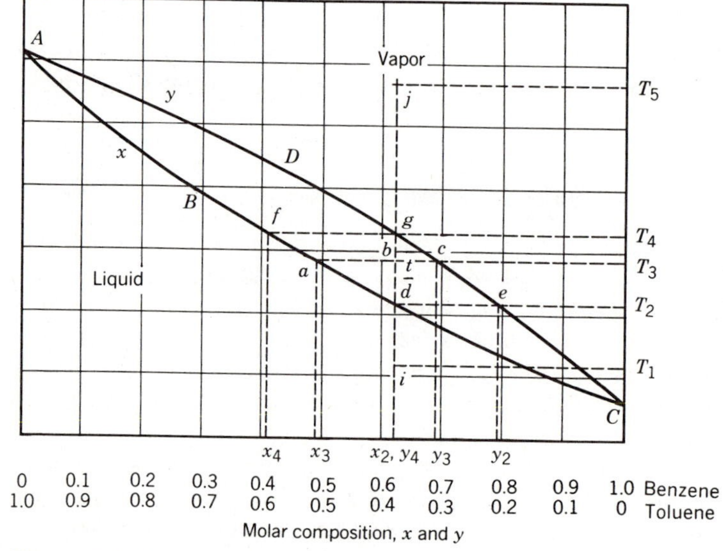

Figure 19·3. Benzene–toluene *Txy* diagram at 1 atm.

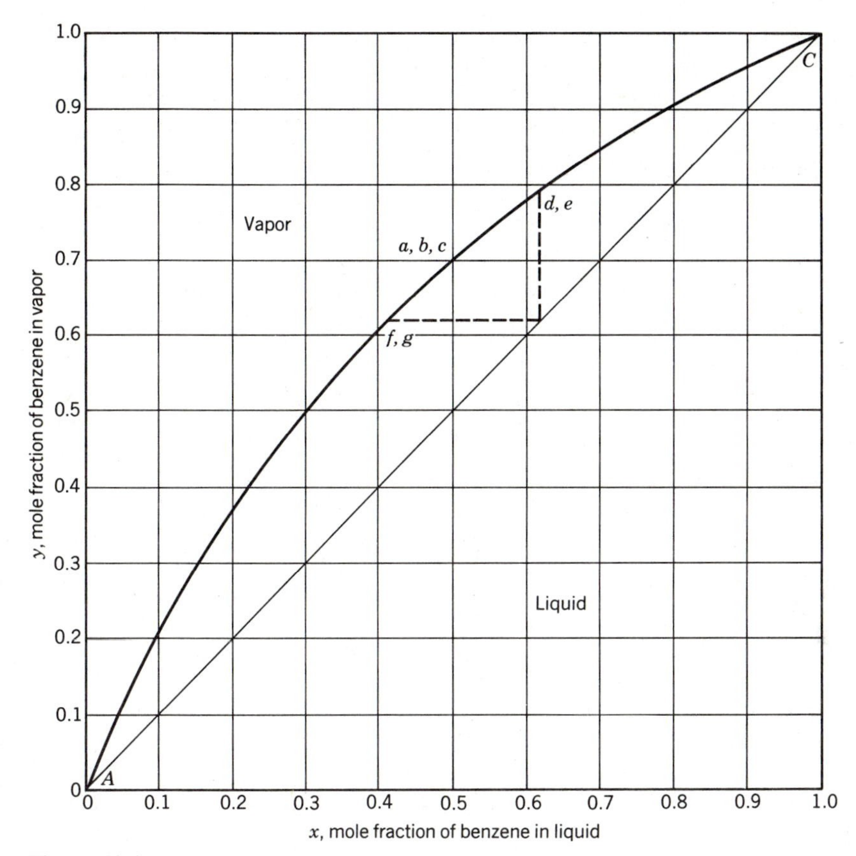

Figure 19·4. Benzene–toluene *yx* diagram at 1 atm.

and *a-b*. For example, if *a-b* and *b-c* are respectively two thirds and one third of the total length *a-c*, then the system is composed of one-third liquid and two-thirds vapor by moles. As the temperature is increased further, the compositions of the two phases change until temperature T_4 is reached, whereupon the liquid phase disappears and the vapor has the same composition as the original liquid mixture ($y_4 = x_2$). Further increase in temperature superheats the vapor with no change in composition.

If a superheated vapor in state *j* is cooled, the temperature decreases until state *g* at T_4 is reached, whereupon liquid of composition x_4 begins to appear. Continued cooling causes the composition of both the liquid and the vapor to change until a temperature T_2 is reached. At T_2 the last trace of vapor condenses, and further cooling simply reduces the temperature of the liquid.

The *yx* diagram of Fig. 19·4 can be plotted from Fig. 19·3. Notice that the liquid–vapor equilibrium *region* of the *Txy* diagram becomes a liquid–vapor equilibrium *line* in the *yx* diagram. Because the saturation temperature varies with composition of the mixture, each point on the equilibrium line is at a different temperature, ranging from the boiling point of toluene at *A* to the boiling point of benzene at *C*. The various states shown on the *Txy* diagram of Fig. 19·3 are also shown on the *yx* diagram. You should trace on the *yx* diagram the processes described above in connection with the *Txy* diagram.

The line of unity slope connecting *A* and *C* is often shown on *yx* diagrams as a convenience in determining the relative locations of states such as *d* and *q* (or *i and j*), where the mole fraction in the liquid in one state equals the mole fraction in the vapor in the other.

A temperature–composition or phase diagram is useful in the study of the separation of volatile miscible liquids. If a mixture of benzene and toluene, for example, is boiled, starting at a point such as *d* in Fig. 19·3, the liquid that remains, as boiling progresses, becomes leaner in benzene. The vapor formed is richer in benzene than the original mixture. *If this vapor is removed and condensed,* the liquid formed from it is therefore richer in benzene than the original mixture. If this liquid is then boiled and the vapor *removed and condensed,* a still richer liquid in benzene will be obtained. By continuing the evaporation and condensation steps it is possible to obtain complete separation of the benzene and toluene. The separation of a mixture into its components is called *rectification.*

Figure 19·5 shows schematically a method of partial rectification. A binary liquid mixture at its boiling point (state *f*) enters the rectifier or rectifying tower at section 1 and flows downward over the trays that are arranged to produce a large liquid surface area for heat and mass transfer. Other physical arrangements to promote contact and the attainment of equilibrium are also used, and some of these replace the continuous process by a series of discrete steps. Heat is added at the bottom of the tower, and only liquid is allowed to leave at section 2. The vapor driven from the liquid flows back up the tower to leave at section 3. Heat transferred from the vapor to the counterflowing liquid causes some of the liquid to evaporate and establishes a temperature gradient throughout the tower. Corresponding to the temperature gradient there are concentration gradients in both the liquid and the vapor streams. Referring to the temperature–composition diagram in Fig. 19·5, if a mixture of *A* and *B* in state *f* enters the tower, the vapor driven from it is in state *g*. This makes the liquid richer in *B* and raises its boiling point. As it flows downward, its temperature rises, evaporation continues, and its *B* concentration increases. The vapor formed lower in the tower is also richer in *B*. Notice that it is possible to withdraw pure *B* at section 2 but the vapor leaving at section 3 can be no richer in *A* than

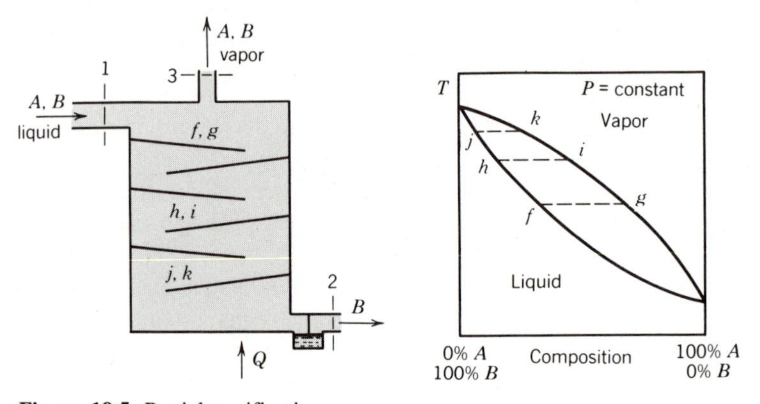

Figure 19·5. Partial rectification.

state g. (In an actual rectifier the liquid and vapor at any section are not in equilibrium with each other because heat and mass are being transferred at a finite rate between the phases. This difference between the actual system and the ideal system we are analyzing should be recognized.)

The maximum yield of pure component B can be determined by means of a mass balance and temperature–composition data. Let x'_A stand for the mass fraction of A in the liquid phase and y'_A stand for the mass fraction of A in the vapor phase. Let $\dot{m}$ represent a total flow rate and $\dot{m}_A$ and $\dot{m}_B$ represent the flow rates of components A and B, respectively. Only B leaves at section 2, so a mass balance is

$$\dot{m}_{B2} = \dot{m}_2 = \dot{m}_1 - \dot{m}_3 = \dot{m}_1 - \frac{\dot{m}_{A3}}{y'_{A3}}$$

All A that enters the system leaves at 3, so $\dot{m}_{A3} = \dot{m}_{A1} = x'_{A3}\,\dot{m}_1$. Also, the liquid entering at 1 and the vapor leaving at 3 are in contact and are assumed to be in equilibrium with each other; so $y'_{A1} = y'_{A3}$. These substitutions in the mass balance and rearrangement lead to

$$\frac{\dot{m}_{B2}}{\dot{m}_1} = 1 - \frac{x'_{A1}}{y'_{A1}}$$

This is the maximum yield. If less liquid is removed at section 2, the vapor in the tower will be richer in B. Its temperature at each section of the tower will be higher, and this is actually required in order to maintain the required heat transfer from the vapor to the liquid; so the yield given by the equation above is not realized by an actual system.

Complete rectification can be accomplished by means of the system shown schematically in Fig. 19·6. Instead of removing vapor at the section where the liquid enters, the tower is extended above the inlet, and a cooler is provided at the top. In the limiting case, pure B liquid leaves at section 2, and pure A vapor leaves at section 3. The condensate from the cooler at the top of the tower flows back down the tower, so that the transfer of heat from the vapor to the liquid and the transfer of mass from the liquid to the vapor

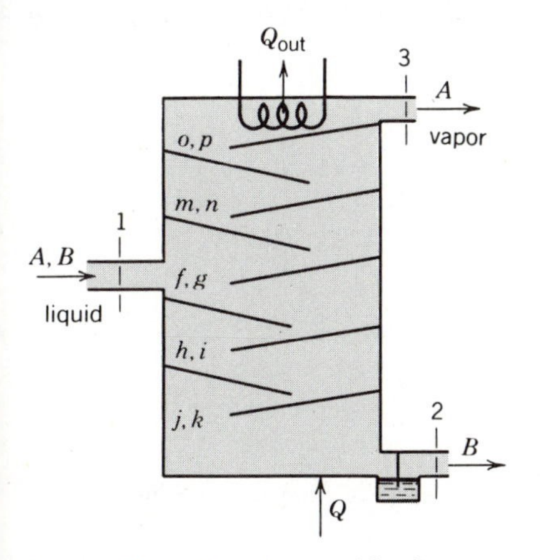

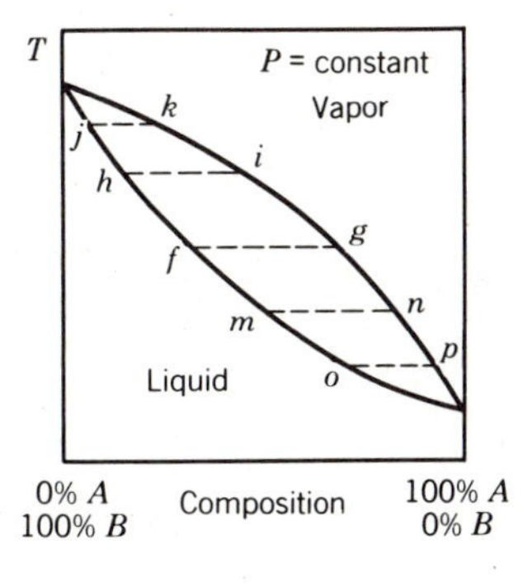

Figure 19·6. Complete rectification.

occurs throughout the tower. The liquid that flows down the tower and mixes with the incoming stream is called the *reflux*. The mass ratio at the mixing point of reflux to incoming flow is a minimum when these two streams are in the same state as they mix near section 1 and when the vapor flowing up the tower past the mixing point at the inlet is in equilibrium with the liquid.

If mixture enthalpy data (such as that for ammonia and water given in Chart A·10) are available, the required heat transfer of a rectification process can be determined by application of the first law.

19·4 Liquid–Vapor Equilibrium: Immiscible and Partially Miscible Liquids

If two liquids are immiscible, each one forms a liquid phase. If they are in equilibrium with a vapor phase, the phase rule shows that $F = 2 + C - P = 2 + 2 - 3 = 1$. There is only one independent intensive property. If the pressure is fixed, then the temperature is fixed, and it is independent of the amount of the phases present. Thus a single saturation pressure-temperature curve holds for all compositions of the liquid mixture. The saturation pressure of the mixture of immiscible liquids A and B is equal to the sum of the saturation pressures of A and B alone at the same temperature as shown in Fig. 19·7.

A temperature-composition diagram for two immiscible liquids A and B and their vapor is shown in Fig. 19·8. The boiling-point temperature for the constant pressure of the diagram is T_b. As pointed out above, it is independent of the composition of the liquid mixture. For a system comprised of two liquid phases A and B coexisting in equilibrium with their vapor there is only one independent intensive property. There is at any specified pressure only one possible composition of the vapor that can exist in equilibrium with a liquid mixture of A and B. This state of the vapor is represented by point a in Fig. 19·8.

As an illustration of the phase relationships of immiscible liquids and their vapor, consider a mixture that is initially in state c as shown in Fig. 19·8. This is actually a mixture of pure B liquid in state q and pure A liquid in state r. The mixture comprises a closed system, so that the total amount of each component remains the same. Also, the

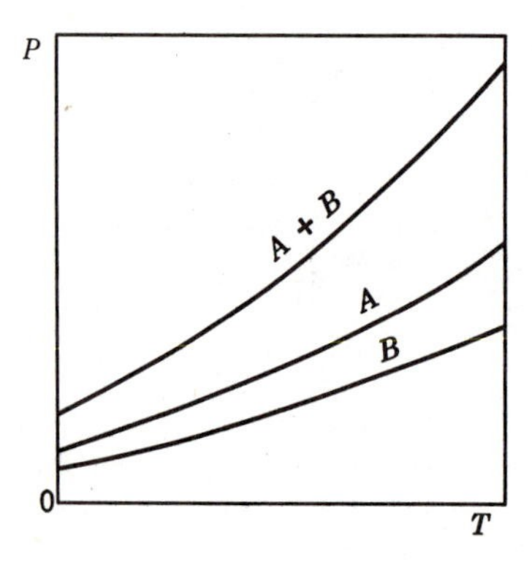

Figure 19·7. Vapor pressures of immiscible liquids A and B.

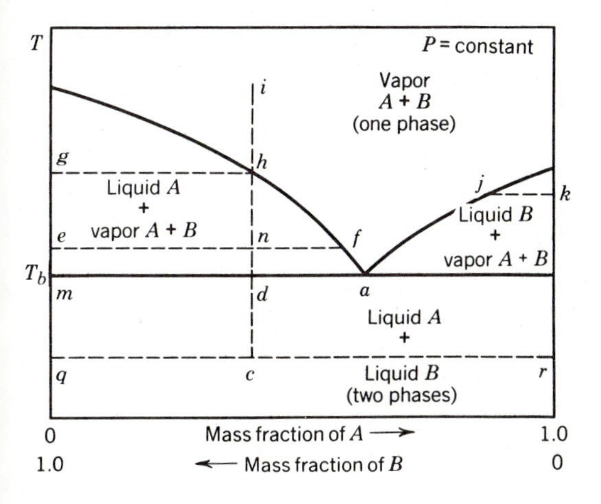

Figure 19·8. Liquid–vapor equilibrium of two immiscible liquids A and B.

pressure is held constant. As the temperature is increased, the mixture goes from state c to state d, always consisting of two liquid phases, one of pure A and the other of pure B. If heat is added to the liquid mixture in state d, evaporation begins, but of course the vapor formed must be in state a as noted previously. Since the vapor formed is richer in A than the system as a whole is, there is an increase in the amount of the liquid phase comprised of pure B, state m. As the last trace of liquid A is evaporated, the system is comprised of two phases: pure B liquid in state m, and a vapor mixture in state a. The relative masses of the two phases are indicated by the lengths of the line segments a-d and m-d. The system is again divariant ($F = 2 + C - P = 2 + 2 - 2 = 2$); so the temperature can be increased at constant pressure. The liquid phase remains pure B, and the vapor phase changes composition as its state point moves from a toward f and h. Since the vapor phase becomes richer in B, the amount of the liquid phase present decreases. When the temperature reaches the value for which the vapor state is h, the pure B liquid phase is represented by point g. The vapor in state h has the same composition

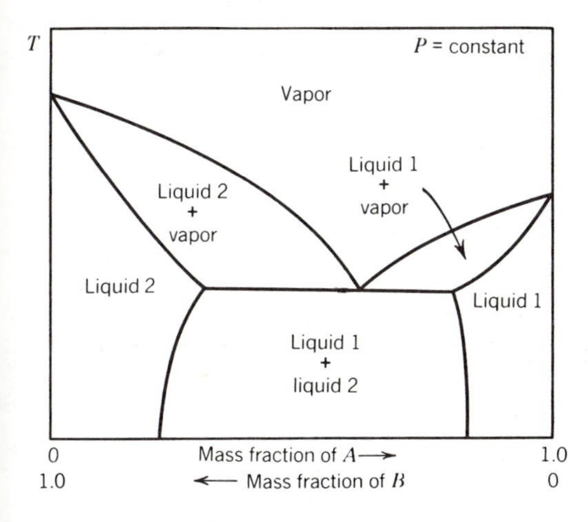

Figure 19·9. Liquid–vapor equilibrium of partially miscible liquids.

as the entire system; so the last trace of liquid evaporates as the vapor reaches state *h*. Further heating superheats the vapor toward state *i*.

Figure 19·8 shows three double-phase regions, one triple-phase point *a*, and one single-phase region. Notice that points *c*, *d*, and *n* represent mixtures of two phases, neither of which has the composition indicated by the single state point. Point *i*, however, represents the composition and temperature of a single phase.

The phase relationships for two partially miscible liquids and their vapor are shown in Fig. 19·9. Liquid 1 is a solution rich in *A*, and liquid 2 is a solution rich in *B*.

19·5 Solid–Liquid Equilibrium: Immiscible Solids and Miscible Liquids

The preceding sections give a general picture of phase relationships in binary mixtures, even though only liquid and vapor phases have been treated. Similar analyses apply to solid-liquid systems. Figure 19·10 is a temperature-composition diagram for a system composed of components *A* and *B* which are immiscible as solids and miscible as liquids.

The only liquid that can be in equilibrium with both solid phases in Fig. 19·10 is that in state *c*, which is called the *eutectic** point. If liquid of any composition other than that of the eutectic mixture is cooled, one of the components will crystallize and cause the liquid remaining to approach the eutectic composition. When the temperature T_m is reached, the system is composed of one single-component solid phase and liquid of the eutectic composition. Further removal of heat causes the liquid to solidify into a mixture of crystals of *A* and crystals of *B*. The two components must crystallize in the ratio of the eutectic mixture because the liquid solution must remain fixed in composition as long as it is in equilibrium with two solid phases. The resulting solid mixture includes all the single-component solid formed as the temperature was lowered toward T_m as well as the solid formed in the eutectic proportion at T_m.

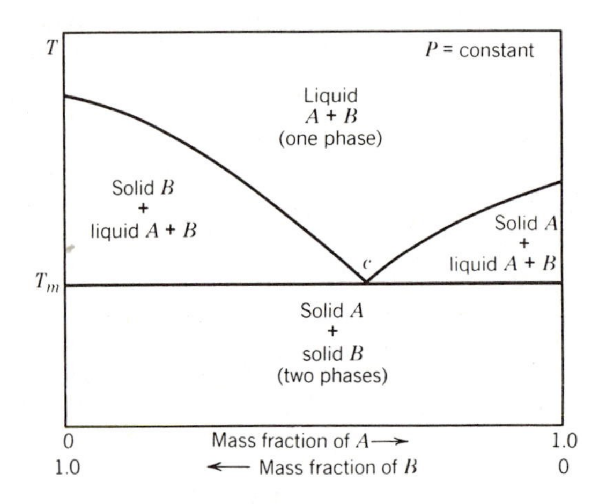

Figure 19·10. Solid–liquid equilibrium of miscible liquids and immiscible solids.

**Eutectic* is derived from two Greek words that mean *easy melting*.

19·6 Summary

The phase rule, which was first formulated by J. Willard Gibbs, states that, for any system that is (1) homogeneous or composed of a finite number of homogeneous parts in contact, (2) in equilibrium, and (3) free of electric, magnetic, gravitational, solid distortion, and surface tension effects, the maximum number of independent intensive properties F is given by

$$F = 2 + C - P \tag{19·1}$$

where C is the number of components and P is the number of phases in the system. The number of components is the least number of independently variable substances in terms of which the composition of any phase can be specified.

If two liquids when mixed in any concentration form a single phase, they are said to be *miscible*. If when mixed they remain in two phases, each identical in composition with one of the liquids before mixing, they are *immiscible*. If the mixing results in a single phase for some concentrations and two phases for the others, the liquids are said to be partially miscible. Solids, like liquids, show varying degrees of miscibility. Gases are always completely miscible.

In the study of phase relationships of multicomponent systems, temperature–composition (Txy) and composition (yx) diagrams plotted for constant pressure are quite useful. In this chapter such diagrams have been used to analyze qualitatively phase relationships for a few types of binary mixtures.

References

19·1 Gosney, W. B., *Principles of Refrigeration,* Cambridge University Press, Cambridge, England, 1982, Sections 6.1-5 and 6.8-13.

19·2 Henley, E. J., and J. D. Seader, *Equilibrium-Stage Separation Operations in Chemical Engineering,* Wiley, New York, 1981.

19·3 McQuiston, Faye C., and Jerald D. Parker, *Heating, Ventilating, and Air Conditioning,* 2nd ed., Wiley, New York, 1982, Section 15-8.

19·4 Smith, J. M., and H. C. Van Ness, *Introduction to Chemical Engineering Thermodynamics,* 2nd ed., McGraw-Hill, New York, 1975, Sections 8-1 through 8-6.

19·5 Wood, Bernard D., *Applications of Thermodynamics,* 2nd ed., Addison-Wesley, Reading, Mass., 1982, Sections 4.6.3, 4.

Problems

19·1 A system consists of solid sodium chloride, a solution of sodium chloride, water, ethyl alcohol, and vapor. Determine the number of phases.

19·2 How many components are present in a gas mixture at room temperature and atmospheric pressure consisting of oxygen, hydrogen, and water vapor?

19·3 Determine the number of phases for a system consisting of sand and solid salt in the presence ed., Wiley, New York, 1982, Section 15-8.

19·4 Smith, J. M., and H. C. Van Ness, *Introduction to Chemical Engineering Thermodynamics*, 2nd ed., McGraw-Hill, New York, 1975, Sections 8-1 through 8-6.

19·5 Saturated sulfur dioxide exists at 100°F. Is the system invariant, univariant, or divariant?

19·6 A system under equilibrium conditions contains a saturated water solution of sodium chloride, excess sodium chloride, ethyl alcohol, and vapor. Determine the number of degrees of freedom.

19·7 For the system in which solid magnesium carbonate dissociates according to the following reaction, determine the number of degrees of freedom: $MgCO_3$(solid) $\rightarrow$ MgO(solid) + CO_2(gas).

19·8 Derive the following expression for the minimum reflux ratio at the mixing point in a tower like that shown in Fig. 19·6:

$$\frac{\dot{m}_{\text{reflux}}}{\dot{m}_1} = \left(\frac{1 - y'_{A1}}{y'_{A1} - x'_{A1}}\right) x'_{A1}$$

19·9 Sketch a yx diagram for Fig. 19·5.

19·10 Sketch a yx diagram for Fig. 19·6.

19·11 A mixture of 20 percent benzene and 80 percent toluene enters a partial rectifier as shown in Fig. 19·5 at a rate of 6000 lbm/h. Compute the maximum yield of pure toluene.

19·12 A continuous rectifying column is used to separate completely 12 000 lbm/h of a solution of 60 percent benzene and 40 percent toluene into its components. Compute the minimum reflux rate.

19·13 An aqua-ammonia solution that is 30 percent ammonia is to be rectified at 100 psia in a tower like that of Fig. 19·6. Determine the reflux ratio, the heat added, and the heat removed per pound of entering solution. What assumptions do you make?

19·14 Refer to Fig. 19·10. A liquid is initially in a state with a composition that is richer in B than the eutectic mixture. Describe the phase changes that occur as this mixture is cooled to a temperature lower than T_m.

CHAPTER TWENTY

Fuel Cells: Direct Energy Conversion

In preceding chapters of this book, several different engines have been analyzed. These engines all convert heat into work by using a working substance—usually a fluid—that alternately absorbs and rejects heat as it undergoes a cycle. A device that would continually and directly convert some part of the heat supplied into work, without involving a cycle of a working substance, would be quite desirable. The efficiency of such a device would still, however, be limited in accordance with the Carnot principle discussed in Chapter 7. Another desirable device would convert chemical energy directly into work, without first converting the chemical energy into heat, because the Carnot principle efficiency limitation would not apply. Both kinds of devices are called direct energy conversion devices.

If the kind of direct energy conversion engines that convert heat directly into work can be operated in reverse, they serve as direct energy conversion refrigerators. Although the Carnot principle limitation on coefficient of performance still applies, the simplicity of such devices makes them valuable in some application.

20·1 Direct Energy Conversion Systems

Direct energy conversion systems continuously convert various forms of thermal, chemical, or nuclear energy into work or electrical energy without the use of working substances that alternately absorb, store, and reject energy. Electrical energy and work are equivalent, because either can be fully converted into the other, at least in the ideal case. That is, there are no inherent limitations on such conversion.

In this chapter we consider one kind of direct energy conversion system, the fuel cell, on the basis of principles presented in earlier chapters. Four other direct energy conversion systems are briefly and qualitatively described. They are:

1. Photovoltaic systems
2. Thermoelectric systems
3. Thermionic systems
4. Magnetohydrodynamic systems

Fuel cells can operate isothermally. They do not involve heat exchange with energy reservoirs at different temperatures, so the Carnot principle does not limit their performance. Other factors may limit performance, even for ideal devices, but there is no limitation based on the Carnot principle.

Photovoltaic devices may operate isothermally, but the input radiant energy comes from a source at a different temperature.

Thermoelectric, thermionic, and magnetohydrodynamic devices all involve the addition of heat from a higher-temperature reservoir and the rejection of heat to a lower-temperature reservoir. Therefore, the Carnot principle applies. Although these devices may not have high efficiencies, they may be attractive because of the simplicity of direct conversion systems and for other reasons. For example, the lack of moving parts may make it possible for these devices to operate at a much higher temperature than conventional power plants. If these devices reject heat at very high temperatures, they can be combined with conventional power plants in ''topping'' arrangements as mentioned in Sec. 17·6 for binary vapor power plants.

In the selection of an energy conversion system for a given application, many factors in addition to efficiency must be considered. Some of these are size, mass, initial cost, reliability, cost of maintenance, environmental pollution, and adaptability to various fuels. Consequently, each of the direct energy conversion devices mentioned here is well suited to certain applications.

20·2 Fuel Cells: Description of Operation

The fuel cell is a steady-flow device in which a fuel and an oxidizer are combined chemically in an isothermal reaction to produce electrical work. This is achieved by dividing the overall reaction into two reactions that occur on separate electrodes. The fuel and oxidizer do not come directly into contact with each other, because direct contact would generally involve a nonisothermal reaction such as a normal combustion process.

One reaction, occurring on the surface of one electrode, ionizes the fuel and sends released electrons into an external electric circuit. On the surface of the other electrode, a reaction occurs that accepts electrons from the external circuit and ionizes the oxidizer. An electrolyte between the electrodes transports ions to complete the overall chemical reaction and also to complete the electric circuit.

One type of hydrogen–oxygen fuel cell is shown schematically in Fig. 20·1 to illustrate the functions of the various parts of the cell. External connections provide for flows of hydrogen and oxygen into the fuel cell and for the flow of the reaction product, water, from the cell. The other connections are to the external electric circuit.

The Electrodes

The *anode* is defined as the electrode that supplies electrons to the external circuit. The *cathode* is defined as the electrode that receives electrons from the external circuit. On the anode surface in Fig. 20·1, hydrogen combines with negative OH ions to produce electrons and water. The electrons enter the external circuit, and the water goes into the electrolyte. On the cathode surface, oxygen combines with water from the electrolyte and

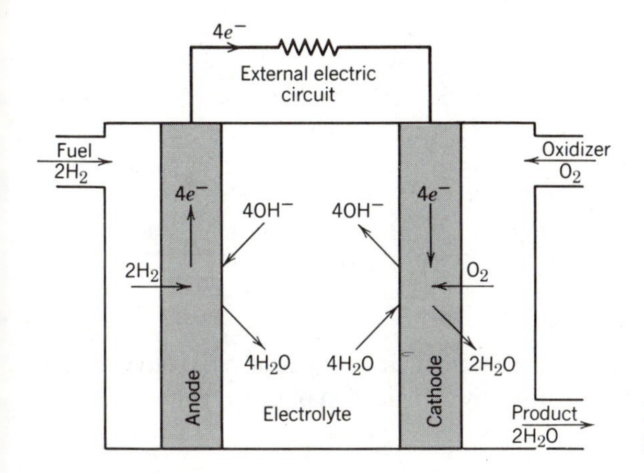

Figure 20·1. Schematic diagram of hydrogen–oxygen fuel cell.

electrons from the external circuit to produce negative OH ions and water. The OH ions are carried away through the electrolyte, and the water is discharged from the cell. In the fuel cell shown, the electrodes also perform the important function of separating the nonionized fuel and oxidizer from the electrolyte.

The Electrolyte

In Fig. 20·1, the electrolyte carries negative ions from the cathode to the anode. This is not always the case. For example, in Fig. 20·2, the electrolyte carries positive ions from the anode to the cathode. In both cases, the electrolyte transports ions to complete the overall chemical reaction and also to complete the electric circuit. Acidic electrolytes may be used to carry positive ions, H^+; alkaline electrolytes carry negative ions, OH^-. Acidic electrolytes are the more widely used.

Although the examples we give here involve aqueous electrolytes, other liquids and also solids are used as electrolytes, and solid polymer acidic electrolytes are the ones most commonly used. Since some fuel cells operate best at elevated temperatures, non-aqueous electrolytes are especially suitable for them.

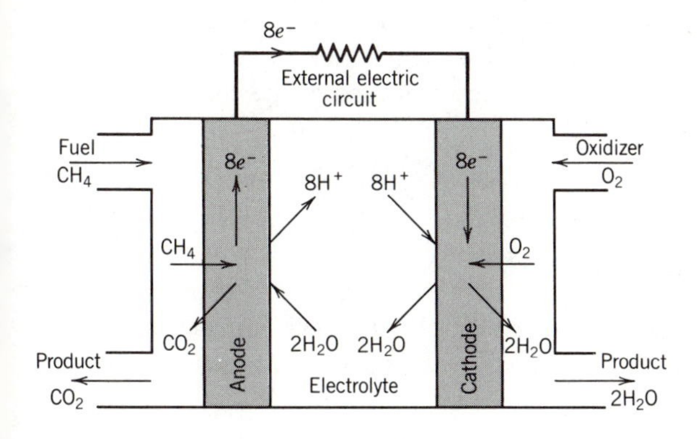

Figure 20·2. Schematic diagram of methane–oxygen fuel cell.

Fuel

A wide variety of fuels may be used in fuel cells. Hydrogen is widely used. In spacecraft applications a hydrogen–oxygen fuel cell provides not only electric power but also water for the crew. In terrestrial applications, hydrocarbon fuels are preferred because they are more readily available and, consequently, less expensive. Liquid as well as gaseous fuels are possible.

Hydrogen can be prepared from a hydrocarbon fuel by a reforming reaction. For example, the reforming of methane is achieved by the reaction

$$CH_4 + H_2O \rightarrow CO + H_2$$

Since CO is always produced by a reforming reaction and CO poisons the catalysts used in fuel cells, the reforming reaction is followed by an oxidation of CO in a reaction called the shift reaction

$$CO + H_2O \rightarrow CO_2 + H_2$$

These separate reactions do not enter ideal cell calculations where catalysis is not considered.

Oxidizer

Although Figs, 20·1 and 20·2 show oxygen as the oxidizer, it is unnecessary to use pure oxygen. Consequently, air is the most common oxidizer.

Fuel cells are often compared with batteries. Indeed, there are similarities in structure and nomenclature. The differences, however, are greater than the similarities. A battery is a closed system in which energy is stored in a chemical form. As energy is taken from the battery, the chemical composition of an electrode, of the electrolyte, or of both changes. In contrast, a fuel cell is a steady-flow system. Fuel and oxidizer are supplied and reaction products must be removed. Ideally, there is no change in the chemical composition of the electrolyte or either electrode.

In comparing the chemical reaction that occurs in a fuel cell with a combustion reaction, notice that in the fuel cell the reaction is isothermal while a combustion reaction usually involves a large temperature change. Energy is removed from the fuel cell as electrical energy, while energy is removed from a combustion reaction as heat or as heat and work together. Because the fuel cell operates isothermally and continually, the extent of its conversion of chemical energy to electrical energy is not limited by the Carnot principle. Whenever a combustion reaction produces heat used as the energy input to a cycle, the Carnot principle does apply.

20·3 Analysis of Fuel-Cell Operation

Section 8·7 shows that for a steady-flow system with a substance both entering and leaving at the temperature of the surrounding atmosphere and exchanging heat with only the atmosphere, the maximum work that can be produced is equal to the decrease in G of the substance

$$W_{max} = -\Delta G \tag{20·1}$$

The value of ΔG for any reaction can be calculated in a manner completely analogous to that for calculating ΔH as given by Eqs. 12·1 and 12·2. That is

$$\Delta G = G_2 - G_1 = \sum_{prod} N(g_2 - g_a) + \Delta G_{R,a} - \sum_{reac} N(g_1 - g_a) \tag{20·2}$$

$$\Delta G_R = \sum_{prod} N\Delta g_f - \sum_{reac} N\Delta g_f \tag{20·3}$$

In Eq. 20·2, the subscript a indicates the temperature for which the ΔG_R is used. In Eq. 20·3, of course, the ΔG_R and all Δg_f values must be taken at the same temperature. Usually, the temperature of a standard reference state is used so that

$$\Delta G = G_2 - G_1 = \sum_{prod} N(g_2 - g^\circ + \Delta g_f^\circ) - \sum_{reac} N(g_1 - g^\circ + \Delta g_f^\circ) \tag{20·4}$$

Values of Δg_f° are often tabulated with other thermodynamic property data. Values can also be calculated from a relationship based on the definition $G \equiv H - TS$. For any element, the Gibbs function of formation is zero; for any compound

$$\Delta G_f^\circ = \Delta H_f^\circ - T^\circ \Delta S_f^\circ$$

and

$$\Delta S_f^\circ = \sum_{compound} Ns^\circ - \sum_{elements} Ns^\circ$$

where all entropies are absolute entropies.

Property values for some substances used in fuel cells are given in Table 20·1.

For a reversible isothermal steady-flow process at T_0 (the temperature of the surrounding atmosphere), during which heat exchange occurs with only the atmosphere, the

TABLE 20·1 Δh_f° and Δg_f° at 1 Atm, 25°C (77°F)*

	Δh_f°		Δg_f°	
Compound	kJ/kmol	B/lbmol	kJ/kmol	B/lbmol
$CO(g)$	−110 530	−47 519	−137 165	−58 970
$CO_2(g)$	−393 510	−169 179	−394 397	−169 560
$CH_4(g)$	−74 850	−32 180	−50 838	−21 856
$C_3H_8(g)$	−104 680	−45 004	−24 465	−10 518
$CH_3OH(l)$	−201 041	−86 432	−162 403	−69 821
$H_2O(l)$	−285 830	−122 885	−237 185	−101 971
$H_2O(g)$	−241 820	−103 964	−228 591	−98 276
$NH_3(g)$	−45 898	−19 733	−16 387	−7 045

*From Table A·8.

work is the maximum work and the heat transfer is given by

$$Q = H_2 - H_1 + W_{max} = H_2 - H_1 + G_1 - G_2 = T_0(S_2 - S_1)$$

The heat transfer may be greater than, equal to, or less than zero, depending on the relative magnitudes of ΔG and ΔH.

Unlike heat engines, fuel cells do not operate cyclically and exchange only heat and work with the surroundings. They diminish the amount of fuel and oxidizer and increase the amount of products in the surroundings. Consequently, no performance parameter such as a thermal efficiency that involves only heat and work transfers is defined for fuel cells.

One performance parameter in use is

$$\varepsilon \equiv \frac{\Delta G}{\Delta H}$$

This parameter ε has been called the "fuel-cell efficiency." Notice that ε is not a thermal efficiency as defined in connection with cycles, and the two cannot be compared directly with each other. Indeed, ε is less than 1 for an ideal hydrogen–oxygen fuel cell, equal to 1 for an ideal methane–oxygen cell, and greater than 1 for a carbon–oxygen cell. A value of ε greater than 1 indicates that W_{max} exceeds the decrease in enthalpy of the stream passing through the cell; therefore, there must be heat transfer from the surroundings to the fuel cell.

The work done by a fuel cell is given by the product of the charge, Q_e, carried by the external circuit and the cell terminal emf or voltage, V, which is the work done per unit charge. The charge delivered is the product of the number of electrons delivered to the external circuit and the charge per electron (1.6022×10^{-19} C/electron). The number of electrons is equal to the product of the number of molecules giving up electrons and the valence j. In turn, the number of molecules is the number of molecules per mole (Avogadro's number, 6.022169×10^{26} molecules/kmol) times the number of moles of reactants giving up electrons. Combining these products gives

$$W_{max} = (6.022169 \times 10^{26})N_j(1.6022 \times 10^{-19})V = \mathfrak{F} N_j V$$

where $\mathfrak{F}$ is the Faraday constant (96.487×10^6 C/kmol of electrons or 96 487 kJ/V·kmol of electrons), N is the number of moles of reactant having a valence j, and V is the cell terminal voltage. Thus, the ideal cell terminal voltage is given by

$$V = \frac{-\Delta G}{\mathfrak{F} N_j} \tag{20·5}$$

Remember that this is the *ideal* cell voltage. Internal cell resistance and other effects prevent this value from being realized in an actual fuel cell.

20·4 Actual Fuel-Cell Performance

As pointed out in Sec. 1·8, we often analyze an ideal system and then adjust our results in predicting an actual system's behavior. In the preceding section we have considered only ideal fuel cells. They operate reversibly, and their performance, including such matters as variation in performance with operating temperature, can be calculated from principles introduced earlier.

Actual fuel cells do not perform as well as ideal ones. Several factors cause the differences, and we now describe briefly the major ones.

We assumed that in the ideal fuel cell, fuel or oxidizer and the electrolyte were always present at the reaction surface of an electrode in proper proportions and that there were no impediments to the reaction. In an actual fuel cell, the transport of fuel, oxidizer, electrolyte, and products to and from the electrode surfaces is complex and introduces several problems. In order to provide a large electrode surface for the three-phase interface of a gaseous fuel or oxidizer, a liquid electrolyte, and a solid electrode, porous electrodes are used. However, if either too much liquid or too much gas enters the electrode and displaces the other fluid phase, the interface surface is reduced. Reducing the interface surface impedes the reaction.

Even if the phases are present in the proper proportions, there is a tendency for the reaction to reduce the concentration of reactants and increase the concentration of products in the reaction zone. This retards the reaction. Agitating the fluids is one method of addressing this problem.

The diffusion process at the electrode surface is quite complex. Some energy is required to cause the chemisorption process at the three-phase interface to occur under any conditions. This is called the activation energy, and however it is supplied, it reduces the work of the cell.

Associated with the diffusion process at the electrode surface is an impedance to electron flow. There is also ohmic resistance elsewhere in the cell. These are irreversible effects that reduce the terminal emf and work of the cell.

Spurious chemical and electrochemical reactions may occur at various places within the cell, and they always impair cell performance.

Most of the losses associated with these various effects increase with increasing current density. Therefore, actual fuel-cell performance approaches ideal cell performance most closely under conditions of very low current density rather than under design conditions. For the same fuel flow rate, an actual cell may provide under low load conditions 80 percent of the maximum work of an ideal cell, whereas under design or rated conditions this ratio may be on the order of 50 percent or lower.

Other losses in actual cells result from leakage of both electrical current and fluids. Also, various auxiliaries such as pumps, fans, and control systems require some energy in an actual fuel-cell plant.

The desire to increase the work done must be balanced by consideration of other factors such as reliability, safety, initial cost, and environmental effects. For example, an electrolyte that is corrosive may be unacceptable for actual fuel cells even though its ideal thermodynamic performance would make it attractive.

Actual fuel cells may be appreciably more complex than the simple ideal ones we have discussed. For example, regenerative fuel cells, in which the reactants are at least partially regenerated by means of an energy input either to the cell or to an external regenerator, are attractive for some applications. The design and performance of actual fuel cells are discussed in some of the references at the end of this chapter as well as in the current periodical literature.

20·5 Other Direct Energy Conversion Systems

Photovoltaic Devices

Photovoltaic energy conversion devices convert energy of light or other electromagnetic radiation directly into work or electrical energy. There is no intermediate step of conversion of other forms of energy to heat. Therefore, thermal efficiency is not defined in connection with photovoltaic devices, and comparisons cannot be made with Carnot cycle performance.

Figure 20·3 is a schematic diagram of a photovoltaic cell. The *n*-type material is a semiconductor such as silicon that contains a trace of an element with one more electron than silicon. This addition makes the silicon conductive. The *p*-type material is formed by adding a trace of an element that contains one fewer electron than silicon. This also causes the silicon to become conductive. The *p*-type layer, especially, is very thin. When light strikes the upper surface of the *p*-type layer, some photons are absorbed near the junction of the two layers. This generates an emf, and if the two layers are connected through an external electric circuit, a current flows.

Losses are caused by the internal ohmic resistance of the cell. This applies both to the external load current and to the shunt current within the cell between the two layers. The ohmic losses tend to increase the cell temperature. Also, some of the incident radiation may be absorbed by the cell material and cause a temperature rise. Consequently, the photovoltaic cell may not operate isothermally. If it does not operate isothermally, the Carnot principle efficiency limitation must be considered.

Although the operation of photovoltaic cells involves many losses so that only a small fraction of the radiant energy incident upon the cell is converted into electrical work, they are simple and reliable. Also, their cost has been reduced considerably in recent years. Consequently, they are the best power source for some applications.

Thermoelectric Devices

Engineers and physical scientists are familiar with *thermoelectric* effects and the underlying *Seebeck, Peltier,* and *Thomson effects.* Figure 20·4 is a schematic diagram of a simple thermoelectric device that can be used in several ways. If the current is nulled, this device could be a thermocouple or temperature-measuring instrument if a suitable potentiometer is used to measure the potential difference across the external circuit connections. By connecting an external electrical load and maintaining a temperature difference between the hot junctions and cold junctions, electrical work can be done on the external circuit.

In selecting the thermoelectric materials, A and B, one is guided by two phenomena.

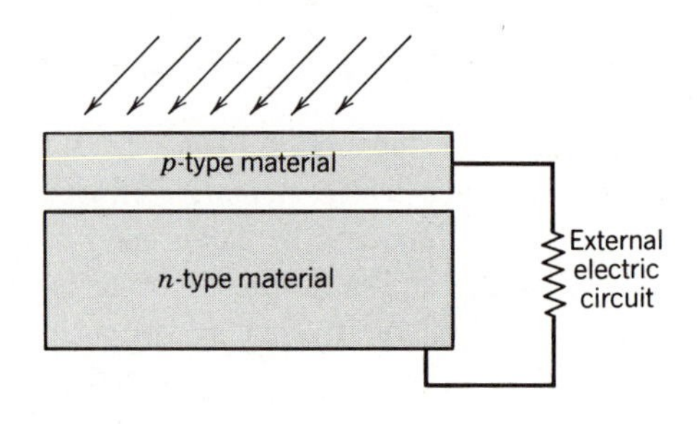

Figure 20·3. Schematic diagram of photovoltaic device.

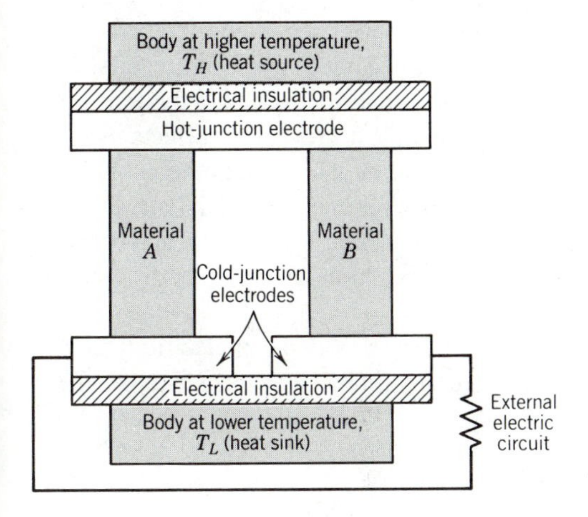

Figure 20·4. Schematic diagram of thermoelectric generator.

The conductivity of a material is directly proportional to the number of free electrons, and the emf generated is inversely proportional to the number of free electrons. Metals have a high density of free electrons and so produce relatively low emfs. However, they have low electrical resistance and therefore permit large current flows for a given emf. This combination results in a low power output. Also, metals have high thermal conductivities so they provide a ready path for wasteful heat flow between the high- and low-temperature ends of each conductor. Insulators have relatively low densities of free electrons and so generate large emfs. However, the high electrical resistance of an insulator keeps the current low, so again little power is generated.

Desirable thermoelectric materials have properties that are between those of metals and insulators. For this reason, semiconductors are used.

Low emfs are generated by thermoelectric materials, so arrangements as shown schematically in Fig. 20·5 are used. The emf between the terminals connected to the external load is the sum of the emfs of the pairs of materials shown.

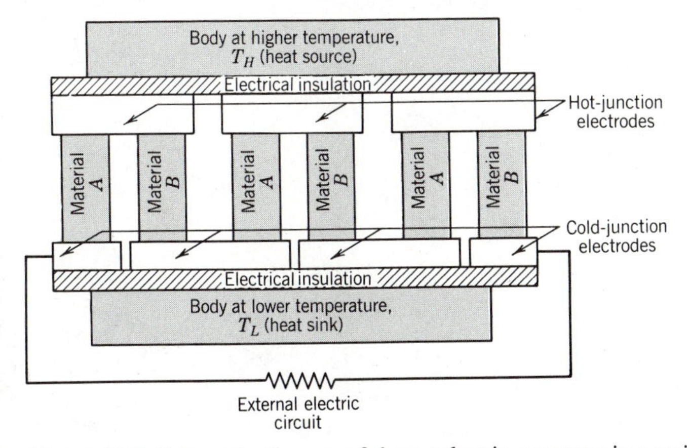

Figure 20·5. Schematic diagram of thermoelectric generator in a series arrangement.

The efficiency of a thermoelectric generator depends on the operating temperatures, the material properties, and the current density. (If the device operated steadily and reversibly, the efficiency would depend only on the temperatures of the bodies the generator exchanges heat with, in accordance with the Carnot principle.) Irreversible effects include heat conduction through the thermoelectric materials from the hot ends to the cold ends, ohmic resistance throughout the device, contact resistance between materials, and current and heat losses. Efficiencies are low, but thermoelectric generators are simple and reliable and can use heat input from various sources: fuels, solar radiation, temperature differences occurring in nature, and nuclear heat sources.

Thermoelectric generators can also be operated in reverse as themoelectric refrigerators. Coefficients of performance are low, but here again other advantages make thermoelectric refrigerators suitable foı some applications.

Thermionic Devices

Figure 20·6 is a schematic diagram of a *thermionic* generator. The two electrodes, here shown as flat plates, are closely spaced and parallel to each other in a sealed enclosure that contains either a vacuum or a plasma. By means of an external heat source, the cathode is raised to such a high temperature that electrons are driven from it. The electrons flow to the anode, which is maintained at a lower temperature by heat transfer to some part of the surroundings. Electrons leave the anode to flow through an external electric circuit and back to the cathode. Cathode temperatures typically exceed 1200 K and may exceed 2000 K. Anode temperatures may be as high as 1000 K. Since heat is rejected at such a high temperature, thermionic generators may best be applied as topping units, with the heat rejected by the thermionic device being the heat input to another power system such as a gas turbine or steam power plant.

Several factors impair the performance of thermionic generators. As electrons pass from the cathode to the anode within the thermionic generator, the mutual repulsion of the electrons in the interelectrode gap inhibits the emission of electrons from the cathode. This is called the space charge effect. One way of reducing this effect is to make the spacing between the electrodes very small. Another is to fill the device with an ionized gas or plasma to replace the vacuum. A plasma permits greater spacing between the

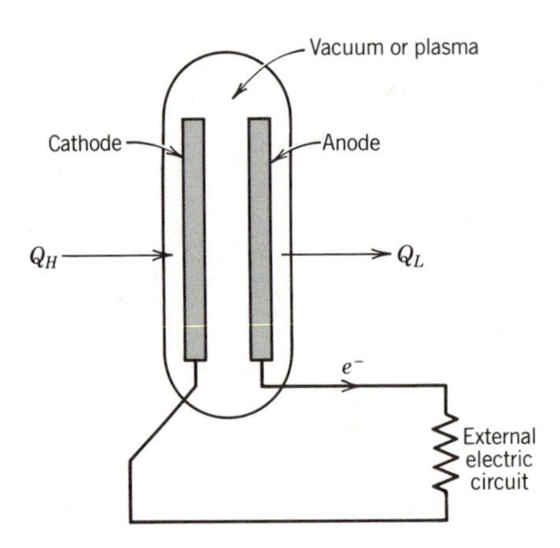

Figure 20·6. Schematic diagram of thermionic generator.

electrodes because of the positive charges in the plasma. Other effects that impair thermionic generator performance are the radiation heat transfer between the electrodes and stray heat losses to the surroundings, including the heat transfer along lead wires. There are also ohmic losses in the electrodes and internal circuitry.

Magnetohydrodynamic Devices

In a conventional electric generator, electric conductors are moved through a magnetic field to induce electric currents in the conductors. The conductors must be moved against a retarding force, and this requires a work input. *Magnetohydrodynamic* (mhd) generators are analogous to conventional electric generators, but a conducting fluid, instead of a solid conductor, is moved through a magnetic field. A current is then induced in the conducting fluid. This current passes through electrodes in the channel walls and through an external electric circuit as shown in Fig. 20·7. Other electromagnetic effects not mentioned here are important in the operation of an mhd generator.

Gases are made electrically conductive either by the addition of a small quantity of an easily ionized substance or by having the gas at a high temperature. The former procedure is called *seeding,* and the substance added is called *seed.*

Work must be done on the fluid in order to move it against the resisting force of the magnetic field. If the fluid is a gas, it must be compressed and heated. The temperature can be raised by burning fuel in the gas. Figure 20·8 is a schematic diagram of an mhd power plant. Clearly, the mhd generator is only a small part of the plant, and much auxiliary equipment is needed. Nevertheless, the mhd generator itself is simple and has no moving parts, so it can withstand higher gas temperatures than turbines. Because the temperature of the gas leaving is so high, mhd generators are considered for use only in topping arrangements, because the energy and availability loss involved in discarding this high-temperature gas would be unacceptable.

Factors that damage the performance of actual mhd generators include space charge effects (similar to those in thermionic converters), ohmic losses in the ionized gas, heat transfer from the gas, fluid friction losses, and some losses associated with the interactions of the electric and magnetic fields in the gas.

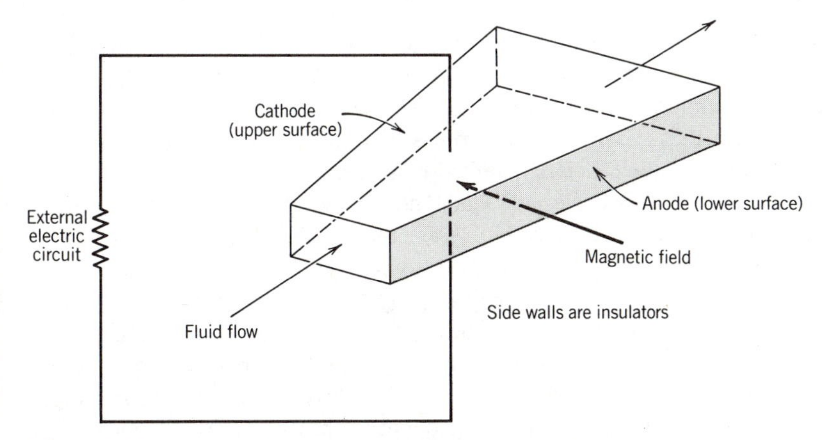

Figure 20·7. Schematic diagram of magnetohydrodynamic generator.

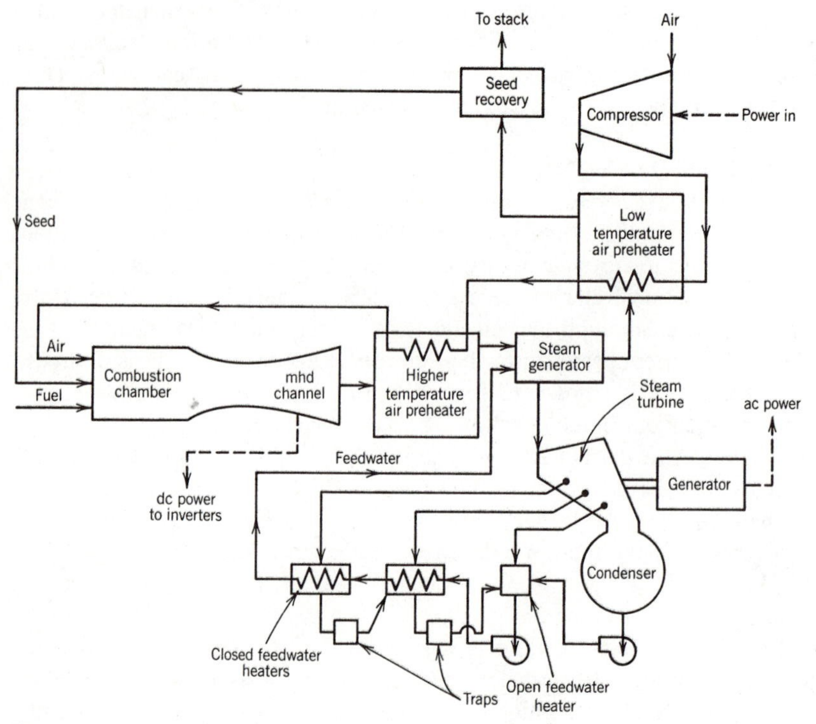

Figure 20·8. Schematic diagram of combined mhd generator and steam cycle.

Remember that an mhd generator involves heat transfer with energy reservoirs at different temperatures, so that the efficiency limitations of the Carnot principle applies.

20·6 Summary

One use of the term *direct energy conversion* refers to the conversion of energy from heat to electric or mechanical work without the interposition of a working substance that undergoes a cycle. Examples are *thermoelectric, thermionic,* and *magnetohydrodynamic* generators. Since these devices absorb heat from a high temperature source and reject heat at a lower temperature, the efficiency limitation imposed by the second law applies. Various effects reduce the efficiencies of actual devices to still lower values. Other advantages, however, make these devices highly suitable for certain applications.

The term *direct energy conversion* refers also to systems such as *photovoltaic* generators that convert light energy into electrical work and *fuel cells* that convert the chemical energy of a fuel into electrical work. Since these devices can operate continually and isothermally, without a direct energy input in the form of heat, thermal efficiency is not defined for them. However, the function of a fuel cell to obtain work from the chemical energy of a fuel is the same as the function of certain types of heat engines, so comparisons between the two must be made. Fuel cells have the advantage of avoiding the step of converting the fuel chemical energy into heat, so their performance is not limited by the Carnot principle.

The fuel cell is a steady-flow device in which a fuel and an oxidizer are combined chemically in an isothermal reaction to produce electrical work. The maximum work obtainable is given by

$$W_{max} = -\Delta G \tag{20·1}$$

For a fuel cell, ΔG can be calculated by

$$\Delta G = G_2 - G_1 = \sum_{prod} N(g_2 - g + g_f) - \sum_{reac} N(g_1 - g_a + g_f) \tag{20·4}$$

The ideal cell terminal voltage is given by

$$V = \frac{-\Delta G}{\mathfrak{F} N_j} \tag{20·5}$$

Several factors cause the performance of actual fuel cells to fall short of ideal cell performance. Most losses increase with increasing current density, so actual fuel-cell performance approaches ideal cell performance most closely under conditions of low current density.

References

20·1 Angrist, S. W., *Direct Energy Conversion,* 4th ed., Allyn and Bacon, Boston, 1982, Chapter 8.

20·2 Holman, J. P., *Thermodynamics,* 2nd ed., McGraw-Hill, New York, 1974, Sections 15-1 and 15-2.

20·3 Sorensen, Harry A., *Energy Conversion Systems,* Wiley, New York, 1983, Articles 12.18-22.

20·4 Wark, K., *Thermodynamics,* 4th ed., McGraw-Hill, New York, 1983, Section 19-1.

20·5 Wood, Bernard D., *Applications of Thermodynamics,* 2nd ed., Addison-Wesley, Reading, Mass., 1982, Chapter 7.

Problems

20·1 Determine the ideal cell voltage at 1 atm, 25°C (77°F) for the following reactions:

(a) $CO + \frac{1}{2} O_2 \rightarrow CO_2$
(b) $C_3H_8 + 5\ O_2 \rightarrow 3\ CO_2 + 4\ H_2O$
(c) $CH_3OH(l) + \frac{3}{2} O_2 \rightarrow CO_2 + 2\ H_2O$
(d) $2\ NH_3 + \frac{3}{2} O_2 \rightarrow N_2 + 3\ H_2O$
(e) $CH_4 + 2\ O_2 \rightarrow CO_2 + 2\ H_2O$

20·2 An ideal fuel cell as shown in Fig. 20·1 operates at 25°C (77°F). Calculate (*a*) the ideal cell voltage, (*b*) ε, and (*c*) the heat transfer per mole of hydrogen entering. (The water leaves as a liquid.)

20·3 An ideal fuel cell as shown in Fig. 20·2 operates at 25°C (77°F). Calculate (*a*) the ideal cell voltage, (*b*) ε, and (*c*) the heat transfer per mole of methane entering.

20·4 An ideal fuel cell using carbon monoxide and oxygen operates at 25°C (77°F). Calculate (*a*) the ideal cell voltage, (*b*) ε, and (*c*) the heat transfer per mole of carbon monoxide entering.

20·5 An ideal fuel cell as shown in Fig. 20·1 operates at 600 K (1080 R). Calculate (*a*) the ideal cell voltage, (*b*) ε, and (*c*) the heat transfer per mole of hydrogen entering.

20·6 An ideal fuel cell as shown in Fig. 20·2 operates at 600 K (1080 R). Calculate (*a*) the ideal cell voltage, (*b*) ε, and (*c*) the heat transfer per mole of methane entering.

20·7 An ideal fuel cell using carbon monoxide and oxygen operates at 500 K (900 R). Calculate (*a*) the ideal cell voltage, (*b*) ε, and (*c*) the heat transfer per mole of carbon monoxide entering.

20·8 Calculate the ideal cell voltage of a carbon-oxygen fuel cell operating at 25°C (77°F).

20·9 Plot the ideal cell voltage and ε against temperature for a carbon monoxide and oxygen cell in the temperature range of 298 to 1200 K.

20·10 Plot the ideal cell voltage and ε against temperature for a hydrogen and oxygen cell in the temperature range of 298 to 1200 K.

20·11 In a bank of fuel cells, each cell has an actual voltage of 60 percent of its ideal voltage when the current is 1.20 A. Twenty percent of the fuel passes through the cell without reacting and is wasted. Determine the fuel flow rate, the power output, and the heat transfer rate of each cell when they are operating at 25°C (77°F) and the reactants are gaseous hydrogen and oxygen.

20·12 In a bank of fuel cells, each cell has an actual voltage of 60 percent of its ideal voltage when the current is 1.20 A. Twenty percent of the fuel passes through the cell without reacting and is wasted. Determine the fuel flow rate, the power output, and the heat transfer rate of each cell when they are operating at 25°C (77°F) and the reactants are liquid methanol (CH_3OH) and gaseous oxygen.

20·13 In a bank of fuel cells, each cell has an actual voltage of 60 percent of its ideal voltage when the current is 1.20 A. Twenty percent of the fuel passes through the cell without reacting and is wasted. Determine the fuel flow rate, the power output, and the heat transfer rate of each cell when they are operating at 600 K (1080 R) and the reactants are gaseous methane and oxygen.

20·14 In a bank of fuel cells, each cell has an actual voltage of 60 percent of its ideal voltage when the current is 1.20 A. Twenty percent of the fuel passes through the cell without reacting and is wasted. Determine the fuel flow rate, the power output, and the heat transfer rate of each cell when they are operating at 25°C (77°F) and the reactants are gaseous propane (C_3H_8) and oxygen.

CHAPTER TWENTY-ONE

Heat Transfer

In many thermodynamic analyses we are concerned with the magnitude of heat transfer, but the mechanisms of heat transfer and the rate at which heat is transferred are not part of thermodynamics. They are the subject of the separate but closely related discipline of heat transfer. Many engineering curricula include a course in heat transfer, but we include a brief chapter on the subject in this thermodynamics textbook because it is often helpful for a student to have knowledge of the modes of heat transfer while studying thermodynamics. The discussion here is limited to the presentation of the basic equations of heat transfer in elementary form, a few definitions, and some simple applications. We restrict ourselves to the conditions of steady state, which means that the temperature at any point is constant with respect to time.

The three methods of heat transfer are conduction, convection, and radiation. We shall see that convection involves the transport of energy by mass transfer, but established usage is to include convection as a mode of heat transfer.

21·1 Steady-State One-Dimensional Conduction

Heat transfer is one-dimensional if the temperature variation can be expressed in terms of a single space coordinate. (Compare one-dimensional heat transfer with one-dimensional fluid flow described in Sec. 14·1.) Conduction is the transfer of energy through a material as a result of a temperature gradient in the material and involves no motion of the material on a macroscopic scale. The basic equation for one-dimensional steady-state heat conduction is

$$\dot{Q} = -kA \frac{dT}{dx} \tag{21·1}$$

where $\dot{Q}$ is the rate of heat transfer, A is the cross-sectional area normal to the direction of heat transfer, x is distance in the direction of heat transfer, T is temperature, and the coefficient k is a property of the material called the *thermal conductivity*. The minus sign indicates that heat is transferred in the direction of decreasing temperature.

Inspection of Eq. 21·1 shows that the dimensions of k are such that typical units are kJ/h·m·°C, W/m·°C, or B/h·ft·°F. k is a function of temperature for most materials, but in many cases the variation of k is small enough so that constant mean values can be used. Values of thermal conductivity of several substances are shown in Fig. 21·1.

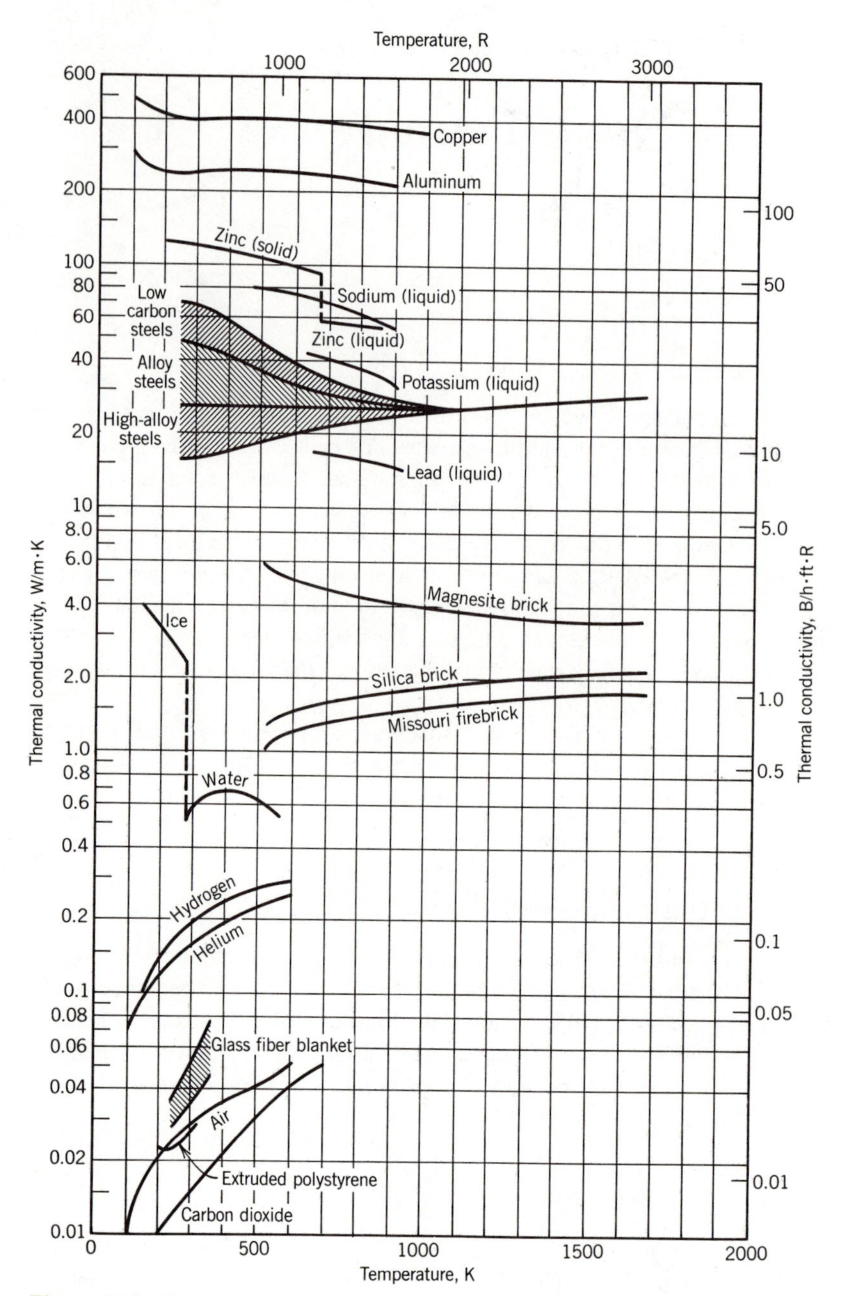

Figure 21·1. Thermal conductivities of several substances.

Consider the integration of the basic equation for two cases of a homogeneous material with constant k value: a plane wall and a cylindrical wall. Considering first the plane wall as shown in Fig. 21·2, we see that A is independent of x. We are treating the case of constant k, so equation 21·1 can be integrated to give

$$\dot{Q} = -kA \frac{T_2 - T_1}{x_2 - x_1}$$

In the field of heat transfer, ΔT is often defined as $(T_1 - T_2)$, where T_1 is the higher temperature; so the last equation can be written as

$$\dot{Q} = kA \frac{\Delta T}{\Delta x} \tag{21·2}$$

For the steady flow of heat in a radial direction through a homogeneous cylindrical wall, the heat transfer is one-dimensional because the temperature at any point is a function of the radius only. The cross-sectional area for heat transfer is $2\pi rL$, where r is the radius, and L is the axial length of the cylinder. Equation 21·1 thus becomes

$$\dot{Q} = -2\pi rLk \frac{dT}{dr}$$

Integrating for the case of constant k and substituting the limits of temperature T_2 at radius r_2 and T_1 at r_1 gives

$$\dot{Q} = \frac{2\pi Lk(T_1 - T_2)}{\ln (r_2/r_1)} \tag{21·3}$$

Now consider the case of composite walls. Figure 21·3 shows a plane wall comprised of three layers. Assume that the k of each layer is constant. Then for steady-state one-dimensional conduction through the wall, Eq. 21·2 may be applied to each section as

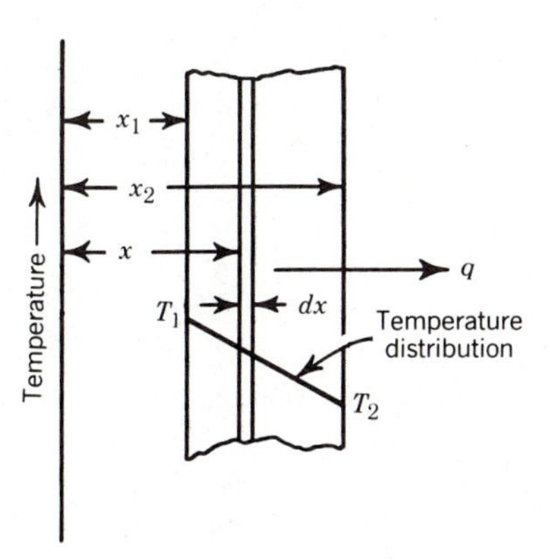

Figure 21·2. Homogeneous plane wall.

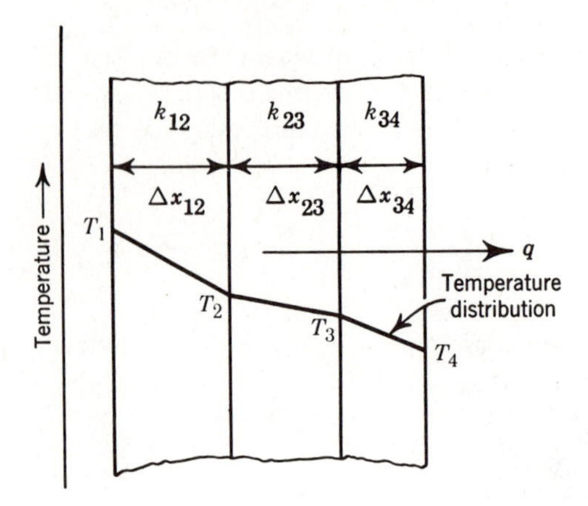

Figure 21·3. Composite plane wall.

follows:

$$\dot{Q} = k_{12} \frac{A(T_1 - T_2)}{\Delta x_{12}}$$

$$\dot{Q} = k_{23} \frac{A(T_2 - T_3)}{\Delta x_{23}}$$

$$\dot{Q} = k_{34} \frac{A(T_3 - T_4)}{\Delta x_{34}}$$

Solving each equation for the temperature difference and adding the expressions gives

$$\dot{Q} = \frac{A(T_1 - T_4)}{\dfrac{\Delta x_{12}}{k_{12}} + \dfrac{\Delta x_{23}}{k_{23}} + \dfrac{\Delta_{34}}{k_{34}}} \tag{21·4}$$

In this equation Δx_{12}, Δx_{23}, and Δx_{34} are the thicknesses of the first, second, and third layers of the wall. T_1 and T_4 are the inner and outer surface temperatures, and k_{12}, k_{23}, and k_{34} are the thermal conductivities of the materials, each of which is assumed to be constant.

A composite cylindrical wall which might be two layers of insulation on a pipe is shown in Fig. 21·4. The steady-state heat conduction in a radial direction is given by the following equation, which can be derived in a manner similar to that used in the development of Eq. 21·4:

$$\dot{Q} = \frac{2\pi L(T_1 - T_3)}{\dfrac{\ln (r_2/r_1)}{k_{12}} + \dfrac{\ln (r_3/r_2)}{k_{23}}} \tag{21·5}$$

where T_1 is the pipe outer surface temperature and T_3 is the outer surface temperature of the outer insulation layer.

The general differential equations representing the current flow in an electric system are similar to those for the heat flow in a thermal system. As a result of the similarity of

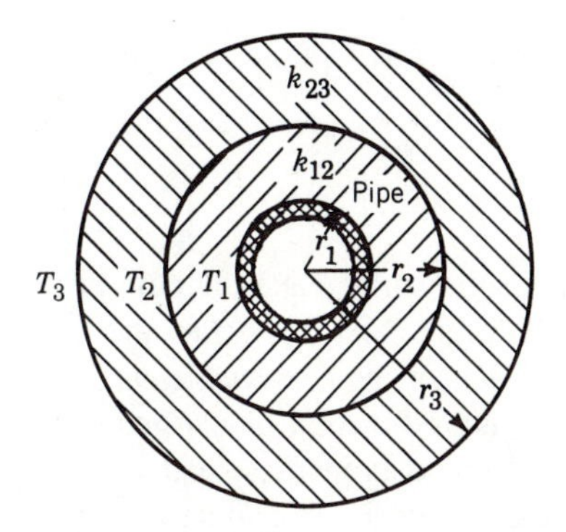

Figure 21·4. Composite cylindrical wall.

the basic equations, a close analogy exists between a thermal circuit and an equivalent electric circuit. The analogy is useful in the study of some complex heat-transfer problems.

The material through which heat flows may be considered as a resistance to the passage of heat. For the plane homogeneous wall shown in Fig. 21·2, the equation for the heat flow may be written

$$\dot{Q} = kA\frac{\Delta T}{\Delta x} = \frac{\Delta T}{\Delta x/kA} = \frac{\Delta T}{R_t}$$

In this expression R_t is the thermal resistance. For the composite plane wall shown in Fig. 21·3, the equation for the heat flow in terms of the thermal resistances is

$$\dot{Q} = \frac{T_1 - T_4}{\dfrac{\Delta x_{12}}{AK_{12}} + \dfrac{\Delta x_{23}}{Ak_{23}} + \dfrac{\Delta x_{34}}{Ak_{34}}} = \frac{T_1 - T_4}{R_{t12} + R_{t23} + R_{t34}}$$

The analogy with a series electric circuit is apparent. Although this illustration is for one-dimensional heat conduction, the analogy is particularly valuable in connection with complex two- and three-dimensional heat-conduction problems.

21·2 Convection

Convection is the transfer of energy by means of a fluid which flows between regions of different temperature. As mentioned in the introduction to this chapter, this involves mass transfer, but it is conventional to classify convection as one of the three modes of heat transfer.

Heat transfer between a solid and a fluid always involves a temperature gradient in the fluid near the solid. The region in which this gradient exists is called the *convective film* or the *thermal boundary layer*. The temperature gradient is accompanied by a density gradient that, as a result of gravity, may cause the fluid to flow. If the fluid flow past the solid surface is caused only by this density gradient, the method of heat transfer is referred to as *free* (or *natural*) *convection*. If the flow past the surface is caused by other means such as a pump, fan, or gravity flow of the entire mass of fluid, then the heat transfer is said to be by *forced convection.*

The basic equation for convective heat transfer is

$$\dot{Q} = hA\,\Delta T \tag{21·6}$$

where ΔT is the difference in temperature between the solid surface and the bulk of the fluid outside the convective film, A is the area of the surface, and h is called the *heat-transfer coefficient,* the *convective heat-transfer coefficient,* or the *film coefficient.*

The convective heat-transfer coefficient h is a complex function of (1) geometry, including the microgeometry of any solid surfaces, (2) fluid properties, and (3) flow characteristics such as velocity, pressure gradients, and turbulence. Accurate prediction of h values is difficult. Data are correlated using dimensionless ratios, and effective application of the correlations usually requires some background knowledge of fluid mechanics.

As one example, consider the flow of a fluid over a single cylinder of circular cross section with the direction of gas flow normal to the cylinder axis. Obviously, the heat-transfer coefficient varies around the circumference of the cylinder, so we use a mean value of h that can be used in the equation

$$Q = hA\,\Delta T$$

where A is the cylinder surface area. Data can best be correlated using the following three dimensionless ratios:

$$\text{Nusselt number} = \text{Nu} = \frac{hD}{k}$$

$$\text{Reynolds number} = \text{Re} = \frac{\rho D V}{\mu}$$

$$\text{Prandtl number} = \text{Pr} = \frac{c_p \mu}{k}$$

where h is the heat-transfer coefficient, D is the cylinder outer diameter, V is the velocity of the fluid approaching the cylinder, and the other symbols represent properties of the fluid: thermal conductivity k, density ρ, dynamic viscosity μ, and the specific heat at constant pressure c_p. A well-known presentation of experimental data that has been published in many books (including References 21·2, 4, 5, and 6) is shown as Fig. 21·5. These data are for air which has a Prandtl number at 1 atm and 25°C of approximately 0.72. Equations of the form

$$\text{Nu} = A(\text{Re})^m(\text{Pr})^{1/3}$$

where A and m are functions of the Reynolds number, can be fitted to the data. A word of caution is in order, however: Although coefficients in correlation equations like the one given here are often published to several significant digits, an accuracy better than about 25 percent is not to be expected in the determination of h values from such equations or from charts like Fig. 21·5. This is indicated by the data scatter on the chart, and the same conclusion can be drawn from similar charts in handbooks.

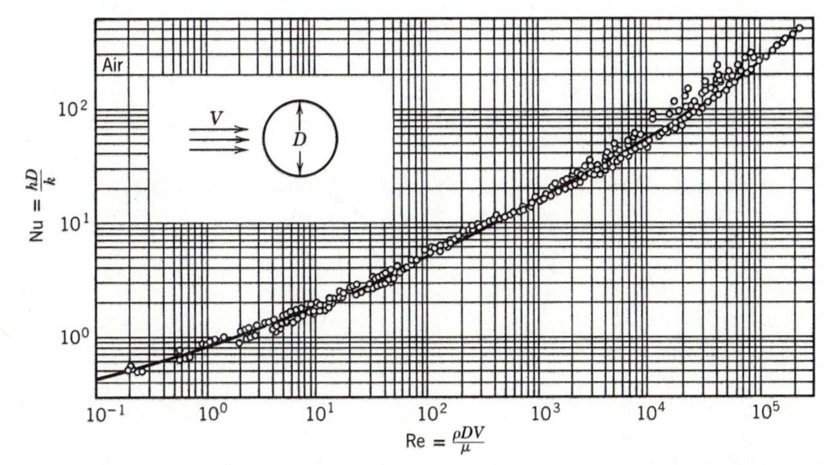

Figure 21·5. Data correlation for flow of air across a circular cylinder. From W. H. McAdams, *Heat Transmission*, 3d ed., McGraw-Hill, New York, 1954. This plot from Reference 21·4.

Correlations are different for each different flow configuration. For example, for fully developed flow (that is, flow in which the velocity distribution across the pipe remains constant from one cross section to another) inside circular tubes, a well-known correlation for a Reynolds number range of approximately 5000 to 500 000 is

$$\text{Nu} = 0.023(\text{Re})^{0.8}(\text{Pr})^{0.4}$$

Again, accuracy of better than 25 percent should not be expected. Nearly all the references listed at the end of this chaper devote much attention to the determination of h values for various situations.

It is often necessary to use two h values as in computing the heat transfer between two fluids separated by a wall as shown in Fig. 21·6. The heat is transferred between the fluids and the wall surfaces by convection and through the wall by conduction. Under these conditions it is convenient to express the transfer of heat in terms of an overall heat-transfer coefficient. The overall coefficient depends on the heat-transfer coefficients and the resistance to heat transfer through the separating wall by conduction. The heat transfer

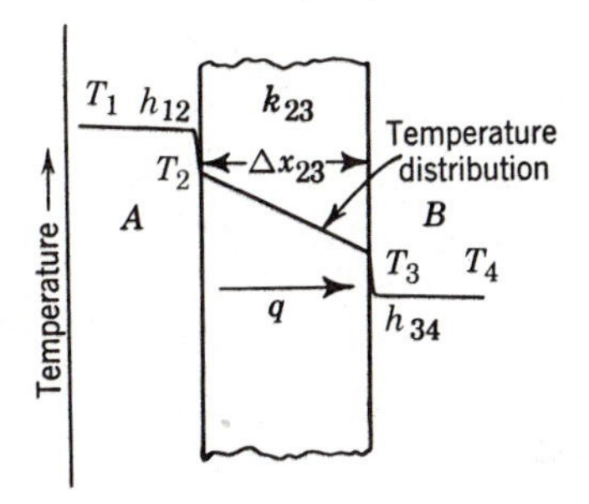

Figure 21·6. Two fluids separated by a wall.

by convection in terms of the overall coefficient may be expressed as

$$\dot{Q} = AU(T_1 - T_4) \tag{21·7}$$

The overall heat-transfer coefficient U has the same units as the heat-transfer coefficient h.

The overall heat-transfer coefficient may be expressed simply in terms of the individual heat-transfer coefficients. Consider the transfer of heat from fluid A to fluid B as in Fig. 21·6. The heat transfer through the fluid and the separating wall is expressed by each of the following:

$$\dot{Q} = Ah_{12}(T_1 - T_2)$$

$$\dot{Q} = Ak_{23}\frac{(T_2 - T_3)}{\Delta x_{23}}$$

$$\dot{Q} = Ah_{34}(T_3 - T_4)$$

Solving each equation for the temperature difference and then adding the equations gives

$$\dot{Q} = \frac{A(T_1 - T_4)}{\dfrac{1}{h_{12}} + \dfrac{\Delta x_{23}}{k_{23}} + \dfrac{1}{h_{34}}} = AU(T_1 - T_4)$$

Solving for U gives

$$U = \frac{1}{\dfrac{1}{h_{12}} + \dfrac{\Delta x_{23}}{k_{23}} + \dfrac{1}{h_{34}}} \tag{21·8}$$

Example 21·1 A furnace wall 28 cm thick is made of two materials bonded together. The "hot side" material is 20 cm thick and has a thermal conductivity of 1.40 W/m·°C; the "cold side" material is 8 cm thick and has a thermal conductivity of 0.50 W/m·°C. The inner surface is exposed to a hot gas at an average temperature of 880°C, and the outer surface is exposed to air at an average temperature of 35°C. If the gas and air convective heat-transfer coefficients are 65 and 6.0 W/m²·°C, respectively determine the heat transfer through a wall area of 2.0 m².

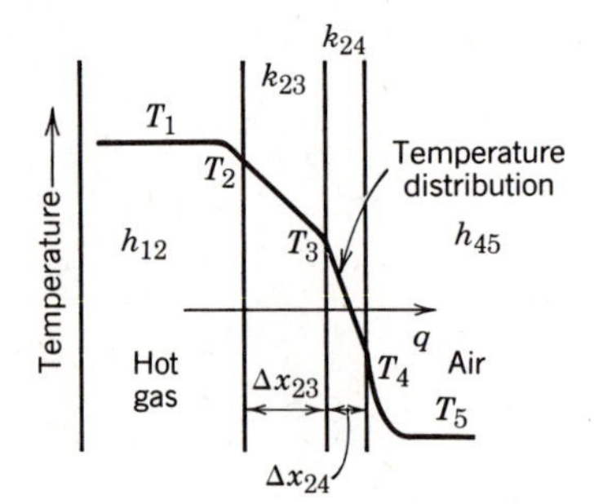

Example 21·1.

Solution. The overall heat-transfer coefficient is given by an extension of Eq. 21·8

$$U = \frac{1}{\dfrac{1}{h_{12}} + \dfrac{\Delta x_{23}}{k_{23}} + \dfrac{\Delta x_{34}}{k_{34}} + \dfrac{1}{h_{45}}}$$

$$= \frac{1}{\dfrac{1}{65} + \dfrac{0.20}{1.40} + \dfrac{0.08}{0.50} + \dfrac{1}{6.0}} = 2.06 \text{ W/m}^2\cdot{}^\circ\text{C}$$

The heat transfer is

$$\dot{Q} = AU\Delta T = 2.0(2.06)(880 - 35) = 3500 \text{ W} = 3.5 \text{ kW}$$

Consider next the heat transfer from the hot fluid to the cold fluid in a tubular heat exchanger as shown in Fig. 21·7. Heat is transferred from the hot fluid to the metal wall by convection, through the wall by conduction, and to the cold fluid by convection. Surface temperatures of the wall are T_2 and T_3, respectively. Fluid temperatures are designated by T_1 and T_4. The outside of the outer pipe is insulated, so no heat flows through it. The rate of heat transfer from the hot fluid to the inner surface of the small pipe is

$$\dot{Q} = h_{12}(2\pi r_2 L)(T_1 - T_2)$$

where $2\pi r_2 L$ is the inner surface area of the pipe and h_{12} the inner fluid heat-transfer coefficient. Similarly the rate of heat transfer from the outer surface of the inner pipe to the cold fluid is

$$\dot{Q} = h_{34}(2\pi r_3 L)(T_3 - T_4)$$

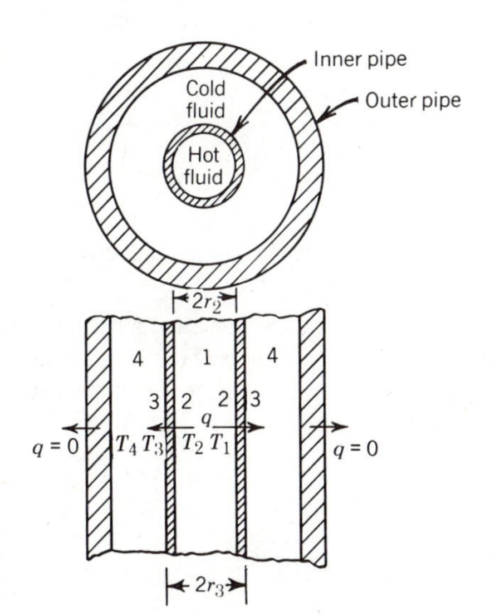

Figure 21·7. Double-pipe heat exchanger.

In this relation $2\pi r_3 L$ is the outer surface area of the inner pipe and h_{34} the heat-transfer coefficient based on this area. The rate of heat transfer through the inner pipe wall may be expressed by Eq. 21·3

$$\dot{Q} = \frac{k_{23} 2\pi L (T_2 - T_3)}{\ln (r_3/r_2)}$$

Solving each of these three relations for the temperature difference and then adding equations gives

$$\dot{Q} = \frac{L(T_1 - T_4)}{\dfrac{1}{h_{12} 2\pi r_2} + \dfrac{\ln (r_3/r_2)}{k_{23} 2\pi} + \dfrac{1}{h_{34} 2\pi r_3}} \tag{21·9}$$

Equation 21·7 may also be employed to express the heat flow in the following manner:

$$\dot{Q} = U_2 A_2 (T_1 - T_4) = U_3 A_3 (T_1 - T_4) \tag{21·10}$$

In this relation U_2 and U_3 are overall heat-transfer coefficients that are related to the inner and outer pipe areas A_2 and A_3, respectively.

Useful equations may be obtained by combining Eqs. 21·9 and 21·10.

$$U_2 = \frac{1}{\dfrac{1}{h_{12}} + \dfrac{A_2 \ln (r_3/r_2)}{k_{23} 2\pi L} + \dfrac{A_2}{h_{34} A_3}} \tag{21·11}$$

$$U_3 = \frac{1}{\dfrac{A_3}{h_{12} A_2} + \dfrac{A_3 \ln (r_3/r_2)}{k_{23} 2\pi L} + \dfrac{1}{h_{34}}} \tag{21·12}$$

Equations 21·11 and 21·12 may be simplified by substituting for the areas the equivalents $2\pi r L$ or $\pi D L$. For example, Eq. 21·11 becomes

$$U_i = \frac{1}{\dfrac{1}{h_i} + \dfrac{r_i}{k}\left(\ln \dfrac{r_o}{r_i}\right) + \dfrac{r_i}{r_o h_o}} \tag{21·13}$$

In this expression the subscripts i and o refer to the inner and outer surfaces, respectively.

Example 21·2. Compute the overall heat-transfer coefficients based on the inner and outer areas for a copper condenser tube of $\frac{3}{4}$-in. outside diameter having a wall thickness of 0.1 in. Assume that the inner heat-transfer coefficient is 280 B/h·ft²·°F, the outer heat-transfer coefficient is 2000 B/h·ft²·°F, and the thermal conductivity of the copper is 200 B/h·ft·°F.

Solution. Substituting into Eqs. 21·11 and 21·12 gives

$$U_i = \frac{1}{\dfrac{1}{280} + \dfrac{\pi(0.55)1 \ln (0.75/0.55)}{12(200)2\pi(1)} + \dfrac{\pi(0.55)1/12}{2000(\pi)0.75(1)/12}}$$
$$= 252 \text{ B/h·ft}^2\text{·°F}$$

$$U_o = \frac{1}{\dfrac{\pi(0.75)1/12}{280(\pi)0.55(1)/12} + \dfrac{\pi(0.75)1 \ln(0.75/0.55)}{12(200)2\pi(1)} + \dfrac{1}{2000}}$$
$$= 184 \text{ B/h·ft}^2\text{·°F}$$

These results show that it is important to use the overall heat-transfer coefficient based on the proper area in any calculation.

In the preceding discussion, the heat exchanger was considered so short that no question was raised regarding the value of ΔT to use in the heat-transfer equation. In most heat exchangers the ΔT varies from section to section, so there is a question as to what value of ΔT to use. We now investigate this point for the two special cases of parallel flow and counterflow. In a parallel-flow heat exchanger the two fluids enter at the same end of the exchanger and flow in the same direction. Heat transfer therefore causes the temperature difference betweeen the fluids to decrease in the direction of flow. The fluids enter a counterflow heat exchanger at opposite ends and flow in opposite directions. The initially hot fluid therefore exchanges heat first with cold fluid that has already been through most of the heat exchanger. Whether the temperature difference increases or decreases in a particular direction depends on the flow rates and the specific heats of the two fluids. Figure 21·8 shows the temperature distributions for parallel-flow and counterflow heat exchangers. The problem at hand is to find the value of mean temperature difference ΔT_m to use in the relationship

$$\dot{Q} = UA\,\Delta T_m$$

for the simple case where U is constant throughout the heat exchanger.

Consider a parallel-flow heat exchanger with a temperature distribution as shown in Fig. 21·8*a*. For an infinitesimal area dA in the heat exchanger, the infinitesimal amount of heat transferred per unit time equals $U(\Delta T)\,dA$ and can also be expressed in terms of the temperature change dT of either fluid that occurs as the fluid passes the area dA. Thus

$$U(\Delta T)\,dA = -\dot{m}_h c_{ph}\,dT_h = \dot{m}_c c_{pc}\,dT_c \qquad \text{(a)}$$

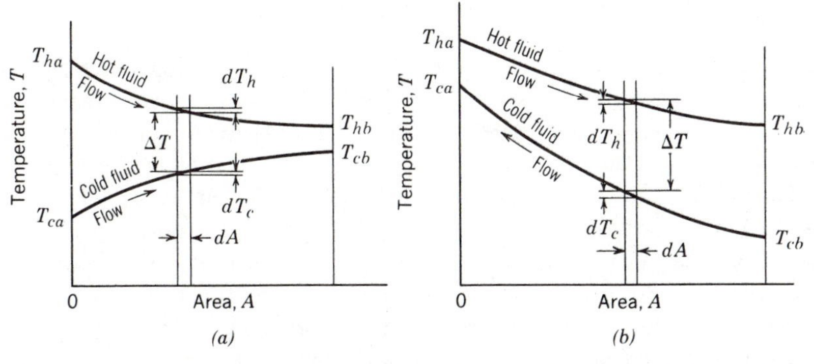

Figure 21·8. Temperature distributions through (*a*) parallel-flow and (*b*) counterflow heat exchangers.

where the subscripts h and c refer to the hot fluid and the cold fluid, respectively, and the minus sign accounts for the fact that we are concerned with heat that leaves the hot fluid. The change in ΔT between the two fluids is

$$d(\Delta T) = d(T_h - T_c) = dT_h - dT_c = -U(\Delta T)\, dA \left(\frac{1}{\dot{m}_h c_{ph}} + \frac{1}{\dot{m}_c c_{pc}} \right) \tag{b}$$

For the entire heat exchanger, if the specific heats are constant,

$$\dot{Q} = -\dot{m}_h c_{ph}(T_{hb} - T_{ha}) = \dot{m}_c c_{pc}(T_{cb} - T_{ca}) \tag{c}$$

where the subscripts a and b denote the end sections of the heat exchanger as shown in Fig. 21·8a. Substituting from Eq. c into Eq. b,

$$d(\Delta T) = -U(\Delta T)\, dA \left(\frac{T_{ha} - T_{hb}}{\dot{Q}} + \frac{T_{cb} - T_{ca}}{\dot{Q}} \right)$$

Integration over the entire heat exchanger (that is, limits of $\Delta T = T_{ha} - T_{ca}$ at $A = 0$ and $\Delta T = T_{hb} - T_{cb}$ at $A = A$) gives

$$\dot{Q} = UA \frac{(T_{hb} - T_{cb}) - (T_{ha} - T_{ca})}{\ln [(T_{hb} - T_{cb})/(T_{ha} - T_{ca})]} = UA \frac{\Delta T_b - \Delta T_a}{\ln (\Delta T_b/\Delta T_a)}$$

The quantity

$$\Delta T_m = \frac{\Delta T_b - \Delta T_a}{\ln (\Delta T_b/\Delta T_a)}$$

is called the *logarithmic mean temperature difference* (lmtd). Although we have derived the lmtd for a parallel-flow heat exchanger, it applies also to counterflow heat exchangers. For crossflow arrangements (as in an automobile-engine radiator) and others the lmtd should not be applied without various correction factors which can be found in the specialized literature.

21·3 Radiation

Even in the absence of an intervening medium, heat can be transferred from one body to another by means of thermal radiation, which is one part of the electromagnetic radiation spectrum that includes radio waves, X rays, and visible light. Thermal radiation covers a broad band of wavelengths adjacent to and partly overlapping that of visible light in the spectrum. Thermal radiation generally has longer wavelengths than visible light. The laws of optics regarding reflection and refraction apply to thermal radiation.

Whenever radiant energy falls on a body, part may be absorbed, part reflected, and the remainder transmitted through the body. If we define *absorptivity* a as the fraction of the incident radiant energy that is absorbed, *reflectivity* ρ as the fraction that is reflected, and *transmissivity* τ as the fraction that is transmitted through the body,

$$a + \rho + \tau = 1 \tag{21·14}$$

For the majority of opaque solid materials encountered in engineering, except for extremely thin bodies, $\tau = 0$; for gases, $\rho = 0$. α, ρ, and τ of a given material depend on the wavelength of the incident radiation. Their values for a given wavelength are called the *monochromatic absorptivity*, etc.

A substance that will absorb all the radiant thermal energy incident upon it at all wavelengths (thus reflecting or transmitting none) is called a *black body* or *black surface*. It should not be inferred from this name that the substance is black in color to the eye. A black body can be simulated by means of a box of uniform temperature with a small hole in it. Radiation incident upon the hole from the outside enters the box and is partially absorbed when it first strikes one of the walls. The part that is reflected from this wall strikes other walls from which only a part is reflected. If the box is large in comparison with the hole in it, a very small fraction of the radiant energy incident upon the hole from the outside is reflected back out through the hole. Therefore, the hole approaches a black body in its behavior.

The hole that acts as a black body not only absorbs radiation from outside the box but also emits radiation from the interior, and the black body is a useful idealization that serves as a standard of comparison for the radiating characteristics of actual bodies as well as for their absorption characteristics.

The Stefan-Boltzmann law* states that the rate of total radiant energy emission from a black body at absolute temperature T is

$$\dot{Q}_b = \sigma A T^4 \tag{21·15}$$

where A is the radiating area and σ is the *Stefan-Boltzmann constant*, 5.670×10^{-8} W/m^2·K^4 or 0.1714×10^{-8} B/h·ft^2·R^4. The subscript b is a reminder that this applies only to black-body radiation. Notice that $\dot{Q}_b$ is the total rate of energy radiation by a black body, not the net rate of heat transfer as commonly used in thermodynamics. $\dot{Q}_b$ is given by Eq. 21·15 for any black body, regardless of the rate at which energy is being received from the surroundings at the same time. The emissivity of a body is defined as

$$\varepsilon \equiv \frac{\dot{Q}/A}{(\dot{Q}/A)_b}$$

where $\dot{Q}/A$ is the rate of thermal radiation per unit area from the body, and $(\dot{Q}/A)_b$ is the rate of thermal radiation per unit area from a black body *at the same temperature*. For a black body, $\varepsilon = 1$; for all other bodies, $\varepsilon < 1$. Thus for a nonblack body,

$$\dot{Q} = \sigma A \varepsilon T^4 \tag{21·16}$$

A few emissivity values for approximately room temperature are given in Table 21·1 for purposes of illustration.

Kirchhoff's law states that for a body *in thermal equilibrium with its surroundings* the emissivity is equal to the absorptivity. This fact can be deduced by considering a body that is in thermal equilibrium with its surroundings and has a total thermal radiation incident upon it of $\dot{Q}$. If the absorptivity of the body is α, the rate of energy absorption

*J. Stefan formulated this law in 1879 from a study of experimental data. Five years later, L. Boltzmann derived the same law from thermodynamic principles.

TABLE 21·1 Approximate Emissivity Values for Several Materials

Material	Emissivity
Iron oxide, carbon, oil	0.8
Rubber (gray, soft), wood (planed), paper	0.85 to 0.90
Roofing paper, enamel, lacquer, porcelain (glazed), fused quartz (rough), brick (red, rough), marble (gray, polished), glass (smooth)	0.91 to 0.94
Asbestos slate (rough), ice, water	0.95 to 0.99
Polished silver	0.01
Polished aluminum	0.04
Polished steel	0.07
Oxidized steel	≈0.8

is $a\dot{Q}$. Since the body is in thermal equilibrium with its surroundings, its rate of energy emission equals its rate of energy absorption:

$$\sigma A \varepsilon T^4 = a\dot{Q}$$

ε and a must be so related that this equality holds for any body. If it is a black body, $\varepsilon = 1$ and $a = 1$. If it is a nonblack body, the fraction of $\dot{Q}$ it absorbs must be the same as the fraction of σAT^4 it emits. Therefore, $a = \varepsilon$ for any body that is in thermal equilibrium with its surroundings. A good absorber of radiant energy is also a good emitter.

The emissivity defined above and used in Eq. 21·16 and in the statement of Kirchhoff's law pertains to the total thermal radiation of all wavelengths from a body. The *monochromatic emissivity* of a body varies with the wavelength. For a black body, the distribution of the radiant energy throughout the wavelength spectrum depends only on the temperature, but for a nonblack body it depends also on the variation of the monochromatic emissivity with wavelength. It can be shown that Kirchhoff's law holds also for monochromatic emissivity and monochromatic absorptivity. A body for which the monochromatic emissivity (and consequently the monochromatic absorptivity) is the same for all wavelengths is called a *gray body*. A gray body is thus a special case of nonblack bodies.

We have so far discussed the rate of energy emission and the rate of energy absorption of a body, but the problem in engineering is usually one of the net interchange of radiant energy between two bodies. Consider first the interchange between two black bodies. If all the radiation from each body falls upon the other body, then the net interchange is simply the difference between $\sigma A_1 T_1^4$ and $\sigma A_2 T_2^4$, where 1 and 2 denote the two bodies. This would be the case for two infinitely large (that is, large enough that edge effects are negligible) parallel black planes. Thus

$$\frac{\dot{Q}}{A} = \sigma(T_1^4 - T_2^4)$$

In general, the radiant heat transfer between two black surfaces is given by

$$\dot{Q} = \sigma F_{A12} A_1 (T_1^4 - T_2^4) = \sigma F_{A21} A_2 (T_1^4 - T_2^4) \tag{21·17}$$

where F_{A12} and F_{A21} are called *shape factors, configuration factors,* or *view factors* based on area A_1 and area A_2, respectively. Shape factors depend only on the geometry of the surfaces. If dA_1 and dA_2 in Fig. 21·9 are infinitesimal parts of two black surfaces, the ϕ's are the angles between the normals to the infinitesimal areas and the line joining them, and r is the distance between them, then

$$F_{A12}A_1 = F_{A21}A_2 = \int_0^{A1}\int_0^{A2} \frac{\cos\phi_1 \cdot \cos\phi_2 \cdot dA_1 \cdot dA_2}{\pi r^2}$$

For infinitely large parallel planes, $F_{A12} = F_{A21} = 1.0$. Also, for any case in which all the radiation from body 1 is incident upon body 2 (as when body 1 is completely enclosed by body 2), $F_{A1} = 1.0$. Figures 21·10 and 21·11 give F_A values for cases of parallel disks and rectangles directly opposed. Most of the references listed at the end of this chapter give data on F_A for various configurations.

For the radiant heat transfer between gray bodies, the fact that not all the radiant energy incident upon either body is absorbed must be considered. Therefore Eq. 21·17 is modified to

$$\dot{Q} = \sigma F_\varepsilon F_{A12} A_1 (T_1^4 - T_2^4) \tag{21·18}$$

where F_ε is the *emissivity factor*, values of which can also be found in the literature for a number of geometries. For very large parallel planes

$$F_\varepsilon = \frac{1}{1/\varepsilon_1 + 1/\varepsilon_2 - 1}$$

and this value applies also to the case of one body completely enclosed by another if the inner body is large compared with the enclosure. If the enclosed body 1 is small compared with the enclosure 2, very little of the reflected radiation from the enclosing surface falls on body 1, so the surface 2 acts as though it is black regardless of its emissivity. Then

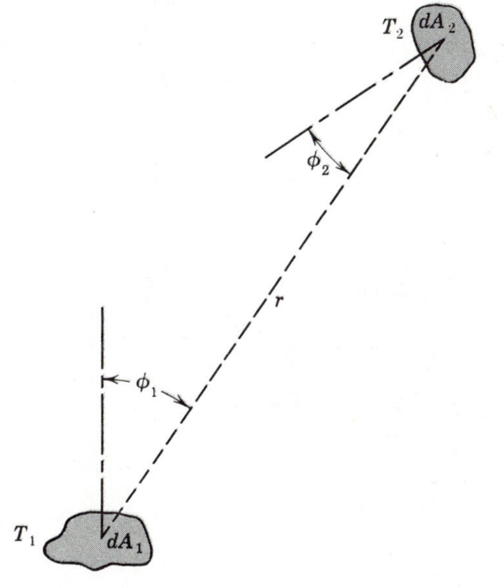

Figure 21·9. Geometrical arrangement of surfaces.

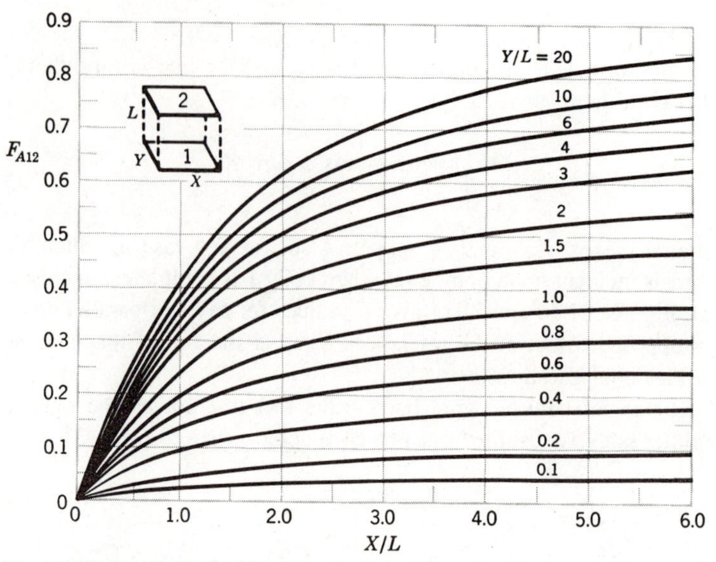

Figure 21·10. Shape factors for directly opposed parallel rectangles. From Reference 21·3.

$F_\varepsilon = \varepsilon_1$. For parallel disks and squares if the distance between them is large compared with their dimensions, $F_\varepsilon = \varepsilon_1\varepsilon_2$.

In this discussion we have ignored the effects of gases between bodies that are exchanging radiant energy. This is a reliable procedure if the gases are transparent ($\tau = 1.0$, $a = 0$, $\rho = 0$) to thermal radiation as nitrogen, oxygen, and hydrogen are. Carbon dioxide, water vapor, and many other gases absorb (and hence emit) appreciable amounts of radiation, however, and this must be accounted for if such gases are present.

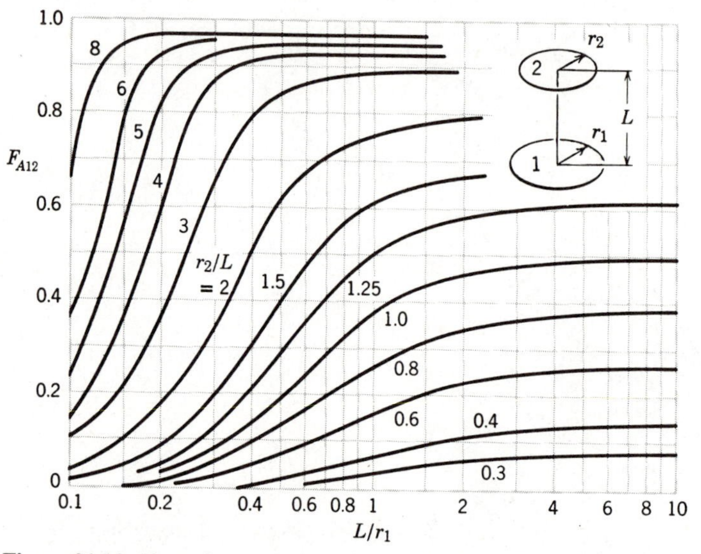

Figure 21·11. Shape factors for coaxial parallel disks. From Reference 21·3.

21·4 Summary

The three methods of heat transfer are *conduction, convection,* and *radiation*. Convection involves the transport of energy by mass transfer, but it is conventionally considered one of the modes of heat transfer.

The basic equation for one-dimensional steady-state heat conduction is

$$\dot{Q} = -kA \frac{dT}{dx} \tag{21·1}$$

where k is a material property called the *thermal conductivity,* x is distance in the direction of heat transfer, and A is area normal to x; k is usually a function of temperature. This equation can be integrated for many cases of heat conduction.

Convection involves the transfer of energy by means of a fluid flowing between regions of different temperature. For the flow of a fluid along a solid surface, the basic equation of convection is

$$\dot{Q} = hA\,\Delta T \tag{21·6}$$

where ΔT is the temperature difference between the surface and the bulk of the fluid far from the surface, A is the surface area, and h is the *convective heat-transfer coefficient,* which is a complex function of fluid properties, geometry, and flow characteristics.

For heat transfer involving both convection and conduction, a convenient formulation is

$$\dot{Q} = AU(T_1 - T_4) \tag{21·7}$$

where U is the *overall heat-transfer coefficient*. It can be expressed in terms of the various convective heat-transfer coefficients, thermal conductivities, and geometries involved.

Heat can be transferred between two bodies that are not in contact with each other, even if there is no intervening medium, by means of thermal *radiation*. Parts of the radiant energy striking any body are absorbed, reflected, or transmitted through a body. These fractions are called the absorptivity α, the reflectivity ρ, and the transmissivity τ, so $\alpha + \rho + \tau = 1$.

A substance that absorbs all of the radiant energy incident upon it at all wavelengths is called a *black body* or *black surface*. The rate of total energy emission from a black body at absolute temperatrue T is given by

$$\dot{Q}_b = \sigma A T^4 \tag{21·15}$$

where A is the radiating area and σ is the Stefan-Boltzmann constant, 5.670×10^{-8} W/m^2·K^4 or 0.714×10^{-8} B/h·ft^2·R^4.

The net radiant energy interchange between two non-black bodies at temperatures T_1 and T_2 is given by

$$\dot{Q} = \sigma F_\varepsilon F_{A12} A_1 (T_1^4 - T_2^4) \tag{21·18}$$

where F_ε and F_{A12} are usually complex functions of the surface characteristics and surface geometries.

References

21·1 Chapman, Alan J., *Heat Transfer,* 3rd ed., Macmillan, New York, 1974.

21·2 Holman, J. P., *Heat Transfer,* 5th ed., McGraw-Hill, New York, 1981.

21·3 Incropera, Frank J., and David P. Dewitt, *Fundamentals of Heat Transfer,* 2nd ed., Wiley, New York, 1985.

21·4 Karlekar, Bhalchandra V., and Robert M. Desmond, *Engineering Heat Transfer,* West Publishing, St. Paul, Minnesota, 1977.

21·5 Krieth, Frank, *Principles of Heat Transfer,* International Publishers, New York, 3rd ed., 1973.

21·6 Krieth, Frank, and William Z. Black, *Basic Heat Transfer,* Harper & Row, New York, 1980.

21·7 Lienhard, John H., *A Heat Transfer Textbook,* Prentice-Hall, Englewood Cliffs, N.J., 1981.

21·8 Schmidt, Frank W., Robert E. Henderson, and Carl H. Wohlgemuth, *Introduction to Thermal Sciences,* Wiley, New York, 1984, chapters 7, 8, 9, and 10.

21·9 Sissom, Leighton E., and Donald R. Pitts, *Elements of Transport Phenomena,* McGraw-Hill, New York, 1972.

21·10 Thomas, Lindon C., *Fundamentals of Heat Transfer,* Prentice-Hall, Englewood Cliffs, N.J., 1980.

Problems

21·1 Calculate the heat transfer through a slab 5.0 cm thick if the thermal conductivity is 0.064 W/m·K and the surface temperatures are 150 and 40°C.

21·2 A wall is made up of 23 cm of refractory material A (k_A = 2.4 W/m·K), 10 cm of insulation B (k_B = 0.087 W/m·K), and 10 cm of brick C (k_C = 1.5 W/m·K). Calculate the heat transfer for outer surface temperatures of materials A and C of 650 and 50°C respectively.

21·3 Calculate the temperatures at the contact surfaces between materials in Problem 21·2.

21·4 Determine the heat transfer through 2.0 m^2 of a wall made up of 23 cm of material A (k_A = 3.5 W/m·K) and 2.5 cm of material B (k_B = 0.50 W/m·K) with outer surface temperatures of materials A and B of 370 and 90°C, respectively.

21·5 Determine the heat loss from 18 m of horizontal pipe, 5 cm in outside diameter, covered with 2.5 cm of insulation (k = 0.070 W/m·K) if the inner and outer surface temperatures of the insulation are 260 and 65°C, respectively.

21·6 Determine the heat loss per linear meter from a pipe, 20 cm in outside diameter, covered with 5 cm of insulation (k = 0.090 W/m·K), if the inner and outer surface temperatures of the insulation are 315 and 50°C, respectively.

21·7 Determine the heat loss per square meter of insulation outer surface from a pipe 10 cm in outside diameter covered with 2.5 cm of insulation A (k_A = 1.60 W/m·K) and 5 cm of insulation B (k_B = 0.10 W/m·K) if the inner and outer surface temperatures of the insulation are 425 and 60°C, respectively.

21·8 Calculate the heat transfer between the two surfaces of a wall 2 in. thick, made of insulation, if the inner and outer surfaces are at 300 and 100°F, respectively. Assume that the thermal conductivity of the material at an average temperature of 200°F is 0.037 B/h·ft·°F.

21·9 A wall is made up of 9 in. of refractory material A (k_A = 1.4 B/h·ft·°F), 4 in. of insulation

B ($k_B = 0.05$ B/h·ft·°F), and 4 in. of brick material C ($k_C = 1.1$ B/h·ft·°F). If the outer surface of materials A and C are at temperatures of 1200 and 120°F, respectively, calculate the heat transfer for each 10 sq ft of surface area.

21·10 Calculate the temperature at the contact surfaces for materials A and B and B and C in Prob. 21.9.

21·11 Determine the heat transmitted per hour through 7 sq ft of a wall made up of 9 in. of material A ($k_A = 2.0$ B/h·ft·°F) and 1 in. of material B ($k_B = 0.3$ B/h·ft·°F). The outer surfaces of materials A and B are at temperatures of 700 and 200°F, respectively.

21·12 Determine the heat loss from 60 ft of horizontal pipe, 2 in. nominal diameter, covered with 1 in. of insulation ($k = 0.04$ B/h·ft·°F), if the inner and outer surface temperatures of the insulation are 500 and 150°F, respectively.

21·13 Calculate the heat loss per linear foot from a pipe 8 in. in nominal diameter covered with 2 in. of insulation ($k = 0.05$ B/h·ft·°F). The inside and outside surface temperatures of the insulation are 600 and 120°F, respectively.

21·14 Determine the heat loss per square foot of insulation surface for a horizontal steam pipe 4 in. in nominal diameter, covered with 1 in. of insulation A ($k_A = 0.9$ B/h·ft·°F) and 2 in. of insulation B ($k_B = 0.06$ B/h·ft·°F), if the inner and outer surface insulation temperatures are 800 and 140°F, respectively.

21·15 Sketch a curve of temperature versus distance through a plane wall in the direction of heat transfer if the thermal conductivity is given by $k = a + bT$, where a and b are positive constants.

21·16 Sketch a curve of temperature versus radius for one-dimensional heat conduction outward through cylindrical wall if the thermal conductivity is constant.

21·17 Would the existence of a negative thermal conductivity violate either the first or the second law? Explain.

21·18 Compute the heat loss by convection from 100 m of bare pipe of 5.0 cm outside diameter to the still surrounding air at 25°C if the pipe surface temperature is 150°C and the film coefficient is 5.7 W/m²·K.

21·19 The wall of a heat exchanger is made of a plate of copper, 2.0 cm thick, ($k = 345$ W/m·K). For film coefficients of 280 W/m²·K and 850 W/m²·K on the two sides of the plate and a temperature difference of 65°C between the fluids, determine the heat transfer per square meter.

21·20 Brass tubes ($k = 100$ W/m·K), 2.3 cm OD × 2.0 cm ID, are used in a heat exchanger. For inner and outer film coefficients of 8500 and 570 W/m²·K, respectively, calculate the overall heat transfer coefficient on the basis of the inner area.

21·21 Solve Problem 21·20 on the basis of the outer tube area.

21·22 A steel tube ($k = 43$ W/m·K), 2.3 cm OD × 2.0 cm ID, is used in a heat exchanger where the inner and outer film coefficients are 8500 and 570 W/m²·K, respectively. Calculate the overall heat transfer coefficient on the basis of the outer surface area.

21·23 Cooling water enters a steam condenser at 20°C and leaves at 30°C. Steam condensing is at 45°C. Calculate the log mean temperature difference.

21·24 A furnace wall 20 cm thick is made of a material having an average thermal conductivity of 1.40 W/m·K. The inner surface is exposed to hot gases at 300°C; the outer surface is in contact with air at 30°C. Heat loss through the wall amounts to 980 W/m². The total wall surface area is 25 m². Calculate the overall heat-transfer coefficient.

21·25 A heat exchanger is to be designed to transmit 525 000 kJ/h. If the maximum and minimum temperature differences between the fluids are to be 17 and 2 Celsius degrees, respectively, and the overall heat transfer coefficient is 1400 W/m^2·K, calculate the amount of heat-transfer surface required for parallel flow.

21·26 Water enters a heat exchanger at 30°C and leaves at 38°C. Steam condenses at 60°C on the outside of tubes carrying the water. For a total heat transfer area of 50 m^2 and an overall heat-transfer coefficient of 1700 W/m^2·K, calculate the heat transfer.

21·27 Water enters a parallel-flow heat exchanger at 70°C and leaves at 130°C. Hot gases enter at 300°C and leave at 150°C. For a total heat-transfer area of 465 m^2 and an overall heat-transfer coefficient of 570 W/m^2·K, calculate the heat transfer.

21·28 Compute the heat loss by convection from 350 ft of bare pipe 2 in. in nominal diameter to the still air of a room. Assume that the pipe surface temperature and air temperature are 300 and 80°F, respectively, and that the film coefficient is 1 B/h·sq ft·°F.

21·29 The wall of a heat exchanger is made of a plate of copper $\frac{3}{4}$ in. thick (k = 200 B/h·ft·°F). If the film coefficients on the two sides of the plate are 50 B/h·sq ft·°F and 150 B/h·sq ft·°F, calculate the heat passing through the wall per day per square foot. The total temperature drop is 150 Fahrenheit degrees.

21·30 Brass tubes, 1 in. OD × 0.92 in. ID, are used in a heat exchanger. If the inner and outer film coefficients are 1500 B/h·sq ft·°F and 1000 B/h·sq ft·°F, calculate the overall heat-transfer coefficient on the basis of the inner surface area. Assume that k for brass is 60 B/h·ft·°F.

21·31 In Prob. 21·30 calculate the overall heat-transfer coefficient based on the outer tube area.

21·32 A steel tube (k = 25 B/h·ft·°F) has an outside diameter of 0.9 in. and an inside diameter of 0.8 in. If the coefficients on the outside and inside are 100 B/h·sq ft·°F and 1500 B/h·sq ft·°F, respectively, calculate the overall heat-transfer coefficient, using the outer surface area as the basis.

21·33 Cooling water enters a steam condenser at 65°F and leaves at 78°F. If the condensing steam is at a temperature of 115°F, calculate the log mean temperature difference.

21·34 A furnace wall 8 in. thick is made of material having an average thermal conductivity of 0.8 B/h·ft·°F. The inner surface is exposed to hot gases at 600°F; the outer surface is in contact with air at 90°F. Heat loss through the wall by conduction and convection amounts to 292 B/h·sq ft. If the total wall surface is 300 sq ft, calculate the overall heat-transfer coefficient U.

21·35 A heat exchanger is to be designed to transmit 500 000 B/h. If the design maximum and minimum temperature differences are 30 and 3 degrees F, respectively, and the overall heat-transfer rate is 250 B/h·sq ft·°F, calculate the total heat-transfer surface required for parallel flow.

21·36 In a proposed design of a liquid-to-liquid heat exchanger, the maximum and minimum temperature differences are 40 and 10 degrees F, respectively. If 100 000 B/h is to be transmitted through 50 sq ft, calculate the overall heat-transfer coefficient required.

21·37 Water enters a heat exchanger at 85°F and leaves at 100°F. Steam at a constant temperature of 140°F condenses on the outside of the tubes carrying the water. If the total heat transfer area is 500 sq ft and the overall heat-transfer coefficient for the unit is 300 B/h·sq ft·°F, calculate the heat transfer.

21·38 Water enters a parallel flow heat exchanger at 160°F and leaves at 270°F. Hot gases enter at 580°F and leave at 300°F. If the total heat-transfer area is 5000 sq ft and the overall heat-transfer coefficient is 100 B/h·sq ft·°F, calculate the heat transfer.

21·39 A house is heated by an oil-burning furnace which is controlled by a thermostat which holds the room temperature within ±2 degrees F of its setting. From the standpoint of fuel saving alone, which procedure is better for winter operation of the heating system: (*a*) leaving the thermostat set at 70°F, or (*b*) leaving the thermostat set at 70°F during the day but setting it at 55°F for 8 h during the night? Explain your reasoning in not more than one page.

21·40 A gearbox driven by a gasoline engine has an efficiency of 92 percent when its output shaft delivers 120 hp at 240 rpm. The gearbox is cooled by means of a fan which blows room air over it. Under these steady operating conditions, how much heat is given off by the gearbox? Sketch a curve of gearbox casing temperature versus gearbox efficiency.

21·41 In order to get a uniform temperature distribution over the bottom of a surface cooking vessel, should the bottom be made (*a*) of high- or low-conductivity material, (*b*) thick or thin?

21·42 Why is grease used in a skillet? Is more grease required in an iron skillet or in a Pyrex one?

21·43 Compute the radiant heat transfer per square meter between two infinite parallel black planes at temperatures of 200 and 40°C.

21·44 Two parallel black disks 0.60 m in diameter are located 0.30 m apart and directly opposite each other. For disk temperatures of 90 and 200°C, compute the radiant heat transfer.

21·45 Two parallel black square surfaces 1.25 m on each side are located 0.30 m apart and directly opposite each other. For surface temperatures of 260 and 400°C, compute the radiant heat transfer.

21·46 Two parallel black rectangular surfaces 4.0 m on one side and 2.0 m on the other side are located 0.50 m apart and directly opposite each other. For surface temperatures of 100 and 300°C, compute the radiant heat transfer.

21·47 Calculate the radiant heat transfer through a hole, approximately 3 cm in diameter, in the wall of an open-hearth furnace, when the average furnace and outside temperatures are 1500 and 40°C, respectively.

21·48 Determine the radiant heat loss through a crack with an area of 2 cm^2 in an oven door when the inside and outside temperatures are 800 and 40°C, respectively.

21·49 Consider the heat transfer by radiation only between two infinitely large parallel black planes, each at a constant temperature. A third infinitely large black plane is inserted between the original two. Determine the ratio of the rate of heat transfer with the third plane present to the original rate.

21·50 Explain why smudge pots are effective in reducing frost damage in fruit groves. Explain how protection is also provided by large fans that circulate air through the groves.

21·51 The outer surface temperature of a gearbox operating steadily at an efficiency (ratio of power output to power input) of 95 percent is 140°F when the surroundings are at 40°F. What will the surface temperature be if the gearbox operates at 90 percent efficiency with the same power input, assuming that the gearbox is always cooled by (*a*) convection alone? (*b*) radiation alone?

21·52 Compute the radiant heat interchange per square foot between two infinite parallel black planes, the temperatures of which are 400 and 100°F, respectively.

21·53 Two parallel black disks 2 ft in diameter are located 1 ft apart and directly opposite each other. If the temperatures of the disks are 400 and 200°F, respectively, compute the net radiant heat transfer per square foot of surface area.

21·54 Two black square surfaces of 4 ft side length are located 1 ft apart and directly opposite each other. If the temperatures are 100 and 500°F, respectively, compute the net radiant heat interchange between the two surfaces.

21·55 Compute the net radiant interchange between two parallel square black surfaces of 6 ft side length located directly opposite each other at a distance of 2 ft. The temperatures of the surfaces are 200 and 600°F, respectively.

21·56 Calculate the radiant heat transfer through a hole, approximately 1 in. in diameter, in the wall of an open-hearth furnace, if the average furnace and outside temperatures are 2800 and 110°F, respectively.

21·57 Determine the radiant heat loss through a crack with an area of 0.3 sq in. in an oven door if the inside and outside temperatures are 1500 and 100°F, respectively.

APPENDIX A

Dimensions and Units

A·1 Dimensions, Units, and Conversion Factors

In this discussion we refer to physical quantities such as length, time, mass, and temperature as *dimensions*. Thus the dimensions of velocity are length/time and the dimensions of pressure are force/(length)^2. We refer to a small group of dimensions from which all others can be formed as *primary dimensions*. These can be selected arbitrarily. If we denote some possible primary dimensions by the symbols L for length, M for mass, F for force, τ for time, θ for temperature, and Q for electric charge, we can write the dimensions of other physical quantities in terms of these as shown in the following list:

Linear velocity	L/τ	Density	M/L^3
Linear acceleration	L/τ^2	Specific volume	L^3/M
Angular velocity*	$1/\tau$	Specific gravity*	—
Angular acceleration*	$1/\tau^2$	Energy	FL
Pressure	F/L^2	Specific heat	$FL/M\theta$
Work	FL	Electric potential	FL/Q
Power	FL/τ	Electric current	Q/τ

Any equation involving physical quantities, whether it is an algebraic, differential, or integral equation, must be dimensionally homogeneous; that is, equal quantities must have the same dimensions and only quantities with the same dimensions can be added to each other. Dimensions are not changed by the performance of any indicated mathematical operation such as differentiation or integration. The dimensional homogeneity of an equation such as

$$Pv = RT$$

is shown by substituting for each quantity its dimensions

$$\left[\frac{F}{L^2}\right]\left[\frac{L^3}{M}\right] = \left[\frac{FL}{M\theta}\right]\theta$$

$$\left[\frac{FL}{M}\right] = \left[\frac{FL}{M}\right]$$

*Angular measure is dimensionless and can be thought of as a ratio of two lengths: an arc length and a radius. Specific gravity is dimensionless because it is a ratio of two densities.

Also, the dimensions of

$$\left(\frac{\partial T}{\partial v}\right)_s = \left(\frac{\partial P}{\partial s}\right)_v$$

are readily checked for homogeneity by

$$\left[\frac{\theta M}{L^3}\right] = \left[\frac{F}{L^2}\right]\left[\frac{M\theta}{FL}\right]$$

$$\left[\frac{\theta M}{L^3}\right] = \left[\frac{\theta M}{L^3}\right]$$

Notice that dimensional homogeneity has nothing to do with numbers or numerical values. If an equation is valid, it must be dimensionally homogeneous. If it is not dimensionally homogeneous, it cannot be made valid simply by a judicious selection of numerical values or the use of conversion factors. The requirement of dimensional homogeneity thus provides a means of checking equations, and it is also useful in determining the general form of physical equations through a process known as dimensional analysis.

Units are arbitrary magnitudes of dimensions which are used for purposes of measurement. Samples of units of the primary dimensions are given in the following table:

Primary Dimension	Sample Units
Length	inch, centimeter, foot, mile, light year
Time	hour, minute, second, day
Force	pound force, poundal, dyne, newton
Mass	pound mass, slug, gram, kilogram
Temperature	rankine, kelvin, degree Celsius, degree Fahrenheit
Electric current	ampere, microampere

As pointed out above, any equation involving physical quantities must be dimensionally homogeneous. For numerical computations, the additional requirements of unitary homogeneity must be met. For example, although the equation

$$v = \frac{RT}{P} \tag{a}$$

is dimensionally homogeneous, substitution of numerical values as

$$v\left[\frac{\text{ft}^3}{\text{lbm}}\right] = \frac{0.0686\left[\dfrac{\text{B}}{\text{lbm}\cdot\text{R}}\right]520[\text{R}]}{14.7\left[\dfrac{\text{lbf}}{\text{in}^2}\right]} \tag{b}$$

does not satisfy the requirement of consistent units and leads to confusion. Consistent units can be provided by means of *conversion factors* or *unitary constants* such as

$$\frac{778\ \text{ft}\cdot\text{lbf}}{\text{B}} = 1$$

$$\frac{144\ \text{in}^2}{\text{ft}^2} = 1$$

that follow from the relations

$$778 \text{ ft·lbf} = 1 \text{ B}$$

$$144 \text{ in}^2 = 1 \text{ ft}^2$$

(Notice that a conversion factor is always dimensionless.) Thus instead of Eq. b we can write

$$v\left[\frac{\text{ft}^3}{\text{lbm}}\right] = \frac{0.0686\left[\dfrac{\text{B}}{\text{lbm·R}}\right]778\left[\dfrac{\text{ft·lbf}}{\text{B}}\right]520[\text{R}]}{14.7\left[\dfrac{\text{lbf}}{\text{in}^2}\right]144\left[\dfrac{\text{in}^2}{\text{ft}^2}\right]}$$

$$= 13.1\,\frac{\text{ft}^3}{\text{lbm}}$$

in which the units are consistent or homogeneous. Notice that in any equation that is dimensionally homogeneous, unitary homogeneity can be obtained by the proper selection of units and the use of conversion factors.

A·2 Systems of Units; Number of Primary Dimensions

Several systems of units are in common use. An important difference among them is the number of primary dimensions on which they are based. All involve primary dimensions such as length, time, electric current, thermodynamic temperature, and luminous intensity. Some systems, however, use both force and mass as primary dimensions while others take only one of these as primary and the other as a derived dimension on the basis of some physical law, often Newton's second law of motion. If both mass and force are taken as primary, any corresponding system of units is called an $FML\tau$ system. If only one of these two is taken as primary, the system of units is called an $FL\tau$ or $ML\tau$ system.

This book uses two systems of units, SI (for le Système International d'Unités) and a common variant of what is usually called the English system. SI is an $ML\tau$ system, and it is clearly so defined. The English system is an $FML\tau$ system, but several variants are in use, and some of these are essentially $ML\tau$ systems.

A·3 International System (SI)

The primary dimensions (and their base units) in SI are mass (kilogram), length (meter), time (seconds), electric current (ampere), thermodynamic temperature (kelvin), luminous intensity (candela), and quantity of matter (mole). The basic equation for defining force and its unit is $F = ma$, and the dimensions of the equation are

$$[F] = \left[\frac{ML}{\tau^2}\right]$$

Newton is the name given to the unit of force defined as 1 kg·m/s^2, so the defining equation is

$$1 \text{ newton} = 1 \text{ kilogram} \times 1 \text{ m/s}^2$$

Consequently, *conversion factors* are

$$1 \frac{\text{N}\cdot\text{s}^2}{\text{kg}\cdot\text{m}} = 1 \qquad \text{and} \qquad 1 \frac{\text{kg}\cdot\text{m}}{\text{N}\cdot\text{s}^2} = 1$$

As is the case with any conversion factor, these must be dimensionless and have a dimensionless value of unity.

The kinetic energy of an object with a mass of 10 kilograms moving in translation at a velocity of 100 meters per second is given by

$$KE = \frac{mV^2}{2} = \frac{10[\text{kg}](100)\ [\text{m}^2/\text{s}^2]}{2} = 50\ 000\ \frac{\text{kg}\cdot\text{m}^2}{\text{s}^2}$$

At first glance these units may not look like units of energy. However, we can convert them to more familiar units by means of a conversion factor because conversion factors always have a value of unity, and multiplication of one side of any equation by unity is always permissible

$$KE = 50\ 000 \left[\frac{\text{kg}\cdot\text{m}^2}{\text{s}^2}\right] \times 1 \left[\frac{\text{N}\cdot\text{s}^2}{\text{kg}\cdot\text{m}}\right] = 50\ 000\ \text{N}\cdot\text{m} = 50\ 000\ \text{joules}$$

Note that in a system that takes as primary dimensions only three of the four dimensions of force, mass, length, and time, the basic equation is $F = ma$, with no proportionality factor. Also, this equation is valid in any consistent set of units, and it is written without conversion factors because we never include conversion factors in symbolic equations. In numerical calculations, conversion factors often must be used to obtain consistency of units.

Some other sets of units for use with primary dimensions of $FL\tau$ or $ML\tau$ are shown in the following table.

F	M	L	τ	Conversion factor
lbf	slug	foot	second	$1 \frac{\text{slug}\cdot\text{ft}}{\text{lbf}\cdot\text{s}^2} = 1$
poundal	lbm	foot	second	$1 \frac{\text{lbm}\cdot\text{ft}}{\text{pdl}\cdot\text{s}^2} = 1$
dyne	gram	centimeter	second	$1 \frac{\text{g}\cdot\text{cm}}{\text{dyne}\cdot\text{s}^2} = 1$

A·4 English System

Force, mass, length, and time are all taken as independent primary dimensions in the English system. A relationship among these four dimensions is given by Newton's second law, $F \propto ma$, but the equation $F = ma$ is invalid because it is not dimensionally homogeneous. We must write instead

$$F = \frac{ma}{g_c}$$

where g_c is a *dimensional constant*. Since the equation above must be dimensionally homogeneous, the dimensions of g_c must be $ML/F\tau^2$.

Since g_c is a dimensional constant, it is usually written in equations to provide dimensional homogeneity. It is unlike unitary constants, such as 100 cm/m or 144 sq in./sq ft, that are necessary in numerical computations but are not written in equations because equations should be valid independently of any set of units or numerical values.*

g_c can be written with various units, and of course its numerical value depends only on the units selected. The relationship among the units of the English system—pound force (lbf), pound mass (lbm), foot, and second—is that a force of 1 lbf accelerates a mass of 1 lbm at a rate of 32.174 ft/s². Therefore

$$F = \frac{ma}{g_c}$$

$$1 \text{ lbf} = \frac{1}{g_c} \times 1 \text{ lbm} \times 32.174 \text{ ft/s}^2$$

In this set of units, therefore, g_c has the value

$$g_c = 32.174 \frac{\text{lbm}\cdot\text{ft}}{\text{lbf}\cdot\text{s}^2}$$

g_c is *dimensional,* not *dimensionless*. It is a dimensional constant and *not* a conversion factor. Conversion factors must always be dimensionless and have a dimensionless value of unity.

(Sometimes, the conversion factor 1 kg·m/N·s² is called g_c, and numerical calculations based on such a procedure can be carried out with success. However, since in SI the newton is *defined* as 1 kg·m/s², and the dimension F is *defined* by

$$[F] = \left[\frac{ML}{\tau^2}\right]$$

1 kg·m/N·s² is *dimensionless* and always has a value of unity. Thus it is a conversion factor, not a dimensional constant like the English system g_c, and need not be written in equations to ensure dimensional homogeneity.)

Many engineers successfully mix an $FL\tau$ or $ML\tau$ system with an $FML\tau$ system, even though the practice has some pitfalls. Some use essentially an $ML\tau$ system but use it with units customarily associated with an $FML\tau$ system by using conversion factors. For example, one may use the units of slug, foot, and second for three primary dimensions and then use implicitly the conversion factor

*If force, mass, length, and time are used as independent primary dimensions, equations such as

$$\gamma = \rho g \qquad KE = \frac{1}{2}mV^2 \qquad \text{Torque} = I\alpha \qquad E = mc^2 \tag{c}$$

are also dimensionally inhomogeneous, and they should be written with a dimensional constant

$$\gamma = \frac{\rho g}{g_c} \qquad KE = \frac{mV^2}{2g_c} \qquad \text{Torque} = \frac{I\alpha}{g_c} \qquad E = \frac{mc^2}{g_c} \tag{d}$$

If, on the other hand, only three primary dimensions are used, Eqs. c are correct, and Eqs. d would include a conversion factor that is superfluous until numerical calculations are made.

of 32.174 lbm/slug perhaps without ever writing the unit *slug*. As another example, one may use the units pound force, pound mass, foot, and second in an $ML\tau$ system where pound force is a unit of the derived dimension force $[F]$ defined in dimensions by $[F] = [ML/\tau^2]$. Then 1 lbf is defined as the unit of force required to accelerate a mass of 1 lbm at a rate of 32.174 ft/s². *In this case,* there is a conversion factor (dimensionless and having a value of unity like all conversion factors) of

$$32.174 \frac{\text{lbm}\cdot\text{ft}}{\text{lbf}\cdot\text{s}^2} = 1$$

(This is not g_c.) Thus the kinetic energy of an object with a mass of 20 lbm moving in translation at a velocity of 100 ft/s is

$$KE = \frac{mV^2}{2} = \frac{20[\text{lbm}](100)^2[\text{ft}^2/\text{s}^2]}{2} = 100\,000 \frac{\text{lbm}\cdot\text{ft}^2}{\text{s}^2}$$

This result is correct, but the units are not customary. To obtain customary units, we use the conversion factor given above so that

$$KE = 100\,000 \left[\frac{\text{lbm}\cdot\text{ft}^2}{\text{s}^2}\right] \times \frac{1}{32.174}\left[\frac{\text{lbf}\cdot\text{s}^2}{\text{lbm}\cdot\text{ft}}\right] = 3108 \text{ ft}\cdot\text{lbf}$$

Clearly, several procedures can be successful. It is essential, however, that you understand fully the procedure you are using.

All equations published in this book are written for $FL\tau$ or $ML\tau$ systems and are dimensionally homogeneous for such systems. Each equation is valid for any consistent set of units. Therefore, *g_c does not appear in the equations in this book.* If you wish to use strictly an $FML\tau$ system, you must remember to include in equations dimensional constants such as g_c as needed for *dimensional* homogeneity.

Recall that g_c has nothing to do with gravity or the acceleration of a freely falling body, even though historically some systems of units were founded on the basis of gravitational force measurements on bodies in certain locations.

Difficulty with units and dimensions in the English system stems largely from the use of the term pound as a unit of force and as a unit of mass. (The unfortunate use of the term *kilogram* as a unit of force promotes the same difficulty in a variant of SI.) The difficulty is compounded by the fact that in some cases, and for some purposes, mass and a particular force—weight, a gravitational force—can apparently be used interchangeably. Let us look into the relationship which involves weight and mass by applying the basic equation for a body of fixed mass

$$F = \frac{ma}{g_c} \tag{e}$$

to a body which is acted upon by a gravitational force and none other. This force we call weight. The acceleration of a body caused by this force alone is called the acceleration of gravity g. Substituting this particular force and the corresponding particular acceleration into Eq. e gives

$$w = \frac{mg}{g_c}$$

Obviously, if g is constant, then weight and mass are in a fixed proportion to each other. For accounting purposes, such as mass balances, mass and weight can actually be used interchangeably. This cannot be done if g varies. A serious danger in this practice is that weight, which is a force, comes to be thought of as mass, and confusion then results between the quantities force and mass that are as different from each other as the quantities length and time are.

Another factor in the confusion between weight and mass is that the operation called *weighing* is usually a determination of mass. This is certainly the case when a balance-type scale is used. It is also the case with a spring scale if the scale has been calibrated by means of standard masses (which are commonly and unfortunately called standard weights).

The exercises below provide a means of checking your understanding of units and dimensions.

Exercises

1. A body weighs 360 N in a location where $g = 8.80\ \text{m/s}^2$. Determine the force required to accelerate this body at a rate of $10.0\ \text{m/s}^2$.

2. Determine the force required to accelerate a body with a mass of 85 kg at a rate of $4.2\ \text{m/s}^2$ in a location where $g = 9.10\ \text{m/s}^2$ if the acceleration is (*a*) horizontal, (*b*) vertically upward.

3. Determine the force in newtons required to accelerate a body with a mass of 68 lbm at a rate of 14 m/s^2 in a location where $g = 930\ \text{cm/s}^2$.

4. A body weighs 100 N in a location where $g = 890\ \text{cm/s}^2$. (*a*) What is its mass in kg? In lbm? In slugs? (*b*) What is its weight in newtons and its mass in kilograms in a location where $g = 500\ \text{m/s}^2$?

5. A liquid has a density of $880\ \text{kg/m}^3$. Determine its specific volume and its specific weight in a location where (*a*) $g = 9.80\ \text{m/s}^2$; (*b*) $g = 920\ \text{m/s}^2$.

6. A body weighs 20 lbf in a location where $g = 31.6\ \text{ft/s}^2$. Determine the force required to accelerate this body at a rate of $30.0\ \text{ft/s}^2$.

7. Determine the force required to accelerate a body with a mass of 500 lbm at a rate of $5.0\ \text{ft/s}^2$ in a location where $g = 31.6\ \text{ft/s}^2$ if the acceleration is (*a*) horizontal, (*b*) vertically upward.

8. Determine the force in pounds required to accelerate a body with a mass of 15 kg at a rate of $10\ \text{ft/s}^2$ in a location where $g = 30.0\ \text{ft/s}^2$.

9. A body weighs 30 lbf in a location where $g = 32.0\ \text{ft/s}^2$. (*a*) What is its mass in lbm? In slugs? (*b*) What are its weight in pounds, mass in pounds, and mass in slugs in a location where $g = 16.0\ \text{ft/s}^2$?

10. A liquid has a density of $55\ \text{lbm/ft}^3$. Determine its specific volume, its specific weight, and its density in slug/ft^3 in a location where (*a*) $g = 32.2\ \text{ft/s}^2$, (*b*) $g = 31.6\ \text{ft/s}^2$.

11. What is the value of g_c in a location where a body with a mass of 270 lbm weighs 195 lbf?

12. A certain liquid has a dynamic viscosity of $6.20 \times 10^{-5}\ \text{slug/ft·s}$ at room temperature. Determine the value of dynamic viscosity in each of the following sets of units: lbf·s/ft^2, N·s/m^2, kg/m·s, and lbm/ft·s.

13. A useful dimensionless ratio in fluid mechanics, Reynolds number, is defined as $N_R = \rho DV/\mu$, where ρ is the density, D is the diameter of a pipe in which the fluid flows, V is the mean velocity of the fluid, and μ is the dynamic viscosity. Determine the value of Reynolds number for a fluid with a density of $52.0\ \text{lbm/ft}^3$ and a dynamic viscosity of $4.00 \times 10^{-5}\ \text{lbf·s/ft}^2$ flowing at a mean velocity of 8.00 ft/s through a pipe of 6.25 in. inside diameter.

APPENDIX B

Note on Partial Derivatives

Partial derivatives are important in thermodynamics because several properties are defined in terms of partial derivatives (for example, specific heats and the Joule-Thomson coefficient), and a knowledge of relationships among partial derivatives is consequently helpful in correlating properties.

B·1 A Graphical Interpretation of Partial Derivatives

If three quantities x, y, and z are functionally related, this fact can be stated in any of the following ways:

$$f(x,y,z) = 0 \qquad x = f(y,z) \qquad y = f(x,z) \qquad z = f(x,y)$$

The differential dz is given by

$$dz = \left(\frac{\partial z}{\partial x}\right)_y dx + \left(\frac{\partial z}{\partial y}\right)_x dy$$

and similar expressions can be written for dx and dy.

Figure B·1 gives a physical picture of this differential dz where a-b-c-d is any very small

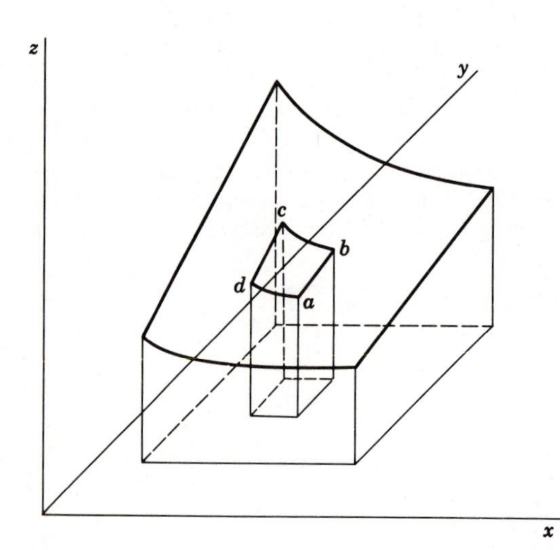

Figure B·1.

element of a surface which satisfies the equation $z = f(x,y)$. The change in z between points a and c can be evaluated as

$$\Delta z_{a\text{-}c} = \Delta z_{a\text{-}b} + \Delta z_{b\text{-}c}$$

Since b and c both lie in a plane of y = constant

$$\begin{aligned}\Delta z_{b\text{-}c} &= \Delta x_{b\text{-}c} \times (\text{slope of } zx \text{ curve in plane of } y = \text{constant}) \\ &= \Delta x_{b\text{-}c}\left(\frac{\Delta z}{\Delta x}\right)_{y=\text{const}}\end{aligned}$$

and similarly

$$\Delta z_{a\text{-}b} = \Delta y_{a\text{-}b}\left(\frac{\Delta z}{\Delta y}\right)_{x=\text{const}}$$

Then

$$\Delta z_{a\text{-}c} = \left(\frac{\Delta z}{\Delta x}\right)_y \Delta x_{b\text{-}c} + \left(\frac{\Delta z}{\Delta y}\right)_x \Delta y_{a\text{-}b}$$

If the surface element a-b-c-d is allowed to become smaller and smaller, Δz approaches dz as a limit, and

$$dz = \left(\frac{\partial z}{\partial x}\right)_y dx + \left(\frac{\partial z}{\partial y}\right)_x dy$$

Notice that, at each point on the surface, $(\partial z/\partial x)_y$ has a single fixed value, because it is the slope of a zx curve along a particular path (which in this case is a path along the intersection of the surface and a y = constant plane). dz/dx, however, has many values at each point on the surface, since it can be evaluated along many different paths through any point. In general

$$\frac{dz}{dx} = \left(\frac{\partial z}{\partial x}\right)_y + \left(\frac{\partial z}{\partial y}\right)_x \frac{dy}{dx}$$

where $(\partial z/\partial x)_y$ and $(\partial z/\partial y)_x$ are point functions which have fixed values at each point, and dz/dx and dy/dx have many possible values at each point, depending on the path along which they are evaluated.

Exercises

Figure B·2 is a contour map with lines of constant elevation H. x is distance measured east of the origin, and y is distance measured north of the origin. At each xy location there is a fixed value of H; so $H = f(x, y)$ or $f(H, x, y) = 0$. The map shows a hill with its summit at 3,5 (i.e., $x = 3$, $y = 5$). A stream flows southwest along a path given approximately by $y = x - 12$.

1. Give the coordinates of a point at which it appears that $(\partial H/\partial y)_x$ is (*a*) a maximum, (*b*) a minimum, and (*c*) zero.
2. Repeat exercise 1 for $(\partial H/\partial x)_y$.
3. Give the coordinates of a point at which it appears that $(\partial y/\partial x)_H$ equals (*a*) zero, (*b*) one, and (*c*) minus one.
4. A path on the ground is given by the relation $y = 2x$. Give the coordinates of points (if any) along this path where dH/dx is (*a*) a maximum, (*b*) a minimum, and (*c*) zero.
5. Repeat exercise 4 for $(\partial H/\partial x)_y$.

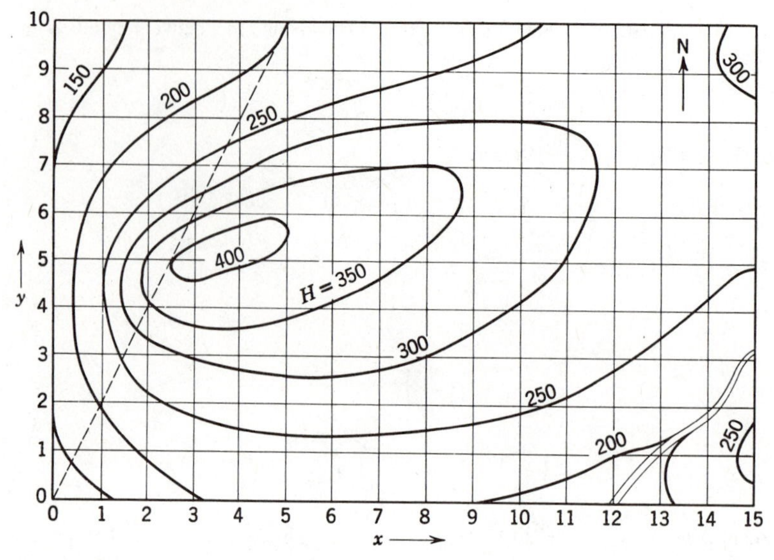

Figure B·2.

B·2 Three Relations among Partial Derivatives

If four quantities w, x, y, and z are related in such a manner that any two, but not more than two, are independent (and this is a common case in thermodynamics), then any one can be considered as a function of any other two:

$$\begin{array}{llll} w = f(x,y) & x = f(y,z) & y = f(x,z) & z = f(x,y) \\ w = f(x,z) & x = f(w,z) & y = f(w,x) & z = f(w,x) \\ w = f(y,z) & x = f(w,y) & y = f(w,z) & z = f(w,y) \end{array}$$

Of course, $f(x,y)$ signifies only *some function of* x *and* y and does not represent any particular function.

Writing the differentials of two of the quantities as expressed above is the first step toward establishing three relationships which are quite useful in thermodynamics. If we take, for example, the following pair from the list above

$$x = f(w,y) \qquad \text{and} \qquad y = f(w,z)$$

the differentials are

$$dx = \left(\frac{\partial x}{\partial w}\right)_y dw + \left(\frac{\partial x}{\partial y}\right)_w dy$$

$$dy = \left(\frac{\partial y}{\partial w}\right)_z dw + \left(\frac{\partial y}{\partial z}\right)_w dz$$

Substituting the value of dy from the second of these two equations into the first one and collecting terms gives

$$dx = \left[\left(\frac{\partial x}{\partial w}\right)_y + \left(\frac{\partial x}{\partial y}\right)_w \left(\frac{\partial y}{\partial w}\right)_z\right] dw + \left[\left(\frac{\partial x}{\partial y}\right)_w \left(\frac{\partial y}{\partial z}\right)_w\right] dz$$

Also,

$$dx = \left(\frac{\partial x}{\partial w}\right)_z dw + \left(\frac{\partial x}{\partial z}\right)_w dz$$

w and z can be varied independently, so the coefficients of dw and dz must be the same in these two equations. Equating the coefficients of dw gives the first useful relationship

$$\left(\frac{\partial x}{\partial w}\right)_z = \left(\frac{\partial x}{\partial w}\right)_y + \left(\frac{\partial x}{\partial y}\right)_w \left(\frac{\partial y}{\partial w}\right)_z \tag{a}$$

Equating the coefficients of dz and rearranging gives

$$\left(\frac{\partial x}{\partial y}\right)_w \left(\frac{\partial y}{\partial z}\right)_w = \left(\frac{\partial x}{\partial z}\right)_w \tag{b}$$

or

$$\left(\frac{\partial x}{\partial y}\right)_w \left(\frac{\partial y}{\partial z}\right)_w \left(\frac{\partial z}{\partial x}\right)_w = 1 \tag{b}$$

A third useful relationship can be obtained by first writing an equation such as

$$dz = \left(\frac{\partial z}{\partial y}\right)_x dy + \left(\frac{\partial z}{\partial x}\right)_y dx$$

in the form

$$\frac{dz}{dy} = \left(\frac{\partial z}{\partial y}\right)_x + \left(\frac{\partial z}{\partial x}\right)_y \frac{dx}{dy}$$

Since any two of the three quantities x, y, and z can be varied independently, we can assign the condition that z is constant. (This means, referring back to the graphical interpretation in Sec. B·1, that we follow a path of constant z. It does not mean that $(\partial z/\partial y)_x$ or $(\partial z/\partial x)_y$ is zero.) Then

$$0 = \left(\frac{\partial z}{\partial y}\right)_x + \left(\frac{\partial z}{\partial x}\right)_y \left(\frac{\partial x}{\partial y}\right)_z$$

This useful relationship is often written as

$$\left(\frac{\partial x}{\partial y}\right)_z \left(\frac{\partial y}{\partial z}\right)_x \left(\frac{\partial z}{\partial x}\right)_y = -1 \tag{c}$$

and is sometimes called the "rotating partials." Equation c shows the absurdity of treating ∂x, ∂y, and ∂z as quantities which can be handled alone and canceled as parts of fractions.

Equations a, b, and c are frequently used in the correlation of thermodynamic property data.

Tables

Appendix charts are listed on page 804.

TABLE A·1.1 (SI) Water: Liquid-Vapor Saturation, Temperature Table (SI)

		Specific Volume m^3/kg		Internal Energy kJ/kg			Enthalpy kJ/kg			Entropy kJ/kg·K		
Temp. °C T	Pressure kPa P	Sat. Liquid v_f	Sat. Vapor v_g	Sat. Liquid u_f	Evap. u_{fg}	Sat. Vapor u_g	Sat. Liquid h_f	Evap. h_{fg}	Sat. Vapor h_g	Sat. Liquid s_f	Evap. s_{fg}	Sat. Vapor s_g
0.01	0.6113	0.001 000	206.14	0.00	2375.3	2375.3	0.01	2501.3	2501.4	.0000	9.1562	9.1562
5	0.8721	0.001 000	147.12	20.97	2361.3	2382.3	20.98	2489.6	2510.6	.0761	8.9496	9.0257
10	1.2276	0.001 000	106.38	42.00	2347.2	2389.2	42.01	2477.7	2519.8	.1510	8.7498	8.9008
15	1.7051	0.001 001	77.93	62.99	2333.1	2396.1	62.99	2465.9	2528.9	.2245	8.5569	8.7814
20	2.339	0.001 002	57.79	83.95	2319.0	2402.9	83.96	2454.1	2538.1	.2966	8.3706	8.6672
25	3.169	0.001 003	43.36	104.88	2304.9	2409.8	104.89	2442.3	2547.2	.3674	8.1905	8.5580
30	4.246	0.001 004	32.89	125.78	2290.8	2416.6	125.79	2430.5	2556.3	.4369	8.0164	8.4533
35	5.628	0.001 006	25.22	146.67	2276.7	2423.4	146.68	2418.6	2565.3	.5053	7.8478	8.3531
40	7.384	0.001 008	19.52	167.56	2262.6	2430.1	167.57	2406.7	2574.3	.5725	7.6845	8.2570
45	9.593	0.001 010	15.26	188.44	2248.4	2436.8	188.45	2394.8	2583.2	.6387	7.5261	8.1648
50	12.349	0.001 012	12.03	209.32	2234.2	2443.5	209.33	2382.7	2592.1	.7038	7.3725	8.0763
55	15.758	0.001 015	9.568	230.21	2219.9	2450.1	230.23	2370.7	2600.9	.7679	7.2234	7.9913
60	19.940	0.001 017	7.671	251.11	2205.5	2456.6	251.13	2358.5	2609.6	.8312	7.0784	7.9096
65	25.03	0.001 020	6.197	272.02	2191.1	2463.1	272.06	2346.2	2618.3	.8935	6.9375	7.8310
70	31.19	0.001 023	5.042	292.95	2176.6	2469.6	292.98	2333.8	2626.8	.9549	6.8004	7.7553
75	38.58	0.001 026	4.131	313.90	2162.0	2475.9	313.93	2321.4	2635.3	1.0155	6.6669	7.6824
80	47.39	0.001 029	3.407	334.86	2147.4	2482.2	334.91	2308.8	2643.7	1.0753	6.5369	7.6122
85	57.83	0.001 033	2.828	355.84	2132.6	2488.4	355.90	2296.0	2651.9	1.1343	6.4102	7.5445
90	70.14	0.001 036	2.361	376.85	2117.7	2494.5	376.92	2283.2	2660.1	1.1925	6.2866	7.4791
95	84.55	0.001 040	1.982	397.88	2102.7	2500.6	397.96	2270.2	2668.1	1.2500	6.1659	7.4159

Source: Adapted from Joseph H. Keenan, Frederick G. Keyes, Philip G. Hill, and Joan G. Moore, *Steam Tables*, Wiley, New York, 1978. Abridgment from Gordon J. Van Wylen and Richard E. Sonntag, *Fundamentals of Classical Thermodynamics*, 3rd ed., Wiley, New York, 1985.

100	0.101 35	0.001 044	1.6729	418.94	2087.6	2506.5	419.04	2257.0	2676.1	1.3069	6.0480	7.3549
105	0.120 82	0.001 048	1.4194	440.02	2072.3	2512.4	440.15	2243.7	2683.8	1.3630	5.9328	7.2958
110	0.143 27	0.001 052	1.2102	461.14	2057.0	2518.1	461.30	2230.2	2691.5	1.4185	5.8202	7.2387
115	0.169 06	0.001 056	1.0366	482.30	2041.4	2523.7	482.48	2216.5	2699.0	1.4734	5.7100	7.1833
120	0.198 53	0.001 060	0.8919	503.50	2025.8	2529.3	503.71	2202.6	2706.3	1.5276	5.6020	7.1296
125	0.2321	0.001 065	0.7706	524.74	2009.9	2534.6	524.99	2188.5	2713.5	1.5813	5.4962	7.0775
130	0.2701	0.001 070	0.6685	546.02	1993.9	2539.9	546.31	2174.2	2720.5	1.6344	5.3925	7.0269
135	0.3130	0.001 075	0.5822	567.35	1977.7	2545.0	567.69	2159.6	2727.3	1.6870	5.2907	6.9777
140	0.3613	0.001 080	0.5089	588.74	1961.3	2550.0	589.13	2144.7	2733.9	1.7391	5.1908	6.9299
145	0.4154	0.001 085	0.4463	610.18	1944.7	2554.9	610.63	2129.6	2740.3	1.7907	5.0926	6.8833
150	0.4758	0.001 091	0.3928	631.68	1927.9	2559.5	632.20	2114.3	2746.5	1.8418	4.9960	6.8379
155	0.5431	0.001 096	0.3468	653.24	1910.8	2564.1	653.84	2098.6	2752.4	1.8925	4.9010	6.7935
160	0.6178	0.001 102	0.3071	674.87	1893.5	2568.4	675.55	2082.6	2758.1	1.9427	4.8075	6.7502
165	0.7005	0.001 108	0.2727	696.56	1876.0	2572.5	697.34	2066.2	2763.5	1.9925	4.7153	6.7078
170	0.7917	0.001 114	0.2428	718.33	1858.1	2576.5	719.21	2049.5	2768.7	2.0419	4.6244	6.6663
175	0.8920	0.001 121	0.2168	740.17	1840.0	2580.2	741.17	2032.4	2773.6	2.0909	4.5347	6.6256
180	1.0021	0.001 127	0.194 05	762.09	1821.6	2583.7	763.22	2015.0	2778.2	2.1396	4.4461	6.5857
185	1.1227	0.001 134	0.174 09	784.10	1802.9	2587.0	785.37	1997.1	2782.4	2.1879	4.3586	6.5465
190	1.2544	0.001 141	0.156 54	806.19	1783.8	2590.0	807.62	1978.8	2786.4	2.2359	4.2720	6.5079
195	1.3978	0.001 149	0.141 05	828.37	1764.4	2592.8	829.98	1960.0	2790.0	2.2835	4.1863	6.4698
200	1.5538	0.001 157	0.127 36	850.65	1744.7	2595.3	852.45	1940.7	2793.2	2.3309	4.1014	6.4323
205	1.7230	0.001 164	0.115 21	873.04	1724.5	2597.5	875.04	1921.0	2796.0	2.3780	4.0172	6.3952
210	1.9062	0.001 173	0.104 41	895.53	1703.9	2599.5	897.76	1900.7	2798.5	2.4248	3.9337	6.3585
215	2.104	0.001 181	0.094 79	918.14	1682.9	2601.1	920.62	1879.9	2800.5	2.4714	3.8507	6.3221
220	2.318	0.001 190	0.086 19	940.87	1661.5	2602.4	943.62	1858.5	2802.1	2.5178	3.7683	6.2861
225	2.548	0.001 199	0.078 49	963.73	1639.6	2603.3	966.78	1836.5	2803.3	2.5639	3.6863	6.2503
230	2.795	0.001 209	0.071 58	986.74	1617.2	2603.9	990.12	1813.8	2804.0	2.6099	3.6047	6.2146
235	3.060	0.001 219	0.065 37	1009.89	1594.2	2604.1	1013.62	1790.5	2804.2	2.6558	3.5233	6.1791
240	3.344	0.001 229	0.059 76	1033.21	1570.8	2604.0	1037.32	1766.5	2803.8	2.7015	3.4422	6.1437
245	3.648	0.001 240	0.054 71	1056.71	1546.7	2603.4	1061.23	1741.7	2803.0	2.7472	3.3612	6.1083

TABLE A·1.1 (SI) *(Concluded)*

		Specific Volume m³/kg		Internal Energy kJ/kg			Enthalpy kJ/kg			Entropy kJ/kg·K		
Temp. °C T	Pressure MPa P	Sat. Liquid v_f	Sat. Vapor v_g	Sat. Liquid u_f	Evap. u_{fg}	Sat. Vapor u_g	Sat. Liquid h_f	Evap. h_{fg}	Sat. Vapor h_g	Sat. Liquid s_f	Evap. s_{fg}	Sat. Vapor s_g
250	3.973	0.001 251	0.050 13	1080.39	1522.0	2602.4	1085.36	1716.2	2801.5	2.7927	3.2802	6.0730
255	4.319	0.001 263	0.045 98	1104.28	1496.7	2600.9	1109.73	1689.8	2799.5	2.8383	3.1992	6.0375
260	4.688	0.001 276	0.042 21	1128.39	1470.6	2599.0	1134.37	1662.5	2796.9	2.8838	3.1181	6.0019
265	5.081	0.001 289	0.038 77	1152.74	1443.9	2596.6	1159.28	1634.4	2793.6	2.9294	3.0368	5.9662
270	5.499	0.001 302	0.035 64	1177.36	1416.3	2593.7	1184.51	1605.2	2789.7	2.9751	2.9551	5.9301
275	5.942	0.001 317	0.032 79	1202.25	1387.9	2590.2	1210.07	1574.9	2785.0	3.0208	2.8730	5.8938
280	6.412	0.001 332	0.030 17	1227.46	1358.7	2586.1	1235.99	1543.6	2779.6	3.0668	2.7903	5.8571
285	6.909	0.001 348	0.027 77	1253.00	1328.4	2581.4	1262.31	1511.0	2773.3	3.1130	2.7070	5.8199
290	7.436	0.001 366	0.025 57	1278.92	1297.1	2576.0	1289.07	1477.1	2766.2	3.1594	2.6227	5.7821
295	7.993	0.001 384	0.023 54	1305.2	1264.7	2569.9	1316.3	1441.8	2758.1	3.2062	2.5375	5.7437
300	8.581	0.001 404	0.021 67	1332.0	1231.0	2563.0	1344.0	1404.9	2749.0	3.2534	2.4511	5.7045
305	9.202	0.001 425	0.019 948	1359.3	1195.9	2555.2	1372.4	1366.4	2738.7	3.3010	2.3633	5.6643
310	9.856	0.001 447	0.018 350	1387.1	1159.4	2546.4	1401.3	1326.0	2727.3	3.3493	2.2737	5.6230
315	10.547	0.001 472	0.016 867	1415.5	1121.1	2536.6	1431.0	1283.5	2714.5	3.3982	2.1821	5.5804
320	11.274	0.001 499	0.015 488	1444.6	1080.9	2525.5	1461.5	1238.6	2700.1	3.4480	2.0882	5.5362
330	12.845	0.001 561	0.012 996	1505.3	993.7	2498.9	1525.3	1140.6	2665.9	3.5507	1.8909	5.4417
340	14.586	0.001 638	0.010 797	1570.3	894.3	2464.6	1594.2	1027.9	2622.0	3.6594	1.6763	5.3357
350	16.513	0.001 740	0.008 813	1641.9	776.6	2418.4	1670.6	893.4	2563.9	3.7777	1.4335	5.2112
360	18.651	0.001 893	0.006 945	1725.2	626.3	2351.5	1760.5	720.5	2481.0	3.9147	1.1379	5.0526
370	21.03	0.002 213	0.004 925	1844.0	384.5	2228.5	1890.5	441.6	2332.1	4.1106	.6865	4.7971
374.14	22.09	0.003 155	0.003 155	2029.6	0	2029.6	2099.3	0	2099.3	4.4298	0	4.4298

TABLE A·1.2 (SI) Water: Liquid-Vapor Saturation, Pressure Table (SI)

		Specific Volume m³/kg		Internal Energy kJ/kg			Enthalpy kJ/kg			Entropy kJ/kg·K		
Pressure kPa P	Temp. °C T	Sat. Liquid v_f	Sat. Vapor v_g	Sat. Liquid u_f	Evap. u_{fg}	Sat. Vapor u_g	Sat. Liquid h_f	Evap. h_{fg}	Sat. Vapor h_g	Sat. Liquid s_f	Evap. s_{fg}	Sat. Vapor s_g
0.6113	0.01	0.001 000	206.14	.00	2375.3	2375.3	.01	2501.3	2501.4	.0000	9.1562	9.1562
1.0	6.98	0.001 000	129.21	29.30	2355.7	2385.0	29.30	2484.9	2514.2	.1059	8.8697	8.9756
1.5	13.03	0.001 001	87.98	54.71	2338.6	2393.3	54.71	2470.6	2525.3	.1957	8.6322	8.8279
2.0	17.50	0.001 001	67.00	73.48	2326.0	2399.5	73.48	2460.0	2533.5	.2607	8.4629	8.7237
2.5	21.08	0.001 002	54.25	88.48	2315.9	2404.4	88.49	2451.6	2540.0	.3120	8.3311	8.6432
3.0	24.08	0.001 003	45.67	101.04	2307.5	2408.5	101.05	2444.5	2545.5	.3545	8.2231	8.5776
4.0	28.96	0.001 004	34.80	121.45	2293.7	2415.2	121.46	2432.9	2554.4	.4226	8.0520	8.4746
5.0	32.88	0.001 005	28.19	137.81	2282.7	2420.5	137.82	2423.7	2561.5	.4764	7.9187	8.3951
7.5	40.29	0.001 008	19.24	168.78	2261.7	2430.5	168.79	2406.0	2574.8	.5764	7.6750	8.2515
10	45.81	0.001 010	14.67	191.82	2246.1	2437.9	191.83	2392.8	2584.7	.6493	7.5009	8.1502
15	53.97	0.001 014	10.02	225.92	2222.8	2448.7	225.94	2373.1	2599.1	.7549	7.2536	8.0085
20	60.06	0.001 017	7.649	251.38	2205.4	2456.7	251.40	2358.3	2609.7	.8320	7.0766	7.9085
25	64.97	0.001 020	6.204	271.90	2191.2	2463.1	271.93	2346.3	2618.2	.8931	6.9383	7.8314
30	69.10	0.001 022	5.229	289.20	2179.2	2468.4	289.23	2336.1	2625.3	.9439	6.8247	7.7686
40	75.87	0.001 027	3.993	317.53	2159.5	2477.0	317.58	2319.2	2636.8	1.0259	6.6441	7.6700
50	81.33	0.001 030	3.240	340.44	2143.4	2483.9	340.49	2305.4	2645.9	1.0910	6.5029	7.5939
75	91.78	0.001 037	2.217	384.31	2112.4	2496.7	384.39	2278.6	2663.0	1.2130	6.2434	7.4564
MPa												
0.100	99.63	0.001 043	1.6940	417.36	2088.7	2506.1	417.46	2258.0	2675.5	1.3026	6.0568	7.3594
0.125	105.99	0.001 048	1.3749	444.19	2069.3	2513.5	444.32	2241.0	2685.4	1.3740	5.9104	7.2844
0.150	111.37	0.001 053	1.1593	466.94	2052.7	2519.7	467.11	2226.5	2693.6	1.4336	5.7897	7.2233
0.175	116.06	0.001 057	1.0036	486.80	2038.1	2524.9	486.99	2213.6	2700.6	1.4849	5.6868	7.1717
0.200	120.23	0.001 061	0.8857	504.49	2025.0	2529.5	504.70	2201.9	2706.7	1.5301	5.5970	7.1271

TABLE A·1.2 (SI) *(Continued)*

Pressure MPa P	Temp. °C T	Specific Volume m³/kg Sat. Liquid v_f	Sat. Vapor v_g	Internal Energy kJ/kg Sat. Liquid u_f	Evap. u_{fg}	Sat. Vapor u_g	Enthalpy kJ/kg Sat. Liquid h_f	Evap. h_{fg}	Sat. Vapor h_g	Entropy kJ/kg·K Sat. Liquid s_f	Evap. s_{fg}	Sat. Vapor s_g
0.225	124.00	0.001 064	0.7933	520.47	2013.1	2533.6	520.72	2191.3	2712.1	1.5706	5.5173	7.0878
0.250	127.44	0.001 067	0.7187	535.10	2002.1	2537.2	535.37	2181.5	2716.9	1.6072	5.4455	7.0527
0.275	130.60	0.001 070	0.6573	548.59	1991.9	2540.5	548.89	2172.4	2721.3	1.6408	5.3801	7.0209
0.300	133.55	0.001 073	0.6058	561.15	1982.4	2543.6	561.47	2163.8	2725.3	1.6718	5.3201	6.9919
0.325	136.30	0.001 076	0.5620	572.90	1973.5	2546.4	573.25	2155.8	2729.0	1.7006	5.2646	6.9652
0.350	138.88	0.001 079	0.5243	583.95	1965.0	2548.9	584.33	2148.1	2732.4	1.7275	5.2130	6.9405
0.375	141.32	0.001 081	0.4914	594.40	1956.9	2551.3	594.81	2140.8	2735.6	1.7528	5.1647	6.9175
0.40	143.63	0.001 084	0.4625	604.31	1949.3	2553.6	604.74	2133.8	2738.6	1.7766	5.1193	6.8959
0.45	147.93	0.001 088	0.4140	622.77	1934.9	2557.6	623.25	2120.7	2743.9	1.8207	5.0359	6.8565
0.50	151.86	0.001 093	0.3749	639.68	1921.6	2561.2	640.23	2108.5	2748.7	1.8607	4.9606	6.8213
0.55	155.48	0.001 097	0.3427	655.32	1909.2	2564.5	655.93	2097.0	2753.0	1.8973	4.8920	6.7893
0.60	158.85	0.001 101	0.3157	669.90	1897.5	2567.4	670.56	2086.3	2756.8	1.9312	4.8288	6.7600
0.65	162.01	0.001 104	0.2927	683.56	1886.5	2570.1	684.28	2076.0	2760.3	1.9627	4.7703	6.7331
0.70	164.97	0.001 108	0.2729	696.44	1876.1	2572.5	697.22	2066.3	2763.5	1.9922	4.7158	6.7080
0.75	167.78	0.001 112	0.2556	708.64	1866.1	2574.7	709.47	2057.0	2766.4	2.0200	4.6647	6.6847
0.80	170.43	0.001 115	0.2404	720.22	1856.6	2576.8	721.11	2048.0	2769.1	2.0462	4.6166	6.6628
0.85	172.96	0.001 118	0.2270	731.27	1847.4	2578.7	732.22	2039.4	2771.6	2.0710	4.5711	6.6421
0.90	175.38	0.001 121	0.2150	741.83	1838.6	2580.5	742.83	2031.1	2773.9	2.0946	4.5280	6.6226
0.95	177.69	0.001 124	0.2042	751.95	1830.2	2582.1	753.02	2023.1	2776.1	2.1172	4.4869	6.6041
1.00	179.91	0.001 127	0.194 44	761.68	1822.0	2583.6	762.81	2015.3	2778.1	2.1387	4.4478	6.5865
1.10	184.09	0.001 133	0.177 53	780.09	1806.3	2586.4	781.34	2000.4	2781.7	2.1792	4.3744	6.5536
1.20	187.99	0.001 139	0.163 33	797.29	1791.5	2588.8	798.65	1986.2	2784.8	2.2166	4.3067	6.5233
1.30	191.64	0.001 144	0.151 25	813.44	1777.5	2591.0	814.93	1972.7	2787.6	2.2515	4.2438	6.4953
1.40	195.07	0.001 149	0.140 84	828.70	1764.1	2592.8	830.30	1959.7	2790.0	2.2842	4.1850	6.4693
1.50	198.32	0.001 154	0.131 77	843.16	1751.3	2594.5	844.89	1947.3	2792.2	2.3150	4.1298	6.4448

1.75	205.76	0.001 166	0.113 49	876.46	1721.4	2597.8	878.50	1917.9	2796.4	2.3851	4.0044	6.3896
2.00	212.42	0.001 177	0.099 63	906.44	1693.8	2600.3	908.79	1890.7	2799.5	2.4474	3.8935	6.3409
2.25	218.45	0.001 187	0.088 75	933.83	1668.2	2602.0	936.49	1865.2	2801.7	2.5035	3.7937	6.2972
2.5	223.99	0.001 197	0.079 98	959.11	1644.0	2603.1	962.11	1841.0	2803.1	2.5547	3.7028	6.2575
3.0	233.90	0.001 217	0.066 68	1004.78	1599.3	2604.1	1008.42	1795.7	2804.2	2.6457	3.5412	6.1869
3.5	242.60	0.001 235	0.057 07	1045.43	1558.3	2603.7	1049.75	1753.7	2803.4	2.7253	3.4000	6.1253
4	250.40	0.001 252	0.049 78	1082.31	1520.0	2602.3	1087.31	1714.1	2801.4	27964	3.2737	6.0701
5	263.99	0.001 286	0.039 44	1147.81	1449.3	2597.1	1154.23	1640.1	2794.3	2.9202	3.0532	5.9734
6	275.64	0.001 319	0.032 44	1205.44	1384.3	2589.7	1213.35	1571.0	2784.3	3.0267	2.8625	5.8892
7	285.88	0.001 351	0.027 37	1257.55	1323.0	2580.5	1267.00	1505.1	2772.1	3.1211	2.6922	5.8133
8	295.06	0.001 384	0.023 52	1305.57	1264.2	2569.8	1316.64	1441.3	2758.0	3.2068	2.5364	5.7432
9	303.40	0.001 418	0.020 48	1350.51	1207.3	2557.8	1363.26	1378.9	2742.1	3.2858	2.3915	5.6772
10	311.06	0.001 452	0.018 026	1393.04	1151.4	2544.4	1407.56	1317.1	2724.7	3.3596	2.2544	5.6141
11	318.15	0.001 489	0.015 987	1433.7	1096.0	2529.8	1450.1	1255.5	2705.6	3.4295	2.1233	5.5527
12	324.75	0.001 527	0.014 263	1473.0	1040.7	2513.7	1491.3	1193.6	2684.9	3.4962	1.9962	5.4924
13	330.93	0.001 567	0.012 780	1511.1	985.0	2496.1	1531.5	1130.7	2662.2	3.5606	1.8718	5.4323
14	336.75	0.001 611	0.011 485	1548.6	928.2	2476.8	1571.1	1066.5	2637.6	3.6232	1.7485	5.3717
15	342.24	0.001 658	0.010 337	1585.6	869.8	2455.5	1610.5	1000.0	2610.5	3.6848	1.6249	5.3098
16	347.44	0.001 711	0.009 306	1622.7	809.0	2431.7	1650.1	930.6	2580.6	3.7461	1.4994	5.2455
17	352.37	0.001 770	0.008 364	1660.2	744.8	2405.0	1690.3	856.9	2547.2	3.8079	1.3698	5.1777
18	357.06	0.001 840	0.007 489	1698.9	675.4	2374.3	1732.0	777.1	2509.1	3.8715	1.2329	5.1044
19	361.54	0.001 924	0.006 657	1739.9	598.1	2338.1	1776.5	688.0	2464.5	3.9388	1.0839	5.0228
20	365.81	0.002 036	0.005 834	1785.6	507.5	2293.0	1826.3	583.4	2409.7	4.0139	.9130	4.9269
21	369.89	0.002 207	0.004 952	1842.1	388.5	2230.6	1888.4	446.2	2334.6	4.1075	.6938	4.8013
22	373.80	0.002 742	0.003 568	1961.9	125.2	2087.1	2022.2	143.4	2165.6	4.3110	.2216	4.5327
22.09	374.14	0.003 155	0.003 155	2029.6	0	2029.6	2099.3	0	2099.3	4.4298	0	4.4298

TABLE A·1.3 (SI) Superheated Steam (Saturation temperature in °C shown in parentheses following each pressure) (SI)

T	v	u	h	s	v	u	h	s	v	u	h	s
	P = 0.010 MPa (45.81)				P = 0.050 MPa (81.33)				P = 0.10 MPa (99.63)			
Sat.	14.674	2437.9	2584.7	8.1502	3.240	2483.9	2645.9	7.5939	1.6940	2506.1	2675.5	7.3594
50	14.869	2443.9	2592.6	8.1749								
100	17.196	2515.5	2687.5	8.4479	3.418	2511.6	2682.5	7.6947	1.6958	2506.7	2676.2	7.3614
150	19.512	2587.9	2783.0	8.6882	3.889	2585.6	2780.1	7.9401	1.9364	2582.8	2776.4	7.6134
200	21.825	2661.3	2879.5	8.9038	4.356	2659.9	2877.7	8.1580	2.172	2658.1	2875.3	7.8343
250	24.136	2736.0	2977.3	9.1002	4.820	2735.0	2976.0	8.3556	2.406	2733.7	2974.3	8.0333
300	26.445	2812.1	3076.5	9.2813	5.284	2811.3	3075.5	8.5373	2.639	2810.4	3074.3	8.2158
400	31.063	2968.9	3279.6	9.6077	6.209	2968.5	3278.9	8.8642	3.103	2967.9	3278.2	8.5435
500	35.679	3132.3	3489.1	9.8978	7.134	3132.0	3488.7	9.1546	3.565	3131.6	3488.1	8.8342
600	40.295	3302.5	3705.4	10.1608	8.057	3302.2	3705.1	9.4178	4.028	3301.9	3704.7	9.0976
700	44.911	3479.6	3928.7	10.4028	8.981	3479.4	3928.5	9.6599	4.490	3479.2	3928.2	9.3398
800	49.526	3663.8	4159.0	10.6281	9.904	3663.6	4158.9	9.8852	4.952	3663.5	4158.6	9.5652
900	54.141	3855.0	4396.4	10.8396	10.828	3854.9	4396.3	10.0967	5.414	3854.8	4396.1	9.7767
1000	58.757	4053.0	4640.6	11.0393	11.751	4052.9	4640.5	10.2964	5.875	4052.8	4640.3	9.9764
1100	63.372	4257.5	4891.2	11.2287	12.674	4257.4	4891.1	10.4859	6.337	4257.3	4891.0	10.1659
1200	67.987	4467.9	5147.8	11.4091	13.597	4467.8	5147.7	10.6662	6.799	4467.7	5147.6	10.3463
1300	72.602	4683.7	5409.7	11.5811	14.521	4683.6	5409.6	10.8382	7.260	4683.5	5409.5	10.5183
	P = 0.20 MPa (120.23)				P = 0.30 MPa (133.55)				P = 0.40 MPa (143.63)			
Sat.	.8857	2529.5	2706.7	7.1272	.6058	2543.6	2725.3	6.9919	.4625	2553.6	2738.6	6.8959
150	.9596	2576.9	2768.8	7.2795	.6339	2570.8	2761.0	7.0778	.4708	2564.5	2752.8	6.9299
200	1.0803	2654.4	2870.5	7.5066	.7163	2650.7	2865.6	7.3115	.5342	2646.8	2860.5	7.1706
250	1.1988	2731.2	2971.0	7.7086	.7964	2728.7	2967.6	7.5166	.5951	2726.1	2964.2	7.3789
300	1.3162	2808.6	3071.8	7.8926	.8753	2806.7	3069.3	7.7022	.6548	2804.8	3066.8	7.5662
400	1.5493	2966.7	3276.6	8.2218	1.0315	2965.6	3275.0	8.0330	.7726	2964.4	3273.4	7.8985

T	v	u	h	s	v	u	h	s	v	u	h	s
500	1.7814	3130.8	3487.1	8.5133	1.1867	3130.0	3486.0	8.3251	.8893	3129.2	3484.9	8.1913
600	2.013	3301.4	3704.0	8.7770	1.3414	3300.8	3703.2	8.5892	1.0055	3300.2	3702.4	8.4558
700	2.244	3478.8	3927.6	9.0194	1.4957	3478.4	3927.1	8.8319	1.1215	3477.9	3926.5	8.6987
800	2.475	3663.1	4158.2	9.2449	1.6499	3662.9	4157.8	9.0576	1.2372	3662.4	4157.3	8.9244
900	2.706	3854.5	4395.8	9.4566	1.8041	3854.2	4395.4	9.2692	1.3529	3853.9	4395.1	9.1362
1000	2.937	4052.5	4640.0	9.6563	1.9581	4052.3	4639.7	9.4690	1.4685	4052.0	4639.4	9.3360
1100	3.168	4257.0	4890.7	9.8458	2.1121	4256.8	4890.4	9.6585	1.5840	4256.5	4890.2	9.5256
1200	3.399	4467.5	5147.3	10.0262	2.2661	4467.2	5147.1	9.8389	1.6996	4467.0	5146.8	9.7060
1300	3.630	4683.2	5409.3	10.1982	2.4201	4683.0	5409.0	10.0110	1.8151	4682.8	5408.8	9.8780

	P = 0.50 MPa (151.86)				P = 0.60 MPa (158.85)				P = 0.80 MPa (170.43)			
Sat.	.3749	2561.2	2748.7	6.8213	.3157	2567.4	2756.8	6.7600	.2404	2576.8	2769.1	6.6628
200	.4249	2642.9	2855.4	7.0592	.3520	2638.9	2850.1	6.9665	.2608	2630.6	2839.3	6.8158
250	.4744	2723.5	2960.7	7.2709	.3938	2720.9	2957.2	7.1816	.2931	2715.5	2950.0	7.0384
300	.5226	2802.9	3064.2	7.4599	.4344	2801.0	3061.6	7.3724	.3241	2797.2	3056.5	7.2328
350	.5701	2882.6	3167.7	7.6329	.4742	2881.2	3165.7	7.5464	.3544	2878.2	3161.7	7.4089
400	.6173	2963.2	3271.9	7.7938	.5137	2962.1	3270.3	7.7079	.3843	2959.7	3267.1	7.5716
500	.7109	3128.4	3483.9	8.0873	.5920	3127.6	3482.8	8.0021	.4433	3126.0	3480.6	7.8673
600	.8041	3299.6	3701.7	8.3522	.6697	3299.1	3700.9	8.2674	.5018	3297.9	3699.4	8.1333
700	.8969	3477.5	3925.9	8.5952	.7472	3477.0	3925.3	8.5107	.5601	3476.2	3924.2	8.3770
800	.9896	3662.1	4156.9	8.8211	.8245	3661.8	4156.5	8.7367	.6181	3661.1	4155.6	8.6033
900	1.0822	3853.6	4394.7	9.0329	.9017	3853.4	4394.4	8.9486	.6761	3852.8	4393.7	8.8153
1000	1.1747	4051.8	4639.1	9.2328	.9788	4051.5	4638.8	9.1485	.7340	4051.0	4638.2	9.0153

TABLE A·1.3 (SI) *(Continued)*

T	v	u	h	s	v	u	h	s	v	u	h	s
	P = 0.50 MPa (151.86)				P = 0.60 MPa (158.85)				P = 0.80 MPa (170.43)			
1100	1.2672	4256.3	4889.9	9.4224	1.0559	4256.1	4889.6	9.3381	.7919	4255.6	4889.1	9.2050
1200	1.3596	4466.8	5146.6	9.6029	1.1330	4466.5	5146.3	9.5185	.8497	4466.1	5145.9	9.3855
1300	1.4521	4682.5	5408.6	9.7749	1.2101	4682.3	5408.3	9.6906	.9076	4681.8	5407.9	9.5575
	P = 1.00 MPa (179.91)				P = 1.20 MPa (187.99)				P = 1.40 MPa (195.07)			
Sat.	.194 44	2583.6	2778.1	6.5865	.163 33	2588.8	2784.8	6.5233	.140 84	2592.8	2790.0	6.4693
200	.2060	2621.9	2827.9	6.6940	.169 30	2612.8	2815.9	6.5898	.143 02	2603.1	2803.3	6.4975
250	.2327	2709.9	2942.6	6.9247	.192 34	2704.2	2935.0	6.8294	.163 50	2698.3	2927.2	6.7467
300	.2579	2793.2	3051.2	7.1229	.2138	2789.2	3045.8	7.0317	.182 28	2785.2	3040.4	6.9534
350	.2825	2875.2	3157.7	7.3011	.2345	2872.2	3153.6	7.2121	.2003	2869.2	3149.5	7.1360
400	.3066	2957.3	3263.9	7.4651	.2548	2954.9	3260.7	7.3774	.2178	2952.5	3257.5	7.3026
500	.3541	3124.4	3478.5	7.7622	.2946	3122.8	3476.3	7.6759	.2521	3121.1	3474.1	7.6027
600	.4011	3296.8	3697.9	8.0290	.3339	3295.6	3696.3	7.9435	.2860	3294.4	3694.8	7.8710
700	.4478	3475.3	3923.1	8.2731	.3729	3474.4	3922.0	8.1881	.3195	3473.6	3920.8	8.1160
800	.4943	3660.4	4154.7	8.4996	.4118	3659.7	4153.8	8.4148	.3528	3659.0	4153.0	8.3431
900	.5407	3852.2	4392.9	8.7118	.4505	3851.6	4392.2	8.6272	.3861	3851.1	4391.5	8.5556
1000	.5871	4050.5	4637.6	8.9119	.4892	4050.0	4637.0	8.8274	.4192	4049.5	4636.4	8.7559
1100	.6335	4255.1	4888.6	9.1017	.5278	4254.6	4888.0	9.0172	.4524	4254.1	4887.5	8.9457
1200	.6798	4465.6	5145.4	9.2822	.5665	4465.1	5144.9	9.1977	.4855	4464.7	5144.4	9.1262
1300	.7261	4681.3	5407.4	9.4543	.6051	4680.9	5407.0	9.3698	.5186	4680.4	5406.5	9.2984
	P = 1.60 MPa (201.41)				P = 1.80 MPa (207.15)				P = 2.00 MPa (212.42)			
Sat.	.123 80	2596.0	2794.0	6.4218	.110 42	2598.4	2797.1	6.3794	.099 63	2600.3	2799.5	6.3409
225	.132 87	2644.7	2857.3	6.5518	.116 73	2636.6	2846.7	6.4808	.103 77	2628.3	2835.8	6.4147
250	.141 84	2692.3	2919.2	6.6732	.124 97	2686.0	2911.0	6.6066	.111 44	2679.6	2902.5	6.5453
300	.158 62	2781.1	3034.8	6.8844	.140 21	2776.9	3029.2	6.8226	.125 47	2772.6	3023.5	6.7664
350	.174 56	2866.1	3145.4	7.0694	.154 57	2863.0	3141.2	7.0100	.138 57	2859.8	3137.0	6.9563
400	.190 05	2950.1	3254.2	7.2374	.168 47	2947.7	3250.9	7.1794	.151 20	2945.2	3247.6	7.1271
500	.2203	3119.5	3472.0	7.5390	.195 50	3117.9	3469.8	7.4825	.175 68	3116.2	3467.6	7.4317
600	.2500	3293.3	3693.2	7.8080	.2220	3292.1	3691.7	7.7523	.199 60	3290.9	3690.1	7.7024
700	.2794	3472.7	3919.7	8.0535	.2482	3471.8	3918.5	7.9983	.2232	3470.9	3917.4	7.9487

800	.3086	3658.3	4152.1	8.2808	.2742	3657.6	4151.2	8.2258	.2467	3657.0	4150.3	8.1765
900	.3377	3850.5	4390.8	8.4935	.3001	3849.9	4390.1	8.4386	.2700	3849.3	4389.4	8.3895
1000	.3668	4049.0	4635.8	8.6938	.3260	4048.5	4635.2	8.6391	.2933	4048.0	4634.6	8.5901
1100	.3958	4253.7	4887.0	8.8837	.3518	4253.2	4886.4	8.8290	.3166	4252.7	4885.9	8.7800
1200	.4248	4464.2	5143.9	9.0643	.3776	4463.7	5143.4	9.0096	.3398	4463.3	5142.9	8.9607
1300	.4538	4679.9	5406.0	9.2364	.4034	4679.5	5405.6	9.1818	.3631	4679.0	5405.1	9.1329
	P = 2.50 MPa (223.99)				P = 3.00 MPa (233.90)				P = 3.50 MPa (242.60)			
Sat.	.079 98	2603.1	2803.1	6.2575	.066 68	2604.1	2804.2	6.1869	.057 07	2603.7	2803.4	6.1253
225	.080 27	2605.6	2806.3	6.2639								
250	.087 00	2662.6	2880.1	6.4085	.070 58	2644.0	2855.8	6.2872	.058 72	2623.7	2829.2	6.1749
300	.098 90	2761.6	3008.8	6.6438	.081 14	2750.1	2993.5	6.5390	.068 42	2738.0	2977.5	6.4461
350	.109 76	2851.9	3126.3	6.8403	.090 53	2843.7	3115.3	6.7428	.076 78	2835.3	3104.0	6.6579
400	.120 10	2939.1	3239.3	7.0148	.099 36	2932.8	3230.9	6.9212	.084 53	2926.4	3222.3	6.8405
450	.130 14	3025.5	3350.8	7.1746	.107 87	3020.4	3344.0	7.0834	.091 96	3015.3	3337.2	7.0052
500	.139 98	3112.1	3462.1	7.3234	.116 19	3108.0	3456.5	7.2338	.099 18	3103.0	3450.9	7.1572
600	.159 30	3288.0	3686.3	7.5960	.132 43	3285.0	3682.3	7.5085	.113 24	3282.1	3678.4	7.4339
700	.178 32	3468.7	3914.5	7.8435	.148 38	3466.5	3911.7	7.7571	.126 99	3464.3	3908.8	7.6837
800	.197 16	3655.3	4148.2	8.0720	.164 14	3653.5	4145.9	7.9862	.140 56	3651.8	4143.7	7.9134
900	.215 90	3847.9	4387.6	8.2853	.179 80	3846.5	4385.9	8.1999	.154 02	3845.0	4384.1	8.1276
1000	.2346	4046.7	4633.1	8.4861	.195 41	4045.4	4631.6	8.4009	.167 43	4044.1	4630.1	8.3288
1100	.2532	4251.5	4884.6	8.6762	.210 98	4250.3	4883.3	8.5912	.180 80	4249.2	4881.9	8.5192
1200	.2718	4462.1	5141.7	8.8569	.226 52	4460.9	5140.5	8.7720	.194 15	4459.8	5139.3	8.7000
1300	.2905	4677.8	5404.0	9.0291	.242 06	4676.6	5402.8	8.9442	.207 49	4675.5	5401.7	8.8723

TABLE A·1.3 (SI) *(Continued)*

T	v	u	h	s	v	u	h	s	v	u	h	s
	P = 4.0 MPa (250.40)				P = 4.5 MPa (257.49)				P = 5.0 MPa (263.99)			
Sat.	.049 78	2602.3	2801.4	6.0701	.044 06	2600.1	2798.3	6.0198	.039 44	2597.1	2794.3	5.9734
275	.054 57	2667.9	2886.2	6.2285	.047 30	2650.3	2863.2	6.1401	.041 41	2631.3	2838.3	6.0544
300	.058 84	2725.3	2960.7	6.3615	.051 35	2712.0	2943.1	6.2828	.045 32	2698.0	2924.5	6.2084
350	.066 45	2826.7	3092.5	6.5821	.058 40	2817.8	3080.6	6.5131	.051 94	2808.7	3068.4	6.4493
400	.073 41	2919.9	3213.6	6.7690	.064 75	2913.3	3204.7	6.7047	.057 81	2906.6	3195.7	6.6459
450	.080 02	3010.2	3330.3	6.9363	.070 74	3005.0	3323.3	6.8746	.063 30	2999.7	3316.2	6.8186
500	.086 43	3099.5	3445.3	7.0901	.076 51	3095.3	3439.6	7.0301	.068 57	3091.0	3433.8	6.9759
600	.098 85	3279.1	3674.4	7.3688	.087 65	3276.0	3670.5	7.3110	.078 69	3273.0	3666.5	7.2589
700	.110 95	3462.1	3905.9	7.6198	.098 47	3459.9	3903.0	7.5631	.088 49	3457.6	3900.1	7.5122
800	.122 87	3650.0	4141.5	7.8502	.109 11	3648.3	4139.3	7.7942	.098 11	3646.6	4137.1	7.7440
900	.134 69	3843.6	4382.3	8.0647	.119 65	3842.2	4380.6	8.0091	.107 62	3840.7	4378.8	7.9593
1000	.146 45	4042.9	4628.7	8.2662	.130 13	4041.6	4627.2	8.2108	.117 07	4040.4	4625.7	8.1612
1100	.158 17	4248.0	4880.6	8.4567	.140 56	4246.8	4879.3	8.4015	.126 48	4245.6	4878.0	8.3520
1200	.169 87	4458.6	5138.1	8.6376	.150 98	4457.5	5136.9	8.5825	.135 87	4456.3	5135.7	8.5331
1300	.181 56	4674.3	5400.5	8.8100	.161 39	4673.1	5399.4	8.7549	.145 26	4672.0	5398.2	8.7055
	P = 6.0 MPa (275.64)				P = 7.0 MPa (285.88)				P = 8.0 MPa (295.06)			
Sat.	.032 44	2589.7	2784.3	5.8892	.027 37	2580.5	2772.1	5.8133	.023 52	2569.8	2758.0	5.7432
300	.036 16	2667.2	2884.2	6.0674	.029 47	2632.2	2838.4	5.9305	.024 26	2590.9	2785.0	5.7906
350	.042 23	2789.6	3043.0	6.3335	.035 24	2769.4	3016.0	6.2283	.029 95	2747.7	2987.3	6.1301
400	.047 39	2892.9	3177.2	6.5408	.039 93	2878.6	3158.1	6.4478	.034 32	2863.8	3138.3	6.3634
450	.052 14	2988.9	3301.8	6.7193	.044 16	2978.0	3287.1	6.6327	.038 17	2966.7	3272.0	6.5551
500	.056 65	3082.2	3422.2	6.8803	.048 14	3073.4	3410.3	6.7975	.041 75	3064.3	3398.3	6.7240
550	.061 01	3174.6	3540.6	7.0288	.051 95	3167.2	3530.9	6.9486	.045 16	3159.8	3521.0	6.8778
600	.065 25	3266.9	3658.4	7.1677	.055 65	3260.7	3650.3	7.0894	.048 45	3254.4	3642.0	7.0206
700	.073 52	3453.1	3894.2	7.4234	.062 83	3448.5	3888.3	7.3476	.054 81	3443.9	3882.4	7.2812
800	.081 60	3643.1	4132.7	7.6566	.069 81	3639.5	4128.2	7.5822	.060 97	3636.0	4123.8	7.5173
900	.089 58	3837.8	4375.3	7.8727	.076 69	3835.0	4371.8	7.7991	.067 02	3832.1	4368.3	7.7351
1000	.097 49	4037.8	4622.7	8.0751	.083 50	4035.3	4619.8	8.0020	.073 01	4032.8	4616.9	7.9384
1100	.105 36	4243.3	4875.4	8.2661	.090 27	4240.9	4872.8	8.1933	.078 96	4238.6	4870.3	8.1300

T	v	u	h	s	v	u	h	s	v	u	h	s
1200	.113 21	4454.0	5133.3	8.4474	.097 03	4451.7	5130.9	8.3747	.084 89	4449.5	5128.5	8.3115
1300	.121 06	4669.6	5396.0	8.6199	.103 77	4667.3	5393.7	8.5473	.090 80	4665.0	5391.5	8.4842
	P = 9.0 MPa (303.40)				P = 10.0 MPa (311.06)				P = 12.5 MPa (327.89)			
Sat.	.020 48	2557.8	2742.1	5.6772	.018 026	2544.4	2724.7	5.6141	.013 495	2505.1	2673.8	5.4624
325	.023 27	2646.6	2856.0	5.8712	.019 861	2610.4	2809.1	5.7568				
350	.025 80	2724.4	2956.6	6.0361	.022 42	2699.2	2923.4	5.9443	.016 126	2624.6	2826.2	5.7118
400	.029 93	2848.4	3117.8	6.2854	.026 41	2832.4	3096.5	6.2120	.020 00	2789.3	3039.3	6.0417
450	.033 50	2955.2	3256.6	6.4844	.029 75	2943.4	3240.9	6.4190	.022 99	2912.5	3199.8	6.2719
500	.036 77	3055.2	3386.1	6.6576	.032 79	3045.8	3373.7	6.5966	.025 60	3021.7	3341.8	6.4618
550	.039 87	3152.2	3511.0	6.8142	.035 64	3144.6	3500.9	6.7561	.028 01	3125.0	3475.2	6.6290
600	.042 85	3248.1	3633.7	6.9589	.038 37	3241.7	3625.3	6.9029	.030 29	3225.4	3604.0	6.7810
650	.045 74	3343.6	3755.3	7.0943	.041 01	3338.2	3748.2	7.0398	.032 48	3324.4	3730.4	6.9218
700	.048 57	3439.3	3876.5	7.2221	.043 58	3434.7	3870.5	7.1687	.034 60	3422.9	3855.3	7.0536
800	.054 09	3632.5	4119.3	7.4596	.048 59	3628.9	4114.8	7.4077	.038 69	3620.0	4103.6	7.2965
900	.059 50	3829.2	4364.8	7.6783	.053 49	3826.3	4361.2	7.6272	.042 67	3819.1	4352.5	7.5182
1000	.064 85	4030.3	4614.0	7.8821	.058 32	4027.8	4611.0	7.8315	.046 58	4021.6	4603.8	7.7237
1100	.070 16	4236.3	4867.7	8.0740	.063 12	4234.0	4865.1	8.0237	.050 45	4228.2	4858.8	7.9165
1200	.075 44	4447.2	5126.2	8.2556	.067 89	4444.9	5123.8	8.2055	.054 30	4439.3	5118.0	8.0987
1300	.080 72	4662.7	5389.2	8.4284	.072 65	4460.5	5387.0	8.3783	.058 13	4654.8	5381.4	8.2717
	P = 15.0 MPa (342.24)				P = 17.5 MPa (354.75)				P = 20.0 MPa (365.81)			
Sat.	.010 337	2455.5	2610.5	5.3098	.007 920	2390.2	2528.8	5.1419	.005 834	2293.0	2409.7	4.9269
350	.011 470	2520.4	2692.4	5.4421								

TABLE A·1.3 (SI) *(Concluded)*

T	v	u	h	s	v	u	h	s	v	u	h	s
	P = 15.0 MPa (342.24)				P = 17.5 MPa (354.75)				P = 20.0 MPa (365.81)			
400	.015 649	2740.7	2975.5	5.8811	.012 447	2685.0	2902.9	5.7213	.009 942	2619.3	2818.1	5.5540
450	.018 445	2879.5	3156.2	6.1404	.015 174	2844.2	3109.7	6.0184	.012 695	2806.2	3060.1	5.9017
500	.020 80	2996.6	3308.6	6.3443	.017 358	2970.3	3274.1	6.2383	.014 768	2942.9	3238.2	6.1401
550	.022 93	3104.7	3448.6	6.5199	.019 288	3083.9	3421.4	6.4230	.016 555	3062.4	3393.5	6.3348
600	.024 91	3208.6	3582.3	6.6776	.021 06	3191.5	3560.1	6.5866	.018 178	3174.0	3537.6	6.5048
650	.026 80	3310.3	3712.3	6.8224	.022 74	3296.0	3693.9	6.7357	.019 693	3281.4	3675.3	6.6582
700	.028 61	3410.9	3840.1	6.9572	.024 34	3398.7	3824.6	6.8736	.021 13	3386.4	3809.0	6.7993
800	.032 10	3610.9	4092.4	7.2040	.027 38	3601.8	4081.1	7.1244	.023 85	3592.7	4069.7	7.0544
900	.035 46	3811.9	4343.8	7.4279	.030 31	3804.7	4335.1	7.3507	.026 45	3797.5	4326.4	7.2830
1000	.038 75	4015.4	4596.6	7.6348	.033 16	4009.3	4589.5	7.5589	.028 97	4003.1	4582.5	7.4925
1100	.042 00	4222.6	4852.6	7.8283	.035 97	4216.9	4846.4	7.7531	.031 45	4211.3	4840.2	7.6874
1200	.045 23	4433.8	5112.3	8.0108	.038 76	4428.3	5106.6	7.9360	.033 91	4422.8	5101.0	7.8707
1300	.048 45	4649.1	5376.0	8.1840	.041 54	4643.5	5370.5	8.1093	.036 36	4638.0	5365.1	8.0442
	P = 25.0 MPa				P = 30.0 MPa				P = 35.0 MPa			
375	.001 973 1	1798.7	1848.0	4.0320	.001 789 2	1737.8	1791.5	3.9305	.001 700 3	1702.9	1762.4	3.8722
400	.006 004	2430.1	2580.2	5.1418	.002 790	2067.4	2151.1	4.4728	.002 100	1914.1	1987.6	4.2126
425	.007 881	2609.2	2806.3	5.4723	.005 303	2455.1	2614.2	5.1504	.003 428	2253.4	2373.4	4.7747
450	.009 162	2720.7	2949.7	5.6744	.006 735	2619.3	2821.4	5.4424	.004 961	2498.7	2672.4	5.1962
500	.011 123	2884.3	3162.4	5.9592	.008 678	2820.7	3081.1	5.7905	.006 927	2751.9	2994.4	5.6282
550	.012 724	3017.5	3335.6	6.1765	.010 168	2970.3	3275.4	6.0342	.008 345	2921.0	3213.0	5.9026
600	.014 137	3137.9	3491.4	6.3602	.011 446	3100.5	3443.9	6.2331	.009 527	3062.0	3395.5	6.1179
650	.015 433	3251.6	3637.4	6.5229	.012 596	3221.0	3598.9	6.4058	.010 575	3189.8	3559.9	6.3010
700	.016 646	3361.3	3777.5	6.6707	.013 661	3335.8	3745.6	6.5606	.011 533	3309.8	3713.5	6.4631
800	.018 912	3574.3	4047.1	6.9345	.015 623	3555.5	4024.2	6.8332	.013 278	3536.7	4001.5	6.7450
900	.021 045	3783.0	4309.1	7.1680	.017 448	3768.5	4291.9	7.0718	.014 883	3754.0	4274.9	6.9886
1000	.023 10	3990.9	4568.5	7.3802	.019 196	3978.8	4554.7	7.2867	.016 410	3966.7	4541.1	7.2064
1100	.025 12	4200.2	4828.2	7.5765	.020 903	4189.2	4816.3	7.4845	.017 895	4178.3	4804.6	7.4057

1200	.027 11	4412.0	5089.9	7.7605	.022 589	4401.3	5079.0	7.6692	.019 360	4390.7	5068.3	7.5910
1300	.029 10	4626.9	5354.4	7.9342	.024 266	4616.0	5344.0	7.8432	.020 815	4605.1	5333.6	7.7653
	P = 40.0 MPa				P = 50.0 MPa				P = 60.0 MPa			
375	.001 640 7	1677.1	1742.8	3.8290	.001 559 4	1638.6	1716.6	3.7639	.001 502 8	1609.4	1699.5	3.7141
400	.001 907 7	1854.6	1930.9	4.1135	.001 730 9	1788.1	1874.6	4.0031	.001 633 5	1745.4	1843.4	3.9318
425	.002 532	2096.9	2198.1	4.5029	.002 007	1959.7	2060.0	4.2734	.001 816 5	1892.7	2001.7	4.1626
450	.003 693	2365.1	2512.8	4.9459	.002 486	2159.6	2284.0	4.5884	.002 085	2053.9	2179.0	4.4121
500	.005 622	2678.4	2903.3	5.4700	.003 892	2525.5	2720.1	5.1726	.002 956	2390.6	2567.9	4.9321
550	.006 984	2869.7	3149.1	5.7785	.005 118	2763.6	3019.5	5.5485	.003 956	2658.8	2896.2	5.3441
600	.008 094	3022.6	3346.4	6.0114	.006 112	2942.0	3247.6	5.8178	.004 834	2861.1	3151.2	5.6452
650	.009 063	3158.0	3520.6	6.2054	.006 966	3093.5	3441.8	6.0342	.005 595	3028.8	3364.5	5.8829
700	.009 941	3283.6	3681.2	6.3750	.007 727	3230.5	3616.8	6.2189	.006 272	3177.2	3553.5	6.0824
800	.011 523	3517.8	3978.7	6.6662	.009 076	3479.8	3933.6	6.5290	.007 459	3441.5	3889.1	6.4109
900	.012 962	3739.4	4257.9	6.9150	.010 283	3710.3	4224.4	6.7882	.008 508	3681.0	4191.5	6.6805
1000	.014 324	3954.6	4527.6	7.1356	.011 411	3930.5	4501.1	7.0146	.009 480	3906.4	4475.2	6.9127
1100	.015 642	4167.4	4793.1	7.3364	.012 496	4145.7	4770.5	7.2184	.010 409	4124.1	4748.6	7.1195
1200	.016 940	4380.1	5057.7	7.5224	.013 561	4359.1	5037.2	7.4058	.011 317	4338.2	5017.2	7.3083
1300	.018 229	4594.3	5323.5	7.6969	.014 616	4572.8	5303.6	7.5808	.012 215	4551.4	5284.3	7.4837

TABLE A·1.4 (SI) Water: Compressed Liquid (SI) v in m³/kg, u and h in kJ/kg, s in kJ/kg·K

T	v	u	h	s	v	u	h	s	v	u	h	s
	P = 5 MPa (263.99)				P = 10 MPa (311.06)				P = 15 MPa (342.24)			
Sat.	.001 285 9	1147.8	1154.2	2.9202	.001 452 4	1393.0	1407.6	3.3596	.001 658 1	1585.6	1610.5	3.6848
0	.000 997 7	.04	5.04	.0001	.000 995 2	.09	10.04	.0002	.000 992 8	.15	15.05	.0004
20	.000 999 5	83.65	88.65	.2956	.000 997 2	83.36	93.33	.2945	.000 995 0	83.06	97.99	.2934
40	.001 005 6	166.95	171.97	.5705	.001 003 4	166.35	176.38	.5686	.001 001 3	165.76	180.78	.5666
60	.001 014 9	250.23	255.30	.8285	.001 012 7	249.36	259.49	.8258	.001 010 5	248.51	263.67	.8232
80	.001 026 8	333.72	338.85	1.0720	.001 024 5	332.59	342.83	1.0688	.001 022 2	331.48	346.81	1.0656
100	.001 041 0	417.52	422.72	1.3030	.001 038 5	416.12	426.50	1.2992	.001 036 1	414.74	430.28	1.2955
120	.001 057 6	501.80	507.09	1.5233	.001 054 9	500.08	510.64	1.5189	.001 052 2	498.40	514.19	1.5145
140	.001 076 8	586.76	592.15	1.7343	.001 073 7	584.68	595.42	1.7292	.001 070 7	582.66	598.72	1.7242
160	.001 098 8	672.62	678.12	1.9375	.001 095 3	670.13	681.08	1.9317	.001 091 8	667.71	684.09	1.9260
180	.001 124 0	759.63	765.25	2.1341	.001 119 9	756.65	767.84	2.1275	.001 115 9	753.76	770.50	2.1210
200	.001 153 0	848.1	853.9	2.3255	.001 148 0	844.5	856.0	2.3178	.001 143 3	841.0	858.2	2.3104
220	.001 186 6	938.4	944.4	2.5128	.001 180 5	934.1	945.9	2.5039	.001 174 8	929.9	947.5	2.4953
240	.001 226 4	1031.4	1037.5	2.6979	.001 218 7	1026.0	1038.1	2.6872	.001 211 4	1020.8	1039.0	2.6771
260	.001 274 9	1127.9	1134.3	2.8830	.001 264 5	1121.1	1133.7	2.8699	.001 255 0	1114.6	1133.4	2.8576
280					.001 321 6	1220.9	1234.1	3.0548	.001 308 4	1212.5	1232.1	3.0393
300					.001 397 2	1328.4	1342.3	3.2469	.001 377 0	1316.6	1337.3	3.2260
320									.001 472 4	1431.1	1453.2	3.4247
340									.001 631 1	1567.5	1591.9	3.6546

	P = 20 MPa (365.81)				P = 30 MPa				P = 50 MPa			
Sat.	.002 036	1785.6	1826.3	4.0139								
0	.000 990 4	.19	20.01	.0004	.000 985 6	.25	29.82	.0001	.000 976 6	.20	49.03	−.0014
20	.000 992 8	82.77	102.62	.2923	.000 988 6	82.17	111.84	.2899	.000 980 4	81.00	130.02	.2848
40	.000 999 2	165.17	185.16	.5646	.000 995 1	164.04	193.89	.5607	.000 987 2	161.86	211.21	.5527
60	.001 008 4	247.68	267.85	.8206	.001 004 2	246.06	276.19	.8154	.000 996 2	242.98	292.79	.8052
80	.001 019 9	330.40	350.80	1.0624	.001 015 6	328.30	358.77	1.0561	.001 007 3	324.34	374.70	1.0440
100	.001 033 7	413.39	434.06	1.2917	.001 029 0	410.78	441.66	1.2844	.001 020 1	405.88	456.89	1.2703
120	.001 049 6	496.76	517.76	1.5102	.001 044 5	493.59	524.93	1.5018	.001 034 8	487.65	539.39	1.4857
140	.001 067 8	580.69	602.04	1.7193	.001 062 1	576.88	608.75	1.7098	.001 051 5	569.77	622.35	1.6915
160	.001 088 5	665.35	687.12	1.9204	.001 082 1	660.82	693.28	1.9096	.001 070 3	652.41	705.92	1.8891
180	.001 112 0	750.95	773.20	2.1147	.001 104 7	745.59	778.73	2.1024	.001 091 2	735.69	790.25	2.0794
200	.001 138 8	837.7	860.5	2.3031	.001 130 2	831.4	865.3	2.2893	.001 114 6	819.7	875.5	2.2634
220	.001 169 3	925.9	949.3	2.4870	.001 159 0	918.3	953.1	2.4711	.001 140 8	904.7	961.7	2.4419
240	.001 204 6	1016.0	1040.0	2.6674	.001 192 0	1006.9	1042.6	2.6490	.001 170 2	990.7	1049.2	2.6158
260	.001 246 2	1108.6	1133.5	2.8459	.001 230 3	1097.4	1134.3	2.8243	.001 203 4	1078.1	1138.2	2.7860
280	.001 296 5	1204.7	1230.6	3.0248	.001 275 5	1190.7	1229.0	2.9986	.001 241 5	1167.2	1229.3	2.9537
300	.001 359 6	1306.1	1333.3	3.2071	.001 330 4	1287.9	1327.8	3.1741	.001 286 0	1258.7	1323.0	3.1200
320	.001 443 7	1415.7	1444.6	3.3979	.001 399 7	1390.7	1432.7	3.3539	.001 338 8	1353.3	1420.2	3.2868
340	.001 568 4	1539.7	1571.0	3.6075	.001 492 0	1501.7	1546.5	3.5426	.001 403 2	1452.0	1522.1	3.4557
360	.001 822 6	1702.8	1739.3	3.8772	.001 626 5	1626.6	1675.4	3.7494	.001 483 8	1556.0	1630.2	3.6291
380					.001 869 1	1781.4	1837.5	4.0012	.001 588 4	1667.2	1746.6	3.8101

TABLE A·1.5 (SI) Water: Solid-Vapor Saturation (SI)

Temp. °C T	Pressure kPa P	Specific Volume m³/kg: Sat. Solid $v_i \times 10^3$	Sat. Vapor v_g	Internal Energy kJ/kg: Sat. Solid u_i	Subl. u_{ig}	Sat. Vapor u_g	Enthalpy kJ/kg: Sat. Solid h_i	Subl. h_{ig}	Sat. Vapor h_g	Entropy kJ/kg·K: Sat. Solid s_i	Subl. s_{ig}	Sat. Vapor s_g
.01	.6113	1.0908	206.1	−333.40	2708.7	2375.3	−333.40	2834.8	2501.4	−1.221	10.378	9.156
0	.6108	1.0908	206.3	−333.43	2708.8	2375.3	−333.43	2834.8	2501.3	−1.221	10.378	9.157
−2	.5176	1.0904	241.7	−337.62	2710.2	2372.6	−337.62	2835.3	2497.7	−1.237	10.456	9.219
−4	.4375	1.0901	283.8	−341.78	2711.6	2369.8	−341.78	2835.7	2494.0	−1.253	10.536	9.283
−6	.3689	1.0898	334.2	−345.91	2712.9	2367.0	−345.91	2836.2	2490.3	−1.268	10.616	9.348
−8	.3102	1.0894	394.4	−350.02	2714.2	2364.2	−350.02	2836.6	2486.6	−1.284	10.698	9.414
−10	.2602	1.0891	466.7	−354.09	2715.5	2361.4	−354.09	2837.0	2482.9	−1.299	10.781	9.481
−12	.2176	1.0888	553.7	−358.14	2716.8	2358.7	−358.14	2837.3	2479.2	−1.315	10.865	9.550
−14	.1815	1.0884	658.8	−362.15	2718.0	2355.9	−362.15	2837.6	2475.5	−1.331	10.950	9.619
−16	.1510	1.0881	786.0	−366.14	2719.2	2353.1	−366.14	2837.9	2471.8	−1.346	11.036	9.690
−18	.1252	1.0878	940.5	−370.10	2720.4	2350.3	−370.10	2838.2	2468.1	−1.362	11.123	9.762
−20	.1035	1.0874	1128.6	−374.03	2721.6	2347.5	−374.03	2838.4	2464.3	−1.377	11.212	9.835
−22	.0853	1.0871	1358.4	−377.93	2722.7	2344.7	−377.93	2838.6	2460.6	−1.393	11.302	9.909
−24	.0701	1.0868	1640.1	−381.80	2723.7	2342.0	−381.80	2838.7	2456.9	−1.408	11.394	9.985
−26	.0574	1.0864	1986.4	−385.64	2724.8	2339.2	−385.64	2838.9	2453.2	−1.424	11.486	10.062
−28	.0469	1.0861	2413.7	−389.45	2725.8	2336.4	−389.45	2839.0	2449.5	−1.439	11.580	10.141
−30	.0381	1.0858	2943	−393.23	2726.8	2333.6	−393.23	2839.0	2445.8	−1.455	11.676	10.221
−32	.0309	1.0854	3600	−396.98	2727.8	2330.8	−396.98	2839.1	2442.1	−1.471	11.773	10.303
−34	.0250	1.0851	4419	−400.71	2728.7	2328.0	−400.71	2839.1	2438.4	−1.486	11.872	10.386
−36	.0201	1.0848	5444	−404.40	2729.6	2325.2	−404.40	2839.1	2434.7	−1.501	11.972	10.470
−38	.0161	1.0844	6731	−408.06	2730.5	2322.4	−408.06	2839.0	2430.9	−1.517	12.073	10.556
−40	.0129	1.0841	8354	−411.70	2731.3	2319.6	−411.70	2838.9	2427.2	−1.532	12.176	10.644

TABLE A·1.1 (E) Water: Liquid-Vapor Saturation, Temperature Table (English units)

	Pressure	Specific Volume ft³/lbm		Internal Energy B/lbm			Enthalpy B/lbm			Entropy B/lbm·R		
Temp. °F T	lbf/ sq in. P	Sat. Liquid v_f	Sat. Vapor v_g	Sat. Liquid u_f	Evap. u_{fg}	Sat. Vapor u_g	Sat. Liquid h_f	Evap. h_{fg}	Sat. Vapor h_g	Sat. Liquid s_f	Evap. s_{fg}	Sat. Vapor s_g
32.018	0.08866	0.016022	3302	0.00	1021.2	1021.2	0.01	1075.4	1075.4	0.00000	2.1869	2.1869
35	0.09992	0.016021	2948	2.99	1019.2	1022.2	3.00	1073.7	1076.7	0.00607	2.1704	2.1764
40	0.12166	0.016020	2445	8.02	1015.8	1023.9	8.02	1070.9	1078.9	0.01617	2.1430	2.1592
45	0.14748	0.016021	2037	13.04	1012.5	1025.5	13.04	1068.1	1081.1	0.02618	2.1162	2.1423
50	0.17803	0.016024	1704.2	18.06	1009.1	1027.2	18.06	1065.2	1083.3	0.03607	2.0899	2.1259
60	0.2563	0.016035	1206.9	28.08	1002.4	1030.4	28.08	1059.6	1087.7	0.05555	2.0388	2.0943
70	0.3632	0.016051	867.7	38.09	995.6	1033.7	38.09	1054.0	1092.0	0.07463	1.9896	2.0642
80	0.5073	0.016073	632.8	48.08	988.9	1037.0	48.09	1048.3	1096.4	0.09332	1.9423	2.0356
90	0.6988	0.016099	467.7	58.07	982.2	1040.2	58.07	1042.7	1100.7	0.11165	1.8966	2.0083
100	0.9503	0.016130	350.0	68.04	975.4	1043.5	68.05	1037.0	1105.0	0.12963	1.8526	1.9822
110	1.2763	0.016166	265.1	78.02	968.7	1046.7	78.02	1031.3	1109.3	0.14730	1.8101	1.9574
120	1.6945	0.016205	203.0	87.99	961.9	1049.9	88.00	1025.5	1113.5	0.16465	1.7690	1.9336
130	2.225	0.016247	157.17	97.97	955.1	1053.0	97.98	1019.8	1117.8	0.18172	1.7292	1.9109
140	2.892	0.016293	122.88	107.95	948.2	1056.2	107.96	1014.0	1121.9	0.19851	1.6907	1.8892
150	3.722	0.016343	96.99	117.95	941.3	1059.3	117.96	1008.1	1126.1	0.21503	1.6533	1.8684
160	4.745	0.016395	77.23	127.94	934.4	1062.3	127.96	1002.2	1130.1	0.23130	1.6171	1.8484
170	5.996	0.016450	62.02	137.95	927.4	1065.4	137.97	996.2	1134.2	0.24732	1.5819	1.8293
180	7.515	0.016509	50.20	147.97	920.4	1068.3	147.99	990.2	1138.2	0.26311	1.5478	1.8109

Source: Abridged from Joseph H. Keenan, Frederick G. Keyes, Philip G. Hill, and Joan G. Moore, *Steam Tables*, Wiley, New York, 1969. Abridgment partially from Gordon J. Van Wylen and Richard E. Sonntag, *Fundamentals of Classical Thermodynamics*, 3rd ed., Wiley, New York, 1985.

TABLE A·1.1 (E) *(Continued)*

Temp. °F T	Pressure lbf/sq in. P	Specific Volume, ft³/lbm Sat. Liquid v_f	Sat. Vapor v_g	Internal Energy, B/lbm Sat. Liquid u_f	Evap. u_{fg}	Sat. Vapor u_g	Enthalpy B/lbm Sat. Liquid h_f	Evap. h_{fg}	Sat. Vapor h_g	Entropy B/lbm·R Sat. Liquid s_f	Evap. s_{fg}	Sat. Vapor s_g
190	9.343	0.016570	40.95	158.00	913.3	1071.3	158.03	984.1	1142.1	0.27866	1.5146	1.7932
200	11.529	0.016634	33.63	168.04	906.2	1074.2	168.07	977.9	1145.9	0.29400	1.4822	1.7762
210	14.125	0.016702	27.82	178.10	898.9	1077.0	178.14	971.6	1149.7	0.30913	1.4508	1.7599
212	14.698	0.016716	26.80	180.11	897.5	1077.6	180.16	970.3	1150.5	0.31213	1.4446	1.7567
220	17.188	0.016772	23.15	188.17	891.7	1079.8	188.22	965.3	1153.5	0.32406	1.4201	1.7441
230	20.78	0.016845	19.386	198.26	884.3	1082.6	198.32	958.8	1157.1	0.33880	1.3901	1.7289
240	24.97	0.016922	16.327	208.36	876.9	1085.3	208.44	952.3	1160.7	0.35335	1.3609	1.7143
250	29.82	0.017001	13.826	218.49	869.4	1087.9	218.59	945.6	1164.2	0.36772	1.3324	1.7001
260	35.42	0.017084	11.768	228.64	861.8	1090.5	228.76	938.8	1167.6	0.38193	1.3044	1.6864
270	41.85	0.017170	10.066	238.82	854.1	1093.0	238.95	932.0	1170.9	0.39597	1.2771	1.6731
280	49.18	0.017259	8.650	249.02	846.3	1095.4	249.18	924.9	1174.1	0.40986	1.2504	1.6602
290	57.53	0.017352	7.467	259.25	838.5	1097.7	259.44	917.8	1177.2	0.42360	1.2241	1.6477
300	66.98	0.017448	6.472	269.52	830.5	1100.0	269.73	910.4	1180.2	0.43720	1.1984	1.6356
310	77.64	0.017548	5.632	279.81	822.3	1102.1	280.06	903.0	1183.0	0.45067	1.1731	1.6238
320	89.60	0.017652	4.919	290.14	814.1	1104.2	290.43	895.3	1185.8	0.46400	1.1483	1.6123
330	103.00	0.017760	4.312	300.51	805.7	1106.2	300.84	887.5	1188.4	0.47722	1.1238	1.6010
340	117.93	0.017872	3.792	310.91	797.1	1108.0	311.30	879.5	1190.8	0.49031	1.0997	1.5901
350	134.53	0.017988	3.346	321.35	788.4	1109.8	321.80	871.3	1193.1	0.50329	1.0760	1.5793

360	152.92	0.018108	2.961	331.84	779.6	1111.4	332.35	862.9	1195.2	0.51617	1.0526	1.5688
370	173.23	0.018233	2.628	342.37	770.6	1112.9	342.96	854.2	1197.2	0.52894	1.0295	1.5585
380	195.60	0.018363	2.339	352.95	761.4	1114.3	353.62	845.4	1199.0	0.54163	1.0067	1.5483
390	220.2	0.018498	2.087	363.58	752.0	1115.6	364.34	836.2	1200.6	0.55422	0.9841	1.5383
400	247.1	0.018638	1.8661	374.27	742.4	1116.6	375.12	826.8	1202.0	0.56672	0.9617	1.5284
410	276.5	0.018784	1.6726	385.01	732.6	1117.6	385.97	817.2	1203.1	0.57916	0.9395	1.5187
420	308.5	0.018936	1.5024	395.81	722.5	1118.3	396.89	807.2	1204.1	0.59152	0.9175	1.5091
430	343.3	0.019094	1.3521	406.68	712.2	1118.9	407.89	796.9	1204.8	0.60381	0.8957	1.4995
440	381.2	0.019260	1.2192	417.62	701.7	1119.3	418.98	786.3	1205.3	0.61605	0.8740	1.4900
450	422.1	0.019433	1.1011	428.6	690.9	1119.5	430.2	775.4	1205.6	0.6282	0.8523	1.4806
460	466.3	0.019614	0.9961	439.7	679.8	1119.6	441.4	764.1	1205.5	0.6404	0.8308	1.4712
470	514.1	0.019803	0.9025	450.9	668.4	1119.4	452.8	752.4	1205.2	0.6525	0.8093	1.4618
480	565.5	0.020002	0.8187	462.2	656.7	1118.9	464.3	740.3	1204.6	0.6646	0.7878	1.4524
490	620.7	0.020211	0.7436	473.6	644.7	1118.3	475.9	727.8	1203.7	0.6767	0.7663	1.4430
500	680.0	0.02043	0.6761	485.1	632.3	1117.4	487.7	714.8	1202.5	0.6888	0.7448	1.4335
520	811.4	0.02091	0.5605	508.5	606.2	1114.8	511.7	687.3	1198.9	0.7130	0.7015	1.4145
540	961.5	0.02145	0.4658	532.6	578.4	1111.0	536.4	657.5	1193.8	0.7374	0.6576	1.3950
560	1131.8	0.02207	0.3877	557.4	548.4	1105.8	562.0	625.0	1187.0	0.7620	0.6129	1.3749
580	1324.3	0.02278	0.3225	583.1	515.9	1098.9	588.6	589.3	1178.0	0.7872	0.5668	1.3540
600	1541.0	0.02363	0.2677	609.9	480.1	1090.0	616.7	549.7	1166.4	0.8130	0.5187	1.3317
620	1784.4	0.02465	0.2209	638.3	440.2	1078.5	646.4	505.0	1151.4	0.8398	0.4677	1.3075
640	2057.1	0.02593	0.1805	668.7	394.5	1063.2	678.6	453.4	1131.9	0.8681	0.4122	1.2803
660	2362	0.02767	0.14459	702.3	340.0	1042.3	714.4	391.1	1105.5	0.8990	0.3493	1.2483
680	2705	0.03032	0.11127	741.7	269.3	1011.0	756.9	309.8	1066.7	0.9350	0.2718	1.2068
700	3090	0.03666	0.07438	801.7	145.9	947.7	822.7	167.5	990.2	0.9902	0.1444	1.1346
705.44	3204	0.05053	0.05053	872.6	0	872.6	902.5	0	902.5	1.0580	0	1.0580

TABLE A·1.2 (E) Water: Liquid-Vapor Saturation, Pressure Table (English units) (ft³/lbm, B/lbm, B/lbm·R)

Pressure lbf/sq in. P	Temp. °F T	Specific Volume: Sat. Liquid v_f	Specific Volume: Sat. Vapor v_g	Internal Energy: Sat. Liquid u_f	Internal Energy: Evap. u_{fg}	Internal Energy: Sat. Vapor u_g	Enthalpy: Sat. Liquid h_f	Enthalpy: Evap. h_{fg}	Enthalpy: Sat. Vapor h_g	Entropy: Sat. Liquid s_f	Entropy: Evap. s_{fg}	Entropy: Sat. Vapor s_g
0.08866	32.02	.016022	3302.	.00	1021.2	1021.2	.01	1075.4	1075.4	.00000	2.1869	2.1869
.10	35.02	.016021	2946.	3.02	1019.2	1022.2	3.02	1073.7	1076.7	.00612	2.1702	2.1764
.20	53.15	.016027	1526.3	21.22	1007.0	1028.2	21.22	1063.5	1084.7	.04225	2.0736	2.1158
.30	64.46	.016041	1039.7	32.55	999.4	1031.9	32.56	1057.1	1089.6	.06411	2.0166	2.0807
.40	72.84	.016056	792.0	40.94	993.7	1034.7	40.94	1052.3	1093.3	.07998	1.9760	2.0559
.50	79.56	.016071	641.5	47.64	989.2	1036.9	47.65	1048.6	1096.2	.09250	1.9443	2.0368
1.0	101.70	.016136	333.6	69.74	974.3	1044.0	69.74	1036.0	1105.8	.13266	1.8453	1.9779
1.5	115.65	.016187	227.7	83.65	964.8	1048.5	83.65	1028.0	1111.7	.15714	1.7867	1.9438
2.0	126.04	.016230	173.75	94.02	957.8	1051.8	94.02	1022.1	1116.1	.17499	1.7448	1.9198
3.0	141.43	.016300	118.72	109.38	947.2	1056.6	109.39	1013.1	1122.5	.20089	1.6852	1.8861
4.0	152.93	.016358	90.64	120.88	939.3	1060.2	120.89	1006.4	1127.3	.21983	1.6426	1.8624
5.0	162.21	.016407	73.53	130.15	932.9	1063.0	130.17	1000.9	1131.0	.23486	1.6093	1.8441
7.5	179.91	.016508	50.30	147.88	920.4	1068.3	147.90	990.2	1138.1	.26297	1.5481	1.8110
10	193.19	.016590	38.42	161.20	911.0	1072.2	161.23	982.1	1143.3	.28358	1.5041	1.7877
14.696	211.99	.016715	26.80	180.10	897.5	1077.6	180.15	970.4	1150.5	.31212	1.4446	1.7567
15	213.03	.016723	26.29	181.14	896.8	1077.9	181.19	969.7	1150.9	.31367	1.4414	1.7551
20	227.96	.016830	20.09	196.19	885.8	1082.0	196.26	960.1	1156.4	.33580	1.3962	1.7320
25	240.08	.016922	16.306	208.44	876.9	1085.3	208.52	952.2	1160.7	.35345	1.3607	1.7142
30	250.34	.017004	13.748	218.84	869.2	1088.0	218.93	945.4	1164.3	.36821	1.3314	1.6996
35	259.30	.017078	11.900	227.93	862.4	1090.3	228.04	939.3	1167.4	.38093	1.3064	1.6873
40	267.26	.017146	10.501	236.03	856.2	1092.3	236.16	933.8	1170.0	.39214	1.2845	1.6767
45	274.46	.017209	9.403	243.37	850.7	1094.0	243.51	928.8	1172.3	.40218	1.2651	1.6673
50	281.03	.017269	8.518	250.08	845.5	1095.6	250.24	924.2	1174.4	.41129	1.2476	1.6589
55	287.10	.017325	7.789	256.28	840.8	1097.0	256.46	919.9	1176.3	.41963	1.2317	1.6513
60	292.73	.017378	7.177	262.06	836.3	1098.3	262.25	915.8	1178.0	.42733	1.2170	1.6444

65	298.00	.017429	6.657	267.46	832.1	1099.5	267.67	911.9	1179.6	.43450	1.2035	1.6380
70	302.96	.017478	6.209	272.56	828.1	1100.6	272.79	908.3	1181.0	.44120	1.1909	1.6321
75	307.63	.017524	5.818	277.37	824.3	1101.6	277.61	904.8	1182.4	.44749	1.1790	1.6265
80	312.07	.017570	5.474	281.95	820.6	1102.6	282.21	901.4	1183.6	.45344	1.1679	1.6214
85	316.29	.017613	5.170	286.30	817.1	1103.5	286.58	898.2	1184.8	.45907	1.1574	1.6165
90	320.31	.017655	4.898	290.46	813.8	1104.3	290.76	895.1	1185.9	.46442	1.1475	1.6119
95	324.16	.017696	4.654	294.45	810.6	1105.0	294.76	892.1	1186.9	.46952	1.1380	1.6076
100	327.86	.017736	4.434	298.28	807.5	1105.8	298.61	889.2	1187.8	.47439	1.1290	1.6034
105	331.41	.017775	4.234	301.97	804.5	1106.5	302.31	886.4	1188.7	.47906	1.1204	1.5995
110	334.82	.017813	4.051	305.52	801.6	1107.1	305.88	883.7	1189.6	.48355	1.1122	1.5957
115	338.12	.017850	3.884	308.95	798.8	1107.7	309.33	881.0	1190.4	.48786	1.1042	1.5921
120	341.30	.017886	3.730	312.27	796.0	1108.3	312.67	878.5	1191.1	.49201	1.0966	1.5886
125	344.39	.017922	3.588	315.49	793.3	1108.8	315.90	875.9	1191.8	.49602	1.0893	1.5853
130	347.37	.017957	3.457	318.61	790.7	1109.4	319.04	873.5	1192.5	.49989	1.0822	1.5821
135	350.27	.017991	3.335	321.64	788.2	1109.8	322.08	871.1	1193.2	.50364	1.0754	1.5790
140	353.08	.018024	3.221	324.58	785.7	1110.3	325.05	868.7	1193.8	.50727	1.0688	1.5761
145	355.82	.018057	3.115	327.45	783.3	1110.8	327.93	866.4	1194.4	.51079	1.0624	1.5732
150	358.48	.018089	3.016	330.24	781.0	1111.2	330.75	864.2	1194.9	.51422	1.0562	1.5704
160	363.60	.018152	2.836	335.63	776.4	1112.0	336.16	859.8	1196.0	.52078	1.0443	1.5651
170	368.47	.018214	2.676	340.76	772.0	1112.7	341.33	855.6	1196.9	.52700	1.0330	1.5600
180	373.13	.018273	2.533	345.68	767.7	1113.4	346.29	851.5	1197.8	.53292	1.0223	1.5553
190	377.59	.018331	2.405	350.39	763.6	1114.0	351.04	847.5	1198.6	.53857	1.0122	1.5507
200	381.86	.018387	2.289	354.9	759.6	1114.6	355.6	843.7	1199.3	.5440	1.0025	1.5464
225	391.87	.018523	2.043	365.6	750.2	1115.8	366.3	834.5	1200.8	.5566	.9799	1.5365
250	401.04	.018653	1.8448	375.4	741.4	1116.7	376.2	825.8	1202.1	.5680	.9594	1.5274
275	409.52	.018777	1.6813	384.5	733.0	1117.5	385.4	817.6	1203.1	.5786	.9406	1.5192
300	417.43	.018896	1.5442	393.0	725.1	1118.2	394.1	809.8	1203.9	.5883	.9232	1.5115
350	431.82	.019124	1.3267	408.7	710.3	1119.0	409.9	795.0	1204.9	.6060	.8917	1.4978
400	444.70	.019340	1.1620	422.8	696.7	1119.5	424.2	781.2	1205.5	.6218	.8638	1.4856
450	456.39	.019547	1.0326	435.7	683.9	1119.6	437.4	768.2	1205.6	.6360	.8385	1.4746

TABLE A·1.2 (E) *(Concluded)*

Pressure lbf/sq in. P	Temp. °F T	Specific Volume		Internal Energy			Enthalpy			Entropy		
		Sat. Liquid v_f	Sat. Vapor v_g	Sat. Liquid u_f	Evap. u_{fg}	Sat. Vapor u_g	Sat. Liquid h_f	Evap. h_{fg}	Sat. Vapor h_g	Sat. Liquid s_f	Evap. s_{fg}	Sat. Vapor s_g
500	467.13	.019748	.9283	447.7	671.7	1119.4	449.5	755.8	1205.3	.6490	.8154	1.4645
550	477.07	.019943	.8423	458.9	660.2	1119.1	460.9	743.9	1204.8	.6611	.7941	1.4551
600	486.33	.02013	.7702	469.4	649.1	1118.6	471.7	732.4	1204.1	.6723	.7742	1.4464
700	503.23	.02051	.6558	488.9	628.2	1117.0	491.5	710.5	1202.0	.6927	.7378	1.4305
800	518.36	.02087	.5691	506.6	608.4	1115.0	509.7	689.6	1199.3	.7110	.7050	1.4160
900	532.12	.02123	.5009	523.0	589.6	1112.6	526.6	669.5	1196.0	.7277	.6750	1.4027
1000	544.75	.02159	.4459	538.4	571.5	1109.9	542.4	650.0	1192.4	.7432	.6471	1.3903
1250	572.56	.02250	.3454	573.4	528.3	1101.7	578.6	603.0	1181.6	.7778	.5841	1.3619
1500	596.39	.02346	.2769	605.0	486.9	1091.8	611.5	557.2	1168.7	.8082	.5276	1.3359
1750	617.31	.02450	.2268	634.4	445.9	1080.2	642.3	511.4	1153.7	.8361	.4748	1.3109
2000	636.00	.02565	.18813	662.4	404.2	1066.6	671.9	464.4	1136.3	.8623	.4238	1.2861
2250	652.90	.02698	.15692	689.9	360.7	1050.6	701.1	414.8	1115.9	.8876	.3728	1.2604
2500	668.31	.02860	.13059	717.7	313.4	1031.0	730.9	360.5	1091.4	.9131	.3196	1.2327
2750	682.46	.03077	.10717	747.3	258.6	1005.9	763.0	297.4	1060.4	.9401	.2604	1.2005
3000	695.52	.03431	.08404	783.4	185.4	968.8	802.5	213.0	1015.5	.9732	.1843	1.1575
3203.6	705.44	.05053	.05053	872.6	0	872.6	902.5	0	902.5	1.0580	0	1.0580

TABLE A·1.3 (E) Water: Superheated Vapor (Saturation temperature in °F shown in parentheses following each pressure in psia) v in ft³/lbm, u and h in B/lbm, s in B/lbm·R

°F	v	u	h	s	v	u	h	s	v	u	h	s
	P = 1.0(101.70)				P = 5.0(162.21)				P = 10.0(193.19)			
Sat	333.6	1044.0	1105.8	1.9779	73.53	1063.0	1131.0	1.8441	38.42	1072.2	1143.3	1.7877
200	392.5	1077.5	1150.1	2.0508	78.15	1076.3	1148.6	1.8715	38.85	1074.7	1146.6	1.7927
240	416.4	1091.2	1168.3	2.0775	83.00	1090.3	1167.1	1.8987	41.32	1089.0	1165.5	1.8205
280	440.3	1105.0	1186.5	2.1028	87.83	1104.3	1185.5	1.9244	43.77	1103.3	1184.3	1.8467
320	464.2	1118.9	1204.8	2.1269	92.64	1118.3	1204.0	1.9487	46.20	1117.6	1203.1	1.8714
360	488.1	1132.9	1223.2	2.1500	97.45	1132.4	1222.6	1.9719	48.62	1131.8	1221.8	1.8948
400	511.9	1147.0	1241.8	2.1720	102.24	1146.6	1241.2	1.9941	51.03	1146.1	1240.5	1.9171
440	535.8	1161.2	1260.4	2.1932	107.03	1160.9	1259.9	2.0154	53.44	1160.5	1259.3	1.9385
500	571.5	1182.8	1288.5	2.2235	114.20	1182.5	1288.2	2.0458	57.04	1182.2	1287.7	1.9690
600	631.1	1219.3	1336.1	2.2706	126.15	1219.1	1335.8	2.0930	63.03	1218.9	1335.5	2.0164
700	690.7	1256.7	1384.5	2.3142	138.08	1256.5	1384.3	2.1367	69.01	1256.3	1384.0	2.0601
800	750.3	1294.9	1433.7	2.3550	150.01	1294.7	1433.5	2.1775	74.98	1294.6	1433.3	2.1009
1000	869.5	1373.9	1534.8	2.4294	173.86	1373.9	1534.7	2.2520	86.91	1373.8	1534.6	2.1755
1200	988.6	1456.7	1639.6	2.4967	197.70	1456.6	1639.5	2.3193	98.84	1456.5	1639.4	2.2428
1400	1107.7	1543.1	1748.1	2.5584	221.54	1543.1	1748.1	2.3810	110.76	1543.0	1748.0	2.3045
	P = 14.696(211.99)				P = 20(227.96)				P = 40(267.26)			
Sat	26.80	1077.6	1150.5	1.7567	20.09	1082.0	1156.4	1.7320	10.501	1092.3	1170.0	1.6767
240	28.00	1087.9	1164.0	1.7764	20.47	1086.5	1162.3	1.7405	...	...	...	...
280	29.69	1102.4	1183.1	1.8030	21.73	1101.4	1181.8	1.7676	10.711	1097.3	1176.6	1.6857
320	31.36	1116.8	1202.1	1.8280	22.98	1116.0	1201.0	1.7930	11.360	1112.8	1196.9	1.7124
360	33.02	1131.2	1221.0	1.8516	24.21	1130.6	1220.1	1.8168	11.996	1128.0	1216.8	1.7373
400	34.67	1145.6	1239.9	1.8741	25.43	1145.1	1239.2	1.8395	12.623	1143.0	1236.4	1.7606
440	36.31	1160.1	1258.8	1.8956	26.64	1159.6	1258.2	1.8611	13.243	1157.8	1255.8	1.7828

TABLE A·1.3 (E) *(Continued)*

°F	v	u	h	s	v	u	h	s	v	u	h	s
500	38.77	1181.8	1287.3	1.9263	28.46	1181.5	1286.8	1.8919	14.164	1180.1	1284.9	1.8140
600	42.86	1218.6	1335.2	1.9737	31.47	1218.4	1334.8	1.9395	15.685	1217.3	1333.4	1.8621
700	46.93	1256.1	1383.8	2.0175	34.47	1255.9	1383.5	1.9834	17.196	1255.1	1382.4	1.9063
800	51.00	1294.4	1433.1	2.0584	37.46	1294.3	1432.9	2.0243	18.701	1293.7	1432.1	1.9474
1000	59.13	1373.7	1534.5	2.1330	43.44	1373.5	1534.3	2.0989	21.70	1373.1	1533.8	2.0223
1200	67.25	1456.5	1639.3	2.2003	49.41	1456.4	1639.2	2.1663	24.69	1456.1	1638.9	2.0897
1400	75.36	1543.0	1747.9	2.2621	55.37	1542.9	1747.9	2.2281	27.68	1542.7	1747.6	2.1515
1600	83.47	1633.2	1860.2	2.3194	61.33	1633.2	1860.1	2.2854	30.66	1633.0	1859.9	2.2089
	P = 60(292.73)				P = 80(312.07)				P = 100(327.86)			
Sat	7.177	1098.3	1178.0	1.6444	5.474	1102.6	1183.6	1.6214	4.434	1105.8	1187.8	1.6034
320	7.485	1109.5	1192.6	1.6634	5.544	1106.0	1188.0	1.6271	...	...	...	...
360	7.924	1125.3	1213.3	1.6893	5.886	1122.5	1209.7	1.6541	4.662	1119.7	1205.9	1.6259
400	8.353	1140.8	1233.5	1.7134	6.217	1138.5	1230.6	1.6790	4.934	1136.2	1227.5	1.6517
440	8.775	1156.0	1253.4	1.7360	6.541	1154.2	1251.0	1.7022	5.199	1152.3	1248.5	1.6755
500	9.399	1178.6	1283.0	1.7678	7.017	1177.2	1281.1	1.7346	5.587	1175.7	1279.1	1.7085
600	10.425	1216.3	1332.1	1.8165	7.794	1215.3	1330.7	1.7838	6.216	1214.2	1329.3	1.7582
700	11.440	1254.4	1381.4	1.8609	8.561	1253.6	1380.3	1.8285	6.834	1252.8	1379.2	1.8033
800	12.448	1293.0	1431.2	1.9022	9.321	1292.4	1430.4	1.8700	7.445	1291.8	1429.6	1.8449
1000	14.454	1372.7	1533.2	1.9773	10.831	1372.3	1532.6	1.9453	8.657	1371.9	1532.1	1.9204
1200	16.452	1455.8	1638.5	2.0448	12.333	1455.5	1638.1	2.0130	9.861	1455.2	1637.7	1.9882
1400	18.445	1542.5	1747.3	2.1067	13.830	1542.3	1747.0	2.0749	11.060	1542.0	1746.7	2.0502
1600	20.44	1632.8	1859.7	2.1641	15.324	1632.6	1859.5	2.1323	12.257	1632.4	1859.3	2.1076
1800	22.43	1726.7	1975.7	2.2179	16.818	1726.5	1975.5	2.1861	13.452	1726.4	1975.3	2.1614
2000	24.41	1824.0	2095.1	2.2685	18.310	1823.9	2094.9	2.2367	14.647	1823.7	2094.8	2.2121
	P = 120(341.30)				P = 140(353.08)				P = 160(363.60)			
Sat	3.730	1108.3	1191.1	1.5886	3.221	1110.3	1193.8	1.5761	2.836	1112.0	1196.0	1.5651
360	3.844	1116.7	1202.0	1.6021	3.259	1113.5	1198.0	1.5812	...	...	...	...
400	4.079	1133.8	1224.4	1.6288	3.466	1131.4	1221.2	1.6088	3.007	1128.8	1217.8	1.5911
450	4.360	1154.3	1251.2	1.6590	3.713	1152.4	1248.6	1.6399	3.228	1150.5	1246.1	1.6230

500	4.633	1174.2	1277.1	1.6868	3.952	1172.7	1275.1	1.6682	3.440	1171.2	1273.0	1.6518
550	4.900	1193.8	1302.6	1.7127	4.184	1192.6	1300.9	1.6944	3.646	1191.3	1299.2	1.6784
600	5.164	1213.2	1327.8	1.7371	4.412	1212.1	1326.4	1.7191	3.848	1211.1	1325.0	1.7034
700	5.682	1252.0	1378.2	1.7825	4.860	1251.2	1377.1	1.7648	4.243	1250.4	1376.0	1.7494
800	6.195	1291.2	1428.7	1.8243	5.301	1290.5	1427.9	1.8068	4.631	1289.9	1427.0	1.7916
1000	7.208	1371.5	1531.5	1.9000	6.173	1371.0	1531.0	1.8827	5.397	1370.6	1530.4	1.8677
1200	8.213	1454.9	1637.3	1.9679	7.036	1454.6	1636.9	1.9507	6.154	1454.3	1636.5	1.9358
1400	9.214	1541.8	1746.4	2.0300	7.895	1541.6	1746.1	2.0129	6.906	1541.4	1745.9	1.9980
1600	10.212	1632.3	1859.0	2.0875	8.752	1632.1	1858.8	2.0704	7.656	1631.9	1858.6	2.0556
1800	11.209	1726.2	1975.1	2.1413	9.607	1726.1	1975.0	2.1242	8.405	1725.9	1974.8	2.1094
2000	12.205	1823.6	2094.6	2.1919	10.461	1823.5	2094.5	2.1749	9.153	1823.3	2094.3	2.1601
	P = 180(373.13)				P = 200(381.86)				P = 225(391.87)			
Sat	2.533	1113.4	1197.8	1.5553	2.289	1114.6	1199.3	1.5464	2.043	1115.8	1200.8	1.5365
400	2.648	1126.2	1214.4	1.5749	2.361	1123.5	1210.8	1.5600	2.073	1119.9	1206.2	1.5427
450	2.850	1148.5	1243.4	1.6078	2.548	1146.4	1240.7	1.5938	2.245	1143.8	1237.3	1.5779
500	3.042	1169.6	1270.9	1.6372	2.724	1168.0	1268.8	1.6239	2.405	1165.9	1266.1	1.6087
550	3.228	1190.0	1297.5	1.6642	2.893	1188.7	1295.7	1.6512	2.558	1187.0	1293.5	1.6366
600	3.409	1210.0	1323.5	1.6893	3.058	1208.9	1322.1	1.6767	2.707	1207.5	1320.2	1.6624
700	3.763	1249.6	1374.9	1.7357	3.379	1248.8	1373.8	1.7234	2.995	1247.7	1372.4	1.7095
800	4.110	1289.3	1426.2	1.7781	3.693	1288.6	1425.3	1.7660	3.276	1287.8	1424.2	1.7523
900	4.453	1329.4	1477.7	1.8175	4.003	1328.9	1477.1	1.8055	3.553	1328.3	1476.2	1.7920
1000	4.793	1370.2	1529.8	1.8545	4.310	1369.8	1529.3	1.8425	3.827	1369.3	1528.6	1.8292
1200	5.467	1454.0	1636.1	1.9227	4.918	1453.7	1635.7	1.9109	4.369	1453.4	1635.3	1.8977

TABLE A·1.3 (E) *(Continued)*

°F	v	u	h	s	v	u	h	s	v	u	h	s
1400	6.137	1541.2	1745.6	1.9849	5.521	1540.9	1745.3	1.9732	4.906	1540.7	1744.9	1.9600
1600	6.804	1631.7	1858.4	2.0425	6.123	1631.6	1858.2	2.0308	5.441	1631.3	1857.9	2.0177
1800	7.470	1725.8	1974.6	2.0964	6.722	1725.6	1974.4	2.0847	5.975	1725.4	1974.2	2.0716
2000	8.135	1823.2	2094.2	2.1470	7.321	1823.0	2094.0	2.1354	6.507	1822.9	2093.8	2.1223
	P = 250(401.04)				P = 275(409.52)				P = 300(417.43)			
Sat	1.8448	1116.7	1202.1	1.5274	1.6813	1117.5	1203.1	1.5192	1.5442	1118.2	1203.9	1.5115
450	2.002	1141.1	1233.7	1.5632	1.8026	1138.3	1230.0	1.5495	1.6361	1135.4	1226.2	1.5365
500	2.150	1163.8	1263.3	1.5948	1.9407	1161.7	1260.4	1.5820	1.7662	1159.5	1257.5	1.5701
550	2.290	1185.3	1291.3	1.6233	2.071	1183.6	1289.0	1.6110	1.8878	1181.9	1286.7	1.5997
600	2.426	1206.1	1318.3	1.6494	2.196	1204.7	1316.4	1.6376	2.004	1203.2	1314.5	1.6266
650	2.558	1226.5	1344.9	1.6739	2.317	1225.3	1343.2	1.6623	2.117	1224.1	1341.6	1.6516
700	2.688	1246.7	1371.1	1.6970	2.436	1245.7	1369.7	1.6856	2.227	1244.6	1368.3	1.6751
800	2.943	1287.0	1423.2	1.7401	2.670	1286.2	1422.1	1.7289	2.442	1285.4	1421.0	1.7187
900	3.193	1327.6	1475.3	1.7799	2.898	1327.0	1474.5	1.7689	2.653	1326.3	1473.6	1.7589
1000	3.440	1368.7	1527.9	1.8172	3.124	1368.2	1527.2	1.8064	2.860	1367.7	1526.5	1.7964
1200	3.929	1453.0	1634.8	1.8858	3.570	1452.6	1634.3	1.8751	3.270	1452.2	1633.8	1.8653
1400	4.414	1540.4	1744.6	1.9483	4.011	1540.1	1744.2	1.9376	3.675	1539.8	1743.8	1.9279
1600	4.896	1631.1	1857.6	2.0060	4.450	1630.9	1857.3	1.9954	4.078	1630.7	1857.0	1.9857
1800	5.376	1725.2	1974.0	2.0599	4.887	1725.0	1973.7	2.0493	4.479	1724.9	1973.5	2.0396
2000	5.856	1822.7	2093.6	2.1106	5.323	1822.5	2093.4	2.1000	4.879	1822.3	2093.2	2.0904
	P = 350(431.82)				P = 400(444.70)				P = 450(456.39)			
Sat	1.3267	1119.0	1204.9	1.4978	1.1620	1119.5	1205.5	1.4856	1.0326	1119.6	1205.6	1.4746
450	1.3733	1129.2	1218.2	1.5125	1.1745	1122.6	1209.6	1.4901	...	...	...	...
500	1.4913	1154.9	1251.5	1.5482	1.2843	1150.1	1245.2	1.5282	1.1226	1145.1	1238.5	1.5097
550	1.5998	1178.3	1281.9	1.5790	1.3833	1174.6	1277.0	1.5605	1.2146	1170.7	1271.9	1.5436
600	1.7025	1200.3	1310.6	1.6068	1.4760	1197.3	1306.6	1.5892	1.2996	1194.3	1302.5	1.5732
650	1.8013	1221.6	1338.3	1.6323	1.5645	1219.1	1334.9	1.6153	1.3803	1216.6	1331.5	1.6000
700	1.8975	1242.5	1365.4	1.6562	1.6503	1240.4	1362.5	1.6397	1.4580	1238.2	1359.6	1.6248
800	2.085	1283.8	1418.8	1.7004	1.8163	1282.1	1416.6	1.6844	1.6077	1280.5	1414.4	1.6701

900	2.267	1325.0	1471.8	1.7409	1.9776	1323.7	1470.1	1.7252	1.7524	1322.4	1468.3	1.7113
1000	2.446	1366.6	1525.0	1.7787	2.136	1365.5	1523.6	1.7632	1.8941	1364.4	1522.2	1.7495
1200	2.799	1451.5	1632.8	1.8478	2.446	1450.7	1631.8	1.8327	2.172	1450.0	1630.8	1.8192
1400	3.148	1539.3	1743.1	1.9106	2.752	1538.7	1742.4	1.8956	2.444	1538.1	1741.7	1.8823
1600	3.494	1630.2	1856.5	1.9685	3.055	1629.8	1855.9	1.9535	2.715	1629.3	1855.4	1.9403
1800	3.838	1724.5	1973.1	2.0225	3.357	1724.1	1972.6	2.0076	2.983	1723.7	1972.1	1.9944
2000	4.182	1822.0	2092.8	2.0733	3.658	1821.6	2092.4	2.0584	3.251	1821.3	2092.0	2.0453
	P = 500(467.13)				P = 600(486.33)				P = 700(503.23)			
Sat	0.9283	1119.4	1205.3	1.4645	0.7702	1118.6	1204.1	1.4464	0.6558	1117.0	1202.0	1.4305
500	0.9924	1139.7	1231.5	1.4923	0.7947	1128.0	1216.2	1.4592	...	...	...	...
550	1.0792	1166.7	1266.6	1.5279	0.8749	1158.2	1255.4	1.4990	0.7275	1149.0	1243.2	1.4723
600	1.1583	1191.1	1298.3	1.5585	0.9456	1184.5	1289.5	1.5320	0.7929	1177.5	1280.2	1.5081
650	1.2327	1214.0	1328.0	1.5860	1.0109	1208.6	1320.9	1.5609	0.8520	1203.1	1313.4	1.5387
700	1.3040	1236.0	1356.7	1.6112	1.0727	1231.5	1350.6	1.5872	0.9073	1226.9	1344.4	1.5661
800	1.4407	1278.8	1412.1	1.6571	1.1900	1275.4	1407.6	1.6343	1.0109	1272.0	1402.9	1.6145
900	1.5723	1321.0	1466.5	1.6987	1.3021	1318.4	1462.9	1.6766	1.1089	1315.6	1459.3	1.6576
1000	1.7008	1363.3	1520.7	1.7371	1.4108	1361.2	1517.8	1.7155	1.2036	1358.9	1514.9	1.6970
1100	1.8271	1406.0	1575.1	1.7731	1.5173	1404.2	1572.7	1.7519	1.2960	1402.4	1570.2	1.7337
1200	1.9518	1449.2	1629.8	1.8072	1.6222	1447.7	1627.8	1.7861	1.3868	1446.2	1625.8	1.7682
1400	2.198	1537.6	1741.0	1.8704	1.8289	1536.5	1739.5	1.8497	1.5652	1535.3	1738.1	1.8321
1600	2.442	1628.9	1854.8	1.9285	2.033	1628.0	1853.7	1.9080	1.7409	1627.1	1852.6	1.8906
1800	2.684	1723.3	1971.7	1.9827	2.236	1722.6	1970.8	1.9622	1.9152	1721.8	1969.9	1.9449
2000	2.926	1820.9	2091.6	2.0335	2.438	1820.2	2090.8	2.0131	2.0887	1819.5	2090.1	1.9958

TABLE A·1.3 (E) *(Concluded)*

°F	v	u	h	s	v	u	h	s	v	u	h	s
	P = 800(518.36)				P = 1000(544.75)				P = 1250(572.56)			
Sat	0.5691	1115.0	1199.3	1.4160	0.4459	1109.9	1192.4	1.3903	0.3454	1101.7	1181.6	1.3619
550	0.6154	1138.8	1229.9	1.4469	0.4534	1114.8	1198.7	1.3966	...	...	...	...
600	0.6776	1170.1	1270.4	1.4861	0.5140	1153.7	1248.8	1.4450	0.3786	1129.0	1216.6	1.3954
650	0.7324	1197.2	1305.6	1.5186	0.5637	1184.7	1289.1	1.4822	0.4267	1167.2	1266.0	1.4410
700	0.7829	1222.1	1338.0	1.5471	0.6080	1212.0	1324.6	1.5135	0.4670	1198.4	1306.4	1.4767
750	0.8306	1245.7	1368.6	1.5730	0.6490	1237.2	1357.3	1.5412	0.5030	1226.1	1342.4	1.5070
800	0.8764	1268.5	1398.2	1.5969	0.6878	1261.2	1388.5	1.5664	0.5364	1251.8	1375.8	1.5341
900	0.9640	1312.9	1455.6	1.6408	0.7610	1307.3	1448.1	1.6120	0.5984	1300.0	1438.4	1.5820
1000	1.0482	1356.7	1511.9	1.6807	0.8305	1352.2	1505.9	1.6530	0.6563	1346.4	1498.2	1.6244
1100	1.1300	1400.5	1567.8	1.7178	0.8976	1396.8	1562.9	1.6908	0.7116	1392.0	1556.6	1.6631
1200	1.2102	1444.6	1623.8	1.7526	0.9630	1441.5	1619.7	1.7261	0.7652	1437.5	1614.5	1.6991
1400	1.3674	1534.2	1736.6	1.8167	1.0905	1531.9	1733.7	1.7909	0.8689	1529.0	1730.0	1.7648
1600	1.5218	1626.2	1851.5	1.8754	1.2152	1624.4	1849.3	1.8499	0.9699	1622.2	1846.5	1.8243
1800	1.6749	1721.0	1969.0	1.9298	1.3384	1719.5	1967.2	1.9046	1.0693	1717.6	1965.0	1.8791
2000	1.8271	1818.8	2089.3	1.9808	1.4608	1817.4	2087.7	1.9557	1.1678	1815.7	2085.8	1.9304
	P = 1500(596.39)				P = 1750(617.31)				P = 2000(636.00)			
Sat	0.2769	1091.8	1168.7	1.3359	0.2268	1080.2	1153.7	1.3109	0.18813	1066.6	1136.3	1.2861
600	0.2816	1096.6	1174.8	1.3416	...	...	...	...	...	...	...	...
650	0.3329	1147.0	1239.4	1.4012	0.2627	1122.5	1207.6	1.3603	0.2057	1091.1	1167.2	1.3141
700	0.3716	1183.4	1286.6	1.4429	0.3022	1166.7	1264.6	1.4106	0.2487	1147.7	1239.8	1.3782
750	0.4049	1214.1	1326.5	1.4767	0.3341	1201.3	1309.5	1.4485	0.2803	1187.3	1291.1	1.4216
800	0.4350	1241.8	1362.5	1.5058	0.3622	1231.3	1348.6	1.4802	0.3071	1220.1	1333.8	1.4562
850	0.4631	1267.7	1396.2	1.5320	0.3878	1258.8	1384.4	1.5081	0.3312	1249.5	1372.0	1.4860
900	0.4897	1292.5	1428.5	1.5562	0.4119	1284.8	1418.2	1.5334	0.3534	1276.8	1407.6	1.5126
1000	0.5400	1340.4	1490.3	1.6001	0.4569	1334.3	1482.3	1.5789	0.3945	1328.1	1474.1	1.5598
1100	0.5876	1387.2	1550.3	1.6399	0.4990	1382.2	1543.8	1.6197	0.4325	1377.2	1537.2	1.6017
1200	0.6334	1433.5	1609.3	1.6765	0.5392	1429.4	1604.0	1.6571	0.4685	1425.2	1598.6	1.6398
1400	0.7213	1526.1	1726.3	1.7431	0.6158	1523.1	1722.6	1.7245	0.5368	1520.2	1718.8	1.7082
1600	0.8064	1619.9	1843.7	1.8031	0.6896	1617.6	1841.0	1.7850	0.6020	1615.4	1838.2	1.7692

TABLE A·1.4 (E) Water: Compressed Liquid (English units) v in ft³/lbm, u and h in B/lbm, s in B/lbm·R

P (T Sat.)	500 psia (467.13°F)				1000 psia (544.75°F)				2000 psia (636.00°F)			
T	v	u	h	s	v	u	h	s	v	u	h	s
Sat.	.019748	447.70	449.53	.64904	.021591	538.39	542.38	.74320	.025649	662.40	671.89	.86227
32	.015994	.00	1.49	.00000	.015967	.03	2.99	.00005	.015912	.06	5.95	.00008
50	.015998	18.02	19.50	.03599	.015972	17.99	20.94	.03592	.015920	17.91	23.81	.03575
100	.016106	67.87	69.36	.12932	.016082	67.70	70.68	.12901	.016034	67.37	73.30	.12839
150	.016318	117.66	119.17	.21457	.016293	117.38	120.40	.21410	.016244	116.83	122.84	.21318
200	.016608	167.65	169.19	.29341	.016580	167.26	170.32	.29281	.016527	166.49	172.60	.29162
250	.016972	217.99	219.56	.36702	.016941	217.47	220.61	.36628	.016880	216.46	222.70	.36482
300	.017416	268.92	270.53	.43641	.017379	268.24	271.46	.43552	.017308	266.93	273.33	.43376
350	.017954	320.71	322.37	.50249	.017909	319.83	323.15	.50140	.017822	318.15	324.74	.49929
400	.018608	373.68	375.40	.56604	.018550	372.55	375.98	.56472	.018439	370.38	377.21	.56216
450	.019420	428.40	430.19	.62798	.019340	426.89	430.47	.62632	.019191	424.04	431.14	.62313
500					.02036	483.8	487.5	.6874	.02014	479.8	487.3	.6832
510					.02060	495.6	499.4	.6997	.02036	491.4	498.9	.6953
520					.02086	507.6	511.5	.7121	.02060	503.1	510.7	.7073
530					.02114	519.9	523.8	.7245	.02085	514.9	522.6	.7195
540					.02144	532.4	536.3	.7372	.02112	527.0	534.8	.7317
550									.02141	539.2	547.2	.7440
560									.02172	551.8	559.8	.7565
570									.02206	564.6	572.8	.7691
580									.02243	577.8	586.1	.7820
590									.02284	591.3	599.8	.7951
600									.02330	605.4	614.0	.8086
610									.02382	620.0	628.8	.8225
620									.02443	635.4	644.5	.8371
630									.02514	651.9	661.2	.8525

TABLE A·1.4 (E) *(Concluded)*

P	4000 psia				6000 psia				8000 psia			
T	v	u	h	s	v	u	h	s	v	u	h	s
32	.015807	.10	11.80	.00005	.015705	.11	17.55	−.00013	.015608	.06	23.17	−.00048
50	.015821	17.76	29.47	.03534	.015726	17.57	35.03	.03480	.015635	17.38	40.53	.03419
100	.015942	66.72	78.52	.12714	.015853	66.09	83.70	.12588	.015767	65.49	88.84	.12461
150	.016150	115.77	127.73	.21136	.016059	114.77	132.60	.20957	.015972	113.82	137.46	.20782
200	.016425	165.02	177.18	.28931	.016328	163.63	181.76	.28707	.016234	162.32	186.35	.28489
250	.016765	214.52	226.93	.36200	.016656	212.70	231.19	.35929	.016553	210.98	235.48	.35668
300	.017174	264.43	277.15	.43038	.017048	262.10	281.03	.42716	.016930	259.91	284.98	.42407
350	.017659	314.98	328.05	.49526	.017510	312.04	331.48	.49147	.017371	309.30	335.01	.48786
400	.018235	366.35	379.85	.55734	.018051	362.66	382.71	.55285	.017883	359.27	385.74	.54865
450	.018924	418.83	432.84	.61725	.018689	414.17	434.92	.61189	.018479	409.95	437.30	.60694
500	.019766	472.9	487.5	.6758	.019451	466.9	488.5	.6692	.019179	461.6	490.0	.6633
520	.020161	495.2	510.1	.6990	.019802	488.4	510.4	.6918	.019495	482.6	511.4	.6854
540	.020600	517.9	533.1	.7223	.020185	510.3	532.7	.7144	.019837	503.8	533.1	.7074
560	.021091	541.2	556.8	.7457	.020606	532.6	555.5	.7369	.020208	525.3	555.2	.7292
580	.021648	565.2	581.2	.7694	.021072	555.3	578.7	.7595	.020613	547.1	577.7	.7510
600	.02229	590.0	606.5	.7936	.02159	578.6	602.6	.7822	.02106	569.4	600.5	.7728
620	.02304	616.0	633.0	.8183	.02218	602.6	627.2	.8052	.02155	592.0	623.9	.7947
640	.02394	643.3	661.1	.8441	.02285	627.3	652.7	.8286	.02209	615.2	647.9	.8167
660	.02506	672.7	691.2	.8712	.02363	653.0	679.2	.8525	.02270	639.0	672.6	.8389
680	.02653	704.9	724.5	.9007	.02454	679.8	707.0	.8771	.02339	663.4	698.0	.8615
700	.02867	742.1	763.4	.9345	.02563	708.1	736.5	.9028	.02418	688.6	724.4	.8844
710	.03026	764.3	786.7	.9545	.02626	722.9	752.1	.9161	.02461	701.5	737.9	.8960

TABLE A·1.5 (E) Water: Solid-Vapor Saturation (English units) v in ft³/lbm, u and h in B/lbm, s in B/lbm·R

		Specific Volume		Internal Energy			Enthalpy			Entropy		
Temp. °F T	Pressure (lbf/in;²) P	Sat. Solid v_i	Sat. Vapor $v_g \times 10^{-3}$	Sat. Solid u_i	Subl. u_{ig}	Sat. Vapor u_g	Sat. Solid h_i	Subl. h_{ig}	Sat. Vapor h_g	Sat. Solid s_i	Subl. s_{ig}	Sat. Vapor s_g
32.018	.0887	.01747	3.302	−143.34	1164.6	1021.2	−143.34	1218.7	1075.4	−.292	2.479	2.187
32	.0886	.01747	3.305	−143.35	1164.6	1021.2	−143.35	1218.7	1075.4	−.292	2.479	2.187
30	.0808	.01747	3.607	−144.35	1164.9	1020.5	−144.35	1218.9	1074.5	−.294	2.489	2.195
25	.0641	.01746	4.506	−146.84	1165.7	1018.9	−146.84	1219.1	1072.3	−.299	2.515	2.216
20	.0505	.01745	5.655	−149.31	1166.5	1017.2	−149.31	1219.4	1070.1	−.304	2.542	2.238
15	.0396	.01745	7.13	−151.75	1167.3	1015.5	−151.75	1219.7	1067.9	−.309	2.569	2.260
10	.0309	.01744	9.04	−154.17	1168.1	1013.9	−154.17	1219.9	1065.7	−.314	2.597	2.283
5	.0240	.01743	11.52	−156.56	1168.8	1012.2	−156.56	1220.1	1063.5	−.320	2.626	2.306
0	.0185	.01743	14.77	−158.93	1169.5	1010.6	−158.93	1220.2	1061.2	−.325	2.655	2.330
−5	.0142	.01742	19.03	−161.27	1170.2	1008.9	−161.27	1220.3	1059.0	−.330	2.684	2.354
−10	.0109	.01741	24.66	−163.59	1170.9	1007.3	−163.59	1220.4	1056.8	−.335	2.714	2.379
−15	.0082	.01740	32.2	−165.89	1171.5	1005.6	−165.89	1220.5	1054.6	−.340	2.745	2.405
−20	.0062	.01740	42.2	−168.16	1172.1	1003.9	−168.16	1220.6	1052.4	−.345	2.776	2.431
−25	.0046	.01739	55.7	−170.40	1172.7	1002.3	−170.40	1220.6	1050.2	−.351	2.808	2.457
−30	.0035	.01738	74.1	−172.63	1173.2	1000.6	−172.63	1220.6	1048.0	−.356	2.841	2.485
−35	.0026	.01737	99.2	−174.82	1173.8	998.9	−174.82	1220.6	1045.8	−.361	2.874	2.513
−40	.0019	.01737	133.8	−177.00	1174.3	997.3	−177.00	1220.6	1043.6	−.366	2.908	2.542

TABLE A·2.1 (SI) Ammonia: Liquid-Vapor Saturation, Temperature Table (SI)

Temp. °C T	Abs. Pressure kPa P	Specific Volume m^3/kg Sat. Liquid v_f	Evap. v_{fg}	Sat. Vapor v_g	Enthalpy kJ/kg Sat. Liquid h_f	Evap. h_{fg}	Sat. Vapor h_g	Entropy kJ/kg·K Sat. Liquid s_f	Evap. s_{fg}	Sat. Vapor s_g
−50	40.88	0.001 424	2.6239	2.6254	−44.3	1416.7	1372.4	−0.1942	6.3502	6.1561
−48	45.96	0.001 429	2.3518	2.3533	−35.5	1411.3	1375.8	−0.1547	6.2696	6.1149
−46	51.55	0.001 434	2.1126	2.1140	−26.6	1405.8	1379.2	−0.1156	6.1902	6.0746
−44	57.69	0.001 439	1.9018	1.9032	−17.8	1400.3	1382.5	−0.0768	6.1120	6.0352
−42	64.42	0.001 444	1.7155	1.7170	−8.9	1394.7	1385.8	−0.0382	6.0349	5.9967
−40	71.77	0.001 449	1.5506	1.5521	0.0	1389.0	1389.0	0.0000	5.9589	5.9589
−38	79.80	0.001 454	1.4043	1.4058	8.9	1383.3	1392.2	0.0380	5.8840	5.9220
−36	88.54	0.001 460	1.2742	1.2757	17.8	1377.6	1395.4	0.0757	5.8101	5.8858
−34	98.05	0.001 465	1.1582	1.1597	26.8	1371.8	1398.5	0.1132	5.7372	5.8504
−32	108.37	0.001 470	1.0547	1.0562	35.7	1365.9	1401.6	0.1504	5.6652	5.8156
−30	119.55	0.001 476	0.9621	0.9635	44.7	1360.0	1404.6	0.1873	5.5942	5.7815
−28	131.64	0.001 481	0.8790	0.8805	53.6	1354.0	1407.6	0.2240	5.5241	5.7481
−26	144.70	0.001 487	0.8044	0.8059	62.6	1347.9	1410.5	0.2605	5.4548	5.7153
−24	158.78	0.001 492	0.7373	0.7388	71.6	1341.8	1413.4	0.2967	5.3864	5.6831
−22	173.93	0.001 498	0.6768	0.6783	80.7	1335.6	1416.2	0.3327	5.3188	5.6515
−20	190.22	0.001 504	0.6222	0.6237	89.7	1329.3	1419.0	0.3684	5.2520	5.6205
−18	207.71	0.001 510	0.5728	0.5743	98.8	1322.9	1421.7	0.4040	5.1860	5.5900
−16	226.45	0.001 515	0.5280	0.5296	107.8	1316.5	1424.4	0.4393	5.1207	5.5600
−14	246.51	0.001 521	0.4874	0.4889	116.9	1310.0	1427.0	0.4744	5.0561	5.5305
−12	267.95	0.001 528	0.4505	0.4520	126.0	1303.5	1429.5	0.5093	4.9922	5.5015

−10	290.85	0.001 534	0.4169	0.4185	135.2	1296.8	1432.0	0.5440	4.9290	5.4730
−8	315.25	0.001 540	0.3863	0.3878	144.3	1290.1	1434.4	0.5785	4.8664	5.4449
−6	341.25	0.001 546	0.3583	0.3599	153.5	1283.3	1436.8	0.6128	4.8045	5.4173
−4	368.90	0.001 553	0.3328	0.3343	162.7	1276.4	1439.1	0.6469	4.7432	5.3901
−2	398.27	0.001 559	0.3094	0.3109	171.9	1269.4	1441.3	0.6808	4.6825	5.3633
0	429.44	0.001 566	0.2879	0.2895	181.1	1262.4	1443.5	0.7145	4.6223	5.3369
2	462.49	0.001 573	0.2683	0.2698	190.4	1255.2	1445.6	0.7481	4.5627	5.3108
4	497.49	0.001 580	0.2502	0.2517	199.6	1248.0	1447.6	0.7815	4.5037	5.2852
6	534.51	0.001 587	0.2335	0.2351	208.9	1240.6	1449.6	0.8148	4.4451	5.2599
8	573.64	0.001 594	0.2182	0.2198	218.3	1233.2	1451.5	0.8479	4.3871	5.2350
10	614.95	0.001 601	0.2040	0.2056	227.6	1225.7	1453.3	0.8808	4.3295	5.2104
12	658.52	0.001 608	0.1910	0.1926	237.0	1218.1	1455.1	0.9136	4.2725	5.1861
14	704.44	0.001 616	0.1789	0.1805	246.4	1210.4	1456.8	0.9463	4.2159	5.1621
16	752.79	0.001 623	0.1677	0.1693	255.9	1202.6	1458.5	0.9788	4.1597	5.1385
18	803.66	0.001 631	0.1574	0.1590	265.4	1194.7	1460.0	1.0112	4.1039	5.1151
20	857.12	0.001 639	0.1477	0.1494	274.9	1186.7	1461.5	1.0434	4.0486	5.0920
22	913.27	0.001 647	0.1388	0.1405	284.4	1178.5	1462.9	1.0755	3.9937	5.0692
24	972.19	0.001 655	0.1305	0.1322	294.0	1170.3	1464.3	1.1075	3.9392	5.0467
26	1033.97	0.001 663	0.1228	0.1245	303.6	1162.0	1465.6	1.1394	3.8850	5.0244
28	1098.71	0.001 671	0.1156	0.1173	313.2	1153.6	1466.8	1.1711	3.8312	5.0023
30	1166.49	0.001 680	0.1089	0.1106	322.9	1145.0	1467.9	1.2028	3.7777	4.9805
32	1237.41	0.001 689	0.1027	0.1044	332.6	1136.4	1469.0	1.2343	3.7246	4.9589
34	1311.55	0.001 698	0.0969	0.0986	342.3	1127.6	1469.9	1.2656	3.6718	4.9374
36	1389.03	0.001 707	0.0914	0.0931	352.1	1118.7	1470.8	1.2969	3.6192	4.9161
38	1469.92	0.001 716	0.0863	0.0880	361.9	1109.7	1471.5	1.3281	3.5669	4.8950
40	1554.33	0.001 726	0.0815	0.0833	371.7	1100.5	1472.2	1.3591	3.5148	4.8740
42	1642.35	0.001 735	0.0771	0.0788	381.6	1091.2	1472.8	1.3901	3.4630	4.8530
44	1734.09	0.001 745	0.0728	0.0746	391.5	1081.7	1473.2	1.4209	3.4112	4.8322
46	1829.65	0.001 756	0.0689	0.0707	401.5	1072.0	1473.5	1.4518	3.3595	4.8113
48	1929.13	0.001 766	0.0652	0.0669	411.5	1062.2	1473.7	1.4826	3.3079	4.7905
50	2032.62	0.001 777	0.0617	0.0635	421.7	1052.0	1473.7	1.5135	3.2561	4.7696

Source: Adapted from National Bureau of Standards Circular No. 142, *Tables of Thermodynamic Properties of Ammonia*. Abridgment from Gordon J. Van Wylen and Richard E. Sonntag, *Fundamentals of Classical Thermodynamics*, 3rd ed., Wiley, New York, 1985.

TABLE A·2.2 (SI) Ammonia: Liquid-Vapor Saturation, Pressure Table (SI)

Abs. Pressure kPa P	Temp. °C T	Specific Volume m^3/kg Sat. Liquid v_f	Evap. v_{fg}	Sat. Vapor v_g	Enthalpy kJ/kg Sat. Liquid h_f	Evap. h_{fg}	Sat. Vapor h_g	Entropy kJ/kg·K Sat. Liquid s_f	Evap. s_{fg}	Sat. Vapor s_g
40	−50.4	.001421	2.683	2.684	−45.9	1417.8	1371.9	−0.201	6.370	6.169
60	−43.3	.001442	1.834	1.836	−14.6	1398.3	1383.6	−0.063	6.086	6.023
80	−37.9	.001456	1.401	1.402	9.1	1383.2	1392.3	0.039	5.883	5.922
100	−33.6	.001467	1.136	1.138	28.4	1370.7	1399.1	0.120	5.725	5.845
150	−25.3	.001489	0.777	0.778	66.1	1345.6	1411.7	0.274	5.432	5.706
200	−18.9	.001507	0.593	0.594	94.9	1325.7	1420.6	0.389	5.217	5.605
300	−9.2	.001535	0.405	0.406	138.7	1294.2	1432.9	0.557	4.906	5.463
400	−1.9	.001559	0.309	0.310	172.4	1269.0	1441.4	0.683	4.681	5.364
500	4.1	.001579	0.249	0.251	200.3	1247.4	1447.7	0.784	4.501	5.285
600	9.3	.001598	0.209	0.210	224.3	1228.3	1452.6	0.869	4.351	5.220
800	17.9	.001631	0.157	0.159	264.7	1195.3	1460.0	1.009	4.109	5.118
1000	24.9	.001660	0.126	0.128	298.3	1166.7	1465.0	1.122	3.916	5.038
1200	31.0	.001686	0.105	0.107	327.5	1141.0	1468.4	1.218	3.754	4.971
1400	36.3	.001710	0.090	0.092	353.4	1117.4	1470.8	1.301	3.613	4.914
1600	41.1	.001732	0.079	0.081	376.9	1095.5	1472.4	1.375	3.488	4.864
1800	45.4	.001752	0.070	0.072	398.5	1074.9	1473.4	1.442	3.376	4.819
2000	49.4	.001771	0.633	0.065	418.4	1055.4	1473.9	1.504	3.274	4.778

TABLE A·2.3 (SI) Ammonia: Superheated Vapor (SI) v in m³/kg, h in kJ/kg, s in kJ/kg·K

Abs. Pressure kPa (Sat. Temp. °C)		Temperature, °C											
		−20	−10	0	10	20	30	40	50	60	70	80	100
	v	2.4474	2.5481	2.6482	2.7479	2.8473	2.9464	3.0453	3.1441	3.2427	3.3413	3.4397	
50	h	1435.8	1457.0	1478.1	1499.2	1520.4	1541.7	1563.0	1584.5	1606.1	1627.8	1649.7	
(−46.54)	s	6.3256	6.4077	6.4865	6.5625	6.6360	6.7073	6.7766	6.8441	6.9099	6.9743	7.0372	
	v	1.6233	1.6915	1.7591	1.8263	1.8932	1.9597	2.0261	2.0923	2.1584	2.2244	2.2903	
75	h	1433.0	1454.7	1476.1	1497.5	1518.9	1540.3	1561.8	1583.4	1605.1	1626.9	1648.9	
(−39.18)	s	6.1190	6.2028	6.2828	6.3597	6.4339	6.5058	6.5756	6.6434	6.7096	6.7742	6.8373	
	v	1.2110	1.2631	1.3145	1.3654	1.4160	1.4664	1.5165	1.5664	1.6163	1.6659	1.7155	1.8145
100	h	1430.1	1452.2	1474.1	1495.7	1517.3	1538.9	1560.5	1582.2	1604.1	1626.0	1648.0	1692.6
(−33.61)	s	5.9695	6.0552	6.1366	6.2144	6.2894	6.3618	6.4321	6.5003	6.5668	6.6316	6.6950	6.8177
	v	0.9635	1.0059	1.0476	1.0889	1.1297	1.1703	1.2107	1.2509	1.2909	1.3309	1.3707	1.4501
125	h	1427.2	1449.8	1472.0	1493.9	1515.7	1537.5	1559.3	1581.1	1603.0	1625.0	1647.2	1691.8
(−29.08)	s	5.8512	5.9389	6.0217	6.1006	6.1763	6.2494	6.3201	6.3887	6.4555	6.5206	6.5842	6.7072
	v	0.7984	0.8344	0.8697	0.9045	0.9388	0.9729	1.0068	1.0405	1.0740	1.1074	1.1408	1.2072
150	h	1424.1	1447.3	1469.8	1492.1	1514.1	1536.1	1558.0	1580.0	1602.0	1624.1	1646.3	1691.1
(−25.23)	s	5.7526	5.8424	5.9266	6.0066	6.0831	6.1568	6.2280	6.2970	6.3641	6.4295	6.4933	6.6167

TABLE A·2.3 (SI) *(Continued)*

Abs. Pressure kPa (Sat. Temp. °C)		Temperature, °C											
		−20	−10	0	10	20	30	40	50	60	70	80	100
	v		0.6199	0.6471	0.6738	0.7001	0.7261	0.7519	0.7774	0.8029	0.8282	0.8533	0.9035
200	h		1442.0	1465.5	1488.4	1510.9	1533.2	1555.5	1577.7	1599.9	1622.2	1644.6	1689.6
(−18.86)	s		5.6863	5.7737	5.8559	5.9342	6.0091	6.0813	6.1512	6.2189	6.2849	6.3491	6.4732
	v		0.4910	0.5135	0.5354	0.5568	0.5780	0.5989	0.6196	0.6401	0.6605	0.6809	0.7212
250	h		1436.6	1461.0	1484.5	1507.6	1530.3	1552.9	1575.4	1597.8	1620.3	1642.8	1688.2
(−13.67)	s		5.5609	5.6517	5.7365	5.8165	5.8928	5.9661	6.0368	6.1052	6.1717	6.2365	6.3613
	v			0.4243	0.4430	0.4613	0.4792	0.4968	0.5143	0.5316	0.5488	0.5658	0.5997
300	h			1456.3	1480.6	1504.2	1527.4	1550.3	1573.0	1595.7	1618.4	1641.1	1686.7
(−9.23)	s			5.5493	5.6366	5.7186	5.7963	5.8707	5.9423	6.0114	6.0785	6.1437	6.2693
	v			0.3605	0.3770	0.3929	0.4086	0.4239	0.4391	0.4541	0.4689	0.4837	0.5129
350	h			1451.5	1476.5	1500.7	1524.4	1547.6	1570.7	1593.6	1616.5	1639.3	1685.2
(−5.35)	s			5.4600	5.5502	5.6342	5.7135	5.7890	5.8615	5.9314	5.9990	6.0647	6.1910
	v			0.3125	0.3274	0.3417	0.3556	0.3692	0.3826	0.3959	0.4090	0.4220	0.4478
400	h			1446.5	1472.4	1497.2	1521.3	1544.9	1568.3	1591.5	1614.5	1637.6	1683.7
(−1.89)	s			5.3803	5.4735	5.5597	5.6405	5.7173	5.7907	5.8613	5.9296	5.9957	6.1228
	v			0.2752	0.2887	0.3017	0.3143	0.3266	0.3387	0.3506	0.3624	0.3740	0.3971
450	h			1441.3	1468.1	1493.6	1518.2	1542.2	1565.9	1589.3	1612.6	1635.8	1682.2
(1.26)	s			5.3078	5.4042	5.4926	5.5752	5.6532	5.7275	5.7989	5.8678	5.9345	6.0623

	v	0.2698	0.2813	0.2926	0.3036	0.3144	0.3251	0.3357	0.3565	0.3771	0.3975	
500	h	1489.9	1515.0	1539.5	1563.4	1587.1	1610.6	1634.0	1680.7	1727.5	1774.7	
(4.14)	s	5.4314	5.5157	5.5950	5.6704	5.7425	5.8120	5.8793	6.0079	6.1301	6.2472	
	v	0.2217	0.2317	0.2414	0.2508	0.2600	0.2691	0.2781	0.2957	0.3130	0.3302	
600	h	1482.4	1508.6	1533.8	1558.5	1582.7	1606.6	1630.4	1677.7	1724.9	1772.4	
(9.29)	s	5.3222	5.4102	5.4923	5.5697	5.6436	5.7144	5.7826	5.9129	6.0363	6.1541	
	v	0.1874	0.1963	0.2048	0.2131	0.2212	0.2291	0.2369	0.2522	0.2672	0.2821	
700	h	1474.5	1501.9	1528.1	1553.4	1578.2	1602.6	1626.8	1674.6	1722.4	1770.2	
(13.81)	s	5.2259	5.3179	5.4029	5.4826	5.5582	5.6303	5.6997	5.8316	5.9562	6.0749	
	v	0.1615	0.1696	0.1773	0.1848	0.1920	0.1991	0.2060	0.2196	0.2329	0.2459	0.2589
800	h	1466.3	1495.0	1522.2	1548.3	1573.7	1598.6	1623.1	1671.6	1719.8	1768.0	1816.4
(17.86)	s	5.1387	5.2351	5.3232	5.4053	5.4827	5.5562	5.6268	5.7603	5.8861	6.0057	6.1202
	v		0.1488	0.1559	0.1627	0.1693	0.1757	0.1820	0.1942	0.2061	0.2178	0.2294
900	h		1488.0	1516.2	1543.0	1569.1	1594.4	1619.4	1668.5	1717.1	1765.7	1814.4
(21.54)	s		5.1593	5.2508	5.3354	5.4147	5.4897	5.5614	5.6968	5.8237	5.9442	6.0594

TABLE A·2.3 (SI) *(Concluded)*

Abs. Pressure kPa (Sat. Temp. °C)		Temperature, °C											
		20	30	40	50	60	70	80	100	120	140	160	180
	v		0.1321	0.1388	0.1450	0.1511	0.1570	0.1627	0.1739	0.1847	0.1954	0.2058	0.2162
1000	h		1480.6	1510.0	1537.7	1564.4	1590.3	1615.6	1665.4	1714.5	1763.4	1812.4	1861.7
(24.91)	s		5.0889	5.1840	5.2713	5.3525	5.4299	5.5021	5.6392	5.7674	5.8888	6.0047	6.1159
	v			0.1129	0.1185	0.1238	0.1289	0.1338	0.1434	0.1526	0.1616	0.1705	0.1792
1200	h			1497.1	1526.6	1554.7	1581.7	1608.0	1659.2	1709.2	1758.9	1808.5	1858.2
(30.96)	s			5.0629	5.1560	5.2416	5.3215	5.3970	5.5379	5.6687	5.7919	5.9091	6.0214
	v			0.0944	0.0995	0.1042	0.1088	0.1132	0.1216	0.1297	0.1376	0.1452	0.1528
1400	h			1483.4	1515.1	1544.7	1573.0	1600.2	1652.8	1703.9	1754.3	1804.5	1854.7
(36.28)	s			4.9534	5.0530	5.1434	5.2270	5.3053	5.4501	5.5836	5.7087	5.8273	5.9406
	v				0.0851	0.0895	0.0937	0.0977	0.1053	0.1125	0.1195	0.1263	0.1330
1600	h				1502.9	1534.4	1564.0	1592.3	1646.4	1698.5	1749.7	1800.5	1851.2
(41.05)	s				4.9584	5.0543	5.1419	5.2232	5.3722	5.5084	5.6355	5.7555	5.8699
	v				0.0739	0.0781	0.0820	0.0856	0.0926	0.0992	0.1055	0.1116	0.1177
1800	h				1490.0	1523.5	1554.6	1584.1	1639.8	1693.1	1745.1	1796.5	1847.7
(45.39)	s				4.8693	4.9715	5.0635	5.1482	5.3018	5.4409	5.5699	5.6914	5.8069
	v				0.0648	0.0688	0.0725	0.0760	0.0824	0.0885	0.0943	0.0999	0.1054
2000	h				1476.1	1512.0	1544.9	1575.6	1633.2	1687.6	1740.4	1792.4	1844.1
(49.38)	s				4.7834	4.8930	4.9902	5.0786	5.2371	5.3793	5.5104	5.6333	5.7499

TABLE A·2.1 (E) Ammonia: Liquid-Vapor Saturation, Temperature Table (English units) v in cu ft/lbm; h in B/lbm; s in B/lbm·R

		Specific Volume		Enthalpy			Entropy		
Temp. °F T	Pressure psia P	Sat. Liquid v_f	Sat. Vapor v_g	Sat. Liquid h_f	Evap. h_{fg}	Sat. Vapor h_g	Sat. Liquid s_f	Evap. s_{fg}	Sat. Vapor s_g
−60	5.55	0.02278	44.73	−21.2	610.8	589.6	−0.0517	1.5286	1.4769
−50	7.67	0.02299	33.08	−10.6	604.3	593.7	−0.0256	1.4753	1.4497
−40	10.41	0.02322	24.86	0.0	597.6	597.6	0.0000	1.4242	1.4242
−30	13.90	0.02345	18.97	10.7	590.7	601.4	0.0250	1.3751	1.4001
−20	18.30	0.02369	14.68	21.4	583.6	605.0	0.0497	1.3277	1.3774
−10	23.74	0.02393	11.50	32.1	576.4	608.5	0.0738	1.2820	1.3558
0	30.42	0.02419	9.116	42.9	568.9	611.8	0.0975	1.2377	1.3352
5	34.27	0.02432	8.150	48.3	565.0	613.3	0.1092	1.2161	1.3253
10	38.51	0.02446	7.304	53.8	561.1	614.9	0.1208	1.1949	1.3157
20	48.21	0.02474	5.910	64.7	553.1	617.8	0.1437	1.1532	1.2969
30	59.74	0.02503	4.825	75.7	544.8	620.5	0.1663	1.1127	1.2790
40	73.32	0.02533	3.971	86.8	536.2	623.0	0.1885	1.0733	1.2618
50	89.19	0.02564	3.294	97.9	527.3	625.2	0.2105	1.0348	1.2453
60	107.6	0.02597	2.751	109.2	518.1	627.3	0.2322	0.9972	1.2294
70	128.8	0.02632	2.312	120.5	508.6	629.1	0.2537	0.9603	1.2140
80	153.0	0.02668	1.955	132.0	498.7	630.7	0.2749	0.9242	1.1991
86	169.2	0.02691	1.772	138.9	492.6	631.5	0.2875	0.9029	1.1904
90	180.6	0.02707	1.661	143.5	488.5	632.0	0.2958	0.8888	1.1846
100	211.9	0.02747	1.419	155.2	477.8	633.0	0.3166	0.8539	1.1705
110	247.0	0.02790	1.217	167.0	466.7	633.7	0.3372	0.8194	1.1566
120	286.4	0.02836	1.047	179.0	455.0	634.0	0.3576	0.7851	1.1427

Source: From *Engineering Thermodynamics*, by H. J. Stoever, Wiley, 1951, as abridged from *Tables of Thermodynamic Properties of Ammonia*, National Bureau of Standards Circular No. 142.

TABLE A·2.2 (E) Ammonia: Liquid-Vapor Saturation, Pressure Table (English units) v in cu ft/lbm; h in B/lbm; s in B/lbm·R

		Specific Volume		Enthalpy			Entropy		
Pressure psia P	Temp. °F T	Sat. Liquid v_f	Sat. Vapor v_g	Sat. Liquid h_f	Evap. h_{fg}	Sat. Vapor h_g	Sat. Liquid s_f	Evap. s_{fg}	Sat. Vapor s_g
5	−63.11	0.02271	49.31	−24.5	612.8	588.3	−0.0599	1.5456	1.4857
10	−41.34	0.02319	25.81	−1.4	598.5	597.1	−0.0034	1.4310	1.4276
15	−27.29	0.02351	17.67	13.6	588.8	602.4	0.0318	1.3620	1.3938
20	−16.64	0.02377	13.50	25.0	581.2	606.2	0.0578	1.3122	1.3700
30	− 0.57	0.02417	9.236	42.3	569.3	611.6	0.0962	1.2402	1.3364
40	11.66	0.02451	7.047	55.6	559.8	615.4	0.1246	1.1879	1.3125
50	21.67	0.02479	5.710	66.5	551.7	618.2	0.1475	1.1464	1.2939
60	30.21	0.02504	4.805	75.9	544.6	620.5	0.1668	1.1119	1.2787
80	44.40	0.02546	3.655	91.7	532.3	624.0	0.1982	1.0563	1.2545
100	56.05	0.02584	2.952	104.7	521.8	626.5	0.2237	1.0119	1.2356
120	66.02	0.02618	2.476	116.0	512.4	628.4	0.2452	0.9749	1.2201
140	74.79	0.02649	2.132	126.0	503.9	629.9	0.2638	0.9430	1.2068
170	86.29	0.02692	1.764	139.3	492.3	631.6	0.2881	0.9019	1.1900
200	96.34	0.02732	1.502	150.9	481.8	632.7	0.3090	0.8666	1.1756
230	105.30	0.02770	1.307	161.4	472.0	633.4	0.3275	0.8356	1.1631
260	113.42	0.02806	1.155	171.1	462.8	633.9	0.3441	0.8077	1.1518

Source: From *Engineering Thermodynamics*, by H. J. Stoever, Wiley, 1951, as abridged from *Tables of Thermodynamic Properties of Ammonia*, National Bureau of Standards Circular No. 142.

TABLE A·2.3 (E) Ammonia: Superheated Vapor (English units) v in cu ft/lbm; h in B/lbm; s in B/lbm·R

Temp. °F	Pressure, psia (saturation temperature in italics)											
	5 *−63.11°*			10 *−41.34°*			15 *−27.29°*			20 *−16.64°*		
	v	h	s	v	h	s	v	h	s	v	h	s
Sat.	*49.31*	*588.3*	*1.4857*	*25.81*	*597.1*	*1.4276*	*17.67*	*602.4*	*1.3938*	*13.50*	*606.2*	*1.3700*
−50	51.05	595.2	1.5025	.....	.....	.....	.....	.....	.....	.....	.....	.....
−40	52.36	600.3	1.5149	25.90	597.8	1.4293	.....	.....	.....	.....	.....	.....
−30	53.67	605.4	1.5269	26.58	603.2	1.4420	.....	.....	.....	.....	.....	.....
−20	54.97	610.4	1.5385	27.26	608.5	1.4542	18.01	606.4	1.4031	.....	.....	.....
−10	56.26	615.4	1.5498	27.92	613.7	1.4659	18.47	611.9	1.4154	13.74	610.0	1.3784
0	57.55	620.4	1.5608	28.58	618.9	1.4773	18.92	617.2	1.4272	14.09	615.5	1.3907
10	58.84	625.4	1.5716	29.24	624.0	1.4884	19.37	622.5	1.4386	14.44	621.0	1.4025
20	60.12	630.4	1.5821	29.90	629.1	1.4992	19.82	627.8	1.4497	14.78	626.4	1.4138
30	61.41	635.4	1.5925	30.55	634.2	1.5097	20.26	633.0	1.4604	15.11	631.7	1.4248
40	62.69	640.4	1.6026	31.20	639.3	1.5200	20.70	638.2	1.4709	15.45	637.0	1.4356
50	63.96	645.5	1.6125	31.85	644.4	1.5301	21.14	643.4	1.4812	15.78	642.3	1.4460
60	65.24	650.5	1.6223	32.49	649.5	1.5400	21.58	648.5	1.4912	16.12	647.5	1.4562
70	66.51	655.5	1.6319	33.14	654.6	1.5497	22.01	653.7	1.5011	16.45	652.8	1.4662
80	67.79	660.6	1.6413	33.78	659.7	1.5593	22.44	658.9	1.5108	16.78	658.0	1.4760
90	69.06	665.6	1.6506	34.42	664.8	1.5687	22.88	664.0	1.5203	17.10	663.2	1.4856
100	70.33	670.7	1.6598	35.07	670.0	1.5779	23.31	669.2	1.5296	17.43	668.5	1.4950
110	71.60	675.8	1.6689	35.71	675.1	1.5870	23.74	674.4	1.5388	17.76	673.7	1.5042
120	72.87	680.9	1.6778	36.35	680.3	1.5960	24.17	679.6	1.5478	18.08	678.9	1.5133
130	74.14	686.1	1.6865	36.99	685.4	1.6049	24.60	684.8	1.5567	18.41	684.2	1.5223
140	75.41	691.2	1.6952	37.62	690.6	1.6136	25.03	690.0	1.5655	18.73	689.4	1.5312
150	76.68	696.4	1.7038	38.26	695.8	1.6222	25.46	695.3	1.5742	19.05	694.7	1.5399
160	77.95	701.6	1.7122	38.90	701.1	1.6307	25.88	700.5	1.5827	19.37	700.0	1.5485
170	79.21	706.8	1.7206	39.54	706.3	1.6391	26.31	705.8	1.5911	19.70	705.3	1.5569
180	80.48	712.1	1.7289	40.17	711.6	1.6474	26.74	711.1	1.5995	20.02	710.6	1.5653
190	.....	.....	.....	40.81	716.9	1.6556	27.16	716.4	1.6077	20.34	715.9	1.5736
200	.....	.....	.....	41.45	722.2	1.6637	27.59	721.7	1.6158	20.66	721.2	1.5817
220	.....	.....	.....	.....	.....	.....	28.44	732.4	1.6318	21.30	732.0	1.5978
240	.....	.....	.....	.....	.....	.....	.....	.....	.....	21.94	742.8	1.6135

TABLE A·2.3 (E) *(Continued)*

Temp. °F	Pressure, psia (saturation temperature in italics)											
	30 (*−0.57°*)			40 (*11.66°*)			50 (*21.67°*)			60 (*30.21°*)		
	v	*h*	*s*	*v*	*h*	*s*	*v*	*h*	*s*	*v*	*h*	*s*
Sat.	*9.236*	*611.6*	*1.3364*	*7.047*	*615.4*	*1.3125*	*5.710*	*618.2*	*1.2939*	*4.805*	*620.5*	*1.2787*
0	9.250	611.9	1.3371	.....	.....	.....	.....	.....	.....	.....	.....	.....
10	9.492	617.8	1.3497	.....	.....	.....	.....	.....	.....	.....	.....	.....
20	9.731	623.5	1.3618	7.203	620.4	1.3231	.....	.....	.....	.....	.....	.....
30	9.966	629.1	1.3733	7.387	626.3	1.3353	5.838	623.4	1.3046	.....	.....	.....
40	10.20	634.6	1.3845	7.568	632.1	1.3470	5.988	629.5	1.3169	4.933	626.8	1.2913
50	10.43	640.1	1.3953	7.746	637.8	1.3583	6.135	635.4	1.3286	5.060	632.9	1.3035
60	10.65	645.5	1.4059	7.922	643.4	1.3692	6.280	641.2	1.3399	5.184	638.0	1.3152
70	10.88	650.9	1.4161	8.096	648.9	1.3797	6.423	646.9	1.3508	5.307	644.9	1.3265
80	11.10	656.2	1.4261	8.268	654.4	1.3900	6.564	652.6	1.3613	5.428	650.7	1.3373
90	11.33	661.6	1.4359	8.439	659.9	1.4000	6.704	658.2	1.3716	5.547	656.4	1.3479
100	11.55	666.9	1.4456	8.609	665.3	1.4098	6.843	663.7	1.3816	5.665	662.1	1.3581
110	11.77	672.2	1.4550	8.777	670.7	1.4194	6.980	669.2	1.3914	5.781	667.7	1.3681
120	11.99	677.5	1.4642	8.945	676.1	1.4288	7.117	674.7	1.4009	5.897	673.3	1.3778
130	12.21	682.9	1.4733	9.112	681.5	1.4381	7.252	680.2	1.4103	6.012	678.9	1.3873
140	12.43	688.2	1.4823	9.278	686.9	1.4471	7.387	685.7	1.4195	6.126	684.4	1.3966
150	12.65	693.5	1.4911	9.444	692.3	1.4561	7.521	691.1	1.4286	6.239	689.9	1.4058
160	12.87	698.8	1.4998	9.609	697.7	1.4648	7.655	696.6	1.4374	6.352	695.5	1.4148
170	13.08	704.2	1.5083	9.774	703.1	1.4735	7.788	702.1	1.4462	6.464	701.0	1.4236
180	13.30	709.6	1.5168	9.938	708.5	1.4820	7.921	707.5	1.4548	6.576	706.5	1.4323
190	13.52	714.9	1.5251	10.10	714.0	1.4904	8.053	713.0	1.4633	6.687	712.0	1.4409
200	13.73	720.3	1.5334	10.27	719.4	1.4987	8.185	718.5	1.4716	6.798	717.5	1.4493
220	14.16	731.1	1.5495	10.59	730.3	1.5150	8.448	729.4	1.4880	7.019	728.6	1.4658
240	14.59	742.0	1.5653	10.92	741.3	1.5309	8.710	740.5	1.5040	7.238	739.7	1.4819
260	15.02	753.0	1.5808	11.24	752.3	1.5465	8.970	751.6	1.5197	7.457	750.9	1.4976
280	.....	.....	.....	11.56	763.4	1.5617	9.230	762.7	1.5350	7.675	762.1	1.5130
300	.....	.....	.....	11.88	774.6	1.5766	9.489	774.0	1.5500	7.892	773.3	1.5281

Source: From *Engineering Thermodynamics*, by H. J. Stoever, Wiley, 1951, as abridged from *Tables of Thermodynamic Properties of Ammonia*, National Bureau of Standards Circular No. 142.

Temp. °F	80 *44.40°*			100 *56.05°*			120 *66.02°*			140 *74.79°*		
	v	*h*	*s*	*v*	*h*	*s*	*v*	*h*	*s*	*v*	*h*	*s*
Sat.	*3.655*	*624.0*	*1.2545*	*2.952*	*626.5*	*1.2356*	*2.476*	*628.4*	*1.2201*	*2.132*	*629.9*	*1.2068*
50	3.712	627.7	1.2619	.....	.....	.....	.....	.....	.....	.....	.....	.....
60	3.812	634.3	1.2745	2.985	629.3	1.2409	.....	.....	.....	.....	.....	.....
70	3.909	640.6	1.2866	3.068	636.0	1.2539	2.505	631.3	1.2255	.....	.....	.....
80	4.005	646.7	1.2981	3.149	642.6	1.2661	2.576	638.3	1.2386	2.166	633.8	1.2140
90	4.098	652.8	1.3092	3.227	649.0	1.2778	2.645	645.0	1.2510	2.228	640.9	1.2272
100	4.190	658.7	1.3199	3.304	655.2	1.2891	2.712	651.6	1.2628	2.288	647.8	1.2396
110	4.281	664.6	1.3303	3.380	661.3	1.2999	2.778	658.0	1.2741	2.347	654.5	1.2515
120	4.371	670.4	1.3404	3.454	667.3	1.3104	2.842	664.2	1.2850	2.404	661.1	1.2628
130	4.460	676.1	1.3502	3.527	673.3	1.3206	2.905	670.4	1.2956	2.460	667.4	1.2738
140	4.548	681.8	1.3598	3.600	679.2	1.3305	2.967	676.5	1.3058	2.515	673.7	1.2843
150	4.635	687.5	1.3692	3.672	685.0	1.3401	3.029	682.5	1.3157	2.569	679.9	1.2945
160	4.722	693.2	1.3784	3.743	690.8	1.3495	3.089	688.4	1.3254	2.622	686.0	1.3045
170	4.808	698.8	1.3874	3.813	696.6	1.3588	3.149	694.3	1.3348	2.675	692.0	1.3141
180	4.893	704.4	1.3963	3.883	702.3	1.3678	3.209	700.2	1.3441	2.727	698.0	1.3236
190	4.978	710.0	1.4050	3.952	708.0	1.3767	3.268	706.0	1.3531	2.779	704.0	1.3328
200	5.063	715.6	1.4136	4.021	713.7	1.3854	3.326	711.8	1.3620	2.830	709.9	1.3418
210	5.147	721.3	1.4220	4.090	719.4	1.3940	3.385	717.6	1.3707	2.880	715.8	1.3507
220	5.231	726.9	1.4304	4.158	725.1	1.4024	3.442	723.4	1.3793	2.931	721.6	1.3594
230	5.315	732.5	1.4386	4.226	730.8	1.4108	3.500	729.2	1.3877	2.981	727.5	1.3679
240	5.398	738.1	1.4467	4.294	736.5	1.4190	3.557	734.9	1.3960	3.030	733.3	1.3763
250	5.482	743.8	1.4547	4.361	742.2	1.4271	3.614	740.7	1.4042	3.080	739.2	1.3846
260	5.565	749.4	1.4626	4.428	747.9	1.4350	3.671	746.5	1.4123	3.129	745.0	1.3928
280	5.730	760.7	1.4781	4.562	759.4	1.4507	3.783	758.0	1.4281	3.227	756.7	1.4088
300	5.894	772.1	1.4933	4.695	770.8	1.4660	3.895	769.6	1.4435	3.323	768.3	1.4243

TABLE A·2.3 (E) *(Continued)*

Temp. °F	Pressure, psia (saturation temperature in italics) 170 *86.29°*			200 *96.34°*			230 *105.30°*			260 *113.42°*		
	v	h	s	v	h	s	v	h	s	v	h	s
Sat.	*1.764*	*631.6*	*1.1900*	*1.502*	*632.7*	*1.1756*	*1.307*	*633.4*	*1.1631*	*1.155*	*633.9*	*1.1518*
90	1.784	634.4	1.1952	.....	.....	.....	.....	.....	.....	.....	.....	.....
100	1.837	641.9	1.2087	1.520	635.6	1.1809	.....	.....	.....	.....	.....	.....
110	1.889	649.1	1.2215	1.567	643.4	1.1947	1.328	637.4	1.1700	.....	.....	.....
120	1.939	656.1	1.2336	1.612	650.9	1.2077	1.370	645.4	1.1840	1.182	639.5	1.1617
130	1.988	662.8	1.2452	1.656	658.1	1.2200	1.410	653.1	1.1971	1.220	647.8	1.1757
140	2.035	669.4	1.2563	1.698	665.0	1.2317	1.449	660.4	1.2095	1.257	655.6	1.1889
150	2.081	675.9	1.2669	1.740	671.8	1.2429	1.487	667.6	1.2213	1.292	663.1	1.2014
160	2.127	682.3	1.2773	1.780	678.4	1.2537	1.524	674.5	1.2325	1.326	670.4	1.2132
170	2.172	688.5	1.2873	1.820	684.9	1.2641	1.559	681.3	1.2434	1.359	677.5	1.2245
180	2.216	694.7	1.2971	1.859	691.3	1.2742	1.594	687.9	1.2538	1.391	684.4	1.2354
190	2.260	700.8	1.3066	1.897	697.7	1.2840	1.629	694.4	1.2640	1.422	691.1	1.2458
200	2.303	706.9	1.3159	1.935	703.9	1.2935	1.663	700.9	1.2738	1.453	697.7	1.2560
210	2.346	713.0	1.3249	1.972	710.1	1.3029	1.696	707.2	1.2834	1.484	704.3	1.2658
220	2.389	719.0	1.3338	2.009	716.3	1.3120	1.729	713.5	1.2927	1.514	710.7	1.2754
230	2.431	724.9	1.3426	2.046	722.4	1.3209	1.762	719.8	1.3018	1.543	717.1	1.2847
240	2.473	730.9	1.3512	2.082	728.4	1.3296	1.794	726.0	1.3107	1.572	723.4	1.2938
250	2.514	736.8	1.3596	2.118	734.5	1.3382	1.826	732.1	1.3195	1.601	729.7	1.3027
260	2.555	742.8	1.3679	2.154	740.5	1.3467	1.857	738.3	1.3281	1.630	736.0	1.3115
270	2.596	748.7	1.3761	2.189	746.5	1.3550	1.889	744.4	1.3365	1.658	742.2	1.3200
280	2.637	754.6	1.3841	2.225	752.5	1.3631	1.920	750.5	1.3448	1.686	748.4	1.3285
290	2.678	760.5	1.3921	2.260	758.5	1.3712	1.951	756.5	1.3530	1.714	754.5	1.3367
300	2.718	766.4	1.3999	2.295	764.5	1.3791	1.982	762.6	1.3610	1.741	760.7	1.3449
320	2.798	778.3	1.4153	2.364	776.5	1.3947	2.043	774.7	1.3767	1.796	772.9	1.3608
340	2.878	790.1	1.4303	2.432	788.5	1.4099	2.103	786.8	1.3921	1.850	785.2	1.3763
360	.....	.....	.....	2.500	800.5	1.4247	2.163	798.9	1.4070	1.904	797.4	1.3914
380	.....	.....	.....	2.568	812.5	1.4392	2.222	811.1	1.4217	1.957	809.6	1.4062

TABLE A·3.1 (SI) Refrigerant 12 (dichlorodifluoromethane): Liquid-Vapor Saturation, Temperature Table (SI)

		Specific Volume m^3/kg			Enthalpy kJ/kg			Entropy kJ/kg·K		
Temp. °C	Abs. Pressure MPa P	Sat. Liquid v_f	Evap. v_{fg}	Sat. Vapor v_g	Sat. Liquid h_f	Evap. h_{fg}	Sat. Vapor h_g	Sat. Liquid s_f	Evap. s_{fg}	Sat. Vapor s_g
−90	0.0028	0.000 608	4.414 937	4.415 545	−43.243	189.618	146.375	−0.2084	1.0352	0.8268
−85	0.0042	0.000 612	3.036 704	3.037 316	−38.968	187.608	148.640	−0.1854	0.9970	0.8116
−80	0.0062	0.000 617	2.137 728	2.138 345	−34.688	185.612	150.924	−0.1630	0.9609	0.7979
−75	0.0088	0.000 622	1.537 030	1.537 651	−30.401	183.625	153.224	−0.1411	0.9266	0.7855
−70	0.0123	0.000 627	1.126 654	1.127 280	−26.103	181.640	155.536	−0.1197	0.8940	0.7744
−65	0.0168	0.000 632	0.840 534	0.841 166	−21.793	179.651	157.857	−0.0987	0.8630	0.7643
−60	0.0226	0.000 637	0.637 274	0.637 910	−17.469	177.653	160.184	−0.0782	0.8334	0.7552
−55	0.0300	0.000 642	0.490 358	0.491 000	−13.129	175.641	162.512	−0.0581	0.8051	0.7470
−50	0.0391	0.000 648	0.382 457	0.383 105	−8.772	173.611	164.840	−0.0384	0.7779	0.7396
−45	0.0504	0.000 654	0.302 029	0.302 682	−4.396	171.558	167.163	−0.0190	0.7519	0.7329
−40	0.0642	0.000 659	0.241 251	0.241 910	−0.000	169.479	169.479	−0.0000	0.7269	0.7269
−35	0.0807	0.000 666	0.194 732	0.195 398	4.416	167.368	171.784	0.0187	0.7027	0.7214
−30	0.1004	0.000 672	0.158 703	0.159 375	8.854	165.222	174.076	0.0371	0.6795	0.7165
−25	0.1237	0.000 679	0.130 487	0.131 166	13.315	163.037	176.352	0.0552	0.6570	0.7121
−20	0.1509	0.000 685	0.108 162	0.108 847	17.800	160.810	178.610	0.0730	0.6352	0.7082
−15	0.1826	0.000 693	0.090 326	0.091 018	22.312	158.534	180.846	0.0906	0.6141	0.7046
−10	0.2191	0.000 700	0.075 946	0.076 646	26.851	156.207	183.058	0.1079	0.5936	0.7014
−5	0.2610	0.000 708	0.064 255	0.064 963	31.420	153.823	185.243	0.1250	0.5736	0.6986

0	0.3086	0.000 716	0.054 673	0.055 389	36.022	151.376	187.397	0.1418	0.5542	0.6960
5	0.3626	0.000 724	0.046 761	0.047 485	40.659	148.859	189.518	0.1585	0.5351	0.6937
10	0.4233	0.000 733	0.040 180	0.040 914	45.337	146.265	191.602	0.1750	0.5165	0.6916
15	0.4914	0.000 743	0.034 671	0.035 413	50.058	143.586	193.644	0.1914	0.4983	0.6897
20	0.5673	0.000 752	0.030 028	0.030 780	54.828	140.812	195.641	0.2076	0.4803	0.6879
25	0.6516	0.000 763	0.026 091	0.026 854	59.653	137.933	197.586	0.2237	0.4626	0.6863
30	0.7449	0.000 774	0.022 734	0.023 508	64.539	134.936	199.475	0.2397	0.4451	0.6848
35	0.8477	0.000 786	0.019 855	0.020 641	69.494	131.805	201.299	0.2557	0.4277	0.6834
40	0.9607	0.000 798	0.017 373	0.018 171	74.527	128.525	203.051	0.2716	0.4104	0.6820
45	1.0843	0.000 811	0.015 220	0.016 032	79.647	125.074	204.722	0.2875	0.3931	0.6806
50	1.2193	0.000 826	0.013 344	0.014 170	84.868	121.430	206.298	0.3034	0.3758	0.6792
55	1.3663	0.000 841	0.011 701	0.012 542	90.201	117.565	207.766	0.3194	0.3582	0.6777
60	1.5259	0.000 858	0.010 253	0.011 111	95.665	113.443	209.109	0.3355	0.3405	0.6760
65	1.6988	0.000 877	0.008 971	0.009 847	101.279	109.024	210.303	0.3518	0.3224	0.6742
70	1.8858	0.000 897	0.007 828	0.008 725	107.067	104.255	211.321	0.3683	0.3038	0.6721
75	2.0874	0.000 920	0.006 802	0.007 723	113.058	99.068	212.126	0.3851	0.2845	0.6697
80	2.3046	0.000 946	0.005 875	0.006 821	119.291	93.373	212.665	0.4023	0.2644	0.6667
85	2.5380	0.000 976	0.005 029	0.006 005	125.818	87.047	212.865	0.4201	0.2430	0.6631
90	2.7885	0.001 012	0.004 246	0.005 258	132.708	79.907	212.614	0.4385	0.2200	0.6585
95	3.0569	0.001 056	0.003 508	0.004 563	140.068	71.658	211.726	0.4579	0.1946	0.6526
100	3.3440	0.001 113	0.002 790	0.003 903	148.076	61.768	209.843	0.4788	0.1655	0.6444
105	3.6509	0.001 197	0.002 045	0.003 242	157.085	49.014	206.099	0.5023	0.1296	0.6319
110	3.9784	0.001 364	0.001 098	0.002 462	168.059	28.425	196.484	0.5322	0.0742	0.6064
112	4.1155	0.001 792	0.000 005	0.001 797	174.920	0.151	175.071	0.5651	0.0004	0.5655

Source: Copyright © 1955 and 1956, E. I. du Pont de Nemours & Company, Inc. Reprinted by permission. Adapted from English units.

TABLE A·3.2 (SI) Refrigerant 12 (dichlorodifluoromethane): Superheated Vapor (SI)

Temp. °C	v m³/kg	h kJ/kg	s kJ/kg·K	v m³/kg	h kJ/kg	s kJ/kg·K	v m³/kg	h kJ/kg	s kJ/kg·K
		0.05 MPa			0.10 MPa			0.15 MPa	
−20.0	0.341 857	181.042	0.7912	0.167 701	179.861	0.7401			
−10.0	0.356 227	186.757	0.8133	0.175 222	185.707	0.7628	0.114 716	184.619	0.7318
0.0	0.370 508	192.567	0.8350	0.182 647	191.628	0.7849	0.119 866	190.660	0.7543
10.0	0.384 716	198.471	0.8562	0.189 994	197.628	0.8064	0.124 932	196.762	0.7763
20.0	0.398 863	204.469	0.8770	0.197 277	203.707	0.8275	0.129 930	202.927	0.7977
30.0	0.412 959	210.557	0.8974	0.204 506	209.866	0.8482	0.134 873	209.160	0.8186
40.0	0.427 012	216.733	0.9175	0.211 691	216.104	0.8684	0.139 768	215.463	0.8390
50.0	0.441 030	222.997	0.9372	0.218 839	222.421	0.8883	0.144 625	221.835	0.8591
60.0	0.455 017	229.344	0.9565	0.225 955	228.815	0.9078	0.149 450	228.277	0.8787
70.0	0.468 978	235.774	0.9755	0.233 044	235.285	0.9269	0.154 247	234.789	0.8980
80.0	0.482 917	242.282	0.9942	0.240 111	241.829	0.9457	0.159 020	241.371	0.9169
90.0	0.496 838	248.868	1.0126	0.247 159	248.446	0.9642	0.163 774	248.020	0.9354
		0.20 MPa			0.25 MPa			0.30 MPa	
0.0	0.088 608	189.669	0.7320	0.069 752	188.644	0.7139	0.057 150	187.583	0.6984
10.0	0.092 550	195.878	0.7543	0.073 024	194.969	0.7366	0.059 984	194.034	0.7216
20.0	0.096 418	202.135	0.7760	0.076 218	201.322	0.7587	0.062 734	200.490	0.7440
30.0	0.100 228	208.446	0.7972	0.079 350	207.715	0.7801	0.065 418	206.969	0.7658
40.0	0.103 989	214.814	0.8178	0.082 431	214.153	0.8010	0.068 049	213.480	0.7869
50.0	0.107 710	221.243	0.8381	0.085 470	220.642	0.8214	0.070 635	220.030	0.8075
60.0	0.111 397	227.735	0.8578	0.088 474	227.185	0.8413	0.073 185	226.627	0.8276
70.0	0.115 055	234.291	0.8772	0.091 449	233.785	0.8608	0.075 705	233.273	0.8473
80.0	0.118 690	240.910	0.8962	0.094 398	240.443	0.8800	0.078 200	239.971	0.8665
90.0	0.122 304	247.593	0.9149	0.097 327	247.160	0.8987	0.080 673	246.723	0.8853
100.0	0.125 901	254.339	0.9332	0.100 238	253.936	0.9171	0.083 127	253.530	0.9038
110.0	0.129 483	261.147	0.9512	0.103 134	260.770	0.9352	0.085 566	260.391	0.9220

	0.40 MPa			0.50 MPa			0.60 MPa		
20.0	0.045 836	198.762	0.7199	0.035 646	196.935	0.6999			
30.0	0.047 971	205.428	0.7423	0.037 464	203.814	0.7230	0.030 422	202.116	0.7063
40.0	0.050 046	212.095	0.7639	0.039 214	210.656	0.7452	0.031 966	209.154	0.7291
50.0	0.052 072	218.779	0.7849	0.040 911	217.484	0.7667	0.033 450	216.141	0.7511
60.0	0.054 059	225.488	0.8054	0.042 565	224.315	0.7875	0.034 887	223.104	0.7723
70.0	0.056 014	232.230	0.8253	0.044 184	231.161	0.8077	0.036 285	230.062	0.7929
80.0	0.057 941	239.012	0.8448	0.045 774	238.031	0.8275	0.037 653	237.027	0.8129
90.0	0.059 846	245.837	0.8638	0.047 340	244.932	0.8467	0.038 995	244.009	0.8324
100.0	0.061 731	252.707	0.8825	0.048 886	251.869	0.8656	0.040 316	251.016	0.8514
110.0	0.063 600	259.624	0.9008	0.050 415	258.845	0.8840	0.041 619	258.053	0.8700
120.0	0.065 455	266.590	0.9187	0.051 929	265.862	0.9021	0.042 907	265.124	0.8882
130.0	0.067 298	273.605	0.9364	0.053 430	272.923	0.9198	0.044 181	272.231	0.9061

	0.70 MPa			0.80 MPa			0.90 MPa		
40.0	0.026 761	207.580	0.7148	0.022 830	205.924	0.7016	0.019 744	204.170	0.6982
50.0	0.028 100	214.745	0.7373	0.024 068	213.290	0.7248	0.020 912	211.765	0.7131
60.0	0.029 387	221.854	0.7590	0.025 247	220.558	0.7469	0.022 012	219.212	0.7358
70.0	0.030 632	228.931	0.7799	0.026 380	227.766	0.7682	0.023 062	226.564	0.7575
80.0	0.031 843	235.997	0.8002	0.027 477	234.941	0.7888	0.024 072	233.856	0.7785
90.0	0.033 027	243.066	0.8199	0.028 545	242.101	0.8088	0.025 051	241.113	0.7987
100.0	0.034 189	250.146	0.8392	0.029 588	249.260	0.8283	0.026 005	248.355	0.8184
110.0	0.035 332	257.247	0.8579	0.030 612	256.428	0.8472	0.026 937	255.593	0.8376
120.0	0.036 458	264.374	0.8763	0.031 619	263.613	0.8657	0.027 851	262.839	0.8562
130.0	0.037 572	271.531	0.8943	0.032 612	270.820	0.8838	0.028 751	270.100	0.8745
140.0	0.038 673	278.720	0.9119	0.033 592	278.055	0.9016	0.029 639	277.381	0.8923
150.0	0.039 764	285.946	0.9292	0.034 563	285.320	0.9189	0.030 515	284.687	0.9098

TABLE A·3.2 (SI) *(Concluded)*

Temp. °C	v m³/kg	h kJ/kg	s kJ/kg·K	v m³/kg	h kJ/kg	s kJ/kg·K	v m³/kg	h kJ/kg	s kJ/kg·K
	1.00 MPa			1.20 MPa			1.40 MPa		
50.0	0.018 366	210.162	0.7021	0.014 483	206.661	0.6812			
60.0	0.019 410	217.810	0.7254	0.015 463	214.805	0.7060	0.012 579	211.457	0.6876
70.0	0.020 397	225.319	0.7476	0.016 368	222.687	0.7293	0.013 448	219.822	0.7123
80.0	0.021 341	232.739	0.7689	0.017 221	230.398	0.7514	0.014 247	227.891	0.7355
90.0	0.022 251	240.101	0.7895	0.018 032	237.995	0.7727	0.014 997	235.766	0.7575
100.0	0.023 133	247.430	0.8094	0.018 812	245.518	0.7931	0.015 710	243.512	0.7785
110.0	0.023 993	254.743	0.8287	0.019 567	252.993	0.8129	0.016 393	251.170	0.7988
120.0	0.024 835	262.053	0.8475	0.020 301	260.441	0.8320	0.017 053	258.770	0.8183
130.0	0.025 661	269.369	0.8659	0.021 018	267.875	0.8507	0.017 695	266.334	0.8373
140.0	0.026 474	276.699	0.8839	0.021 721	275.307	0.8689	0.018 321	273.877	0.8558
150.0	0.027 275	284.047	0.9015	0.022 412	282.745	0.8867	0.018 934	281.411	0.8738
160.0	0.028 068	291.419	0.9187	0.023 093	290.195	0.9041	0.019 535	288.946	0.8914
	1.60 MPa			1.80 MPa			2.00 MPa		
70.0	0.011 208	216.650	0.6959	0.009 406	213.049	0.6794			
80.0	0.011 984	225.177	0.7204	0.010 187	222.198	0.7057	0.008 704	218.859	0.6909
90.0	0.012 698	233.390	0.7433	0.010 884	230.835	0.7298	0.009 406	228.056	0.7166
100.0	0.013 366	241.397	0.7651	0.011 526	239.155	0.7524	0.010 035	236.760	0.7402
110.0	0.014 000	249.264	0.7859	0.012 126	247.264	0.7739	0.010 615	245.154	0.7624
120.0	0.014 608	257.035	0.8059	0.012 697	255.228	0.7944	0.011 159	253.341	0.7835
130.0	0.015 195	264.742	0.8253	0.013 244	263.094	0.8141	0.011 676	261.384	0.8037
140.0	0.015 765	272.406	0.8440	0.013 772	270.891	0.8332	0.012 172	269.327	0.8232
150.0	0.016 320	280.044	0.8623	0.014 284	278.642	0.8518	0.012 651	277.201	0.8420
160.0	0.016 864	287.669	0.8801	0.014 784	286.364	0.8698	0.013 116	285.027	0.8603
170.9	0.017 398	295.290	0.8975	0.015 272	294.069	0.8874	0.013 570	292.822	0.8781
180.0	0.017 923	302.914	0.9145	0.015 752	301.767	0.9046	0.014 013	300.598	0.8955

	2.50 MPa			3.00 MPa			3.50 MPa		
90.0	0.006 595	219.562	0.6823						
100.0	0.007 264	229.852	0.7103	0.005 231	220.529	0.6770			
110.0	0.007 837	239.271	0.7352	0.005 886	232.068	0.7075	0.004 324	222.121	0.6750
120.0	0.008 351	248.192	0.7582	0.006 419	242.208	0.7336	0.004 959	234.875	0.7078
130.0	0.008 827	256.794	0.7798	0.006 887	251.632	0.7573	0.005 456	245.661	0.7349
140.0	0.009 273	265.180	0.8003	0.007 313	260.620	0.7793	0.005 884	255.524	0.7591
150.0	0.009 697	273.414	0.8200	0.007 709	269.319	0.8001	0.006 270	264.846	0.7814
160.0	0.010 104	281.540	0.8390	0.008 083	277.817	0.8200	0.006 626	273.817	0.8023
170.0	0.010 497	289.589	0.8574	0.008 439	286.171	0.8391	0.006 961	282.545	0.8222
180.0	0.010 879	297.583	0.8752	0.008 782	294.422	0.8575	0.007 279	291.100	0.8413
190.0	0.011 250	305.540	0.8926	0.009 114	302.597	0.8753	0.007 584	299.528	0.8597
200.0	0.011 614	313.472	0.9095	0.009 436	310.718	0.8927	0.007 878	307.864	0.8775

	4.00 MPa		
120.0	0.003 736	224.863	0.6771
130.0	0.004 325	238.443	0.7111
140.0	0.004 781	249.703	0.7386
150.0	0.005 172	259.904	0.7630
160.0	0.005 522	269.492	0.7854
170.0	0.005 845	278.684	0.8063
180.0	0.006 147	287.602	0.8262
190.0	0.006 434	296.326	0.8453
200.0	0.006 708	304.906	0.8636
210.0	0.006 972	313.380	0.8813
220.0	0.007 228	321.774	0.8985
230.0	0.007 477	330.108	0.9152

TABLE A·3.1 (E) Refrigerant 12 (dichlorodifluoromethane): Liquid-Vapor Saturation, Temperature Table (English units)

		Specific Volume ft³/lbm			Enthalpy B/lbm			Entropy B/lbm·R		
Temp. °F T	Abs. Pressure lbf/in.² P	Sat. Liquid v_f	Evap. v_{fg}	Sat. Vapor v_g	Sat. Liquid h_f	Evap. h_{fg}	Sat. Vapor h_g	Sat. Liquid s_f	Evap. s_{fg}	Sat. Vapor s_g
−130	0.41224	0.009736	70.7203	70.730	−18.609	81.577	62.968	−0.04983	0.24743	0.19760
−120	0.64190	0.009816	46.7312	46.741	−16.565	80.617	64.052	−0.04372	0.23731	0.19359
−110	0.97034	0.009899	31.7671	31.777	−14.518	79.663	65.145	−0.03779	0.22780	0.19002
−100	1.4280	0.009985	21.1541	22.164	−12.466	78.714	66.248	−0.03200	0.21883	0.18683
−90	2.0509	0.010073	15.8109	15.821	−10.409	77.764	67.355	−0.02637	0.21034	0.18398
−80	2.8807	0.010164	11.5228	11.533	−8.3451	76.812	68.467	−0.02086	0.20229	0.18143
−70	3.9651	0.010259	8.5584	8.5687	−6.2730	75.853	69.580	−0.01548	0.19464	0.17916
−60	5.3575	0.010357	6.4670	6.4774	−4.1919	74.885	70.693	−0.01021	0.18716	0.17714
−50	7.1168	0.010459	4.9637	4.9742	−2.1011	73.906	71.805	−0.00506	0.18038	0.17533
−40	9.3076	0.010564	3.8644	3.8750	0	72.913	72.913	0	0.17373	0.17373
−30	11.999	0.010674	3.0478	3.0585	2.1120	71.903	74.015	0.00496	0.16733	0.17229
−20	15.267	0.010788	2.4321	2.4429	4.2357	70.874	75.110	0.00983	0.16119	0.17102
−10	19.189	0.010906	1.9628	1.9727	6.3716	69.824	76.196	0.01462	0.15527	0.16989
0	23.849	0.011030	1.5979	1.6089	8.5207	68.750	77.271	0.01932	0.14956	0.16888
10	29.335	0.011160	1.3129	1.3241	10.684	67.651	78.335	0.02395	0.14403	0.16798
20	35.736	0.011296	1.0875	1.0988	12.863	66.522	79.385	0.02852	0.13867	0.16719
30	43.148	0.011438	0.90736	0.91880	15.058	65.361	80.419	0.03301	0.13347	0.16648
40	51.667	0.011588	0.76198	0.77357	17.273	64.163	81.436	0.03745	0.12841	0.16586
50	61.394	0.011746	0.64362	0.65537	19.507	62.926	82.433	0.04184	0.12346	0.16530
60	72.433	0.011913	0.54648	0.55839	21.766	61.643	83.409	0.04618	0.11861	0.16479
70	84.888	0.012089	0.46609	0.47818	24.050	60.309	84.359	0.05048	0.11386	0.16434
80	98.870	0.012277	0.39907	0.41135	26.365	58.917	85.282	0.05475	0.10917	0.16392
90	114.49	0.012478	0.34281	0.35529	28.713	57.461	86.174	0.05900	0.10453	0.16353
100	131.86	0.012693	0.29525	0.30794	31.100	55.929	87.029	0.06323	0.09992	0.16315

110	151.11	0.012924	0.25577	0.26769	33.531	54.313	87.844	0.06745	0.09534	0.16279
120	172.35	0.013174	0.22019	0.23326	36.013	52.597	88.610	0.07168	0.09073	0.16241
130	195.71	0.013447	0.19019	0.20364	38.553	50.768	89.321	0.07583	0.08609	0.16202
140	221.32	0.013746	0.16424	0.17799	41.162	48.805	89.967	0.08021	0.08138	0.16159
150	249.31	0.014078	0.14156	0.15564	43.850	46.684	90.534	0.08453	0.07657	0.16110
160	279.82	0.014449	0.12159	0.13604	46.633	44.373	91.006	0.08893	0.07260	0.16053
170	313.00	0.014871	0.10386	0.11873	49.529	41.830	91.359	0.09342	0.06643	0.15985
180	349.00	0.015360	0.08794	0.10330	52.562	38.999	91.561	0.09804	0.06096	0.15900
190	387.98	0.015942	0.073476	0.089418	55.769	35.792	91.561	0.10284	0.05511	0.15793
200	430.09	0.016659	0.060069	0.076728	59.203	32.075	91.278	0.10789	0.04862	0.15651
210	475.52	0.017601	0.047242	0.064843	62.959	27.599	90.558	0.11332	0.03921	0.15453
220	524.43	0.018986	0.035154	0.053140	67.246	21.790	89.036	0.11943	0.03206	0.15149
230	577.03	0.021854	0.017581	0.039435	72.893	12.229	85.122	0.12739	0.01773	0.14512
233.6 (critical)	596.9	0.02870	0	0.2870	78.86	0	78.86	0.1359	0	0.1359

Source: Copyright 1955 and 1956, E. I. du Pont de Nemours & Company, Inc. Reprinted by permission.

TABLE A·3.2 (E) Refrigerant 12 (dichlorodifluoromethane): Superheated Vapor (English units) v in ft³/lbm, h in B/lbm, s in B/lbm·R

Temp. °F	v	h	s	v	h	s	v	h	s
		5 lbf/in.²			10 lbf/in.²			15 lbf/in.²	
0	8.0611	78.582	0.19663	3.9809	78.246	0.18471	2.6201	77.902	0.17751
20	8.4265	81.309	0.20244	4.1691	81.014	0.19061	2.7494	80.712	0.18349
40	8.7903	84.090	0.20812	4.3556	83.828	0.19635	2.8770	83.561	0.18931
60	9.1528	86.922	0.21367	4.5408	86.689	0.20197	3.0031	86.451	0.19498
80	9.5142	89.806	0.21912	4.7248	89.596	0.20746	3.1281	89.383	0.20051
100	9.8747	92.738	0.22445	4.9079	92.548	0.21283	3.2521	92.357	0.20593
120	10.234	95.717	0.22968	5.0903	95.546	0.21809	3.3754	95.373	0.21122
140	10.594	98.743	0.23481	5.2720	98.586	0.22325	3.4981	98.429	0.21640
160	10.952	101.812	0.23985	5.4533	101.669	0.22830	3.6202	101.525	0.22148
180	11.311	104.925	0.24479	5.6341	104.793	0.23326	3.7419	104.661	0.22646
200	11.668	108.079	0.24964	5.8145	107.957	0.23813	3.8632	107.835	0.23135
220	12.026	111.272	0.25441	5.9946	111.159	0.24291	3.9841	111.046	0.23614
		20 lbf/in.²			25 lbf/in.²			30 lbf/in.²	
20	2.0391	80.403	0.17829	1.6125	80.088	0.17414	1.3278	79.765	0.17065
40	2.1373	83.289	0.18419	1.6932	83.012	0.18012	1.3969	82.730	0.17671
60	2.2340	86.210	0.18992	1.7723	85.965	0.18591	1.4644	85.716	0.18257
80	2.3295	89.168	0.19550	1.8502	88.950	0.19155	1.5306	88.729	0.18826
100	2.4241	92.164	0.20095	1.9271	91.968	0.19704	1.5957	91.770	0.19379
120	2.5179	95.198	0.20628	2.0032	95.021	0.20240	1.6600	94.843	0.19918
140	2.6110	98.270	0.21149	2.0786	98.110	0.20763	1.7237	97.948	0.20445
160	2.7036	101.380	0.21659	2.1535	101.234	0.21276	1.7868	101.086	0.20960
180	2.7957	104.528	0.22159	2.2279	104.393	0.21778	1.8494	104.258	0.21463
200	2.8874	107.712	0.22649	2.3019	107.588	0.22269	1.9116	107.464	0.21957
220	2.9789	110.932	0.23130	2.3756	110.817	0.22752	1.9735	110.702	0.22440
240	3.0700	114.186	0.23602	2.4491	114.080	0.23225	2.0351	113.973	0.22915

	35 lbf/in.²			40 lbf/in.²			50 lbf/in.²		
40	1.1850	82.442	0.17375	1.0258	82.148	0.17112	0.80248	81.540	0.16655
60	1.2442	85.463	0.17968	1.0789	85.206	0.17712	0.84713	84.676	0.17271
80	1.3021	88.504	0.18542	1.1306	88.277	0.18292	0.89025	87.811	0.17862
100	1.3589	91.570	0.19100	1.1812	91.367	0.18854	0.93216	90.953	0.18434
120	1.4148	94.663	0.19643	1.2309	94.480	0.19401	0.97313	94.110	0.18988
140	1.4701	97.785	0.20172	1.2798	97.620	0.19933	1.0133	97.286	0.19527
160	1.5248	100.938	0.20689	1.3282	100.788	0.20453	1.0529	100.485	0.20051
180	1.5789	104.122	0.21195	1.3761	103.985	0.20961	1.0920	103.708	0.20563
200	1.6327	107.338	0.21690	1.4236	107.212	0.21457	1.1307	106.958	0.21064
220	1.6862	110.586	0.22175	1.4707	110.469	0.21944	1.1690	110.235	0.21553
240	1.7394	113.865	0.22651	1.5176	113.757	0.22420	1.2070	113.539	0.22032
260	1.7923	117.175	0.23117	1.5642	117.074	0.22888	1.2447	116.871	0.22502

	60 lbf/in.²			70 lbf/in.²			80 lbf/in.²		
60	0.69210	84.126	0.16892	0.58088	83.552	0.16556	...	...	...
80	0.72964	87.330	0.17497	0.61458	86.832	0.17175	0.52795	86.316	0.16885
100	0.76588	90.528	0.18079	0.64685	90.091	0.17768	0.55734	89.640	0.17489
120	0.80110	93.731	0.18641	0.67803	93.343	0.18339	0.58556	92.945	0.18070
140	0.83551	96.945	0.19186	0.70836	96.597	0.18891	0.61286	96.242	0.18629
160	0.86928	100.776	0.19716	0.73800	99.862	0.19427	0.63943	99.542	0.19170
180	0.90252	103.427	0.20233	0.76708	103.141	0.19948	0.66543	102.851	0.19696
200	0.93531	106.700	0.20736	0.79571	106.439	0.20455	0.69095	106.174	0.20207
220	0.96775	109.997	0.21229	0.82397	109.756	0.20951	0.71609	109.513	0.20706
240	0.99988	113.319	0.21710	0.85191	113.096	0.21435	0.74090	112.872	0.21193
260	1.0318	116.666	0.22182	0.87959	116.459	0.21909	0.76544	116.251	0.21669
280	1.0634	120.039	0.22644	0.90705	119.846	0.22373	0.78975	119.652	0.22135

TABLE A·3.2 (E) *(Continued)*

Temp. °F	v	h	s	v	h	s	v	h	s
	90 lbf/in.^{2}			100 lbf/in.^{2}			125 lbf/in.^{2}		
100	0.48749	89.175	0.17234	0.43138	88.694	0.16996	0.32943	87.407	0.16455
120	0.51346	92.536	0.17824	0.45562	92.116	0.17597	0.35086	91.008	0.17087
140	0.53845	95.879	0.18391	0.47881	95.507	0.18172	0.37098	94.537	0.17686
160	0.56268	99.216	0.18938	0.50118	98.884	0.18726	0.39015	98.023	0.18258
180	0.58629	102.557	0.19469	0.52291	102.257	0.19262	0.40857	101.484	0.18807
200	0.60941	105.905	0.19984	0.54413	105.633	0.19782	0.42642	104.934	0.19338
220	0.63213	109.267	0.20486	0.56492	109.018	0.20287	0.44380	108.380	0.19853
240	0.65451	112.644	0.20976	0.58538	112.415	0.20780	0.46081	111.829	0.20353
260	0.67662	116.040	0.21455	0.60554	115.828	0.21261	0.47750	115.287	0.20840
280	0.69849	119.456	0.21923	0.62546	119.258	0.21731	0.49394	118.756	0.21316
300	0.72016	122.892	0.22381	0.64518	122.707	0.22191	0.51016	122.238	0.21780
320	0.74166	126.349	0.22830	0.66472	126.176	0.22641	0.52619	125.737	0.22235
	150 lbf/in.^{2}			175 lbf/in.^{2}			200 lbf/in.^{2}		
120	0.28007	89.800	0.16629	...	...	...	...	...	...
140	0.29845	93.498	0.17256	0.24595	92.373	0.16859	0.20579	91.137	0.16480
160	0.31566	97.112	0.17849	0.26198	96.142	0.17478	0.22121	95.100	0.17130
180	0.33200	100.675	0.18415	0.27697	99.823	0.18062	0.23535	98.921	0.17737
200	0.34769	104.206	0.18958	0.29120	103.447	0.18620	0.24860	102.652	0.18311
220	0.36285	107.720	0.19483	0.30485	107.036	0.19156	0.26117	106.325	0.18860
240	0.37761	111.226	0.19992	0.31804	110.605	0.19674	0.27323	109.962	0.19387
260	0.39203	114.732	0.20485	0.33087	114.162	0.20175	0.28489	113.576	0.19896
280	0.40617	118.242	0.20967	0.34339	117.717	0.20662	0.29623	117.178	0.20390
300	0.42008	121.761	0.21436	0.35567	121.273	0.21137	0.30730	120.775	0.20870
320	0.43379	125.290	0.21894	0.36773	124.835	0.21599	0.31815	124.373	0.21337
340	0.44733	128.833	0.22343	0.37963	128.407	0.22052	0.32881	127.974	0.21793

	250 lbf/in.²			300 lbf/in.²			400 lbf/in.²		
160	0.16249	92.717	0.16462	...	...	...	...	...	...
180	0.17605	96.925	0.17130	0.13482	94.556	0.16537	...	...	...
200	0.18824	100.930	0.17747	0.14697	98.975	0.17217	0.091005	93.718	0.16092
220	0.19952	104.809	0.18326	0.15774	103.136	0.17838	0.10316	99.046	0.16888
240	0.21014	108.607	0.18877	0.16761	107.140	0.18419	0.11300	103.735	0.17568
260	0.22027	112.351	0.19404	0.17685	111.043	0.18969	0.12163	108.105	0.18183
280	0.23001	116.060	0.19913	0.18562	114.879	0.19495	0.12949	112.286	0.18756
300	0.23944	119.747	0.20405	0.19402	118.670	0.20000	0.13680	116.343	0.19298
320	0.24862	123.420	0.20882	0.20214	122.430	0.20489	0.14372	120.318	0.19814
340	0.25759	127.088	0.21346	0.21002	126.171	0.20963	0.15032	124.235	0.20310
360	0.26639	130.754	0.21799	0.21770	129.900	0.21423	0.15668	128.112	0.20789
380	0.27504	134.423	0.22241	0.22522	133.624	0.21872	0.16285	131.961	0.21258

	500 lbf/in.²			600 lbf/in.²		
220	0.064207	92.397	0.15683	...	...	...
240	0.077620	99.218	0.16672	0.047488	91.024	0.15335
260	0.087054	104.526	0.17421	0.061922	99.741	0.16566
280	0.094923	109.277	0.18072	0.070859	105.637	0.17374
300	0.10190	113.729	0.18666	0.078059	110.729	0.18053
320	0.10829	117.997	0.19221	0.084333	115.420	0.18663
340	0.11426	122.143	0.19746	0.090017	119.871	0.19227
360	0.11992	126.205	0.20247	0.095289	124.167	0.19757
380	0.12533	130.207	0.20730	0.10025	128.355	0.20262
400	0.13054	134.166	0.21196	0.10498	132.466	0.20746
420	0.13559	138.096	0.21648	0.10952	136.523	0.21213
440	0.14051	142.004	0.22087	0.11391	140.539	0.21664

TABLE A·4 (SI) Properties of Gases (SI)

Gas	Molar Mass M $\frac{kg}{kmol}$	Gas Constant R $\frac{kJ}{kg \cdot K}$	Specific Heats at 25°C c_p $\frac{kJ}{kg \cdot K}$	Specific Heats at 25°C c_v $\frac{kJ}{kg \cdot K}$	Specific Heats at 25°C k	c_p/R Equation Coefficients for $c_p/R = C_p/R_u = a + bT + cT^2 + dT^3 + eT^4$: T range	a	$b \times 10^3$	$c \times 10^6$	$d \times 10^9$	$e \times 10^{12}$	Critical State Properties P_c MPa	T_c K	v_{N_c} $\frac{m^3}{kmol}$	Z_c	van der Waals' Constants a $\frac{MPa \cdot m^6}{kmol^2}$	b $\frac{m^3}{kmol}$	Gas
Acetylene, C_2H_2	26.04	0.319	1.69	1.37	1.232	1000–5000K	4.751	5.124	−1.745	0.2867	−0.01795	6.242	310	0.113	0.274	0.4472	0.0515	C_2H_2
						300–1000K	1.410	19.06	−24.50	16.39	−4.135							
Air	28.97	0.287	1.005	0.718	1.400	1000–5000K	3.045	1.337	−0.4879	0.08548	−0.005696	3.769	132			0.1352	0.0365	Air
						300–1000K	3.653	−1.337	3.294	−1.913	0.2763							
Butane, C_4H_{10}	58.12	0.143	1.71	1.56	1.091	300–1500K	$\frac{c_p}{R} = 0.4756 + 0.04465T - 22.04 \times 10^{-6}T^2 + 4.207 \times 10^{-9}T^3$					3.780	425	0.255	0.274	1.3858	0.1162	C_4H_{10}
Carbon dioxide, CO_2	44.01	0.189	0.844	0.655	1.289	1000–5000K	4.461	3.098	−1.239	0.2274	−0.01553	7.387	304	0.094	0.275	0.3658	0.0428	CO_2
						300–1000K	2.401	8.735	−6.607	2.002	6.327×10^{-4}							
Carbon monoxide, CO	28.01	0.297	1.04	0.744	1.399	1000–5000K	2.984	1.489	−0.5790	0.1036	−0.006935	3.496	133	0.093	0.294	0.1470	0.0395	CO
						300–1000K	3.710	−1.619	3.692	−2.032	0.2395							
Ethane, C_2H_6	30.07	0.276	1.75	1.48	1.187	300–1500K	$\frac{c_p}{R} = 0.8293 + 0.02075T - 7.704 \times 10^{-6}T^2 + 0.8756 \times 10^{-9}T^3$					4.884	305	0.148	0.285	0.5573	0.06502	C_2H_6
Ethene, C_2H_4	28.05	0.296	1.53	1.23	1.240	1000–5000K	3.455	11.49	−4.365	0.7616	−0.05012	5.117	283	0.124	0.270	0.4573	0.0575	C_2H_4
						300–1000K	1.426	11.38	7.989	−16.25	6.749							
Hydrogen, H	2.016	4.124	14.3	10.2	1.405	1000–5000K	3.100	0.5112	0.05264	−0.03491	0.003694	1.297	33.2	0.065	0.305	0.02481	0.0266	H_2
						300–1000K	3.057	2.677	−5.810	5.521	−1.812							
Methane, CH_4	16.04	0.518	2.22	1.70	1.304	1000–5000K	1.503	10.42	−3.918	0.6778	−0.04428	4.641	191	0.099	0.290	0.2281	0.0427	CH_4
						300–1000K	3.826	−3.979	24.56	−22.73	6.963							
Nitrogen, N_2	28.01	0.297	1.04	0.743	1.400	1000–5000K	2.896	1.515	−0.5724	0.09981	−0.006522	3.394	126	0.090	0.291	0.1366	0.0386	N_2
						300–1000K	3.675	−1.208	2.324	−0.6322	−0.2258							
Oxygen, O_2	32.00	0.260	0.919	0.659	1.395	1000–5000K	3.622	0.7362	−0.1965	0.03620	−0.002895	5.042	155	0.074	0.288	0.1380	0.0317	O_2
						300–1000K	3.626	−1.878	7.055	−6.764	2.156							
Propane, C_3H_8	44.10	0.189	1.67	1.48	1.127	300–1500K	$\frac{c_p}{R} = -0.4861 + 0.03663T - 18.90 \times 10^{-6}T^2 + 3.814 \times 10^{-9}T^3$					4.266	370	0.200	0.277	0.9355	0.0901	C_3H_8
Water, H_2O	18.02	0.461	1.86	1.40	1.329	1000–5000K	2.717	2.945	−0.8022	0.1023	−0.004847	22.09	647	0.056	0.230	0.5528	0.0305	H_2O
						300–1000K	4.070	−1.108	4.152	−2.964	0.8070							

Source: c_p/R equations and coefficients from NASA SP-273, 1971, except for butane, ethane, and propane for which the source is Kobe, K. A., and E. G. Long,

TABLE A·4 (E) Properties of Gases (English units)

Gas	Molar Mass M $\frac{\text{lbm}}{\text{lbmol}}$	Gas Constant R $\frac{\text{ft}\cdot\text{lbf}}{\text{lbm}\cdot\text{R}}$	Specific heats at 77°F: c_p $\frac{\text{B}}{\text{lbm}\cdot\text{R}}$	Specific heats at 77°F: c_v $\frac{\text{B}}{\text{lbm}\cdot\text{R}}$	k	c_p/R Equation Coefficients for $c_p/R = C_p/R_u = a + bT + cT^2 + dT^3 + eT^4$: T range	a	$b \times 10^3$	$c \times 10^6$	$d \times 10^9$	$e \times 10^{12}$	Critical State Properties: P_c psia	T_c R	v_{N_c} $\frac{\text{ft}^3}{\text{lbmol}}$	Z_c	van der Waals' Constants: a $\frac{\text{psi}\cdot\text{ft}^6}{\text{lbmol}^2}$	b $\frac{\text{ft}^3}{\text{lbmol}}$	Gas
Acetylene, C_2H_2	26.04	59.3	0.404	0.328	1.232	1800–9000 R 540–1800 R	4.751 1.410	2.847 10.59	−0.5387 −7.562	0.04917 2.811	−0.001710 −0.3939	906	557	1.81	0.274	16 650	0.825	C_2H_2
Air	28.97	53.3	0.240	0.171	1.400	1800–9000 R 540–1800 R	3.045 3.653	0.7428 −0.7428	−0.1506 1.017	0.01466 −0.3280	-5.426×10^{-4} 0.02632	547	238			5032	0.584	Air
Butane, C_4H_{10}	58.12	26.6	0.408	0.373	1.091	540–2700 R	$\frac{c_p}{R} = 0.4755 + 0.02481T - 6.806\ 10^{-6}T^2 + 0.7215 \times 10^{-9}T^3$					551	765	4.08	0.274	51 580	1.862	C_4H_{10}
Carbon dioxide, CO_2	44.01	35.1	0.2016	0.1564	1.289	1800–9000 R 540–1800 R	4.461 2.401	1.721 4.853	−0.3825 −2.039	0.03899 0.3433	−0.001479 6.027×10^{-5}	1072	548	1.51	0.275	13 610	0.686	CO_2
Carbon monoxide, CO	28.01	55.2	0.249	0.178	1.399	1800–9000 R 540–1800 R	2.984 3.710	0.8273 −0.08995	−0.1787 1.140	0.01777 −0.3484	-6.607×10^{-4} 0.02282	507	239	1.49	0.294	5472	0.632	CO
Ethane, C_2H_6	30.07	51.4	0.419	0.353	1.187	540–2700 R	$\frac{c_p}{R} = 0.8293 + 0.01153T - 2.378 \times 10^{-6}T^2 + 0.1501 \times 10^{-3}T^3$					709	550	2.37	0.285	20 740	1.041	C_2H_6
Ethene, C_2H_4	28.07	55.1	0.365	0.294	1.240	1800–9000 R 540–1800 R	3.455 1.426	6.384 6.324	−1.347 2.466	0.1306 −2.787	−0.004775 0.6429	742	510	1.99	0.270	17 020	0.922	C_2H_4
Hydrogen, H_2	2.016	776.	3.42	2.43	1.405	1800–9000 R 540–1800 R	3.100 3.057	0.2840 1.487	0.01625 −1.793	−0.005986 0.9467	3.519×10^{-4} −0.1726	188	59.8	1.04	0.305	923.3	0.426	H_2
Methane, CH_4	16.04	96.3	0.531	0.407	1.304	1800–9000 R 540–1800 R	1.503 3.826	5.787 −2.211	−1.209 7.580	0.1162 −3.898	−0.004218 0.6633	673	343	1.59	0.290	8490	0.684	CH_4
Nitrogen, N_2	28.01	55.2	0.2485	0.1775	1.400	1800–9000 R 540–1800 R	2.896 3.675	0.8419 −0.6712	−0.1767 0.7173	0.01711 −0.1084	-6.213×10^{-4} −0.02151	492	227	1.44	0.291	5084	0.618	N_2
Oxygen, O_2	32.00	48.3	0.219	0.157	1.395	1800–9000 R 540–1800 R	3.622 3.626	0.4090 −1.043	−0.06066 2.178	0.006207 −1.160	$-2.757\ 10^{-4}$ 0.2053	731	278	1.18	0.288	5135	0.508	O_2
Propane, C_3H_8	44.10	35.0	0.400	0.355	1.127	540–2700 R	$\frac{c_p}{R} = -0.4861 + 0.02035T - 5.833 \times 10^{-6}T^2 + 0.6540 \times 10^{-9}T^3$					619	666	3.20	0.277	34 820	1.444	C_3H_8
Water, H_2O	18.02	85.8	0.445	0.335	1.329	1800–9000 R 540–1800 R	2.717 4.070	1.636 −0.6158	−0.2476 1.282	0.01754 −0.5082	-4.618×10^{-4} 0.07688	3205	1165	0.897	0.230	20 580	0.488	H_2O

Source: c_p/R equations and coefficients from NASA SP-273, 1971, except for butane, ethane, and propane for which the source is Kobe, K. A., and E. G. Long, "Thermochemistry for the Petrochemical Industry, Part II—Paraffinic Hydrocarbons, C_1–C_6," *Petroleum Refiner,* Vol. 28, No. 2, 1949, pp. 113–116.

TABLE A·5(SI) Air at Low Pressures (SI) T in K, t in °C, h and u in kJ/kg, and ϕ in kJ/kg·K

T	t	h	p_r	u	v_r	ϕ	T	t	h	p_r	u	v_r	ϕ
200	−73.15	200.13	.33468	142.72	171.52	5.2950	700	426.85	713.51	28.679	512.59	7.0058	6.5725
210	−63.15	210.15	.39684	149.88	151.89	5.3439	710	436.85	724.27	30.245	520.47	6.7380	6.5877
220	−53.15	220.18	.46684	157.03	135.26	5.3905	720	446.85	735.05	31.876	528.39	6.4832	6.6028
230	−43.15	230.20	.54524	164.18	121.08	5.4350	730	456.85	745.85	33.575	536.31	6.2407	6.6177
240	−33.15	240.22	.63263	171.34	108.89	5.4777	740	466.85	756.68	35.344	544.28	6.0097	6.6324
250	−23.15	250.25	.7296	178.49	98.353	5.5186	750	476.85	767.53	37.184	552.26	5.7894	6.6470
260	−13.15	260.28	.8368	185.65	89.188	5.5580	760	486.85	778.41	39.098	560.27	5.5795	6.6614
270	−3.15	270.31	.9547	192.81	81.173	5.5958	770	496.85	789.31	41.088	568.30	5.3791	6.6757
280	6.85	280.35	1.0842	199.98	74.129	5.6323	780	506.85	800.23	43.156	576.35	5.1878	6.6898
290	16.85	290.39	1.2258	207.15	67.909	5.6676	790	516.85	811.18	45.304	584.43	5.0052	6.7037
300	26.85	300.43	1.3801	214.32	62.393	5.7016	800	526.85	822.15	47.535	592.53	4.8306	6.7175
310	36.85	310.48	1.5480	221.50	57.481	5.7346	810	536.85	833.15	49.852	600.65	4.6637	6.7312
320	46.85	320.53	1.7301	228.68	53.091	5.7665	820	546.85	844.16	52.256	608.80	4.5041	6.7447
330	56.85	330.59	1.9271	235.87	49.152	5.7974	830	556.85	855.20	54.749	616.97	4.3514	6.7581
340	66.85	340.66	2.1398	243.07	45.608	5.8275	840	566.85	866.26	57.336	625.16	4.2052	6.7713
350	76.85	350.73	2.3689	250.27	42.407	5.8567	850	576.85	877.35	60.017	633.37	4.0652	6.7844
360	86.85	360.81	2.6154	257.48	39.509	5.8851	860	586.85	888.45	62.795	641.61	3.9310	6.7974
370	96.85	370.91	2.8799	264.71	36.877	5.9127	870	596.85	899.58	65.674	649.87	3.8024	6.8103
380	106.85	381.01	3.1633	271.94	34.481	5.9397	880	606.85	910.73	68.655	658.15	3.6791	6.8230
390	116.85	391.12	3.4664	279.18	32.293	5.9660	890	616.85	921.90	71.742	666.45	3.5608	6.8356
400	126.85	401.25	3.7902	286.43	30.292	5.9916	900	626.85	933.10	74.937	674.77	3.4473	6.8482
410	136.85	411.38	4.1356	293.70	28.456	6.0166	910	636.85	944.31	78.243	683.11	3.3383	6.8605
420	146.85	421.54	4.5035	300.98	26.769	6.0411	920	646.85	955.55	81.663	691.48	3.2336	6.8728
430	156.85	431.70	4.8949	308.28	25.215	6.0650	930	656.85	966.80	85.200	699.87	3.1331	6.8850
440	166.85	441.88	5.3106	315.59	23.781	6.0884	940	666.85	978.08	88.856	708.27	3.0365	6.8971

450	176.85	452.07	5.7519	322.91	22.456	6.1113	950	676.85	989.38	92.63	716.70	2.9436	6.9090
460	186.85	462.28	6.2197	330.25	21.228	6.1338	960	686.85	1000.69	96.54	725.14	2.8543	6.9209
470	196.85	472.51	6.7150	337.61	20.090	6.1557	970	696.85	1012.03	100.57	733.61	2.7683	6.9326
480	206.85	482.76	7.2391	344.98	19.032	6.1773	980	706.85	1023.39	104.74	742.10	2.6857	6.9443
490	216.85	493.02	7.7930	352.38	18.048	6.1985	990	716.85	1034.76	109.04	750.60	2.6061	6.9558
500	226.85	503.30	8.378	359.79	17.130	6.2193	1000	726.85	1046.16	113.48	759.13	2.5294	6.9673
510	236.85	513.60	8.995	367.22	16.274	6.2396	1010	736.85	1057.57	118.06	767.67	2.4556	6.9786
520	246.85	523.93	9.645	374.67	15.474	6.2597	1020	746.85	1069.01	122.78	776.23	2.3845	6.9899
530	256.85	534.27	10.331	338.14	14.726	6.2794	1030	756.85	1080.46	127.65	784.81	2.3160	7.0010
540	266.85	544.63	11.052	389.63	14.025	6.2988	1040	766.85	1091.93	132.68	793.41	2.2499	7.0121
550	276.85	555.01	11.810	397.15	13.367	6.3178	1050	776.85	1103.41	137.86	802.03	2.1862	7.0231
560	286.85	565.42	12.608	404.68	12.749	6.3366	1060	786.85	1114.92	143.20	810.66	2.1247	7.0340
570	296.85	575.84	13.445	412.24	12.169	6.3550	1070	796.85	1126.44	148.70	819.32	2.0654	7.0448
580	306.85	586.29	14.324	419.82	11.623	6.3732	1080	806.85	1137.98	154.37	827.99	2.0082	7.0556
590	316.85	596.77	15.245	427.42	11.108	6.3911	1090	816.85	1149.54	160.20	836.67	1.9529	7.0661
600	326.85	607.26	16.212	435.04	10.623	6.4087	1100	826.85	1161.11	166.21	845.38	1.8996	7.0768
610	336.85	617.78	17.224	442.69	10.165	6.4261	1110	836.85	1172.70	172.40	854.10	1.8481	7.0873
620	346.85	628.32	18.284	450.36	9.733	6.4433	1120	846.85	1184.31	178.76	862.83	1.7983	7.0977
630	356.85	638.89	19.393	458.06	9.324	6.4602	1130	856.85	1195.93	185.32	871.59	1.7502	7.1080
640	366.85	649.47	20.553	465.77	8.938	6.4768	1140	866.85	1207.57	192.06	880.35	1.7038	7.1183
650	376.85	660.09	21.766	473.52	8.5718	6.4933	1150	876.85	1219.23	198.99	889.14	1.6588	7.1285
660	386.85	670.72	23.033	481.28	8.2249	6.5095	1160	886.85	1230.90	206.12	897.94	1.6154	7.1386
670	396.85	681.38	24.356	489.07	7.8960	6.5256	1170	896.85	1242.58	213.45	906.76	1.5733	7.1486
680	406.85	692.07	25.736	496.89	7.5839	6.5414	1180	906.85	1254.28	220.99	915.59	1.5327	7.1586
690	416.85	702.78	27.177	504.73	7.2875	6.5570	1190	916.85	1266.00	228.73	924.43	1.4933	7.1684

Source: Abridged from Joseph H. Keenan, Jing Chao, and Joseph Kaye, *Gas Tables*, 2nd ed., SI, Wiley, New York, 1983.

TABLE A·5 (SI) *(Continued)*

T	t	h	p_r	u	v_r	ϕ	T	t	h	p_r	u	v_r	ϕ
1200	926.85	1277.73	236.69	933.29	1.4552	7.1783	1700	1426.85	1879.58	1017.9	1391.63	.47937	7.5970
1210	936.85	1289.48	244.86	942.17	1.4184	7.1880	1725	1451.85	1910.31	1083.6	1415.18	.45694	7.6149
1220	946.85	1301.24	253.26	951.06	1.3827	7.1977	1750	1476.85	1941.09	1152.6	1438.78	.43582	7.6326
1230	956.85	1313.01	261.89	959.96	1.3481	7.2073	1775	1501.85	1971.91	1225.0	1462.43	.41592	7.6501
1240	966.85	1324.80	270.74	968.88	1.3146	7.2168	1800	1526.85	2002.78	1300.9	1486.12	.39714	7.6674
1250	976.85	1336.60	279.83	977.81	1.2822	7.2263	1825	1551.85	2033.69	1380.6	1509.86	.37943	7.6844
1260	986.85	1348.42	289.16	986.76	1.2507	7.2357	1850	1576.85	2064.65	1464.0	1533.65	.36270	7.7013
1270	996.85	1360.25	298.74	995.72	1.2202	7.2451	1875	1601.85	2095.66	1551.5	1557.47	.34689	7.7179
1280	1006.85	1372.09	308.57	1004.69	1.1907	7.2544	1900	1626.85	2126.70	1643.0	1581.34	.33194	7.7344
1290	1016.85	1383.95	318.65	1013.68	1.1620	7.2636	1925	1651.85	2157.79	1738.7	1605.25	.31779	7.7506
1300	1026.85	1395.81	328.98	1022.67	1.1342	7.2728	1950	1676.85	2188.91	1838.8	1629.20	.30439	7.7667
1310	1036.85	1407.70	339.59	1031.69	1.1073	7.2819	1975	1701.85	2220.08	1943.4	1653.19	.29170	7.7826
1320	1046.85	1419.59	350.46	1040.71	1.0811	7.2909	2000	1726.85	2251.28	2052.6	1677.22	.27967	7.7983
1330	1056.85	1431.50	361.61	1049.75	1.0557	7.2999	2025	1751.85	2282.52	2166.7	1701.28	.26826	7.8138
1340	1066.85	1443.42	373.03	1058.79	1.0311	7.3088	2050	1776.85	2313.80	2285.7	1725.38	.25743	7.8292
1350	1076.85	1455.35	384.74	1067.86	1.00716	7.3177	2075	1801.85	2345.11	2409.9	1749.52	.24714	7.8443
1360	1086.85	1467.29	396.74	1076.93	.98393	7.3265	2100	1826.85	2376.46	2539.3	1773.70	.23737	7.8594
1370	1096.85	1479.25	409.03	1086.01	.96138	7.3353	2125	1851.85	2407.85	2674.2	1797.90	.22808	7.8742
1380	1106.85	1491.21	421.62	1095.11	.93947	7.3440	2150	1876.85	2439.26	2814.7	1822.15	.21924	7.8889
1390	1116.85	1503.19	434.52	1104.22	.91819	7.3526	2175	1901.85	2470.71	2961.0	1846.42	.21084	7.9035
1400	1126.85	1515.18	447.73	1113.34	.89752	7.3612	2200	1926.85	2502.20	3113.3	1870.73	.20283	7.9179
1410	1136.85	1527.18	461.25	1122.47	.87742	7.3698	2225	1951.85	2533.71	3271.7	1895.07	.19520	7.9321
1420	1146.85	1539.20	475.10	1131.61	.85789	7.3783	2250	1976.85	2565.26	3436.4	1919.44	.18793	7.9462
1430	1156.85	1551.22	489.27	1140.77	.83891	7.3867	2275	2001.85	2596.84	3607.7	1943.84	.18100	7.9602
1440	1166.85	1563.26	503.78	1149.93	.82045	7.3951	2300	2026.85	2628.45	3785.6	1968.27	.17439	7.9740

1450	1176.85	1575.30	518.63	1159.11	.80250	7.4034	2325	2051.85	2660.08	3970.4	1992.74	.16808	7.9877
1460	1186.85	1587.36	533.81	1168.29	.78504	7.4117	2350	2076.85	2691.75	4162.3	2017.23	.16206	8.0012
1470	1196.85	1599.42	549.35	1177.49	.76806	7.4199	2375	2101.85	2723.45	4361.4	2041.75	.15630	8.0146
1480	1206.85	1611.50	565.25	1186.69	.75154	7.4281	2400	2126.85	2755.17	4568.1	2066.29	.15080	8.0279
1490	1216.85	1623.59	581.51	1195.91	.73546	7.4363	2425	2151.85	2786.92	4782.4	2090.87	.14554	8.0411
1500	1226.85	1635.68	598.14	1205.14	.71982	7.4444	2450	2176.85	2818.70	5004.7	2115.47	.14051	8.0541
1510	1236.85	1647.79	615.14	1214.37	.70459	7.4524	2475	2201.85	2850.50	5235.0	2140.10	.13570	8.0670
1520	1246.85	1659.91	632.52	1223.62	.68976	7.4604	2500	2226.85	2882.34	5473.7	2164.76	.13110	8.0798
1530	1256.85	1672.03	650.29	1232.88	.67533	7.4684	2525	2251.85	2914.19	5720.9	2189.44	.12669	8.0925
1540	1266.85	1684.17	668.45	1242.14	.66128	7.4763	2550	2276.85	2946.07	5976.9	2214.15	.12246	8.1051
1550	1276.85	1696.32	687.01	1251.42	.64759	7.4841	2575	2301.85	2977.98	6241.9	2238.88	.11841	8.1175
1560	1286.85	1708.47	705.98	1260.70	.63425	7.4919	2600	2326.85	3009.91	6516.1	2263.63	.11453	8.1299
1570	1296.85	1720.64	725.36	1270.00	.62127	7.4997	2625	2351.85	3041.87	6799.8	2288.41	.11081	8.1421
1580	1306.85	1732.81	745.16	1279.30	.60861	7.5074	2650	2376.85	3073.85	7093.2	2313.22	.10723	8.1542
1590	1316.85	1744.99	765.38	1288.61	.59628	7.5151	2675	2401.85	3105.85	7396.5	2338.05	.10381	8.1662
1600	1326.85	1757.19	786.04	1297.94	.58426	7.5228	2700	2426.85	3137.88	7710.1	2362.90	.10052	8.1781
1610	1336.85	1769.39	807.13	1307.27	.57255	7.5304	2725	2451.85	3169.93	8034.1	2387.77	.09736	8.1900
1620	1346.85	1781.60	828.67	1316.61	.56113	7.5379	2750	2476.85	3202.00	8368.8	2412.67	.09432	8.2017
1630	1356.85	1793.82	850.67	1325.95	.54999	7.5455	2775	2501.85	3234.09	8714.5	2437.58	.09140	8.2133
1640	1366.85	1806.04	873.12	1335.31	.53913	7.5529	2800	2526.85	3266.21	9071.4	2462.52	.08860	8.2248
1650	1376.85	1818.28	896.05	1344.68	.52855	7.5604	2825	2551.85	3298.35	9439.8	2487.48	.08590	8.2362
1660	1386.85	1830.52	929.44	1354.05	.51822	7.5678	2850	2576.85	3330.50	9820	2512.46	.08330	8.2476
1670	1396.85	1842.78	943.32	1363.43	.50814	7.5751	2875	2601.85	3362.68	10212	2537.47	.08081	8.2588
1680	1406.85	1855.04	967.69	1372.82	.49832	7.5825	2900	2626.85	3394.88	10617	2562.49	.07840	8.2700
1690	1416.85	1867.30	992.55	1382.22	.48873	7.5897	2925	2651.85	3427.10	11034	2587.53	.07609	8.2810

TABLE A·5(E) Air at Low Pressures (English units) T in R, t in °F, h and u in B/lbm, and ϕ in B/lbm·R

T	t	h	p_r	u	v_r	ϕ	T	t	h	p_r	u	v_r	ϕ
400	−59.67	95.62	.4835	68.19	306.5	.52875	1400	940.33	342.99	42.69	247.02	12.150	.83592
420	−39.67	100.40	.5733	71.61	271.4	.54043	1420	960.33	348.22	45.06	250.87	11.675	.83963
440	−19.67	105.19	.6745	75.03	241.7	.55157	1440	980.33	353.46	47.54	254.74	11.223	.84329
460	.33	109.98	.7878	78.45	216.31	.56222	1460	1000.33	358.71	50.11	258.62	10.793	.84691
480	20.33	114.78	.9142	81.87	194.52	.57241	1480	1020.33	363.98	52.80	262.51	10.384	.85050
500	40.33	119.57	1.0544	85.29	175.68	.58220	1500	1040.33	369.25	55.60	266.42	9.995	.85404
520	60.33	124.36	1.2094	88.71	159.29	.59160	1520	1060.33	374.54	58.52	270.34	9.623	.85754
540	80.33	129.16	1.3801	92.14	144.96	.60065	1540	1080.33	379.84	61.55	274.27	9.270	.86100
560	100.33	133.96	1.5675	95.57	132.35	.60938	1560	1100.33	385.15	64.70	278.21	8.932	.86443
580	120.33	138.76	1.7725	99.00	121.23	.61781	1580	1120.33	390.48	67.98	282.16	8.610	.86782
600	140.33	143.57	1.996	102.44	111.35	.62595	1600	1140.33	395.81	71.39	286.12	8.303	.87118
620	160.33	148.38	2.240	105.88	102.56	.63384	1620	1160.33	401.16	74.94	290.10	8.009	.87450
640	180.33	153.20	2.504	109.32	94.70	.64148	1640	1180.33	406.52	78.62	294.09	7.728	.87778
660	200.33	158.01	2.790	112.77	87.65	.64890	1660	1200.33	411.89	82.44	298.08	7.460	.88104
680	220.33	162.84	3.099	116.22	81.30	.65610	1680	1220.33	417.27	86.41	302.09	7.203	.88426
700	240.33	167.67	3.432	119.68	75.57	.66310	1700	1240.33	422.66	90.52	306.11	6.958	.88745
720	260.33	172.51	3.790	123.14	70.38	.66991	1720	1260.33	428.06	94.79	310.14	6.723	.89061
740	280.33	177.35	4.175	126.62	65.66	.67655	1740	1280.33	433.47	99.21	314.18	6.497	.89374
760	300.33	182.20	4.588	130.10	61.36	.68301	1760	1300.33	438.89	103.80	318.23	6.282	.89684
780	320.33	187.06	5.031	133.58	57.44	.68932	1780	1320.33	444.32	108.29	322.29	6.075	.89990
800	340.33	192.92	5.504	137.08	53.85	.69548	1800	1340.33	449.77	113.48	326.37	5.877	.90294
820	360.33	196.79	6.008	140.58	50.56	.70150	1820	1360.33	455.22	118.57	330.45	5.686	.90596
840	380.33	201.68	6.547	144.09	47.53	.70738	1840	1380.33	460.68	123.85	334.54	5.504	.90894
860	400.33	206.57	7.120	147.61	44.75	.71314	1860	1400.33	466.16	129.31	338.64	5.329	.91190
880	420.33	211.47	7.730	151.14	42.18	.71877	1880	1420.33	471.64	134.96	342.96	5.161	.91483

900	440.33	216.38	8.378	154.68	39.80	.72429	1900	1440.33	477.13	140.81	346.87	4.999	.91774
920	460.33	221.30	9.066	158.23	37.60	.72970	1920	1460.33	482.63	146.85	351.00	4.844	.92062
940	480.33	226.23	9.795	161.79	35.55	.73500	1940	1480.33	488.14	153.09	355.14	4.695	.92347
960	500.33	231.18	10.567	165.36	33.66	.74020	1960	1500.33	493.66	159.54	359.29	4.551	.92630
980	520.33	236.13	11.384	168.95	31.89	.74531	1980	1520.33	499.19	166.21	363.45	4.413	.92911
1000	540.33	241.10	12.248	172.54	30.25	.75032	2000	1540.33	504.72	173.10	367.61	4.281	.93189
1020	560.33	246.07	13.161	176.15	28.71	.75525	2020	1560.33	510.27	180.20	371.79	4.153	.93465
1040	580.33	251.06	14.125	179.76	27.28	.76010	2040	1580.33	515.82	187.54	375.97	4.030	.93739
1060	600.33	256.06	15.141	183.39	25.94	.76486	2060	1600.33	521.39	195.11	380.16	3.911	.94010
1080	620.33	261.08	16.212	187.03	24.68	.76954	2080	1620.33	526.96	202.93	384.36	3.797	.94279
1100	640.33	266.10	17.339	190.69	23.50	.77415	2100	1640.33	532.54	211.0	388.57	3.687	.94546
1120	660.33	271.14	18.526	194.35	22.40	.77869	2120	1660.33	538.13	219.3	392.79	3.582	.94811
1140	680.33	276.19	19.774	198.03	21.36	.78316	2140	1680.33	543.72	227.9	397.01	3.479	.95074
1160	700.33	281.25	21.09	201.72	20.381	.78756	2160	1700.33	549.32	236.7	401.24	3.381	.95334
1180	720.33	286.33	22.46	205.43	19.462	.79190	2180	1720.33	554.94	245.8	405.48	3.286	.95593
1200	740.33	291.41	23.91	209.15	18.595	.79618	2200	1740.33	560.56	255.2	409.73	3.194	.95850
1220	760.33	296.51	25.42	212.88	17.777	.80039	2220	1760.33	566.18	264.8	413.99	3.106	.96104
1240	780.33	301.63	27.01	216.62	17.006	.80455	2240	1780.33	571.82	274.8	418.25	3.020	.96357
1260	800.33	306.75	28.68	220.37	16.276	.80865	2260	1800.33	577.46	285.0	422.52	2.938	.96607
1280	820.33	311.89	30.42	244.14	15.587	.81270	2280	1820.33	583.10	295.5	426.80	2.858	.96856
1300	840.33	317.04	32.25	227.92	14.935	.81669	2300	1840.33	588.76	306.4	431.08	2.781	.97103
1320	860.33	322.21	34.16	231.72	14.317	.82063	2320	1860.33	594.42	317.5	435.37	2.707	.97348
1340	880.33	327.39	36.15	235.52	13.732	.82453	2340	1880.33	600.09	329.0	439.67	2.635	.97592
1360	900.33	332.58	38.24	239.34	13.177	.82837	2360	1900.33	605.77	340.8	443.98	2.566	.97833
1380	920.33	337.78	40.42	243.17	12.650	.83217	2380	1920.33	611.45	352.9	448.29	2.498	.98073

Source: Abridged from Joseph H. Keenan, Jing Chao, and Joseph Kaye, *Gas Tables*, 2nd ed., English units, Wiley, New York, 1980.

TABLE A·5 (E) *(Continued)*

T	t	h	p_r	u	v_r	ϕ	T	t	h	p_r	u	v_r	ϕ
2400	1940.33	617.14	365.4	452.61	2.433	.98311	3400	2940.33	908.38	1601.8	675.29	.7864	1.08443
2420	1960.33	622.84	378.2	456.93	2.371	.98547	3450	2990.33	923.22	1706.3	686.70	.7491	1.08877
2440	1980.33	628.54	391.4	461.26	2.310	.98782	3500	3040.33	938.09	1816.1	698.14	.7140	1.09304
2460	2000.33	634.25	404.9	465.60	2.251	.99015	3550	3090.33	952.97	1931.5	709.60	.6809	1.09727
2480	2020.33	639.96	418.8	469.94	2.194	.99247	3600	3140.33	967.88	2052.6	721.07	.6498	1.10143
2500	2040.33	645.68	433.1	474.29	2.1386	.99476	3650	3190.33	982.80	2179.7	732.57	.6204	1.10555
2520	2060.33	651.41	447.7	478.65	2.0852	.99704	3700	3240.33	997.74	2312.9	744.09	.5927	1.10962
2540	2080.33	657.14	462.8	483.01	2.0334	.99931	3750	3290.33	1012.71	2452.4	755.62	.5665	1.11363
2560	2100.33	662.88	478.2	487.38	1.9832	1.00156	3800	3340.33	1027.69	2598.6	767.18	.5418	1.11760
2580	2120.33	668.63	494.1	491.75	1.9346	1.00380	3850	3390.33	1042.69	2742.6	778.75	.5184	1.12153
2600	2140.33	674.38	510.3	496.13	1.8875	1.00602	3900	3440.33	1057.71	2911.6	790.34	.4962	1.12540
2620	2160.33	680.14	527.0	500.52	1.8418	1.00822	3950	3490.33	1072.74	3078.9	801.94	.4753	1.12923
2640	2180.33	685.90	544.1	504.91	1.7974	1.01041	4000	3540.33	1087.79	3254	813.57	.4554	1.13302
2660	2200.33	691.67	561.7	509.31	1.7545	1.01259	4050	3590.33	1102.86	3436	825.21	.4366	1.13676
2680	2220.33	697.44	579.7	513.71	1.7128	1.01475	4100	3640.33	1117.95	3627	836.87	.4188	1.14046
2700	2240.33	703.22	598.1	518.12	1.6723	1.01690	4150	3690.33	1133.05	3826	848.54	.4018	1.14412
2720	2260.33	709.00	617.0	522.53	1.6331	1.01903	4200	3740.33	1148.17	4034	860.23	.3858	1.14775
2740	2280.33	714.79	636.4	526.95	1.5950	1.02116	4250	3790.33	1163.30	4250	871.93	.3705	1.15133
2760	2300.33	720.58	656.3	531.37	1.5580	1.02326	4300	3840.33	1178.45	4475	883.65	.3560	1.15487
2780	2320.33	726.38	676.6	535.80	1.5221	1.02536	4350	3890.33	1193.61	4710	895.39	.3421	1.15838
2800	2340.33	732.19	697.5	540.23	1.4872	1.02744	4400	3940.33	1208.78	4955	907.14	.3290	1.16185
2820	2360.33	738.00	718.9	544.67	1.4533	1.02950	4450	3990.33	1223.98	5209	918.90	.3165	1.16528
2840	2380.33	743.81	740.7	549.11	1.4204	1.03156	4500	4040.33	1239.18	5474	930.68	.3046	1.16868
2860	2400.33	749.63	763.1	553.56	1.3885	1.03360	4550	4090.33	1254.40	5749	942.47	.2932	1.17204
2880	2420.33	755.45	786.0	558.01	1.3574	1.03563	4600	4140.33	1269.63	6035	954.27	.2824	1.17537

2900	2440.33	761.28	809.5	562.47	1.3272	1.03765	4650	4190.33	1284.88	6332	966.09	.2721	1.17867
2920	2460.33	767.12	833.5	566.93	1.2978	1.03965	4700	4240.33	1300.13	6641	977.92	.2622	1.18193
2940	2480.33	772.95	858.1	571.40	1.2693	1.04164	4750	4290.33	1315.41	6962	989.76	.2528	1.18516
2960	2500.33	778.80	883.3	575.87	1.2416	1.04362	4800	4340.33	1330.69	7294	1001.62	.2438	1.18836
2980	2520.33	784.64	909.0	580.35	1.2146	1.04559	4850	4390.33	1345.98	7640	1013.49	.2352	1.19153
3000	2540.33	790.49	935.3	584.83	1.1883	1.04755	4900	4440.33	1361.29	7998	1025.37	.2270	1.19467
3020	2560.33	796.35	962.2	589.31	1.1628	1.04949	4950	4490.33	1376.61	8369	1037.26	.2191	1.19778
3040	2580.33	802.21	989.8	593.80	1.1379	1.05143	5000	4540.33	1391.94	8754	1049.16	.21161	1.20086
3060	2600.33	808.07	1017.9	598.29	1.1137	1.05335	5050	4590.33	1407.29	9152	1061.08	.20442	1.20392
3080	2620.33	813.94	1046.7	602.79	1.0902	1.05526	5100	4640.33	1422.64	9565	1073.00	.19753	1.20694
3100	2640.33	819.82	1076.1	607.29	1.0672	1.05716	5150	4690.33	1438.01	9993	1084.94	.19093	1.20994
3120	2660.33	825.69	1106.2	611.80	1.0449	1.05905	5200	4740.33	1453.38	10435	1096.89	.18461	1.21291
3140	2680.33	831.57	1136.9	616.31	1.0232	1.06093	5250	4790.33	1468.77	10893	1108.85	.17855	1.21586
3160	2700.33	837.46	1168.3	620.82	1.0020	1.06280	5300	4840.33	1484.17	11367	1120.82	.17273	1.21878
3180	2720.33	843.35	1200.4	625.34	.9814	1.06466	5350	4890.33	1499.57	11857	1132.80	.16716	1.22167
3200	2740.33	849.24	1233.2	629.86	.9613	1.06651	5400	4940.33	1514.99	12364	1144.79	.16181	1.22454
3220	2760.33	855.14	1266.7	634.39	.9417	1.06834	5450	4990.33	1530.42	12888	1156.79	.15667	1.22738
3240	2780.33	862.04	1300.9	638.92	.9227	1.07017	5500	5040.33	1545.86	13429	1168.80	.15173	1.23020
3260	2800.33	866.94	1335.9	643.45	.9041	1.07199	5550	5090.33	1561.31	13988	1180.82	.14700	1.23300
3280	2820.33	872.85	1371.5	647.99	.8860	1.07379	5600	5140.33	1576.76	14565	1192.85	.14244	1.23577
3300	2840.33	878.77	1408.0	652.53	.8683	1.07559	5650	5190.33	1592.23	15161	1204.89	.13806	1.23852
3320	2860.33	884.68	1445.2	657.07	.8511	1.07738	5700	5240.33	1607.71	15776	1216.94	.13385	1.24125
3340	2880.33	890.60	1483.1	661.62	.8343	1.07916	5750	5290.33	1623.19	16411	1228.99	.12980	1.24395
3360	2900.33	896.52	1521.9	666.18	.8179	1.08092	5800	5340.33	1638.68	17066	1241.06	.12591	1.24663
3380	2920.33	902.45	1561.4	670.73	.8020	1.08268	5850	5390.33	1654.19	17742	1253.13	.12216	1.24930

TABLE A·6(SI) Properties of Ideal Gases (SI) h in kJ/kmol, ϕ in kJ/kmol·K

	CO		CO_2		H_2		H_2O		N_2		O_2		OH	
T	h	ϕ	h	ϕ	h	ϕ	h	ϕ	h	ϕ	h	ϕ	h	ϕ
0	0		0		0		0		0		0		0	0
200	5813.2	185.891	5952.3	199.859	5697.2	119.328	6621.8	175.369	5812.8	179.842	5815.7	193.330	6196.5	171.481
298	8666.7	197.499	9359.7	213.657	8463.1	130.558	9899.0	188.697	8665.8	191.448	8676.9	204.975	9171.3	183.594
300	8725.0	197.697	9433.7	213.908	8520.7	130.752	9966.1	188.925	8724.1	191.646	8737.6	205.174	9225.7	183.782
400	11647.1	206.102	13368.5	225.199	11426.7	139.117	13355.6	198.672	11641.3	200.037	11708.8	213.717	12204.7	192.355
500	14601.7	212.693	17671.7	234.788	14349.2	145.635	16828.4	206.417	14581.0	206.596	14767.7	220.538	15162.8	198.953
600	17612.1	218.180	22273.3	243.170	17278.3	150.973	20405.3	212.935	17564.1	212.033	17927.3	226.296	18112.5	204.334
700	20692.3	222.927	27120.9	250.639	20216.1	155.504	24095.7	218.622	20606.8	216.722	21182.0	231.312	21074.8	208.895
800	23845.9	227.137	32173.5	257.382	23168.1	159.445	27906.2	223.708	23716.2	220.873	24518.8	235.766	24049.6	212.911
900	27069.9	230.933	37397.4	263.533	26143.3	162.946	31841.2	228.342	26892.6	224.614	27924.2	239.776	27057.9	216.413
1000	30357.9	234.397	42764.6	269.186	29148.1	166.105	35904.1	232.621	30132.4	228.027	31386.2	243.423	30103.9	219.622
1100	33702.6	237.584	48250.8	274.417	31286.9	169.007	40094.4	236.614	33429.7	231.169	34895.2	246.768	33195.9	222.568
1200	37096.7	240.537	53840.2	279.272	35264.3	171.684	44410.1	240.368	36778.3	234.082	38443.8	249.855	36329.7	225.296
1300	40533.7	243.288	59515.8	283.820	38384.9	174.178	48846.1	243.918	40172.0	236.798	42026.6	252.723	39513.7	227.840
1400	44007.6	245.862	65261.8	288.086	41548.6	176.531	53396.6	247.290	43605.3	239.343	45639.6	255.400	42739.6	230.233
1500	47513.7	248.281	71071.5	292.093	44757.8	178.743	58054.5	250.503	47073.2	241.735	49280.1	257.912	46011.4	232.488
1600	51047.6	250.562	76935.6	295.868	48010.1	180.838	62811.9	253.573	50571.4	243.993	52946.2	260.278	49321.0	234.626
1700	54605.8	252.719	82847.5	299.452	51303.3	182.834	67662.0	259.513	54096.3	246.129	56636.7	262.515	52672.4	236.660
1800	58185.4	254.765	88797.9	302.852	54635.1	184.738	72596.8	259.334	57644.5	248.158	60350.6	264.638	56061.4	238.593
1900	61783.7	256.710	94784.7	306.095	58007.8	186.559	77609.5	262.044	61213.5	250.087	64087.5	266.658	59479.7	240.442
2000	65398.6	258.565	100805.5	309.179	61419.0	188.313	82693.7	264.652	64800.9	251.927	67846.8	268.586	62931.5	242.216
2100	69028.3	260.336	106853.6	312.131	64864.4	189.992	87843.6	267.164	68404.6	253.685	71628.4	270.431	66412.6	243.910
2200	72671.3	262.030	112929.0	314.958	68343.8	191.614	93053.8	269.588	72023.1	255.369	75431.8	272.201	69923.0	245.542
2300	76326.2	263.655	119024.8	317.668	71855.0	193.169	98319.5	271.928	75654.7	256.983	79256.9	273.901	73454.3	247.115
2400	79991.9	265.215	125143.3	320.271	75395.8	194.682	103636.5	274.191	79298.3	258.534	83103.1	275.538	77010.7	248.630
2500	83667.3	266.715	131282.3	322.782	78966.0	196.137	109001.3	276.381	82952.7	260.025	86970.2	277.116	80588.0	250.090
2600	87351.7	268.160	137437.1	325.193	82563.6	197.550	114410.3	278.502	86617.0	261.463	90857.6	278.641	84186.3	251.500
2700	91044.2	269.554	143610.1	327.521	86186.1	198.914	119860.8	280.559	90290.3	262.849	94764.9	280.116	87805.4	252.864
2800	94744.3	270.900	149796.8	329.774	89833.6	200.244	125349.8	282.556	93971.8	264.188	98691.6	281.544	91440.0	254.186
2900	98451.3	272.200	155999.3	331.953	93506.0	201.533	130875.5	284.495	97661.0	265.482	102637.0	282.928	95089.8	255.467
3000	102164.8	273.459	162215.5	334.056	97201.2	202.780	136436.7	286.380	101357.3	266.735	106600.4	284.272	98755.0	256.709

Source: Abridged from Joseph H. Keenan, Jing Chao, and Joseph Kaye, *Gas Tables*, 2nd ed., SI, Wiley, New York, 1983, except data for OH from *JANAF Thermochemical Tables*, 2nd ed., National Bureau of Standards, NSRDS-NBS 37, 1971.

TABLE A·6(E) Properties of Ideal Gases (English units) h in B/lbmol, ϕ in B/lbmol·R

	CO		CO_2		H_2		H_2O		N_2		O_2		OH	
T	h	ϕ	h	ϕ	h	ϕ	h	ϕ	h	ϕ	h	ϕ	h	ϕ
0	0		0		0		0		0		0		0	
300	2082.1	43.132	2106.3	46.361	2066.2	27.337	2369.1	40.434	2081.9	41.687	2083.0	44.908	2227	39.545
400	2777.3	45.132	2873.3	48.563	2712.6	29.194	3165.7	42.726	2777.1	43.687	2778.6	46.909	2952	41.636
500	3472.7	46.684	3705.1	50.417	3388.5	30.701	3964.2	44.507	3472.5	45.238	3476.2	48.466	3674	43.259
537	3730.2	47.180	4029.1	51.041	3642.6	31.192	4260.5	45.079	3729.8	45.735	3735.4	48.966	3940	43.778
600	4169.0	47.953	4599.8	52.046	4078.0	31.958	4767.4	45.971	4168.2	46.507	4178.9	49.747	4392	44.584
700	4867.3	49.029	5551.0	53.511	4773.1	33.031	5578.5	47.222	4865.2	47.581	4890.3	50.843	5108	45.705
800	5569.5	49.967	6552.1	54.847	5470.6	33.960	6400.4	48.319	5564.8	48.516	5613.1	51.808	5821	46.676
900	6277.6	50.801	7597.5	56.078	6169.0	34.784	7234.9	49.302	6268.7	49.344	6349.0	52.675	6533	47.528
1000	6993.2	51.555	8682.3	57.221	6868.4	35.521	8083.4	50.196	6978.4	50.092	7098.3	53.464	7243	48.285
1100	7717.4	52.245	9802.5	58.288	7568.5	36.188	8946.5	51.018	7695.2	50.775	7860.9	54.191	7952	48.970
1200	8451.2	52.883	10954.6	59.290	8269.7	36.798	9824.9	51.782	8420.2	51.406	8636.0	54.865	8660	49.595
1300	9194.7	53.478	12135.6	60.236	8972.7	37.360	10718.8	52.498	9153.9	51.993	9422.6	55.495	9366	50.170
1400	9947.9	54.037	13342.5	61.130	9678.0	37.884	11628.9	53.172	9896.5	52.544	10219.6	56.085	10071	50.702
1500	10710.7	54.563	14573.2	61.979	10386.2	38.371	12555.3	53.811	10648.2	53.062	11026.2	56.642	10775	51.198
1600	11482.5	55.061	15825.2	62.787	11097.1	38.830	13498.7	54.420	11408.6	53.553	11841.3	57.168	11478	51.661
1700	12263.0	55.534	17096.6	63.557	11812.8	39.264	14458.8	55.002	12177.6	54.019	12664.0	57.666	12181	52.097
1800	13051.6	55.985	18385.1	64.294	12531.4	39.674	15436.0	55.561	12954.6	54.463	13493.6	58.141	12882	52.508
1900	13847.6	56.415	19689.6	64.999	13254.9	40.065	16430.1	56.098	13739.3	54.887	14329.5	58.593	13608	52.896
2000	14650.7	56.827	21008.7	65.676	13984.3	40.440	17441.0	56.616	14531.1	55.294	15170.9	59.024	14376	53.265
2200	16275.6	57.601	23686.1	66.951	15457.8	41.141	19512.3	57.603	16134.4	56.058	16868.9	59.833	15921	53.949
2400	17922.5	58.318	26407.4	68.137	16953.6	41.794	21646.8	58.532	17761.1	56.765	18584.6	60.580	17476	54.574
2600	19588.2	58.984	29164.8	69.239	18473.9	42.400	23840.8	59.410	19407.8	57.424	20315.7	61.272	19040	55.149
2800	21269.9	59.607	31953.5	70.272	20017.7	42.974	26090.1	60.243	21071.9	58.041	22060.9	61.919	20612	55.777
3000	22965.3	60.192	34768.5	71.243	21582.9	43.512	28390.1	61.036	22750.9	58.620	23819.4	62.526	22193	56.354

3200	24672.5	60.743	37606.0	72.158	23168.6	44.025	30736.4	61.793	24442.9	59.166	25590.4	63.097	23781	56.894
3400	26389.9	61.264	40463.0	73.026	24776.7	44.511	33125.0	62.517	26146.2	59.682	27373.6	63.638	25376	57.401
3600	28116.3	61.757	43338.6	73.846	26405.4	44.978	35551.9	63.211	27859.3	60.172	29168.9	64.151	27033	57.879
3800	29850.6	62.226	46228.8	74.629	28052.6	45.423	38013.5	63.876	29581.2	60.637	30975.9	64.639	28709	58.332
4000	31592.8	62.672	49132.7	75.373	29716.5	45.850	40507.0	64.516	31310.8	61.081	32794.5	65.106	30396	58.761
4200	33339.2	63.099	52047.3	76.084	31397.9	46.261	43029.3	65.131	33047.3	61.504	34624.5	65.552	32093	59.169
4400	35092.1	63.506	54973.6	76.765	33095.9	46.654	45578.4	65.724	34789.8	61.910	36465.7	65.980	33798	59.558
4600	36850.0	63.897	57909.7	77.417	34806.6	47.035	48151.7	66.296	36537.9	62.298	38317.9	66.392	35513	59.930
4800	38612.4	64.272	60855.6	78.044	36532.9	47.402	50748.0	66.848	38291.0	62.671	40180.7	66.788	37236	60.286
5000	40378.9	64.633	63809.3	78.648	38271.9	47.758	53364.9	67.382	40048.6	63.030	42053.9	67.171	38967	60.627

Source: Abridged from Joseph H. Keenan, Jing Chao, and Joseph Kaye, *Gas Tables*, 2nd ed., English units, Wiley, New York, 1980, except data for OH from *JANAF Thermochemical Tables*, 2nd ed., National Bureau of Standards, NSRDS-NBS 37, 1971.

TABLE A·7(SI) Enthalpy of Eight Gases at Low Pressure (SI) (in kJ/kmol)

T, K	Methane CH_4	Ethane C_2H_6	Propane C_3H_8	n-Butane C_4H_{10}	n-Octane C_8H_{18}	Methanol CH_3OH	Ammonia NH_3	Hydrazine N_2H_4	T, K
0	0	0	0	0	0	0	0	0	0
200	6644	7259	8414	10744	21581		6665	7067	200
298	10016	11874	14740	19276	37782	11427	10058	11510	298
300	10083	11975	14874	19460	38116	11510	10125	11602	300
400	13887	17874	23276	30648	59580	16263	13836	17113	400
500	18238	25058	33623	44350	85939	21811	17870	23510	500
600	13192	33430	45689	60250	116650	28146	22230	30648	600
700	28727	42844	59287	78115	151084	35192	26907	38397	700
800	34819	53220	74182	97613	188698	42865	31878	46677	800
900	41417	64434	90207	118533	228466	51087	37129	55421	900
1000	48492	76358	107194	140666	270705	59810	42639	64584	1000
1100	55982	88910	125102	163929			48396	74124	1100

Source: Data for all gases but ammonia from *Selected Values of Properties of Hydrocarbons and Related Compounds,* Thermodynamics Research Center Data Project, Thermodynamics Research Center, Texas A & M University, College Station, Texas 77843 (looseleaf data sheets extant 1984); data for ammonia from *JANAF Thermochemical Tables,* National Bureau of Standards, 2nd ed., National Bureau of Standards, NSRDS-NBS 37, 1971.

TABLE A·7(E) Enthalpy of Eight Gases at Low Pressure (English units) (in B/lbmol)

T, R	Methane CH_4	Ethane C_2H_6	Propane C_3H_8	*n*-Butane C_4H_{10}	*n*-Octane C_8H_{18}	Methanol CH_3OH	Ammonia NH_3	Hydrazine N_2H_4	*T*, R
0	0	0	0	0	0	0	0	0	0
537	4306	5105	6337	8287	16243	4913	4324	4948	537
600	4853	5929	7499	9841	19215	5589	4866	5733	600
800	6782	9005	11903	15693	30446	8011	6700	8542	800
1000	8999	12717	17261	22769	44087	10853	8708	11776	1000
1200	11525	17018	23466	30927	59839	14085	10884	15367	1200
1400	14359	21850	30412	40028	77374	17671	13219	19257	1400
1600	17482	27148	37993	49932	96340	21564	15705	23402	1600
1800	20857	32837	46105	60501	116354	25712	18332	27767	1800
2000	24427	38831	54646	71594	137009	30050	21088	32332	2000

Source: Data for all gases but ammonia from *Selected Values of Properties of Hydrocarbons and Related Compounds*, Thermodynamics Research Center Data Project, Thermodynamics Research Center, Texas A & M University, College Station, Texas 77843 (looseleaf data sheets extant 1984); data for ammonia from *JANAF Thermochemical Tables*, National Bureau of Standards, 2nd ed., National Bureau of Standards, NSRDS-NBS 37, 1971.

TABLE A·8(SI) Properties of Substances at 25°C and Low Pressure (SI)

Name	Molar Mass kg/kmol	Phase	h_{fg} kJ/kmol	Enthalpy of Combustion Δh_c at 25°C, 1 atm: H_2O (l) kJ/kg	H_2O (l) kJ/kmol	H_2O (g) kJ/kg	H_2O (g) kJ/kmol	Enthalpy of Formation Δh_f° kJ/kmol	Absolute Entropy S° (1 atm) kJ/kmol·K	Formula
Hydrogen, H_2	2.016	gas	…	−141784	−285836	−119954	−241826	0	130.57	H_2
Water, H_2O	18.02	gas	44010	…	…	…	…	−241820	188.71	H_2O
Carbon, C	12.01	solid	…	−32765	−393510	−32765	−393510	0	5.734	C
Carbon monixide, CO	28.01	gas	…	−10103	−282980	−10103	−282980	−110530	197.56	CO
Carbon dioxide, CO_2	44.01	gas	…	…	…	…	…	−393510	213.68	CO_2
Nitrogen, N_2	28.01	gas	…	…	…	…	…	0	191.46	N_2
Oxygen, O_2	32.00	gas	…	…	…	…	…	0	205.04	O_2
Methane, CH_4	16.04	gas	…	−55497	−890320	−50010	−802300	−74850	186.27	CH_4
Ethane, C_2H_8	30.07	gas	5021	−51902	−1560660	−47511	−1428630	−83850	229.12	C_2H_6
Propane, C_3H_8	44.10	gas	14820	−50326	−2219160	−46333	−2043120	−104680	270.20	C_3H_8
n-Butane, C_4H_{10}	58.12	gas	21066	−49508	−2877540	−45722	−2657480	−125650	309.91	C_4H_{10}
n-Pentane, C_5H_{12}	72.15	gas	26426	−49007	−3535820	−45347	−3271750	−146710	349.45	C_5H_{12}
n-Hexane, C_6H_{14}	86.18	gas	31552	−48681	−4194753	−45104	−3886685	−166940	388.74	C_6H_{14}
n-Heptane, C_7H_{16}	100.2	gas	36547	−48438	−4853482	−44924	−4501398	−187650	427.98	C_7H_{16}
n-Octane, C_8H_{18}	114.2	gas	41484	−48258	−5512210	−44790	−5116112	−208447	466.73	C_8H_{18}
Ethene, C_2H_4	28.05	gas	…	−50303	−1411200	−47166	−1323200	52283	219.45	C_2H_4
Propene, C_3H_6	42.08	gas	…	−48900	−2057700	−45762	−1925700	20414	266.94	C_3H_6
Acetylene, C_2H_2	26.04	gas	…	−49915	−1299634	−48255	−1255618	266748	200.82	C_2H_2
Benzene, C_6H_6	78.11	gas	33849	−42265	−3301500	−40575	−3169400	82927	269.20	C_6H_6
Methanol, CH_3OH	32.04	gas	37430	−23850	−764153	−21103	−676133	−201041	239.70	CH_3OH
Ethanol, C_2H_5OH	46.07	gas	42309	−30608	−1410122	−27742	−1278092	−234430	282.59	C_2H_5OH
Hydrazine, N_2H_4	32.05	liquid	44727	−19420	−622277	−16670	−534280	50626	88.63	N_2H_4

Source: Data from *Selected Values of Properties of Chemical Compounds*, Thermodynamics Research Center Project, Thermodynamics Research Center, Texas A & M University, College Station, Texas 77843 (looseleaf data sheets extant 1984).

TABLE A·8(E) Properties of Substances at 77°F and Low Pressure (English units)

Name	Molar Mass lbm/lbmol	Phase	h_{fg} B/lbmol	Enthalpy of Combustion Δh_c at 77°F and 1 atm, H_2O (l) B/lbm	H_2O (l) B/lbmol	H_2O (g) B/lbm	H_2O (g) B/lbmol	Enthalpy of Formation Δh_f° B/lbmol	Absolute Entropy S° (1 atm) B/lbmol·R	Formula
Hydrogen, H_2	2.016	gas	…	−60956	−122887	−51571	−103966	0	31.186	H_2
Water, H_2O	18.02	gas	18921	…	…	…	…	−103963	45.073	H_2O
Carbon, C	12.01	solid	…	−14087	−169179	−14087	−169179	0	1.370	C
Carbon monoxide, CO	28.01	gas	…	−4343	−121660	−4344	−121660	−47519	47.186	CO
Carbon dioxide, CO_2	44.01	gas	…	…	…	…	…	−169179	51.037	CO_2
Nitrogen, N_2	28.01	gas	…	…	…	…	…	0	45.730	N_2
Oxygen, O_2	32.00	gas	…	…	…	…	…	0	48.972	O_2
Methane, CH_4	16.04	gas	…	−23859	−382769	−21500	−344927	−32180	44.490	CH_4
Ethane, C_2H_6	30.07	gas	2159	−22314	−670963	−20426	−614200	−36049	54.724	C_2H_6
Propane, C_3H_8	44.10	gas	6371	−21636	−954067	−19920	−878383	−45004	64.536	C_3H_8
n-Butane, C_4H_{10}	58.12	gas	9057	−21285	−1237120	−19657	−1142511	−54020	74.021	C_4H_{10}
n-Pentane, C_5H_{12}	72.15	gas	11361	−21069	−1520129	−19496	−1406599	−63074	83.465	C_5H_{12}
n-Hexane, C_6H_{14}	86.18	gas	13565	−20928	−1803419	−19391	−1670974	−71771	92.849	C_6H_{14}
n-Heptane, C_7H_{16}	100.2	gas	15712	−20825	−2086622	−19314	−1935253	−80675	102.22	C_7H_{16}
n-Octane, C_8H_{18}	114.2	gas	17835	−20747	−2369824	−19256	−2199532	−89616	111.48	C_8H_{18}
Ethene, C_2H_4	28.05	gas	…	−21626	−606707	−20278	−568874	22478	52.415	C_2H_4
Propene, C_3H_6	42.08	gas	…	−21023	−884652	−19674	−827902	8776	63.758	C_3H_6
Acetylene, C_2H_2	26.04	gas	…	−21460	−558818	−20734	−539913	97484	47.965	C_2H_2
Benzene, C_6H_6	78.11	gas	14552	−18171	−1419390	−17444	−1362597	35652	64.297	C_6H_6
Methanol, CH_3OH	32.04	gas	16092	−10254	−328527	−9073	−290685	−86432	57.251	CH_3OH
Ethanol, C_2H_5OH	46.07	gas	18190	−13159	−606243	−11927	−549481	−100787	67.495	C_2H_5OH
Hydrazine, N_2H_4	32.05	liquid	19229	−8349	−267531	−7167	−229699	21765	21.17	N_2H_4

Source: Data from *Selected Values of Properties of Chemical Compounds*, Thermodynamics Research Center Project, Thermodynamics Research Center, Texas A & M University, College Station, Texas 77843 (looseleaf data sheets extant 1984).

TABLE A·9 Log_{10} K_p for Eight Ideal Gas Reactions $P° = 1$ atm

T, K	$H_2 \rightarrow 2H$	$N_2 \rightarrow 2N$	$O_2 \rightarrow 2O$	$CO_2 \rightarrow CO + \frac{1}{2}O_2$	$H_2O \rightarrow OH + \frac{1}{2}H_2$	$H_2O \rightarrow H_2 + \frac{1}{2}O_2$	$\frac{1}{2}N_2 + \frac{1}{2}O_2 \rightarrow NO$	$CO_2 + H_2 \rightarrow CO + H_2O$	T, R
298	−71.23	−159.6	−81.20	−45.07	−46.05	−40.05	−15.17	−5.018	537
500	−40.32	−92.69	−45.88	−25.03	−26.13	−22.89	−8.783	−2.139	900
1000	−17.29	−43.06	−19.61	−10.02	−11.28	−10.06	−4.062	−0.159	1800
1200	−18.41	−34.76	−15.21	−7.764	−8.789	−7.899	−3.275	0.135	2160
1400	−10.63	−28.82	−12.05	−6.014	−7.003	−6.347	−2.712	0.333	2520
1500	−9.514	−26.44	−10.79	−5.316	−6.288	−5.725	−2.487	0.409	2700
1600	−8.534	−24.36	−9.682	−4.706	−5.662	−5.180	−2.290	0.474	2880
1700	−7.668	−22.52	−8.706	−4.169	−5.109	−4.699	−2.116	0.530	3060
1800	−6.896	−20.88	−7.836	−3.693	−4.617	−4.270	−1.962	0.577	3240
1900	−6.206	−19.42	−7.056	−3.267	−4.177	−3.886	−1.823	0.619	3420
2000	−5.582	−18.10	−6.354	−2.884	−3.780	−3.540	−1.699	0.656	3600
2100	−5.018	−16.90	−5.720	−2.539	−3.422	−3.227	−1.586	0.688	3780
2200	−4.504	−15.82	−5.142	−2.226	−3.095	−2.942	−1.484	0.716	3960
2300	−4.032	−14.82	−4.614	−1.940	−2.798	−2.682	−1.391	0.742	4140
2400	−3.602	−13.91	−4.130	−1.679	−2.525	−2.443	−1.305	0.764	4320
2500	−3.204	−13.08	−3.684	−1.440	−2.274	−2.224	−1.227	0.784	4500
2600	−2.836	−12.30	−3.272	−1.219	−2.042	−2.021	−1.154	0.802	4680
2700	−2.496	−11.59	−2.890	−1.015	−1.828	−1.833	−1.087	0.818	4860
2800	−2.178	−10.92	−2.536	−0.825	−1.628	−1.658	−1.025	0.833	5040
2900	−1.884	−10.30	−2.206	−0.649	−1.442	−1.495	−0.967	0.846	5220
3000	−1.608	−9.720	−1.898	−0.485	−1.269	−1.343	−0.913	0.858	5400
3100	−1.350	−9.178	−1.610	−0.332	−1.107	−1.201	−0.863	0.869	5580
3200	−1.108	−8.668	−1.338	−0.189	−0.955	−1.067	−0.815	0.878	5760
3300	−0.880	−8.190	−1.084	−0.054	−0.813	−0.942	−0.771	0.888	5940
3400	−0.664	−7.740	−0.844	0.071	−0.679	−0.824	−0.729	0.895	6120
3500	−0.462	−7.316	−0.620	0.190	−0.552	−0.712	−0.690	0.902	6300

Source: Based on data of *JANAF Thermochemical Tables*, 2nd ed., National Bureau of Standards, NSRDS-NBS 37, 1971, and revisions published in *Journal of Physical and Chemical Data* through 1982.

TABLE A·10 One-Dimensional Ideal Gas Isentropic Compressible-Flow Functions, $k = 1.4$

M	$\frac{A}{A^*}$	$\frac{P}{P_t}$	$\frac{\rho}{\rho_t}$	$\frac{T}{T_t}$	M	$\frac{A}{A^*}$	$\frac{P}{P_t}$	$\frac{\rho}{\rho_t}$	$\frac{T}{T_t}$
0	∞	1.00000	1.00000	1.00000	2.00	1.6875	.12780	.23005	.55556
.10	5.8218	.99303	.99502	.99800	2.10	1.8369	.10935	.20580	.53135
.20	2.9635	.97250	.98027	.99206	2.20	2.0050	.09352	.18405	.50813
.30	2.0351	.93947	.95638	.98232	2.30	2.1931	.07997	.16458	.48591
.40	1.5901	.89562	.92428	.96899	2.40	2.4031	.06840	.14720	.46468
.50	1.3398	.84302	.88517	.95238	2.50	2.6367	.05853	.13169	.44444
.60	1.1882	.78400	.84045	.93284	2.60	2.8960	.05012	.11787	.42517
.70	1.09437	.72092	.79158	.91075	2.70	3.1830	.04295	.10557	.40684
.80	1.03823	.65602	.74000	.88652	2.80	3.5001	.03685	.09462	.38941
.90	1.00886	.59126	.68704	.86058	2.90	3.8498	.03165	.08489	.37286
1.00	1.00000	.52828	.63394	.83333	3.00	4.2346	.02722	.07623	.35714
1.10	1.00793	.46835	.58169	.80515	3.50	6.7896	.01311	.04523	.28986
1.20	1.03044	.41238	.53114	.77640	4.00	10.719	.00658	.02766	.23810
1.30	1.06631	.36092	.48291	.74738	4.50	16.562	.00346	.01745	.19802
1.40	1.1149	.31424	.43742	.71839	5.00	25.000	$189(10)^{-5}$	.01134	.16667
1.50	1.1762	.27240	.39498	.68965	6.00	53.180	$633(10)^{-6}$	.00519	.12195
1.60	1.2502	.23527	.35573	.66138	7.00	104.143	$242(10)^{-6}$	.00261	.09259
1.70	1.3376	.20259	.31969	.63372	8.00	190.109	$102(10)^{-6}$	.00141	.07246
1.80	1.4390	.17404	.28682	.60680	9.00	327.189	$474(10)^{-7}$	.000815	.05814
1.90	1.5552	.14924	.25699	.58072	10.00	535.938	$236(10)^{-7}$	.000495	.04762

Source: Abridged by permission from Table 30 of *Gas Tables* by Joseph H. Keenan, Jing Chao, and Joseph Kaye, Wiley, New York, 1980 (English Units) and 1983 (SI Units).

M_x	M_y	$\frac{P_y}{P_x}$	$\frac{\rho_y}{\rho_x}$	$\frac{T_y}{T_x}$	$\frac{P_{0y}}{P_{0x}}$	$\frac{P_{0y}}{P_x}$
1.00	1.00000	1.00000	1.00000	1.00000	1.00000	1.8929
1.10	.91177	1.2450	1.1691	1.06494	.99892	2.1328
1.20	.84217	1.5133	1.3416	1.1280	.99280	2.4075
1.30	.78596	1.8050	1.5157	1.1909	.97935	2.7135
1.40	.73971	2.1200	1.6896	1.2547	.95819	3.0493
1.50	.70109	2.4583	1.8621	1.3202	.92978	3.4133
1.60	.66844	2.8201	2.0317	1.3880	.89520	3.8049
1.70	.64055	3.2050	2.1977	1.4583	.85573	4.2238
1.80	.61650	3.6133	2.3592	1.5316	.81268	4.6695
1.90	.59562	4.0450	2.5157	1.6079	.76735	5.1417
2.00	.57735	4.5000	2.6666	1.6875	.72088	5.6405
2.10	.56128	4.9784	2.8119	1.7704	.67422	6.1655
2.20	.54706	5.4800	2.9512	1.8569	.62812	6.7163
2.30	.53441	6.0050	3.0846	1.9468	.58331	7.2937
2.40	.52312	6.5533	3.2119	2.0403	.54015	7.8969
2.50	.51299	7.1250	3.3333	2.1375	.49902	8.5262
2.60	.50387	7.7200	3.4489	2.2383	.46012	9.1813
2.70	.49563	8.3383	3.5590	2.3429	.42359	9.8625
2.80	.48817	8.9800	3.6635	2.4512	.38946	10.569
2.90	.48138	9.6450	3.7629	2.5632	.35773	11.302
3.00	.47519	10.333	3.8571	2.6790	.32834	12.061
3.50	.45115	14.125	4.2608	3.3150	.21295	16.242
4.00	.43496	18.500	4.5714	4.0469	.13876	21.068
4.50	.42355	23.458	4.8119	4.8751	.09170	26.539
5.00	.41523	29.000	5.0000	5.8000	.06172	32.654
6.00	.40416	41.833	5.2683	7.941	.02965	46.815
7.00	.39736	57.000	5.4444	10.469	.01535	63.552
8.00	.39289	74.500	5.5652	13.387	.00849	82.865
9.00	.38980	94.333	5.6512	16.693	.00496	104.753
10.00	.38757	116.50	5.7143	20.388	.00304	129.217

Source: Abridged by permission from Table 48 of *Gas Tables* by Joseph H. Keenan, Jing Chao, and Joseph Kaye, Wiley, New York, 1980 (English Units) and 1983 (SI Units).

Charts

Appendix tables are listed on pages 722–723.

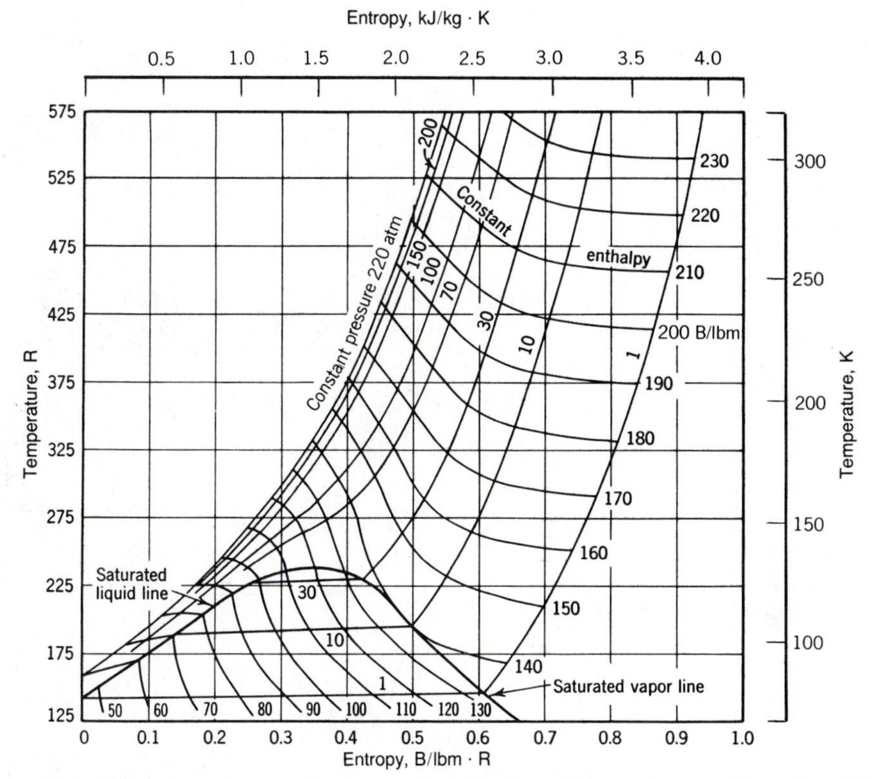

Chart A·1. Temperature–entropy diagram for air. Data from "Thermodynamic Properties of Air at Low Temperatures," by V. C. Williams, *Transactions, AIChE,* vol. 39, no. 1, Feb. 1943.

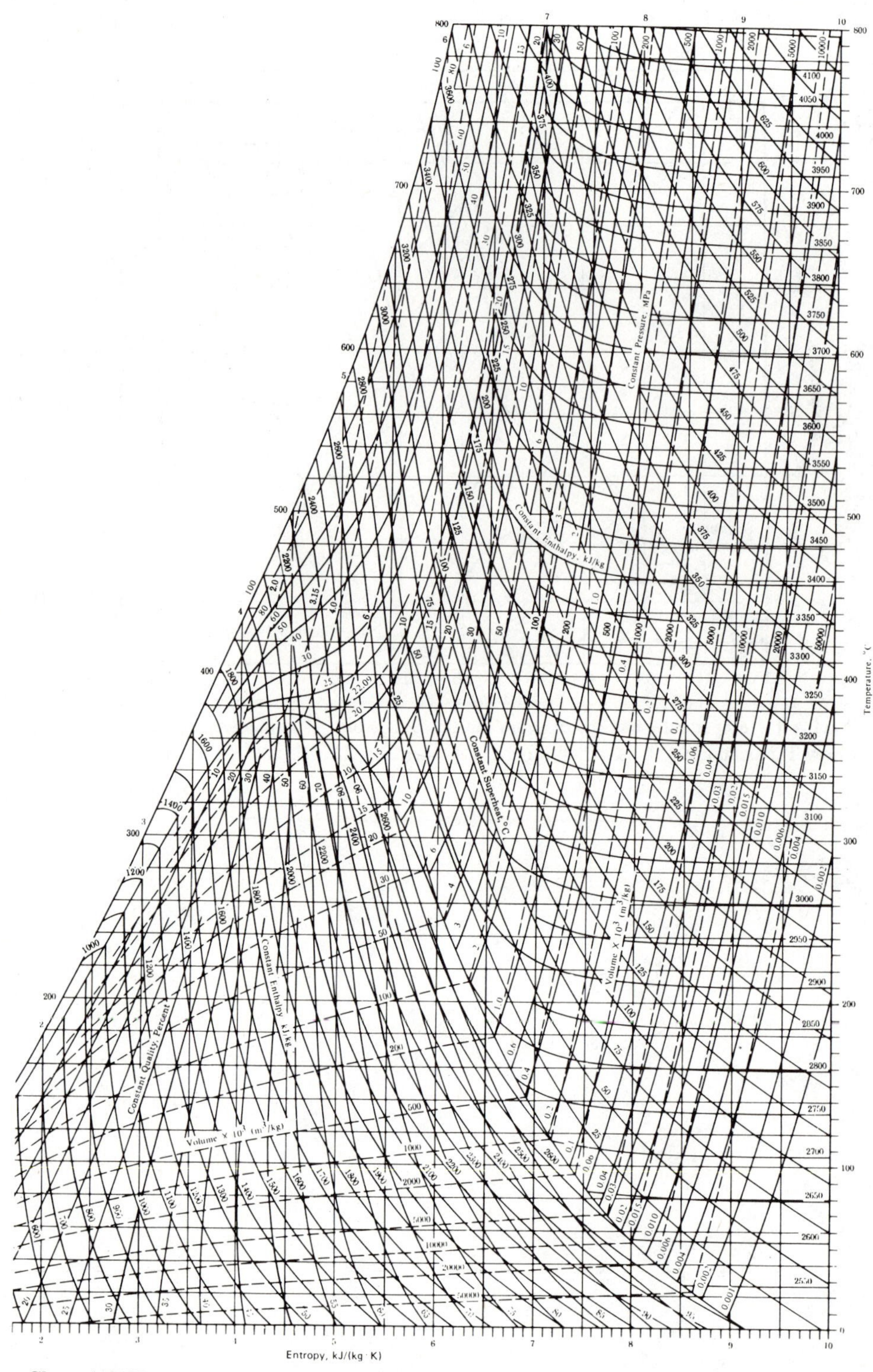

Chart A·2(SI). Temperature–entropy chart for water (SI). Reprinted from Joseph H. Keenan, Frederick G. Keyes, Philip G. Hill, and Joan G. Moore, *Steam Tables*, SI Units, Wiley, New York, 1978.

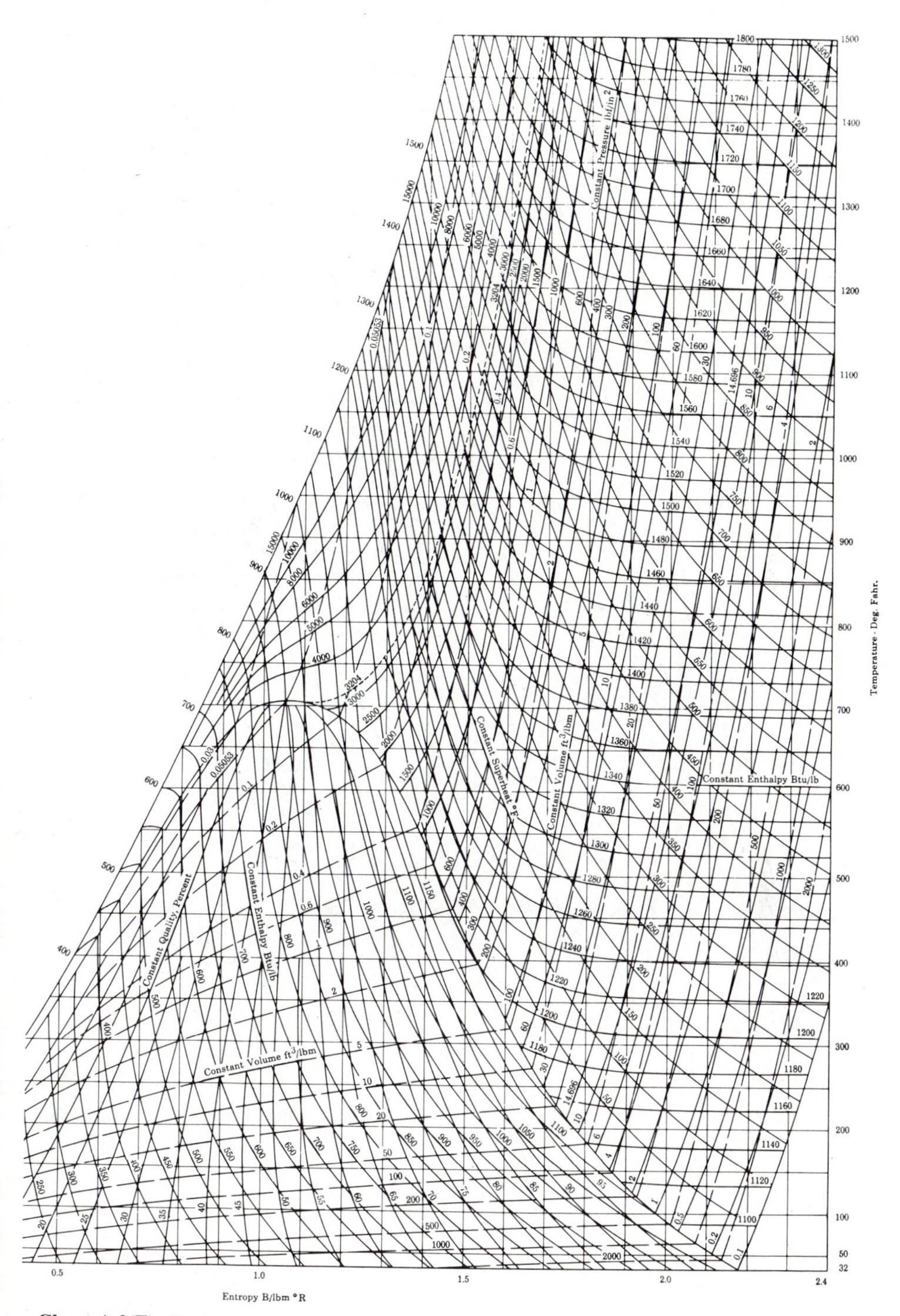

Chart A·2(E). Temperature–entropy chart for water (English units). Reprinted from Joseph H. Keenan, Frederick G. Keyes, Philip G. Hill, and Joan G. Moore, *Steam Tables*, English Units, Wiley, New York, 1969.

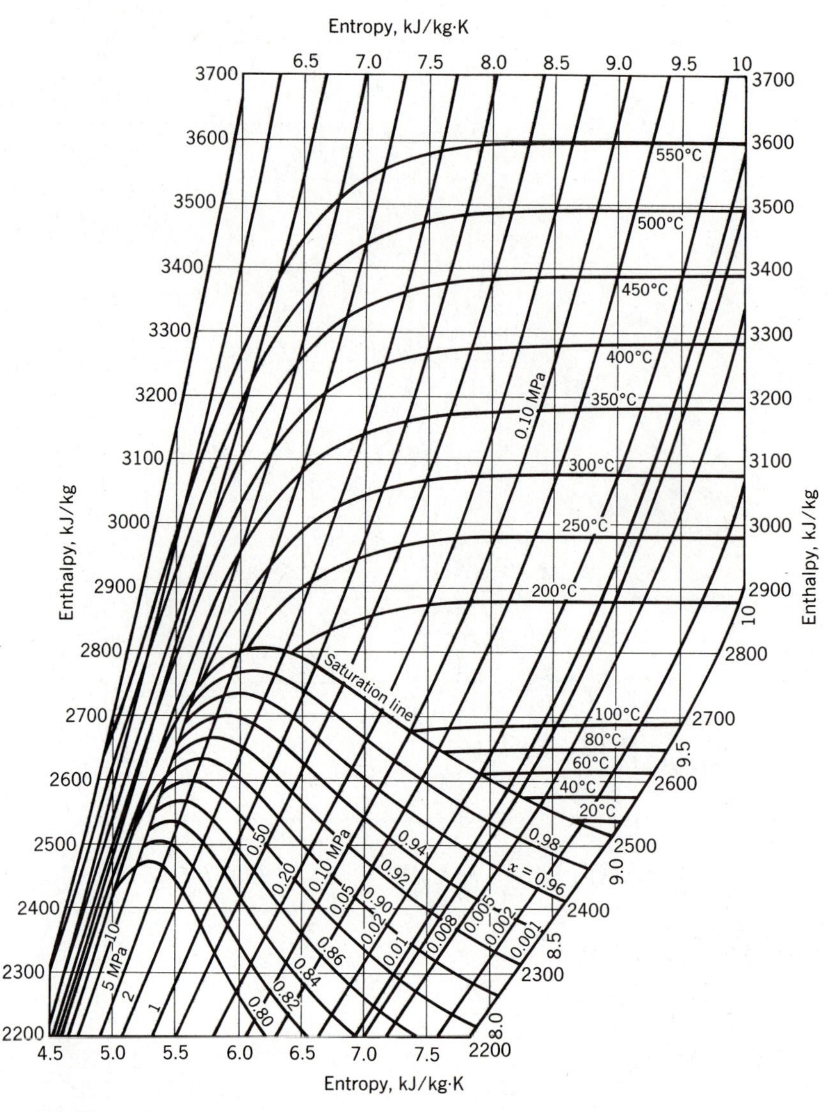

Chart A·3(SI). Mollier (hs) chart for water (SI). Adapted from Joseph H. Keenan, Frederick G. Keyes, Philip G. Hill, and Joan G. Moore, *Steam Tables*, SI Units, Wiley, New York, 1978.

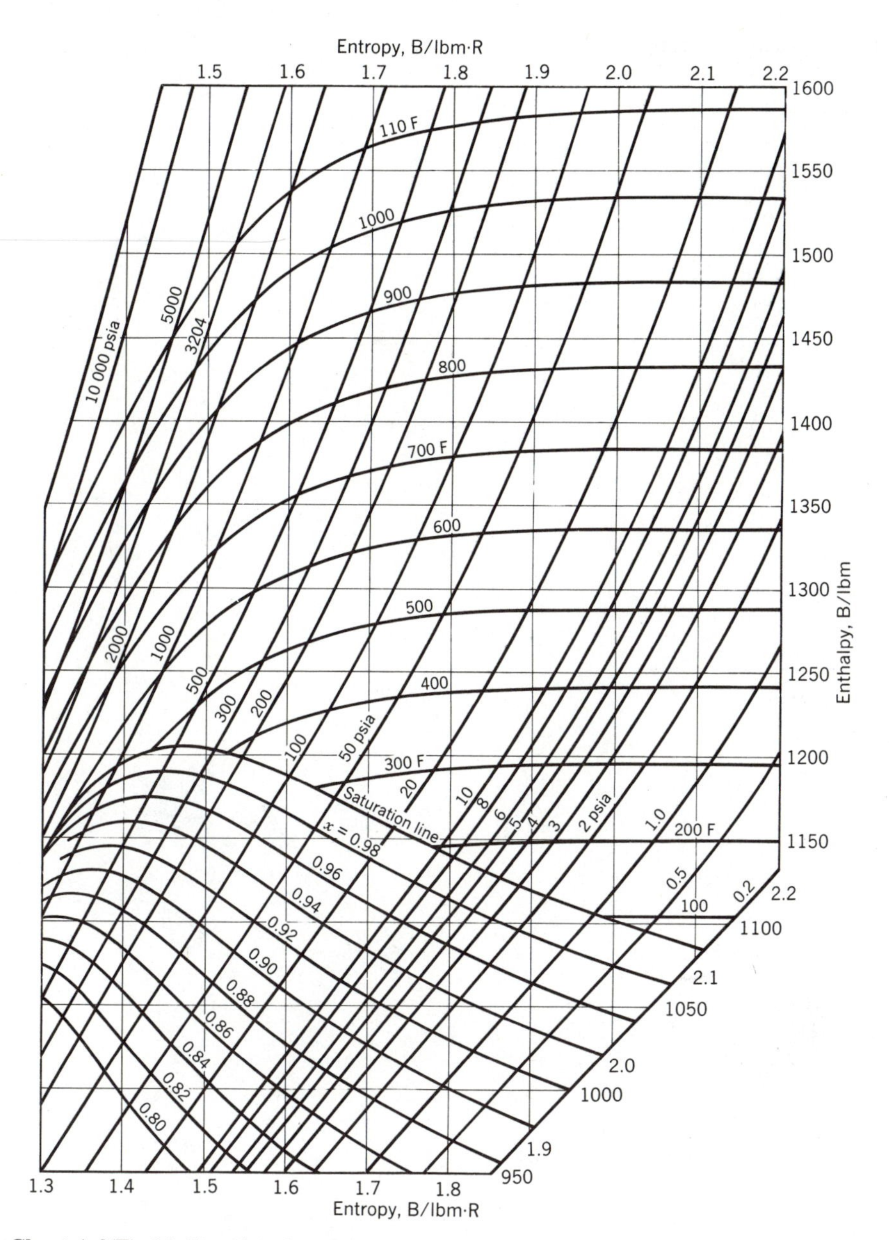

Chart A·3(E). Mollier (*hs*) chart for water (English units). Adapted from Joseph H. Keenan, Frederick G. Keyes, Philip G. Hill, and Joan G. Moore, *Steam Tables,* English Units, Wiley, New York, 1969.

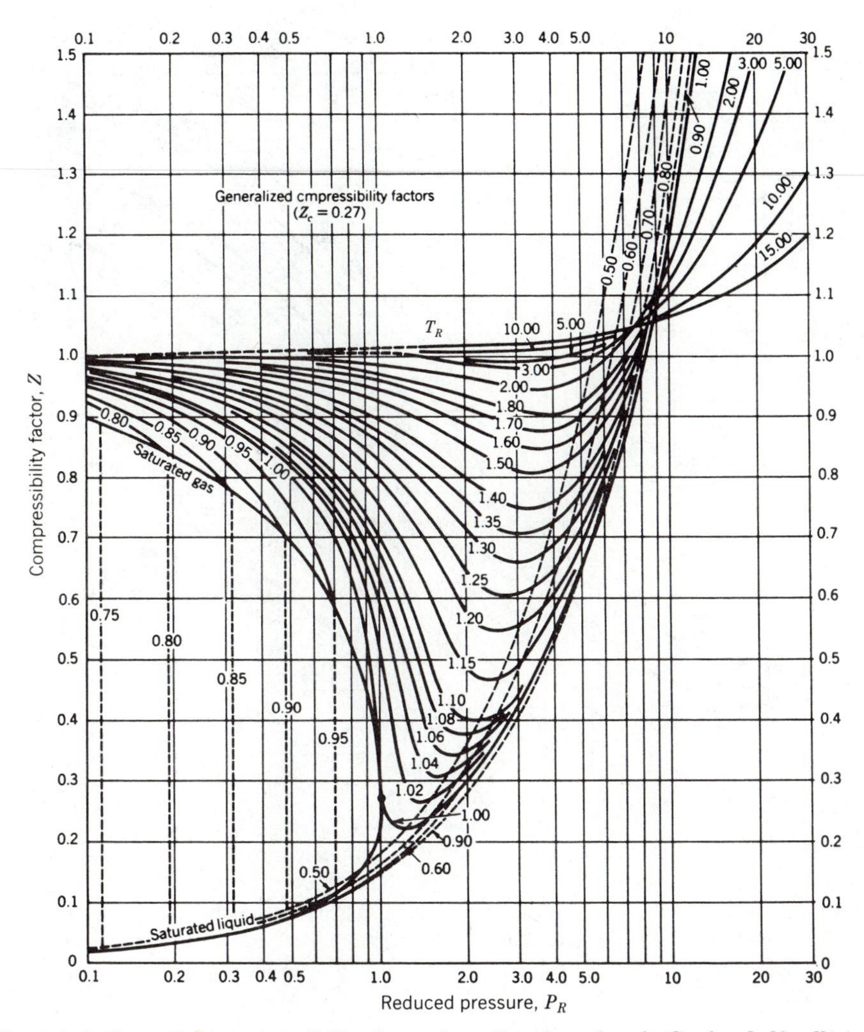

Chart A·4. Generalized compressibility factor chart. Based on chart in Gordon J. Van Wylen and Richard E. Sonntag, *Fundamentals of Classical Thermodynamics,* 3rd ed., SI Version, Wiley, New York, 1985.

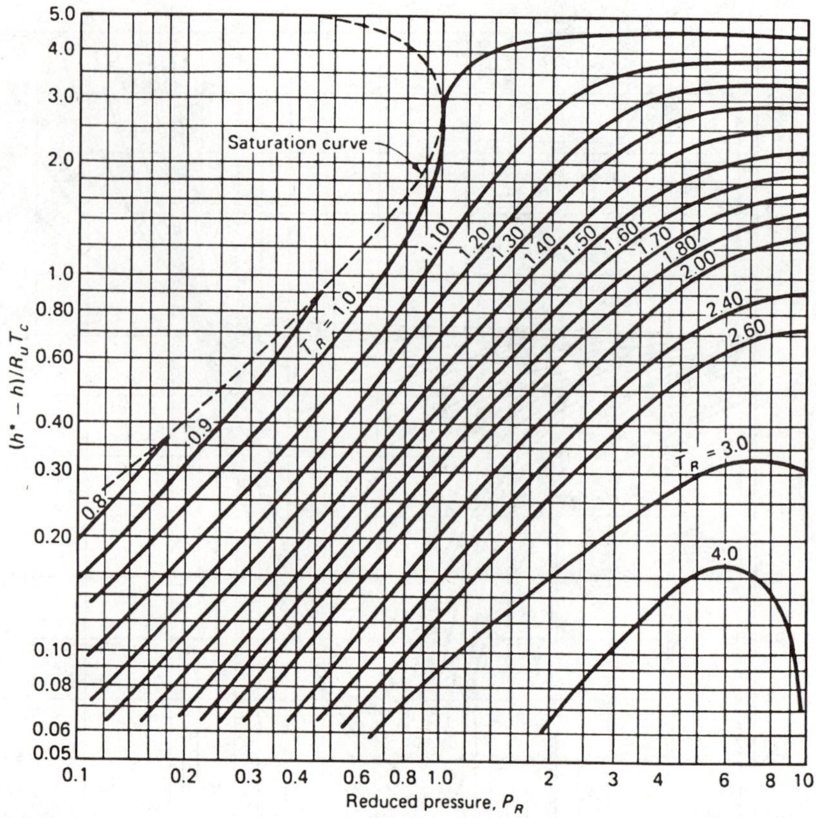

Chart A·5. Generalized enthalpy chart. Reprinted by permission from Kenneth Wark, *Thermodynamics,* 4th ed., McGraw-Hill, New York, 1983. Based on data from A. L. Lyderson, R. A. Greenkorn, and O. A. Hougen, "Engineering Experiment Station Report No. 4," University of Wisconsin, 1955. $T_R = T/T_c$ = reduced temperature, $P_R = P/P_c$ = reduced pressure, T_c = critical temperature, P_c = critical pressure, $h*$ = enthalpy of an ideal gas, h = enthalpy of an actual gas.

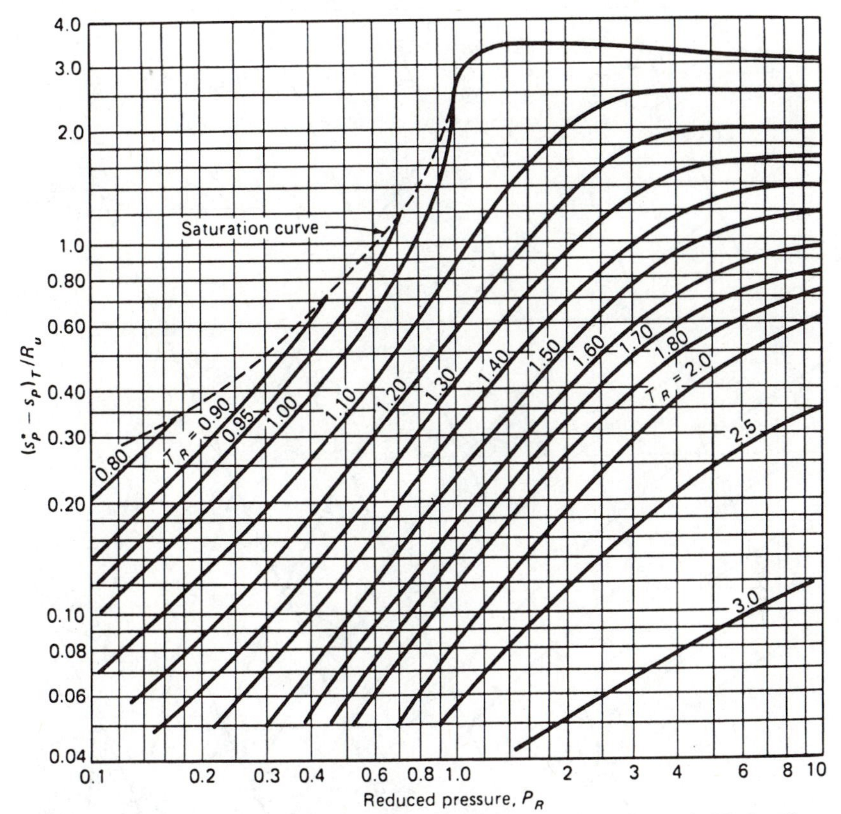

Chart A·6. Generalized entropy chart. Reprinted by permission from Kenneth Wark, *Thermodynamics*, 4th ed., McGraw-Hill, New York, 1983. Based on data from A. L. Lyderson, R. A. Greenkorn, and O. A. Hougen, "Engineering Experiment Station Report No. 4," University of Wisconsin, 1955. $T_R = T/T_c$ = reduced temperature, $P_R = P/P_c$ = reduced pressure, T_c = critical temperature, P_c = critical pressure, h^* = enthalpy of an ideal gas, h = enthalpy of an actual gas.

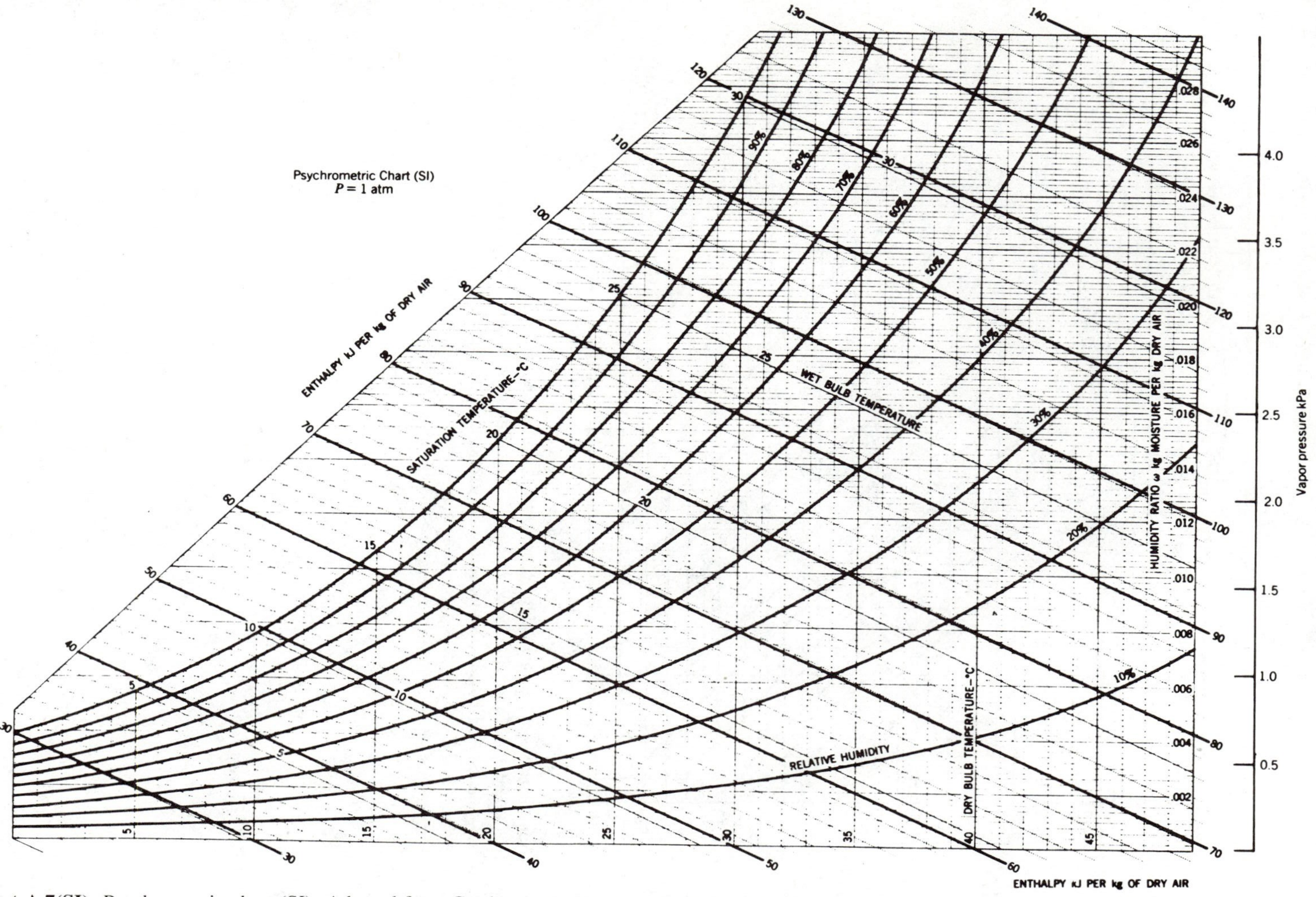

Chart A·7(SI). Psychrometric chart (SI). Adapted from Gordon J. Van Wylen and Richard E. Sonntag, *Fundamentals of Classical Thermodynamics*, 3rd ed., SI Version, Wiley, New York, 1985.

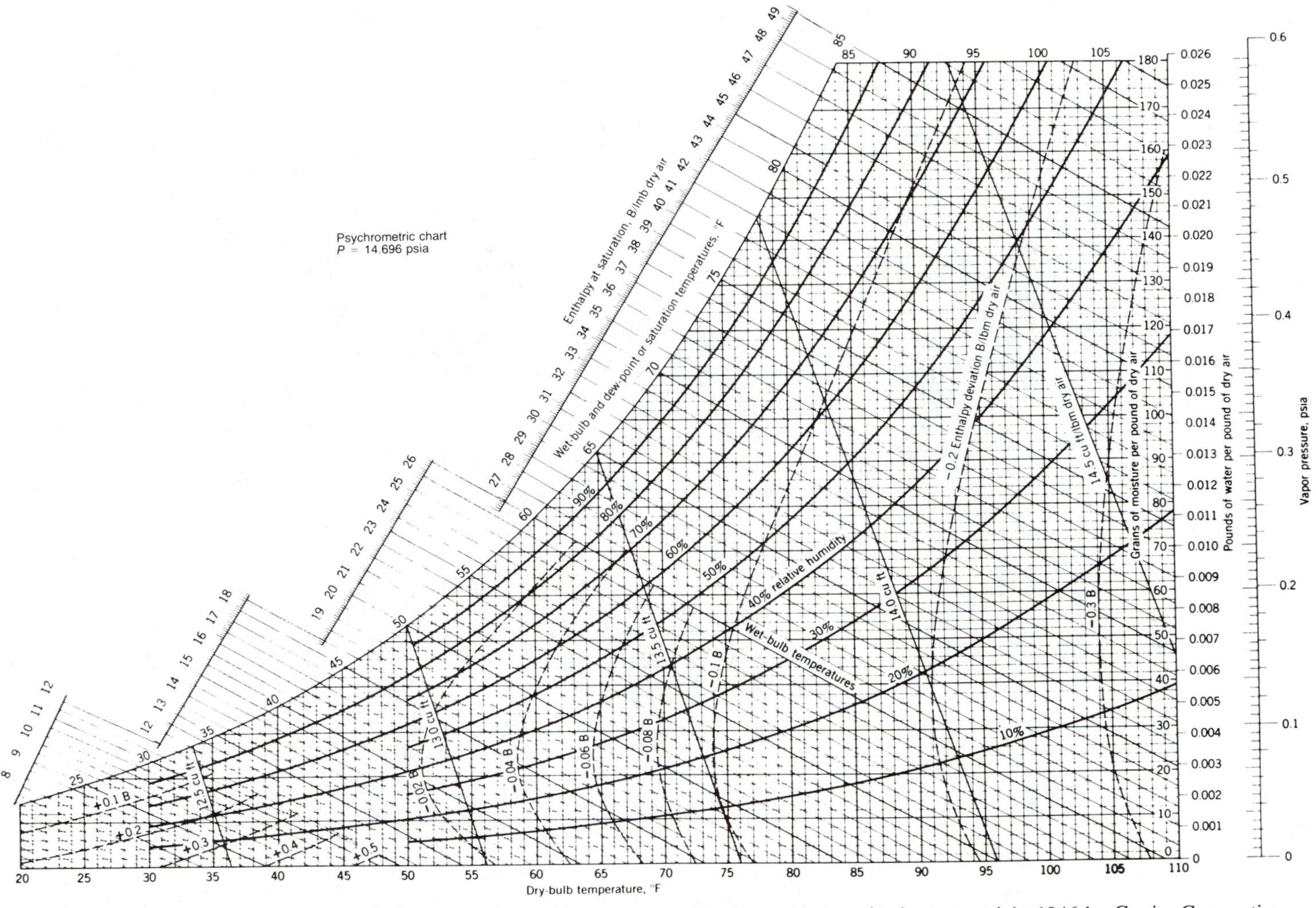

Chart A·7(E). Psychrometric chart (English units). Adapted by permission from Carrier psychrometric chart, copyright 1946 by Carrier Corporation.

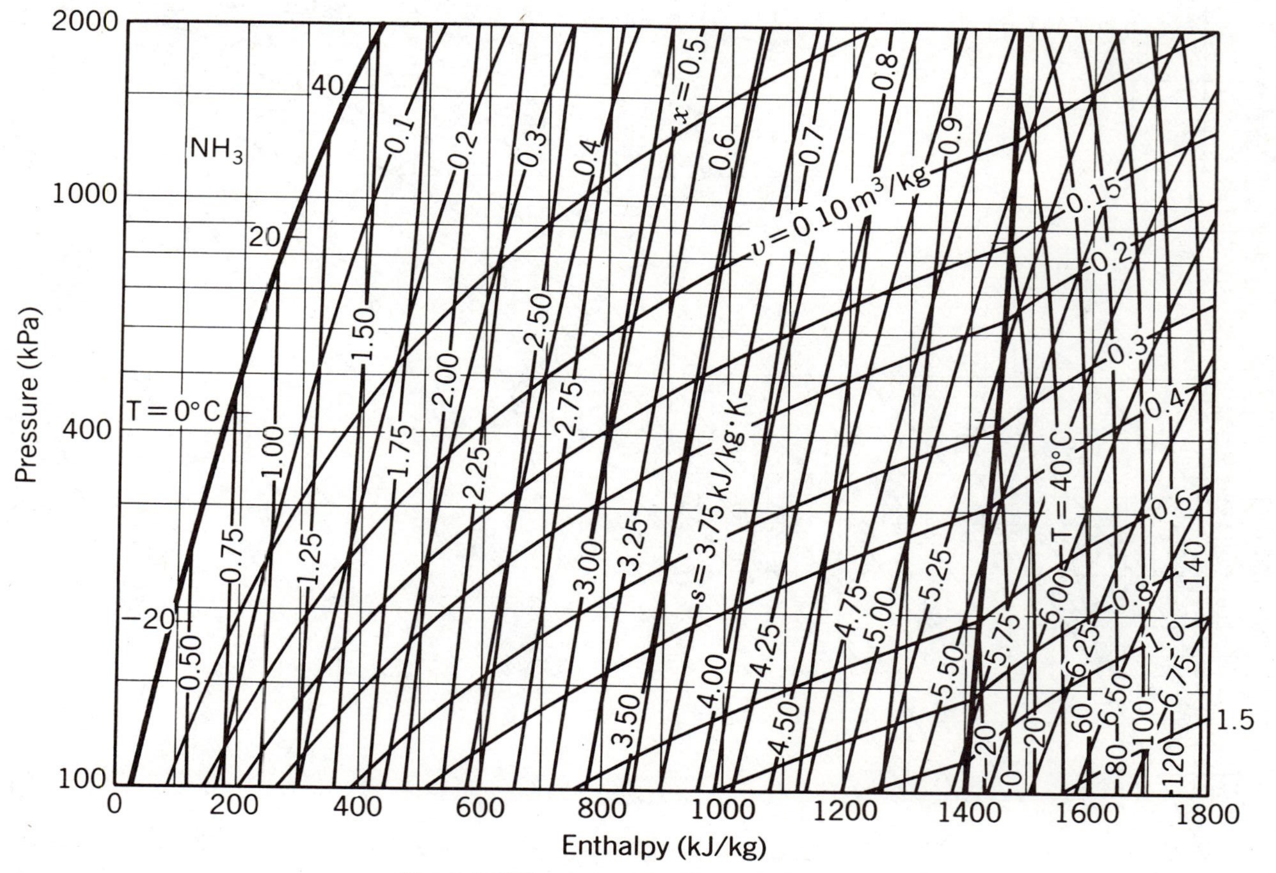

Chart A·8(SI). Mollier (*Ph*) chart for ammonia (SI).

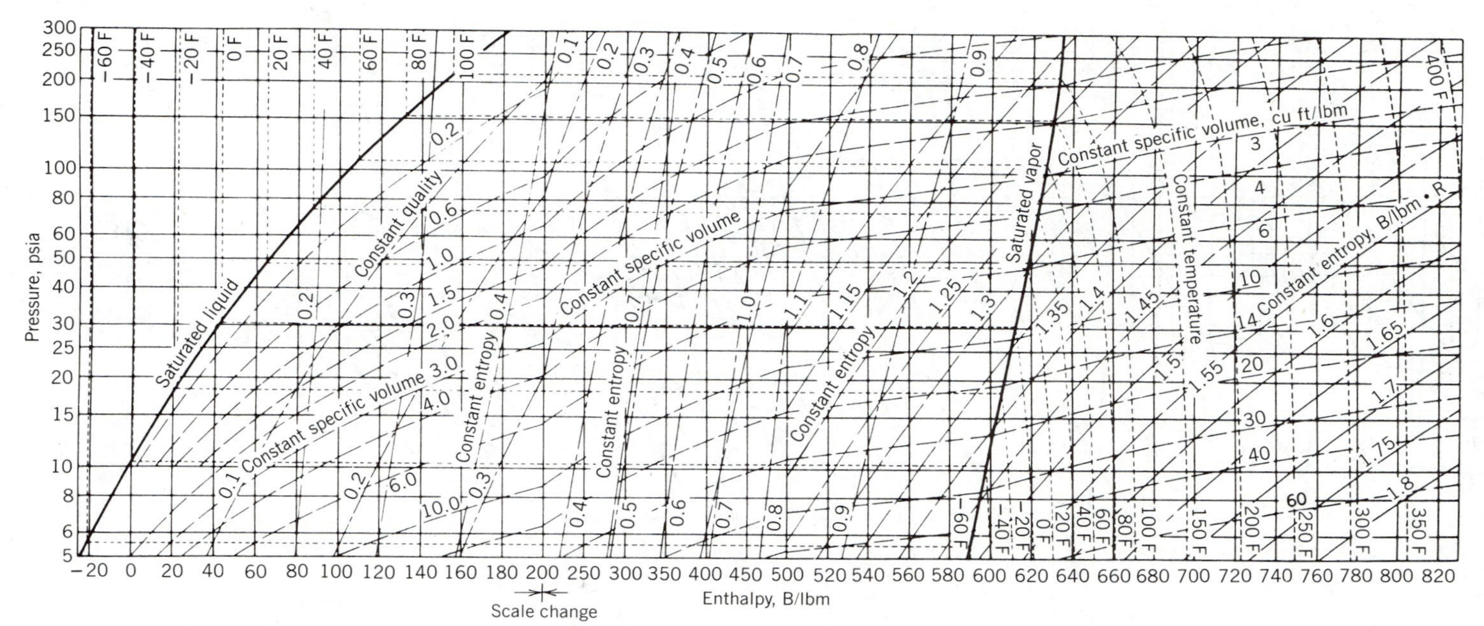

Chart A·8(E). Mollier (*Ph*) chart for ammonia (English units). Data from *Tables of Thermodynamic Properties of Ammonia*, National Bureau of Standards Circular 142.

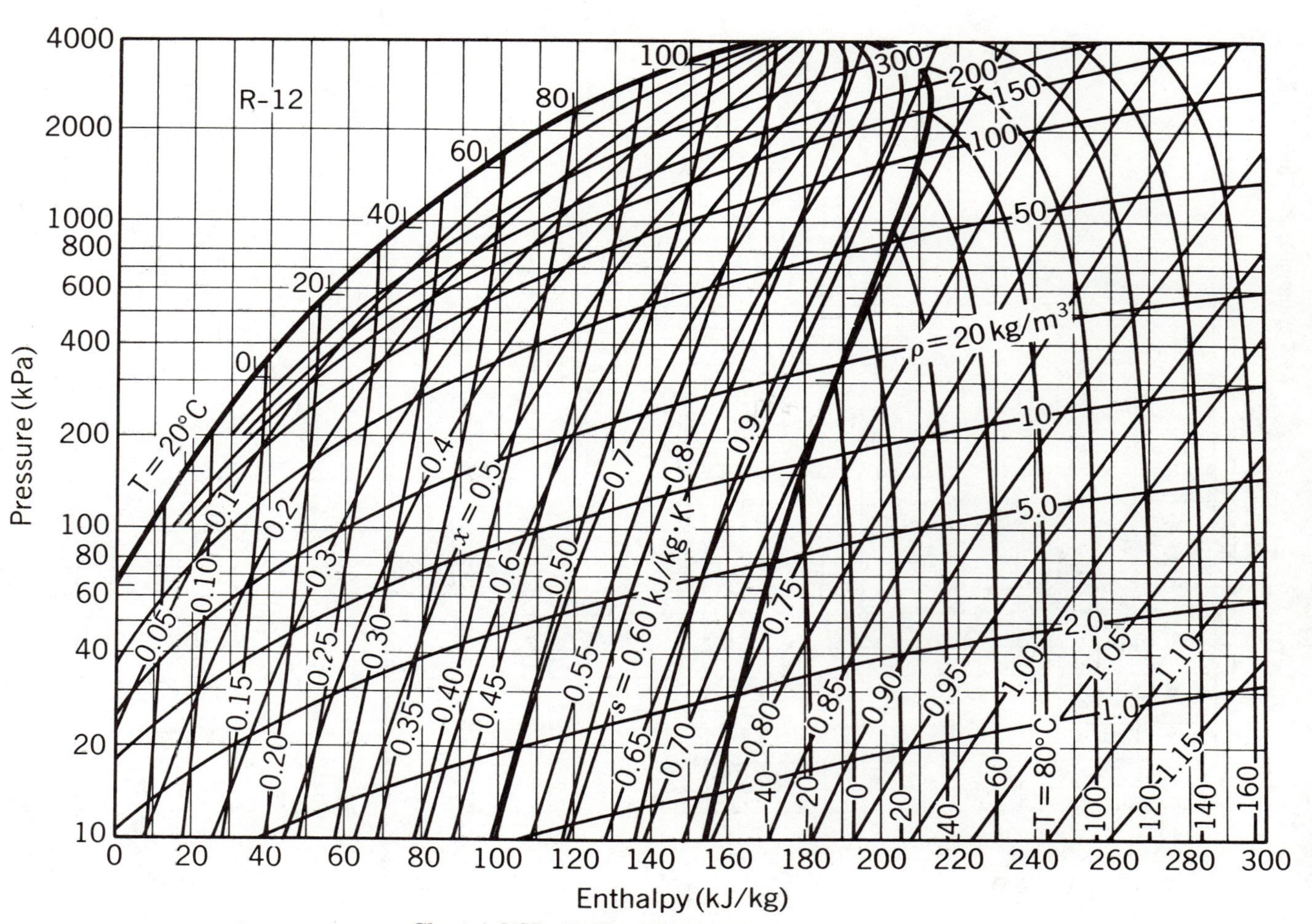

Chart A·9(SI). Mollier (*Ph*) chart for refrigerant 12 (SI).

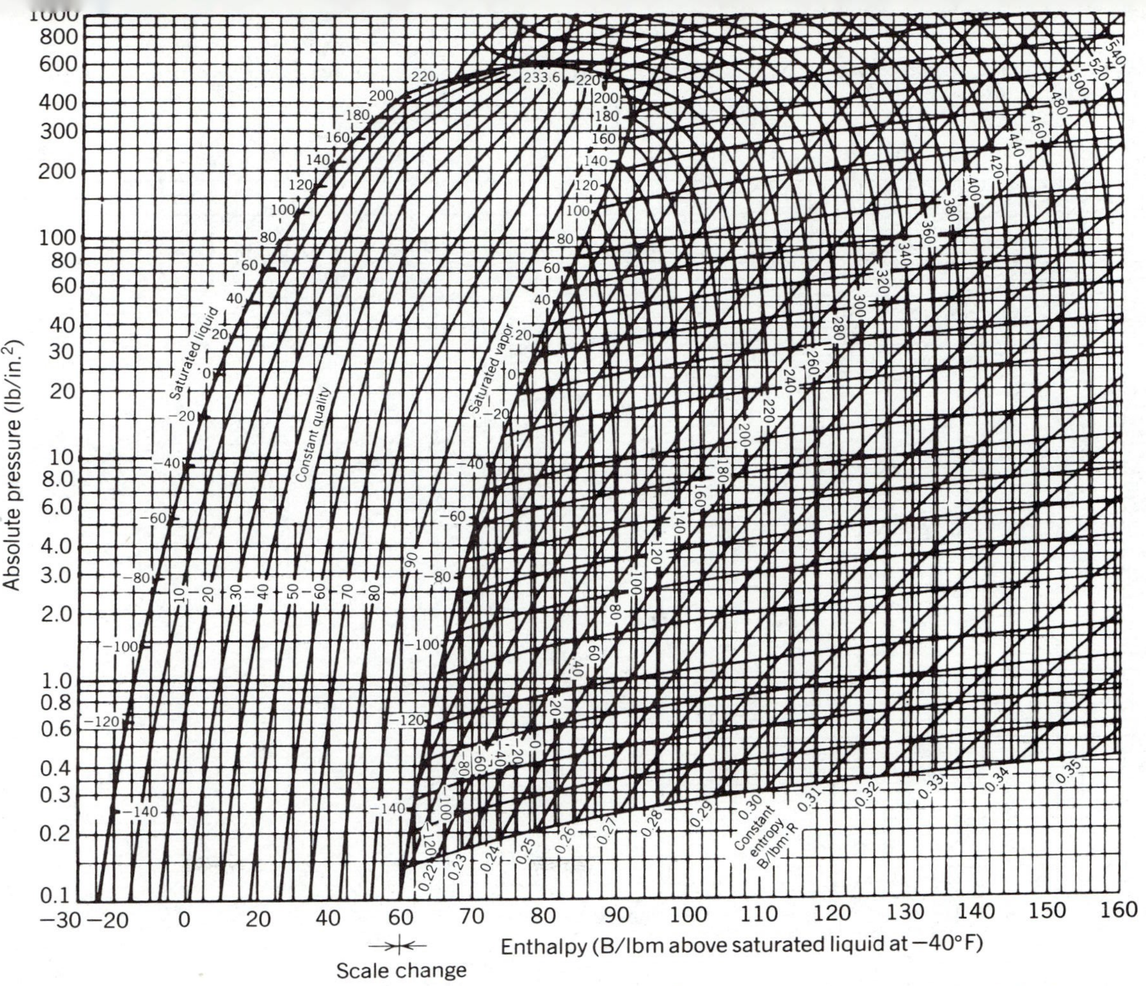

Chart A·9(E). Mollier (*Ph*) chart for refrigerant 12 (English units). Reprinted by permission from *ASHRAE Handbook—1985 Fundamentals*, American Society of Heating, Refrigerating and Air-Conditioning Engineers, Atlanta, 1985.

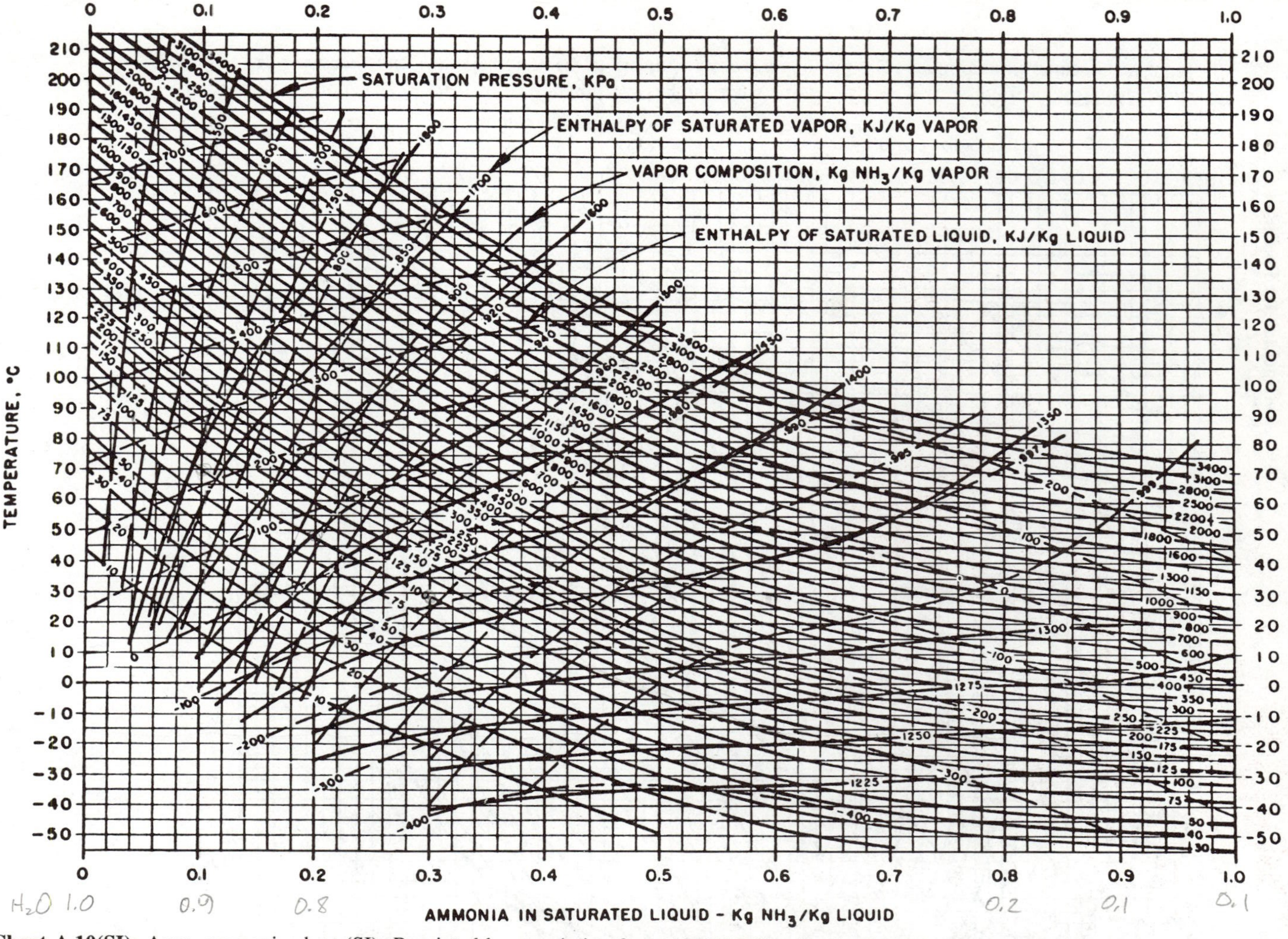

Chart A·10(SI). Aqua–ammonia chart (SI). Reprinted by permission from *ASHRAE Handbook—1985 Fundamentals,* American Society of Heating, Refrigerating and Air-Conditioning Engineers, Atlanta, 1985.

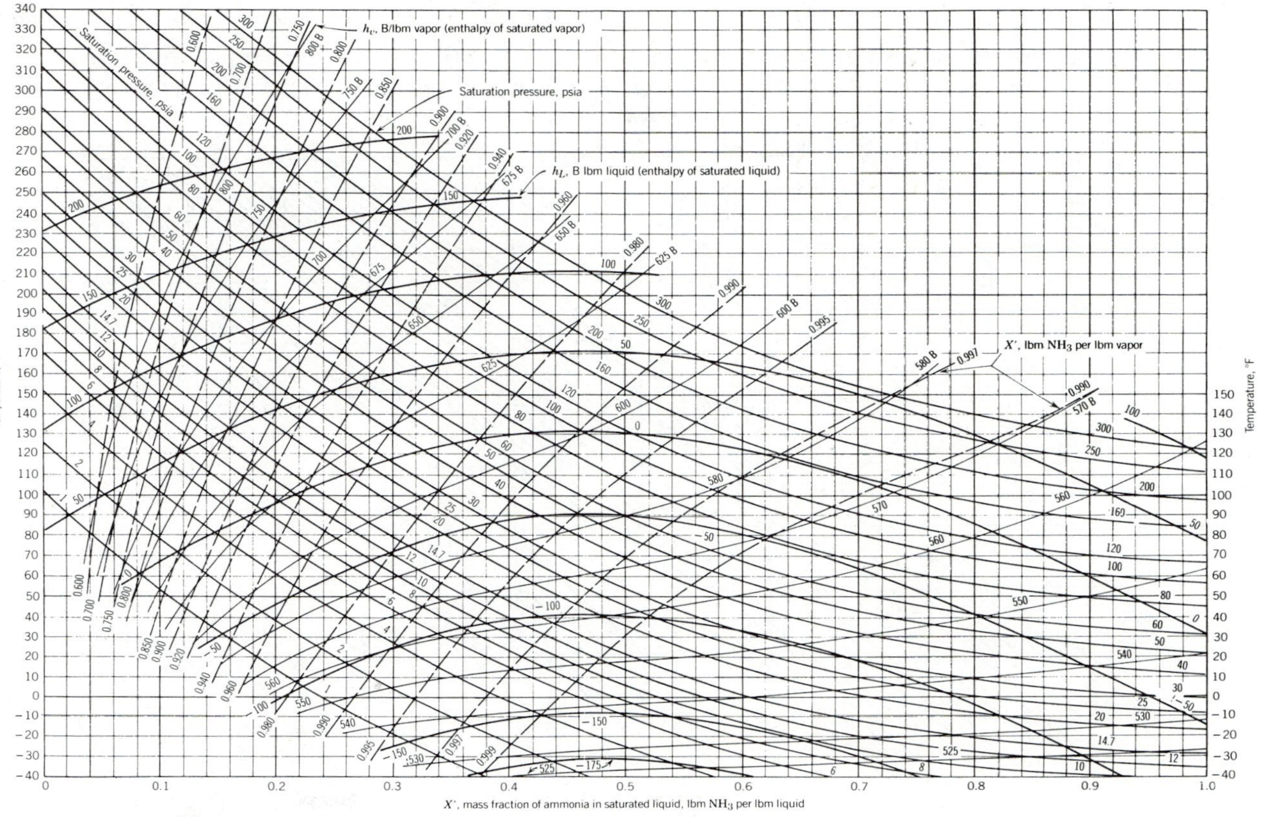

Chart A·10(E). Aqua–ammonia Chart (English units). Adapted from ''Equilibrium Properties of Aqua–ammonia in Chart Form,'' by F. H. Kohloss, Jr. and G. L. Scott, *Refrigeration Engineering,* vol. 58, no. 10, Oct. 1950, p. 970.

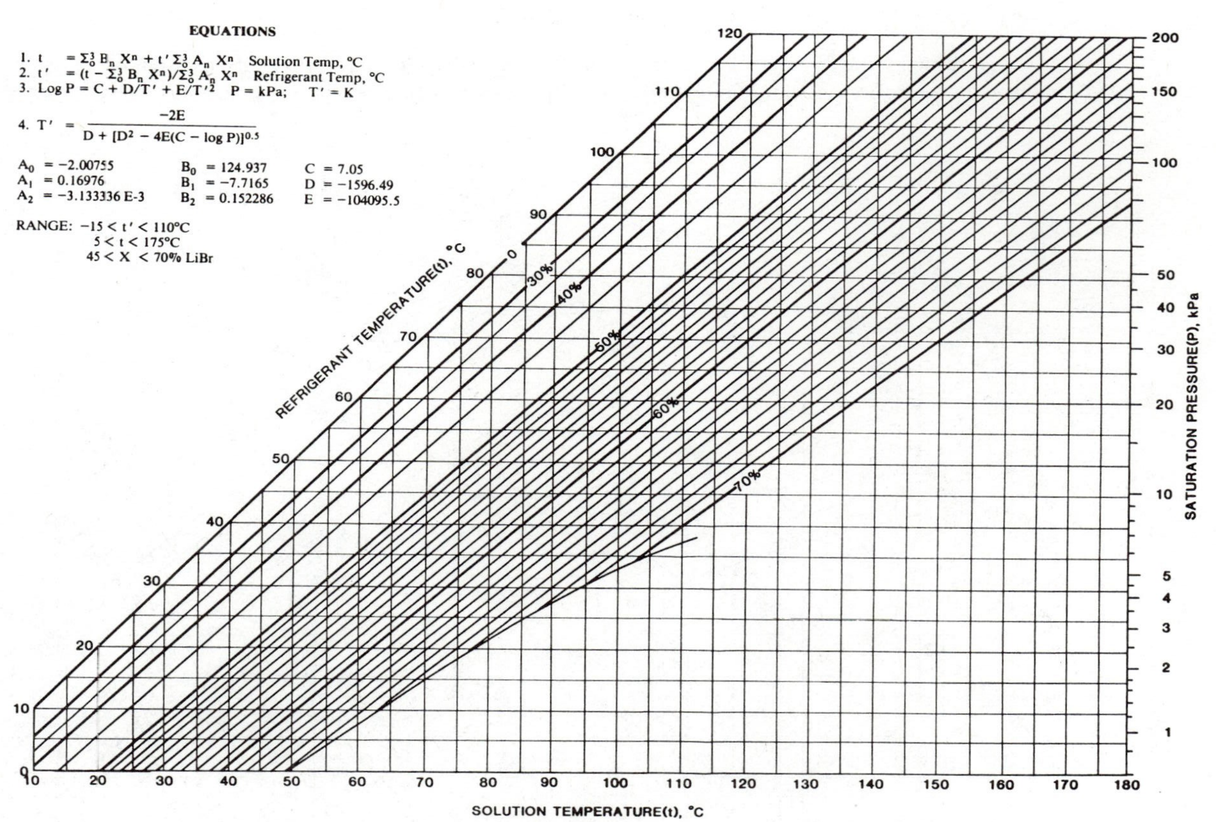

Chart A·11. Aqueous lithium bromide equilibrium chart. Reprinted by permission from *ASHRAE Handbook—1985 Fundamentals,* American Society of Heating, Refrigerating and Air-Conditioning Engineers, Atlanta, 1985.

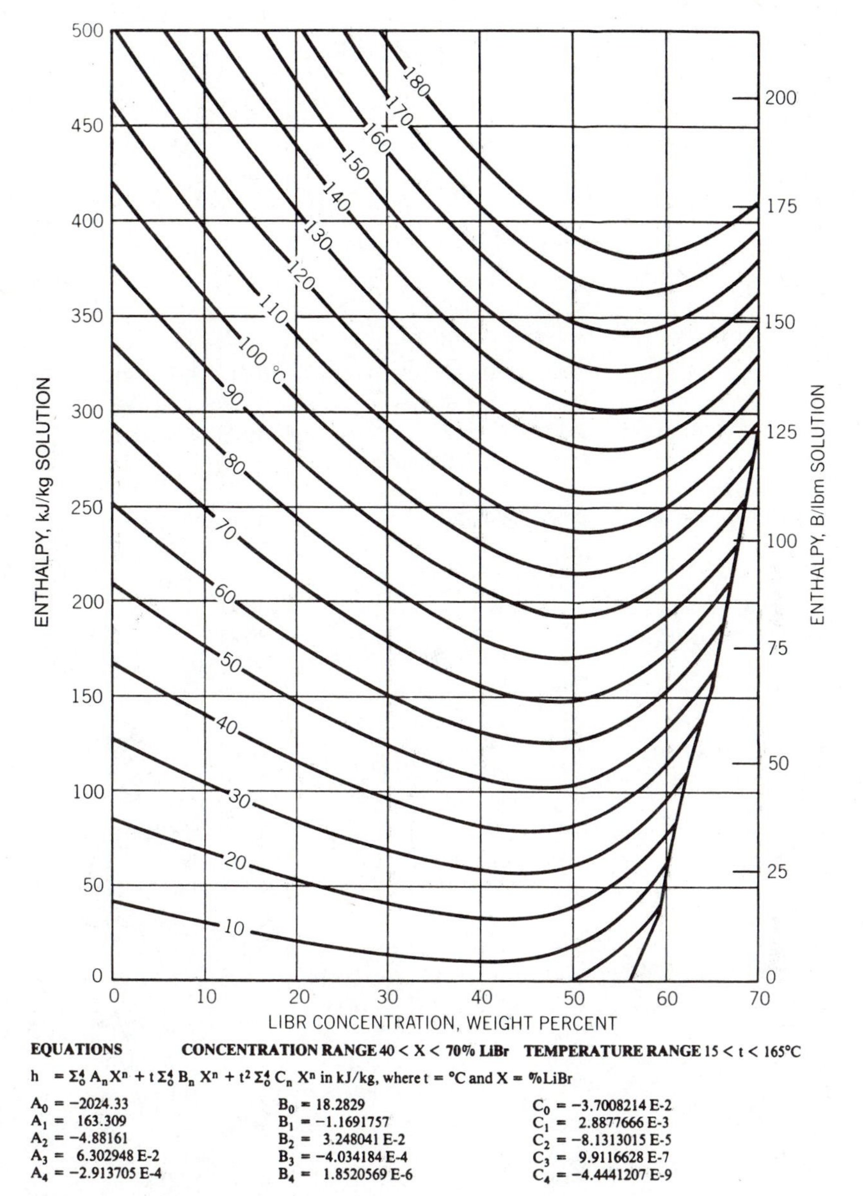

EQUATIONS **CONCENTRATION RANGE** $40 < X < 70\%$ **LiBr** **TEMPERATURE RANGE** $15 < t < 165°C$

$$h = \Sigma_0^4 A_n X^n + t\,\Sigma_0^4 B_n X^n + t^2\,\Sigma_0^4 C_n X^n \text{ in kJ/kg, where } t = °C \text{ and } X = \%\text{LiBr}$$

$A_0 = -2024.33$	$B_0 = 18.2829$	$C_0 = -3.7008214$ E-2
$A_1 = 163.309$	$B_1 = -1.1691757$	$C_1 = 2.8877666$ E-3
$A_2 = -4.88161$	$B_2 = 3.248041$ E-2	$C_2 = -8.1313015$ E-5
$A_3 = 6.302948$ E-2	$B_3 = -4.034184$ E-4	$C_3 = 9.9116628$ E-7
$A_4 = -2.913705$ E-4	$B_4 = 1.8520569$ E-6	$C_4 = -4.4441207$ E-9

Chart A·12. Aqueous lithium bromide enthalpy-concentration chart. Reprinted by permission from *ASHRAE Handbook—1985 Fundamentals,* American Society of Heating, Refrigerating and Air-Conditioning Engineers, Atlanta, 1985.

PROBLEM ANSWERS

1·3 5.1 ft^3
1·6 437 cm^3/g
1·9 5400 kPa, 72.4 cm Hg vac.
1·15 −218 Réaumur
1·18 15 000 lbf, 2400 hp
1·21 1620 kW
1·24 58 970 ft·lbf
1·36 (*a*) 1.11 J
1·39 12.5 kJ, 5.0 kJ, 7.5 kJ, 7650 kg
1·42 368 kJ
1·48 −4.29 kJ
1·51 33.7 ft/s, 0.589 m/s
1·54 0.158 kg/s
1·60 27.5 kJ/kg

2·3 −20 kJ/kg
2·6 −70 kJ/kg
2·9 80 kJ/kg
2·12 2020 cal
2·15 2.40 kJ
2·18 −10 B
2·21 −800 B
2·24 −230 B
2·27 0
2·30 0
2·33 −184 W
2·39 20 kJ
2·42 99 kJ
2·54 1.33
2·57 −315 B/lbm
2·60 447 m/s
2·63 −82.2 B/lbm
2·66 969 B/lbm
2·69 255 kW
2·72 1850 kW
2·75 80.7 kW
2·78 20.9 kJ/kg
2·81 520 kJ/kg
2·84 −6.1 B/lbm
2·87 21.3 kW
2·90 −6.7 B/lbm
2·93 41 400 kJ/kg
2·96 299 kJ/kg
2·108 0.223
2·111 (*a*) 5.5 kW, (*b*) 15.5 kW

3·3 94 K, 104 K, 116 K
3·6 No
3·9 21°F
3·21 6.48 × 10^{12}
3·30 (*a*) 179.9°C, (*b*) 2.57 kg
3·33 2790 kJ/kg
3·36 (*a*) 361°F, (*b*) 6.84 lbf
3·39 244 B/lbm, 0.7404 ft^3/lbm
3·54 107 m
3·60 (*a*) 135°C, (*b*) 70.14 kPa, (*c*) 0.753, (*d*) 0
3·66 7.18 lbm/lbm
3·69 2.89 kJ/kg·K, 2.49 kJ/kg·K, 7.17 kPa/K
3·75 113.2 kJ/kg
3·78 84.2 kJ
3·81 52.5 kJ
3·84 345 kJ/kg
3·87 7.37 B
3·90 38.5 B
3·93 −0.3 B/lbm
3·96 8 kJ/kg
3·99 0.948
3·102 2.79 m^2
3·105 209°C
3·108 −29.5 B/lbm
3·111 27.1 ft^2
3·114 351 B/lbm
3·120 1160 kW
3·123 0.382 kg/s
3·126 142 kJ/kg
3·129 17.4 B/lbm
3·132 −29.4 × 10^6 B/h
3·135 2438 kg/h
3·138 −35°C, 93 kPa
3·141 288°C
3·144 95 700 B
3·147 9.54 psia, 191°F
3·150 118°F

4·3 1.05 m^3, 1350 kPa
4·6 25.9 kg
4·9 0.646 N
4·12 568 lbm
4·15 39.2 ft^3, 288 psi
4·18 0.0326 lbmol
4·21 3.6 K
4·24 (*a*) −85.1 kJ, (*b*) −85.1 kJ, (*c*) 0
4·27 (*a*) −87.3 kJ, (*b*) −87.3 kJ, (*c*) 0
4·30 5.11 lbf/ft^2
4·33 224°F
4·36 4.7 B
4·48 22.5°C
4·51 69.0 kJ/kg
4·54 1.53
4·57 0.073 B/lbm·F
4·66 478 kJ
4·69 235 B
4·72 21.1 B
4·75 (*a*) −0.409 kJ, (*b*) 0.571 kJ
4·78 (*a*) −0.629 kJ
4·81 18.4 B
4·84 0.128 kJ

4·87 (*a*) −1.15 B
4·90 (*a*) 9.4, (*b*) −1.2 B
4·93 −108 kJ/kg
4·96 125 kW
4·99 337 hp
4·102 937 hp
4·108 47.7 kJ/kg, 66.6 kJ/kg, −47.7 kJ/kg
4·111 1.18
4·114 (*a*) 343 m/s, (*b*) −58.8 kJ/kg
4·117 17.6 kW
4·120 49.5 kJ/kg
4·123 32.6 kJ
4·126 172 B, 240 B, −172 B
4·129 1.143
4·132 (*a*) 1140 ft/s, (*b*) −25.9 B
4·135 26.2 hp
4·138 21.3 B/lbm
4·141 −24.200 ft·lbf, 0, 31.1 B
4·147 0.00262 kg/s
4·150 212 ft^3/min
4·165 96 200 kJ
4·168 114 kJ out
4·171 9.9 min
4·174 436 B
4·177 287°F
4·180 57.5 psia

6·9 207 kJ, 176 kJ
6·12 324 K
6·15 −36.4 B/lbm, −54.6 B/lbm
6·18 0.50, 78.8 kJ, 21.4 kPa
6·21 1.61, 25.2 B/lbm
6·30 36 700 B/h

7·18 602 kJ/min
7·21 238 K
7·24 0.76 kw
7·27 4.35
7·30 0.23 kw
7·33 7.11 kw, 12.8 kJ/s
7·36 104 B/s, 33.1 B/s, 0.681
7·39 0.51, 661 R
7·42 37.0 B, 63.0 B
7·45 0.48
7·48 23.3 hp
7·51 2.8 hp
7·54 8.3
7·57 150 B/h
7·60 207 B/h
7·63 0.22, 35.1 kJ

8·6 350 kw
8·9 510 kPa
8·12 91.5 kJ/kg
8·15 1036 kJ/kg·K
8·18 18.5 B/lbm
8·21 82 B/lbm
8·24 10.3 B/lbm
8·27 32.0 B/lbm
8·36 680 kJ, 406 kJ, 274 kJ, 0.403, 680 kJ, 394 kJ, 286 kJ, 0.421
8·39 (*b*) 3.4 kJ, (*c*) 0.17 kJ/K
8·42 0.189 B/R
8·45 1.36 hp
8·51 (*a*) −725 kJ, (*b*) 0.011 m^3
8·54 0.0389 kJ/K
8·57 16.0 kJ/kg
8·60 (*a*) 2.2 kJ, (*b*) 0.016 kJ/K
8·63 (*a*) 292 B, (*b*) 2.09 ft^3
8·66 0.0248 B/R
8·69 6.9 B/lbm
8·78 274 K, 333 K
8·81 484 K
8·84 3.3 kJ/kg
8·87 0.93, >0.93, <0.93
8·93 59.3 hp
8·99 21.7 kJ/kg
8·102 0.84
8·105 (*a*) 171 B, (*b*) 85.5 B, 85.5 B, (*c*) 48 B, 123 B
8·108 70 B; −0.733 B/R, 0.124 B/R; 0
8·111 (*a*) 0.35, (*b*) 0.20

9·3 512 N/m^2·K
9·6 6.1 psf/°R, 7.2 psf/°R
9·9 2060 kJ/kg
9·12 2.1 Pa
9·15 887 B/lbm
9·18 19°F
9·21 (*a*) 0, (*b*) 1.6 K/atm
9·39 3.29×10^{-3} K^{-1}
9·42 6.9670 kJ/kg·K, 1046.0 kJ/kg
9·45 170°C
9·48 283 kJ/kg
9·51 780 K
9·54 1010 R
9·57 120.5 B/lbm
9·60 0.598
9·63 2.32×10^{-3} m^3/kg
9·66 58.7 MPa
9·72 0.875
9·75 955 R
9·78 1.07 lbm/ft^3
9·87 (*a*) 2.77×10^{-3} m^3/kg, (*b*) 3.56×10^{-3} m^3/kg, (*c*) 3.43×10^{-3} m^3/kg, (*d*) 3.55×10^{-3} m^3/kg
9·90 38.5 kPa
9·93 (*a*) 2.23×10^{-3} m^3/kg, (*b*) 2.39×10^{-3} m^3/kg
9·96 1720 psia
9·99 1680 psia
9·102 68.4%
9·105 25 hp, 6 B/s, 0.052 in^2
9·108 0.0031 kg/s
9·111 32.4 B/lbm, 31.1 B/lbm
9·114 143°C
9·117 2.9 kJ

10·3 398 kJ
10·6 16.3 kJ
10·9 24.0 kJ
10·12 2.59 B; 1.66 B
10·15 455 B
10·18 15.0 B
10·21 215 kJ/kg, 0.31
10·24 63 600 kJ/h
10·30 (*a*) 270 B/lbm, (*b*) 158 B/lbm
10·33 59 900 B/h
10·36 14.7 kJ; −8.7 kJ; 4.8 kJ
10·39 12.8 kJ/kg, 720 kJ/K
10·42 26.0 kJ/kg
10·45 94.5 B/lbm, 94.5 B/lbm, 22.6 B/lbm
10·48 130 B
10·51 63.6 B/lbm
10·54 13.1 B/lbm

11·3 P_{O_2} = 0.32 atm, P_{N_2} = 0.48 atm, 40% O_2, 60% N_2
11·6 0.0542 kJ/kg·K
11·9 36% O_2, 64% CO_2, 39% O_2, 61% CO_2
11·12 0.0542 kJ/kg·K
11·15 329 K; 407 kPa
11·18 42.6 kJ/kg
11·21 −24.2 kJ
11·27 0.0281 B/lbm·R
11·30 39.3 B
11·33 (*a*) 600 R, (*b*) 31.5, (*c*) 0.213 B/lbm·F, (*d*) 49 ft·lb/lb·R, (*e*) 63.8 psia, (*f*) 24.0, 31.4, 8.4 psia
11·36 0.311 lbm
11·42 0.85 MPa
11·45 25°C, 1.9 kPa
11·48 0.0163
11·54 24.5 psia; 11.53, 11.29, 1.65 psia
11·57 0.0125 lbm/lbm, 0.29 psia
11·60 0.57, 72.8°F
11·63 0.0208 lbm/lbm, 78°F
11·66 7.57×10^{-4}
11·69 0.28
11·75 90°F, 0.0308 lbm/lbm da
11·78 100.7°F
11·90 25.7 kJ/kg
11·93 (a) 0.0076 kg/kg da, (b) 47.8 kJ/kg da, (c) 18.2 kJ/kg da
11·96 2810 kg/min, 100 kg/min
11·99 61.2 kJ/kg
11·102 644 kJ/kg
11·108 (*a*) 80°F, (*b*) 0.507 psia, (*c*) 0.0223 lbm v/lbm da, (*d*) 0.0223 lbm v/lbm da
11·111 0.54 hp
11·114 139°F
11·117 64.7 B/lbm

12·3 36°C (96°F), 13.8 kg/kg
12·6 (*a*) 11.1, 0.8, 12.7, 11.9, 12.7, 17.5, 33.3 percent; (*b*) 6.3, 31.7, 28.6, 33.4 percent; (*c*) 16.2 cu ft/lbm; (*d*) 4.65; (*e*) 12.5, 4.7, 82.8 percent; (*f*) 6.35; (*g*) 0.57; (*h*) 5.35 kg/kg; (*i*) 53°C (128°F); (*j*) 58°C (136°F)
12·9 (*a*) 11.7 kg/kg f; (*b*) 4.9, 14.3, 80.6, 0.2 percent; (*c*) 12.6 kg/kg f, (d) 0.45 kg/kg f, (e) 11.6 kg/kg f, (f) 36°C, (g) 4.7°C
12·12 17.2 mass units
12·15 (*a*) 16.5 mass units, (*b*) 1.85 mass units
12·18 0.79 lbm/lbm
12·24 52 700 B/lbm
12·27 18 000 B/lbm
12·36 −20 800 B/lbm
12·39 1650 K
12·42 10 300 B/lbm

13·3 (a) 0.00168, (b) 595
13·6 Percentages: (a) 30.4 CO_2, 3.7 CO, 1.8 O_2, 64.1 N_2; (b) 31.2 CO_2, 3.0 CO, 1.5 O_2, 64.3 N_2
13·9 Percentages: (*a*) 30.4 CO_2, 3.7 CO, 1.8 O_2, 64.1 N_2; (*b*) 31.2 CO_2, 3.0 CO, 1.5 O_2, 64.3 N_2
13·12 Percentages: (*a*) 46.8 CO_2, 48.4 H_2O, 2.4 CO, 1.3 O_2, 0.6 OH, 0.5 H_2
13·15 2180 K
13·18 4220 kJ/kg
13·21 6010 kJ/kg
13·24 4730 kJ/kg, 4230 kJ/kg
13·27 8330 kJ/kg
13·30 3020 K
13·33 2400 K
13·36 2370 K
13·39 2280 K
13·42 2270 K
13·45 2.10 kg/kg CH_4
13·48 −250 000 kJ/kg
13·51 4.4 atm

14·6 276°C, 263.6°C
14·9 543 R; 8.9 psia
14·12 31 kN
14·15 7.2 kg/s; 1.68 m
14·18 772 lbf
14·21 0.59
14·24 0.59
14·30 0.127 m^2
14·33 (*a*) 0.134 m^3, (*b*) 0.90
14·36 0.314
14·39 1.58 ft^2
14·42 1.41 ft^2; 0.907
14·48 0.876 kg/s
14·51 1.51 lbm/s
14·54 16.9 lbm/s
14·57 795 kPa, 278°C
14·60 0.193 m^2; 0.0291 m^2
14·63 0.0026 m^2
14·66 1100 kPa, 278°C, 26.9 kJ/kg
14·72 0.535 ft^2; 0.296 ft^2
14·75 0.51, 0.45 ft^3
14·78 13.8 psia, 191.5 psia
14·81 5.45 lbm/s, 0.59 B/lbm
14·84 159 psia, 539°F, 11.5 B/lbm
14·96 4.72 cm^2, 14.7 cm^2
14·102 3.105 in^2
14·108 80 kPa
14·114 0.385
14·117 18.4 psia; 1045 ft/s
14·120 14.2 psia; $b \rightarrow a$
14·126 (a) 450 kPa, (b) 620 kJ/kg
14·129 (a) 145 kPa, (b) 48.4 kPa
14·132 (a) 51.3 psia, (b) 25.4 psia

15·3 0.048, 1.00
15·12 q_{out} = 27.3 kJ/kg
15·15 56.6 kW

15·18 289 kW
15·21 0.817, 130.6 kJ/kg
15·30 1.4 hp
15·33 359 hp
15·36 1058 hp
15·39 13.2%
15·42 3.2 kg/s
15·48 29 psia, 53 psia, 0.92, 0.91, 0.90
15·51 15.3 B/lbm
15·54 4.88 m^3
15·60 (*a*) 10.5 kg/kW·h, (*b*) 145 kJ/kg, (c) 0.702
15·63 (*a*) 23.3 lbm/kW·h, (*b*) 62.6 B/lbm, (c) 0.700
15·66 0.273
15·69 5.54 hp
15·72 (*a*) 4.43 B/lbm, (*b*) 108 hp
15·75 0.0153 m^3/min
15·78 26 hp
15·81 89 hp, 5.73 ft^3

16·3 4.67 MW and 0.257 for 300 kPa
16·12 9290 hp, 0.345
16·15 Impossible data
16·18 0.131
16·21 (*a*) 615 K, (*b*) 0.902
16·30 (*a*) 0.507, (*b*) 0.616, (*c*) impossible
16·33 0.481
16·39 0.58
16·42 (*a*) kg/s, 12.7 kg/s; (*b*) 208.2 kJ/kg
16·45 49.2 kg/s, 0.213
16·48 1150 B/s, 1170 B/s
16·51 152 lbm/s, 82.3 lbm/s
16·54 19 200 N
16·57 2350 fps
16·60 5.43
16·63 43 600 lbf
16·69 (*a*) 0.464, (*b*) 6.1 kJ, (*c*) 1.4 kJ
16·72 3680 kPa, 493 kJ/kg, 0.699
16·75 (*a*) 0.482, (*b*) 6.19 B, (*c*) 1.36 B
16·78 (a) 411%, (b) 0.621
16·81 10.9 kJ
16·87 (a) 820 kJ, (b) 0.504, (c) 1306 kJ, (d) 486 kJ

17·3 0.432, 0.0068
17·12 0.224
17·15 (*a*) 0.290, (*b*) 0.0129
17·18 3800 psia
17·21 145 psia, 340 psia, 875 psia, 2000 psia
17·30 0.397
17·33 0.432
17·36 450 kPa
17·42 0.395
17·45 0.424
17·48 90 psia
17·51 0.372, 2143 kJ/kg, 0.880, 0.854
17·54 0.497, 87.6 kg/s
17·57 1.107×10^6 lbm/h, 0.424
17·60 0.457, 644 000 lbm/h
17·63 Cannot meet all conditions
17·66 0.96 h
17·69 0.77 h
17·72 8.51 kg/s, 2340 kW
17·78 (*a*) 437 kW, (*b*) 1400 lbm/h

18·3 4.57
18·6 1.85 hp
18·9 4.57
18·15 (a) 2.16 kW
18·21 (a) 25.7 kW, (b) No
18·24 21.1 kW
18·27 2.83 hp
18·33 (*a*) 35 hp
18·36 27.8 hp
18·39 4.72
18·45 1180 lbm/h
18·51 (a) 2.81 kg/kg NH_3, (b) 2460 kJ/kg
18·54 (*a*) 5.67 lbm/lbm, (*b*) 1400 B/lbm
18·57 (*a*) 884 kg/h, (*b*) 23 600 kg/h, (*c*) 31 400 kg/h
18·60 (a) 2000 lbm/h, (b) 61 900 lbm/h, (c) 96 100 lbm/h
18·63 −70°F

19·3 4
19·6 2
19·12 0.711

20·3 1.06 V
20·6 (*a*) 1.2 V, (*b*) 1.22, (*c*) 170 000 kJ/kmol
20·12 2.07×10^{-9} kmol/s, 0.70 W, −0.56 W

21·3 606°C, 80°C
21·6 370 W/m
21·9 1440 B/h
21·12 8650 B/h
21·18 11.2 kW
21·21 525 W/m^2·K
21·24 3.63 W/m^2·K
21·27 22 800 kW
21·30 449 B/h·ft^2·°F
21·33 43.2°F
21·36 92.4 B/h·ft^2·°F
21·45 7.1 kW
21·48 14.9 W
21·51 (*a*) 240°F, (*b*) 206°F
21·54 53 200 B/h
21·57 218 B/h

Appendix A exercises

3. 432 N
6. 19.0 lbf
9. (*a*) 30.2 lbm, 0.938 slug; (*b*) 15 lbf, 30.2 lbm, 0.938 slug
12. 6.20×10^{-5} lbf·s/ft^2, 0.002 97 N·s/m^2, 0.002 97 kg/m·s, 0.001 99 lbm/ft·s

INDEX

CONVERSION FACTORS

LENGTH

1 ft = 0.3048* m	1 m = 3.281 ft
1 in. = 0.0254* m	1 cm = 0.3937 in.
1 mi = 5280 ft	1 km = 0.6214 mi

AREA

1 ft^2 = 0.0929 m^2	1 m^2 = 10.76 ft^2
1 $in.^2$ = 645.16 mm^2	1 cm^2 = 0.1550 $in.^2$

VOLUME

1 ft^3 = 0.028 317 m^3	1 m^3 = 35.32 ft^3
1 $in.^3$ = 1.639×10^{-5} m^3	1 cm^3 = 0.061 02 $in.^3$
1 gal = 0.003 785 4 m^3	1 gal = 231* $in.^3$
1 l = 0.001 m^3	1 gal/min = 0.000 063 1 m^3/s
1 gal/min = 0.002 228 ft^3/s	

MASS

1 lbm = 0.453 592 kg	1 kg = 2.204 62 lbm
1 slug = 14.594 kg	1 tonne = 1000 kg
1 ton = 2000 lbm	

PRESSURE

1 psi = 6.894 757 kPa	1 kPa = 0.145 038 psi
1 in. Hg = 3.387 kPa	1 in. Hg = 0.4912 psi
1 bar = 100* kPa	1 mm Hg = 0.1333 kPa

1 atm = 101.325 kPa = 14.696 psi = 760 mm Hg = 29.92 in. Hg

FORCE

1 lbf = 4.448 222 N	1 N = 0.224 809 lbf
1 dyne = 1×10^{-5} N	

*Exact value